Autonomous Mobile Robots

Emerging Methodologies and
Applications in Modelling,
Identification and Control

Autonomous Mobile
Robots

Planning, Navigation and Simulation

Rahul Kala
Centre of Intelligent Robotics, Indian Institute of Information
Technology, Allahabad, India

Series Editor

Quan Min Zhu

ELSEVIER

ACADEMIC PRESS
An imprint of Elsevier

Academic Press is an imprint of Elsevier
125 London Wall, London EC2Y 5AS, United Kingdom
525 B Street, Suite 1650, San Diego, CA 92101, United States
50 Hampshire Street, 5th Floor, Cambridge, MA 02139, United States
The Boulevard, Langford Lane, Kidlington, Oxford OX5 1GB, United Kingdom

Notices
Knowledge and best practice in this field are constantly changing. As new research and experience broaden our understanding, changes in research methods, professional practices, or medical treatment may become necessary.

Practitioners and researchers must always rely on their own experience and knowledge in evaluating and using any information, methods, compounds, or experiments described herein. In using such information or methods they should be mindful of their own safety and the safety of others, including parties for whom they have a professional responsibility.

To the fullest extent of the law, neither the Publisher nor the authors, contributors, or editors, assume any liability for any injury and/or damage to persons or property as a matter of products liability, negligence or otherwise, or from any use or operation of any methods, products, instructions, or ideas contained in the material herein.

ISBN 978-0-443-18908-1

For information on all Academic Press publications
visit our website at https://www.elsevier.com/books-and-journals

Publisher: Mara E. Conner
Acquisitions Editor: Sonnini R. Yura
Editorial Project Manager: Tom Mearns
Production Project Manager: Erragounta Saibabu Rao
Cover Designer: Mark Rogers

Typeset by STRAIVE, India

Working together
to grow libraries in
developing countries

www.elsevier.com • www.bookaid.org

Contents

4. Intelligent graph search basics

5. Graph search-based motion planning

6. Configuration space and collision checking

7. Roadmap and cell decomposition-based motion planning

11. Geometric and fuzzy logic-based motion planning

12. An introduction to machine learning and deep learning

13. Learning from demonstrations for robotics

17. Hybrid planning techniques

18. Multi-robot motion planning

19. Task planning approaches

20. Swarm and evolutionary robotics

21. Simulation systems and case studies

Preface

The idea of this book came from the teaching experience of the author at their parent institution, which runs a postgraduate-level degree programme in robotics and machine intelligence. Being a premium institution in artificial intelligence that attracts the best students interested in this field, the institute has several robotics courses. However, no current book is useful for these courses, due to three main reasons. First, giving a good robotics background to a student who otherwise has only a theoretical background in artificial intelligence is not an easy task, and is often neglected in the books. One could spend a lot of time in the complexities of the algorithms, which will have little utility if not put in the context of robotics and an overall explanation of robotics issues. Second, the domain of robotics is too wide. Even the specific problem of robot planning can be solved by a multitude of methods. Covering only one school of learning does not convey the diversity that is necessary for robotics courses. Third, in robotics particularly, it is important to cover the applied and the practical perspectives without an assumption of the possession of expensive robots by the readers. The coverage also needs to iteratively build up an aptitude to develop robotic systems in a simulation framework.

The incidence of specialized robotics projects is steadily increasing at the undergraduate level. A typical learning requirement for undergraduates is to be able to get to the introductory robotics context as quickly as possible and to delve deeper into the specific topic. An understanding from the experience of teaching the related courses and guiding student projects for around 8 years has made it clear that the current books have only partial coverage: beautifully covering the depth of some specific topics, however, not providing enough context, not providing exhaustive coverage of the topics, and only offering a limited encouragement to programming.

In the same vein, this book recognizes that a significant proportion of the audience studying robotics are from noncomputer science domains; in particular, many are from the mechanical engineering domain, who often have a weaker background in computing courses than those widely known to students from computer science backgrounds. Without a suitable background, the coverage becomes particularly hard for such students. This book covers all the basic concepts that are required for readers coming from different streams and backgrounds.

Learning modalities have drastically changed in the last few years due to increased use of massively open online courses, blogs, multimedia, and the greater availability of open-source software. The material included in books such as this one needs, therefore, to adapt to the changes to co-exist in the same ecosystem. Therefore, the writing is more graphic, further supplemented with illustrations, and lots of references to available software for implementations are made.

The book primarily covers the problem of motion planning of robots. Numerous algorithms solve the same problem, which are appropriately discussed in separate chapters. The book extends to the other issues involved with robot navigation, including avoiding dynamic objects and people in the way. The book also puts all this in the perspective of a simulation system to eliminate the need for any real robot. These problems largely rely upon other problems in robotics, such as perception, simultaneous localization and mapping, control, etc. These are also covered in the book to enable understanding of the overall domain.

The writing of this book has been performed rigorously, based on this background. There are three unique writing aspects of the book. First, a lot of emphasis is placed on getting the basics of robotics clear. Introductory chapters are specially written to clarify the background before focussing on the core topics. Second, more emphasis is put on developing intuitions using multiple modalities. The ideas are conveyed using engaging examples, illustrations, summary figures, boxes, algorithms, and exercises. A good part of the book can be given a quick and casual read to get abreast of the technology, and the illustrations further help the reader to stay on track. The book is written in a manner enabling the reader to grasp most of the concepts using illustrations and the text surrounding the illustrations alone. Third, more emphasis is put on implementation. The illustrations are made using programmes that are miniaturized implementations of the actual algorithms; and the exercises encourage readers to iteratively build the same programmes. A separate chapter is devoted to the use of simulation for solving robotics problems. Therefore, the book has an unusually large number of illustrations and typically has more substantial introductions to establish each chapter in the current context.

The book is suitable for course adoption. At the parent institution of the author, the coverage spans across three courses in two semesters. Teaching resources shall be continuously populated and made available on a dedicated book website. Numerous examples, exercises, illustrations, etc., are taken from the teaching materials and student projects that have successfully worked at the author's parent institution. The exercises are made to encourage the thought process, rather than arriving at a benchmark standard answer. At the parent university, the teaching is blended with implementations using small simulations made by the students themselves as well as developed using standard simulators. Programming exercises are provided with every chapter, which the students can attempt in order to get hands-on experience. The questions are typically open-ended and can be extended to projects. If you are doing a course adoption, an updated list of exercises, projects, and teaching materials can be requested from the author at any time.

Chapters 1–3 give an overall introduction to the problem of robotics. Chapter 1 discusses robotics from both hardware and software perspectives. The chapter models the robots from a kinematic perspective and introduces the basics of vision. Chapters 2 and 3 advance the discussions to the problems of localization (enabling the robot to know its position), mapping (enabling the robot to make an environment representation), simultaneous localization and mapping for practical robotics, and controlling the robot. Chapter 3 also introduces the planning aspects of robotics with the simplest algorithm imitating a bug.

Chapters 4–9 present the deliberative planning approaches in robotics. First, Chapters 4 and 5 give different graph search algorithms, presenting all issues that come with robotics. Then, Chapter 6 presents the concept of configuration space, in which the planning problem is solved. The free configuration space identifies all configurations that do not incur a collision. Therefore, collision-checking is also discussed in the chapter. Chapter 7 computes an off-line roadmap for online motion planning in a static environment. The chapter discusses the visibility roadmap technique that connects every corner with the other corners visible, Voronoi graphs that capture the topology by computing the deformation retract with the maximum separation, and cell decomposition that fills the free space with multiple cells of selected shapes. Chapter 8 uses random vertices and edges to sample a roadmap, while Chapter 9 iteratively grows a random tree representing the roadmap for a fixed source and goal.

Chapters 10–14 represent the reactive techniques to plan the robot. Chapter 10 models an electrostatic potential where a robot is attracted towards the goal and is repelled by obstacles. The forces can be modelled for social compliance and for modifying a trajectory modelled as an elastic strip. Chapter 11 moves a robot based on a geometric understanding of the obstacles and based on fuzzy rules. The navigation problem is solved from a learning perspective in Chapters 12 and 13 using the supervised learning techniques, and in Chapter 14, using the Reinforcement Learning techniques.

Chapters 15 and 16 solve the planning problem from an evolutionary perspective, discussing all the robotics-specific issues. Chapter 17 notes the limitations of the individual approaches and makes hybrid architectures, creating possibilities for an ensemble of robot behaviours. Chapter 18 discusses the planning approaches for multiple robots in a centralized (jointly planning all robots) or decentralized (planning robots separately and rectifying errors) architecture. Chapter 19 presents the task planning approaches, modelling the environment using Boolean facts, and solving the problem using classic Artificial Intelligence approaches.

Chapter 20 extends the discussions to the use of many robots as a swarm displaying complex behaviours. The chapter also discusses an evolutionary mechanism to evolve robot controllers for swarms. Finally, Chapter 21 presents the different challenges in robotics, including presenting a brief introduction to the popular robotics simulators.

Acknowledgements

The book started with an ambitious plan to fill the void of a good textbook for the robotics courses at the Indian Institute of Information Technology, Allahabad. The book is largely a result of teaching the robotics courses and guiding the student projects at the Institute. Therefore, the entire robotics community deserves a special appreciation for the work that led to this book. Specifically, the author thanks Prof. G.C. Nandi, as the book is a result of an endless number of motivational and technical discussions with him. The author also thanks Prof. Pavan Chakraborty for always motivating him.

The book is a result of the experiences gained at different institutions, where the author spent time as part of the degree programmes. The author especially thanks Prof. Kevin Warwick, for shaping the researcher in the author, and the University of Reading for the research environment. The author also thanks Prof. Anupam Shukla and Dr. Ritu Tiwari for introducing the domain of robotics and book writing to him and the Indian Institute of Information Technology and Management, Gwalior, for catalyzing the research.

The author thanks Vaibhav Malviya, Abhinav Malviya, Arun Kumar Reddy K., and Himanshu Vimal for their work on social robotics that provided a new perspective to enrich the book. Many components of SLAM discussed in the book are a result of the works of Lhilo Kenye, Rohit Yadav, Shubham Singh, Utkarsh, Mehul Arora, Dhruv Agarwal, Ritik Mahajan, Ambuj Agrawal, Shivansh Beohar, Vishal Pani, Naman Tiwari, Arpit Mishra, Prashant Rawat, Rahul Gautam, Harsh Jain, Mayank Poply, Rajkumar Jain, and Mukul Anand, for which these students are especially thanked. In the same vein, the author also thanks the team at navAjna Technologies Pvt. Ltd. for the technical and financial assistance provided for the SLAM projects. The author also thanks the team at Mercedes-Benz Research and Development India for enriching discussions that helped the book.

The author also thanks Sayed Haider Mohammad Jafri, Manav Vallecha, Shivansh Beohar, Arun Kumar Reddy K., Naveen Kumar Mittal, Abhinav Khare, Ritika Motwani, S. Akash, Jayesh Patil, Rahul Birendra Jha, Anant Rai, Ishneet Sethi, Alka Trivedi, Pranav Singhal, Mrinal Bhave, and Rishika Agarwal for the work on learning-based motion planning that is also summarized in the book. Haider also contributed to the costmap-based planning research. The swarm and evolutionary robotics concepts were enriched by discussions with Utkarsh, Adrish Banerjee, Rahul Gautam, Apoorva, Shourya

Munim, Ashu Sharma, and Adarsh Suresan. The author also thanks Lavu Vignan, Amol Upreti, Ashwin Kannan, Prashant Gupta, Rishabh Tiwari, Shubham Prasad, and Apurv Khatri for their work on sampling-based motion planning. The author also thanks Chandra Shekhar Pati for his work on motion control, the experience of which enriched the book. The book discusses vision aspects that are attributed to the rich discussions with Ayush Goyal, Abhinav Malviya, and Aashna Sharma. The author also thanks Lhilo Kenye for discussions on his work with pick and place robots. The author had an enriching experience of working with robots for which a lot of domain knowledge was gained from the works of Priya Shukla, Vandana Kushwaha, and Soumyarka Mondal.

The mission planning problem discussed in the book is partly inspired by the works of Akanksha Bharadwaj, Anil Kumar, Sugandha Dumka, Smiti Maheshwari, Abeer Khan, D. Diksha, S. Shelly, Surabhi Sinha, and Ishant Wadhwa. Many hardware perspectives mentioned in the book are attributed to the works of Gaurav Kumar Pandey, Tara Prasad Tripathy, Bishnu Kishor Sahu, Heena Meena, and Priyanshu Dhiman. The author thanks the Science and Engineering Research Board, Department of Science and Technology, Government of India, for the financial support in the same domain, vide grant number ECR/2015/000406.

The author also thanks the I-Hub Foundation for Cobotics, Indian Institute of Technology Delhi, and the Department of Science and Technology, Government of India, for the financial support towards his research in mobile manipulation and warehousing technologies that helped him include relevant examples in the book. The author also conveys his thanks to Garvit Suri, Hardik Bhati, Rishi Gupta, Shubhang Singh, and Surya Kant for their work in the same domain and to Addverb Technologies Pvt. Ltd. for enriching discussions. Finally, thanks also go out to Hardik Bhati, Rishi Gupta, and Garvit Suri, who were excellent students of the motion planning course; the experience gained from teaching this course helped perfect the book.

Chapter 1

An introduction to robotics

1.1 Introduction

The field of robotics has grown exponentially with robots coming into industrial and personal use rather than just being concepts to be found in books. The increase in robotics is not only on the application front but also backed by a sudden surge in the robotics literature. The book is an attempt to capture the hardest problems in robotics and to present an intuitive and simple understanding of the solutions that power sophisticated robots.

Robotics has always been an interdisciplinary field of study that has elements from all aspects of science and technology. It gets the kinematics, dynamics, and design principles from mechanical engineering; the circuitry, motors, and controls from electrical engineering; the sensing rooted into physics; the embedded systems and digital communication from electronics engineering; the intelligence and programming from the schools of computing; and numerous other schools of learning. Therefore, the domain of robotics is not far from any single field of study, and examples of robotics decorate the curricula of all schools of learning. The aim here is not to cover every single detail but to reason and solve the hardest problems in robotics.

The most challenging aspect of robotics nowadays is in *decision-making* (Tiwari et al., 2012). The robot should be able to autonomously navigate, execute higher-order tasks, partner with other robots, and deal with humans in the vicinity. The robot alone or as a team should be able to drive as a self-driving car, get your favourite book from the shelf, clean up the mess on the floor, deliver your parcel from the post office, etc. All these require a lot of decision-making at every step, and all aspects will be discussed in this book.

Making a real robot to solve a real-life problem requires numerous programming modules. The choice of programming modules and sensors is also dependent on the context of the operation of the robot and the robotic environment. This book discusses in great breadth and depth the different algorithms that enable the decision-making of the robot. The context and the larger picture in which such decisions are made are shown in this introductory chapter. The introduction chapter discusses all common modules that make up an autonomous robot and explains the relation between the modules. The subsequent

Autonomous Mobile Robots. https://doi.org/10.1016/B978-0-443-18908-1.00010-8

chapters delve into the motion planning aspects of robotics. That said, the first three chapters cover several topics that aid in motion planning and decision-making.

1.2 Application areas

Nowadays, robots are found in a wide variety of applications. *Industries* witnessed one of the first applications of robots, wherein robots could be used for automation and manipulation of industrial processes, which would take too long to be done by a human. Robots are faster, more efficient, and highly accurate. Unlike automation machines, robots could analyse the scenario and adapt their movements while at the same time being easily re-programmable. The current transformation in Industry 4.0 calls for greater interest in robotics in industries, along with Internet-of-Things-based sensors and augmented reality for human diagnosis. The *agricultural* sector is another similar sector where robots provide automated services such as ploughing, seeding, spraying pesticides, and harvesting. Robots can be highly precise, which can lead to more accurate use of pesticides for a better yield and fewer health hazards. According to research findings, the entire field can be more efficiently used to space plants, and robots can complete the processes a lot earlier.

The *defence* is another application, wherein the drones are used to combat terrorism in remote areas without risking human life; the humanoids are used to enter dangerous areas for rescue, and mobile robots are used for surveying. *Space* is another common application. The space shuttles are robots that hover over the earth. The Mars rover was a robot that walked on Mars to get data for earth. Robots are also used for other space explorations. Robots can enter hazardous areas and therefore find a lot of application in areas like searching for mines, searching for toxins, identifying gas leakages, surveying and rescuing in earthquake and landslide-affected areas, and doing long-range surveys with mobile robots and drones.

Self-driving cars (Kala, 2016) have recently been a big buzzword in all automobile companies demonstrating their prototypes. The self-driving cars are mobile robots on roads. These cars are safer for humans, reduce accidents caused by human error, can park themselves and are therefore best used as automated taxis, can use all traffic information for better decision-making, and the vehicles can even talk and coordinate with each other over the Internet. The *warehouses* are increasingly getting robots to assemble, pack, and ship items of interest, where again the chances of packing the wrong item or delivering the item at the wrong place are smaller. With the surge in the e-commerce industry, people are increasingly shopping online, leading to larger orders to be shipped. Robots add to efficiency. The *drones* themselves are excellent agents to ship the items quickly, which otherwise takes days even for small distances. The transport of items within environments, like hospitals, factories, offices, etc., is also now increasingly taking place with the help of robots.

Robots are also entering personal space. *Entertainment robotics* covers robots that entertain and take care of human subjects. Robotic humanoids can narrate stories and answer queries. Robotic dogs are used for affection. These robots are especially useful in hospital environments as aids to patients. Autistic patients can get a therapy session using such robots. *Assistive robotics* covers robots that assist the elderly and the disabled. They can monitor the subject and alarm for events like the subject falling, feeling ill, or becoming weak. They can get items of interest for the ones with lower mobility. They can also be used as exoskeletons or prosthetics for hands, legs, or other body parts. Surgical robots can be used to perform an entire surgery. These robots are teleoperated by the surgeon, and the microcontrol algorithms are used for precision in the surgery in the most minimally invasive way. Robots can now clean homes with robotic vacuum cleaners that ensure they visit all corners of the room and detect the surfaces. Robotics is also common in *sports*, with robots playing table tennis, chess, football, and other games.

1.3 Hardware perspectives

The first image of every robot is a piece of hardware. Let us, therefore, decipher robotic hardware. A summary is also given in Box 1.1. The hardware consists of the *sensors* through which the robot looks at the entire world. The *actuators* act on the physical world to make some manipulation or movement in the world. The *computing system* enables information processing and taking decisions. The computing system can be a combination of CPUs and GPUs on-board. The GPUs process information in parallel and are therefore faster for vision operations. Different computing units require networking with a fast protocol. The smaller robots rely on embedded boards, like Raspberry Pi or Arduino.

Box 1.1 Different hardware components in robots.

Sensors (vision):
    Cameras: Should have high shutter rate, not blur with robot in high motion
    Stereo Cameras: Can estimate depth using 2 cameras like the human eye
    3D cameras: Combine a depth image with colour image. Have low resolution; only work indoor, e.g. Intel RealSense, Microsoft Kinect, etc.
    Omnidirectional (360 degree) cameras
    Event-driven cameras: Report only difference between consecutive images and capture fast-moving objects.
    Optical flow sensors
    Thermal and infrared cameras: Take heat images, useful for seeing at night, sensing motion by heat, and can detect specific colours.
    Pan-tilt unit for cameras: Turn the camera to focus on the object of interest and expand the field of view.
Sensors (Proximity): Detect distance to object (and type of object) by hitting waves.
    SONAR: Use sound wave, low resolution, and short range.

Continued

Box 1.1 Different hardware components in robots—cont'd

Ultrasonic: Use high-frequency ultrasonic wave, large range, and reasonable resolution.

Radar: Use radio waves, short-range and long-range variants, and reasonable resolution.

Lidar: Use light waves. High resolution and expensive. 2D and 3D variants.

Sensors (Others):

Tactile: Bump (press of a button), Touch (measured by a change in capacitance/resistance/thermal sensing), and Force sensors (change of conductance under stress).

GPS: Latitude, longitude, and altitude measured using geostationary satellites, errors of a few metres. Differential GPS uses the difference in GPS reading with another base station GPS for better accuracy.

Inertial Measurement Unit (IMU): 3-axis accelerometer (measures acceleration), 3-axis gyroscope (measures angular speed), and 3-axis magnetometer (measures orientation with respect to the earth's poles).

Actuators:

Motors: DC motors, brushless DC motors (BLDC) motors, high-torque DC motors, servo motors (control position using feedback), micro servo motors, stepper motors, and motor gears.

Wheels: Castor wheel (spherical for support), omnidirectional wheels (can move in any direction), mecanum wheel (omnidirectional using differential speed of opposing wheels), wheel belts, and brakes.

Manipulators: Joints (with motors), links (arms), grippers for picking (vacuum, friction pads, and electrical fingers), and underactuated fingers (1 motor moves multiple fingers by common thread or gears).

Others: Buzzers and LEDs to communicate with other robots and indicate status to humans.

Computing system: CPUs and GPUs.

Communication system: Bluetooth, Wi-Fi, Zigbee, etc.

Power (e.g. Lithium-Ion batteries).

Human-Robot Interaction: push buttons, touch screens, and emergency stop.

The *communication system* enables robots to talk to other robots or the off-board computing system (computing cluster or cloud). The communication could be using Bluetooth, Wi-Fi, Zigbee, or any other channel. The range, speed, and cost are the major indicators of the communication protocol. Wi-Fi is generally preferred as most spaces typically have campus-wide Wi-Fi in place. The *power system* supplies power to all resources. Lithium-ion (Li-ion) batteries are a typical choice of power. Lithium-Polymer (Li-Po) is another type that has a limited application due to its exothermic nature and its risks of igniting. A battery bearing more weight will enable the robot to operate for a longer duration of time without requiring charging, which, however, increases the cost, size, and payload of the robot. The user interacts with the robot using systems like push buttons or touch screens, which are also parts of the robot hardware. An emergency stop button using both hardware and software is a common requirement. The sensors and actuators are the most interesting. Let us discuss those in greater detail.

1.3.1 Actuators

The actuators are used by the robot to cause a change in the physical world. The mobile robots are driven by wheels, which are powered by motors. The control systems operate the motors to navigate the robot. The *direct current (DC) motors* are commonly used with the current supplied by the batteries. When current is supplied, the motors rotate at a constant speed. Pulse-width modulation is used to switch on or off the motors to control the speed. A motor driver is used to control the current and voltage given to the motor to control the motor. The technique sends signals to the motors; for example, to operate at half the speed, alternating on and off signals are given. The motors can work in a bidirectional mode to make the robot move forwards and backwards. The *brushless DC* (BLDC) motors are more efficient than the native brushed motor. The brushless motors alternate the current supplied into the coils located outside to rotate a permanent magnet (and thus opposite in functioning to the brushed motor). The alternating ensures that the motor generates torque in the same direction. The motors have higher efficiency and power while requiring less maintenance. The high-torque motors are used in case the application requires carrying excess payloads, requiring more power. The motors may be *geared* to improve the output torque or speed in contrast to the input speed as per the application. The gear consists of teeth between two rotating devices, while the teeth interfere with each other. It translates the rotation motion of one device to the other with respect to the ratio of the number of teeth or the gear ratio.

The motors are fitted with *encoders* to determine how much the motor has turned. Knowing how much the wheels of all motors have turned enables to compute how much the robot has moved and ultimately the position of the robot. Most budget encoders consist of spikes as shown in Fig. 1.1. A transmitter is placed at one end (under the paper in the figure) of the spikes, and a receiver is at the other end (over the paper in the figure). As the wheel rotates, therefore, the receiver gets pulses. Counting the number of pulses, multiplied by the resolution of the encoders, gives the angular rotation of the wheel, which can be used to compute the linear movement of the wheel. These are called *incremental binary encoders*. The dual encoders, quad encoders, and higher-order encoders use multiple encoding schemes at multiple rims of the wheel, above and below each rim, for a high degree of accuracy. The *absolute encoders* are also possible. Here, every spike is of a different length. The receiver calculates the length of the spike from the time duration for which the signal was received (correspondingly, not received). Every length corresponds to a unique position of the wheel. This is especially good for manipulators where the task is to reach a specific position. Another manner of making encoders is by using *potentiometers* that fall under the category of the continuous and absolute encoding schemes. One end of an electrical probe is fixed outside the wheel, and another electrical probe is fixed on the wheel. As the wheel turns, the potential changes as per the position of the wheel, which is used to calculate the wheel position. The problem with encoders for a mobile robot is *slippage*. In case of a slip, there is a translation motion of the mobile robot, but no rotation of the wheel. The receiver does not receive pulses and therefore reports that no motion took place, while the robot slipped without causing a rotation of the wheel. Therefore, the encoders cannot be used to calculate the absolute position of the mobile robot.

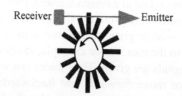

Receiver ← → Emitter

FIG. 1.1 Wheel Encoder. The position of the robot can be calculated from the rotation of every wheel/joint in the manipulator. The receiver receives spikes, and a count of spikes gives rotation. The wheels can also be fitted with an electrical probe, whose potential value changes as the wheel moves, useful for finding absolute rotation of a manipulator joint.

The other type of motor is the *servo motor*. The servo motors are used for precise positioning. Imagine a robotic hand or a robotic manipulator, where the aim is to place it at an exact angle to pick up an item of interest. The servo motors use encoders (potentiometers or incremental binary encoders) to calculate the precise position, which is used in a closed-loop control system to reduce the error and get to the specified position. Many times, the motors control fine parts of the robot, like the small fingers, eyelids of a humanoid, etc. The *micro servo motors* are the miniature version that provides power but uses less space. Servo motors are a better form of another type of motor, *stepper motor*, that is rarely used in robotic applications. The stepper motors are driven by electrical pulses, and the motors incrementally move based on the input pulses. Therefore, they should be ideal for positioning, like the manipulators, but the absence of any feedback mechanism renders them impractical.

The wheel is another characteristic of the robot. The mobile robot wheels should never slip or skid. Therefore, the wheels typically use high-quality rubber. The wheels are further ruggedized to handle harsh outdoor terrains, without showing wear. A *castor wheel* is a spherical wheel that is typically used for support. The wheel can be used to turn in any direction at any time and therefore does not inhibit motion. The omnidirectional wheels are increasingly getting popular. Imagine a mobile robot with a left and right wheel. It can go ahead, back, and turn clockwise and anticlockwise. It cannot go sideways. As the name suggests, the *omnidirectional wheels* remove this constraint and the wheels do not skid even if the motion is perpendicular to the wheel (or that the robot is going sideways). This happens as the rim of the wheel has passive wheels all over the rim, which can rotate perpendicular to the main wheel. Normally, these wheels are passive, but when the motion is perpendicular to the orientation of the wheel, these passive wheels show rotation to allow motion. Robots have multiple wheels in multiple orientations to allow motion at all angles. The *mecanum wheel* is also used for omnidirectional motion, except that the working principle for sideways motion is based on the differential of the opposing wheels, which causes the passive wheels to be active in the opposite direction. Another type of wheel uses belts for a greater hold in gravel and similar terrain.

The chassis is the physical base of a robot. Inside the base, the shaft is the physical rod over which the wheels are placed. The wheels are placed over a wheel bearing attached to the shaft, over which the wheel rotates. Standard mounts exist to fit the motors at the wheelbase, connected via screws. The different sensors and body parts are attached to the chassis through different brackets, which are custom designed as per the type and requirements of the robot. The designs are typically done in CAD before manufacturing. These days, it is equally common to have the designs 3D printed. An external brake may sometimes be useful to immediately stop the robot, just like in automobiles, except that the braking is automatically done as per the input speed profile.

The manipulators have similar motors at joints, and every joint is separated by links. The motors rotate to a specific angle to get the robotic hand in place. The *grippers* are a similar motor arrangement, imitating fingers. A finger may have many links connected by joints, and there may be many fingers in a hand. Ideally, each joint needs a motor of its own. Typically, the fingers are *under-actuated*, meaning that the number of motors is less than the number of joints. The same motor turns all links in a finger, or multiple fingers simultaneously. This can be done by passing the same thread pulled by the motor to move all fingers at once. While closing a fist, the different links in the fingers move at different angles. The geared mechanism can be used to move one finger link as a fraction of another.

Vacuum suction and friction pads are also used as an alternative to fingers. Vacuum suction can suck in objects and therefore eliminate the need to accurately place the hand. The friction pads hold the object against two pads, providing high friction. Therefore, the chances of an object slipping from the hand (requiring a re-grasp) are small. Other actuators include LED lights which indicate the status of humans or other robots, buzzers, etc.

1.3.2 Vision sensors

The sensors enable the robot to look at and decipher the physical world. The *camera* is one of the foremost sensors used. The cameras work on a pinhole mechanism, wherein the rich 3D world is projected onto a 2D surface and the result is in the form of an image. The cameras used for video conferencing and mobile phone cameras may not be ideal for robotics, as the camera itself may move with the robot and therefore the images may get blurry. The robotic cameras need to have a high shutter rate to eliminate motion blur. The depth information is, however, lost, which is important since the robot needs to reason out the depth of an object to pick up and the depth of an obstacle to decide how to avoid it.

The *stereo cameras* use two cameras to get the depth. The two cameras are kept parallel to each other, separated by a distance called the baseline. The depth is estimated using the principle of triangulation. The other way to get 3D vision is by using *3D cameras* like Microsoft Kinect, Intel RealSense, etc., which use a depth image camera along with a monocular camera and bind the two images. Even though they are affordable, the depth precision is low, and the cameras work for a limited range in indoor environments without sunlight.

A camera needs to sense the entire image and give it to the software, which can take a long time and limit the frequency of operation of the camera. The *event-driven cameras* are suited to capture fast motion. They only report the difference from the previous image, which can be done at the hardware level. Similarly, *optical flow vision sensors* are suited to report optical flow, which is computed based on the live feed at the sensor itself. The vision sensors are often put in a pan-tilt unit so that the camera base can be rotated to make the camera look at a specific point of interest. This is an alternate arrangement to having multiple cameras and stitching the images to create a full 360-degree panorama. The 360-degree vision can also be created by making 1 camera look vertically up at a convex mirror attached to the robot, while the mirror looks 360 degrees around the robot. This is only possible for smaller robots as the convex mirror should be high for far-end vision. The technology behind the 360-degree vision is getting increasingly sophisticated day by day.

The *thermal cameras* take thermal or heat images and can thus be used in the absence of visible light. There are application areas where thermal images are more informative. The *infrared camera* is another common option, which is more affordable and takes a response from the near infrared rather than visible light or thermal images (far infrared). The sensor can even be used to detect specific colours and characteristic objects based on the response to infrared. The *passive infrared sensor* does not emit any waves (unlike an active infrared camera) but only detects infrared energy from the environment. It can therefore be used to detect motion, where the body in motion will emit heat. This is used in automatic doors and lights.

1.3.3 Proximity and other sensors

The proximity sensors detect the distance from the sensor to the nearest object in the direction of the sensor. These sensors are therefore widely used to compute the distance from the obstacle, to rationally avoid it, and to know about the world. The proximity sensors send an electromagnetic beam that hits the object and comes back. The time of flight is used to measure the distance. The secondary wave and the wave characteristics can also be used to detect the type of material that the wave hit. The sound is one of the common carrier waves, called sound navigation and ranging or *SONAR sensor*. The problem with sound is that some materials absorb sound. Some others cause an echo with an angle that produces a secondary wave that can be hard to filter. The sound wave diffracts by some angle, and any obstacle within that angular range is reported. The sensor cannot detect objects far away. The *ultrasonic sensor* uses a separate frequency which makes it suitable for larger range obstacles. However, accuracy is still small. Most self-driving cars use ultrasonics to detect vehicles in front and back, used for automatic cruise control and collision avoidance. Radio waves are also common, giving rise to radio detection and ranging or the *RADAR sensor*. The wavelength can be varied for short-range or long-range detection. The radar transmitter and detector can spin on their axes to take a profile of the obstacles around them. The radar source can even be moved in all solid angles to get a complete depth image. Microwaves, radiowaves, or ultrasonic waves using the Doppler effect are also used for the detection of motion as motion sensors.

The most accurate (and costly) sensing is using light as the sensor, which makes the light detection and ranging or the *LiDAR sensor*. The laser transmitter can even rotate 360 degrees to get the entire 2D data of the obstacles around it. This is called a 2D laser scan. Having multiple such laser beams at multiple elevations can give a 3D map as shown in Fig. 1.2. This is referred to as the 3D lidar. The depth cameras combine a monocular RGB camera with depth, making depth images. Typically, they are low budget but give poor depth resolution. An image from such a camera is also given in Fig. 1.2.

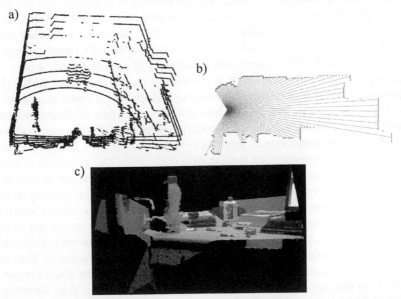

FIG. 1.2 Depth Images produced by (A) a low-resolution 3D laser scanner, (B) a 2D laser scanner, and (C) an RGB-Depth (RGBD) Camera. In a 2D laser scan, the laser rays are also shown.

The tactile sensors work based on touch. The simplest case is a button that is pressed when the robot collides and completes the circuit, called the *bump sensor*. The motors can then be switched off. The *touch sensor* is another variant. The touch is easy to detect as the human body emits heat that can be detected by thermal sensing. The fingers can even cause a change in the capacitance or resistance in the electrical field of the sensor, which is used to detect touch. The *force sensors* can be used to detect the force or pressure being exerted on the sensor. The sensor uses materials whose conductance changes in proportion to the stress exerted. Unlike other sensors, they give a continuous value of the force applied and not a binary signal. They are therefore commonly used to properly grasp the objects, neither too hard nor too soft.

The *Global Positioning System* (GPS) is used to give the latitude, longitude, and altitude of the body with respect to the earth's coordinate axis system. The sensor makes use of geostationary satellites whose position relative to the earth is fixed and known. The receiver receives signals and uses the same to calculate the distance from the satellites. By knowing the distances from 3 or more such satellites, the position can be estimated using triangulation. Construct spheres around the geostationary satellites with a radius equal to the estimated distance. The intersection of all spheres gives the position. The GPS gives errors of a few metres and does not work with thick walls that can block the GPS signals. The accuracy can be improved by using a *differential GPS* (DGPS) system, where 2 or more GPS receivers are used. The calculation is done based on the difference in readings from a stationary GPS fixed at a known position. The noises, therefore, get subtracted by taking a differential, and the position is calculated with respect to the static GPS base station. Similarly, the *real-time kinematic GPS* (RTK GPS) system uses a fixed base station along with a GPS receiver to calculate the position, while the fixed base station is used to correct the GPS signal readings. This gives a relative GPS system with an accuracy of around a centimetre.

The last commonly used sensor is the *inertial measurement unit* (IMU). It consists of a 3-axis accelerometer, a 3-axis gyroscope, and a 3-axis magnetometer. The *accelerometer* measures the acceleration in the 3 axes of the sensor. The sensor uses the property that the acceleration is a result of a force that causes stress that changes the conductance properties of some materials. This electrical response is used to measure the force at each of the axes. The *gyroscope* is used to measure the angular velocity along each of the three axes. The sensor uses the principle of conservation of angular momentum. Imagine a wheel rotating on its axis, which is surrounded by a mounting plane. The wheel keeps rotating until no external torque is applied. However, on the application of an external torque to the outer mounting, the principle of conservation of angular momentum shifts the orientation of the wheel in the opposite direction, which can be measured. The last component is a *magnetometer*. The instrument uses a magnet that always orients itself in the direction of the earth's pole. This can be used to compute the orientation of the robot. These sensors can get a lot of noise from the environment, due to which the sensor reading may not be reliable.

1.4 Kinematics

Robotics deals with multiple coordinate frames. The sensors and actuators have their coordinate frames, and they all share data and information. *Transformation* deals with converting data as observed from one coordinate frame into another coordinate frame (Craig, 2005; Fu et al., 1987). Imagine an external camera at some location seeing an object and reporting the same. It is important to know where the object is with respect to the centre of the robot so that the robot can avoid it, and this is done by using transformations.

The motion also is seen as migration from one coordinate frame to the other by the robot. The *kinematics* tells us the new pose of the robot and all its parts, given the previous pose of the robot and the control signals applied. Say a mobile robot is given a known linear and angular speed, the kinematics enables computation of the new pose of the robot. Similarly, with the manipulator. If each joint is rotated by a known amount, the new pose of the tool in the hand of the manipulator is determined by using kinematics.

1.4.1 Transformations

First let us consider a 3D world, in which a point p is located at position $(p_x, p_y, p_z)^T$ with respect to the XYZ coordinate axis system. Here the transpose is applied because the points are represented as column vectors. Suppose the point translates by an amount $\Delta p = (\Delta p_x, \Delta p_y, \Delta p_z)^T$ with respect to the same coordinate axis system, the new position p' is given by Eqs (1.1), (1.2). The translation is shown in Fig. 1.3A.

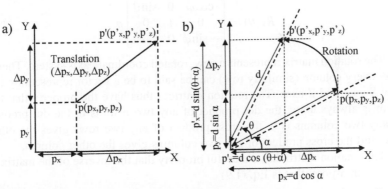

FIG. 1.3 Transformation of a point in 3D (A) Translation (B) Rotation. The new point coordinates can be derived using the fact that the distance from origin remains constant.

$$p' = p + \Delta p \tag{1.1}$$

$$\begin{bmatrix} p_x' \\ p_y' \\ p_z' \end{bmatrix} = \begin{bmatrix} p_x \\ p_x \\ p_x \end{bmatrix} + \begin{bmatrix} \Delta p_x \\ \Delta p_y \\ \Delta p_z \end{bmatrix} \tag{1.2}$$

Now consider that the point rotates by an amount θ along the Z-axis. As the rotation is only in the XY plane, the rotation is conveniently shown in Fig. 1.3B. From the figure, it can be calculated that the new values of X and Y coordinate values are given by Eqs (1.3), (1.4), while the Z-axis value remains unchanged due to this rotation.

$$p'_x = p_x \cos\theta - p_y \sin\theta \qquad (1.3)$$

$$p'_y = p_x \sin\theta + p_y \cos\theta \qquad (1.4)$$

The equations can be combined and written as Eqs (1.5), (1.6)

$$p' = R_Z(\theta)p = \begin{bmatrix} \cos\theta & -\sin\theta & 0 \\ \sin\theta & \cos\theta & 0 \\ 0 & 0 & 1 \end{bmatrix} p \qquad (1.5)$$

$$\begin{bmatrix} p_x' \\ p_y' \\ p_z' \end{bmatrix} = \begin{bmatrix} \cos\theta & -\sin\theta & 0 \\ \sin\theta & \cos\theta & 0 \\ 0 & 0 & 1 \end{bmatrix} \begin{bmatrix} p_x \\ p_y \\ p_z \end{bmatrix} \qquad (1.6)$$

$R_Z(\theta)$ is called the *rotation matrix*, representing rotation along the Z-axis by an amount θ. The axes are symmetric, and therefore rotation along the X-axis is represented by $R_X(\alpha)$ and given by Eq. (1.7), while the rotation along Y-axis is represented by $R_Y(\varphi)$ and given by Eq. (1.8):

$$R_X(\theta) = \begin{bmatrix} 1 & 0 & 0 \\ 0 & \cos\alpha & -\sin\alpha \\ 0 & \sin\alpha & \cos\alpha \end{bmatrix} p \qquad (1.7)$$

$$R_Y(\theta) = \begin{bmatrix} \cos\phi & 0 & \sin\theta \\ 0 & 1 & 0 \\ -\sin\phi & 0 & \cos\theta \end{bmatrix} p \qquad (1.8)$$

The rotation matrix represents a new, rotated coordinate axis system. Therefore, every column (similarly row) can be said to be a vector representing the new X-, Y-, and Z-axes. The rotation matrices thus have the property that the norm is always unity, the dot product of any two rows is 0, the dot product of any two columns is 0, the cross product of any two rows gives the other row, and the cross product of any two columns gives the other column.

The rotation matrix has the useful property that the inverse of the matrix is just the transpose, that is Eq. (1.9)

$$R^{-1} = R^T \qquad (1.9)$$

This property is particularly useful in situations where the inverse needs to be calculated. Image a point p' after rotation by an amount θ along the Z-axis gives a known point p. The point p' can now be computed as given in Eqs (1.10), (1.11)

$$p = R_Z(\theta)p' \qquad (1.10)$$

$$p' = R_Z^{-1}(\theta)p = R_Z^T(\theta)p \qquad (1.11)$$

In general, a body may not be constrained to rotate around the X-, Y-, or Z-axis only, but along any arbitrary axis. Let us aim to represent any general rotation as multiple rotations applied along with either of the X-, Y-, or Z-axis sequentially. These are called *composite rotations* or making multiple rotations to the same point. Imagine a point p first rotates by $R_X(\alpha)$. The new point is given by $R_X(\alpha)p$, which is then subjected to a rotation $R_Z(\theta)$. The point is now at $R_Z(\theta)R_X(\alpha)p$. Suppose after both rotations, another rotation is made of $R_Y(\varphi)$, giving the final point as Eq. (1.12).

$$p' = R_Y(\varphi)R_Z(\theta)R_X(\alpha)p \tag{1.12}$$

Let us now talk about physical bodies (like a robot) instead of points. The point p is a characteristic point in the body of the robot, which also has its coordinate axis system. Imagine a rectangular robot. Let the heading direction of the robot be the X' axis, and the Y' and Z' axes be perpendicular to the robot. $X'Y'Z'$ axis is called the *body coordinate frame* because the coordinate system is attached to the robot body that will move (translate and rotate) as the robot moves (translates and rotates). So far, we have only considered the cases where the rotations happen with respect to the *world coordinate axis system (XYZ)*. Many robot specifications may have the robot rotations specified in terms of its coordinate axis system, called the body coordinate axis system $(X'Y'Z')$. The two coordinate axis systems are shown in Fig. 1.4. A rotation of θ along the Z-axis (world coordinate frame) is not the same as a rotation of θ along Z' axis (body coordinate frame). We now study rotations specified in the body coordinate frame. Suppose the orientation of the body coordinate axis system with respect to the world coordinate axis system is represented by the rotation matrix R. Suppose the body rotates by θ along the Z' axis (body coordinate frame). The new body coordinate axis system is given by $R \times R_{Z'}(\theta)$, or a *postmultiplication* of the rotation. Extending to composite transforms, starting from

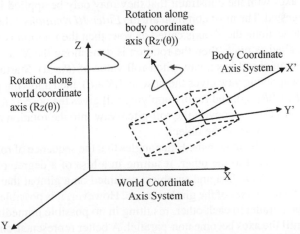

FIG. 1.4 Coordinate Systems. The orientation of the body is given by a 3×3 rotation matrix (R). Rotations specified in the world coordinate system are pre-multiplied (new pose $= R_Z(\theta)R$), and rotations specified in the body coordinate system are postmultiplied (new pose $= RR_{Z'}(\theta)$).

identity (I_3), if the body first rotates by $R_{X'}(\alpha)$ with respect to its own body (giving pose $I_3 R_{X'}(\alpha)$), then rotates by $R_{Z'}(\theta)$ with respect to its own rotated body (giving pose $I_3 R_{X'}(\alpha) R_{Z'}(\theta)$), and further rotates by $R_{Y'}(\varphi)$ with respect to its own body, the new point will be given by Eq. (1.13).

$$p' = I_3 R_{X'}(\alpha) R_{Z'}(\theta) R_{Y'}(\varphi) \tag{1.13}$$

Here $X'Y'Z'$ denotes the body coordinate frame at any time, which keeps rotating. I_3 is the 3×3 identity matrix denoting the initial orientation of the body with respect to the global coordinate system (or that the body is aligned to the world coordinate axis frame). The output is a rotation matrix as we are interested in the orientation of the body.

An interesting observation is about the representation of the position and orientation of a body. Suppose the body has a position (p_x, p_y, p_z). It means first translation in X by p_x, then in Y by p_y, and then in Z by p_z to get the position. Alternatively, translation in Z, followed by X, and then followed by Y, or any other order, gives the same point. This is because the translation uses the addition of the translation matrices, which follows the commutative property. However, the rotation of the body cannot be represented by rotations along 3 axes irrespective of the order, because the rotation uses multiplication of matrices, which is not commutative. Take an object and rotate it along the X-axis by 90 degrees (by holding the object with both fingers along the X-axis and rotating) and then rotate it along the Y-axis by 90 degrees. Now again reset the object to its original position, and this time it rotates along Y and then X. The appearance of the object shall be different.

This gives rise to the problem of *representation of orientation*. A mechanism is obviously needed to represent the orientation of bodies as a 3×3 *rotation matrix*, which is used in all calculations. This is computationally reasonable, however, not intuitive for humans. An alternate is to represent orientations along the 3 axes with the constraint that they may only be applied in the same order as suggested. The most common are the *Euler-III rotations*, where the first rotation is done along the Z'-axis called as *yaw*, then the rotation is done along the Y'-axis called as *pitch*, then the rotation is done along the X'-axis called as *roll*. Suppose an object has a yaw-pitch-roll value of (θ, φ, α). Starting from an identity, first applying yaw gives the pose as $I_3 R_{Z'}(\theta)$, then applying pitch gives the pose as $R_{Z'}(\theta) R_{Y'}(\varphi)$, and finally applying roll gives the pose as $R_{Z'}(\theta) R_{Y'}(\varphi) R_{X'}(\alpha)$, which is used to convert the roll-pitch-yaw into the rotation matrix representation of the pose of the object.

There is one problem with the Euler angles that the sequence of rotations can make an axis parallel to the other, resulting in a loss of a degree of freedom, called the gimbal lock. Imagine an object mounted on a gimbal that can rotate as either of the three axes of the gimbal rotate. However, it is possible for 2 gimbal axes to be parallel to each other, resulting in no possible immediate motion at an axis, until the axes become non-parallel. A better representation technique is by using 4 entities for representation rather than 3, called the *quaternions*.

Consider a quaternion representing the orientation given by $(q_0, q_1, q_2, q_3)^T$. The quaternion can be represented more visually as Eq. (1.14).

$$q = q_0 + q_1 i + q_2 j + q_3 k \qquad (1.14)$$

As an extension to the theory of complex numbers, the fundamental law of quaternion arithmetic has the following relations: $i^2 = j^2 = k^2 = ijk = -1$. Further, $ij = -ji = k$. The three together represent a 3 axis coordinate system with the axes as $i, j,$ and k. Consider a vector from the origin to the point (q_1, q_2, q_3) in this coordinate axis system and rotate it by an angular value of q_0. The rotation obtained can be used to visualize the representation of the quaternion, shown in Fig. 1.5.

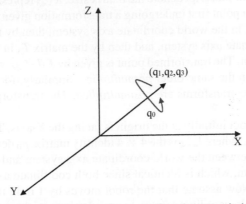

FIG. 1.5 Representation of orientation using quaternions (q_0, q_1, q_2, q_3). (q_1, q_2, q_3) represents a vector around which a rotation of q_0 is applied.

Currently, translation is represented as a matrix addition, while rotation is represented as a matrix multiplication. Any sensor reading in one coordinate axis system must be *transformed* into another coordinate axis system by using both translation and rotation. Similarly, a robot is transformed from one coordinate axis system to another as the robot moves, consisting of both translation and rotation. Let us therefore represent translation as a matrix multiplication instead of a matrix addition by increasing the dimensionality by 1 called *homogenization*. Eq. (1.2) can now be homogenized and written as Eq. (1.15).

$$\begin{bmatrix} p_x' \\ p_y' \\ p_z' \\ 1 \end{bmatrix} = \begin{bmatrix} 1 & 0 & 0 & \Delta p_x \\ 0 & 1 & 0 & \Delta p_y \\ 0 & 0 & 1 & \Delta p_z \\ 0 & 0 & 0 & 1 \end{bmatrix} \begin{bmatrix} p_x \\ p_x \\ p_x \\ 1 \end{bmatrix} \qquad (1.15)$$

Similarly, to match dimensionality, Eq. (1.6) can also be homogenized by adding 1 dimension, given by Eq. (1.16).

$$\begin{bmatrix} p_x' \\ p_y' \\ p_z' \\ 1 \end{bmatrix} = \begin{bmatrix} \cos\theta & -\sin\theta & 0 & 0 \\ \sin\theta & \cos\theta & 0 & 0 \\ 0 & 0 & 1 & 0 \\ 0 & 0 & 0 & 1 \end{bmatrix} \begin{bmatrix} p_x \\ p_y \\ p_z \\ 1 \end{bmatrix} \qquad (1.16)$$

The rotation and translation can now be combined as one transformation matrix by increasing the dimensionality by 1 or homogenization. The resultant relation is given by Eqs (1.17), (1.18).

$$p'_{4\times 1} = T_{4\times 4} p_{4\times 1} \qquad (1.17)$$

$$
\begin{bmatrix} p_x' \\ p_y' \\ p_z' \\ 1 \end{bmatrix} = \begin{bmatrix} R_{3\times 3} & \Delta p_{3\times 1} \\ 0_{1\times 3} & 1 \end{bmatrix} \begin{bmatrix} p_x \\ p_y \\ p_z \\ 1 \end{bmatrix} = \begin{bmatrix} & R_{3\times 3} & & \Delta p_x \\ & & & \Delta p_y \\ & & & \Delta p_z \\ 0 & 0 & 0 & 1 \end{bmatrix} \begin{bmatrix} p_x \\ p_y \\ p_z \\ 1 \end{bmatrix} \qquad (1.18)
$$

The numbers in subscript denote the matrix size. $R_{3\times 3}$ represents the rotation matrix. Imagine a point first undergoing a transformation given by the transformation matrix T_1 in the world coordinate axis system, then by the matrix T_2 in the world coordinate axis system, and then by the matrix T_3 in the world coordinate axis system. The transformed point is given by $T_3 T_2 T_1 p$, which has all the transformations in the same order *premultiplied*. Similarly, for the body coordinate system, the transforms are *postmultiplied*. The transforms are summarized in Box 1.2.

Consider a robot initially at the origin is facing the X-axis. The pose is thus given by $p_0 = I_{4\times 4}$, where $I_{4\times 4}$ is the 4 × 4 identity matrix. p_0 denotes the initial transformation between the world coordinate axis system and the body coordinate axis system, which is identical since both coordinate axis systems initially coincide. Now assume that the robot moves by 1 m in the X'-axis with respect to the body coordinate frame (currently same as the world coordinate frame). The pose will now be given by $p_1 = p_0 [I_{3\times 3} [1\,0\,0]^T | 0\,0\,0\,1]$. Here $I_{3\times 3}$ is the rotation matrix (identity in the example) and $[1\,0\,0]^T$ is the translation vector. All transformation matrices have a [0 0 0 1] added as the last row, which will be represented as "| 0 0 0 1" to represent the transformation matrix as a single line.

The robot now rotates by $-\pi/2$ along the Z' axis with respect to the body coordinate frame, giving the new pose as $p_2 = p_1 R_Z(-\pi/2)$, where $R_Z(-\pi/2)$ is the 4 × 4 homogenized rotation matrix. Now the robot goes ahead by 1 m in the X' axis with respect to the body coordinate frame. The new position is given by $p_3 = p_2 [I_{3\times 3} [1\,0\,0]^T | 0\,0\,0\,1]$. Now there is a wind blowing in the X-axis of the world coordinate frame, which pushes the robot by 1 m along the X-axis, that gives the new pose of the robot as $p_4 = [I_{3\times 3} [1\,0\,0]^T | 0\,0\,0\,1] p_3$. Since the transformation is with respect to the world coordinate frame, it will be pre-multiplied. The robot now turns by $\pi/2$ along the Z'-axis with respect to the body coordinate frame, giving a new pose as $p_5 = p_4 R_Z(\pi/2)$. Finally, the robot moves sideways along the Y'-axis with respect to the body coordinate frame, giving the new pose as $p_6 = p_5 [I_{3\times 3} [0\,1\,0]^T | 0\,0\,0\,1]$, which is the new pose of the robot in the world coordinate frames shown in Fig. 1.6.

Box 1.2 Different notations used with transforms.

Transforms

T_a^b: Transformation from coordinate axis system a to coordinate axis system b. Equivalently, coordinate axis system a as observed from the coordinate axis system b (4×4 matrix)

$T_a^b = (T_b^a)^{-1}$

R_a^b: Rotation from coordinate axis system a to coordinate axis system b (3×3 matrix). Equivalently, coordinate axis system b after rotation by R_a^b becomes coordinate axis system a.

$t_a^b(t_{a,x}^b, t_{a,y}^b, t_{a,z}^b, 1)^T$: Translation from coordinate axis system a to coordinate axis system b (3×1 matrix). Equivalently, coordinate axis system b after rotation by t_a^b becomes coordinate axis system a.

$$T_a^b = \begin{bmatrix} R_a^b 3 \times 3 & t_a^b 3 \times 1 \\ 0_{1 \times 3} & 1 \end{bmatrix} = \begin{bmatrix} & & & t_{a,x}^b \\ & R_a^b 3 \times 3 & & t_{a,y}^b \\ & & & t_{a,z}^b \\ 0 & 0 & 0 & 1 \end{bmatrix}$$

$p(p_x, p_y, p_z, 1)^T$: Point in the world coordinate frame (4×1 matrix)
$p_a(p_{a,x}, p_{a,y}, p_{a,z}, 1)^T$: Point in the coordinate frame a (4×1 matrix)
$p_b(p_{b,x}, p_{b,y}, p_{b,z}, 1)^T$: Point in the coordinate frame b (4×1 matrix)

$p_b = T_a^b p_a$

P: Pose (position and orientation) in the world coordinate frame (4×4 matrix)
P_a: Pose in the coordinate frame a (4×4 matrix)
P_b: Pose in the coordinate frame b (4×4 matrix)

$P_b = T_a^b P_a$

Image

T_W^{Cam}: Transformation from world to camera coordinate frame (in camera coordinate frame, Z points in the line of sight, X is horizontal, and Y is vertically down).

$T_{Cam}^W = (T_W^{Cam})^{-1}$ is called as the pose of the camera in the world.

$p_{Cam} = T_W^{Cam} p$: Point p in camera coordinate frame.

$$C = \begin{bmatrix} f_x & 0 & c_x \\ 0 & f_y & c_y \\ 0 & 0 & 1 \end{bmatrix}: \text{Camera Calibration Matrix}$$

$$\begin{bmatrix} u'_l \\ v'_l \\ w \end{bmatrix} = C p_{Cam} = C T_W^{Cam} p$$

$\pi(p_{Cam}) = (u_l, v_l)^T = (u'_l/w, v_l = v'_l/w)$: **point viewed in the image.** u_l is the horizontal (left to right) X value, v_l is the vertical (top to bottom) Y value, and top-left corner is the origin.

Calibration

Extrinsic calibration: Find T_b^a

Let point p_a observed by sensor a be observed as p_b^{obs} by sensor b (with noise)
$T_a^b = \text{argmin} T_a^b \sum_{p_a} \|p_b^{obs} - T_a^b p_a\|_2^2$, $\|.\|_2$ is the Euclidian norm

Intrinsic Calibration: Find C (Chessboard calibration method)

Show chessboard images. The detected corners make a local coordinate frame with known coordinate values. Find C such that the projected corners and observed corners in the image are as close as possible.

Continued

Box 1.2 Different notations used with transforms—cont'd

Extrinsic Calibration for camera: Find T_W^C (or find the pose of the camera in the world T_C^W)

Let a known point p be observed at $(u_{l,p}^{obs}, v_{l,p}^{obs})$ in the image

$$T_W^C = argmin \sum_p \left\| \left(u_{l,p}^{obs}, v_{l,p}^{obs}\right)^T - \pi\left(T_W^{Cam}p\right) \right\|_2^2$$

$$p_{Cam} = T_W^{Cam}p, \begin{pmatrix} u_l' \\ v_l' \\ w \\ 1 \end{pmatrix} = Cp_{Cam}, \begin{pmatrix} u_{l,p}^{proj} \\ v_{l,p}^{proj} \end{pmatrix} = \begin{pmatrix} u_l'/w \\ v_l'/w \end{pmatrix}, \pi(p_{Cam}) = \begin{pmatrix} u_{l,p}^{proj} \\ v_{l,p}^{proj} \end{pmatrix}$$

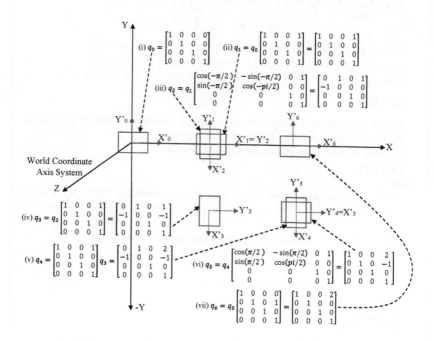

FIG. 1.6 Transformations. The robot is moved by giving numerous transformations. X_i' denotes the X-axis of the robot after transformation I, and q_i denotes the pose after transformation i.

1.4.2 TF tree

Imagine a multirobotic system, where each robot has multiple sensors and actuators. All the sensors and actuators have their coordinate axis systems. Every coordinate axis system is related to every other coordinate axis system using a transformation matrix denoted by T_a^b. T_a^b is the transformation matrix from the coordinate axis system a to the coordinate axis system b. Therefore, a point with the pose (position and orientation as a 4×4 matrix) p_a in coordinate axis system a will be seen at pose $p_b = T^b_a p_a$ in coordinate axis system b. Correspondingly, $p_a = T^a_b p_b$. Alternatively, T^b_a represents the pose of the coordinate axis system a when seen in the frame of reference of coordinate axis system b.

From the definition, $T^b_a = (T^a_b)^{-1}$, and hence only one of the transformations (T^a_b or T^b_a) may be stored for facilitating both-way conversions. There is a trick in the calculation of the inverse of such coordinate axis system transforms. The transform T^a_b has a rotation component (R^a_b) and a translation component (t^a_b). The rotation matrix has the property that $R^b_a = (R^a_b)^{-1} = (R^a_b)^T$. The translation component needs to be calculated $t^b_a = (t^a_b)^{-1}$, which is the origin of coordinate axis system a as seen from the coordinate axis system b. The inversion of the position t^a_b to account a change in observing a from b instead of b from a as previously is $-t^a_b$, assuming both coordinate axis systems do not have a rotation between them. Further adding rotation makes the inversion as $-R^b_a t^a_b = -(R^a_b)^T t^a_b$. The inversion of a transformation matrix is hence given by Eqs (1.19), (1.20)

$$T^a_b = \begin{bmatrix} R^a_{b3\times3} & t^a_{b3\times1} \\ 0 \quad 0 \quad 0 & 1 \end{bmatrix} \tag{1.19}$$

$$T^b_a = \begin{bmatrix} R^b_{a3\times3} & t^b_{a3\times1} \\ 0 \quad 0 \quad 0 & 1 \end{bmatrix} = (T^a_b)^{-1}$$

$$= \begin{bmatrix} (R^a_{b3\times3})^T & -(R^a_{b3\times3})^T t^a_{b3\times1} \\ 0 \quad 0 \quad 0 & 1 \end{bmatrix} \tag{1.20}$$

The transformation may be static. As an example, the transformation between the location of the centre of rotation of the robot and the lidar sensor is static. This can be computed by using the CAD model of the robot used to manufacture the robot. Similarly, the transformation between the centre of rotation of the robot and the wheels is static. The transformations may also be dynamic as a mobile robot moves, its transformation with the world keeps changing.

The transformations may be computed by a process of *calibration*. Say the transformation between a lidar sensor and a stereo camera, both attached to a robot, needs to be done. Show a point object to both sensors and compute the position of the object as computed by the 2 sensors. The object is at one position only in the world, which can be used to compute the transformation. In other words, if $p = [p_x, p_y, p_z, 1]^T$ is the actual position of the object in the 3D world in the world coordinate axis system, its position as seen by the lidar sensor will be given by $p^{proj}_L = T^L_W p$, where the subscript in point denotes its coordinate axis system (unless it is a world coordinate axis system), T^L_W denotes the transformation matrix from the world to the lidar, W is used to denote the world coordinate frame, and *proj* means that the point is projected. T_L^W is the pose of the lidar in the world (which is the same as the transformation from lidar to world) that one can measure with a measuring device, noting the distance from the origin along the 3 axes, and the rotation. $T_W^L = (T_L^W)^{-1}$ is the matrix used here. Since such a measurement in the world is rarely possible, the same matrix is obtained by calibration. Suppose the point is observed at a position p_L^{obs} (which may be different due to noise). The error $\|p_L^{obs} - p_L^{proj}\|_2^2 = \|p_L^{obs} - T_W^L p\|_2^2$ should be simultaneously minimized at many points to get the transformation between the world and lidar T_W^L, or Eq. (1.21)

$$T_W^L = \arg\min_{T_W^L} \sum_p \left\| p_L^{obs} - p_L^{proj} \right\|_2^2 = \arg\min_{T_W^L} \sum_p \left\| p_L^{obs} - T_W^L p \right\|_2^2 \quad (1.21)$$

The inverse T_L^W is the true pose of the lidar in the world. Here $\|.\|_2$ is the Euclidean norm of the vector. Unfortunately, the position of such points p in the world is not known, which is an input in the current equation. Therefore, let us transform from the lidar to the stereo. Let p_L be the position of the point as seen by the lidar (which will be the true position in the lidar coordinate axis system with some noise). Let p_S^{proj} be the position of the same point in the stereo coordinate axis system given by $p_S^{proj} = T_L^S p_L$, where *proj* means a projection. The same point as observed by the stereo is given by p_S^{obs} (which will be the true position in the stereo coordinate axis system with some noise). This means the error $\|p_S^{obs} - T_L^S p_L\|_2^2$ should be minimized simultaneously for many points p_L. The value of T_L^S that gives the smallest error for all observed point pairs in the stereo and lidar images is stored as the true transformation from the lidar to the stereo sensor.

The multirobot, multisensor, and multiactuator system will have such a transformation relationship between every pair of sensors, actuators, and robots. The transforms are related to each other as a composite transform obtained by multiplication. So, a transform from sensor to world (T_{sensor}^W) is the transform from robot to world (T_{robot}^W), times the transform from sensor to robot (T_{sensor}^{robot}) or $T_{sensor}^W = T_{robot}^W T_{sensor}^{robot}$. Hence out of the three transforms, only two need to be stored, while the third can be derived from the relation. Therefore, a tree is enough to store all transforms. The *transformation tree* (or *TF tree*) has all the coordinate systems as the nodes, and an edge existing between two nodes stores the transformation matrix between the nodes (Foote, 2013). Suppose a user wants the transform between the coordinate system a (say 1st robot's lidar) and b (say 2nd robot's centre of rotation). A single path exists between the two nodes a and b in the tree, and multiplication of all transforms in the edges of the path gives the transformation between the two queried nodes. It must be noted that even though a graph is possible to store all transforms, however, a graph can have multiple paths and thus multiple transforms between the queried nodes, thus leading to an anomaly if all transforms from different paths do not match. A synthetic TF tree is shown in Fig. 1.7.

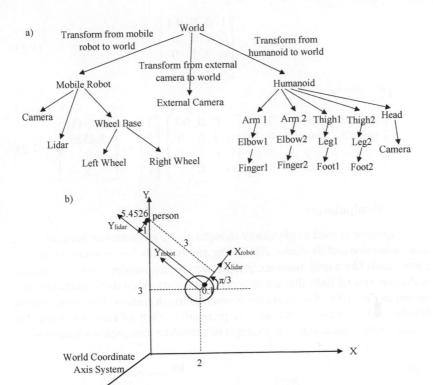

FIG. 1.7 Transformation Tree. (A) Every vertex represents a coordinate axis system, and edges store the transformation between the systems (not all shown in the figure). To get transform between two vertices, multiply all transformation matrices on the path between the vertices. (B) Example to calculate the position of the person in the real world based on known robot pose and lidar position.

Let us take a small example shown in Fig. 1.7B. Suppose a mobile robot is at position (2,3) oriented by $\pi/3$ rad (produced by a rotation of $\pi/3$ rad with respect to the Z-axis), using the centre of the differential wheel drive system as the characteristic point. The robot has a lidar scanner placed ahead of the characteristic point along the robot's heading direction (X-axis of the robot coordinate system) by 0.1 m. To make matters simple, assume no rotation between the robot and lidar coordinate axis systems. The lidar scanner sees a person at the location (1,3,2) in its coordinate axis system. This can be represented as a TF tree with nodes world (W), robot, and lidar. The position of the person in the real world is given by Eqs (1.22), (1.24).

$$T_{robot}^{W} = \begin{bmatrix} \cos(\pi/3) & -\sin(\pi/3) & 0 & 2 \\ \sin(\pi/3) & \cos(\pi/3) & 0 & 3 \\ 0 & 0 & 1 & 0 \\ 0 & 0 & 0 & 1 \end{bmatrix} \qquad (1.22)$$

$$T^{robot}_{lidar} = \begin{bmatrix} 1 & 0 & 0 & 0.1 \\ 0 & 1 & 0 & 0 \\ 0 & 0 & 1 & 0 \\ 0 & 0 & 0 & 1 \end{bmatrix} \tag{1.23}$$

$$p_W = T^W_{robot} T^{robot}_{lidar} p_{lidar}$$

$$= \begin{bmatrix} \cos\pi/3 & -\sin\pi/3 & 0 & 2 \\ \sin\pi/3 & \cos\pi/3 & 0 & 3 \\ 0 & 0 & 1 & 0 \\ 0 & 0 & 0 & 1 \end{bmatrix} \begin{bmatrix} 1 & 0 & 0 & 0.1 \\ 0 & 1 & 0 & 0 \\ 0 & 0 & 1 & 0 \\ 0 & 0 & 0 & 1 \end{bmatrix} \begin{bmatrix} 1 \\ 3 \\ 2 \\ 1 \end{bmatrix} = \begin{bmatrix} -0.0481 \\ 5.4526 \\ 2 \\ 1 \end{bmatrix} \tag{1.24}$$

1.4.3 Manipulators

A manipulator is used to physically manipulate the environment because of the arm (Sciavicco and Siciliano, 2000). The arm typically has an end effector that carries tools like a drill, hammer, gripper, etc., for the manipulation. The manipulator has several *links* that are connected by *joints*. A synthetic manipulator is shown in Fig. 1.8A. The joints are actuated through motors. The joints may be *revolute*, which rotate on the axis, or *prismatic*, which go back and forth. The turning of the human arm is an example of a revolute joint, while a beam coming

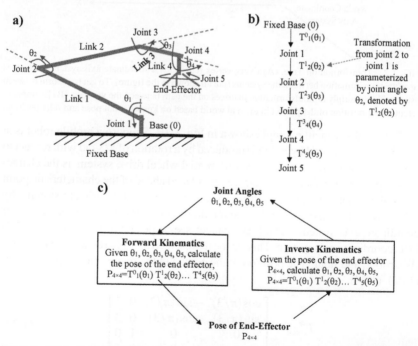

FIG. 1.8 Manipulator (A) Links and joints (B) Kinematic Chain (C) Forward and Inverse Kinematics. Inverse kinematics enables to place the motors at desired joint angles to grasp an object at a known pose.

out of a machine and hitting the object is an example of a prismatic joint. The prismatic joints are also actuated by geared motors, where the gear mechanism changes rotation into translation. The revolute joints are currently more common.

First let us compute the transformation matrix between every pair of consecutive links. Ideally, there are 3 translations and 3 rotations that can happen between every pair of consecutive links. However, the transformation is stated in terms associated with the manipulators specifically. Every consecutive link is connected by a joint that has one motor. As the motor moves, the following link moves, which changes the transformation. The transformation is thus a function of the current motor position or the joint angle. To compute the transformations, we need to fix the coordinate axis systems for all links. The commonly followed norms are the *Denavit Hartenberg parameters* (DH parameters) that standardize and metricize the transformations. Let $T_i^{i-1}(\theta_i)$ denote the transformation matrix from the links i to the link $i - 1$, which depends on the joint angle θ_i (considering all revolute joints only).

The *forward kinematics* of a manipulator is used to compute the pose of the end effector of the manipulator, given a knowledge of all the joint angles. The joint angles are computable by using encoders or potentiometers on the motors. The operation is intuitive. The base of the manipulator is assumed to be in a known pose. Given knowledge of the first joint, the transformation matrix between the base and the first link (T_1^0) can be computed. Given knowledge of the second joint, the transformation matrix between the first link and the second link (T_2^1) can be computed, which can be used to get the transform between the base and the second link by postmultiplication as $T_2^0 = T_1^0 T_2^1$. Given a knowledge of the third joint, the transformation between the second and the third link (T_3^2) can be computed, which further gives the pose of the third link relative to the base as $T_3^0 = T_2^0 T_3^2 = T_1^0 T_2^1 T_3^2$. In this way, we can get the pose of the end effector with respect to the base (T_n^0) by transforming from one link to the other by postmultiplication. T_n^0 represents a transformation from the end effort to the manipulator base. For an n-link manipulator, the pose of the end effector with respect to the base (T_n^0) is given by Eqs (1.25), (1.26)

$$T_n^0 = T_1^0(\theta_1)T_2^1(\theta_2)T_3^2(\theta_3)\dots T_n^{n-1}(\theta_n) = \Pi_{i=1}^n T_i^{i-1}(\theta_i) \qquad (1.25)$$

$$\begin{bmatrix} R_{3\times3} & (p_n^0)_{3\times1} \\ 0_{1\times3} & 1 \end{bmatrix} = T_1^0(\theta_1)T_2^1(\theta_2)T_3^2(\theta_3)\dots T_n^{n-1}(\theta_n) = \Pi_{i=1}^n T_i^{i-1}(\theta_i) \qquad (1.26)$$

Note that the pose is a transformation matrix from which the position (p_n^0) and quaternion angles can be extracted. The *kinematic chain* of the manipulator is like the TF tree for the specific problem of manipulation and specifies how the joints are connected by transforms. The kinematic chain is shown in Fig. 1.8B. The vector $(\theta_0, \theta_1, \dots \theta_n)^T$ is known as the *joint configuration* of the manipulator since it encodes the condition of every joint. The *forward kinematics* hence converts a joint configuration into the pose of the end effector in the real world or Cartesian coordinate axis system, shown in Fig. 1.8C.

A derivative of the transformation with respect to time converts the joint velocity (rate of change of every joint angle) into a linear velocity (a derivative of position) and angular velocity (a derivative of rotation) of the manipulator end effector in the real world (Cartesian coordinate system), which is also known as the *manipulator Jacobian*. The Cartesian speed (derivative of the position of the end effector) is Jacobian times the joint speed (angular speed of each joint). Similarly, the joint speed is inverse of Jacobian times the Cartesian speed. Therefore, if the determinant of the manipulator Jacobian becomes zero at a joint configuration, the joint velocity tends to infinity for a finite Cartesian speed, and the situation is called a *singularity*. The planning of manipulators is done to avoid such joint configurations. Many times, people hold the hand of the manipulator and move it, which gives a Cartesian speed at every point, but the motors should move the end effector to give the same speed. If the motion goes through a singularity point, the joint speeds will have to go infinite for a continuous motion, which is not possible. Similarly, if planning the trajectory of the end effector to navigate the object in hand in a characteristic motion, the planned motion of the end effector should be reflected by a motion of the joints, which must not pass through a point of singularity.

A common problem in robotics is solving the inverse problem. The robot may need to pick up an object like a ball. Image sensors detect the ball. Proximity sensors enable them to determine the location of the ball. Computer vision algorithms state which places to grasp the ball, which is all used to compute the pose (position and rotation) that the end effector should reach to grasp the ball. However, the joint angles need to be known that make the manipulator reach the ball physically. The desired joint angles can then be reached by the control algorithms for each motor. Similarly, to drill, the pose (position and orientation) of the drilling site is known, but not the joint angles. The forward kinematics converts the joint angles into the pose of the end effector. On the contrary, *inverse kinematics* converts the pose of the end effector into joint angles. The problem is hard, and there may not be a closed-form solution. The hardest aspect is that there may be multiple valid solutions known as *kinematic redundancy*. Imagine a 3-link planar link that needs to reach a point. It could be done elbow up or elbow down, and both are valid solutions as shown in Fig. 1.9. The end effector pose has 6 degrees of freedom (3 for position and 3 for rotation), and therefore if the manipulator has more than 6 joints, there will be redundancy in the solution. The inverse kinematics is solved using geometrical approaches, optimization approaches, and sometimes learning approaches. A small intuitive mechanism is that the human hand uses the initial links (arms) to reach the object and the later ones (wrist) to orient as per need. Many manipulators are human inspired and have the same principle, which can be used as a heuristic to get solutions. The initial links are used to get the position, and the last ones to get the orientation of the end effector.

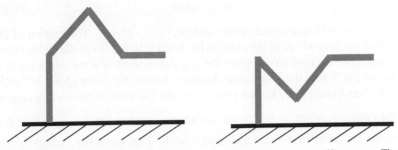

FIG. 1.9 Kinematic Redundancy. Multiple joint angles lead to the same end-effector pose. Therefore, inverse kinematics can have many solutions.

1.4.4 Mobile robots

Let us now repeat our understanding of kinematics for wheeled mobile robots. Let the mobile robot be like a car (Siegwart et al., 2004). Suppose the position of the centre of the rear wheels is given by (x,y) while the orientation of the vehicle is itself given by θ with respect to the world coordinate axis system. Suppose the vehicle rotates with the help of the steering which is directly connected to the front wheels that rotate. Suppose λ is the steering angle relative to the heading of the vehicle. The distance between the rear and the front wheels is given by L. Let $v_x = dx/dt$ and $v_y = dy/dt$ denote the linear speeds in the 2 axes. Let $\omega = d\theta/dt$ denote the angular speed. The notations are shown in Fig. 1.10.

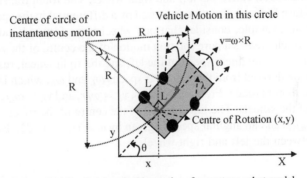

FIG. 1.10 Notations used for deriving the kinematics of a car-type robot model.

The vehicle cannot move sideways, which is a fundamental constraint in such robots given by Eq. (1.27)

$$\frac{v_y}{v_x} = \tan\theta \tag{1.27}$$

Therefore, the system is modelled as Eqs (1.28), (1.29)

$$v_x = v\cos\theta \tag{1.28}$$

$$v_y = v \sin \theta \tag{1.29}$$

Here v is the linear speed of the vehicle, $v = \sqrt{v_x^2 + v_y^2}$. The motion of the centre of the rear wheel of the vehicle for fixed steering is circular with radius R (suppose). The time to complete the angular motion of a complete circle is $2\pi/\omega$, which is also the total linear distance divided by linear speed or $2\pi R/v$, which gives a commonly known result $v = \omega R$. The vehicle travels as a tangent to the circle, considering which we can approximate as $\tan \lambda = \dfrac{L}{R}$, where L is the distance between the rear and front wheels. This gives Eq. (1.30)

$$\omega = \frac{v}{R} = \frac{v}{L} \tan \lambda \tag{1.30}$$

The vehicle is controlled using the linear speed v, which is generally bound to the interval $[-v_{min}, v_{max}]$, and the steering, which is also bound to an interval $[-\lambda_{max}, -\lambda_{max}]$. The equations are the same for a car, which is technically known as the *dual bicycle model* and obtained for a *tricycle model*. The *Reeds-Shepp Car* is a special type of car where the linear speed is restricted to only 3 values $-v_{max}$, 0, and v_{max}. This helps in planning as shall be seen in the later chapters. The *Dubins Car* is an even simpler model where the vehicle can only move forward with a speed of v_{max} (or stay stationary with speed 0).

A similar model is the *differential wheel drive model*, where there are two wheels on the robot, called the left and right wheel. The robot translates when both robots turn forward. Rotation is caused by a differential in the speeds of the wheels. As an example, consider that one wheel turns forward and the other backward, which creates a pure rotation motion of the centre of the robot. Suppose v_L and v_R are the linear speed of the left and the right wheel, respectively. The speeds are derived from their angular speeds ω_L and ω_R, which is the speed at which the motors rotate. From kinematics, $v_L = \omega_L r$, and $v_L = \omega_L r$, where r is the radius of the wheels. The linear speed of the centre of the robot is given by $v = (v_L + v_R)/2$, and an angular speed is given by $\omega = (v_R - v_L)/2l$. Here l is the distance between the left and right wheels.

1.5 Computer vision

The primary sensor of interest is a camera, and therefore let us have a quick primer on the vision that will help us understand the rest of the book. The image is a 2-dimensional array, and extracting as much information out of it as possible constitutes the problem of vision (Szeliski, 2011). The overall problem is described in Fig. 1.11. A summary of different terms is given in Box 1.3.

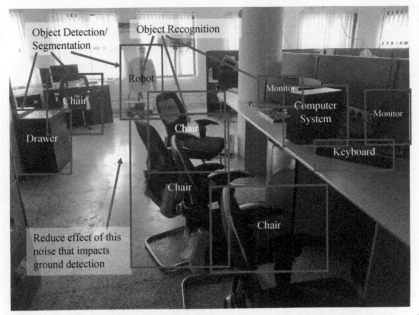

FIG. 1.11 Computer vision: A robot removes noise, detects, and recognizes different objects of common use. Stereo correspondence can be used to also extract 3D information.

Box 1.3 Summary of vision terminologies.

Pre-processing

    Colour Spaces: For light invariance, the light should preferably be an independent component neglected in vision. Common spaces: red green blue (RGB), hue saturation value (HSV), Luma Component blue Component red (YCbCr), and Luma-AB (LAB).

    Noise Removal: Median filters, mean filters, Gaussian filters, statistical filters, and histogram equalization.

Segmentation and Detection:

    Edge based: Different objects separate by visible edges (difference in colours), e.g. Canny edge detector, Sobel edge detector, and grow cut (flood fill object till colour is same).

    Clustering: Colour-based clustering and spatial clustering.

    Morphological Analysis: Erosion (shrink to remove small objects) and dilation (expand to fill gaps).

    Recognition based: Place multiple candidate windows in X, Y, and scale. If an object is recognized in the window, the object exists.

Machine Learning:

    Supervised Learning learns a model to match input to known targets. Binary classifiers have 2 targets only (0 or 1). Multiclass classifiers use multiple binary classifiers.

Continued

Box 1.3 Summary of vision terminologies—cont'd

Learning: Change model parameters to best match prediction to targets.

Models: Artificial neural networks (ANN), support vector machines (SVM), decision trees, bagging (multiple models learnt on different data), adaptive boosting (multiple models, a model learns on data on which other models perform worst to increase the diversity of models), and deep learning (ANN with many layers can automatically extract features and take high dimensional input).

Recognition

Factors of Variation: Translation, scaling, rotation, light, occlusion, etc.

Feature Extraction: Detection (detect corners), description (represent a local image patch around corners invariant to factors of variation), e.g. SIFT, SURF, and oriented gradients.

Feature Matching: Match feature descriptors between a test image and a template image. A feature is robust if the best matching feature is good and the second-best matching feature is bad.

Object Recognition using local features: Test image and template images have the same features, and the same transformation maps every feature in the test image into the template image (geometric consistency of features).

Object Recognition using global features: Convert objects into a fixed-size global feature vector (e.g. histogram of local features with fixed bins, histogram of oriented gradients, etc.; dimensionality reduction of images like PCA, LDA, etc.), using supervised machine learning to map features to class labels.

Deep Learning solutions can take an image as an input and generate bounding boxes and labels across all objects as an output as end-to-end learning.

Tracking

Track objects in image/world. Reduces effects of noises, predicts position when not in view (occlusion), predicts future position for decision-making (catching object and avoiding obstacle), and predicts region of interest for faster recognition/detection.

Human-Robot Interaction

Hand Gestures: Static (recognition similar to image) and dynamic (may use frequency features, e.g. signal represented by X/Y position of hand with time).

Speech Recognition: Detection/recognition of words similar to image. Use signal frequency features, time features, or both (spectrum image).

Multimodal: Using multiple modalities, e.g. use speech to say pick up an object and hand to point to the object to pick up.

1.5.1 Calibration

The first decision is always to place the sensor in an appropriate place to capture the desired data. Thereafter, it is important to calibrate the camera. The vision camera works on the pinhole camera principle, which is implemented using a camera lens. The lens develops errors, and sometimes the light does not bend as desired within a lens leading to *distortion*. Because of this effect, the straight lines appear curved. Furthermore, the images may have a skew, which is another kind of distortion due to manufacturing errors where the lens is not completely in alignment with the image plane. The distortion correction happens by placing chessboard images and detecting lines. The image is passed through models whose coefficients are regressed to straighten the lines.

Imagine a point in the real world is at $p\,(p_x, p_y, p_z)^T$, which if seen from the camera coordinate frame of reference will be at $p_{Cam} = T_W^{Cam} p$. Here T_W^{Cam} is the transformation matrix from the world to the camera coordinate axis system, and p_{Cam} denotes the point as observed in the camera coordinate system. Note that in the camera, the Z-axis is the one that goes out from the centre of the lens in the line of sight of the camera, the X-axis is typically horizontal in the plane of the camera, and the Y-axis is vertical in the plane of the camera. This is given by Eq. (1.31).

$$p_{Cam} = T_W^{Cam} p = [R_{3 \times 3}\ t_{3 \times 1}]p_{4 \times 1} = \begin{bmatrix} r_{11} & r_{12} & r_{13} & t_x \\ r_{21} & r_{22} & r_{23} & t_y \\ r_{31} & r_{32} & r_{33} & t_z \end{bmatrix} \begin{bmatrix} p_x \\ p_y \\ p_z \\ 1 \end{bmatrix} \tag{1.31}$$

The typicality of the equation is that the last row $[0\ 0\ 0\ 1]$ is omitted without a loss of generality as it is not needed in the subsequent calculations. The camera projects all such points in the world on its optical lens, which takes the form of an image using a projection through the camera calibration matrix C. The inverted image is formed at focal length behind the camera lens. The image as seen is equivalent to the projection at focal length distance in front of the camera. Using the principle of similar triangles, the projection onto the 2D image can be calculated. Suppose $p_I\,(u'_I, v'_I, w)^T$ is the point in the 2D image formed by the camera, given by Eqs (1.32), (1.33)

$$p_I = C p_{Cam} \tag{1.32}$$

$$\begin{bmatrix} u'_I \\ v'_I \\ w \end{bmatrix} = \begin{bmatrix} f_x & 0 & c_x \\ 0 & f_y & c_y \\ 0 & 0 & 1 \end{bmatrix} p_{Cam} \tag{1.33}$$

Here f_x and f_y are the focal lengths in the 2 axes, and (c_x, c_y) are the coordinates of the centre of the image. This gives Eq. (1.34)

$$\begin{bmatrix} u'_I \\ v'_I \\ w \end{bmatrix} = \begin{bmatrix} f_x & 0 & c_x \\ 0 & f_y & c_y \\ 0 & 0 & 1 \end{bmatrix} T_W^{Cam} p = \begin{bmatrix} f_x & 0 & c_x \\ 0 & f_y & c_y \\ 0 & 0 & 1 \end{bmatrix} \begin{bmatrix} r_{11} & r_{12} & r_{13} & t_x \\ r_{21} & r_{22} & r_{23} & t_y \\ r_{31} & r_{32} & r_{33} & t_z \end{bmatrix} \begin{bmatrix} p_x \\ p_y \\ p_z \\ 1 \end{bmatrix} \tag{1.34}$$

The equations have an extra parameter w representing scaling. In the projection on the image plane, the scale information is lost because the camera sees in 2D. The vector $(u'_I, v'_I, w)^T$ is hence converted into $(u_I, v_I, 1)^T = (u'_I/w, v'_I/w, 1)^T$ as per homogenization to get the image coordinates. Suppose the point in the camera coordinate axis system is $(p_{xCam}, p_{yCam}, p_{zCam}, 1)^T$. This gives the coordinates in the image as given by Eqs (1.35), (1.36)

$$\begin{bmatrix} u'_I \\ v'_I \\ w \end{bmatrix} = \begin{bmatrix} f_x & 0 & c_x \\ 0 & f_y & c_y \\ 0 & 0 & 1 \end{bmatrix} \begin{bmatrix} p_{xCam} \\ p_{yCam} \\ p_{zCam} \end{bmatrix} = \begin{bmatrix} f_x p_{xCam} + c_x p_{zCam} \\ f_y p_{yCam} + c_y p_{zCam} \\ p_{zCam} \end{bmatrix} \quad (1.35)$$

$$\begin{bmatrix} u_I \\ v_I \end{bmatrix} = \pi \left(\begin{bmatrix} p_{xCam} \\ p_{yCam} \\ p_{zCam} \end{bmatrix} \right) = \begin{bmatrix} (f_x p_{xCam} + c_x p_{zCam})/p_{zCam} \\ (f_y p_{yCam} + c_y p_{zCam})/p_{zCam} \end{bmatrix} = \begin{bmatrix} f_x p_{xCam}/p_{zCam} + c_x \\ f_y p_{yCam}/p_{zCam} + c_y \end{bmatrix}$$

$$(1.36)$$

Here π is called the projection function that takes the position of any point in the camera coordinate axis system and computes its projection in the 2D image plane. Note that the image coordinates here have an origin at the top-left corner of the image. The X-axis of the image (u_I-value) is horizontal from left to right, and the Y-axis of the image (v_I-value) is vertical from top to bottom. Some image processing and vision libraries differently interpret the image where either the origin or the axes system may be different. Suitable conversions may therefore be needed before plugging the image coordinates into image processing and vision libraries. Similarly, the input is a point in the camera coordinate system with the Z-axis of the camera looking out of the camera, the X-axis of the camera horizontal to the camera plane, and the Y-axis of the camera looking vertically down.

As an example, consider the image in Fig. 1.12 captured by a camera. Consider the top-left corner of the alphabet A. The real-world coordinates of the top-left corner of A are computed (most approximately) as $p = (-13.235, 0.70, -1.851)$, where the motion of the robot happens in the XZ plane and the Y-axis looks vertically down. The map is defined in the XZ plane rather than the

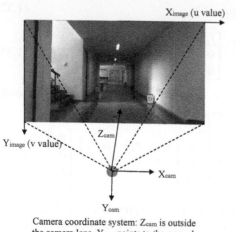

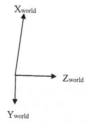

Camera coordinate system: Z_{cam} is outside the camera lens, Y_{cam} points to the ground, X_{cam} looks at the right

World coordinate system: Y_{world} points to the ground, the robot translates and rotates in the X_{world}-Z_{world} plane

FIG. 1.12 Image used for the understanding of projections.

conventionally used *XY* plane because the starting pose of the camera at the home location is taken as the origin, wherein the Z-axis is the one that goes outside the lens of the camera in the line of sight. Suppose the position of the camera at which this image was taken was (most approximately) measured as $(x_{Cam}, y_{Cam}, z_{Cam}) = (-5.801, 0, 1.805)$ while the camera at the time of taking the image had an orientation along the Y-axis of $\theta_{Ycam} = -1.885$ rad. The pose of the camera is as given by Eq. (1.37).

$$pose\ of\ camera = T_{Cam}^{W} = \begin{bmatrix} \cos\theta_{Ycam} & 0 & \sin\theta_{Ycam} & x_{cam} \\ 0 & 1 & 0 & y_{cam} \\ -\sin\theta_{Ycam} & 0 & \cos\theta_{Ycam} & z_{cam} \\ 0 & 0 & 0 & 1 \end{bmatrix}$$

$$= \begin{bmatrix} -0.309 & 0 & -0.951 & -5.801 \\ 0 & 1 & 0 & 0 \\ 0.951 & 0 & -0.309 & 1.805 \\ 0 & 0 & 0 & 1 \end{bmatrix} \quad (1.37)$$

The top-left corner of the alphabet *A* will thus be in the coordinates $(p_{xCam}, p_{yCam}, p_{zCam}, 1)^{T}$ in the camera's coordinate frame given by Eq. (1.38), where the inverse is taken to enable conversion from the world to the camera coordinate axis while the pose of the camera does the conversion the other way around.

$$\begin{bmatrix} p_{xCam} \\ p_{yCam} \\ p_{zCam} \\ 1 \end{bmatrix} = T_{W}^{Cam} p = \left(T_{Cam}^{W} \right)^{-1} p$$

$$\begin{bmatrix} -0.309 & 0 & -0.9511 & -5.8014 \\ 0 & 1 & 0 & 0 \\ 0.9511 & 0 & -0.309 & 1.805 \\ 0 & 0 & 0 & 1 \end{bmatrix}^{-1} \begin{bmatrix} -13.235 \\ 0.7 \\ -1.851 \\ 1 \end{bmatrix} = \begin{bmatrix} -1.180 \\ 0.7 \\ 8.2 \\ 1 \end{bmatrix} \quad (1.38)$$

The image coordinates are thus given by Eqs (1.39), (1.40), where the calibration matrix *C* is assumed to be known and is directly fed.

$$\begin{bmatrix} u_I' \\ v_I' \\ w \end{bmatrix} = C \begin{bmatrix} p_{xCam} \\ p_{yCam} \\ p_{zCam} \end{bmatrix} = \begin{bmatrix} f_x & 0 & c_x \\ 0 & f_y & c_y \\ 0 & 0 & 1 \end{bmatrix} \begin{bmatrix} p_{xCam} \\ p_{yCam} \\ p_{zCam} \end{bmatrix}$$

$$= \begin{bmatrix} 1421.28 & 0 & 865.24 \\ 0 & 1420.64 & 528.13 \\ 0 & 0 & 1 \end{bmatrix} \begin{bmatrix} -1.180 \\ 0.7 \\ 8.2 \end{bmatrix} = \begin{bmatrix} 5418.4 \\ 5325.3 \\ 8.2 \end{bmatrix} \quad (1.39)$$

$$\begin{bmatrix} u \\ v \end{bmatrix} = \begin{bmatrix} u_I'/w \\ v_I'/w \end{bmatrix} = \begin{bmatrix} 5418.4/8.2 \\ 5325.3/8.2 \end{bmatrix} = \begin{bmatrix} 661 \\ 649 \end{bmatrix} \quad (1.40)$$

The coordinates are approximately the same as seen in the image, subject to approximation errors in measuring the position of the top-left corner in the ground truth, the pose of the robot, and the calibration matrix. Many software systems use a different origin and coordinate axis system, for which the results can be transformed.

Camera calibration is the process of finding all the parameters. It is composed of two parts, an *extrinsic calibration* which calculates the parameters of T_W^{Cam} and an *intrinsic calibration* which computes the parameters of the calibration matrix C. Typically this is done by showing multiple images of a chessboard to the camera at multiple viewing angles. Each chessboard makes a local planar coordinate frame, and every corner is given a coordinate value in this local frame. With some parameter estimates, the corners can then be projected onto the image. The algorithm attempts to get parameter estimates such that the corner projection on the image using the projection formula and the observed coordinates of the same corners on the image are as close as possible. This is used to compute both extrinsic and intrinsic calibration matrices. The extrinsic is not useful as it uses the chessboard as the temporary world coordinate frame. The results are taken from multiple such chessboards to get the intrinsic calibration matrix. The extrinsic calibration is doable only when objects with known positions in the world coordinate axis system exist.

1.5.2 Pre-processing

The images are shown in the Red-Green-Blue (RGB) colour space; however, the colour space is not very useful for working with images since the images look very different in different lighting conditions that cause many vision algorithms to fail. The hue-saturation-value (HSV) is a better colour space, which resembles the way humans perceive, where light intensity is the 3rd component (value), which may be neglected for vision algorithms. The hue stores the colour information and saturation stores the greyness, which are both (theoretically) light invariant. The YCbCr is another colour space that stores light as the first component (Y or luma), while the Cb stores the blue colour component minus luma, and similarly Cr stores the red colour component minus luma. A similar space is the LAB space, where L stands for lightness and denotes the light component, while a and b store the red and blue colour components only.

The given input image is additionally denoised by passing through *noise removal* filters. This removes the noises such as bright reflection from a light source, spots in the image, etc. The noise may be salt and pepper, Gaussian, or from other complex noise models. The noise is removed by passing it through appropriate filters. The median filters, mean filters, and Gaussian filters are common. The statistical analysis for noise rejection and histogram equalization of colour is also used to improve the image quality.

1.5.3 Segmentation and detection

Given a denoised image, a common requirement of robotic applications is to understand the objects to be found inside the image. The first step for this is to separate the different objects from each other known as the problem of *segmentation* or *object detection*. The classical approaches for segmentation use the fact that different objects will be of distinct colours and thus will have a visible boundary between them. *Edge detection* techniques use a gradient of the image along the X- and Y-axis of the image, and edges are marked by a high gradient showing a colour change. The *Canny edge detection* method and the *Sobel edge detection* method are specialized techniques to cut out such objects. The boundary can be traced, and any pixel within the closed boundary is the object. A similar technique is the *grow-cut method*. The method assumes that a part of the object can be detected using colour or similar information. Using this pixel as seed, the region growing is used where anything neighbouring an object pixel is believed to be an object pixel and the object pixel region is iteratively expanded. The growth stops when a significant change in a colour denoting a boundary is reached. Such detection schemes annotate the pixels corresponding to the object. Many regions may be detected incorrectly due to noise. These will typically be small. Therefore, *erosion* is applied, wherein the identified object regions are shrunk. This eliminates small objects. Thereafter, *dilation* is applied, wherein the objects are grown to eliminate small holes in them. Together, the methods constitute the morphological operations in computer vision. Alternatively, if the objects are of different colours, the colour-based *clustering* algorithms may be used. Otherwise, the spatial clustering algorithms can extract the objects if the objects in the foreground are known distinctly in contrast to the background.

The objects of interest now come in complex geometries with multiple colours, and the classic techniques are therefore no longer relevant. A modern way of detecting objects is to make a system that identifies objects contained fully within the image (the recognition problem is discussed in the next subsection). The image captured may have multiple objects at any place and of any size. Therefore, multiple *candidate windows* or region proposals are placed in different X and Y locations and with different scales. The subimage within the candidate window is given to the recognition system that identifies the object within it. If the recognition fails or succeeds with a low probability, it is believed that no object was contained in that candidate window. On the contrary, if the recognition system can identify an object within the candidate window, the candidate window is used to denote the segmented object. A local search on the X-axis, Y-axis, or size of the candidate window may be done to increase the accuracy of predicting the bounding box around the object.

Similarly, classic techniques find bounding boxes around the object of interest. A modern requirement is a semantic recognition or finding the object class to which a pixel belongs. The self-driving cars need pixels of roads for navigation and not a bounding box of the road. Similarly, algorithms may want to remove dynamic people from the image at the pixel level to study the background, rather than removing a big patch of bounding box in which the human lies. Machine learning algorithms take local and global windows around the pixel to predict the object class to which the pixel belongs.

1.5.4 Machine learning

Machine learning deals with the summarization of information contained in the historic data collected in data repositories, which is generalized to reason about new data in the system that may have never been seen before (Haykin, 2009; Shukla et al., 2010). The *supervised machine learning* systems are provided with training input data and a desired target, and they try to learn a function that maps the inputs to the targets. Thereafter, if a new input is provided, they guess the output using the same learned model. Classification is one of the prominent problems where machine learning is used. The models take input and predict the probability that the output belongs to a specific class. Many systems model a binary classifier where the output can be either of two classes. The output is thus the probability that the input belongs to the first (or second) class.

The multiclass classifier systems are based on the binary classifier systems using either of two methodologies. For C classes, the *one-vs-all methodology* trains C classifiers, each trying to separate one class from all other inputs. The ith classifier learns a binary classification problem with output as the probability that the input belongs to class C_i. The *one-vs-one classifier* trains $C(C-1)/2$ classifiers, one for every pair of classes. Thereafter, the input is given to all classes and a majority vote of the output is used to determine the output class. The class label that gets most of the votes is the predicted label. In a one-vs-all methodology, the binary classifiers can have a significantly low balance between the positive and negative class training data for a large number of classes, and the binary classifiers perform poorly when the positive and negative examples are not balanced. Typically, the class having a larger number of examples starts to dominate, which biases the classifier to predict the label corresponding to the class that has a larger number of examples. The one-vs-one methodology, however, leads to too many classifiers that may be impractical to train for too many classes. A middle way is giving every class a unique binary encoding called *output coding*. Every bit represents a classifier, and the value represents the target value for the specific class for the classifier. The input is thus passed to all classifiers, and the classifier outputs (0 or 1) are noted in the same order. This gives a code corresponding to the input. The class that has the closest output code is searched using the Hamming distance metric, which is the predicted label. The rule is that the Hamming distance between every pair of output codes should be the largest so that the classifier cannot confuse one class with the other even if a few classifiers give the wrong output, while the bit length should be the least to have a manageable number of classifiers that are easy to learn.

The machine learning systems differ on the parameterized mathematical equation by which they model the input to output mapping. The parameters are typically called the weights. *Learning* is the process of guessing the correct weights that give the best input to the output mapping function. The learning algorithms adjust the weights automatically to iteratively reduce the error.

Neural network is a common classifier that imitates the human brain and models the relationship using a weighted addition of inputs and layered processing to make outputs. The *Support Vector Machine* projects the data into higher-dimensional space and learns a linear hyperplane that best separates the positive and negative classes. The *decision trees* instead recursively partition the data in a tree-like manner based on learnable conditions. A traversal of the tree leads to the leaves, which have the class information.

It is common to use multiple classifiers instead of a single classifier making a classifier *ensemble*. Multiple classifiers redundantly solve the same problem. The input is given to all the classifiers, and the outputs are integrated. The principle in using a multiclassifier system is that the different classifiers must be good individually while having high diversity. If the diversity is small, it is observed that all classifiers end up giving nearly the same output, and the integration of output is not much better than any one single classifier. The best way to increase diversity among classifiers is to give each classifier a separate segment of data to learn. The *bagging* techniques divide the data among the classifiers. The *boosting* techniques instead prefer to give data to a classifier that has been classified as wrong by most of the previously trained classifiers, thus incentivizing learning on data on which no classifier has performed well. This makes the new classifier better on rare data while compromising on other data, thus increasing diversity. Adaptive boosting (Adaboost) is a common implementation.

The classical machine learning techniques can only take a limited amount of input attributes, and thus feature engineering is applied to limit the input dimension. The *deep learning* techniques (Goodfellow et al., 2016) can take a lot of data and learn from the same, without compromising on generality or producing results on unknown inputs. These techniques thus provide end-to-end learning for problems. They extract features automatically in the process, which is free of any human bias. They can take images directly as input to provide the class label, bounding box, or multiple labels and bounding boxes as output.

1.5.5 Recognition

Recognition deals with identifying the object in the image or giving a label to the object contained in the image. The challenge is that the test image will have all *factors of variation,* including rotation of the object, translation of the object, scaling of the object, illumination effects, etc., and the recognition must be invariant to these changes. The recognition is done by using local and global techniques. The local techniques study the local patterns in the image for recognizing the object. The image has too many pixels, and therefore *feature extraction* is done to limit the input size. In an image, corners are the most interesting things. Humans can detect objects based on corners alone. The lines are the next most interesting. Therefore, humans can make out the object using line art alone, which neither has colour nor texture. Nothing is interesting in completely coloured regions of the image. The feature extraction techniques thus mine out the corners. Each feature extraction technique gives the position of the feature called the problem of *feature detection*, and a unique signature that locally identifies the feature called the problem of *feature description*.

A common feature extraction technique is the *Scale Invariant Feature Transform* (SIFT). The technique uses a difference in Gaussian filter responses at multiple scales. The corners have a different response to a difference in Gaussians, and such an intensity of response can be used to extract the feature. The signature is also computed using bits of the local windows in the difference of Gaussians. The *Speeded Up Robust Transform* (SURF) is a similar feature that is faster than SIFT as it uses integral images that enable quick computation of the gradients in the actual processing. The image is used to compute the Hessian matrix quickly at multiple scales, and response to Haar wavelets is used to detect the strong features. The *oriented gradients* simply detect local edges by measuring the gradient of the image in the X and Y directions and computing the magnitude and direction of the local edge. Other features include Features from Accelerated Segment Test (FAST), Adaptive and Generic Accelerated Segment Test (AGAST), Binary Robust Independent Elementary Features (BRIEF), Binary Robust Invariant Scalable Keypoints (BRISK), Oriented FAST and Rotated BRIEF (ORB), Good Features to Track (GFTT), KAZE, Accelerated KAZE (A-KAZE), Fast Retina Keypoint (FREAK), Maximally Stable Extremal Regions (MSER), etc.

For recognition using local techniques, a template of every object is available in the library. The matching using local features happens by matching the signature of every feature in the template with the nearest feature in the image. The robustness of the feature can be checked by computing a ratio of the best matching to the second-best matching feature. Ideally, if the feature signature is unique, it would be found once, and the second-best match will be poor. Matching of all features between the template and test image is done. For an ideal match, all features would be found between the test image and the template. The algorithms may further use *geometrical consistency*. The features between the template and test image must have the same transformation (scaling, translation, and rotation). If a transformation matrix can be computed such that all features agree to be witnessing the same transformation between the test image and the template, the object is said to be recognized. The Hough transform finds out straight lines, which is useful to see if all transforms are matching for geometric consistency. It may be noted that this methodology exists for simple objects only, which are not deformable. The human body, as an example, is a complex deformable body since the hands and legs can move. The approaches of the *deformable part model* first detect simple shapes and thereafter all possible geometric arrangements that the simple shapes can take. Like hands and legs in a human body can move. So, a technique may separately detect the hands, legs, and body and additionally check that the hands and legs are attached to the body using the transformations permissible for the human body to deform to. The test object can hence match any of the possible leaned geometries.

The other manner of object recognition is to do it globally, typically using machine learning techniques. A machine takes the *global features* of a fixed-size object as input and gives the probability distribution of different classes as output. The global features can be derived from the local features using the technique of a histogram. Every feature signature is quantized to be within some discrete set of values. A histogram of such values is made, which acts as the feature descriptor. As an example, the *Histogram of Oriented Gradients* (HOG) is a global feature. The local edges may be between $-\pi$ and π, which are broken down into k bins. Each edge is added to its bin, and counting of all edges is done based on their intensity. This, after normalization, makes a k-feature vector.

The complete image cannot be given to the classifier due to its massive size. The technique of *dimensionality reduction* measures the correlation between the input features to create new features as a combination of existing features which are highly uncorrelated. This can be used to reduce the number of dimensions for any problem, attacking redundancy among the features. The common techniques are principle component analysis (PCA), independent component analysis (ICA), and linear discriminant analysis (LDA). Hence every pixel can be a feature, while these techniques can severely reduce the dimensionality. The wavelet analysis is also useful for breaking an image and intensifying the informative parts, which can be used for dimensionality reduction.

1.5.6   Pose and depth estimation

The recognition was done so far using a 2D image. However, a robot needs to know the 3D location of all entities, including obstacles for decision-making. This is typically facilitated using *stereo cameras*. The use of 2 cameras can facilitate the estimation of the depth of the object of interest. Let us consider the most common case where the two cameras are kept parallel to each other. The separation between the cameras is called the *baseline*. The depth is estimated from the principle of *triangulation*. Imagine seeing the same point (e.g. point of a pen) in both the left and right images. The specific position in the left and right images at which the point in 3D is projected will be different in the 2 images. Identifying the same point in both images is called *correspondence*, which itself is a hard problem. The difference in position between the left and right camera images is called *disparity*. Imagine an object at infinity, say, the moon. There will be barely any difference in the position between the left and right images where the moon is projected. However, imagine a near object, say, your finger directly ahead of the stereo camera. In the left camera, it would be seen at the far right and vice versa. So, the distance is inversely proportional to the disparity. The notations are shown in Fig. 1.13.

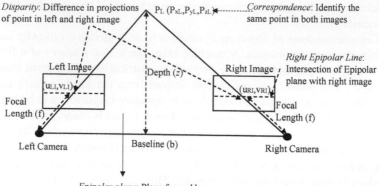

FIG. 1.13 Calculation of depth using a stereo camera set-up. The depth is inversely proportional to the disparity.

Consider the point $P_L(p_{xL}, p_{yL}, p_{zL})$ in the left camera coordinate axis system. The plane joining the point P_L with the centres of the two cameras is called the *epipolar plane*. The point P_L is projected into the left camera image at a point $P_{LI} = (u_{LI}, v_{LI})^T$, where the subscript *LI* stands for left-image coordinate axis system. The Z-axis points to the object, and thus p_{zL} is the depth. The X-axis is pointing towards the right camera. The Y-axis is perpendicular to both the X and Z axes. Using the projection formula of Eq. (1.36), we get Eq. (1.41).

$$\begin{bmatrix} u_{LI} \\ v_{LI} \end{bmatrix} = \begin{bmatrix} f_x p_{xL}/p_{zL} + c_x \\ f_y p_{yL}/p_{zL} + c_y \end{bmatrix} \tag{1.41}$$

The projection at the right camera will be given by $P_{RI} = \pi(T_L^R P_L)$, where T_R^L is the extrinsic camera calibration matrix that converts the point from the left camera's coordinate axis system to the right camera's coordinate axis system and π is the projection function. Suppose the two cameras are placed in parallel and separated by a baseline b (distance between the left and right cameras). They are both on the same Z-axis. The transformation hence only has a translation of baseline (b) along the X-axis, or Eq. (1.42)

$$T_L^R = \begin{bmatrix} 1 & 0 & 0 & -b \\ 0 & 1 & 0 & 0 \\ 0 & 0 & 1 & 0 \\ 0 & 0 & 0 & 1 \end{bmatrix} \tag{1.42}$$

This means that the point will lie in the same row in the corresponding images. This is called the *epipolar constraint*. More precisely, the epipolar plane intersects the image plane (at focal length distance in front of the camera plane as per the pinhole camera principle) at the *epipolar line*. The point in both images needs to be at the corresponding epipolar lines. Suppose (u_{LI}, v_{LI}) and (u_{RI}, v_{RI}) are the projection of the point in the left and right images, $v_{LI} = v_{RI}$ due to the epipolar constraint. The difference between the projection in the left and right images is called as the disparity (d), that is $d = u_{LI} - u_{RI}$. The projection on the right camera is given by Eqs (1.43)–(1.47)

$$P_R = T_L^R P_L \tag{1.43}$$

$$\begin{bmatrix} u'_{RI} \\ v'_{RI} \\ w \end{bmatrix} = CT_L^R P_L = \begin{bmatrix} f_x & 0 & c_x \\ 0 & f_y & c_y \\ 0 & 0 & 1 \end{bmatrix} \begin{bmatrix} 1 & 0 & 0 & -b \\ 0 & 1 & 0 & 0 \\ 0 & 0 & 1 & 0 \end{bmatrix} \begin{bmatrix} p_{xL} \\ p_{yL} \\ p_{zL} \\ 1 \end{bmatrix} \tag{1.44}$$

$$\begin{bmatrix} u'_{RI} \\ v'_{RI} \\ w \end{bmatrix} = \begin{bmatrix} f_x & 0 & c_x \\ 0 & f_y & c_y \\ 0 & 0 & 1 \end{bmatrix} \begin{bmatrix} p_{xL} - b \\ p_{yL} \\ p_{zL} \end{bmatrix} \tag{1.45}$$

$$\begin{bmatrix} u'_{RI} \\ v'_{RI} \\ w \end{bmatrix} = \begin{bmatrix} f_x(p_{xL} - b) + c_x p_{zL} \\ f_y p_{yL} + c_y p_{zL} \\ p_{zL} \end{bmatrix} \tag{1.46}$$

$$\begin{bmatrix} u_{RI} \\ v_{RI} \end{bmatrix} = \begin{bmatrix} u'_{RI}/w \\ v'_{RI}/w \end{bmatrix} = \begin{bmatrix} (f_x p_{xL} - f_x b + c_x p_{zL})/p_{zL} \\ (f_y p_{yL} + c_y p_{zL})/p_{zL} \end{bmatrix} \tag{1.47}$$

Comparing the projections along X-axis and using (u_L, v_L) as the projection of the left camera using Eq. (1.41) gives Eqs (1.48)–(1.51)

$$u_{RI} = (f_x p_{xL} - f_x b + c_x p_{zL})/p_{zL} = (f_x p_{xL} + c_x p_{zL})/p_{zL} - f_x b/p_{zL} \tag{1.48}$$

$$u_{RI} = u_{LI} - f_x b/p_{zL} \tag{1.49}$$

$$u_{LI} - u_{RI} = d = f_x b/p_{zL} \tag{1.50}$$

$$p_{zL} = \frac{f_x b}{d} \tag{1.51}$$

This gives the depth p_{zL} as an inverse of the disparity (d) for a given focal length (f_x) and baseline (b). Using the projection formula of Eq. (1.41), the coordinates of the point in the left camera coordinate frame can be calculated as given by Eq. (1.52).

$$\begin{bmatrix} p_{xL} \\ p_{yL} \\ p_{zL} \end{bmatrix} = \begin{bmatrix} (u_{LI} - c_x)p_{zL}/f_x \\ (v_{LI} - c_y)p_{zL}/f_y \\ p_{zL} \end{bmatrix} \tag{1.52}$$

The hard problem is still finding the correspondence or the same point in the left and right cameras. The epipolar restriction means that only the corresponding row in both images needs to be searched for. A small image block around the point is taken. The same block is searched for along the same row in the other image for correspondence. The centre of the best matching block gives the corresponding point.

Consider the same image as shown in Fig. 1.12 taken by the left camera and a similar image taken by the right camera. Suppose the coordinates of the top-left corner of the alphabet A in the left camera frame are $(u_{LI}, v_{LI}) = (753, 690)$ and the same corner in the right image is seen at coordinates $(u_{RI}, v_{RI}) = (726, 690)$. The coordinates take the top to down as the positive Y-axis with the origin at the top-left corner in the image. The disparity is $753-726 = 27$, which gives a depth of $p_{zL} = f_x b/d = 1421.28 \times 0.12/27 = 6.3168$. Here focal length $f_x = 1421.28$ and baseline $b = 0.12$ m are assumed to be known. The top-left corner of the alphabet A thus has a position in the left camera's coordinate frame given by Eq. (1.53).

$$\begin{bmatrix} p_{xL} \\ p_{yL} \\ p_{zL} \end{bmatrix} = \begin{bmatrix} (u_{LI} - c_x)p_{zL}/f_x \\ (v_{LI} - c_y)p_{zL}/f_y \\ p_{zL} \end{bmatrix}$$

$$\begin{bmatrix} (753 - 865.24)6.3168/1421.28 \\ (690 - 528.13)6.3168/1420.64 \\ 6.3168 \end{bmatrix} = \begin{bmatrix} -0.4988 \\ 0.7197 \\ 6.3168 \end{bmatrix} \tag{1.53}$$

It is assumed that $c_x = 865.24$ and $c_y = 528.13$ are known from the calibration procedure. Suppose the position of the left camera at which this image was taken was (most approximately) measured as $(x_L, y_L, z_L) = (-5.801, 0, 1.805)$, while the left camera at the time of taking the image had an orientation along the Y-axis of $\theta_{YL} = -1.885$ rad. The pose of the camera is as given by Eq. (1.54), and the position of the point in the world coordinates will be given by Eq. (1.55).

$$pose\ of\ camera = T_L^W = \begin{bmatrix} \cos\theta_{YL} & 0 & \sin\theta_{Ycam} & x_L \\ 0 & 1 & 0 & y_L \\ -\sin\theta_{YL} & 0 & \cos\theta_{Ycam} & z_L \\ 0 & 0 & 0 & 1 \end{bmatrix}$$

$$= \begin{bmatrix} -0.309 & 0 & -0.951 & -5.801 \\ 0 & 1 & 0 & 0 \\ 0.951 & 0 & -0.309 & 1.805 \\ 0 & 0 & 0 & 1 \end{bmatrix} \tag{1.54}$$

$$p_{world} = T_L^W p_L = \begin{bmatrix} -0.309 & 0 & -0.951 & -5.801 \\ 0 & 1 & 0 & 0 \\ 0.951 & 0 & -0.309 & 1.805 \\ 0 & 0 & 0 & 1 \end{bmatrix} \begin{bmatrix} -0.4988 \\ 0.7197 \\ 6.3168 \\ 1 \end{bmatrix} = \begin{bmatrix} -11.654 \\ 0.718 \\ -0.621 \\ 1 \end{bmatrix}$$

$$\tag{1.55}$$

Again, the calculations depend upon the approximations in the computation of the disparity, pose of the robot, and the calibration matrix. Averaging across multiple observations reduces the effects of approximations.

Knowledge of depth completes the position of the object in 3D with respect to the camera. Some applications like grasping additionally need the *object's pose* (position and orientation) in 3D. Assume that an object template is available. The pose is the transformation to be applied to the object template, which gives the object as per the current appearance. You have a picture of an upright bottle in mind, and therefore if the bottle is lying flat on the table with an orientation of 30 degrees, you can spot the pose (rotations) by comparing it with the picture of the upright bottle. Assume that some characteristic feature points (corners) in the recognized object are triangulated in 3D. The corresponding points in the object template are known. Correspondence matching by looking at a corner with a strong signature in the test point cloud and matching the same in the template point cloud gives a list of feature points in the test point cloud and the corresponding points in the template point cloud. Pose is the transformation of the template point cloud that aligns with the test point cloud. Apply multiple candidate transformations to the template point cloud to see if the transformed template exactly matches the test point cloud object. This reduces pose identification to an optimization problem that finds the transformation that most closely aligns the template with the test object. We will see more details on the mathematical front in Chapter 2.

1.5.7 Point clouds and depth images

Modern robots use 3D vision instead of 2D images. There are two common sources for 3D images: first using a 3D lidar where every point has an *XYZ* value. A collection of such points is called a *point cloud* (Guo et al., 2014; Rusu and Cousins, 2011). The other is using a stereo camera or depth camera. These cameras take 2D images and append depths over them. The point cloud taken by a single lidar scan and a depth image of a stereo camera are more correctly 2.5 dimensions. For the stereo camera, the data is a 2D array representing the RGB image and another 2D array representing the depth of every pixel. For lidar, the data is a 2D array representing the depth at every solid angle. In both cases, the data is hence 2D, which is projected into 3D for visualization. The 3rd dimension is not complete as both sources of 3D images cannot handle occlusion. For an occluded segment of space, both sensors have no information about the presence or absence of objects. The point cloud data structure, as an array of points, can integrate multiple laser scans from multiple lidars and readings from the same lidar placed at different locations with time to produce a full 3D depth map. The stereo has RGB information in addition to the depth, and the images are

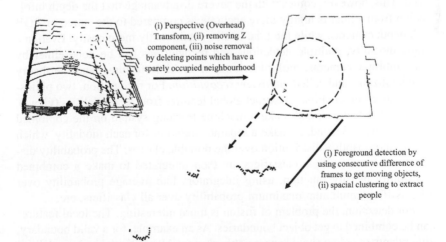

FIG. 1.14 Recognizing moving people using a low-resolution 3D lidar.

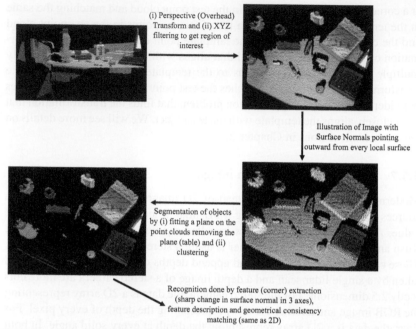

FIG. 1.15 Vision with RGB-Depth (RGBD) images.

more accurately called the RGB-depth (RGBD) images. The vision is illustrated with the help of real-world examples. The vision for the problem of detecting moving people using a 3D lidar is shown in Fig. 1.14, and the vision to detect and recognize objects on a tabletop using RGBD images is shown in Fig. 1.15.

The advantage of RGBD is that the segmentation and recognition can be done in both colour space and depth space, and thereafter the results can be fused. This, however, comes with the severe disadvantage that the depth information from a lidar is highly more accurate as compared to the stereo or RGB with depth cameras, while the lidar is also significantly more costly. Even in a single modality, multiple detections and recognitions can be done either by using multiple cameras, multiple views, or multiple features, and the theory is generalizable and called *multiview recognition*. For recognition, two philosophies are common. First, to extract global features from all modalities (views), combine the features, and make a machine learning system on the combined feature vectors. Second, to make a separate classifier for each modality, which outputs a probability distribution over the possible classes. The probability distributions from multiple classifiers are then integrated to make a combined probability density function, using integrators like average probability over all classifiers, minimum/maximum probability over all classifiers, etc.

For detection, the problem of fusion is more interesting. The local features can be combined to get object boundaries. As an example, for a valid boundary, either depth or colour should appreciably change. Alternatively, both modalities can give a list of objects. The final set of objects can be a union/intersection of the lists. An object detected in one modality is the same as that detected in another modality if the intersection over union is sufficiently high. Common errors in detection are having 1 object split into two and having 2 objects combined into 1, which can be used as a heuristic to improve fusion.

Dealing with a point cloud is like dealing with monocular images with the addition of depth in the case of stereo images. The lidar images have only depth and no RGB information. The pipeline remains the same. The point clouds may be externally *transformed* to have the best view since they represent 3D images. A bird's eye view from an otherwise oblique camera view is a common transformation. The *noise removal* or *filtering* in terms of depth is done based on the proximity of every point with the neighbouring points. From every point in the point cloud, draw a sphere in 3D with a small radius. If there are very few points in the region, the point is mostly an outlier and can be deleted. Statistical measures of mean and variance are used to adaptively compute the threshold that depends on the resolution of the sensor.

The equivalents of pixels in the 2D images are local surfaces in a 3D depth image. The *local surfaces* are metricized by the surface normal that uniquely identifies the local surface. There will be a surface normal pointing into the object and another one pointing outside the object; the one outside is used as a convention. The object can be assumed to be locally convex for finding the same. The surface normal is computed by finding all points within a neighbourhood radius and computing the eigenvector corresponding to the smallest eigenvalue, which is the surface normal. Thereafter, the computationally heavy processes start. It is reasonable to *downsample* the point cloud for computational reasons, equivalent to resizing images.

The features shall be corners, but the definition needs to be adapted. In an image, a strong gradient along 2 axes constitutes a corner. In a 3D image, a point is a corner if the surface normal shows a drastic change for a small change in all the 3 axes. This only detects the corner. The encoding needs to be done to make the rotation and scale-invariant encoding of the local patch around the corner. The Normal Aligned Radial Features (NARF) is a technique that uses depth. The 3D equivalents of 2D descriptors also exist. The other descriptors that specifically represent the depth component include the Signature of Histograms of Orientations (SHOT), Camera Roll Histogram (CRH), Viewpoint Feature Histogram (VFH), Clustered Viewpoint Feature Histogram (CVFH), Oriented Unique and Repeatable Clustered Viewpoint Feature Histogram (OUR-CVFH), Ensemble of Shape Features (ESF), Point Feature Histogram (PFH), Fast Point Feature Histogram (FPFH), Global Fast Point Feature Histogram (GFPFH), Globally Aligned Spatial Distribution (GASD), Radius based Surface Descriptor (RSD), Global Radius based Surface Descriptor (GRSD), Rotation Invariant Feature Transform (RIFT), etc.

The segmentation can be done using any of the 2D image processes, which can be extended to account for the depth. Like the grow-cut method, it not only grows based on the colour information (if available) but also on the difference in depth. A significant difference in depth indicates a different object. Similarly, the spatial clustering could be extended to all 3 axes. The recognition using local features may again be done based on geometric consistency like the images from monocular vision. Alternatively using a histogram of global features can be fed into a machine learning library.

1.5.8 Tracking

Vision in the case of robots is more interesting because the camera will continuously see objects and do the decision-making rather than a single frame alone. This may be because the robotic hand continuously approaches the object to grasp, the robot continuously walks towards the human of interest, or the object is being searched for by exploration. Video adds the time dimension to vision, which is interesting because the results may be reused for already seen objects and noises may be corrected. This is done by using *tracking* along with recognition (Malviya and Kala, 2018, 2022). Tracking is useful for several reasons. First, assume that an object was seen at the top left part of the image and is not to be recognized in the next frame. It is highly unlikely that the object will suddenly disappear. This is an error in recognition that can be rectified by tracking.

Further, assume that the object went behind a table and hence was invisible due to occlusion. The tracker can extrapolate the object trajectory to guess its location, even if it is not in a visible region. Further, assume that the robotic hand needs to intercept the object in motion. For the same, the current position of the object is not important, but the position afterwards at the intercept time is important. This is facilitated by first tracking and then predicting the future motion. The robotic applications of avoiding walking humans, catching a thrown object, approaching a walking human, etc., all use tracking and predict the motion of the object of interest. The tracking also makes it computationally feasible to work with large images. Imagine an object was detected at a location (x,y). To detect the same object, only a small window around $(x + v_x, y + v_y)$ is enough where (v_x, v_y) is the guessed speed of the object relative to the image. This reduces the burden of searching the entire image. The algorithms used in tracking are particle filters and Kalman filters, which are discussed in the Chapter 2.

1.5.9 Human-robot interface

Robots often interact with humans, and therefore the *human-robot interface* is an important vision problem. The most interesting application is when humans instruct robots using gestures. The *gestures* may be visual, by speech, or a combination of both. The visual gestures are commonly done with the help of hand gestures. The gestures may be static, where the hand posture determines the instruction or dynamic where the complete hand motion determines the instruction. The problem has the same object recognition pipeline, where different hand postures make different classes. The identification of the foreground needs to be done additionally. If only humans are dynamic, it could be done by a difference in consecutive frames, by the characteristic colour of the hand, or by the characteristic shape of the hand. The dynamic gestures have an additional time component. This makes it possible to extract frequency-domain features. The motion of the tip of the hand over time is a signal in X and a signal in Y, which will have frequency-level characteristics. Optical flow-based features are also possible. Alternatively, recurrent models of a machine learning system that take temporal sequences (of features) can be used.

Speech recognition is an old problem, and sophisticated speech recognition systems exist. Our discussion of images was primarily from a signal processing perspective and therefore the same pipeline of noise removal, segmentation, feature extraction, and recognition work. The features may be over the frequency domain or time domain as the same signal can be viewed in both time and frequency. Both domains can be simultaneously used by viewing the image as a spectrum with time and frequency as the axes, over which amplitude intensity is derived, making a *spectrum image*. It may be noted that humans do not pause in between two words, and therefore pause is not a valid segmentation technique. The approaches use candidate windows and increase the width that locally maximizes the class probability as adjudged by the classifier. A dynamic programming-based maximization of cuts is also possible. For assistive robotics associated with the disabled, the signals may instead come directly from the brain, specifically called the *brain robot interface*. Such amputees typically require training to give signals for the desired motion of the prosthetics in place, while the machines try to learn to interpret such signals.

Many approaches complement both speech and visual gestures, making a fused system called a *multimodal human-robot interface*. Some systems use the two gestures redundantly, and the best classification with the highest probability of recognition out of the two is used. As an example, the user could ask the robot to come towards him/her while also gesturing the same. Some instead reserve one modality for some aspects of the instruction and another modality for another aspect of the instruction. As an example, the user could ask the robot to switch off the light and point to which room's light is in question.

Questions

1. Fix a table and mark X, Y, and Z global coordinate axes as per your liking, keeping the origin as one corner. Take a cuboid object (like a pencil box) and mark the $X'Y'Z'$ body coordinate axis keeping the origin as one corner. Initially let both coordinate systems be coinciding ($I_{4\times4}$ pose). Now perform the following operations in the same order: (i) Translate by $(3,1,2)$ along with the world coordinate frame. (ii) Rotate by $\pi/3$ along the Z-axis (world coordinate frame). (iii) Rotate by $\pi/2$ along the X' axis (body coordinate frame). (iv) Rotate by $\pi/3$ along the Y' axis (body coordinate frame). (v) Rotate by $\pi/3$ along the Z-axis (world coordinate frame). (vi) Translate by $(1,0,1)$ along the body coordinate frame. Convert the resultant transformation matrix into Euler angles (using any library). Physically perform the

same operation using sequence given and translation/Euler angles obtained. Confirm that the results match.

2. Take a planar 2 link manipulator with all revolute joints. The base is at the origin. Assuming all joints are at $\pi/6$ degrees with respect to the previous link (refer to Fig. 1.8) and all links are of 1 m, calculate the position of the end effector. Calculate and plot all positions that the end effector of the manipulator can take using forward kinematics. For any general (x,y) position of the end effector, geometrically calculate the joint angles that can reach the position (inverse kinematics). Repeat the calculations for a 3-link manipulator. Include the case when the end effector should have a specific desired orientation.

3. Mark a coordinate axis system on a table. Assume the camera is initially at the origin, looking towards the Z-axis. The camera coordinate frame does coincide with the world coordinate frame. Translate the camera by $(1,1,2)$ with respect to the world coordinate frame and rotate along $\pi/2$ along the X' axis (body coordinate frame). Comment on whether the camera can view the points $(\pm 3, \pm 3, \pm 3)$, and if yes, estimate the position on the image. Take an actual calibrated camera and verify the results.

4. Assume that the error in correspondence is 5 pixels, which causes an error in the disparity of 5 pixels. Comment on the error in depth at points nearby and far away. Comment on how (and why) the resolution of the camera affect the accuracy of depth estimation in stereo vision. Also, comment on what factors affect how far a stereo-vision camera can see sufficiently accurately.

5. Identify different popular machine learning approaches. Note down how they represent the input to output mapping and the possible pros and cons of such a representation.

6. Enlist all popular robots and get an estimate of the price, hardware, and software features. Go to any popular hardware store's website to get the price of all the hardware involved. Using this study, reason out what software and hardware features increase the price of robots.

7. [Programming] Make a stereo pair using 2 webcams and measure the baseline. Calibrate both webcams. Take a chessboard at a known depth from the cameras. Estimate the depth of the chessboard using the stereo pair. Repeat with objects of everyday utility for the estimation of depth.

8. [Programming] Fix two web cameras at an arbitrary position in a room, such that a reasonably large area of the room is seen by both cameras. Use a laser pointer and detect the laser point in the images taken from both cameras. Using multiple such readings at a known distance from either camera, estimate the transformation from the first camera to the second.

9. [Programming] Take a web camera and make it look at a flat surface. Draw a visible coordinate frame at the ground. Now convert the image into an orthographic image, as would be seen by a camera from a bird's eye view.

10. [Programming] Take pictures of common indoor and outdoor environments. Use different feature detection techniques from any computer vision library. Draw the features detected to understand how computer vision algorithms work.
11. [Programming] Take some data sets from the UCI Machine Learning Repository (archive.ics.uci.edu). Use popular machine learning algorithms to learn the data and get results for both classification and regression problems.
12. [Programming] Take some synthetic 3D object templates available from the Point Cloud Library (pointclouds.org). Segment and recognize these objects from common scenes found in the same library.
13. [Programming] Make a simple animation which reads the arrow keys from the keyboard and moves a point robot on the display screen. The up/down arrow keys should control the linear speed, and the left/right arrow keys should control the angular speed. Using arrow keys, move this point robot on a synthetic map so that the robot avoids all obstacles.
14. [Programming] Make a simple language to teleoperate the robot using gestures recognized from a webcam attached to your laptop/computer. You should be able to increase/decrease linear and angular speeds like the steering and brake/throttle of a car. Detect these gestures as static and dynamic gestures. Assuming the robot is at a point in a simulated game with obstacles, note all the problems in controlling the robot using these gestures.
15. [Programming] Go around your home. Note the different objects that are of interest to the robot, like a water bottle that the robot may like to bring, the furniture that the robot would like to avoid, etc. Make a video while walking around the house. Extract the different frames. Annotate a few of the objects of interest noted earlier. Using popular object recognition libraries, how many of these objects are you appreciably able to detect and recognize?

References

Craig, J.J., 2005. Introduction to Robotics: Mechanics and Control. Pearson, Upper Saddle River, NJ.

Foote, T, 2013. The transform library. In: IEEE Conference on Technologies for Practical Robot Applications. IEEE, Woburn, MA, pp. 1–6.

Fu, K.S., Gonzalez, R., Lee, C.S.G., 1987. Robotics: Control Sensing, Vision and Intelligence. Tata McGraw-Hill, Delhi, India.

Goodfellow, I., Bengio, Y., Courville, A., 2016. Deep Learning. MIT Press, Cambridge, MA.

Guo, Y., Bennamoun, M., Sohel, F., Lu, M., Wan, J., 2014. 3D object recognition in cluttered scenes with local surface features: a survey. IEEE Trans. Pattern Anal. Mach. Intell. 36 (11), 2270–2287.

Haykin, S., 2009. Neural Networks and Learning Machines. Pearson Education, Upper Saddle River, NJ.

Kala, R., 2016. On-Road Intelligent Vehicles: Motion Planning for Intelligent Transportation Systems. Elsevier, Waltham, MA.

Malviya, V., Kala, R., 2018. Tracking vehicle and faces: towards socialistic assessment of human behaviour. In: 2018 Conference on Information and Communication Technology (CICT). IEEE, Jabalpur, India, pp. 1–6.

Malviya, V., Kala, R., 2022. Trajectory prediction and tracking using a multi-behaviour social particle filter. Appl. Intell. 52 (7), 7158–7200.

Rusu, R.B., Cousins, S., 2011. 3D is here: point cloud library (PCL). In: 2011 IEEE International Conference on Robotics and Automation. IEEE, Shanghai, pp. 1–4.

Sciavicco, L., Siciliano, B., 2000. Modelling and Control of Robot Manipulators. Springer-Verlag, London.

Shukla, A., Tiwari, R., Kala, R., 2010. Towards Hybrid and Adaptive Computing: A Perspective. Springer-Verlag, Berlin, Heidelberg.

Siegwart, R., Nourbakhsh, I.R., Scaramuzza, D., 2004. Introduction to Autonomous Mobile Robots. MIT Press, Cambridge, MA.

Szeliski, R., 2011. Computer Vision: Algorithms and Applications. Springer-Verlag, London.

Tiwari, R., Shukla, A., Kala, R., 2012. Intelligent Planning for Mobile Robotics: Algorithmic Approaches. IGI Global, Hershey, PA.

Chapter 2

Localization and mapping

2.1 Introduction

From a software perspective, the robot needs to make a lot of decisions at every point in time. The robot should decide how to avoid the obstacle in front, the route to take while going to the lab from the classroom, how to pick up the pen from the table while avoiding a collision with the other items, etc. This constitutes the problem of *motion planning* in which the planner plans for the robot to get to the desired goal without colliding. Motion planning will be taken through multiple algorithms throughout the book. Planning robotic systems need relevant inputs, and getting them is an integral part of the problem of robotics.

The first input needed for decision-making is a map of the environment that states every piece of information about the obstacles and items of interest. The creation of a map is called the problem of *mapping*. If asked to get a pen from the living room, you need to at least know the map of the relevant part of the house and where the living room is. The other major input is knowing where one is on the map, which is referred to as the problem of *localization*. Many times, you may have a map of a city, but you may not be able to navigate until and unless you know your pose on the map. Similarly, robots need to know their pose at every time. Navigation within office spaces is only possible because the global map of the place is known, the local map can be seen by humans, and the position inside the map is equally firmly known. The problems are related and often solved together as the problem of *Simultaneous Localization and Mapping* (*SLAM*).

The motion planning algorithm only lays out a plan for the robot. The motion planning algorithms provide a trajectory that the robotic system should follow. The execution of the plan or following the planned trajectory is another problem that physically moves the robot, which is done by using a *controller*. The controller makes the robot follow the planned trajectory as precisely as possible. The robot may often deviate from the trajectory due to errors in sensing, robot slipping, errors in actuation, etc. The controller constantly adapts the control of the motors to minimize such errors.

Autonomous Mobile Robots. https://doi.org/10.1016/B978-0-443-18908-1.00012-1

This chapter provides an overall view of robotics from a software perspective and gives an in-depth discussion of the specific problems of localization, mapping, and SLAM. Chapter 3 furthers the discussion by giving an in-depth discussion on the problems of planning and control and making a practical SLAM solution.

2.2 AI primer

Largely, the software design primitives are the same as that of *Artificial Intelligence* (AI) programs. The *robotic agent* (Russell and Norvig, 2010) takes the sensor readings from the environment and processes them to calculate an action, which is used to physically alter the environment through the actuators. The robot looks at the world through the sensors, manipulates the world through the actuators, and the process of mapping the sensor readings to actuator outputs is called the *problem of robotics* (Murphy, 2000). This happens in a loop shown in Fig. 2.1. This figure is also called the *sense-plan-act* loop of the robot, where the information processing relies on the notion to plan. The plan could be replaced with a *react* module, where the robot merely reacts to what it sees. The plan could also be replaced with a *learn* module, where the robot learns as it travels and makes decisions.

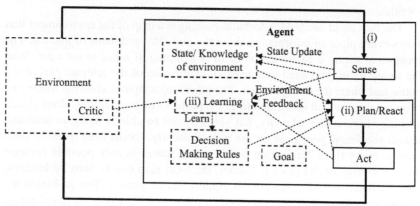

FIG. 2.1 The AI Agent. (i) The agent looks at the world using its sensors, plans its decision, and acts using the actuators. The knowledge of the world (state) is accumulated using sensing and actions. The rule base is used for making decisions, based on a known goal. (ii) Planning takes time and memory once; however, execution is real-time; reaction is fast but not optimal. (iii) The rule base may be learned using feedback from environment from the last action (e.g., fell down and felt pain, reinforcement learning), or using a dedicated critic in the environment (e.g., a human teacher, supervised learning), or by finding frequently observed patterns (whose output may be given by humans, unsupervised learning).

There is a difference between the notions of planning, reaction, and learning. In *planning*, initially, the agent acquires all the information that is needed for decision-making. This information is summarized as a *knowledge model* in the memory of the agent. The knowledge model has information like places where there is an obstacle, current robot position, facts about the environment like whether the master is at home. Thereafter, the agent makes a plan from the current state to the desired *goal state*. Once the plan is made, the agent executes the plan in multiple cycles of the loop. When the plan is invalidated due to a change in the external world, or the robot's actions did not do what they were supposed to do, or a change of goal; the robot can adapt the plan or replan.

Reaction rarely requires a big knowledge model. The sensors are near directly mapped to the actuators by arithmetic equations, functions, or rules. Some help from the knowledge model may be taken though. It can be as rules like stop when there is an obstacle directly ahead, turn right if there is an obstacle on the left, etc. The reaction is fast but lacks optimality. It can get the robot stuck. The planning is optimal but presses on having a high compute and memory and thus is an offline process.

The reactive agent moves by rules. However, the rules themselves cannot be static and predefined. If a human being falls while walking, the human being learns a new rule about how to walk. This *learning* is entirely driven by feedback from the environment and constitutes *reinforcement learning*. Sometimes, a programmer or user tells the robot the correct action to take, which is called *supervised learning*. Robots constantly observe inputs from the environment, which are often repetitive or contain frequent patterns, and *unsupervised learning* mines such frequent patterns. The robot must make decisions based on such patterns and therefore an expert must provide correct decisions corresponding to each input or pattern identified by the robot. Since only a few patterns recur, getting target output is easy without requiring a human-intensive annotation of a large part of the dataset.

A robotic workplace may deploy many robots together, referred to as the *multi-agent systems*. In multi-agent settings, the agents could be *cooperative* like a team of robots doing the same job of security; or *competitive* (also called adversarial), when the agents in the environment act opposite to the robotic agent, like a robot playing a game with a human. Another classification is based on communication. The agents may have *direct communication*, either wired or wireless; or the agents may only be able to see each other through the effects produced in the environment, called *indirect communication*. As an example, a robot can avoid another robot coming directly at it, just by looking at it.

2.3 Localization primer

Let us assume that the problem is to make the robot do one of the hardest jobs possible, which is to go to the home library and get this book for you. Let us discuss the different components that are needed to solve the problem. The problem can be fundamentally broken into a navigation problem to go to the home library and a manipulation problem, to pick up the book. Returning and giving the book to you is another navigation and manipulation problem. Let us attempt to solve the problem at a higher level of abstraction. The different problems are shown in Fig. 2.2 for the mobile base only. The problems are also summarized in Box 2.1.

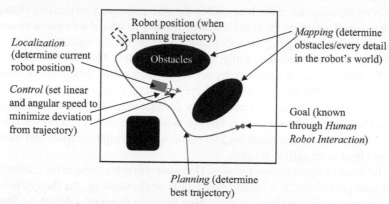

FIG. 2.2   Various subproblems in robotics. Mapping gives the details of the environment and localization gives the current robot pose, which is used to plan a trajectory and for control. A controller instructs the robot's motors to execute the planned trajectory.

Box 2.1 Different problems in robotics

Localization: Where is the robot on the map?

    Global Level: GPS, multiple wifi transmitters with the known location

    Local Level using motion model: Integrate encoders (cannot be used when the robot skids/slips causing translation without wheel rotation) or IMU signals (have large noise). Both integrate noise and errors increase with time.

    Local Level using measurement model: Landmark (say an LED light) has a known position l. Landmark is seen at pose o with respect to the robot. This is enough to calculate the pose r of the robot. However, how to determine l?

    Manipulators: Absolute encoders give joint angles

Mapping: What does the world look like?

    Global Level: Overall road-network map, room floor map, etc.

    Local Level: Specific positions of moving people, open door, etc.

    Construction: At time t, the robot pose is known to be r. The robot sees a landmark (say LED light) at position o with respect to it. So, the position of landmark with respect to world l can be calculated and registered. However, how to determine r?

Simultaneous Localization and Mapping:

    Mapping (l) assumes knowledge of localization (r); localization (r) assumes knowledge of map (l). Solve both simultaneously.

    EKF-SLAM: Simultaneously track the landmarks and robot pose. Motion is only applied to mean/variance pose variables corresponding to the robot. Observation of landmark i only corrects the mean/covariance between the robot and landmark i.

Box 2.1 Different problems in robotics—cont'd

Fast-SLAM: EKF-SLAM represents covariance between every two landmarks, which is wasteful. Fast-SLAM only represents covariance between the robot and every landmark, thus breaking the SLAM problem into k problems for k landmarks.

Planning: Make a short, smooth, and collision-free trajectory around obstacles
  Global Level: Optimal, time-consuming
  Local Level: Not-optimal, efficient, can handle moving objects

Control:
- Gives speeds to motors to trace the planned trajectory
- Handle motor noise (slip, skid), sensor noise, modelling errors
- Use localization to get the actual pose. Use a computed motion plan to get the desired pose. Feedback error between the desired and actual pose to the motors to drive the robot close to the desired pose.

The robot must always know its pose in the map of the environment, which in this example is the home. The problem of *localization* (Thrun et al., 2006) tells the robot its pose on the map and is an answer to the question *Where am I?* The localization may be the coordinates and orientation in the room for the mobile base or the specific pose of the manipulator's hand. Imagine being asked to roam a new city with a city map and you want to go to a museum. It is impossible to navigate until someone tells you where you are on the map (and your current orientation). Similarly, you can move in your house freely, only because at each instant of time you know where you are within the house. Imagine roaming a museum where all halls look alike and there is no signage. It would be impossible to locate yourself on the floor map and thus impossible to find out where to go.

The localization is broadly solved at two levels of abstraction, a global level and a local level. Examples of global level include finding the road of navigation in the road network map or knowing the room where one is at a home or office environment. At the local level, localization refers to finding the precise point in the coordinate system used, relative to the objects around. It enables one to answer for precise distance from the doors, tables, and other objects in an indoor scenario, while precisely answering the distances from cars, signals, and intersections around in an outdoor scenario.

Let us take intuitive mechanisms of localizing the robot. The GPS solves the localization problem at the global level in the outdoor environment. The problem is that the sensor has a few metres of error. The equivalent of an indoor environment is to either use multiple wifi transmitters, whose position is known. This makes a system like GPS. Both systems know the position of the transmitter and use *triangulation* from multiple such transmitters to get the position of the robot. The signal strength is a measure of the distance from the transmitter. Cast a circle of the same radius around each transmitter, and the intersection of the circles is the position of the robot. In the case of a manipulator, the encoders or potentiometer give the exact position of each joint (motor) and thus the position of the robotic hand can be precisely calculated.

2.3.1 Localization using motion models

The localization at the local level is more involved. One of the means is to assume that the robot starts from the origin (which is the charger as the robot always starts its journey from the charger) and to integrate encoder readings to know the position. The encoders tell how much the left and right wheels have turned, which can be used to calculate the position and orientation of the robot using kinematic equations. However, there are a few problems here. The encoders have a poor resolution, which limits the resolution of position calculation. When the robot slips, there is a translation motion and no rotation, which messes up the formulae. Moreover, the encoders always will have noise, and integration of signal to get position means an integration of noise, which can quickly become substantially large. The Inertial Measurement Unit (IMU) can report signals even when the robot slips, but again the IMU has a large amount of noise, which will also be integrated to know the position, thus leading to large errors. A *motion model* calculates the position of the robot at every point in time by an integration of the control sequence estimated by using suitable sensors. Encoders and IMU can both be used as motion models for the robot. The position calculated in this mechanism is only reliable for the starting few minutes as eventually the constant addition of noise with time makes the position estimate highly unreliable.

2.3.2 Localization using observation models: Camera

Another mechanism to calculate the positions does not integrate and is, therefore, more reliable. Imagine the coordinates of every object are known (specifically the coordinates of every feature in every object are known). The objects (features) are now called *landmarks*. So, you know the coordinates of (every feature contained in) tables, chairs, doors, fans, lights, trees, traffic lights, signboards, etc. Now when you observe a landmark whose coordinates in the map are precisely known; and you know the distance, angles, or coordinates at which you see the landmark from you, you can infer your coordinates. The only noise is from the current sensors observing the landmarks. However, the challenge is to know the coordinates of the landmarks.

Suppose there is a point landmark i whose position is precisely known as l_i. Suppose the transformation from the robot to the world (pose of the robot to be determined) is T_R^W, meaning the landmark should be observed at $T_W^R l_i = (T_R^W)^{-1} l_i$ in the robot's coordinate frame. Suppose the sensor S is used to observe the landmark and the transformation from the robot to the sensor is T_R^S, meaning that the sensor will see the landmark at $o_i^{proj} = T_R^S T_W^R l_i = T_R^S (T_R^W)^{-1} l_i$, where *proj* means a projection. Suppose the sensor was a 3D lidar, which observed the object at position o_i^{obs}. This means that Eq. (2.1) should be minimized by varying T_R^W, which gives the estimated pose of the robot T_R^{W*}, which is the output of localization that can be computed from any optimization method.

$$
\begin{aligned}
T_R^{W*} &= \operatorname{argmin}_{T_R^W} \sum_{o_i} \left\| o_i^{\text{proj}} - o_i^{\text{obs}} \right\|_2^2 = \operatorname{argmin}_{T_W^R} \sum_{o_i} \left\| T_R^S T_W^R l_i - o_i^{\text{obs}} \right\|_2^2 \\
&= \operatorname{argmin}_{T_R^W} \sum_{o_i} \left\| T_R^S (T_R^W)^{-1} l_i - o_i^{\text{obs}} \right\|_2^2
\end{aligned}
\tag{2.1}
$$

Finding the location by using Eq. (2.1), constitutes the *observation model* since the location is derived from the observations made by the robot's sensors alone. The equation is also called as minimizing the *re-projection error* since the equation re-projects the landmarks into the sensor plane and minimizes the error between the expected re-projection of the landmark with a known pose with the observed projection of the landmark.

Suppose a camera was used and the landmark was observed at $(u_i^{\text{obs}}, v_i^{\text{obs}})^T$ in the image, the pose can now be computed similarly as Eq. (2.2), where C is the camera calibration matrix (homogenized to 4×4 to match dimensionality) and $\pi()$ is the projection function that projects a point in the camera coordinate frame (*cam*) onto the image plane, which uses the scale (w) for normalization.

$$
\begin{aligned}
T_R^{W*} &= \operatorname{argmin}_{T_R^W} \Sigma_i \left\| \left(u_i^{\text{proj}} v_i^{\text{proj}} \right)^T - \left(u_i^{\text{obs}} v_i^{\text{obs}} \right)^T \right\|_2^2 \\
&= \operatorname{argmin}_{T_R^W} \Sigma_i \left\| \pi \left(T_R^{\text{cam}} T_W^R l_i \right) - \left(u_i^{\text{obs}} v_i^{\text{obs}} \right)^T \right\|_2^2
\end{aligned}
$$

$$
\left(u_i^{\text{proj}\prime}, v_i^{\text{proj}\prime}, w, 1 \right)^T = C T_R^{\text{cam}} T_W^R l_i
$$

$$
\left(u_i^{\text{proj}}, v_i^{\text{proj}} \right)^T = \pi \left(T_R^{\text{cam}} T_W^R l_i \right) = \left(\frac{u_i^{\text{proj}\prime}}{w}, \frac{v_i^{\text{proj}\prime}}{w} \right)^T
\tag{2.2}
$$

As an intuitive example, consider that the map contains several alphabets placed manually, doors, and some other commonly occurring objects, shown in Fig. 2.3A. Consider that a live image taken by the robot is as shown in Fig. 2.3B. For simplicity, let the characteristic point of the robot used for localization be the same as the camera, which gives $T_R^{\text{cam}} = I_{4 \times 4}$ as an identity matrix. Let the semantic objects be detected along with the four corners. The number of corners is taken as four instead of the eight corners for a cuboid since the objects being considered are planar. Even though the object recognition tells the specific object that the camera is looking at, still there is a correspondence matching problem since an object can occur multiple times in the map. The problem of *correspondence matching* is to tell, which object is being seen out of several such objects in the map. Suppose one knows an approximate robot pose and the pose of all occurrences of the detected object. All the object occurrences can be projected on the image plane as further detailed in the example. The occurrence that lies closest to the observed position of the object is the one that is being seen.

FIG. 2.3 (A) Semantic map used for localization. Each planar object has four corners, top-left, top-right, bottom-left, and bottom-right whose coordinates are approximately measured. Only two corners are visible since the 3D map is plotted as 2D. (B) Image used for the understanding of localization and mapping.

Consider the camera detects four objects, alphabet A, alphabet B, alphabet D, and alphabet E. The positions of all four corners (landmarks) of all four objects in the image $\{(u_i^{\text{obs}}, v_i^{\text{obs}})^T\}$ are noted and are given in Table 2.1. Suppose the positions of the corners in the world coordinates are known (by approximately measuring them) as $\{l_i\}$, which is also given in Table 2.1. Let us assume a random pose of the robot given by Eq. (2.3) and the inverse used in the calculations given by Eq. (2.4).

$$\text{guessed pose of robot} = T_R^W = \begin{bmatrix} -0.3536 & 0 & -0.9354 & -5.5404 \\ 0 & 1 & 0 & -0.1486 \\ 0.9354 & 0 & -0.3536 & 1.7396 \\ 0 & 0 & 0 & 1 \end{bmatrix} \quad (2.3)$$

$$T_W^R = \left(T_R^W\right)^{-1} = \begin{bmatrix} -0.3536 & 0 & 0.9354 & -3.5863 \\ 0 & 1 & 0 & 0.1486 \\ -0.9354 & 0 & -0.3536 & -4.5673 \\ 0 & 0 & 0 & 1 \end{bmatrix} \quad (2.4)$$

TABLE 2.1 Localization using objects detected in a known Semantic Map.

Semantic object	Corner	Observed image coordinates $(u_i^{obs}, v_i^{obs})^{Ta}$	Homogenized world coordinates from semantic database $(l_i)^b$	Unnormalized image coordinates $(u_i^{proj'}, v_i^{proj'}, w, 1)^T = CT_R^{cam} T_W^R l_i^c$	Predicted image coordinates $(u_i^{proj}, v_i^{proj})^T = \left(\frac{u_i^{proj'}}{w}, \frac{v_i^{proj'}}{w}\right)^T$	Squared error $\left(u_i^{obs} - u_i^{proj}\right)^2 + \left(v_i^{obs} - v_i^{proj}\right)^2$
A	Top-left	(752, 685)	(−13.236, 0.7, −1.8512, 1)	(6420.65, 5677.82, 8.47, 1)	(758.21, 670.49)	249.1042
	Top-right	(802, 685)	(−13.316, 0.7, −1.6039, 1)	(6778.72, 5671.16, 8.45, 1)	(801.69, 670.71)	204.3002
	Bottom-right	(802, 752)	(−13.316, 1.05, −1.6039, 1)	(6778.72, 6168.39, 8.45, 1)	(801.69, 729.51)	505.8962
	Bottom-left	(752, 752)	(−13.236, 1.05, −1.8512, 1)	(6420.65, 6175.05, 8.47, 1)	(758.21, 729.21)	557.9482
B	Top-left	(806, 652)	(−16.564, 0.7, −2.9327, 1)	(9679.74, 7523.85, 11.96, 1)	(809.10, 628.90)	543.22
	Top-right	(842, 652)	(−16.645, 0.7, −2.6855, 1)	(10039.01, 7517.70, 11.95, 1)	(839.95, 629.00)	533.2025
	Bottom-right	(842, 700)	(−16.645, 1.05, −2.6855, 1)	(10039.01, 8014.93, 11.95, 1)	(839.95, 670.60)	868.5625
	Bottom-left	(806, 700)	(−16.564, 1.05, −2.9327, 1)	(9679.74, 8021.08, 11.96, 1)	(809.10, 670.46)	882.2216

Continued

TABLE 2.1 Localization using objects detected in a known Semantic Map—cont'd

Semantic object	Corner	Observed image coordinates $(u_i^{obs}, v_i^{obs})^T$	Homogenized world coordinates from semantic database (l_i)	Unnormalized image coordinates $(u_i^{proj'}, v_i^{proj'}, w, 1)^T = CT_R^{cam} T_{WI} l_i$	Predicted image coordinates $(u_i^{proj}, v_i^{proj})^T = \left(\frac{u_i^{proj'}}{w}, \frac{v_i^{proj'}}{w}\right)^T$	Squared error $(u_i^{obs} - u_i^{proj})^2 + (v_i^{obs} - v_i^{proj})^2$
D	Top-left	(1302, 762)	(−11.127, 0.7, 1.3679, 1)	(6948.63, 4034.80, 5.36, 1)	(1297.09, 753.17)	102.077
	Top-right	(1378, 762)	(−11.204, 0.7, 1.6057, 1)	(7293.04, 4028.43, 5.34, 1)	(1364.45, 753.68)	252.8249
	Bottom-right	(1378, 861)	(−11.204, 1.05, 1.6057, 1)	(7293.04, 4525.66, 5.34, 1)	(1364.44, 846.70)	388.3636
	Bottom-left	(1302, 861)	(−11.127, 1.05, 1.3679, 1)	(6948.63, 4532.03, 5.36, 1)	(1297.09, 845.98)	249.7085
E	Top-left	(1058, 639)	(−19.591, 0.7, −1.3823, 1)	(15237.71, 8729.69, 14.25, 1)	(1069.58, 612.75)	823.1589
	Top-right	(1090, 639)	(−19.669, 0.7, −1.1446, 1)	(15583.33, 8723.83, 14.24, 1)	(1094.67, 612.82)	707.2013
	Bottom-right	(1090, 678)	(−19.669, 1.05, −1.1446, 1)	(15583.33, 9221.06, 14.24, 1)	(1094.67, 647.74)	937.4765
	Bottom-left	(1058, 678)	(−19.591, 1.05, −1.3823, 1)	(15237.71, 9226.91, 14.27, 1)	(1069.56, 647.65)	1054.756
Root mean square error						23.53

Calculation of error for a single pose estimate is shown. Optimization is used to get the best pose estimate (localization).

[a]Image: Y-axis is positive in the top-down direction, X-axis is positive in the left-right direction, and origin at top-left.

[b]World: Y-axis is positive facing the ground. The robot moves at the XZ plane.

[c]guessed pose of robot $= T_R^W = \begin{bmatrix} -0.3536 & 0 & -0.9354 & -5.5404 \\ 0 & 1 & 0 & -0.1486 \\ 0.9354 & 0 & -0.3536 & 1.7396 \\ 0 & 0 & 0 & 1 \end{bmatrix}$ (this guess is optimized as the problem of localization).

$$T_W^R = (T_R^W)^{-1} = \begin{bmatrix} -0.3536 & 0 & 0.9354 & -3.5863 \\ 0 & 1 & 0 & 0.1486 \\ -0.9354 & 0 & -0.3536 & -4.5673 \\ 0 & 0 & 0 & 1 \end{bmatrix}$$

$T_R^{cam} = I_{4 \times 4} = \begin{bmatrix} 1 & 0 & 0 & 0 \\ 0 & 1 & 0 & 0 \\ 0 & 0 & 1 & 0 \\ 0 & 0 & 0 & 1 \end{bmatrix}$ (camera's coordinate system is used as the robot's coordinate system).

$C = \begin{bmatrix} 1421.28 & 0 & 865.24 & 0 \\ 0 & 1420.64 & 528.13 & 0 \\ 0 & 0 & 1 & 0 \\ 0 & 0 & 0 & 1 \end{bmatrix}$ (known Calibration matrix, extended to 4×4).

Let us project all corners of all objects in the image plane assuming the guessed robot pose, with a known calibration matrix. Let us also measure the error between the observed corners and the projected corners, shown in Table 2.1. If the guess in Eq. (2.3) was good, the projected corners should coincide with the observed corners with some margin of error. Now an optimization library can be used to modify the guessed pose, such that the error is minimal, which constitutes the problem of localization.

The example is not realistic since a robot would require many features to effectively localize while the example has very few of them. Furthermore, a little error in the assumed position of the object corners close to the camera translates to a large change in the projected position of the object on the image, and the optimization library will be sensitive to such objects. Imagine moving the alphabet D by a metre, which produces a large change in its projected position on the image. Imagine moving the door at the far end by a metre. That hardly produces a change in the image. The positions of the nearby objects should thus be highly accurate. If the object positions are known with good accuracy, the far objects produce a small change in the image for a large change of the robot position and thus do not enable an accurate localization.

2.3.3   Localization using observation models: Lidar

Let us rediscuss the same example assuming a laser scanner. A 3D laser scanner produces 3D images, while a 2D laser scanner produces 2D images. Both images can be processed to get landmarks (features). Suppose the map is perfectly known, so the position of all landmarks (l_i) is known in 2D or 3D. Suppose the live laser scanner detects the landmark l_i, whose position can be fetched from the database. This requires a solution to the correspondence problem. From all the landmarks in the database, the landmark with the closest signature (feature descriptor) is used. Additionally, if a prospective pose of the robot is known (from computed pose at previous steps), the projection of landmark should be at the same spatial region of the lidar image as actually observed. A 2D lidar measures the distances (called *range*) of obstacles at all angles (called *bearing*) with respect to the robot heading. A 3D lidar measures the distances (called range) of obstacles at all solid angles (called *bearing* and *elevation*) with respect to the robot heading.

Consider a 2D lidar mounted at the centre of a planar robot. Suppose the landmark l_i is observed at a range d_i^{obs} and bearing α_i^{obs} (similarly, range-bearing-elevation of d_i^{obs}-α_i^{obs}-ϕ_i^{obs} in 3D). This means that the ray of the laser scan that hit the landmark was at an angle α_i^{obs} and reported a distance of d_i^{obs}. Suppose the robot position is given by (x_t, y_t) and orientation by θ_t at time t, which is assumed (and actually needs to be calculated). Suppose the transformation between the lidar and robot is an identity matrix. The landmark l_i (l_{ix}, l_{iy}) would be seen at a range d_i^{proj} and bearing α_i^{proj}, where the quantities are given by Eq. (2.5) and shown in Fig. 2.4.

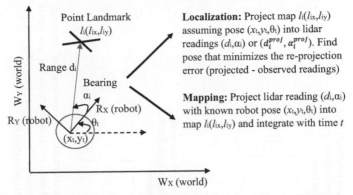

FIG. 2.4 Measurement model using lidar. The sensor reports the distance to the landmark (d_i or range) and the angle to the landmark (α_i or bearing).

$$\begin{bmatrix} d_i^{\text{proj}}(x_t, y_t, \theta_t) \\ \alpha_i^{\text{proj}}(x_t, y_t, \theta_t) \end{bmatrix} = \begin{bmatrix} \sqrt{(l_{ix} - x_t)^2 - (l_{iy} - y_t)^2} \\ \text{atan2}(l_{iy} - y_t, l_{ix} - x_t) - \theta_t \end{bmatrix} \quad (2.5)$$

The optimization algorithm optimizes the robot pose (x_t, y_t, θ_t) to minimize the error, or Eq. (2.6)

$$(x_t, y_t, \theta_t)^T = \text{argmin}_{(x_t, y_t, \theta_t)^T} \Sigma_i \left\| \left(d_i^{\text{proj}}(x_t, y_t, \theta_t), \alpha_i^{\text{proj}}(x_t, y_t, \theta_t) \right)^T - \left(d_i^{\text{obs}}, \alpha_i^{\text{obs}} \right)^T \right\|_2^2$$

$$(2.6)$$

Both previous techniques use features or landmarks. Let us take another example of localization using completely dense maps called *dense localization*. Consider that the complete map as a binary image is known as a grid. A laser scan image is essentially a number of laser rays that report the distance to the obstacles. Let the laser ray at angle α (*bearing*) with respect to the robot's heading for a 2D lidar scanner (α or bearing and φ or elevation for a 3D lidar) have an observed distance to obstacle value of d_α^{obs} (called *range*). Assume that a live laser scan is made at time t. Suppose the robot's pose is guessed to be at a position (x_t, y_t) and orientation by θ_t. Place a virtual robot with translation and rotation (pose) given by (x_t, y_t, θ_t). Project a virtual laser at the same angle α (α and φ for 3D) to get an expected reading if the robot was placed at (x_t, y_t, θ_t). Say the laser should record a distance value of $d_\alpha^{\text{proj}}(x_t, y_t, \theta_t)$. The error is $\Sigma_\alpha (d_\alpha^{\text{proj}}(x_t, y_t, \theta_t) - d_\alpha^{\text{obs}})^2$, which is minimized over the robot pose (x_t, y_t, θ_t). The robot pose (x_t, y_t, θ_t) that gives the least error is the solution to the localization problem.

An advantage of lasers is that the laser images can be stitched with time just like the camera images can be stitched, which is very common in photography to produce a panoramic view. Consider the laser scan image produced by the laser at the centre of the robot shown in Fig. 2.5A. Suppose, the first image

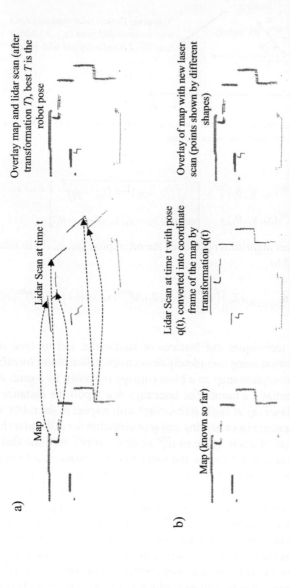

FIG. 2.5 (A) Localization. Assume map is known. Compute transformation T that minimizes the difference between the lidar scan (after transformation T) and map, where the best T is the robot pose. *Arrows* show corresponding points. (B) Mapping. The positions are assumed to be known and therefore the second lidar scan is converted into the frame of reference of the map and overlaid to extend the map. Similarly, all scans are merged. Localization assumes a known map, while mapping assumes localization (pose) is always known. Simultaneously solving them is called Simultaneous Localization and Mapping (SLAM).

is taken when the robot starts its journey at the origin, or the pose of the robot at time 0 with respect to the world is identity ($T_0^W = I_{4\times4}$). The second laser scan image that is captured live is transformed in multiple trials until it best matches the first image, and the transformation gives the new pose of the robot (T_1^0), given that the environment remained static and the change between images was only due to the motion of the robot. Imagine the original laser image in one colour. Apply an unknown (guessed) transform T_1^0 to the live (second) image. Plot the transformed 2nd image by another colour. If the guess was good, the two images will overlap to the maximum extent. Correlation and occurrence of the same pixels in a grid are good metrics to use to measure the similarity. The optimizer finds the best transformation that minimizes the error. Similarly, subsequent laser scan images can be stitched by transforming the subsequent images that best match the previous ones, while the transformation that minimizes the error between the transformed new image and the old image is the motion of the robot between the successive frames (T_t^{t-1}) at time t. The transforms can be continuously integrated to give the localization or the robot pose ($T_t^W = T_0^W \, T_1^0 \, T_2^1 \dots T_t^{t-1}$). Alternatively, the transformation (T_t^W) that minimizes the error between the transformed new image with the stitched image map so far is the solution to the localization problem, only considering the segment of the map that is likely visible in the new lidar scan image.

2.4 Mapping primer

The problem of mapping is to represent the complete environment in a manner to aid in decision-making. The *mapping* is an answer to the question *What does the world around me look like*? Knowing the position alone, humans cannot navigate in any environment. You know how to reach the library because the map of the home is embedded in the mind. It is possible to drive in a city because the navigation system shows the map of the place. It is also possible to roam around unknown museums because the floor map tells what is to be found and where. You can lift books because the location of the book and its vicinity can be seen. Mapping itself may be at a global level, which is seen as a road network graph of a city, a room guide of homes, or floor plans of museums. It may even be at a detailed local level, wherein it additionally mentions moving (dynamic) people and cars in the vicinity, the specific positions of chairs and tables that may have moved, etc.

Let us also look at mapping using some intuitive means. Maps may sometimes be seen like a visual maps drawn on paper as a 2D or 3D grid. Let us attempt to make the same map. Imagine a mobile robot that has a 2D (equivalently 3D) lidar that reports distances to obstacles around. Assume that the robot starts from the origin. Suppose ith ray $l(0)$ at bearing α_i observes a range of d_i. Here 0 refers to time. This means in the sensor's coordinate frame, the laser ray has detected a point obstacle at position $l(0) = [d_i \cos \alpha_i, d_i \sin \alpha_i]^T$, where there are several such rays giving a collection of obstacles $\{l(0)\}$. This can be projected as an image shown in Fig. 2.5B. The point obstacles in the sensor's coordinate frame are at $\{l(0)\}$, or $T_S^R\{l(0)\}$ with respect to the robot. Here 0 refers to time. Let the robot initially be at the origin or $T_R^W(0)$ is an identity matrix at time 0. Now after some time, the robot moves to a new position, which is known from the localization system. This creates another new image. The robot sees a new set of points $\{l(1) = [d_i \cos \alpha_i, d_i \sin \alpha_i]^T\}$ with respect to the sensor, which is $T_S^R\{l(1)\}$ with respect to the robot and $T_R^W(1)T_S^R\{l(1)\}$ with respect to the world, where 1 represents the time. $T_R^W(1)$ is known from the localization system (it is the robot pose). These points are just appended over the previous points making the map. The map thus contains the points $T_R^W(0)T_S^R\{l(0)\} \cup T_R^W(1) T_S^R\{l(1)\}$. The map iteratively acquires points $\cup_t T_R^W(t)T_S^R\{l(t)\}$.

Suppose a planar robot is used with a 2D lidar sensor. The robot is assumed to be at position (x_t, y_t) with orientation θ_t, where t refers to the time. The pose (x_t, y_t, θ_t) is accurately known from the localization system. Furthermore, suppose the transformation between the robot and the sensor is identity. The new point obstacles are $\{l(t) = \cup_i (d_i \cos \alpha_i, d_i \sin \alpha_i)^T\}$ in the robot's coordinate system, which transforms to $\{\cup_i (x_t + d_i \cos(\theta_t + \alpha_i), y_t + d_i \sin(\theta_t + \alpha_i))^T\}$ in the world coordinate system. The conversion is shown in Fig. 2.4. The map is thus a collection of point obstacles $\{\cup_t \cup_i (x_t + d_i \cos(\theta_t + \alpha_i), y_t + d_i \sin(\theta_t + \alpha_i))^T\}$. For storage purposes, points remarkably close to each other may be merged to limit the map size or the map may be stored in the form of a grid with a fixed resolution, with only 1 point inside the grid denoting an obstacle. Both techniques limit the map by the resolution due to possibly a very large number of continuous points; however, incur a discretization loss. To limit the discretization loss, the resolution may be kept very high. However, a very large resolution of a grid in 3D will require a massively high memory that may not be available. The technique has practical issues since false positives will have to be handled. In other words, a laser scanner ray may occasionally give a wrong reading, projecting a point obstacle that does not exist. This causes a false point obstacle to be projected into the map. The approach should, hence, only consider cells as obstacles if they are consistently reported as obstacles.

Similarly, consider Fig. 2.3B. Assume that the door at the end of the corridor is detected, that is not in the current semantic map. Now the robot detects the door and all four corners. The detection of the door corners on the image plane is shown in Table 2.2 for both the left and the right cameras. This is used to compute the disparity, through which the coordinates in the camera coordinate frame are computed (l_{cam}), where the subscript cam refers to the left camera coordinate frame. As previously, assume that $T_{cam}^R = I_{4\times4}$ or the robot's coordinate frame is the same as the left camera coordinate frame. Assume that the pose of the robot at the time of taking the image is the same as Eq. (2.3). This gives the position of the four coordinates in the world as $T_R^W T_{cam}^R l_{cam}$. The computation is shown in Table 2.2. This object is now added on top of the objects in the semantic map and can be used for localization in the future.

TABLE 2.2 Mapping the objects detected from the stereo camera.

Semantic object	Corner	Observed left image coordinates $(u_{Ll}, v_{Ll})^a$	Observed right image coordinates $(u_{Rl}, v_{Rl})^a$	Disparity $(d = u_{Ll} - u_{Rl})$	Depth $I_{z,cam} = bf/d$, b = known baseline, f = known focal length	Position in sensor (left camera)^b coordinates $I_{cam} = \begin{bmatrix} (u_{Ll}-c_x)p_{z,cam}/f_x \\ (v_{Ll}-c_y)p_{z,cam}/f_y \\ p_{z,cam} \\ 1 \end{bmatrix}$	Position in world coordinates $T_R^W T_{cam}^R I_{cam}^{\,c}$
Door	Top-left	(910, 513)	(895, 513)	15	11.3702	(0.35808, −0.12109, 11.3702, 1)	(−16.3027, −0.2697, −1.9460, 1)
	Top-right	(1008, 513)	(994, 513)	14	12.1824	(1.223657, −0.12974, 12.1824, 1)	(−17.3685, −0.2783, −1.4235, 1)
	Bottom-right	(1008, 630)	(994, 630)	14	12.1824	(1.223657, 0.873565, 12.1824, 1)	(−17.3685, 0.7250, −1.4235, 1)
	Bottom-left	(910, 630)	(895, 630)	15	11.3702	(0.35808, −0.815327, 11.3702, 1)	(−16.3027, 0.6667, −1.9460, 1)

Assuming that the robot knows its pose, the new objects are added to the semantic database.

^{a}Image: Y-axis is positive in the top-down direction, X-axis is positive in the left-right direction, and origin at top-left.

$^bC = \begin{bmatrix} f_x & 0 & c_x \\ 0 & f_y & c_y \\ 0 & 0 & 1 \end{bmatrix} = \begin{bmatrix} 1421.28 & 0 & 865.24 \\ 0 & 1420.64 & 528.13 \\ 0 & 0 & 1 \end{bmatrix}$, known as calibration matrix.

^{c}World: Y-axis is positive facing the ground. The robot moves at the XZ plane.

$T_{cam}^R = I_{4\times4} = \begin{bmatrix} 1 & 0 & 0 & 0 \\ 0 & 1 & 0 & 0 \\ 0 & 0 & 1 & 0 \\ 0 & 0 & 0 & 1 \end{bmatrix}$ (left camera's coordinate system is used as the robot's coordinate system).

Suppose pose of robot $= T_R^W = \begin{bmatrix} -0.3536 & 0 & -0.9354 & -5.5404 \\ 0 & 1 & 0 & -0.1486 \\ 0.9354 & 0 & -0.3536 & 1.7396 \\ 0 & 0 & 0 & 1 \end{bmatrix}$ (assumed to be known for solving the mapping problem). Obtained from the localization module in Table 2.1

on prior-registered landmarks for the SLAM problem).

The physical location of the door is reasonably further away from the one calculated in Table 2.1. The maximum depth that a camera can sense depends upon the resolution of the camera, the precision of the camera (internal unmodelled errors), and the precision of the calibration parameters. The depth is the inverse of disparity. Therefore, a few pixel errors in disparity does not lead to a significant error for nearby objects that have a large disparity, while the errors are extremely high for far objects that have a low disparity. Hence, stereovision cannot be used to measure the distance of far objects like the moon and stars since the disparity will come out to be zero. The advantage is that the robot will see the same door multiple times, sometimes nearby while sometimes far away. The door's position is taken as the one largely computed when the door was close enough by using a weighted averaging.

It may be interesting to note that the robot pose was always calculated with respect to the map. So, a map with a known coordinate axis system was assumed. The coordinates of everything in the map, including (all the features in) the doors, tables, chairs, paintings, traffic signals, traffic lights, trees, etc., were assumed to be known. The robot observed these landmarks (map) to calculate the robot's pose (localization) with respect to these landmarks (map). At the same time, the map was calculated with knowledge of the robot's pose. The robot saw landmarks around and (knowing its pose or localization, and the position of landmarks with respect to itself) calculated the coordinates of the landmarks (map). The localization (position), therefore, needs the map (landmarks) as input, while mapping, in turn, needs the localization (pose) as input. This is the chicken and egg problem in robotics called *Simultaneous Localization and Mapping* (SLAM).

Reconsider the example in Table 2.1. First assuming the map, the pose was computed in Table 2.1 subjected to optimization. Thereafter, the pose computed from Table 2.1 after optimization was used to make an expanded map that included the door. As the robot moves, it can localize itself with respect to the known location of the door calculated from Table 2.2. The computed pose through an optimization like Table 2.1 while taking the door as the feature can be used to compute the location of new features detected and update the ones known previously (like computing a better position of the door). This goes on as the robot moves.

Previously, it was stated that the nearby objects need to be registered accurately, otherwise, a small error in registering them can make a large error in the projected image plane, which translates to the need of having an accurate robot position during registration. Say, the robot registers a feature (object corner), moves a small distance, and uses the registered feature (object corner) to estimate the motion, after which the registered feature is updated. The relative motion (odometry) computation is not a function of the robot's position to initially register the feature, and therefore, an error in the initial pose does not translate into an error in measuring the relative motion between frames. If there was an error T in the pose of the robot when the feature was registered and (suppose) the computation of the corner's position relative to the robot pose was error-free, then there will be an error of T in the registration of the feature. If the projection on the observed image is error-free, the only error will be from the error T in registering the object. The odometry or the computation of transformation between two frames is invariant to the magnitude of error present at the first frame. Practically, there will be an additional error due to the computation of the feature pose relative to the robot and projection on the Image plane. The robot picks up errors along with time causing a gradual drift in the trajectory, but there is no fatal localization error due to a small initial localization error while registering the nearby features.

2.5 Planning

Given a map of a new place and your pose on the map, numerous ways make you reach your goal. The best way will have to be found. Similarly, knowing the map of your campus, sitting at an office, the route to the coffee shop has numerous options, out of which the best route must be computed. If you see an obstacle directly ahead like a chair, it could be avoided from the left or the right, a decision about the same needs to be made. If you see numerous people approaching opposite to where you are going, a decision whether to intersect the group, avoid the group from the left or avoid the group from the right needs to be made. All these are *planning decisions* to be made (Choset et al., 2005; Tiwari et al., 2012). Even manipulators need to reason on how to reach the obstacle to grasp, without causing a collision at any of the links. Planning happens at multiple levels of hierarchy from route planning at the highest level in indoor or outdoor scenarios; to deciding the distance to keep from the obstacle while avoiding it. In all cases, the aim is to make the navigation collision-free, while optimizing on the objectives of path length, smoothness, safety, etc.

The fundamental manner of dealing with the problem is a graph search at a global level, called deliberative planning. The actual environment is continuous and therefore multiple theories exist to convert it into a discrete and finite set of vertices and edges. The graph search can take a prolonged time and many times a quick solution is required to avoid a rapidly moving obstacle. The other type of approach is at the local level, which is reactive. These approaches make quick decisions from the sensor percept based on some rules. Most of the book is devoted to the problem of motion planning. The dynamic obstacles like moving people cause a problem at all levels of problem-solving with robotics. Since the people are dynamically moving, the algorithm must be quick to decide at least the immediate moves (if not the complete plan) before the people move further and invalidate the plan. Similarly, the local maps need to be made to include such dynamic people for motion planning while the map will be constantly change as people move. Localization must never use these people as features who keep moving (and the landmarks were assumed to be static) while localizing at a local level (knowing the pose relative to such people) is needed for decision-making on how to avoid them.

2.6 Control

The planning algorithms take decisions from the highest level on the route to take, to the lowest level on how to avoid the obstacles. The actual execution of the decisions is done by the *control algorithms* (Lynch and Park, 2017). The problem is interesting because ultimately the robots are operated by motors that can be given reference speeds to operate. The task of calculating good reference speeds is done by the control algorithm. The control algorithms will incur errors as they go because neither the motors are ideal, nor do the sensors report the correct data, nor are the robot models fail-proof, nor are the modules in complete synchronization. The robot may skid, or slip or the robot may de-orient along with the motion. The controller corrects all such errors as the motion happens.

The control problem is to get the current pose from the localization system directly or from the other sensors indirectly, to get the desired pose from the planning system, and to then reason on what control signal should be given to the motors to have the actual pose as close as possible to the desired one. The errors are continuously computed and rectified in the process. The easiest case is that of the manipulator where every motor is responsible to control a joint whose desired angular position is known. It can get more complex. As an example, a controller may want to make a car go sideways, but no motor can make the car go sideways. In other words, the degree of freedom requiring correction is not actuated. The controller must then make the car move and turn to get the desired configuration. There are three configurations (x, y, θ) controlled by using only two control inputs (linear and angular speed). Similar is the case of a drone, four motors control six degrees of freedom. The control of a robot is discussed in detail in Chapter 3.

2.7 Tracking

The problem of determining one's position can also be inferred as tracking of one's position. *Tracking algorithms* gained popularity for the need for ground stations to track moving objects, vehicles, airplanes, etc. The tracking is done by using *filtering algorithms*, that filter out noise from the signal to compute the position along with time. There are always two ways to track an object. Imagine that the pose of the object was known at time t and further that the controls applied, speed of the object, and all kinematics are known precisely. This means that the kinematics can be used to compute the pose at time $t+1$ and thereafter. This is called the *motion model*. Furthermore, suppose that an observation is made by a sensor that depends upon the pose of the object. With a knowledge of the world, the expected (projected) sensor reading for a given pose can be computed, while the actually observed readings are known. The pose is thus the one that minimizes the error between the expected and observed values. Knowing the observation, it is, therefore, possible to recover the pose of the object at any time, called the *observation model*. While the separate models are already discussed, let us fuse both models into one system.

The state at time t (say q_t) is represented by a Probability Density Function called *belief*. Let $u_{1:t} = u_1, u_2 \ldots u_t$ be the control signals till time t (linear speed, angular speed, etc., for a robot or object of study). Similarly, let $z_{1:t} = z_1, z_2 \ldots z_t$ be the observations till time t (landmarks observed by the robot or object in the study observed from the base). The belief is given by Eq. (2.7):

$$bel(q_t) = P(q_t|u_{1:t}, z_{1:t}) = P(q_t|u_1, z_1, u_2, z_2 \ldots u_t, z_t) \tag{2.7}$$

Let us assume that the robot is given one control step, makes one observation, takes another control step, and so on. The belief may even be calculated before the application of the observation, given by a slightly different notation shown in Eq. (2.8). The only difference with Eq. (2.7) is the absence of z_t as the belief is measured before observation.

$$\overline{bel}(q_t) = P(q_t|u_{1:t}, z_{1:t-1}) = P(q_t|u_1, z_1, u_2, z_2 \ldots u_t) \tag{2.8}$$

The bar denotes that the belief computed is the preobservation estimate. The terms can be recursively stated. First, apply the Bayes algorithm to get the relation as shown in Eq. (2.9).

$$bel(q_t) = P(q_t|u_{1:t}, z_{1:t-1}, z_t) = \frac{P(z_t|q_t, u_{1:t}z_{1:t-1})P(q_t|u_{1:t}z_{1:t-1})}{\displaystyle\int_{q_t} P(z_t|q_t, u_{1:t}z_{1:t-1})P(q_t|u_{1:t}z_{1:t-1})dq_t} \tag{2.9}$$

Using the Markovian assumption, the current observation (z_t) only depends upon the current state (q_t) and not on the sequence of observations and actions of the past. Furthermore, the denominator can always be calculated as a normalization factor of the numerator to make the summation 1, and thus using η to represent the denominator gives Eqs (2.10) and (2.11):

$$bel(q_t) = \eta P(z_t|q_t)P(q_t|u_{1:t}z_{1:t-1}) \tag{2.10}$$

$$bel(q_t) = \eta P(z_t|q_t)\overline{bel}(q_t) \tag{2.11}$$

Similarly, consider Eq. (2.8), in which let q_{t-1} be added to give Eq. (2.12)

$$\overline{bel}(q_t) = P(q_t|u_{1:t}, z_{1:t-1}) = \int_q P(q_t|q_{t-1}u_{1:t}, z_{1:t-1})P(q_{t-1}|u_{1:t}, z_{1:t-1})dq_{t-1} \tag{2.12}$$

Again, use the Markovian assumption that the state at time t (say q_t) only depends upon the last state (q_{t-1}) and last control input (u_t) and not the complete history of inputs. Similarly, note that u_t does not affect the probability for q_{t-1} and can be conditionally neglected. This gives Eqs (2.13)–(2.15):

$$\overline{bel}(q_t) = \int_q P(q_t|q_{t-1}u_t)P(q_{t-1}|u_{1:t}, z_{1:t-1})dq_{t-1} \tag{2.13}$$

$$\overline{bel}(q_t) = \int_q P(q_t|q_{t-1}u_t)P(q_{t-1}|u_{1:t-1}, z_{1:t-1})dq_{t-1} \tag{2.14}$$

$$\overline{bel}(q_t) = \int_q P(q_t|q_{t-1}u_t)bel(q_t)dq_{t-1} \tag{2.15}$$

2.7.1   Kalman Filters

The recursive representation of the belief gives a generic mechanism to compute the state, however, the actual calculation depends a lot on how the probability density function is represented. Different filtering algorithms have different modelling of the same recursive representation of belief. Let us now assume that the probabilities are Gaussian multivariate, which makes the Kalman Filter. This means that the probability density function can be represented by a mean $(\mu_t)_{n\times 1}$ and covariance $(\Sigma)_{n\times n}$ where the numbers in subscript show the matrix size for n state variables. To ease calculations, let us further assume that the motion model and observation model are both linear, as given by Eqs (2.16) and (2.17):

$$q_t = Aq_{t-1} + Bu_t + \epsilon_t^{motion} \tag{2.16}$$

$$z_t = Cq_t + \epsilon_t^{obs} \tag{2.17}$$

Here, A and B are the coefficients of the linear motion model system, q_t is the state, u_t is the control input, z_t is the observation, C is the coefficient of the observation system, and ϵ_t^{motion} and ϵ_t^{obs} are the Gaussian noises of the motion and observation processes, respectively. As an example, consider that a two-link manipulator can be controlled independently for both joints by giving suitable rotation speeds, and encoders at both joints enable one to see the position of the motor at any instant. The model is given by Eqs (2.18) and (2.19):

$$\begin{bmatrix} \theta_{1,t} \\ \theta_{2,t} \end{bmatrix} = \begin{bmatrix} 1 & 0 \\ 0 & 1 \end{bmatrix} \begin{bmatrix} \theta_{1,t-1} \\ \theta_{2,t-1} \end{bmatrix} + \begin{bmatrix} \Delta t & 0 \\ 0 & \Delta t \end{bmatrix} \begin{bmatrix} \omega_{1,t} \\ \omega_{2,t} \end{bmatrix} + \begin{bmatrix} \epsilon_{t,2}^{motion} \\ \epsilon_{t,1}^{motion} \end{bmatrix} \tag{2.18}$$

$$\begin{bmatrix} \theta_{1,t}^{obs} \\ \theta_{2,t}^{obs} \end{bmatrix} = \begin{bmatrix} 1 & 0 \\ 0 & 1 \end{bmatrix} \begin{bmatrix} \theta_{1,t} \\ \theta_{2,t} \end{bmatrix} + \begin{bmatrix} \epsilon_{1,t}^{obs} \\ \epsilon_{2,t}^{obs} \end{bmatrix} \tag{2.19}$$

Here, $q_t = [\theta_{1,t}, \theta_{2,t}]^T$ represents the state at time t, which is observed as $z_t = [\theta_{1,t}^{obs}, \theta_{2,t}^{obs}]^T$ in the form of both joint angles due to observation noise. Δt is the difference in times between consecutive iterations, and

$u_t = [\omega_{1,\ t}, \omega_{2,\ t}]^T$ are the angular speeds of the two joints. From the equations $A = \begin{bmatrix} 1 & 0 \\ 0 & 1 \end{bmatrix}$, $B = \begin{bmatrix} \Delta t & 0 \\ 0 & \Delta t \end{bmatrix}$, and $C = \begin{bmatrix} 1 & 0 \\ 0 & 1 \end{bmatrix}$.

The task is now to guess the state, given that the noises are unknown and not measurable. Let μ_t be the belief at time t. The robot makes a move using a control u_t and some noise. The noise is assumed to be Gaussian. The belief after the application of the control step involves the multiplication of the Gaussian probabilities of motion and previous belief using Eq. (2.15). However, multiplication of two Gaussian distributions to get another Gaussian distribution is an intensive step that will be skipped. Luckily, the result is very intuitive. The expected mean, if the noise has a zero mean, is given by Eq. (2.20), while the covariance after the application of the control is given by Eq. (2.21).

$$\bar{\mu}_t = A\mu_{t-1} + Bu_t \tag{2.20}$$

$$\bar{\Sigma}_t = A\Sigma_{t-1}A^T + R \tag{2.21}$$

This is also called the *prediction step*. Here, R is the covariance of the motion noise. The bar is applied to both mean and covariance in satisfaction of the definition that they represent the belief before correction by the observation. The equation states that the expected mean is the one without noise, while if only controls are applied, the noise will continue to be added to the covariance, making it even noisier. This means motion alone causes the continuous addition of errors.

Let us, therefore, correct the same using an *observation*. The observation noise is also assumed to be Gaussian in nature. We again need to multiply the current belief from Eqs (2.20) and (2.21) with the Gaussian distribution of observation to get the updated belief as per Eq. (2.11), which is again an involved step and will be skipped. However, the results are again intuitive. The model will correct the belief of Eqs (2.20) and (2.21) in the proportionality of the error measured in the observation. Therefore, after making an observation, a weighted correction will be applied, and the weight is called the *Kalman Gain* (K_t) given by Eq. (2.22):

$$K_t = \bar{\Sigma}_t C^T \left(C\bar{\Sigma}_t C^T + Q \right)^{-1} \tag{2.22}$$

Here, Q is the covariance of the observation noise. Given a gain factor (K_t), the correction is K_t times the error in observation for the mean. z_t^{obs} was the observation, which using Eq. (2.17) should have been $z_t^{\text{proj}} = C\bar{\mu}_t$ in the absence of noise and for the best estimate of state so far. This gives the error as $z_t^{\text{proj}} - z_t^{\text{obs}} = z_t^{\text{proj}} - C\bar{\mu}_t$. Similarly, for the covariance, where the correction is over the old value $\bar{\Sigma}_t$ and is multiplicative. The new mean and covariance are thus given by Eqs (2.23) and (2.24):

$$\mu_t = \bar{\mu}_t + K_t\left(z_t^{\text{obs}} - z_t^{\text{proj}} \right) = \bar{\mu}_t + K_t\left(z_t^{\text{obs}} - C\bar{\mu}_t \right) \tag{2.23}$$

$$\Sigma_t = (I - K_t C)\bar{\Sigma}_t \tag{2.24}$$

Informally, the *Kalman gain* is how much the system believes in the observation in contrast to the motion model. Consider that the observation noise (Q) is nearly infinite, in which case the inverse becomes nearly 0, drawing an analogy to the real numbers and the Kalman gain is 0, indicating nothing should be done using the observation model. Neither mean nor covariance shall change. On the contrary, if the observation noise (Q) is 0, the Kalman gain (informally) becomes C^{-1}, meaning that correction covariance will become 0 while the mean will be (informally) $C^{-1}z_t^{obs}$, which is the ideal mean with a zero-observation error. These discussions are done informally, as the observation matrix C may not be invertible.

Unlike the motion model, the observation model may increase or reduce uncertainty, which depends upon the quantity K_tC relative to the identity matrix. Typically, if the measurement uncertainty is larger than the one due to motion (rarely the case), the uncertainty may increase. However, mostly there is a substantial reduction in uncertainty due to the observation, which ensures that tracking never grows uncertain due to motion. The general methodology is summarized in Fig. 2.6.

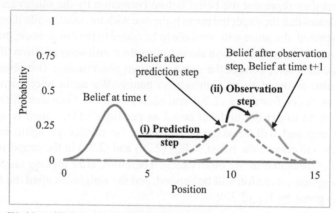

FIG. 2.6 Working of Kalman Filters. Belief is represented by a Gaussian. (i) Prediction step: Mean is moved by motion model/kinematics (e.g., manipulator's joints are rotated with known speeds) and variance increases as per motion noise covariance. (ii) Observation step: Robot observes a landmark (e.g., laser) at a position z, which as per belief mean should be at z'. Error $z - z'$ is used to correct belief (mean and covariance). Nonlinear models are locally linearized (Extended Kalman Filter).

2.7.2   Extended Kalman Filters

The major problem with the Kalman Filter is the assumption that the system has a linear motion model and a linear observation model. This will rarely be the case. The kinematics of most mobile robots has a state as $(x, y, \theta)^T$, while there is nonlinearity due to the $\sin \theta$ and $\cos \theta$ terms. Similarly, with drones and other robots. Similarly, the cameras may have an extrinsic calibration matrix that again has cosine and sine terms causing nonlinearity. However, it is possible to locally assume the system as linear by taking the linear term coefficients as the first derivatives of the motion and observation functions.

Let the motion model be given by a generic function f and the observation model be given by another generic function g, shown as Eqs (2.25) and (2.26)

$$q_t = f(q_{t-1}, u_t) + \epsilon_t^{\text{motion}} \tag{2.25}$$

$$z_t = g(q_t) + \epsilon_t^{\text{obs}} \tag{2.26}$$

Let us use the Taylor Series Expansion to the first term only to give the same function across the operating point μ_{t-1} for the motion model and $\bar{\mu}_t$ for the observation model, given by Eqs (2.27) and (2.28).

$$q_t = f(\mu_{t-1}, u_t) + (f(q_{t-1}, u_t) - f(\mu_{t-1}, u_t)) \frac{\partial f(q_{t-1}, u_t)}{\partial q_{t-1}} + \epsilon_t^{\text{motion}} \tag{2.27}$$

$$z_t = g(\bar{\mu}_t) + (g(q_t) - g(\bar{\mu}_t)) \frac{\partial g(q_t)}{\partial q_t} + \epsilon_t^{\text{obs}} \tag{2.28}$$

At any time, t, thus using $A_t = \frac{\partial f(q_t, u_t)}{\partial q_t}$ for $q_t = \mu_{t-1}$ and $C_t = \frac{\partial g(q_t)}{\partial q_t}$ for $q_t = \bar{\mu}_t$, gives the corresponding equations for the locally linearized version of the Kalman Filter, known as the *Extended Kalman Filter* (EKF). The filter is given by Eqs (2.29)–(2.33)

$$\bar{\mu}_t = f(\mu_{t-1}) \tag{2.29}$$

$$\bar{\Sigma}_t = A_t \Sigma_{t-1} A_t^T + R, A_t = \frac{\partial f(q_t, u_t)}{\partial q_t} \tag{2.30}$$

$$K_t = \bar{\Sigma}_t C_t^T (C_t \bar{\Sigma}_t C_t^T + Q)^{-1}, C_t = \frac{\partial g(q_t)}{\partial q_t} \tag{2.31}$$

$$\mu_t = \bar{\mu}_t + K_t \left(z_t^{\text{obs}} - z_t^{\text{proj}} \right) = \bar{\mu}_t + K_t (z_t^{\text{obs}} - g(\bar{\mu}_t)) \tag{2.32}$$

$$\Sigma_t = (I - K_t C_t) \bar{\Sigma}_t \tag{2.33}$$

2.7.3 Particle Filters

The Kalman Filter assumes that the probability is Gaussian. However, imagine a situation where a self-driving car does not have any indication about its current orientation but knows its current position well. The city is perfectly symmetric and therefore there are traffic lights, intersections, roundabouts, and houses in every direction at the same distance. Now the vehicle, because of its motion, can estimate the distance it travels, but it will not know whether it is travelling north, south, east, or west. The probability distribution, hence, has four maxima, and the probability distribution is multimodal. This becomes even more relevant in the context of a *kidnapped robot problem*. Imagine you believe you are going left in the corridor, while you are going on the right. Assume that the distinctive landmarks are symmetric. Now the more you go ahead, the more you are confident about your position because the observations match the expectations, while you are going in the opposite direction. Hence, a distribution that can represent the other extreme possibility is of value.

The aim is now to make a filter that can imitate any probability density function. Such a representation shall be nonparametric as no mean and covariance are used to parametrize the generic probability density function. Consider any general probability density function over all possible states. Let us discretize the states into a finite number of discrete bins. The probability density function $bel(q_t)$ representing the belief at time t is now a collection of a finite number of probabilities, one for every bin. The probabilities can be directly computed from Eqs (2.11) and (2.15) replacing the integral with a summation due to the discrete nature of states (bin). This is called as the *histogram filter* that makes the states discrete into bins and represents the probabilities as a histogram distribution over the bins. The poses must be known to an accuracy of a few centimetres or fractions of radians, which means placing a high number of bins (called *resolution*) for every state variable. The total number of bins is $(resolution)^n$, where n is the number of state variables. One bin is reserved for every combination of state variables. After the motion model, the new probability of a bin i is the summation of probabilities of all ways to reach the bin or the summation of probability of going from bin j to bin i (for a given control input) times the probability of being at bin j at the previous time step (Eq. 2.15). This needs to be calculated for all bins at every time. This is a computationally heavy step for a sufficiently large number of bins. These operations need to be done in a real-time as the robot moves.

Let us, therefore, improve the histogram filter and eliminate discretization. It is imperative that most bins will have a near-zero probability and the probability will only be sufficiently high in bins around a few areas. The areas where the probability is high will, however, keep changing. The near-zero probabilities and hence their bins need not be represented as the results of both motion and observation models retain the probabilities to near-zero and near-zero bins do not contribute to any calculation.

To represent a generic probability density function, let us now place a few *particles* at areas (states) with a reasonably high probability. A particle is a sample from the probability density function that represents a possible state. Only a few particles are used to limit computation and the particles together nearly represent the complete probability density function. The representation is adaptive and when the actual probability density function changes, the samples redistribute themselves to imitate the same representation with a small number of particles.

Let p_t^i represent the ith particle, which believes that the current state of the system at time t is p_t^i. The particle is associated with weight w_t^i representing the degree of confidence of the particle. This can be used to represent a probability density function with the probability of sampled state p_t^i given by $P(p_t^i) = w_t^i / \sum_i w_t^i$. This is shown in Fig. 2.7.

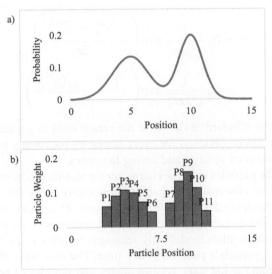

FIG. 2.7 Representation of belief using particles. (A) A sample multimodal belief. (B) Using 11 particles P1–P11 to represent the belief. Each particle has a position and weight. Particles are only kept in areas with a promising probability.

First, let us apply a control input u_t based on which every particle updates its belief as given by Eq. (2.34).

$$p_t^i = f\left(p_{t-1}^i, u_t\right) + \epsilon_t^{\text{motion}}, \epsilon_t^{\text{motion}} \sim N(0, R) \tag{2.34}$$

Here, $f()$ is the motion model. The noise is unknown for the particle and therefore the particle guesses the noise by applying sampling over the known probability density function of the noise model. Assume that the noise is Gaussian with mean 0 and covariance R. Each particle draws a random sample from this noise. The system has many such particles.

Now the particle guesses its weight by using observation. Suppose the sensors make an observation z_t^{obs}. If the observation matches the expected observation with the particle at p_t^i, the weight is high, and vice versa. This is given by Eq. (2.35):

$$w_t^i = P\left(z_t^{\text{obs}} | p_t^i\right) \tag{2.35}$$

Here $P(z_t^{\text{obs}}|p_t^i)$ is the probability of observing z_t^{obs} given the state of the system is p_t^i. If there are multiple observations $z_t^{\text{obs}} = [z_{t,1}^{\text{obs}}, z_{t,2}^{\text{obs}}, \ldots]^T$, assuming the observations are independent and uncorrelated, gives Eq. (2.36):

$$w_t^i = \Pi_k P\left(z_{t,k}^{\text{obs}}|p_t^i\right) \tag{2.36}$$

Let the probability be modelled as a normal curve around the expected value $z_{t,k}^{\text{proj}} = g_k(p_t^i)$, given by Eq. (2.37).

$$w_t^i = \Pi_k \frac{1}{\sqrt{2\pi\sigma_k^2}} \exp\left(-\frac{\left(z_{t,k}^{\text{obs}} - z_{t,k}^{\text{proj}}\right)^2}{2\sigma_k^2}\right)$$

$$= \Pi_k \frac{1}{\sqrt{2\pi\sigma_k^2}} \exp\left(-\frac{\left(z_{t,k}^{\text{obs}} - g_k\left(p_t^i\right)\right)^2}{2\sigma_k^2}\right) \tag{2.37}$$

Here σ_k is the standard deviation of the sensor used to get the observation. The parameter can be obtained for every sensor by measuring many observations under controlled settings and noting the errors.

Finally, if the particles keep moving as per the motion model and adjust their weights as per the observation probability, it is highly likely that very soon the particles will physically lie at low probability areas. This is like the incremental building up of the errors in the motion model, which was not a good way to localize. The observation model only reassigns weights to particles, it does not determine a particle's position at any time. The new state of the particles is calculated from the old state, control inputs, and samples of noise. The particles, therefore, keep scattering in space along with time due to incremental addition of noise and eventually have a near-zero weight.

Therefore, the redistribution of particles is desirable. The principle is to have more particles in areas that are currently more promising, and lesser at places where the probability appears to be negligible currently. This is done by *resampling*. The resampling algorithm draws, with replacement, from the existing population of particles, with the selection chance of a particle proportional to its weight. The higher-weight particles thus typically get selected multiple times, while the lower-weight particles are more likely to get eliminated or extinct due to no selection. The resampled particles make the next set of particles for the next iteration. For resampling, assume that the particles are spread along the circumference of a Roulette Wheel. The circumference coverage of every particle is proportional to its weight. The Roulette Wheel is rotated n times, and a winning particle is selected every time. All selected n particles make the new particles for the next iteration.

The Particle Filter faces problem when the motion noise does not represent the true nature of noise. Under such circumstances, the particles ultimately scatter at regions far from the true state and none of the particles have a small observation error or a large weight. Since the weights are normalized, all particles get nearly equal probability. This resists the incentivization of particles being close to the system state and an eventual correction. Similarly, if the observation model has a noise different from the one used in the algorithm, either all particles will get a large weight (in case of a too-high observation σ_k) or none of them will get a large weight (in case of a too low observation σ_k), which again makes all particles share similar weights and the tracker is likely to lose track with time.

2.8 Localization

Let us use the knowledge of the filters to localize the robot (Thrun et al., 2006). This section builds specific modules to complete the localization problem. The different terminologies are also summarized in Box 2.2.

Box 2.2 Different terminologies used with localization and mapping

Localization:
- **Tracking**: Assume a belief (probability distribution on states), (i) predict belief with knowledge of the applied control signals, (ii) make an observation, the error is the projected observation from belief minus the actual observation, and (iii) correct belief based on the observed error.
- **Correspondence**: Suppose the map has 15 landmarks and 4 landmarks are currently seen. Find corresponding pairs (i,j) such that landmark (at index) i in the map is landmark (at index) j in the current observation (to compute the error).
- **Common observations**: Error is the projected observation from belief minus actual observation by the sensor. Here observation could mean:
  - 2D lidar: Range-bearing (distance and angle to landmark, landmarks are obstacle corners).
  - 3D lidar/stereo: Range-bearing-elevation-signature (a generalization of range-bearing to 3D. Feature detector gives landmarks; feature descriptor gives signature).
  - 2D/3D lidar: Range reading for every bearing that the lidar measures (bearing and elevation for a 3D lidar).
  - 2D/3D lidar: Correlation of local and global map. A local map is made using the current sensor percept. Correlation is an error metric.
  - Image: X, Y coordinates of the landmark in the image and signature (feature detector gives landmarks and feature descriptor gives signature).

Mapping:
- **General concept**: Given the pose of the robot, take a laser scan, convert the laser scan to world coordinate, and plot the laser readings as obstacles.
- **Occupancy map**: Probability (log-odds) that a cell is occupied by obstacles. Each laser ray (equivalently stereo) illustrates (i) cell is occupied by an obstacle, adding to log-odds, (ii) cell is not occupied by an obstacle, reducing log-odds, and (iii) no information about the cell due to occlusion, cell outside the field of view.
- **Semantic map**: List of all landmarks (objects/shapes with pose) along with semantics (object label). A landmark is traced for some time before adding to

---

Continued

Box 2.2 Different terminologies used with localization and mapping—cont'd

the map (reduces false positives), updated for the pose by averaging across all observations, and deleted when not seen for a long time.
- **Feature map**: Only used for localization. Store a list of all point features, their position on the map, along with their signatures.
- **Using different maps to get desired observation for localization**: Error is the projected observation from belief minus actual observation by the sensor. Projected observation for a given pose q is obtained from the map.
  - **Occupancy**: For a virtual robot at q, use virtual sensors to get the expected range-bearing for all bearings.
  - **Semantic (camera)**: Project recognized object polygons on a virtual camera placed at q to get the synthetic projected image/features.
  - **Semantic (lidar)**: For a virtual robot at q, use a ray-tracing algorithm to get range-bearing-elevation of all local corners.
  - **Feature (camera)**: Project 3D position of features on the calibrated virtual camera placed at q, to get their position in 2D image. Use the registered signature.

2.8.1   Motion model

The motion model takes the pose of the robot and the control signal is applied as input and produces the state at the next time instance as an output. Let us build a motion model $f()$ for a differential wheel drive robot that takes the previous pose $(x_{t-1}, y_{t-1}, \theta_{t-1})^T$ and the current control input (v_t, ω_t) to produce the next pose as output $(x_t, y_t, \theta_t)^T$. Assuming a known constant linear speed (v_t) and angular speed (ω_t), the robot traces a large circle with centre ICC $(ICC_x, ICC_y)^T$ for time Δt. The kinematics is thus transforming the coordinate system with ICC as the centre with the robot position given as $p = (x_{t-1} - ICC_x, y_{t-1} - ICC_y)^T$, performing a Z-axis rotation of $\omega_t \Delta t$ $(p' = R_Z(\omega_t \Delta t)p)$ to this position, and transforming the coordinate system back from ICC to origin $(p' + (ICC_x, ICC_y)^T)$. Additionally, accounting for noises, the kinematics is given by Eqs (2.38)–(2.44).

$$\begin{bmatrix} \dot{x} \\ \dot{y} \\ \dot{\theta} \end{bmatrix} = \begin{bmatrix} v_t \cos(\theta_t) \\ v_t \sin(\theta_t) \\ \omega_t \end{bmatrix} \tag{2.38}$$

$$\begin{bmatrix} x_t \\ y_t \\ \theta_t \end{bmatrix} = f\left(\begin{bmatrix} x_{t-1} \\ y_{t-1} \\ \theta_{t-1} \end{bmatrix}, \begin{bmatrix} v'_t \\ \omega'_t \end{bmatrix} \right)$$

$$= \begin{bmatrix} \cos(\omega'_t \Delta t) & -\sin(\omega'_t \Delta t) & 0 \\ \sin(\omega'_t \Delta t) & \cos(\omega'_t \Delta t) & 0 \\ 0 & 0 & 1 \end{bmatrix} \begin{bmatrix} x_{t-1} - ICC_{x,t} \\ y_{t-1} - ICC_{y,t} \\ \theta_{t-1} \end{bmatrix} + \begin{bmatrix} ICC_{x,t} \\ ICC_{y,t} \\ \omega'_t \Delta t + \epsilon_\theta \Delta t \end{bmatrix}$$

$$\tag{2.39}$$

$$\epsilon_\theta \sim N(0, \alpha_{v1} abs(v_t) + \alpha_{\omega 1} abs(\omega_t)) \tag{2.40}$$

$$ICC = \begin{bmatrix} x_{t-1} - L\sin(\theta_{t-1}) \\ y_{t-1} + L\cos(\theta_{t-1}) \end{bmatrix} \tag{2.41}$$

$$L = \frac{v'_t}{\omega'_t} \tag{2.42}$$

$$v'_t = v_t + \epsilon_v, \epsilon_v \sim N(0, \alpha_{v2}abs(v_t) + \alpha_{\omega2}abs(\omega_t)) \tag{2.43}$$

$$\omega'_t = \omega_t + \epsilon_\omega, \epsilon_\omega \sim N(0, \alpha_{v3}abs(v_t) + \alpha_{\omega3}abs(\omega_t)) \tag{2.44}$$

Here, $(x_t, y_t, \theta_t)^T$ represents the state of the system at time t, where (x_t, y_t) is the position and θ_t is the orientation. The two controls (u_t) are the linear velocity (v_t) and angular velocity (ω_t) or $u_t = (v_t, \omega_t)^T$. The kinematic relation is defined for a robot turning on a circle with centre ICC for Δt unit of time with a radius L. The actual speeds v'_t, and ω'_t will be affected by samples from the noise ϵ_v, and ϵ_ω. The noise is typically linearly proportional to the speeds. A robot moving fast is likely to have a larger noise as compared to a slow-moving robot. A normal expectation is that the noise in the linear speed ϵ_v is proportional to the linear speed $abs(v_t)$, and the noise in the angular speed ϵ_ω is proportional to the angular speed $abs(\omega_t)$, where $abs()$ is the absolute value function. However, practically both speeds affect each other's noise in different proportions. Therefore, ϵ_v is taken from a normal distribution with variance $\alpha_{v2}abs(v_t) + \alpha_{\omega2}abs(\omega_t)$ and ϵ_ω is taken from another normal distribution with variance $\alpha_{v3}abs(v_t) + \alpha_{\omega3}abs(\omega_t)$. Here $\alpha_{v1}, \alpha_{v2}, \alpha_{v3}, \alpha_{\omega1}, \alpha_{\omega2}$, and $\alpha_{\omega3}$ are the proportionality constants. The orientation is subjected to another such noise (ϵ_θ) by the joint effect of the linear and angular speeds with proportionality constants α_{v1} and $\alpha_{\omega1}$. $N()$ denotes the normal distribution.

The implementation of the motion model is slightly different for the two filters discussed. For the application of the Extended Kalman Filter, a derivative of the model with respect to the state variables is to be done. Similarly, sampling noise and adding the same can be used for the Particle Filters, which are shown in the equations. The notations are shown in Fig. 2.8. Let us expand the motion model of Eq. (2.39) by eliminating the matrices as Eq. (2.45).

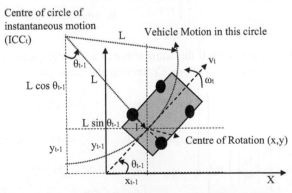

FIG. 2.8 Notations used for deriving the kinematics (motion model) of a car-type robot model. A derivative of kinematics gives the matrices used in Kalman Filter. Noised version of kinematics is used for the Particle Filter.

$$\begin{bmatrix} x_t \\ y_t \\ \theta_t \end{bmatrix} = f\left(\begin{bmatrix} x_{t-1} \\ y_{t-1} \\ \theta_{t-1} \end{bmatrix}, \begin{bmatrix} v_t \\ \omega_t \end{bmatrix} \right)$$

$$= \begin{bmatrix} (x_{t-1} - ICC_{x,t})\cos(\omega'_t \Delta t) - (y_{t-1} - ICC_{y,t})\sin(\omega'_t \Delta t) + ICC_{x,t} \\ (x_{t-1} - ICC_{x,t})\sin(\omega'_t \Delta t) + (y_{t-1} - ICC_{y,t})\cos(\omega'_t \Delta t) + ICC_{y,t} \\ \theta_{t-1} + \omega'_t \Delta t + \epsilon_\theta \Delta t \end{bmatrix} \quad (2.45)$$

Substituting for ICC_t $(ICC_{x,t}, ICC_{y,t})^T$ using Eq. (2.41) gives Eq. (2.46):

$$\begin{bmatrix} x_t \\ y_t \\ \theta_t \end{bmatrix} = f\begin{bmatrix} x_{t-1} \\ y_{t-1} \\ \theta_{t-1} \end{bmatrix}\begin{bmatrix} v_t \\ \omega_t \end{bmatrix}$$

$$\quad (2.46)$$

$$= \begin{bmatrix} L\sin\theta_{t-1}\cos\omega'_t\Delta t + L\cos\theta_{t-1}\sin\omega'_t\Delta t + x_{t-1} - L\sin\theta_{t-1} \\ L\sin\theta_{t-1}\sin\omega'_t\Delta t - L\cos\theta_{t-1}\cos\omega'_t\Delta t + y_{t-1} + L\cos\theta_{t-1} \\ \theta_{t-1} + \omega'_t\Delta t + \epsilon_\theta\Delta t \end{bmatrix}$$

Further substituting L from Eq. (2.42) and rearranging the terms gives Eqs (2.47) and (2.48).

$$\begin{bmatrix} x_t \\ y_t \\ \theta_t \end{bmatrix} = f\begin{bmatrix} x_{t-1} \\ y_{t-1} \\ \theta_{t-1} \end{bmatrix}\begin{bmatrix} v_t \\ \omega_t \end{bmatrix}$$

$$= \begin{bmatrix} x_{t-1} + \dfrac{v'_t}{\omega'_t}\sin\theta_{t-1}\cos\omega'_t\Delta t + \dfrac{v'_t}{\omega'_t}\cos\theta_{t-1}\sin\omega'_t\Delta t - \dfrac{v'_t}{\omega'_t}\sin\theta_{t-1} \\ y_{t-1} + \dfrac{v'_t}{\omega'_t}\sin\theta_{t-1}\sin\omega'_t\Delta t - \dfrac{v'_t}{\omega'_t}\cos\theta_{t-1}\cos\omega'_t\Delta t + \dfrac{v'_t}{\omega'_t}\cos\theta_{t-1} \\ \theta_{t-1} + \omega'_t\Delta t + \epsilon_\theta\Delta t \end{bmatrix} \quad (2.47)$$

$$\begin{bmatrix} x_t \\ y_t \\ \theta_t \end{bmatrix} = f\left(\begin{bmatrix} x_{t-1} \\ y_{t-1} \\ \theta_{t-1} \end{bmatrix}, \begin{bmatrix} v_t \\ \omega_t \end{bmatrix} \right) = \begin{bmatrix} x_{t-1} - \dfrac{v'_t}{\omega'_t}\sin(\theta_{t-1}) + \dfrac{v'_t}{\omega'_t}\sin(\theta_{t-1} + \omega'_t\Delta t) \\ y_{t-1} + \dfrac{v'_t}{\omega'_t}\cos(\theta_{t-1}) - \dfrac{v'_t}{\omega'_t}\cos(\theta_{t-1} + \omega'_t\Delta t) \\ \theta_{t-1} + \omega'_t\Delta t + \epsilon_\theta\Delta t \end{bmatrix}$$

$$\quad (2.48)$$

Eq. (2.48) will not work if the angular speed is (nearly) zero or the vehicle moves by a linear speed (v_t) alone. Correcting the same gives Eq. (2.49).

$$\begin{bmatrix} x_t \\ y_t \\ \theta_t \end{bmatrix} = f\left(\begin{bmatrix} x_{t-1} \\ y_{t-1} \\ \theta_{t-1} \end{bmatrix}, \begin{bmatrix} v_t \\ \omega_t \end{bmatrix} \right) = \begin{cases} \begin{bmatrix} x_{t-1} - \dfrac{v'_t}{\omega'_t}\sin(\theta_{t-1}) + \dfrac{v'_t}{\omega'_t}\sin(\theta_{t-1} + \omega'_t\Delta t) \\ y_{t-1} + \dfrac{v'_t}{\omega'_t}\cos(\theta_{t-1}) - \dfrac{v'_t}{\omega'_t}\cos(\theta_{t-1} + \omega'_t\Delta t) \\ \theta_{t-1} + \omega'_t\Delta t + \epsilon_\theta\Delta t \end{bmatrix} & \text{if } \omega'_t \neq 0 \\[3em] \begin{bmatrix} x_{t-1} + v'_t\cos(\theta_{t-1}) \\ y_{t-1} + v'_t\sin(\theta_{t-1}) \\ \theta_{t-1} + \epsilon_\theta\Delta t \end{bmatrix} & \text{if } \omega'_t \approx 0 \end{cases}$$

$$\quad (2.49)$$

Taking a derivative of Eq. (2.49) with respect to the state variables $(x_t, y_t, \theta_t)^T$ gives Eq. (2.50)

$$A_t = \frac{\partial f}{\partial (x_t, y_t, \theta_t)^T} = \begin{bmatrix} \dfrac{\partial f(x)}{\partial x_t} & \dfrac{\partial f(x)}{\partial y_t} & \dfrac{\partial f(x)}{\partial \theta_t} \\[2mm] \dfrac{\partial f(y)}{\partial x_t} & \dfrac{\partial f(y)}{\partial y_t} & \dfrac{\partial f(y)}{\partial \theta_t} \\[2mm] \dfrac{\partial f(\theta)}{\partial x_t} & \dfrac{\partial f(\theta)}{\partial y_t} & \dfrac{\partial f(\theta)}{\partial \theta_t} \end{bmatrix}$$

$$= \begin{cases} \begin{bmatrix} 1 & 0 & -\dfrac{v'_t}{\omega'_t}\cos(\theta_{t-1}) + \dfrac{v'_t}{\omega'_t}\cos(\theta_{t-1} + \omega'_t \Delta t) \\[2mm] 0 & 1 & -\dfrac{v'_t}{\omega'_t}\sin(\theta_{t-1}) + \dfrac{v'_t}{\omega'_t}\sin(\theta_{t-1} + \omega'_t \Delta t) \\[2mm] 0 & 0 & 1 \end{bmatrix} & \text{if } \omega'_t \neq 0 \\[10mm] \begin{bmatrix} 1 & 0 & -v'_t \sin(\theta_{t-1}) \\[2mm] 0 & 1 & v'_t \cos(\theta_{t-1}) \\[2mm] 0 & 0 & 1 \end{bmatrix} & \text{if } \omega'_t \approx 0 \end{cases} \qquad (2.50)$$

2.8.2 Motion model for encoders

The current robots do not always rely on the velocities. It is hard to synchronize the velocity given to the motors with the localization module. The models are more sensitive to drift. The motors are more nonideal in terms of adherence to the reference velocities. The *encoders* or, in general, the *odometry-based models* sometimes have a better performance as a motion model. The advantage is that the robots can internally calculate the new positions (or change in position with previous reading) to be used in localization, which is a capability given by most libraries. The problem, however, is that the position will never be the same as reported, requiring fusion with observation models using filters. The filtering requires a formal motion model, which is based on velocities and not the reported positions by the odometry model.

Consider that the robot is reported by the sensor to have moved from pose $q'_{t-1}(x'_{t-1}, y'_{t-1}, \theta'_{t-1})$ to pose $q'_t(x'_t, y'_t, \theta'_t)$, where the prime represents that the terms are observed by the encoder and not the true positions. The intent is to use the terms as a motion model and not an observation model. To use this as a motion model, we will need to add new terms representing a control signal. Therefore, let us consider the process between q'_{t-1} and q'_t involved an initial rotation from θ'_{t-1} by a magnitude of $\Delta\theta'_{1,t}$ followed by a traversal of $\Delta d'_t$ and finally a rotation by $\Delta\theta'_{2,t}$ to finally get an orientation of θ'_t. The three control signals added are thus $\Delta\theta'_{1,t}$, $\Delta d'_t$, and $\Delta\theta'_{2,t}$. The terms can be calculated as given by Eqs (2.51)–(2.53). The notations are shown in Fig. 2.9.

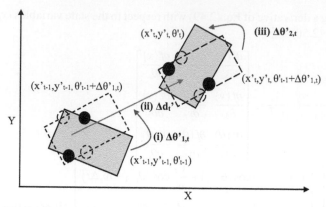

FIG. 2.9 Notations used for deriving the kinematics (motion model) of a car-type robot model from odometry sensors like encoders that report the (noisy) pose at every time. The vehicle motion is approximated by (i) initial rotation by $\Delta\theta'_{1,t}$, (ii) linear translation by $\Delta d'_t$, and (iii) final rotation by $\Delta\theta'_{2,t}$ to attain the pose reported by the odometry sensors. ($\Delta\theta'_{1,t}$, $\Delta d'_t$, and $\Delta\theta'_{2,t}$) makes the control input that is noised to represent the real control.

$$\Delta\theta'_{1,t} = \operatorname{atan2}(y'_t - y'_{t-1}, x'_t - x'_{t-1}) - \theta'_{t-1} \qquad (2.51)$$

$$\Delta d'_t = \sqrt{\left(y'_t - y'_{t-1}\right)^2 + \left(x'_t - x'_{t-1}\right)^2} \qquad (2.52)$$

$$\Delta\theta'_{2,t} = \theta'_t - \left(\theta'_{t-1} + \Delta\theta'_{1,t}\right) \qquad (2.53)$$

Eqs (2.41)–(2.53) may seem conceptually unusual. Previously we used previous state and control inputs to calculate the new state. However, now both the previous state and new state (ideal and noiseless) are given from which the control signal ($\Delta\theta'_{1,t}$, $\Delta d'_t$, and $\Delta\theta'_{2,t}$) is being derived. The aim here is to first model the control inputs and then to derive a new kinematic equation of the robot with these inputs (incorporating noise) as a motion model so that the filtering techniques can be used.

The actual motion will not be the ideal one as reported by the odometry system. The three variables $\Delta\theta'_{1,t}$, $\Delta d'_t$, and $\Delta\theta'_{2,t}$ are assumed to be the ideal control inputs, and therefore, noise is added to them giving the actual control inputs as $\Delta\theta_{1,t}$, Δd_t, and $\Delta\theta_{2,t}$, while the noise is sampled. After adding for noise, the controls become as given by Eqs (2.54)–(2.56).

$$\Delta\theta_{1,t} = \Delta\theta'_{1,t} + \epsilon_{\theta1}, \epsilon_{\theta1} \sim N(0, \alpha_{\theta1}\Delta\theta'_{1,t} + \alpha_{d1}\Delta d'_t) \qquad (2.54)$$

$$\Delta d_t = \Delta d'_t + \epsilon_d, \epsilon_d \sim N(0, \alpha_{\theta2}\Delta\theta'_{1,t} + \alpha_{d2}\Delta d'_t + \alpha_{\theta3}\Delta\theta'_{2,t}) \qquad (2.55)$$

$$\Delta\theta_{2,t} = \Delta\theta'_{2,t} + \epsilon_{\theta2}, \epsilon_{\theta2} \sim N(0, \alpha_{\theta4}\Delta\theta'_{2,t} + \alpha_{d3}\Delta d'_t) \qquad (2.56)$$

Here, all α are the coefficients of the noise model. The more is the magnitude of the control inputs, the larger is the noise representing by a larger variance of the noise model. This can now be used to calculate the new position with noise, which also gives the motion model. The new position is given by Eq. (2.57).

$$\begin{bmatrix} x_t \\ y_t \\ \theta_t \end{bmatrix} = f\left(\begin{bmatrix} x_{t-1} \\ y_{t-1} \\ \theta_{t-1} \end{bmatrix}, \begin{bmatrix} \Delta\theta_{1,t} \\ \Delta d_t \\ \Delta\theta_{2,t} \end{bmatrix} \right) = \begin{bmatrix} x_{t-1} + \Delta d_t \cos(\theta_{t-1} + \Delta\theta_{1,t}) \\ y_{t-1} + \Delta d_t \sin(\theta_{t-1} + \Delta\theta_{1,t}) \\ \theta_{t-1} + \Delta\theta_{1,t} + \Delta\theta_{2,t} \end{bmatrix} \quad (2.57)$$

The notations are similar to the case with velocities and therefore self-explanatory.

2.8.3 Measurement model using landmarks and camera

The measurement model takes as input the expected pose of the robot and gives as output the expected sensor readings if the robot is at the input pose. The measurement model is dependent upon the *map*. This can be used to compute the error as the difference between the expected and observed sensor readings.

Projecting a rich map into the sensor plate can be a computationally intensive step. *Feature engineering* is a popular mechanism to reduce the input to a few features. This can be done for 2D or 3D lidars as well as stereo or monocular camera images. The complete input (RGB image or depth image) is processed through the local feature extraction technique, which typically computes corners in the 2D RGB image or the 2D or 3D depth image. A feature is called a *landmark*. Since the map is assumed to be known, the position of the landmark (feature) in the world $(x_j, y_j, z_j)^T$ and its signature (feature descriptor that encodes the local region around the feature) as per the feature encoding technique s_j are both accurately known. The map is now a collection of features or landmarks creating a *feature map* $\{(x_j, y_j, z_j, s_j)^T\}$. First let us assume a monocular camera, which should observe the landmark at $\{(u_{j,t}^{\text{proj}}, v_{j,t}^{\text{proj}})\}$ with signature s_j at time t as given by Eq. (2.58), given that landmark is within the field of view, which also constitutes the observation model g.

$$g\left(T_{R,t}^{W}, x_j, y_j, z_j, s_j \right) = \left(u_{j,t}^{\text{proj}}, v_{j,t}^{\text{proj}}, s_j \right) \quad (2.58)$$
$$\left(u_{j,t}^{\text{proj}}, v_{j,t}^{\text{proj}} \right) = \pi\left(T_R^{\text{cam}} T_{W,t}^{R} \left(x_j, y_j, z_j, 1 \right)^T \right) = \pi\left(T_R^{\text{cam}} \left(T_{R,t}^{W} \right)^{-1} \left(x_j, y_j, z_j, 1 \right)^T \right)$$

Here g is the observation model that takes the estimated robot pose ($T_{R,t}^{W}$) at time t and landmark position/signature as the input to give the sensor readings as an output. π is the projection function that takes a point in the camera coordinate frame as input to produce the projection of the point in the image as output, T_R^{cam} is the transformation from the robot to the camera, and $T_{W,t}^{R}$ is the transformation from the world to the robot (whose inverse $T_{R,t}^{W}$ is the pose of the robot).

A problem is that the image observes many features or landmarks, while many others are already in the data set. An observed landmark k hence needs to be searched in terms of space and signature both to the best landmark in the map (say j). Suppose you observed the top-left corner of a door (landmark), and the observation model requires the ground truth position and signature of the top-left corner of the same door (map). The problem is that there are three doors with different ground truth positions. For the application of the observation model, you need to make sure that the correct corner's ground truth position is supplied. Suppose an indicative pose of the robot is already known as $T_{R,t}^{W\sim}$. In the Extended Kalman Filter, it is the preobservation state estimate; while in the Particle Filter, it could be taken from the motion model applied to the last iteration's state estimate. Project all landmarks into the image plane for this pose and filter out the landmarks, which will not be observed (lie behind the robot at this pose or are projected outside the size of the image plane). Signatures (feature descriptors) are invariant to most factors of variation. For the observed landmark k search for the closest signature out of all filtered landmarks. The stability may be checked by comparing the best and 2nd best matches, wherein the observed corner should have a good signature match to one landmark in the map, while it should have a poor second-best matching landmark in the map. Sometimes another constraint can be added that the projection of the feature assuming the robot pose $T_{R,t}^{W\sim}$ should be close to the observed location of the feature on the image ($u_{k,t}^{\text{obs}}, v_{k,t}^{\text{obs}}$). Suppose kth observed landmark matches with jth landmark in the map database, or $j = c(k)$, where c is the correspondence function. This is called the *correspondence problem*. It is not necessary that every observed feature k may match to one landmark j in the database; or that every filtered landmark j that should be visible in the image is actually observed. Possible reasons include occlusions, dynamic people walking, false positives due to illumination effect, lightning conditions suppressing a landmark, wrong feature registered in the map, etc.

The derivative of the observation function (Eq. 2.58) is used in the Extended Kalman Filter. The error function is the difference between the expected (projected) observation given by the model and the observed position of the landmarks, used to compute the weight of a particle in a Particle Filter. The probability is thus given by Eq. (2.59):

$$P\left(\left(u_t^{\text{obs}}, v_t^{\text{obs}} s_t^{\text{obs}}\right) \mid T_{R,t}^W, \text{map}\right) =$$

$$\Pi_k \frac{1}{\sqrt{2\pi\sigma_u^2}} \exp\left(-\frac{\left(u_{k,t}^{\text{obs}} - u_{c(k),t}^{\text{proj}}\right)^2}{2\sigma_u^2}\right) \frac{1}{\sqrt{2\pi\sigma_v^2}} \exp\left(-\frac{\left(v_{k,t}^{\text{obs}} - v_{c(k),t}^{\text{proj}}\right)^2}{2\sigma_v^2}\right)$$

$$\frac{1}{\sqrt{2\pi\sigma_s^2}} \exp\left(-\frac{\left\|s_{k,t}^{\text{obs}} - s_{c(k)}\right\|_2^2}{2\sigma_s^2}\right)$$

$$(2.59)$$

$$\left(u_{c(k),t}^{\text{proj}}, v_{c(k),t}^{\text{proj}}\right) = \pi\left(T_R^{\text{cam}} T_{W,t}^R \left(x_{c(k)}, y_{c(k)}, z_{c(k)}, 1\right)^T\right)$$

$$= \pi\left(T_R^{\text{cam}} \left(T_{R,t}^W\right)^{-1} \left(x_{c(k)}, y_{c(k)}, z_{c(k)}, 1\right)^T\right)$$

Here $c(k)$ is the correspondence function that locates a feature k observed in the image to the one in the map and ($u_{k,t}^{\text{obs}}, v_{k,t}^{\text{obs}}, s_{k,t}^{\text{obs}}$) is the kth landmark's observed position and signature.

2.8.4 Measurement model using landmarks and Lidar

In the case of a 3D laser, the observation of the landmark is modified to the distance reported by the laser ray called the range ($d_{k,t}^{obs}$), the angle with respect to the Z-axis called as *bearing* ($\alpha_{k,t}^{obs}$), the *elevation angle* ($\phi_{k,t}^{obs}$), and the *signature* ($s_{k,t}^{obs}$). The landmark is first located in the frame of reference of the laser, $T_R^S T_{W,t}^R (x_{c(i)}, y_{c(i)}, z_{c(i)}, 1)^T$, and then the corresponding angles and range are calculated as the observation model. The same correspondence problem needs to be solved for localization.

For simplicity, let us consider the case of a 2D laser scanner instead. Suppose the landmark is at the known position (x_j, y_j). The observation vector consists of the distance to the landmark (range or $d_{k,t}^{obs}$), angle to landmark in the heading direction of the robot (bearing or $\alpha_{k,t}^{obs}$) and the signature of the landmark (feature descriptor or $s_{k,t}^{obs}$). The observation function takes a robot pose (say, x_t, y_t, and θ_t) as input along with the landmark's position and signature, and returns the projected observation vector, given by Eq. (2.60), which makes the observation model with notations similar to Fig. 2.4.

$$
\begin{bmatrix} d_{j,t}^{proj} \\ \alpha_{j,t}^{proj} \\ s_j \end{bmatrix} = g(x_t, y_t, \theta_t, x_j, y_j, s_j) = \begin{bmatrix} \sqrt{(x_j - x_t)^2 + (y_j - y_t)^2} \\ \mathrm{atan2}(y_j - y_t, x_j - x_t) - \theta_t \\ s_j \end{bmatrix} \tag{2.60}
$$

Its derivative is used for the EKF. Taking a derivation of Eq. (2.60) with respect to the robot pose $(x_t, y_t, \theta_t)^T$, gives Eq. (2.61).

$$
C_{j,t} = \frac{\partial g}{\partial (x_t, y_t, \theta_t)^T} = \begin{bmatrix} \dfrac{\partial g(d_{j,t})}{\partial x_t} & \dfrac{\partial g(d_{j,t})}{\partial y_t} & \dfrac{\partial g(d_{j,t})}{\partial \theta_t} \\[2mm] \dfrac{\partial g(\alpha_{j,t})}{\partial x_t} & \dfrac{\partial g(\alpha_{j,t})}{\partial y_t} & \dfrac{\partial g(\alpha_{j,t})}{\partial \theta_t} \\[2mm] \dfrac{\partial g(s_j)}{\partial x_t} & \dfrac{\partial g(s_j)}{\partial y_t} & \dfrac{\partial g(s_j)}{\partial \theta_t} \end{bmatrix}
$$

$$
C_{j,t} = \begin{bmatrix} \dfrac{-(x_j - x_t)}{\sqrt{(x_j - x_t)^2 + (y_j - y_t)^2}} & \dfrac{-(y_j - y_t)}{\sqrt{(x_j - x_t)^2 + (y_j - y_t)^2}} & 0 \\[3mm] \dfrac{y_j - y_t}{(x_j - x_t)^2 + (y_j - y_t)^2} & \dfrac{-(x_j - x_t)}{(x_j - x_t)^2 + (y_j - y_t)^2} & -1 \\[3mm] 0 & 0 & 0 \end{bmatrix} \tag{2.61}
$$

Let $D = (x_j - x_t)^2 + (y_j - y_t)^2$ to simply Eq. (2.61) as Eq. (2.62).

$$
C_{j,t} = \frac{1}{D} \begin{bmatrix} -\sqrt{D}(x_j - x_t) & -\sqrt{D}(y_j - y_t) & 0 \\ y_j - y_t & -(x_j - x_t) & -1 \\ 0 & 0 & 0 \end{bmatrix} \tag{2.62}
$$

The Particle Filter uses the same equation to compute the probabilities as the difference between the expected (projected) observation from a particle at $p_t(x_t, y_t, \theta_t)$ and the actual observation as physically observed from the sensor. Suppose a laser scan is given that can be made as a local depth image. The features are extracted from the same image giving the landmark position and the signature ($s_{k,t}^{obs}$). Trace the ray that hit the landmark. Assume it is at a bearing $\alpha_{k,t}^{obs}$ and reports a range of $d_{k,t}^{obs}$. To apply the model, we need to locate the same landmark on the global map and fetch its coordinates. Assume that the landmark with the closest signature map in the database is $j = c(k)$, where c is the correspondence function, which for the currently estimated pose is also at a similar range-bearing as per the observation. The probability is calculated independently for range (d), bearing (α), and signature, given as Eqs (2.63) and (2.64).

$$P\left((d_t^{obs}, \alpha_t^{obs} s_t^{obs}) \mid (x_t, y_t \theta_t), \text{map}\right)$$

$$= \Pi_k \frac{1}{\sqrt{2\pi\sigma_d^2}} \exp\left(-\frac{\left(d_{k,t}^{obs} - d_{c(k),t}^{proj}\right)^2}{2\sigma_u^2}\right) \frac{1}{\sqrt{2\pi\sigma_\alpha^2}} \exp\left(-\frac{\left(\alpha_{k,t}^{obs} - \alpha_{c(k),t}^{proj}\right)^2}{2\sigma_v^2}\right)$$

$$\frac{1}{\sqrt{2\pi\sigma_s^2}} \exp\left(-\frac{\left\|s_{k,t}^{obs} - s_{c(k)}\right\|_2^2}{2\sigma_s^2}\right)$$

$$\tag{2.63}$$

$$\begin{bmatrix} d_{c(k),t}^{proj} \\ \alpha_{c(k),t}^{proj} \\ s_{c(k)}^{proj} \end{bmatrix} = \begin{bmatrix} \sqrt{\left(x_{c(k)} - x_t\right)^2 + \left(y_{c(k)} - y_t\right)^2} \\ \text{atan2}\left(y_{c(k)} - y_t, x_{c(k)} - x_t\right) - \theta_t \\ s_{c(k)} \end{bmatrix} \tag{2.64}$$

2.8.5   Measurement model for dense maps

The observation techniques used previously assumed a good feature detector and carried localization using the selected features only. It is possible to perform localization on the dense map without features as well. Let us make an observation model for the lidar sensor in a map with some (large number of) point obstacles only, like a grid map. Let the actual reading of the lidar be d_α^{obs}, which records the distance to the obstacle for several bearings α. Let the known position of the point obstacle be o and let the hypothesis be that the robot is placed at $T_{R,t}^W$ at time t, which is the robot pose. The obstacle o which will be seen by the robot at $T_{W,t}^R o = (T_{R,t}^W)^{-1} o$. The laser scanner will see the obstacle at $T_R^S T_{W,t}^R o$, where T_R^S is the transformation from robot to sensor known by the extrinsic calibration. The laser ray at an angle α (bearing) of the 2D laser scanner will be another coordinate transform T_S^α from the laser scanner coordinate frame to the coordinate frame of the specific laser ray. The transform is a rotation by α along the Z-axis (world coordinate system). $T_S^\alpha = R_Z(\alpha)$. The coordinate axis system is given in Fig. 2.10. Let the obstacle in the laser's coordinate frame be given by $o_t^{proj}(o_{t,x}^{proj}, o_{t,y}^{proj}, o_{t,z}^{proj}) = T_S^\alpha T_R^S T_{W,t}^R o$. Suppose the local coordinate system of the laser is such that the direction of the laser ray is the X-axis while the Y-axis is horizontal and the Z-axis points from the ground to the roof. The laser hits the obstacle if its Y coordinate value is 0 ($o_{t,y}^{\text{proj}}=0$) with some tolerance and the laser reading ($d_\alpha^{\text{proj}}(T_{R,t}^W, o)$) is recorded in the positive X-axis ($o_{t,x}^{\text{proj}}$). $d_\alpha^{\text{proj}}(T_{R,t}^W, o) = o_{t,x}^{\text{proj}} : o_{t,y}^{\text{proj}} = 0, o_{t,x}^{\text{proj}} > 0$. The laser may intercept multiple obstacles at the specific angle α and thus the minimum positive value is taken for all such obstacles, given by Eq. (2.65), which also constitutes the observation model for the lidar system that takes the robot pose and map as input to produce the sensor readings as an output.

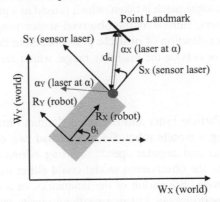

FIG. 2.10 Observation model for dense maps. The sensor reports the distance to the obstacle (d_α^{obs} or range) for every angle to the obstacle (α or bearing). Observation $z^{\text{obs}} = \{\cup_\alpha d_\alpha^{\text{obs}}\}$. The observation model (z^{proj}) is the expected reading with the robot at (x_t, y_t, θ_t) with a known map for every bearing. A derivative is used for the Extended Kalman Filter. Particle Filter uses probability (weight) proportional to error (abs($z^{\text{proj}} - z^{\text{obs}}$)).

$$d_\alpha^{\text{proj}}\left(T_{R,t}^W, o\right) = \min_{o_t^{\text{proj}} : o_{t,y}^{\text{proj}} = 0, o_{t,x}^{\text{proj}} > 0,} o_{t,x}^{\text{proj}}, o_t^{\text{proj}} = T_S^\alpha T_R^S T_{W,t}^R o \qquad (2.65)$$

Alternatively, assume that a grid map is given. Place a robot at an assumed pose (x, y, θ). Propagate a virtual laser ray at a bearing angle α. Compute the distance (d_α^{proj}) when the ray strikes an obstacle in the grid map, which gives the same range value making the observation model. For polygon objects, the same formulation may be obtained from the ray-tracing algorithms. The implementation of the observation model is different for the Particle Filter and Extended Kalman Filter. Observation $d_\alpha^{\text{proj}}(T_{R,t}^W, o)$ is known as a function of the robot pose (observation model), whose derivative gives the observation matrix for the Extended Kalman Filter, with a strong assumption that the map (obstacle positions) is perfectly known. $d_\alpha^{\text{proj}}(T_{R,t}^W, o)$ can be computed for any input robot pose represented by the particle while the observed distance d_α^{obs} from the laser for all bearings is known, whose error is inversely proportional to the particle weight for a Particle Filter. Currently, the probabilities of detecting a new obstacle, not detecting an old obstacle or a deviation in the angle are not accounted for by simplicity.

Another observation model is commonly used. Here, we first make a local map in the robot coordinate frame of reference. Let the actual observation of laser ray at angle α be o'. Making a local map is an inverse transformation from the one earlier, given by $\{T_S^R T_\alpha^S o'\}$. Every laser ray observes an obstacle o' in its coordinate frame, which is transformed into the sensor's and thereafter the robot's coordinate frame. The transformation between the laser ray at α to the robot can be calculated for every sensed point o'. The hypothesis is that the robot is at pose $T_{R,t}^W$ and the map in the world coordinate frame is known as a collection of obstacles o. The registered map is transformed to the robot's frame of reference as per the expected robot pose, giving the expected map as $\{T_{W,t}^R o\}$. The *correlation* between the observed map ($\{T_\alpha^R o'\}$) and the expected map ($\{T_{W,t}^R o\}$) in the same patch is taken, which is used as a probability measure for the Particle Filters. Here o' is the observed sensor readings and o is the registered map as a collection of obstacle positions. Note that some obstacles o will be occluded or outside the visibility range, which are excluded.

2.8.6 Examples

So far two filters (Particle Filter and Extended Kalman Filter) have been discussed for localizing a mobile robot. Each filter had two options for motion models (using linear and angular speed; or using odometry reported from encoders). Similarly, the observation model could either use landmark-based lidar reading, a camera observation of the landmarks, or a dense localization on the lidar (or camera) image. Let us specifically write the pseudo-code for three cases. As the first case let us use the Extended Kalman Filter on a robot whose motion commands (linear and angular speeds) are known. The observation model uses the lidar sensor on features or landmarks. The algorithm is shown as Algorithm 2.1. Specifically observe that since there are multiple observations (one from each landmark), there are multiple Kalman gains, and correction of mean (line 19) and covariance (line 20) uses a summation of all Kalman gains. Line 6 applies the motion model to compute the pre-observation mean and Lines 7-8 apply a derivative of the motion model to compute the pre-observation covariance for the case when $\omega_t \neq 0$. More formally Eqs. 2.49 and 2.50 should be used.

In another case consider a Particle Filter is used to localize the same robot. However, the robot uses a vision camera instead of a lidar, while using point features for localization. The pseudo-code is given as Algorithm 2.2. Line 13 is for the case when $\omega_t \neq 0$, while more formally Eq. 2.49 should be used. For this case, the robot pose was as a state vector, while the observation needs a pose matrix which is computed from the state vector in Line 14. Since there are multiple observations (from different landmarks), the probabilities are multiplied to get the particle weight. In the same example let us replace the camera with a lidar and perform a dense localization. Now the observation model places a virtual lidar at the pose predicted by the particle and gets the projected reading from this virtual lidar. The error in the range readings of this virtual lidar sensor and the observed lidar sensor reading are used to set the particle weight. The odometry-based motion model is used. The algorithm is given as Algorithm 2.3.

Algorithm 2.1 Localization using Extended Kalman Filter with lidar sensor

---

1   $\mu_0 = [\mu_{0,x}, \mu_{0,y}, \mu_{0,\theta}]^T$;   ▷ Initial known pose estimate (x, y, θ)

2   $\Sigma_0 = \begin{bmatrix} \sigma_{0,x} & 0 & 0 \\ 0 & \sigma_{0,y} & 0 \\ 0 & 0 & \sigma_{0,\theta} \end{bmatrix}$;   ▷ Initial covariance

3   $t \leftarrow 0$;   ▷ time

4   **while** *end of sequence* **do**

5     $t \leftarrow t + 1$, Get control $u_t = [v_t, \omega_t]$;   ▷ linear, angular speed
      ▷ Prediction Step using Motion Model ;

6     $\bar{\mu}_t = \begin{bmatrix} \bar{\mu}_{t,x} \\ \bar{\mu}_{t,y} \\ \bar{\mu}_{t,\theta} \end{bmatrix} = \begin{bmatrix} \mu_{t-1,x} - \frac{v_t}{\omega_t} sin(\mu_{t-1,\theta}) + \frac{v_t}{\omega_t} sin(\mu_{t-1,\theta} + \omega_{t-1}\Delta t) \\ \mu_{t-1,y} + \frac{v_t}{\omega_t} cos(\mu_{t-1,\theta}) - \frac{v_t}{\omega_t} cos(\mu_{t-1,\theta} + \omega_t\Delta t) \\ \mu_{t-1,\theta} + \omega_{t-1}\Delta t \end{bmatrix}$;

7     $A_t = \begin{bmatrix} 1 & 0 & -\frac{v_t}{\omega_t} cos(\mu_{t-1,\theta}) + \frac{v_t}{\omega_t} cos(\mu_{t-1,\theta} + \omega_{t-1}\Delta t) \\ 0 & 1 & -\frac{v_t}{\omega_t} sin(\mu_{t-1,\theta}) + \frac{v_t}{\omega_t} sin(\mu_{t-1,\theta} + \omega_t \Delta t) \\ 0 & 0 & 1 \end{bmatrix}$;

8     $\bar{\Sigma}_t = A_t \Sigma_{t-1} A_t^T + R_t$;

9     $Q_t = \begin{bmatrix} \sigma_d & 0 & 0 \\ 0 & \sigma_\alpha & 0 \\ 0 & 0 & \sigma_s \end{bmatrix}$;   ▷ Correction by observation step

10   Observe lidar as an array of (range-bearing) $\{(d_t^{obs}, \alpha_t^{obs})\}$;

11   Compute landmarks (features, L) as an array of range-bearing-signature using feature detection and feature description algorithms on the depth image, $L = \{(d_{k,t}^{obs}, \alpha_{k,t}^{obs}, s_{k,t}^{obs})\}$;

12   **for** *all observed landmarks* $z_{k,t}^{obs}(d_{k,t}^{obs}, \alpha_{k,t}^{obs}, s_{k,t}^{obs}) \in L$ **do**

13   Locate the best signature matching landmark in the map, say $j = c(k)$, such that landmark would be visible to robot at $(\bar{\mu}_{t-1,x}, \bar{\mu}_{t-1,y}, \bar{\mu}_{t-1,\theta})$ near observed range-bearing ;

14   Get landmark position $(x_{c(k)}, y_{c(k)})$ and signature $s_{c(k)}$ from map ;

15      $D = (x_{c(k)} - \bar{\mu}_{t,x})^2 + (y_{c(k)} - \bar{\mu}_{t,x})^2$;

16      $\begin{bmatrix} d_{c(k),t}^{proj} \\ \alpha_{c(k),t}^{proj} \\ s_{c(k)}^{proj} \end{bmatrix} = \begin{bmatrix} \sqrt{D} \\ atan2(y_{c(k)} - \bar{\mu}_{t,y}, x_{c(k)} - \bar{\mu}_{t,x}) - \bar{\mu}_{t,\theta} \\ s_{c(k)} \end{bmatrix}$;
      ▷ Observation model (project on lidar)

17      $C_{k,t} = \frac{1}{D} \begin{bmatrix} -\sqrt{D}(x_{c(k)} - \bar{\mu}_{t,x}) & -\sqrt{D}(y_{c(k)} - \bar{\mu}_{t,y}) & 0 \\ y_{c(k)} - \bar{\mu}_{t,y} & -(x_{c(k)} - \bar{\mu}_{t,x}) & -1 \\ 0 & 0 & 0 \end{bmatrix}$;

18      $K_{k,t} = \bar{\Sigma}_t C_{k,t}^T (C_{k,t} \bar{\Sigma}_t C_{k,t}^T + Q_t)^{-1}$;   ▷ Kalman Gain

19    $\mu_t = \bar{\mu}_t + \Sigma_k K_{k,t} [(d_{k,t}^{obs}, \alpha_{k,t}^{obs}, s_{k,t}^{obs})^T - (d_{c(k),t}^{proj}, \alpha_{c(k),t}^{proj}, s_{c(k)}^{proj})^T]$;

20    $\Sigma_t = (I - \Sigma_k K_{k,t} C_{k,t}) \bar{\Sigma}_t$;   ▷ Updated mean and covariance

21   publish μ_t, Σ_t ;

Algorithm 2.2 Localization using Particle Filter with camera

1 $\mathcal{P}$ = initial Probability Density Function of pose, $t \leftarrow 0$;
▷ Initialize Particles. Note: As per camera conventions, robot moves in ZX plane, Y axis points down into the floor. θ is rotation along Y-axis. ;

2 **for** *i from 1 to n* **do** $p_0^i = (p_{0,z}^i, p_{0,x}^i, p_{0,\theta}^i)^T \sim \mathcal{P}$; ▷ sample particles

3 **while** *end of sequence* **do**

4 $\quad$ $t \leftarrow t + 1$, Get control $u_t = [v_t, \omega_t]$; ▷ linear, angular speed

5 $\quad$ Observe camera image. Compute landmarks (features, L) as an array of image coordinates-signature using feature detection and feature description algorithms, $L = \{(u_k^{obs}, v_k^{obs}, s_k^{obs})\}$;

6 $\quad$ **for** *all observed landmarks* $z_{k,t}^{obs}(u_{k,t}^{obs}, v_{k,t}^{obs}, s_{k,t}^{obs}) \in L$ **do**

7 $\quad\quad$ Correspondence Matching: Locate best (as per signature) landmark $c(k)$ in the map that would be visible to the robot at expected pose near observed image coordinates, where the expected robot pose is motion model applied to the previous best estimate ;

8 $\quad\quad$ Get landmark position $(x_{c(k)}, y_{c(k)}, z_{c(k)})$, signature $s_{c(k)}$ from map;

9 $\quad$ **for** *all particles i* **do**

10 $\quad\quad$ $\epsilon_\theta \sim \mathcal{N}(0, \alpha_{v1} abs(v_t) + \alpha_{\omega 1} abs(\omega_t))$; ▷ Prediction Step

11 $\quad\quad$ $v_t' = v_t + \epsilon_v, \epsilon_v \sim \mathcal{N}(0, \alpha_{v2} abs(v_t) + \alpha_{\omega 2} abs(\omega_t))$;

12 $\quad\quad$ $\omega_t' = \omega_t + \epsilon_\omega, \epsilon_\omega \sim \mathcal{N}(0, \alpha_{v3} abs(v_t) + \alpha_{\omega 3} abs(\omega_t))$;

13 $\quad\quad$ $p_t^i = \begin{bmatrix} p_{t,z}^i \\ p_{t,x}^i \\ p_{t,\theta}^i \end{bmatrix} = \begin{bmatrix} p_{t-1,z}^i - \frac{v_t'}{\omega_t'} sin(p_{t-1,\theta}^i) + \frac{v_t'}{\omega_t'} sin(p_{t-1,\theta}^i + \omega_{t-1}'\Delta t) \\ p_{t-1,x}^i + \frac{v_t'}{\omega_t'} cos(p_{t-1,\theta}^i) - \frac{v_t'}{\omega_t'} cos(p_{t-1,\theta}^i + \omega_t'\Delta t) \\ p_{t-1,\theta}^i + \omega_{t-1}'\Delta t + \epsilon_\theta \Delta t \end{bmatrix}$

14 $\quad\quad$ $T_{R,t}^W = \begin{bmatrix} cos(p_{t,\theta}^i) & 0 & sin(p_{t,\theta}^i) & p_{t,x}^i \\ 0 & 1 & 0 & 0 \\ -sin(p_{t,\theta}^i) & 0 & cos(p_{t,\theta}^i) & p_{t,z}^i \\ 0 & 0 & 0 & 1 \end{bmatrix}$; ▷ Particle pose as

$\quad\quad$ a transformation matrix

15 $\quad\quad$ **for** *all observed landmarks* $z_{k,t}^{obs}(u_{k,t}^{obs}, v_{k,t}^{obs}, s_{k,t}^{obs}) \in L$ **do**

$\quad\quad\quad$ ▷ Observation: Project $(x_{c(k)}, y_{c(k)}, z_{c(k)})$ on image;

16 $\quad\quad\quad$ $l_{cam} = \begin{bmatrix} l_{cam,x} \\ l_{cam,y} \\ l_{cam,z} \\ 1 \end{bmatrix} = T_{R,t}^{cam}(T_{R,t}^W)^{-1} \begin{bmatrix} x_{c(k)} \\ y_{c(k)} \\ z_{c(k)} \\ 1 \end{bmatrix}$

17 $\quad\quad\quad$ $\begin{bmatrix} u_{c(k)}'^{obs} \\ v_{c(k)}'^{obs} \\ w \end{bmatrix} = C \begin{bmatrix} l_{cam,x} \\ l_{cam,y} \\ l_{cam,z} \end{bmatrix} = \begin{bmatrix} f_x & 0 & c_x \\ 0 & f_y & c_y \\ 0 & 0 & 1 \end{bmatrix} \begin{bmatrix} l_{cam,x} \\ l_{cam,y} \\ l_{cam,z} \end{bmatrix}$;

18 $\quad\quad\quad$ $(u_{c(k)}^{obs}, v_{c(k)}^{obs})^T = (u_{c(k)}'^{obs}/w, v_{c(k)}'^{obs}/w)^T$;

19 $\quad\quad$ $w_t^i = \Pi_k exp(-\frac{(u_k^{obs} - u_{c(k)}^{proj})^2}{\sigma_u^2}) exp(-\frac{(v_k^{obs} - v_{c(k)}^{proj})^2}{\sigma_v^2}) exp(-\frac{||s_k^{obs} - s_{c(k)}||_2^2}{\sigma_s^2})$;

20 $\quad$ publish $\frac{\Sigma_i w_t^i p_t^i}{\Sigma_i w_t^i}$, Re-sample particles ;

Algorithm 2.3 Localization using Particle Filter with Lidar

---

1 $\mathcal{P}$ = initial Probability Density Function of pose, e.g. known mean with a small covariance (high confidence) ;

2 **for** i *from 1 to n* **do**   ▷ initialize n particles

3 $\quad\lfloor\ p_0^i = (p_{0,x}^i, p_{0,y}^i, p_{0,\theta}^i)^T \sim$ sample taken from $\mathcal{P}$

4 Get initial pose according to encoders $(x_0', y_0', \theta_0')^T$;

5 $t \leftarrow 0$;   ▷ time

6 **while** *end of sequence* **do**

7 $\quad t \leftarrow t + 1$;

8 $\quad$ Get current encoder readings $(x_t', y_t', \theta_t')^T$;

$\quad$ ▷ Convert encoder readings into control input ;

9 $\quad u_t' = \begin{bmatrix} \Delta\theta_{1,t}' \\ \Delta d_t' \\ \Delta\theta_{2,t}' \end{bmatrix} = \begin{bmatrix} atan2(y_t' - y_{t-1}', x_t' - x_{t-1}') - \theta_{t-1}' \\ \sqrt{(x_t' - x_{t-1}')^2 + (y_t' - y_{t-1}')^2} \\ \theta_t' - (\theta_{t-1}' + \Delta\theta_{1,t}') \end{bmatrix}$;

10 $\quad$ Observe lidar (array of range-bearing) $z_t = \{(d_\alpha, \alpha)\}$;

11 $\quad$ **for** *all particles i* **do**

$\quad\quad$ ▷ Prediction Step ;

12 $\quad\quad \epsilon_{\theta 1} \sim \mathcal{N}(0, \alpha_{\theta 1}\Delta\theta_{1,t}' + \alpha_{d1}\Delta d_t')$;

13 $\quad\quad \epsilon_d \sim \mathcal{N}(0, \alpha_{\theta 2}\Delta\theta_{2,t}' + \alpha_{d2}\Delta d_t' + \alpha_{\theta 3}\Delta\theta_{1,t}')$;

14 $\quad\quad \epsilon_{\theta 2} \sim \mathcal{N}(0, \alpha_{\theta 4}\Delta\theta_{2,t}' + \alpha_{d3}\Delta d_t')$;

$\quad\quad$ ▷ noise input ;

15 $\quad\quad u_t = (\Delta\theta_{1,t}, \Delta d_t, \Delta\theta_{2,t})^T = (\Delta\theta_{1,t}' + \epsilon_{\theta 1}, \Delta d_t' + \epsilon_d, \Delta\theta_{2,t}' + \epsilon_{\theta 2})^T$

16 $\quad\quad p_t^i = \begin{bmatrix} p_{t,x}^i \\ p_{t,y}^i \\ p_{t,\theta}^i \end{bmatrix} = \begin{bmatrix} p_{t-1,x}^i + \Delta d_t cos(p_{t-1,\theta}^i + \Delta_{theta1,t}) \\ p_{t-1,y}^i + \Delta d_t sin(p_{t-1,\theta}^i + \Delta_{theta1,t}) \\ p_{t-1,\theta}^i + \Delta\theta_{1,t} + \Delta\theta_{2,t} \end{bmatrix}$

$\quad\quad$ ▷ Correction by Observation Step ;

17 $\quad\quad$ Place virtual robot at $(p_{t,x}^i, p_{t,y}^i)$, orientation $p_{t,\theta}^i$ in known map ;

18 $\quad\quad z_{proj} = \emptyset$;

19 $\quad\quad$ **for** *all bearing α in observed lidar image z_t* **do**

20 $\quad\quad\quad$ use ray tracing algorithm to compute the range of the virtual robot at bearing α, say $d_{proj,\alpha}$;

21 $\quad\quad\quad$ add $(d_{proj,\alpha}, \alpha)$ to z_{proj} ;

22 $\quad\quad w_t^i = exp(-\frac{||d_{proj,\alpha} - d_\alpha||_2^2}{\sigma^2})$;

23 $\quad$ publish $\frac{\Sigma_i w_t^i p_t^i}{\Sigma_i w_t^i}$;

24 $\quad$ Re-sample particles ;

Assume that the robot moves using a known linear and angular speed. The observation model is the reading for one landmark (range and bearing) from a lidar sensor whose position is accurately known. The solution for one iteration of the algorithm using the Extended Kalman Filter is given in Fig. 2.11 and using the Particle Filter is given in Fig. 2.12.

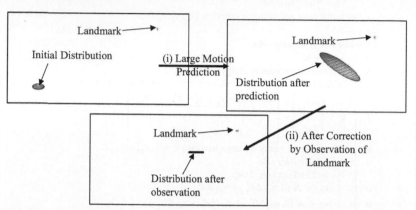

FIG. 2.11 Localization using an Extended Kalman Filter. (i) A robot is given a large motion (linear and angular speed), which occurs as a large increase in covariance. (ii) After observation (distance to the landmark, angle to the landmark), the covariance quickly converges back. The X-axis and Y-axis are visible, but the orientation axis is not visible in the figure.

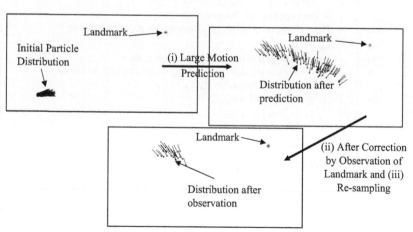

FIG. 2.12 Localization using a Particle Filter. (i) All particles (x, y, θ) are subjected to a motion model (linear and angular speed) with different sampled noises. (ii) The particles calculate weights (not shown) based on the projected and actual observation (angle/distance to landmark). (iii) Particles are resampled proportional to weight.

2.9 Mapping

The next problem is to make a map of the world. The problem is relatively easier to solve given the true pose of the robot at every time step.

2.9.1 Occupancy maps

Suppose the robot at a time t is at q_t and a lidar scanner observes an obstacle at a distance (range) $d_{\alpha,t}$ with the laser ray at angle α (bearing). This means the obstacle is at position $(d_{\alpha,t}, 0, 0)^T$ with respect to the ray (ray's direction is the X-axis), $T_\alpha^S(d_{\alpha,t}, 0, 0, 1)^T$ with respect to the laser scanner, $T_S^R T_\alpha^S(d_{\alpha,t}, 0, 0, 1)^T$ with respect to the robot and finally $T_{R,t}^W T_S^R T_\alpha^S(d_{\alpha,t}, 0, 0, 1)^T$ with respect to the world. The last transformation is known as the pose of the robot which is assumed to be known. Suppose the robot is at (x_t, y_t, θ_t) at time t and that the robot frame of reference and lidar frame of reference are the same ($T_S^R = I_4$, identity matrix). The point obstacle detected by the lidar will be at $(x_t + d_{\alpha,t}\cos(\theta_t + \alpha), y_t + d_{\alpha,t}\sin(\theta_t + \alpha))^T$ in the world coordinate frame. This reading also tells us that there is no obstacle in the direction of the laser ray till (and excluding) a distance of $d_{\alpha,t}$, otherwise the laser ray would have detected that obstacle and reported a range less than $d_{\alpha,t}$. The cells $\{(x_t + d\cos(\theta_t + \alpha), y_t + d\sin(\theta_t + \alpha))^T, 0 < d < d_{\alpha,t}\}$ are thus obstacle-free. The detected point obstacles can be added to the list of obstacles making a map for every laser scan at time t and for all time t that the robot moves, or map is $\{\cup_t \cup_\alpha (x_t + d_{\alpha,t}\cos(\theta_t + \alpha), y_t + d_{\alpha,t}\sin(\theta_t + \alpha))^T\}$. The same calculation may be done for the case of a 3D laser scanner. For a stereo camera, the depth will now have to be triangulated and found out, to get the position of the landmark in 3D, which can then be transformed to get the map with respect to the world.

However, the actual problem is a little more involved since multiple laser scanners can at multiple times see multiple obstacles and obstacle-free regions and report anomalies. One laser ray may state an obstacle at a specific position in the world, and another might negate the possibility of an obstacle at that position. In such a case of the presence of multiple readings supporting or (not supporting) the presence of obstacle at a grid cell, there is a need to fuse the different observations with time into one decision regarding the presence of obstacle at the grid. Similarly, in case of the use of multiple sensors, some sensors may support the presence of an obstacle while some other sensors may not support the presence of obstacle at a grid, while the decisions of the sensors need to be fused into one decision. The fusion may be *optimistic*, in which case an obstacle is said to be at a place only when all readings support the presence of the obstacle, otherwise, it is assumed that no obstacle lies. Alternatively, a more common implementation is to have a *pessimistic fusion*, where the obstacle is said to be at a place even if one sensor states the position as an obstacle. Alternatively, *probabilistic fusion* may be used. The mapping in this manner produces a binary image as an output, which can be passed through noise removal, straightening of the bent lines, regression of polygons, and similar techniques. One such map produced by a Pioneer LX robot is shown in Fig. 2.13.

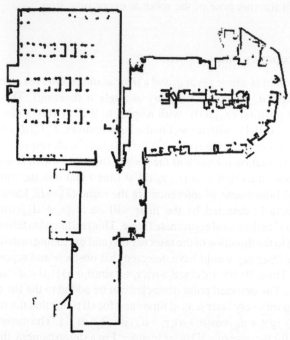

FIG. 2.13 A sample grid map. The map is produced by integrating laser scans as the robot moves. Postprocessing corrects lines and corners and removes outliers.

More generally, let the map be divided into grids and let $P(m_{(x,y)}| z_{0:t}, q_{0:t})$ be the probability that the cell (x, y) of the map m is an obstacle, given all the past poses $(q_{0:t})$ and sensor observations $(z_{0:t})$. The technique is called *occupancy grid mapping*. The map $m = \{m_{(x,y)}\}$ is a collection of all such cells, and mapping can be conveniently defined as the problem of computing the probability of every cell being an obstacle, given all past states and observations, $P(m| z_{0:t}, q_{0:t})$. The computation of probabilities will involve multiplication of several probabilities. Sometimes it is better to work in the logarithmic domain where the log probabilities shall become additive in nature. Let the *log-odds* of a grid $m_{(x,y)}$ being an obstacle at time t be given by Eq. (2.66):

$$l_{(x,y),t} = \log \frac{P\left(m_{(x,y)}|z_{0:t}, q_{0:t}\right)}{P\left(\overline{m}_{(x,y)}|z_{0:t}, q_{0:t}\right)} = \log \frac{P\left(m_{(x,y)}|z_{0:t}, q_{0:t}\right)}{1 - P\left(m_{(x,y)}|z_{0:t}, q_{0:t}\right)} \tag{2.66}$$

Let us assume that the log-odds at time $t-1$ are $l_{(x,y),t-1}$. Suppose a sensor reading is made that reports presence of an obstacle with a probability $P(m_{(x,y)}|z_t, q_t)$, and reports an absence of an obstacle with probability

$P\left(\overline{m_{(x,y)}}|z_t, q_t\right)$. Multiplying the new probability of observations gives the updated log-odds as given by Eq. (2.67).

$$l_{(xy),t} = \log \frac{P\left(m_{(xy)}|z_{0:t-1}, q_{0:t-1}\right)P\left(m_{(xy)}|z_t, q_t\right)}{P\left(\overline{m_{(x,y)}}|z_{0:t-1}, q_{0:t-1}\right)P\left(\overline{m_{(x,y)}}|z_t, q_t\right)}$$

$$l_{(xy),t} = \log \frac{P\left(m_{(xy)}|z_{0:t-1}, q_{0:t-1}\right)}{P\left(\overline{m_{(x,y)}}|z_{0:t-1}, q_{0:t-1}\right)} + \log \frac{P\left(m_{(xy)}|z_t, q_t\right)}{P\left(\overline{m_{(x,y)}}|z_t, q_t\right)}$$

$$l_{(x,y),t} = l_{(x,y),t-1} + \log \frac{P\left(m_{(x,y)}|z_t, q_t\right)}{1 - P\left(m_{(x,y)}|z_t, q_t\right)} \tag{2.67}$$

The notion is now to keep adding log-odds of the cell $m_{(x,y)}$ being an obstacle along with time for the sensor readings while modelling the log-odds of every sensor reading. If there were multiple sensors, the reading at time t would be $z_t = \{z_{k,t}\}$, where $z_{k,t}$ is the reading of the kth sensor, giving the log-odds as Eq. (2.68):

$$l_{(x,y),t} = l_{(x,y),t-1} + \log \frac{\Pi_k P\left(m_{(x,y)}|z_{k,t}, q_t\right)}{\Pi_k P\left(\overline{m_{(x,y)}}|z_{k,t}, q_t\right)}$$

$$l_{(x,y),t} = l_{(x,y),t-1} + \sum_k \log \frac{P\left(m_{(x,y)}|z_{k,t}, q_t\right)}{P\left(\overline{m_{(x,y)}}|z_{k,t}, q_t\right)}$$

$$l_{(x,y),t} = l_{(x,y),t-1} + \sum_k \log \frac{P\left(m_{(x,y)}|z_{k,t}, q_t\right)}{1 - P\left(m_{(x,y)}|z_{k,t}, q_t\right)} \tag{2.68}$$

Here $P(m_{(x,y)}|z_{k,t}, q_t)$ is the probability that the grid $m_{(x,y)}$ is an obstacle as per the observation of the kth sensor at time t with the robot at pose q_t. Let us denote the same quantity as the log-odds of the observation given by Eq. (2.69):

$$l_{(x,t)}(z_{k,t}, q_t) = \log \frac{P\left(m_{(x,y)}|z_{k,t}, q_t\right)}{P\left(\overline{m_{(x,y)}}|z_{k,t}, q_t\right)} = \log \frac{P\left(m_{(x,y)}|z_{k,t}, q_t\right)}{1 - P\left(m_{(x,y)}|z_{k,t}, q_t\right)} \tag{2.69}$$

The updates of log-odds of a cell are thus given by Eq. (2.70):

$$l_{(x,y),t} = l_{(x,y),t-1} + \sum_k l_{(x,t)}(z_{k,t}, q_t) \tag{2.70}$$

Let us do a simple modelling of $l_{(x,t)}(z_{k,t}, q_t)$. Let l_0 (say 0) be the prior of the log-odds of observing an obstacle, which is the log-odds of the probability of the cell being an obstacle without any information. An observation $z_{k,t}$ at any instance of time may report either of three things about every cell:

(i) The cell $m_{(x,y)}$ is an obstacle, in which case $P\left(m_{(x,y)}|z_{k,t}, q_t\right) > P\left(\overline{m_{(x,y)}}|z_{k,t}, q_t\right)$ or $l_{(x,t)}(z_{k,t}, q_t) > l_0$. In other words, new evidence for the presence of obstacle is known and is added to the log-odds of the obstacle (equivalent to multiplication of probabilities), which appreciably increases the log-odds. Practically, lasers are highly accurate. Let us assume that a high value of l_{max} is added in case an obstacle is detected.

(ii) The cell $m_{(x,y)}$ is not an obstacle, in which case $P(m_{(x,y)}|z_{k,t}, q_t) < P(\overline{m_{(x,y)}}|z_{k,t}, q_t)$ or $l_{(x,t)}(z_{k,t}, q_t) < l_0$. In other words, new evidence for the absence of the obstacle is known and is subtracted from the log-odds of the presence of the obstacle (numerator is less than denominator and the logarithmic term is negative). Attributed to the accuracy of lidars, let us assume that a low (negative) value of l_{min} is added in such a case.

(iii) The observation has no information about the presence or absence of the obstacle at $m_{(x,y)}$ as $m_{(x,y)}$ is out of the field of view of the sensor or there is an obstacle that occludes the cell $m_{(x,y)}$ from being detected. In this case, the log-odds remain unaffected or are affected by just the prior l_0 (say 0). The value l_0 is added in this case.

Let us assume that the sensor has a small radial dispersion of ε_α. This means that the laser ray at bearing α could be anywhere between the range $[\alpha - \varepsilon_\alpha, \alpha + \varepsilon_\alpha]$. Let the lidar have a small radial dispersion of ε_d, which means that for the true range d, the reported range could be anywhere in $[d - \varepsilon_d, d + \varepsilon_d]$. Apart from the sensor accuracy, ε_d and ε_α must deal with the discretization loss incurred by the conversion of a continuous map into discrete grids. We need to calculate $P(m_{(x,y)}|z_{k,t}, q_t)$ or whether the cell (x,y) in map m is occupied by an obstacle as reported by the laser reading $z_{k,t} = \{(d_\alpha, \alpha)\}$ with the robot at pose $q_t(x_t, y_t, \theta_t)$. The laser reading is assumed to be an array of range bearings of the lidar, with a range value for every bearing. If the cell was an obstacle, it would be seen at a range given by Eq. (2.71) and bearing given by Eq. (2.72).

$$d(x, y) = \sqrt{(x - x_t)^2 + (y - y_t)^2} \tag{2.71}$$

$$\alpha(x, y) = \text{atan2}(y - y_t, x - x_t) - \theta_t \tag{2.72}$$

The specific laser ray in $z_{k,t}$ that would detect the obstacle would be the one at the closest bearing in the observation, say at a bearing α' given by Eq. (2.73):

$$\alpha' = \arg\min_{\alpha' \in z_{k,t}} abs(\alpha' - \alpha(x, y)) \tag{2.73}$$

Here $abs()$ is the absolute value function. Say the laser ray reports a range of d'_α. If $abs(\alpha' - \alpha(x, y)) < \varepsilon_\alpha$ (condition A), the cell is covered by the chosen laser ray in terms of bearing. Similarly, if $abs(d' - d(x, y)) < \varepsilon_d$ (condition B), the cell is covered by the chosen laser ray in terms of range. If both A and B hold, the cell (x, y) is an obstacle, and l_{max} is added as per point (i). If A holds and the cell is (by range) before the selected obstacle with threshold ε_d ($d(x,y) < d' - \varepsilon_d$), then the cell (x, y) is free and l_{min} is added as per point (ii). If A does not hold, the cell (x, y) is outside the field of view of the sensor, and l_0 is added as per point (iii). If A holds and the cell is (by range) after the selected obstacle with threshold ε_d ($d(x,y) > d' + \varepsilon_d$), then the cell (x, y) is occluded by the obstacle detected by the laser, and l_0 is added as per point (iv). Let d_{max} be the maximum range that the sensor can detect. A common technique is to exclude all readings of d_{max} (or larger), where the laser does not strike an obstacle. The computation of the updated log-odds of a cell is shown in Algorithm 2.4. In the algorithm, the assumption is a single observation, and hence, the subscript k is dropped. Every step that the robot moves and takes a laser scan, the updated log-odds are calculated for the potentially affected cells. The notations used are given in Fig. 2.14.

Algorithm 2.4 Mapping

> **Input** : Cell (x, y) with log-odds $l_{(x,y),t-1}$ whose updated log-odds
> needs to be computed. Lidar reading z_t as an array of
> range-bearings $\{d_\alpha, \alpha\}$. Robot pose $q_t(x_t, y_t, \theta_t)$
>
> **Output:** Updated log-odds $l_{(x,y),t-1}$

1   $d(x, y) = \sqrt{(x - x_t)^2 + (y - y_t)^2}$;   ▷ range of (x, y)
2   $\alpha(x, y) = atan2(y - y_t, x - x_t) - \theta_t$;   ▷ bearing of (x, y)
3   $\alpha' = argmin_{\alpha' \in z_t} abs(\alpha' - \alpha(x, y))$; ▷ bearing that detects (x, y)
4   d'_α = range of laser at bearing α' ;   ▷ range that detects (x, y)
5   **if** $abs(\alpha' - \alpha(x, y)) < \epsilon_\alpha \land abs(d'_\alpha - d(x, y)) < \epsilon_d \land d'_\alpha < d_{max}$ **then**
6      $l_{(x,y),t} = l_{(x,y),t-1} + l_{max}$;   ▷ (x, y) is obstacle
7   **else if** $abs(\alpha' - \alpha(x, y)) < \epsilon_\alpha \land d(x, y) < d'_\alpha - \epsilon_d$ **then**
8      $l_{(x,y),t} = l_{(x,y),t-1} + l_{min}$;   ▷ (x, y) is free
9   **else if** $abs(\alpha' - \alpha(x, y)) > \epsilon_\alpha \lor d(x, y) > d'_\alpha + \epsilon_d \lor d(x, y) > d_{max}$ **then**
10      $l_{(x,y),t} = l_{(x,y),t-1} + l_0$; ▷ (x, y) is outside field of view or occluded
11   **return** $l_{(x,y),t}$;

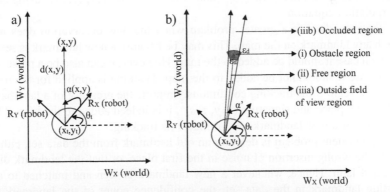

FIG. 2.14 Calculating log-odds of a cell (x, y). (A) Range $d(x, y)$ and bearing $\alpha(x, y)$ of the laser ray that will detect obstacles at (x, y), if any. (B) The log-odds of (x, y) is updated based on the region (i–iii) at which it lies. (d', α') is the range-bearing of the closest laser ray (in terms of bearing, minimizes $abs(\alpha' - \alpha(x, y))$) that serves (x, y). The readings are relaxed by ε_α (bearing) and ε_d (range).

2.9.2   Geometric, semantic, and feature maps

The maps may also be *geometric* or *semantic*, wherein the map is a collection of the type of object, the size of the object, the pose of the object, and other semantic definitions of the object (like a door can be opened, a pen can be picked up, etc.). As an example, walk around the house and register every item minutely that you see. A detailed list of all such objects makes the semantic map. This requires a detector for every possible object. Alternatively, look around for primitive shapes and regress simple shapes around every object that you see. A collection of the pose, size, and shape of every such object makes a geometric map. The maps are ultimately used for decision-making or planning.

Some maps are made for localization alone called as the *feature map*. The feature map only stores point features (or landmarks, given by feature detector), their signature (given by the feature descriptor), and the position of the point feature (obtained by stereo pair or lidar at the time of map creation) in the world coordinate frame. The feature map is used as a reference for features in localization. In the case of stereo cameras and 3D laser scanners where the landmark position and signature constitute the map, the approach is similar.

Let us discuss the maintenance of the feature maps as a special case, which can be generalized to geometric and semantic maps as well. The feature map is a collection of landmarks $m = \{(x_i, y_i, z_i, s_i)^T\}$. Upon observing a new landmark at $(x, y, z)^T$ in the real world (after necessary transformations) which has a signature s, a decision needs to be made whether this is a landmark already on the map or a new landmark. If it closely matches an existing landmark in the map, with respect to position and signature, the task is now to update the old landmark in the map. Say the landmark $(x_i, y_i, z_i, s_i)^T$ in the map has a very similar signature s_i to the observed signature s, and the position of the landmark $(x_i, y_i, z_i)^T$ is also very similar to the observed position $(x, y, z)^T$. The updated position and signature are taken from an interpolation of the old observation $(x_i, y_i, z_i, s_i)^T$ and new observation $(x, y, z, s)^T$ using any interpolation equation.

However, there is a greater problem when the new observation does not match any landmark on the map. This may be because a new landmark is seen, in which case it should be added to the map. However, it can also be a noise, in which case it should not be added to the map. Tracking is applied for the new landmark seen. If the tracking continuously reports the presence of a landmark at the same place and signature $(x, y, z, s)^T$, it is indeed added to the map. The first sight of a new landmark only initiates its tracking.

A prominent problem is deleting an old landmark from the data set, either due to the wrong insertion of noise in the first place, or that the landmark disappeared later. Hence, whenever a new landmark is seen and matched to an existing landmark in the data set, the confidence count of the landmark is increased. Gradually as the robot moves, every landmark is seen multiple times, and the confidence count increases. If a landmark maintains a poor confidence count, while nearby landmarks maintain a significantly larger confidence count, the landmark with a low confidence count is bound to be deleted.

2.10 Simultaneous Localization and Mapping

So far, we have solved two problems separately. First, to compute location given the map or the problem of localization, given by $P(q_t|z_{0:t}, u_{0:t}, m)$ with q_t as the pose, u_t as the control input, and z_t as the observation at time t. The map (m) was used as an input for the observation model. The second problem was to calculate the map, or the problem of mapping, given by $P(m|z_{0:t}, u_{0:t}, q_t)$, where the location (pose) was the input. So, the map is derived from the pose, and in turn, the pose is derived from the map, which is the chicken and egg problem in robotics. We need pose to calculate the map and the map to calculate the pose. The input to localization (map) is obtained from mapping, while the input to mapping (robot pose) is obtained from localization. The problems are, hence, solved together as the problem of *Simultaneous Localization and Mapping* (SLAM), given by $P(q_{0:t}, m|z_{0:t}, u_{0:t})$.

Let us first consider EKF to solve the problem with the algorithm known as EKF-SLAM. The notion is to embed both the robot pose (say $(x, y, \theta)^T$) and the landmarks (say $(x_1, y_1, z_1, s_1, x_2, y_2, z_2, s_2 \ldots)^T$) as one state vector $(x, y, \theta, x_1, y_1, z_1, s_1, x_2, y_2, z_2, s_2, \ldots)^T$ of size $(3+4l) \times 1$, where l is the number of landmarks, each consisting of three variables for position and 1 for signature. The covariance matrix will be of the size $(3+4l) \times (3+4l)$. The motion model only deals with the state variables of the robot $(x, y, \theta)^T$ since the landmarks are stationary. The observation adapts both landmarks as well as the robot state vector. However, not all landmarks are within the field of view and hence the submatrix representing the visible landmarks is considered. The major problem with the EKF-SLAM approach is that the covariance matrix size is extremely large for practical problems with many landmarks. The problem is that the covariance matrix models covariance between landmarks, which can be conveniently assumed to be independent of each.

Let us, therefore, decompose the problem into a localization and mapping problem using the probability basics, specifically known here as the Rao-Blackwellization equation given by Eqs (2.74) and (2.75):

$$P(q_{1:t}, m|z_{1:t}, u_{1:t}) = P(q_{1:t}|z_{1:t}, u_{1:t}) P(m|q_{1:t} z_{1:t}, u_{1:t}) \qquad (2.74)$$

$$P(q_{1:t}, m|z_{1:t}, u_{1:t}) = P(q_{1:t}|z_{1:t}, u_{1:t}) \Pi_k P(m_k|q_{1:t} z_{1:t}) \qquad (2.75)$$

Here m_k is the kth landmark and the decomposition assumes that all landmarks are conditionally independent of each other. The first part represents the localization problem, while the second part represents the mapping problem. The *Fast-SLAM* benefits from the decomposition of the problem into two parts. In this algorithm, a Particle Filter is used to solve the overall problem. Given a location, the mapping problem is solved by using EKF. Each particle, hence, has a robot's position and the EKF parameters associated with all landmarks. Given a position, the EKF of a landmark is a 4×4 covariance vector and 4×1 state vector only. The particle has l such EKFs for l landmarks. The EKF part is a conventional tracking algorithm, given a robot's pose. The Particle Filter as well can be easily implemented. The motion is moving the robot with noise. The measurement sets the particle weights based on the probability of observations made. This technique breaks a large covariance matrix into a larger number of matrices of smaller dimensions and is thus more usable.

Questions

1. A point robot is at the centre of the room. The room has no obstacles other than the walls. Therefore, the proximity sensors observe the room corners. The room corners are at $(0\,m, 0\,m)$, $(3\,m, 3\,m)$, $(0\,m, 3\,m)$, and $(3\,m, 3\,m)$. The following is the motion and observation of the robot.
   (a) Initially, the robot is at $(2\,m, 2\,m)$ oriented at 0 degree
   (b) $v = 10\,cm/s$ and $\omega = 5$ degrees/s for 3 s
   (c) Observation of landmarks: (i) One landmark at range $= 1.28\,m$ and bearing $= 38$ degrees; (ii) One landmark at range $= 2.45\,m$ and bearing $= -60$ degrees; (iii) One landmark at range $= 2.13\,m$ and bearing $= 165$ degrees; and (iv) One landmark at range $= 3.07\,m$ and bearing $= -134$ degrees
   The linear speed has a motion noise with a variance of $0.05\,m$ and the angular speed has a motion noise with a variance of 10 degrees. The observation noise has a variance of $0.1\,m$ for range and 30 degrees for bearing. Solve for localization using Extended Kalman Filter and Particle Filter (five particles).

2. Consider a humanoid robot at the origin oriented on the X-axis
   (a) Humanoid walks without turning in the positive X-axis. The humanoid walks by taking steps and every step is "approximately" 10 cm (mean 10 cm and variance 1 cm). The robot takes three steps in the positive X-axis. Give the mean and variance of the position using the Extended Kalman Filter after every step for three steps.
   (b) Humanoid is attempting to turn on its feet, which it can do without moving forward. It starts from the origin with an orientation of $0.3\,rad$ anticlockwise (positive), with a variance of $0.1\,rad$. Give the mean and variance of the orientation using Extended Kalman Filter after every step for three steps.
   (c) Assume that the humanoid can either turn or can move forward and not both at the same time. Consider the following motion commands given

one after the other in the same order: 1 step forward, 1 step turn, 1 step forward, and 1 step turn. Using the Extended Kalman Filter, give the posterior on the pose after every command.

3. Assume the robot workspace has a camera that gives an aerial view of the workspace. Machine learning algorithms are used to convert the map into an obstacle map. The machine learning algorithm gives a posterior over the obstacle, that is the probability that a grid is an obstacle and the probability that the grid is not an obstacle. The same overhead map detects the robot and gives the position and orientation of the robot, with the same probability given by mean and variance. The robot has 10 sonar rings. You may assume the relevant parameters. Given the readings of the sonar sensors, make a function that gives the map. How do you account for the fact that the overhead camera may be more accurate than the sonar?

4. Enlist all popular robots and get some estimates about the price, hardware, and software features. Go to any popular hardware store website to get the price of all hardware involved. Identify how do the robots localize themselves and the sensors that they use for the same. Hence, reason how does localization accuracy affects the choice of sensors and robot cost.

5. [Programming] Make a simulator where you move a simulated robot on a map using arrow keys. The up/down keys control the linear speed and left/right arrow keys control the angular speed. Add some error on the simulated robot, so that the expected speeds are different from the actual. The map has no obstacles except for five points. The robot has a virtual sensor that reports the distance to all five points and angle to all five points (measured with respect to the robot's heading direction and bearing). Move the robot on a random track and print the observations and controls (linear and angular speed) at every point of time.

6. [Programming] Using the observations and controls from question 5, track the robot using the Extended Kalman Filter and Particle Filter. For Extended Kalman Filter assume a large initial covariance and a moderate error in the mean at the start of the simulation. For Particle Filter spread the initial particles at a large distance, denoting a poor knowledge of the initial position.

7. [Programming] Extend the solution to the case when the initial position of the five landmarks is not known, which should also be computed.

8. [Programming] Extend the solution of question 6 when the environment consists of all polygon obstacles only, which are detected using a 2D lidar. Use the obstacle corners as features. Use ray tracing to determine if a corner (feature) will be visible.

9. [Programming] Assume that the map is a grid map in the form of a binary 2D array. Give the laser readings of a virtual 2D lidar that scans from -120 degrees to 120 degrees in steps of 1-degree resolution. Add some noise to

readings. Log the robot pose, control, and laser readings. Then using the logs:

(a) Construct the map using the robot pose and laser readings only.

(b) Calculate the robot pose using the observation function as the lidar readings (range value for all bearings).

(c) Calculate the robot pose by using a correlation of the local and global map using laser readings and known binary maps only.

(d) Calculate the robot pose using corners as features and using the observation as the range and bearing values corresponding to the detected features.

Use Particle Filter and Extended Kalman Filter to get positions. Furthermore, use local optimization to compute the transform between any two scans separated by some time. *Hint*: To make a virtual laser scanner, use the parametric equation of line as $[x, y]^T = [x_r, y_r]^T + d[\cos(\theta_r + \alpha)$ $\sin(\theta_r + \alpha)]^T$ where (x_r, y_r) is the position of the robot, θ_r is the orientation of the robot and α is the angle of the laser ray (bearing). By iteratively increasing d, find the smallest d such that the resultant point (x, y) is on an obstacle in the image. The image coordinates will not be the same as the world coordinates. Derive how the (x, y) coordinates of the robot relate to the pixels in the bitmap.

10. [Programming] Take a stereo camera setup. Take two images from the stereo camera, after moving a small step and making a small turn in between the images. Use any feature detection technique to register the features as landmarks. Using any feature-matching techniques locate the same features in the 2nd image. Make a function that takes a prospective transform between two stereo images and calculates the re-projection error. Using any optimization technique, calculate the best transformation, thus solving for visual odometry. Note that you would have used both images of the first stereo image, but only the left image of the 2nd stereo image.

References

Choset, H., Lynch, K.M., Hutchinson, S., Kantor, G., Burgard, W., Kavraki, L.E., Thrun, S., 2005. Principles of Robot Motion: Theory, Algorithms, and Implementations. MIT Press, Cambridge, MA.

Lynch, K.M., Park, F.C., 2017. Modern Robotics: Mechanics, Planning, and Control. Cambridge University Press, Cambridge, UK.

Murphy, R.R., 2000. Introduction to AI Robotics. MIT Press, Cambridge, MA.

Russell, S., Norvig, P., 2010. Artificial Intelligence: A Modern Approach. Pearson, Upper Saddle River, NJ.

Thrun, S., Burgard, W., Fox, D., 2006. Probabilistic Robotics. MIT Press, Cambridge, MA.

Tiwari, R., Shukla, A., Kala, R., 2012. Intelligent Planning for Mobile Robotics: Algorithmic Approaches. IGI Global, Hershey, PA.

Chapter 3

Visual SLAM, planning, and control

3.1 Introduction

Modern-day robots can perform many sophisticated actions autonomously and with great precision. This is facilitated by sophisticated algorithms that continuously assess the situation and make all the decisions without human intervention. The problem of *robot motion planning* (Choset et al., 2005; Tiwari et al., 2012) is to construct a trajectory that takes the robot from the current state, called the source, to some desired state, called the goal. The robot may need to find a trajectory that optimizes some optimization criteria like travelling time and risk of colliding with an obstacle. The knowledge of the world as gathered by the sensors from time to time is already summarized in the form of a map. Typically, to render any service, the robot needs to physically navigate to the desired state to carry out the necessary operation. The problem of motion planning guides the robot regarding the route to follow, the turns to make, and broadly the path to follow. The indoor robots and self-driving cars move around by using motion planning to decide the path to follow.

The planning may not be restricted to a mobile robot or a drone. Imagine a robotic manipulator who needs to fix a screw at a place. Carrying the screw using the hands of the robot is also a motion planning problem. Here as well, the map refers to the information about all the obstacles which need to be always avoided. Hence, picking and placing items in a robotic warehouse is a hard motion planning problem. Opening the door by hand or operating an appliance by hand is also a motion planning problem.

The examples so far are more precisely the *geometric motion planning* problems (Kala et al., 2010a) wherein the aim is to find a trajectory in a continuous geometrical space. To accomplish a larger mission, the robot needs to schedule many operations, selecting which operations need to be done and their relative ordering. This is also a planning problem often referred to as the *task planning* or *mission planning* problem. The map here represents the known facts about the position of different items and their operations. The solution is typically a sequence of operations, each of them being independent and different

Autonomous Mobile Robots. https://doi.org/10.1016/B978-0-443-18908-1.00007-8

103

geometric motion planning problems. As an example, asking the robot to prepare a dish needs the robot to plan out the recipe, decide how and from where to get the ingredients, and plan to put all the ingredients in a desired order. Similarly, asking a robot to clean a desk requires the robot to schedule a pickup of all the items from the place and keep them in the appropriate place, with the constraints that an item under another item cannot be picked up directly. Similarly, the assembly of an entire vehicle by a robot needs planning and scheduling and to be further able to deal with uncertainties that may arise with time.

The problem of planning is typically solved by taking a map or the world model as an input and knowing the position in the map along with any information that may be needed for decision-making. In this implementation, the robot needs to first see and explore the world, making a world model, which gives information about the world for planning. Chapter 2 described making a map using known positions, while the positions were computed using the map as an input. The problems of localization and mapping are solved together as Simultaneous Localization and Mapping (SLAM), where a robot travels in an unknown place and simultaneously computes its pose and maps the environment. This is the first step to be performed by any new robot on being introduced to a new environment, while the robot is tele-operated for this purpose by a human. This map is used for geometric motion planning. Similarly, for task and mission planning problems, the robot must be fed information about everything on the map which may be operated upon. A robot preparing a dish needs to know the location of all the ingredients and the objects around the ingredients.

The exploration and planning can be combined so that the robot makes the least number of movements to get the desired actions done. In such a working philosophy, planning is an integrated part of mapping and localization, which is referred to as the problem of *active SLAM*. The geometric motion planner in such a case works in a partially known environment and makes some decisions regarding the initial navigation. As the robot moves, more information is available which is incorporated in the map or knowledge base and the plans are rectified or extended. So, if a robot is asked to visit an in-campus stationery shop, by knowing only the current location and the approximate location of the stationary shop, it would not go wondering the entire campus premises. It would try to go by the shortest route and reroute if the map does not allow going through certain areas as discovered while navigating. As humans, when we are asked to go to new places on the campus, we also tend to follow the same algorithmics, not exhaustively exploring the entire campus.

For the task or mission problems as well, the robot starts with some, largely unknown, *belief* of the world and takes actions that increase the knowledge of the world or clarify the belief about the important constructs, while optimizing the overall time to explore and solve the problem. Any new information observed is updated in the knowledge base as the new belief of the world. The robot makes the best decisions with limited knowledge, while continuously updating the beliefs. So, the robot assumes all ingredients at their respective obvious places, which is a belief. If the robot does not find an ingredient at a place, it tries to look around the best possible places and also notes or gathers the other ingredients in the process. It does not exhaustively explore the entire kitchen as the start of the cooking exercise. The different nomenclatures and definitions are summarized in Box 3.1.

Box 3.1 Different terms used in motion planning.

Generic definitions
 Path planning: continuous curve that connects source to goal avoiding obstacles.
 Motion planning: Path planning with speed information.
 Path: Continuous function from source to goal (without speed), parametrized by relative distance from the source.
 Trajectory: Continuous function from source to goal with speed, parametrized by time.
Types of problems:
 Geometric motion planning: Continuous trajectory from source to goal avoiding obstacles.
 Task planning/mission planning: Discrete actions that must be performed in sequence/schedule (e.g. robot assembling a car from parts, delivering letters to offices).
 Integration of task and geometric planning: Task planning problem where each action is a geometric planning problem.
Simultaneous localization and mapping (SLAM): Mapping assumes knowledge of localization; localization assumes knowledge of map. Solve both simultaneously.
 Sparse SLAM using visual odometry: Given image at frame k (keyframe), compute XYZ position of all detected features (landmarks). Pose after a short time t is calculated by minimizing the error between the projection of the landmarks on the image and the actual position where the landmarks are observed on the image.
 Dense SLAM using visual odometry: Replace features by all points in the working of sparse SLAM.
Planning in partially known environments: Map/facts needed for decision making are not available, hence,
 Belief spaces: (probabilistically) assume missing facts, plan and act; rectify plans as more information rectifies beliefs (e.g. assume the lift is working, reroute to stairs if not).
 SLAM methodology: Explore and gather facts/make map (SLAM problem), and then plan (e.g. explore the campus and make map, then use the map for motion planning).
 Active SLAM problem: Motion planning alongside SLAM. Take actions that get the most important facts/parts of the map; take actions that increase relevant knowledge and get towards the goal (e.g. walk towards the target museum exhibit, if stuck, explore a new route).

Continued

Box 3.1 Different terms used in motion planning—cont'd

Problem modelling definitions

Ambient space: World as we see it; **Workspace**: Parts of the world reachable by end-effector of the manipulator (both terms used interchangeably in this book).

Configuration: Encoding of the problem scenario (static only, nonconstants) into variables. E.g. (position X, position Y, orientation θ) for a mobile robot.

Free configuration space (C^{free}): Collection of configurations where robot (in the workspace) does not collide with an obstacle or itself.

Obstacle-prone configuration space (C^{obs}): Collection of configurations where robot (in the workspace) collides with an obstacle or itself.

Holonomic constraints: Constraints that depend on configuration (position) only. E.g. robot cannot collide with an obstacle.

Nonholonomic constraints: Constraints that depend on configuration (position) and its derivative (speed). E.g. for many robots sideways speed is always 0.

Metrics

Completeness: Property that an algorithm returns a solution if one exists.

Optimality: Property that an algorithm returns the best solution if one exists.

Resolution completeness (optimality): Property that an algorithm returns a (the best) solution, if one exists, limited by the current resolution. Higher resolutions may result in a (better) solution.

Probabilistic completeness (optimality): Probability that an algorithm returns a (the best) solution (if one exists) tends to 1 as time tends to infinity.

Solution classification definitions

Real-time/online solutions (reactive algorithms): Can be calculated by the robot on the fly (near-real-time can be calculated with some smaller frequency). Can handle suddenly appearing obstacles, dynamic obstacles, etc. Typically, neither optimal nor complete.

Offline planning (deliberative algorithms): Assuming the map is static, the robot waits at the source, calculates a trajectory, and executes the complete trajectory. Can handle only static obstacles. Typically, optimal and complete.

Replanning: For offline algorithms, in case of a change of map during planning/execution, the old trajectory is *recomputed* for the new map, typically attempting to reuse all past calculations. Used for suddenly appearing static obstacles, very slowly moving dynamic obstacles (e.g. travelling to the office by car, if a road is blocked, an alternate route is available from memory).

Trajectory correction: For offline algorithms, in case of a change of map during planning/execution, the old trajectory is *locally corrected* by small modifications for the new map by a local algorithm. Useful for dynamic obstacles unless they are very fast. E.g. while moving in a corridor, if you spot an oncoming person, you constantly drift at the side correcting the occupancy at the middle of the corridor.

A fusion of deliberative and reactive techniques: Deliberative component handles static obstacles (walls, furniture, etc.), reactive handles dynamic obstacles (moving people, suddenly dropped flower pot, etc.)

Anytime algorithms: A solution is always available and improves with time.

The planning systems only give future suggestive motions of the robot. The physical motion is to be performed by the motors of the robot. The control algorithms assess the current state of the robot and the desired state as per the planning algorithm's output, and the algorithms give the speeds to the individual motors to make the future robot state close to the desired state. If the planning algorithms give a path as output for a mobile robot, the control algorithms generate the linear and angular speeds to make the robot closely trace the same path.

Even though the notions of geometric planning, task planning, mission planning, planning in uncertainties, and belief spaces are discussed, this chapter only motivates and introduces geometric robot motion planning. Other concepts shall be taken in the subsequent chapters. The chapter also makes numerous assumptions and presents the first algorithm for motion planning, which is the Bug algorithm. The algorithm imitates the navigation principles used by the insects as they go to reach their goals.

3.2 Visual simultaneous localization and mapping

Chapter 2 presents the basics of the SLAM algorithm for a limited number of preknown landmarks. Consider a self-driving car is moving with a high-definition stereo camera or a high-definition 3D laser scanner. The images can be processed by a feature detection technique and every feature is a landmark. This means that the number of landmarks is extremely high to start. As the vehicle moves, the same landmark would be seen only for a short period till it is in the field of view of the sensor. However, the vehicle will accumulate a lot of features as it moves. Hence, the SLAM problem is solved for such scenarios where the number of landmarks is extremely large.

3.2.1 Visual odometry

Let us first assume that the pose at some time t is well known as a transformation matrix $q_t = T_t^W$. The aim is to find the pose at time $t + \Delta t$. Let us estimate the pose based on the images from a stereo pair (or 3D LiDAR) alone. The problem is also known as *visual odometry*, or extraction of odometry information using vision only (Pire et al., 2017). Features are extracted from both the images. Let us consider the robot (vehicle) coordinate axis system at time t with the vehicle at q_t. The features from the image at time t can be triangulated since a stereo camera is used to get the XYZ location along with the signature. This makes a map at time t in the robot (vehicle) coordinate axis system at time t without which localization cannot happen. To localize the vehicle at time $t + \Delta t$ in this robot coordinate axis system, we need to first match features between the frames. While matching features between time t and $t + \Delta t$, there is a correspondence problem and let the ith observed feature at location $(u_{i,t+\Delta t}^{\text{obs}}, v_{i,t+\Delta t}^{\text{obs}})$ in the image with signature $s_i^{t+\Delta t}$ at time $t + \Delta t$ match with mapped location $(x_j^t, y_j^t, z_j^t, s_j^t)$ at time t in the robot coordinate axis system at time t. This makes the correspondence function $j = c(i)$. Correspondence matching between frames can be done by matching signature and/or location. It is assumed that $(x_{i,t}, y_{i,t}, z_{i,t})$ is with respect to the robot pose at time t. Features common to both frames and hence the ones that appear in the correspondence function are also called the *covisible features*. The motion between the times can thus be estimated in Eq. (3.1).

$$T_t^{t+\Delta t} = \mathrm{argmin}_{T_t^{t+\Delta t}} \sum_{co-\text{visible features } i} \left\| \begin{bmatrix} u_{i,t+\Delta t}^{\mathrm{obs}} \\ v_{i,t+\Delta t}^{\mathrm{obs}} \end{bmatrix} - \pi \left(T_R^S T_t^{t+\Delta t} \begin{bmatrix} x_{c(i),t} \\ y_{c(i),t} \\ z_{c(i),t} \\ 1 \end{bmatrix} \right) \right\|_2^2 \quad (3.1)$$

Here, π is the projection function that projects a point in the camera coordinate frame into the image plane. $T_{t+\Delta t}^t$ denotes the motion of the robot from time t to $t+\Delta t$ as a transformation matrix. $T_R^S T_t^{t+\Delta t}$ is the transformation from the coordinate frame at time t to the camera's current frame (denoted by S). The equation is also called the minimization of the *reprojection error*, which is projecting the 3D features into 2D image place to see the observation error. The pose at time $t+\Delta t$ is thus given by $q_{t+\Delta t} = T_{t+\Delta t}^W = T_t^W T_{t+\Delta t}^t = q_t(T_t^{t+\Delta t})^{-1}$. This means if the vehicle starts from the origin, or $T_{t=0}^W = I_{4\times4}$, the pose at any point in time can be calculated by iteratively postmultiplying the visual odometry output. The concept is shown in Fig. 3.1.

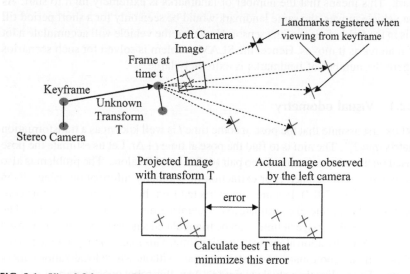

FIG. 3.1 **Visual Odometry.** The stereo camera registers landmarks in the map at the keyframe. For any other frame, the transformation from keyframe (to calculate pose) is the one that minimizes the projected position of landmarks (with the assumed transformation) and the one actually observed in the image. The keyframe changes when there are very few common landmarks left.

The problem is also known as *Perspective n Point* problem (PnP), which means to calculate the 6D relative motion between two poses of the camera, given some points in 3D as captured in the image. Ideally, the error should be exactly 0, which means it is possible to estimate 6 unknown parameters of the transformation using 3 points alone (also called a P3P problem). For the chosen 3 points observed at $(u_{1,t+\Delta t}^{obs}, v_{1,t+\Delta t}^{obs})$, $(u_{2,t+\Delta t}^{obs}, v_{2,t+\Delta t}^{obs})$, and $(u_{3,t+\Delta t}^{obs}, v_{3,t+\Delta t}^{obs})$, consider Eq. (3.2).

$$\begin{bmatrix} u_{i,t+\Delta t}^{obs} \\ v_{i,t+\Delta t}^{obs} \end{bmatrix} = \pi \left(T_R^S T_t^{t+\Delta t} \begin{bmatrix} x_{c(i),t} \\ y_{c(i),t} \\ z_{c(i),t} \\ 1 \end{bmatrix} \right), i = 1, 2, 3 \tag{3.2}$$

Expanding for $i = 1,2,3$ gives 6 equations (3 for u-value and 3 for v-value), using which 6 unknown parameters of $T_t^{t+\Delta t}$ (3 translation terms and 3 Euler rotation terms of roll, pitch, and yaw) can be calculated. However, the equations are not linear and cannot be solved by using a linear equation solver. Geometric solvers can use geometric properties of the points to get a solution. However, since there may be more points, it is possible to use them to calculate a better pose. A technique is to use *optimization* (like gradient descent) to minimize the error that gives the best transformation. The optimization can be done in real time with the modern computing infrastructure.

Another technique to solve the same equation is using the *RANdom SAmpling Consensus* (RANSAC) algorithm (Fischler and Bolles, 1981). There are more points than the minimum (3) that are needed to solve the equation, assuming 0 error. Therefore, randomly sample 3 points with correspondences and calculate the transformation using P3P. The transformation is now called a hypothesis. Now this transformation is checked for all landmarks which are covisible among the two images. If the error using the transformation on a landmark is appreciably low, it is said that the landmark supports the hypothesis or is an inlier. Iterating through all landmarks gives a count of the total number of landmarks that support the hypothesis or the total number of inliers, which is called the value of the hypothesis. This is repeated for multiple sampled combinations of 3 points to solve the P3P problem to get a new hypothesis transformation. If the value of a hypothesis is large enough, it is said to be the correct solution, and the algorithm breaks. Otherwise, the algorithm randomly keeps searching for a better hypothesis and within the time limit, the best hypothesis is returned. The concept is shown in Fig. 3.2. RANSAC is widely used for solving the problem because of its ability to reject outliers. Imagine the correspondence problem was solved and the ith observed feature was found to be best matching the jth landmark, while in reality the ith feature was a noise which should not be matched to any map feature, and the correspondence matching is wrong. The machines will try to optimize the transformations to minimize the error including this wrongly matched pair, which will lead to an erroneous result. Even though it may be said that the impact of the error will be reduced due to abundance of other features. However, practically there may not be several features in the image and the large impact due to one bad correspondence matching will be too large in scale as compared to the other features. RANSAC filters out such outliers and fits a model using inliers alone.

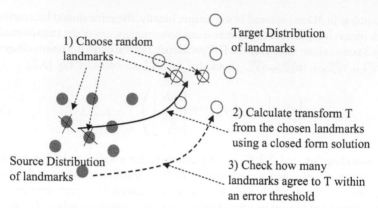

FIG. 3.2 Working of RANSAC algorithm to calculate the pose of the robot. Algorithm iterates in steps 1, 2, and 3 to iteratively improve the guessed transform. This is used to solve the problem given in Fig. 3.1.

A problem with this approach is that if Δt tends to 0 due to the high processing rate, the features would have barely moved. The effect of noise and discretization due to pixelization will be large making the algorithm sensitive. Therefore, the motion is estimated from one fixed frame called the *keyframe* with another frame at a distance, rather than between consecutive frames. As the robot moves, the covisible features between the keyframe and the new frame will reduce, up to 0. This means there will be fewer points to calculate the relative motion. Hence the keyframe is changed when the similarity between a frame and keyframe drops below a threshold, and the next frame is taken as the new keyframe. The localization happens at any time measuring the odometry with respect to the last keyframe, while a new keyframe is added if the similarity (percentage of covisible features) between the current frame and the last keyframe drops below a threshold (say 90%).

Consider a path for which the poses are to be estimated. Mark the keyframes as *pose vertices* (vertices that store the robot pose) and the connection between the vertices as *motion edges* (edges that store the relative motion information). Sometimes an IMU or wheel encoders are available to give an indicative odometry between the frames that can be written on top of the motion edges. Suppose a keyframe K_i (now called pose vertex) sees a landmark L_j (x_i, y_i, z_i, s_j) in the image at a point $(u_{i,j}^{obs}, v_{i,j}^{obs})^T$. The landmark position is called the *landmark vertex* (which stores the fixed landmark position and signature). If the landmark $L_j(x_i, y_i, z_i, s_j)$ is seen by the keyframe K_i at $(u_{i,j}^{obs}, v_{i,j}^{obs})^T$, it is said that an *observation edge* exists between the landmark and the keyframe storing the observation value $(u_{i,j}^{obs}, v_{i,j}^{obs})^T$. This makes a *pose graph* (currently the pose-vertices form a path and not a graph).

The problem of SLAM is thus easily decomposed separately into localization and mapping. Assume that initially the vehicle starts from the origin which is the 1st keyframe. Make a map using features visible at the 1st keyframe, each feature computing and adding a landmark vertex. After some time, at the 2nd keyframe, use the map made at the 1st keyframe to localize. Every observed landmark vertex is connected to the 2nd keyframe pose vertex using an observation edge. Each observation edge represents an observation constraint or a term of the re-projection error, while the solver must compute the keyframe pose (associated with the pose vertex) to minimize the observation edge constraint error. This gives the pose of the 2nd keyframe. Using the computed pose of the 2nd keyframe make a new map (adding newly observed features on top of the old map) at the 2nd keyframe, each added landmark making a new landmark vertex and the associated observation edge. A problem is that a newly observed point may be a noise, which if admitted into the map can cause erroneous future localizations. Hence, every landmark before being admitted into the map should be checked for consistency. *Tracking* over the past few frames is used to see if the feature is strong and not a noise. If the feature can be consistently tracked in the past few frames, it is admitted into the map. Use this map made at the 2nd keyframe, localize at the 3rd keyframe, and so on.

Let us very briefly consider a similar case of using 3D LiDARs. In this case, let us observe a few points in the 3D space in the robot's coordinate frame along with the signature in the depth image, out of the points in the registered map, given by Eq. (3.3).

$$
T_t^{t+\Delta t} = \mathrm{argmin}_{T_t^{t+\Delta t}} \sum_{\text{co-visible features } i} \left\| \begin{bmatrix} x_{i,t+\Delta t}^{\mathrm{obs}} \\ y_{i,t+\Delta t}^{\mathrm{obs}} \\ z_{i,t+\Delta t}^{\mathrm{obs}} \\ 1 \end{bmatrix} - T_R^S T_t^{t+\Delta t} \begin{bmatrix} x_{c(i),t} \\ y_{c(i),t} \\ z_{c(i),t} \\ 1 \end{bmatrix} \right\|_2^2 \tag{3.3}
$$

Here $(x_{i,t+\Delta t}^{\mathrm{obs}}, y_{i,t+\Delta t}^{\mathrm{obs}}, z_{i,t+\Delta t}^{\mathrm{obs}})$ is the place where the landmark is observed at time $t + \Delta t$. A solution can be by using similar techniques. A solution to this equation is also called the *Iterative Closest Point* algorithm. The advantage of the algorithm is that it can deal with unknown correspondences or where $c(i)$ is not known. The algorithm iteratively improves the matching of the original landmarks with the new landmarks by calculating a transformation. If a better transformation is computed, it is accepted iteratively. The algorithm requires a good initial guess of the transform, which is available for the problem since the best transform for the previous frame is already calculated, or one can be obtained by using the IMU or encoder readings. The algorithm then matches every point to the nearest point in the new image, using which a new hypothesis transform is calculated. By projecting all the points, the cumulative error is calculated as the value.

The discussions currently using both image and LiDAR were assuming features. The *dense SLAM* techniques perform the same operations with the same set of equations, however, do it for all pixels (image) and points in the point cloud (LiDARs) without any feature extraction. The approaches are typically more computationally intensive but do not suffer the loss of information due to improper features.

3.2.2 Bundle adjustment

Let us now shift the interest from the localization aspect to the *mapping aspect* of SLAM. To reduce computation, let us only focus upon the keyframes and leave the data recorded in-between keyframes. Consider a path with keyframes as $\{K_i\}$. Each keyframe observes several landmarks $\{L_j\}$. If a keyframe K_i does observe a landmark L_j, the observation is given as $(u_{i,j}^{obs}, v_{i,j}^{obs})^T$. Now the problem in mapping is that the same landmark will appear at slightly different XYZ positions at different keyframes after triangulation and the mapping needs to be able to optimize the exact location. Further, so far it was stated that the landmark can be easily registered once the position of the robot is known. However, there may be noise in seeing the landmark at a particular keyframe and if the same gets registered, it shall never be corrected later. This can be fatal as a wrong landmark registration shall give wrong future poses of the subsequent keyframes where the landmark is observed, and in-turn the wrong future poses will register wrong subsequent landmarks, resulting in large future errors in both localization and mapping thereafter. So far, only the positions were corrected by optimization based on several landmarks. It should also be possible to correct the landmarks based on several positions as the mapping aspect of the problem.

Consider the robot iteratively travels the keyframes and makes the observations. The problem is now to simultaneously calculate the poses of the robot as well as the landmarks in the world, given the constraint that the landmark positions are constants. Consider the pose graph with the keyframes as pose vertices, landmarks as vertices, motion edges between the keyframes, and an edge between a landmark and the pose vertex if the associated keyframe observes the landmark. We now simultaneously optimize the complete graph, optimizing for the landmark vertices (landmark positions) as well as the pose vertices (keyframe poses), such that the constraint represented by the edges has the least errors. The observation edge represents a term of the reprojection error. Sometimes, relative motion information between keyframes is available from the IMU or wheel encoders and the optimizer must attempt to compute the relative poses that minimize the error with such readings as well. The equation tries to compute robot poses and landmark positions such that the projected observations match as closely as possible with the actual observations and (not shown here) the odometry available from IMU or wheel encoders matches the relative odometry between keyframes. This is done by solving the two quantities jointly, as given by Eq. (3.4)

$$T_W^t, L_j = \arg \min_{T_W^t, L_j} \sum_{i \in \text{Keyframes}} \sum_{\substack{j \in \text{Landmarks,} \\ \text{keyframe } i \\ \text{observes landmark } j}} \left\| \begin{bmatrix} u_{i,j}^{obs} \\ v_{i,j}^{obs} \end{bmatrix} - \pi\left(T_R^S T_W^t L_j\right) \right\|_2^2 \quad (3.4)$$

Here, T_W^t is the transform from the world coordinate frame to the robot's pose at time t (while T_t^W is the robot pose at time t). T_R^S is a constant transformation from the robot to the camera sensor. π is the projection function to project the point from the camera coordinate frame onto the image. $L_j(x_i,y_i,z_i,1)$ is the homogenized landmark position. A difference from the earlier formulation is that now the landmarks are registered in the world coordinate system.

The problem is that since several quantities are being simultaneously optimized, it will take a long time and may not be possible to do so as the vehicle runs. The first constraint hence is that the graph size is limited, also called the *local bundle adjustment* to stress that only the recent few segments of the graph are considered for optimization. The second constraint is that this optimization happens as a separate thread with a smaller frequency (because of a higher computation time) so as not to directly affect the localization thread that calculates the poses using the keyframes alone. The equation is similar when using LiDARs. The concept is shown in Fig. 3.3.

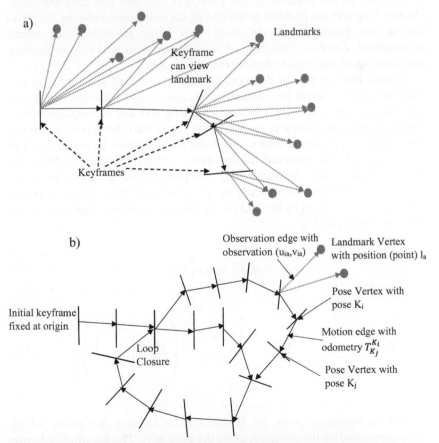

FIG. 3.3 (A) Local bundle adjustment. (B) Pose graph optimization. Each keyframe captures an image with viewable landmarks shown by *dotted arrows*. Motion between keyframes may be available from odometry. The technique jointly optimizes the position of landmarks and poses of keyframes (the SLAM problem), such that the landmarks are visible as viewed in the actual images taken by the robot at the keyframes. Loop closure causes cycles in the pose graph.

3.2.3 Loop closure and loop closure detection

The localization problem so far built from the previous keyframe's pose to calculate the new keyframe's pose and thus incrementally solved the problem. Unfortunately, this means that the errors shall also grow up incrementally with time, making the actual pose poor after some time. The vehicle soon starts to drift from the true position, while the drift continuously increases. Bundle adjustment eliminates some mapping errors to reduce the drift but does not eliminate the problem of incremental drift. This problem is solved by the concept of *loop closure*. Imagine roaming around a new city before discovering that you are lost as none of the roads seem familiar. However, after some time, you see your accommodation location which you know with absolute certainty. Now you can relocalize yourself and correct your beliefs on the positions in the past. It is said that you took and completed a loop and the position at the end of the loop should also be the position at the beginning of the loop. This helps to correct the errors accumulated. Correcting the map of the past positions will help to localize next time when the vehicle travels the same route.

Assume that the pose of the robot is known fairly well at a time t_1 as q_{t1}, while the robot observes the landmarks $\{(x_{j,t1}, y_{j,t1}, z_{j,t1})\}$ with signature $\{s_{j,t1}\}$ in its frame of reference. The robot travels a long way and accumulates errors. However, after some time at t_2, the sensor percept of the robot reports seeing the same thing as was seen at time t_1, when the robot believes its pose is q_{t2}. Let $\{(u_{i,t_2}^{obs}, v_{i,t_2}^{obs})^T\}$ be the observed features along with their signatures. Since the robot perceives nearly the same thing, the pose should be nearly the same. Specifically, the robot pose should be q'_{t2}, calculated from visual odometry between t_1 and t_2, given by Eq. (3.5) using the correspondence function $j = c(i)$ between times t_1 and t_2.

$$q'_{t2} = q_{t_1} T_{t_2}^{t_1} = q_{t1} \left(T_{t_1}^{t_2} \right)^{-1},$$

$$T_{t_1}^{t_2} = \arg\min_{T_{t_1}^{t_2}} \sum_{\text{co-visible features } i} \left\| \begin{bmatrix} u_{i,t_2}^{obs} \\ v_{i,t_2}^{obs} \end{bmatrix} - \pi \left(T_R^S T_{t_1}^{t_2} \begin{bmatrix} x_{c(i),t} \\ y_{c(i),t} \\ z_{c(i),t} \\ 1 \end{bmatrix} \right) \right\|_2^2 \quad (3.5)$$

The minimization gives the transformation between the poses, which when multiplied by the old pose, gives the pose at t_2. The accumulated error is thus known to be $(q_{t_2})^{-1} q'_{t_2}$, which needs to be corrected, as shown in Fig. 3.4A.

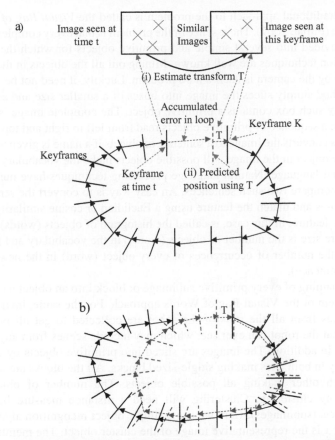

FIG. 3.4 (A) Loop closure detection. Assume the robot sees an image that is very similar to the one seen at keyframe K, and therefore the current pose is corrected to be just next to K (transformation can be estimated using visual odometry between two frames) and loop closure is said to have occurred. (B) Loop closure correction. The last node is pulled to its desired place as a correction, which in turn pulls all the keyframes backward. The correction is largest at the last keyframe and 0 at the first, proportionately at all the middle ones.

This gives rise to two subproblems, loop closure detection and loop closure correction. Let us first solve the problem of loop closure detection. *Loop closure detection* is a difficult problem because of two reasons. First, for loop closure, the two images should be nearly the same. This is a more complex problem than object recognition because object recognition problems typically solve for many simple objects than images where a variety of objects are embedded together as a scene. Second, every new frame can have a loop closure with every other frame in the data set. Some space-based pruning can be done, however the numbers are still large.

The traditional approach to the problem is called the *Visual Bag of Words* approach (Filliat, 2007). The essence of the problem is that every complex scene can be divided into several simple and primitive objects for which the object recognition techniques are well known. Mining out all the objects in the scene captured by the camera is itself a hard problem. Luckily, it need not be solved. The method simply slices the image into boxes of a smaller size and assumes that every such box contains a primitive object. The complete image is hence said to be a sequence of primitive objects read from left to right and top to bottom. This converts the image to a sentence of objects if a name is given to every primitive object in the image. All possible objects make the vocabulary of this new visual language. Natural language processing techniques have numerous ways to compare two such sentences. An easy way is to convert the sentences into features and match the feature using a Euclidian or cosine similarity measure. The feature, in this case, is called the histogram of objects (words) where the feature size is the number of objects (words) in the vocabulary and feature value is the number of occurrences of every object (word) in the new scene image (sentence).

The naming of every primitive subimage or block into an object name is a subproblem of the Visual Bag of Words approach. For the same, imagine all the images from all the runs of the robot are collected to get all types of scenes that the robot can ever see, which may include scenes from any scene database in addition. The images are sliced into primitive objects by cutting uniformly in both axes making single-sized blocks. All the blocks are stacked over each other making all possible objects. The number of objects is reduced by clustering. Clustering will need a distance measure between two objects (subimages). Here we use a simple object recognition algorithm. Consider x is the representative image of the cluster object. The membership of a new object y to the cluster represented by x will be based on the number of feature matches between x and y. Feature detectors and descriptors are used in both the images. The feature descriptors give signatures that are used to match features between images, while the matching features are filtered for geometric consistency. The higher the number of matched features, the larger is the affinity of y towards x. Every cluster is assigned a random name. The new test image is also divided into blocks and for every block, the closest cluster to which it matches is seen. This is a simple object recognition problem. The same name is used for the block. Once all blocks get their names, the histogram of words features is extracted to match the test image with a candidate image for loop closure.

The technique described converts the raw images into global feature vectors. The local features as used previously detected and described pixels. The global features describe the complete image into a fixed size of feature descriptor. The images can now be plotted into a *feature space*, each image being a point in the feature space. As the vehicle travels, each keyframe image is additionally registered in the feature space. A live image is converted into features and the closest matching image in the feature space (if any) is used for loop closure. The proximity query in the feature space can be done in a logarithmic time using the k-d tree data structure. The images are further verified for a loop closure by matching the local features.

3.2.4 Loop closure correction

The detection of loop closure is not enough, and corrections need to mend the map once the errors are known. Imagine the pose graph as shown in Fig. 3.4B, with the keyframes, explicitly marked. Since there is an explicit error, the last keyframe marking the loop closure would be corrected to its true calculated position. This means the keyframe behind it will also proportionately move, and the one still at the back will also be moved. The first keyframe making the start of the loop closure will however not be moved. The keyframes are hence moved in proportional to the error. The first keyframe moves by the largest magnitude and the last one by the least. Suppose n keyframes make a loop with error $e = (q_{t_2})^{-1} q'_{t_2}$. The last keyframe moves by e, the first by 0, and any intermediate by an amount $(i/n)e$. The interpolation however is a little more involved since poses are transformation matrices.

Let us revisit pose *graph optimization (Kümmerle et al., 2011)*. Let every keyframe be a *pose vertex* of the graph. The pose vertex stores the pose as a transformation matrix of the corresponding keyframe (say, K_i). If a vehicle directly goes from keyframe K_i to a keyframe K_j, then a *motion edge* is added between the vertices i and j. The motion edges store the odometry information between the frames (say $T^{K_i}_{K_j,\text{odom}}$), which could be measured by any odometry system. If the vehicle goes from keyframe K_k and completes a loop closure with keyframe K_j, then also a motion edge is added between the vertices K_k and K_j. Loop closures complete the cycle to make this a graph instead of a path. Further, every landmark is also a *landmark vertex* in this graph. The landmark vertices store the position in XYZ of the corresponding landmark (say L_a). An *observation edge* is added between a landmark vertex L_a and a pose (keyframe) vertex K_i if the keyframe K_i observes the landmark L_a. The observation edges store the image coordinates $(u^{\text{obs}}_{i,a}, v^{\text{obs}}_{i,a})^T$, where the landmark L_a was observed in the image taken from the keyframe K_i. Every edge stores a constraint between the information stored at the connecting vertices and the edge itself. Consider the motion edge, $T^{K_i}_{K_j}$ between keyframes K_i and K_j, where the error in information at any time is $(K_i - T^{K_i}_{K_j,\text{odom}} K_j)$. Similarly, the measurement edge $(u^{\text{obs}}_{i,a}, v^{\text{obs}}_{i,a})^T$ between the landmark L_a and keyframe K_i represents the error $(u^{\text{obs}}_{i,a}, v^{\text{obs}}_{i,a})^T - \pi(T^S_R T^{K_i}_W L_a)$. Both errors may not be ideally zero due to errors in the associated sensors. The graph optimization technique now attempts to set the pose values of all keyframes in the loop and position values of all landmarks to minimize the cumulative error represented in the graph, shown by Eq. (3.6) and Fig. 3.3B.

$$K_i, L_a = \arg \min_{K_i, L_a} \left(\sum_{K_i} \sum_{\substack{L_a, \\ K_i \\ \text{observes } L_a}} \left\| \begin{bmatrix} u_{i,a}^{\text{obs}} \\ v_{i,a}^{\text{obs}} \end{bmatrix} - \pi T_R^S T_W^{K_i} L_a \right\|_2^2 \right. \tag{3.6}$$

$$\left. + \sum_{K_i} \sum_{K_j, \text{motion edge } K_i, K_j} \left\| K_i - T_{K_j, \text{odom}}^{K_i} K_j \right\|_2^2 \right)$$

There is a small difference between bundle adjustment and loop closure. Bundle adjustment looks at a small graph around the current keyframe to locally compute the landmark positions and keyframe poses, while the loop closure does it for a loop of keyframes and the associated landmarks. The benefit of moving in loops is that all subsequent keyframes are already registered and will trigger loop closures as the vehicle visits them again on taking the same loop. The second visit of the loop can correct the errors of the map. The map produced from a stereo camera is shown in Fig. 3.5.

FIG. 3.5 Map produced using a stereo camera. Dots represent the landmarks, and the pose graph is shown using lines.

3.3 Configuration space and problem formulation

The map of the area of operation is always made by the internal and external sensors, whose sensor readings are summarized in the form of known information of the obstacles. Imagine such a space with open areas and obstacles in 2D or 3D environments. This space is called the *ambient space* or *workspace W*. So, the references to the map previously are more precisely known as ambient space or workspace. Note that workspace specifically means the volume of space reachable by the end effector of the manipulator rather than the complete space and therefore the term ambient space is technically more correct. However, throughout the book, the terms workspace and ambient space shall be used interchangeably, since most of the focus shall be on robots with a mobile base which can reach nearly the entire ambient space.

The workspace (W) consists of two sets of points. The first set of points denote the obstacles (W^{obs}) and the second set of points denote the free space (W^{free}). Naturally, the superset W is made of these two sets of points and hence $W^{free} = W \setminus W^{obs}$, where $\setminus$ denotes the set complement of W^{obs} with the superset W.

Imagine a virtual robot traversing this space as per its constraints. The virtual robot needs to be encoded by the minimum number of variables; such that, given the encoding, a user may visualize the complete robot in the workspace. Typically for a robot that flies, the XYZ coordinate values and the Euler angles perfectly encode the robot. The size of the robot, colour, length of the bounding box, etc. are not encoded since they remain constant as the robot moves. This encoding is called the *configuration* of the robot. This is different from a complete state of the robot that also encodes the time derivatives like the speed of every variable.

A virtual robot at a configuration (q) may collide with an obstacle (or in some cases with itself like a manipulator arm can hit itself). If a robot cannot be placed at a configuration q because of such a collision, the configuration is said to belong to the obstacle-prone configuration space or C^{obs}. The *obstacle prone configuration space* is the set of all configurations q such that a robot placed at the configuration q collides with a static obstacle or itself. Assuming no self-collisions, the space may be conveniently denoted by Eq. (3.7).

$$C^{obs} = \{q : R(q) \cap W^{obs} \neq \varnothing\} \tag{3.7}$$

$R(q)$ denotes the set of all points occupied by the robot in the workspace when placed at the configuration q. The Eq. states that if there are points common between those occupied by the robot and the obstacles as known in the workspace, a collision is said to have occurred. The collision-checking algorithms are used to test if a configuration is free or not.

The free configuration space (C^{free}) is a set of all configurations which do not incur any collision with a static obstacle or the robot with itself. Like the workspace, the configuration space consisting of all configurations is a superset of the free and obstacle-prone configuration spaces, giving $C^{free} = C \setminus C^{obs}$. The notions of workspace and configuration space with examples are taken in detail in Chapter 6.

A path $\tau:[0,1] \rightarrow C^{free}$ is defined as a continuum of configurations that start from $\tau(0) = source$ and end at $\tau(1) = goal$, such that all intermediate configurations $\tau(s)$ are collision-free, or $\tau(s) \in C^{free} \forall 0 \leq s \leq 1$. Assume the configuration space given in Fig. 3.6. The path that the algorithm should find is also shown in the same Fig. A *trajectory* is the same as the path; however, it is parametrized by time, τ: $\mathbb{R}_{\geq 0} \to C^{\text{free}}$, with the configuration of the robot at time t given by $\tau(t)$. Typically, the robot starts at the source $\tau(0) = source$, ends at the goal $\tau(T) = goal$, where T is the time to reach the goal, and stays at the same place thereafter, $\tau(t) = goal \,\forall\, t \geq T$. Parameterizing by time has an advantage that the derivatives can also be used to give the speeds and accelerations to control the robot. However, trajectory computation implies that the speed assignment needs to be done. The path only states the generic path to follow with no information of the speed. In some cases, the robots may disappear upon reaching the goal, like while planning for a road-segment, a robot that reaches the end of the segment ceases to exist; or a robot that reaches the exit point (goal) ceases to exist, assuming that it has left the exit.

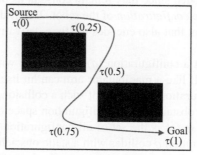

FIG. 3.6 Sample configuration space and path. The path is collision-free and is parametrized by relative distance from the source.

The pursuit of finding the path or trajectory of a robot may imply constraints. Typically, it is desired that the path or trajectory has a derivative so that a control algorithm can be used as a realization of the path or trajectory. The control algorithms cannot typically be used to trace sharp corners. A mobile robot with a differential wheel drive cannot travel side-ways, which is an additional constraint that the algorithms may like to optimize for. This constraint depends upon both the configuration and the speed of the robot and is called the *nonholonomic constraint*. A constraint that can be specified by just the configuration is called the *holonomic constraint*.

3.4 Planning objectives

The motion planning algorithms typically try to compute the best possible trajectory. The notion is hence to assign a cost factor that illustrates metrically how good or bad a trajectory is. One of the most used metrics is called the *path length*, given by Eq. (3.8).

$$\text{length}(\tau) = \int \tau(s)ds = \sum_{s=0,\text{ step size } \Delta}^{1-\Delta} d(\tau(s), \tau(s + \Delta)) \tag{3.8}$$

For trajectory planning, alternatively, the time to reach the goal is a metric. This accounts for the fact that many places on the path may have a steep turn.

For mobile robots, the allowable speed is inversely proportional to the curvature, given by Eq. (3.9).

$$v(s) = \min\left(\sqrt{\frac{\rho}{\kappa(s)}}, v^{\max}\right) \tag{3.9}$$

Here, $v(s)$ is the maximum allowed speed at s, ρ is a constant, $\kappa(s)$ is the curvature, and $v^{\max}$ is the maximum permissible speed of the robot at any time because of its limitations. The curvature (in a discrete sense, discretized by δ) is given by Eq. (3.10).

$$\kappa(s) = \|\tau(s + \delta) + \tau(s - \delta) - 2\tau(s)\| \tag{3.10}$$

The curvature is the norm of the vector shown in Fig. 3.7A. It models the second derivative of the curve equation with respect to the parameterization variable. The curvature is calculated in a discrete sense here but could also be calculated by the derivative.

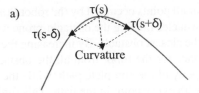

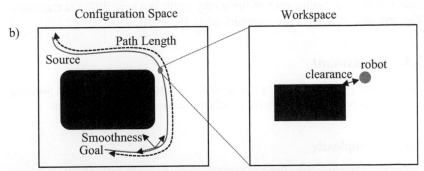

FIG. 3.7 Commonly used objectives for motion planning. **(A)** Curvature. The sharper the turns, the higher is the curvature. A straight line as a 0 curvature. **(B)** Planning objectives: Path length, time to reach goal (not shown), clearance in workspace (minimum or average over path), smoothness (speed is inverse of curvature, minimum or average over path).

As per this modelling, one may go at very large speeds on straight roads, but on encountering an intersection, one must slow down. To make a U-turn, one travels at a near-zero speed. In all these cases, as curvature increases, the speed of the operation reduces. The time to reach the destination accounts for the same factor. Hence, a popular mechanism of trajectory planning in mobile robots is to first calculate the path and then to assign speeds by knowing the curvature.

Smoothness is another metric that tries to find smooth trajectories such that the operating speed can be maximized. The smoothness of a path at the parametrized point s is given by its curvature $\kappa(s)$, with a smaller value meaning a better path.

The stated objectives may take the robot too close to the obstacle which is hazardous since the obstacles are known only by a limited precision and the control algorithms can move the robot only to some limited precision. The map may have a few centimetres error due to mapping errors while performing SLAM. The control algorithms may have a few centimetres errors between the desired pose on the trajectory and the actual pose of the robot. Keeping the robot too close to the obstacle to increase the smoothness or decrease the path length unnecessarily risks a collision. Hence, another objective takes the robot far from the obstacles. *Clearance* at a configuration q is the minimum distance between the robot and any other obstacle, given by Eq. (3.11).

$$\text{clearance}(q) = \min_{o \in W^{\text{obs}}, r \in R(q)} d(r, o) \tag{3.11}$$

The equation iterates over all points occupied by the robot $R(q)$ when placed at configuration q and the obstacles making the obstacle-prone workspace. The clearance is the distance to the closest obstacle. By increasing the clearance, the algorithm keeps a safe distance for the robot from all the obstacles.

The path length is a property of the complete path, while the clearance and smoothness are associated with every point in the path. This further makes the objective functions as minimum clearance over path, average clearance over path, minimum smoothness over the path, and average smoothness over the path. Optimizing against each of these may result in a very different trajectory. The different objectives are summarized in Fig. 3.7B.

3.5 Assessment

This section presents the categorization to understand which motion planning algorithm is better in what metrics and can be used in what situations.

3.5.1 Complexity

Computational complexity denotes how the computation time increases with input size, in the worst case. The exact time can be improved with hardware, therefore the order of change is studied. Assume in the worst case the time taken by an algorithm is given by $T(n) = 6n^2 + 12n + 30$, where n is the input size. As $n \to \infty$,

$6n^2 >> 12n >> 30$, giving effective time as nearly $6n^2$. The constant 6 can be improved by getting a better hardware. The order of change is conveniently given by $O(n^2)$, where $O()$ refers to the worst-case asymptotic complexity analysis.

The input size n varies along with the algorithms, which may be the hyper-volume of the configuration space, length of the optimal trajectory, number of samples drawn in space, number of distinct obstacles, etc. The *memory complexity* is similarly the additional memory required for the computation of the trajectory. Some very small robots may not have the memory to store a lot of information. For larger robots, a larger memory complexity typically also implies a larger computational complexity. The memory complexity is still important to consider because many times memory may exhaust faster than the available computation time.

The absolute computation time is also used as a factor to assess the working of the algorithms. An algorithm claiming to be *real-time* in nature for the real-time operation of the robot must guarantee a computation time within a few milliseconds. The criterion may be relaxed for the algorithms that claim to be *near real-time*, meaning that they guarantee to miss the real-time requirements by a small constant factor. The near-real-time algorithms can be operated at reasonably high frequencies for planning and controlling the robot. The *offline algorithms* suggest they take a large computation time. Typically, the offline algorithms make a trajectory while the robot is waiting for the trajectory to be supplied assuming a static map. Once a trajectory is constructed and given to the robot, the robot works by using *online control algorithms*.

3.5.2 Completeness, optimality, and soundness

Completeness is a property of the algorithm that guarantees to find at least one solution, provided one exists. Complete algorithms find at least one trajectory that connects the source and the goal. There may be situations when the source and goal are not connected in the first place, in which case the algorithm needs to graciously exit as no trajectory is possible. *Optimality* is the property that the algorithm finds the best trajectory, provided one exists. The objective measure for defining optimality may be any generic objective defined by the user or may be specific for the algorithm (like path length).

Many times, a resolution factor creeps into the algorithm working. This may be the resolution of representing the map, searching for a solution, collision-checking, or any other resolution parameter which may be specific to the algorithm. A higher resolution generally implies better results; however, the algorithm also takes a longer computation time to get the results. *Resolution completeness* means that the algorithm guarantees to find a trajectory if one exists, subject to the constraint of resolution setting. In other words, the algorithm guarantees to find a trajectory, if one exists, for a reasonably large resolution. An algorithm is a *resolution optimal* if it guarantees to find the optimal trajectory subjected to the constraints of the resolution. Typically, the quality of the trajectory keeps increasing as the resolution increases. The algorithm finds the optimal trajectory as the resolution tends to infinity.

Many algorithms start with a solution and improve the solution along with time. As time increases, the solutions keep getting better and better. The *probabilistically complete* algorithms are the ones in which the probability of finding a solution, if one exists, tends to one as time tends to infinity. Practically, the algorithms return at least one solution soon enough than infinity. Similarly, *probabilistically optimal* algorithms are the ones in which the probability of returning the optimal solution tends to one as time tends to infinity. Practically, the algorithms give a good enough solution after a moderate time, although the guarantee of the best solution comes as time tends to infinity. The solutions typically improve with time.

3.6 Terminologies

There are a few terminologies that will be used repeatedly in the book. The planning algorithms typically take a map and plan a trajectory from the source to the goal for the robot. The trajectory is used by the control algorithms to trace the trajectory. This however implies that the map remains static from the time the trajectory planning started to the time the robot completely traces the trajectory. If anything (apart from the robot) moves or anything changes in the map, the complete trajectory may get invalidated. This style of working constitutes *deliberative algorithms*. The algorithms assume a static map, deliberate by taking time, and give a trajectory valid for the static map only.

The *reactive algorithms* on the other hand assess the situation including the information of the goal relative to the robot and information of the nearby obstacles relative to the robot. The algorithms just compute the immediate motion of the robot. They typically do not construct the entire trajectory. In the next time instance, another calculation is made to calculate the next move. Typically, these algorithms take lesser computation time, but lack optimality and completeness. The map may change with time in such algorithms, which is reactively taken care of. A typical manner of working is therefore to use deliberative algorithms for the static components of the map and reactive algorithms for the dynamic components. The fusion of the two algorithms gives a good navigation algorithm (Kala et al., 2010b).

Suppose a deliberative algorithm is used to navigate a robot as the map largely remains static. However, after the algorithm has computed a path and the robot starts its journey, the map changes by the introduction of some new obstacles, motion of some obstacles, or by deletion of some obstacles. A deliberative algorithm typically stops the robot and calculates a new trajectory. However, in many algorithms, it may be possible to reuse any of the computations previously, including reuse of metadata generated during the initial search. This fastens the algorithm, instead of running a new algorithm from scratch. Such algorithms are called *replanning algorithms*. Alternatively, the *trajectory correction algorithms* (Kala, 2014) take a trajectory that has been invalidated due to the changed map and locally repair and correct the trajectory in a small computation time. With sufficient resources on-board, this can be done dynamically as the robot moves. This avoids the robot to wait and replan. This is also better as it provides near-completeness and local-optimality, unlike reactive algorithms. The trajectory correction algorithms are thus good choices to handle slowly moving obstacles and suddenly appearing obstacles. Very fast-moving obstacles are best handled by the reactive algorithms since the dynamics changes faster than the algorithmic time required to correct the trajectory. If using trajectory correction algorithms for very fast-moving obstacles, before the trajectory can be corrected for the observed motion of an obstacle, the obstacle further moves invalidating the (yet to be corrected) trajectory.

Many algorithms plan iteratively, wherein the solution keeps improving with time. Such algorithms are referred to as the *anytime algorithm*, where the name is derived from the observation that a solution is available at any time, and therefore the robot can make a decision to stop the algorithm and start motion from source to goal at any time. The longer the robot waits, the better the solution as the solution typically improves with time. These algorithms are suited for robots that need a guarantee to make a motion within a few milliseconds (or seconds) as per the application requirements. The best available solution within the time limit is used for the motion.

3.7 Control

Given a map, position on the map, and a reference trajectory, the final task is to give control signals to the motors of the robot to trace the desired trajectory. The motors already have an embedded controller in them, very similar to the controller that we would be discussing. These controllers can be given a reference speed, and the embedded controllers ensure that the actual speed of the motor is as close as possible to the reference speed. The control problem is thus to generate speeds of operation of the different motors. Here, we will only study the kinematics-based controller. It is also possible to calculate the forces and torques in the system and hence calculate the desired motor torque which is a dynamics-based controller and derives its modelling from the system dynamics.

3.7.1 PID controller

Let us consider the simplest problem, which is to control the manipulator using kinematics alone. Suppose that the arena is obstacle-free, and the manipulator needs to reach a specific object, for which the desired position of the joints is known by inverse kinematics. The problem is easy because each joint is operated by an independent motor, and the desired angular position is known by inverse kinematics. Thus, the whole problem can be decomposed into solving for 1 motor alone.

Let $\theta(0)$ be the current position of the motor at time 0 and let θ_d be the desired motor position. One solution to the control problem is to apply a fixed angular speed of ω_{max} for $(\theta_d - \theta(0))/\omega_{max}$ seconds. This is called the *open-loop control system*. However, this solution has problems, for example maintaining the exact speed profile may not be possible, instantaneously gaining a speed ω_{max} from 0 may not be possible, stopping after the specific time suddenly may not be possible, and the motor may have losses due to induction, noise, and other reasons, due to which perfectly maintaining ω_{max} may also be not possible.

To overcome all these problems, the aim is to take *feedback* regularly from the system. Now, the supplied speed operates the motor and physically changes the position and produces a change, which is measured and supplied back to the control algorithm to rectify the next control signal. This completes the loop, and the controller is called the *closed-loop feedback controller*. At any general time, t, let the angular position of the joint be $\theta(t)$ measured by using motor encoders. Using an analogy of the way humans walk and drive, the aim is to gently stop at the desired position, so as not to risk moving beyond the desired position by a large margin of momentum. Furthermore, the aim is to move as quickly as possible. So, when close to the desired position, the braking must happen gently and finish when the desired position is reached. When far away from the desired position, the speed can be as large as possible. This gives rise to the *control law*, which relates the input error to the output control signal. The control scheme is summarized in Fig. 3.8. The control law is given by Eqs (3.12)–(3.13).

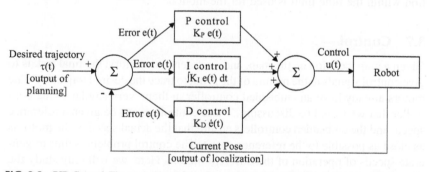

FIG. 3.8   PID Control. The control input (say, linear and angular speed) is derived from the current error (P term), integral of error since the start (I term), and derivative of the error (D term). The control is given to the physical robot to move, whose feedback is obtained from localization. The integral term removes the steady-state error even with constant external noise, thus removing long term problems. The derivative term anticipates error and dampens overshoot.

$$u(t) = K_P \, e(t) \tag{3.12}$$

$$\omega(t) = K_P \, (\theta_d - \theta(t)) \tag{3.13}$$

Here, $u(t)$ is the control input, which for the specific example is the angular speed $u(t) = \omega(t)$. $e(t)$ is the error, which for the specific example is $e(t) = (\theta_d - \theta(t))$. K_P is a proportionality constant.

This controller is called the P-controller or the *proportional controller*, as the control signal is proportional to the error. Ideally, the bound checks are applied to limit the control signal to the permissible limits. The controller meets the design expectation with a large control input for large deviation. As soon as the desired angle is set, the controller stops.

The speed will be large initially when the deviation from the desired value is larger. However, speed will gradually reduce as the desired value is reached. This is called the *rise phase*. The rise time is the time required to reach 90% of the steady-state or reference value. Even though the control signal becomes zero when the desired value is reached, due to momentum and kinematic constraints, there will be an *overshoot*, when the value will go beyond the desired value. This will then change the polarity of the error and the control signal (angular speed) will invert direction to reduce the error. The overshoot will still be there but be lesser due to the smaller operating speeds. This will go on with the system oscillating around the reference value. The oscillations will typically dampen for some time. Finally, the system will settle with very small oscillations within a permissible error limit. The time required is called the *setting time*. The final oscillations observed are said to constitute the *steady-state error*. The notations are shown in Fig. 3.9. The controller runs as quickly as possible and recomputes the control signal with a frequency known as the control frequency. Note that the error is the difference between the actual pose and desired pose, while the former is the localization problem.

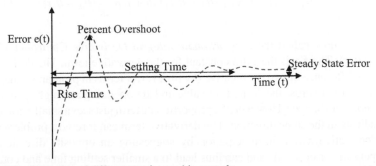

FIG. 3.9 Illustration of different metrics used in controls.

A problem with the P-controller is when the system has a constant external noise that gets added. Imagine a manipulator gets a small positive noise at every instance. So even if the error is zero, the small positive noise will drag the system further away. Of course, when the error becomes sufficiently large, the correction is applied in the other direction. However, in the steady state, the errors are always positive. Similarly, assume that the steering of the car-like robot is slightly bent, or the wheels of the differential wheel drive robot are loose and keep angling a little away. Since at the steady state, the error is always positive (or negative), the area under the curve is a strong indicator of a long-term accumulated problem in control. The integrator is the mathematical formula that keeps adding such errors and is applied as the correction uses the same. The *Proportionate-Integral controller* (PI controller) uses the control law given by Eqs (3.14)–(3.15).

$$u(t) = K_P\, e(t) + \int_t K_I\, e(t)\, dt \qquad\qquad (3.14)$$

$$\omega(t) = K_P(\theta_d - \theta(t)) + K_I \int_t (\theta_d - \theta(t))\, dt \qquad\qquad (3.15)$$

Here K_I is the integral term constant. The controller applies corrections based on all past errors accumulated and thus corrects the long-term historic problems. This brings the steady-state error even with a constant external noise to 0. However, the integral term may often oppose the proportionate term and thus increase overshoot and settling time. In programs, the integral term is just a constant adder (summation) that, in a discrete sense, keeps adding all the error terms. Integrating arbitrary functions may not be possible.

Another factor often added in controller design is to anticipate the problems that may come later and act accordingly in anticipation. This is done by using a derivative term also, which makes the control law as given by Eqs (3.16)–(3.17).

$$u(t) = K_P\, e(t) + \int_t K_I\, e(t)\, dt + K_D\, \dot{e}(t) \qquad\qquad (3.16)$$

$$\omega(t) = K_P(\theta_d - \theta(t)) + K_I \int_t (\theta_d - \theta(t))\, dt + K_D\, \frac{d}{dt}(\theta_d - \theta(t)) \qquad (3.17)$$

This is now called the *Proportionate-Integral-Derivative Controller (PID Controller)*. The K_D is the extra constant for the derivative term. The derivative is also a difference of two consecutive (by time) error terms in programs. Imagine you are about to cross the desired set value and are operating at a high speed. An overshoot will happen. However, the proportionate term just sees a small error term and adjusts in the same direction. The derivative term can foresee a problem and take corrective action in anticipation by suggesting an opposite direction. It dampens the error profile and can thus lead to a smaller settling time and continuously dampens the overshoot. If one is at a highly positive error, the proportionality term gives rise to a large positive control signal. At the next iteration, the error would reduce to a smaller value. The derivative becomes negative and thus the derivative term dampens the error correction by preferring negative control signals opposing the proportionate term. Similarly, if the error is highly negative. In case the system is oscillating, an oscillation from a negative error to a positive error would make a positive control signal, supporting the proportional term to reduce the error, and vice versa for the oscillation from a positive error to a negative error.

3.7.2 Mobile robot control

Let us now focus on wheeled mobile robots, which can be conveniently represented by an XY coordinate value and an orientation. The robot is controlled by using a linear and angular velocity. This is now an interesting problem because 3 variables have the desired value, but only 2 control terms. The control terms also do not independently operate on isolated variables and thus a decomposition is not possible.

Consider the problem is to control the robot over a path planned by using any planning algorithm, as shown in Fig. 3.10A. While attempting to control the robot over this path, the robot may not necessarily always be on the path and may deviate at times. Let the current position of the robot be as shown in the same figure. The *crosstrack error* is defined as the signed distance perpendicular to the path. Above the path, the error is taken to be negative and below the path, the error is taken to be positive. Imagine being under the path. Now you need to have a positive angular speed to attempt going towards the path. The angular speed is higher when the crosstrack error is larger. At the trajectory, the error is zero, and therefore there is no need to immediately steer (which will have to be done eventually due to overshoot). The PID controller hence uses the same generic equation, except that the signed crosstrack error is used. The linear speed may be reduced if the error is larger and vice versa.

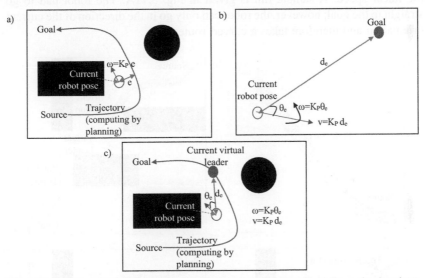

FIG. 3.10 Error terms used in controlling a mobile robot. (A) Cross Track Error, the signed perpendicular distance to the planned trajectory, used to set linear (v) and angular (ω) speed. (B) Error from a goal, consisting of distance to goal (to set linear speed) and angular deviation to the goal (used to set angular speed). (C) Following a trajectory by using a virtual leader as the goal. The leader travels at a constant speed in the planned trajectory.

The crosstrack error keeps the robot as close as possible to the trajectory. Sometimes, the robot needs to go to a specific point. Consider an obstacle-free map and a goal $G(G_x, G_y)$ in the map that the robot wishes to attain. The current position of the robot is $(x(t), y(t))$ with an orientation $\theta(t)$, both obtained from the localization module. The steering now aims to make the robot constantly face the goal, while the speed is set to stop exactly at the goal. The terms are shown in Fig. 3.10B. The angular error (to control angular velocity) and linear error (to control the linear velocity) are given by Eqs (3.18)–(3.22).

$$\theta_e(t) = \text{atan2}\left(G_{y-y(t)}, G_{x-x(t)}\right) - \theta(t) \tag{3.18}$$

$$\theta'_e(t) = \text{atan2}(\sin(\theta_e(t)), \cos(\theta_e(t))) \tag{3.19}$$

$$d_e(t) = \sqrt{(x(t) - G_x)^2 + (y(t) - G_y)^2} \tag{3.20}$$

$$e(t) = \begin{bmatrix} \theta'_e(t) \\ d_e(t) \end{bmatrix} \tag{3.21}$$

$$\begin{bmatrix} \omega(t) \\ v(t) \end{bmatrix} = K_P e(t) + K_I \int_t e(t)\, dt + \frac{d}{dt} e(t) \tag{3.22}$$

Equation (3.18) denotes the angular error as the difference between the desired and the current orientation, while Eq. (3.19) limits the difference to the interval $(-\pi, \pi]$. This mechanism is assumed to be used for all angular differences throughout the book. The linear error is the distance to goal used to set the linear speed. A sample run is given in Fig. 3.11A. The robot had to go straight to the goal; however, the robot can only go in the direction of the current orientation and therefore takes a curved route.

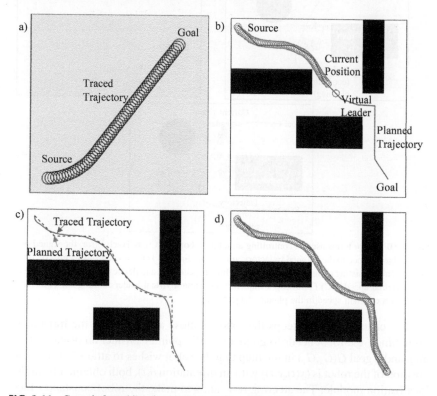

FIG. 3.11 Control of a mobile robot (A) Control of a mobile robot to reach a goal, (B) Virtual leader kept at a certain distance ahead of the robot in the planned trajectory (c-d) Traced trajectory and planned trajectories. Note that the corners are smoothened adhering to kinematic constraints.

The problem is that there are obstacles and therefore the robot cannot assume that a straight-line motion to the goal will be collision-free. Consider that a path is already planned by the path planning algorithm, which the robot must trace as closely as possible. The path may not be perfectly traced due to noises and adherence to constraints. Let a *ghost robot* trace the path, also called the *virtual leader*, which can be perfectly done since the robot is virtual, ideal, and noise-free. The ghost robot now acts as a moving goal for the real robot, which is controlled by the same set of equations. The control law for the real robot has already been set.

Let us discuss the motion of the ghost robot. Suppose the ghost robot travels faster than the maximum speed of the actual robot. The actual robot shall never be able to catch up and will lag, thus risking a collision. Contrary, if the ghost robot is too slow, the actual robot shall quickly catch it and will have to wait for the ghost robot to go ahead. In this case, the real robot waits for the ghost robot. Therefore, the ghost robot is always kept sufficiently ahead of the actual robot. The ghost robot starts its journey at the maximum speed; however, if the separation with the real robot becomes more than a threshold, the ghost robot waits for the actual robot to catch up. Once the real robot catches up and is again close to the ghost robot, the ghost robot shall move further away on the planned trajectory. Eventually, the ghost robot reaches the final goal in the planned trajectory and stops. The approach is called the *virtual leader following* approach. The terms are shown in Fig. 3.11C. A sample run is shown in Fig. 3.11B–D.

3.7.3 Other control laws

Imagine that the robot is given a reference trajectory of a road-network map of a city, and the next step is to make a sharp right turn at the intersection. The drivers can foresee the sharp right turn and start making a right turn in anticipation, while also start applying the brakes in anticipation. The current control laws are based upon the immediate error only and therefore do not foresee such problems in the reference trajectory beforehand. The *model predictive control* is a strategy based on predicting the building up of the error and taking corrective control actions in anticipation of the same. The control strategy applies a PID kind of control not to the actual robot but a ghost robot whose kinematics is well known. The error with time is noted. The controller in the actual robot, therefore, tries to mitigate the long-term error seen in the ghost robot. In the example, the ghost robot will report a sudden increase in error on crossing the intersection, and the control law of the actual robot will derive a control sequence to mitigate the same.

Let $q(0)$ be the current pose of the vehicle and $\tau(t)$ be the desired future trajectory. Note that here the desired trajectory is time-stamped, while previously it was just a path in space with no restriction of time. Suppose $q(t+\Delta t) = K(q(t), u(t), \Delta(t))$ be the known kinematic function (K) of the robot which takes the robot state $q(t)$ and a constant control input $u(t)$ applied for Δt time, to produce as output the state after the control input is over, or $q(t+\Delta t)$. The aim is to minimize the error to get the best control sequence, or $u^* = \mathrm{argmin}_u \sum_{t=0}^{T} \|q(t) - \tau(t)\|_2^2$. Here, T is the time horizon to which the controller foresees. This calculation is performed at every step, so only $u(0)$ is used. A variety of local optimization techniques, some to be discussed in the subsequent chapters, can aid an online optimization.

Imagine that the robot model is perfectly known, and everything behaves ideally. In such a case, the time horizon can be kept as infinity to get the best control sequence, which can be directly fed into the robot to perfectly trace the trajectory. The assumption is that the environment is static, and the application is offline. So, there is time to do a global optimization, which can be time-taking. This is called *optimal control*. The term optimal control can also be used in a local sense, to optimize a sequence of control inputs to be given, which ensures disturbance rejection online when the robot does not behave as anticipated.

Other control mechanisms are good to mention. The *neural control* takes the error (sometimes the change in error and sometimes integral error as well) as input to produce the control signal as output. The training of the neural controllers can be done using supervised or reinforcement learning techniques. Later chapters delve deeper into the use of neural networks for navigation. Similarly, the *fuzzy controller* enables specifying lingual rules to relate the error (including change in error and error integral, if desired) into the corresponding control output. The fuzzy arithmetic is used to calculate the actual output. The *sliding mode controller* can also take multiple controllers and switch the control based on the sliding rule, like switching of fuzzy rules. Finally, the *robust controllers* model the uncertainties in their working and take control parameters that are robust to uncertainties and noises.

3.8 Bug algorithms

Let us now construct a very simple algorithm that solves the problem of motion planning. The inspiration is obtained by the bugs which beautifully navigate with the least amount of sensing and intelligence. Assume that the robot is point sized and hence the workspace and configuration space are the same. It is assumed that the robot always knows its location and the direction to the goal.

3.8.1 Bug 0 algorithm

The most primitive behaviour shown by the robot is to walk straight towards the goal. This is called the *motion to goal* behaviour and can be defined as making a step towards the goal. The motion to goal behaviour is not possible when an obstacle comes in the way of the robot and the goal, and the robot collides with the obstacle in pursuit of reaching the goal. Now the robot displays another behaviour called the *boundary following* behaviour. The robot needs to avoid the obstacle and an appropriate mechanism of doing the same is to follow the obstacle boundary, till it is not possible to move towards the goal. As soon as a step towards the goal is possible, the robot will leave the boundary following behaviour and start the goal-seeking behaviour. The working of the algorithm for a synthetic scenario is shown in Fig. 3.12.

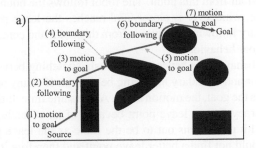

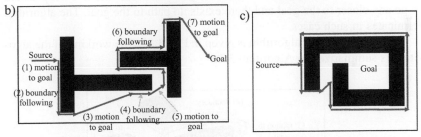

FIG. 3.12 Working of the Bug 0 algorithm. The robot moves straight to the goal if possible, otherwise circumvents the obstacle boundary till a straight-line motion is possible. The boundary following may randomly be clockwise or anticlockwise. (A) Example 1, (B) Example 2, (C) shows the algorithm is neither optimal nor complete.

The algorithm is not complete. Consider the situation as shown in Fig. 3.12C. There is a path from source to goal, however the algorithm fails to spot it and ends up going into a loop. The algorithm is not optimal as well. The optimal path makes the robot take tangents to the obstacles to get to the goal with the smallest path length.

3.8.2 Bug 1 algorithm

The Bug 0 is not complete and therefore the Bug 1 algorithm extends the Bug 0 algorithm by adding a little memory to get completeness. The robot again starts with the motion to goal behaviour by making steps towards the goal. When it is not possible to move any further due to collision with the obstacles, the robot invokes the boundary following behaviour in either clockwise or anticlockwise direction. The point at which the robot collides with the obstacle is called the *hit point.* However, now the robot traverses the complete obstacle boundary till it again reaches the same hit point. As the robot travels, it keeps an account of the closest distance from the obstacle boundary point to the goal. After the robot has traversed the complete boundary, it knows which point in the boundary is closest to the goal and is therefore of value for the continuation of the outwards journey. This point is called the *leave point* as the robot leaves the boundary following behaviour from this point. The robot follows the boundary clockwise or anticlockwise, whichever is closer, and reaches the leave point. The robot stops the boundary following behaviour from this point and continues to display the motion to goal behaviour.

The overall behaviour is hence continuously switching between the motion to goal behaviour and boundary following behaviour. If at any point in time the robot encounters the goal, the motion stops. At the same time, if the robot cannot continue its journey from a leave point because of an obstacle, no solution is possible. If the hit point turns out to be the same as the leave point, it means that the robot could not find a better leave point and therefore it is not possible to go any closer to the goal and there exists no path to the goal. The algorithm terminates in such cases.

The working of the algorithm is given in Fig. 3.13. The working of the algorithm is summarized in Algorithm 3.1.

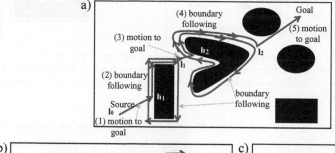

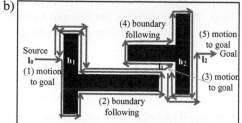
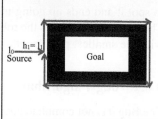

FIG. 3.13 Working of the Bug 1 algorithm. The robot moves straight to the goal if possible until it hits a 'hit point' denoted h_i. Thereafter, the boundary of the complete obstacle is travelled, to compute the nearest point to the goal called 'leave point' denoted l_i. The robot goes to the leave point and starts the motion to goal behaviour. (A) Example 1, (B) Example 2, (C) The algorithm is complete and detects that there is no path to the goal.

Algorithm 3.1 Bug 1 algorithm.

Input: map (sensed by the robot) and goal. It is assumed that the
position is always known.
Output: Motion of the robot to the goal

1 $l_0 \leftarrow source$;   ▷ 0th leave point
2 $i \leftarrow 1$;
3 get current position q ;
4 **while** $q \neq Goal$ **do**
▷ Motion to goal behaviour ;
5   **while** *robot not hits an obstacle* **do**
6   move a step towards goal and update position q ;
7   **if** $q=Goal$ **then** return true;
8      $h_i \leftarrow q$;   ▷ i^{th} hit point
9      $m_i \leftarrow q$;   ▷ i^{th} leave point computed in next loop
10   move a step clockwise (or anti-clockwise) around obstacle ;
11   update position q ;
▷ Boundary following behaviour ;
12   **while** $q \neq h_i$ **do**
13   move a step clockwise (or anti-clockwise) around obstacle ;
14   update position q ;
15   **if** $q = Goal$ **then** return true;
16   **if** $d(q, Goal) < d(m_i, Goal)$ **then** $m_i \leftarrow q$;
17      $l_i \leftarrow m_i$;   ▷ i^{th} leave point
18   **if** *robot cannot make a step towards goal from* l_i **then** return false;
19   move to l_i following the obstacle boundary clockwise or
anti-clockwise (whichever is smaller) ;
20      $i \leftarrow i + 1$;
21 return true;

---

h_i is used to denote the ith hit point and similarly, l_i is used to denote the ith
leave point. The position of the robot q is always assumed to be available from
the localization system. The source is called the 0th leave point for the working
of the algorithm.

Suppose the robot hits the ith hit point (h_i) when it encounters an obstacle.
The robot goes all around the obstacle to find the closest point to the goal as the

leave point (l_i). Since l_i is the closest point in the boundary to the goal, $d(h_i, G) \geq d(l_i, G)$, where $d(x, G)$ denotes the distance of point x to the goal (G). Thereafter, the robot goes towards the goal, meeting the next hit point (h_{i+1}). Since the motion is obtained by moving towards the goal, $d(l_i, G) \geq d(h_{i+1}, G)$. In this way, the distance to the goal keeps getting decreased and the robot keeps moving towards the goal till the goal is eventually reached. Imagine it is not possible to reach the goal, in which case the hit point and the leave point will turn out to be the same, or $h_i = l_i$. In such a case, there is no path to the goal, otherwise, some point in the boundary would have been closer to the goal. The situation is shown in Fig. 3.13C. This shows that the algorithm is *complete*. However, the algorithm is *not optimal*, in which case the path should take tangents to the obstacles. The path can be easily shortened by drawing straight lines to obstacle corners.

Suppose the straight line distance from the source (S) to the goal (G) is $d(S, G)$. Suppose there are n obstacles with the perimeter of obstacle i as p_i. In the worst case, the algorithm traverses all obstacles and traverses the obstacle i by a distance of $1.5p_i$. Here the robot must first traverse the entire perimeter to find the best leave point and additionally go from the hit point to the leave point by the shortest (clockwise or anticlockwise) direction, which will guarantee to reach the leave point in $0.5p_i$ distance. The robot additionally travels straight-line distances between source and goal, which is under a threshold of $d(S, G)$. The maximum path length produced by the algorithm is thus given by Eq. (3.23).

$$\text{Maximum Path Length} = d(S, G) + \Sigma_i 1.5p_i \tag{3.23}$$

The complexity is thus $O(n)$ where n is the number of obstacles.

3.8.3 Bug 2 algorithm

The Bug 1 algorithm wastes a significant amount of time in traversing the entire perimeter of each obstacle encountered from source to goal. An attempt is thus made to enable the robot to make a greedy decision wherein it decides to leave the obstacle boundary without following the complete boundary of the obstacle (Lumelsky and Stepanov, 1987). An m-line is defined as a straight line from the source to the goal. The line is constant and does not change as the robot moves. The source here is the position at the start of the journey of the robot.

The robot essentially uses the m-line to go straight from the source to the goal, which constitutes the *motion to goal behaviour* of the robot. However, the m-line will, at many points, be intercepted by the obstacles and the robot cannot surpass the obstacles. Hence, whenever the robot meets an obstacle, while navigating through the *m*-line, the robot invokes the *boundary following behaviour* in which it follows the obstacle boundary. The robot keeps following the obstacle boundary till it again meets the *m*-line at a position that is better (nearer to the goal) than the position where it left the *m*-line. Now the robot is in a better position to the goal and continues following the *m*-line. The nomenclature used is the same as the Bug 1 algorithm. The robot initially leaves the 0th leave point or source, following the *m*-line. The robot on encountering the obstacle is said to have hit the *i*th hit point, denoted by h_i. In such a case, the boundary following behaviour starts till a nearer to goal point on the *m*-line is encountered by the robot, called the *i*th leave point, l_i. At any point in time, if the goal is reached, the robot terminates the loop with success.

The working of the algorithm is shown in Fig. 3.14A–B for the case when a path is possible and found and Fig. 3.14C for the case when no path exists. In case no path exists, like the Bug 1 algorithm, it is not possible to proceed towards the goal and the robot cannot find a better leave point or cannot move towards the goal from the leave point. The algorithm is summarized as Algorithm 3.2.

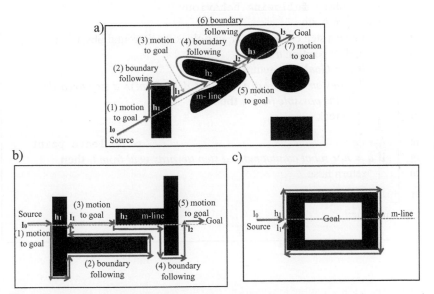

FIG. 3.14 Working of the Bug 2 algorithm. The line joining source to the goal is constant and called the *m*-line. The robot travels to the goal until it hits the obstacle at the hit point (h_i). Thereafter the robot circumvents the obstacle until it gets to a position nearer to the goal at the *m*-line. The robot leaves at this leave point (l_i) and continues with the motion to the goal. (A) Example 1, (B) Example 2, (C) The algorithm is complete and detects that there is no path to the goal.

Algorithm 3.2 Bug 2 algorithm.

Input: map (sensed by the robot) and goal. It is assumed that the
position is always known.

Output: Motion of the robot to the goal

1   $l_0 \leftarrow$ *source* ;   ▷ 0th `leave point`

2   $i \leftarrow 1$;

3   get current position q ;

4   $m - line = \{\lambda Goal + (1 - \lambda)Source = 0, 0 \leq \lambda \leq 1\}$;   ▷ `Line from`
      Source `to` *Goal*

5   **while** $q \neq Goal$ **do**

         ▷ `Motion to goal behaviour` ;

6   **while** *robot not hits an obstacle* **do**

7   move a step towards goal following m-line, update position q ;

8   **if** *q=Goal* **then** return true;

9      $h_i \leftarrow q$;   ▷ i^{th} `hit point`

10   move a step clockwise (or anti-clockwise) around obstacle ;

11   update position q ;

         ▷ `Boundary following behaviour` ;

12   **while** $q \neq h_i$ **do**

13   move a step clockwise (or anti-clockwise) around obstacle ;

14   update position q ;

15   **if** $q = Goal$ **then** return true;

16   **if** *q lies on m-line* $\wedge d(q, Goal) < d(h_i, Goal) \wedge$ *a step towards*
          goal is possible from q **then**

17   break

18     $l_i \leftarrow q$;   ▷ i^{th} `leave point`

19   **if** $q = h_i \vee$ *robot cannot make a step towards goal from* l_i **then**

20   return false

21     $i \leftarrow i + 1$;

22 return true;

In the first instance, it appears that the greedy selection of the leave point is valuable as the robot need not traverse the complete perimeter like Bug 1, however this algorithm may not always produce such short paths. Consider the situation given in Fig. 3.15. Even though there is a single obstacle, the robot will meet the same obstacle numerous times, once at every spike of the obstacle that cuts the *m*-line. The decision of whether the robot should take a clockwise or anticlockwise turn is not hard defined in the algorithm. In the worst case, the robot could take all incorrect decisions and end up taking the entire perimeter of the obstacle for every spike encountered.

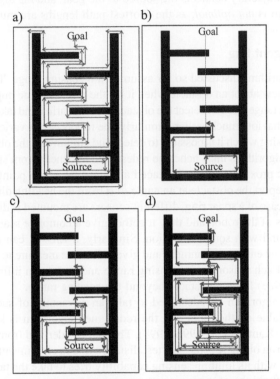

FIG. 3.15 Better output of Bug 1 as compared to Bug 2 on an object with spikes. (A) Bug 1 algorithm (B) Bug 2 algorithm output till 1st leave point (C) Bug 2 algorithm output till 2nd leave point (D) Bug 2 algorithm output till 3rd leave point.

Suppose there are n obstacles, which in the worst case will all lie on the m-line. Suppose an obstacle i has n_i spikes and in the worst case, a boundary following behaviour is invoked at every such spike that traverses nearly the complete obstacle perimeter p_i. This means the total path length is $d(S,G)$ to traverse the m-line and an additional path length of $n_i p_i$ for an obstacle i. The maximum path length is given by Eq. (3.24).

$$\text{Maximum Path Length} = d(S,G) + \Sigma_i n_i p_i \qquad (3.24)$$

The path length and hence the computation time is theoretically larger than the Bug 1 algorithm as also shown by the example. The complexity is $O(n)$ where n is the total number of spikes in all obstacles.

The algorithm is *complete* with the same discussions as in Bug 1. Every leave point is closer than the previous hit point, and since the robot follows the *m*-line, every hit point is closer than the previous leave point. The algorithm thus makes the robot keep getting closer and closer to the goal. In case this is not possible, no trajectory connects the source to the goal, and the algorithm exits. The algorithm is *not optimal*, as the shortest path lengths are much smaller.

3.8.4 Tangent Bug

The Bug algorithms discussed so far assumed a 0-sensing range. Therefore, the robot only knew about an obstacle when it hit the obstacle. The bugs have some finite sensing range and can detect the obstacles much earlier and take the obstacle avoidance behaviour much earlier. Similarly, the robots have a variety of proximity sensors, using which it is possible to detect and circumvent the obstacles much earlier. The algorithms must hence be redesigned to benefit from such sensors.

A generic proximity sensor is placed at the centre of the point robot and is mounted to report the obstacle at an angle of α from the heading direction of the robot. In the case of a sonar ring, different sensors are mounted at different heading directions of the robot, and they together give a complete scan of the environment around with some resolution. Similarly, a LiDAR can make a laser beam sweep the entire environment and give a distance measure to all the obstacles around. Each sensor has its sensing range and returns an infinite value (or maximum value) for any obstacle beyond the sensing range.

The Bug algorithm is redesigned to take an advantage of such proximity sensing available on the robot and the redesigned algorithm is known as the *Tangent Bug* (Kamon et al., 1996, 1998). The name is derived from the fact that the tangent to an obstacle is the shortest distance line to circumvent the obstacle and the robot with its limiting sensing capability tries to take the tangents to the obstacles to avoid the obstacles.

First, suppose no obstacle is detected by the proximity sensors in the direction of the goal. In such a case, the robot simply invokes the *motion to goal behaviour* and takes steps towards the goal. As the robot goes nearer to the goal, many obstacles may now be within the sensing range of the robot, including those in the straight line joining the source to the goal. The robot must now change its motion to avoid the obstacles. The first thing to be done is to know the corners of the obstacles which make the tangents to the obstacles. Suppose the robot has an infinite distance range sensor. Suppose the laser starts scanning from 0 degrees and completes 360 degrees as it sweeps radially. At the surface of the obstacle, the laser shows a very small change in value between every two consecutive radial scans. However, as it transits from one obstacle to another; or one spike of a concave obstacle to another, there is a dramatic change in the distance value. This marks an *obstacle corner*. A complete scan thus gives all such obstacle corners visible to the robot. Since the robot has a finite sensing range, any time when the value changes from a finite to infinity (maximum value of the sensor) or from infinity (maximum value of the sensor) to a finite value, it also suggests a guessed corner due to the limited sensing range of the robot. The actual corner may lie beyond the range of the proximity sensor, but the robot can see only as far as the sensor range, and therefore the extreme distance values are also taken as corners. Both conditions are given by Eq. (3.25) and shown in Fig. 3.16.

$$O = \left\{ d_\alpha [\cos \alpha \quad \sin \alpha]^T : \text{abs}(d_\alpha - d_{\alpha-\delta}) > \epsilon \lor \text{abs}(d_\alpha - d_{\alpha+\delta}) \right.$$

$$\left. > \epsilon \lor (d_\alpha < d_{max} \land d_{\alpha-\delta} = d_{max}) \lor (d_\alpha < d_{max} \land d_{\alpha+\delta} = d_{max}) \right\}$$

(3.25)

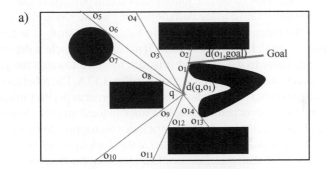

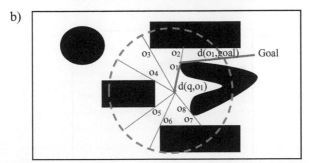

FIG. 3.16 (A) Detection of corners from high-resolution proximity sensor scan. The proximity sensor scans in all angles and a discontinuity in proximity value denotes corners at both points of discontinuity. (B) For finite sensing, a value from finite to infinity or vice versa is also a corner. The corner o indicating least value of $d(q,o) + d(o,goal)$ is selected as the immediate goal, where q is the current robot position.

Here d_α is the proximity sensor's reading at a direction α (relative to the heading direction of the robot, assuming the proximity sensor is aligned in the heading direction of the robot), $\alpha + \delta$ is the immediately next reading of the sensor, while $\alpha - \delta$ is the immediately previous reading. ε is the distance threshold and d_{max} is the maximum proximity sensor's reading. The obstacle positions are relative to the robot.

The robot at any point may see multiple corners and needs to select one corner which acts as the immediate goal to circumvent the obstacle sensed.

The entire map can now be summarized by the obstacle corners alone. Therefore, if the robot chooses to avoid the obstacle with corner o, assuming that there are no further obstacles, it will first go from the current position to o and thereafter from o to the goal, incurring a total cost of $d(q,o) + d(o, goal)$, $o \in O$. Out of all the obstacles corners, the one that minimizes this heuristic cost is the one that is taken as the immediate goal, and the robot moves towards the same. The immediate goal for the robot to follow is thus given by (Eq. 3.26).

$$
o = \begin{cases} q + d_{\max}[\cos\theta_G \sin\theta_G]^T & \text{if } q + d_{\max}[\cos\theta_G \sin\theta_G]^T \in W^{\text{free}} \\ \arg\min_{o\in O} d(q, o) + q(o, goal) & \text{otherwise} \end{cases} \tag{3.26}
$$

Here, θ_G is the slope of the line joining the current position of the robot (q) to the goal, measured relative to the current heading of the robot. d_{max} is the sensing radius of the range sensor. The point $q + d_{\max}[\cos\theta_G \sin\theta_G]^T$ is the point at d_{max} towards the goal. If the point is valid, it means no obstacle is detected in the straight-line distance to the goal and it is safe to move towards the goal.

However, consider the situation as shown in Fig. 3.17A. The robot moves as per the motion to goal behaviour by following the best corner as per the limited sensing range. The robot however realizes that it is stuck in local minima. At point x, if it takes a step clockwise, the corner at the anticlockwise happens to be the best and if it then takes a step anticlockwise, the corner at the clockwise direction turns out to be the best. The robot shall therefore oscillate endlessly. Such an obstacle is called the *blocking obstacle* since it blocks the way of the robot.

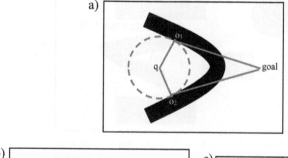

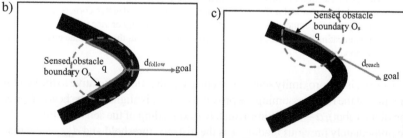

FIG. 3.17   Local minima. (A) The robot can no longer minimize the heuristic value of distance to the goal. As soon as the robot dips towards o_2, the new shifted position of o_1 (as sensed by onboard sensors) becomes better and vice versa. The boundary following behaviour is invoked. (B) Followed distance (C) Reach distance used to switch to motion to goal behaviour.

Whenever a robot identifies a local minimum of the heuristic value noted above, it knows that there is no way to improve the value by greedy selections noted above. Hence, the robot switches to a *boundary following* behaviour instead. In this behaviour, the robot selects a direction to circumvent the obstacle, clockwise or anticlockwise. The robot only considers the obstacle corners that lie in the same direction for motion. This is called the *following obstacle*, which represents the obstacle whose boundary is being followed. Initially, both the blocking obstacle and the following obstacle are the same.

The transition back to the motion to goal behaviour is a more critical condition. The notion is the same that if the robot believes that it is in a better position and nearer to the goal as compared to the last hit point (local minima), it must start invoking the motion to goal behaviour. However, now the robot has a sensor, and the range sensing may be incorporated so that the robot may sense reaching a better position in anticipation and start invoking the motion to goal behaviour.

To write the transition condition, let us define two distance measures. The *followed distance* or d_{follow} is defined as the shortest distance between the sensed obstacle boundary and the goal. Consider the workspace at any point of time as shown in Fig. 3.17B–C. The sensed points by the robot are given by Eq. (3.27).

$$O_s = \left\{ o \in \partial W^{\text{obs}} : \forall_{0 \le \lambda \le 1} \lambda q + (1 - \lambda)o \in W^{\text{free}} \right\} \tag{3.27}$$

The equation takes all points o such that the straight line from q to o, represented by $\lambda q + (1 - \lambda)o$, is collision-free (and therefore sensed by the range sensor) and that the point o is at the obstacle boundary (∂W^{obs}). Out of all such points, the distance to the closest point is given by Eq. (3.28).

$$d_{\text{follow}} = \min_{o \in Os} d(o, \text{goal}) \tag{3.28}$$

This distance is easy to measure as the range sensors directly give the range readings which can be projected on the map and the closest point to the goal is ultimately selected.

The other distance measure is called d_{reach}, which is the distance to the goal that the robot is guaranteed to reach as it travels. The robot senses obstacles and all sensor values report points in the obstacle boundary which the robot can reach by travelling in a straight line. The closest such point from the goal (o) is selected and its distance to the goal is taken as d_{reach}. As the robot travels, it detects a point o at obstacle boundary $\partial O_{\text{following}}$, to which it can travel in a straight line. The closest such point to the goal means that the robot will eventually reach o and thereafter can have a small distance to the goal. The modelling is same as d_{follow}, however only points on the following obstacle are considered, given by Eqs (3.29)–(3.30).

$$O_s = \left\{ o \in \partial O_{\text{following}} : \forall_{0 \le \lambda \le 1} \lambda q + (1 - \lambda)o \in W^{\text{free}} \right\} \tag{3.29}$$

$$d_{\text{reach}} = \min_{o \in Os} d(o, \text{goal}) \tag{3.30}$$

The transition condition is that the robot believes it will reach a better position as previously known or $d_{\text{reach}} < d_{\text{follow}}$.

The working of the algorithm is shown in Fig. 3.18. An interesting thing to note is how the algorithm changes as the proximity range increases. Hence, the same map is shown for the case of 0-sensing range in Fig. 3.19A, and infinite sensing range in Fig. 3.19B. For a 0-sensing range, the robot can only detect the obstacle once it collides at it, in which case it keeps going in the direction (clockwise or anticlockwise) that is closer to the goal. On encountering a local optimum, the boundary following behaviour is invoked. On the contrary, in case of infinite sensing radius, the local optimum was not even a corner as per the definition (difference between two consecutive distance range sensor readings being very high) and the robot did not enter the region. The working of the algorithm is given by Algorithm 3.3.

FIG. 3.18 Working of the Tangent Bug algorithm. The robot moves straight to the goal till its proximity sensors detect the obstacle, in which case the best corner is used for navigation. The robot can detect local optima and invokes boundary following behaviour till $d_{\text{reach}} < d_{\text{follow}}$.

a)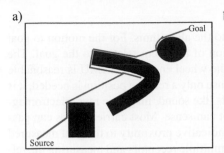
b)

FIG. 3.19   Tangent Bug Algorithm (A) 0 sensing range (B) infinite sensing range.

Algorithm 3.3 Tangent Bug algorithm.

Input: map (sensed by the robot) and goal. It is assumed that the
position is always known and range sensors sense obstacles till R
Output: Motion of the robot to the goal

1   $l_0 \leftarrow source$;   ▷ 0th leave point
2   $i \leftarrow 1$;
3   get current position q ;
4   **while** $q \neq Goal$ **do**
        ▷ Motion to goal behaviour ;
5   **while** *no local minima encountered* **do**
6   select sub-goal o by using Eq. (3.9) ;
7   move a step towards o and update position q ;
8   **if** $q=Goal$ **then** return true;
9       $h_i \leftarrow q$;   ▷ i^{th} hit point
10   select direction (clockwise or anti-clockwise) and mark the
        following obstacle $o_{following}$;
        ▷ Boundary following behaviour ;
11   **while** *robot not traversed entire obstacle perimeter* **do**
12   select sub-goal o restricted to the selected obstacle/direction ;
13   move a step towards o and update position q ;
14   **if** $q = Goal$ **then** return true;
15   **if** $d_{reach} < d_{follow}$ **then** break;
16      $l_i \leftarrow q$;   ▷ i^{th} leave point
17   **if** $d_{reach} \geq d_{follow} \lor$ *robot cannot make a step towards goal from* l_i
        then return false;
18      $i \leftarrow i + 1$;
19 return true;

3.8.5 Practical implementations

This section adds some practicalities to the algorithms. For the motion to goal behaviour, the only input is the position of the robot relative to the goal. The approximate position available from the wheel encoders or IMU is reasonable to use, while the goal is constant. Because only a relative position is needed, it is also possible that the goal emits signals like sound, magnetic rays, electromagnetic waves, light, etc. which the robot can sense. Most carrier waves can pass through regular obstacles. This gives indicative proximity to the goal measured by the strength of the signal. By using multiple receptors and measuring the relative intensity, the direction can also be guessed. Alternatively, the change in signal intensity as the robot moves also gives a relative direction. This enables practically implementing the algorithm without the need to invest in localization systems. The concepts are discussed in detail in Chapter 20 on swarm robotics. That said motion to the goal is the primary guiding behaviour in the Bug algorithms and these readings are approximate and thus the robot practically deviates.

The boundary following behaviour is also simple. Imagine that the robot has a range of proximity sensors, either a high-resolution LiDAR or a ring of sonars. The sonar or the laser ray that reports the least distance to the obstacle is the one that gives the direction of the obstacle. Since the direction of all laser rays and all sonar sensors are known, the robot may simply attempt motion perpendicular to this direction. This is shown in Fig. 3.20A. The boundary following behaviour makes the robot move away from the obstacle if the sensed distance is too small. Similarly, the robot will be made to move towards the obstacle, if the sensed distance is too large. The aim is to control the robot such that the robot maintains a constant distance from the obstacle (as reported by the least value recoded by range sensing), while in general moving perpendicular to the direction of the least distance value.

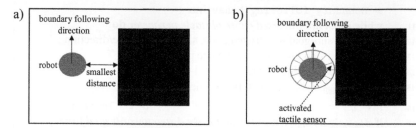

FIG. 3.20 Practical implementation of the Bug Algorithm. (A) In the case of a proximity sensor, the sensor corresponding to the smallest proximity reading gives direction to the boundary, and the boundary following is keeping a perpendicular direction (B) In the case of a tactile sensor, the robot should move perpendicular to the invoked tactile sensor.

If the robot is traversing using the motion to goal behaviour, a small range sensing value indicates that an obstacle is near and a shift of behaviour to obstacle avoidance is suggestive. Similarly, if the obstacle does not strictly lie at the line of sight of the goal, it indicates that it is worthwhile to shift to the motion to goal behaviour.

Another manner of getting the algorithm working is by using a circular array of tactile sensors. The robot must hit the obstacle for it to be detected. The robot displays the motion to goal behaviour as usual, until it collides with an obstacle. The tactile sensor which reports the collision gives the direction information and the robot must move perpendicular. The robot will, again and again, try to collide with the obstacle by moving towards the goal, and on reporting of a collision take a step away. This is shown in Fig. 3.20B.

The algorithms additionally need to be able to sense when traversing the entire perimeter of the obstacle is over (Bug 1) or when the m-line is encountered (Bug 2), which can be approximately done by using wheel encoders or IMU. If the goal emits signals, the strength of the signal is a measure of the proximity to the goal indicating the position. For boundary following using tactile sensors, the robot must aim repeatedly to collide with the obstacle to get an estimate of the boundary, otherwise, it is impossible to guess the boundary. The reliance on the approximate position due to the need to measure a perimeter trace is more involved.

Questions

1. Write a control law such that the robot reaches the desired goal (G_x, G_y) from the current position, however also reaches the goal with a specific orientation θ_G.
2. A differential drive robot is at $(0\,m, 0\,m)$ and is oriented at the X-axis. The robot wants to go to the location $(0\,m, 2\,m)$ and it should be oriented at the X-axis when it reaches $(0\,m, 2\,m)$. In other words, the robot intends to go $2\,m$ sideways. Explain any 1 control law and draw the robot's trajectory using the control law. Does it differ from how a human would drive the vehicle for the same problem?
3. Is it possible to extend the Bug algorithm for 3D cases, while still making a complete solution? Why or why not?
4. Is it possible to extend the Bug algorithm for the case of a 2-revolute link planar manipulator with a fixed base, while still making a complete solution? Why or why not? Answer the same question for a 3 revolute-link manipulator. The goal is defined in joint space (θ_1, θ_2) for 2-link and $(\theta_1, \theta_2, \theta_3)$ for 3-link.
5. Give the output of the Bug 0 algorithm, Bug 1 algorithm, Bug 2 algorithm, Tangent Bug algorithm with 0 range sensing, Tangent Bug algorithm with infinite range sensing, and Tangent Bug algorithm with finite range sensing, on the maps shown in Fig. 3.21.

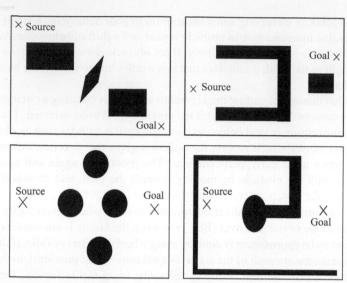

FIG. 3.21 Maps for question 5.

6. Does the Tangent Bug with infinite sensing radius produce optimal results? Does the algorithm guarantee smoothness of the path? If yes, can you prove the same? If no, give counter-examples. What is the effect of increasing the sensing range to the optimality and completeness of the Tangent Bug algorithm?

7. Modify the Tangent Bug algorithm to account for the radius of the robot (R) and maintain a safety distance of ε. (*Hint*: It is enough to maintain a safety distance of $R + \varepsilon$ from the centre of the robot).

8. Why does the Tangent Bug algorithm stop displaying the boundary following behaviour on reaching somewhat nearer to the goal; while the Bug 2 algorithm, like Tangent Bug, stops displaying the behaviour on visiting the m-line? Can the condition used in the Tangent Bug algorithm be also used in the Bug 2 algorithm, thereby making a better Bug 2 algorithm?

9. Establish clear trade-offs between the deliberative and reactive planning techniques, clearly explaining which one is more applicable in which situations. Also, suggest all means of fusing the two techniques if the deliberative planner returns the best trajectory, and the reactive planner returns the immediate move in pursuit of the goal. Which technique (or fusion) is more relevant for the scenarios (i) Going from one room in your house to another room assuming nobody else at home, (ii) further assuming that even though the room plan is known, the exact placement of all furniture is not known, (iii) further assuming that other housemates are moving around, (iv) further assuming that some kids are running very quickly, (v) further if someone accidentally drops a box just in front of you? Repeat

the same questions for the situation of a manipulator operating to pick up items. You may want to repeat your solutions after reading all algorithms in this book.

10. A robot uses wheel encoders to localize itself, however the error keeps on increasing. Discuss the ways to relocalize the robot when the error is large enough. Consider the complete system and discuss how to account for this uncertainty in the motion planning algorithms.

11. Construct some synthetic maps. Using different trajectories in each of the maps, state if making trajectories better as per the path length objective also makes them simultaneously better (or worse) as per the clearance objective. Do the same for all pairs of the objectives of path length, smoothness, and clearance.

12. When walking around at different places, observe how do you adjust your trajectory to account for other people. In which situations do you cooperate and move to ease their walking; when do they move to ease your walking; what happens you see someone going to the same spot as yours (simultaneously attempting entering through a door, when only one at a time can go in; or taking a right from an intersection that blocks someone going straight from the opposite direction); what happens when a person charges by moving swiftly towards you? Can you make simple reactive rules that accommodate all of your observations?

13. [Programming] Use popular SLAM libraries (SPTAM, RTABMAP, ORB-SLAM, Cartographer, etc.) and public data sets as documented in these libraries to perform SLAM.

14. [Programming] Create a basic robotic simulator that adds noises to the control inputs. Using the simulator implement a controller for a robot to go in the following trajectories (i) Straight line to a known goal (ii) Sinusoidal wave (iii) Go indefinitely on an elliptical track (iv) Go indefinitely on a general polygon track with all known vertices.

15. [Programming] In the implementation of the control algorithm, you assumed the robot's position is known. Combine the solutions to localization (from Chapter 2) and control problems. The localization algorithm uses the applied controls and landmarks to give the position, while the control algorithm uses the same position to calculate the error and next control input.

16. [Programming] Assume a 2D world. Each obstacle is defined as a polygon with known vertices defined in a counter-clockwise manner. There are a few such obstacles. Implement Bug 0, Bug 1, Bug 2, and Tangent Bug algorithms. (*Hint*: In case of a collision, the point robot will nearly lie on the polygon line. For Tangent Bug, can you calculate the distance between a point and line at a certain heading direction representing laser rays using geometry/equation of lines? The laser ray may not strike the polygon).

17. [Programming] Repeat the programs for the case when the map is given as a bitmap image. (*Hint*: Derive how do the (x,y) coordinates of the robot relate to the pixels in the bitmap. Make a virtual laser scanner to scan the map in each direction for the Tangent Bug).

18. [Programming] Mount a camera that (approximately) looks downwards on a 2D plane. The plane should have a clear floor with a black-coloured planar object with a small height representing the obstacles. Correct the image to transform the resultant image into an orthographic projection by finding the correct projection matrix from a fixed set of calibration points. Now, classify all image points as black (obstacles) and nonblack (free) to make the workspace. Take a toy robot and determine its position and orientation in the image for localization. Take a red flag that marks the goal and detect that as well. Use the Bug algorithm solution from the previous question on this map to get the trajectory. Thereafter, program the motion of the actual robot to make the robot go towards the goal using the Bug algorithms.

19. [Programming] Use popular robotic simulators (CoppeliaSim, Player-Stage, etc.) to control the robot at preknown trajectories.

References

Choset, H., Lynch, K.M., Hutchinson, S., Kantor, G., Burgard, W., Kavraki, L.E., Thrun, S., 2005. Principles of Robot Motion: Theory, Algorithms, and Implementations. MIT Press, Cambridge, MA.

Filliat, D., 2007. A visual bag of words method for interactive qualitative localization and mapping. In: Proceedings 2007 IEEE International Conference on Robotics and Automation. IEEE, Roma, pp. 3921–3926.

Fischler, M.A., Bolles, R.C., 1981. Random sample consensus: a paradigm for model fitting with applications to image analysis and automated cartography. Commun. ACM 24 (6), 381–395.

Kala, R., 2014. Coordination in navigation of multiple mobile robots. Cybern. Syst. 45 (1), 1–24.

Kala, R., Shukla, A., Tiwari, R., 2010a. Dynamic environment robot path planning using hierarchical evolutionary algorithms. Cybern. Syst. 41 (6), 435–454.

Kala, R., Shukla, A., Tiwari, R., 2010b. Fusion of probabilistic A* algorithm and fuzzy inference system for robotic path planning. Artif. Intell. Rev. 33 (4), 275–306.

Kamon, I., Rivlin, E., Rimon, E., 1996. A new range-sensor based globally convergent navigation for mobile robots. In: IEEE International Conference on Robotics and Automation. IEEE, Minneapolis, MN, pp. 429–435.

Kamon, I., Rimon, E., Rivlin, E., 1998. Tangentbug: a range-sensor based navigation algorithm. Int. J. Robot. Res. 17 (9), 934–953.

Kümmerle, R., Grisetti, G., Strasdat, H., Konolige, K., Burgard, W., 2011. G2o: a general framework for graph optimization. In: 2011 IEEE International Conference on Robotics and Automation. IEEE, Shanghai, pp. 3607–3613.

Lumelsky, V., Stepanov, A., 1987. Path planning strategies for point mobile automaton moving amidst unknown obstacles of arbitrary shape. Algorithmica 2, 403–430.

Pire, T., Fischer, T., Castro, G., Cristoforis, P.D., Civera, J., Berlles, J.J., 2017. S-PTAM: stereo parallel tracking and mapping. Robot. Auton. Syst. 93, 27–42.

Tiwari, R., Shukla, A., Kala, R., 2012. Intelligent Planning for Mobile Robotics: Algorithmic Approaches. IGI Global, Hershey, PA.

Chapter 4

Intelligent graph search basics

4.1 Introduction

Imagine a general space, called the *configuration space* (C), in which you need to search for a path from a predefined point called a source (S) to another predefined point called the goal (G). Space is filled with some regions which are nontraversable called obstacles. The space occupied by all obstacles is called the *collision-prone configuration space* (C^{obs}). The rest of the space is free and traversable, called the *free-configuration space* (C^{free}), in which the current position of the robot is called the *source* and the destination to which the robot intends to travel is known as the *goal*, with the robot taken as a point (Choset et al., 2005; Tiwari et al., 2012). A detailed account of the configuration space is discussed in Chapter 6. Informally, the problem is encoded by the identification of all the variables called the *robot configuration*. The configuration space is a space that has all such variables as the axis. A configuration, which is a point in this space, is black (obstacle) if a virtual robot placed at the same configuration intersects or collides with some obstacle (or self-collides with itself). Alternatively, the configuration is white (free), denoting that the virtual robot placed at the configuration will be collision-free. The *collision-checking algorithms* are used to classify the configurations as collision-prone or collision-free.

Robotics is one of the most common application areas of artificial intelligence (AI). Hence, to solve the problem of making a path from the source to the goal, while avoiding obstacles, it is imperative to look at the *classic AI* techniques of *Graph Search*. For any problem to be solved by using a graph search, some assumptions apply. The environment must be *static*. If the planner takes a few seconds to return a solution, a lot can change in a few seconds. There may be moving people, suddenly fallen objects, etc. The solution generated does not know about these changes, and therefore the solution may not be valid. After the calculation of the path, the robot is still physically at the source. The physical motion from the source to the goal will take a few minutes. If the environment changes in these few minutes, the path that the robot is following will still become invalid. The movement of people or change of map is handled by *reactive approaches, trajectory correction,* or *replanning algorithms*. This way of breaking the problem is called *abstraction*, wherein the dynamic elements of the

Autonomous Mobile Robots. https://doi.org/10.1016/B978-0-443-18908-1.00005-4

problem are abstracted out at a higher level of planning. While deciding the route to take to your favourite holiday destination while travelling by car, you may not consider the finer details like how to avoid the vehicles that you may encounter while moving.

The other assumption is that the environment must be *fully observable*. So, every piece of information that could potentially be useful in decision-making should be available. We have already implemented this assumption by considering that the map is available from the SLAM algorithm. However, there are things that the robot will not know of. Consider that the robot is asked to go from one room to the other and there are two ways possible. One way has a door in the middle that may be open or closed, while the other way is a freeway. The robot cannot operate doors. The robot may not know the status of the door, which affects its decision-making. If the door is mostly open (and similarly closed), one may plan accordingly and replan if the fact happens to be contradictory. Alternatively, the search could probabilistically account for all possibilities and take the action which has the least expected cost based on partial knowledge. Such decision-making is discussed in Chapter 14. While deciding the route to the office, it is possible that your favourite route had an accident and is blocked. However, you do not account for such facts and reroute only when you reach the road and see it blocked. If a road is often blocked, it is best to permanently avoid it.

The next assumption made is that the model is *deterministic*. This means that if the robot takes one step of 10 cm right, it does land at exactly one step right at a distance of 10 cm. More formally, it means that from any state s, if the robot takes an action a and lands at a state s', then every time when the same state-action pair is executed, the result will be the same. This is also a big assumption in robotics, where everything is stochastic, or all actions have uncertainties. The motors may jam, the robot may slip, the friction may change, etc., giving errors of a few centimetres. The mechanism for handling stochasticity is to assume that the robot's operations are deterministic and make plans accordingly. When the robot is unable to trace the plans due to stochasticity, the *controller* will try to give corrective measures in real time.

Another assumption is that the world is *discrete* in nature. A graph is a finite collection of discrete vertices and edges. The robots operate in continuous worlds, and the assumption also does not naturally hold. The states are made discrete by filling the configuration space with *cells* of various sizes. This leads to a loss of information as some cells may be partially free and partially occupied by obstacles; however, for the purpose of working, the complete cell will have to be classified as either free or obstacle. Similarly, the robot will only be given some finite options of actions to choose from in every state, typically only the option to choose from the neighbouring cells.

The assumption is that the system is a *single agent*. Hence the only entity operative in the entire world is the robot. Broadly, the multiagent systems may have direct communication (e.g. Wi-Fi) or no direct communication.

If the agents can communicate directly, it is possible to take the different agents as entities of one major robot and use the same algorithms as discussed here or in the rest of the book. Such a mechanism for dealing with the problem is dealt with in Chapter 18. If the agents cannot communicate directly, they become dynamic obstacles that are not admissible as per the static clause. An exception is that the robots can have rules on how to act when seeing each other, and they all have the same set of rules, unlike dynamic obstacles, wherein humans are not strictly governed by rules or navigation etiquette. Based on the rules, the future motion of every robot can be accurately guessed and accounted for while planning.

In AI, systems are classified as sequential and episodic. In *sequential* systems, one plans for a sequence of decisions to be taken, while in episodic systems, the intent is only to make the best current actions, which are neither governed by the past actions nor affect the future actions. Episodic systems look at the world and make an immediate decision without concerning the past or future actions. For example, if a robot needs to pick up all green balls going through a conveyer belt at a low speed, picking up the current ball does not depend on the past actions and does not affect future actions. Robot motion planning gives a sequence of actions that the robot must make, and therefore, the problem is sequential. Sequential systems are more generalized systems, and the search problems are for sequential decision-making. The generic assumptions are summarized in Box 4.1.

Box 4.1 General assumptions for problem solving using graph search.

Static: Environment does not change.
- Static from the start of planning (~few seconds) till the end of execution (~minutes)
- This planning handles only static obstacles, and dynamic obstacles are handled by a reactive planner, or trajectory correction algorithms and larger errors by replanning
- For example, plan a route as if no other vehicles exist; on seeing one, reactively avoid/overtake it. If a road is blocked, reroute.

Fully observable: All information is available
- Use SLAM to make a map.
- Some facts can be assumed, e.g. a lift that is often operative is operative, replan if not.
- Some facts can be based on probabilistically expected values, e.g. expected time to wait for a lift.

Deterministic: Action consequences are error-free
- Some noise can be handled by control algorithms.
- For example, if the robot moves an extra centimetre, control will mitigate the error.

Continued

Box 4.1 General assumptions for problem solving using graph search—cont'd

Discrete: Environment has a finite number of states
- Every continuous scale can be made discrete by dividing into tiles.
- More tiles imply a higher resolution and better results, with more computation time.

Single agent: No other agent can move
- Multiagent systems without communication: Other agents equivalent to dynamic obstacles. If they all obey the same rules, their motion can be guessed, useful for motion planning.
- Multiagent systems with communication: Can be planned as one complex system controlling all agents.

Sequential: Current action affects the future actions and is affected by the past actions.
- Plan a sequence of moves.
- On the contrary episodic systems, only see the current input to take the current best action (e.g. classification).

It may be interesting to note that practically the problem of robot motion planning is dynamic, partially observable, stochastic, and continuous. However, it is forcibly taken to be exactly the contrary: static, fully observable, deterministic, and discrete. This is an application of the concept of *abstraction*. The problem is solved at a higher level of abstraction, wherein the finer details are all ignored. The problem does not even consider the electric voltage to apply at every motor to move the robot, which constitutes an abstracted detail. The abstract solution is found with assumptions, and the assumptions are rectified at the finer stages. So, small changes in the environment are reactively dealt with as the robot moves. If an assumed detail is found to be wrong, the robot can completely replan at a later stage. Errors due to stochasticity are rectified by the controller. Similarly, while the trajectory may be planned in a discrete space, the control algorithm is suitable for continuous space. The control phase also accounts for the low-level signals to be given to the motors to roughly follow the planned path.

4.2 An overview of graphs

A *graph* $G < V, E >$ is defined as a collection of vertices V and edges E (Cormen et al., 2009). An edge $\langle u, v \rangle \in E$ connects two vertices u and v and may optionally have a weight $w(u,v)$. Here $w : V^2 \rightarrow \mathbb{R}$ denotes the weight function. If edges have weights associated with them, the graph is called *weighted*; if the edges do not have weights, the graph is called unweighted, and the edges are said to have a unity weight. In *undirected graphs,* the edges can be travelled in both directions; that is, an edge between a vertex u and a vertex v implies that an inverse edge between the vertex v and the vertex u shall also exist. Conversely, if the edge cannot be travelled in both directions, it is called a *directed graph*. Fig. 4.1 shows the different graph types diagrammatically. Edges make it possible to travel the graph by visiting vertices in a sequence, and the traversal is called a *path* (τ). The path is hence an ordered sequence $\tau = [v_1, v_2, v_3....]$, such that an edge exists between every two consecutive edges, that is $\forall_i$, $\langle v_i, v_{i+1} \rangle \in E$. Consider the graph in Fig. 4.1C, [1,2,4,5] is a path. The cost of the path is typically defined as the sum of all the edge costs, that is $\text{cost}(\tau) = \sum_i w(v_i, v_{i+1})$. The cost of the path [1,2,4,5] in Fig. 4.1C is $2+4+2=8$. Graph search aims to find the least cost path from a specific vertex called source (S) to any vertex that satisfies a prespecified goal criterion (say G), that is, to find $\tau^* = \text{argmin}_\tau \ \text{cost}(\tau)$, $v_1 = S$, $v_{\text{end}} = G$. Here * is a typical way of denoting optimality or the smallest cost.

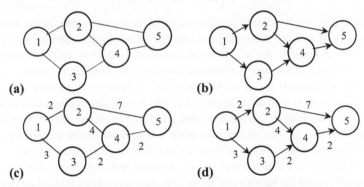

FIG. 4.1 Graph types (A) undirected and unweighted graph, (B) directed graph and unweighted graph, (C) undirected and weighted graph, (D) directed and weighted graph.

Graph search techniques constitute the earliest ways to solve most real-life problems. Routing yourself from your home to office through traffic is a graph search, with intersections representing the vertices and roads representing the edges (weighted, generally bidirectional, however unidirectional for one-ways). Similarly, solving a Rubik's cube is also a graph search, with different patterns (or states) of the cube denoting the vertices and the movements of the blocks denoting the edges (unweighted, bidirectional).

The *adjacency matrix* is a matrix ($A:V^2 \rightarrow [0,1]$) of size $n \times n$, where n is the number of vertices of the graph. The cell $A(u,v)$ stores 1 if $<u,v>$ is an edge and 0 otherwise, that is Eq. (4.1):

$$A(u, v) = \begin{cases} 1 & \text{if } (u, v) \in E \\ 0 & \text{if } (u, v) \notin E \end{cases} \tag{4.1}$$

For a weighted graph, one may instead store the weight matrix $W:V^2 \rightarrow \mathbb{R}$, where $w(u,v)$ denotes the weight of the edge if one exists and ∞ otherwise. For undirected graphs, the matrix is symmetric, that is $W=W^T$, and hence only the upper triangle or the lower triangle matrix may be stored. The matrix representation of the graph in Fig. 4.1A and Fig. 4.1C is given as Eqs (4.2)–(4.3), assuming all vertices are integers with no metadata.

$$A = \begin{bmatrix} 0 & 1 & 1 & 0 & 0 \\ 1 & 0 & 0 & 1 & 1 \\ 1 & 0 & 0 & 1 & 0 \\ 0 & 1 & 1 & 0 & 1 \\ 0 & 1 & 0 & 1 & 0 \end{bmatrix} \tag{4.2}$$

$$W = \begin{bmatrix} \infty & 2 & 3 & \infty & \infty \\ 2 & \infty & \infty & 4 & 7 \\ 3 & \infty & \infty & 2 & \infty \\ \infty & 4 & 2 & \infty & 2 \\ \infty & 7 & 0 & 2 & \infty \end{bmatrix} \tag{4.3}$$

The major problem with the adjacency matrix representation is the memory complexity of $O(n^2)$. Practically, the number of edges is far less than the theoretical limit possible for n^2. As an example, if a city has 1000 intersections and every intersection has 4 roads on average, the number of roads is 4000 (2000 for undirected) and is far less than 10^6. A better way to store the graph in such a context is with an *adjacency list*, which only stores edges that actually exist. The technique is the same as that used by the sparse matrix technique, which is the mechanism to store a matrix where most of the entries are zero. The easiest implementation is to store the graph as an array of linked lists, where every linked list is a collection of edges outgoing from the vertex. For the edge u-v, the uth linked list stores v and its weight. Any associated meta-data can also be stored, such as the name of the road, traffic volume on the road, etc. The adjacency list representation for the graph in Fig. 4.1B is shown in Fig. 4.2A and for the graph in Fig. 4.1D is shown in Fig. 4.2B.

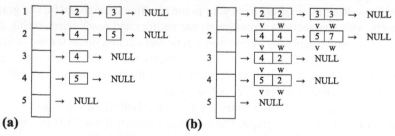

(a)　　　　　　　　　　　　　**(b)**

FIG. 4.2 Adjacency list representation (A) Unweighted graph (corresponding to Fig. 4.1B) (B) Weighted graph (corresponding to Fig. 4.1D).

An assumption taken so far was that the vertices are integers, which is not the usual case. Like in the case of a Rubik's cube, the vertices are the states or the patterns that cannot be taken as integers. The weight (or adjacency) matrix hence becomes a map (or hash map) that takes a pair of vertices as the key and the weight of the edge (or its existential status) as a value. Similarly, the adjacency list is also a map (or hash map) that takes the name of the vertex as the key and returns the list of edges (names, metadata, and weights) as a value. In such cases, the list of vertices may be stored additionally as an array or map.

4.3   Breadth-first search

First, let us solve for a general graph for the shortest path. In this section, it is assumed that the graph is *unweighted* (all weights as equal/unity). Numerous problems have unweighted graphs, including games wherein each move is equivalent in terms of cost or the motion of a robot that can only move up, down, right, and left.

4.3.1   General graph search working

Before we delve specifically into the *breadth-first search* (BFS) algorithm, let us first discuss the general philosophy of search that is common to all search algorithms. The basic intent is to traverse or walk through the entire graph. While doing so, we would keep bookkeeping variables that enable us to compute the path from the source. In all searches, we use a *fringe* or *frontier*, which is a collection of vertices that we have witnessed but not fully explored. So, the traversal phenomenon is to take out vertices from the fringe one after the other and to explore them up. Once a vertex is fully explored, it is no longer in the fringe. As we explore the vertices, we come to know about many more vertices, which are added into the fringe. During search, it is possible to accidentally visit a vertex twice, and therefore every time a vertex is fully explored, it is coloured. The fringe is a collection of unvisited vertices, and at every point of time, any one of the vertices may be selected for exploration. The selection of vertices is given by the *search strategy*, which differentiates between the different search algorithms.

This principle is intuitive. Imagine you have shifted to a new housing society, and you are excited to visit the complete complex. Initially you only know the entrance once you step in. Upon entering, you see three roads leading to three different intersections. You memorize (add in fringe) all the places and select any one to walk into. Upon walking on the first road, you see 2 more new places and memorize (add to fringe) them as well. Once you have completely visited a place, you remove it from your memory and colour it black as an indicator. On visiting a new place, if you see that the place is coloured back, you know you have already been here. This stops you from visiting places repeatedly.

The search technique uses the *principle of optimality*, which states that every subpath within a shortest path is also the shortest. This means that if $S \to a \to b \to c \to d$ is the shortest path from S to d, then the subpaths $S \to a \to b$, $b \to c \to d, a \to b \to c$, etc., are also the shortest, which can be proved by a contradiction. If there was a shorter path between S and b in $S \to a \to b \to c \to d$, replacing $S \to a \to b$ with it would have led to a better result for the path between S and d.

Now, let us discuss the bookkeeping variables that we will maintain. $d(u)$ denotes the distance of the vertex u from the source, making the distance function $d:V \rightarrow \mathbb{R}_{\geq 0}$. Parent $\pi(u)$ is the penultimate (last but one) vertex in the shortest path from source to the vertex u, just before the vertex u, making the parent function $\pi:V \rightarrow V \cup \{NIL\}$. So, the shortest path from source to u will start from source to $\pi(u)$ and then the edge $<\pi(u),u>$. In other words, if $S \rightarrow a \rightarrow b \rightarrow c \rightarrow d$ is the shortest path from S to d, then $\pi(d) = c$, $\pi(c) = b$, $\pi(b) = a$, $\pi(a) = S$, and $\pi(S) = NIL$. While searching, one may wish to avoid the vertices being visited twice, for which reason they are coloured. $Colour(u)$ denotes the state of any vertex. Every vertex is initially white, meaning the vertex is unseen and unexplored. When a vertex is first seen and added to the fringe, it becomes grey. So, all vertices inside the fringe are always grey. When a vertex is fully explored, it is taken outside the fringe, and the colour changes to black.

Consider that at any given time the vertex taken out of the fringe is u. All outgoing edges from u are processed one after the other, and when all such edges get processed, the vertex u is said to have been fully explored. The vertex is coloured black, and the vertex is removed from the fringe. Consider the edge $<u,v>$ with u being taken from the fringe, for which the values of $d(u)$ and $\pi(u)$ are known, and hence there exists a path from source (S) to u with a cost of $d(u)$ as shown in Fig. 4.3. Assume v is currently coloured white. One of the possible ways to go from source to v is to first go to u and then take the $<u,v>$ edge. This gives a total distance of $d(v) = d(u) + 1$, and in the path so formed, the parent of v is u, or $\pi(v) = u$, since it is the penultimate vertex in the path from source to v as shown in Fig. 4.3. The vertex v is added to the fringe and coloured grey.

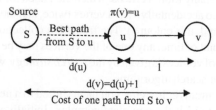

FIG. 4.3 Expansion of an edge (u,v) using BFS.

4.3.2 Algorithm working

With this background, let us now discuss the specific principle used by *breadth-first search*. In this search, the fringe is implemented with a *queue* data structure, maintaining a first-come first-serve order. A vertex with a lower d value enters the queue first and therefore leaves the queue first, starting from the source initially with a d value of 0. This gives a level-by-level traversal of the graph. First, the source is visited; then all vertices at a distance of 1 from the source are visited; later all vertices at a distance of 2 from the source are visited, and so on. The same principle applies to proving the optimality of the algorithm as well. If the goal vertex is found at a distance of $d(G)$, it means $d(G)$ is the optimal cost. If a goal existed at a smaller level, it would have been found earlier as per the level-by-level traversal strategy.

We put all the pieces of the algorithm together and present it in the form of a pseudo-code given as Algorithm 4.1. The main logic behind the pseudo-code (lines 10–21) is to extract a vertex from the fringe (lines 11–14) maintained in the form of a queue, colour the vertex black (line 21), and process all outgoing edges from that vertex (lines 15–20). The loop is a normal traversal of the adjacency list typically stored as a linked list. The traversal gives all outgoing edges and ending vertices. The processing of every vertex involves computation of the variables (lines 17–19), after which the vertex is added to the queue (line 20). To avoid processing a vertex twice, the colour is checked before processing (line 16).

Algorithm 4.1 Breadth-First Search (BFS).

Input: Graph G with vertices V and edges E, source (S) and goal test criterion ($GoalTest$)

Output: path cost (d) and shortest path (τ)

▷ Initialization ;

1 **for** $v \in V$ **do**   ▷ for all vertices
2 | $colour(v) \leftarrow White$;   ▷ White=Unseen, Grey=seen/in queue, Black=processed
3 | $d(v) \leftarrow \infty$;   ▷ distance from source
4 | $\pi(v) \leftarrow NIL$;   ▷ parent
5 $d(S) \leftarrow 0$;   ▷ Initialize source
6 $colour(S) \leftarrow grey$;
7 $Q \leftarrow$ An Empty Queue ;
8 $Enqueue(Q, S)$;
9 $found \leftarrow false$;
10 **while** $\neg Empty(Q)$ **do**   ▷ processing all vertices
11 | $u \leftarrow Front(Q)$;
12 | **if** $GoalTest(u)$ **then**
13 | | $found \leftarrow true$, break
14 | $dequeue(Q)$;
15 | **for** $v \in V : < u, v > \in E$ **do**   ▷ for all outgoing edges
16 | | **if** $colour(v) = White$ **then**   ▷ never seen before
17 | | | $d(v) \leftarrow d(u) + 1$;
18 | | | $\pi(v) \leftarrow u$;
19 | | | $colour(v) \leftarrow grey$;
20 | | | $enqueue(Q, v)$;
21 | $colour(u) \leftarrow black$;
22 **if** $\neg found$ **then** return ∞, NIL;
23 $\tau \leftarrow calculatePath(\pi, v)$;
24 return $d(v), \tau$;

The terminal criteria are encountering a goal vertex (lines 12–13) and a goal not found. Instead of taking a specific goal vertex, a generic case of a goal test is used. This allows for solving cases where there are multiple goal vertices (e.g. the robot needs to go to the refrigerator in the kitchen or the one in the living room), or for cases where a specific condition is met (e.g. the goal position with any orientation needs to be achieved). The initialization of all vertices but the source (lines 1–4) involves setting their bookkeeping variables. The source is separately initialized (lines 5–6) and added to an empty queue (lines 8).

Let us discuss the complexity of the algorithm. Informally, let V be the number of vertices and E be the number of edges. Lines 1–4 are initializations, which have a complexity of $O(V)$. Lines 11–14 and 21 are repeated once for every vertex and therefore come with a total complexity of $O(V)$. Lines 16–20 are executed once for every edge. For a vertex u, lines 16–20 are executed for all outgoing edges from u, repeated for all vertices u, giving a complexity of $O(E)$, including the effect of the outer loop. The overall complexity of the algorithm is hence $O(V+E)$.

Consider a graph $G' < V',E' >$ with all vertices as in the original graph $V' = V$ but edges consisting of only edges between vertices and their parents $E' = \{<\pi(v),v>\}$. Since every vertex will have only one parent, the graph is actually a tree called the *search tree*. The tree stores all possible shortest paths. A bottom-up traversal is made to find the shortest path. The path is reversed to get the path from source to goal. The algorithm gets its name since the search tree is traversed in a breadth-first manner. The algorithm to compute the path is given as Algorithm 4.2.

Algorithm 4.2 Calculatepath(π,v).

    Input: parent tree (π) and goal (v)
    Output: path (τ)
1 $\tau \leftarrow$ empty ordered list ;
2 **while** $u \neq NIL$ **do**
3   add u to the end of τ ;
4     $u \leftarrow \pi(u)$;   ▷ travel bottom to up in parent tree π
5 *reverse*(τ) ;   ▷ path from root (source) to goal
6 return τ ;

4.3.3 Example

Let us look into the working of the BFS algorithm with a suitable example. Consider the graph given in Fig. 4.1A. The steps involved in BFS are shown in

Fig. 4.4. The source is vertex 1, and the goal is vertex 5. Initially, the fringe only contains vertex 1 with cost 0 and parent NIL. At the first iteration, vertex 1 is taken out of the fringe (and turned black). The vertex produces children 2 and 3. 1 also serves as the parent of 2 and 3. Both vertices 2 and 3 are added in the fringe with grey colour.

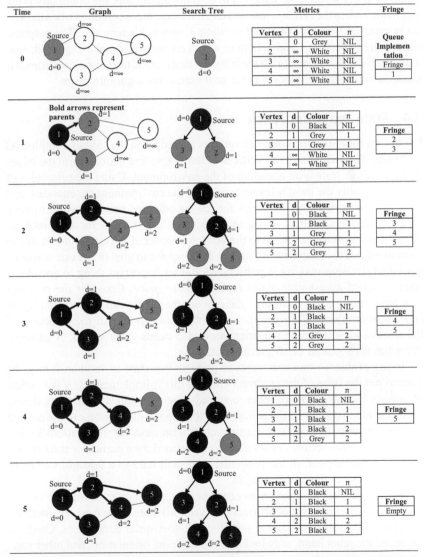

FIG. 4.4 Working of Breadth-First Search. At every iteration, one vertex is extracted from the fringe; all outgoing edges are used to calculate new values of *d* and π, while white vertices are added in the fringe with colour updated to *grey*. Vertices taken out of the fringe are changed to *black*.

The next vertex to be taken out of the fringe as per queue methodology is 2. The vertex 2 is connected to the vertices 1, 4, and 5. Since 1 is already black, it is ignored. The costs of the vertices 4 and 5 are calculated as $d(4) = d(2) + 1 = 1 + 1 = 2$. The two new vertices 4 and 5 are added to the fringe with grey colour. It may be noted that vertex 5 is the goal vertex, but the search does not stop when the goal is inserted into the fringe. However, the search stops when the *goal is removed from the fringe*. It may not have an effect in BFS, but in all other algorithms, the optimality is not ascertained when the vertex is inserted into the fringe but ascertained when the vertex is removed from the fringe. At iteration 3, vertex 3 is removed. The outgoing edges point to vertices 1 and 4, both of which are nonwhite and are ignored; similarly, with the next vertex (4) in the fringe. Finally, at the next iteration, vertex 5 is taken out of the fringe. Since it is the goal vertex, the algorithm stops.

4.4 State space approach

Let us first formalize the notion of vertices and edges in the case of complex AI problems. The states form the vertices of the graph, and actions form the edges of the graph. A *state* is an encoding of the environment. Using the principles of abstraction, only the most relevant aspects of the environment are encoded, and the aim is to have as few variables for the encoding as possible, so as to have efficient searches, without losing many details. Specifically, for any problem, imagine going from the source to the goal. An aspect of the environment that is guaranteed not to change during the run from source to goal or the one that does not affect decision-making is particularly not an important thing to encode as state. A set of all possible states makes a *state space*. Consider the problem of a point robot going from one place to another. This can be encoded by its (x,y) location alone. Consider that the robot only moves in the direction of the wheels, in which case the orientation also needs to be stored. A game of chess has the location of all pieces as its state.

In any problem when one wants to go from one state to another, it must be done as per the rules of the environment, and this is implemented by the notion of *actions*. To move a point robot, one may give commands such as up, down, right, and left that become the actions. For the robot with orientation, the motion may be by moving forward, turning clockwise, and turning anticlockwise, which become the actions. *Feasible actions* defined for a particular state ensure that the action applied to the particular state will lead to a feasible successor state, valid as per the problem logic.

The chapter has used the terms configuration and state to denote the vertex. Both terms differ in terms of abstraction. A robot's state more formally includes position, orientation, speed, and sometimes higher-order derivatives. Configuration, on the other hand, only includes the position, orientation, and other variables independent of time (without derivatives). When discussing robotics, the specific word configuration space is used, while for generic AI perspectives the word state is used.

Consider that the problem is to solve a Rubik's cube. Line 1 of Algorithm 4.1 loops through all the vertices or all states possible. The states are typically exponential to the number of variables needed to encode the state, and therefore initialization cannot be done in a short time and will also be out of available system memory. Hence, the same algorithm must be written so that the graph is not explicitly stored. While the graph is infinitely large, one does not need to store the entire graph a priori. In any game, given a state, it is possible to compute the next state by considering all possible actions. Assume that the necessary variables d, π, and *colour* are hash-maps with a constant time lookup and update. The size of the data structures is the number of vertices discovered, which is significantly less than the total number of possible vertices. Now, the actions have names (up, down, turn, etc.) and may be stored as additional information, just like *cost*.

A change introduced here is the *colour* bookkeeping variable. A state is white when it is unseen. It is obviously not possible to store all states as white initially. A state is white if it is unseen, grey if it is inside the fringe, and black if it is outside the fringe. Instead of storing colour, all states processed from the queue are stored as a data structure called *closed*. So, if a state is neither in the fringe nor in the closed, it is white; if a state is in the fringe and not in the closed, it is grey; and if it is in the closed, it is black.

Another variant of the algorithm neither stores colour nor is closed and is called the *tree-search* implementation of the algorithm. For this implementation, it is important that any information like cost, depth, etc., be stored inside the node only and not as maps, as was done so far. The name is derived from the fact that a tree is a special case of graphs where cycles are not admissible. Imagine a search tree expanding from the source onwards, exploring all states. Since we do not prune already visited states, such states will be visited again, making the same search subtree as was made by the same vertex when visited for the first time. The vertices are repeated in the search tree. On the contrary, the implementation with *closed* is called as the graph search implementation that prunes the latter vertex and stops the same calculations from being done again. Suppose a vertex u produces a vertex v, which is also found somewhere else in the search tree as the child of the vertex x. Since the children of vertex u and x are the same vertex v, one could represent it as a graph with one vertex v as the child of both u and x. From a search perspective, of course, the parent tree will remain as a tree, and only the better of the two v vertices will be retained and the other pruned. The tree search technique will visit the vertex v both times and generate the same subtree twice, once as the child of x and once as the child of u. In a tree search, after visiting the vertex v as a child of x, if the search revisits v as a child of u, there is no data structure that tells the algorithm that v has already been visited. The vertex is revisited as if it were a new vertex. In case of a *graph search* strategy, however, there is a *closed* data structure that tells the algorithm that the vertex v was already visited and need not be visited again.

Considering the discussions, let us modify the pseudo-code of BFS, which is presented as Algorithm 4.3 for a graph search. In terms of implementations, using hash maps for d and *closed* is suggestive. The only difference is the absence of colour and the use of *closed* instead. Further all actions are applied in lines 12–13, and only feasible ones are used for further processing of lines 15–19. The pseudo-code assumes that $v \notin Q$ in line 15 can be calculated in a constant time, which is only possible if a mapping between state v and fringe exists, that may not be the case if using a vanilla library implementation. The use of *colour* instead is one option, or the search may be done on the d values. Alternatively, even if this condition is not checked, the result is correct with extra computation time (and same complexity). The tree search implementation does not take maps for *closed*, d, *parent,* and *action* and instead stores the data for distance, parent, and action in the state node.

Algorithm 4.3 Breadth-First Search (BFS).

Input: action function (A), feasibility function, source (S) and goal test
criterion ($GoalTest$)
Output: path cost (d) and shortest path (τ)
   ▷ Initialization ;
1 $d(S) \leftarrow 0$;
2 $\pi(S) \leftarrow NIL$;
3 $Q \leftarrow$ An Empty Queue ;
4 $Enqueue(Q, S)$;
5 $closed \leftarrow$ An empty hash map;
6 $found \leftarrow false$;
7 **while** $\neg Empty(Q)$ **do**   ▷ processing all states
8     $u \leftarrow Front(Q)$;
9   **if** $GoalTest(u)$ **then**
10        $found \leftarrow true$, break
11     $dequeue(Q)$;
12   **for** *all actions a* **do**   ▷ equivalent to edges
13        $v \leftarrow$ state on application of action a on u ;
14   **if** $\neg Feasible(v)$ **then** continue;
15   **if** $v \notin Q \wedge v \notin closed$ **then**   ▷ never seen before
16           $d(v) \leftarrow d(u) + 1$;   ▷ distance from source
17           $\pi(v) \leftarrow u$;   ▷ parent state
18           $action(v) \leftarrow a$;   ▷ action from parent to v
19           $enqueue(Q, v)$;
20   Add u to $closed$;
21 **if** $\neg found$ **then** return ∞, NIL;
22 $\tau \leftarrow calculatePath(\pi, v)$;
23 return $d(v), \tau$;

Let us take a small example to study the approach. Imagine a differential wheel drive robot being driven with two wheels. The robot can go forward by rotating both wheels and turn $\pi/2$ degrees clockwise or anticlockwise by turning one wheel forward and the other backward. The action set is continuous; however, it needs to be discretized in this mechanism to enable graph search. The state consists of the (x,y) coordinates and the orientation (θ). The problem is shown in Fig. 4.5A. Assume that the world is discretized into blocks, and the world is limited to a size of 10×10 blocks, with X coordinates in the range $[0,10]$, Y coordinates in the range $[0,10]$, and orientation in the range $[0,2\pi)$. More correctly, because of discretization, X coordinates possible are $\{0, 1, 2, 3, 4, 5, 6, 7, 8, 9, 10\}$, Y coordinates possible are $\{0, 1, 2, 3, 4, 5, 6, 7, 8, 9, 10\}$, and orientations possible are $\{0, \pi/2, \pi, 3\pi/2\}$. The robot can have any X and Y coordinates and rotation from this set. Hence the state space is the Cartesian product of the sets, or $\{0, 1, 2, 3, 4, 5, 6, 7, 8, 9, 10\}^2 \times \{0, \pi/2, \pi, 3\pi/2\}$. The maximum size of the space is hence $10 \times 10 \times 4 = 400$. It may be noted that this is the total state space. Not all states will be possible due to the feasibility of collisions with obstacles. As an example, standing at $(0,0)$ and at an orientation of π, the robot on being moved forward will go to (-1.0), which is not possible since there is a world boundary at $X = 0$. The goal test is simply to reach a position irrespective of orientation.

It is interesting to observe that if the turns were kept at $\pi/4$ instead of $\pi/2$, moving 1 unit forward at an angle of $\pi/4$ from $(0,0)$ leads to the position $(1/\sqrt{2}, 1/\sqrt{2})$. However, the states are discretized, and there is no such position as $(1/\sqrt{2}, 1/\sqrt{2})$. One way is to map this to the closest position in the discretized state space $(1,1)$, with the control algorithm responsible for handling the small error.

Interestingly, graph search in the state-space approach never assumed the states as discrete, while the actions need to be compulsorily discrete. The tree-search strategy does not implement closed but admits continuous states. The graph-search variant implements a closed set of states (to prune rediscovered states to be visited again); however, for continuous states, the chances of two doubles matching for equality is small and therefore rediscovering a state in closed is unlikely.

The problem is summarized in Box 4.2. Assume starting from position $(0,0)$ with an orientation of 0 rad. The search tree for a few iterations is given in Fig. 4.5B. It is assumed that the search tree is formed using a graph search.

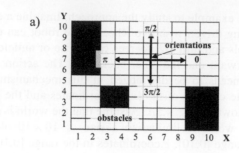

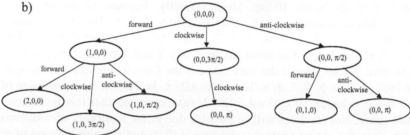

FIG. 4.5 Robot with orientation motion problem (A) Representation (B) Search Tree (using state represented as (x,y,θ)).

Box 4.2 Robot with orientation motion problem.

State: (x,y,θ) [vertex of the graph].

State space: $\{0,1,2,3,4,5,6,7,8,9,10\}^2 \times \{0, \pi/2, \pi, 3\pi/2\}$.

State space size: $10 \times 10 \times 4 = 400$ (considering no obstacles, set of all vertices of the graph).

> Actual space excludes states where the robot collides with a static obstacle (or itself).
>
> Imagine a ghost robot kept at (x,y,θ). If it intersects an obstacle, exclude such states.

Actions: {forward, clockwise, anticlockwise} [edges of the graph].

Action functions:

$$\text{forward}(x,y,\theta)=\begin{cases}(x+1,y,\theta) & \theta=0\\(x,y+1,\theta) & \theta=\pi/2\\(x-1,y,\theta) & \theta=\pi\\(x,y-1,\theta) & \theta=3\pi/2.\end{cases}$$

$\text{clockwise}(x,y,\theta)=(x,y,(\theta-\pi/2+2\pi) \bmod 2\pi)$

$\text{anticlockwise}(x,y,\theta)=(x,y,(\theta+\pi/2) \bmod 2\pi)$

Branching factor: 3

Feasible actions: Actions such that the state after the application of action is feasible. As an example, for a point robot:

$A=\{\text{action}\in\{\text{forward, clockwise, anticlockwise}\}: (x',y',\theta')=\text{action}(x,y,\theta), 0\leq x' \leq 10, 0\leq y' \leq 10, \text{map}(x',y')=\text{free}\}$

Goal test: $x=G_x \wedge y=G_y$

The complexity of the algorithms again needs to be redefined as $O(V+E)$ implies a very large number since the number of states (and hence total possible actions) for some problems may be infinite. Assume that the solution is found at a distance of d from the source. Further assume that every state can have a maximum of b actions possible, where b is called the *branching factor*. So, if a robot is moved in the four orthogonal directions, b is 4. The term branching factor comes from the fact that in the search tree, any vertex can have a maximum number of children corresponding to the branching factor. Both solution depth (d) and branching factor (b) are finite for any realistic problem (although some problems may have the infinite branching factor not solvable by BFS). The computation cost is the total number of nodes in the search tree. At 0th depth, there is only the source or 1 node. At depth 1, the source can have a maximum of b nodes. Each of the b nodes can have b children, and hence b parents can have b^2 children at depth 2. Similarly, b^2 parents can have b^3 children for the next depth. The total number of nodes till a maximum depth of d is hence $1+b+b^2+b^3+\dots$, which by a geometric progression addition has a complexity $O(b^d)$. The same is also shown in Fig. 4.6. The use of the term *node* instead of state is deliberative here. The state only stores the associated encoded variables, while the node can store any additional information as needed for the implementation such as parent, depth, etc. The information is assumed to be stored as maps and not inside the node; however, it is better to use the term node to have the possibility of any metainformation along with the state.

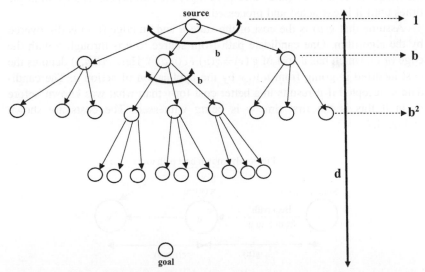

FIG. 4.6 Computation of complexity of BFS. If a solution exists at depth d, the complexity is the number of nodes, $O(b^d)$.

4.5 Uniform cost search

The BFS has a major assumption that the edges are unweighted (or have the same weight). However, most real-life problems have a different weight associated with every action. Therefore, the BFS needs to be generalized to the use of weighted graphs; however, for *uniform cost search* (UCS), the weights should be strictly nonnegative only. This is practical in all AI problems, where taking actions always incurs some kind of cost. The UCS algorithm is the same as the more commonly known Dijkstra's algorithm with the small difference that the Dijkstra's algorithm asks for adding all states to the fringe at the very start, and therefore it cannot be done in a state space-based approach where the number of states could be infinite.

4.5.1 Algorithm

The major change in the use of UCS in contrast to the BFS is that the fringe is implemented by using a *priority queue*, where the priority value of a node is given by the distance of the node from the source. The distance is termed g here instead of d as used in BFS, since the term d is also commonly used to denote depth. The other basics are exactly the same as the generic search principles discussed with BFS. The nodes may not have optimal costs when inserted into the fringe (also called as the fringe); however, they will have optimal costs when extracted from the fringe (or frontier). This means that while a state is in the fringe, its cost value (and correspondingly the parent) can change multiple times until it is extracted and processed.

Assume that $g(u)$ is the cost of the state u and an edge (u,v) is discovered by the algorithm. One candidate path from source to v is through u with the edge (u,v), which has a cost of $g'(v) = g(u) + c(u,a,v)$. Here $c(u,a,v)$ denotes the cost incurred in going from u to v by the application of action a. The candidate is accepted if it results in a better cost for v than what was known before or in if this is the first time v is being witnessed. The costs are shown in Fig. 4.7.

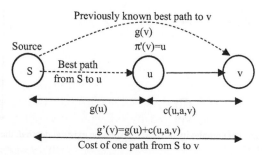

FIG. 4.7 Expansion of an edge (u,v) using UCS.

For now, assume that the algorithm is optimal. If the state v is neither in fringe nor in closed (white), the candidate cost is accepted since no prior cost is available. If v is in fringe but not in closed (grey), the candidate is accepted only if it gets a better cost value than previously known. If v is in closed (black), since the algorithm is optimal, a state in closed can never take a better value, and such a case is rejected. Although when working with *suboptimal searches* and when the *graph itself can change* with time, it is suggestive to account for this case as well. Assuming $g(v)$ as the estimated cost from source at any point in time, the updated $g(v)$ value after expansion of edge (u,v) with the assumption of optimal search is given by Eq. (4.4) and the updated parent by Eq. (4.5):

$$g(v) \leftarrow \begin{cases} g(v) & \text{if } v \in \text{fringe} \wedge g(v) \leq g(u) + c(u,a,v) \vee v \in \text{closed} \\ g(u) + c(u,a,v) & \text{if } v \notin \text{fringe} \vee v \in \text{fringe} \wedge g(u) + c(u,a,v) < g(v) \end{cases}$$

$$(4.4)$$

$$\pi(v) \leftarrow \begin{cases} \pi(v) & \text{if } v \in \text{fringe} \wedge g(v) \leq g(u) + c(u,a,v) \vee v \in \text{closed} \\ u & \text{if } v \notin \text{fringe} \vee v \in \text{fringe} \wedge g(u) + c(u,a,v) < g(v) \end{cases} \quad (4.5)$$

The search principle is derived from BFS. In BFS, we explored nodes at a depth of 0, then 1, then 2, and so on so that the algorithm is optimal. However, in the case of UCS, the costs are real numbers. The same principle in UCS may be read as follows: first explore the nearest state to the source (or the source itself), then the next farthest away state from the source, then the still next farthest away state from the source, and so on till the goal is reached. The search principle needs to select one of the vertices in the fringe, and the node with the smallest g value is the best candidate as per the principle.

The UCS algorithm for the state-based approach is given as Algorithm 4.4. The pseudo-code is the same as BFS with the exceptions discussed. The focus should be given to lines 15–24. Lines 15–19 are the cases when the state is neither in fringe nor in closed and should hence have a normal insertion. Lines 20–24 check whether the state was already in fringe and is rediscovered with a better cost, in which case the better cost (and parent) is accepted and correspondingly the priority queue is updated. The cases when the state is already in closed or in fringe with a better value are not dealt with since there is no computation involved and the edge should just be ignored. From a programming perspective, a heap is a good mechanism for implementing the priority queue. The pseudo-code in lines 15 ($v \notin Q$) and 20 ($v \in Q$) assumes it is possible to have a constant-time search on the fringe, which will require the use of colour as an alternative, or g implemented as a map can be searched. More importantly, line 24 assumes that the state u can be mapped to a heap node to update its priority, which will not be possible in a vanilla library implementation, although pointers may be used and maintained otherwise. Even if the state is reinserted instead of updating the priority, the algorithm will work with more computation time (and same complexity). Therefore, a popular implementation reinserts a better rediscovered node in the priority queue while the old one with a worse cost is already in the priority queue.

Algorithm 4.4 Uniform Cost Search (UCS).

Input: action function (A), feasibility function, cost function $c(u, a, v)$
(cost from u to v by action a), source (S), goal test $(GoalTest)$
Output: path cost (d) and shortest path (τ)
   ▷ Initialization ;
1 $g(S) \leftarrow 0$;   ▷ cost from source
2 $\pi(S) \leftarrow NIL$;
3 $Q \leftarrow$ An Empty Priority Queue prioritized by g values;
4 $Enqueue(Q, S)$ with priority $g(S)$;
5 $closed \leftarrow$ An empty hash map;
6 $found \leftarrow false$;
7 **while** $\neg Empty(Q)$ **do**   ▷ processing all states
8      $u \leftarrow Front(Q)$;
9   **if** $GoalTest(u)$ **then**
10        $\lfloor$ $found \leftarrow true$, break
11    $dequeue(Q)$;
12   **for** $all\ actions\ a$ **do**   ▷ equivalent to edges
13        $v \leftarrow$ state on application of action a on u ;
14   **if** $\neg Feasible(v)$ **then** continue;
15   **if** $v \notin Q \wedge v \notin closed$ **then**   ▷ never seen before
16            $g(v) \leftarrow g(u) + c(u, a, v)$;   ▷ distance from source
17            $\pi(v) \leftarrow u$;   ▷ parent state
18            $action(v) \leftarrow a$;   ▷ action from parent to v
19            $enqueue(Q, v)$ with priority $g(v)$;
20   **else if** $v \notin closed \wedge v \in Q \wedge g(v) > g(u) + c(u, a, v)$ **then**
               ▷ re-seen v with a better cost
21            $g(v) \leftarrow g(u) + c(u, a, v)$;
22            $\pi(v) \leftarrow u$;
23            $action(v) \leftarrow a$;
24   update priority of v in Q to $g(v)$
25   Add u to $closed$;
26 **if** $\neg found$ **then** return ∞, NIL;
27 $\tau \leftarrow calculatePath(\pi, v)$;
28 return $d(v), \tau$;

The complexity of UCS is intuitively derived similar to the BFS. The cost will be the same $O(b^d)$, with the problem that the solution depth is not a representative. So, let us have a worst-case estimate of the depth. Suppose the optimal cost from source to goal is C^*. Further suppose the minimum edge cost is ε. The computation time is the total time in expanding all nodes with a cost of less than or equal to C^*. Out of all such nodes, let us consider the node with the maximum depth. The depth of any node is given by the cost of the node divided by the cost of an average edge in the path from source to the node, $d(u) = g(u)/(\text{average edge cost})$, where $d(u)$ is the depth of node u. To find maximum possible depth, we find maximum $g(u)$, which is C^* among all processed nodes, and minimize average edge cost, which cannot be lower than ε. So, the complexity is given by $O(b^{1+\lfloor C^*/\varepsilon \rfloor})$.

The optimality can also be established on the same lines as the BFS. First, all nodes with a g value of 0 are processed, then all nodes with a little higher g value are processed, then the ones with a still higher g value, and so on. The search expands with the contours of cost. At a given time if the node with cost $g(u)$ is being expanded, it represents a cost contour of value $g(u)$, signifying all nodes with a cost less than $g(u)$ have already been explored by an earlier cost contour while the ones with a cost more than $g(u)$ will be explored by future contours. The expansion happens at an increasing order of cost. Let us study this in contradiction. Suppose a node u with a smaller cost $g(u)$ is processed after a node v with a higher cost $g(v)$, or $g(u) < g(v)$. u would have been nondiscovered at the time of processing of v; otherwise the priority queue would have prioritized u. However, some ancestor of u would have been processed, say x, before v was processed, $g(x) < g(v)$. Let $x \rightarrow s_1 \rightarrow s_2 \rightarrow \ldots u$ be the optimal path from x to u. Since $g(u) < g(v)$ and all edge costs are nonnegative, $g(x) < g(s_1) < g(s_2) < \ldots < g(u) < g(v)$. x would have given a lower cost to a child s_1 in the optimal path from x to u, which would have been inserted and processed before v (since $g(s_1) < g(v)$), which again would have given a lower g value to a node s_2 in the optimal path to u that would also have been inserted and processed in the queue before v (since $g(s_1) < g(v)$). This goes on till u gets a lower cost than v and is inserted and processed before v in the priority queue (since $g(u) < g(v)$), thus disproving the contradiction. However, a node v being inserted into the priority queue by the processing of a node u may not be optimal because even if $g(u)$ is the smallest cost from source to u, $c(u,a,v)$ may not be the smallest cost from u to v or the optimal path may not go through u at all. Hence nodes dequeued from the priority queue have an optimal cost, but the nodes being enqueued into the priority queue may not have an optimal cost.

4.5.2 Example

Let us study the working of the algorithm with a simple example. The graph used for the problem is the same as that in Fig. 4.1D. In the algorithm, first only the source vertex is known, and the same is added to the fringe with a priority of 0. At the first iteration, the source is extracted and processed. As a result, vertices 2 and 3 are discovered with costs (and priorities) 2 and 3, and both are added to the fringe. Since 1 is processed, it is added to closed. At the next iteration, the least cost (best priority) vertex is 2, so it is processed and added to closed. Processing produces the vertices 4 (with cost $g(4) = g(2) + 4 = 2 + 4 = 6$) and 5 (with cost $g(5) = g(2) + 7 = 2 + 7 = 9$). Both are added to the fringe. It may be seen that even though the goal vertex (5) is seen, the search does not stop. If the search stops, the current path from source (1) to goal (5) would be suboptimal.

At the next iteration, out of 3, 4, and 5, the least cost is 3, which is hence extracted and processed. The vertex rediscovers the vertex 4 with a better cost, and so are the g values and priorities updated. Similarly, the next iteration extracts 4 from the fringe and produces vertex 5 with a better cost. Lastly, vertex 5 is extracted. Therefore, the goal test is passed and the search stops. The steps are shown in Fig. 4.8.

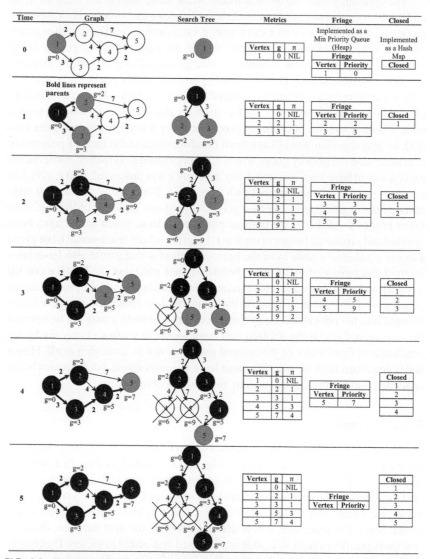

FIG. 4.8 Working of Uniform Cost Search. At every iteration, one vertex is extracted from the fringe, all outgoing edges are used to calculate new values of g and π, costs are updated (if better), and new vertices are added to the fringe (if necessary). The processed vertices are added in closed.

4.5.3   Linear memory complexity searches

The difference between tree search (without *closed*) and graph search (with *closed*) strategies may not seem to be major in BFS and UCS. However, consider a similar search mechanism, the *depth-first search* (DFS) which searches the depth of the tree first. If the variables *colour, parent, action,* and *d* are maintained in the node (along with state) instead of global maps, the only memory needed is the fringe, which for DFS is implemented as a stack. In case of DFS, at any time, the fringe (stack) stores the children of all nodes in one candidate path from source to the current node, which is limited to the length of the deepest path. The algorithm has a memory complexity of $O(bd_{max})$, where b is the branching factor and d_{max} is the maximum depth to which the DFS may go (may be infinite). The backtracking implementation of DFS (which does not create all children at once but rather recurses on children one by one) only stores the path from source to the current node, giving the memory complexity $O(d_{max})$. Note that costs or any other data is stored in the node (along with the state) and not as global maps in the tree variants of search algorithms. The graph search variant stores the costs, parents, and closed status of all nodes ever processed; however, the tree search variant stores the costs and parents of only the nodes in the current path. The memory of the search becomes linear instead of exponential in the case of a tree search implementation without *closed*. The cost and path can still be traced by passing relevant information to the child in its creation by the parent, or by storing the path from parent to child, which also has a linear complexity. The linear memory complexity DFS algorithm is neither optimal nor complete as compared to the exponential memory complexity BFS and UCS; however, practically one runs out of memory a lot earlier than running out of time for static maps. The linear-memory complexity of the DFS makes it an interesting algorithm.

A typical problem with the DFS is that the algorithm gets stuck exploring cycles. Consider from source the algorithm gets to the state a, from which it goes to b; from b it goes to c; from c the next node to expand is a, from which it goes into b and gets into the cycle between a, b, and c. The algorithm indefinitely traverses the path *source, a, b, c, a, b, c,* and so on. Any graph search variant with *closed* would have pruned node a from being revisited, thus eliminating the cycle. Local measures such as not visiting a node in the current path can break this cycle; however, DFS can have cycles where a node is closed and reopened indefinitely in the tree search implementation (with the absence of a closed set of nodes stored explicitly).

A solution to the problem is *depth-limited DFS* where expansion happens until a maximum specified limit. If a node has a depth greater than a *limit*, it is not expanded. Thus, one could estimate d_{max} as the maximum depth at which the goal may exist and only expand the tree within this limit. If the estimate d_{max} is correct, the resultant algorithm would be complete (but not optimal) with a linear memory complexity and an exponential time complexity; however, it may be practically impossible to guess such a limit.

Since an estimate d_{max} cannot be made, let us make another algorithm that iteratively guesses the estimate d_{max}. A previous estimate being wrong initiates another run of the algorithm with a new estimate. The resultant algorithm is called iterative deepening. *Iterative deepening* calls a depth-limited DFS multiple times with increasing depth limits. From the source, the algorithm first calls depth-limited DFS with a depth limit of 0, which only visits the source. The next time it calls DFS with a depth limit of 1, which visits all the children of the source. Thereafter, it calls DFS with a limit of 2 to also visit the grandchildren of the source. If the goal lies at a depth of d, then the depth-limited DFS at the depth limit of d results in the optimal path to the goal. The algorithm is optimal for an unweighted graph. Surprisingly, the computation time is not much different from the BFS since in an exponentially increasing series representing the total number of nodes visited by the DFS, most of the time is devoted only to the last iteration of the depth limit of d, and the cumulative total of the earlier computations (terms in the exponential series) is comparatively small. The total computation time is given by $1 + b + b^2 + b^3 + \ldots + b^{d+1}$, where b is the branching factor and the ith term denotes the number of nodes expanded at depth limit i. Since $b^{d+1} \gg b^d \gg \ldots \gg 1$, the summation is nearly the last term alone. After every iteration, the memory is cleared, and a new memory is initialized for the next run. Therefore, the memory complexity is $O(bd)$, which is only from the last iteration. The iterative deepening algorithm is optimal since it imitates the BFS with a level-by-level traversal in every iteration, while it has a linear memory complexity since the implementation at every iteration is that of a DFS. Iterative deepening takes the advantages of both BFS (optimality) and DFS (linear memory complexity) and represents an algorithm better than both individual algorithms in the case where the state space is too large to be stored in the memory. It can be used for algorithms that take too long and exhaust the system memory, while as per the problem one may be in a position to devote more time. The time complexity remains exponential.

The discussions can be extended to weighted graphs as well. *Cost-limited DFS* is an implementation of DFS that limits the total cost of paths searched by DFS. If a node has a cost greater than a set *limit*, the expansion of that node does not take place. Calling cost-limited DFS multiple times with increasing cost limits gives an algorithm that is optimal with linear memory complexity (and still exponential time complexity). The algorithm runs at any time with a cost limit C, expanding all possible paths such that the maximum path cost does not cross C. In doing so, the algorithm also peeks into the nodes at the next depth to get the next highest cost limit

C'. C' is the smallest cost considering all paths possible up to 1 depth more than the current pruning limit C. This becomes the cost limit for the next iteration. The algorithm keeps increasing the cost limits until the goal is found. The algorithm is called *iterative lengthening* and is a good solution if the problem runs out of memory before time.

4.6   A* algorithm

The uniform cost search is not an intelligent manner of search. The search technique searches the entire space from the source in an outward direction and stops when the goal is found. The *A* algorithm* adds *intelligence* to the search mechanism using *heuristics* (Hart et al., 1968; Russell and Norvig, 2010). The aim is to tell the search procedure about the good places to explore and bias the search towards the same good areas. This directs the search towards more relevant areas in the search space, because of which the search becomes faster. We first define the notion of heuristics and then discuss mechanisms to add heuristics in the search procedure.

4.6.1   Heuristics

In general, *heuristics* is defined as a rule of thumb that one believes to be correct. Heuristics serve as rules that guides decision-making in any intelligent system. The rules are usually guesses injected into the system to get the desired behaviour. In the A* algorithm, heuristics are taken by implementing a function called as a *heuristic function* $h{:}\zeta \to \mathbb{R}_{\geq 0}$ that *guesses* the cost of a state to the goal. Here ζ is the state space. The heuristic value of a state u, given by $h(u)$, is the estimated cost incurred in going from u to the goal. The heuristic value of the goal is always 0. It may be noted here that the heuristic value is just an estimate and comes with no guarantees of being correct. The actual values can only be calculated by an optimal search algorithm, which is not computationally efficient. Hence the aim is to have a function that guesses the cost to the goal without taking much computation time.

The heuristic value is defined for every state, and it does not depend up the search process. This makes it different from the cost from the source (g value) also called the *historic cost* of a state, which is calculated in the search process and represents the actual values and not the guesses. The calculation of g value is computationally expensive, while that of h value is a quick guess. The estimated total cost of a path through a state u is the cost incurred in going from source to u and the estimated cost of the path in going from u to the goal. Let $f{:}\zeta \to \mathbb{R}_{\geq 0}$ be a function that estimates the total cost of the path compulsorily through a state u, which by definition is given by $f(u) = g(u) + h(u)$, explained graphically in Fig. 4.9. The prime used in the figure denotes a candidate that may be accepted if better than the current best.

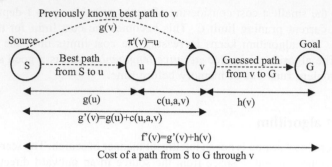

FIG. 4.9 Expansion of an edge (u,v) in the A* algorithm.

Here the first component or g is an exact number, and the second component or h is an approximation. Assume the entire graph is explored and the f, g, and h values for all the states are computed. Let S be the source. For the source $g(S)=0$ and $h(S) \geq 0$, while for the goal $g(G) \geq 0$ and $h(G)=0$. Both $f(S) = g(S)+h(S)=h(S)$ and $f(G)=g(G)+h(G)=g(G)$ are estimates of the total path costs. For source, $f(S)$ contains an estimated component only ($h(S)$), while for goal $f(G)$ contains an exact component only ($g(G)$). This means that as one travels away from the source towards the goal, the approximate part or h reduces and the exact part or g increases, until at the goal where there is no approximate part, and hence, we get the exact distance between source and goal.

4.6.2 Algorithm

The search technique used by the A* algorithm is to use the total estimated cost of the path or f value for the prioritization of the fringe. Nodes with a small f value are processed first, and the nodes with a larger f value are processed at the end. The process of expanding an edge results in some new vertices being discovered and a better cost of some new vertex being computed. Given a lot of path estimates from source to goal, it is intuitive to refine the best possible path estimate found so far in search of the optimal path from source to goal. The algorithm is presented as Algorithm 4.5. It may be seen that the algorithm is the same as the UCS with the only difference that the prioritization is on f instead of the g values. The heuristic function is specific to the problem, and therefore the algorithm assumes that such a function is already available.

Algorithm 4.5 A* Algorithm.

Input: action function (A), feasibility function, cost function $c(u, a, v)$
(cost from u to v by action a), heuristic function $h(u)$ (heuristic
value for state u), source (S) and goal test criterion $(GoalTest)$

Output: path cost (d) and shortest path (τ)

▷ Initialization ;

1 $g(S) \leftarrow 0$;   ▷ cost from source

2 $f(S) \leftarrow g(S) + h(S)$;   ▷ total cost

3 $\pi(S) \leftarrow NIL$;

4 $Q \leftarrow$ An Empty Priority Queue, prioritized on f values ;

5 $Enqueue(Q, S)$ with priority $f(S)$;

6 $closed \leftarrow$ An empty hash map;

7 $found \leftarrow false$;

8 **while** $\neg Empty(Q)$ **do**   ▷ processing all states

9 | $u \leftarrow Front(Q)$;

10 | **if** $GoalTest(u)$ **then**

11 | | $found \leftarrow true$, break

12 | $dequeue(Q)$;

13 | **for** *all actions* a **do**   ▷ equivalent to edges

14 | | $v \leftarrow$ state on application of action a on u ;

15 | | **if** $\neg Feasible(v)$ **then** continue;

16 | | **if** $v \notin Q \wedge v \notin closed$ **then**   ▷ never seen before

17 | | | $g(v) \leftarrow g(u) + c(u, a, v)$;   ▷ distance from source

18 | | | $f(v) \leftarrow g(v) + h(v)$;   ▷ total cost

19 | | | $\pi(v) \leftarrow u$;   ▷ parent state

20 | | | $action(v) \leftarrow a$;   ▷ action from parent to v

21 | | | $enqueue(Q, v)$ with priority $f(v)$;

22 | | **else if** $v \notin closed \wedge v \in Q \wedge g(v) > g(u) + c(u, a, v)$ **then**
       ▷ re-seen v with a better cost

23 | | | $g(v) \leftarrow g(u) + c(u, a, v)$;

24 | | | $f(v) \leftarrow g(v) + h(v)$;

25 | | | $\pi(v) \leftarrow u$;

26 | | | $action(v) \leftarrow a$;

27 | | | update priority of v in Q to $f(v)$

28 | Add u to $closed$;

29 **if** $\neg found$ **then** return ∞, NIL;

30 $\tau \leftarrow calculatePath(\pi, v)$;

31 return $d(v), \tau$;

4.6.3 Example

Let us apply the A* algorithm to the graph shown in Fig. 4.1B. To apply the algorithm, the heuristic function needs to be additionally supplied, which is given in Table 4.1. The actual estimates come from the specific problem whose design principles are discussed later. The run of the A* algorithm is shown in Fig. 4.10.

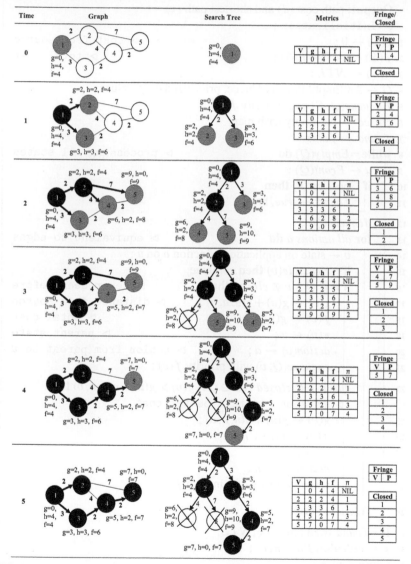

FIG. 4.10 Working of A* algorithm. The working and notations are the same as UCS, with the difference that the priority queue is on f values ($f = g + h$), where h values are supplied in Table 4.1 (heuristic value). V denotes vertex and P denotes priority.

TABLE 4.1 Heuristic values.

Vertex	h
1	4
2	2
3	3
4	2
5	0

Initially, only the source vertex is at the fringe. The g value is 0, while the h value is directly taken from Table 4.1. The f value is the sum of the g and h component values. The source is used to produce the two adjoining vertices, 2 and 3, which are also added to the fringe with priorities taken as their f values. The g values are the same as UCS, while the h values are as supplied in Table 4.1. At the next iteration, the best node is 2 with a priority value of 4. The node is extracted and processed, which produces the other two vertices, 4 and 5. The search does not stop on encountering the goal vertex 5, since it is being inserted into the fringe. The next best priority value is vertex 3 with a value of 6. On being expanded, it results in a better g value and hence a better f value for vertex 4. Therefore, the costs, parents, and priorities are updated. Later, 4 is extracted as the next best, which further reduces the estimated cost of the goal to a value of 7. In the end, the goal is extracted from the fringe, and therefore the search stops.

4.6.4 Admissibility and consistency

The next major discussion concerns the optimality of the algorithm. We first define two terms, admissibility and consistency, concerning the heuristic function. A heuristic function is called *admissible*, if it never overestimates the best cost to the goal, that is Eq. (4.6):

$$\forall_u \, h(u) \leq h^*(u) \tag{4.6}$$

Here $h^*(u)$ is the optimal cost from u to the goal. The heuristic function may underestimate the cost to the goal or return the exact cost to the goal, but it must never overestimate for any state u. This is achieved by making the heuristic function *optimistic*, wherein the estimate is typically better (lesser) than the actual cost. To assess the admissibility of the heuristic function used in Table 4.1, let us calculate the h^* values and check for admissibility. The same is shown in Table 4.2. The h^* values are simple to calculate manually since the graph is small. It may be noted that the values are being calculated for analysis only, and the calculation does not form a computational part of the algorithm, which is infeasible since there may be infinite states. Since for all vertices $h(u) \leq h^*(u)$, the heuristic function is admissible.

TABLE 4.2 Checking for admissibility. Since for all vertices h^* (shortest path cost to goal) $\leq h$ (Table 4.1), the heuristic function is admissible.

Vertex	h	h^*	$h \leq h^*$
1	4	7	Yes
2	2	6	Yes
3	3	4	Yes
4	2	2	Yes
5	0	0	Yes

The other term is consistency. A heuristic function is called *consistent* if

$$\forall_{(u,v) \in E} \; h(u) \leq c(u, a, v) + h(v) \tag{4.7}$$

The inequality is also known as the triangle inequality. Imagine a triangle with the vertices u, v, and goal (G); a directed edge between u and v denoted by the action a with a cost of $c(u,a,v)$; an estimated path from u to G that potentially has a cost of $h(u)$ with a direction from u to G; and similarly an estimated path from v to G that potentially has a cost of $h(v)$ directed from v to G. The consistency equation states that the sum of two sides of the triangle is greater than the side. In other words, the estimated cost of reaching u to G should be smaller than the one incurred in going from u to v and then v to G. The consistency check for the heuristic function of Table 4.1 is given in Table 4.3. The function is consistent. Again, the calculations are for analysis only and do not constitute an algorithmic procedure.

TABLE 4.3 Checking for consistency, h taken from Table 4.1 and edge weights from Fig. 4.1B.

Edge (u,v)	$h(u)$	$c(u,a,v)$	$h(v)$	$h(u) \leq c(u,a,v) + h(v)$
(1,2)	4	2	2	Yes
(1,3)	4	3	3	Yes
(2,1)	2	2	4	Yes
(2,4)	2	4	2	Yes
(2,5)	2	7	0	Yes
(3,1)	3	3	4	Yes
(3,4)	3	2	2	Yes
(4,2)	2	4	2	Yes
(4,3)	2	2	3	Yes
(4,5)	2	2	0	Yes
(5,2)	0	7	2	Yes
(5,4)	0	2	2	Yes

Since for all edges the inequality holds, the heuristic function is consistent.

The A* algorithm is *optimal if the heuristic function is admissible for a tree search*. Since $h(u) \leq h^*(u)$ for all nodes, this means $g(u) + h(u) \leq g(u) + h^*(u)$ or the estimated cost from source to goal for any node is always either lesser or equal to the optimal cost to goal through u. This implies that as we expand the nodes, the f values continue to increase, and no value other than the goal can have an f value greater than C^*, the optimal cost from source to goal, until the goal is expanded with a cost equal to C^*. The A* algorithm is *optimal if the heuristic function is admissible and consistent for a graph search*. Consistency ensures that a node placed in closed cannot be rediscovered with a better cost as a result of heuristic estimate errors. This forms the basic genesis behind the design of a heuristic function. Consistency is a stronger criterion than admissibility. A consistent heuristic function is admissible, but the vice versa may not be true. Consider a consistent heuristic function. The heuristic value of the goal is 0. Put this in the consistency equation of all edges inbound to the goal to get new inequalities, which are inserted at other edges that have goal as the grandchild. Iteratively solve all inequalities of consistency that will simply lead to the admissibility criterion, implying that admissibility can be derived from consistency.

The computation time of the A* algorithm is of another concern. Imagine that, somehow, the best-known heuristic function h^* is known for the problem. The assumption is for understanding only; since if the best heuristic function h^* was known for a problem, and for every vertex, the cost to goal was known precisely and one could just follow a greedy mechanism of selecting the closest next vertex to the goal to compute the shortest path. In such a case, the solution would behave greedy. If such a function is known, for any node in the search, the g value is optimal for the vertex being expanded, while the h value is also optimal as per assumption. Meaning, the A* algorithm takes a linear path from source to goal and the *complexity is linear*. In the other extreme case, the worst heuristic function is when we do not know anything about the problem and therefore use a constant heuristic function $h = 0$. In such a case, the A* algorithm becomes the same as the UCS with *exponential computational complexity*. The average case lies in between the two extremes, and the complexity depends on the error in the heuristic function defined as the difference between the estimate and the actual heuristic value or $h^*(u) - h(u)$. The absence of an absolute function in the error is with the understanding that the heuristic function is admissible and hence the error is strictly positive ($h^*(u) \geq h(u)$ as per admissibility criterion). Typically, the heuristic error is very large, and the algorithm tends towards the exponential case, although it can be ascertained that if the heuristic error is bounded within the logarithmic of the optimal heuristic, the A* algorithm can maintain a linear complexity. This is practically difficult to achieve for real-life problems.

h^* is the highest value that the algorithm can accommodate since any higher value will contradict the admissibility assumption and make the algorithm suboptimal, while the higher the heuristic value, the smaller is the error. The aim in the design of a heuristic function is hence to produce as large estimates as possible, while still guaranteeing admissibility (not overestimating). In all discussions here, it is assumed that the heuristic function calculation takes a constant (and small) computation time. For many problems, it may be possible to design very good heuristic functions that make the search time (minus time spent in heuristic calculations) extremely small; however, the calculation of the heuristic function may require a very long time that does not result in an overall computational benefit.

The A* algorithm is *optimally efficient*, meaning that for a given heuristic function and no other information about the problem, no algorithm can guarantee the expansion of fewer nodes than the A* algorithm while guaranteeing optimality. Some algorithms may happen to expand fewer nodes only by tie breaking for specific cases with no overall guarantee. This makes the A* algorithm the best algorithm to use for any optimal search problem already represented as a graph.

4.6.5 Heuristic function design

Given a problem, the challenge is the design of a good heuristic function that maintains as little heuristic error as possible while still being admissible (and consistent in the case of a graph search implementation). The easiest way to design such a heuristic function is called *relaxation*. The method suggests that a greedy design of a solution to a search problem is not possible due to some constraints in the problem; however, if a few constraints are relaxed, the problem may be naively solvable in a constant time. The problem is called the relaxed problem, and the cost of the solution is used as the heuristic function. A constrained solution always has a higher cost than an unconstrained solution, and therefore relaxation results in an admissible heuristic. Similarly, relaxation obeys the consistency criterion as all the solutions represent simpler problems in natural spaces where the law of triangle inequality holds.

Consider the example of computing the path from source to goal for a point robot in some grid map. A relaxed problem does not consider obstacles, for which a simple solution is to go in a straight line from the node to goal with solution cost as the straight-line distance. Going to a place with obstacles generates solutions of higher cost, and hence heuristic is admissible. Consistency is a natural phenomenon while walking in obstacle-free spaces. The *Euclidean distance* is hence a good heuristic function.

A typical AI example is the 8-piece puzzle in which one is expected to arrange pieces numbered 1 to 8 with a blank space in a 3×3 board by only sliding the pieces next to the blank. The relaxed problem considers that a piece can be moved on top of the other pieces (while in the actual problem a piece can only be moved by sliding it into the blank). Considering the relaxed problem where every piece can be directly moved to its destination, the sum of Manhattan distances of all pieces to their desired destination is a good heuristic function. Another relaxation for the same problem is to assume that a unit move is picking up a tile and directly keeping it at the desired place, assuming that the board is broken with no constraints on which piece can move to what place. For this relaxation, the number of misplaced tiles is a valid heuristic function. However, the sum of Manhattan distance shall always be larger than the number of misplaced tiles, and thus the former heuristic dominates the latter, while both heuristics are admissible and consistent. Hence the sum of Manhattan distances is the chosen heuristic function for the problem.

Relaxation does give admissible heuristics and optimal searches; however, in many problems the heuristic error may be large. This can be seen from the example of searching the path of a robot, wherein the path with obstacles will be very long compared to the one without obstacles. A popular mechanism, with the widespread popularity of the machine learning algorithms, is to *learn* a heuristic function instead. Solutions are recorded from numerous sources, including solutions generated by humans from visual perception, searching for solutions from random states for a prolonged time, solutions generated by greedy mechanisms, etc. The states are encoded as a set of features. A mapping of the feature to the solution cost is done using machine learning algorithms. This results in more realistic heuristic functions, resulting in faster searches. However, admissibility and hence optimality cannot be ascertained. The heuristic values can be discounted by some factor depending on the uncertainties in learning, error in learning, uncertainties in data set, possible error margins in data set, etc., to induce some optimality.

The notion of heuristics in the A* algorithm is studied from another perspective. Imagine going from a source to a goal with some obstacles in between. The UCS first expands all nodes with a small cost, then higher cost, and so on. Therefore, if the closed nodes are shaded for visual representation, the search tree appears as circles around the source and the search happens circularly. The concept is studied for the problem of searching a path in a grid for a point robot, with every vertex connected to the 8 neighbours (up, down, right, left, top-right, top-left, bottom-right, and bottom-left). The step cost metric is the Euclidean distance, and hence four straight actions (up, down, right, and left) have a unit cost, while the other four diagonal actions (top-right, top-left, bottom-right, and bottom-left) have a cost of $\sqrt{2}$. The exploration using UCS is shown in Figs 4.11 and 4.12. The closed nodes almost appear as a circle and not a circle because the distance metric used is the cost from the source. Take the outermost points in the search circle. They all have the same distance to the source (with the restriction of 8 possible actions and accounting for obstacles that make the path longer than straight-line distances).

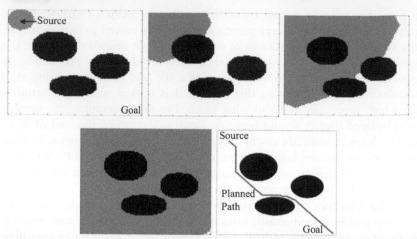

FIG. 4.11 Searching using Uniform Cost Search for Convex Obstacles. The white coloured nodes are unvisited; the grey nodes are in the closed. The search happens as a circle around the source.

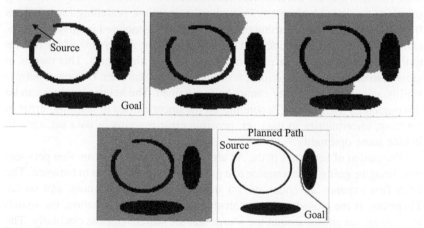

FIG. 4.12 Searching using Uniform Cost Search for a Concave Obstacle.

If one were to search the space with no information about the obstacles, it is imperative to be biased towards the goal, assuming some knowledge is available about the nodes closer and further apart from the goal. This means that the search is elliptical, rather than circular around the goal as a mark of bias towards the goal. This happens in the use of A* algorithm, with the search shown in Figs 4.13 and 4.14. At any time, the width of the ellipse is proportional to the current heuristic error. An error of 0 makes the width of the ellipse 0 with a straight path from source towards the goal, while the highest heuristic error makes the ellipse with the highest width, making it the same as the circle. The notion of optimality is to maintain the width of the ellipse or bias the search

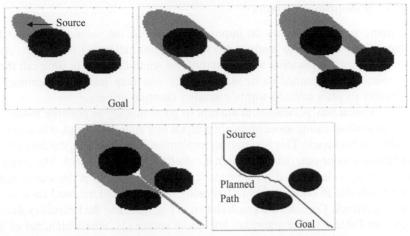

FIG. 4.13 Searching using A* algorithm for Convex Obstacles. The notations are the same as the UCS. The search is more pointed towards the goal, forming an ellipse instead of a circle.

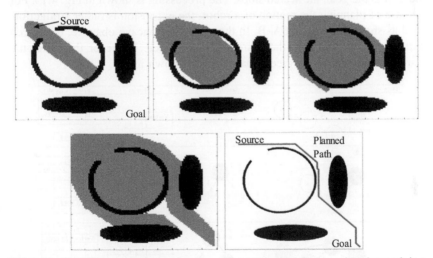

FIG. 4.14 Searching using A* algorithm for a concave obstacle. Concave obstacles result in a significantly higher heuristic error, and therefore the ellipse expands to nearly a circle.

just enough, to keep the search optimal. Fig. 4.13 shows the expansions in case of a convex obstacle wherein the error in the heuristic is small, or in other words, there is a limited difference between the straight-line distance to the goal and the path length avoiding obstacles. However, in the case of a concave obstacle, the heuristic error is exceptionally large as shown in Fig. 4.14. Inside the concave obstacle, the straight-line distance to the goal is small; however, the actual path involves going first outside the obstacle and then following the obstacle boundary until the goal. Hence the search ellipse almost becomes a circle in the region.

4.6.6 Suboptimal variants

For many problems, it may be impossible to search the solution in a small computation time due to many states or small time allocated for search. As an example, when planning a robot, only for a small duration of time can the environment be assumed to be static. Hence one may instead compromise optimality to get a solution within a smaller duration of time.

The fastest way possible is to attempt to go as quickly as possible towards the goal without caring about optimality. Upon reaching a dead end, it is always possible to backtrack. This is simple to implement by prioritizing the fringe by the heuristic value only, and the search is called the *heuristic search*. The search being greedy is fast in reality, although it is still exponential in computational complexity due to critical cases of too many dead ends and the need for a frequent backtrack. Consider the graph shown in Fig. 4.1C and the heuristic values shown in Table 4.1. The algorithm goes from source (1), after which, out of 2 and 3, 2 has a better heuristic value and is therefore selected. Then 2 is processed, and out of 4 and 5, 5 has a better heuristic value. Then 5 is processed, and as it is the goal, the search stops. The processing is shown in Fig. 4.15. For the case of planning the motion of a robot, the expansion is rather straightforward as shown in Fig. 4.16A–B. In the case of a concave obstacle, the search starts to ultimately explore the entire concave obstacle. However, thereafter, the search quickly follows a greedy paradigm as shown in Fig. 4.16C–F.

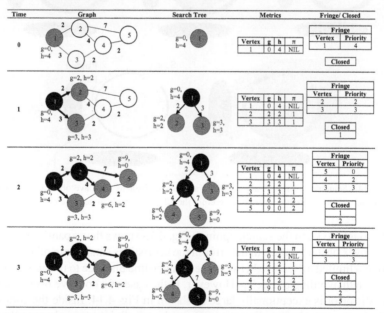

FIG. 4.15 Working of heuristic search. The working and notations are the same as the A* algorithm, with the difference that the priority queue is on the heuristic (*h*) values. The *h* values are supplied as a part of the problem definition (Table 4.1).

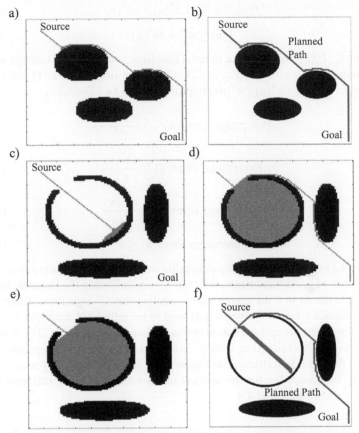

FIG. 4.16   Searching using heuristic search. The notations are the same as the A* algorithm. The search is highly pointed, but sub-optimal. (A–B) Convex Obstacles (C–F) Concave Obstacles. Concave obstacles result in a dead end for all states inside the obstacle and hence an increase in exploration.

The A* algorithm (with a prioritization on $f(u) = g(u) + h(u)$) is slow and optimal, while the heuristic search (with a prioritization on $f(u) = h(u)$) is fast and highly sub-optimal. A middle way out is to use a search with a prioritization in-between the two techniques, called as *ε-heuristic search*. The prioritization is given by Eq. (4.8)

$$f(u) = g(u) + (1 + \epsilon)h(u) \tag{4.8}$$

Here ε controls the heuristic contribution, $\varepsilon \geq 0$. As per the admissibility criterion, $h(u) \leq h^*(u)$. Injecting this in Eq. (4.8) gives Eqs (4.9)–(4.11).

$$f(u) \leq g(u) + (1 + \epsilon)h^*(u) \tag{4.9}$$

$$f(u) \leq g(u) + h^*(u) + \epsilon h^*(u) \tag{4.10}$$

$$f(u) \leq C^* + \epsilon h^*(u) \tag{4.11}$$

Here C^* $(g(u)+h^*(u))$ is the optimal cost from source to goal through u. The percentage error is given by Eq. (4.12), which after using Eq. (4.11) and knowing that $C^* = g(u)+h^*(u) \geq h^*(u)$ simplifies to Eqs (4.13)–(4.15)

$$\text{Percentage Error}(u) = \frac{f(u) - C^*}{C^*} \tag{4.12}$$

$$\text{Percentage Error}(u) \leq \frac{\epsilon h^*(u)}{C^*} \tag{4.13}$$

$$\text{Percentage Error}(u) \leq \frac{\epsilon h^*(u)}{h^*(u)} \tag{4.14}$$

$$\text{Percentage Error}(u) \leq \epsilon \tag{4.15}$$

Eq. (4.15) states that the sub-optimality at every time is limited to a maximum factor of ϵ. As ϵ increases, the search starts imitating the heuristic behaviour, while closer to 0 the search starts imitating the A* algorithm. There is no need to reduce ϵ below 0 since it will tend to become UCS from the A* algorithm. Say a user accepts a solution that is deviated by a factor of 20% from the optimal cost. In this situation, setting $\epsilon = 0.2$ can significantly improve the computation time, while restricting deviation from the optimal cost by a mere 20% that may be acceptable for the application. A summary of the different techniques is shown in Table 4.4.

4.7 Adversarial search

Sometimes an agent must act in adversarial settings, where it is in the environment with another agent that opposes the actions of the agent being planned. The classic example of such a system is games. In games, if the first player wins, the second player loses and vice versa. Such games are called *zero-sum games* because the person winning gets 1 (or the actual score), the person losing gets − 1 (or the negative of the actual score), and the total cumulative return of both players is 0. In case of a tie, each player gets a 0, making the cumulative return 0. Zero-sum games require the players to be adversarial to each other as any bit of cooperation by one player to increase the returns of the other player will reciprocally result in a drop in the returns of the cooperating player by an equal magnitude. The interest here is in sequential games where the players take turns making moves one after the other.

So far, the output of the search was a path, say $\tau = source \rightarrow s_1 \rightarrow s_2 \ldots \rightarrow goal$. However, now after the agent lands at s_1, the adversarial agent will act and affect the environment, which governs the next action of the agent being

TABLE 4.4 A summary of different search techniques.

S. No.	Algorithm	Fringe operating policy	Complete	Optimal	Time Complexity	Memory Complexity
1.	Breadth-First Search (BFS)	Queue	Yes	Yes, for unweighted graphs	$O(b^d)$	$O(b^d)$
2.	Depth First Search (DCS) as a tree search	Stack	No	No	$O(b^D)$	$O(bD)$
3.	Depth First Search (DCS) as a graph search	Stack	Yes for finite state space	No	$O(b^D)$	$O(b^D)$
4.	Uniform Cost Search (UCS)	Prioritized on g	Yes	Yes	$O(b^{1+\lfloor C^*/\varepsilon \rfloor})$	$O(b^{1+\lfloor C^*/\varepsilon \rfloor})$
5.	Depth-limited DFS	In DFS, only nodes with depths less than or equal to limit l are expanded	No (Yes if $l \geq d$, which typically cannot be guaranteed)	No	$O(b^l)$	$O(bl)$
6.	Iterative Deepening	Depth-limited DFS is run for increasing depth limits (l=0,1,2,3,...)	Yes	Yes, for unweighted graphs	$O(b^d)$	$O(bd)$
7.	Iterative Lengthening	DFS is run for increasing cost limits (in DFS, only nodes with costs less than or equal to limit are expanded)	Yes	Yes	$O(b^{1+\lfloor C^*/\varepsilon \rfloor})$	$O(b(1 + \lfloor C^*/\varepsilon \rfloor))$
8.	A* algorithm	Prioritized on $g+h$	Yes	Yes, if the heuristic is admissible (tree search) and consistent (graph search)	Proportional to the heuristic error (cost to goal minus heuristic estimate), linear for best heuristic ($h(s)=h^*(s)$), exponential for worst heuristic ($h(s)=0$)	
9.	Heuristic	Prioritized on h	No as a tree search, Yes as a graph search for a finite state space	No	$O(b^D)$	$O(b^D)$
10.	ε-A* algorithm	Prioritized on $g+(1+\varepsilon)h$, $\varepsilon \geq 0$	No as a tree search, Yes as a graph search for a finite state space	Sub-optimality limited to ε% deviation from the optimal cost	Intermediate between A* and heuristic search controlled by factor ε	

Notations: g (cost from source), h (heuristic estimate to goal), b (branching factor), d (solution depth), C^* (optimal solution cost), D (maximum depth in the problem, maybe infinite), ε (smallest step cost), graph search (use closed that stores all processed states to stop a state from being visited twice; the data structure takes exponential memory), tree search (do not use closed, or any other global bookkeeping variable; only variables in the node are stored).

planned that may no longer be going to s_2. The next action of the adversarial agent will only be known at runtime. Hence, instead of a path, let us say that we compute the best action for every possible state. The output is thus a policy π, which is the best action $\pi(s)$ for any general state s. In other words, adversarial search deals with computation of policy function π.

4.7.1 Game trees

Solving for such games involves making a *game tree*, where the root of the tree is the initial game condition and every layer expands the possible moves that the player has. The *utility* of a node in a game tree is the expected return of the game to both players in that specific state. The agents chose actions that maximize their returns from the game. If the game is zero-sum, the sum of the expected utility of both players is 0 and therefore the utility of only one player may be stored (and that of the other player will be negative, the same value). Therefore, let us call the two players MAX and MIN. The player called MAX attempts to maximize the expected utility. The player is graphically denoted by a triangle. The player called MIN minimizes the expected utility of MAX (thereby maximizing its own expected utility). The player is graphically denoted by an inverted triangle. More specifically, adversarial planning involves planning an agent's behaviour against an opponent. The agent being planned is called MAX, while the opponent is called MIN. The algorithm is correspondingly called the *MINIMAX* algorithm.

The expansion of the game tree involves making all possible moves from a state. The depth-first search is a suited technique considering memory complexity. While expanding by making all possible combination of moves, ultimately a state would be reached where the game terminates as per the game logic. Such a state is called the *terminal state* and the utility of such a state is known as per the game logic called as the *terminal utility*. For a zero-sum game, it is typically 1 if MAX wins, -1 if MAX loses, and 0 in case of a tie, although games could have any range of pay-offs. For the nonzero-sum game, the utility of each playing agent needs to be explicitly noted.

The utility of the terminal states is used to calculate the utility of all other states. Assume that both players play rationally or make the best possible moves. Consider a state s where MAX needs to make a move. Consider the possible actions $A = \{a\}$, such that an action a leads to the state $s' = a(s)$. Let the utilities for all such target states be known (say $U(s')$, where U is the utility function). The agent MAX wants to maximize its utility and therefore selects an action that gets it the best utility from state s. The MIN will similarly try to minimize the utility and can be conversely modelled. This gives the utility of state s as Eq. (4.16):

$$U(s) = \begin{cases} \max_{a \in A} U(a(s)) & \text{if player}(s) = \text{MAX} \\ \min_{a \in A} U(a(s)) & \text{if player}(s) = \text{MIN} \\ U_{\text{terminal}}(s) & \text{if terminal}(s) \end{cases} \quad (4.16)$$

$U_{\text{terminal}}(s)$ is the terminal utility as per the game logic. $\pi(s)$ denotes the best action to take at state s and is given as Eq. (4.17):

$$\pi(s) = \begin{cases} \underset{a \in A}{\text{argmax}}\, U(a(s)) & \text{if player}(s) = \text{MAX} \\ \underset{a \in A}{\text{argmin}}\, U(a(s)) & \text{if player}(s) = \text{MIN} \\ \text{NIL} & \text{if terminal}(s) \end{cases} \quad (4.17)$$

The formulas are recursive, where a larger problem of finding the utility of s is defined as a smaller problem of finding the utility of $s' = a(s)$, knowing the utility of the terminals. The expansion happens in a depth-first search manner to limit the memory complexity, like any recursive implementation. The process resembles human decision-making in games. Humans think, "if I make this move and the opponent makes that move, then I select this as my next move," and so on. They consider all possible combinations of actions along with their outcomes to select a move. An assumption made was that MIN plays optimally. If MIN instead often erred or played sub-optimally, the returns to MAX would be at least the same, or sometimes higher than what is calculated by the algorithm.

The policy may be too large to store in memory and may therefore be calculated when the game is played, with the current state as the root of the game tree. The player calculates a policy for the current state and makes a new move. The opponent makes a move, in which time, the player calculates a policy for the next move, starting from the current state. This eliminates the need to store the policy to limit memory. This also resembles human thinking in games where upon seeing the player's move, humans take a new thinking process only to compute the next move.

However, for most games, the tree is too deep, and calculating the complete tree may be impossible for time factors as well, both offline and online. Therefore, instead of going deep until the terminal state, the search stops at a depth satisfying a qualifying criterion. A typical criterion to use is a maximum *cut-off depth*. The utility of such states must be known, which is not available from the game logic since the game was terminated as premature. The heuristics are used to estimate the winning chances of the player, which serves as the utility. The heuristics may be *logic-based* defining rules of what are good winning patterns and vice versa, or they may be *machine learning based* where the state is the input and winning chance is the output (learned from games played in online matches). It is important that the cut-off happens when the game is in a relatively stable state. If the next move can result in a significant improvement in the score of MAX or MIN (e.g., a major monster shot down and a chess piece acquired), the heuristic utility estimation function may not detect the importance of the move and may incur a huge error in estimation.

4.7.2 Example

Let us take a small example. Practical games are too long and therefore cannot be represented in a short amount of space. As an example, consider the problem of tic-tac-toe, which is played on a 3×3 board with the players making X or O. The first player to completely occupy any row, any column, or any diagonal wins. To save space, let us play the game on a 1×4 board, and getting any two consecutive spaces occupied is called a win. The game tree is shown in Fig. 4.17. At every level, the information shown is the state (denoted by drawing

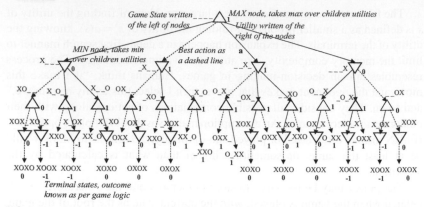

FIG. 4.17 Game Tree for a synthetic problem. A 1×4 tic-tac-toe, 2 consecutive X or O wins. Max node takes a maximum over all children. Min node takes a minimum over all children. Terminal nodes have a utility value known as per the game logic.

the game state), the player who is making a turn (denoted by the triangle or inverted triangle), the utility of MAX at the state (written as a numeric under the state), and the best action to take from the state (drawn as a dashed line).

The ties are arbitrarily handled. A good heuristic in real life is that if you are winning, try to take the shortest route to reduce the chance of making a mistake, and if you are losing, try to take the deepest route to maximize the chance of MIN making a mistake. The *depth* is hence used as a *tiebreaker*.

If at any instant the state reads 1, it means that MAX is guaranteed to win if it plays optimally, irrespective of how MIN plays. If a state reads 0, it means that MAX is at least guaranteed a tie if it plays optimally. MAX may still win if MIN makes an error. The presence of −1 means that MIN is guaranteed to win if MIN plays optimally. If MIN makes an error, anything may happen. At any state, if MAX is asked to make a move, it selects the successor that has the maximum utility value.

4.7.3 Pruning

It may be interesting to observe that in the game tree, many times a branch was expanded even though it appeared to be meaningless for expansion. While the computer programs cannot use common sense, cutting down some of the branches is possible by a technique called *pruning*. Pruning prohibits expansion of certain branches if it is guaranteed that they cannot produce a better result for MAX. As an example, consider branch '*a*' in Fig. 4.18A. The tree is computed in a depth-first paradigm. The root MAX node already has the best value possible by an earlier branch or has discovered the winning formula. It is immaterial to expand any further. The subsequent branches shall hence be pruned.

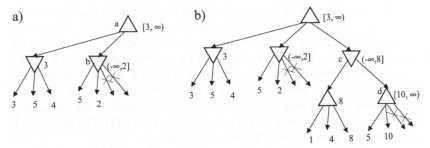

FIG. 4.18 $\alpha\beta$ Pruning. (A) Expand a game tree noting the expectation of every node as children expand. b expects a value less than 2, while a already has a larger value (3). Branches crossed are pruned. (B) d expects values greater than 10, while c already has a lower value (8). Branches crossed are pruned.

A popular means of pruning is the *Alpha-Beta pruning* ($\alpha\beta$ pruning). At every node, keep a track of the range of values expected for a node. Specifically, the α value is the maximum value seen so far at any node in that branch for a MAX. The MAX is trying to maximize the utility and is therefore not interested in anything lesser than α, which has already been guaranteed. Similarly, β is the least value seen in any node in that branch for MIN. MIN tries to minimize the utility and is therefore not interested in anything higher than the current best value of β. Consider the situation shown in Fig. 4.18A. The numbers in bracket denote the range of expected values as per the search so far. The root 'a' is already guaranteed a value of at least 3. However, the node 'b' will produce a value of less than 2. So, there is no point in expanding 'b' further, and all subsequent branches are pruned. Similarly, for Fig. 4.18B, on the expansion of the first child fully, the node 'c' expects a value of 8 or lower, while the second child expansion of node 'd' clarifies that the returned value will be 10 or higher. Hence MIN is never going to pick the node and subsequent values are pruned.

The $\alpha\beta$ pruning is lossless and does not affect optimality. Suppose a heuristic is available that guesses the utility of a node. If a node is known to be poor for a player and the heuristic is confident about the guess, it may be pruned. Pruning should be avoided if the heuristic guess may fluctuate within the next few moves. Alternatively, $\alpha\beta$ pruning depends on the order in which the children are processed. It is better to try better-looking moves first that give a tight value for the bounds ($\alpha\beta$ values), which helps to prune more branches. Even a human player rarely accounts for poor future moves of the player or the opponent.

4.7.4 Stochastic games

Some games are stochastic involving a toss of a coin or a roll of a dice. This is relevant to our problem since even a single robot with slipping items is *playing a game against nature*, which is a stochastic game with nature (slippage or stochasticity) acting as the opponent. The agent knows its moves but does not know the move of nature, which sometimes allows the correct action and sometimes results in a slip. The modelling of the moves of the players remains the same. For the stochastic games, there is a new player, nature, that controls the stochastic elements. The player is denoted by a circle and is called a *chance node*. Say the probability that nature's move leads to a state s' from state s is given by $P(s'|s)$. The utility of a state is called the *total expected utility* and is a weighted addition of all target state's utility, weighted by the probability. The expected utility of the state s is now given by Eq. (4.18). The algorithm is now referred to as the *ExpectiMiniMax* algorithm.

$$U(s) = \begin{cases} \sum_{s'} P(s'|s)U(s') & \text{if player}(s) = \text{CHANCE} \wedge \neg \text{cutoff}(s) \\ \max_{a \in A} U(a(s)) & \text{if player}(s) = \text{MAX} \wedge \neg \text{cutoff}(s) \\ \min_{a \in A} U(a(s)) & \text{if player}(s) = \text{MIN} \wedge \neg \text{cutoff}(s) \\ U_{\text{terminal}}(s) & \text{if terminal}(s) \\ U_{\text{guess}}(s) & \text{if cutoff}(s) \wedge \neg \text{terminal}(s) \end{cases} \quad (4.18)$$

Here *cut-off* defines the criterion to terminate the depth-first search and return the heuristic guess utility or $U_{\text{guess}}(s)$. The terminal utility $U_{\text{terminal}}(s)$ is as per the game logic if s is terminal.

The algorithm is shown as Algorithm 4.6 for the motion of MAX, which is the only function that will be called during the gameplay. The utility value is given by Algorithm 4.7.

Algorithm 4.6 PlayMAX (s).

Input: current game state (s)
Output: best move to make for MAX $(bestMove)$
1 $best \leftarrow -\infty$;   ▷ value of best move so far
2 $bestMove \leftarrow NULL$;   ▷ best move so far
3 **for** *all actions a* **do**
4     $s' \leftarrow a(s)$;   ▷ state after application of move a
5     $v \leftarrow Solve(s', -\infty, \infty, MIN)$;   ▷ utility of s' using DFS
6   **if** $v > best$ **then** $best \leftarrow v, bestMove \leftarrow a$;
7 **return** $bestMove$

Algorithm 4.7 Solves (s,α,β,player).

   Input: current game state (s), best value for MAX player in the current
          branch ($α$), best value for MIN player in the current branch ($β$),
          current $player$ (MAX or MIN)
   Output: utility of the current state for MAX

1 **if** $isTerminal(s)$ **then**   ▷ base condition of recursion
2   ⌊ return score of MAX as per game logic

3 **if** $cutoff(s)$ **then**   ▷ pruning to limit complexity
4   ⌊ return score of MAX as per heuristics/guess/learnt utility function

5 **if** $player{=}MAX$ **then**   ▷ find move with the maximum utility
6    $best \leftarrow -\infty$;
7   **for** $all\ actions\ a$ **do**
8        $s' \leftarrow a(s)$;
9        $v \leftarrow Solve(s', α, β, MIN)$;   ▷ assumed MIN moves next
10   **if** $v > best$ **then** $best \leftarrow v$;
11   **if** $v \geq β$ **then** return $best$;   ▷ $αβ$ pruning
12   ⌊ **if** $v > α$ **then** $α \leftarrow v$;

13 **else**   ▷ player=MIN, move with the least utility value
14    $best \leftarrow \infty$;
15   **for** $all\ actions\ a$ **do**
16        $s' \leftarrow a(s)$;
17        $v \leftarrow Solve(s', α, β, MAX)$;
18   **if** $v < best$ **then** $best \leftarrow v$;
19   **if** $v \leq α$ **then** return $best$;
20   ⌊ **if** $v < β$ **then** $β \leftarrow v$;

21 return $best$

Again, as an example consider a small variant of the tic-tac-toe game with a board size of 1×2. Imagine that a coin is tossed before every move and the outcome decides who makes the next move. If head comes, MAX makes a move, and if tail comes, MIN makes a move. The coin is biased for the head, the probability of which is 0.6. The probabilities are written on the arrows associated with the chance nodes. The game tree because of such modelling is shown in Fig. 4.19. The root has a value of +0.19. Simulate any game any number of

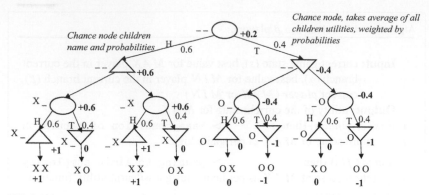

FIG. 4.19 Stochastic Game Tree for a synthetic problem. A 1×2 tic-tac-toe, 2 consecutive X or O wins. 1×2. A biased toss to decide who moves next. The probability of heads (for X to play) is 0.6 and tails (for O to play) is 0.4.

times. The returns for MAX will be $+1$, -1, or 0. It will never be $+0.19$. The value denotes that if the game is played an infinite number of times and the results are averaged, the average returns to MAX will be $+0.19$ as long as MAX and MIN both play optimally, and nature plays as per the assumed probabilities.

It is obvious that if the game is not zero-sum, the utility is an array of returns of both players, and every player maximizes its utility. An interesting behaviour in such games is *cooperation*. Normally in walking one is happy to give a little way to someone, sacrificing his/her objective by a little to make a bigger difference in the objective of another person. So, a player may take an action using which it is at a little loss, but that helps the other player significantly.

Similarly, if the game is multiplayer, the utility is the pay-off for every player and the players maximize their objectives. However, it is common to see *alliances*, where a few players unite to play against the others. The players play to maximize the returns of one or a few of the alliance members, even if those move give the player a very poor individual return. The players also attempt to minimize the returns of the players not in the alliance. Once and if the others are practically out of the game, then the alliances are broken, and the alliance members play against each other. Such social behaviours are hard to model in programs.

Questions

1. Solve the graphs given in Fig. 4.20 using (a) breadth-first search, (b) uniform cost search, (c) heuristic search, and (d) A* Algorithm. The heuristic values for each graph are given aside the graph. Also, find out whether the heuristic is admissible and consistent. Hence comment on which of these algorithms are optimal.

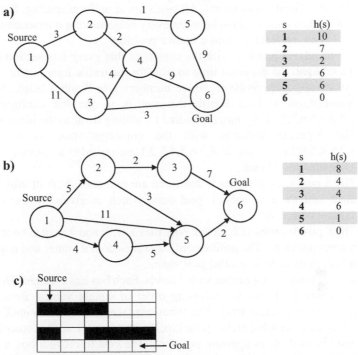

a)

s	h(s)
1	10
2	7
3	2
4	6
5	6
6	0

b)

s	h(s)
1	8
2	4
3	4
4	6
5	1
6	0

c)

FIG. 4.20 Sample maps for the calculation of the shortest path. (A) Problem 1, (B) Problem 2, (C) Problem 3. For (C), assume that the robot can take 8 neighbouring actions with a Euclidian step cost function, and design a good heuristic function.

2. Give the heuristic function for the following cases:
   - A robot needs to go to the goal (G_x, G_y), which is located on the same floor of the building at which the robot is standing.
   - A robot is asked to go to any coffee shop. There are two coffee shops on the current floor of the building located at (s_{1x}, s_{1y}) and (s_{2x}, s_{2y}).
   - A robot needs to go from the 2nd floor to a location (G_x, G_y) at the ground floor of a building, for which the robot may take a lift at the location (l_x, l_y). The lift takes 10 s to go from the 2nd floor to the ground floor. The robot moves at a uniform speed of v m/s.
   - A robot needs to go from the 2nd floor to a location (G_x, G_y) at the ground floor of a building, for which the robot may take either lift at the location (l_x, l_y) or stairs at the location (s_x, s_y). The lift takes 10 s to go from the 2nd floor to the ground floor with an average waiting time of 30 s. Taking stairs takes 40 s to climb down to the ground floor from the 2nd floor. The robot moves at a uniform speed of v m/s.
   - Driving around a well-planned city where all roads are either parallel or perpendicular to each other, and perpendicular roads intersect each other

at well-defined intersections. All roads are at uniform spacing. Suppose that it takes 1 min to drive between every neighbouring intersection. The goal is an intersection with a known position.

- For the above problem with the constraint that every intersection has a traffic light, the expected time to wait for the traffic light is 10 s.
- For the 8-puzzle problem, where numbers 1 to 8 and a blank (b) are arranged on a 3×3 board, the aim is to get the configuration 1,2,3,4,5,6,7,8,'b' by moving pieces by sliding them to the blank only.
- The 8-puzzle problem with the constraint that arrangements 1,2,3,4,5,6,7,8,'b' and 'b',8,7,6,5,4,3,2,1 are regarded as correct, where 'b' stands for a blank.
- The 8-puzzle problem such that there are two ones, both of which are alike and there is only 1 goal state which is given by the order: 1,1,2,3,4,5,6,7,'b'.
- The 8-puzzle problem such that only tiles 1, 3, 5, and 7 need to be at their correct positions. The positions of other tiles do not matter and any permutation of these is a valid goal state.
- Sorting boxes using a robot with 2 hands. Each box has a weight. The aim is to sort the boxes in increasing order of weight values. Boxes are arranged as a linear array. The robot can take commands $swap(X,Y)$ in which case the robot picks up the box X with 1 hand, Y with another hand, and places them in opposite positions. The cost to move a box b from index i to index j is given by $w_b(c + \text{abs}(i-j))$, where w_b is the weight of the box, c is a known constant, and $abs()$ is the absolute value function. When moving 2 boxes simultaneously, the cost of both boxes is added.

3. State whether the following problems can be solved optimally by the named algorithm or not. If the problem can be solved by the named algorithm, also propose a solution in one or two lines of text. If the problem cannot be solved by the named algorithm, propose a new algorithm in one or two lines of text, which may not be optimal.

- A* algorithm to make a robot go from point a to b in a grid map. The search takes 10 s to complete; the navigation from a to b by the shortest route takes 10 s, while the map changes by small amounts every 5 s.
- A* algorithm to make a robot go from point a to b in a grid map. The search takes 2 s to complete; the navigation from a to b by the shortest route takes 10 s, while the map changes by small amounts every 5 s.
- A* algorithm to make an agent go from point a to b in a grid map. The search takes 10 s to complete; the navigation from a to b by the shortest route takes 10 s; the map is static; however, there is one person who keeps encircling the place at a fixed speed.

- Breadth-first search for the problem of planning a route from home to office, given the road network map.
- Breadth-first search for the problem with a deep solution and with a large search space that will exhaust the system memory.

4. Consider that the A* algorithm is used for problem-solving over a reasonably sized graph. Suppose the heuristic function h is known to be admissible and consistent. Sort the following heuristic functions in increasing order of their likely computational time. Which of the following heuristic functions are likely to provide optimal paths? (a) $h/2$, (b) $h \times 2$, (c) h, (d) 0 (heuristic $= 0$ for all states).

5. Suppose the heuristic functions h_1 and h_2 are known to be admissible and consistent. Sort the following heuristic functions in increasing order of their likely computational time. Which of the following heuristic functions are likely to provide optimal paths? (a) $h_1 + h_2$, (b) $\max(h_1 - h_2, 0)$, (c) $\max(-h_1, h_2)$, (d) $\min(h_1, h_2)$, (e) $h_1/2$, (e) $h_1 \times 2$.

6. Solve the game tree shown in Fig. 4.21. Also, encircle the nodes which will be pruned when solved using $\alpha\beta$ pruning.

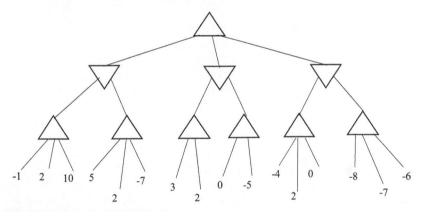

FIG. 4.21 Game tree for question 6.

7. [Programming] Solve the following classic AI problems using the A* algorithm. Search the web for problem descriptions. (a) 8 tile problem, (b) Missionaries and cannibals problem, (c) Water jug problem.

8. [Programming] Make a black and white bitmap image and read it in a programming language. The black represents the obstacle and white represents the free areas. Every pixel is a grid for a point robot. Every pixel is connected to the 8 neighbours with a Euclidean step cost. Write a program using the A* algorithm to go from a prespecified source to a prespecified goal, avoiding obstacles.

References

Choset, H., Lynch, K.M., Hutchinson, S., Kantor, G., Burgard, W., Kavraki, L.E., Thrun, S., 2005. Principles of Robot Motion: Theory, Algorithms, and Implementations. MIT Press, Cambridge, MA.

Cormen, T.H., Leiserson, C.E., Rivest, R.L., Stein, C., 2009. Introduction to Algorithms, third ed. MIT Press, Cambridge, MA.

Hart, P.E., Nilsson, N.J., Raphael, B., 1968. A formal basis for the heuristic determination of minimum cost paths. IEEE Trans. Syst. Sci. Cybern. 4 (2), 100–107.

Russell, S., Norvig, P., 2010. Artificial Intelligence: A Modern Approach. Pearson, Upper Saddle River, NJ.

Tiwari, R., Shukla, A., Kala, R., 2012. Intelligent Planning for Mobile Robotics: Algorithmic Approaches. IGI Global, Hershey, PA.

Chapter 5

Graph search-based motion planning

5.1 Introduction

The problem of robotics can be solved by the classic search paradigm of artificial intelligence, and a convenient way to do so is to encode the problem as a configuration with the configuration space as the set of all possible configurations (Choset et al., 2005; Tiwari et al., 2012). A robot configuration may be collision-prone, making the obstacle-prone configuration space to be avoided by the robot. The free-configuration space consists of all configurations where the robot collides with neither an obstacle nor itself. The configuration space is discretized into grids, and every neighbouring grid is said to be connected. This transforms the problem into a graph search problem.

However, the search here has two major problems. The first is that the map is assumed to be static, while the map may change with time. If the map changes during the searching or execution time, the path of the robot is invalidated and the search needs to be re-performed. In such cases, it is important to reuse as much past information as possible to make the re-planning efficient.

Furthermore, the search itself over the complete configuration space can be extremely computationally expensive, making it impossible to plan in near-real-time for the robot to solve realistic problems. The search is therefore made more efficient, primarily by the use of good heuristics. This chapter is devoted to underlying all practical problems that hinder the application of the search methodology for practical motion planning and solving the same.

5.2 Motion planning using A* algorithm

This section applies the basic concepts of the A* algorithm (Hart et al., 1968, 1972; Russell and Norvig, 2010) to the problem of motion planning. This section highlights a few important issues not discussed elsewhere (Kala et al., 2009, 2010, 2011).

Autonomous Mobile Robots. https://doi.org/10.1016/B978-0-443-18908-1.00008-X

5.2.1 Vertices

Before solving the problem, the first issue is to convert the problem into a graph. It is already assumed that the environment is static and fully observable, the problem is for a single agent, and the robot follows the commands reasonably accurately. It is assumed that the problem is already available as a configuration space with a source and goal specified in which the solution needs to be searched for.

The first issue is to convert a continuous configuration space into a discrete space, which is done by packing cells throughout the space. Such a conversion is natural in information technology, wherein a continuous visible world is discretized into pixels by taking a picture, or a continuous sound signal is discretized by sampling, and likewise. All the more, the specification of objects in the real world constituting the world map is discretized and stored in the system.

The world map with spatial coordinates will not have the same axes as the configuration space that includes the rotations along with the positions of the robot, with the configuration space being typically higher dimensional in contrast to the workspace. Even if it is a point robot whose world map is the same as the configuration space, the map may be made by rich cameras with high resolutions, while it may not be possible to compute the path in such high resolutions in small computation times. Therefore, a discrete map may need to be reduced in resolution so that a search algorithm can compute a path within the permissible time limit.

It may be noted that the environment can only be assumed to be static for a limited period and the algorithm must return a path within the same duration. The best way out is to compute fast, make short plans, and execute short plans frequently. This requires a small computation time. In any case, it is not possible to have the robot wait for a few minutes to compute the path.

After the map is discretized, all the discrete cells become the vertices of the map. Only the cells that are outside obstacles are considered to be vertices. Sometimes a cell may be partly within an obstacle and partly outside the obstacle. In such a case, one may take an *optimistic approach*, where a cell is considered obstacle-free if at least some parts of the cell are collision-free; or one may take a *pessimistic approach*, where a cell is said to be obstacle-prone if any segment within the cell has an obstacle. For safety reasons, a pessimistic approach is more desirable. A middle-way out is to consider the *ratio of obstacle occupancy* within the cell and to search for a path that is short and simultaneously has a low obstacle occupancy ratio. This makes the search bi-objective in nature that can be converted into a single-objective search by a weighted addition. The problem is worsened if the resolution is small. Smaller resolutions result in faster searches, however, result in suboptimal paths and ambiguity regarding the occupancy status of cells.

A greater problem associated when the resolution is poor is associated with *narrow corridors*. Imagine a narrow corridor as shown in Fig. 5.1. A finer resolution sees the corridor and finds a way in-between. However, a coarser resolution has cells that overlap both ends of the narrow corridor and therefore the cell is taken as obstacle-prone. This results in a loss of completeness; as even if a path exists, the algorithm for the given resolution will not be able to find it. It may appear that the case of such narrow corridors should be unlikely in a real-world setting. However, imagine a modest size robot going in a corridor with some obstacles. The size of the corridor in the configuration space is the available corridor width between the obstacles minus the size of the robot, which can be reasonably small. It is common to have buildings with narrow doors in which the robots can just seep in, or to have narrow lanes in which self-driving cars need to perfectly align and get in, or to have a manipulator pick up an object cluttered with other objects (obstacles) all around. In all such cases, the resolution loss can lead to a lack of completeness.

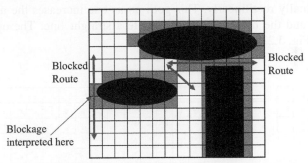

FIG. 5.1 Narrow corridor problem. Continuous space is discretized into cells. Grey cells (partly occupied by an obstacle) are taken as an obstacle. Due to resolution narrow corridors are blocked.

In the worst case, the search explores all vertices or the complete search space. Assume that the search space has d dimensions. Each dimension is normalized to have values between 0 and 1 and divided into r cells. So, the time and space complexities are the total number of vertices which is $O(r^d)$. Higher resolutions result in a much slower search but yield better paths. Smaller resolutions are faster, however, yield poorer paths.

5.2.2 Edges

A cell is connected to all neighbourhood cells to form the edges. Only those edges are added that are completely collision-free between the connecting vertices. Here, as well there is a big design issue about how large the *neighbourhood function* should be. For simplicity consider a 2-dimensional space. The smallest neighbourhood function is to connect the 4 surrounding cells (up, down, right, and left). A little larger neighbourhood function connects the diagonal cells as well with 8 ($3^2 - 1$) neighbours. It is also possible to have a vertex connect to all 24 ($5^2 - 1$) neighbours that include all cells in the 5×5 block. Higher options also include 48 ($7^2 - 1$), 80 ($9^2 - 1$), and so on. The different neighbourhood functions are graphically shown in Fig. 5.2A. The number of neighbours constitutes the number of edges per vertex or the branching factor and constitutes the *action resolution* of the algorithm. A neighbourhood function of 4 can only travel rectilinearly, along the 2 axes, and therefore even if the map has no obstacles, the robot can only take the Manhattan route. In the case of 8 neighbours, the robot can travel along the axis and diagonals, meaning all angles in multiples of $\pi/4$. So, in an obstacle-free map, if the goal is not an angle of $\pi/4$, the robot will move diagonally approaching it, and walk rectilinearly for the rest of the journey. A neighbourhood function of 24 is more flexible allowing all multiples of $\tan^{-1}(1,2)$, $\pi/4$, and rectilinear motion. The robot may still not reach the goal in a single line, in which case it will travel first by the closest angle to $\tan^{-1}(1,2)$, then by the closest angle to $\pi/4$, and finally rectilinearly. This goes on as one increases the neighbourhood size and the motion starts to imitate a straight line. The options are shown in Fig. 5.2B.

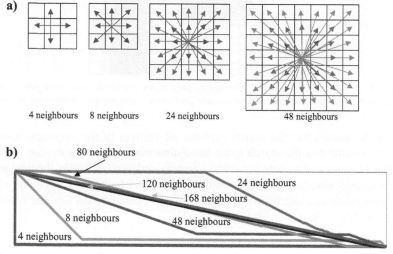

FIG. 5.2 Different neighbourhood functions for a 2D planning problem. (A) Graph. An edge is feasible if it is entirely away from the obstacle. (B) As the number of neighbours increase, path length reduces, and computation time increases. Experiments are done on 4 and $(2n+1)^2 - 1$, $n = 1, 2, 3, 4, 5$ and 6 neighbourhood functions.

The computational complexity of the UCS is $O(b^{1+\lfloor C^*/\varepsilon \rfloor})$. As the neighbourhood size increases there is an increase in b and a reduction in C^*. Initially, the reduction in C^* is significant, but soon the action resolution is rich enough and therefore the branching factor is of a larger concern. This does depend upon the complexity of the environment, as in simple environments the heuristics can be very reliable and the search can be fast with a large neighbourhood function as well. The same is studied by experiments by increasing the neighbourhood size. The paths are shown in Fig. 5.3, while computation time initially reduces and then increases.

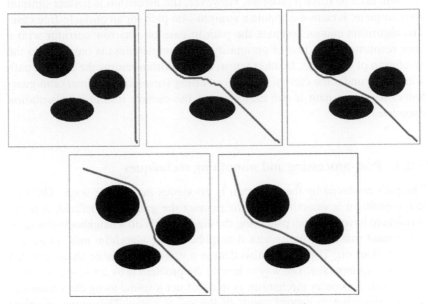

FIG. 5.3   The paths become shorter on increasing the action resolution.

An edge connecting a vertex u to a vertex v is collision-free if the complete straight line connecting u to v does not collide with an obstacle. It may be possible that both vertices u and v are collision-free, however, some intermediate cell in the straight line between u and v incurs a collision. For the graph, the weight of the edge may be taken as the Euclidian distance between the vertices. In such a case, the heuristic function then is the Euclidian distance from the node to the goal. For the example of a 2D configuration space with neighbourhood connectivity, better heuristic functions are possible. For the 4-neighbour case, the Manhattan distance is a valid heuristic function since the robot can only travel rectilinearly. For the 8-neighbour case, the heuristic function is made by removing obstacles, making the robot reach goal by using only the diagonals, and then travelling the rest of the distance remaining rectilinearly, which has a simple closed-form solution. Similarly, other neighbourhood functions can be dealt with. The cost function can be amended to include the clearance terms, in which case the algorithm does not get too close to the obstacles. Clearance of a point is the distance to the closest obstacle and the term attempts to make the robot stay further away from obstacles. Since graph search cannot handle multi-objective searches, using a weighted addition is a feasible criterion. The configuration space-specific distance functions may be used as well.

It may be counter-intuitive that an optimal and complete search algorithm, A*, was used to solve a problem. However, the algorithm is neither optimal nor complete. It cannot compute a straight-line path for an obstacle-free case. The algorithm cannot compute the path in case of a narrow corridor with a poor resolution. The loss of optimality and completeness is only due to the resolution of operation. In other terms, the algorithm returns the optimal path admissible under the current resolution setting (*resolution optimal*) and guarantees to give a path if one exists as per the current resolution (*resolution complete*).

5.2.3 Post-processing and smoothing techniques

The path produced by the algorithm is erroneous in multiple ways. The first major problem is smoothness. In the manner the graph was defined, it is not possible to have a smooth path using the algorithm. With 4 neighbours the algorithm must make $\pi/2$ turns, with 8 neighbours the algorithm must make $\pi/4$ turns, and so on. These turns also denote a non-differentiable sharpness that the robots cannot take. The way to handle the problem is by using *smoothening* operations. A popular mechanism is to construct a spline using the vertices of the A* algorithm as control points of the *spline* curve. The output of spline smoothing on a sample trajectory is shown in Fig. 5.4. When using a high resolution, the control points of the spline can be sampled down from the set of all points on the trajectory, since using all points will result in sharp trajectories. The splines smooth the trajectory by distorting the original trajectory. There needs to be a balance between the two-opposing criterion of keeping the transformed trajectory as close as possible to the original trajectory and smoothing the trajectory as much as possible (which may make the trajectory different from the original one). The balance between the objectives can be controlled by the degree of the spline. During the search, it must be ensured that enough clearance is left to keep the trajectory collision-free on being subjected to small deformations. Clearance denotes the distance between the points in the trajectory to the obstacles which must be maximized. A way of achieving this is by specifying a minimum clearance C_{min} and only accepting states that have this clearance. This further restricts the robot from entering a narrow corridor, and hence, results in a lack of completeness. Taking weighted addition with clearance is an option, which again makes the weight parameter a little hard to set.

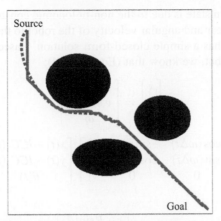

FIG. 5.4 Smoothing of the trajectory. The sampled A* output is smoothened by using a spline curve.

A better way of dealing with smoothness is to take an objective function and locally optimize the trajectory, taking the A* computed trajectory as the initial seed. At every iteration, the local optimizer moves the control points of the spline by a small amount, deletes some points, or adds some points in the middle of the existing points to create a new temporary trajectory. The better trajectory of the two trajectories is retained for the next iteration, judged as per the objective criterion. The optimization-based approach is discussed in Chapters 14 and 15.

5.3 Planning with robot's kinematics

A problem associated with the approach so far is neglecting the kinematics. Imagine a wheeled mobile robot tracing the trajectory. The algorithm may conveniently ask the robot to travel sideways, which is not possible for the robot. The infeasibility may be corrected by smoothening to some extent; however, the algorithm does not guarantee that it will be feasible to trace the trajectory. A solution to the problem is, hence, to compute trajectories that are feasible during the search process. The same can be done by incorporating the robot kinematics in the search process. The search happens in the *state space* and not the configuration space. Consider a very small round robot that moves by a differential wheel drive mechanism. The configuration only consists of the (x,y) variables and so the configuration space is 2D in nature that appears very similar to the workspace. However, the state space consists of the kinematics or the derivative of the variables as well (or the velocities as well). Consider a differential wheel drive robot that moves by creating a differential of the rotations of the left and the right wheel. For the same robot, the state is given by $(x, y, \theta, \dot{x}, \dot{y}, \dot{\theta})$. However, since the robot is constrained to only move forward instantaneously, it is given by (x,y,θ,v,ω), where v is the linear velocity, ω is the angular velocity, and the reduction in state is due to the non-holonomic constraint $tan\theta = \frac{\dot{y}}{\dot{x}}$. The mapping to the linear and angular velocity of the robot to the rotation speed of the robot's wheel has a simple closed-form solution. Based on the kinematic Equation of the robot, we know that (Eqs 5.1–5.4):

$$\begin{bmatrix} \dot{x} \\ \dot{y} \\ \dot{\theta} \end{bmatrix} = \begin{bmatrix} v\cos(\theta) \\ v\sin(\theta) \\ \omega \end{bmatrix} \tag{5.1}$$

$$\begin{bmatrix} x(t+\Delta t) \\ x(t+\Delta t) \\ \theta(t+\Delta t) \end{bmatrix} = \begin{bmatrix} \cos(\omega\Delta t) & -\sin(\omega\Delta t) & 0 \\ \sin(\omega\Delta t) & \cos(\omega\Delta t) & 0 \\ 0 & 0 & 1 \end{bmatrix} \begin{bmatrix} x(t) - ICC_x(t) \\ y(t) - ICC_y(t) \\ \theta(t) \end{bmatrix} + \begin{bmatrix} ICC_x(t) \\ ICC_y(t) \\ \omega\Delta t \end{bmatrix} \tag{5.2}$$

$$ICC = \begin{bmatrix} x - R\sin(\theta) \\ y + R\cos(\theta) \end{bmatrix} \tag{5.3}$$

$$R = \frac{v}{\omega}, v = \frac{v_L + v_R}{2}, \omega = \frac{v_R - v_L}{l} \tag{5.4}$$

Here, the two controls are based on the left and right wheels of the robot with velocities v_L and v_R. The kinematic relation is defined for a robot turning on a circle with centre ICC for Δt unit of time with a radius R. l is the distance between the wheels. Let the equation be summarized as $s(t+\Delta t) = f(s(t), u, \Delta t)$, where s is the state vector, $u = (v_L, v_R)$ is the control input and Δt is the time for which the control input is applied. The number of control inputs possible is $[-v_{max}, v_{max}]$ for the left wheel and $[-v_{max}, v_{max}]$ for the right wheel, leading to a continuous control speed of $[-v_{max}, v_{max}]^2$. However, the space or the set of controls needs to be discretized for search, and let us sample some control inputs out given by $u_1, u_2, u_3, u_4 \dots u_m$, where m is the maximum number of actions and is a resolution parameter. As an example, if $m=8$, the sampled controls represented by (v_L, v_R) are $u_1 = (v_{max}, v_{max})$, $u_2 = (v_{max}, 0)$, $u_3 = (v_{max}, -v_{max})$, $u_4 = (0, v_{max})$, $u_5 = (0, -v_{max})$, $u_6 = (-v_{max}, v_{max})$, $u_7 = (-v_{max}, 0)$, and $u_8 = (-v_{max}, -v_{max})$.

The search happens without discretizing the state and the tree-search variant of the search paradigm is used wherein a node is not checked for a revisit. This means no *closed* map checks for a revisit. The search tree is rooted at the source state. The expansion of any state s produces m children (not all feasible), given by $s_1 = f(s, u_1, \Delta t)$, $s_2 = f(s, u_2, \Delta t)$, $s_3 = f(s, u_3, \Delta t)$, ... $s_m = f(s, u_m, \Delta t)$. Δt is another resolution parameter. The feasible children are added into the search tree with an action cost taken as per the requirements of the problem. Typical action costs can be the path length, smoothness, jerk, acceleration required, fuel cost, etc., all of which are computable easily since the kinematics of the vehicle from source to the child are known. When searching for a solution directly in the configuration space there were limited options of path cost as the kinematics of the robot was not encountered. Therefore, any reference to velocity and cost terms involving velocity could not be accounted for. The search only carried out path planning and the assignment of velocities was leftover. While planning in state space, the kinematics is known and therefore any kinematic-based cost function can be designed. The collision checking can no longer be done by interpolating a straight line and needs simulation of the vehicle from the parent to the child, which can be more quickly done by modern collision checking libraries that account for moving bodies. Since the state is not discretized, the goal test criterion is relaxed to the robot reaching a state close enough, rather than precisely moving onto the goal. The search tree so produced is given in Fig. 5.5. Fig. 5.5A shows a dummy tree while the search tree over the robotic grid map is shown in Fig. 5.5B.

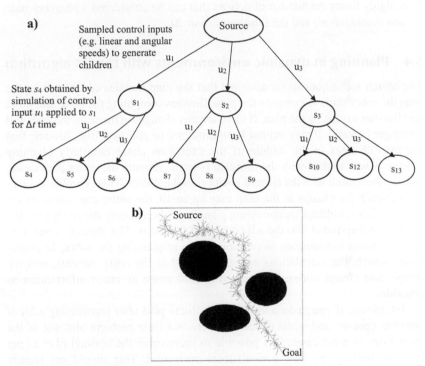

FIG. 5.5 Search tree using search with robot kinematics. (A) Search Tree. (B) Visualization on a workspace. Edges are naturally smooth and the path obeys kinematic constraints.

The search tree edges are already annotated with the actions taken (control signal or speeds used), which incorporates the robot kinematics. Therefore, in principle, the robot can be given control signals from source to goal as mentioned in each edge in the same order, and the robot shall reach the goal. The speed constraints are already incorporated. However, there are factors not accounted for, for example, maintenance of exact speed, acceleration constraints, friction, dynamics, etc. This makes a deviation between the actual trajectory and the computed trajectory. The trajectory computed is open-loop control, while a feedback-based closed-loop control is needed to account for such uncertainties. However, the state-based search produces feasible trajectories that do not require post-processing.

The problem with the search technique is adding dimensions into a search that is already exponential in terms of the number of dimensions with complexity $O(r^d)$, where d is the number of dimensions and r is the resolution. This means that adding variables to the search can highly increase the computation cost. Assuming that the algorithm minimizes travel time, the optimal cost as T (time to reach the goal), and every action is applied for Δt time. In terms of the search tree depth, the computation complexity is given by $O(m^{1+\lfloor T/\Delta t \rfloor})$. This highly limits the number of actions that can be considered with every state (action resolution m) and the time resolution Δt.

5.4 Planning in dynamic environments with the D* algorithm

The search techniques so far assumed that the map remains constant from the time the robot starts to compute the plan, finishes computing the plan, and starts and finishes executing the plan. If there are any changes in the map, the path will no longer be valid. It may neither be optimal nor be collision-free. Assume that the map changes in the middle of the execution phase (similarly planning phase). A way to deal with the problem is to re-compute the entire path again, starting the search with the current position as the source. This is problematic as even though the change in the map may be small, the entire computations are redone. This is unlike a human doing planning, wherein only the changes in the map are incorporated into the adjustment of the plans. The attempt is hence to reuse as many calculations as possible while re-planning the robot. In graph-based search, the calculations are summarized as the *costs*, *parents*, *actions*, *fringe*, and *closed* nodes. The search should reuse as much information as possible.

Intuitively, if you make a very good holiday plan after considering a lot of possible options, and while executing there is a little problem and one of the steps fails, in most cases, it is possible to reconstruct the optimal plan as per the new findings by only a small local correction. This should not require the weeks-long planning done initially. Even in the context of robotics, if there is a small obstacle in the way of the robot from the source to the goal; mostly only a small correction needs to be made to rectify the plan.

The concepts introduced here will be only applicable to small changes in the map. Large changes in the map can be re-computed by the algorithms discussed here; however, the performance will be nearly as poor as a complete re-planning. A robot on the run should not be asked to wait for a prolonged time for re-calculations due to changed maps. If the map is changing by a small amount at a high frequency, this philosophy will again not work as the map would have further changed by the time the previous changes are updated. The re-planning must happen faster than the rate at which changes are made in the environment. Very quick changes in the environment can only be handled by the reactive motion planning paradigm. A summary of the different algorithms is given in Box 5.1.

Box 5.1 A summary of different algorithms.

Re-planning: Adjust plan if the map changes from the start of the planning phase to the end of the execution phase, *reusing* information (costs, parents, fringe, closed) from the planning algorithm

Backward Search. Can only be applied when:
- Actions are invertible (up becomes down), which may not be possible when planning with robot kinematics
- There is a single goal (or finitely small number numbers of goals), which is not possible when the robot should reach a goal position with any orientation.
- Cannot be used when the map/cost function depends on time, e.g. obstacles with known dynamics (another robot) can be forward simulated and forward search is possible but backward is not.

D* Algorithm
- Corrects costs and parents for changed maps
- Can handle small changes only (re-planning time is proportional to changes) and changes with a very small frequency (re-planning should be complete till the map further changes)
- Highly dynamic elements are handled by reactive algorithms, moderate dynamics are handled by trajectory local correction algorithms, small dynamics by D* like algorithms.
- Uses a backward search (goal to source) so that the metric *cost from goal* is invariant to changing robot positions (source) enroute.
- It is possible to revisit a node with a better cost in closed, which is processed and re-inserted into the fringe.
- *h*-value: cost to the goal (computed exhaustively by search and not a heuristic estimate).
- *k*-value: smallest cost from goal seen by a node.
- Fringe is prioritized on *k*-values.
- Focussed D* adds a heuristic function to D*.

Lifelong Planning A*
- Find edges affected by map changes and correct indefinitely.
- *g*-value: cost from goal (backward search)
- *rhs*-value: 1-step lookahead cost from goal ($rhs(v) = \min_{u,a}(g(u) + c(v,a,u))$).
- *k*-value, $k(v) = [\min(g(v), rhs(v)) + h(v),\ \min(g(v), rhs(v))] = $ [primary priority criterion, secondary priority criterion]
- If $g(u) = rhs(u)$, consistent, nothing to do

Continued

Box 5.1 A summary of different algorithms—cont'd

- If $g(u) > rhs(u)$, over-consistent, update g-value, which may make neighbours inconsistent, added to queue
- If $g(u) < rhs(u)$, under-consistent, invalidate g-value (set to ∞), and reprocess u and neighbours, added to queue
- Queue (of inconsistent nodes) prioritized on k-value

D*-Lite

- Extends Lifelong Planning A* for robot motion
- In a backward search, the heuristic estimate is the cost from the source (current robot position), and the historic cost is the cost to the goal.
- If the robot moves by distance s, the source changes, which reduces heuristic components of all items in the priority queue by a maximum of s.
- Instead of reducing s to all elements of the priority queue, s is added to all future insertions.

Anytime A*

- A solution available anytime, more time means a better solution
- $f(u) = g(u) + h(u)$: cost from source to goal through u [same as A*]
- $f'(u) = g(u) + (1 + \varepsilon) h(u)$, $\varepsilon > 0$ [priority of queue]
- Since priority is on f', a suboptimal goal is seen very early.
- Revisiting a node (even in closed) with a better f-value, reprocesses it by insertion into the fringe. So, each node, including the goal, is revisited with better cost with time.
- A node with an f-value larger than the current best cost to the goal is pruned from search.

Theta*

- Paths in the grid are not discretized to any angle
- In expansion of v from u, also try a direct connection from parent of u or $\pi(u)$ to v

LRTA*:

- Learning Real-Time A*: The robot moves as it calculates every step of the search. Can be used for immediate motion and in partially known maps.
- Move from u to v by greedy selection of 1-step lookahead value ($\arg\min_a c(u, a, v) + h(v)$)
- $h'(u) = \min_a(c(u, a, v) + h(v))$ is a valid heuristic estimate for u, learn the estimate $h(u) \leftarrow \max(h(u), h'(u))$
- For a tree search, the algorithm may visit a node several times, updating (increasing) the heuristics of the nodes, eventually leading to the goal.
- For a graph search, prune revisited nodes. Greedy action selection may result in a deadlock. Perform backtracking by a physical motion to the next node.

ALT Heuristics

- Multi-query (multiple source-goal pairs for the same map) problem
- Select landmarks (l) and compute distance from landmark to everywhere offline
- Heuristic of node u to goal G, is given by $h(u, G) = \max_l d(l, G) - d(l, u)$, where $d(l, u)$ is the pre-calculated optimal distance from landmark l to u.

5.4.1 Using a backward search

The *D* algorithm* (Stentz, 1994), also called the *Dynamic A* algorithm*, adapts the changes in the map while working in the same philosophy as the A* algorithm. To do so, let us first introduce a basic change in the algorithm, which is to search a path *backward* from the goal to the source instead of the other way around. The reversal does not make any difference as the trajectory can always be reversed. In a forward search from state s, all possible actions a are applied to get states s', or $s' = a(s)$. In a backward search, for the expansion of a state s, we need to find all states s', which on the application of an action a produces s, or $s = a(s')$, $s' = a^{-1}(s)$, implying that the actions must be invertible. Up becomes down, right becomes left, and so on. An exception where this does not hold well is the state-based search of a robot (Section 5.3), which uses a forward kinematic equation of motion. The forward motion cannot be reversed to get the same reverse motion, or the kinematics cannot be inverted. Knowing sampled linear (v) and angular (ω) speeds applied at a state s, one can get s'; however, it is unknown what controls (v, ω) should be applied on what states s', such that the state s is achieved. Strictly speaking, Eqs (5.1)–(5.4) did not apply acceleration constraints and the kinematics in the presented form can be inverted, although a general case in such robotic systems assumes that the kinematics is non-invertible.

Another example of when backward search cannot be applied is when there are multiple goals. Imagine only goal position (and not orientation is important) when planning a rectangular robot. Now, there are infinitely many goals with the given position and any orientation. Another example where backward search cannot be applied is when the map changes with time with a known function. Imagine a person is moving in the robotic environment, but the person's position at every instant of time can be accurately predicted. Consider the time from source to a state s is $T(s)$ and the time to reach a state s' using an edge s-s' is $T(s')$, which can both be computed for a forward search. The feasibility of expansion of the edge s-s' can be done as the map (with the person) is known at every instant of time. Simultaneously, simulating both robot traversing from s to s' and the person traversing as per known plan from time $T(s)$ to $T(s')$, and checking if this simulation succeeds without a collision is the feasibility check. However, the time of reaching the goal or $T(goal)$, and hence the map at the root in a backward search is not known, due to which no feasibility check can take place.

The problem with the D* algorithm is re-computing the costs and parents when the map changes while the robot moves from the source to the goal. The metric *cost from the source* changes when the robot moves from the source towards the goal, with no change in the map, as the source (robot's current position) changes. However, the metric *cost from goal* does not change when the robot moves as long as the map remains static. Hence a backward search is used.

5.4.2   Raised and lowered waves

The other change introduced in the algorithm deals with the philosophy of optimality of the A* algorithm. In the A* algorithm, a node in *closed* is optimal and therefore no changes must be made in its cost. However, in the D* algorithm, a node in *closed* may become suboptimal due to the changes in the environment. Hence, it is acceptable to find a better value of a node otherwise in *closed*, to update its value, and to reintroduce the node in the fringe so that the cost may be propagated to the other nodes, including some in *closed*.

To understand the algorithm, let us consider 2 situations. In the first situation, an obstacle slightly clears. Now, it is possible to have a smaller cost to the goal. The changes are reported by the mapping systems to the algorithm. The node in the surrounding of the cleared obstacle realizes that its cost to the goal has now been reduced. It gets re-inserted into the fringe at a better cost. The processing of this node results in a reduction of the costs of the neighbouring nodes, which are re-inserted into the fringe. This flows like a wave from the affected area backward and is called the *lowered wave*. The wave results in a reduction in costs as the wave proceeds. This is also shown in Fig. 5.6A. Fig. 5.6B shows the regeneration of the search tree with the updated calculations for a synthetic test case when no node can find a better parent due to the changed map.

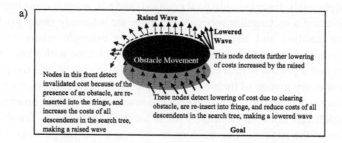

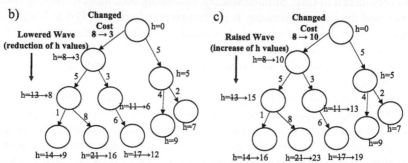

FIG. 5.6   Lowered and raised wave in D* algorithm. (A) Indicative scenario. (B) Example of lowered wave. (C) Example of raised wave.

In the other case imagine that the obstacle moves towards the optimal path to the goal, in which case the robot will have to now take a longer path. The mapping system notifies the algorithm. The node now inside the obstacle realizes that it has turned from being feasible to infeasible and lets all the neighbouring nodes know about this. The costs of the neighbouring nodes increase as a result. The neighbouring nodes try to find a better path by finding a better parent, as the previous parent is now collision-prone with a very large cost. Ultimately, some feasible nodes will get a feasible neighbour. However, the costs would have increased from the prior path (which is now infeasible) to the new path (which is feasible). This propagates a similar wave from the node towards the source called the *raise wave*. The raise wave increases the cost to goal for the affected nodes. This is shown in Fig. 5.6A. Fig. 5.6C specifically shows the regeneration of the search tree with the updated costs, with a (bad) assumption that no node can find a better parent. Typically, the waves may die as the remaining nodes do not experience a changed cost; or the two waves may be produced simultaneously at different regions and may merge.

There is a change in the naming convention in the discussion of the algorithm. Since the reference to cost here is the cost from the node to the goal, let us denote the cost as h, even though it is not an estimate and is being calculated by an exhaustive search. Previously g-value was used as the cost from source for a forward search. Now the same term for a backward search or the cost to goal is denoted by h.

The raise wave is particularly interesting. If the old value of the node is better than the new raised value, it is unnatural for a node to accept the new and worse value. The worse cost is otherwise correct since the occupancy of the obstacle will raise the costs and the optimal cost to goal will be higher than the one previously. This necessitates to intimate a node that the older value, even though smaller, is not valid and should be re-computed. Consider a simple problem shown in Fig. 5.7. An edge was previously valid, however, is now invalid shown by a very large cost. A popular mechanism to re-calculate the costs is that every node (that considers itself affected) looks into the neighbours and re-calculates the cost. Here, 2 nodes A and B are in a racing condition. A offers a better cost to B who accepts it; while B offers a better cost to A who accepts it. Iteratively both nodes increase each other's cost to the correct (very high) value. However, the re-planning takes more time than a complete re-planning. Hence, a better means to re-calculate the cost is to tell A that it cannot use the current cost of B for re-calculating its value, since B is fundamentally after A unless B has re-computed its updated value. So, if A chooses a new parent to account for the changed costs, B is not considered. However, a normal expansion of B due to the changed costs (raised wave) may update the costs of A. This immediately first updates the cost of C, followed by A and B as a raised wave, giving the corrected costs to all affected nodes.

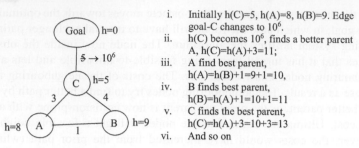

i. Initially h(C)=5, h(A)=8, h(B)=9. Edge goal-C changes to 10^6.
ii. h(C) becomes 10^6, finds a better parent A, h(C)=h(A)+3=11;
iii. A find best parent, h(A)=h(B)+1=9+1=10,
iv. B finds best parent, h(B)=h(A)+1=10+1=11
v. C finds the best parent, h(C)=h(A)+3=10+3=13
vi. And so on

FIG. 5.7 Updating costs because of a change of map. At each iteration, an affected node selects the best parent to calculate the updated cost. The re-planning takes significantly more time than planning a graph with 4 nodes, necessitating stopping A from using B as a parent, and similarly for others.

5.4.3   D* algorithm principle

To make matters simple, let us consider the Uniform Cost Search (UCS) operating principle. The basic concept of the D* algorithm is to maintain a priority queue of nodes that may be affected by a change of map. The affected nodes are inserted into the priority queue and processed to give them the updated cost. While doing so more affected nodes are discovered that are inserted. The priority queue is prioritized by a new criterion called the *k-value*. The value is stored in addition to the other costs. Consider a node v is currently inside *closed* with cost to goal as $h(v)$. Consider that the node v discovers a *lowered cost* as a result of a change of map with the reduced (new) cost as $h'(v)$. This node must be processed before any other node for which v can act as a parent. Hence, the new value $h'(v)$ with a reduced cost is a better indicator of priority. In such a case, it may help reduce the cost of an existing node (say a) in the priority queue by acting as a parent and reducing the cost. If the old (larger) value was taken as the priority and the old cost ($h(v)$) was larger than the current priority of a, first the node a would have been dequeued and processed giving rise to a wave, and later v would have been processed, ultimately reducing the cost of the node a resulting in the same nodes being reprocessed.

Consider the opposite case where the node v is informed about an *increase* in its cost from the old value $h(v)$ to a higher (new) value of $h'(v)$. Since v has an irregularity (change in cost), it will be inserted into the priority queue and subsequently processed to propagate the new values to the other nodes, while some other nodes with irregularities are already in the priority queue. Search techniques rarely visit the complete space and stop when the source (or current pose of the robot, considering a backward search) is dequeued. The new value $h'(v)$ may be larger than the updated cost for the source node and using priority as the new value $h'(v)$ as a priority may never get processed as it will be scheduled for processing after the source is dequeued. Therefore, the old value $h(v)$ is a better candidate for priority. Furthermore, a node in the priority queue whose current cost may be computed from v should be processed after v in the priority queue, and hence, the old value $h(v)$ is used as the priority. Consider a change in cost puts 2 nodes in the priority queue (say v and a), while currently, a is a descendent of v (before change of map). Since a's cost is computed from v, it is important to first process v and then process a in the priority queue. Hence, the old value $h(v)$ is a better indication of the priority. Since v has a smaller cost to goal before update as compared to a, it will have a more favourable priority as compared to a.

Combining both cases using priority as the minimum of the old and the new cost, or $k(v) = min(h(v), h'(v))$ is a good choice of priority for the case when v is in closed. If v is neither in fringe nor in closed, there is no known old cost value, and hence, the priority is given by the new (and only) estimate $h'(v)$. A more complicated case is when a node v is rediscovered while it is in the fringe. The node v has a previous priority $k(v)$ computed as per the discussed mechanism that needs to be updated. Before being rediscovered, the estimated cost to goal was $h(v)$, while the new cost estimate on being rediscovered is $h'(v)$. Using the same ideology as per the previous case, $k(v)$ acts as a minimizer that keeps storing the minimum of all estimates as node v is repeatedly rediscovered. Hence, the priority is updated to $min(k(v), h'(v))$. The priority of the node v when it is (re)discovered from an old estimate of $h(v)$ to a new estimate of $h'(v)$ with the old priority as $k(v)$ is given by Eq. (5.5), where $\leftarrow$ is used to denote the update of priority from some old value to the new value.

$$k(v) \leftarrow \begin{cases} min(h(v), h'(v)) & v \in closed \\ h'(v) & v \notin fringe \land v \notin closed \\ min(k(v), h'(v)) & v \in fringe \end{cases} \tag{5.5}$$

Initially, the k-value and h-value are the same and remain to be the same until the costs change due to a change in the map. If the h-value of a node is larger than the k-value ($h(v) > k(v)$), it denotes a raised state or the propagation of a raised wave. If an edge increases its cost, the h-value increases, while the k-value continues to be the old smaller value. If the h-value of a node is equal to the k-value, it may denote a lowered state (propagation of a lowered wave) or a no change of cost. Since the D* algorithm only calculates the new costs for nodes that change costs, the condition $h(v) = k(v)$ denotes a lowered stated for all nodes in the priority queue in the replanning stage.

Consider a node u is dequeued from the priority queue and is being processed. If u is in a raised state $(h(u) > k(u))$, then it searches for a new parent (say v). This is unlike any search technique discussed so far where parents produce children. Because of a change of map, it is valid for a child to search for the best parent. However, it is important for the parent v to be conceptually before the child u. $h(v)$ is the guaranteed and updated cost of v. $k(u)$ is the smallest value that u has seen since it was last made consistent (put in closed). If $h(v) \leq k(u)$, it means that there is no way that v has calculated its cost from u and can serve as the parent of u. Thereafter, a normal processing of the UCS continues. In case of a lowered value (or no change), the processing is similar to a standard UCS involving updating the parents and costs, with an additional step of computing the k-value as the priority.

The bound $h(v) \leq k(u)$ may appear tight and there may be cases where v is a valid candidate as a parent for u which does not obey the constraints $h(v) \leq k(u)$. In general, while processing an edge u-v it is possible that v may appear to be the candidate that may generate a lowered wave, which is important to process by inserting it into the priority queue. The algorithm keeps track of such a condition. Equivalently, if a vertex u is found to be lowered (like because of finding a better parent), it is also re-inserted into the priority queue and reprocessed.

In all cases if an edge u-v is being processed, while u is the current parent of v (or $\pi(v) = u$), it is mandatory to update the cost of v. The old cost of v was calculated from the old value of u which is changed and v needs to be updated irrespective of whether it reduces or increases the cost. In other words, a child always derives its cost from the parent. If any parent changes its cost, it is important for the parent to also change the corresponding cost of the child and to insert it into the priority queue. When the child is dequeued it becomes the parent and compulsorily updates the cost of all its immediate children.

5.4.4   D* algorithm

For simplicity, we assume that the UCS has already computed a path from the goal to the source with costs given by the h-values along with parents, fringe, and closed. The robot starts from some source and moves forward, until the cost of some edge (u, v) changes, which needs to be incorporated in the algorithm. The affected node is inserted into the fringe, from which state the re-calculation process starts. The algorithm for the changes in costs and parents are given by Algorithm 5.1 and Algorithm 5.2. Even though the algorithm assumes that a single edge has a changed cost, in reality, if multiple edges have costs changed, all such affected vertices could similarly be inserted into the fringe for processing.

Algorithm 5.1 D^*.

Input: Modified edge cost $c(v, a, u)$; fringe (Q) prioritized on k, *closed*,
   cost from goal (h), parent (π) from UCS. Assume $k = h$ initially.

1   $Enqueue(Q, u)$ with priority $k(u)$, $found \leftarrow false$;

2   **while** $\neg Empty(Q)$ **do**

3     $u \leftarrow Front(Q), dequeue(Q)$;

4   **if** $u = currentPosition$ **then** $found \leftarrow true$, break;

5   **if** $k(u) < h(u)$ **then**   ▷ cost of u is raised

6   **for** *all actions a* **do**   ▷ find a better parent v for u

7         $v \leftarrow$ state such that application of action a on u produces v ;
         ▷ $h(v) < k(u)$ means v exists before u ;

8   **if** $feasible(v) \wedge (v \in closed \vee v \in Q) \wedge h(v) \leq k(u) \wedge h(u) > h(v) + c(u, a, v)$ **then** $\pi(u) \leftarrow v, h(u) \leftarrow h(v) + c(u, a, v)$;

9   **if** $k(u) = h(u)$ **then**   ▷ Lowered condition

10   **for** *all actions a* **do**   ▷ conventional backward search

11         $v \leftarrow$ state such that application of action a on v produces u ;
         ▷ if unseen or parent u/edge $u - v$ propagating
         lowered wave or v has a better cost ;

12   **if** $feasible \wedge ((v \notin closed \wedge \notin Q) \vee (\pi(v) = u \wedge h(v) \neq h(u) + c(v, a, u)) \vee (\pi(v) \neq u \wedge h(v) > h(u) + c(v, a, u)))$ **then**

13           $h'(v) \leftarrow h(u) + c(v, a, u), k(v) \leftarrow calculateK(h'(v))$;

14           $h(v) \leftarrow h'(v), \pi(v) \leftarrow u, action(v) \leftarrow a$;

15   **if** $v \notin Q$ **then** $enqueue(Q, v)$ with priority $k(v)$;

16   **else** update priority of v to $k(v)$;

17   **else**   ▷ Raised Condition

18   **for** *all actions a* **do**   ▷ conventional backward search

19         $v \leftarrow$ state such that application of action a on v produces u ;
         ▷ if unseen or parent u/edge $u - v$ propagating a
         raised wave ;

20   **if** $feasible(v) \wedge ((v \notin closed \wedge v \notin frontier) \vee (\pi(v) = u \wedge h(v) \neq h(u) + c(v, a, u)))$ **then**

21           $h'(v) \leftarrow h(u) + c(v, a, u), k(v) \leftarrow calculateK(h'(v))$;

22           $h(v) \leftarrow h'(v), \pi(v) \leftarrow u, action(v) \leftarrow a$;

23   **if** $v \notin Q$ **then** $enqueue(Q, v)$ with priority $k(v)$;

24   **else** update priority of v to $k(v)$;

          ▷ lowered/inconsistent u, enqueue and process ;

25   **else if** $feasible(v) \wedge \pi(v) \neq u \wedge h(v) > h(u) + c(v, a, u)$ **then**

26           $k(u) \leftarrow calculateK(h(u)), enqueue(Q, u)$ priority $k(u)$;

          ▷ lowered/inconsistent v, enqueue and process ;

27   **else if** $feasible(v) \wedge \pi(v) \neq u \wedge h(u) > h(v) + c(u, a, v) \wedge v \in closed \wedge h(v) > k(u)$ **then**

28           $k(v) \leftarrow calculateK(h(v))$;

29   **if** $v \notin Q$ **then** $enqueue(Q, v)$ with priority $k(v)$;

30   **else** update priority of v to $k(v)$;

31   **If** $found$, return $d(v), calculatePath(\pi, v)$, **else** return ∞, NIL ;

Algorithm 5.2 Calculate $K(h'(v))$.

---

1 **if** $v \notin closed \wedge v \notin frontier$ **then** return $h'(v)$;
2 **if** $v \in frontier$ **then** return $min(h'(v), k(v))$;
3 **if** $v \in closed$ **then** return $min(h(v), h'(v))$;

It is assumed that the initial search itself happens in the same principle of a backward search with k-values, and hence, the fringe and closed are properly initialized. Furthermore, it is assumed that the affected nodes are already inserted into the fringe. Due to the changes in the map, it is possible for a node to get a better parent (Lines 5–8), like in the case of an increase in cost or a raised wave, a node may get a reduced cost due to a change of parent. The rest of the pseudo-code is fundamentally bifurcated into the cases of lowered (Lines 9–16) and raised (Lines 17–30). The lowered condition also holds good for a no change in cost condition for new nodes that are expanded. The cost computation for the cases is shown in Fig. 5.8. For the processing of any node, the first task is to search for a new parent for the node which can offer a better cost (Lines 5–8), which is important since a node may be experiencing a raised wave and an alternate parent might offer a better cost. This is unlike a normal search wherein a parent is expanded to create children, here a child searches for a better parent, admissible due to the dynamic environment. A check on the k-value of the eligible parent is performed ($h(v) \leq k(u)$) to ensure that the parent primarily exists before the child. The lowered case of Lines 9–16 is the same as a conventional UCS with the difference that even a closed node can be reinserted into the fringe and that a cost may no longer be valid due to a changed parent cost and therefore must be updated. So $\pi(v) = u$ and $h(v) \neq h(u) + c(v, a, u)$, which means that the calculation of $h(v)$ using $h(u)$ is no longer consistent (due to a change cost of the parent u) and consistency must therefore be added by re-calculating the cost and adding the same to the fringe.

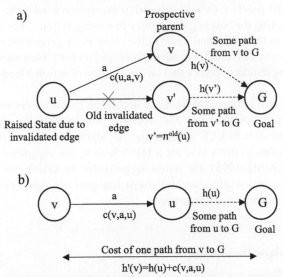

a)

Prospective parent

Some path from v to G

v

h(v)

a

c(u,a,v)

u

Old invalidated edge

h(v')

v'

h(v')

G

Some path from v' to G   Goal

Raised State due to invalidated edge

v'=π^old(u)

b)

v

a

c(v,a,u)

u

h(u)

G

Some path from u to G   Goal

Cost of one path from v to G

h'(v)=h(u)+c(v,a,u)

FIG. 5.8 Calculation of costs for D* algorithm (A) A raised state u searching for a better parent v (B) Expansion of an edge u-v to find a better cost for v or to correct for changed cost $c(v,a,u)$. The diagram is a mirror image of the ones used in Chapter 4 since it represents a backward search.

The more interesting case is the raised condition of Lines 17–30, which also fundamentally looks the same as the default UCS with two exceptional cases. The first exception is if a node v is rediscovered with a better cost (lowered u, Line 25–26), the vertex u is inserted back into the fringe to process the vertex v as per normal UCS working philosophy. The line detects that u is lowered (inconsistent) and therefore a candidate for reprocessing. In the other exceptional condition, if a vertex u can get a better cost due to a vertex v in closed (lowered v, Lines 27–30) then the vertex v is re-inserted into the fringe from closed as it is now found to be a possible source of a lowered wave. The check $h(v) > k(u)$ is converse of the one used for finding a better parent since $h(v) \leq k(u)$ condition is already checked.

The notations may seem inverted in some places, which is because the algorithm uses the notion of going from goal to source rather than the other way around. $c(u, a, v)$ means cost in going from u to v by application of an action a. In the forward search, u denotes the parent and v denotes the child. For the backward version, v denotes the child and u denotes the parent. Since the search is backward instead of using this expression to denote the expansion of parent u producing a child v, the expression denotes the expansion of parent v to produce the child u. Alternatively $c(u, a, v)$ denotes the quest of the child

u to find a better parent v. Consider the other expression used in the pseudo-code $c(v, a, u)$, denoting the cost in going from v to u using action a. For the backward variant u acts as the parent and v acts as the child. For a generation of children of parent u, the cost of a child v is taken as $c(v, a, u)$ for a backward search signifying finding children states v that on application of action a lead to the parent state u.

The algorithm is based on the UCS algorithm that does not use heuristics. Heuristics can be inserted into the working principles of the algorithm like the use of heuristics in UCS to make the A* algorithm. If heuristics are used in all calculations to get a bias for a faster search, the algorithm is known as *focussed D** (Stentz, 1995). The name suggests that the search process is biased towards the source while searching from the goal for path correction upon dynamic changes in the environment.

5.4.5   D* algorithm example

The working of the D* algorithm is studied by a very simple example. Take the graph given in Fig. 5.9. The robot first plans a path with the only difference that a backward Uniform Cost Search is used. The same is rather trivial and not shown in the example. Thereafter the robot starts its journey from the source to the goal. However, upon reaching vertex 4, it discovers that the edge (4,5) is no longer feasible. This can be visualized as the robot planned to assume the door connecting two rooms will be open, however, upon traversal, the robot found that the door was locked. Thereafter, D* algorithm is used to rewire the parent tree and get a new path for the robot. The calculations are shown in Fig. 5.9. The cost of the edge (4,5) is made significantly large to signify that the edge does not exist and the same shall not appear in the shortest path. First, vertex 4 computes that the edge is no longer feasible and therefore its cost is significantly high. This propagates a raised wave in the graph. The wave travels from 4, outwards, raising the costs of all edges. As the vertices increase their values, they are added into the fringe to propagate the increased costs to the vertices ahead. The expansions are in the order of k-values.

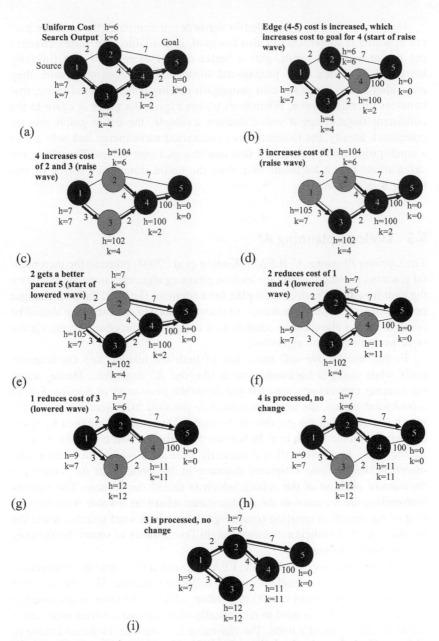

FIG. 5.9 Working of D* algorithm. The *black* nodes are in the closed, *grey* in the fringe and the bold edges denote the parent-child relation (since it is a backward search, the *arrows* point from child to the parent). (A)–(I) show the working steps.

However, when 2 is expanded in the order, it computes that a better parent 5, which is coincidentally also the goal, is available. The vertex changes the parent and as a result, gets a better cost. This reduced cost (initially increased by vertex 4) is propagated further. As vertices are found, they are added to the fringe for cost propagation. This is the lowered wave, that balances the raised wave. When both waves expire, the graph is again in the consistent stage. Since it was a dummy example, the entire graph was recomputed. Ideally, the lowered wave and raised wave merge and only affect a small proportion of the graph that has changed cost values. Therefore, the calculations take a lot lesser time than the entire state space search or initiating a new search.

5.5 Lifelong planning A*

The *Lifelong Planning A** (LPA*) (Koenig et al., 2004) presents the incremental planning paradigm where the motion planning algorithm must always be on the hunt for a better path or must plan for a lifetime. Hence, if the map changes because of any motion, appearance, or disappearance of an obstacle, it should be possible for the algorithm to adapt it as it computes. The general notion is the same as that of the D* algorithm.

In this section, we will make use of both the historic and the heuristic costs while doing a backward search like the A* algorithm. Hence, again the naming conventions are changed from the ones used in Section 5.4. In a backward search, the aim is to calculate the cost from goal for any node, which is now called the g-value of the node. This cost is computed by using deliberation in the search tree. In Section 5.4, this cost was called the h-value. At the same time, we will use heuristic estimates, which for a backward search is the heuristic estimated distance of the node from the source (or the current position of the robot), which is called the h-value. The naming conventions are the same as the A* algorithm where the g-value was the exact cost of the search in question (cost to goal for a backward search), while the h-value was the heuristic estimate (straight line distance to source for the considered backward search).

The basic philosophy behind the LPA* algorithm is to store the information about the cost to goal redundantly in two different variables. The first one is the cost from goal (backward search) or g-value, which is the same as discussed so far. Another variable is used to redundantly store the same information and is called the *rhs*-value of a node. The *rhs*-value is a one-step lookahead value of the node and is calculated by a node by looking at the g values of its neighbours.

The *rhs*(v) for a vertex v is defined by Eq. (5.6) in the context of a backward search.

$$rhs(v) = \begin{cases} \min_{u,a:a \text{ applied on } v \text{ produces } u} g(u) + c(v,a,u) & \text{if } v \neq Goal \\ 0 & \text{if } v = Goal \end{cases} \quad (5.6)$$

The *rhs*(v) considers the cost in going from v to u with a cost of c(v, a, u) using action a and then from u to goal with a cost of g(u), while the u that gives the least cost is selected as the predecessor of v (for a backward search, equivalent to a successor for a forward search). The equation does not hold good for the root of the backward search (goal) for which a value of 0 for cost to goal is explicitly set.

Normally, the *rhs*-value of the node should be the same as the g-value. However, imagine that due to some changes in the map, the edge cost c(u, a, v) changes to a different value, say $c_{new}(u, a, v)$. The changed edge cost can be used to re-compute the new value of *rhs*(v). In such a situation, *rhs*(v) using the new edge cost is not the same as the g-value using the old edge cost, and there is an *inconsistency*. This consistency may be resolved by re-computing g(v) and/or *rhs*(v), which causes inconsistency. However, doing so also propagates an inconsistency somewhere else, which needs to be resolved. This generates a wave of inconsistency that flows in the graph, similar to the raised and lowered waves of the D* algorithm. So, a fringe of inconsistent nodes is made and processed one by one at every stage taking a vertex and making it consistent. Once all the vertices are made consistent, the costs point out the values after updates.

The priority of the fringe is given by a *key* comprising the minimum expected distance from the source to goal through the vertex (f value of A*) and cost from goal for a backward search (g value of A*), given by Eq. (5.7).

$$k(v) = [k_1(v), k_2(v)] = [\min(g(v), rhs(v)) + h(v), \min(g(v), rhs(v))] \quad (5.7)$$

The priority is given by the first value of the key which is $k_1(v)$ or the expected distance from source to goal through v like the A* algorithm. However, in the case of a tie in the primary criterion, the second value $k_2(v)$ or the cost from goal is used as a tie-breaker. In the A* algorithm, processing two nodes with the same f-value could be done in any order, both yielding the optimal solution. However, here the question is resolution of inconsistencies and two inconsistent nodes having the same $k_1(v)$ value in the priority queue could also have a parent and child relationship necessitating the processing of the parent before the child given by the secondary measure. In the A* algorithm, every node only once got in and out of the priority queue, and therefore processing a child before the parent was not possible at any stage.

The comparator between the vertices u and v is hence given by Eq. (5.8), with $u < v$ denoting that v is preferred over u because of the priority of a max priority queue.

$$u < v, iff \; k_1(v) < k_1(u) \lor (k_1(v) = k_1(u) \land k_1(v) < k_2(u)) \qquad (5.8)$$

The same is used to make the priority comparison operations of the priority queue.

Similar to the D* algorithm, the smaller of the $g(v)$ and $rhs(v)$ values are taken to determine the priority. Consider that the weight of an edge is reduced from $c(v, a, u)$ to $c_{new}(v, a, u)$, potentially giving rise to a lowered wave, in which case $rhs(v)$ with the updated edge cost is smaller than $g(v)$ with the previous (higher) edge cost. The node v may now serve as a parent to a node in the priority queue and should be processed first indicating the use of $rhs(v)$ with a smaller cost for the priority calculation. However, consider the edge cost was increased from $c(v, a, u)$ to $c_{new}(v, a, u)$, causing a raised wave. $rhs(v)$ updates the new (raised) cost while $g(v)$ has the old (lower) cost, giving $rhs(v) > g(v)$. Now v must first update all nodes in the priority queue for which v may be an ancestor in the path, and hence, using old $g(v)$ value, which is also smaller, is a better option. Combining the two concepts, the priority is given by taking the minimum of $rhs(v)$ and $g(v)$ values for the historic cost component.

The resolution of the conflicts is done by setting the correct g-value, which makes the computation consistent with the rhs-value. Typically, this will result in setting another vertex potentially inconsistent and insertion into the fringe. Let us study this as two different test cases. In the first case is when a vertex v is *over-consistent* given by $g(v) > rhs(v)$, meaning that an edge cost was decreased. In this case, the vertex v can accept the lowered cost as the new cost to make it consistent. This may make the neighbours inconsistent, whose rhs-values are re-calculated and checked for consistency. In case a change of cost of v makes a neighbour inconsistent, it is added into the priority queue of inconsistent vertices.

Consider the opposite case where a node v is *under-consistent*, given by $g(v) < rhs(v)$, potentially giving rise to a raised wave. Intuitively, the best solution may appear to update the cost $g(v)$ to the $rhs(v)$ that has the updated cost. However, it is possible that in this process v calculates its cost from its child and the parent and child get into a race condition to increase each other's cost up to infinity for cases where a feasible path is no longer possible. Hence, a different strategy is adopted. If a vertex v is found to be under-consistent or $g(v) > rhs(v)$, then its cost is invalidated or raised to the maximum value (∞) such that no node can take v as the parent. All neighbours re-calculate the *rhs*-value using the new cost of $g(v) = \infty$ to propagate the invalidity, which makes it likely for the neighbouring nodes not to take v as a parent in the *rhs*-value calculations. $g(v)$ also re-calculates its *rhs*-value, which will not be the same as the *g*-value (∞) and v will be re-inserted into the queue. Eventually, a lowered wave would propagate after which v will have the correct *rhs*-value. Thereafter, as per priority, v will be dequeued. On processing of the vertex v, the $g(v) = \infty$ will be lowered to a finite value (updated) cost, further propagating the lowered wave. The processing of an under-consistent vertex happens in this mechanism by invalidating it or giving it an infinite value first, and then letting the lowered wave reduce the cost to the correct finite value.

The algorithm is given as Algorithm 5.3. The processing is also shown in Fig. 5.10. Since for D* the search was assumed to be backward from goal to source. Hence the same terminology has been used in the pseudo-code of LPA* as well, even though the algorithm can work in both directions subjected to the reversibility of the actions. For the same reasons the notations will appear reversed when dealing with traversing edges and edge costs. The pseudo-code initializes just like any other search with the addition of the *k*-values and *rhs*-values, shown in Lines 1–4. The infinite loop in Line 5 is the lifelong learning component that implies that the costs will keep getting updated. The loop has two very similar codes one after the other. One part (Lines 7–19) does the normal processing of the queue for resolving inconsistencies like the A* algorithm, while the other part (Lines 20–29) catches all changed edge costs and resolves the same. Since the procedure for processing an edge or resolving inconsistencies between g and *rhs*-values is the same for both parts, a similar code is replicated twice. When processing a queue, the inconsistency can be of two types. First is over-consistent wherein the *rhs*-value is smaller than the g value ($g(u) > rhs(u)$), meaning a lowered wave is being propagated (Lines 9, 11–19). The resolution of consistency, in such a case, is by updating the *g*-value to the *rhs*-value as the *rhs*-value is more updated due to a 1-step lookahead (Line 9). In such a case, the *rhs*-values of the neighbours may be impacted and the same are re-calculated for all neighbouring nodes (Lines 12–14). If any vertex v is found to be inconsistent ($g(v) \neq rhs(v)$, Line 16), then it is either inserted into the queue (Line 17) or its priority is updated (Line 18) depending upon its presence or absence from the queue. If a vertex v however is consistent ($g(v) = rhs(v)$), it may be removed from the queue if already present (Line 19).

Algorithm 5.3 Lifelong planning A^*.

Input: action function (A), feasibility function, cost function (c),
   heuristic function (h), source (S) and goal (G)
Output: Updated vertex costs as per changing edge costs, computed
   infinitely waiting for change

1 $g(G) \leftarrow 0$;   ▷ backward search, cost from goal
2 $\pi(G) \leftarrow NIL, action(G) \leftarrow NIL$;
3 $k(G) \leftarrow [g(G) + h(G),\ g(G)]$;   ▷ priority of fringe [primary
   criterion, secondary criterion] or [total cost,
   historic cost (cost from goal)]
4 $Q \leftarrow$ empty priority queue, $enqueue(Q, G)$ with priority $k(G)$;
5 **while** *forever* **do**
6   **while** $\neg empty(Q)$ **do**   ▷ remove known inconsistencies
7          $u \leftarrow front(Q), dequeue(Q)$;
           ▷ if all nodes before source by priority
   processed and source is consistent ;
8   **if** $k(u) \geq k(S) \wedge rhs(S) = g(S)$ **then** break;
9   **if** $g(u) > rhs(u)$ **then** $g(u) \leftarrow rhs(u)$;   ▷ over-consistent
10   **else** $g(u) \leftarrow \infty$; ▷ under-consistent, invalidate old
11   **for** (*state v : any action a applied on v produces u*) $\cup \{u\}$ **do**
           ▷ re-calculate *rhs* values
12   **if** $v \neq G$ **then**   ▷ rhs(G) is fixed to 0
             ▷ $a'(v) = u'$: action a' on v gives u' ;
13           $rhs(v) \leftarrow min_{a'u' : a'(v)=u'} g(u') + c(v, a, u')$;
14           $[action(v), \pi(v)] \leftarrow argmin_{a'u' : a'(v)=u'} g(u') + c(v, a', u')$;
15           $k(v) \leftarrow [min(g(v), rhs(v)) + h(v),\ min(g(v), rhs(v))]$;
16   **if** $g(v) \neq rhs(v)$ **then**   ▷ add inconsistent
17   **if** $v \notin Q$ **then** $Enqueue(Q, v)$;
18   **else** update priority of v in Q;
19   **else** remove v from Q, if present;
20   **for** *all edges v, a, u, with changed cost $c_{new}(v, a, u)$* **do**
21         $c(v, a, u) \leftarrow c_{new}(v, a, u)$;
22   **if** $v \neq G$ **then**   ▷ same as previous block
23           $rhs(v) \leftarrow min_{a'u' : a'(v)=u'} g(u') + c(v, a, u')$;
24           $[action(v), \pi(v)] \leftarrow argmin_{a'u' : a'(v)=u'} g(u') + c(v, a', u')$;
25         $k(v) \leftarrow [min(g(v), rhs(v)) + h(v),\ min(g(v), rhs(v))]$;
26   **if** $g(v) \neq rhs(v)$ **then**
27   **if** $v \notin Q$ **then** $Enqueue(Q, v)$;
28   **else** update priority of v in Q;
29   **else** remove v from Q, if present;

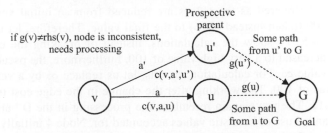

FIG. 5.10 Processing of a vertex u using Lifelong Planning A*. If u is over-consistent $(g(u) > rhs(u))$, inconsistency is rectified $(g(u) = rhs(u))$ and inconsistency is propagated to all v, by re-calculating their $rhs(v)$ value $(rhs(v) = min_{u',a'} g(v) + c(v,a', u'))$. If u is under-consistent $(g(u) < rhs(u))$, $g(u)$ is invalidated $(g(u) = \infty)$. Inconsistent nodes are re-inserted into the queue.

On the contrary, an under-consistent edge potentially generating a raised wave is characterized by a higher rhs-value as compared to a g value (Line 10, 11–18). It means the rhs-value did a one-step lookahead to sense increased values, while the g value is yet to be updated. Here, the problem is that the $g(u)$ value of a vertex u is not yet valid as it represents the cost before the increase in the edge cost. Therefore, the same is first invalidated by setting it to infinity in Line 10, stopping any node to use u as a parent. The $rhs(u)$ computed in Lines 12–14 may still be finite through an alternate parent chosen by u. A change of parent will make u inconsistent, which will be enqueued and dequeued again (as per priority), however upon dequeue, it will be a lowered node. The neighbouring nodes of u also re-calculate the rhs-value and the inconsistent ones are added to the priority queue (Lines 12–19).

Finally, the breaking condition of the loop (Line 8) is that the priority of the vertex being processed is higher than the source (equivalent to goal in case of a forward search). The priority increases as one keeps processing vertices, and a value higher than source means that the source has either been processed or it is not necessary to process the source since it is not affected by the changes. Another clause added in the condition is $rhs(S) = g(S)$, which means that the source (equivalent to goal in the forward search) should be consistent when exiting the loop. Lines 20–29 have a very similar implementation and modify all edges (u, v) with changed costs with the new cost given by $c_{new}(u, v)$. The loop iterates on all edges and modifies v as per the changed costs of the edge. The vertex is added to the queue for processing, similar to the first part. The algorithm casually accesses any vertex v, while in implementation, it will need to be ascertained that the same exists, and if not, its cost must be taken as infinite.

The working of the algorithm on a sample graph is shown in Fig. 5.11 for no pre-calculated values, starting from the goal. The heuristic function values are given in Table 5.1. The wave resembles the one that would be seen in a backward A* algorithm. In this example, only over-consistent nodes are discovered as the values are reduced from an initial very large value of 100 (taken instead of ∞) to the final value. Therefore, let us break the edge 4–5. To show the calculations, instead of breaking the edge, its value is increased to a very high value of 100. Furthermore, the pseudo-code uses the value ∞. For calculation reasons, let us replace ∞ by a very large finite value of 100. The working after the change in the edge cost is shown in Fig. 5.12. The wave again resembles the processing in the D* algorithm with the added use of heuristic values accounted for. Node 4 initially starts a raised wave. Later, node 2 starts the lowered wave through which the other nodes get their updated values.

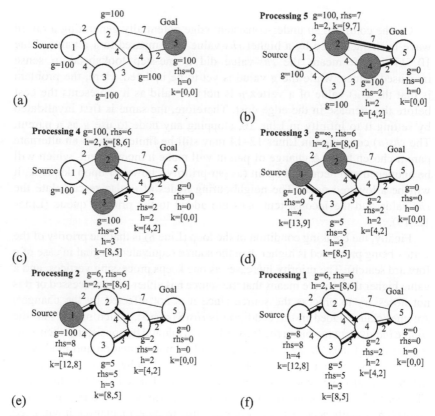

FIG. 5.11 Working of the LPA* algorithm. The *grey* nodes are in the queue for processing, processed as per the *k*-values. The bold edges denote the parent-child relation (child to parent). For a node, if the *rhs*-value is lower than the *g*-value, the *g*-value of the node and the *rhs*-value of the neighbours are updated. Inconsistent nodes are enqueued. (A)–(F) show the working steps.

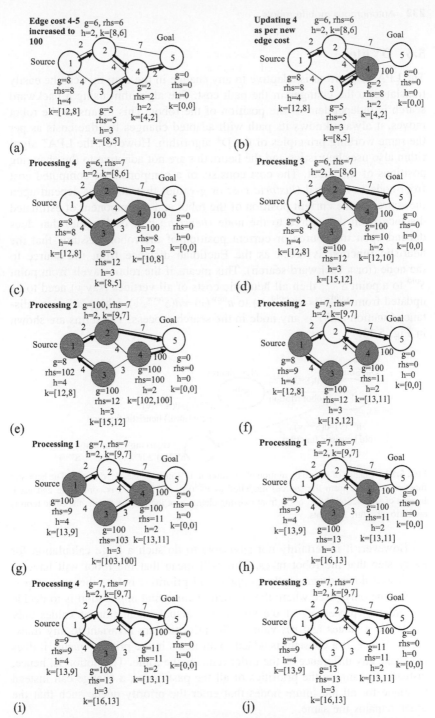

FIG. 5.12 Working of the LPA* algorithm when an edge cost is increased. For a node, if the *rhs*-value is larger than the *g*-value, the *g*-value is set to ∞ (here 100). The *rhs*-value of the node and neighbours are updated. Inconsistent nodes are enqueued. (A)–(J) show the working steps.

5.6   D* lite

The LPA* algorithm is adaptive to any change in the map that can be easily translated to the changes in the path cost. The algorithm using a backward search is independent of the position of the robot. This means as the robot moves, it always knows its path with adapted changes to edge costs as per the same working principles of the D* algorithm. However, the LPA* algorithm also uses heuristics and the heuristics are not adaptive to the changing positions of the robot. The cost consists of a component of computed cost from node to the goal (*historic cost* or g-value) that does not depend upon the source or the current position of the robot and a component of estimated distance from the source to the node (*heuristic cost* or h-value) that does depend upon the source or current position of the robot. Assume that the heuristic function is taken as the Euclidian distance from the source to the node (for a backward search). This means if the robot travels from point S^{old} to a point S^{new}, then all heuristic costs of all vertices (say v) need to be updated from $h^{old}(v) = d(S^{old}, v)$ to $h^{new}(v) = d(S^{new}, v)$, where $d()$ is the distance function and v is any node in the search process. The terms are shown in Fig. 5.13.

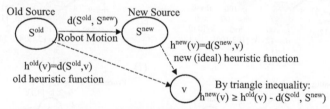

FIG. 5.13   Change of source by motion of the robot in D*Lite on a backward search: The heuristic function (distance from the source) modelled as $h^{old}(v) - d(S^{old}, S^{new})$ is admissible and used. Instead of subtracting $d(S^{old}, S^{new})$ from existing elements of the priority queue, the same term is added to all future elements.

However, it is certainly not advisable to do such a large calculation for every step that the robot takes. This will mean that the robot will have to be slowed down dramatically to update all priorities before the next step is taken. The only place where the heuristic values find application is to decide the order of processing of the nodes. The actual calculation of values only needs the g values and *rhs*-values. The priorities are invariant to any translation. Hence, if a constant is added to all items in the priority queue, it does not make any difference as the order remains the same. The notion is, hence, instead of changing the priorities of all the past nodes, a change can instead be made for all the future nodes that enter the priority queue such that the order remains the same.

The heuristic function is also modified to ease computation. The main constraint is that the heuristic function should be admissible (heuristic estimate should always be smaller than the optimal solution cost or $h(v) \leq h^*(v)$). If the robot travels from point S^{old} to point S^{new}, and Euclidian distance from the source to nodes $h^{old}(v) = d(S^{old}, v)$ is the (ideal) heuristic function, the new (ideal) heuristic value after motion will be given by $h^{new}(v) = d(S^{new}, v)$. Based on the triangle inequality shown in Eqs (5.9)–(5.11),

$$d\left(S^{old}, v\right) \leq d\left(S^{old}, S^{new}\right) + d\left(S^{new}, v\right) \tag{5.9}$$

$$d\left(S^{new}, v\right) \geq d\left(S^{old}, v\right) - d\left(S^{old}, S^{new}\right) \tag{5.10}$$

$$h^{new}(v) \geq h^{old}(v) - d\left(S^{old}, S^{new}\right) \tag{5.11}$$

This means if the cost $d(S^{old}, S^{new})$ is subtracted from the old admissible heuristic value of a node, the resultant heuristic function, which is not the ideal one, will still be admissible. The heuristic value being set is lesser than the ideal ones, which affects the computation time of the search, but being admissible does not affect the optimality.

Let us summarize two key deductions made so far. (i) If $d(S^{old}, S^{new})$ is subtracted from all heuristic values signifying priority, the resultant heuristic values are still admissible even if the robot moves by a distance of $d(S^{old}, S^{new})$. (ii) Furthermore, if any constant, say $d(S^{old}, S^{new})$ is subtracted from all priorities of a priority queue, the order remains the same and there is no effect. Combining both deductions, no change should be required in the priority queue. However, a new node (say v) being inserted into the priority queue will use the new heuristic function $h^{new}(v) = d(S^{new}, v)$ which needs to be adjusted for a lack of uniform subtraction of $d(S^{old}, S^{new})$ from all existing nodes in the priority queue. Hence, any new node being inserted into the priority queue should have $d(S^{old}, S^{new})$ added to maintain the order of nodes in the priority queue.

To sum this up, instead of subtracting $d(S^{old}, S^{new})$ from all nodes, $d(S^{old}, S^{new})$ is additionally added to the heuristic cost of the new node and any future node that may be added to the priority queue. This can be extended for the subsequent motion of the robot as well. Suppose the robot further moves from S^{new} to S_2^{new}, now $d(S^{old}, S^{new})$ needs to be added to any future node because of the calculations so far, and similarly, $d(S^{new}, S_2^{new})$ also needs to be added due to the subsequent motion, and hence, the overhead cost to add in all future insertions into the priority queue will be $d(S^{old}, S^{new}) + d(S^{new}, S_2^{new})$. This can be generalized to any number of motions of the robot and only needs to be done when there is a change in the cost of an edge.

So, the algorithm keeps adding the costs as the robot moves to a variable, say s, and the k-values are modified as given by Eqs (5.12), (5.13).

$$k(v) = [k_1(v), k_2(v)]$$
$$= [\min(g(v), rhs(v)) + h(v) + s, \ \min(g(v), rhs(v))] \tag{5.12}$$

$$s = \sum_i s_i = \sum_i d(q_{i-1}, q_i) \tag{5.13}$$

Where s is the total distance travelled by the robot, s_i is the distance travelled at ith change of any edge cost, q_i is the state at the ith time when a change in edge cost is detected ($q_0 = source$) and $d()$ is the distance function. The result is important because it enables the search to be invariant to the changing source of the robot, without iterating through all the nodes of the priority queue every time there is a motion of the robot. This is the first algorithm where re-planning is benefitted by the use of heuristics, making a realistic solution to the problem of re-planning for small changes in the map as the robot moves.

The other change introduced in the algorithm is relaxing the breaking condition. The typical breaking condition is $rhs(S) = g(S)$ for source S. Typically, the source S can be in any one of the three conditions, consistent ($g(S) = rhs(S)$), under-consistent ($g(S) < rhs(S)$) or over-consistent ($g(S) > rhs(S)$). The consistent state is suitable to break. The motion of the robot happens by using the rhs-values and not the g values, even though both the values are redundant in a consistent state. This is just because the rhs-values have a one-step lookahead and are more updated to a minute extent in case of changes in the edge costs. The over-consistent condition ($g(S) > rhs(S)$) emphasizes a lowered state and given that the priorities lower than the source are already processed, it is acceptable to terminate. $rhs(S)$ already has the updated value and that will be used for calculations. However, in case of being under-consistent ($g(S) < rhs(S)$), a raised event is said to have occurred and the new value of $g(S)$ will have to be calculated. $rhs(S)$ is itself based on the g values of the surrounding nodes and is also unstable.

The algorithm after making the modifications is shown as Algorithm 5.4, called the D*-Lite algorithm (Koenig and Likhachev, 2005). The algorithm assumes $\tau(i)$ as the position of the robot at step I, which is the position of the real robot and can also be obtained from the localization system. S is taken as a cumulative sum of the distance between the current position and the position at the last change (q) and is added to all future priorities. The changes to the LPA* are highlighted.

Algorithm 5.4 D^* lite.

Input: action function (A), feasibility function, cost function (c),
heuristic function (h), source (S) and goal (G)
Output: Path traversed by the robot

1   $g(G) \leftarrow 0, \pi(G) \leftarrow NIL, action(G) \leftarrow NIL$;
2   $k(G) \leftarrow [g(G) + h(G), \ g(G)]$;
3   $i \leftarrow 0, \tau(i) \leftarrow Source$;   ▷ Trajectory traced
4   $q \leftarrow Source$;   ▷ state at last step
5   $s \leftarrow 0$;   ▷ path length to be added to priorities
6   $Q \leftarrow$ empty priority queue, $enqueue(Q, G)$ with priority $k(G)$;
7   **while** $\tau(i) \neq G$ **do**   ▷ not reached goal
8   **while** $\neg empty(Q)$ **do**
9          $u \leftarrow front(Q), dequeue(Q)$;
10   **if** $k(u) \geq k(\tau(i)) \wedge rhs(\tau(i)) \leq g(\tau(i))$ **then** break;
             ▷ under-consistent $rhs(\tau(i)) < g(\tau(i))$ is added
11   **if** $g(u) > rhs(u)$ **then** $g(u) \leftarrow rhs(u)$;
12   **else** $g(u) \leftarrow \infty$;
13   **for** *(state v : any action a applied on v produces u)* $\cup \{u\}$ **do**
14   **if** $v \neq G$ **then**
15                $rhs(v) \leftarrow min_{a'u':a'(v)=u'} g(u') + c(v, a, u')$;
16                $[action(v), \pi(v)] \leftarrow argmin_{a'u':a'(v)=u'} g(u') + c(v, a', u')$;
17             $k(v) \leftarrow [min(g(v), rhs(v)) + h(v) + s, \ min(g(v), rhs(v))]$;
             ▷ added s to priorities
18   **if** $g(v) \neq rhs(v)$ **then**
19   **if** $v \notin Q$ **then** $Enqueue(Q, v)$;
20   **else** update priority of v in Q;
21   **else** remove v from Q, if present;
22   **if** *changed cost with any edge* **then**
23          $s \leftarrow s + d(q, \tau(i)), q \leftarrow \tau(i)$;   ▷ update path length
24   **for** *all edges v, a, u, with changed cost $c_{new}(v, a, u)$* **do**
25          $c(v, a, u) \leftarrow c_{new}(v, a, u)$;
26   **if** $v \neq G$ **then**
27             $rhs(v) \leftarrow min_{a'u':a'(v)=u'} g(u') + c(v, a, u')$;
28             $[action(v), \pi(v)] \leftarrow argmin_{a'u':a'(v)=u'} g(u') + c(v, a', u')$;
29          $k(v) \leftarrow [min(g(v), rhs(v)) + h(v) + s, \ min(g(v), rhs(v))]$;
30   **if** $g(v) \neq rhs(v)$ **then**
31   **if** $v \notin Q$ **then** $Enqueue(Q, v)$;
32   **else** update priority of v in Q;
33   **else** remove v from Q, if present;
34      $i \leftarrow i + 1$;
35      $a \leftarrow action(\tau(i)), \tau(i) \leftarrow \pi(\tau(i))$, Go to $\tau(i)$ by a ;   ▷ Next step
36   return τ

5.7 Other variants

There are many other issues related to the use of the A* algorithm over the problem of motion planning for robotics. This section is devoted to all such issues.

5.7.1 Anytime A*

A major problem with the A* algorithm is still the computational complexity. Even if the map is static, the robot can stand at the source and compute a path for a prolonged time and not make a move. This can be detrimental as the robot may waste a few seconds to get a smaller path of the order of a few milliseconds. The heuristic search was much faster but could be extremely suboptimal in numerous cases. A middle ground was the ε-A* algorithm in which the historic and the heuristic component were added in a weighted manner and the weight could tradeoff between computational time and optimality. The problem with the search is that this parameter is difficult to set during the design phase or even before the planning starts. The results highly depend upon the complexity of the problem and the map both. Therefore, it is unnatural to supply a robot with a set parameter that also meets the user's requirements. Also, it is hard for the end-user to set the parameter since it highly depends upon the map, which will change along with time.

The solution to the problem is to make the algorithm *iterative*. In motion planning, the same concept is known by the name of *anytime algorithms*. The iterative or anytime mechanism of solving a problem states that a solution can be extracted at any time of the solution searching phase; while the more time is given to the solver, the better will the results be. In such a case, the user can even halt the search process at runtime and the robot shall follow the best path known so far.

The A* and heuristic algorithms have a constraint that every node is processed at most once. So, in the worst case, every node gets into the fringe and out of fringe when it is a part of closed. Thereafter, even if the node is revisited, the node cannot be re-inserted back into the fringe. Since the search is iterative or anytime, the intent is to let nodes keep improving their costs. Therefore, if a node gets a better cost while closed, it may be re-inserted back into the fringe for propagating the lower cost. The concept is the same as D* in which re-insertions were instead due to changing the map.

The aim is to start from an ε-A* with a high ε making it effectively a heuristic search while ending at the A* algorithm. The only difference between the two algorithms is the cost function, and hence, let the algorithm have both cost functions, that is $f(u) = g(u) + h(u)$ and $f'(u) = g(u) + (1 + \varepsilon)h(u)$. The costs are maintained in parallel. Since the initial search should be fast, the processing is done by prioritizing the fringe based on the $f'(u)$ values. This gives a solution reasonably early. However, unlike a conventional search, the algorithm does not stop and waits for the solutions to improve. The improvement is done by letting the search process continue by taking nodes out of the fringe and processing them.

Imagine that the search continues to happen indefinitely until the fringe is empty. Since the graph is connected, every vertex gets into the fringe at least once. Imagine the optimal path to the goal as S, u_1, u_2, u_3, and so on. The vertex u_1 was directly produced from the source and therefore has an optimal value. Imagine u_2 has been already processed with a suboptimal cost because of a better heuristic component value, and u_1 is still in the fringe. At some point in time, the vertex u_1 will be extracted from the fringe, which will process the edge (u_1, u_2) and get the optimal cost to u_2. Then u_2 will be forced to get into the fringe again, even though it was already processed. Since u_2 is into the fringe, it will eventually get out and optimize the cost of u_3. This goes on and on until the optimal path is completely discovered. In worst cases, a vertex could get into and out of the fringe multiple times.

It is important to note that the initial search effort with a big heuristic component may first find a suboptimal path including processing of the goal. Hence, removal of goal from the fringe, as in the A* algorithm, is not a valid termination criterion. Two changes are desirable in the algorithm. In the A* algorithm it is known that, for an admissible heuristic, f-values of nodes keep increasing up to the f-value of the goal, when the search stops. Hence, if the f-value of the node being processed exceeds the best-known f-value of the goal, then it is a good condition to prune the node from being processed, as the subsequent successor nodes will go towards optimizing areas beyond the space between the source and goal as per the A* algorithm search ellipse. The other change in the algorithm is very similar. Any node with an f-value greater than that of the best-known goal is irrelevant to search and therefore such nodes may as well be pruned to be inserted into the fringe or processed at any time.

The algorithm is shown in Algorithm 5.5 and is very similar to the A* algorithm. The major differences are highlighted. The major difference is calculating both f- and f'-values while prioritizing on f'-values alone. Lines 8 and 11 suppress the processing of any vertex with a higher f-value (and not f' value) than the best-known goal. The only major change is Lines 12–16, wherein, if a goal is obtained, the best-known goal is updated. Since the search is iterative, better goals or the goal with a better cost can be found multiple times. The other major change is Lines 27–29, wherein a node may be re-inserted from closed into the fringe if a better cost is found. The terminal criterion based on goal is removed since the goal would be found multiple times. The search does not go indefinitely for infinite search spaces since there is a bound in the f-value when inserting or processing from a fringe.

Algorithm 5.5 Anytime A^*.

Input: action function (A), feasibility function, cost function (c),
      heuristic function (h), source (S) and goal (G)

Output: Path cost (d), shortest path (τ)

1   $g(S) \leftarrow 0, \pi(S) \leftarrow NIL, f(S) \leftarrow g(S) + h(S)$;

2   $f'(S) \leftarrow g(S) + (1 + \epsilon)h(S)$; ▷ priority: weighted heuristic

3   $Q \leftarrow$ empty priority queue, prioritized on f' values;

4   $enqueue(Q, S)$ with priority $f'(S)$, $closed \leftarrow empty$;

5   $best \leftarrow nil$;   ▷ best cost iteratively improves

6   **while** $\neg empty(Q) \wedge$ *not interrupted by user* **do**

7       $u \leftarrow Front(Q), dequeue(Q)$;
        ▷ pruning condition: u's best cost estimate $f(u)$
        should be less than the current best cost $f(best)$;

8   **if** $best = nil \vee f(u) < f(best)$ **then**

9   **for** *all actions a* **do**

10              $v \leftarrow$ state on application of action a on u ;

11   **if** $\neg feasible(v) \vee best \neq nil \wedge g(v) + h(v) \geq f(best)$ **then**
              continue;

12   **if** *GoalTest(v)* **then**   ▷ found better path to goal

13                 $g(v) \leftarrow g(u) + c(u, a, v), f(v) \leftarrow g(v) + h(v)$;

14                 $f'(v) \leftarrow g(v) + (1 + \epsilon)h(v)$;

15                 $\pi(v) \leftarrow u, action(v) \leftarrow a$;

16                 $best \leftarrow v$;

17   **if** $v \notin closed \wedge v \notin frontier$ **then**

18                 $g(v) \leftarrow g(u) + c(u, a, v), f(v) \leftarrow g(v) + h(v)$;

19                 $f'(v) \leftarrow g(v) + (1 + \epsilon)h(v)$;

20                 $\pi(v) \leftarrow u, action(v) \leftarrow a$;

21                 $enqueue(Q, v)$ with priority $f'(v)$;

22   **else if** $g(v) > g(u) + c(u, a, v)$ **then**

23                 $g(v) \leftarrow g(u) + c(u, a, v), f(v) \leftarrow g(v) + h(v)$;

24                 $f'(v) \leftarrow g(v) + (1 + \epsilon)h(v)$;

25                 $\pi(v) \leftarrow u, action(v) \leftarrow a$;

26   **if** $v \in frontier$ **then** update priority of v in Q to $f'(v)$;

27   **else if** $v \in closed$ **then**

28   Delete v from closed ;

29                   $enqueue(Q, v)$ with priority $f'(v)$;

30 **if** $best = nil$ **then** return ∞, NIL;

31 $\tau \leftarrow calculatePath(\pi, best)$;

32 return $d(best), \tau$

The algorithm is used to solve a small graph to explain the working and the process is shown in Fig. 5.14. The heuristic values are given in Table 5.1. In order to get the initial results in the fastest possible time, the priority is taken as the heuristic value or $f'(u) = h(u)$. At every iteration, the node with the smallest heuristic value is dequeued and processed. It can be seen that the first path was readily available, while it improved twice with time, giving the final path.

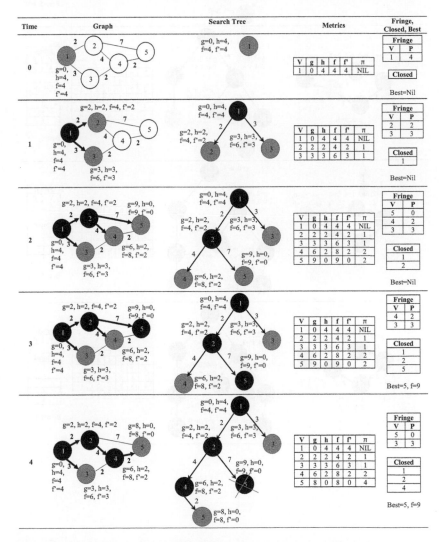

FIG. 5.14 Working of Anytime A* algorithm. Two cost functions are simultaneously used, $f(u) = g(u) + h(u)$ and $f'(u) = g(u) + (1 + \varepsilon) h(u)$. In this example $f'(u) = h(u)$ or $\varepsilon = \infty$ The fringe is prioritized by f' values.

(Continued)

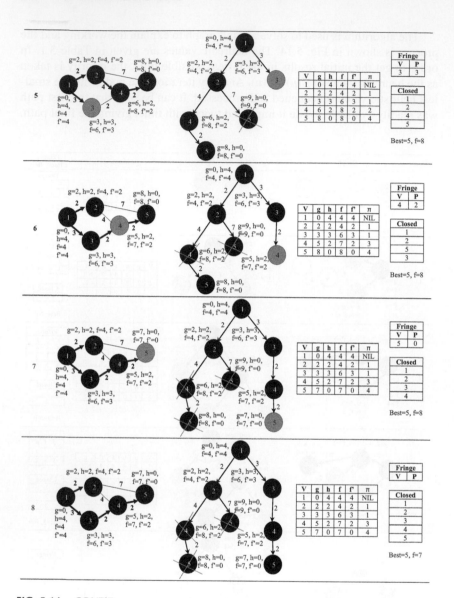

FIG. 5.14,   CONT'D

TABLE 5.1 Heuristic values.

Vertex	h
1	4
2	2
3	3
4	2
5	0

5.7.2   Theta*

Any Theta A* or *Theta** (Daniel et al., 2010) is an algorithm that removes the constraint of the fixed turning radius of the robot as discussed in Section 5.2.2, wherein the restriction was due to the restricted action resolution. The Theta* algorithm is a generalization of the A* algorithm covering the motion planning of a mobile robot in a grid, that checks for some additional paths that are neglected by the A* algorithm due to resolution. This ultimately removes the restriction of fixed angles imposed by the A* algorithm.

Imagine expanding a vertex u to produce a vertex v, which is done by the expansion of the edge (u,v), while the parent of the vertex u is $\pi(u)$. The sub-path produced as a result is $\pi(u)$ to u to v. However, A* ignores the possibility of going straight from $\pi(u)$ to v without passing via u, which if feasible will be a shorter path unless the three vertices are in a straight line. The situation is shown in Fig. 5.15A. Hence, in the Theta* algorithm, a child not only checks for a straight-line connection with the parent but also from the grandparent, to reduce the path length. Since u is already processed by a non-angle constraining algorithm, it means that $\pi(u)$ may not be the immediate neighbour of u but may as well be the result of such a shortening. This makes *Theta** free of any discretization in terms of actions (although the states are still discretized). In a grid, Theta* results in a path touching obstacle corners. The reduction is shown in Fig. 5.15B. Fig. 5.16 shows the results of the algorithm on a synthetic map.

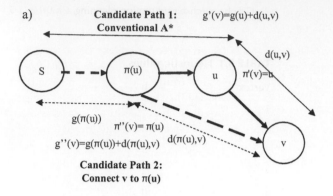

a) **Candidate Path 1:** $g'(v)=g(u)+d(u,v)$
Conventional A*

$d(u,v)$

S $\pi(u)$ u $\pi'(v)=u$

$g(\pi(u))$ $\pi''(v)= \pi(u)$

$g''(v)=g(\pi(u))+d(\pi(u),v)$ $d(\pi(u),v)$ v

Candidate Path 2:
Connect v to $\pi(u)$

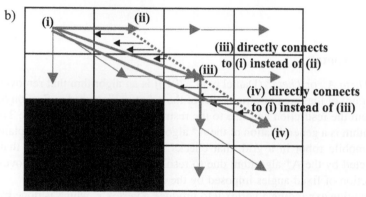

b)   (i)   (ii)

(iii) directly connects
to (i) instead of (ii)

(iii)

(iv) directly connects
to (i) instead of (iii)

(iv)

FIG. 5.15 Theta* Algorithm. (A) The algorithm accepts candidate path 2, if feasible, else candidate path 1. (B) Representation on a grid map.

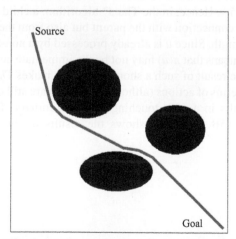

Source

Goal

FIG. 5.16 Results of Theta* algorithm on a grid map. The space between the path and obstacle is due to discretization. The path almost touches the obstacle without any discretization of turning angle.

The *Theta** algorithm is summarized as Algorithm 5.6. It can be seen that there is no change in the A* algorithm with the difference that if $\pi(u)$ to v is collision-free then a direct connection from $\pi(u)$ to v is made. The term action from classic AI does not have applicability here as the search is focussed upon grids, and hence the same is removed. The algorithm discussed here is a basic version of *Theta**. Another version of *Theta** instead propagates the angles and angle constraints from the parent vertex to the child vertex, which results in a faster implementation.

Algorithm 5.6 *Theta*.*

Input: action function (A), feasibility function, cost function (c),
heuristic function (h), source (S) and goal (G)

Output: Path cost (d), shortest path (τ)

1   $g(S) \leftarrow 0, f(S) \leftarrow g(S) + h(S), \pi(S) \leftarrow NIL$;
2   $Q \leftarrow$ empty priority queue, prioritized on f values ;
3   *enqueue*(Q, S) with priority $f(S)$;
4   *closed* $\leftarrow$ *empty* ;
5   *found* $\leftarrow$ *false* ;
6   **while** $\neg empty(Q)$ **do**
7      $u \leftarrow Front(Q)$;
8   **if** *GoalTest*(u) **then** *found* $\leftarrow$ *true*, *break*;
9   *dequeue*(Q) ;
10   **for** *all directions a* **do**
11        $v \leftarrow$ state on moving in direction a from u ;
12   **if** $\neg feasible(v)$ **then** continue;
         ▷ Try directly connecting $\pi(u)$ to v ;
13   **if** $\pi(u) \neq null \wedge (\pi(u)$ *to* v *is collision-free*) **then** $u' \leftarrow \pi(u)$;
14   **else** $u' \leftarrow u$;
15   **if** $v \notin closed \wedge v \notin frontier$ **then**
16           $g(v) \leftarrow g(u') + c(u', v)$;
17           $f(v) \leftarrow g(v) + h(v)$;
18           $\pi(v) \leftarrow u'$;
19   *enqueue*(Q, v) with priority $f(v)$;
20   **else if** $v \notin closed \wedge v \in frontier \wedge g(v) > g(u') + c(u', v)$ **then**
21           $g(v) \leftarrow g(u') + c(u', v)$;
22           $f(v) \leftarrow g(v) + h(v)$;
23           $\pi(v) \leftarrow u'$;
24   update priority of v in Q to $f(v)$;

25   **if** $\neg found$ **then** return ∞, NIL;
26   $\tau \leftarrow calculatePath(\pi, v)$;
27   return $d(v), \tau$

5.7.3 Learning heuristics

So far the search assumed a fixed source and a fixed goal, while the aim was to compute the shortest path between the source and the goal. We now divert the attention to multi-query problems wherein the robot raises search queries multiple times. This may be because the robot may slip and need a new path from the updated source or multiple robots need to reach the same place. The assumption here is that the goal remains unchanged. Because of the same, the results of the initial search may be re-utilized by some means by all future searches for the same goal. This introduces the notion of *learning*, wherein the experience of finding the shortest path to a goal by one robot could be used by the same robot again or by the other robots.

Suppose a search is made which produces a path τ with a cost C^* and a closed list (*closed*). Suppose a vertex u inside is closed. $g(u)$ is the optimal cost from source to u, which is known after the search. Let $h^*(u)$ be the best-known heuristic function which may not be the initially given heuristic function $h(u)$. This means $f(u) = g(u) + h^*(u) \geq C^*$, or $h^*(u) \geq C^* - g(u)$. It may be intuitive that the equality is enough considering the optimal cost C^* is known; however, the possibility of $f(u) > C^*$ is taken for the case considering the cost through u may be larger when the optimal path giving C^* does not pass-through u. The best estimate that can be taken for the heuristic function for learning is $H(u) = C^* - g(u)$. These learned values $H(u)$ are better than the initial estimate $h(u)$ and are stored separately. For now, mere storage of these values externally is the learning, which may not coincide with the definition of learning as per popular machine learning libraries.

The algorithm is hence simple, for every query, we learn the heuristic function $H(u)$ for every u in closed. The A* algorithm should itself benefit from the learned heuristics and hence uses $H(u)$ instead of the given heuristic function $h(u)$. The heuristic function points to $h(u)$ values if nothing is learned. The general structure of the algorithm is given by Algorithm 5.7.

Algorithm 5.7 Learning heuristics.

1 $[\tau, C^*, closed, g] \leftarrow A^*()$;
   ▷ C^* : optimal cost from source to goal;
   ▷ H : learnt heuristic function;
2 **for** $u \in closed$ **do**
3 | **if** $u \in H$ **then** $H(u) \leftarrow max(H(u), C^* - g(u))$;
4 | **else** $H(u) \leftarrow C^* - g(u)$;
5 **Function** getHeuristic(s):
6 | **if** $u \in H$ **then** return $H(u)$
7 | **else** return $h(u)$ ▷ conventional heuristic function
8 |

5.7.4 Learning real-time A*

Another major problem associated with the A* algorithm is that the robot waits for a very long time searching for a plan, before even making a single move. This can be counter-intuitive as the robot can take a few seconds to optimize a plan, which after optimization is only marginally better than a plan computed in a small time. Furthermore, it is not very appealing for robots taking the time to make the initial moves. This section, hence, adds a real-time element to the search wherein, the robot makes immediate moves based on the currently known calculations and the calculations of only the surrounding few vertices. The algorithm is called the *Learning Real-Time A*** (LRTA*) algorithm.

Another motivation is when the robot operates at a partially known map. The robot can use the onboard sensors to look at the current vertex and its neighbours, however, it cannot look at the distant vertices which may be occluded. Many times, the robot only has limited proximity sensors that can only sense obstacles in the vicinity. In such a case, the immediate decision making needs to be made based on the known partial map alone and the complete map is not available to perform a global search.

The robot moves in a greedy mechanism, taking the best available action out of the available successor options, or chooses the successor that has the minimum expected cost to the goal. Suppose the robot is at vertex u, the action taken by the robot will be given by Eq. (5.14).

$$a(u) = \arg\min_a c(u, a, v) + H(v) \tag{5.14}$$

The first term is the cost to move from u to v and the second term is the estimated distance from v to the goal. $H(v)$ denotes the heuristic function being learned. If a large part of the map is visible from the current position and a large time is available for the immediate move, it is possible to take a larger neighbourhood or consider more target vertices v. In such a case, the robot plans for all nodes till a certain *planning horizon* or till a certain cost from the current position (assuming heuristic estimates beyond the planning horizon). The algorithm selects the best vertex v at the horizon that promises the best total cost considering the cost from the current position and the heuristic estimates. The first move of the plan is used for the motion, and the process is iteratively repeated with new such planning at every step.

While making moves, the robot learns the heuristic function from every possible move. Let us define the heuristic function as the distance from the vertex (u) to the goal. Assume the shortest distance from u to goal has v as the next vertex, making the heuristic function as the cost from u to v and v to goal, while selecting the best out of all options (v). The heuristic function for the vertex u can be conveniently defined as given by Eq. (5.15)

$$H(u) \leftarrow \max(\min_a(c(u, a, v) + H(v)), H(u)) \tag{5.15}$$

This is called a one-step lookahead heuristic estimate, wherein the heuristic function peeks into the heuristic values of all successors to calculate a better heuristic value. The arrow indicates the heuristic values continuously adapting themselves, starting from the initial heuristic function designed by the relaxed problem method.

The pseudo-code is given by Algorithm 5.8 and is a straight implementation of a greedy decision-making to take the least expected cost and a one-step lookahead-based heuristic learning (Korf, 2006; Bulitko and Lee, 2006). The algorithm may be called multiple times by multiple robots or by the same robot, and the heuristic function keeps updating itself to better values.

Algorithm 5.8 *LRTA*.*

Input: action function (A), feasibility function, cost function (c),
  heuristic function, current state (S), goal test ($GoalTest$)
Output: Traversed Path (τ), whether goal was reached

1   $u \leftarrow S$;   ▷ current state
2   $\tau = \{S\}$;
3   $H(u) = h(u)$;   ▷ initialize learnt heuristic by the given
  heuristic function
4   **while** $\neg GoalTest(u)$ **do**
5      $a_{min} \leftarrow NIL, c_{min} \leftarrow \infty$;
6   **for** *all directions a from u* **do**
7          $v \leftarrow$ state on moving in direction a from u;
8   **if** $\neg feasible(v)$ **then** continue;
9   **if** $v \notin H$ **then** $H(v) = h(v)$;
10   **if** $c(u, a, v) + h(v) < c_{min}$ **then**
11          $c_{min} \leftarrow c(u, a, v) + H(v), a_{min} \leftarrow a$
12   **if** $a_{min} \neq NIL$ **then**
13        $H(u) \leftarrow max(c_{min}, H(u))$;   ▷ learning heuristic
14   Execute action a_{min} to move to state v ;   ▷ greedy action
15        $u \leftarrow v$;
16   add u to τ ;
17   **else** return $\tau, false$;
18   return $\tau, true$;

Let us discuss Algorithm 5.8 in the context of a tree search variant of the A* algorithm rather than a greedy path to goal. Assume all edges are bidirectional. The reason for this assumption is that a unidirectional edge from u to v may be taken due to a very good heuristic value of v; however, v may have no outgoing edges resulting in a dead-end that the robot cannot get out of. Having all edges as bidirectional ensures that there is no trap. A general heuristic search as a tree variant is not complete. However, in this case, upon every step, the heuristic function is learned, typically increased to a more realistic estimate. The algorithm may get stuck at a local optima and oscillate between a vertex sequence in a loop by constantly visiting them; however, upon every visit the heuristic function will be increased for the vertices in the sequence, and eventually the edge cost plus heuristic estimate will exceed an out of loop sequence node, and the algorithm will prioritize the out of loop sequence node, rendering the algorithm complete for the case of a bidirectional graph. If the algorithm visits the nodes for sufficient number of time, it can learn the ideal heuristic function (h^*) with optimal cost to goal in which a greedy search leads to the goal.

Sometimes a graph search variant may be better instead of a tree search, where the exploration of an already visited vertex again is pruned. This can be built on top of Algorithm 5.8 by adding the fringe nodes and closed nodes on top of the search process. The fringe in this case is only a list of nodes that require processing. The next node to be selected (or the fringe prioritization policy) is from the immediate successors using Eq. (5.14). This can be understood as the source is always the current node being processed where the robot is physically located. The cost from source is, therefore, the cost from this node, that for the successor nodes is the edge cost for a grid-map kind of environment with all step costs as 1. A successor node typically has the least cost, unless the heuristics suddenly becomes very high.

A problem with this implementation is that the robot will get stuck in a local optimum, wherein it is not possible to move to a successor state since all successor states may have been already processed and are in *closed*. It is still possible to find the next node in the fringe with the smallest cost with the updated source, which typically would be a sibling of the current node (parent's best-unprocessed child excluding the current node). In such a case, the A* algorithm would have used backtracking to go to the node and continue the search. In A* the robot is at the source (at the start of the planning) and the planning is done ahead of time. Hence, as the algorithm backtracks, no physical navigation of the robot is required. However, in the case of LRTA*, the robot physically navigates, and therefore, backtracking will require the robot to physically navigate to the backtracked node. The backtracking always leads to going back to a parent state. Assuming that the edges are bidirectional, the robot starts to retrace the path for backtracking. The robot backtracks to the parent whose all actions have not yet been tried, and the untried actions (in the priority as per cost) are now attempted as per the set heuristic. If the edges were not bidirectional, the robot will have to perform an exhaustive search to backtrack and a path may not always be possible, leading to the robot being stuck. Luckily, in most cases the edges are bidirectional. Further, if the motion for the LRTA* was a partially known map, and the robot during its navigation performed mapping near the visited areas, it would be possible to do a global path search to the next backtracked node. Taking fringe and closed nodes eliminate the possibility of the robot unnecessarily visiting places multiple times. The robot does not revisit a place, in which case it is added to *closed*. Upon visiting a vertex only those actions are tried which have never been tried previously.

5.7.5 ALT heuristics

So far, we have seen the mechanism of learning a heuristic function when the goal is fixed. Imagine that the goal moves, or some other robots operate in the same environment that needs to plan between a separate pair of source and goal. Such problems are common with robotics under the category of multi-query approaches wherein the same environment is queried multiple times for a new pair of source and goal. Navigating within the home and office environment is an appropriate example. One navigates numerous times within the campus or society and every such navigation is a query on the same map.

This section presents a mechanism to construct very good heuristic functions for such multi-query approaches where both the source and goal are variable. The approach is known as *A* Landmark Triangulation* (ALT) and makes use of a large number of *landmarks* (*l*). Let the landmarks be given by $L = \{l\}$. The landmarks are spread around the search space. The distance between the landmark and every other point in the search space can be computed in an offline manner. The assumption is that the map is static and therefore a lot of time is available to compute any metadata associated with search. However, the source and goal are not available. When the robot comes into the environment and queries for a path between a pair of source and goal, the path should be readily available by using any precomputed metadata.

Suppose $d(l,u)$ is the pre-calculated distance between every landmark *l* and every other vertex *u*. This is the exact distance calculated through an exhaustive search technique. Suppose *G* is the goal. The heuristic function is thus $h^*(u) = d(u, G)$. It can be conveniently assumed that the distances $d(l, G)$ and $d(l, u)$ are known. Consider the triangle formed by *u*, *l*, and *G*, as shown in Fig. 5.17. Using triangle inequality, we get $d(l,G) \leq d(l,u) + d(u,G)$ or $d(l,G) \leq d(l,u) + h^*(u)$. Hence, $h^*(u) \geq d(l,G) - d(l,u)$. This means if a heuristic function is taken as $d(l, G)-d(l, u)$, it underestimates the optimal cost to the goal and is a good candidate for a heuristic function. Such a heuristic function can be made for all possible landmarks, while the best heuristic function will be the maximum of all of them. The heuristic function taken is hence given by Eq. (5.16):

$$h(u) = \max_l d(l,G) - d(l,u) \tag{5.16}$$

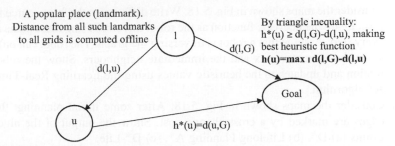

FIG. 5.17 ALT Triangulation heuristic. Works for any source/goal using landmarks pre-computed offline.

The heuristic function so designed assumes that the heuristic will be taken through a landmark and neglects the straight-line distance from the source to goal. This may sometimes lead to a non-admissible heuristic will a minute loss of optimality. A little discount to the heuristic component when running the A* algorithm is sufficient to account for the small loss of optimality.

Questions

1. Consider the maps shown in Fig. 5.18. Plan the maps by using Anytime A* algorithm.

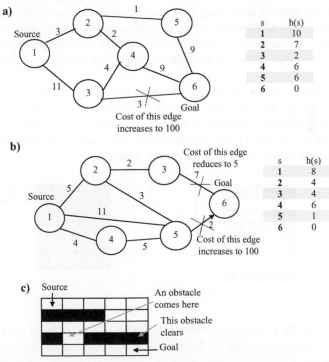

FIG. 5.18 Sample maps for calculation of shortest path. (A) Problem 1. (B) Problem 2. (C) Problem 3.

2. Consider the maps shown in Fig. 5.18. Write different paths from source to goal. Learn the heuristic function as the robot traverses using these paths at every step. Assume that a robot is traversing the graph. The robot can only sense the current vertex and the immediate neighbours. Show the robot motion and updates to the heuristic values using the Learning Real-Time A* algorithm.

3. Consider the maps shown in Fig. 5.18. After some initial planning, the edges are marked by a cross-change cost. Show the output of the algorithms (a) D*, (b) Lifelong Planning A*, (c) D*-Lite.

4. Compare and contrast between the A* algorithm and the anytime A* algorithm for different real-life scenarios.

5. Suppose that the robot is planned using the A* algorithm. Would the algorithm be optimal and complete if (a) obstacles are added while A* algorithm is computing the results, (b) obstacles are deleted when A* algorithm is computing the results, (c) some obstacle is moved while A* algorithm is computing the results, and (d) some obstacles are altered after the A* algorithm computes the results and while the robot is tracing the trajectory computed.

6. State whether the D* algorithm will work in the following situations or not: (a) the goal is changed to a new location, (b) the entire map suddenly changes, (c) the robot slips and reaches a new location nearby which was not the one initially desired, (d) an enemy surrounds the robot from all sides and traps the robot with no other way to go, (e) a small obstacle suddenly appears in the map, (f) a very large obstacle suddenly appears in the map, (g) an obstacle moves at a small speed in the map, (h) large changes are applied to an otherwise sparsely occupied map, while all the changes lie at the map boundaries, and (i) the cost function is changed dynamically.

7. Consider the robot operating in maps shown in Fig. 5.19. Show the path traversed by the robot for different resolutions. Does the robot travel through different corridors in different resolutions? Comment upon the change in optimality and completeness with resolution. Resolutions of states and actions should be discussed separately.

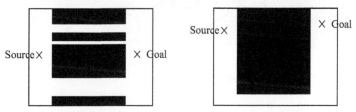

FIG. 5.19 Maps to study the effect of resolution.

8. Which of the following problems cannot be solved by using (a) Forward Search Algorithms, and (b) Backward Search Algorithms?

- In a grid map, finding the shortest path from a given source to either of the given k goals.
- In a grid map, finding a path from a given source to a given goal, with a moving obstacle walking on a known trajectory with a constant speed.
- In a grid map, finding a path from a given source to a given goal, with a moving obstacle walking on an unknown trajectory.
- Playing a deterministic Atari game with a game simulator, for which the effects of the action can be seen on the screen, but the internal working of the game is unknown. A programmable simulator is available.
- In a map moving with 4 actions up, down, right, and left. However, the distance moved in either of the actions is $1 + \log(1 + ct)$, where c is a known constant and t is the number of times the same action has been used in the immediate past consecutively without being interrupted by any other action.

9. [Programming] Make a high-resolution bitmap. Select a source and goal in the bitmap. Solve the following questions:
   - Plan the robot from source to goal.
   - Plan the robot from source to goal using *Theta** algorithm
   - Plan the robot such that a minimum clearance of C_{min} is guaranteed.
   - Plan the robot such that a minimum clearance of C_{min} is guaranteed and a penalty of $C_{max} - C$ per unit move is given if the clearance is less than C_{max} (and greater than C_{min}).
   - Plan the robot such that it reaches either of two goals, separately marked.
   - Plan the robot, such that it reaches both goals in any order.
   - Plan the robot such that it reaches the first goal and then the second goal.
   - Add a robot R that traverses diagonally in the map with a known speed. Plan a robot that avoids static obstacles and known robot R.
   - Assume that the robot can push small obstacles of size less than A. The push happens by banging on the opposite side of the obstacle. The moves and push can only happen in the 4 rectilinear directions. Plan the robot.
   - Add a small obstacle in the path of the robot and re-plan using D*, Life-long Planning A*, D*-Lite algorithms.

10. [Programming] A robot may not always have a binary map. Sometimes robots have a non-binary *cost map*, where every grid denotes the relative difficulty in moving through the grid (e.g. slippery floor, rough terrain, etc.).
    - Paint 5 different terrains with 5 different colours and associate every terrain with a cost. Plan a robot from a predetermined source to a predetermined goal.
    - Suppose the cost map is the height of obstacles. Suppose the robot can jump with a cost C for a maximum s steps. It may be better to jump over an object rather than climbing up and down an obstacle. Plan the robot

under the settings. Modify the search such that the robot can only make a maximum of 1 jump in T time.

11. [Programming] Make a high-resolution bitmap. Select a source and goal in the bitmap. Make a function that downgrades the map to a lower resolution, plans a path, and projects the path up to the higher resolution map. Study the effect of (a) state resolution and (b) action resolution on path length, computation time, and completeness for different maps.

12. [Programming] In the maps used in previous questions, assume that a robot wishes to traverse from a source to a goal; however, the robot can only sense the current grid and the immediately neighbouring grids. Show the path traced by the robot when using the Learning Real-Time A* (LRTA*) algorithm. Show how the path changes as the neighbourhood that the robot can sense is increased.

13. [Programming] In the maps used in previous questions, study the effect of increasing the number of landmarks of ALT heuristics on the same metrics. Also, write a program to teleoperate a synthetic robot from different sources and learn the heuristics as the robot moves. Study if the planning with leaned heuristics improves the performance of the future runs, and on what metrics.

14. [Programming] In the maps used in previous questions, for different combinations of source and goal, show how the path length varies with time using Anytime A* algorithm.

References

Bulitko, V., Lee, G., 2006. Learning in real-time search: a unifying framework. J. Artif. Intell. Res. 25, 119–157.

Choset, H., Lynch, K.M., Hutchinson, S., Kantor, G., Burgard, W., Kavraki, L.E., Thrun, S., 2005. Principles of Robot Motion: Theory, Algorithms, and Implementations. MIT Press, Cambridge, MA.

Daniel, K., Nash, A., Koenig, S., 2010. Theta*: any-angle path planning on grids. J. Artif. Intell. Res. 39, 533–579.

Hart, P.E., Nilsson, N.J., Raphael, B., 1968. A formal basis for the heuristic determination of minimum cost paths. IEEE Trans. Syst. Sci. Cybern. 4 (2), 100–107.

Hart, P.E., Nilsson, N.J., Raphael, B., 1972. Correction to 'a formal basis for the heuristic determination of minimum cost paths'. ACM SIGART Bull. 37, 28–29.

Kala, R., Shukla, A., Tiwari, R., Rungta, S., Janghel, R.R., 2009. Mobile robot navigation control in moving obstacle environment using genetic algorithm, artificial neural networks and A* algorithm. In: Proceedings of the IEEE World Congress on Computer Science and Information Engineering. IEEE, Los Angeles/Anaheim, pp. 705–713.

Kala, R., Shukla, A., Tiwari, R., 2010. Fusion of probabilistic A* algorithm and fuzzy inference system for robotic path planning. Artif. Intell. Rev. 33 (4), 275–306.

Kala, R., Shukla, A., Tiwari, R., 2011. Robotic path planning in static environment using hierarchical multi-neuron heuristic search and probability based fitness. Neurocomputing 74 (14–15), 2314–2335.

Koenig, S., Likhachev, M., 2005. Fast replanning for navigation in unknown terrain. IEEE Trans. Robot. 21 (3), 354–363.

Koenig, S., Likhachev, M., Furcy, D., 2004. Lifelong Planning A*. Artif. Intell. 155 (1–2), 93–146.

Korf, R., 2006. Real-time heuristic search. Artif. Intell. 42 (2–3), 189–211.

Russell, S., Norvig, P., 2010. Artificial Intelligence: A Modern Approach. Pearson, Upper Saddle River, NJ.

Stentz, A., 1994. Optimal and efficient path planning for partially-known environments. In: Proceedings of the 1994 IEEE International Conference on Robotics and Automation. IEEE, San Diego, CA, pp. 3310–3317.

Stentz, A., 1995. The focussed D* algorithm for real-time replanning. In: Proceedings of the 1994 IJCAI'95 Proceedings of the 14th International Joint Conference on Artificial Intelligence – Volume 2. IEEE, Montreal, Quebec, Canada, pp. 1652–1659.

Tiwari, R., Shukla, A., Kala, R., 2012. Intelligent Planning for Mobile Robotics: Algorithmic Approaches. IGI Global, Hershey, PA.

Chapter 6

Configuration space and collision checking

6.1 Introduction

The problem of motion planning of a robot is to enable the robot to go from the current pose (or source) to the desired pose (or goal) without colliding with any obstacle on the way. Examples include navigating a car in dense roads, navigating to parallel park, and solving a maze. In such cases, a human visualizes different steps or actions, for example, while driving a car on needing a tight manoeuvre at an intersection, a person visualizes whether the move will be possible or will incur a collision. Similarly, in solving a maze, a human visualizes different routes to see if they reach the goal without collision.

Motion planning algorithms thus visualize the state of the moving entity in the entire environment and check whether the same is feasible or incurs a collision. This is done by encoding the possible situation into a configuration and thereafter checking if the configurations are feasible or not. This chapter is devoted to this exercise that forms the backbone of all motion planning algorithms.

6.2 Configuration and configuration space

Let us encode a moving entity, typically a robot, using a few variables. In the case of a car, the position (x, y) and orientation (θ) of the car completely describe the situation for a static scenario. The other variables such as the dimension of the car and positions of obstacles are constant and therefore can be removed from the description of the situation. The variables like weather conditions, lightning, and time of the day do not contribute to decision-making regarding navigation, despite being constant for the manoeuvre, and can be left in the description as well.

The minimal representation that completely describes the situation at any point of time, needed for decision-making, is called the *configuration* (Choset et al., 2005; LaValle, 2006), and all variables that participate in the

Autonomous Mobile Robots. https://doi.org/10.1016/B978-0-443-18908-1.00021-2

configuration are called the *configuration variables*. The configuration encodes the problem into a set of real numbers that can be used by computer programs to solve the problem. For the example of a car, the triplicate (x, y, θ) encodes the configuration. For the maze-solving problem, the aim is to only move a point which can be encoded by the coordinates (x, y).

The physical robot or entity cannot be moved around to check if it incurs a collision or not; however, a virtual ghost robot or an avatar of the moving entity can be moved around in a computer program to check for collisions without damaging the real robot. While operating a car manually, the human typically imagines a ghost car, and if the ghost car makes a move sequence without incurring collisions, the real car is allowed to make the moves. To incur the same functionality in the problem of motion planning, it is important to know whether a configuration will incur collisions or not.

Collision checking algorithms take several polygons as input and check if the polygons intersect with each other or not. A set of polygons denotes the map of the environment, while another set of polygons denotes the robot. If the robot intersects with any obstacle or with itself, a collision can be detected by using these algorithms. A *collision checking query* can thus be made, giving a configuration of the robot in a known map, to determine if the configuration is free or incurs a collision with the obstacle or the robot collides with itself.

The map of the world constitutes the *ambient space* of the robot. The subspace which is reachable by a characteristic point of the robot (or the end-effector in case of manipulators) is called *workspace* (W). Typically, for mobile robots, no distinction is made between the workspace and the ambient space because the robot can move to most of the areas and operate. Hence, the terms may seldom be used interchangeably. In case of a manipulator with a fixed base, the ambient space consists of the entire world, out of which only a subset of the space can be reached by the end-effector of the manipulator which lies within the maximum reach of the manipulator. Thus, the workspace (set of points reachable by the manipulator end-effector) is a subset of the ambient space (entire world). The set of points in the real world, which is free of any obstacle, constitutes the *free workspace* or W^{free}. Similarly, the set of points in the real world that is occupied by obstacles constitutes the obstacle-prone workspace or W^{obs}. Thus, $W^{\text{free}} = W \backslash W^{\text{obs}}$ means W^{free} is the compliment of W^{obs} on the superset W.

It is imperative to thus classify configurations into two categories, one which is free of collisions and the second which either incur collisions with the environment, or incurs collisions of the robot with itself, or the configuration is not possible for the robot to take for other reason. It is imperative to visualize the configurations in space. A collection of all configurations is called the *configuration space* (C). Imagine a car with configuration (x, y, θ).

The new space has every configuration variable as an axis. In the example, there will be three axes for the three variables, 1 denoting the X-axis with the x value, 1 denoting the Y-axis with the y value, and the last denoting the Θ axis with the θ value.

The set of all configurations without any collisions constitutes the *free configuration space* (C^{free}), while the configurations which incur collisions or are otherwise not possible constitute the forbidden configuration space or the collision-prone configuration space (C^{obs}). Thus, $C^{\text{free}} = C \backslash C^{\text{obs}}$, which means C^{free}, is the compliment of C^{obs} on the superset C. Imagine any general configuration (x, y, θ) of a car. This configuration is a point in the configuration space. Now imagine a ghost car placed at (x, y) with the orientation θ. This ghost car may intersect some obstacle, in which case the point (x, y, θ) in the configuration space is coloured black and the point belongs to C^{obs}. On the contrary, if the ghost car does not collide with any obstacle, then the point (x, y, θ) in the configuration space is coloured white and the point belongs to C^{free}. This makes a black and white space, where the collection of all black points is C^{obs}, collection of all white points is C^{free}, and the union of the two is C.

It is important to note the differences between the workspace (more concretely the ambient space) and configuration space. The workspace is the physical space where the robot moves. It is sensed by the sensors and so constructed. The robot is a real mass with size in the workspace. However, configuration space is constructed to aid and define motion planning. The complete robot is just a point, as it is in the encoded format in this space. Every point may be collision-prone or collision-free depending upon the results of collision checking. A point in the configuration space can be visualized as a ghost robot occupying a volume in the workspace placed at the configuration of the chosen point in the configuration space. Similarly, every ghost robot in the workspace can be mapped into a point in the configuration space by noting the values of all the configuration variables of the ghost robot.

More formally, let the robot be placed at the configuration q. Let the set of all points occupied by the robot in the workspace is given by $R(q)$. There is an easy way to define the function $R(q)$. Let $R(0)$ be the set of all points in the frame of reference of the characteristic point of the robot called the robot placed at an absolute zero. As the robot is transformed into the current configuration q, all points undergo some transformation, say $T(q)$. The transformed set of points is hence given by Eq. (6.1):

$$R(q) = T(q)R(0) \tag{6.1}$$

A configuration q is called collision-free, if the robot collides neither with an obstacle nor with itself, given by Eq. (6.2):

$$q \in C^{\text{free}} \text{ iff } R(q) \cap W^{\text{obs}} = \emptyset \wedge \neg \text{selfCollisions}(R(q)) \tag{6.2}$$

In other words, a configuration q belongs to the free configuration space (C^{free}) if and only if the intersection of all points ($R(q)$) occupied by the robot in the workspace at configuration q, with all obstacles in the workspace (W^{obs}) is a null set. Assuming that the robot is made of smaller primitive objects $R(q) = \{r(q)\}$ and the workspace itself is made up of smaller objects $W^{obs} = \{a\}$, this simplifies to Eq. (6.3):

$$q \in C^{free} \textit{iff} \left(\forall_{r(q) \in R(q), a \in W^{obs}} r(q) \cap a = \emptyset \right)$$
$$\wedge \left(\forall_{r_1(q) \in R(q), r_2(q) \in R(q), r_1(q) \neq r_2(q)} r_1(q) \cap r_2(q) = \emptyset \right) \tag{6.3}$$

The equation states that a configuration q belongs to free configuration space (C^{free}) if and only if the intersection between every primitive object ($r(q)$) of the robot and every primitive object of the workspace (a) is null, and so must be the intersection between every two distinct primitive objects constituting the robot (say $r_1(q)$ and $r_2(q)$). The free configuration space is a set of all free configurations, or Eq. (6.4):

$$C^{free} = \left\{ q: \left(\forall_{r(q) \in R(q), a \in W^{obs}} r(q) \cap a = \emptyset \right) \right.$$
$$\left. \wedge \left(\forall_{r_1(q) \in R(q), r_2(2) \in R(q), r_1(q) \neq r_2(q)} r_1(q) \cap r_2(q) = \emptyset \right) \right\} \tag{6.4}$$

Let us study this by a very small example. A very small room in a house has a size of 300 cm × 300 cm. One corner of the room is used as the origin with the axis system as shown in Fig. 6.1A. There are two obstacles in the room as shown in the same figure. The workspace of the robot is hence $W = [0,300] \times [0,300] = [0,300]^2$ and denotes all possible areas in question. The first obstacle has an occupancy of $W_1^{obs} = [75,125] \times [180,230]$ and the second obstacle has an occupancy of $W_2^{obs} = [180,230] \times [80,130]$. The two obstacles together make the obstacle-prone workspace given by $W^{obs} = [75,125] \times [180,230] \cup [180,230] \times [80,130]$. The free workspace is the leftover space, given by $W^{free} = [0,300]^2 / ([175,125] \times [180,230] \cup [180,230] \times [80,130])$. The complementation (/) can also be replaced with a set difference.

Imagine the robot is a square of size length (l) × width (w) = 20 cm × 20 cm, with the centre of the square used to denote its position. Consider that the robot can be translated but not rotated. In such a case, the configuration of the robot is given by (x, y). Since the robot must always be in the room, x and y can take values in the range [0,300]. This means the configuration space is given by $C = [0,300] \times [0,300] = [0,300]^2$. By rolling a ghost robot in the entire space, the configurations that incur collisions are noted. The normal occupancy of the robot is given by $R(0) = [-10,10] \times [-10,10] = [-10,10]^2$, which denotes the robot with the centre at the origin. If the robot is translated by (x, y) denoting the robot's configuration, the occupancy of the robot is given by $R(q\langle x,y \rangle) = [x - 10, x + 10] \times [y - 10, y + 10]$. In terms of transformations, the same function is also given by Eq. (6.5), where the additional 1 is to add homogenization.

$$R(q\langle x, y, 1 \rangle) = T(q)R(0) = \begin{bmatrix} 1 & 0 & x \\ 0 & 1 & y \\ 0 & 0 & 1 \end{bmatrix} \times \begin{bmatrix} R_X(0) \\ R_Y(0) \\ 1 \end{bmatrix},$$

$$R_X(0) = [-10, 10], R_Y(0) = [-10, 10] \tag{6.5}$$

Here, $R_X(0) = [-10,10]$ (and similarly $R_Y(0)$) is the set of all X values (similarly Y values) occupied by the robot when placed at an absolute 0. Each combination of X and Y values is transformed by $T(q)$ to get the same point with the robot placed at q.

The ghost robot is shown in different configurations as a lined box. If the ghost robot intersects the obstacle, the configuration (centre position) is collision-prone and vice versa. The limiting cases when the robot touches the obstacles are of interest since they form the boundary of the free and obstacle-prone configuration space. For the obstacle $[75,125] \times [180,230]$, the robot cannot occupy the space $C_1^{obs} = [75 - l/2, 125 + l/2] \times [180 - w/2, 230 + w/2] = [65,135] \times [170,240]$, since the robot kept at such a centre shall collide with the obstacle. Similarly, the second obstacle $[180,230] \times [80,130]$ gives the obstacle-prone configurations as $C_2^{obs} = [170,240] \times [70,140]$. The workspace boundary also acts as an obstacle, giving $C_{boundary}^{obs} = [0,300]^2/(10,290)^2$. If the robot is placed with centre at $x \leq l/2$ or $x \leq 10$, the left part of the robot body shall intersect the map boundary making a collision and similarly for $x \geq 300 - l/2$ or $x \geq 290$. The robot will not collide with any boundary when its centre is at $x \in (10,290)$ and $y \in (10,290)$, meaning $(10,290)^2$ is the free configuration space with respect to the map boundary. Correspondingly, its compliment, $C_{boundary}^{obs} = [0,300]^2/(10,290)^2$, is collision-prone with respect to the boundary. The obstacle-prone configuration space is thus given by $C^{obs} = C_1^{obs} \cup C_2^{obs} \cup C_{boundary}^{obs} = [65,135] \times [170,240] \cup [170,240] \times [70,140] \cup ([0,300]^2/(10,290)^2)$. The remaining space is the free configuration space ($C^{free} = C/C^{obs}$). The limiting cases are drawn as lined boxes. The centre of the box denotes the position in the configuration space, which marks the boundary. Correspondingly, the configuration space is shown in Fig. 6.1B and appears as all boundaries inflated by the size of the robot.

Imagine a similar workspace shown in Fig. 6.1C and D with the space between the obstacles lesser than the size of the robot. Upon doing the same calculations, it turns up that there is no configuration free in-between the two obstacles after rolling the ghost robot around the obstacles and inflating the obstacles. This is true because there is no way that the big robot could have fitted in the small space between the two obstacles.

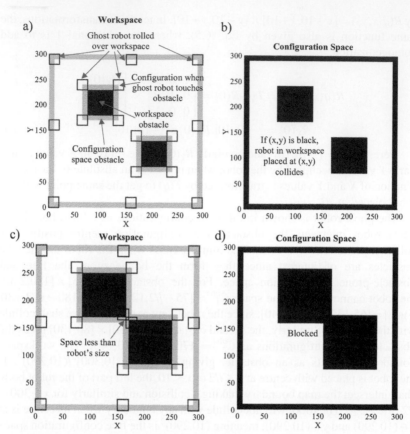

FIG. 6.1 Construction of configuration space. If the ghost robot centred at (x, y) configuration collides in the workspace (A), then (x, y) is marked as an obstacle in the configuration space (B). The configuration space appears as the obstacles are inflated by the size of the robot. (C) shows the workspace for the case where the space is too small for the robot to pass, which appears blocked in the configuration space shown in (D).

The source and goal are special configurations, which are supplied as a part of the problem. Typically, the appearance of the robot at source and goal is given, whose configuration can be computed. The path $\tau:[0,1] \to C^{\text{free}}$ is thus a continuum of configurations, starting from the source ($\tau(0) = S$) and ending at the goal ($\tau(0) = G$), such that all intermediate configurations are collision-free, or $\tau(s) \in C^{\text{free}} \forall 0 \leq s \leq 1$. A sample path is given in Fig. 6.2. The path is parametrized by a variable s starting from a value 0 at the source, 1 at the goal, 0.5 in the middle, and so on.

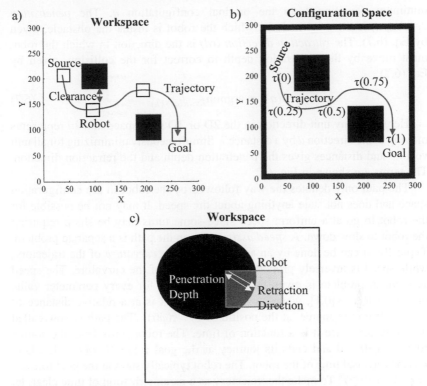

FIG. 6.2 Notation used in motion planning (A) physical path traced by the robot in workspace and clearance (the shortest distance between robot and obstacle), (B) robot trajectory in configuration space, (C) penetration depth and retraction direction.

Many times, the robot may not collide with an obstacle, but the configuration may be risky as it may be too close to the obstacle. *Clearance* measures the proximity of the robot to the obstacle and is the least distance between the robot and the obstacle, given by Eq. (6.6).

$$C(q) = \min_{r \in R(q),\, a \in W^{obs}} \|r - a\| \tag{6.6}$$

Here, $\|.\|$ denotes the Euclidian norm. In other words, clearance is the minimum distance between any point in the robot (r), to any point (a) in the obstacle-prone workspace. The clearance is referred to as the *proximity query* as it measures the proximity between the robot and obstacles.

Suppose a collision occurs with the robot at configuration q in a prospective robot path. Many algorithms require that such collisions be corrected by minimum deviation from the original configuration q. The *penetration depth* (pd) is the magnitude by which the robot is inside the obstacle, given by Eq. (6.7). The *retraction direction* (rd) is the direction in which the robot must move by the penetration depth to correct for the collision, given by Eq. (6.7)

$$[pd(q), rd(q)] = \text{argmin}_{\lambda, u:(R(q)+\lambda.u)\cap W^{obs}=\emptyset} \lambda \qquad (6.7)$$

Here, u is any unit direction in the 2D or 3D workspace and λu represents moving in the direction u by a distance λ. Simultaneously minimizing for all unit vectors and distances gives the penetration depth and the retraction direction. The terms are shown in Fig. 6.2.

The path only denotes the way followed by the robot in the configuration space and does not state anything about the speed. It may not be possible for the robot to go at a uniform speed since some turns may be sharp requiring the robot to slow down. A *speed assignment* on the path is a separate problem. Typically, it can be done by an assessment of the *curvature* of the trajectory, while speed is inversely proportional to the root of the curvature. The speed assignment problem is given by assigning a time for every parameter value of s, or $\varphi: R_{\geq 0} \to [0,1]$, where at time t, the robot is at a relative distance of $s = \varphi(t)$ from the source, at the position $\tau(s) = \tau(\varphi(t))$. The path is now called the *trajectory* since it is a function of time. The robot starts from the source $\tau(0) = S$, $\varphi(0) = 0$ and ends its journey at the goal $\tau(1) = G$, $\varphi(T) = 1$, where T is the traversal time of the robot. The robot typically stays at the goal forever, or $\varphi(t) = 1 \forall t \geq T$. To make the trajectory as a parametrization of time clear, let us re-formulate the trajectory as $\tau': R_{\geq 0} \to C^{free}$, where τ' is the trajectory as a parametrization of time and $\tau'(t)$ is the configuration of the robot at time t. The robot starts from source, or $\tau'(0) = S$; and ends at the goal $\tau'(t) = G \forall t \geq T$. The functions are related as $\tau'(t) = \tau(\varphi(t))$. While the path only gives the way followed by the robot, the *trajectory* additionally gives the time information, speed profile (as the first derivative), acceleration profile (as the second derivative), and so on.

This now gives the ability to send different ghost robots to take actions and move strictly in the free configuration space to reach the goal while minimizing the objective function. At any time if even a part motion of the ghost robot seems to be in a collision, the motion is abandoned and a new way of tackling the obstacle is sought. The configuration space gives an easy means to plan a point in a space, which corresponds to moving a robot with size and dimensions in real workspace. The configuration space enables working with a few configuration variables as an easy means to encode the problem, and to visualize the motion in a convenient space. A summary of the terms is given in Box 6.1.

Box 6.1 A summary of terms

Need of configuration space

Planning: Imagine a ghost robot traversing a trajectory/trajectory segment. Motion planning algorithms join multiple such segments or compute one trajectory, such that the ghost robot traversal is collision-free and optimal.

Collision-detection: Check if a ghost robot collides with obstacles.

Configuration space: Tells the collision status of all configurations that the robot can take. All collision-prone configurations are visualized as black and noncollision prone as white. Motion planning is planning a point-robot in this space.

Definitions

Ambient space: Physical world as seen and constructed by the sensors.

Workspace: Points reachable by the end-effector of the manipulator or mobile robot (used synonymously as ambient space in this book).

Free workspace: Points free of obstacles in the physical world.

Obstacle-prone workspace: Points with obstacles in the physical world.

Configuration (q): Minimum variables used to encode the robot at any time.

R(q): Set of all points occupied by the robot at configuration q in the workspace.

Configuration space: Set of all configurations a robot can take, visualized as a space with every configuration variable as an independent axis.

Obstacle-prone configuration space: All configurations where a ghost robot collides with an obstacle or itself.

Free-configuration space: All configurations which are not obstacle-prone.

State space/control space: Space using configuration variables, their derivatives (speeds), and other variables used to encode the kinematics/dynamics of the robot.

Path: A function that gives continuous collision-free configurations of the robot from source to goal, function parametrized by relative distance from the source ($s \in [0,1]$).

Trajectory: Path along with time (speed) information, function parameterized by time.

Motion/trajectory planning: Finding trajectory, typically solved by computing a path and then assigning speed at every point (speed assignment problem).

Differences in spaces

- Workspace is the physical space with a physical robot with size and mass, which is a point in the configuration space.
- One can take any point in the configuration space and place a ghost robot with known mass/size in the physical workspace at pose given by the configuration.
- One can note all configuration variables of a ghost robot in the workspace to get its equivalent point in the configuration space.
- Workspace may store geometric obstacles or grid-maps; however, a geometric workspace may have a nongeometric configuration space and thus cannot be used by algorithms that rely on obstacle geometries.

Continued

Box 6.1 A summary of terms—cont'd

- State space has speed information and can be used by planners to model non-holonomic (speed-dependent) constraints. For example, speed limits the curvature that the robots can take.

Construction of configuration space

- Need not be explicitly constructed, motion planning algorithms construct only parts of configuration space as per need
- Identify configuration variables
- Make overall configuration space with every variable as an axis
- For every point in this space, put a ghost robot at the same configuration in the workspace, where it will have a size
  - If the ghost robot collides, mark the point as black (obstacle-prone configuration space)
  - If the ghost robot does not collide, mark the point as white (free configuration space)
- Collision-checking libraries perform efficient collision checking

Configuration space examples

Planar circular robot: Configuration (x, y), Configuration Space: $C = \mathbb{R}^2$

Spherical robots: Configuration (x, y, z), Configuration Space: $C = \mathbb{R}^3$

Planar rectangular robot: Configuration (x, y, θ), Configuration Space: $C = \mathbb{R}^2 \times SO(2) = SE(2)$

Cuboidal robot: Configuration $(x, y, z,$ Euler Angles or Quaternions), Configuration Space: $C = \mathbb{R}^2 \times SO(3) = SE(3)$

Manipulator: Configuration $(\theta_1, \theta_2 \ldots \theta_n)$, Configuration Space: $C = T^n$ (if joint angles can turn infinitely), $\mathbb{R}^n$ (if joint angles have limits)

Mobile manipulator (mobile robot + manipulator): Configuration $(x, y, \theta_{Robot}, \theta_1, \theta_2 \ldots \theta_n)$, Configuration Space: $C = SE(2) \times T^n$

Composite systems: Multiple robot/robot parts as one system (e.g. multirobotics, mobile manipulator). The configuration is a join of all configuration variables; configuration space is the cross product of configuration space of individual parts.

6.3 Collision detection and proximity query primer

The basic problem of construction of configuration space is thus collision checking. The problem of collision checking is dealt with in a bottom-up manner, starting from the most primitive case to the more complex ones.

6.3.1 Collision checking with spheres

The simplest case is when the robot and the obstacles are spheres. Two spheres (or circles in a 2D world) do not collide if the distance between their centres is more than the sum of their radii. Let q be the configuration of the robot (X, Y, Z position) which has a radius ρ. Let the obstacle-prone workspace be a collection of spheres $\{o\}$ at centre q_o and with radius ρ_o, or $W^{obs} = \{o \langle q_o, \rho_o \rangle\}$. The collision checking is given by Eq. (6.8), where d is the distance function.

$$collides(q) \; iff \; \exists_o \, d(q, q_o) \leq \rho + \rho_o \tag{6.8}$$

A positive collision detection states that the configuration belongs to C^{obs} and vice versa, or $q \in C^{obs} \; iff \; \exists_o \, d(q, q_o) \leq \rho + \rho_o$. The clearance can also be measured easily in this case, given by Eq. (6.9).

$$C(q) = \min_o \, d(q, q_o) - (\rho + \rho_o) \tag{6.9}$$

A negative clearance denotes a collision. Similarly, the penetration is given by Eq. (6.10) and is applicable if the robot collides.

$$pd(q) = \max_{o:d(q, q_o) \leq \rho + \rho_o} \rho + \rho_o - d(q, q_o) \tag{6.10}$$

In case of a penetration, the retraction direction is simply away from the obstacle, or $(q - q_o)/\|q - q_o\|$. Here, $(q - q_o)$ denotes the vector moving in which direction the penetration distance will reduce by the largest amount, while $(q - q_o)/\|q - q_o\|$ normalizes the vector to make a unit vector.

6.3.2   Collision checking with polygons for a point robot

Now consider that the robot is point-sized and the obstacle is any general convex polygon for a 2D case and a polyhedral for a 3D case. This is shown in Fig. 6.3. Consider a line $L_{o_i(1)}(x) = 0$ (plane for 3D) between two corners ($o_1^{(1)}$ and $o_2^{(1)}$) of the obstacle O_1. Every line divides the workspace into two half-spaces, $L_{o_i(1)}(x) > 0$ (positive half-space) and $L_{o_i(1)}(x) < 0$ (negative half-space). The obstacle corners, as a convention, are numbered in the counter-clockwise direction. The equation of the line is written in a manner such that the obstacle lies completely in the negative half-space of the line. The obstacle O_j may now be given by the intersection of all the negative half-spaces, shown in Fig. 6.3 and given by Eqs (6.11) and (6.12).

$$O_j = \cap_i \left(L_{o_i^{(j)}}(x) \leq 0 \right) \tag{6.11}$$

$$O_j = \left\{ x : \forall_i \, L_{o_i^{(j)}}(x) \leq 0 \right\} \tag{6.12}$$

If there are multiple such obstacles (O_j), the obstacle-prone workspace W^{obs} is the collection of all such obstacles and hence a union over all such obstacles may be taken, given by Eqs (6.13) and (6.14).

$$W^{obs} = \cup_j O_j \tag{6.13}$$

$$W^{obs} = \left\{ x : \exists_{O_j} \forall_{o_i^{(j)} \in O_j} L_{o_i^{(j)}}(x) \leq 0 \right\} \tag{6.14}$$

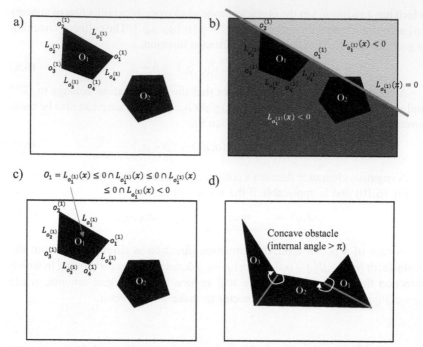

FIG. 6.3 Collision detection between a polygon obstacle and a point robot. (A) The obstacles are numbered counter-clockwise. (B) Every line divides the space into two halves. (C) The obstacle forms the intersection of negative half-spaces, used for collision checking. (D) Concave obstacle can be broken into multiple convex obstacles.

The small letter $o_i^{(j)}$ denotes the ith corner of obstacle j, while the capital letter O_j denotes a single obstacle. A point robot is completely described by its coordinates, or the configuration (q) is the same as the position (x), that is $q = x$. This means that the workspace and the configuration space are the same, which means that the collision-checking function is given by Eq. (6.15).

$$\text{collides}(q) \, iff \, \exists_{O_j} \forall_{o_i \in O_j} L_{o_i}(q) \leq 0 \qquad (6.15)$$

A problem happens when the obstacle is concave, in which case the assumption that an obstacle is made up of all negative half spaces does not hold good. However, every concave obstacle can be easily decomposed into multiple convex obstacles. For the case of 2D, the concave obstacles have at least one internal angle which is more than π. Detecting whether an obstacle is convex or not is hence an easy problem. The obstacle is characterized by the corners alone. All corners are iterated till the obstacle remains convex, and the corner where the convex nature is dropped is used to cut the obstacle into a convex part and the leftover obstacle. Iteratively, the obstacle breaking point can be set as a corner that has a larger internal angle, from which a new convex obstacle can be cut out. The convex hull uses a similar approach. Consider the example in Fig. 6.3D. Now the obstacle is simply split into three parts as if they were three independent obstacles.

The clearance (in case of an obstacle) and penetration depth (in case of a collision) can be obtained by computing the normal distance of a point from the closest side of the obstacle. The retraction direction is always making the point move in the normal direction to the closest side of the obstacle. The calculations are harder in case of a concave obstacle which was decomposed into a convex obstacle, in which case the line where the obstacle was decomposed cannot be used as an obstacle exists on the other side.

6.3.3 Collision checking between polygons

A more practical case is when the robot itself is a polygon (or a collection of polygons, polyhedral in a 3D case). The problem in principality reduces to an intersection test between two polygons (or polyhedral for a 3D case). The line of the obstacle (and the robot) shall now be more generally denoted by its *normals* (to generalize with 3D surface normals). The normal is taken to be outgoing or pointing out of the robot or the obstacle. The general notion remains the same, that the collision is said to have incurred if the robot lies in all the negative half-spaces of the obstacle; however since the robot is not point-sized, it is important to determine the shape of the new obstacle, such that the robot can be assumed to be point-sized. The new obstacle will have a different shape and different sides as compared to the original obstacle polygon. To visualize the effect, let us keep the orientation of the robot as a constant and sweep it along the plane, especially keeping it at the boundaries of the obstacle and not making it intersect the obstacle. The effect is shown in Fig. 6.4. The centre of mass of the robot is used to denote the configuration. Sweeping around the obstacle makes a new obstacle, such that the robot can now be reduced to a point at its centre.

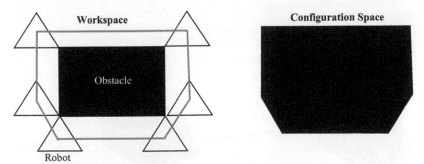

FIG. 6.4 Making a polygon configuration space obstacle for polygon robot and polygon workspace obstacle, for a fixed orientation. The robot is slid across all boundaries keeping the orientation fixed. All transitions are marked as edges. The collision checking now can be done assuming a point robot for the fixed orientation in the new polygon obstacle in the configuration space.

All corners of the obstacle polygon and the robot polygon are numbered in a counter-clockwise direction as shown in Fig. 6.5A. The limiting cases are of interest which defines the new obstacle, which happens in two cases. In the first case, a corner of the obstacle touches a line of the robot (shown in Fig. 6.5B). In the second case, a line of the obstacle touches a corner of the robot (shown in Fig. 6.5C). The other case involves a corner of the obstacle touching the corner of the robot, which however includes characteristics of both the other cases without any new characteristic. Similarly, in another limiting case, the line of the obstacle can touch another line of the robot, in which case either corner touches as well as incurring the characteristics.

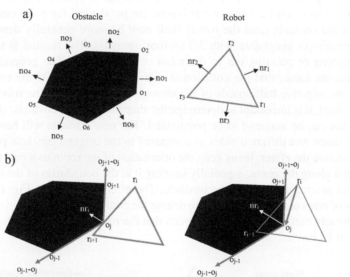

Obstacle corner/robot edge: for all applicable pairs satisfying $(o_{j-1} - o_j)$. $nr_i \geq 0$ and $(o_{j+1} - o_j)$. $nr_i \geq 0$, robot collides if it lies in all negative half spaces $nr_i \cdot (o_j - r_i) \leq 0$

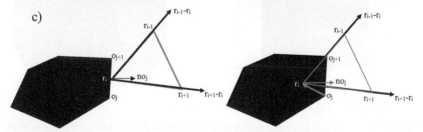

FIG. 6.5 Collision detection with polygons: (A) naming convention, (B) boundary case when obstacle corner touches robot boundary, (C) boundary case when obstacle corner touches robot boundary (replace robot by obstacle and vice versa in (B)).

Let us discuss the first case when a corner of the obstacle corner (o_j) touches the line of the robot (with normal nr_i). The first condition is called the *applicability condition* which checks if the obstacle corner and the robot normal come into contact in this sweeping process. If not, the pair does not make any edge of the new obstacle and hence may be neglected. If the pair is applicable, the pair makes an edge of the new obstacle. The condition is given by Eqs (6.16) and (6.17).

$$(o_{j-1} - o_j) \cdot nr_i \geq 0 \qquad (6.16)$$

$$(o_{j+1} - o_j) \cdot nr_i \geq 0 \qquad (6.17)$$

The equations state that the projection of both edges making the obstacle corner should have a positive projection on the robot's normal. If the obstacle is rotated so that this condition does not hold good, some other corner instead becomes applicable. If the robot is rotated such that the conditions do not hold good, some other edge becomes applicable instead, which will be the participating edge in the sweeping process.

This is repeated over all the pairs of obstacle and robot gives the pair of participating obstacle corner and robot edges that make the obstacle edges for collision checking. The equations only depend upon the normal of the robot, which is a function of the orientation of the robot and not the actual position of the robot. The edge of the half-plane is defined for every pair satisfying the applicability condition. The half-plane is given by Eq. (6.18).

$$nr_i \cdot (o_j - r_i) \leq 0 \qquad (6.18)$$

Imagine the obstacle corner o_j touches the line $r_{i+1} - r_i$ with the normal nr_i. Since o_j is a part of the line (r_i, r_{i+1}), the projection of the same line $o_j - r_i$ with its normal nr_i will be exactly zero. However, as o_j penetrates deeper into the robot, the angle starts to increase more than $\pi/2$, giving a negative projection.

The other case of the robot corner touching the obstacle line is the same and can be simply derived by changing the roles of the robot and the obstacle, giving the applicability conditions as Eqs (6.19) and (6.20) and the equation of the half-space as Eq. (6.21).

$$(r_{i-1} - r_i) \cdot no_j \geq 0 \qquad (6.19)$$

$$(r_{i+1} - r_i) \cdot no_j \geq 0 \qquad (6.20)$$

$$no_j \cdot (r_i - o_j) \leq 0 \qquad (6.21)$$

Given the half-plane lines over all pairs of corners and edges between the robot and the obstacle, a collision is said to have occurred if the robot lies in the negative half-space of every such edge. All applicability conditions are checked to get the pairs of applicable edges and corners, and if the robot qualifies by lying in the negative half-space of every such pair, the collision is said to have occurred. The equation defining the half-space also gives the clearance or penetration depth by calculating the projection of the normal with the line joining the corners. The retraction direction is towards the normal.

6.3.4 Intersection checks with the GJK algorithm

The *Gilbert–Johnson–Keerthi* (GJK, Gilbert and Foo, 1990; Gilbert et al., 1988) algorithm is a quick mechanism to determine if two polygons in higher dimensions intersect or not. Let O be the set of all points inside the obstacle polyhedron and R be the set of all points inside the robot polyhedron in 3D. Let us compute the minimum distance between any point in O and any point in R, given by Eq. (6.22).

$$C(q) = \min_{r \in R(q), o \in O} \|o - r\| \tag{6.22}$$

Here $\|.\|$ is the Euclidian distance. If the distance is zero, the robot and obstacle intersect, while a positive distance reports the clearance. It is eminent that the difference of all points between the robot and the obstacle is an interesting entity, which can be conveniently represented by a Minkowski difference operator, given by Eqs (6.23) and (6.24).

$$Z(q) = O \ominus R(q) \tag{6.23}$$

$$Z(q) = O \oplus (-R(q)) \tag{6.24}$$

Here, $\ominus$ is the Minkowski difference operator. Given two sets of points O and R, the operator creates a new set of points created by taking a difference between every point in O with every point in R. The $\oplus$ is correspondingly the Minkowski addition operator. To understand the Minkowski's difference, first let us take the robot to be point sized, placed at point x in space. Now the Minkowski difference simply takes the obstacle and translates it by $-x$ in space, keeping the shape of the obstacle as constant. However, if the robot has a size, then the Minkowski difference can be visualized by first translating the obstacle by the negative of the centre of the robot as previously, and then re-sizing the obstacle appropriately as per the size of the robot. This is shown in Fig. 6.6.

In case Z contains the origin, then the robot and the obstacle indeed intersect, because some $q \in O$ and $q \in R(q)$ cancel each other out giving a 0 difference. Furthermore, the minimum distance between the robot and the obstacle (clearance) is the distance of the closest point in Z from the origin. The GJK algorithm does the same, which is to try to get as close as possible to the origin, travelling through the points in Z, iteratively.

Computing the complete Minkowski difference Z and thereafter finding the closest point to the origin can be computationally impossible considering that it contains infinite continuous points. We first assume that both the robot and obstacle polyhedral are convex in nature, and further that the Minkowski difference of them (set Z) is also convex in nature. Previously, the set of all obstacle points making a convex obstacle was summarized by the vertices alone for

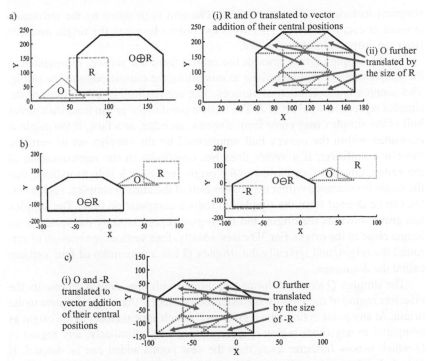

FIG. 6.6 (A) Calculating the Minkowski addition $O \oplus R$. The addition makes a new figure by adding all points inside O and R. (B) and (C) Calculating the Minkowski difference $O \ominus R$.

collision checking. Similarly, a set of points representing a convex Z is used to represent the infinite points within the convex hull. The *convex hull* represented as a set of vertices is enough to represent a region, given that the original surface was convex. The notion of a convex hull is shown in Fig. 6.7A for a 2D case. Given a convex hull representing Z consisting of vertices, edges, and faces, the problem is to compute the point closest to the origin. If the convex hull contains the origin in it, there is a case of collision. If the origin is outside the convex hull, the closest distance of any vertex, edge, or face from the origin gives the clearance.

However, a problem is that computing all vertices, edges, and faces of the convex hull of Z is itself not feasible. Therefore, we start with a small sample of the convex hull representing a part of Z, represented by a set of vertices (Q) called the *simplex*. The simplex represents a set of vertices whose convex hull models a subset of Z. Now the aim is to slowly grow the represented convex hull (simplex) towards the origin, till no further growth is possible. At any iteration, if it appears that a part of Z not currently contained in the simplex Q may be closer to the origin, the simplex is expanded by adding such points. Several such points may be candidates to add. The simplex will always be a convex hull. Therefore, given a collection of points in Z closer to the origin than the current simplex, only such points need to be added whose convex hull (with the current simplex) includes all the other points. The aim is to move by the maximum amount in every iteration. Therefore, the point closest to the origin assessed from the current simplex is added.

As the simplex grows towards the origin, there are vertices represented in the simplex that do not contribute to modelling the closest point to the origin. The simplex is a collection of vertices. The convex hull of all vertices of the simplex represents a subset of Z. The closest point to the origin from the convex hull of the simplex may come from a vertex, an edge, or a face. If the origin is contained within the convex hull represented by the simplex set of vertices, there is a collision. If a vertex does not contribute to the representation of the vertex, edge, or face closest to the origin, nor does it help to indicate that the origin is contained within the convex hull of the current simplex, such a vertex can be deleted from the simplex to reduce computation time. The simplex has grown towards the origin, and the region represented by the vertex is no longer close to the origin. For 3D cases, ideally, four vertices are enough to surround the origin, and typically the simplex Q has a maximum of four vertices called the 3-simplex.

The simplex Q slowly deforms and moves with time until it represents the effective region of the Minkowski difference surface Z that may be closest to the origin. At any point in time, the regions in Z which appear closer to the origin as compared to any other region in Q are added in Q. Similarly, any region in Q which is now furtherer away than the new points added can be deleted. If Q includes the vertices all around the origin, whose closest point to the origin is the origin itself, a collision is said to have occurred.

Let us now discuss the growing of the simplex more formally. Given a simplex Q, the intention is to grow Q towards the origin and to find the closest point towards the origin. Let u be the point that is currently closest to the origin by considering all points within the simplex Q. A traversal towards the origin will mean that we travel as much as possible in the direction $-u$ (vector $-u$ points from u towards the origin) to potentially reach the origin. The *support* gives the furthest away point in a specific direction while traversing along with Z. The notion is shown in Fig. 6.7B. Therefore, $v = support(-u)$ is potentially a step closer towards the origin, considering complete Z. v is now of interest and is therefore added to the simplex Q, making a new simplex, from which some old points may be removed as they are not required to make the convex hull containing v. Since the simplex Q has changed, the closest point in the simplex Q to the origin, u, will also change. The process goes on and on, till an improvement can be made in the closest point.

The algorithm starts with a random point, u, which is a vertex in Z and makes the simplex using this point only, $Q = \{u\}$. The point is called the 0 simplex, which is also the current closest point to the origin. Taking the support in the direction $-u$ gives a new vertex v, which can be added to the simplex Q. This makes the simplex to have two points called the 1-simplex. A new closest vertex in simplex to the origin is computed, which is used to get a new direction and support in that direction. The algorithm is given by Algorithm 6.1.

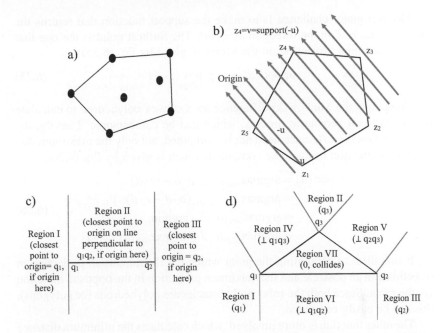

FIG. 6.7 Steps of GJK algorithm. (A) Convex hull (the least number of vertices fully covering all points), is enough for collision and proximity queries. (B) Support (furthermost point in a direction $-u$). (C) Finding closest point to the origin for 1-simplex. (D) Finding the closest point to the origin for 2-simplex. Below every region is an indicator of the closest point to the origin, provided origin lies in the same region.

Algorithm 6.1 GJK algorithm

Input: Minkowski difference Z between obstacle O and robot R (need not be explicitly calculated)

Output: distance of the closest point in Z to origin (clearance)

1   $u \leftarrow$ random vertex in Z ;
2   $Q \leftarrow \{u\}$;   ▷ `simplex`
3   **while** *true* **do**
4      $v \leftarrow support(-u)$;   ▷ `Fig. 6.7(b)`
5   **if** $||v|| \geq ||u||$ **then** break;
6   add v to Q, maintaining Q as the smallest convex hull that contains v by deleting extra points in Q ;
7   **if** *volume(Q) contains origin* **then** return 0;   ▷ `collision`
         ▷ `u: closest point to origin, Fig. 6.7(c-d)` ;
8      $u \leftarrow argmin_{u \in volume(Q)}||u||$;
9   return $||u||$;

The first major challenge is to make the support function that returns the furthest point in Z in the direction $d(=-u)$. The furthest point is the one that has the maximum projection on the vector d, given by Eq. (6.25).

$$\text{support}(d) = \arg\max_{z \in Z(q)} (z \cdot d) \qquad (6.25)$$

Computing the Minkowski difference for a convex polyhedron to calculate the support is a geometric process, which can be cumbersome. Luckily, the entire Minkowski difference need not be computed, but only the maximum distance in a direction needs to be computed, which is given by Eq. (6.26).

$$
\begin{aligned}
\text{support}(d) &= \text{argmax}_{r \in R(q), o \in O}((o - r) \cdot d) \\
&= \text{argmax}_{r \in R(q), o \in O}(o \cdot d - r \cdot d) \\
&= \text{argmax}_{o \in O} o \cdot d - \text{argmin}_{r \in R(q)} r \cdot d \\
&= \text{argmax}_{o \in O} o \cdot d + \text{argmax}_{r \in R(q)} r \cdot (-d)
\end{aligned}
\qquad (6.26)
$$

It simplifies to computing the point which has the maximum projection in direction d in an obstacle and the maximum projection in the opposite direction for the robot. Since both the robot and obstacles are polyhedrons (or polygons), this can be easily computed.

The other function is more involved, which calculates the minimum distance of a point in the volume represented by the simplex Q, to the origin. Let us decompose the problem and solve it for simple cases. The simplest case is for the 0-simplex when Q contains just 1 point, which is itself closest to the origin. The next simplest case is when Q is 1-simplex, or Q contains 2 points. In this case, the closest point to the origin may lie at either of the three regions shown in Fig. 6.7C. The closest points for the two extreme regions are the vertices themselves, while the closest point for the central region is given by the perpendicular on the line, which can be geometrically calculated. Projections can be used to find which of the three regions are applicable. A trick that can be used is that the point which was added later to the simplex which is known (say q_2). This means we moved from the first point of the simplex (say q_1) to reach the second point (say q_2) in an attempt of moving towards the origin. This negates out any possibility to have the origin in either region (region I in the example).

The problem is more complex when Q is 2-simplex or has three vertices, making a triangle. The regions are shown in Fig. 6.7D. The division of space in this manner constitutes the Voronoi decomposition, grouping areas that are closer to one point than any other point. The closest point will continue to be a vertex or the perpendicular to an edge, depending upon the condition or region. Again, knowing the order of the insertion of vertices rules out some of the regions. Each region that said, can be geometrically calculated. In the case of a 3-simplex, when four vertices lie in a space making a three-dimensional structure cuts out more regions in 3D space. This is difficult to visualize in 2D paper and is thus left out. This constitutes the geometric way of finding the least distances. Alternatively, it is also possible to compute the same terms algebraically by modelling as a minimization equation and optimizing the point such that the point lies inside the space while minimizing the distance.

The working of the algorithm for a synthetic scenario is shown in Fig. 6.8.

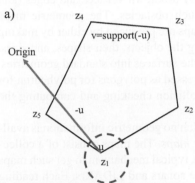 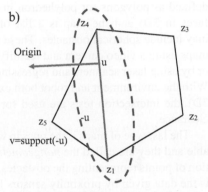

0-simplex Q, selected randomly. Out of all points in simplex (currently only 1, z_1), $u=z_1$ is closest to origin, $z_4=\text{support}(-u)$ (out of all vertices, largest projection on $-u$), added to simplex.

1-simplex Q. Out of all points in simplex, u is closest to origin, $z_5=\text{support}(-u)$ (out of all vertices, largest projection on $-u$), added to simplex

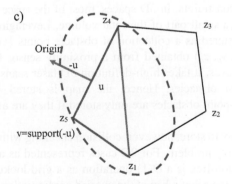

2-simplex Q. Out of all points in simplex, u is closest to origin, $z_5=\text{support}(-u)$ (out of all vertices, largest projection on $-u$). Algorithm converged. u is closest to origin with no collision.

FIG. 6.8 Working of the GJK algorithm. $Z=O\ominus R$ for a synthetic map is plotted. Note that the Minkowski difference need not be explicitly computed.

6.4 Types of maps

We shall be typically dealing with three kinds of maps. The first type of map is the *topological map*, as it represents the topology of the space and is available as a graph of vertices and edges. The most common example is the satellite navigation or the navigation utilities in smartphones, which operates on the road-network graph available for every place. The road-network graph is an abstraction of the big roads with multiple-lanes and big intersections.

The other kind of map is called the *geometric map*, wherein the obstacles are defined as polygons (or polyhedron in 3D) of known vertices and edges (and faces in 3D) and the map is a list of such obstacles. The geometric maps may include spherical obstacles. These maps are produced by either by making maps using a video camera and identifying the objects, their shapes, and sizes, or by using laser scanners and regressing the surfaces into standard geometries. With the environment and robot both expressed as polygons (or polyhedron for 3D), the intersection tests are used for collision checking and computing the clearance.

The last type of maps is the ones in which no geometric information is available and they are called the *nongeometric maps*. The maps consist of a collection of points representing the obstacles. A typical mechanism to get such maps is the data given by proximity sensors like sonars and LiDARrs. Each reading gives a point in space that is occupied by the obstacle. The grid map (2D or 3D) is a common representation, wherein the space is fitted with cells of some resolution. There is a discretization loss due to the conversion of the continuous map into discrete cells. It is possible to store 2D maps as a grid bitmap with a reasonable resolution. However, for 3D cases storing such bitmaps can be infeasible for memory constraints. In 3D spaces, most of the space is free and obstacles consume only a small part of the total volume. Leveraging this fact, the 3D maps are often stored as a collection of obstacle points $\{(x_o, y_o, z_o)\}$, each obstacle point (x_o, y_o, z_o) obtained from a proximity sensor like a laser scanner. As the robot moves, it takes high-definition 3D laser scans which can overpopulate the list of obstacles. Hence, the map is stored with some resolution R, where two-point obstacles are only stored if they are at a distance of R or more.

The grid maps are easy to store, however, collision checking with grid-maps is a computationally heavy problem. For 2D cases represented as a grid, querying if a point is collision-free is a $O(1)$ operation as a grid lookup. The set of all points occupied by the robot when at the queried configuration q, denoted by $R(q)$, is obtained and checked for collision. Being continuous, $R(q)$ contains infinite points that can be sampled at a high resolution. Consider a rectangular robot as an example. Place the robot at the queried configuration $q(x, y, \theta)$. Sample all points $R(q)$ in the grid that the robot occupies and check for the validity of all sampled points. If there are no obstacles smaller than the size of the robot, only checking for the perimeter of the rectangular robot instead of the complete area suffices. For lower resolution initial motion planning, only checking the four corners is a good approximation to get an initial approximate plan, while the final plan will require a higher resolution collision checking by checking for the complete area.

Collision checking for 3D maps stored as an array of point obstacles is the same as the 2D case, except that checking every point sampled from $R(q)$ is no longer a constant time operation. Ideally, one must iterate through all the obstacle points in the 3D map to check if the sampled point in $R(q)$ lies very close to any of the obstacle points. A technique discussed in Section 6.5 will make this an operation with a complexity logarithmic to the number of points in the map.

The three types of maps are given in Fig. 6.9. A special case is when the obstacles have splines, Bezier, and similar curves as their geometries. Since the geometries are known, they may be taken as geometric, but it is difficult to use the geometrical information for collision checking and heuristics in motion planning. Therefore, the geometrical information may not be in a usable format, and these are called the *semigeometric maps*.

a) Topological Map (vertices and edges)

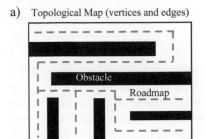

b) Geometric Map (shapes)

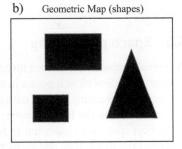

c) Non-Geometric Map (bitmap, gridmap)

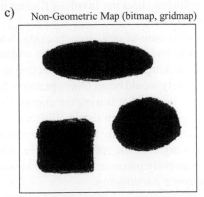

FIG. 6.9 Types of maps. Geometric and nongeometric maps are interconvertible. Standard shapes can be regressed to known geometries from nongeometric maps. Geometries in a geometric map can be painted to make nongeometric maps.

It is possible to convert a type of map into the another. The obstacles in a grid with a nongeometric representation may be regressed over by geometric surfaces, yielding a geometric map. The Bounding Volume Hierarchy Technique of Section 6.6 is one way to do the same. Similarly, the geometric obstacles can be converted into nongeometric grids by drawing on a bitmap canvas in the same way as geometries are drawn as images on the screen. The conversion of nongeometric and geometric maps into a topological map uses techniques like cell decomposition and probabilistic roadmaps, discussed in separate chapters.

A major misunderstanding when dealing with the type of maps for motion planning is to appreciate the difference between the workspace and configuration space. The deliberative motion planning algorithms typically use the configuration space for motion planning. The entire algorithm operates on the configuration space. A geometric workspace constructed by the sensors of a manipulator produces a configuration space that does not have a regular geometry (or that the geometry cannot be derived). Hence, the workspace may be a geometric map, but the configuration space may not be geometric. Hence, many pure geometric techniques cannot be used for planning a manipulator, operating in a geometric workspace.

6.5 Space partitioning

A typical problem that comes throughout robotics, machine learning, and AI is to query for a few closest points in space. These are called the *proximity queries*. For machine learning, the test data output is assumed to be like the nearest training set of data in the feature space. In robotics, the neighbouring points in configuration space are connected to form a graph to traverse. Furthermore, every robot makes immediate decisions based on the neighbouring set of robots and obstacles only. Even collision checking involves finding the nearest obstacles and checking for intersections.

Let us take the simplest case of collision checking, wherein the robot is a sphere with a small radius and there are n obstacles which are spheres with a small radius. Collision checking by doing an intersection test between the robot and every obstacle separately would be an $O(n)$ operation. Since the spheres are small, it may alternatively appear that storing the map as a hash table (grid map) is a good option, which gives a constant time collision checking. However, the memory requirements can be arbitrarily large for 3D maps. The better way to store the map is hence to organize the spheres as a tree, where every node represents a region and recursively partitions the region into smaller regions. The concept used is called *space partitioning*.

6.5.1   *k*-d tree

The first implementation of space partitioning is the *k*-d tree which divides the space into halves at every level. Now to search for a node, one just needs to reach the correct segment of space. We will discuss one of the many implementation variants possible. Consider several points scattered in the space, shown in Fig. 6.10. The need is to construct a *k*-d tree data structure by recursively partitioning the space into halves. Let us first partition the *X*-axis. The point which lies in the middle of all the points considering the *X*-axis values is taken. Let the point be given by *q*. This means *q* is such that $q[X]$ is the median out of all $t[X]$ points in the collection. Here $q[X]$ denotes the *X* coordinate of *q*. The node stores the point *q* along with the splitting value $q[X]$ (which is anyways present in *q*),

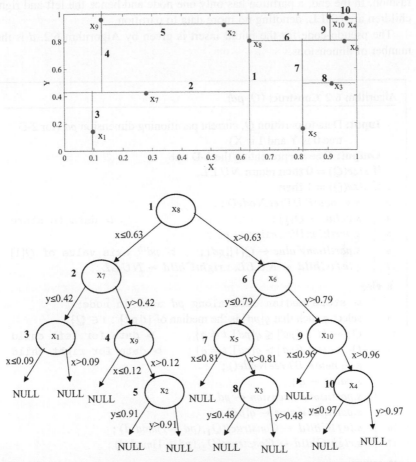

FIG. 6.10 Construction of *k*-d tree. Each node divides the space into two halves across the median, starting from the root (*X*-axis), and then *Y*-axis (depth 1), storing the median data.

the splitting axis (X), and the pointers to the left and the right child. The collection of points can be partitioned into two parts, one with all points t such that $t[X] \leq q[X]$ (or which lie in the left space of the new partition) and the second with all points t such that $t[X] > q[X]$ (or which lie in the right space of the new partition). Now we can recursively partition the left collection of points and the right collection of points by a similar process.

The left partition contains a subset of points. We will now use the Y-axis to partition. Let q be the point taken, which is such that $q[Y]$ becomes the median out of all points $t[Y]$ in the subset. This means in the left partition, we create a new partition over q, partitioning the Y-axis into two. This further divides the set of points into two partitions. The node stores the value (q), partitioning dimension (Y), and the partitioning value ($q[Y]$), along with the two children. Similarly, with the right partition. The partitioning will recurse in the same fashion. In the end, a partition has only one node and hence the left and right children are NULL, denoting no more data to partition.

The pseudo-code for the initial insert is given by Algorithm 6.2. d is the number of dimensions.

Algorithm 6.2 Construct (Q, pd)

Input: Data to partition Q, current partitioning dimension pd (for 2-D tree 0 is Y and 1 is X)

Output: Tree s representing the k-D tree

1 **if** $size(Q) = 0$ **then** return $NULL$;
2 **if** $size(Q) = 1$ **then**
3      $s \leftarrow newKDTreeNode()$;
4      $s.value \leftarrow Q[1]$;   ▷ data to store
5      $s.partitionDimension \leftarrow pd$;
6      $s.partitionValue \leftarrow Q[1][pd]$;   ▷ pd^{th} axis value of $Q[1]$
7      $s.leftChild \leftarrow NULL, s.rightChild \leftarrow NULL$;
8 **else**
     ▷ store median data along pd axis in node;
9   select q such that $q[pd]$ is the median of $\{t[pd] : t \in Q\}$;
10     $Q_1 \leftarrow \{t : t[pd] \leq q[pd], t \neq q\}$;   ▷ data for left child
11     $Q_2 \leftarrow \{t : t[pd] > q[pd], t \neq q\}$;   ▷ data for right child
12     $s \leftarrow newKDTreeNode()$;
13     $s.value \leftarrow q$;
14     $s.partitionDimension \leftarrow pd$;
15     $s.partitionValue \leftarrow q[pd]$;
16     $s.leftChild \leftarrow Construct(Q_1, (pd + 1)mod\ d)$;
17     $s.rightChild \leftarrow Construct(Q_2, (pd + 1)mod\ d)$;
18 **return** s

If data is added after the construction of the *k*-d tree, the addition of a new node is also a similar exercise. The addition neither ensures that the partitioning dimensions will be medians nor that the complete tree will be balanced. Imagine a data *q* is to be added. The addition takes place by first traversing the tree to find the subspace that it belongs to. If the *X*-axis value is larger than the root, the search happens to the right child. At a tree node with *Y*-partitioning, if the *Y*-axis value of the new node is lesser than the *Y*-axis value of the current node in the tree, the search proceeds to the left child. Later the tree is added as the appropriate child of the previous leaf node. The working is similar to a Binary Search Tree. In the worst case, the tree will be unbalanced giving $O(dn)$ time complexity, however, practically the complexities are close to $O(d\log n)$. Here *n* is the number of points and *d* is the dimensionality. The addition process is shown in Fig. 6.11. The addition is given by Algorithm 6.3.

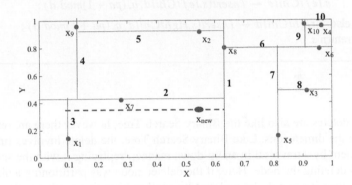

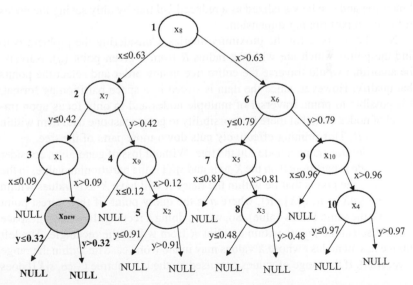

FIG. 6.11 Insertion of a node in *k*-d tree. The tree is traversed, and the node is added to a leaf by maintaining the partitioning.

Algorithm 6.3 Insert (*s, q, pd*)

Input: current node *s*, data to insert *q*, partitioning dimension of the
current node *pd*

Output: Updated tree *s*

1 **if** $s = NULL$ **then**
2 | $s \leftarrow newKDTreeNode()$;
3 | $s.value \leftarrow q$;
4 | $s.partitionDimension \leftarrow pd$;
5 | $s.partitionValue \leftarrow q[pd]$;
6 | $s.leftChild \leftarrow NULL, s.rightChild \leftarrow NULL$;
7 | return *s* ;

8 **else if** $q[pd] \leq s.partitionValue$ **then**
9 | $s.leftChild \leftarrow Insert(s.leftChild, q, (pd + 1) mod\ d)$;

10 **else** $s.rightChild \leftarrow Insert(s.rightChild, q, (pd + 1) mod\ d)$;
11 return *s*

The deletes are also like the Binary Search Tree, however there are restrictions on the dimensions. Like Binary Search Tree, the delete involves finding the element by a traversal in an up-down manner, replacing it with the successor, and deleting the node. Here, if the deleted node was partitioning a dimension *pd*, then it can only be replaced by another node that partitions the same dimension and can be visualized as a reduced *k*-d tree by only seeing the nodes of the same partitioning dimension.

Now let us solve for the proximity queries. Considering the problem is to find the points which are within a radius *R* from a given point (*q*). Naively, the algorithm would traverse the entire tree in any order and select the points that qualify. However, since the data is stored in a space partitioning format, it is possible to prune traversal of multiple nodes and to only focus upon traversal of nodes such that there is a possibility to get at least one data item within the radius *R*. This pruning effectively cuts down most parts of the tree.

Consider traversing a node *s* in the tree. Without loss of generality, consider that the node is partitioned on the *X*-axis and *s*[*X*] is the partitioning value. In the worst case, the points that lie within the range *R* will have *X*-axis values within the range $[q[X] - R, q[X] + R]$, where *q* is the query point (if the nearest point will have the same *Y*-axis value). So, we are doing a rectangular search in space and will select the points within a radius *R* from within the rectangle. The left subtree has all nodes *t* whose *X* values may in the worst case lie within the range $(-\infty, s[X]]$. If the range of values promised by the left subtree $(-\infty, s[X]]$ does not coincide with the range of values needed by the search $[q[X] - R, q[X] + R]$, there is no need to visit the left subtree. The two ranges intersect if $s[X] \geq q$ $[X] - R$. Similarly, for the right tree. The range of expected values is $(s[X], \infty)$, while the range being searched for is $[q[X] - R, q[X] + R]$. This means it is unnecessary to search for the right subtree if the two ranges do not intersect. The two ranges intersect if $s[X] < q[X] + R$. The working of the search is shown in Fig. 6.12. The algorithm is shown in Algorithm 6.4.

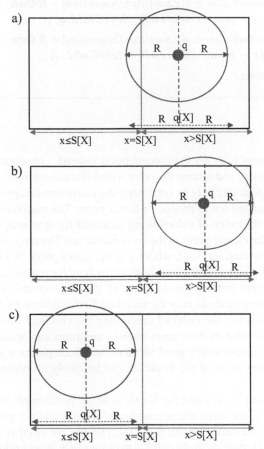

FIG. 6.12 Pruning for computing the nearest R in a k-d tree. Query is to get all nodes within a distance R from q in the subtree rooted at S, partitioned in the X-axis at a value $S[X]$. (A) In the first case both children subtrees need to be checked for the nearest neighbours $(S[X] \geq q[X] - R$ and S $[X] \leq q[X] + R)$. (B) In the second case the left subtree can be pruned $(S[X] < q[X] - R)$. (C) In the third case, the right subtree can be pruned $(S[X] > q[X] + R)$.

Algorithm 6.4 Nearest $R(s, q)$

Input: k-D tree rooted at s, query point q
Output: List of points within radius R from query point q

1   $points \leftarrow \emptyset$;
2   **if** $s = NULL$ **then** return $points$;
3   **if** $||q - s.value|| \leq R$ **then** $points \leftarrow points \cup \{s\}$;
    ▷ pruning condition: if, along pd axis, left child's coverage is no further than R from q, then prune;
4   **if** $s.partitionValue \geq q[s.partitionDimension] - R$ **then**
5     $points \leftarrow points \cup nearestR(s.leftChild, q)$;
6   **if** $s.partitionValue < q[s.partitionDimension] + R$ **then**
7     $points \leftarrow points \cup nearestR(s.rightChild, q)$;
8   return $points$

A similar query is to get the nearest point instead. The algorithm searches within the entire tree and returns the point which lies closest to the query point q. The search happens similarly by traversing the entire tree and pruning all nodes which are guaranteed not to produce a better point. The search differs from the nearest R since the range of values being searched for is now dynamic. At any point, let m be the closest point in the traversal so far. The range being searched for is strictly less than $||q - m||$, where q is the query point. If a better point is found, the best point m updates and shortens the search even further. The shorter the range, the more pruning can be done. In the tree, several nodes have two children and either of them may be searched first followed by the other. We use heuristics to select the order of processing the children. The aim is to traverse the tree such that the best areas where a minimum can occur are traversed first. This gives a reasonably good value of m, which places a small range of search. Thereafter, most of the branches can be easily pruned out because of a small range.

Consider at any node s and the X-axis is being partitioned. Now if q lies to the left side of the partition, the nearest node is likely, and not guaranteed to be on the left and vice versa. Hence if $q[X] \leq s[X]$, the left child is explored first, and if $q[X] > s[X]$, then the right child is explored first. Both children may have to be explored. Pruning is checked as per the derivations made in the earlier section, with the range R altered to $||m - q||$. The working of the algorithm is shown in Fig. 6.13. The algorithm is given as Algorithm 6.5.

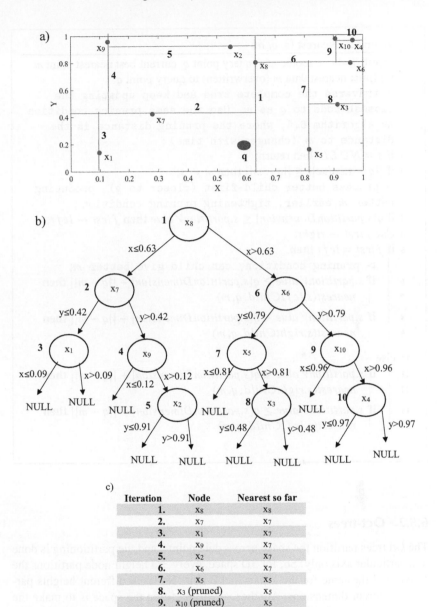

FIG. 6.13 Searching for the nearest node to q in the k-d tree. The search is similar to the nearest R case of Fig. 6.12, with the exception that R is always the distance to the closest node so far. For tree traversal, a child more likely to store the closest node is traversed first. (A) Search space. (B) k-d tree. (C) Search iterations.

Algorithm 6.5 Nearest (s, q, m)

Input: k-d tree rooted at s, query point q, current best nearest point m
Output: nearest data m (overwritten) to query point q
▷ traverse the complete tree and keep updating the closest node to q as m. Use the same pruning condition as Algorithm 6.4, where the pruning distance is the distance to m (changes with time) ;
1 **if** $s = NULL$ **then** return;
2 **if** $||q - s.value|| \leq ||q - m||$ **then** $m \leftarrow s$;
   ▷ process better child first (closer to q), producing better m earlier, tightening pruning condition ;
3 **if** $q[s.partitionDimension] \leq s.partitionValue$ **then** $first \leftarrow left$;
4 **else** $first \leftarrow right$;
5 **if** $first = left$ **then**
   | ▷ pruning condition, can child give better m;
6 | **if** $s.partitionValue \geq q[s.partitionDimension] - ||q - m||$ **then**
7 | | $nearest(s.leftChild, q, m)$
8 | **if** $s.partitionValue < q[s.partitionDimension] + ||q - m||$ **then**
9 | | $nearest(s.rightChild, q, m)$
10 **else**
11 | **if** $s.partitionValue < q[s.partitionDimension] + ||q - m||$ **then**
12 | | $nearest(s.rightChild, q, m)$
13 | **if** $s.partitionValue \geq q[s.partitionDimension] - ||q - m||$ **then**
14 | | $nearest(s.leftChild, q, m)$

6.5.2 Oct-trees

The k-d trees partition the space across data points, and the partitioning is done in a particular axis only. So, for 3D space, every 3rd height node partitions the X-axis and the same for the Y-axis and Z-axis. Nodes at different heights partition different dimensions. Another way to partition the space is to make the partitions across all dimensions at every single node of the tree (Meagher, 1982; Hornung et al., 2013).

Consider a 3D space. Every node of the tree partitions the space into the X-axis, Y-axis, and Z-axis simultaneously, creating 2^3 or 8 partitions. These 8 partitions become the children of the node. In a k-d tree, the partitioning is done at the median value, which means that the balance of the tree changes as more nodes are added or deleted. Here, in oct-trees, the partitioning is done at the

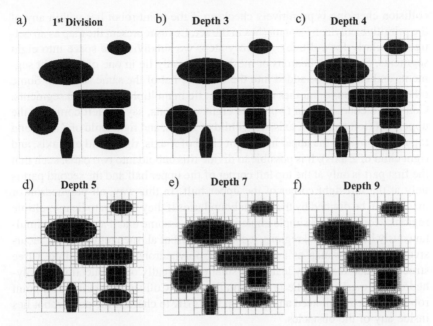

FIG. 6.14 Quad-tree representation. At every level, the cell is divided into four parts, except completely free cells and completely obstacle-occupied cells. (A) First division, (B) Depth 3, (C) Depth 4 (D) Depth 5, (E) Depth 7, (F) Depth 9.

middle of the space, dividing a space into eight equal partitions. It does not consider the data when deciding the partitions. However, space is only partitioned if it has some data. Empty spaces are not partitioned. Similarly, for rich point cloud data, there may be a threshold onto the minimum size of a cell. Otherwise, one may indefinitely go on dividing the cells. If v is the volume of the cell, all the children of the cell have a volume of $(v/8)$. This means children at depth d have a volume of $(v/8)^d$, which should be limited. This constitutes the resolution of representation of the oct-tree. All leaf nodes hence are taken as completely obstacle-free or filled with obstacles.

Further, the notion of the same concept in 2D spaces is called the *quad-tree*, where every node has 2^2 or four children. The quad-tree decomposition is shown in Fig. 6.14.

6.5.3   Collision checking with oct-trees

The discussions so far assumed the problem of collision detection for a point-sized robot with the map represented as a set of point obstacles or a grid of obstacles. Another possibility is to have a geometric map, where every obstacle is a known geometry like a polyhedron. Imagine a general obstacle that can have a complex shape. For simplicity, assume that the obstacle is already broken down into a set of primitive polyhedrons or a set of primitive meshes. The collision checking is primitively checking if the point robot lies within any of the primitive polyhedron or meshes in the set. Let us represent the map as an oct-tree (similarly for k-d tree). At every step, we subdivide the space into eight segments. Some polyhedrons or meshes shall only lie in one of the eight segments and can thus be added into the obstacle list of the same segment. Some other polyhedrons or meshes may however lie in multiple of the eight segments. Consider a cuboid lies in two of the eight segments, say top left corner of the upper half (left child of X axis, right child of Y axis, and right child of Z axis) and top right corner of the upper half (right child of X axis, right child of Y axis, and right child of Z axis). It is possible to divide this cuboid into two parts, such that the first part is only at the top left corner of the upper half and the second part is only at the top right corner of the upper half. In this way, all polyhedron or meshes can be divided by the region boundaries if they happen to lie in multiple regions. However, as a simple implementation, assume that if an obstacle simultaneously lies in several regions, it shall be a part of all such regions. This recursively breaks the map as a collection of polyhedrons or meshes, into a tree structure, such that every leaf has a smaller (typically one) collection of polyhedrons or meshes that lie in that small region. Collision checking of a point robot at the leaf will need an algorithm like GJK to check if the point robot lies inside any of the obstacles.

Now let us assume that the robot is itself represented as a polyhedron or that the robot itself has some shape and size. As a simple implementation, consider the tightest angle-aligned bounding box that encloses the robot. Recursively travel the tree. At every node, the robot may be colliding with one or several children nodes. Consider the volume occupied by each of the children node and the volume occupied by the robot (bounding box). If the volumes intersect, there is a possible collision, and the child will be recursively traversed. If there is no possibility of a collision or that the volumes do not intersect, the child is pruned. At the leaves, the exact robot as a set of polyhedrons is considered for an exact collision checking with algorithms like the GJK with every polyhedron or mesh obstacle contained in the leaf. The proximity checks may similarly be performed. This is shown for a sample 2D map in Fig. 6.15.

As a more complex representation, assume that the robot is a set of polyhedrons or meshes. Let us recursively divide the robot and store the same as an oct-tree like the map. If a polyhedron or mesh of the robot lies in more than one region, the same may be broken into separate polygons, one for each region; or may be redundantly stored. Now the test is whether the robot oct-tree intersects with a the oct-tree representing obstacles or map. This can be achieved by traversing all combinations of both trees in the worst case. Imagine that at any time of the search, a prospective collision needs to be checked with the node o of the obstacle oct-tree and the node r of the robot oct-tree. The o node of the obstacle oct-tree has eight children $o_1, o_2, o_3, \ldots o_8$. A few of these children (o_i) may not have any possible collision with the robot node r, or the volume of a few o_i may not intersect the volume of r. For the children of o that possibly intersect

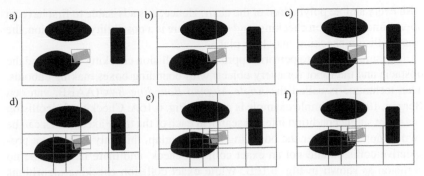

FIG. 6.15 Collision checking using the quad-tree approach. The robot is enclosed to a bounding box which is used for approximate collision checking (to get the prospective colliding obstacle parts with which an exact collision checking follows). The obstacles are hierarchically decomposed using the quad-tree approach, only decomposing a cell if that intersects the robot bounding box and is neither completely obstacle-free nor completely occupied by an obstacle. The collision checking happens using a Depth-First Search on the quad-tree. (A) 1st level, (B) Depth 1, (C) Depth 2, (D) Depth 3, (E) Depth 4, (F) Depth 5.

r (say o_i), the most relevant children of r (say r_j) are found such that the volume of o_i and r_j intersect. This gives new subproblems of smaller sizes that are recursively checked for collision. The leaves contain only a few polyhedrons or meshes (typically 1) of the robot and the obstacles that are checked for collision by an exact algorithm like the GJK. The all-combination traversals are thus severely pruned to get the most prominent colliding pairs of parts of the robot and obstacles. This also gives the penetration depth/retraction direction or clearance. If the map was instead represented as a set of point obstacles, the leaves will contain a few point obstacles which is checked for a collision with the robot polyhedron or mesh. Similarly, if the robot is represented as a set of points, the leaves check for collision between the points of the robot and the points of the obstacle, where two points collide if the proximity between the points is less than the resolution of the map and robot representation.

6.6 Bounding volume hierarchy

The space partitioning techniques aim to model the complete space by recursive hierarchical partitioning. Another methodology is to recursively subdivide the obstacles by bounding them within simpler shapes. Imagine a beautiful chair-shaped obstacle which has a rich curvature that makes the different parts of the chair. Collision checking with this chair can take a significantly long time, checking for possible collisions with each small local face which is a primitive shape polyhedron. Similarly, the robot itself may have rich faces that can be decomposed into small local faces with each small local face being a polyhedron. However, the robot may be far away from the chair and there may be no need of checking for collisions between every simple polyhedron of the robot with every simple polyhedron of the chair. Hence, it is advisable to first have an approximate collision checking, and only if there is a possibility of collision, the finer collision checking may take place.

The simplest way to perform approximate collision checking is to bound the obstacle and the robot (or every object) with bounding boxes making cuboids. This technique is called the *Angle Aligned Bounding Box* (AABB, van den Bergen, 1997). A simple example is given in Fig. 6.16A. Checking for collision is a naïve single condition interval search at each of the three axes, which can be swiftly done. Consider the bounding boxes overlap, this only indicates a prospective collision and not an exact collision check, and there may still be no collision as shown in Fig. 6.16B. While exact collision checking algorithms may still be applied, iterating over all faces of the objects; it is possible to recursively divide the problem.

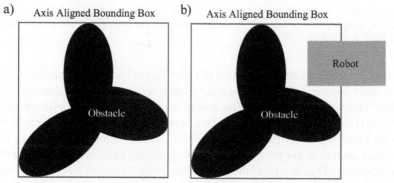

FIG. 6.16 Angle Aligned Bounding Box. (A) Fast collision checking is done for the bounding box only. (B) The robot does not collide with the obstacle, but a collision would be reported at this resolution.

Now consider that the obstacle is broken down into two obstacles. The partitioning can be done by partitioning across the longest axis, partitioning such that the volumes are equally distributed, partitioning across the axis with the highest variance, or any similar heuristic measure. This produces two obstacles as candidates for collision checking. The obstacles may be even further divided into finer obstacles. This constitutes the top-down approach of making an object hierarchy. The tree is constructed in a top-down manner, each node having two children. It is also possible to make a bottom-up tree, wherein unit obstacles are grouped into one to create a parent node of the tree. The breaking down in a tree is shown in Fig. 6.17. The modelling error of the bounding box is the volume of free space inside the bounding box. When an obstacle is broken down into two obstacles, the modelling error reduces, while increasing the complexity since more checks will have to be performed with each of the obstacles.

Another figure which also conveniently models the object while making easy collision checks is a sphere. The collision checking between two spheres

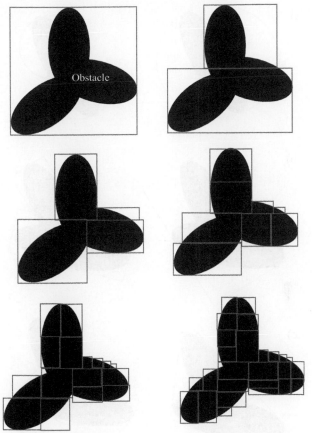

FIG. 6.17 Angle Aligned Bounding Box Tree: At every depth, the cells (not entirely black or white or smaller than the smallest size) are subdivided into two parts maintained as a tree. With increasing depth, the modelling error reduces, and collision-checking time increases.

requires just one condition check (that the distance between the centres is less than the sum of the radii) and can be quickly done. The *spherical bounding box* technique bounds the complete object into a sphere, which can be recursively subdivided. The bounding box is shown in Fig. 6.18A.

The problem with the AABB is that the restriction of the axis of the bounding box being rectilinear is too restrictive, which can sometimes result in a significant loss as shown in Fig. 6.18B. The *Oriented Bounding Box* (OBB, Gottschalk et al., 1996) approach removes this restriction and allows the bounding boxes to be aligned at any general axis, with the only restriction that it takes a rectangular shape. Given a shape (typically a polyhedron with known vertices, edges, and faces), the eigenvalues and eigenvectors help to get the principal axis where the polyhedron gives the maximum variation, taken as the principal axis

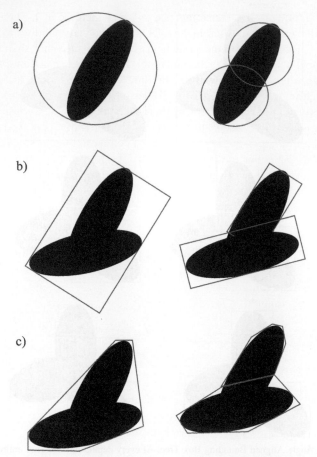

FIG. 6.18 (A) Spherical Bounding Box and its 1st level hierarchical decomposition, (B) Oriented Bounding Box and its 1st level hierarchical decomposition, (C) Discrete Oriented Polytope of degree k (k-DOP) and its 1st level hierarchical decomposition ($k=6$). The smallest degree polygon ($k=6$) is fitted.

of the bounding box. The other axes are orthogonal to the principal axis. The OBB is also recursively subdivided into smaller parts to make finer modelling of the obstacles, like the AABB.

Another generalization is to remove the restriction that the bounding box will be rectangular and to allow any convex polygon to represent the object. This modelling is called the *Discrete Oriented Polytope of degree k* (*k*-DOP, Klosowski et al., 1998). The degree k controls the number of vertices that can be used in the representation of the object. More is the degree of the *k*-DOP, lesser is the modelling errors and better is the representation. This is shown in Fig. 6.18C. However, higher degrees also require more conditions in the collision checking, thus requiring more computation time. The construction time of k-DOP also increases with an increase in k. The representation is still hierarchical, wherein the k-DOP is recursively subdivided into typically two objects based on the same heuristics as for the other shapes.

Even though the term obstacles and robots have been discussed for long, it is good to see where they come from. In simulation systems, all obstacles are robots which are models created using 3D design and CAD-like software, which can be imported into the simulator. Even though the designer may use rich curvatures to make the models, they can be geometrically broken down into *meshes*. The *triangular meshes* are the most common, though any shape may be used. The triangular mesh may be visualized as a mesh wrapped around the entire surface of the obstacle and robot, where each cell of the mesh is a triangle. Powerful meshing algorithms are available for all formats of objects. A collection of unorganized meshes in the naïve format is also called a *mesh soup*. In the real world, the robot will use 3D range scanners like a 3D LiDAR to get point clouds, which may then be regressed into surfaces. Alternatively, the video cameras may be used to detect the object whose shapes and sizes may be computed to place a standard model. These objects are also similarly decomposable into meshes forming a mesh soup. The triangular meshes are also discussed in Chapter 7, with the difference that the intent in Chapter 7 is to model the free space while the intent here is to model the obstacle.

The problem of collision checking is to check if the robot represented as a set of polyhedrons intersects any obstacle in the map. The collision checking algorithm behaves like the case of space partitioning or the oct-tree approach. The map always has a few such obstacles like a few tables, chairs, doors, etc. However, each obstacle can have a very complex shape. Consider that the Bounding Volume Hierarchy is used to represent each obstacle. Further consider that the robot itself is described as a Bounding Volume Hierarchy. Both robot and obstacle hierarchies have leaves that are completely occupied by an obstacle while the shape of the leaves may be an angle-aligned bounding box, oriented bounding box, spherical bounding box, k-DOP, etc. as per the modelling technique used. The collision detection could in the worst case check for intersection between all leaves of the obstacle hierarchy with all leaves of the robot hierarchy. However, it will be possible to prune most of the leaves that are unlikely to collide. Both trees are traversed in a top-down manner. At any time, let the obstacle tree be at a node o and the robot tree be at the node r. If the volumes occupied by o and r do not intersect, no collision is possible. However, if the volumes intersect, a collision may be possible which needs to be checked at the deeper levels. If neither of these is a leaf node, the obstacle node has children o_1 and o_2, while the robot node has children r_1 and r_2. If the volumes occupied by o_1 and r_1 intersect, they are checked for a collision at the next level, else the pair is pruned. Similarly, with the other possibilities of $o_1 - r_2$, $o_2 - r_1$, and $o_1 - r_2$. The leaves are completely occupied by the obstacles. Hence, intersection of two leaves confirms a collision. Consider the robot and obstacle as shown in Fig. 6.19. Recursive division of the potentially colliding obstacle

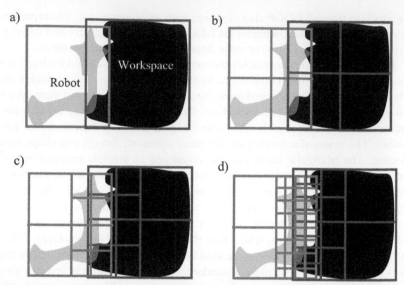

FIG. 6.19 Collision checking using AABB. The robot and obstacles are hierarchically decomposed using AABB, only decomposing a cell of the robot at a level if that intersects another obstacle cell at the current operating level, and vice versa. The collision checking happens using a simultaneous Depth-First Search traversal of both trees. (A) 1st level, (B) Depth 1, (C) subdividing top-right and bottom right robot in (B) due to intersection with AABB of the obstacle at the same level. Similarly, subdividing top left and bottom left obstacle. (D) Subdividing the 16 cells on the right of the robot, and the four leftmost cells of the obstacle, the only possible candidates for a collision.

and robot pairs at different depths is shown in the same figure. Even though all nodes of the same depth are shown in the same figure for clarity, the algorithm will work as a Depth First Search over the obstacle and robot hierarchical tree representations.

This checks for a possible collision with one obstacle, while collision with multiple obstacles needs to be iteratively checked. It is possible to take the robot as a primitive shape and to check for collision of the same with the obstacle hierarchy by doing an intersection test of the robot polyhedron with the obstacle hierarchy nodes, or to have the robot and obstacles represented at different depths.

6.7 Continuous collision detection

Many times, the requirement is to see if a collision would happen if the robot traversed a particular path or moves ahead by a little distance in some direction, and so on. The trajectory computation of a robot is itself doing collision detection on small segments which join to form a complete trajectory. The discussions so far were limited to a single time instant, checking if the robot can take a particular configuration by doing intersection tests. If the robot moves along with time and collision checking need to be done for the entire anticipated motion, it is called the problem of *continuous collision detection*.

The fundamental way to do continuous collision detection is to break the continuous motion into several discrete small steps, where the number of steps determines the resolution of collision checking. Collision checking is done for the configuration at every step. However, this adds an unnecessary additional parameter of resolution into the working of the algorithm. *Swept volume* (Kim et al., 2004) is an easy mechanism to avoid this parameter. The total volume that a robot sweeps can be described as a shape by knowing the motion of the robot. The collision checker checks for a collision of this swept volume shape with all other obstacles in space. The method however does not hold good when there are multiple moving obstacles, when the robot is displaying complex dynamics (for which the shape of the swept volume is hard to determine), or the robot has multiple moving parts like both hands or several links simultaneously moving. Sometimes approximations may be used to determine if a collision is likely to occur and the potential areas of the collision, after which an exact collision checking algorithm, must be used. As an example, the complete robot trajectory including all its moving parts can be bounded to a bounding box for approximate collision checking. Only if this bounding box intersects an obstacle, an exact collision checking is used.

Imagine a rectangular robot that operates by traversing straight as shown in Fig. 6.20A. The volume swept is shown in the same figure. The volume swept is a convenient convex shape for collision checking. The robot may be an object hierarchy forming a volume swept which may also be defined by a similar object hierarchy. If the shape is rectangular, the technique is called the rectangular swept volume, and similarly if the shape is spherical, it is called the swept spherical volume.

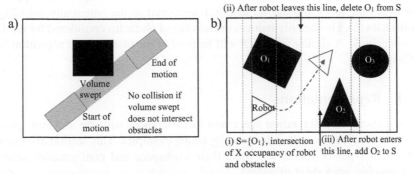

FIG. 6.20 Continuous collision detection: (A) swept volume (B) sweep and prune. S is a set of potentially colliding obstacles as per the X-axis occupancy only, maintained as the robot moves. Exact collision checking happens with only obstacles in S.

If the swept volume is somehow not applicable or easy to use due to different reasons, discretising the trajectory into a few sampled configurations and collision checking on the sampled configuration is the only option. However, the motion of the robot has a *temporal cohesion*, which means the robot can only move a small distance in a small point of time. Knowing the robot and obstacle parts which were close to each other at a point in time, at the next instance it is known that only the same parts are likely to collide. This information can be cashed to speed up some collision checks at a later instance. So, assume the robot has a small proximity with a set of obstacles (S). At the next time instance, the collision is only possible with the obstacles in S and their neighbours in the motion direction of the robot. The set of obstacles in S that the robot has crossed can be further deleted from the set S.

A technique using the same properties is called the *Sweep and Prune* (Tracy et al., 2009). The algorithm sweeps the robot as per its motion and maintains a list of potentially colliding obstacles (S), thus pruning the other obstacles from an exact collision-check. As the robot sweeps its motion, if it is potentially colliding with a new obstacle, it is added to S, while if it left the potentially colliding area of an obstacle, the obstacle is deleted from S. At every step only obstacles in S are considered for an exact collision check. The obstacles are stored in a manner such that the inclusion and deletion conditions can be checked without iterating through all obstacles at every step, like storing the obstacles in a sorted order along the major motion axis of the robot.

As a simple implementation consider that the robot is a triangle that moves along the X-axis in a map shown in Fig. 6.20B. It is imperative to store all the obstacles as per the starting X-axis value. As the robot moves, the moment it encounters the starting X-axis value of an obstacle, the obstacle is a likely candidate for collision check and is added to the set of potentially colliding obstacles (S). As it leaves the last X-axis vertex of the obstacle, it cannot have a collision and the obstacle is no longer considered in the set of potentially colliding obstacles (S). At any time, the exact collision check only happens between the robot and the set of obstacles stored in the potentially colliding obstacles (S). This heavily reduces the number of obstacles considered for a collision check. A similar approach will be used in the plane sweep algorithm in Chapter 7.

6.8 Representations

The notion of configuration space and its differences with the ambient space (workspace) can be best understood by taking examples. This section discusses different types of robots and plots their workspace and configuration space (if possible, on a sheet of paper).

6.8.1 Point and spherical robots

The simplest example is a point-sized robot operating in a 2D space. The workspace is all points attainable by the robot, which is the complete ambient space.

The configuration is given by $q = (x, y)$. The configuration space is the set of all values attainable by the robot. If the map has no limits, each x and y variable can attain any real number as its range, meaning the configuration space is given by $C = \mathbb{R} \times \mathbb{R} = \mathbb{R}^2$. Even though the map has limits, the planning algorithms and their properties do not depend upon the bounds on the map, which means the bounds can be arbitrarily taken as $\mathbb{R}$ to define the class of the configuration space. A point robot standing at the workspace position which is obstacle-prone is deemed to be in a collision and vice versa. Hence, the free configuration space C^{free} is the same as the free workspace W^{free}, and C^{obs} is the same as W^{obs}. The workspace and configuration for a synthetic map are shown in Fig. 6.21A and B. Similarly, a point robot operating in a 3D workspace has a configuration (x, y, z). The workspace and configuration space for a 3D robot is the same, given by $C = \mathbb{R}^3$.

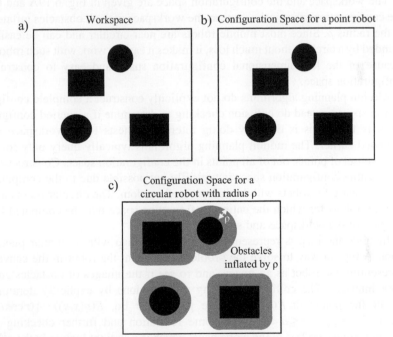

a) Workspace
b) Configuration Space for a point robot
c) Configuration Space for a circular robot with radius ρ
Obstacles inflated by ρ

FIG. 6.21 Workspace and configuration space: (A) workspace space for a point robot, (B) configuration space for a point robot (same as workspace), (C) configuration space for a circular robot.

An interesting case is that of a circular planar robot operating in a 2D space. To encode the configuration of the robot, we only need the x and y coordinate values making the configuration as $q = (x, y)$ and the configuration space as $C = \mathbb{R}^2$. Rotation is not used because even if the robot is rotated, it will look the same without changing anything in the collision-checking algorithm.

A configuration (x, y) is collision prone if the robot of radius ρ centred at $q(x, y)$ collides with an obstacle. C^{obs} is a collection of all such points. The function $R(q)$ that returns all points in the workspace occupied by the robot at configuration $q(x, y)$ is given by Eq. (6.27), where $R(q(x, y))$ is the set of all points (t_x, t_y) that lies at a distance of less than radius ρ from the centre (x, y) of the robot.

$$R(q(x,y)) = \left\{ (t_x, t_y) \colon (t_x - x)^2 + (t_y - y)^2 \le \rho^2 \right\} \tag{6.27}$$

Using this definition of $R(q)$, C^{obs} and C^{free} are given by Eqs (6.28)–(6.29).

$$C^{\text{obs}} = \left\{ q(x,y) \colon \exists_{(t_x, t_y) \in W^{\text{obs}}} (t_x - x)^2 + (t_y - y)^2 \le \rho^2 \right\} \tag{6.28}$$

$$C^{\text{free}} = \left\{ q \colon \left(R(q) \cap W^{\text{obs}} = \emptyset \right) \right\} \tag{6.29}$$

The workspace and the configuration space are given in Fig. 6.19A and C. The configuration space is the same as the workspace with all obstacles inflated by the radius ρ. Since most mobile robots are near-circular and can be easily bounded by a circle without much loss, it makes it easy to work with such robots because of the low-dimensional configuration space and easy to construct configuration space.

Motion planning algorithms do not explicitly construct a complete configuration space, instead do collision checking to determine if a queried configuration is free. This is done by doing intersection tests for a workspace as discussed earlier. The motion planning algorithms typically query only for a small subset of points, out of all points in the configuration space. Construction of the entire configuration space is typically not possible due to the computational expense for robots with high degree of freedom. The circular robots are the only robots for which the entire configuration space may be computed by inflation of the workspace and stored explicitly.

Imagine the map is represented as a 2D grid map with a circular planar robot. A typical way to solve the problem is to paint the robot in the canvas representing the robot as an image, and to see if the images of obstacles and robot intersect. The collision checking may be done by explicitly iterating for all the points in $R(q)$ using the parametric Eq. $R(q(x,y)) = \{(r\cos\alpha, r\sin\alpha) \colon 0 \le r \le \rho, -\pi \le \alpha < \pi\}$ with some resolution and further checking if the same points are free in the workspace, which is a straight lookup in the grid map. This involves looping through r and α to get all points in $R(q)$ followed by a look-up at the grid map.

In case the map is geometrically represented as a collection of polygons, a typical collision checking involves checking if a circle intersects any of the polygons. It is acceptable to store the map as a quad-tree hierarchy for a more efficient collision checking, approximating the circular robot as a square for getting the initial set of possibly colliding objects, between which an exact collision checking is performed.

6.8.2 Representation of orientation

The subsequent examples also incorporate the orientation of the robot. It may initially appear that the angles can be in the range $[-\pi, \pi)$ and since the bounds do not matter taken as $\mathbb{R}$. The problem with this representation is that it does not account for the fact that the angles are *circular*. Hence, the robot could either rotate $\pi/2$ to face the goal or turn $-3\pi/2$ to look at the goal, both of which are valid options and should be available to the planner. The representation is elevated to two dimensions.

Let $S^1 = \{(x,y): x^2 + y^2 = 1\}$ be a collection of points at a unit distance from the origin. Superscript 1 denotes the degree which is 1 or a circle. The set of points is a convenient mechanism to represent an angle θ, conveying the circular properties. Mapping between the two systems is possible as given by Eqs (6.30) and (6.31):

$$\theta = \operatorname{atan2}(y, x) \tag{6.30}$$

$$[x, y]^T = [\cos\theta, \ \sin\theta]^T \tag{6.31}$$

It should be noted that the dimensionality of the space is the number of parameters needed for the representation minus the number of independent constraints. Here, the representation involves two parameters (x, y) and one constraint keeping the original dimensionality of one to represent an angle. The constraint is $x^2 + y^2 = 1$ or that the representation can only be made at the unit circle.

6.8.3 Mobile robots

A common example is to have a planar rectangular robot that moves on a 2D plane. The configuration variables are the location and orientation, or $q = (x, y, \theta)$. The configuration space allows all real numbers for the x and y variables, while S^1 for the orientation. The group representing the orientation of a planar robot has a specific name called SO(2) or a *special orthogonal group of order 2*. The name is derived from the Lie Group theory. Similarly, the configuration space $C = \mathbb{R} \times \mathbb{R} \times SO(2) = \mathbb{R}^2 \times SO(2) = SE(2)$ has a special name SE(2) called the *special Euclidian Group of order 2*.

Let us plot the workspace and the SE(2) configuration space. SE(2) is a three-dimensional configuration space. However, since 1 of the variables is an orientation, it needs to be mapped to 1 higher dimensional space making a 4D figure which cannot be plotted on paper. However, let us informally represent the orientation by an independent axis and plot it as a 3D figure. The workspace and the configuration space are shown in Fig. 6.22A and B. The configuration space is displayed by stacking the cases of different orientations on top of each other. Even if the workspace is geometric, the configuration space is non-geometric. To understand the configuration space, imagine a robot at any orientation (say 0) and move it around. When it is even partly inside the obstacle, it is taken as black in the configuration space around the characteristic point. This forms a slice of the configuration space. As the orientation is changed, the robot will be able to come further to the obstacle as shown in Fig. 6.22C. As the orientation changes, the robot comes furtherer to the obstacle up to an angle of $\pi/2$ and can later come closer to the obstacle from $\pi/2$ to π. The configuration space due to symmetry is the same for orientations π to 2π. Because of the cyclic nature, the configuration space at an orientation of 0 and 2π looks the same.

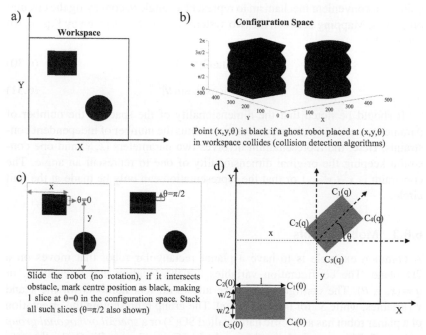

FIG. 6.22 Configuration space for a planar mobile robot. (A) Workspace, (B) configuration space. Angles assumed noncircular for plotting. A point is black if the robot placed in the same configuration at the workspace incurs collision. (C) Visualizing a slice of configuration space at $\theta = 0$ and $\theta = \pi/2$. (D) Notations used for collision checking in a grid map. Iterate all points inside the robot and none should be in the obstacle.

For collision checking with geometric maps, intersection tests may be used. Alternatively, for grid maps, we need to paint the robot over the workspace available as a grid to see if the robot intersects any obstacle. This requires iterating over all points of the robot using the parametric equation for the rectangle given by Eqs (6.32)–(6.35):

$$R(q(x,y,\theta)) = \Big\{(\lambda a + (1 - a)b) : a = \mu C_1(q) + (1 - \mu)C_2(q), b = \mu C_4(q)$$

$$+ (1 - \mu)C_3(q) : 0 \le \lambda \le 1, 0, \ \le \mu \le 1\Big\} \tag{6.32}$$

$$C_1(q) = T(q)C_1(0), C_2(q) = T(q)C_2(0),$$
$$C_3(q) = T(q)C_3(0), C_4(q) = T(q)C_4(0) \tag{6.33}$$

$$T(q(x,y,\theta)) = \begin{bmatrix} \cos\theta & -\sin\theta & x \\ \sin\theta & \cos\theta & y \\ 0 & 0 & 1 \end{bmatrix} \tag{6.34}$$

$$C_1(0) = [l \ \ w/2 \ \ 1]^T, C_2(0) = [0 \ \ w/2 \ \ 1]^T \tag{6.35}$$

$$C_3(0) = [0 \ \ -w/2 \ \ 1]^T, C_4(0) = [l \ \ -w/2 \ \ 1]^T$$

Here, C_1, C_2, C_3, and C_4 are the corners of the robot. (a, b) represents a line interpolated at μ between the corners along the length, and the point inside the rectangle is an interpolation across λ on the line (a, b). The procedure involves looping through λ and μ to get a point in $R(q)$ followed by a look-up at the grid map. l represents the robot length, and w represents the robot width. The characteristic point used to denote the configuration is in the middle of the rear axis of the robot, which is the axis on which the robot turns. The notations are shown in Fig. 6.22D.

While the configuration space for the cyclic case cannot be visualized for the 2D robot, let us put one more constraint to visualize the space. Let us state that the robot can only move in the X-axis and rotate, which makes the configuration $q = (x, \theta)$ and can be plotted in 3D. It needs 1 straight and 1 circular axis, which can be represented by a cylinder. The means to get the relation is to take a 2D figure showing the X-axis and the orientation axis and making a sign to state that the orientation axis is cyclic as shown in Fig. 6.23. The two axis lines with the sign show the circular property, or that they need to be joined. Hence, modify the plot such that the sides with the sign join, which gives a cylinder.

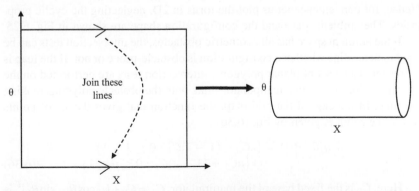

FIG. 6.23 Visualizing configuration spaces with angles. A robot translates in X-axis and changes orientation which is plotted as a 2D configuration space with one circular axis. Joining the circular axis lines gives a cylindrical configuration space. Obstacles are not shown.

The case of flying drones involves the robot to translate in three dimensions as well as to rotate in three dimensions. To specify the rotation, a minimum of three parameters are needed which are typically the Euler's angles denoting the pitch, roll, and yaw. Alternatively, the same orientation can also be denoted by the quaternions with four variables and one constraint giving a dimensionality of 3. Similarly, a 3×3 rotation matrix also denotes the rotation. In general, the rotation representations are called SO(3), meaning *Special Orthogonal group of order 3*, where the name is from the Lie groups. Correspondingly, the configuration space has additional three terms for translation, giving a configuration space $C = \mathbb{R}^3 \times SO(3) = SE(3)$. The configuration space, as stated, has a name SE(3) or *special Euclidian group of order 3*.

6.8.4 Manipulators

Consider a planar manipulator with two links. The ambient space of the manipulator is made by the sensors, or the map supplied externally. The manipulator has two revolute joints. Assume both the joints can rotate in a complete circle any number of times. Since there are two links, and the base is fixed, the joint angles are sufficient to denote the configuration. The configuration becomes $q = (\theta_1, \theta_2)$. The angles are taken as per the Denavit–Hartenberg (DH) parameter principles and shown in Fig. 6.24A. Since each joint angle is S^1, it means the configuration space is $C = S^1 \times S^1 = T^2$. T^2 is called the torus of degree 2. Similarly, for n links, T^n is called a torus of degree n. To visualize the torus T^2, let us first plot the two angles independently on two axes, which makes a rectangle. Let us mark the two sides which should be cyclic as shown in Fig. 6.24B. First one side is rolled, which makes a cylinder. Then the second side is rolled which makes the torus a doughnut shape. It is important to see that in this case $S^1 \times S^1 \neq S^2$, where S^2 is a sphere. The torus and the sphere are not equivalent to each other. The spherical configuration shape is produced by a spherical joint, which can simultaneously rotate on two axes.

The torus is not easy to visualize on paper since half of it shall be hidden. Hence, for convenience let us plot the torus in 2D, neglecting the cyclic properties. The ambient space and the configuration shape are shown in Fig. 6.25.

If the ambient space has all geometric obstacles, the intersection tests can be used to determine whether a configuration is obstacle-prone or not. If the map is represented as a set of planar polygons, intersection tests are performed on the rectangles formed by the manipulator links with the obstacle polygons making the map. In the case of the grid map, the function that gives the set of points inside the robot is given by Eq. (6.36).

$$R(q(\theta_1, \theta_2)) = \{\lambda C_0 + (1 - \lambda)C_1 : 0 \leq \lambda \leq 1\}$$
$$\cup \{\mu C_1 + (1 - \mu)C_2 : 0 \leq \mu \leq 1\} \qquad (6.36)$$

Here, C_0 is the fixed base of the manipulator. $C_1 = C_0 + l_1[\cos\theta_1 \quad \sin\theta_1]^T$ is the endpoint of the first link and $C_2 = C_1 + l_2[\cos(\theta_1 + \theta_2) \quad \sin(\theta_1 + \theta_2)]^T$ is the

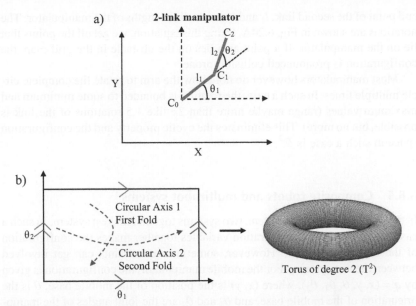

FIG. 6.24 Two link manipulator: (A) notations used for collision checking, (B) visualizing the configuration space as a torus. Joining the circular axis gives a torus.

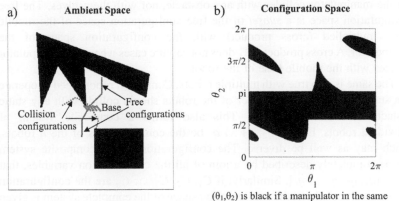

(θ_1, θ_2) is black if a manipulator in the same
configuration in the ambient space does not collide.

FIG. 6.25 Configuration space for a two-link manipulator: (A) ambient space, (B) configuration space plotted in 2D and not torus for easier visibility. A point black in the configuration space denotes that the two-link manipulator kept at the same configuration in the ambient space will incur collisions.

end point of the second link. l_1 and l_2 are the link lengths of the manipulator. The notations are shown in Fig. 6.24A. Using this equation, we get all the points that lie on the manipulator. If a point also lies on the obstacle in the grid-map, the configuration is pronounced collision-prone.

Most manipulators however do not allow the arm to rotate the complete circle multiple times. In such a case, the rotation is bounded to some minimum and maximum values (range maybe more than 2π like 1.5 rotations of the link is possible, but no more). This eliminates the cyclic property and the configuration space in such a case is $\mathbb{R}^2$.

6.8.5 Composite robots and multirobot systems

In the case of a *composite system*, two systems together make a system. In such a case, a join of all the configuration variables together makes the configuration of the composite system. However, sometimes constraints can get involved between the two systems. For the mobile manipulator, the configuration is given by $q = (x, y, \theta, \theta_1, \theta_2)$, where (x, y) is the position of the mobile base, θ is the orientation of the mobile base, and θ_1 and θ_2 are the joint angles of the manipulator arm. The configuration space of the composite system is given by the cross product of all the individual systems. The configuration space in such a case is given by $C = \text{SE}(2) \times T^2$. This assumes that there are no constraints. The free configuration space has all configurations such that neither the robot, nor the manipulator collides with any obstacle, nor with themselves. The free configuration space is a *subset* of the free configuration space of the mobile robot multiplied (cross product) with the configuration space of the manipulator. A cross product only does not capture cases where the manipulator collides with the mobile base of the robot.

The same is also true with multiple robots. Consider n robots need to operate in a static workspace and neither of the robots should collide with any static obstacle or with each other. This also forms a composite system of individual robots. Let $q_1, q_2, q_3 \dots q_n$ be the configurations of the n robots, which may as well be diverse. The configuration of the composite system can be completely described by a join of all the configuration variables, that is $q = [q_1, q_2, q_3 \dots q_n]$. Similarly, if $C_1, C_2, C_3, \dots C_n$ are the configuration spaces of all the robots, the configuration space of the complete system is given by $C = C_1 \times C_2 \times C_3 \dots C_n$.

In this case, the free configuration space is such that no robot collides with any static obstacle of the workspace, and no two robots collide with each other. The free configuration space can be defined by the cross product of the free configuration spaces for all robots, excluding configurations such that any two robots collide. This is given by Eq. (6.37).

$$C^{\text{free}} = C_1^{\text{free}} \times C_2^{\text{free}} \times C_3^{\text{free}} \dots \times C_n^{\text{free}} \setminus$$
$$\{[q_1, q_2, q_3, \dots q_n] : R_i(q_i) \cap R_j(q_j) \neq \emptyset, i \neq j\} \qquad (6.37)$$

The equation considers the joint configurations where any pair of distinct robots i and j collide, or that the occupancy of the robot i (given by $R_i(q_i)$) and robot j (given by $R_j(q_j)$) have a non-null ($\varnothing$) intersection. All such cases are complimented from the cross product of configuration spaces of the individual robots $C_1^{\text{free}} \times C_2^{\text{free}} \times C_3^{\text{free}} \ldots \times C_n^{\text{free}}$ to make the free configuration system for the composite system.

6.9 State spaces

The configuration space only encodes the robot in a static sense as per the spatial occupancy of the robot. The state on the other hand also encodes velocities and (if needed) other higher-order derivatives. A robot may be planned in both configuration space as well as the state space. The state space has more dimensions making planning more complex. The configuration space, however, does not have all variables and therefore it may be hard to specify all constraints.

The robots have two types of constraints. The first type of constraints are called the *holonomic constraints* represented by $g(q,t)=0$ where $g()$ is the constraint function, q is the configuration variable, and t denotes the time. The constraint states that the robot cannot take a few configurations at some time. This is possible to easily specify in the configuration space planners. As an example, the robot can never cross the boundary of the map.

The other type of constraint is called *nonholonomic constraint*, specified as $g(q, \dot{q}, t)$. Here, q denotes the configuration and $\dot{q}$ denotes the derivative of the configuration or speed of every configuration variable. The nonholonomic constraints cannot be modelled in configuration space as they depend upon the speed. However, it is easy to account for them in the state space.

Consider a rectangular differential wheel drive robot. The state consists of all configuration variables and their derivatives $(x, y, \theta, \dot{x}, \dot{y}, \dot{\theta})$. However, the robot cannot walk sideways, which is a constraint, meaning $\tan\theta = \frac{\dot{y}}{\dot{x}}$. Taking $\dot{y}$ as $v\cos\theta$, $\dot{x}$ as $v\sin\theta$, $\dot{\theta}$ as ω, and removing the constraint gives the representation $(x, y, \theta, v, \omega)$, where θ is the orientation, v is the linear speed and ω is the angular speed. The constraint can be incorporated in state space. This means a configuration space planner may ask a robot to travel sideways and the trajectory will have to be corrected. In another example, the curvature of the turn that the robot can take is given by its speed, with larger speed requiring smoother trajectory. This is another nonholonomic constraint. To incorporate this in planning, it is essential that the planner knows the speeds and hence the allowable curvature of turn.

The configuration space planners often yield trajectories that are not traceable. Due to this, there is now an inclination to plan in state spaces, even if the dimensionality is much higher. Consider that the robot has to circumvent a tight turn at a corner. Now one cannot plan a path as a sequence of configurations and then use a control algorithm to trace the path, as the path may have turns for which no control sequence exists. A better approach therefore is to directly calculate the control sequence that given to the robot makes the robot reach the desired goal configuration. The motion of robots is typically governed by kinematics equation of the form $s(t+\Delta t) = K(s(t), u(t), \Delta t)$, which computes the new state $s(t+\Delta t)$ using the inputs as the state at time t, the control signal applied ($u(t)$), and the duration of time (Δt) for which the control signal is applied. The planner hence in the state space finds the sequence of control signals to be applied that makes the robot reach the goal. The state space is also called the *control space* since it is the space in which the control signals act.

Questions

1. Show the k-D tree and quad-tree formed on addition of the following points: (7,2), (7,1), (4,8), (2,4), (4,8), (2,7), (5,7), (2,5), (6,8), (1,4). Also, show the mechanism to get the closest point and all points within two units to the query points (1,6) and (4,2).

2. Show the decomposition of the objects in Fig. 6.26 till depth 2 using Angle-Aligned Bounding Box, Oriented Bounding Box, spherical bounding box, and k-DOP.

a)   b)   c)

FIG. 6.26 Sample maps for question 2.

3. Describe the configuration space for the following robots (a) a three-link manipulator with a two-finger gripper, (b) a rectangular planar robot with a three-link manipulator and a two-finger gripper mounted on top, (c) two robots of type (b), (d) a six-link manipulator with each joint limited to a minimum and a maximum threshold.

4. A car is operating in an obstacle-free region, for which the shortest distance is always a straight line between the source and goal. However, the car is not looking at the goal but is perpendicular to the straight-line distance to the goal. So practically the car cannot travel sideways to reach the goal. Explain whether motion planning fails in such a context and if there is an algorithm that can plan the path of the robotic car.

5. [Programming] Take a bitmap image representing the workspace. You can use any drawing tool to make a bitmap. For each of the following cases, make a function that checks if the robot at a particular configuration is free

or not. Check at extremely low resolutions and only check at robot boundary/corner points to get early results (if necessary). If the configuration is free, plot it in the workspace as blue. If the configuration is not free, plot it in the workspace as red. Check for randomly created configurations. Also, draw the configuration space as an image/3D plot for whichever case the dimensionality of space is 3 or less.

- A two-link revolute manipulator with a fixed base
- A three-link revolute manipulator with a fixed base (also check for self-collisions by deriving an intersection test for lines. The lines will intersect if a solution to the simultaneous equations gives a point in-between the points making the lines).
- A three-link revolute manipulator with a fixed base and a two-finger gripper (also check for self-collisions)
- A three-link manipulator with the second joint nonprogrammable and fixed to 30 degrees
- A robot with one revolute and one prismatic joint with a fixed base
- A circular planar disc of radius R
- A rectangular robot

6. [Programming] For each of the cases in question 5, take two configurations (say A and B) and check if a continuous motion from A to B is collision-prone or collision-free. Draw both configurations in the workspace in red (in case of a collision) and blue (if no collision). You may check for a sampled configuration in-between A and B.

7. [Programming] For each of the cases in question 5, take four configurations (say A, B, C, and D) and check if a continuous motion from A to B for one robot and simultaneous a continuous motion from C to D for another robot (both robots with the same speed) is collision-prone or collision-free. Simulate the motion on the screen to verify the results.

8. [Programming] Consider the environment as a collection of convex polygon obstacles with vertices numbered anticlockwise instead of a bitmap. Repeat the solution to questions 5–7. For the circular robot case, assume all obstacles as circles.

9. [Programming] Given a 3D space as a large list of point obstacles (point clouds), represent the workspace using the k-D tree and oct-tree representations. Hence, print the closest points to a given query point within a distance D.

References

Choset, H., Lynch, K.M., Hutchinson, S., Kantor, G., Burgard, W., Kavraki, L.E., Thrun, S., 2005. Principles of Robot Motion: Theory, Algorithms, and Implementations. MIT Press, Cambridge, MA.

Gilbert, E.G., Johnson, D.W., Keerthi, S.S., 1988. A fast procedure for computing the distance between complex objects in three-dimensional space. IEEE J. Rob. Autom. 4 (2), 193–203. https://doi.org/10.1109/56.2083.

Gilbert, E.G., Foo, C., 1990. Computing the distance between general convex objects in three-dimensional space. IEEE Trans. Robot. Autom. 6 (1), 53–61.

Gottschalk, S., Lin, M., Manocha, D., 1996. OBBTree: a hierarchical structure for rapid interference detection. In: Proceedings of the of ACM Siggraph Symposium on Interactive 3D Graphics and Games. ACM, New York, NY, pp. 171–180.

Hornung, A., Wurm, K.M., Bennewitz, M., Stachniss, C., Burgard, W., 2013. OctoMap: an efficient probabilistic 3D mapping framework based on octrees. Auton. Robot. 34, 189–206.

Kim, Y.J., Varadhan, G., Lin, M.C., Manocha, D., 2004. Fast swept volume approximation of complex polyhedral models. Comput. Aided Des. 36 (11), 1013–1027.

Klosowski, J., Held, M., Mitchell, J., Sowizral, H., Zikan, K., 1998. Efficient collision detection using bounding volume hierarchies of k-dops. IEEE Trans. Vis. Comput. Graph. 4 (1), 21–37.

LaValle, S., 2006. Planning Algorithms. Cambridge University Press, Cambridge, NY.

Meagher, D., 1982. Geometric modeling using octree encoding. Comput. Graph. Image Process. 19 (2), 129–147.

Tracy, D.J., Buss, S.R., Woods, B.M., 2009. Efficient large-scale sweep and prune methods with AABB insertion and removal. In: 2009 IEEE Virtual Reality Conference. IEEE, Lafayette, LA, pp. 191–198.

van den Bergen, G., 1997. Efficient collision detection of complex deformable models using AABB trees. J. Graph. Tools 2 (4), 1–14.

Chapter 7

Roadmap and cell decomposition-based motion planning

7.1 Introduction

The problem of robot motion planning is to compute a robot's trajectory from a given source to a given goal such that the robot does not collide with any static or dynamic obstacle, including self-collisions with itself. The two methods studied so far are the A* algorithm working over discretized cells and the Bug algorithms. The A* algorithm had the problem of exponential complexity in terms of the degrees of freedom, high computation time for a high-resolution map, inability to detect narrow corridors in a low-resolution map leading to a lack of completeness and optimality, and difficulty in fixing the resolution parameter. So, the algorithm is only practically usable under high-resolution settings, for which it consumes too much time. The Bug algorithms are highly suboptimal and not recommendable for any practical robotic applications.

On the contrary, with modern technology, self-driving cars can easily plan and navigate for endless miles; humans can easily plan complex motions from one place in the office to another and navigate themselves even at new campuses or shopping complexes with ease. This makes it inquisitive to exploit any heuristic or trick being used by these systems to make the overall navigation a much simpler problem, solvable in near real-time.

The major underlying mechanism in all the examples is that there is a lot of precomputation to store the rich map information into a structure that can be used for online planning. This summarized structure is called the *roadmap*. The conversion of any general map into a roadmap is an offline method; however, this converts any general map into a *topological map*, making a graph with vertices V and edges E. The topological map has limited states and edges, and therefore it is easy to use for online planning.

In the analogy of the self-driving car, the road network graph (roadmap) of the city is already known and can be directly used by any application. The graph search can be used to give a *route* for the vehicle, which specifies all roads,

Autonomous Mobile Robots. https://doi.org/10.1016/B978-0-443-18908-1.00015-7

intersections, and exits to take while driving. This reduces the problem of navigating in a city to following the route or taking control actions to keep oneself within the roads and intersections. Satellite imagery gives a high-definition aerial view of the world. The process of identifying roads and connecting them in a graph format is an offline process. That said, there are other ways to make maps, both human assisted and autonomous, including plotting the route taken by a test vehicle, SLAM using on-board vision sensors, using local human expertise, etc.

Similarly, for home and office scenarios, one knows the rooms, hallways, corridors, stairs, etc. available on the map and the way they connect. Therefore, the problem of navigating through the office reduces to going from one root to the next room, which is a much simpler navigation problem. The static obstacle networks within the rooms are also summarized as topological maps in the human brain, and therefore it is much easier for humans to make their way out.

In all the problems, there is an offline approach to convert the map in its native format into a roadmap. On the roadmap, there are numerous queries to solve. For example, satellite navigation systems for driving get numerous queries every second for routing vehicles and all visitors to museums frequently refer to the same map for navigation information. All roadmaps abstract space into a few points representing vertices and edges. Strictly speaking, to move using roadmaps only, one needs to first access the roadmap by reaching the nearest abstracted configuration in the space, then travel as per the abstracted vertices and edges, and finally depart the roadmap or the abstracted configurations to connect to the goal. One's room is not a vertex in the satellite navigation system. To use the satellite navigation system, one must reach the nearest vertex or edge (intersection or road in the transportation network). Similarly, the satellite navigation system gives a route till the last vertex (intersection or road in the transportation network), after which one drives into the home garage or parking using the local knowledge not represented in the roadmap.

The difference between the A* algorithm over a grid map and the roadmap becomes more interesting when you visualize the graph created by the A* algorithm as a roadmap. The purpose behind the roadmap-based approaches is to capture some structure in the given map and to exquisitely summarize the same in a much smaller number of vertices and edges. The graph overlay on grids does not fully capture this notion. However, the bigger problem with grids was controlling the number of vertices in the graph or the resolution parameter.

Another concept is used to decompose the map into smaller cells, representing only the free space, called *cell decomposition*. The aim is to subdivide the *free space* continuously up to the level of unit cells. The unit cells constitute the vertices of the map while any two neighbouring cells sharing any part of a cell edge are said to be connected by an edge. In cell decomposition, all free configurations are a part of some cell, while all cells together form the roadmap, unlike roadmap-based approaches where the roadmap is a small subset of all free cells, represented in a vertex and edge manner. Since a cell or an area denotes a vertex, rather than a single configuration, it is not naturally clear how to define distances between two big cells; a command to go to a cell means going to which characteristic configuration inside the cell, etc. All these issues must be explicitly defined by the implementing algorithms or applications. This chapter is devoted to motion planning using both roadmap-based approaches and cell decomposition-based approaches. Some of the key concepts are summarized in Box 7.1.

Box 7.1 A summary of terms.

Roadmap approaches:
- Summarize static space as a graph using an offline approach.
- The same roadmap can be used to solve multiple motion planning queries online.
- Given a query:
  - First connect source to roadmap (*access*)
  - Then connect goal to roadmap (*depart*)
  - Then do an online graph search (*connect*)

Visibility graph:
- Assumes robot is point sized (can inflate obstacles for circular robots).
- Assumes geometric maps (polygon/polyhedral obstacles).
- Connect every obstacle corner with other visible obstacle corners making the roadmap.
- Complete and optimal using path length metric for 2D cases only.

Voronoi:
- Captures the topology of the space (as a *deformation retract*).
- **Voronoi diagram**: Points equidistant to two control points, making a graph.
- **Generalized Voronoi diagram**: Taking sized obstacles as control points, points equidistant to two obstacles form edges, and points equidistant to three obstacles form edges of a roadmap.
- **Generalized Voronoi graph**: Extended to 3D spaces. Edges are equidistant to three obstacles, vertices are equidistant to four obstacles, and additional edges are added to get connectivity.
- Can be generalized to higher dimensions, workable till 3D spaces only.
- Planning maximizes clearance and is complete and not optimal.

Cell decomposition:
- Fill cells in the continuous space-forming vertices. Neighbouring cells form edges.
- **Single resolution**: Fill rectangles, triangles, hexagons, etc. Chosen shape may be constrained to be completely in free space or may be allowed to be partly in obstacle as such shapes are uniformly filled in the complete space.
- **Multiresolution**: Use grids of different sizes to represent free cell and recursively subdivide space as quadtree. Can model higher resolutions near obstacles/narrow corridors and hence better. During the algorithmic search, resolution can be dynamically set, as per the current area/need.

Continued

Box 7.1 A summary of terms—cont'd

- **Trapezoidal:** Place trapezoids by a single sweep of a vertical line in space.
- **Boustrophedon:** Merges some neighbouring cells of trapezoidal decomposition, suitable for complete coverage problem, and produces fewer cells.
- **Navigation mesh:** Produce a mesh representing navigable corridors within which the robot is free to move.

7.2 Roadmaps

Consider the free configuration space of the robot given by C^{free}. The roadmap aims to represent the same information in the form of a graph RM with vertices V and edges E, which represents a significant reduction in size from the hyper-volume of the free configuration space to a mere small collection of vertices and edges. The vertices are the samples taken from the configuration space ($V = \{v\} \subset C^{\text{free}}$), while the edges ($E = \{e < v_i, v_j >\}$) connect any two vertices (v_i and v_j) such that a defined path ($\tau_e(v_i, v_j)$) between them is collision-free. The manner of deciding the vertices and edges is a characteristic of every algorithm. The roadmap RM is defined as all sets of points, either vertices or edges that are abstracted from the free configuration space and are a part of the graph, given by $RM = V \cup E$. The conversion of the free configuration space into the roadmap is an offline process. Roadmap-based motion planning is a *multiquery approach* wherein multiple motion planning queries are issued that need to be solved for a path in near real-time using the same roadmap. The multiple queries may be generated from multiple robots or multiple times from the same robot. The query $q(S,G)$ consists of a source (S) and goal (G), and the aim is to compute the path $\tau:[0,1] \to C^{\text{free}}$ such that the path starts from the source $\tau(0) = S$, ends at the goal $\tau(1) = G$, and is collision-free at all times $\tau(s) \in C^{\text{free}}$, $0 \le s \le 1$.

Since space is already abstracted to the roadmap RM, it is expected that the path τ shall only pass through the points in the roadmap, which however is not possible since the source and goal are not a part of the roadmap. The solution to the query $q(S,G)$ therefore consists of three subparts:

- *Access*: Going from source to any point in the roadmap, which gives a trajectory $\tau_{\text{access}}:[0,1] \to C^{\text{free}}$ from $\tau_{\text{access}}(0) = S$ to $\tau_{\text{access}}(1) = q_1 \in RM$.
- *Connect*: Going from q_1 in the roadmap to q_2 in roadmap such that the robot is always in the roadmap, giving the trajectory $\tau_{\text{connect}}:[0,1] \to RM$ from $\tau_{\text{connect}}(0) = q_1 \in RM$ to $\tau_{\text{connect}}(1) = q_2 \in RM$, passing through the roadmap all the way $\tau_{\text{connect}}(s) \in RM$, $0 \le s \le 1$.
- *Depart*: Going from q_2 to the goal, which gives a trajectory $\tau_{\text{depart}}:[0,1] \to C^{\text{free}}$ from $\tau_{\text{depart}}(0) = q_2 \in RM$ to $\tau_{\text{depart}}(1) = G$.

The resultant trajectory is a concatenation of the three parts that are τ_{access}, $\tau_{connect}$, and τ_{depart}. Consider the example shown in Fig. 7.1. From the source, first access the roadmap to point q_1. If there are multiple access points (q_1), the planner will try a path through all of them and then select the best. Then the point q_1 is connected to departure point q_2, while planning using a graph search. If multiple departure points (q_2) are possible, all such points would be tried, and the trajectory corresponding to the best is retained. Finally, the last segment is departing the roadmap at q_2 and reaching the goal.

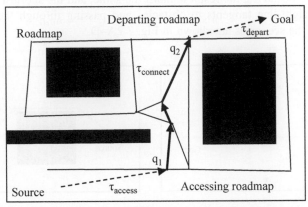

FIG. 7.1   Roadmaps. The continuous configuration space is abstracted into a graph called roadmap. Planning consists of accessing the roadmap (dashed, τ_{access}, not a part of the roadmap), going through the roadmap (bold lines, $\tau_{connect}$, part of the roadmap), and departing the roadmap (dashed, τ_{depart}, not a part of the roadmap).

For this chapter, it will be assumed that the robot is a point robot and the workspace and configuration space are the same. All approaches discussed in this chapter need explicit knowledge of the entire space for different purposes. It is not computationally possible to compute the high-dimensional configuration space for a high-dimensional robot in totality. Furthermore, some methods assume the geometry of the space or a *geometrical knowledge* of the obstacles. The assumption is that all obstacles are polygons in 2D or polyhedrons in 3D. Geometric shapes like polygons lose their geometry in the configuration space. A rectangular robot operating in a workspace with all rectangles also has a nonrectangular configuration space in SE(2), since the orientation changes the shape of the obstacles along the orientation axis. Hence geometry in configuration space is both unknown and indeterminable.

7.3   Visibility graphs

The first approach is called the *visibility graph* and uses the notion of *visibility* to define a roadmap (Janet et al., 1995; Oommen et al., 1987). We are given space with polygon obstacles. The approach fits into a roadmap in the same space to facilitate the motion of the robot.

7.3.1 Concepts

To understand the notion, let us first restrict ourselves to 2D maps. Imagine a 2D obstacle-free map with some sources and goals. The shortest path is the straight line joining the source to goal. Now insert an obstacle in the middle of the source and goal. The shortest path now is composed of tangents to the obstacle from source to goal and traversing the obstacle boundary in between. In the case of a polygon obstacle, the shortest path is a traversal to the corner of the obstacle, travel through the obstacle boundary, and departure from another corner. On adding more obstacles, the result remains the same, and the shortest path consists of straight-line tangents to obstacles and passing through the obstacle boundary. The scenarios are shown in Fig. 7.2A–D.

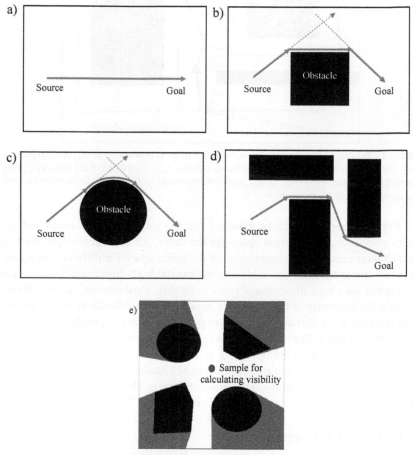

FIG. 7.2 Conceptualizing visibility roadmap. Shortest paths constitute a mixture of the straight-line source to goal connections (shown in (A)), tangents to obstacles for both polygons (shown in (B)) and nonpolygon (shown in (C)) cases, and traversal on obstacle surfaces (shown in (B), (C), and (D)). (E) shows regions visible to a point (white) and not visible (grey).

We first make a naïve implementation of a roadmap for polygon obstacles, knowing that the shortest path always consists of straight lines passing through obstacle corners. Obstacle corners (in the case of polygon maps) are interesting since they facilitate the shortest paths. So, take all obstacle corners as vertices on the roadmap. Two vertices (v_i, v_j) are said to be connected if they are visible to each other; that is, a straight line from v_i to v_j does not intersect any obstacle given by Eq. (7.1).

$$E = \left\{ (v_i, v_j) : \lambda v_i + (1 - \lambda)v_j \in C^{free}, 0 \leq \lambda \leq 1, v_i \in V, v_j \in V, v_i \neq v_j \right\} \quad (7.1)$$

This is obtained from the same intuition that the shortest path has straight lines between obstacle corners in a polygon environment. Eq. (7.1) uses the parametric equation of a line from v_i to v_j to check that all intermediate points on the line $(\lambda v_i + (1 - \lambda)v_j)$ are collision-free.

The name *visibility graph* for the approach is based on the notion that every pair of vertices *visible* to each other is connected by an edge. Correspondingly, the *visibility* of any state q is defined as the set of all states that are visible to the state q or the set of all states x such that a straight line from q to the state x is collision-free, given by Eq. (7.2).

$$V(q) = \left\{ x : \lambda x + (1 - \lambda)q \in C^{free}, 0 \leq \lambda \leq 1 \right\} \quad (7.2)$$

The visibility of a sample point on a synthetic map is shown in Fig. 7.2E. The visibility is more intuitively obtained by imagining an omnidirectional light source at q. The visibility is all the points, excluding obstacles that are directly illuminated by the light source. Taking all obstacle corners as vertices and connecting every pair of visible vertices by an edge, the roadmap so produced is given in Fig. 7.3.

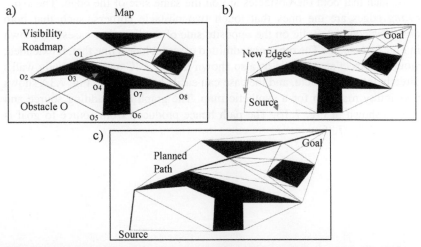

FIG. 7.3 Planning using visibility roadmap. (A) The initial roadmap construction (all corners are vertices, and all possible pairs of visible vertices are edges). (B) Adding source and goal temporarily to the roadmap for solving planning query. (C) Path produced by the graph search. The initial roadmap is shown in red (fine line), the new edges on adding source and goal in green (fine line), and the planned path in blue (bold line).

Once the roadmap is offline, it has to be used for the online motion planning of the robot solving the query $q(S,G)$. Since the source (S) and goal (G) are not part of the roadmap, a graph search algorithm cannot be used directly. Therefore, first, make a new temporary roadmap over the original roadmap by adding in the source (and goal) as the new vertices and connecting the source (and goal) to any other vertex in the roadmap that is in the visibility of the source (and goal). The roadmap (RM') with vertices (V') and edges (E') is given by Eqs. (7.3), (7.4)

$$V' = V \cup \{S, G\} \tag{7.3}$$

$$E' = E \cup \left\{ (S, v_j) : \lambda S + (1 - \lambda) v_j \in C^{\text{free}}, 0 \leq \lambda \leq 1, v_j \in V \cup \{G\} \right\}$$
$$\cup \left\{ (v_i, G) : \lambda v_i + (1 - \lambda) G \in C^{\text{free}}, 0 \leq \lambda \leq 1, v_i \in V \cup \{S\} \right\} \tag{7.4}$$

This solves the problem of *accessibility* or connecting to the roadmap from the source, and *departability* or leaving the roadmap at the goal. The path is a graph search over the new temporary roadmap. Note that the source, goal, and all edges associated with them are deleted once the query is processed.

The execution time of the online part of the problem is of prime concern as it facilitates online planning of the motion planning query. To facilitate faster searches, it is advisable to have as few edges as possible without affecting the quality. Hence all edges that are guaranteed not to come in the path of any motion planning query are pruned from the graph.

The *restricted visibility graphs* only admit two types of edges: supporting and separating. The *supporting* edges are the ones that touch two obstacle corners, such that both the obstacles are on the same side of the edge. The *separating* edges are the ones that touch two obstacle corners, such that both obstacles completely lie on the opposite side of the line. The cases are shown in Fig. 7.4A and B. The other nonadmitted cases include a line that, if extended, would intersect an obstacle as also shown in Fig. 7.4C and D. Nobody walks directly into an obstacle, and the case can easily be neglected. Fig. 7.3A shows a visibility roadmap for a synthetic map. The graph after adding source and goal is shown in Fig. 7.3B. The path hence produced from source to goal is shown in Fig. 7.3C.

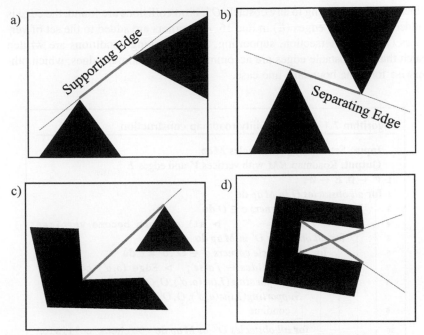

FIG. 7.4 Different types of edges in a visibility graph. (A) Supporting, both obstacles on the same side. (B) Separating, both obstacles on opposite sides. (C) and (D) are neither supporting nor separating. Only supporting (A) and separating (B) edges are admitted.

7.3.2 Construction

A naïve implementation of a visibility graph involves taking all corners as vertices and doing collision checking between any two pairs of corners. Assume that the map is 2D and is available in the form of a collection of obstacles $Map = \{O\}$, with every obstacle defined as a sequence of vertices in a counterclockwise manner, $O = [o_1, o_2, o_3, \ldots]$. The notations are shown in Fig. 7.3A. The collision checking (line intersections) between an edge (a,b) and an edge (c,d) solves the simultaneous pair of equations representing lines of the two edges (say, $\lambda a + (1 - \lambda)a = 0$ and $\mu c + (1 - \mu)d = 0$) to see if a unique solution exists (λ, μ) and the solution is in the interior of the two lines ($0 \leq \lambda \leq 1$ and $0 \leq \mu \leq 1$), which can be done in a constant unit of time.

The implementation is shown in Algorithm 7.1. The algorithm iterates through all possible edges between any pair of vertices (o, o') in lines 2–6 and checks possible collisions (line intersections) with any edge (o_i'', o_{i+1}'') in lines 10–12, belonging to any obstacle O''. If no collisions are found, the edge is added to the set of edges (E) in line 16. All corners are added to the set of vertices (V). The intersection, supporting, and separating conditions are written such that the obstacle edges are accommodated as roadmap edges, which otherwise form the boundary line case.

Algorithm 7.1 Naïve visibility roadmap construction.

Input: Set of polygon obstacles Map
Output: Roadmap RM with vertices V and edges E

1 $V \leftarrow \emptyset, E \leftarrow \emptyset$;
2 **for** *all obstacles O in Map* **do**
3   **for** *all obstacle corners* $o \in O$ **do**
4          $V \leftarrow V \cup \{o\}$;   ▷ all corners become vertices
5   **for** *all obstacles O' in Map* **do**
6   **for** *all obstacle corners* $o' \in O', o' \neq o$ **do**
7              $edgeCollides \leftarrow false$;   ▷ Edge (o, o') collides?
8   **if** $\neg(Separating(Line(o, o'), O, O') \vee$
                  $Supporting(Line(o, o'), O, O'))$ **then**
9   continue
10   **for** *all obstacles O'' in Map* **do** ▷ check collision
                 with all obstacle edges
11   **for** *all obstacle edges* $(o''_i, o''_{i+1}) \in O''$ **do**
12   **if** $Line(o, o')$ *intersects* $Line(o''_i, o''_{i+1})$ **then**
13                         $edgeCollides \leftarrow true, break$
14   **if** $edgeCollides$ **then** break;
15   **if** $\neg edgeCollides$ **then**
16                 $E \leftarrow E \cup <o, o'>$

17 return $RM <V, E>$

Typicality is Line 12. Imagine a square obstacle. The two diagonals are obstacle prone; however, the algorithm as a naïve implementation will not detect any collision. Hence the previous and next edges are taken, and if (o, o') is strictly in-between, it means that the line goes inside the obstacle and should be neglected. For the same, the obstacle vertices must be numbered counterclockwise. The implementation of supporting and separating concepts is simple. Each line divides the space into two half-planes, and to identify the half-plane to which a point belongs, put the point in the equation of the line.

For supporting case, the previous and next obstacle corners in the same obstacle lie on the same side for both obstacles, while for the separating case both previous and next obstacle corners lie on one side for the first obstacle and both previous and next obstacle corners lie on the other side for the second obstacle.

The complexity of the algorithm is defined in terms of the number of vertices of all obstacles combined (n). For any polygon obstacle, the number of vertices is always equal to the number of edges. The complexity of the algorithm is thus $O(n^3)$ because each loop of lines 2 and 3, 5 and 6, and 10 and 11 iterates through all vertices or edges.

For any modestly sized map, the complexity is too high, and therefore some optimization in the algorithm is done. The optimized algorithm is known as the *rotational sweep* algorithm and uses the analogy that to check for edge connections emanating from the vertex o, a light beam is rotated from o in an anticlockwise (or clockwise) direction. Any vertex (o') that the light beam touches corresponds to an acceptable edge (o,o'), while if the light beam cannot touch a vertex due to an obstacle, it is not added as an edge. The problem with Algorithm 7.1 is the need for collision checking. The rotational sweep algorithm is selective on the obstacle edges to be selected for every candidate edge that wishes to be on the roadmap. Since not all edges are checked for collision, the complexity is significantly reduced.

To understand the concept, let us say that collision checking of an edge $<o,o'>$ emanating from a vertex o will only happen for the obstacle edges in the candidate set S. No other edge outside S will be checked for collision. The issue is maintaining (and initializing) such a candidate set of obstacle edges S. Rotations are harder to understand in contrast to translations, so let us study the concept assuming a vertical line L ($X =$ constant) is translated across the horizontal (X) axis from $X = -\infty$ to $X = \infty$. Suppose an obstacle edge is between obstacle corners (x_1, y_1) and (x_2, y_2). As the vertical line L travels, it will intersect the line at $X = \min(x_1, x_2)$ and will keep intersecting the line until it reaches $X = \max(x_1, x_2)$. There will be neither collisions before $X = \min(x_1, x_2)$ nor after $X = \max(x_1, x_2)$. Therefore for collision checking of L with any line, the line with obstacle corners (x_1, y_1) and (x_2, y_2) should be inserted into S at $\min(x_1, x_2)$ and extracted out of S at $\max(x_1, x_2)$. Generalizing to when there are many edges whose lateral coverage along the X-axis overlaps, each edge is inserted at its minimum X coordinate and extracted at its own maximum X coordinate value. The edges may therefore be made directional from a lower X coordinate value to a higher X coordinate value. When L meets the tail of the edge, it is added in S, and when L meets the head of the edge, it is removed from S. The notations are shown in Fig. 7.5.

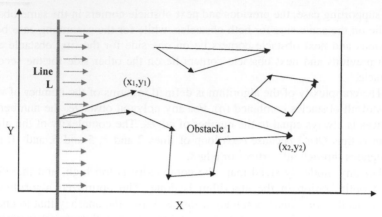

FIG. 7.5 Understanding the plane sweep. The line L goes from left to right. The edges are added in set S on encountering the tail (lower X coordinate) and deleted on encountering the head (higher X coordinate). Set S contains all edges that L intersects.

Now replace the translating line L with a rotating line L that rotates from $\theta = 0$ to $\theta = 2\pi$, where θ is the slope of the line. The choice of angles is arbitrary, and any starting point can be chosen in the circular range while ending at the same point. Similarly, all candidate obstacle edges have a circular coverage, which is the range of slopes of L within which they will lie. The circular coverage is the angle subtended by the two corners at the rotating point (x_o, y_o). Assuming the obstacle corners are (x_1, y_1) and (x_2, y_2), the angles are given by $\alpha_1 = \text{atan} 2(y_1 - y_o, x_1 - x_o)$ and $\alpha_2 = \text{atan} 2(y_2 - y_o, x_1 - x_o)$. Conceptually, the angular range is given by $\min(\alpha_1, \alpha_2)$ to $\max(\alpha_1, \alpha_2)$; however, the angles cannot be directly compared because of the circular property that 0 is not less than 2π. The comparison metric here is strictly in the counterclockwise direction, the direction of rotation of the sweeping line. Imagine the obstacle edge and the rotating line, and see the first corner that the line strikes on the obstacle edge, such that thereafter it is always on the obstacle edge till it leaves the obstacle. This becomes the tail. The other obstacle corner, after which the rotation line leaves the obstacle edge, becomes the head. When the rotating line touches the tail of the obstacle edge, the corresponding edge is added to S, and when the rotating line touches the head of the obstacle edge, the edge is removed from S.

Further, let us assume that S (the set of candidate obstacle edges) is always sorted using the metric of distance between the pivot point o (start point of the rotating line) and the intersection point of the obstacle edge with the rotating line. This can be done by implementing the data structure S as a balanced binary search tree using the distance metric. Only the first edge in S is a candidate for collision checking, which is the nearest to the rotating line.

The algorithm is now given by Algorithm 7.2. The algorithm checks for all possible edges (o, o') (lines 2–3 and 9) for collisions with the nearest (first) edge in S (lines 10–12). The difference with a naïve implementation is that o' is taken as per the angles $\alpha_{o'}$ as an implementation of rotation sweep. Initially, the rotation line is at $\theta = 0$, and therefore all edges that cut the horizontal line are candidates (line 4). As the line rotates, it stops at $\theta = \alpha_1$, then $\theta = \alpha_2$, then $\theta = \alpha_3$, and so on. At any point, if it is at the tail of an edge, the same is added (lines 13–14), while if the line is at the head of an edge, it is deleted (lines 15–16). There are two loops that go through all vertices in lines 2–3 and line 9, giving a complexity of $O(n^2)$. At each point, edges are either added or deleted in the set S maintained as a balanced binary tree, giving a complexity of $O(n^2 \log n)$.

Algorithm 7.2 Visibility roadmap construction with rotation sweep algorithm.

Input: Set of polygon obstacles Map

Output: Roadmap RM with vertices V and edges E

1   $V \leftarrow \emptyset, E \leftarrow \emptyset$;

2   **for** *all obstacles O in Map* **do**

3   **for** *all obstacle corners $o(x_o, y_o) \in O$* **do**

      ▷ S : set of obstacle edges colliding with rotation line centred at o at 0 radians ;

4       $S \leftarrow \{< o_i, o_{i+1} >$ intersects line $y = y_o\}$;   ▷ all obstacle edges intersecting horizontal line at o

5   sort S using distance of intersection point of line $y = y_o$ from o ;

6       $V \leftarrow V \cup \{o\}$;

7   **for** *all obstacle corners $o'(x_{o'}, y_{o'})$ except o* **do**

8        $\alpha_{o'} = (atan2(y_{o'} - y_o, x_{o'} - x_o) + 2\pi)mod\,2\pi$;

9   **for** *all obstacle corners o' except o in increasing order of $\alpha_{o'}$ in counter-clockwise direction from 0* **do**

       ▷ rotation line stop at o' in order $\alpha_{o'}$. Only first edge in S may cause a collision;

10   **if** *Line(o, o') intersects first edge e in S* **then**

11         $edgeCollides \leftarrow true$

12   **else** $edgeCollides \leftarrow false$;

13   **for** *all obstacle edges (o', o'') emanating from o'* **do**

14         $S \leftarrow S \cup < o', o'' >$;   ▷ maintaining sorted order

15   **for** *all obstacle edges (o'', o') ending at o'* **do**

16         $S \leftarrow S- < o'', o' >$;   ▷ maintaining sorted order

17   **if** $\neg edgeCollides \wedge (Separating(Line(o, o'), O, O') \vee Supporting(Line(o, o'), O, O')$ **then**

18         $E \leftarrow E \cup < o, o' >$

19   return $RM < V, E >$

It was previously stated that the edges that go inside an obstacle are not candidate edges to be inserted in the roadmap, irrespective of the collision status, like the diagonals of a square obstacle. Therefore, for considering prospective roadmap edges from an obstacle corner o_i, the circular region between the slope of edges $<o_{i-1},o_i>$ and $<o_{i+1},o_i>$ will not result in any edge irrespective of the collision status. Any circular region outside these edges will not incur collisions with obstacle edges $<o_{i-1},o_i>$ and $<o_{i+1},o_i>$. Therefore, the edges $<o_{i-1},o_i>$ and $<o_{i+1},o_i>$ may be excluded from the calculations.

To understand the construction of the roadmap, let us take an example while doing it for only one obstacle corner taken as a vertex (o_{13}) as a representative, shown in Fig. 7.6. α values are in the range $[0,2\pi)$. Edges are made directed in nature from a smaller α value to a higher α value, and all vertices are sorted as per the α value. The sweeping line starts from 0 degrees, and thus initially S is all lines that intersect the horizontal line at o_{13}. The rotation line stops at every vertex as per the α values. If it stops at the start of an edge, the edge is added in S. Correspondingly, if it stops at the head of an edge, it is deleted in S. The set S is sorted as per the distance between o_{13} and the intersecting point of the edge. The detailed calculations are shown in Table 7.1. Thereafter for collision checking, only the first edge in S is considered, which is not shown in the calculations.

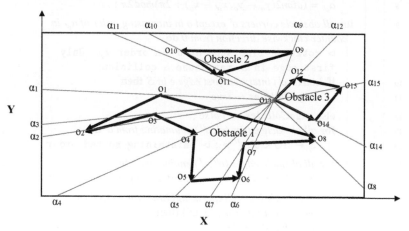

FIG. 7.6 Rotation sweep algorithm. The calculations showing maintenance of candidate set S are given in Table 7.1. Edges are directed from a lower slope using o_{13} as the origin to a higher slope in the counterclockwise direction.

TABLE 7.1 Processing of the map in Fig. 7.6 for rotation sweep algorithm for vertex o_{13} excluding the directly connected edges (o_{13},o_{14}) and (o_{13},o_{12}). Vertices are sorted as per the angle to the rotation point. The rotation line stops at all the vertices in the same order. On encountering a tail, the vertex is added to S, and on encountering the head, the edge is deleted from S. Since S is sorted by distance from the rotating line start and the intersection point on edge, only the first (nearest to the line) edge in S is a candidate for collision check.

S. no.	Angle of line	Vertex	Operation	S	Edges added in roadmap
1.	0		Initial, Add (o_{14},o_{15})	(o_{14},o_{15})	
2.	α_{15}	o_{15}	Delete (o_{14},o_{15}), Add (o_{15},o_{12})	(o_{15},o_{12})	Rotating line between circular regions of o_{14}-o_{13}-o_{12}, no edge is added
3.	α_{12}	o_{12}	Delete (o_{15},o_{12})	Empty	(o_{13},o_{12}) added
4.	α_9	o_9	Add (o_9,o_{11}), (o_9,o_{10})	(o_9,o_{11}), (o_9,o_{10})	(o_{13},o_9) added
5.	α_{10}	o_{10}	Delete (o_9,o_{10}), Add (o_{10},o_{11})	(o_9,o_{11}), (o_{10},o_{11})	(o_{13},o_{10}) collides with (o_9,o_{11})
6.	α_{11}	o_{11}	Delete (o_{10},o_{11}), (o_9,o_{11})	Empty	(o_{13},o_{11}) added
7.	α_1	o_1	Add (o_1,o_8), (o_1,o_2)	(o_1,o_8), (o_1,o_2)	(o_{13},o_1) added
8.	α_3	o_3	Add (o_3,o_2), (o_3,o_4)	(o_1,o_8), (o_1,o_2), (o_3,o_4), (o_3,o_2),	(o_{13},o_3) collides with (o_1,o_8)
9.	α_2	o_2	Delete (o_1,o_2), (o_3,o_2)	(o_1,o_8), (o_3,o_4)	(o_{13},o_2) collides with (o_1,o_8)
10.	α_4	o_4	Delete (o_3,o_4), Add (o_4,o_5)	(o_1,o_8), (o_4,o_5)	(o_{13},o_4) collides with (o_1,o_8)
11.	α_5	o_5	Delete (o_4,o_5), Add (o_5,o_6)	(o_1,o_8), (o_5,o_6)	(o_{13},o_5) collides with (o_1,o_8)
12.	α_7	o_7	Add (o_7,o_6), (o_7,o_8)	(o_1,o_8), (o_7,o_8), (o_7,o_6), (o_5,o_6),	(o_{13},o_7) collides with (o_1,o_8)

Continued

TABLE 7.1 Processing of the map in Fig. 7.6 for rotation sweep algorithm for vertex o_{13} excluding the directly connected edges (o_{13},o_{14}) and (o_{13},o_{12}). Vertices are sorted as per the angle to the rotation point. The rotation line stops at all the vertices in the same order. On encountering a tail, the vertex is added to S, and on encountering the head, the edge is deleted from S. Since S is sorted by distance from the rotating line start and the intersection point on edge, only the first (nearest to the line) edge in S is a candidate for collision check—cont'd

S. no.	Angle of line	Vertex	Operation	S	Edges added in roadmap
13.	α_6	o_6	Delete (o_5,o_6), (o_7,o_6)	(o_1,o_8), (o_7,o_8)	(o_{13},o_6) collides with (o_1,o_8)
14.	α_8	o_8	Delete (o_1,o_8), (o_7,o_8)	Empty	(o_{13},o_8) added
15.	α_{14}	o_{14}		Empty	(o_{13},o_{14}) added

7.3.3 Completeness and optimality

The algorithm completeness depends on the properties of *accessibility* or the ability to connect to a roadmap at any time; *connectivity* or the ability to navigate within the roadmap; and *departability* or the ability to leave the roadmap and join the goal. For completeness, let us assume that at least one path τ exists that connects the source to the goal. Accessibility is easy to establish since if there are no obstacles around the source, it will directly connect to the goal by a straight line, and if there is at least one obstacle in front of the source, it would connect to its corner, making a tangent to the obstacle. Departability is the opposite of connectivity and can be inferred by replacing the source with the goal. For connectivity, imagine a path τ goes from source to goal. Through discussions similar to accessibility and departability, we know that every point in the path $\tau(s)$ is visible to at least one vertex in the roadmap, which is an obstacle corner. The specific vertex or set of vertices will change as we travel from source to goal. While travelling from source to goal using τ, if visibility to a vertex u is lost, it means visibility to some neighbouring vertex of u (say v) will be obtained. If there was an obstacle O blocking the visibility of (u,v), then a way around the obstacle O would be found as at least one corner of O is visible to both u and v, and all obstacle edges are part of the roadmap. In short, the visibility map represents all *homotopic groups,* and hence a path homotopic to τ must be present in the visibility graph. This proves completeness. A discussion over homotopic groups is deferred until the very end of the chapter.

The algorithm is *optimal using path length* as a metric only for *2D maps*. This is derived directly from the intuition that all possible paths covering all corners have been encoded into the roadmap, and the shortest path in case of 2D only passes through obstacle corners. For 3D, the shortest path still comprises tangents to obstacles; however, the shortest path edges may pass through an edge of the obstacle. Imagine a room, and the task is to go from one extreme (0,0,0) to the other extreme (1,1,1). Further, imagine a large cube right at the centre that inhibits a straight-line motion from source to goal. The optimal path passes through the obstacle edge that is formed between the top face and side face of the cubical obstacle. It does not pass through any of the 8 corners of the cubical obstacle.

A problem with the visibility graph-based approach is that the robot kisses the obstacle boundaries while navigating. This can be problematic as sensing and control errors may cause the robot to collide. Further, the robot was assumed to be a point even if planning in the workspace, which may occupy space. The two problems can be resolved to a good degree by inflating the obstacles to the size of the robot and an additional minimum clearance that the robot may require. However, it is hard if the robot is a square in which case the magnitude to inflate is half the diagonal length, which can incur a large modelling loss. The geometric (polygon) obstacles in the workspace will not be geometric (polyhedron) in the 3D configuration space. The algorithm requires too much computation time for high-degree polygons. The algorithm assumes geometric obstacles, and to model a curved obstacle, one may need to use too many vertices, thus making the algorithm take too long to compute the roadmap (which is an offline process).

7.4 Voronoi

The visibility roadmap attempts to keep the robot close to the obstacle corners, thus producing the shortest paths in 2D maps. The Voronoi has the exact opposite motivation and attempts to keep the robots as far away from the obstacles as possible (Lee, 1982; Dong et al., 2010). Therefore, the roadmap models paths that have the highest clearance possible, which is another acceptable metric for motion planning. The notion behind Voronoi-based motion planning is to connect the map, with very few vertices and edges while capturing the *topology* of the map, with the general intent of keeping the roadmap vertices and edges far away from the obstacles. First, the concept is studied strictly in a 2D environment, and the resulting roadmap is called the *generalized Voronoi diagram* (GVD). Thereafter, the concept is further generalized to higher dimensions called the *generalized Voronoi graph* (GVG).

7.4.1 Deformation retracts

In a 2D map, the GVD plays the role of a *deformation retract*. Imagine a torus (doughnut-shaped) space. In this doughnut, scrape off the material from the entire surface of the doughnut and keep doing this until a very thin line is left. No material can be scraped off any further because it will break the connectivity. Hence a thin ring will be left, which is called the deformation retract of the doughnut. The notations are shown in Fig. 7.7A. Similarly, imagine a general 2D map with some obstacles. Imagine the white space in between obstacles filled with a solid material different from the obstacles. At every point in time, keep scraping the material until you get to the situation where any further scraping will affect connectivity. It will come earlier for some regions and later for some other regions. The shape leftover is called the GVD, which is a deformation retract of the original map.

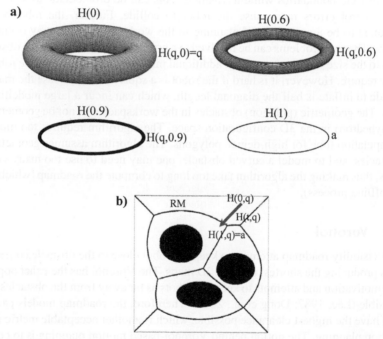

FIG. 7.7 Deformation retract. (A) A torus is retracted by scraping off the boundaries till it becomes a ring. (B) Obstacle-free space is retracted to the roadmap (*RM*) by moving every point using a deformation function *H*. *H*(0) is the obstacle-free space, *H*(1) is the roadmap.

The scraping process maps all points $q \in C^{\text{free}}$ in the free configuration space to a point $a \in RM$ in the roadmap (*RM*). Let the mapping function be given by H: $[0,1] \times C^{\text{free}} \rightarrow RM$, where $RM \subseteq C^{\text{free}}$. The first argument denotes the time step in the transformation starting for q at $t = 0$ to a at $t = 1$, and as t goes from 0 to 1, the point $H(t,q)$ travels as a continuous function without intersecting with obstacles. Here *RM* denotes the deformation retract, while *H* denotes the retraction function. For deformation retraction:

- Every point starts at its initial position in the original space at $t = 0$, or $H(0, q) = q \forall q \in C^{\text{free}}$.
- Every point ends at the roadmap at $t = 1$, or $H(1, q) \in RM \subset C^{\text{free}} \forall q \in C^{\text{free}}$.
- The points in RM should not move at any time, or $H(t, a) = a \forall a \in RM$, $0 \leq t \leq 1$.
- All deformations must be always collision-free, or $H(t, q) \in C^{\text{free}} \forall q \in C^{\text{free}}$, $0 \leq t \leq 1$.

Numerous functions can enable the deformation, and any one of them may be used. If one such function exists, it denotes a valid deformation retract. The concept is shown in Fig. 7.7.

7.4.2 Generalized Voronoi diagram

The *Voronoi diagram* is a mechanism to partition space into a set of discrete regions called the Voronoi regions. Let $\{O_1, O_2, O_3, \ldots\}$ be the set of control points in the same plane. Alternatively, assume that $\{O_1, O_2, O_3, \ldots\}$ are point obstacles that are scattered in the workspace for which the Voronoi diagram is being constructed. Each control point has its Voronoi region of space. A point in space is said to belong to the region of the closest control point. The set of points that lie in the *Voronoi region* of the control point o_i are given by Eq. (7.5).

$$R(O_i) = \{q : d(q, O_i) \leq d(q, O_j) \forall O_j \neq O_i\} \qquad (7.5)$$

The Voronoi diagrams partition space and therefore form a graph wherein the partition boundaries are taken as the edges (E) and the place where any two edges meet is taken as the vertices (V). The edges are defined in a slightly different manner unlike the previous approaches and constitute a collection of points (path) lying on the physical space, and the edges are not characterized by the vertex pair alone (using a generalization made soon after, edges may not be straight lines). The edges are defined for every pair of control points (O_i and O_j), denoted by e_{ij}, as the set of points *equidistant* to the two closest control points, given by Eq. (7.6).

$$E = \cup e_{ij} = \{q : d(q, O_i) = d(q, O_j) \wedge d(q, O_i) < d(q, O_k) \forall O_i \neq O_j \neq O_k\} \qquad (7.6)$$

The equation should be read as follows: The edge e_{ij} is the set of all points q that are equidistant to O_i and O_j ($d(q, O_i) = d(q, O_j)$), and O_i and O_j are the two closest control points to q. An empty set of points denotes the absence of the edge e_{ij}.

The vertices are the meeting points of edges and the places where the distance between the three closest obstacles is the same. Suppose a vertex v_{ijk} lies on the edges e_{ij} and e_{ik}. Since v_{ijk} lies on the edge e_{ij}, its distance between o_i and o_j is equal, and since it lies on e_{ik}, its distance between o_i and o_k is also the same. Since $d(v_{ijk}, O_i) = d(v_{ijk}, O_j)$ and $d(v_{ijk}, O_i) = d(v_{ijk}, O_k)$, it means

$d(v_{ijk},O_j) = d(v_{ijk},O_k)$, or v_{ijk} is on the edge e_{jk} as well. The set of vertices are given by Eq. (7.7) and is similar to Eq. (7.6) with the notion of closest 2 control points replaced by closest 3 control points.

$$V = \cup v_{ijk} = \left\{ q : d(q,O_i) = d(q,O_j) = d(q,O_k) \wedge d(q,O_i) < d(q,O_l) \right.$$
$$\left. \forall O_i \neq O_j \neq O_k \neq O_l \right\}$$

(7.7)

The Voronoi diagram for a set of points is given in Fig. 7.8A. Take any point on any edge and verify that it is equidistant to two closest control points. It is not possible to locate any point equidistant to two closest control points that is not on an edge. Similarly, it can be verified that every vertex is equidistant to three closest control points, and all such points are vertices.

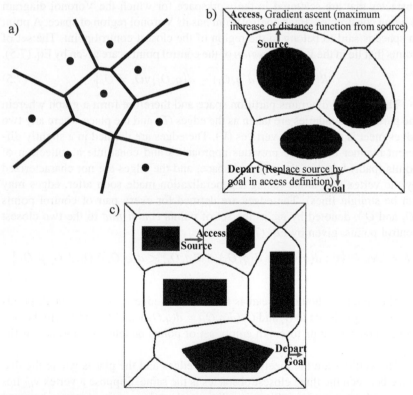

FIG. 7.8 (A) Voronoi diagram. Set of points equidistant to two control points. (B and C) Generalized Voronoi diagram. Each edge is equidistant to 2 nearest obstacles, and the vertex is equidistant to three equidistant obstacles.

For motion planning, the obstacles act as control points. To accommodate nonpoint control points (obstacles), the distance function is re-formulated. Formally, the distance between point q and obstacle O_i is defined by the least distance between any point in the obstacle with q, or $d(q, O_i) = \min_{o \in O_i} d(q, o)$, and is the same as clearance. Another alternative definition defines the distance between an obstacle and a point q, as the least distance between any points on the obstacle boundaries with q while the distance function from obstacle is defined as positive for points outside the obstacle and negative for points inside the obstacle.

The Voronoi diagrams are generalized using the new distance function for obstacles (control points) with size and called the GVD. Since the definitions remain the same, the discussions are not repeated here. The GVD for a synthetic map is shown in Fig. 7.8B and C. It may be noted that the edges may not be straight lines anymore, however may be curves. The exact shape depends on the underlying shape of the obstacle since the shape of the obstacle boundary plays a major role in determining the shape of the edges. It can also be seen that the graph very beautifully captures the topology of the space. If only the graph was given, it would be reasonably easy to visualize the actual environment. This is exactly how the route planning in vehicles works, with the satellite imagery converted into a GVD that can be used for the route planning.

There is a problem associated with the concave obstacles. Every concave obstacle can be naively decomposed into a set of convex obstacles. For a 2D case, start traversing the edges until convexity is lost. The same can be defined by measuring the internal angles, which for convex obstacles need to be strictly less than π. This gets one convex slice out of the obstacle, leaving another similar concave obstacle that can be iteratively decomposed. This is not the only mechanism, and different starting points can generate different results. The decomposition of a concave obstacle into multiple convex obstacles is nonunique. The different decompositions unfortunately result in the generation of different roadmaps for the same space, meaning for the same input there can be many outputs depending on the decomposition technique. Consider the concave obstacle in Fig. 7.8. There are two ways to decompose the obstacle as shown in Fig. 7.9A and B. The obstacle has an 'inside' region and an 'outside' region. Consider the point x in the inside region shown at the same place on the two decompositions. To maximize the distance from the first obstacle, one needs to move x in the horizontal axis and from the second obstacle on the vertical axis. The gradients from the two obstacles are thus different. However, consider the point y_1 in the outside region of the first map and the point y_2 in the second map. In the first map, to maximize the distance from both obstacles, one would need to go diagonally away. Similarly, for y_2 in the second map, for both obstacles going horizontally maximizes the distance. To have the same roadmap for any decomposition, a constraint is added that the derivatives should be different. This forces the edges that can change with the type of decomposition to be eliminated from being admitted into the roadmap. This changes the edge function to Eq. (7.8).

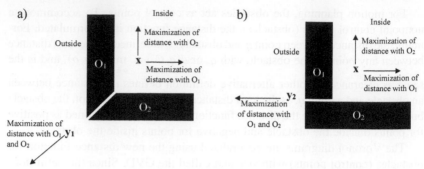

FIG. 7.9 Problems with concave obstacles. A concave obstacle can be decomposed into convex obstacles in many ways. The gradient of the distance function for inside and outside obstacles for different decompositions of the concave obstacle is shown. Only edges that result in different orientations of the gradient are added in the roadmap (inside region only). (A) Example 1, (B) Example 2.

$$E = \cup e_{ij} = \Big\{ q : d(q, O_i) = d(q, O_j) \wedge d(q, O_i) < d(q, O_k) \\ \wedge \nabla d(q, O_i) \neq \nabla d(q, O_j) \forall O_i \neq O_j \neq O_k \Big\}$$

(7.8)

The construction of GVD denotes the offline roadmap construction phase, which is used by the robot for the online motion planning of the robot when a query involving source and goal is given. Given a query, the source and goal are temporarily added to the roadmap. The connection of the source and goal to the roadmap may be done at any of the vertices or edges. The first problem is *connectivity* and requires an edge from the source to the roadmap. Let q be any point on the edge starting from the source to the roadmap. Let $d(q)$ be the distance of q from the nearest obstacle, defined as $d(q) = \min_{O_i} d(q, O_i)$. The motion of q is done in the direction that maximizes the distance by the greatest amount, and thus q moves with the principle $\dot{q} = \nabla d(q)$. This is called the *gradient ascent* and attempts to maximize the distance of q from obstacles as quickly as possible. For any bounded space, the distance cannot reach infinity, and therefore q stops at a local maxima or when it is equidistant to two obstacles. This is either an edge or a vertex as per definition, and hence q ends its journey at the roadmap that solves the connectivity problem. The problem of *departability* is the opposite of connectivity and can be solved by moving from goal to roadmap and adding the inverse of the path as an edge.

The same principles can also be used to show the completeness of the algorithm. From every point in space, one could start a gradient ascent on the distance function and be guaranteed to come to the roadmap. Since departability is the opposite of connectivity, the same can be ascertained as well. Connectivity is again like the visibility roadmap. The GVD is also capable generating all homotopic groups of paths because it is primarily a deformation retract that does not lose any connectivity. This states that the algorithm is complete.

Optimality is harder to illustrate. The paths do have the local maximum clearance. However, clearance is alone rarely the sole optimality criterion in motion planning. Humans have a habit of walking in the centre of pathways, the centre of the lanes while driving, and the centre of the entrance gates on entering/leaving, etc. Even if they avoid obstacles, they tend to take the centre of the obstacle-free space. However, humans do not leave a short path for a longer one that guarantees a higher clearance. The GVDs imitate the same concepts in the case of the robots.

7.4.3   Construction of GVD: Polygon maps

The easiest way to construct the GVD is by using a polygon map. The edges are defined as the equidistant function between two polygon obstacles. The point contributing to the equidistant edge of the roadmap may come from an obstacle vertex or an obstacle edge for either of the two obstacles. Consider the construction of an edge between the obstacles O_i and O_j. The equidistant edge may be made by the following modalities:

- Case I: The shortest distance is contributed by a vertex of O_i (say o_i) and another vertex of O_j (say o_j). In this case, the equidistant edge of the roadmap is a line midway between the corners o_i and o_j, and perpendicular to the line (o_i, o_j). The line is shown in Fig. 7.10A.

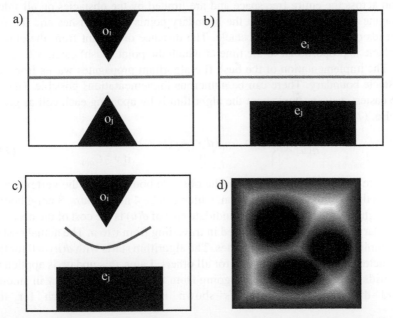

FIG. 7.10   Generation of GVD. (A) Curve maintaining equal distance between two obstacle corners is a line, (B) between two obstacle edges is a line, and (C) between a vertex and an edge is a parabola. GVD is a mixture of these cases. (D) For nonpolygon maps, compute the distance from the nearest obstacle. The GVD is the local maxima.

- Case II: The shortest distance is contributed by an edge of O_i (say e_i) and another edge of O_j (say e_j). In this case, the equidistant edge of the roadmap is a line midway between the edges e_i and e_j acting as an angle bisector of the two lines. The line is shown in Fig. 7.10B.
- Case III: The shortest distance is contributed by a vertex of O_i (say o_i) and an edge of O_j (say e_j). The geometric curve that represents the relationship between a set of points whose distance from a point is equal to the line is a parabola. Given the equations of o_i and line e_j, it is easy to compute the parabola, as shown in Fig. 7.10C.
- Case IV: The shortest distance is contributed by an edge of O_i (say e_i) and a vertex of O_j (say o_j). The case is the same as Case III with the roles of the two obstacles reversed.

Given the equations of the lines and parabolas and knowing the geometric equations to get the distance between every pair of entities, the edges and thus the vertices can be computed.

7.4.4 Construction of GVD: Grid maps

The other format that the map can be given is a nongeometric map that is specified as a grid of some resolution $m \times n$, also called the occupancy map. The *Bushfire algorithm* is commonly used to calculate the distance between the obstacle and any point in the free region. Imagine the free areas as bushes that span across the entire free space and are limited by the obstacles on all sides. Imagine that there is a fire at the boundary points of the bushes and the fire spreads outwards from the obstacle. The distance of a point from the closest obstacle is proportional to the time at which the point (bush) catches fire.

The implementation of the bush fire algorithm propagates waves from the obstacle boundary. There can be numerous implementations possible. One of the easiest ways to implement the algorithm is by updating each cell as given by Eq. (7.9)

$$d(u) \leftarrow \begin{cases} \min_{v \in \delta(u)}(d(v) + c(v, u)) & \text{if } v \notin C_{obs} \\ 0 & \text{if } v \in C_{obs} \end{cases} \tag{7.9}$$

Here $d(u)$ is the distance from the obstacle boundary for the vertex u and $\delta(u)$ is the neighbourhood function, which can be 4 neighbours, 8 neighbours, 16 neighbours, and so on. So, a candidate cost of $d(u)$ is the cost of the obstacle boundary to v and the cost incurred in travelling from v to u. The actual cost is the minimum for all such candidates. The algorithm starts with $d(u) = 0$ for the obstacle boundary and $d(u) = \infty$ for all others. Later, this update is applied to all grids and simulates the wave going from the obstacle boundary in an outward direction. The trade-offs shown in the graph search for the neighbourhood function size hold. For 4 neighbours, the wave grows in a square mechanism, due to the restricted resolution. For 8 neighbours, it also accounts for the diagonal movements. The larger neighbourhood sizes enable more flexible movements. The other mechanism for implementing the algorithm is to use the uniform cost search algorithm. However, initialize it with all obstacle boundaries rather than a single source. Just like the source-initialized uniform cost search represents a wave moving from the source; when initialized with all obstacle boundaries, the algorithm represents multiple waves moving outwards. The distance values for a synthetic map are shown in Fig. 7.10D. The Voronoi consists of the local maxima and is visible in the same figure.

7.4.5 Construction of GVD: By a sensor-based robot

One of the most interesting applications of the GVD is to capture the topology of any place where robots and humans operate (Choset and Nagatani, 2001). The GVD itself forms a map of the world, and therefore there is a curiosity about whether the robot can itself carry out mapping and autonomously make the map of the area of operation. This corresponds to the construction of GVD by a real robot. Assume that the robot is equipped with a high-definition lidar that gives distances to all obstacles at any point in time and that the robot has high-resolution sensors to accurately know its position at any point in time. To make the GVD, the robot is initially asked to move to maximize its distances from all obstacles, for which the robot selects the nearest obstacle and moves in an opposing direction (π more than the angle of the laser ray that records the minimum distance). This is the connect phase and ends with the robot at in equidistant region forming a Voronoi vertex or edge. Thereafter, the task is to move the robot keeping itself at the Voronoi edge. The same can be applied by making perturbations to the robot and asking it to connect again to the roadmap while moving in a general direction of motion. Control algorithms working over the error function to keep the distances equal will themselves cause the robot to move forward while steering the robot to keep the robot on the equidistant edge. The robot can be monitored to indicate when it reaches a point such that the distance to the nearest three distinct obstacles is equal; that will come as three distant laser rays with discontinuity in between that record the same distance. This is registered as a Voronoi vertex.

From the vertex, the robot will have multiple options, and any one option may be taken by the robot. This makes the robot traverse a new edge and registers the same in the roadmap. The outgoing edges are hard to compute given that one only has sensory information. The robot has a local map produced by the current sensor (lidar) percept. The map is assessed to determine the obstacles and the corridors in between the obstacles. Such computations are detailed later in Chapter 11 on geometric motion planning. The robot selects one of the corridors, subsequently going midway between the two obstacles. However, the vertices must be remembered since the robot will have to return to the vertices to compute other outgoing edges. This completes the building of the GVD by autonomous robots.

The construction so described unfortunately faces too many issues. The location may not be perfectly known to the robot without requiring complex sensing. False edges may be detected that deteriorate the quality of the roadmap formed. It is also possible to miss some edges. Hence modern-day simultaneous localization and mapping (SLAM) is more actively used.

7.4.6 Generalized Voronoi graph

The notion of GVD is generalizable to the higher dimensions as well, and under such a case, the concept is known by the name of a GVG and captures the intent to fit in a graph in a general higher-dimensional space. Practically, the GVG applies to a maximum of three dimensions, as the assumption is a polyhedral space, while high-dimensional configuration spaces are rarely polyhedral and cannot be traversed in totality for the building of the roadmap. The definitions used earlier cannot be directly applied here because the set of points equidistant to two obstacles forms a plane in 3D and not an edge. Imagine the two obstacles as two opposite walls of a room. An equidistant surface is a plane in between the two walls. This can also be understood by the fact that in 2D cases, the 2D points had one equidistant constraint, which reduced the dimensionality by one and hence formed an edge.

In 3D cases, we will need to put two constraints to get the edges, and therefore the edges are defined as the set of points whose distance from the three nearest obstacles is the same. Correspondingly, the vertices are defined as the set of points whose distance from the four nearest obstacles is the same. The set of edges and vertices are thus given by Eqs. (7.10), (7.11).

$$
\begin{aligned}
E = \cup e_{ijk} \\
= \Big\{ q : d(q, O_i) = d(q, O_j) = d(q, O_k) \wedge d(q, O_i) < d(q, O_l) \\
\wedge \nabla d(q, O_i) \neq \nabla d(q, O_j) \neq \nabla d(q, O_k) \forall O_i \neq O_j \neq O_k \neq O_l \Big\}
\end{aligned}
\tag{7.10}
$$

$$
\begin{aligned}
V = \cup v_{ijkl} \\
= \Big\{ q : d(q, O_i) = d(q, O_j) = d(q, O_k) = d(q, O_l) \\
\wedge d(q, O_i) < d(q, O_m) \forall O_i \neq O_j \neq O_k \neq O_l \neq O_m \Big\}
\end{aligned}
\tag{7.11}
$$

The issue with the constraints suggested is that they do not necessarily always form connected structures. This creates a problem as the individual components represent a graph, but the graphs are disconnected from each other, which poses a threat to the completeness of the algorithm. Therefore, the edges are added from the GVD defining the set of points equidistant to two obstacles, or to facilitate connections, a dimension is dropped to get a plane equidistant to two obstacles, from which an edge is taken for connections. This completes the graph.

7.5 Cell decomposition

The section presents another interesting paradigm for solving the problem of motion planning for robotics by decomposing the available map into cells, forming a graph with a limited number of vertices and edges. The decomposition may be done offline for a static map. The graph is used to process the query through an online mechanism. Unlike the construction of roadmaps using visibility graph and Voronoi, here the intention is just the grouping up of the free space into discrete cells in a manner that the decomposition can be done quickly to produce coarser-level graphs with reasonably sized regions abstracted to a cell. The cells then become the vertices of the graph, and any two cells sharing a boundary or vertex are said to be connected by an edge.

7.5.1 Single-resolution cell decomposition

There are different types of cell decompositions possible merely by the shape of the cell. The most basic cell decomposition possible is with a fixed cell size and using rectangular cells. This decomposition is already covered when dealing with graph search and converting a high-resolution map into a lower-resolution map. The reduction in resolution is the same as placing rectangular grids and taking any grid even partly occupied by an obstacle as an obstacle grid. Such a decomposition is shown in Fig. 7.11A. However, the decomposition results in a significant free space being modelled as an obstacle, as shown in Fig. 7.11B. To control this effect, another mechanism more intelligently traverses through only the free space to fit in as many rectangular cells as possible. Not all free space can be covered in this mechanism, and there is a loss incurred, particularly in the narrow corridor regions. The decomposition is shown in Fig. 7.11C.

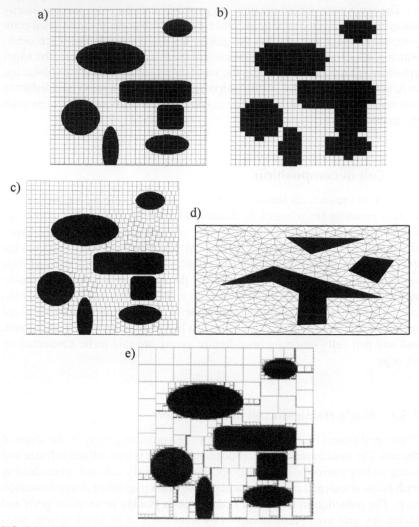

FIG. 7.11 Cell decomposition (A) using rectangular grids of a single size, (B) loss incurred by taking all partly obstacle-occupied cells as obstacles, (C) cell decomposition by only modelling the free space, resulting in unmodelled narrow corridors, (D) cell decomposition using triangular cells (similarly for hexagon, not shown), which are constraint to connect to neighbours by paths feasible by kinematics, and (E) multiresolution cell decomposition with narrower cells in narrow corridors and larger otherwise (filled iteratively).

The shape of the cell may even be *triangular*, in which case space is filled with triangles instead. When discussing the rectangular cells, we did not account for the fact that the robot may be asked to move sideways because sideways cells are neighbours; however, the motion may not be possible by kinematics. A solution is to plan in the state space instead wherein the control signals are edges while the state incorporates the velocities also. However, this complicates the state space. In triangular cells, it is possible to control the size of the cell such that the motions are possible by kinematics for a given speed. Such triangulation is also called *constrained triangulation*. The triangulation for a generic map is shown in Fig. 7.11D.

The *hexagons* as cells are popular in the wireless community because they maximize the area while being close to circular. The same is also true in the case of the robotic community as well, wherein the hexagonal cells find it useful to have high coverage of any general area while making it possible to model for feasible connections by kinematics to the neighbouring cells. This again happens by controlling the neighbouring cells to which a cell is connected and only accounting for feasible turns.

7.5.2 Multiresolution decomposition

The single-cell size has problems in the narrow corridor, where the cells must be fine to get a path inside the narrow corridor, which, however, also increases the number of cells. The other problem is that the size of the cell becomes a difficult parameter to fix as smaller values result in too many cells while larger values result in difficulty travelling through regions requiring dextrous manoeuvers. In general, one requires small cells around obstacles and narrow corridors to be able to avoid them; however, in open spaces, the size of the cells may be much larger instead to save on the number of cells. This is facilitated by the *multiresolution techniques* that do not require a specific size of the cell to be placed on the map.

The easiest way to carry out multiresolution cell placement is to start placement with some rectangular cells as previously discussed. This will cover most of the areas with cells, however, leaving behind modestly large spaces between the cells and obstacles, which can be covered by taking cells of a smaller size. The still-leftover space is filled up by cells of a much smaller size. Caution must be given though an attempt to cover the entire free space will result in too many such cells. On the contrary, the benefits of having extremely small cells are slim since they marginally add to optimality and completeness. Too large cells are also problematic since they lead to the loss of optimality, making the robot go far away from the obstacle boundaries, enabling only limited turning angles, and visiting the centre of any large cell means travelling a large distance from the cell boundary to the centre, which increases the distance from obstacles and may have been otherwise avoidable. Such a greedy assignment for a synthetic map is shown in Fig. 7.11D.

The edges are defined as any two neighbouring cells that share a cell corner or a part of an edge. A typical imagination is to take the vertices as the centre of the rectangle rather than the complete cells. The edges then are imagined as the straight lines joining the centres of the rectangles. This, however, may not always be the case with multiresolution settings as straight lines may pass through neighbouring obstacle cells. In cell decomposition, one needs to travel from anywhere in one cell to the common cell edge (or common part of the cell edge) and then from the common cell edge to anywhere in the other cell. This makes the roadmap edge. The path computed by a planner is a sequence of cells to be visited by the robot. The robot notes the next cell to reach and travels to the part of the cell edge that is common between the current and the next cell. On reaching the cell boundary, it selects the next cell in the path and travels to the common cell edge with the next cell, and so on.

7.5.3 Quadtree approach

The oct-tree divides the 3D space into eight regions, and similarly, the quadtree recursively divides the 2D space into four regions using a X-axis bisector and a Y-axis bisector, until regions are completely obstacle-free (Kambhampati and Davis, 1986). Let the four children be called the north-west cell (NW), the north-east (NE), the south-west (SW), and south-east (SE). Let a cell be white, denoting the absence of any obstacle; black denoting the presence of an obstacle; or grey denoting some regions of the cell are obstacles, and some others are not. The quadtree decomposition for a synthetic map at different levels is shown in Fig. 7.12. The loss incurred due to discretization at different levels is shown in Fig. 7.13. The greyness indicates the density of obstacles in the region. The map at small depths only is shown in Fig. 7.14.

The beauty of the quadtree is that if the tree is cut at any nonconstant depth for different branches, it represents the complete map at some nonuniform resolution. So, it is possible to represent every multiresolution map in the quadtree by the manner of cutting the quadtree at different depths at different branches. The complete tree is not made a priori, which will be a very cumbersome approach, but is made while the algorithm traverses. Let us assume that the search traverses the tree at any generally fixed depth like in the case with a fixed cell size. However, during the search, it needs a finer idea to see if the robot can traverse through a narrow region or not. In such a case, a query about the connection may be made to the lower depth for reporting the connectivity, with the lower depth having a better idea because of operating at a better resolution. The lower depth could further query its children at a much lower level, and so on. It is up to the algorithm to decide the level of operation at any stage, and whether to go deeper or continue the search at the same level. Typically, far away from obstacles where connectivity is ascertained, it is acceptable to not go to deeper levels. Near obstacles and narrow corridors, fine manoeuvers may be needed, and the algorithm may query lower levels for connection checks.

There are other alternative ways in which a search can be strategized while working around a quadtree approach. Searches may be done iteratively from lower depths to higher depths while increasing the resolution only at the critical regions in the lower depth paths (Kala et al., 2011). Alternatively, some

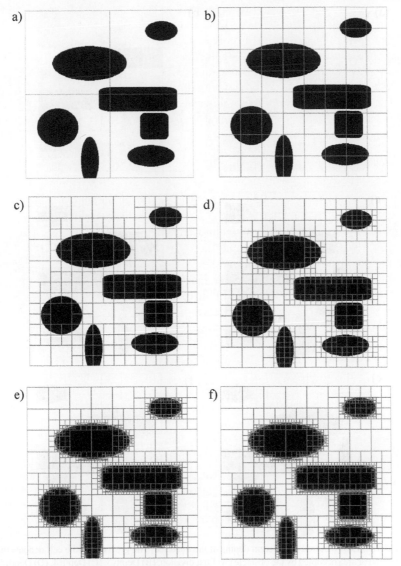

FIG. 7.12 Quadtree representation. At every level, the cell is divided into four parts, except completely free cells and completely obstacle occupied cells. (A) First division, (B) Depth 3, (C) Depth 4, (D) Depth 5, (E) Depth 7, and (F) Depth 9.

acceptable threshold on grey level may be set at both minimum and maximum levels, say 10%–90% (Kala et al., 2010). Anything below the greyness level is assumed to be obstacle-free, and anything above the greyness level is assumed to be obstacle prone. This produces a binary map for search. The quadtree represents every possible multiresolution setting that the algorithm may exploit.

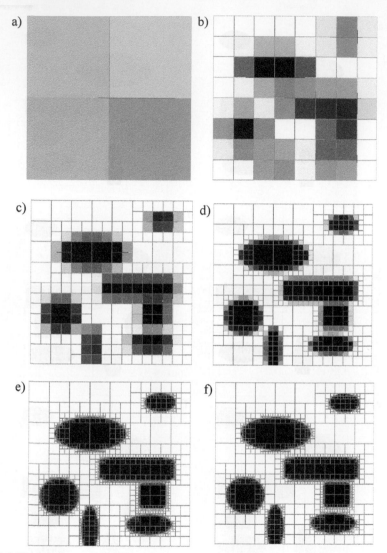

FIG. 7.13 Loss incurred at different depths of the quadtree. The greyness of the cell denotes the proportion of area occupied by the obstacle (A) First division, (B) Depth 3, (C) Depth 4, (D) Depth 5, (E) Depth 7, and (F) Depth 9.

A mechanism to address the problem of optimality and the robot going into the centre of a large cell thus far away from the obstacle is called the *framed quadtree* (Yahja et al., 1998). In a framed quadtree approach, the cells are divided exactly like the quadtree; however, the borders of the cells are added with additional cells of the smallest size, thus forming a frame around the cell. The robots may not go to the centre of the main cell but may as well navigate by

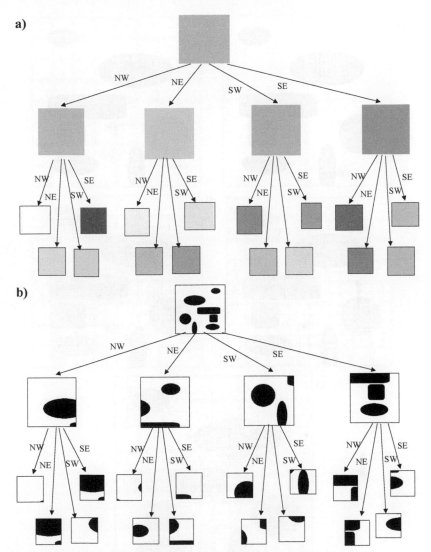

FIG. 7.14 Quadtree decomposition. The node is divided into four regions: north-west (NW), north-east (NE), south-west (SW), and south-east (SE). (A) The greyness denotes the proportion of obstacle occupancy. (B) The part map instead of greyness for a better understanding.

taking the bordering cells that add to optimality. Similarly, it is not necessary to go from one cell centre to the other cell centre. One may as well go to a frame cell, thus taking a shortcut around the main cells. The framed quadtrees thus result in shorter paths, but do not guarantee the shortest paths and do not completely overcome the loss due to discretization. The decomposition is shown in Fig. 7.15.

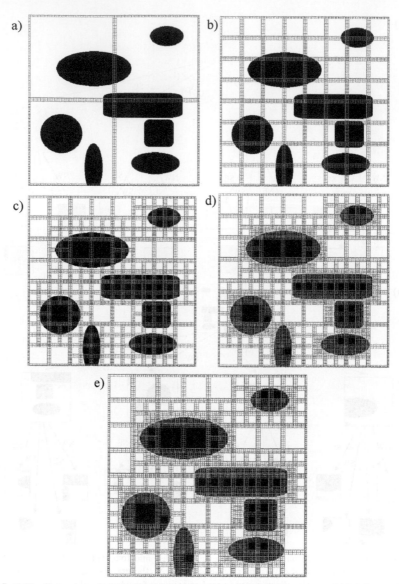

FIG. 7.15 Framed quadtree decomposition. The decomposition is the same as the quadtree, with the difference that a frame is added at the boundary of cells to enable shorter paths not going through centre of cells (A) First division, (B) Depth 3, (C) Depth 4, (D) Depth 5, and (E) Depth 6.

7.6   Trapezoids

The section fills up space with *trapezoidal*-shaped cells (Seidel, 1991). Foremost, it is another shape that can be used to decompose the map. Furthermore, the choice of the specific shape is to enable the construction of the roadmap in a single read of the map by merely sweeping a line and storing the results in a summarized structure of a roadmap.

7.6.1   Construction

First, we place an assumption that the map is polygon in nature, and therefore all obstacles are defined by a collection of vertices numbered in a counterclockwise direction. The decomposition is carried out by traversing the map from one extreme to the other by using a sweeping line. For convenience, let us assume that the sweeping line is vertical. At any point in time, the sweeping line is stopped when it touches some obstacle corner o_i. The sweeping line cuts the obstacles at some points and passes through free regions at the other places. Consider two rays emanating from o_i (ith stop of the sweeping line at obstacle corner) and travelling upwards and downwards, respectively; let us call them the *upper* and *lower* rays. The rays will intersect obstacles (or map boundaries) at points $v_{i\uparrow}$ (intersecting with obstacle $o_{i\uparrow}$) and $v_{i\downarrow}$ (intersecting with obstacle $o_{i\downarrow}$). The *lower* and *upper* rays may not be simultaneously possible as the corner may be such that travelling above or below is not possible. In such a case, the corner (o_i) coincides with one of the vertices ($v_{i\uparrow}$ or $v_{i\downarrow}$). The sweeping line stops at all the obstacle corners. The space from $v_{i\uparrow}$ to $v_{i\downarrow}$ is collision-free and forms the base of the trapezoid. A similar base will be formed by the line from $v_{j\uparrow}$ to $v_{j\downarrow}$ at some previous stop of the line, wherein the stop refers to the space between the same obstacles $o_{i\uparrow}$ and $o_{i\downarrow}$. The obstacle boundaries $o_{i\downarrow}$ and $o_{i\uparrow}$ form the two sides of the trapezoids. In some cases, triangles may instead be formed, which is a special case of trapezoids.

Consider a synthetic map as shown in Fig. 7.16A. The sweeping line needs to stop at every obstacle corner as it traverses from one extremity to the other. While the line may traverse in any direction, for convenience, let us take the sweeping line as a vertical that traverses across the X-axis. Let $O = \{o\} = \{(o_x, o_y)\}$ be the set of obstacle corners for all obstacles. The sweeping line starts from the line $X = \min_o(o_x)$ and goes on to the end of the sweep at $X = \max_o(o_x)$. The line stops at all obstacle corners one after the other in increasing order of X coordinate value. There is a problem when two vertices have the same X-coordinate value, which can be easily resolved by adding a minutely small noise to either of the vertices. Hence a rectangular boundary of the map is distorted to make the four corners have a different X-coordinate value. When the sweeping line stops at a corner, the upper and lower rays are drawn to calculate the intersecting points that give one base of the trapezoid. The other base is obtained as the line traverses, while the obstacle boundaries form the sides.

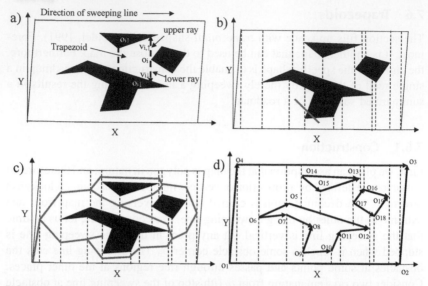

FIG. 7.16 Trapezoidal cell decomposition. (A) One trapezoid with the base at obstacle corner o_i (vertical sweeping lines form the base and obstacle edges form the sides). (B) Trapezoidal decomposition. Taking straight lines joining the cell centre does not result in collision-free edges. (C) The graph is formed by taking the centre of the baseline as vertices. (D) Construction of the trapezoidal decomposition using a sweeping line. The lines are directed from a lesser X-coordinate value to a higher X-coordinate value.

The trapezoids form the vertices of the graph, while any two trapezoids that share an edge are said to be connected by an edge. It is easier to imagine the graphs with points as vertices and paths as edges. In such a case, the most common intuition is to take the centre of the trapezoids as the vertex and the straight-line distance between the centres as edges. This is wrong as the edges of collision-free trapezoidal vertices may still be collision-prone (Fig. 7.16B). Therefore, the centre of the base of the trapezoids is taken as the characteristic point to denote the vertices. Two neighbouring trapezoids being connected means an edge from the centre of the left base of the first trapezoid to the centre of the left base of the second trapezoid. As trapezoids are completely obstacle-free, going from one vertex to a neighbouring vertex means going from one base to the other base of the trapezoid, which is the same as traversing a trapezoid that guarantees an edge to be collision-free. This is shown in Fig. 7.16C.

The complexity of the algorithm is given by the need to sort all the vertices in time $O(n \log n)$, where n is the number of corners. Thereafter, the collision checking of both upper and lower rays is done, which translates to checking the collision of the vertical line at every corner with all obstacle edges, to find the closest colliding edge above and below the corner, which can be done in $O(n)$ time for one corner, or $O(n^2)$ time for all corners, resulting in a complexity of $O(n^2)$.

The search can be made efficient by using the *sweeping line* formulation as done in a visibility roadmap. In this case, the sweeping line translates. The notion is to keep an active set S that stores edges that collide with the sweeping line. Let every line be made directed in nature from a lower X-coordinate value to a higher X-coordinate value. Vertical edges may be assigned direction by adding a minute noise to either of the vertices. As the sweeping line stops at any obstacle corner, all edges emanating from the corner are added to an active set S. Similarly, all edges that end at the corner are deleted from S. In such a case, only set S is considered for collisions. The working of the algorithm for a synthetic map is shown in Fig. 7.16D and Table 7.2.

TABLE 7.2 Processing of the map in Fig. 7.16D for translating the sweep line algorithm. The line makes a stop at all the vertices in the order of X-coordinate values. On encountering a tail, the vertex is added in S, and on encountering the head, the edge is deleted from S. Collision checking only happens with edges in S to make the trapezoid.

S. no.	Vertex	Operation	S
1.		Initial	Empty
2.	1	Add (o_1,o_4), Add (o_1,o_2)	(o_1,o_4), (o_1,o_2)
3.	4	Delete (o_1,o_4), Add (o_4,o_3)	(o_1,o_2), (o_4,o_3)
4.	6	Add (o_6,o_5), Add (o_6,o_7)	(o_1,o_2), (o_4,o_3), (o_6,o_5), (o_6,o_7)
5.	7	Delete (o_6,o_7), Add (o_7,o_8)	(o_1,o_2), (o_4,o_3), (o_6,o_5), (o_7,o_8)
6.	5	Delete (o_6,o_5), Add (o_5,o_{12})	(o_1,o_2), (o_4,o_3), (o_7,o_8), (o_5,o_{12})
7.	14	Add (o_{14},o_{13}), Add (o_{14},o_{15})	(o_1,o_2), (o_4,o_3), (o_7,o_8), (o_5,o_{12}), (o_{14},o_{13}), (o_{14},o_{15})
8.	9	Add (o_9,o_8), Add (o_9,o_{10})	(o_1,o_2), (o_4,o_3), (o_7,o_8), (o_5,o_{12}), (o_{14},o_{13}), (o_{14},o_{15}), (o_9,o_8), (o_9,o_{10})
9.	8	Delete (o_7,o_8), Delete (o_9,o_8)	(o_1,o_2), (o_4,o_3), (o_5,o_{12}), (o_{14},o_{13}), (o_{14},o_{15}), (o_9,o_{10})
10.	15	Delete (o_{14},o_{15}), Add (o_{15},o_{13})	(o_1,o_2), (o_4,o_3), (o_5,o_{12}), (o_{14},o_{13}), (o_9,o_{10}), (o_{15},o_{13})
11.	10	Delete (o_9,o_{10}), Add (o_{10},o_{11})	(o_1,o_2), (o_4,o_3), (o_5,o_{12}), (o_{14},o_{13}), (o_{15},o_{13}), (o_{10},o_{11})
12.	11	Delete (o_{10},o_{11}), Add (o_{11},o_{12})	(o_1,o_2), (o_4,o_3), (o_5,o_{12}), (o_{14},o_{13}), (o_{15},o_{13}), (o_{11},o_{12})

Continued

TABLE 7.2 Processing of the map in Fig. 7.16D for translating the sweep line algorithm. The line makes a stop at all the vertices in the order of X-coordinate values. On encountering a tail, the vertex is added in S, and on encountering the head, the edge is deleted from S. Collision checking only happens with edges in S to make the trapezoid—cont'd

S. no.	Vertex	Operation	S
13.	17	Add (o_{17},o_{16}), Add (o_{17},o_{18})	(o_1,o_2), (o_4,o_3), (o_5,o_{12}), (o_{14},o_{13}), (o_{15},o_{13}), (o_{11},o_{12}), (o_{17},o_{16}), (o_{17},o_{18})
14.	13	Delete (o_{14},o_{13}), Delete (o_{15},o_{13})	(o_1,o_2), (o_4,o_3), (o_5,o_{12}), (o_{11},o_{12}), (o_{17},o_{16}), (o_{17},o_{18})
15.	16	Delete (o_{17},o_{16}), Add (o_{16},o_{19})	(o_1,o_2), (o_4,o_3), (o_5,o_{12}), (o_{11},o_{12}), (o_{17},o_{18}), (o_{16},o_{19})
16.	12	Delete (o_5,o_{12}), Delete (o_{11},o_{12})	(o_1,o_2), (o_4,o_3), (o_{17},o_{18}), (o_{16},o_{19})
17.	18	Delete (o_{17},o_{18}), Add (o_{18},o_{19})	(o_1,o_2), (o_4,o_3), (o_{16},o_{19}), (o_{18},o_{19})
18.	19	Delete (o_{16},o_{19}), Delete (o_{18},o_{19})	(o_1,o_2), (o_4,o_3)
19.	2	Delete (o_1,o_2), Add (o_2,o_3)	(o_4,o_3), (o_2,o_3)
20.	3	Delete (o_4,o_3), Delete (o_2,o_3)	Empty

7.6.2   Complete coverage

The complete coverage problem aims at a robot completely covering an area. This is needed for applications such as household cleaning, inspection of land-mines, cutting grass on lawns, etc. In all such applications, all cells need to be completely traversed. Therefore, the problem can be decomposed into two aspects: The first is to decide the order to visit all cells, and the second is to completely cover a cell. In all algorithms, the transition from one cell to the other has overhead costs so it is important to have only a few cells.

In such a case, many cells of the trapezoidal decomposition may be merged to produce a smaller number of cells, while still maintaining connectivity between the cells. The rule adopted is that the sweeping line stops only at the obstacle corners such that both lower and upper rays are possible to construct; the rays are not immediately blocked because of the obstacle, resulting in either an upper ($v_{i,\uparrow}$) or lower ($v_{i,\downarrow}$) ray to coincide with the obstacle corner (o_i).

The decomposition thus produced is called the *boustrophedon decomposition* (Choset, 2000). The trapezoidal decomposition has more stops in the sweeping lines and thus produces more cells. The boustrophedon decomposition merges two cells where the sweeping line is at an obstacle corner where either the upper or lower ray is not possible. If both rays are possible, topologically different paths merge or diverge. If only one of the upper or lower rays is possible, the deformation retract changes the direction of the line denoting the topology without adding topologically different paths. The decomposition for a synthetic map is shown in Fig. 7.17A. The graph made with the vertices denoting the regions is shown in Fig. 7.17B.

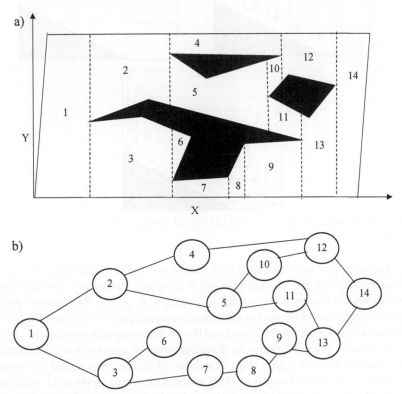

FIG. 7.17 Boustrophedon decomposition (A) construction, The construction is the same as the trapezoidal decomposition with only those lines admitted such that extensions above and below the corner are both simultaneously possible. (B) Conversion into a graph.

For solving the complete area coverage problem, let us first make an algorithm that traverses a cell in its entirety. Three common strategies can be used. The first is the *vertical coverage* in which a robot traverses like the vertical sweeping line throughout the cell, going up and down as the line progresses. The other strategy is making the sweeping line horizontal and asking the robot to go left and right instead, called *horizontal coverage*. The last coverage possible is *circular coverage*, where the robot covers the cell circularly, starting at the outer boundary of the cell and slowly converging into the centre of the cell. The different coverage mechanisms are given in Fig. 7.18.

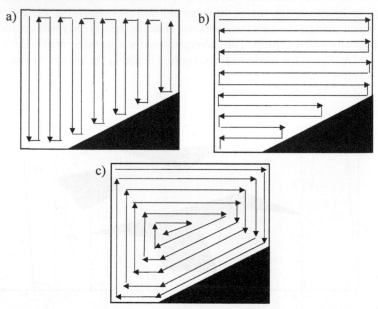

FIG. 7.18 Complete traversal of a cell (A) vertical coverage, (B) horizontal coverage, and (C) circular coverage.

The other problem associated with complete coverage is deciding the order of visits to every cell. A common graph traversal technique is breadth-first search, which traverses the complete graph, also giving the shortest path. The technique cannot be used in a complete coverage problem because the order of visits to vertices is level by level and the problem requires the robot to physically traverse from one vertex to the other vertex. Consider that the breadth-first search visits cell j after cell i at a depth of 4. The sequence of cell visits contains consecutive elements i and j. The robot after visiting the cell i will have to plan the shortest path route from i to a possibly distant j and continue its navigation. This traversal will happen between most consecutive cells in the output, as they are unlikely to be consecutive geographically. The algorithm frequently backtracks, which must be avoided for the choice of the application.

Therefore, the *depth-first search* algorithm is used, which backtracks only when necessary. The algorithm goes to the maximum depth possible and backtracks only if more cells cannot be reached from the vertex. When backtracking, a physical motion of the robot through a motion planning algorithm will be required. The algorithm reduces the number of backtracks and is hence desirable. The complete coverage for a synthetic map is shown in Fig. 7.19.

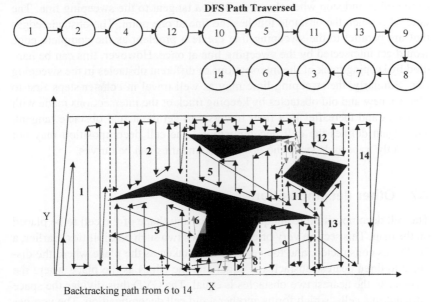

FIG. 7.19 Complete area coverage. First depth first search is called, and the traversal order is taken as the global path. The robot travels cells in the same order. The path traversed due to switching between cells is given by a dotted line.

Alternatively, building an exhaustive search solution is also possible. The problem is to compute a path such that all vertices are visited at least once while the total path cost is minimized. When visiting a cell for the first time, the robot travels the complete cell area and hence the cost. Subsequently (like when backtracking), the distance between the cells is used as the cost of the edge. The complexity is exponential to the number of cells. However, the problem can also be solved by optimization algorithms like the genetic algorithm. In such a case, it represents a combinatorial optimization problem while making queries to the classic motion planning problems.

7.6.3 Occupancy grid maps

The approach so far assumed that the map consists of polygon obstacles. Assume that the map is specified as a grid of some size, with every cell denoting the presence or absence of an obstacle. In general, it is possible to convert the grid map into a polygon map using a variety of methods. The simplest mechanism is to regress polygons around every obstacle. Since the map is assumed to be two dimensional, it is easy to segment all obstacles and get the obstacle boundaries. The obstacle boundaries can be regressed to polygons of sufficient degree to get polygon obstacles.

However, in this case, an explicit conversion into a polygon map is not required. The sweeping line may as well be made to traverse from one extreme to the other and stop when the obstacle is a tangent to the sweeping line. The tangent is when the obstacle touches the line at one point. The obstacle edges that are parallel to the sweeping line create a problem as the complete edge would get intersected by the sweeping line at once. However, this can be handled by keeping track of the intersections of different obstacles in the sweeping line. Similarly, the sweeping lime may as well travel in coarser steps first to identify new and old obstacles by keeping track of the intersections made with obstacles and thereafter in a finer traversal to get the actual obstacle tangent. This again defines the cell decomposition. The cell decomposition may not be strictly trapezoids since the obstacle boundaries can be curves.

7.7 Other decompositions

The cell decomposition technique allows for any shape of the cell to be placed on the map. This section very briefly describes two such possibilities. Earlier, a GVD was constructed, and the vertices were taken as the place where the distance to three nearest obstacles is equal and the edges as the place where the distance to the nearest two obstacles is equal. The GVD thus divides the space into multiple cells, which forms another valid cell decomposition. The vertices are taken as complete regions of space bounded by obstacles or the GVD edges. The edges are taken whenever two cells share a part of an edge. The formulation can be easily used for a complete area coverage problem as well. It may be observed that this is the only example wherein there is an obstacle in the centre of the cell, unlike all other examples where the cells were obstacle-free. However, every cell will have a maximum of one obstacle inside it, and this can be used as a direct heuristic for complete area coverage using the spiral methodology or for travelling between cells wherein one may easily encircle the obstacle as only one exists. The graph hence formed is shown in Fig. 7.20A.

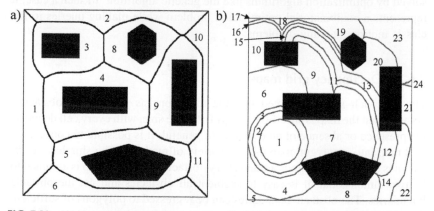

FIG. 7.20 (A) Voronoi-based cell decomposition. Each Voronoi region is a cell. (B) Cell decomposition by wavefront algorithm. Each contour has the same distance from the source (1) accounting for obstacles (and hence the irregular shape). Every time the wave strikes or leaves the obstacle, a critical point is added, which divides the space into cells.

The other mechanism of decomposition is by using a circular sweeping curve instead of a vertical line. The use of the term curve instead of a line is intentional since the circular arc is the notional sweeping line. The circular sweeping curve is obtained by an algorithm called the *wave-front algorithm*. The algorithm is like the uniform cost search in the sense that it represents a wave going from a source in an outward direction. At any point in time, the vertices in the fringe denote the wave points. Since the fringe represents the circular sweeping curve, again tangents to obstacles are taken as the points of stoppage for the sweeping curve. Thereafter, the sweeping curve edges along with the obstacle edges around the stopping points form the cells. This is shown in Fig. 7.20B.

7.8 Navigation mesh

A related representation of the environment is the navigation mesh. The navigation of humans in most situations happens in well-defined corridors and similar structures, like a corridor that leads you from one room to the other; a corridor is made by a flight of stairs, a walking corridor around the park, etc. The corridors may be bound by walls, nonnavigable terrains, etc. The human can move anywhere within the corridor. The corridors represent the navigable areas on the static map and are therefore enough to represent the environment.

The *navigation mesh* is a technique to represent free space that knits a mesh around the free space. Each cell of the mesh is a convex polygon. The convex constraint enforces that the robot can move collision-free between any two points inside the same cell of the mesh. Furthermore, a path from source to goal specifies the mesh cells and represents the area wherein the robot can go collision-free, instead of just a path. Every two cells sharing an edge are said to be connected, making a search graph. The navigation mesh for a synthetic scenario is shown in Fig. 7.21. The advantage of having a navigation mesh

FIG. 7.21 Navigation Mesh. Placing a mesh of convex polygons within free space defines a corridor within which the robot can navigate, useful especially to locally avoid dynamic obstacles where the geometry of the corridor is known rather than just a path.

instead of a roadmap is that the navigation mesh confines the robot to a complete area where it can freely move or be planned to move, unlike a roadmap that represents the edge where the agent should move. This is especially useful when multiple robots are occupying the same space, in which case all of them would use the roadmap edges and get head-on. The navigation mesh represents the area within which the robots can move to avoid each other.

The navigation meshes originated in game play, where the meshes are handcrafted as a part of the game scene specification. Even for the buildings within which the robots move, the navigation meshes are workable from the architecture design. There are algorithmic ways to automatically construct a navigation mesh. The primary mechanism assumes that the workspace specification uses geometrical information about obstacles and map boundaries. The representation needs to be inverted to get the geometries of free movable spaces, which is done by using the primitives of triangulation and other cell decomposition approaches representing free spaces. The constraint now is to have a few cells only to make distinct corridors. Alternatively, the cells from a finer triangulation could be merged with the neighbours to make polygons around the obstacles.

The path finding for a given mesh also has interesting problems. So far, we have taken characteristic points within cells to get the distance function using A* algorithm for graph search. This forms one way of converting the mesh into a roadmap, taking the mesh cell polygon corners for the shortest path or the centre of cell polygon edges that join to the neighbouring cells. In both cases, the output is a path that confines the robot and is suitable for single-robot applications. For multirobot applications, the other robots are dynamic obstacles and are reactively avoided. To enable the robot to benefit from the mesh knowledge, the robot strives to move next to anywhere in the common polygon edge between the current cell and the next cell given by the path-finding algorithm (called the *subgoal*), rather than striving to reach the corner or centre of the next cell in the case of a fixed path. There are hence infinite subgoals represented in the common polygon edge between the current cell and the next cell, making the robot free to move around than focusing on a specific subgoal point while avoiding the other dynamic obstacles. The fusion between such a deliberative path and local obstacle avoidance is dealt with in greater detail in later chapters.

7.9   Homotopy and homology

Homotopy and homology are not algorithms, but terms used in the assessment of spaces and algorithms from a topological perspective. Two paths τ_1 and τ_2 are said to be in the same *homotopic group* if τ_2 can be obtained from τ_1 by multiple infinitely small deformations using the deformation function H, such that all the intermediate trajectories are always collision-free (Aubry, 1995; Kala, 2016a, b). If any deformation function can deform τ_1 to τ_2, both trajectories are said to be in the same homotopic group. Hence the homotopic groups divide the set of all trajectories into groups. The deformation function $H:[0,1] \times [0,1] \rightarrow C^{free}$ starts from the first trajectory $H(0) = \tau_1$ and ends at the second trajectory $H(1) = \tau_2$, with intermediate trajectories $H(s)$, $0 \leq s \leq 1$. All trajectories start at the source $H(s,0) = S$, $0 \leq s \leq 1$, and end at the goal $H(s,1) = G$, $0 \leq s \leq 1$. All the trajectories must always be collision-free $H(s,t) \in C^{free}$, $0 \leq s \leq 1$ and $0 \leq t \leq 1$. The first parameter of the deformation parameterizes the iteration of deformation from the first trajectory to the other, and the second parameter is the parameterization of a trajectory from source to goal. The set of trajectories in the same and different homotopic groups are given in Fig. 7.22A.

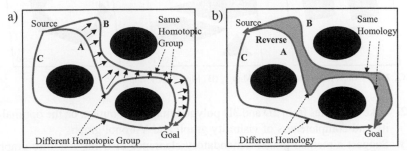

FIG. 7.22 (A) Homotopic groups. *A* can be deformed without collision to *B* (keeping source and goal fixed) and therefore is in the same homotopic group. (B) Homologies. *A* and *B* are in the same homology as the area occupied by the closed surface is collision-free.

The number of homotopic groups between a source and goal is ideally infinite since any obstacle can be encircled by the trajectory any number of times, and every encircling of the obstacle is a trajectory in a different homotopic group. However, the trajectory rarely needs to have circles, and therefore the number of practically useful homotopic groups is finite and exponential to the number of obstacles (every obstacle can be avoided in a clockwise or anti-clockwise manner). Global motion planning in robotics is primarily finding out the homotopic group that the robot must take. Once a homotopic group is known, it is easy to locally optimize the trajectory. Local and reactive planning is about avoiding dynamic obstacles and other uncertainties while moving with a predetermined homotopic group.

Homology is a similar concept and divides the trajectories into multiple groups. Two trajectories τ_1 and τ_2 are said to be *homologous* to each other if the area occupied by the closed curve formed by the first trajectory and the inverse of the second trajectory is collision-free. To determine the homology, one of the trajectories needs to be reversed to form a closed surface from source to source; the first going from source to goal and the reversed second going from goal to source. The concept is shown in Fig. 7.22B. It may appear that both homotopy and homology point to the same concept; however, this is not true, and homology does not imply homotopy.

Questions

1. Using the following approaches, solve the robot motion planning problems in the scenarios shown in Fig. 7.23. (a) Visibility graph, (b) Voronoi graph, (c) quadtree, (d) trapezoidal cell decomposition, (e) and A* algorithm on unit grids. Which of the algorithms are optimal for 2D maps? For the algorithms that are optimal for 2D maps, are their outputs different? If the optimal outputs are different, can the algorithms be called optimal?

a)  b) 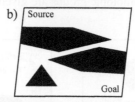 c)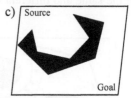

FIG. 7.23   Maps for question 1. (A) Map 1, (B) Map 2, (C) Map 3.

2. Take examples from 2D and 3D polygon maps to comment on the optimality and completeness of visibility graph and Voronoi graph.

3. Suggest a method to add a mandatory clearance of C for visibility graph and Voronoi graph for 2D polygon maps.

4. Both tangent bug and visibility roadmaps take tangents to obstacles in 2D maps. Give one example where the outputs of the two algorithms differ. Explain the pros and cons of the two approaches if they differ in functionality.

5. Consider a map with a narrow corridor. State which of the following algorithms can effectively solve the problem: (a) A* algorithm on unit grids, (b) A* algorithm using a quadtree representation, (c) visibility graph (d) Voronoi graph, and (e) trapezoidal decomposition. Compare and contrast the working of all mentioned algorithms, clearly denoting in which context which algorithm shall perform best and with what assumptions.

6. Assume that the map changes with time. Suggest a re-planning algorithm for visibility roadmap.

7. A workspace is perfectly rectangular with all obstacles being rectangles. A 2-link planar manipulator operates on this workspace. Can the visibility roadmap be used in such an operational scenario?

8. Explain the difference between geometric, nongeometric, and topological maps. In each case, explain how can such a map be practically obtained by using the robot. Categorize between cases where robots use a stereo camera and 2D and 3D lidar sensors.

9. Explain the mechanism for solving the complete area coverage problem using a trapezoidal cell decomposition with multiple robots.

10. [Programming] Make a program that takes a 2D geometrical map as input and makes (a) visibility graph, (b) Voronoi graph, (c) quadtree,

(d) trapezoidal cell decomposition, and (e) boustrophedon cell decomposition. Extend (a–d) for solving a motion planning problem, and extend (e) to solve a complete area coverage problem.

11. [Programming] Extend question 10 to work on nongeometric 2D grid maps represented as a bitmap image. The SLAM algorithms of mobile robots typically use a 2D lidar sensor and store the map as a bitmap image that represents a typical input for motion planning algorithms. In some cases, you may extract the geometry of the obstacle by making suitable algorithms.

12. [Programming] Given a bitmap image representing a workspace, show all points that are visible to a specific input point. Repeat for the case of a geometric map with polygon obstacles.

13. [Programming] Taking a grid map as input, solve the motion planning problem using each grid as a cell and A* algorithm, with the constraint, that the path constructed should always have at least k visible landmarks (a list of all point landmarks is prespecified); otherwise the robot shall not be able to localize itself. Extend the solution to the case that at least k landmarks should be visible within every D distance.

References

Aubry, M., 1995. Homotopy Theory and Models. Birkhäuser, Boston, MA.

Choset, H., 2000. Coverage of known spaces: the boustrophedon cellular decomposition. Auton. Robot. 9 (3), 247–253.

Choset, H., Nagatani, K., 2001. Topological simultaneous localization and mapping (SLAM): toward exact localization without explicit localization. IEEE Trans. Robot. Autom. 17 (2), 125–137.

Dong, H., Li, W., Zhu, J., Duan, S., 2010. The path planning for mobile robot based on Voronoi diagram. In: 2010 Third International Conference on Intelligent Networks and Intelligent Systems, pp. 446–449.

Janet, J.A., Luo, R.C., Kay, M.G., 1995. The essential visibility graph: an approach to global motion planning for autonomous mobile robots. In: Proceedings of 1995 IEEE International Conference on Robotics and Automation, vol. 2, pp. 1958–1963.

Kala, R., 2016a. Homotopic roadmap generation for robot motion planning. J. Intell. Robot. Syst. 82 (3), 555–575.

Kala, R., 2016b. Homotopy conscious roadmap construction by fast sampling of narrow corridors. Appl. Intell. 45 (4), 1089–1102.

Kala, R., Shukla, A., Tiwari, R., 2010. Fusion of probabilistic A* algorithm and fuzzy inference system for robotic path planning. Artif. Intell. Rev. 33 (4), 275–306.

Kala, R., Shukla, A., Tiwari, R., 2011. Robotic path planning in static environment using hierarchical multi-neuron heuristic search and probability based fitness. Neurocomputing 74 (14–15), 2314–2335.

Kambhampati, S., Davis, L.S., 1986. Multiresolution path planning for mobile robots. IEEE J. Robot. Autom. RA-2 (3), 135–145.

Lee, D.T., 1982. Medial axis transformation of a planar shape. IEEE Trans. Pattern Anal. Mach. Intell. PAMI-4 (4), 363–369.

Oommen, B., Iyengar, S., Rao, N., Kashyap, R., 1987. Robot navigation in unknown terrains using learned visibility graphs. Part I. The disjoint convex obstacle case. IEEE J. Robot. Automat. 3 (6), 672–681.

Seidel, R., 1991. A simple and fast incremental randomized algorithm for computing trapezoidal decompositions and for triangulating polygons. Comput. Geom. 1 (1), 51–64.

Yahja, A., Stentz, A., Singh, S., Brumitt, B.L., 1998. Framed-quadtree path planning for mobile robots operating in sparse environments. In: Proceedings of the 1998 IEEE International Conference on Robotics and Automation. IEEE, Leuven, Belgium, pp. 650–655.

Chapter 8

Probabilistic roadmap

8.1 Introduction to sampling-based motion planning

So far, we have seen two methods for the motion planning of robots. The first method was a graph search, which was too complex for online planning in high dimensional spaces further requiring a discretization resulting in a resolution parameter that is hard to set. The second method was roadmap/cell-decomposition, which had a significant computation time for the offline construction of the roadmap rendering them useful only in 2D/3D scenarios with explicit knowledge of the complete space. Although the roadmap construction or the cell decomposition is an offline process, it cannot take indefinitely long. There is only a limited time within which the map can be assumed to be static, for example the furnishings of buildings move with time, the locations of chairs and tables are frequently changed, etc. This invalidates the old computations and requires new ones.

Another problem associated with the approaches is dimensionality. Previously, the problems were solved typically in two dimensions; however, the problems were extendable to higher dimensions as well. Humanoids may have 48–60 degrees of freedom and a drone has 6 degrees of freedom, while a manipulator alone can have 6–9 degrees of freedom. The configuration space is highly unstructured and nongeometric. A geometric workspace for any of these robots also leads to nongeometric configuration space. So, a visibility graph that uses obstacle corners cannot be used. For the Generalized Voronoi Graph using a wavefront planner and cell decomposition approaches, the entire space needs to be traversed. The number of discretized cells is r^d, where r is the number of cells per dimension and d is the number of dimensions, which is workable only for robots with 2–3 degrees of freedom. A high degree of freedom robot will make the system run out of time and memory.

More appropriately, the cell decomposition methods of Chapter 7 are called *exact cell decomposition* methods as the methods return a solution that guarantees completeness; however, the computation of the solution might take severely long. The exact solution is available at the end of the computation and not even an approximate solution is available before that time.

Autonomous Mobile Robots. https://doi.org/10.1016/B978-0-443-18908-1.00003-0

In any branch of computing, when the exact solutions fail, there is an instinct to use *approximate methods*. Many examples exist from different fields of computing. The inability to search the complete optimization space using classic methods led to the development of Genetic Algorithms; the difficulty in handling complex systems by probabilistic reasoning alone led to Fuzzy-based systems; the full posterior is hard to compute by Bayesian methods keeping all assumptions intact, which is otherwise approximated using neural networks. Approximate solutions do not guarantee providing the exact correct answer like the exact methods; however, approximations enable doing computations that were otherwise not possible.

The approximation considered in this chapter (and the next) is called *sampling*. The term is common in marketing research and statistics, wherein if it is not possible to research a complete market space extensively and exhaustively, the research is done only on a chosen number of samples, and the results of limited samples are generalized to the entire population. The results are typically good if the number of samples chosen is large enough to be significant and are chosen by an unbiased and intelligent technique to make the samples as close as possible to the target population.

The same notion is used in this chapter. The free configuration space (C^{free}) is too large, and it is impossible to analyse the same. It is impossible to construct a roadmap within a limited time. Therefore, instead of analysing the complete space, this chapter first draws strategized selected samples from C^{free} and then use the selected samples only to make the roadmap and answer the motion planning queries. Since the number of samples is limited, the computation time will also be limited. The goodness of the solution in terms of optimality and completeness, hence, depends upon both the quality of samples drawn and the number of samples drawn. Typically, more samples yield a better solution. More samples also increase the chances of getting a solution if one exists. This is called *Probabilistic Completeness*, meaning that the probability of computing a solution, if one exists tends to one as time (or the number of samples drawn) tends to infinity. Similarly, the solution is *Probabilistically Optimal*, meaning that the probability of getting the optimal solution will tend to be one, as time (or the number of samples drawn) tends to infinity. The notion of Probabilistic Optimality is placed with concern as many methods will have constraints that guarantee good enough, but not necessarily optimal solutions when executed for a long time. This means that after sufficient time, subject to the complexity of the problem, the solutions are good enough for use, which may not be strictly optimal though. This is a great boon as the exact methods would not even have managed to return a single solution.

The sampling-based motion planning comes in two variants. The first is called the *multiquery approach* wherein the solver is asked to solve multiple queries at different points of time. The queries may come from the same robot during its need to devise a further plan, replan, handle a change of goal, replan due to a major localization error, etc. The queries may also come from different robots operating in the same space. The methodology is to first construct a road-map offline and then solve the motion planning queries in an online manner using the roadmap. The multiquery approaches will be discussed in this chapter.

The other variant is called the *single-query approach*. Here, the solver just needs to solve a single query. If subsequent queries arrive, new similar searches shall be made. The notion is to sample out configurations and to try to construct a trajectory by the same samples, increasing the number of samples with time as per the need. When a good enough trajectory from the source to the goal is computed, the same is returned and not summarized for future use as a roadmap.

The choice between single and multiquery approaches is particularly inter-esting. The choice depends upon whether the map remains static for a prolonged time, the admissible online planning time, the number of motion planning queries to be answered, and the complexity of the scenario. The single-query approaches know the source and the goal and, therefore, need to effectively search in a small region around the source and goal for simple scenarios, which is a significantly small part of the entire search space. However, the multiquery approaches have no information about the source and the goal and need to pre-pare a roadmap for the entire space. In contrast, if the online planning time is too low for a static map, it may be better to summarize the space as a roadmap using the multiquery approaches.

Let us discuss two specific cases. Consider a humanoid robot operating in a kitchen along with humans. Items in the kitchen are often placed here and there and, therefore, space is not largely static and will change every time the robot operates, many times by the human. The total space for a humanoid is too large. The robot can think for a while like humans first think and then find the ingre-dients before putting them in the dish. So, single query-based planning is appro-priate. However, consider a warehouse where the mobile robots navigate from one place to the other carrying items between stations. The scenario may be eas-ily summarized by a roadmap, and hence, the multiquery approach is suitable. The discussions here are summarized as Box 8.1.

Box 8.1 Tradeoffs and concepts in the use of probabilistic roadmaps.

Grid-based graph search
- Have high computation time
- Resolution (size of continuous space that makes a grid) parameter is hard to specify
- Work in low dimensions only

Visibility graph and Voronoi
- Work faster for geometric maps only (with polygon obstacles)
- Require explicit knowledge of the complete space
- A high degree of freedom robot in a geometric workspace has a nongeometric configuration space

---

Continued

Box 8.1 Tradeoffs and concepts in the use of probabilistic roadmaps.—cont'd

Concepts
- Need to get approximate solutions quickly
- Approximate by drawing samples from space
- Probabilistically complete solutions
- Some implementations are Probabilistically Optimal

Multiquery problem solving
- Numerous source-goal pairs (queries)
- Construct a roadmap offline and solve motion planning queries online
- Online query planning possible due to roadmap
- Requires: large offline planning time, static map, ideal for a large number of queries

Single query problem solving
- Solve for 1 source/goal pair
- No roadmap is constructed for reuse
- Since source and goal are known, can strategize search in free configuration space

---

8.2 Probabilistic roadmaps

The *Probabilistic Roadmap* (PRM) (Kavraki et al., 1996a, b) is a multiquery approach that first constructs an offline roadmap in the form of a graph $RM < V, E >$ with vertices V and edges E. The same roadmap is used for answering multiple queries, each query specifying the source and the goal. The aim is to construct a collision-free trajectory from the source to the goal using the roadmap in a small computation time.

8.2.1 Vertices and edges

The vertices of the roadmap $V = \{q\}$ are a set of samples drawn from the free configuration space, $q \sim Sampling\ Strategy(C^{free})$. Here *sampling strategy* governs, which samples should be taken, given that an infinite number of samples exist in C^{free}. The simplest strategy is to take a uniform sampling strategy, wherein every sample in C^{free} has the same probability of being selected. This produces a roadmap whose vertices are spread all over the free configuration space, which for a synthetic 2D map is shown in Fig. 8.1A.

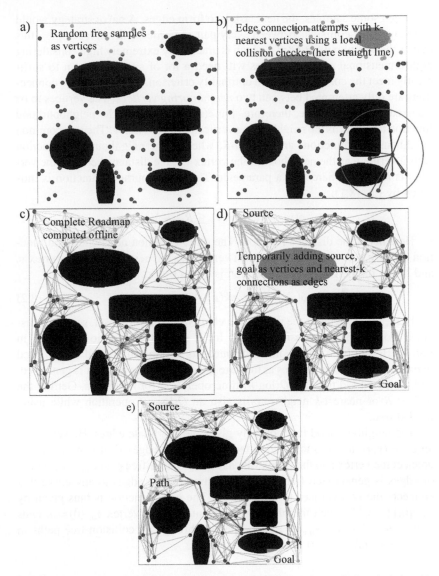

FIG. 8.1 The working of probabilistic roadmap. Actual implementation works iteratively by adding a sample, checking for connections with neighbours, adding edges, and then adding the next sample. (A) Samples, (B) *k*-nearest neighbours, (C) Constructed roadmap, (D) Temporarily adding source and goal, (E) Planned path.

Given some samples, the next task is to form edges. A naïve strategy would be to check for collisions between every pair of vertices. Given n samples, this makes the complexity as $O(n^2)$, which is practically extremely high considering high dimensional complex spaces will need a lot of samples drawn to sufficiently meet the completeness and optimality criterion. Therefore, edge connections are only attempted between two *neighbouring vertices*. The vertices to be checked for collision are, therefore, $\{\langle v, \delta(v) \rangle\}$ where δ is the neighbourhood function that returns the neighbouring vertices to any vertex v. The two common neighbourhood functions are *nearest-R*, wherein the neighbourhood function calls a vertex u as the neighbour of the vertex v if the distance between the vertices is R or less, here R is a parameter. The neighbourhood function is thus given by Eq. (8.1):

$$\delta(v) = \{u : d(u, v) \leq R, u \in V\} \tag{8.1}$$

Here $d()$ is the distance function. The other common neighbourhood function is *nearest-k*, which calls k nearest vertices as the neighbour of the vertex v, and hence the neighbourhood function is given by Eq. (8.2):

$$\delta(v) = \{u : rank(u) \leq k, u \in V\} \tag{8.2}$$

Here, $rank()$ denotes the rank of the element in the sorted set, sorted by distance from v. The *nearest-k* vertices for a synthetic scenario are given in Fig. 8.1B. Both the neighbourhood functions assume that a k-d tree is used in parallel to store the vertices as they are generated. Once a vertex is generated, adding it to a k-d tree is a $O(\log n)$ time operation for n vertices. Getting the nearest-R or nearest-k neighbours is also an $O(\log n)$ operation while using the k-d tree.

The neighbourhood function only gives the candidate edges. However, for an edge (v, u), $u = \delta(v)$ to be admitted in the set of edges E, it is necessary to connect the vertex u to the vertex v by a local collision-free path $e_{u,v}$. The notion of edges is generalized from the strictly straight-line edges to any curve that connects the two vertices to be the edge. The edge function is thus given by $e_{u,v}:[0,1] \to C^{\text{free}}$, such that the edge starts at the first vertex $e_{u,v}(0) = u$, ends at the second vertex $e_{u,v}(1) = v$, and passes through all collision-free paths in between $e_{u,v}(s) \in C^{\text{free}}$, $0 \leq s \leq 1$.

8.2.2 Local planner

The computation of a collision-free edge ($e_{u,v}$) is another motion planning problem solved by a motion planning algorithm called the *local planner* (Δ). The local planning algorithm is a quick way to solve a local motion planning problem between 2 close vertices, v and u separated by an obstacle part or two. The attempt is to quickly solve the problem while not necessarily guaranteeing optimality and completeness that can only be done by time-consuming global planning algorithms. The local planning algorithms use simple heuristic means to quickly travel between vertices in a few attempts. If such attempts result in a path, the same is returned, otherwise, a failure is returned even, if it was possible to connect the vertices had more time been devoted.

The simplest and fastest local planner is the straight-line edge connection that returns a straight line between the vertices v and u if the straight line is collision-free, NIL otherwise, and given by Eq. (8.3):

$$\Delta(v, u) = \begin{cases} \{\lambda v + (1-\lambda)u, 0 \le \lambda \le 1\} & \lambda v + (1-\lambda)u \in C^{free} \forall 0 \le \lambda \le 1 \\ \text{NIL} & \text{otherwise} \end{cases} \quad (8.3)$$

The Equation $\lambda u + (1 - \lambda)v$ is the interpolation equation of a straight line between vertices v and u, and the equation checks that all interpolations are obstacle-free. This way of writing the equation is not completely correct since not all spaces can be interpolated by such addition, and especially the case of angles is interesting since the circular property needs to be maintained. In angles, it needs to be decided whether to travel clockwise or anticlockwise, and whether to cross from π radians and enter from $-\pi$ radians (or vice versa), and the interpolation so adjusted. Consider going from an angle $-3\pi/4$ to $3\pi/4$ for a robot. Now instead of going from $-3\pi/4$ to 0 and from 0 to $3\pi/4$ (what a normal interpolation would do), it may be better to go from $-3\pi/4$ to $-\pi$ and from π to $3\pi/4$ that is valid as per the circular nature of the angles. The interpolation from θ_1 ($-\pi < \theta_1 \le \pi$) to θ_2 ($-\pi < \theta_2 \le \pi$) without crossing the boundary at π has a distance $|\theta_2-\theta_1|$, while the same distance upon choosing to cross the boundary is $\min(|2\pi+\theta_1-\theta_2|, |2\pi+\theta_2-\theta_1|)$. The strategy corresponds to the smaller of the two cases that can be used for the shortest distances.

The collision checking may itself be done in two mechanisms. The first is iterating over the edge in small steps (of size *resolution*) till an obstacle or target vertex is achieved. This is called *incremental collision checking*. It may be observed that so far, the collision checking was exact and now it is itself being done at some *resolution* denoted by the size of the step. It is further stressed that every step is a complex collision checking algorithm that may take time. It is not a unit computation. Continuous collision detection techniques perform collision for a complete edge at one go without discretizing into steps; however, they require the geometrical information of the obstacles in workspace that may or may not be available. The collision checking is shown in Fig. 8.2A. The algorithm is given by Algorithm 8.1. Line 1 calculates the increment in λ using the resolution and distance between v and u ($d(v,u)$). Line 3 is a simplified interpolation and may be replaced by any interpolation as per space. Line 4 is a call to the collision checker, otherwise, no explicit set C^{free} is provided.

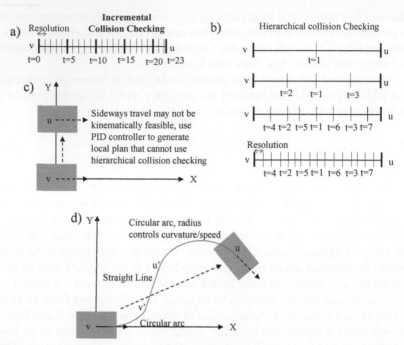

FIG. 8.2 Collision checking of a prospective edge *v-u*. (A) Incremental collision checking by moving from *v* to *u* (*vertical lines* represent points of collision check). (B) Hierarchical collision checking, at every level, the middle points of the previous level are checked. Hierarchical collision checking works better but may not be possible in all mechanisms. (C) A case where hierarchical collision checking cannot be used, (D) edges considering robot's kinematic constraints by adapting *circular arches* and *straight lines* as per samples *v* and *u*.

Algorithm 8.1 Incremental collision check (*v,u*).

> **Input:** Vertices *v* and *u* denoting prospective edge from *v* to *u*
> **Output:** *true* if the straight line edge (v, u) is collission free, *false* otherwise
>
> 1 $\alpha \leftarrow resolution/d(v,u)$;
> 2 **for** λ *from 0 to 1 in small steps of* α **do**
> 3 $\quad q \leftarrow \lambda u + (1 - \lambda)v$;
> 4 $\quad$ **if** $q \notin C^{free}$ **then** return *false*;
> 5 return *true*

The other mechanism of collision-checking is *hierarchical collision-checking,* and the collision checking is done by first checking the midpoint of the line segment (*v,u*) for a collision. If free, this divides the lines into 2 segments, whose midpoints are similarly checked. This divides the line into 4 segments now. By checking their midpoints, the line is divided into 8 segments now. At every iteration, the midpoint of the segment is checked for a collision that produces 2 new segments. The real space is continuous, and the subdivisions cannot happen endlessly. The search must stop at some depth or resolution distance. The collision-checking in this format is shown in Fig. 8.2B and Algorithm 8.2.

Algorithm 8.2 Hierarchical collision check (v,u).

Input: Vertices v and u denoting prospective edge from v to u
Output: *true* if the straight line edge (v, u) is collision free, *false*
        otherwise

1 $Q \leftarrow$ Empty Queue ;   ▷ segments for collision checking
2 $depth(v, u) \leftarrow 1$;   ▷ depth of current segment
3 Add segment (v, u) to Q ;
4 **while** $\neg Empty(Q)$ **do**
5      $(v, u) \leftarrow Q.top(), Q.dequeue()$;
6   **if** $d(v, u) < resolution$ **then**   ▷ threshold distance reached
7   break;
8      $q \leftarrow 0.5u + 0.5v$;
9   **if** $q \notin C^{free}$ **then** return *false*;
10    $depth(v, q) \leftarrow depth(v, u) + 1$; ▷ depth of child segment 1
11    $depth(q, u) \leftarrow depth(v, u) + 1$; ▷ depth of child segment 2
12   Enqueue (v, q) and (q, u) to Q
13 return *true*

The roadmap formed after a straight-line local planner is shown in Fig. 8.1C.

The worst-case complexities of both the algorithms are the same, which happens at the end of the line segment in the incremental and when the last point is checked for collisions records a small obstacle in the hierarchy. However, practically obstacles are large and cover a larger space and therefore are more easily identified in the hierarchical search mechanism. Hence, practically hierarchically subdividing space works better. There may be cases like taking feedback-driven steps oriented towards a goal. In such a mechanism a subsequent step can only be calculated by a knowledge of the prior step, and therefore, incremental search makes sense as hierarchical collision-checking is not possible. Similarly, when moving with the kinematics of the robot in control spaces by simulating the path of the robot, one can only do a stepwise forward simulation, and hence hierarchical collision-checking is not possible.

Consider a planar rectangular robot with 2 samples $v(v_x, v_y, v_\theta)$ and $u(u_x, u_y, u_\theta)$, each storing the position and orientation of the robot. It may be inquisitive to apply the interpolation of Eq. (8.3) as a local planar. However, if v is a robot at the origin facing the X-axis and u is the robot vertically above v, still facing the

X-axis, the interpolation equation asks the robot to go sideways, which violates a nonholonomic constraint that some robots have, shown in Fig. 8.2C. A natural local planner is thus to use a PID control algorithm to drive the robot from v to u that gives the local plan for incremental collision checking that being feedback-driven cannot be done in a hierarchical paradigm. That said, an alternative is to use specialized curves (Fig. 8.2D) that take the robot from v to a point v' using a circular arc whose radius controls the maximum or desirable curvature (or speed), followed by directly connecting v' to a point u' by a straight line, thereafter connecting u' to u by a similar circular arc setting radius as per the curvature (or speed). The curve computation has a solution subjected to conditions, and if such a curve can be generated, the same is added as an edge after collision checking.

8.2.3 The algorithm

Another aspect of the roadmap is that the algorithm should be *anytime* or iterative. This means, starting from an empty set of vertices and edges, the roadmap must incrementally grow. The major advantage of this is that the maximum number of vertices n, which is an important algorithm parameter, need not be specified. Instead, any generic stopping criterion may be used like maximum construction time, the maximum number of vertices, the maximum number of edges, testing for completeness, and optimality measures on a test suite of queries, etc. This can be naively done by incrementally adding vertices and connecting the same to the neighbouring vertices by edges as and when they are added. The pseudocode of the algorithm is given by Algorithm 8.3. Here, Δ

Algorithm 8.3 PRM-build roadmap.

Output: Roadmap $RM < V, E >$ consisting of vertices V and edges E
1   $V \leftarrow \emptyset, E \leftarrow \emptyset$;   ▷ vertices and edges
2   $\kappa \leftarrow \emptyset$;   ▷ Disjoint set to check vertex connectivity
3   $T \leftarrow \emptyset$;   ▷ k-D tree for neighbourhood queries
4 **while** *stopping criterion* **do**
5     $q \leftarrow SamplingStrategy(C^{free})$;
6   add q to V ;
7   add q to $T, MakeSet(q, \kappa)$;
8   **for** *all* $v \in \delta(V) - q$ **do**   ▷ all neighbours
9   **if** $\Delta(q, v)$ *is not null* **then**   ▷ path by a local planner
10   add $\Delta(q, v)$ to E ;
11          $Union(\kappa(q), \kappa(v))$;

12 return $RM < V, E >$

is the local planner. It returns a trajectory between queried configurations, if possible, *null* otherwise. Two data structures are used in parallel. The first is the *k*-d tree denoted by T used by the neighbourhood function. The second is a disjoint set of data structured denoted by κ, which is used to count the total number of subgraphs that are not connected and whether two vertices have a path between them or not.

After the roadmap is built, the next step is to process each query and to compute a trajectory for the query. The source and goal are first added as temporary vertices in the roadmap and corresponding edges are added as per the neighbourhood function and local planner. Thereafter, the new graph can be directly searched by any graph search algorithm like the A* or Uniform Cost Search. The pseudocode is shown in Algorithm 8.4. The processing of a query is shown in Fig. 8.1D and E. Fig. 8.1D shows the roadmap after adding the source and goal, while Fig. 8.1E gives the path produced. The trajectory as computed will have sharp turns that are not realizable by the robot. The trajectory is hence *post-processed* and *smoothened*, which is the same process as discussed with graph search approaches.

Algorithm 8.4 PRM-process query ($q < S,G >,RM < V,E >$).

    Input: query $q < S, G >$ consisting of source S and goal G, Roadmap
             $RM < V, E >$ consisting of vertices V and edges E
    Output: Path τ between source (S) and Goal (G)
1 $V' \leftarrow V \cup \{S, G\}, E' \leftarrow E$; ▷ add S, G to temporary roadmap
2 **for** $q \in \{S, G\}$ **do**   ▷ edges for S and G
3   **for** *all* $v \in (\delta(V') - q)$ **do**
4   **if** $\Delta(q, v)$ **then**
5   add (q, v) to E' ;

6 $\tau \leftarrow GraphSearch(RM' < V', E' >, S, G)$;
7 return τ

8.2.4 Completeness and optimality

As time tends to infinity, virtually all the samples in the free configuration space are sampled out, which will be connected to their neighbouring samples. Therefore, if a path exists, it is bound to be found for sufficiently large neighbourhoods. This means that the algorithm is *probabilistically complete*, or the probability that the solution is found tends to be one as time tends to infinity. The PRM is not probabilistically optimal. Even if all configurations are discovered, the edges may not be connected in a mechanism to enable the shortest paths. The optimality hence depends upon the neighbourhood function. The neighbourhood function should be large enough to get optimality and completeness (Karaman and Frazzoli, 2011). If using the nearest-R methodology, wherein all vertices in a radius of R are called as neighbours, the condition in Eq. (8.4) must hold.

$$R > 2\left(1 + \frac{1}{d}\right)^{\frac{1}{d}}\left(\frac{V\left(C^{\text{free}}\right)}{u_d}\right)^{\frac{1}{d}}\left(\frac{\log n}{n}\right)^{\frac{1}{d}} \tag{8.4}$$

Here, d is the dimensionality, $V()$ is the volume of space, and u_d is the volume of a unit ball. This means as more proportion of the volume of the space is filled with vertices, the lesser the requirement from the radius. Correspondingly, for nearest-k, the value of the parameter k must follow Eq. (8.5).

$$k > e\left(1 + \frac{1}{d}\right)\log n \tag{8.5}$$

In general, the algorithm is probabilistically optimal and probabilistically complete for large values of neighbourhood functions.

The *complexity* of the algorithm is $O(n \log n)$, where finding the nearest k neighbours or neighbouring vertices within the radius R requires $O(\log n)$ complexity worth of time when implemented using a k-d tree data structure that is repeated for n vertices. The sampling in the current form is a unit time algorithm and so is edge detection with neighbours and adding vertices and edges to the roadmap.

8.3 Sampling techniques

The sampling technique governs the strategy to sample the configurations in free space C^{free} and the choice of the sampling strategy largely governs the performance of the algorithm. To make it easy to simultaneously understand all techniques and their tradeoffs, the techniques are summarized in Box 8.2.

Box 8.2 Tradeoffs and summary of different methods in PRM.

Sampling techniques

Uniform: No strategized placement, necessitates too many samples

Obstacle: Attempts modelling for optimal paths by sampling near obstacles. Retraction from obstacle to free space is computationally heavy, and has a high success rate.

Gaussian: Samples near obstacles. Allows some samples away from obstacles for enhanced connectivity.

Bridge test: Samples only in narrow corridors. Other samplers may overpopulate the space to get one sample inside the narrow corridor. Cannot be used for other areas of the space. Small success rate.

Maximum clearance: The high clearance samples are easier to connect and result in the completeness property with a smaller number of samples, optimality is somewhat compromised.

Medial axis: Generation of samples at the medial axis more intelligently increases clearance, also enabling generation of samples in narrow regions.

Grid sampling: Generates at least 1 sample per low-resolution grid. Ensures judicious spread of samples all over space, and boosts samples in poorly represented areas.

Toggle: Maintains a roadmap in obstacles. A small additional time to maintain roadmap in obstacle enables the generation of more samples in narrow corridor-like situations that are difficult otherwise.

Hybrid sampling

Takes advantage of all methods, adapts contributions.

Time-bound: Increase the selection probability of a sampler with more success. Success is connecting two subroadmaps that are not path-connected, adding a sample invisible to all others, adding a sample with very few edges (or difficult region)

Space-bound: The probability of selection of a sampler in the region depends upon obstacle density, average edges, successful connection attempts in that region.

Edge connection methods

Connect connected components: Join disjoint subgraphs by a better local planning algorithm, which by straight-line connections will require too many samples.

Expanding roadmap: Random walk from samples at difficult areas, to walk out of the narrow corridor or generate easy-to-connect samples. Not strategized between 2 samples like *Connect Connected Components*.

Cycle-free roadmaps: Eliminate cycles to reduce computation time, compromising optimality, useful cycles may be added.

Visibility PRM: Cycle-free roadmap, greatly limits the number of samples using heuristics, and compromises optimality adversely.

Lazy PRM: Defers collision checking to the online query phase for a rapid roadmap construction, slowing down the query phase.

8.3.1 Uniform sampling

A typical mechanism to sample configurations is *uniform sampling*, which states that all free configurations have the sample probability of getting selected at every point in time. Generating a random sample is easy for any configuration space and requires the generation of random numbers for every configuration variable involved. The free configuration sampling only accepts the sample if it lies in the free configuration space, which is a call to the collision-checking algorithm. The pseudocode is, hence given by Algorithm 8.5. Line 3 is a call to the collision-checking algorithm. Line 4 is a rare condition when a sample cannot be generated, wherein the algorithm stops after some time to eliminate getting into an infinite loop, especially when there are better sampling strategies available. A synthetic roadmap is shown in Fig. 8.3.

Algorithm 8.5 Uniform sampling.

Output: random sample $q \in C^{free}$
1 **for** *some maximum number of attempts* **do**
2     $q \sim U(C)$;   ▷ `random sample`
3   **if** $q \in C^{free}$ **then** return q;   ▷ `collission detection`
4 return $NULL$

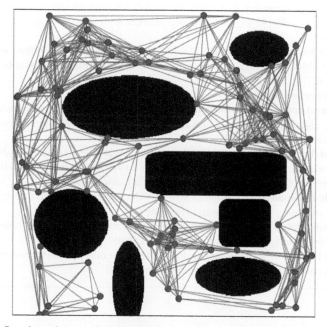

FIG. 8.3 Sample roadmap produced by a uniform sampler.

8.3.2 Obstacle-based sampling

The problem with the uniform sampling is that the search happens in the entire free configuration space, which may result in exploring a significant part of the search space. The shortest path between any two points is always a straight line connecting the two points. If an obstacle or an obstacle framework comes in between the obstacles, the shortest path instead takes the tangent to the obstacle corners in case of a 2D space, and the obstacle faces in cases of higher dimensions. The shortest paths between any two points separated by obstacles make the robot take turns around the obstacle boundary only and travel straight otherwise. The areas near obstacles are hence important in motion planning. Every sample in the case of a PRM models a potential turn of the robot. The samples become the vertices of the roadmap, which makes the points when the robot turns from one edge to the other. Hence, if the obstacle boundaries are sampled out, all shortest paths may be constructed subjected to a sufficiently large neighbourhood function.

Another problem associated with uniform sampling is that of *narrow corridors*. For a narrow corridor to be discovered, it is important to generate at least one sample inside the narrow corridor. The probability of generation of a sample inside the narrow corridor is the volume of the narrow corridor upon the volume of the entire configuration space. As per the definition of the narrow corridor, the numerator is too small in comparison to the denominator, meaning the sample inside the narrow corridor is extremely unlikely to get generated, and it will take extremely long to discover the narrow corridor and connect to the roadmap. This is shown in Fig. 8.4A.

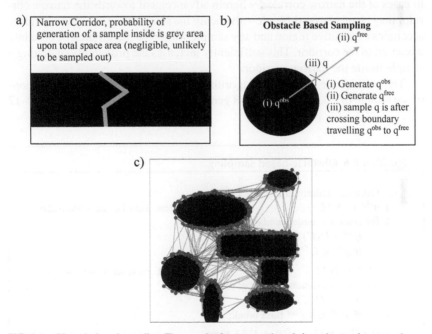

FIG. 8.4 Obstacle-based sampling. The sampler focuses on obstacle boundary as shortest paths are tangents to obstacles (A) narrow corridor problem, (B) obstacle-based sampling concept, and (C) sample roadmap.

Therefore, a sampling strategy is used that boosts samples near the obstacle boundary. This is called *obstacle-based sampling* (Amato et al., 1998; Yeh et al., 2012). The sampling technique first generates a sample inside the obstacle-based configuration space C^{obs}. The technique then advances the sample to the obstacle boundary by making the sample travel in a random direction. In the traversal from inside the obstacle to anywhere else, as soon as the sample crosses the obstacle boundary, that same position is used as the sample. This ensures that the sample is generated near the obstacle boundary. Some space from the boundary may be kept for facilitating easier connectivity to the nearby samples, otherwise, a sample at the boundary is hard to connect as any movement may easily go into the obstacle. The motion from inside the obstacle to the obstacle boundary is called a *retraction*. The traversal to a random direction may be done by generating a free sample and moving towards the free sample.

Another variant advances the sample to the *nearest* obstacle boundary instead. The retraction direction information provided by the collision checker is used and the sample is moved in the retraction direction to get to the nearest boundary. As an example, for a circular robot (random sample) intersecting a circular obstacle, the retraction direction and also the direction of motion of the robot is away from the centre of the circular obstacle. This is especially helpful in cases of the narrow corridor, wherein advancement towards the narrow corridor may be otherwise difficult. Advancing the sample to the nearest boundary ascertains a definitive region and any sample within the region shall be certainly advanced to the corridor. This sufficiently increases the probability of getting a sample inside the narrow corridor.

The pseudocode is given as Algorithm 8.6. Lines 1–4 generate a collision-prone sample (q^{obs}), while Lines 5–8 generate a free sample (q^{free}). Lines 10–12

Algorithm 8.6 Obstacle-based sampling.

    Output: random sample $q \in C^{free}$
1 $q^{obs} \leftarrow NIL$;   ▷ find random sample in obstacle
2 **for** *some maximum number of attempts* **do**
3 | $q^{obs} \sim U(C)$;
4 |__ **if** $q^{obs} \notin C^{free}$ **then** break;
5 $q^{free} \leftarrow NIL$;   ▷ find random free sample
6 **for** *some maximum number of attempts* **do**
7 | $q^{free} \sim U(C)$;
8 |__ **if** $q^{free} \in C^{free}$ **then** break;
9 **if** $q^{obs} \neq NIL \wedge q^{free} \neq NIL$ **then**
10 | **for** λ *from 0 to 1 in small steps* **do**   ▷ travel q^{obs} to q^{free}
11 | | $q \leftarrow \lambda q^{free} + (1 - \lambda)q^{obs}$;
12 |__|__ **if** $q \in C^{free}$ **then** return q;   ▷ crossed boundary
13 return $NULL$

advances the sample from q^{obs} to q^{free} in small steps. As soon as the first free sample is found, the same is returned. The working of the algorithm is also shown in Fig. 8.4B. Fig. 8.4C, further shows a synthetic roadmap generated by using obstacle-based sampling. It may be observed that most of the samples are near obstacle boundaries.

8.3.3 Gaussian sampling

The obstacle-based sampling only generates samples at the obstacle boundary, which however creates a problem. Imagine an obstacle framework with rich sample density across the obstacles and every sample connected to the nearest k samples. For every sample near the obstacle boundary, the nearest k samples would be found adjacent to it near the same obstacle boundary. This will produce a framework, where the obstacle boundaries are nicely covered by roadmaps; however, no two obstacle boundaries have connections between themselves. The situation is shown in Fig. 8.5A. Therefore, it is important to keep some samples far away from the obstacles as well for connectivity, while

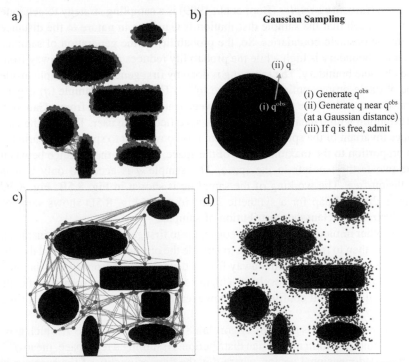

FIG. 8.5 Gaussian sampling (A) Generating samples near obstacles and connecting to nearest k-neighbours produces a nonconnected roadmap. The sampler should also add samples in free areas. (B) Gaussian Sampling, more samples near boundaries, lesser further away, (C) sample roadmap, and (D) samples at a Gaussian distribution from distance from obstacle boundaries.

keeping most samples near the obstacle boundaries. This means the sample density near the obstacle boundary should be high while that far away from the obstacle boundary should be reasonably low. Gaussian sampling is one mechanism to achieve this distribution. That said, Gaussian sampling is not the best way to solve the specific problem and it is better to use a hybrid of obstacle sampling and uniform sampling. Say, 90% of samples are generated by obstacle-based sampling, while 10% of samples are generated by using uniform sampling. It may also be noted that the roadmap is incrementally constructed and in low-density situations, the nearest k samples point to the ones at the neighbouring obstacle boundaries, so it is not that obstacle-based sampling cannot solve problems on its own as was shown in Fig. 8.4C.

The other problem with obstacle-based sampling is that if the sample in obstacle-based configuration space is inside a very deep obstacle, the retraction of the same sample to free configuration space can take a long time. It may instead be desirable sometimes to discard that sample and instead use the time to generate more samples outside obstacle boundaries. The aim is to get the same effect of getting samples near obstacle boundaries, but without using the time-consuming retraction of samples.

The *Gaussian sampler* (Boor et al., 1999) is a sampling strategy to generate samples such that the sample distribution is Gaussian in nature to the distance from the obstacle boundaries. So, the probability of the generation of samples near the boundary is high while the probability reduces as we move away from the obstacle boundary. The sampling is done by first generating a sample inside the obstacle-prone configuration space (q^{obs}). Then another sample (q) is generated near q^{obs}, where the distance between q and q^{obs} is taken from a Gaussian distribution with 0 mean and variance as specified by the user. To keep the variance invariant of the space of operation, it is convenient to specify the variance in proportion to the maximum size of the space, which is the distance between the two farthest points in the space. The sample q is accepted only if it is collision-free. The working of the sampling is shown in Fig. 8.5B. Fig. 8.5C shows the roadmap for a synthetic scenario while Fig. 8.5D shows samples to illustrate the Gaussian distribution of samples.

Another implementation of the concept is to first generate a sample inside an obstacle, then to retract the sample to outside the obstacle boundary, and then to generate a free sample further away from a distance taken from a Gaussian distribution. This technique results in a higher *success rate*; the percentage of calls that result in a valid configuration, however, has a higher computational time as well.

The pseudocode is given by Algorithm 8.7. Typicality is Line 7, which generates samples within a Gaussian distribution. Assuming a real space, mean q^{obs} is a $d \times 1$ dimensional vector, where d is the dimensionality of the space. σ is interpreted as a $d \times 1$ vector, to produce a $d \times 1$ target vector. The Gaussian distribution-based random number generation in such a manner is an available utility in every library. In the case of orientations and other complex spaces, the space property and constraints will have to be worked out. The Gaussian

---

Algorithm 8.7 Gaussian sampling.

Output: random sample $q \in C^{free}$
1 **for** *some maximum number of iterations* **do**
2 $\quad q^{obs} \leftarrow NIL$; $\qquad\qquad\quad$ ▷ find random sample in obstacle
3 $\quad$ **for** *some maximum number of attempts* **do**
4 $\quad\quad q^{obs} \sim U(C)$;
5 $\quad\quad$ **if** $q^{obs} \notin C^{free}$ **then** break;
6 $\quad$ **if** $q^{obs} \neq NIL$ **then**
7 $\quad\quad q \sim G(q^{obs}, \sigma)$; $\qquad\qquad$ ▷ For each axis sample from a
$\qquad\qquad\qquad$ Gaussian distribution (mean q^{obs}, deviation σ)
8 $\quad\quad$ **if** *q is within bounds* $\wedge q \in C^{free}$ **then** return q;
9 return $NULL$

---

sampler practically generates samples close to the boundary and therefore mixing it with a Uniform sampler to generate a few samples for enhanced connectivity is typically recommendable.

8.3.4 Bridge test sampling

The mechanisms of generating samples near the obstacle boundary cannot completely solve the problem of sampling inside narrow corridors. It is unlikely that a sample within an obstacle will retract in the correct direction to lie inside the narrow corridor. During the time in which a sampler adds a sample inside a narrow corridor, too many samples get added in the space, which harms both roadmap construction and the online query phase. Edge collision checking takes the maximum time, and therefore, a generated sample must be assessed for utility before admitting into the roadmap.

The *bridge test* (Hsu et al., 2003; Sun et al., 2005) is a test that checks if a sample is inside the narrow corridor or not. Alternatively, bridge test sampling is a mechanism of generating samples only inside the narrow corridors. To generate such samples, first, a sample q^{obs} is generated inside obstacle-prone configuration space. Then a sample q_1 is generated at a distance $2d_C$ from q^{obs} in a random direction, that is $q_1 = q^{obs} + 2d_C u$, u is a random unit vector that illustrates a random direction. Here d_C is the width of the narrow corridor that the sampler targets. Any corridor larger than d_C is not tagged as a narrow corridor. d_C is an algorithm parameter and is conveniently defined in a proportion to the maximum distance between any two samples in space. A sample q is also generated at the midpoint of q^{obs} and q_1, that is Eq. (8.6):

$$q = 0.5q^{obs} + 0.5q_1 \tag{8.6}$$

This means q is at d_C from q^{obs} and q_1 is at $2d_C$ from q^{obs} in the same direction.

The sample q is said to be inside the narrow corridor if q is free while both the neighbours (q_1 and q^{obs}) are inside obstacles. In other words, q is inside a narrow corridor, if a traversal from q^{obs} to q marks a transition from obstacle to free, and a subsequent traversal in the same direction from q to q_1 marks a transition from free to obstacle.

Typicality is the generation of q_1 that in the proposed mechanism is only possible for real-valued spaces as orientations and spaces with complex constraints do not allow moving in any general random direction. Unlike the Gaussian sampling, the process cannot be decomposed into solving for every axis independently. Furthermore, since orientations and real vectors are both involved simultaneously, the distance function may be complex. An easy way to do so is to use interpolation to move from q^{obs} towards a random sample $q^{rand} \sim U(C)$ by a distance of $2d_C$. The random sample must hence be far away, at least after a distance of $2d_C$. The sample q_1 can be obtained by using the interpolation, that is Eq. (8.7).

$$q_1 \leftarrow \lambda q^{rand} + (1 - \lambda)q^{obs}, \lambda = \frac{2d_C}{d(q^{obs}, q^{rand})}, d(q^{obs}, q^{rand}) \geq 2d_C \quad (8.7)$$

Again, the mechanism of writing the interpolation equation is only for real numbered space and the equation may require adjustment for other spaces like orientations which are cyclic.

The mechanism of working of the algorithm is shown in Fig. 8.6A. The samples generated by the process and the roadmap so produced are shown in Fig. 8.6B. The map boundaries are also taken as obstacles. The algorithm is shown as Algorithm 8.8. The bridge test sampler only generates samples inside the narrow corridor and therefore is unlikely to produce good results in itself, which will require the generation of samples around obstacle boundaries and other places. Therefore, it is suggestive to hybridize the sampler with a few calls

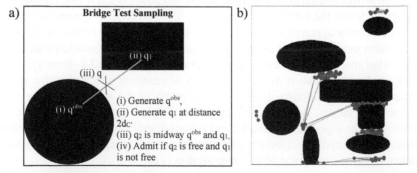

a) **Bridge Test Sampling**

(ii) q_1

(iii) q

(i) q^{obs}

(i) Generate q^{obs},
(ii) Generate q_1 at distance $2d_C$,
(iii) q_2 is midway q^{obs} and q_1,
(iv) Admit if q_2 is free and q_1 is not free

b)

FIG. 8.6 Bridge test sampling (A) concept (B) sample roadmap.

Algorithm 8.8 Bridge test sampling.

    Output: random sample $q \in C^{free}$
1 **for** *some maximum number of iterations* **do**
2     $q^{obs} \leftarrow NIL$;   ▷ find random sample in obstacle
3   **for** *some maximum number of attempts* **do**
4         $q^{obs} \sim U(C)$;
5   **if** $q^{obs} \notin C^{free}$ **then** break;
6   **if** $q^{obs} \neq NIL$ **then**
        ▷ Generate a random sample at distance $2d_C$ from
        q^{obs} in a random direction u;
7         $q_1 \leftarrow q^{obs} + 2d_C.u, u \sim U$ (possible unit directions) ;
8         $q \leftarrow 0.5q^{obs} + 0.5q_1$;
9   **if** $q \in C^{free} \wedge q_1 \notin C^{obs}$ **then** return q;

10 return $NULL$

---

for a uniform sampler, in which case the bridge test sampler generates samples in difficult regions that are connected to the overall framework using a uniform sampler.

8.3.5 Maximum clearance sampling

The complexity of the algorithm is given by $O(n \log n)$ for roadmap construction and $O(n)$ for the online query. Therefore, if the attempt is to reduce the computation time, one of the possibilities is to have a few samples only. This is especially important for cases involving challenging situations and narrow corridors, as ensuring completeness between every part of the space to every other part of the space can effectively mean injecting too many samples.

First, let us assume that the edge connections are made with every possible sample (disregarding neighbourhood function). This means a sample generated q will be connected to all other samples that lie within its visibility region. The visibility region of a sample q is defined as $Visible(q)$ and consists of all regions q' of the configuration space such that the straight line from q to q' is collision-free, or Eq. (8.8):

$$Visible(q) = \{q' : \lambda q' + (1 - \lambda)q \in C^{free} \forall 0 \leq \lambda \leq 1\} \qquad (8.8)$$

Intuitively, the larger the clearance of the sample, the farther away it is from the obstacle boundary, and therefore, larger its visibility. This effect is shown in Fig. 8.7. Completeness can be ascertained for any space if accessibility, departability, and connectivity are all ascertained. *Accessibility* is guaranteed if the union of all visibility regions covers the entire free configuration space, or $\cup_{q \in V} Visible(q) = C^{free}$. In such a case, every possible source is visible to at least one sample. Imagine having omnidirectional light bulbs at the places of the samples, while visibility region of a sample denotes the region illuminated by the bulb or sample. The union of visibility regions of all samples covering the entire configuration space means that the entire space is illuminated by at least one bulb. Now, wherever one implants a source, it will be in the illuminated region. The bulb giving illumination at the region will have a straight-line connection to the implanted source, guaranteeing accessibility. This is shown in Fig. 8.8A–D. The *departability* is the opposite of accessibility and is similarly ascertained. Hence, to reduce the number of samples one of the mechanisms is to increase the visibility of every sample, which is like maximizing clearance. The *connectivity* requires samples on top of those that meet the visibility condition; however since every sample has large visibility, there is a larger area within which the connectivity maintenance samples may lie to aid in getting all samples connected in a roadmap, minimizing the number of samples needed for connectivity.

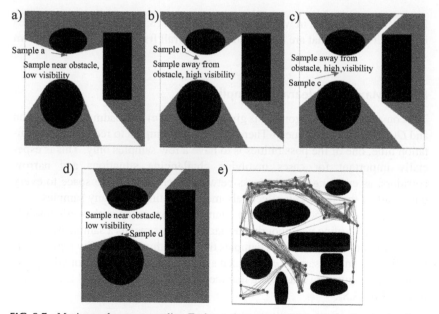

FIG. 8.7 Maximum clearance sampling. Each sample maximizes clearance, by choosing best from a few samples randomly generated. (A–D) As the clearance of the sample increases, the visibility also typically increases. *Grey* regions are invisible while *white* regions are visible to the sample shown. (E) Sample roadmap.

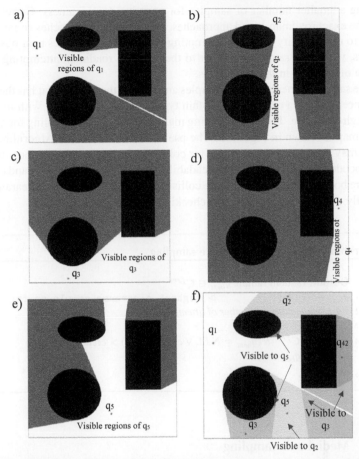

FIG. 8.8 Maximizing clearances results in the completeness property with a lesser number of samples. The visibility of five large clearance samples is shown in (A)–(E). (F) combines the visibility of all samples (choosing a lower indexed sample in case a region is visible to many samples). The entire map is covered meaning any sample generated anywhere can connect to at least one sample, ensuring accessibility and departability properties for completeness.

It may be seen that any attempt to increase the clearance does lead to a *loss of optimality* as the robot now needs to go far away from the obstacles, while the optimal paths would be close to the obstacle boundaries. However, typically the greatest problem of motion planning at the deliberative or global level in complex environments is the identification of the homotopic group of the solution or a strategy for whether to avoid obstacles in a clockwise or anticlockwise manner. If a reasonable homotopic group is taken, the paths may be locally optimized for performance within the same homotopic group, which is a relatively simpler problem. Hence the sampler has a reasonably good performance while needing a smaller number of samples for ensuring completeness. Correlating with the exact roadmap-based approaches, the sampling approaches so far were closer to the visibility roadmap attempting generation of samples near obstacle corners, while this sampler is closer to the Voronoi roadmap, attempting generation of samples in the open areas.

The sampler generates a few samples and returns the sample that has the best clearance. In such a case, the algorithm typically returns samples with reasonably high clearance. The effect of using maximum clearance sampling for a synthetic map is shown in Fig. 8.7E. The pseudocode is shown in Algorithm 8.9. Stress may be put on separate calls for collision checking and clearance. In the pseudocode, this is done for easier readability. The collision checking and clearance are both calculated by the same collision checking algorithms. Clearance is typically a by-product of collision checking.

Algorithm 8.9 Maximum clearance sampling.

Output: random sample $q_{best} \in C^{free}$

1 $q_{best} \leftarrow NIL$;   ▷ highest clearance sample so far
2 **for** *some maximum number of attempts* **do**
3      $q \sim U(C)$;
4   **if** $q \in C^{free} \wedge (q_{best} = NIL \vee clearance(q) > clearance(q_{best}))$
        then $q_{best} \leftarrow q$;
5 return q_{best}

8.3.6 Medial axis sampling

The intention behind the medial axis sampling is the same as the maximum clearance sampling, which is to generate samples with as high clearance as possible (Wilmarth et al., 1999; Holleman and Kavraki, 2000; Lien et al., 2003; Yang and Brock, 2004; Kala, 2018). However, a more intelligent mechanism is adopted to generate samples with maximum local sampling. This is done by sampling at the axis that separates two obstacles in C^{obs} with the largest distance, called the *medial axis*. In simpler terms, the attempt is to generate samples in the Voronoi graph; however, the Voronoi cannot be computed in high dimensional spaces. Therefore, the aim is to generate a sample in free space and to promote it gradually and approximately to the *maximum local clearance*. This is like the ascend of the Voronoi graphs to connect a source to the Voronoi graphs.

The sampling is, hence, intuitively simple; however, the practical implementations are not straightforward until enough approximations are made. As the first implementation, consider the generation of a sample q_1 in C^{free}.

The clearance of q_1 is computed and the closest point in the obstacle boundary in C^{free} is computed (say q_{near}). Correspondingly, the sample q is moved away from q_{near}, such that its clearance increases. As q moves away from q_{near} in the same direction, the clearance keeps increasing, until the sample is at the maximum clearance. The problem with the approach is however the requirement of computationally expensive algorithms to compute the nearest sample at the obstacle boundary, which is not a natural product of collision checking algorithms. Since the obstacle and robot geometries are known in the workspace, the same may be exploited to compute the nearest configuration at the boundary; however, it requires intensive algorithmic efforts. The concept is shown in Fig. 8.9A–B.

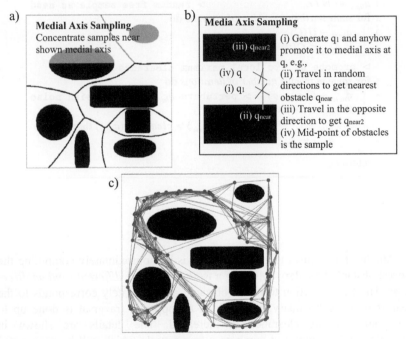

FIG. 8.9 Medial axis sampling. Sampler is used to maximize connectivity (A) concept (B) working (C) sample roadmap.

A simpler mechanism is to work in the configuration space only. So, a sample q in C^{free} is generated and is constantly perturbed. The *perturbation* can be done by using the Gaussian sampling technique for the generation of a sample near q. If the resultant sample has a better clearance, it is accepted. The subsequent attempts further perform perturbation of the original sample q. In such a mechanism, the sample q constantly travels to have a larger clearance. The implementation is shown by Algorithm 8.10. The number of attempts can be difficult to set in the implementation. A mechanism may be to set minimum improvement in a few iterations or to have stall iterations till when no improvement is made denoting a local optimum. The parameter is important since every sample will incur a computation time and therefore the effects are multiplied by the number of samples.

Algorithm 8.10 Medial axis sampling by perturbation.

Output: random sample $q_{best} \in C^{free}$

1 $q_{best} \leftarrow NIL$;   ▷ random free sample as seed
2 **for** *some maximum number of attempts* **do**
3 | $q_{best} \sim U(C)$;
4 | **if** $q_{best} \in C^{free}$ **then** break;

   ▷ promote q_{best} to local maximum clearance ;
5 **for** *some maximum number of iterations* **do**
6 | $q_{new} \sim G(q_{best}, \sigma)$;   ▷ pertubate q_{best} by a Gaussian noise
7 | **if** q_{new} *is within bounds*
   | $\wedge q_{new} \in C^{free} \wedge clearance(q_{new}) > clearance(q_{best})$ **then**
8 | | $q_{best} \leftarrow q_{new}$

9 return q_{best}

Another mechanism to get the sample is by approximately computing the closest obstacle boundary by taking a *walk in multiple different random directions*. The first direction to hit an obstacle approximately corresponds to the nearest obstacle boundary, from which an opposite traversal is done up to the point till the clearance is increased. The details are shown in Algorithm 8.11. Lines 1–4 compute a free sample which will be propagated to the medial axis eventually. Lines 5–6 sample some random directions as unit vectors. Lines 7–16 take a random walk in all random directions until one of the directions meets an obstacle denoting the nearest obstacle in direction u_{on} with last free sample is q_{near}. From q_{near} a walk is taken back towards q in the opposite direction $-u_{on}$, to get the obstacle at the other end at a distance of d. The corridor width is d and therefore the medial axis is at a distance of $d/2$ from q_{near}, which is the sample returned. The results for a synthetic map are shown in Fig. 8.9C.

Algorithm 8.11 Medial axis sampling by approximation.

      Output: random sample $q \in C^{free}$
1 $q \leftarrow NIL$;
2 **for** *some maximum number of attempts* **do**
3      $q \sim U(C)$;
4   **if** $q \in C^{free}$ **then** break;
5 **for** *o from 1 to N* **do**
6      $u_o \sim U($possible unit directions$)$
7 $q_{near} \leftarrow NIL$;   ▷ sample nearest to q in obstacle
8 $d_1 \leftarrow 0$;   ▷ distance to nearest sample in obstacle
9 **while** *true* **do**   ▷ iteratively increase d_1
10   **for** *o from 1 to N* **do**
11        $q_{new} \leftarrow q + d_1.u_o$; ▷ sample at distance d_1 in chosen
             sampled directions
12   **if** q_{new} *is not within bounds* $\vee q_{new} \notin C^{free}$ **then**
13            $q_{near} \leftarrow q + (d_1 - \epsilon).u_o, on \leftarrow o$;
14   break ;
15   **if** $q_{near} \neq NIL$ **then** break;
16    $d_1 \leftarrow d_1 + \epsilon$;
17 **for** $d = 0$ *to* ∞ *in steps* ϵ **do**
      ▷ travel in the opposite direction to reach the
      other end of the corridor ;
18    $q_{new} \leftarrow q_{near} + d.(-u_{on})$;
19   **if** q_{new} *is not within bounds* $\vee q_{new} \notin C^{free}$ **then**
20        $d \leftarrow d - \epsilon$, break ;   ▷ width of the corrdor
21 return $q_{near} + (d/2).(-u_{on})$

8.3.7 Grid sampling

The discussions over motion planning started with representing the problem as a graph by packing the configuration space with *grids* of some uniform size. A problem therein is that it is hard to fix a resolution. Moderately low resolutions are preferable for computational speedup, but risk the narrow corridors being unnoticed, shown in Fig. 8.10A.

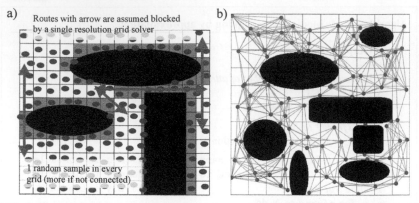

FIG. 8.10 Grid Sampling. Sampler initially places one sample per grid that are iteratively increased in perceived unconnected/difficult areas. (A) Concept and (B) sample roadmap.

The *grid sampling* (LaValle et al., 2004; Lindemann et al., 2005) mechanism solves this problem by applying a sampling framework of PRM over the grids. In grid-based sampling, even though space is filled with grids, a random sample is chosen inside every grid which becomes a vertex. The sample may neither be the centre of the grid nor be any other characteristic point in the grid. The sample should be collision-free, while parts of the grid may have obstacles and other parts may be obstacle-free. This is unlike working with graph searches when the emphasis was on the entire (or at least most part) grid to be obstacle-free. The samples inside every grid are connected to the neighbours by using edges as per PRM functionality.

It is eminent that the samples in narrow corridors will not get connected to their neighbours in such a mechanism. However, there is no restriction of every grid having only one sample. The additional time is devoted to randomly add more samples to other grids. Heuristics may be used in the selection of grids that get more samples. If a sample in a grid is sparsely connected to the samples in the neighbouring grids, it denotes a difficult region and therefore is a good candidate for adding samples inside such grids. The number of samples inside difficult grids is frequently increased to ultimately sample out and connect the narrow corridors. The initial resolution is an important parameter. However, the resolution iteratively increases by the placement of more samples inside grids, notionally subdividing the grids. The sampling done for a synthetic map is shown in Fig. 8.10B.

In higher dimensions, it is impossible to place initial grids due to the exponential number of grids required. Even though the resolution may be kept manageably low to facilitate a workable number of grids, low resolutions do not capture the essence of the algorithm. An approach of using a hash function to map the high dimensional configuration space into a low dimensional space for visualization and grid placement is an alternative way out. The grids may be placed in the hashed space which is of a lower dimension. However, this requires the hash function to be invertible so that a random sample in the hashed space can be mapped to a sample in C^{free}, which can alternatively be achieved by taking samples and accepting only those which follow the target sample distribution given by the hashed space. For visualizing the concept, imagine moving a piano from one room to the other in your house. It is convenient to look at the house from a bird's eye view and see the easy and difficult regions. This corresponds to a hash function that ignores the Z-axis and the orientations.

8.3.8 Toggle PRM

A typical sampler accepts samples in C^{free} and rejects those in C^{obs}. In *Toggle PRM* (Denny and Amato, 2011, 2013; Denny et al., 2013; Kala, 2019) both kinds of samples are accepted, and two roadmaps are made instead of just one. The first roadmap is the conventional roadmap with vertices and edges in the free space. At the same time, the second roadmap is the one that has C^{free} replaced with C^{obs}, meaning it has vertices and edges inside obstacles. Two samples are connected if the straight line between them is completely in the obstacle.

The advantage of constructing and maintaining a roadmap in C^{obs} is to boost samples inside the narrow corridor. Imagine a sample q^{obs} in C^{obs} is generated and is attempted to be connected to the nearest few samples. The situation is shown in Fig. 8.11. Consider that the attempted connectivity between q^{obs} and a q_1 in obstacle roadmap fails, as there is a free space between the two samples. However, the failed attempt discovers a free space between two obstacles, which is a narrow corridor. A sample is thus added in the middle of the narrow corridor.

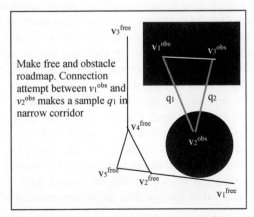

FIG. 8.11 Toggle PRM. Sampler adds new samples in narrow corridors using obstacle roadmap with vertices and edges in obstacles.

8.3.9 Hybrid sampling

All samplers discussed so far have some or the other problem. The uniform sampler performs poorly for focussing near the obstacle boundaries and narrow corridors, while the obstacle-based sampling and bridge test sampling need the assistance of uniform sampling for maintaining connectivity. The maximum clearance and medial axis sampling work on the notion of increasing clearance to get maximum connectivity in minimum time for completeness, while the obstacle and Gaussian sampling aim at optimality by generating samples near obstacle boundaries. Each sampler has its advantage and disadvantage that is complimented by some other sampler.

Hence, instead of using a single sampler, a combination of samplers is used (Hsu et al., 2005; Kala, 2016a, b; Kannan et al., 2016). Therefore, the roadmap has a *diverse set* of samples generated from different sampling methods. Instead of one sampler, several samplers $S = \{s_i\}$ is used. Each sampler has a probability P_i to be used for the generation of a sample, $0 \leq P_i \leq 1$, $\sum P_i = 1$. When a hybrid sampler is called, a sampler is selected probabilistically and the same is used for the generation of the next sample. This mechanism of hybrid sampling generates samples in proportion to their probability. Some samplers take a long time in sampling and some little. Therefore, the time distribution between the sampling algorithms will be different even if they have an equal probability of selection. Alternatively, it is also possible to distribute computation time between the samplers as per their probabilities, in which case they generate a distinct number of samples.

The probabilities of selection of samplers is an algorithm parameter that is hard to set. The parameter can be made time-varying in nature, such that the probabilities become a function of time, $P_i(t)$. This gives the ability to change the behaviour of the algorithm at different times. The rule to increase or decrease the sampling probability may be based on the type of sampler, problem, and solution complexity. As an example, the initial iterations may focus on completeness, thereafter the next iterations may focus more on optimality. The narrow corridor sampling may be initially stressed to discover all such corridors; however, in the long run, the uniform sampling may be stressed assuming the narrow corridors have been discovered. There is no single rule that can be generalized for all problems and scenarios.

A better way to handle the algorithm is to make the same parameter of probability *adaptive* in nature, such that the algorithm can determine the current performance of the sampler and adjust the probability correspondingly. The metrics that assess the success of the sampler are of vital importance. A sampler is said to be *successful* if it results in connecting two subroadmaps that are not connected or instead adds a sample that is not connected to any other sample thus exploring an otherwise unexplored area. Both aspects are discussed in detail later in the chapter under the *Visibility PRM* algorithm. If a newly generated sample is unable to connect to most of the neighbouring samples or has a lesser number of neighbouring samples, it means that the sample is in a difficult region. The difficulty is defined as the number of edges (successful connections) to the total number of attempted edge connections (typically the number of samples in the neighbourhood function). The sampler generating a sample in a difficult region must be boosted by increasing its probability. The probability of a sampler may not be reduced to an exceedingly small value. A sampler can have initial failures, which can dramatically reduce the probability after which it is unlikely to get selected. However, a sampler may be good for the later iterations. If more chances were given, the sampler would have succeeded to increase the probability. The selection may also account for the computation time which can be different from the different sampling algorithms.

Here the sampling was primarily *time-bound* where the sampling probabilities adapted to the changing time. The sampling may as well be *space-bound* (Morales et al., 2005; Kurniawati and Hsu, 2008; Rodriguez et al., 2008) where the sampling probabilities change with space. Different regions of the free configuration space have different properties and may therefore require different sampling techniques. Regions may be randomly taken by sampling. For every region identified, properties like obstacle density can be measured by sampling. If a roadmap lies in the region, the average edges, successful connection attempts, number of samples inside, etc., can also be used as features. Based on all the features, rules may be made to assign probabilities to the samplers. The generic structure of the sampler is shown in Algorithm 8.12. The concept is summarized in Fig. 8.12A. Using a few samplers, the roadmap produced by using hybrid sampling is given in Fig. 8.12B.

Algorithm 8.12 Hybrid sampling.

    Input: samples $S_1, S_2, S_3, \ldots S_N$, N is the number of samplers
    Output: random sample $q \in C^{free}$
1 **if** *time* $= 0$ **then**
2    $\lfloor\; P_i = 1/N \forall i$;   ▷ selection probability of sampler i
3 Select sampler s with probability distribution $\{S_i : P_i\}$;
4 Generate sample q using sampler s ;
5 Adapt P_i based on performance/space ;
6 **return** q ;

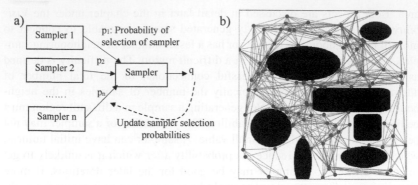

FIG. 8.12 Hybrid sampling. Different samplers are called with different probabilities that are adapted with time (A) concept and (B) sample roadmap.

8.4 Edge connection strategies

Any *local planning algorithm* can be used for edge connection between a vertex with the neighbouring vertices. A typical choice is a straight-line connection, which is the fastest local planning algorithm. Typically, any motion planning algorithm can be used for edge connectivity. There is a tradeoff between the *computation time* taken by the local planning algorithm and the *completeness* of the algorithm. Fast algorithms can miss out on very simple connections, while the algorithms ensuring some completeness can take a very long time. This section discusses some issues related to the edge connections.

8.4.1 Connecting connected components

Completeness is a major problem in sampling-based motion planning for higher dimensions. A typical roadmap produced in such scenarios is given in Fig. 8.13A. There are some connected roadmaps (components), which are all disjoint to each other. Because the individual components are disjoint, it is impossible to compute a solution when the source is near one disjoint graph (component) and the goal is near the other. It is also difficult to connect the two components since it takes more than a straight-line connection unless a few samples are placed in precise locations. The probability of such accurate sample placements is nearly zero for any sampler, which makes it extremely time-consuming to connect the components. Within that time the sampler will add a large number of samples in the other parts of the configuration space that increases the online planning time. Alternatively, consider that a stronger local search algorithm is used for every pair of neighbouring vertices, rather than a straight-line connection, it will waste too much time and the roadmap can thus have a very few samples only.

The intuition behind the *connecting connected roadmaps* (Morales et al., 2003) strategy is to identify the connected components, which are otherwise disjoint from each other after sufficient steps have been taken. Hence, the initial

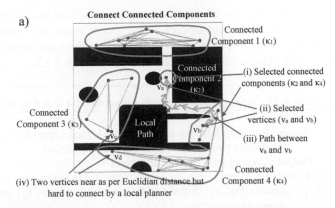

a) **Connect Connected Components**

Connected
Component 1 (κ_1)

Connected
Component 2
(κ_2)

v_a

(i) Selected connected
components (κ_2 and κ_4)

Connected
Component 3 (κ_3)

Local
Path

v_c

v_b

(ii) Selected
vertices (v_a and v_b)

(iii) Path between
v_a and v_b

v_d

Connected
Component 4 (κ_4)

(iv) Two vertices near as per Euclidian distance but
hard to connect by a local planner

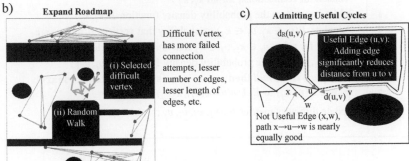

b) **Expand Roadmap**

(i) Selected
difficult
vertex

(ii) Random
Walk

Difficult Vertex
has more failed
connection
attempts, lesser
number of edges,
lesser length of
edges, etc.

c) **Admitting Useful Cycles**

$d_R(u,v)$

Useful Edge (u,v):
Adding edge
significantly reduces
distance from u to v

x
u
$d(u,v)$
v
w

Not Useful Edge (x,w),
path x→u→w is nearly
equally good

FIG. 8.13 Edge connection mechanisms (A) connect connected components strategy. Out of connected components κ_1, κ_2, κ_3, and κ_4, a pair is connected by a better local search algorithm. (B) Expand roadmap strategy. (C) Admitting useful cycles only in PRM.

roadmap may be constructed by a simple local planning algorithm; however, once it is identified that 2 or more disjoint subgraphs (components) exist, efforts are focussed towards connecting the 2 components by using a better local planning algorithm. Typical choices include Rapidly-exploring Random Trees (Chapter 9) and optimization (Chapters 15–16). The nearest pair of vertices may be chosen from the nearest components and subsequent attempts may be made to connect them by a better algorithm as shown in Fig. 8.13. However, sometimes the nearest pair of vertices in the closest components may be separated by a thin wall or similar obstacle and may not be connectible as shown in Fig. 8.13. This leads to wastage of time in attempting a connection. The same vertices will get selected every time the disjoint graph is found and continuously waste time. Hence, the selections of connected components and vertices are done probabilistically wherein the nearer are the components and vertices, the more likely are they to get selected. The failed attempts in every pair of components and vertices are also noted so as not to select the components and vertices that have incurred a lot of past failures. In such a case, even if two components are hard to connect, the algorithm consistently devotes time to make connections by selecting better candidate vertices for connections. This strategy is only applicable when the complete roadmap has at least two connected components disjoint to each other. Hence at the later iterations when the roadmap is available as one chunk, the strategy does not waste any computing time. The generic structure is shown in Algorithm 8.13.

Algorithm 8.13 Connect connected components ($RM < V,E >$, κ).

Output: Roadmap RM as a set of vertices (V) and edges (E). A
disjoint graph data structure (κ) storing vertices
Output: Updated set of vertices and edges

1 **if** $size(\kappa) \geq 2$ **then**   ▷ more than 2 disjoint sub-graphs
2   Calculate $P(\kappa_i, \kappa_j)$, probability that disjoint graphs (κ_i, κ_j) are
      chosen for connections, for all κ_i, κ_j ;
3   Select κ_i, κ_j from the probability density function $P(\kappa_i, \kappa_j)$;
4   Calculate $P(v_i \in \kappa_i, v_j \in \kappa_j)$, probability that vertices v_i and v_j are
      chosen for a few $v_i \in \kappa_i, v_j \in \kappa_j$ that are relatively close;
5   Select v_i, v_j from the probability density function $P(v_i, v_j)$;
6   **if** $\Delta(v_i, v_j)$ *is not null* **then**   ▷ stronger local search
7   add $\Delta(v_i, v_j)$ to E, $Union(\kappa(v_i), \kappa(v_j))$
8   Update difficulty metrics for $\kappa_i, \kappa_j, v_i, v_j$;

8.4.2 Disjoint forest

The assessment of the roadmaps in terms of the number of disjoint components is a simple implementation of the *disjoint forest* data structure (Cormen et al., 2009), that gives the number of disjoint components, the members of each component, and whether two nodes belong to the same subgraph (component) or not.

The disjoint set data structure is a mechanism to store elements of sets such that every two pairs of sets are disjoint or their intersection is null, say the sets are $A = \{1,4,5\}$, $B = \{2,3\}$ and $C = \{6\}$. The three sets together denote a valid disjoint set because of $A \cap B = \varnothing$, $A \cap C = \varnothing$, $B \cap C = \varnothing$. The valid operations are to make a new set with one element v, called *Make-Set(v)*; and to take a union over two sets called as $union(v_1, v_2)$. As an example, *Make-Set(7)* will make the sets as $A = \{1,4,5\}$, $B = \{2,3\}$, $C = \{6\}$ and $D = \{7\}$. Further, taking union of sets A and B ($E = A \cup B$) makes the resultant sets as, $E = \{1,4,5,2,3\}$, $C = \{6\}$ and $D = \{7\}$. Let us drop the set names A, B, C, etc. In every set, one of the elements is called the *representative element* and the set name is taken to be the name of the representative. The sets hence become, $\{\underline{1},4,5\}$, $\{\underline{2},3\}$, $\{\underline{6}\}$, and $\{\underline{7}\}$, where the name of the representative is underlined.

The implementation is done by making a tree for every set. The tree is stored in a manner that the child stores the pointer (index) of the parent (same as the parent tree of graph search). The root of the tree is called the representative of the tree.

Every element directly or indirectly points to the root or the representative. Therefore, the root of the tree (being its representative) also points to itself. The disjoint set forest is stored as an array of vertices, with each entry storing the index of the parent (same as parent tree in graph search). For non-integer vertices, a hash function may be used, and the disjoint forest is stored as a hash table.

In the specific implementation, the vertices become the elements of the set, and every pair of vertices connected by a path are in the same set. The implementation is given by Algorithm 8.14. The *Make-Set* is called whenever a new vertex is generated, which corresponds to the insert operation of a regular data structure. When a vertex is generated, it is unconnected and therefore a tree of its own with just one element that points to itself. The *rank* of a node approximates the depth of the tree.

Algorithm 8.14 Disjoint forest.

---

1 **Function** Make-Set (u):
2     $\pi(u) \leftarrow u$ ▷ parent
3     $rank(u) \leftarrow 1$ ▷ approximated depth

4 **Function** Find-Representative (u):
5   **if** $\pi(u) = u$ **then** return u
6     $\pi(u) \leftarrow$ Find-Representative($\pi(u)$)   ▷ compression (every node traversed directly points to root)
7   return $\pi(u)$

8 **Function** Union (u, v):
9     $u_{root} \leftarrow$ Find-Representative(u)
10    $v_{root} \leftarrow$ Find-Representative(v)
11   **if** $u_{root} \neq v_{root}$ **then**
          ▷ attach smaller tree under larger tree
12   **if** $rank(u_{root}) > rank(v_{root})$ **then**
13          $\pi(v_{root}) \leftarrow u_{root}$
14   **else if** $rank(v_{root}) > rank(u_{root})$ **then**
15          $\pi(u_{root}) \leftarrow v_{root}$
16   **else if** $rank(v_{root}) = rank(u_{root})$ **then**
17          $\pi(u_{root}) \leftarrow v_{root}, rank(u_{root}) \leftarrow rank(u_{root}) + 1$

If the root of two nodes (vertices) is the same, they belong to the same component and are path-connected. Finding the root or the representative is a simple traversal from the node to the root. The solution is by recursion. The root is the only node that is its parent, which acts as the base condition. If $\pi(u)$ in lines 6 and 7 is replaced with any general variable, it represents a recursion. Using $\pi(u)$ to store the representative links the node u to the root directly and thus compresses the path. When executed, the path from u to root will be flattened and every node visited will be directly connected to the root. The algorithm works by first traversing from *u* to the root. Once the root is found, the algorithm traverses the path back, linking every node to the root on the way. This is known as the *path compression* heuristic.

The other operation is a *union*, which is called when edges are added from one vertex to the other vertices. Addition of an edge (*u*,*v*) corresponds to *union* (*u*,*v*) operation in disjoint forest. In a union, the two trees are merged into one, by adding one of the trees as a child of the root of the other tree. Hence, first the two roots are computed and then the roots are *linked* to each other. The complexity of find-representative, which is the backbone for all operations, depends only on the height of the tree. Hence, the height should be as low as possible. Hence, a lower height (rank) tree is added to a higher height (rank) tree. The heuristic is called the *union by the rank* heuristic. The usual algorithm is problematic when taking the union of two elements of the same set, which does not create any difference since the union of a set with itself results in the same set. The situation is hence handled by the condition of Line 11 of Algorithm 8.14. Any traversal from any node to the root to check if they belong to the same disjoint graph (which is also done before a union is taken to avoid union of a tree with itself) also flattens the tree by making all nodes visited from the node to the root as the first-level children of the root. Note that the term rank is not exactly the same as height, rank is only an estimate of the height of the tree. When a path is compressed, the height of the tree may or may not change depending upon whether (or not) the deepest path that was compressed. In case the deepest path was compressed, the height of the tree would reduce, but it is not possible to compute the same in a small computation time. In such a case, the rank is left unchanged, which differs from the height of the tree.

The process while generating a roadmap is shown in Fig. 8.14. Fig. 8.14A shows the initial few steps in the generation of a roadmap. Fig. 8.14B shows the disjoint forest produced at every step.

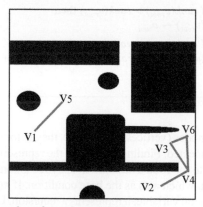

FIG. 8.14 (A) Sample roadmap for studying disjoint set.

(Continued)

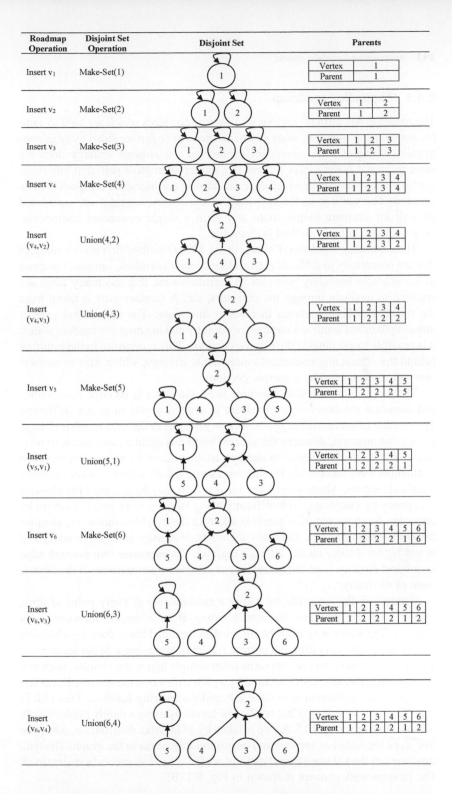

FIG. 8.14, CONT'D (B) Disjoint forest produced during the different operations in the construction of the roadmap. Every make-set or vertex insertion adds a new tree linked to itself. Every union links the root of the smaller tree to a larger tree. The tree is stored in the form of parent pointers, also displayed.

8.4.3 Expanding roadmap

The roadmap has difficult regions like narrow corridors which are hard to sample out by using sampling alone and hence need to be further aided by edge connection strategies. The *connecting connected components* strategy does the same thing but is not useful if there is a longer alternative path that otherwise connects the roadmap, in which case the narrow corridor does not qualify to be selected. The vertices at the two ends of the narrow corridor are connected through an alternate longer route and form a single connected component. Hence a more generic method is desirable.

The *expanding roadmap* (Kavraki et al., 1996a) method first selects samples that are potentially in difficult regions like narrow corridors, samples too close to the obstacle boundary, samples in corridors such that too many turns are required to navigate through the corridors, etc. A *random walk* is taken from the selected difficult samples in random directions. The aim is that even if one sample exists inside the narrow corridor, by making multiple random walks, it is possible to get outside the narrow corridor. This is opposite to the intuition behind the connecting connected components strategy, which tries to connect two regions separated by a narrow corridor.

The more difficult region a sample is in, the lesser is its edge connections and therefore the connectivity measure is a good means to assess difficulty. The number of successful edge connection attempts to the total number of edge connection attempts, denoting the success metric, is again a valid metric to take. Say a vertex v made connection attempts to the nearest-k vertices, out of which k_1 attempts were successful. The success metric is k_1/k, which if lower indicates a difficult region. Alternatively, the local region may be assessed for obstacle occupancy by samplings, with difficulty being larger if more area is covered by obstacles. Suppose out of m samples taken in the neighbourhood, m_1 samples are collision prone. This indicates the obstacle-density is approximately m_1/m and higher density means difficult regions. The measures like average edge length and distance from the nearest set of samples are also useful in the assessment of difficulty.

Random walks are made by taking a random step at every point of time, starting from the previous location. Landing at an infeasible location is not counted. The walks may even be directed in nature. This is done by choosing a difficult sample (q_1) from the current roadmap and using a better local planning algorithm to move towards some other sample (q_2) in the vicinity, such that q_1 and q_2 are not currently connected by a path with a cost close to the Euclidian cost. A good mechanism is to have a Rapidly-exploring Random Tree (RRT) grow from q_1 towards q_2 (Chapter 9). The travel by using a purely random walk is shown in Algorithm 8.15. A step is taken by a Gaussian distribution, while the free steps are added as vertices and corresponding edges in the graph. The data structures of the k-d tree (T) and disjoint set (κ) are simultaneously maintained. The random walk concept is shown in Fig. 8.13B.

Algorithm 8.15 Expand roadmap (*RM < V,E >*).

> **Input:** Roadmap RM as a set of vertices (V) and edges (E)
> **Output:** Updated set of vertices and edges
> 1 Calculate $P(v)$, probability of selection of vertex v based on difficulty metrics ;
> 2 Select vertex v from Discrete Probability Density Function $P(v)$;
> 3 *last* ← v ;   ▷ last valid free sample in walk
> 4 **for** *some number of steps* **do**
> 5   **for** *some number of attempts* **do**
> 6 $q_{new} \sim G(last, \sigma)$;   ▷ Apply pertubation at each axis
> 7   **if** *q is within bounds* $\land q \in C^{free} \land \Delta(last, q)$ *is not null* **then**
> 8   add q_{new} to V, add $\Delta(last, q)$ to E ;
> 9   add q_{new} to T ;   ▷ add q_{new} to k-d tree
> 10 $MakeSet(v, \kappa)$;   ▷ add q_{new} to disjoint set κ
> 11 $union(\kappa(q_{new}), \kappa(last))$;
> 12   *last* ← q_{new}, *break* ;

8.4.4 Generating cycle-free roadmaps

In the construction of the roadmap, the maximum efforts are incurred towards collision checking, within which the maximum efforts are incurred towards collision checking of edges. Therefore, to reduce the computation time, edge connectivity should be limited. Furthermore, a larger number of edges means a larger online query time as well. Here the focus is only on the *completeness* of the algorithm and not the optimality. For complex problems, it is hard to get even a single solution and therefore optimality must be compromised for completeness. The efforts are to get a solution to the problem quickly that the robot can follow, instead of spending a significant amount of additional time in optimality. Some optimality can be later added as postprocessing of the obtained paths.

Since optimality is not a criterion, connectivity is the only measure. Therefore, let us remove any excess edges in the graph without breaking connectivity. This reduces the graph to a *tree*. Adding any new edge would give alternate ways of adding to optimality without affecting completeness. For a tree with n vertices, there are $n - 1$ edges, which limits the edge collision checking time. The search time is proportional to the number of vertices in the path since a tree has only one path between any pair of vertices.

Therefore, the edge connections are restricted to make the roadmap a tree instead of a graph (Algorithm 8.16). The only change involved is Line 9 wherein the disjoint set data structure is referred to before making any edge connection. Only when both the vertices of a prospective edge belong to two different disjoint graphs, a connection is attempted, and on being feasible, an edge is added.

Algorithm 8.16 PRM-build roadmap without cycles.

> **Output:** Roadmap $RM < V, E >$ consisting of vertices V and edges E
> 1 $V \leftarrow \emptyset, E \leftarrow \emptyset$;
> 2 $\kappa \leftarrow \emptyset, T \leftarrow \emptyset$;
> 3   **while** *stopping criterion* **do**
> 4 $q \sim SamplingStrategy(C^{free})$;
> 5   add q to V ;
> 6   add q to $T, MakeSet(v, \kappa)$;
> 7   **for** *all* $v \in \delta(V) - q$ **do**
> 8   ▷ if Find-Representative different ;
> 9   **if** $\kappa(q) \neq \kappa(v) \wedge \Delta(q, v)$ *is not null* **then**
> 10   └ add $\Delta(q, v)$ to $E \, Union(\kappa(q), \kappa(v))$
>
> 11   return $RM < V, E >$

The roadmap without cycles can many times be far too suboptimal, wherein the robot takes a long path around all obstacles only because the cycles were restricted. This affects the optimality metric, and sometimes optimality may also be a matter of concern. The without-cycle restriction may hence be relaxed to have some *useful cycles* (Nieuwenhuisen and Overmars, 2004). This is shown in Fig. 8.13C. Not adding an edge shall poorly affect the optimality as the robot will have to unnecessarily take the longer route across the obstacle. Adding the edge significantly reduces the cost. The *utility* of an edge (u,v) is defined as the percentage improvement it can give in reducing the path cost between u and v. Consider the roadmap $RM < V, E >$ without the edge (u,v) and the path as per graph search between u and v gives a cost of $d_{RM}(u,v)$ where the subscript RM denotes the roadmap. On adding the edge, the distance would reduce to $d(u,v)$, which is the straight line distance between the two vertices. The utility (U) of the edge, which is the relative reduction on inserting the edge, is given by Eq. (8.9).

$$U(e\langle u, v \rangle) = \frac{d_{RM}(u, v) - d(u, v)}{d(u, v)} \tag{8.9}$$

If the improvement is more than a threshold, the edge may instead be added with the acceptability condition given by Eq. (8.10).

$$U(e\langle u, v \rangle) > \eta, \tag{8.10}$$

$$\frac{d_{RM}(u, v) - d(u, v)}{d(u, v)} > \eta$$

where η is the acceptance threshold.

Hence, if an alternative path is available with nearly the straight-line cost, the acceptance test will fail, while if the only alternative way is around the obstacle, the acceptance test will pass. The pseudocode is obtained by changing Line 9 in Algorithm 8.16 to

$$if \left(\kappa(u) \neq \kappa(v) \vee \frac{d_{RM}(u,v) - d(u,v)}{d(u,v)} > \eta \right) \wedge \Delta(q,v) \text{ is not null.}$$

An alternative approach is called *Roadmap Spanning*. Even if a roadmap is produced with a large number of vertices and edges, it is advisable to prune some of the edges to have smaller online query times. The more edges are pruned, the more online computation time is saved. Hence, every edge (u,v) is assessed for utility as per the same formula, $(d_{RM'}(u,v) - d_{RM}(u,v))/d_{RM}(u,v)$. Here, $d_{RM'}(u,v)$ is the distance between the vertices u and v in the roadmap RM' produced from the roadmap RM by deleting the edge (u,v). $d_{RM}(u,v)$ is the Euclidean distance between the vertices u and v in the roadmap RM. The edges which do not have a utility more than the threshold are deleted.

8.4.5 Visibility PRM

For small roadmap construction and online query times, the roadmap should have as few vertices and edges as possible. The notion is to cater to completeness with a compromise to optimality. This is important since for several problems, finding a feasible solution in the first place is extremely challenging. Once done, optimality can be induced by postprocessing or local optimization of the path. During sampling, the concept of *visibility* was used to generate samples with the highest visibility and largest clearance, which resulted in a reduction in the number of samples. This section takes forward the same ideas. The essence is to restrict samples (and edges) that will not be useful for the completeness of the algorithm, and therefore those are pruned to be inserted into the roadmap. A sample q can be inserted into the roadmap only when either of the two conditions holds (Nissoux et al., 1999):

- q is not visible to any other sample in the roadmap. In such a case, the sample q represents an unexplored area and therefore should be retained. This will help in the completeness property due to solving *accessibility* and *departability* properties, wherein a sample in a nearby region may only be visible to q. The condition is shown in Fig. 8.15A.
- q enables the connection between two connected components (subgraphs) that are not connected to each other. The condition is shown in Fig. 8.15B, wherein the sample q is of utility since otherwise there is no way to connect the two regions. This helps in the *connectivity* property of completeness and ensures that the roadmap is one tree and not disconnected trees.

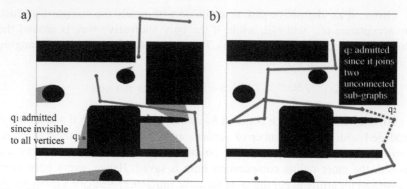

FIG. 8.15 Visibility PRM. The technique generates roadmaps with the least connectivity by only accepting samples that follow either of the two conditions shown in (A) and (B).

The Visibility PRM is *probabilistically complete*. Suppose with time all possible samples have been generated in the roadmap construction, while a very few of them shall naturally be used for the roadmap construction that qualifies for either of the two conditions. When a query is made to the roadmap specifying the source and goal, it is evident that the source and goal shall be visible to at least one sample in the roadmap. If this is not the case, a similar sample during roadmap construction would be added to the roadmap. This proves accessibility and departability. Similarly, the samples connecting the source and goal would be in the same connected component. If this is not the case, the vertices and edges making a path ensuring connectivity would have got added to the roadmap when sampled.

The working of the algorithm for a synthetic scenario is shown in Fig. 8.16. The pseudocode of the algorithm is shown in Algorithm 8.17. Lines 5–7 check if the sample generated (q) is visible to any other sample in the roadmap. In case the sample is invisible to all other samples, the sample is accepted in the roadmap as per the first criterion and added in lines 8–10. However, if the sample is visible to at least one other sample, criterion 2 is checked. If sample q results in connecting two roadmaps with vertices v_1 and v_2 respectively, which are otherwise disconnected or $\kappa(v_1) \neq \kappa(v_2)$, the sample is accepted. For the sample to be accepted q must be visible to v_1 (Eq. 8.11) and v_2 (Eq. 8.12).

$$\lambda v_1 + (1 - \lambda)q \in C^{\text{free}} \forall 0 \leq \lambda \leq 1 \tag{8.11}$$

$$\lambda v_2 + (1-\lambda)q \in C^{\text{free}} \forall 0 \le \lambda \le 1 \qquad (8.12)$$

Lines 12–17 check for the same criterion and on passing the criterion, the sample insertion is done in Lines 15–16. Lines 18–21 check if the sample was added to the roadmap, the edge connections should be made until the edge connections retain the tree structure.

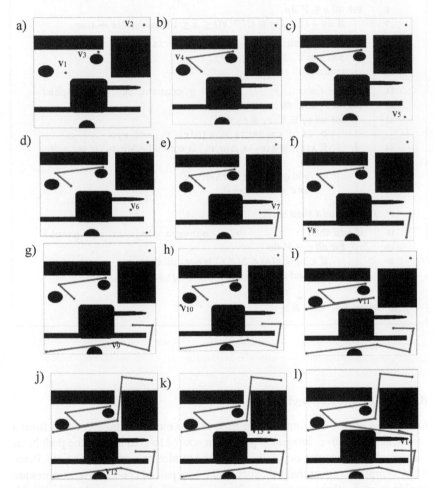

FIG. 8.16 Working of visibility PRM. Each vertex is added if it is not visible to any other vertex or enables connections of two disconnected roadmaps. (A–L) show the algorithm working steps.

Algorithm 8.17 Visibility PRM.

Output: Roadmap $RM < V, E >$ consisting of vertices V and edges E

1 $V \leftarrow \emptyset, E \leftarrow \emptyset$;

2 $\kappa \leftarrow \emptyset$;

3 **while** *stopping criterion* **do**

4      $q \sim SamplingStrategy(C^{free})$;

5      $visible \leftarrow false$;   ▷ Is q visible to any $v \in V$?

6   **for** *all* $v \in V$ **do**

7   **if** $\lambda v + (1 - \lambda)q \in C^{free} \forall 0 \le \lambda \le 1$ **then** $visible \leftarrow true$;

8   **if** $\neg visible$ **then**   ▷ q not visible to any vertex

9   add q to V, $MakeSet(v, \kappa)$;

10   continue ;

11     $add \leftarrow false$;   ▷ Does q connect two sub-graphs?

12   **for** *all* $v_1 \in V$ **do**

13   **for** *all* $v_2 \in V, v_1 \ne v_2$ **do**

            ▷ if q connects disjoint v_1 and v_2 ;

14   **if** $\kappa(v_1) \ne \kappa(v_2) \wedge \Delta(q, v_1)$ *is not null* $\wedge \Delta(q, v_2)$ *is not null* **then**

15   add q to V, $MakeSet(q, \kappa)$;

16                 $add \leftarrow true$, break ;

17   **if** add **then** break;

18   **if** add **then**   ▷ q connects disjoint sub-graphs

19   **for** *all* $v \in V$ **do**

20   **if** $\kappa(v) \ne \kappa(q) \wedge \Delta(q, v)$ *is not null* **then**

21   Add $\Delta(q, v)$ to E, $Union(\kappa(q), \kappa(v))$

22 return $RM < V, E >$;

8.4.6 Complexity revisited

Let us now assess the computation time requirements of the algorithm from a practical perspective. Imagine a path from source to goal and let the path be as far away from obstacles as possible to get the minimum computation time. Place a few samples to cover the path, such that every pair of consecutive samples are visible to each other, generally minimizing the number of samples. Every pair of samples will have a hypercube of length ρ, different for each sample, such that the motion of the sample within the hypercube does not affect the connectivity to the previous and the next sample. Let us select the most difficult sample in this path, which will have the least length ρ. The probability of generation of samples within this ball is $(\rho/R)^d$, where d is the number of dimensions and R is the length of an axis of the configuration space, assuming that every axis is normalized to have the same length and each degree of freedom has an independent axis with no constraints (other than obstacles). The expected number of samples to be drawn to get a sample in the region is $(R/\rho)^d$, which is indicative of the computation time. If there are f such difficult samples, each will have to be discovered with a probability of $(\rho/R)^{fd}$ with an expectation of $(R/\rho)^{fd}$ number of samples. Some samplers increase the probability of generation of samples, while others minimize the number of samples required to solve the problem. Practically, the computation time is exponential to the dimensionality of the space and the difficulty of the problem. This puts a practical limitation on the complexity of the robots that can be planned using the technique.

8.5 Lazy PRM

A significant amount of computation time is spent in the collision checking in the working of the PRM. The collision-checking efforts are incurred in the edge connectivity as well as during sampling. The notion of Lazy PRM (Bohlin and Kavraki, 2000, 2001; Song et al., 2001; Bohlin, 2001; Hauser, 2015) is to be *lazy* in the collision checking in the functioning of the PRM and to delay the collision checking to as late as possible. The delay in collision checking enables identifying the useful vertices and edges, and the ones which are not important since they are unlikely to be in the shortest path for a query. Only the useful vertices and edges are checked for connectivity, which prunes a significant number of vertices and edges for collision checking, thus saving a lot of overall time.

Hence, the principle is to construct a roadmap *without* incurring any collision checking. So, the vertices are added directly at random points in the roadmap. Many of them will be in the obstacles and hence should otherwise not be a part of the roadmap. Similarly, every vertex is connected to the nearest k samples directly without collision checking. This completes the roadmap construction.

Now, a query is raised supplying a source and goal. While searching the roadmap is traversed. Since the roadmap was constructed without any collision checks, it is imperative that the path returned may be invalid and infeasible. Therefore, collision checking on the path is done by doing it first on the vertices and then on the edges. Any infeasible vertex or edge invalidates the path. The infeasible vertices and edges are removed from the roadmap and a fresh search is done using a graph search algorithm. The new graph has lesser infeasible vertices and edges. The new search gives a new path. The vertices in the new path are checked for infeasibility, followed by the edges. The infeasible vertices and edges are deleted, and in the improved roadmap with lesser infeasible vertices and edges, a new search is done. This happens till a feasible path is found.

In an alternate implementation, collision checking is done when the vertices are processed in the graph search technique. First, the vertex collision checking is done leaving the lazy paradigm, while the edge connectivity is still lazy whereby the edges are assumed to be collision-free. A typical mechanism is using the A* algorithm, in which a priority queue is maintained called the fringe. First, the source is added to the fringe. At any point, a vertex is dequeued from the fringe, and its successors are inserted into the fringe if they are either unvisited or discovered at a better cost. The A* algorithm only discovers a subset of all the vertices in the search process, which are added in the fringe. Once, a candidate vertex is discovered to be added into the fringe, only then it is checked for collision. If the vertex is collision-prone, it is deleted from the roadmap including all edges associated with the vertex. When the search completes, a path is found such that no vertex lies on C^{obs}.

However, the edges are still not checked for collisions, and hence, in the next step, the lazy component is removed from the edges. The search is repeated from the source. This time edges (u,v) are checked for collision upon processing. Suppose dequeuing a vertex (u) produces a successor (v). Before inserting v into the fringe, the collision checking is done on edge (u,v). v is inserted into the fringe only if (u,v) is feasible. This produces a path that is collision-free from source to goal. In the process, a huge number of vertices and edges are saved from being checked for collision.

When a vertex or edge is checked for collision, it is *tagged* as having passed the collision check. Therefore, if any other query is raised mentioning a source and goal, the vertices and edges that have already passed the collision check are not checked for collision again. A possibility is that after all collision checking, no path is found from source to goal. In such a case, more samples and edges can be added, which again do not undergo collision checking until called during the query time or graph search.

An alternative implementation does not completely avoid collision checking, but performs collision checking with a *poor resolution*. The computation time and the accuracy of collision checking entirely depend upon the resolution. The path is computed during the online query. The path is then subjected to a high-resolution collision checking. If the path turns out to be infeasible, two things can be done. The first is to apply local perturbation to the path to make it collision-free. The second is to delete the obstacle-prone edges from the roadmap and to repeat the search, adding more samples and edges if necessary.

Sometimes it may be undesirable to delay collision checking till the query time because the lazy collision checking reduces roadmap construction time, but increases the query time. A manner is to initially construct the roadmap using a poor resolution, which is hence fast. Iteratively the resolution is increased, and the edges are rechecked for collisions. So, at every time, the edges are associated with a probability that they are collision-free, with the probability being proportional to the resolution. The resolution may as well be adapted later based on the need for the edge. If an edge connects two samples between which there are too many other edges through the roadmap, its collision checking may be deferred. However, if an edge appears to be joining two vital regions, being the only such edge, its collision checking may be promoted to be done as soon as possible. To keep matters simple, the concepts are summarized in Box 8.3.

Box 8.3 Summary of Lazy PRM.

Lazy PRM
- Construct a roadmap without collision check
- In the online query, compute a path in the roadmap
- Check vertices for collision in path (alternatively, check for collision during search)
- Check edges for collision in path (alternatively, check for collision during search)
- Remove invalid vertices/edges
- Search a new path if previous invalid
- If no valid path found, add new vertices and edges in a lazy manner
- Tag vertices/edges if they have passed collision check to avoid a recheck

Similar concept 1
- The initial roadmap is with a poor-resolution collision check
- At query carry a finer collision check
- An infeasible path is first attempted to be locally improved, then a new path is searched

Similar concept 2
- Increase resolution of collision checking during roadmap construction iteratively
- Increase resolution depending upon the need like availability of alternate edges

---

Questions

1. Consider the maps shown in Fig. 8.17. Show the working of every sampler discussed. For different scenarios also show the working of every edge connection strategy and PRM variants discussed.
2. Visibility Roadmap and Visibility Probabilistic Roadmaps have similar names and can be often confused with each other. Compare and contrast their work, clearly stating which algorithm is better in what context.
3. Obstacle-based (and related) sampling generates samples with the minimum clearance while maximum clearance (and related) sampling generates samples with the maximum clearance. Explain the pros and cons of

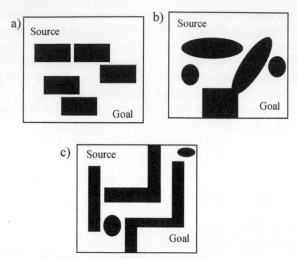

FIG. 8.17 Maps for Question 1. (A) Problem 1. (B) Problem 2. (C) Problem 3.

the two techniques while also noting in which context which sampler works better.

4. Compare and contrast between the following algorithm pairs, explicitly showing situations where each one of them is better than the other (a) - Lazy-PRM and PRM, (b) Visibility Roadmap and PRM

5. Assume edge connection in PRM is done at a low resolution to avoid wastage of time. However, the path that comes, as a result, is invalid. Suggest strategies to correct the path or roadmap for navigation.

6. Explain the metrics based on which a sample can be said to be in a difficult region. Explain the metrics by which a region in space can be called difficult for sampling. Suggest sampling strategies and edge connection mechanisms for such difficult regions.

7. PRM is executed for a few seconds for producing an offline roadmap that has several vertices and edges. The PRM is found to be too dense for online planning by popular graph search algorithms. Propose a technique to delete (a) edges only and (b) vertices and edges, from the roadmap to make a resultant roadmap that has a smaller number of vertices and edges.

8. A robot is operating in a home environment. The robot works throughout the day. For every task that is asked to be done by the robot, the robot makes a motion planning query using a PRM technique. The source and goals are added *permanently* (and not temporarily) to the roadmap, and the plan is returned and executed using the other navigation modules. Explain if this is a correct implementation. If not, suggest problems with the scheme.

9. [Programming] Assume planning for a point robot in a 2D workspace. Read the map as bitmap images. Write a program to make a roadmap using every sampler, every edge connection strategy and PRM variant discussed.

10. [Programming] Repeat the solution to the above question assuming the map as polygons instead of a bitmap.

11. [Programming] Instead of a point robot, solve for (a) a 2-link revolute manipulator, (b) a circular robot, (c) a rectangular robot, and (d) two robots as a circular disk of radius R. Solve for both polygon and bitmap workspace maps.

12. [Programming] Make a Lazy PRM implementation for a point robot in a 2-D workspace. Subject the roadmap for many randomly generated queries. Show how does the online computation time varies as the number of queries increase. Mark the vertices and edges that were never checked for collision.

13. [Programming] Go through the Open Motion Planning Library and MoveIt Library. Simulate different scenarios as available in these libraries. Benchmark different robots for different scenarios. If a scenario takes too long to compute a plan, explain what is challenging about the scenario and how can the algorithms be made more efficient in such scenarios.

14. [Programming] Install the Open Motion Planning Library from the source. Study the implementation of PRM. (a) Modify the code such that the collision checking only happens between two vertices from different connected components/disjoint sets. (b) Implement a new sampling strategy. The strategy should sample a uniform sample and an obstacle-based sample. The new sample must be inserted at a relative distance of 0.25 from the obstacle and 0.75 from the uniform. (c) Implement a rule that checks for edge connectivity with nearest k neighbours and with vertices that are not in nearest k neighbours but nearest $2k$ neighbours only if the vertex is from a different disjoint set component.

References

Amato, N.M., Bayazit, O.B., Dale, L.K., Jones, C., Vallejo, D., 1998. OBPRM: An obstacle-based PRM for 3D workspaces. In: Proceedings of the Third Workshop on the Algorithmic Foundations of Robotics. CRC Press, Boca Raton, FL, pp. 155–168.

Bohlin, R., 2001. Path planning in practice; lazy evaluation on a multi-resolution grid. In: Proceedings of the 2001 IEEE/RSJ International Conference on Intelligent Robots and Systems. IEEE, Maui, HI, pp. 49–54.

Bohlin, R., Kavraki, L.E., 2000. Path planning using lazy PRM. In: Proceedings of the 2000 IEEE International Conference on Robotics and Automation. IEEE, San Francisco, CA, pp. 521–528.

Bohlin, R., Kavraki, L.E., 2001. A randomized algorithm for robot path planning based on lazy evaluation. In: Handbook on Randomized Computing. Kluwer, Dordrecht, Netherlands, pp. 221–249.

Boor, V., Overmars, M.H., van der Stappen, A.F., 1999. The Gaussian sampling strategy for probabilistic roadmap planners. In: Proceedings of the 1999 IEEE International Conference on Robotics and Automation. IEEE, Detroit, MI, pp. 1018–1023.

Cormen, T.H., Leiserson, C.E., Rivest, R.L., Stein, C., 2009. Data structures for disjoint sets. In: Introduction to Algorithms, third ed. MIT Press, Cambridge, MA, pp. 561–586.

Denny, J., Amato, N.M., 2011. Toggle PRM: Simultaneous mapping of C-free and C-obstacle - a study in 2D. In: Proceedings of the 2011 IEEE/RSJ International Conference on Intelligent Robots and Systems. IEEE, San Francisco, CA, pp. 2632–2639.

Denny, J., Amato, N.M., 2013. Toggle PRM: A coordinated mapping of C-free and C-obstacle in arbitrary dimension. In: Algorithmic Foundations of Robotics X. Springer, Berlin, Heidelberg, pp. 297–312.

Denny, J., Shi, K., Amato, N.M., 2013. Lazy toggle PRM: A single-query approach to motion planning. In: Proceedings of the 2013 IEEE International Conference on Robotics and Automation. IEEE, Karlsruhe, pp. 2407–2414.

Hauser, K., 2015. Lazy collision checking in asymptotically-optimal motion planning. In: Proceedings of the 2015 IEEE International Conference on Robotics and Automation. IEEE, Seattle, WA, pp. 2951–2957.

Holleman, C., Kavraki, L.E., 2000. A framework for using the workspace medial axis in PRM planners. In: Proceedings of the 2000 IEEE International Conference on Robotics and Automation. IEEE, San Francisco, CA, pp. 1408–1413.

Hsu, D., Jiang, T., Reif, J., Sun, Z., 2003. The bridge test for sampling narrow passages with probabilistic roadmap planners. In: Proceedings of the 2003 IEEE International Conference on Robotics and Automation. IEEE, Taipei, Taiwan, pp. 4420–4426.

Hsu, D., Sanchez-Ante, G., Sun, Z., 2005. Hybrid PRM sampling with a cost-sensitive adaptive strategy. In: Proceedings of the 2005 IEEE International Conference on Robotics and Automation. IEEE, Barcelona, Spain, pp. 3874–3880.

Kala, R., 2016a. Homotopic roadmap generation for robot motion planning. J. Intell. Robot. Syst. 82 (3), 555–575.

Kala, R., 2016b. Homotopy conscious roadmap construction by fast sampling of narrow corridors. Appl. Intell. 45 (4), 1089–1102.

Kala, R., 2018. Increased visibility sampling for probabilistic roadmaps. In: Proceedings of the IEEE Conference on Simulation, Modelling and Programming for Autonomous Robots. IEEE, Brisbane, Australia, pp. 87–92.

Kala, R., 2019. On sampling inside obstacles for boosted sampling of narrow corridors. Comput. Intell. 35 (2), 430–447.

Kannan, A., Gupta, P., Tiwari, R., Prasad, S., Khatri, A., Kala, R., 2016. Robot motion planning using adaptive hybrid sampling in probabilistic roadmaps. Electronics 5 (2), 16.

Karaman, S., Frazzoli, E., 2011. Sampling-based algorithms for optimal motion planning. The International Journal of Robotics Research 30 (7), 846–894.

Kavraki, L.E., Svestka, P., Latombe, J.C., Overmars, M.H., 1996a. Probabilistic roadmaps for path planning in high-dimensional configuration spaces. IEEE Trans. Robot. Autom. 12 (4), 566–580.

Kavraki, L.E., Kolountzakis, M.N., Latombe, J.C., 1996b. Analysis of probabilistic roadmaps for path planning. In: Proceedings of the IEEE International Conference on Robotics and Automation. IEEE, Minneapolis, MN, pp. 3020–3025.

Kurniawati, H., Hsu, D., 2008. Workspace-based connectivity Oracle: An adaptive sampling strategy for PRM planning. In: Algorithmic Foundation of Robotics VII. Springer Tracts in Advanced Robotics. Vol. 47. Springer, Berlin, Heidelberg, pp. 35–51.

LaValle, S.M., Branicky, M.S., Lindemann, S.R., 2004. On the relationship between classical grid search and probabilistic roadmaps. The International Journal of Robotics Research 23 (7–8), 673–692.

Lien, J.M., Thomas, S.L., Amato, N.M., 2003. A general framework for sampling on the medial axis of the free space. In: Proceedings of the 2003 IEEE International Conference on Robotics and Automation. IEEE, Taipei, Taiwan, pp. 4439–4444.

Lindemann, S.R., Yershova, A., LaValle, S.M., 2005. Incremental grid sampling strategies in robotics. In: Algorithmic Foundations of Robotics. Vol. VI. Springer, Berlin, Heidelberg, pp. 313–328.

Morales, M., Rodriguez, S., Amato, N.M., 2003. Improving the connectivity of PRM roadmaps. In: Proceedings of the IEEE International Conference on Robotics and Automation. IEEE, Taipei, Taiwan, pp. 4427–4432.

Morales, M., Tapia, L., Pearce, R., Rodriguez, S., Amato, N.M., 2005. A machine learning approach for feature-sensitive motion planning. In: Algorithmic Foundations of Robotics VI, Springer Tracts in Advanced Robotics. Vol. 17. Springer, Berlin, Heidelberg, pp. 361–376.

Nieuwenhuisen, D., Overmars, M.H., 2004. Useful cycles in probabilistic roadmap graphs. In: In: *Proceedings of the IEEE International Conference on Robotics and Automation*, 446–452. IEEE, New Orleans, LA.

Nissoux, C., Simeon, T., Laumond, J., 1999. Visibility based probabilistic roadmaps. In: Proceedings of the 1999 IEEE/RSJ International Conference on Intelligent Robots and Systems. IEEE, Kyongju, South Korea, pp. 1316–1321.

Rodriguez, S., Thomas, S., Pearce, R., Amato, N.M., 2008. RESAMPL: A region-sensitive adaptive motion planner. In: Algorithmic Foundation of Robotics VII, Springer Tracts in Advanced Robotics. Vol. 47. Springer, Berlin, Heidelberg, pp. 285–300.

Song, G., Miller, S., Amato, N.M., 2001. Customizing PRM roadmaps at query time. In: Proceedings of the 2001 IEEE International Conference on Robotics and Automation. IEEE, Seoul, South Korea, pp. 1500–1505.

Sun, Z., Hsu, D., Jiang, T., Kurniawati, H., Reif, J.H., 2005. Narrow passage sampling for probabilistic roadmap planning. IEEE Trans. Robot. 21 (6), 1105–1115.

Wilmarth, S.A., Amato, N.M., Stiller, P.F., 1999. MAPRM: A probabilistic roadmap planner with sampling on the medial axis of the free space. In: Proceedings of the 1999 IEEE International Conference on Robotics and Automation. IEEE, Detroit, MI, pp. 1024–1031.

Yang, Y., Brock, O., 2004. Adapting the sampling distribution in PRM planners based on an approximated medial axis. In: Proceedings of the IEEE International Conference on Robotics and Automation. IEEE, New Orleans, LA, pp. 4405–4410.

Yeh, H., Thomas, S., Eppstein, D., Amato, N.M., 2012. UOBPRM: A uniformly distributed obstacle-based PRM. In: Proceedings of the 2012 IEEE/RSJ International Conference on Intelligent Robots and Systems. IEEE, Vilamoura, pp. 2655–2662.

Chapter 9

Rapidly-exploring random trees

9.1 Introduction

The problem of motion planning is to find a trajectory from a given source to a given goal. The Rapidly-exploring Random Trees (RRTs) (LaValle and Kuffner, 1999, 2001) use the concept of *sampling* to generate a few samples from the configuration space and search the trajectory in the configuration space represented by the samples only. Unlike probabilistic roadmaps, the sampling here is done for a *single query* only, and therefore the source and goal are known a priori. This can help in focussing the search towards areas intermediate between source and goal for searching a path, rather than diversifying into the entire configuration space. At the same time, there is now a limited computation time that can be invested because the search does not benefit from a roadmap constructed offline, and the entire planning is done online.

The representation in this chapter is in the form of a *tree* rather than a graph. Because the source is fixed, for every sample, storing the shortest path from the source to the sample is enough because of the *principle of optimality*, which states that every sub-path in the optimal path is also optimal. This means that if a good enough path from source (S) to a configuration q is known, it can be used to construct a good enough path from source to a configuration q_2, provided q comes near the shortest route from source to q_2. Graphs have at least one redundant edge causing a cycle that will not be in the shortest path from the source to any other node.

The tree closely resembles the search tree used in the *graph search approaches*. Previously, the tree was discretized in terms of both the states and the actions by taking a state resolution and an action resolution. The concept of the search tree remains the same, while the discretization is replaced by sampling. While the graph search technique pre-discretized the space into discrete cells, the sampling techniques instead will generate samples that represent the cells. The samples will typically be at any general angle, and therefore the resolution restricting the angle of turn of the robot is not true with the RRT approach. In the same vein, in RRT, the sample density can be increased at

Autonomous Mobile Robots. https://doi.org/10.1016/B978-0-443-18908-1.00001-7

409

the later iterations, especially in the difficult regions, resembling an iterative increase of resolution in graph search.

The RRT-based search can therefore be visualized as a tree initially rooted at the source that along with time expands in all directions. The search can be directed towards the goal as well. At every step, the tree grows as per the strategy adopted. When the tree reaches the goal, the search may be made to stop as a trajectory is obtained or further continued to attain a better path. The search, if continued indefinitely, will have the tree cover the entire space, thus computing a path from source to every configuration in space. The concept is hence *probabilistically complete*, meaning that the probability of finding a solution tends to one as time tends to infinity. A resolution in terms of the minimum length of the edges adds a notion of resolution completeness.

Typically, the concept is not optimal. A* algorithm had optimality because of the strategized mechanism of search, which resulted in far too many unnecessary nodes being visited before the goal was discovered optimally. Changing the search strategy resulted in a loss of optimality. The notion here is to obtain probabilistic completeness, meaning that the solution will be attempted to be discovered quickly, which may be later enhanced. Hence, a sub-optimal path may get discovered first, which may be retained by the algorithm. However, a variety of mechanisms and concepts can be adopted for adding optimality to the generic concept. As an example, if the algorithm is indefinitely run while re-wiring the best paths obtained with time, the algorithm will eventually improve on the path. This philosophy was also studied in the *Anytime A** approach. A naïve mechanism may be to search for solutions multiple times to take the best solution out of all runs. Alternatively, the best sub-paths from multiple runs may be intelligently knit to make the path.

An important metric associated with the algorithms is their ability to *parallelize*. This enables them to take advantage of the multi-core processors, clusters with multiple computing nodes, GPUs, and the cloud framework. The ability to compute in parallel is another advantage of the sampling-based approaches, both probabilistic roadmaps and RRTs.

This will be the last chapter covering the deliberative motion planning algorithms on a configuration space. With a greater understanding of the configuration space through several algorithms, we finally present two important discussions related to the construction of a distance function and the topological aspects of the configuration spaces. The discussions are summarized in Box 9.1.

Box 9.1 A summary of different algorithms.

RRT

- Single query approach, no roadmap is made
- Can strategize search and use heuristics as source and goal are known
- Optimal paths from source to any point can be conveniently stored as a tree as per the principle of optimality.
- Tree rooted at the source, progresses like cost wave of graph search, however sampling is used in place of discretizations
- Cannot be used for online planning
- Probabilistically Complete, not optimal
- Easy to parallelize

RRT Variants

Rapidly-exploring Random Trees (RRT):
Tree grows from source to all directions

Goal-Biased RRT:
Tree growth is biased towards goal

Bidirectional RRT:
Two trees are grown from source and goal. Typically, needs lesser expansions. A narrow corridor may be approached by two trees from two directions and hence easy to tackle. Not possible in control spaces.

RRT-Connect:
Expansion moves multiple steps towards a random sample, results in a faster growth. If the straight-line path to the goal is obstacle-free, taking the goal as the sample results in the complete path being computed.

RRT*:
Adds optimality. Algorithm keeps running. A new node considers other parents that give it a better cost. A new node attempts to reduce the cost of neighbouring nodes by acting as a parent and re-wiring the tree. Paths to all nodes improve with time.

Lazy RRT:
The tree is constructed without a collision check. If the path is collision prone, the collision prone nodes are deleted, and tree growth continues.

Anytime RRT:
Run RRT multiple times, every iteration targets a shorter path to the one computed in the last iteration. The cost from source and heuristics (estimated cost to goal) is used to select high-quality (lower expected cost from source to goal) samples for the tree growth. Samples that are unlikely to get costs better than the current target are pruned to be added to the tree.

Kinodynamic Planning using RRT:
Planning in control space, wherein edges represent control signals. Only forward simulation/uni-directional RRT is possible. Heuristics or local optimizations are used to grow a node towards a random node.

Continued

Box 9.1 A summary of different algorithms—cont'd

Expansive Search Trees (EST):
A random node in the tree is grown in a random direction.
Single-Query Bi-directional Lazy Collision Checking (SBL):
Implemented using EST

KPIECE:
Identifies *outer* cells in the tree and grows them *outward*. A multi-resolution grid is used to store sample density and classify interior/exterior cells. Does not require a distance function.

Sampling-based Roadmap of Trees (SRT):
RRT as a local search for PRM in the Connect Connected Components (disjoint subgraphs) methodology. Stopping criterion of RRT can make the algorithm PRM at one extreme setting and RRT at another extreme setting.

Parallel RRT

OR Parallelization:
Each processor runs an RRT, RRT that completes first is accepted.

AND Parallelization:
Each processor runs an RRT*. After all processors find a path, the best path is taken.

Fused Parallelization:
Each processor runs an RRT*. After all processors find a path, the best path is constructed by knitting paths by all processors.

Multi-Tree RRT
Fusing from multiple trees can approach narrow corridors in multiple directions/ strategies, aids optimality, enables parallel implementation

Local Trees:
Root trees at source, goal and identified difficult areas. The trees eventually merge

Rapidly-exploring Random Graphs (RRG):
A graph grows randomly, initially taking source, goal, and difficult vertices as samples, enabling cycles and redundant connections. Focus is made to join disconnected sub-graphs (especially the one with source and goal), grow at random areas and under-represented areas.

CForest:
Each processor runs an RRT*. The best path computed at one processor is added in the RRT*s of all other processors by adding nodes (if not already existent) and re-wiring. Heuristics are used to prune the tree and bound new samples.

Box 9.1 A summary of different algorithms—cont'd

Distance Functions

Planar Robot:
$$d(q_1, q_2) = \sqrt{(x_1 - x_2)^2 + (y_1 - y_2)^2 + w.AngleDiff(\theta_1 - \theta_2)^2} -\ \text{hard to set}$$
weight parameter, models rotating truck is difficult than a bot by using different weights

Any Robot:
Volume Swept - parameter-free

Manipulator:
$$d(q_1, q_2) = \sqrt{(\theta_{11} - \theta_{21})^2 + w(\theta_{12} - \theta_{22})^2}, \text{ link 1 may be harder to move than}$$
link 2 and hence a different weight

9.2 The algorithm

The intuition behind the algorithm is that the search process is maintained in the form of a tree rooted at the source. A tree has a unique path from the root to any node, which represents a candidate path from the source to the configuration represented by the node. The search is hence expanding a tree in a strategized manner, such that the tree goes outwards, starting from the source, as time proceeds. When the growing tree hits the goal, a path is found, and therefore the search can stop.

9.2.1 RRT

Let T be the tree at any general point in time. A sample q_{rand} is generated as per the sampling technique. The sample pulls the tree towards itself which results in an expansion of the tree. This *pulling* the tree towards a random sample is the main concept behind the algorithm. Like a natural tree grows in all directions along with time, the RRT grows in a random direction at every time step, where the random direction of growth is governed by q_{rand}. q_{rand} pulls the nearest node in RRT towards itself causing an expansion. Let q_{near} be the sample in T that is nearest to the generated sample q_{rand}, as shown in Eq. (9.1).

$$q_{near} = \arg \min_{q_t \in T} d(q_t, q_{rand}) \tag{9.1}$$

Here $d()$ is the distance function. The nearest sample q_{near} is attracted towards the sample q_{rand} and grows in the same direction, resulting in a new sample q_{new}. The sample q_{near} grows by taking a step S towards q_{rand}; however, it cannot jump over the sample q_{rand}. Therefore, the step moved is given by

$\min(S, d(q_{near}, q_{rand}))$. This generates a new sample in the direction from q_{near} to q_{rand}, given by Eq. (9.2):

$$q_{new} = q_{near} + \min(S, d(q_{near}, q_{rand})).u(q_{rand} - q_{near}) \tag{9.2}$$

where $u(q_{rand} - q_{near})$ is the unit vector in the direction of $q_{rand} - q_{near}$. The equation may be re-written in the same format as used throughout the book in the form of an interpolation of a straight line, where q_{new} is at a relative distance of $\frac{\min(S, d(q_{near}, q_{rand}))}{d(q_{near}, q_{rand})}$ from q_{near}. Therefore, q_{new} is given by Eqs (9.3), (9.4).

$$\lambda = \frac{\min(S, d(q_{near}, q_{rand}))}{d(q_{near}, q_{rand})} \tag{9.3}$$

$$q_{new} = (1 - \lambda)q_{near} + \lambda q_{new} \tag{9.4}$$

As previously, the interpolation is for real spaces only and will need adaptation to deal with angles and other complex spaces. The sample q_{new} is accepted as a node of the tree only if a collision-free path between q_{near} and q_{new} exists or $\Delta(q_{near}, q_{new})$ is not null. Here Δ is the local collision checking algorithm that is typically a straight-line collision checking. If the sample is accepted, (q_{near}, q_{new}) is added as an edge of the tree, and q_{near} becomes the parent of the node q_{new} or $\pi(q_{new}) = q_{near}$. Here, π is the data structure used to store all the parents. Further, $E(q_{new})$ is taken as the trajectory that connects q_{near} (parent) to q_{new}. The data structures π and E together constitute the tree (T). Typically, in trees, the parent stores the pointers to the children; however, like the graph search techniques, the children store the pointer to the parent. Like the graph search, the pointer here may even refer to the index of the element in the data structure used to store the vertices.

The notations are shown in Fig. 9.1A. The tree starts from the source and slowly fills up the complete space. When the search reaches near enough to the goal, the search process is stopped. It may be seen that the tree may never land up exactly at the goal as the search is not discretized. The search will be stopped when the tree reaches near enough to the goal. A threshold of the step size (S) is typically acceptable. A sample tree produced because of the same search principles is shown in Fig. 9.1. The pseudo-code is given in Algorithm 9.1 for RRT expansion and Algorithm 9.2 to get the path after the search. Lines 1–4 initialize a tree at the root. Lines 6–9 generate a new sample, which is tested for feasibility in line 10 and added to the tree in lines 11–12. Line 13 is the goal checking criterion.

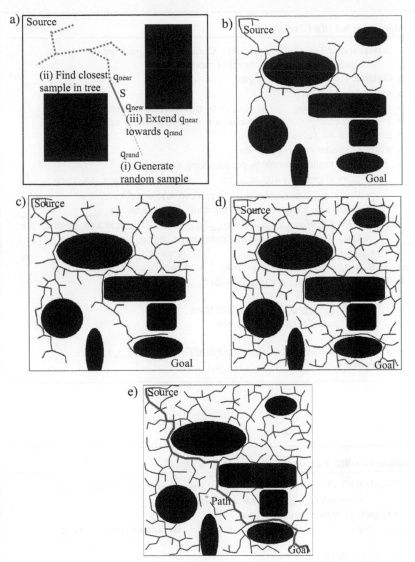

FIG. 9.1   RRT (A) Notations; (B)–(E) Expansion. The tree grows in all directions randomly and stops when it meets the goal.

Algorithm 9.1 RRT (*source,G*).

Input: source (*source*) and goal (*G*). Configuration space definitions
including a collision checking mechanism

Output: Path from source to goal

1 $\pi(source) \leftarrow NIL$;   ▷ parent
2 $E(source) \leftarrow NIL$;   ▷ path from parent to node
3 $T \leftarrow source$;   ▷ tree as a collection of vertices
4 $pathFound \leftarrow false$;
5 **while** *maximum running time* **do**
6      $q_{rand} \leftarrow$ random sample as per sampling strategy ;
7      $q_{near} \leftarrow argmin_{q_t \in T} d(q_t, q_{rand})$;   ▷ closest node in tree
    ▷ Grow q_{near} towards q_{rand} by step size (*S*). *S* is
    inversely proportional to the clearance at q_{near} ;
8      $\lambda \leftarrow min(S, d(q_{near}, q_{rand}))/d(q_{near}, q_{rand})$;   ▷ relative step
    size, threshold avoids overshooting q_{rand}
9      $q_{new} \leftarrow (1 - \lambda)q_{near} + \lambda q_{new}$;   ▷ grow q_{near} towards q_{rand}
10   **if** $\Delta(q_{near}, q_{new})$ *is not null* **then**   ▷ check local connection
11         $\pi(q_{new}) \leftarrow q_{near}, E(q_{new}) \leftarrow \Delta(q_{near}, q_{new})$;
12   add q_{new} to T ;
13   **if** $d(q_{new}, goal) < Threshold$ **then**
14             $pathFound \leftarrow true$, break

15 **if** *pathFound* **then** return *calculatePath*($T < \pi, E >, q_{new}$);
16 return NIL

Algorithm 9.2 CalculatePath($T < \Pi, E >,$ *goal*).

Input: RRT ($T < \pi, E >$) in the form of parents (π), edges from parent
to vertex (E), goal vertex in tree (*goal*)

Output: Path from source to goal

1 $q \leftarrow goal$;   ▷ last node to be added in the path
2 $\tau \leftarrow \emptyset$;   ▷ path
3 **while** $q \neq NIL$ **do**
4   add $E(q)$ to the beginning of τ ;
5      $q \leftarrow \pi(q)$;
6 return τ ;

Line 7 calculates the nearest neighbour to q_{rand}, which is done using the k-d tree data structure using which the search becomes $O(\log n)$, where n is the number of vertices in the tree. Hence every vertex is simultaneously added in the k-d tree data structure as well, which is not shown in the algorithm for clarity. The overall complexity in terms of the number of samples in the tree is $O(n \log n)$.

9.2.2   Goal biased RRT

A problem with the algorithm is that it explores the free space completely and randomly in all directions. This can significantly increase the search time. It may be wise to be more directed towards the goal, which results in quicker growth towards the goal and thereafter a quicker goal discovery. This was observed in the graph search techniques as well wherein the A* algorithm directed the search efforts towards the goal in contrast to the Uniform Cost Search.

To *bias* the search towards the goal, sample the goal node instead of a random sample. In the case of multiple goals or a big goal region, any sample satisfying the goal criterion may be sampled. The goal hence attracts the tree and results in the growth of the tree towards the goal. However, the problem here is that of a failure. Imagine the nearest node to the goal in the tree (q_{near}) and a nearby obstacle in the straight-line distance to the goal. The same goal will get sampled every time and the same q_{near} will get selected at every instant, which will attempt the same move to produce the same q_{new}, which will fail every time. During the heuristic search in graphs, this could be avoided because the number of states was discrete, and the same node was stopped from getting produced every time. Now since the states are continuous, the repetition in the generation of a state cannot be checked. The other problem associated with the approach is that of optimality. Stressing too much on heuristics can make the algorithm take an initial straight motion to the goal which may later become too long a path. Even though RRT does not guarantee optimality, some emphasis on exploring enough around different areas should be made to continuously discover diverse paths eventually leading to the goal.

Therefore, the efforts to be directed towards the goal are only made for a small factor of times called the goal bias or simply *bias*. In all other times, the random sample is used to grow the tree. The general concept is shown in Fig. 9.2A. The tree produced because of this approach is shown in Fig. 9.2. The modified algorithm is given in Algorithm 9.3. The threshold is suppressed from the goal condition since now eventually the goal will be sampled and connected, and thus the exact goal will be reached. The path is the branch from the root to the goal node, shown in Fig. 9.2. The path will have too many unnecessary turns and will need to be smoothened, locally optimized or post-processed.

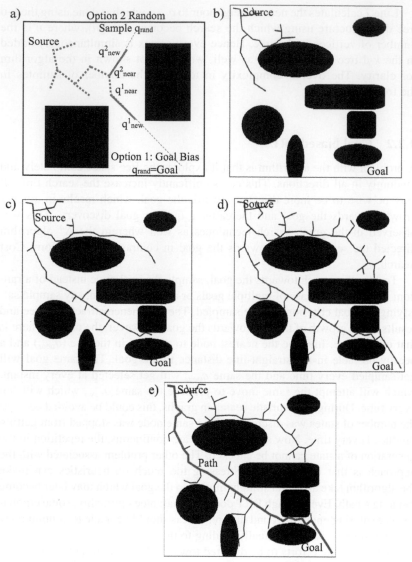

FIG. 9.2 Goal-biased RRT: (A) Notations. The random sample (q_{rand}) is chosen as the goal for some *bias* proportion of time (option 1 in the figure). (B)–(D) Expansion.

Algorithm 9.3 Goal-biased RRT (*source,G*).

Input: Source (*source*) and set of goals (*G*). Configuration space
        definitions including a collision checking mechanism
Output: Path from source to goal

1 $\pi(source) \leftarrow NIL, E(source) \leftarrow NIL$;
2 $T \leftarrow source$;
3 $pathFound \leftarrow false$;
4 **while** *maximum running time* **do**
5    $r \sim U[0,1]$;
6   **if** $r < bias$ **then** $q_{rand} \leftarrow$ sample from G;   ▷ goal bias
7   **else** $q_{rand} \leftarrow$ random sample as per sampling strategy;
8    $q_{near} \leftarrow argmin_{q_t \in T} d(q_t, q_{rand})$;
9    $\lambda \leftarrow min(S, d(q_{near}, q_{rand})) / d(q_{near}, q_{rand})$;
10   $q_{new} \leftarrow (1 - \lambda)q_{near} + \lambda q_{new}$;
11   **if** $\Delta(q_{near}, q_{new})$ *is not null* **then**
12     $\pi(qnew) \leftarrow q_{near}, E(q_{new}) \leftarrow \Delta(q_{near}, q_{new})$;
13   add q_{new} to T ;
14   **if** $q_{new} \in G$ **then** $pathFound \leftarrow true$, break;

15 **if** $pathFound$ **then** return $calculatePath(T < \pi, E >, q_{new})$;
16 return NIL ;

9.2.3 Parameters and optimality

The first major factor introduced in the algorithm is the *step size* (*S*). The
effect of the parameter is like the *resolution* parameter in different
approaches. A larger value of the step size (poorer resolution) means that the
algorithm makes the tree take big steps and thus advance very quickly in different directions including towards the goal. The tree itself has a lesser number
of nodes which results in a speedup. However, larger factors of step size restrict
the motion of the robot, which cannot take fine turns. Fine turns can only be
made by small step sizes. This makes it difficult to enter narrow corridors
and similar places requiring high dexterity. Imagine having to move in a narrow
region, but you can only make big steps. Entering the region would require that
you are perfectly oriented, which is hard to happen randomly. This is shown
in Fig. 9.3.

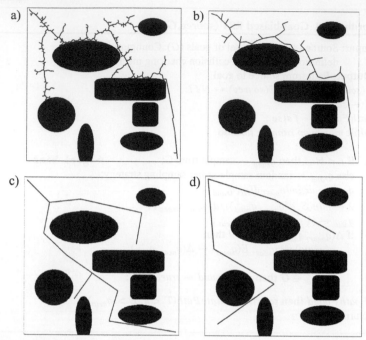

FIG. 9.3 Effect of increasing step size is shown in (A–D). Small step size makes the tree grow slowly, however can make sharp turns. In (D), the step size is too large, and the algorithm fails.

This motivates the need to have an *adaptive step size*, like the notion of having multi-resolution cell-decomposition. The problem with a large step size was to easily manoeuvre around the obstacles, otherwise, a large step size was better for a better computation time. The step size is large away from the obstacle and small near the obstacle. In other words, the step size is inversely proportional to the clearance ($S \propto 1/clearance(q_{near})$). This makes the algorithm reduce the step size to enter a narrow corridor or other regions requiring high dexterity. This is like the notion wherein a car driver drives at high speeds on roads, while reduces speeds to turn at intersections, or to enter narrow garage gates.

The other factor is *bias*, which is used to give some bias to the search algorithm to be directed towards the goal. A very high value will certainly make the algorithm initially march quickly towards the goal, however, higher iterations will result in too many failures, since the nearest sample in the tree may be collision prone. This will waste significant time. In general, optimality will also be affected. At the same time, a low value of the factor will lead to exploring virtually the entire configuration space. This is also wasteful as the tree is not directed towards the goal. A low bias leads to higher *exploration*, wherein a significant portion of space is explored with the cost of time; while a high bias leads to high *exploitation*, wherein the algorithm attempts to exploit the straight line to the goal. The effect is shown in Fig. 9.4.

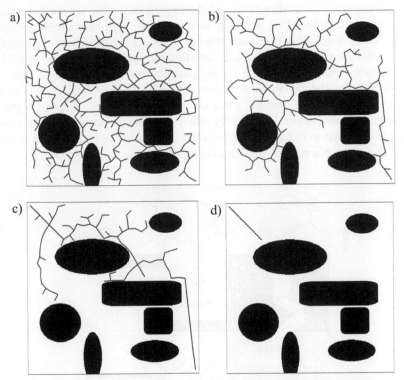

FIG. 9.4 Effect of increasing bias is shown in (A–D). As bias is increased, the tree becomes more focused towards the goal resulting in a smaller number of expansions. In (D), bias is 1 and the same nearest node is attempted which fails. Larger bias may result in too many failures if the nearest sample to the goal cannot be extended due to obstacles.

The algorithm is not optimal as there is no guarantee that the search tree produces optimal paths first. It may be inquisitive to state that the A^* algorithm was optimal unlike heuristic search as just enough bias was given to the Uniform Cost Search. The A^* algorithm searched through all the discrete nodes giving optimality, while the RRT uses sampling, and the absence of a bias still does not guarantee optimality. The sampling faces a severe bottleneck that it becomes biased towards open areas, which is called the *Voronoi bias*. The probability of generation of a sample as per the Uniform sampling strategy is the volume of the region upon the volume of the total space. Open areas have more free volume and hence more likely to get samples in contrast to narrow areas. Therefore, the tree grows quickly towards open areas in contrast to narrow regions.

Suppose there are two paths (homotopic groups of paths) from source to goal, one longer with open areas and the other shorter with the obstacle-congested area. If the sampling were bias-free, the tree would have grown uniformly in all directions and the branch following the shorter route would have travelled first resulting in the shorter path (homotopic group) being discovered. However, the number of samples generated in longer path/wider area space is more, and therefore the growth of the tree towards this region is more. Hence the longer path instead gets discovered due to Voronoi bias. The situation is shown in Fig. 9.5. The Voronoi bias is a source of an immense problem when using RRT. Typically, if the path goes through a narrow region, it is likely that a significant part of the free configuration space gets discovered because the narrow corridor has a small rate of growth because of the Voronoi bias.

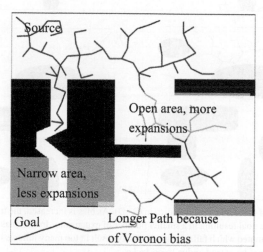

FIG. 9.5 Voronoi bias in RRT. Expansion happens faster in wide-open areas and longer path going through wide-open areas will be returned.

9.3 RRT variants

The basic algorithm can perform well; however, a few useful modifications can help in making the approach better towards the generation of results faster and being more favourable to handle special situations.

9.3.1 Bi-directional RRT

The *bi-directional RRT* (Lavalle and Kuffner, 2000; Kuffner and LaValle, 2000) consists of two trees instead of just one. One of the trees is rooted at the source and grows towards the goal. The other tree is rooted at the goal and grows towards the source. The two trees hence grow (as a bias) towards opposite ends. The search stops when the two trees meet anywhere mid-way. The search technique is called bi-directional since the search happens in two directions by the two trees in parallel. In the algorithmic implementation, both RRT expansions happen one after the other in a loop. Since one of the trees searches backward, the usual constraints of backward search hold including that the approach cannot be used for control spaces where the robots cannot be back simulated, the goals should be explicitly known and finite in numbers, the map should not change by known dynamics, etc.

Let the two trees be T_{source} which is rooted at the source (*source*) and T_{goal}, which is rooted in the goal (*G*). First, a sampled configuration is used to attract the nearest node in T_{source}. Then another sampled configuration is used to attract the nearest node in T_{goal}. Sampling goal is another interesting phenomenon in bi-directional search. The tree rooted at the source (T_{source}) can take any node in the other tree as the goal ($q_{rand} \in T_{goal}$) and vice versa. This is because the intent is only to merge the trees and not to fully grow the two trees. As soon as both trees merge at a sample q_{merge}, the search stops because a path from *source* to q_{merge} is known in T_{source} as well as a path from q_{merge} to the goal is known in T_{goal}. The two paths can thus be merged to produce a path from source to goal.

The concept is shown in Fig. 9.6A. The algorithm is shown in Algorithm 9.4. The working for a synthetic map is shown in Fig. 9.6.

The algorithm is effective from an analysis point of view. Assume that the Voronoi bias is limited and the bias factor is kept low (worst case). Suppose the optimal path has a length of C^* and step size is S. So, the number of RRT steps required in the optimal path is $\frac{C^*}{S}$. A unidirectional RRT will move in the correct way towards the goal for $\frac{C^*}{S}$ steps, however, while doing so will also move in all other directions for roughly the same steps resulting in a total of $O\left(\left(\frac{C^*}{S}\right)^d\right)$ samples, where d is the dimensionality of the space. As an example, in 2D, all samples will lie in a circle of radius $\frac{C^*}{S}$. However, in a bi-directional search, each tree will explore potentially in all directions producing paths of length $\frac{C^*}{2S}$ for the first tree and $\frac{C^*}{2S}$ for the second tree. After this time, the mid-point of the optimal path will be discovered in both search trees and the expansions can stop. This will mean samples of the order of $O\left(\left(\frac{C^*}{2S}\right)^d\right)$, which means a significant reduction.

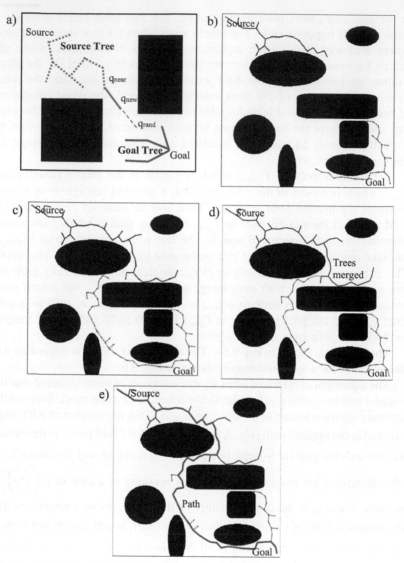

FIG. 9.6 Bi-directional RRT. (A) Notations. Two trees are grown rooted at source and goal respectively. Two trees reduce computation time and approach a narrow corridor from different sides, for which any tree succeeding tackles the narrow corridor. (B)–(E) Expansion.

Algorithm 9.4 Bi-directional RRT (*source,G*).

Input: Source (*source*) and Goal (*G*). Configuration space definitions
   including a collision checking mechanism
Output: Path from source to goal

1 $\pi_{source}(source) \leftarrow NIL, T_{source} \leftarrow source$;   ▷ tree from source
2 $E_{source}(source) \leftarrow NIL$;
3 $\pi_{goal}(G) \leftarrow NIL, T_{goal} \leftarrow G$;   ▷ tree from goal
4 $E_{goal}(G) \leftarrow NIL$;
5 $pathFound \leftarrow false$;
6 $swapped \leftarrow false$;   ▷ source and goal trees swapped?
7 **while** *maximum running time* **do**
8 $\quad$ $r \sim U[0, 1]$;
9 $\quad$ **if** $r < bias$ **then** $q_{rand} \leftarrow$ sample from T_{goal};
10 $\quad$ **else** $q_{rand} \leftarrow$ random sample as per sampling strategy;
11 $\quad$ $q_{near} \leftarrow argmin_{q_t \in T_{source}} d(q_t, q_{rand})$;
12 $\quad$ $\lambda \leftarrow min(S, d(q_{near}, q_{rand}))/d(q_{near}, q_{rand})$;
13 $\quad$ $q_{new} \leftarrow (1 - \lambda)q_{near} + \lambda q_{new}$;
14 $\quad$ **if** $\Delta(q_{near}, q_{new})$ *is not null* **then**
15 $\quad\quad$ $\pi(q_{new}) \leftarrow q_{near}, E_{source}(q_{new}) \leftarrow \Delta(q_{near}, q_{new})$;
16 $\quad\quad$ add q_{new} to T_{source} ;
17 $\quad\quad$ **if** $q_{new} \in T_{goal}$ **then**   ▷ q_{new} exists in both trees
18 $\quad\quad\quad$ $pathFound \leftarrow true, q_{merge} \leftarrow q_{new}$, break

$\quad$ ▷ to expand the goal tree, swap the trees
$\quad$ $swap(T_{source} < \pi_{source}, E_{source} >, T_{goal} < \pi_{goal}, E_{goal} >)$;
19 $\quad$ $swapped \leftarrow \neg swapped$;
20 **if** $swapped$ **then**
$\quad$ $swap(T_{source} < \pi_{source}, E_{source} >, T_{goal} < \pi_{goal}, E_{goal} >)$;
21 **if** $pathFound$ **then**
22 $\quad$ $\tau_1 \leftarrow calculatePath(T_{source} < \pi_{source}, E_{source} >, q_{merge})$;
23 $\quad$ $\tau_2 \leftarrow calculatePath(T_{goal} < \pi_{goal}, E_{goal} >, q_{merge})$;
24 $\quad$ return $join(\tau_1, reverse(\tau_2))$;
25 return NIL;

The Voronoi bias adds another essence behind the use of 2 trees. It may be difficult to enter a narrow region from one end, but it may be easy to *enter* the same narrow region from the other end of the goal and to *exit* through the first end to meet the tree rooted at the source. The problem is simplified, since either of the trees needs to succeed in entering/exiting the narrow region, which makes the algorithm successful.

9.3.2 RRT-connect

In the RRT algorithm, a sample is generated by the sampling technique, however only one step is made towards the sample. This significantly slows down the search as the tree expands by many single steps in different directions. The RRT-Connect algorithm (Lavalle and Kuffner, 2000; Kuffner and LaValle, 2000) accelerates this process by expanding the tree by *multiple steps*, if feasible, at a single random sample. Therefore, the expansion process may now be viewed as taking multiple steps towards the generated sample to accelerate directed growth.

Let T be the tree and q_{rand} be the random sample generated. Let the nearest sample in T be q_{near}. An RRT approach would just take a single step of size S towards q_{rand}. However, the RRT-Connect approach keeps walking in steps of S till either the sample is reached, or an obstacle is reached. In both cases, there is no basis for further expansion and therefore the progression must stop. The sample may even be the goal in which case the progression may happen till the goal, in which case the search stops. This is an interesting property. Many times, the algorithms are given rather simple maps with simple obstacles and design principles may need complex deliberate algorithms for problem-solving. The ease by which the complex algorithms can handle simple maps is also of interest in motion planning. The RRT and PRM algorithms may generate many nodes in such cases. In the RRT-Connect algorithm, when the nearest sample to the goal has a straight-line collision-free connection to the goal, the search immediately stops. This is because of the goal bias when the goal will be sampled as q_{rand}, and the nearest node will start progressing towards the goal till the goal is connected in the same move.

The concept behind the algorithm is shown in Fig. 9.7A. The algorithm as a pseudo-code is shown in Algorithm 9.5. The working on a synthetic map is shown in Fig. 9.7.

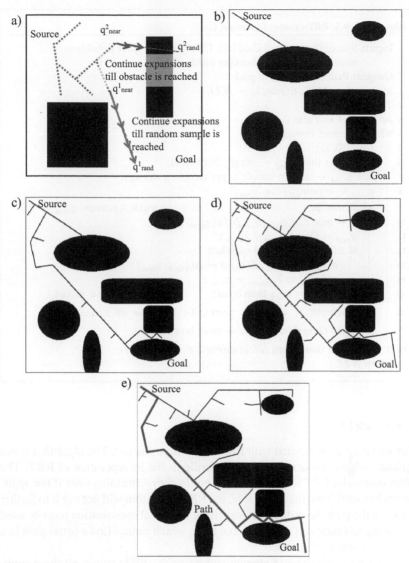

FIG. 9.7 RRT-Connect. (A) Notations. (B–D) Working. Multiple steps are made and hence longer edges. (E) The algorithm stops on finding a path.

Algorithm 9.5 RRT-connect (*source,G*).

Input: Source (*source*) and Goal (*G*). Configuration space definitions
 including a collision checking mechanism
Output: Path from source to goal
1 $\pi(source) \leftarrow NIL, E(source) \leftarrow NIL$;
2 $T \leftarrow source$;
3 $pathFound \leftarrow false$;
4 **while** *maximum running time* **do**
5    $r \sim U[0,1]$;
6   **if** $r < bias$ **then** $q_{rand} \leftarrow$ sample from Goal;
7   **else** $q_{rand} \leftarrow$ random sample as per sampling strategy;
8    $q_{near} \leftarrow argmin_{q_t \in T} d(q_t, q_{rand})$;
9   **while** *true* **do** $\triangleright$ walk towards q_{rand}
10     $\lambda \leftarrow min(S, d(q_{near}, q_{rand}))/d(q_{near}, q_{rand})$;
11     $q_{new} \leftarrow (1 - \lambda)q_{near} + \lambda q_{new}$;
12   **if** $\Delta(q_{near}, q_{new})$ *is not null* **then**
13      $\pi(q_{new}) \leftarrow q_{near}, E(q_{new}) \leftarrow \Delta(q_{near}, q_{new})$;
14   add q_{new} to $T, q_{near} \leftarrow q_{new}$;
15   **if** $q_{new} = q_{rand}$ **then** break; $\triangleright$ walk end point
16   **else** break; $\triangleright$ walk not possible because of collision
17   **if** $q_{new} = G$ **then** $pathFound \leftarrow true$, break;
18 **if** $pathFound$ **then** return $calculatePath(T < \pi, E >, G)$;
19 return NIL ;

9.3.3   RRT*

The major issue associated with RRT so far is *optimality*. The algorithm is not optimal, which is otherwise a major hurdle in the incorporation of RRT. The other issue with RRT so far is that it is not *iterative*, meaning even if the application has additional time available, the RRT algorithm will not use it to further improve the path. So, a different algorithm like local optimization must be used for taking advantage of the additional time, which cannot find a better path in a different homotopic group.

The *RRT* algorithm* (Karaman and Frazzoli, 2011) solves all these problems associated with the RRT algorithm. The RRT* algorithm continues the search process even after a path has been found by continuing to generate new samples and using them to extend the tree. However, in the process, it is possible to get a better path to any node than the one used previously. In such a case, the algorithm accepts the new path. The path improves with time. By running the algorithm indefinitely, the best path can be found, and hence the algorithm is probabilistically optimal.

Suppose $C(q)$ is the cost from the source for a tree node q. The cost is computable easily since there is only one path from the root to any node in a tree and therefore the cost is given by the cost of the parent and additional cost to go from parent to the node, or Eq. (9.5).

$$C(q) = \begin{cases} C(\pi(q)) + d(\pi(q), \, q) & if \; q \neq source \\ 0 & if \; q = source \end{cases} \qquad (9.5)$$

The costs are computed and maintained during the insertion of a node in the tree. Consider that generation of a random sample q_{rand} results in the nearest node q_{near} and taking a step towards q_{rand} results in a sample q_{new}; which can be connected as a straight-line path $\Delta(q_{near}, q_{new})$ and is added to the tree. The RRT* algorithm has the responsibility of improvement of the tree and hence an additional step is performed. All samples in the tree (T) near q_{new} are scanned for possible improvements. Let the set of candidate samples for improvements be all samples in the tree at a maximum distance of R from q_{new}, given by Eq. (9.6).

$$\delta(q_{new}) = \{q \in T : d(q_{new}, q) \leq R\} \qquad (9.6)$$

Here R is an acceptance threshold of neighbours. Consider a candidate sample $q_C \in \delta(q_{new})$. If the straight-line path from q_{new} to q_C is collision-free, it means that q_C has two candidate paths, one as previously through the current parent $\pi(q_C)$ with cost $C(q_C)$ and the other going from source through q_{near}, q_{new} to q_C with cost $C(q_{new}) + d(q_{new}, q_C)$. The better of the two costs is taken. The notations are shown in Fig. 9.8.

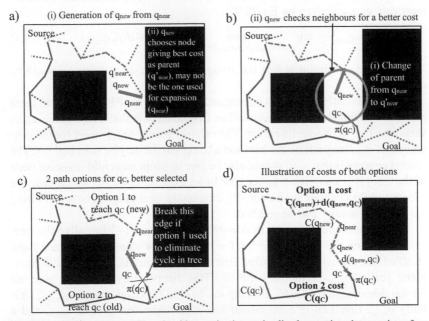

FIG. 9.8 RRT* Notations: The algorithm maintains optimality by continued expansion after reaching the goal and admitting better paths. (A) q_{new} searches for a better parent q'_{near} that offers q_{new} a better cost from source (B) q_{new} changes its original parent (q_{near}) to q'_{near} and q_{new} checks if it can give better paths for neighbouring nodes. (C) and (D) Admitting a better path to a node q_C and re-wiring the tree. $C(q)$ denotes the cost from source in the tree.

If the latter cost is better, it cannot be directly accepted since the addition of an edge between q_{new} and q_C will make a cycle between source, q_{new}, and q_C and the tree will become a graph. Therefore, the tree is instead *re-wired* by a change of parent, such that the tree properties are intact. This happens by removing the old edge (between $\pi(q_C)$ and q_C) and making the new one (between q_{new} and q_C). Alternatively, q_C changes its parent from the old one to the new one. This is shown in Fig. 9.8B. This was previously done in the graph-search approaches where nodes often changed their parents on finding a better cost through some other parent. Note that as q_C changes its parent, q_C itself has descendants (including direct children) whose costs are computed by $C(q_C)$. A change of parent of q_C changes $C(q_C)$ and all decedents must re-compute the costs.

Another change needs to be made because the notion of optimality is added. The RRT algorithm generates a path from source to q_{new} through q_{near}, however, when there are enough samples, there is no guarantee that q_{near} is the best intermediate node to have the shortest costs for q_{new}. Imagine after many iterations, the configuration space is reasonably densely filled with samples. q_{new} may as well be behind (opposite to the direction from q_{near} to goal), in such a case there may be a nearby sample that better connects q_{new} to the tree giving a better cost. Hence q_{near} is seen as a candidate to connect q_{new}, while better candidates are also searched for in the tree in the neighbourhood $\delta(q_{new})$. If a better candidate is found, the same is used instead. The candidate is given by the one that gives the least cost to q_{new}, given by Eq. (9.7).

$$q'_{near} = \arg \min_{q \in \delta(q_{new}):\Delta(q, q_{new}) \text{ is not null}} C(q) + d(q, q_{new}) \tag{9.7}$$

An added constraint is that the straight line from q'_{near} to q_{new} should be collision-free.

RRT* is intuitively easy to understand. On top of the RRT algorithm, there are only two fundamental concepts. The first concept is that a new child (q_{new}) should be free to choose its parent. The original parent (q_{near}) used for the generation of q_{new} may not be the best and a better parent may be searched for. The second concept is that q_{new} should also be free to adopt any child (existing node of the tree) if it can give a better cost to an existing child, in which case the adopted child changes its parent. By constantly changing parents (re-wiring the tree), optimality can be reached.

The pseudo-code is given in Algorithm 9.6. Lines 1–3 initialize the tree. Lines 5–11 are the same as the RRT wherein a new sample is generated and checked for collision. Thereafter first a better candidate for q_{near} is computed (called q'_{near}) in lines 12–19. This happens by iterating from all the neighbours (line 14–16), keeping the best so far neighbour as q'_{near}, (initially q_{near}, line 12). If a neighbour has a collision-free path to q_{new} and a better cost (line 15), it is accepted as the best (line 16). The other part is to re-wire the path if a better cost of any neighbouring node is computed (lines 20–24). For this again all candidate paths are iterated (line 20) in the set of neighbours. If a candidate path has a collision-free motion from q_{new} and with a better cost (line 21), the path is rewired by the change of parent and cost (lines 22–24). The results for a synthetic map are given in Fig. 9.9 for RRT expansions and Fig. 9.10 for the path improvements as RRT expands.

Algorithm 9.6 *RRT* (source,G)*.

Input: Source (*source*) and Goal (*G*). Configuration space definitions
including a collision checking mechanism
Output: Path from source to goal

1   $\pi(source) \leftarrow NIL, E(source) \leftarrow NIL, T \leftarrow source$;
2   $C(source) \leftarrow 0$;   ▷ cost from source
3   $pathFound \leftarrow false$;
4   **while** *maximum running time* **do**
5     $r \sim U[0, 1]$;
6   **if** $r < bias$ **then** $q_{rand} \leftarrow$ sample from Goal;
7   **else** $q_{rand} \leftarrow$ random sample as per sampling strategy;
8     $q_{near} \leftarrow argmin_{q_t \in T} d(q_t, q_{rand})$;
9     $\lambda \leftarrow min(S, d(q_{near}, q_{rand}))/d(q_{near}, q_{rand})$;
10    $q_{new} \leftarrow (1 - \lambda)q_{near} + \lambda q_{new}$;
11   **if** $\Delta(q_{near}, q_{new})$ *is not null* **then**
        ▷ Part I: Find a better parent(q_{near}) for q_{new} ;
12      $q'_{near} \leftarrow q_{near}$;   ▷ best parent so far
13      $\delta(q_{new}) \leftarrow \{q \in T : d(q_{new}, q) < R)\}$;   ▷ all neighbours
14   **for** $q_C \in \delta(q_{new})$ **do**
          ▷ if candidate parent q_C to q_{new} is
          collision-free and serves a better cost ;
15   **if** $\Delta(q_C, q_{new})$ *is not null*
          $\wedge C(q_C) + d(q_C, q_{new}) < C(q'_{near}) + d(q'_{near}, q_{new})$ **then**
16          $q'_{near} \leftarrow q_C$
17      $\pi(q_{new}) \leftarrow q'_{near}, E(q_{new}) \leftarrow \Delta(q'_{near}, q_{new})$;
18      $C(q_{new}) \leftarrow C(q'_{near}) + d(q'_{near}, q_{new})$;
19   add q_{new} to T ;
        ▷ Part II: Find better costs for all neighbours
        of q_{new}, using q_{new} as parent ;
20   **for** $q_C \in \delta(q_{new})$ **do**
21   **if** $\Delta(q_{new}, q_C)$ *is not null* $\wedge C(q_{new}) + d(q_{new}, q_C) < C(q_C)$
         then
22          $\pi(q_C) \leftarrow q_{new}, E(q_C) \leftarrow \Delta(q_{new}, q_C)$;
23          $C(q_C) \leftarrow C(q_{new}) + d(q_{new}, q_C)$;   ▷ re-wire path
24   Re-calculate the cost of all descendants of q_C

25   **if** *pathFound* **then** return *calculatePath*($T < \pi, E >, G$);
26   return NIL

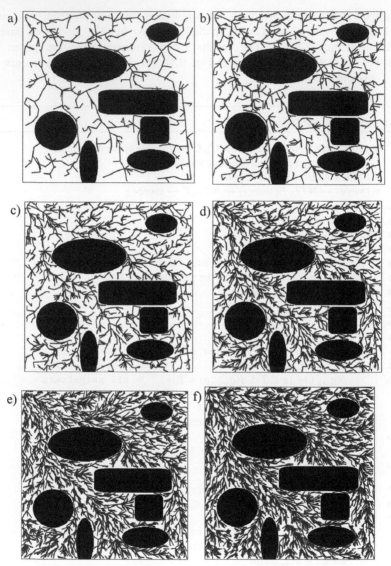

FIG. 9.9 RRT* Expansion along with time is shown in (A–F). The tree initially grows randomly in all directions and soon thereafter better paths are found which is visible by the tree being directed away from the source, enabling short connections to any node from the source.

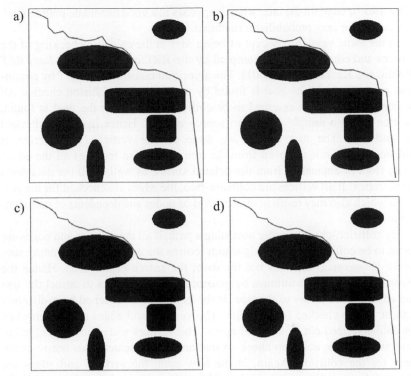

FIG. 9.10 Path improvement in RRT* Expansion along with time is shown in (A–D). The path continuously gets shortened, up to the end when it nearly represents the optimal path.

A catch in the approach is that the optimality depends upon the neighbourhood threshold R. The algorithm will not be optimal for small thresholds, while the algorithm may take too long if the acceptance threshold is set to be infinitely large. The algorithm is only optimal if the neighbourhood threshold R is taken as large enough.

9.3.4 Lazy RRT

The maximum time in the RRT implementation goes in the collision checks which need to be performed for all the vertices and all the edges of the tree. Therefore, collision checking of a vertex or edge must only be done when the collision-check is important. In the probabilistic roadmap technique, it was observed that by delaying the collision checking to as late as the query time, numerous vertices and edges of the roadmap were pruned and did not require any collision-checking as they did not come into good candidate paths from source to goal, which saved a lot of time. The delaying of collision-checking of a vertex or edge till the late, when it appears in a good candidate path to goal, was called the lazy probabilistic roadmap.

In the same vein, the concept of being *lazy* in the collision-checking of the vertices and edges of the tree sampled by the RRT is termed as the *Lazy RRT* (Bohlin and Kavraki, 2000, 2001). The algorithm constructs an RRT by continuous expansions till the goal is found by not doing any collision checks. All vertices and edges are assumed to be collision-free. When the goal is found, the tree stores a sample path from source to goal. Hence it is now checked for collision. The first checks are done on the vertices. If a vertex is collision-prone, it is removed from the tree which also removes all the edges going from or emanating from the deleted vertex, as well as all the decedents of the vertex. If all vertices are collision-free, the edges are checked for the collision which also may result in the removal of nodes and decedents of collision-prone edges.

By collision-checking after obtaining a path, if all the vertices and edges are found to be collision-free, the algorithm returns the path, and the planning succeeds. However, if this was not the case, the search cannot stop. Hence the expansion of the tree continues by generating more samples to attract the tree and adding more nodes to the tree. If the goal is again reached by a different path, it is also checked for collision. The vertices and edges already checked for collision need not be checked again. The vertices and edges are tagged to have survived the collision check so that any future search does not re-check them for the collision. Again, in the new path, the vertices and edges are checked for a collision. The infeasible vertices and edges, along with the descendants, are removed. This continues till a feasible path to the goal is found.

9.3.5 Anytime RRT

Typically, the RRT algorithm lacks the notion of optimality (except the RRT* algorithm). The anytime RRT algorithm is an *iterative* algorithm that keeps improving the solution along with time such that the initial solution is readily available and thereafter better solutions need longer runs, while a solution would be available anytime. Unlike the RRT* algorithm, here the heuristics is also used in assessing the quality of a prospective node to be inserted in the tree.

Before discussing the anytime RRT algorithm, let us incorporate *heuristics* into the working of the RRT algorithm (Urmson and Simmons, 2003). The heuristic value of a node is an estimate made by the algorithm about the solution quality for paths going through the node. In the A* algorithm the heuristic total cost $f(v)$ was defined as the sum of cost from source $g(v)$ and estimated cost to the goal $h(v)$, and a higher f value meant a poorer node whose expansion was deferred till all high-quality (low total cost) nodes were expanded.

Similarly, the *quality* in the case of an RRT node q_{near} is defined as Eq. (9.8).

$$Q(q_{near}) = w_1 C(q_{near}) + w_2 h_G(q_{near}) \tag{9.8}$$

Here $Q(q_{near})$ is the quality of a candidate q_{near}. $C(q_{near})$ is the cost from source to q_{new} and $h_G(q_{near})$ is the estimated (heuristic) cost from q_{near} to the goal. The sub-script G is used to denote the heuristic distance to goal (G). A lower $Q(q_{new})$ value means a better node. Here w_1 and w_2 are the weights associated with the historic and heuristic components, respectively.

The weights w_1 and w_2 are adapted as the algorithm runs. Initially, a result should be quickly obtained and therefore the heuristic value weight is large (high w_2) and the cost from source has a small weight value (low w_1). This biases the search to expand nodes nearer to the goal to get a result quickly. As the algorithm proceeds, the weights are adjusted to have a high cost from source weight (high w_1) and a low heuristic weight (low w_2). This biases the search to explore more in different areas, than to focus on the goal. Hence, w_1 increases at a rate of α_1 and w_2 reduce at a rate of α_2.

The selection of a q_{near} near q_{rand} is now a bi-objective problem. The nodes nearer to q_{rand} should be preferred as per logic, and simultaneously the better-quality nodes should also be preferred due to optimality. There are numerous ways to implement this logic with logarithmic complexity. After selecting a random q_{rand}, the quality measure is additionally used in the selection of the nearest sample q_{near}. The RRT selects the nearest neighbour by distance from q_{rand}, that is Eq. (9.9).

$$q_{near} \leftarrow \arg\min_{q_t \in T} d(q_t, q_{rand}) \tag{9.9}$$

However, in anytime RRT, the *acceptance probability* is proportional to its quality. In case q_{near} is not accepted, the process will be repeated with a new q_{rand}. Note that the quality measure will need to be normalized and inverted because as per the current metric a lower value is better quality. A poor-quality node is likely to be rejected with a large probability of rejection, while a high-quality node is likely to be accepted having a large probability of acceptance.

Alternatively, one may select the nearest k neighbours by the distance metric from q_{rand}, that is $\delta(q_{rand})$; however, use only the best quality sample among the nearest k neighbours that is, Eq. (9.10).

$$q_{near} = \arg\min_{q \in \delta(q_{rand})} Q(q) \tag{9.10}$$

This is additionally passed through an acceptance test to meet a minimum quality level. Another mechanism is to consider samples in $\delta(q_{rand})$ in the order or quality value and probabilistically accept them as per the quality measure. The first few samples are of high quality and more likely to be accepted; however, with a small probability, it is possible that the initial samples get rejected and the lesser quality samples are instead selected.

It may be inquisitive that the A* algorithm uses the quality (cost) measure only, while the Anytime RRT algorithm attempts to balance between the quality measure and the distance from q_{rand}. The two algorithms are conceptually different. The A* algorithm discretizes the search space into a few cells and a cell can only be visited once. The algorithm therefore prioritizes the expansion order of the cells. The Anytime RRT algorithm on the other hand has no discretization and a sample can be selected infinite times to generate infinite (maybe very similar) children. Because of a lack of discretization, no node in the tree is ever permanently closed for future expansions. Therefore, the algorithm cannot repeatedly select the best quality (lowest cost) node, whose expansion towards goal giving a better child may not be feasible.

The concept of quality beautifully injects the heuristic measure and is used in the design of the *Anytime RRT* algorithm (Ferguson and Stentz, 2006, 2007; Karaman et al., 2011). The algorithm proceeds in iterations wherein at the first iteration a path with some cost is computed by a run of the RRT, in the second iteration, a better path than the previous iteration is obtained, and so on, with every iteration giving a better path. Every iteration starts from a new and blank tree and therefore the tree is deleted after every iteration, taking only the best path found so far. When the allocated time to the algorithm (or stopping criterion) is over, the algorithm breaks and returns the best solution found so far.

Since the solution will be continuously improving, the algorithm always has a prospective path of cost $C^*(t)$, where t denotes the time (except for the first run). Let us assume that we aim to find a better path, which should hence have a cost of at maximum $C_{lim} = (1 - \epsilon)C^*(t)$. Here ϵ is the minimum improvement between iterations. The next iteration will keep trying until it gets a path with a cost better than ϵ times or less. Assume q_{rand} to be a random sample which the tree intends to expand to with the restrictions of the step size. $h_S(q_{rand}) + h_G(q_{rand})$ is the least cost of the path possible through q_{rand}, where $h_S(q_{rand})$ is the heuristic cost from the source and $h_G(q_{rand})$ is the heuristic cost from the goal. Both costs are taken as straight-line distances. If $h_S(q_{rand}) + h_G(q_{rand}) > C_{lim}$, it is evident that q_{rand} cannot result in a better cost and therefore should not be used. The acceptability condition of q_{rand} is hence Eq. (9.11).

$$h_S(q_{rand}) + h_G(q_{rand}) \leq C_{lim} \qquad (9.11)$$

The same criterion of cost threshold-based acceptance is also used when a new node q_{new} is generated by extension of a node q_{near}. Since q_{near} produces q_{new}, the cost of q_{new} will be the one obtained through q_{near}, or Eq. (9.12).

$$C(q_{new}) = C(q_{near}) + d(q_{near}, q_{new}) \qquad (9.12)$$

Here $C(q_{new})$ is the cost from the source. The best cost that can be produced with q_{new} is $C(q_{new}) + h_G(q_{new})$ and therefore the node q_{new} must only be accepted if Eq. (9.13) holds,

$$C(q_{new}) + h_G(q_{new}) \leq C_{lim} \qquad (9.13)$$

The algorithm generates multiple prospective samples q_{new} instead of just one and takes the best of them to be inserted into the tree.

Before putting all the concepts into a formal algorithm let us summarize the three major concepts in the design of the Anytime RRT algorithm. (i) The first major concept is to use a quality measure along with the distance from q_{rand} as a measure to select q_{near}, with an aim to be biased towards selection of samples that can potentially give shorter distances from source to goal. The quality was given as the weighted sum of cost from source and the heuristic cost to goal. (ii) The second major concept was to prune any q_{rand} and q_{new} if it is unlikely to be in the optimal path from source to goal. If the historic cost and the heuristic cost became more than the current cost being aimed for, the node was pruned. (iii) Finally, the third major concept was the anytime nature where the solution should improve with time. This was done first by setting the aimed path cost limit of the next iteration a little lower than the previously obtained cost. Thereafter, this was done by setting the quality function weights to prefer better heuristic cost nodes at the start for exploitation, while selecting better historic cost nodes later for more exploration. A small additional concept was also to generate several q_{new} and to select the best among them.

The algorithm is illustrated as Algorithm 9.7 for the overall loop and Algorithm 9.8 for the specific RRT generation. The concepts are shown in Fig. 9.11A and B. In Algorithm 9.7, at every instant of time a path is returned (Line 6), and a better path is aimed at by the algorithm by setting the limit (Line 7). The weights corresponding to the ordering of nodes for taking q_{near} are adapted (Lines 8–9), with the initial setting as the highest weight to the heuristic component (Lines 3–4). Algorithm 9.8 initializes the tree with source as the root (Line 1–2). Lines 3–10 select a random sample q_{rand}, for which it is necessary to pass the limiting condition in Line 9 and hence attempt to generate such a sample is done in a few attempts. Lines 12–21 select the k nearest samples q_{near} (Line 12) which are ordered as per the quality metric (Line 13) and processed in the same order till a sample is added.

Algorithm 9.7 Anytime RRT (*source,G*).

Input: Source (*source*) and Goal (*G*). Configuration space definitions
including a collision checking mechanism
Output: Path from source to goal

1 $C_{lim} \leftarrow \infty$;   ▷ computed path should have a cost $\leq C_{lim}$
2 $\tau_{best} \leftarrow NIL$;   ▷ best path so far
3 $w_1 \leftarrow 0$;   ▷ historic component weight
4 $w_2 \leftarrow 1$;   ▷ heuristic component weight
5 **while** *stopping criterion* **do**
6 | $[\tau_{best}, C_{best}] \leftarrow AnytimeRRTGrow(S, G, C_{lim}, w_1, w_2)$;
7 | $C_{lim} \leftarrow (1 - \epsilon)C_{best}$;   ▷ next path be better by ϵ
8 | $w_1 \leftarrow min(w_1 + \alpha_1, 1)$;   ▷ slowly increase w_1, explore
9 | $w_2 \leftarrow max(w_2 - \alpha_2, 0)$;   ▷ decrease w_2, less greedy
10 return τ_{best}

Algorithm 9.8 Anytime RRT grow ($source, G, C_{lim}, w_1, w_2$).

Input: Source ($source$) and Goal (G), least path length aimed C_{lim}, historic weight (w_1), heuristic weight (w_2), Configuration space definitions including a collision checking mechanism

Output: Path from source to goal

1   $\pi(source) \leftarrow NIL, E(source) \leftarrow NIL, T \leftarrow source$;
2   $C(source) \leftarrow 0$;   ▷ cost from source
3   **while** *maximum running time* **do**
4      $r \sim U[0,1]$;
5   **if** $r < bias$ **then** $q_{rand} \leftarrow$ sample from Goal;
6   **else**
7   **for** *a few attempts* **do**
8              $q_{rand} \leftarrow$ random sample as per sampling strategy ;
              ▷ Can best path via q_{rand} be better than C_{lim}? ;
9   **if** $h_S(q_{rand}) + h_G(q_{rand}) \leq C_{lim}$ **then** break;
10   **else** $q_{rand} \leftarrow NIL$;
11   **if** $q_{rand} = NIL$ **then** continue;
12      $Q_{near} \leftarrow k$ nearest neighbours to q_{rand} ;   ▷ q_{near} candidates
     ▷ process candidates in the order of quality ;
13   **for** *all* $q_{near} \in Q_{near}$ *as per order* $w_1 C(q_{near}) + w_2 h_G(q_{near})$ **do**
14          $Q_{new} \leftarrow Extensions(q_{near}, q_{rand})$;   ▷ a number of collision-free extensions
15          $q_{new} \leftarrow argmin_{q \in Q_{new}} C(q_{near}) + d(q_{near}, q)$; ▷ best in Q_{new}
16          $C(q_{new}) \leftarrow C(q_{near}) + d(q_{near}, q_{new})$;   ▷ cost from source
         ▷ Can best path from q_{new} be better than C_{lim}? ;
17   **if** $C(q_{new}) + h_G(q_{new}) \leq C_{lim}$ **then**
18              $\pi(q_{new}) \leftarrow q_{near}, E(q_{new}) \leftarrow \Delta(q_{near}, q_{new})$;
19   add q_{new} to T ;
20   break ;
21   **if** $q_{new} = G$ **then** return $calculatePath(T < \tau, E >, G)$;
22   **return** NIL ;

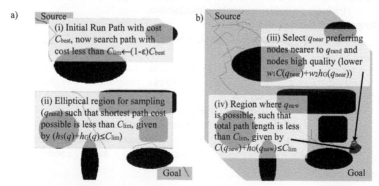

a)
Source

(i) Initial Run Path with cost C_{best}, now search path with cost less than $C_{lim} \leftarrow (1-\varepsilon)C_{best}$

(ii) Elliptical region for sampling (q_{rand}) such that shortest path cost possible is less than C_{lim}, given by $(h_S(q) + h_G(q) \leq C_{lim})$

Goal

b)
Source

(iii) Select q_{near} preferring nodes nearer to q_{rand} and nodes high quality (lower $w_1 C(q_{near}) + w_2 h_G(q_{near})$)

(iv) Region where q_{new} is possible, such that total path length is less than C_{lim}, given by $C(q_{new}) + h_G(q_{new}) \leq C_{lim}$

Goal

FIG. 9.11 Anytime RRT working. RRT runs multiple times, improving the next path by at least a factor of ε (A) Elliptical region *(grey)* within which q_{rand} must be generated (B) Selection of q_{near} and region within which q_{new} may be accepted. C_{lim}: minimum path length of prospective solution, $h_S(q)$: heuristic cost from source to q, $h_G(q)$: heuristic cost from q to goal, $C(q)$: cost from source to q through RRT.

Line 14 generates several possible valid extensions, where validity is taken as the ability to make straight-line collision-free connections with q_{near}. The best sample is used for adding in the tree (Lines 18–20). It is important that the sample aids in getting costs within the threshold for which the acceptability condition is checked (Line 17). Lines 19–20 add the sample. Line 21 checks for the goal condition.

9.3.6 Kinodynamic planning using RRT

The kinodynamic planning is not a variant of RRT like the other sub-sections, but this exposes one of the reasons for the high applicability of RRTs in real life motion planning which is to directly plan in the state (control) space wherein the velocities and maybe higher-level terms like acceleration can be incorporated as state variables and used for planning. The term *kinodynamic* specifically implies that the trajectory generated will adhere to the velocity, acceleration, and other kinematic constraints that are imposed upon the vehicle. This is possible since all the variables are either in the state representation or control signals which are being planned. In such spaces, the manner to move from one state to the other is via control signals. Let f be the kinematic equation of the robotic system. Given the system is at state s_1 at a point of time and is subjected to a control signal u for Δt period after which it moves to the state s_2. Using the kinematic equation, s_2 will be given by $s_2 = f(s_1, u, \Delta t)$. This makes the vertices s_1 and s_2 be connected by a directed edge $(u, \Delta t)$ from s_1 to s_2.

The RRT extends q_{near} towards q_{rand} to generate a new sample q_{new}. This was possible so far since space was interpolated between q_{near} and q_{rand} to generate an intermediate point q_{new}. However, now such interpolation cannot be done since the kinematic and other constraints may not be adhered to by the interpolated point. In other words, no u and Δt may exist such that $q_{new} = f(q_{near}, u, \Delta t)$, or even if they exist it may be impossible to compute them since the general function $f()$ may not have a closed-form solution for u. Hence, an alternative approach is suggestive to compute q_{new}.

The problem is generating a q_{new} from q_{near} towards q_{rand}, such that $q_{new} = f(q_{near}, u, S)$ and q_{new} is in the route from q_{near} to q_{rand}, where S is the step size (in terms of time). The constraint that q_{new} should be between q_{rand} and q_{new} for a fixed time step S means that out of all possibilities of navigation within the time step, the one closest to q_{rand} is the most appropriate, given by Eq. (9.14).

$$q_{new} = \arg \min_{u \in U} d(f(q_{near}, u, S), q_{rand}) \tag{9.14}$$

Here $d()$ is the distance function. The set of controls are infinite and therefore it is impossible to find q_{new}, unless the function $f()$ is invertible like in the case of a straight line. A sampling-based solution is used to find u which most closely places q_{new} near q_{rand} by taking a few samples u from U (set of all possible/infinite control signals) and accepting the best one among the samples. The pseudo-code undergoes a minimum modification to account for this, which is represented as Algorithm 9.9 in entirety for completeness. The addition is Lines 10–14, wherein the best sample is assumed to be q_{near}, and the same is iteratively improved for a small number of attempts.

Algorithm 9.9 Kinodynamic RRT (*source,G*).

Input: Source (*source*) and Goal (*G*). Configuration space definitions
        including a collision checking mechanism
Output: Path from source to goal

1   $\pi(source) \leftarrow NIL, E(source) \leftarrow NIL$;
2   $T \leftarrow source$;
3   $pathFound \leftarrow false$;
4   **while** *maximum running time* **do**
5      $r \sim Uniform[0,1]$;
6   **if** $r < bias$ **then** $q_{rand} \leftarrow$ sample from Goal;
7   **else** $q_{rand} \leftarrow$ random sample as per sampling strategy;
8      $q_{near} \leftarrow argmin_{q_t \in T} d(q_t, q_{rand})$;
      ▷ Find q_{new} and control u such that q_{new} is closest
      to q_{rand} in a unit step time S. This can be done by
      several methods - the simplest strategy is used ;
9      $q_{new} \leftarrow NIL$;   ▷ best candidate so far
10   **for** *some attempts* **do**
11       $u \sim U$;   ▷ take a random control signal
12       $q'_{new} \leftarrow f(q_{near}, u, S)$;   ▷ robot's kinematic equation
13   **if** $q_{new} = NIL \vee d(q'_{new}, q_{rand}) < d(q_{new}, q_{rand})$ **then**
14         $q_{new} \leftarrow q'_{new}, u_{sel} \leftarrow u$
15   **if** $f(q_{near}, u_{sel}, s)$ *is not null* $\forall 0 \leq s \leq S$ **then**   ▷ simulation of
      the edge to check collision
16       $\pi(q_{new}) \leftarrow q_{near}, E(q_{new}) \leftarrow f(q_{near}, u_{sel}, s), 0 \leq s \leq S$;
17   add q_{new} to T ;
18   **if** $q_{new} = G$ **then** $pathFound \leftarrow true$, break;

19   **if** *pathFound* **then** return $calculatePath(T < \pi, E >, G)$;
20   return NIL ;

This is not the only way to generate q_{new}. Imagine a control problem to go from q_{near} to q_{rand}. The control systems already generate control signals for the same problem and the first (or first few) control signals are valid control signals for edges, and the robot state after the application of such control signals gives q_{new}. Alternatively, the kinematic equation $f()$ is known and therefore a gradient-descent based algorithm can be used to compute to quickly find a solution to Eq. (9.14), which is also popularly done in control literatures.

To illustrate the working of the algorithm, consider a planar differential wheel drive robot whose configuration is (x,y,θ) and additionally has speed associated with it consisting of the linear speed (v) and angular speed (ω), giving the state vector (x,y,θ,v,ω). The robot moves by turning the two wheels independently by speeds v_L (left wheel speed) and v_R (right wheel speed), which translates to the linear and angular speeds. The non-holonomic constraint here is that the robot cannot move sideways, or $tan\theta = \frac{\dot{y}}{\dot{x}}$. This is already incorporated by taking 2 speeds for 3 configurations. The kinematics of the robot is given by Eqs (9.15)–(9.18).

$$\begin{bmatrix} \dot{x} \\ \dot{y} \\ \dot{\theta} \end{bmatrix} = \begin{bmatrix} v\cos(\theta) \\ v\sin(\theta) \\ \omega \end{bmatrix} \tag{9.15}$$

$$\begin{bmatrix} x(t+S) \\ y(t+S) \\ \theta(t+S) \end{bmatrix} = f\left(\begin{bmatrix} x(t) \\ y(t) \\ \theta(t) \end{bmatrix}, \begin{bmatrix} v_L \\ v_R \end{bmatrix}, S \right)$$

$$= \begin{bmatrix} \cos(\omega S) & -\sin(\omega S) & 0 \\ \sin(\omega S) & \cos(\omega S) & 0 \\ 0 & 0 & 1 \end{bmatrix} \begin{bmatrix} x(t) - ICC_x(t) \\ y(t) - ICC_y(t) \\ \theta(t) \end{bmatrix} + \begin{bmatrix} ICC_x(t) \\ ICC_y(t) \\ \omega S \end{bmatrix} \tag{9.16}$$

$$ICC = \begin{bmatrix} x - R\sin(\theta) \\ y + R\cos(\theta) \end{bmatrix} \tag{9.17}$$

$$R = \frac{v}{\omega}, v = \frac{v_L + v_R}{2}, \omega = \frac{v_R - v_L}{l} \tag{9.18}$$

ICC is the centre of the circle around which the robot turns, which has a radius of R. The robot merely turns by an angular speed of ω for S duration of time. l is the distance between the wheels. The two control signals are for the left and the right wheel, or $u = [v_L \ v_R]^T$. Both the controls are constrained to a minimum and maximum limit, or $-v_{max} \le v_L \le v_{max}$ and $-v_{max} \le v_R \le v_{max}$. The RRT generated for a synthetic map is shown in Fig. 9.12. It may be seen that there are only smooth curves because the trajectories are now in control space which is smooth because of the constraints on the vehicle which are obeyed by the search process.

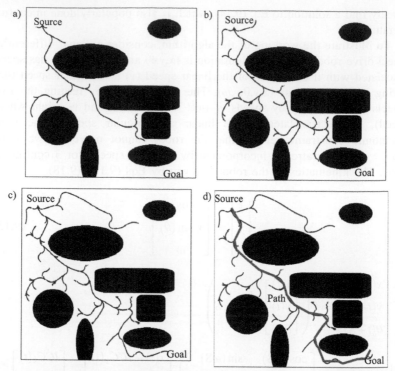

FIG. 9.12 RRT Expansion under kinodynamic planning along with time is shown in (A–D). After selecting a random sample (q_{rand}) and the nearest sample in the tree (q_{near}), numerous random control signals are applied to q_{near} to get candidate new samples (q_{new}), and the one closest to q_{rand} is accepted and added. Due to the incorporation of kinematics, the path is smooth.

The typicality of the RRT as discussed here is that the kinematics used may only allow for a forward simulation and not backward simulation. This means that from a state s if a control signal u is given for S period, we may simulate and calculate in advance the next state s_2; however, given a state s_2 and a control signal u applied for S amount of time, it is not possible to get the parent state s or backward simulation is not possible. If the backward simulation is not possible, the bi-directional RRT algorithm cannot be used since the tree rooted at the goal cannot generate backward steps.

9.3.7 Expansive search trees

The *Expansive Search Trees* (EST) (Hsu et al., 1999, 2002) are not a usual variant of the RRT but are discussed here because they very closely resemble the working of the RRT. EST also maintains a tree rooted at the source, at every step expands the tree and the search stops when the tree spans to cover the goal region.

The only difference between RRT and EST is in the mechanism used to *expand* the tree. In EST, a sample q_{rand} is first generated randomly out of all available samples in the tree. q_{rand} in EST resembles q_{near} in RRT, except for in EST it is directly randomly sampled. Heuristics may be used as well in the selection of the sample. The sample q_{rand} is then used for propagation by the generation of a sample q_{new} in the neighbourhood. The sample q_{new} is accepted if a collision-free path can be computed by the local planner $\Delta(q_{rand}, q_{new})$. The concept is thus to randomly grow a sample q_{rand} in a random direction taken from within the tree. The working of the algorithm is shown in Fig. 9.13A. The pseudo-code is given by Algorithm 9.10. The tree generated for

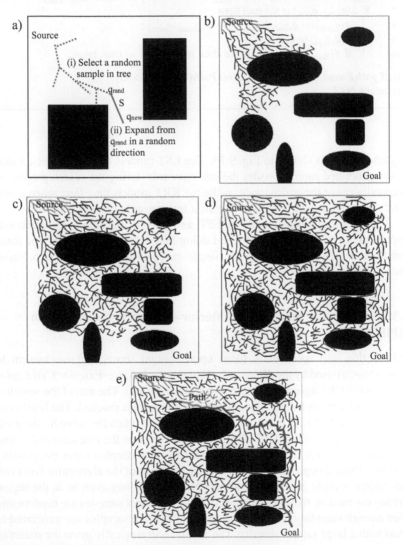

FIG. 9.13   EST Expansion (A) Notations; (B)–(E) Expansion.

Algorithm 9.10 EST (*source,G*).

Input: Source (*source*) and Goal (*G*). Configuration space definitions
        including a collision checking mechanism
Output: Path from source to goal

1 $\pi(source) \leftarrow NIL, E(source) \leftarrow NIL$;
2 $T \leftarrow source$;
3 $pathFound \leftarrow false$;
4 **while** *maximum running time* **do**
5      $q_{rand} \leftarrow$ random sample from T ;
6      $q_{new} \leftarrow$ random sample near q_{near} ;   ▷ by perturbating q_{near}
7   **if** $\Delta(q_{rand}, q_{new})$ *is not null* **then**
8         $\pi(q_{new}) \leftarrow q_{rand}, E(q_{new}) \leftarrow \Delta(q_{rand}, q_{new})$;
9   add q_{new} to T ;
10   **if** $d(q_{new}, G) < Threshold$ **then** $pathFound \leftarrow true$, break;

11 **if** $pathFound$ **then** return $calculatePath(T < \pi, E >, G)$;
12 return NIL

---

a synthetic map is shown in Fig. 9.13. The EST models the effect when a randomly generated sample *pushes* the tree outwards during the search process to eventually cover the entire space, while the RRT models the effect when a randomly generated sample is used to *pull* a tree towards strategized areas.

A good variant constitutes the EST as the base algorithm, bi-directional implementation of the algorithm, and doing lazy collision checking. The three constituents together constitute the *Single-query Bi-directional Lazy collision checking algorithm* (*SBL*).

9.3.8 Kinematic planning by interior-exterior cell exploration (KPIECE)

This section will specifically focus upon another strategized mechanism to grow a tree, as used by the *Kinematic Planning by Interior-Exterior Cell Exploration* (KPIECE) algorithm (Şucan and Kavraki, 2009). The aim of the search is primarily to grow outwards along with time till a goal is reached. The frontier of the graph search always grew outwards and stopped when the wave hit the goal. Similarly, the aim of the RRT algorithm is also to grow the tree *outwards* along with time, such that the algorithm appears to be a sampled wave progressing outward. Even though this is exactly the appearance of the algorithm; however, the random sample generated to attract the roadmap may even be in the region already covered by the search process. Typically, most samples are random and after enough sampling, a large proportion of random samples are generated in areas with a large sample density. The aim is to strategically grow the roadmap outwards. In the graph search algorithms, the implementation did not need any specific technique due to discretization as all the interior cells were expanded after which the closed list of cells stopped any further expansion in the interior region resulting in a constant propagation of the wavefront in an outward direction. However, in RRT, due to the absence of any discretization of space into a finite number of cells, the interior region can have infinite samples drawn that cannot be exhaustively visited, needing a specialized technique facilitating the outward growth.

The other problem associated with the RRT is that it is difficult to keep track of what areas are covered and whatnot, a knowledge of which can help to direct the growth of the tree towards the areas that are largely uncovered. This strategizes the search to grow in difficult areas in which the coverage is small. This can be done by generating more samples in the difficult region or giving more importance to tree nodes near areas that appear uncovered.

In extremely complex and hybrid spaces, it is difficult to define *distance functions*, which highly affects the performance of the RRT algorithm for the choice of the nearest sample. A generic goal test function may be available, but a goal is explicitly not given, and thus a goal bias is not possible.

The fundamental behind KPIECE is to generate a cell decomposition of the area. Since the space is extremely complex, the cell decomposition may be done with *coarse resolutions* by filling in cells in the entire configuration space, covering both C^{obs} and C^{free}. The cells are maintained at *multiple resolutions* as a tree, to avoid the problems associated with a single resolution, with every cell of a lower resolution connected to a bunch of finer cells of a higher resolution.

The working philosophy requires the identification of samples at the *exterior* of the tree, so that their progression may be boosted. Fig. 9.14A explains the concepts of interior and exterior cells, while Fig. 9.14B extends the idea to multiple resolutions. Initially, leaves maybe the exterior nodes at the boundary of a progressing tree, but this is not true because many late expansions will also result in leaf nodes in the interior. Hence cells are used to identify samples at the exterior of the progressing tree. A cell is called exterior if it has less than threshold neighbours which are reasonably densely occupied by samples. Any internal cell will have high density of samples inside the cell and the neighbouring cells, while at the exterior there will be lesser density among the neighbouring cells. There is a bias towards choosing an exterior cell while the internal cells are still chosen for progression. The decision of the cell being interior, or exterior is worked out in a multi-resolution setting. The same map may be decomposed into multiple resolutions. This is especially important when the sample density in general increases, in which case multiple resolutions help in the identification of specific regions which can be conveniently called as exterior, and the growth of the tree be so directed.

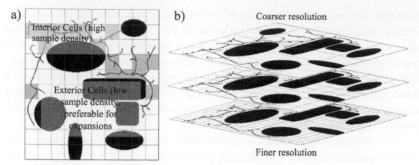

FIG. 9.14 KPIECE Algorithm (A) Interior and Exterior Cells. Expansions through exterior cells, unless low success rate in past. The algorithm assumes no distance function and uses random expansion. (B) Multi-resolution cell representation. Computations are done in multiple resolutions to get the regional sample density for classification and expansions and to identify specific areas laden with samples.

The algorithm cannot generate a random sample and find the nearest node because of the absence of the distance function. Hence a sample is selected and propagated randomly. The progression itself is by applying a random control signal for a random time step to generate a new node. The selection is biased towards the external cells. Inside the cells, because the cells are stored in a multi-resolution manner, the higher resolutions may be used to find the best area when enough samples exist in the tree. The samples are prioritized for selection based on their success metrics, coverage, number of times they are already selected, and so on. The selected sample on application of a random control signal for a random time makes the new node. The coverage metrics summarized as cells are updated whenever any motion is performed for progression.

9.4 Sampling-based roadmap of trees

The previous chapter discussed the probabilistic roadmap (PRM) technique of making an offline roadmap and using the same to solve the online query problem. The PRM uses a local collision-checking algorithm denoted by $\Delta(q_1, q_2)$ to compute the shortest path from q_1 to q_2. This chapter discussed the mechanism of solving single query problems between two configurations.

The *Sampling-based Roadmap of Trees* (SRT) (Plaku et al., 2005; Bekris et al., 2003; Akin et al., 2005) is a concept that uses the *RRT algorithm as the local planning algorithm* for PRM. Before discussing the SRT algorithm, it is wise to see if RRT is a good choice as the local planning algorithm of PRM, in contrast to the other choices. The global planning algorithms are the ones that are computationally heavy and time-consuming, however give fair guarantees to optimality and completeness. In contrast, the local planning algorithms are computationally less intensive, however, they neither guarantee completeness nor optimality. The typical use case of local planning algorithms is when the source and goal are separated by simple obstacles, they can quickly compute a fair way out. The straight-line collision checking was a poor local search algorithm because even though it was fast, it could not even tackle a simple obstacle in the way from source to goal. Doing exhaustive searches between source and goal was again a bad choice as the local algorithm because of a high computational time.

The RRT in such a context has a uniqueness. In simple cases, the RRT is extremely fast in computing a few quick turns around obstacles to connect two samples. This feature can even be scaled to complex scenarios in which RRT takes a long time like any other global search technique. Consider the simplest case wherein the two configurations are collision-free. The goal bias constantly moves the tree towards the goal to connect the source and goal quickly. The RRT-connect may succeed in doing so nearly as quickly as the straight-line collision checking. Unlike PRM and cell decomposition techniques, the RRT in its growing style explores the local area first, before expanding into further away areas, which also gives it the local properties desirable to be used as the local search for the PRM.

Even though RRT makes a good local search algorithm, it is not better than a straight-line collision checker for the initial roadmap generation wherein a straight line itself results in numerous good edges. The notion is hence to use a straight-line collision checking initially for the simpler scenarios and to only use RRT as the local collision avoidance algorithm for the harder scenarios later.

The initial roadmap generation is done in the same manner as the PRM, using a straight-line local collision checking algorithm. This results in the generation of vertices and edges that make in the initial roadmap. In this roadmap, the easier parts of the configuration space are connected, and the regions separated by narrow corridors are disconnected. The RRT algorithm selects the vertices belonging to different components and makes a tree to join the components. The concept is illustrated in Fig. 9.15.

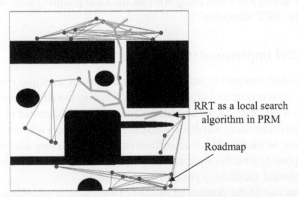

FIG. 9.15   Sampling-based Roadmap of Trees. The initial PRM consists of some disconnected sub-graphs (connected components) and RRT is used as a local search algorithm to get connections. RRT being a better local search algorithm connects through the difficult areas faster than sampling alone.

An important parameter chosen here is the stopping criterion of the RRT algorithm, or in other words, the maximum number of samples (n_{RRT}) that may be a part of the tree before the tree is discarded, and the local collision checking algorithm is deemed to have failed. The failure may happen because the two samples were separated by obstacles with no simple local path connecting them. If the stopping criterion is too small, even simple obstacle corners may not be avoidable by RRT as the local planning algorithm. This reduces the RRT to nearly a straight-line collision checker. On the other hand, if the stopping criterion is too large, the algorithm may invest too much time trying to connect two samples that are not locally connectable, thus wasting a significant amount of time.

For further discussions, let us assume that n_{PRM} is the number of (difficult) vertices in the PRM, each of which is checked for collisions with its neighbours using RRT. The complexity of the algorithm thus becomes $O(n_{PRM}(\log(n_{PRM}) + n_{RRT}\log(n_{RRT})))$. For 1 PRM sample it takes $O(\log n_{PRM})$ time to compute the nearest neighbours and in a maximum of $O(n_{RRT}\log n_{RRT})$ time the n_{RRT} vertices are generated where the logarithmic term is for finding the nearest sample in the tree. For n_{PRM} samples, the time hence becomes $O(n_{PRM}(\log(n_{PRM}) + n_{RRT}\log(n_{RRT})))$. If the total time is restricted by the stopping criterion of the entire algorithm, an increase of n_{RRT} will result in a reduction of n_{PRM} and vice versa. Let us discuss the two extreme cases. If n_{PRM} is too high and n_{RRT} is too low, RRT has just enough time to act as a straight-line collision checker and thus the resultant algorithm is merely a PRM. Similarly, if n_{PRM} is too low and n_{RRT} is too high, there are very few PRM samples that are intensely aimed for connection by RRT, in which case the algorithm primarily behaves like the RRT.

The PRM technique used a *Connect Connected Components* strategy of edge connection that identified vertices that belong to different roadmaps that are disconnected and stated the use of a more deliberative algorithm as the local planner. RRT is a good choice for such an algorithm. The connect connected components strategy of PRM using RRT as the local planning algorithm closely resembles the SRT algorithm.

9.5 Parallel implementations of RRT

The modern-day computing infrastructure consists of multiple cores and the modern-day data processing algorithms are primarily parallel in nature. Therefore, there is a big demand for algorithms that can best make use of the parallel computing infrastructure. Robotics is no exception wherein it is anticipated that the robots will be backed by cloud infrastructure, thus having access to a lot of computing power, with multiple computing nodes at the access. However, the algorithms should facilitate a parallel computation.

This gives rise to the domain of *parallel robot motion planning*. So far, the book discussed graph search techniques that could be parallelized by making the different processors work on different nodes, generate, and merge the search trees; but this requires a good amount of information exchange between processors. Parallelism is naive and useful for the roadmap and cell decomposition approaches. The RRT has a clear advantage here (Raveh et al., 2011; Ichnowski and Alterovitz, 2012; Devaurs et al., 2011, 2013). In the parallel implementation of the algorithms, the problem needs to be divided and distributed to be solved in parallel, and the part solutions at different stages need to be communicated by the nodes and integrated. The process of division, communication, coordination, and integration may happen multiple times. The easier it is to decompose the problem into multiple problems that can be solved without communication, the more effective is the computation. A problem that requires a huge exchange of data at every instant can make the communication cost more than the one saved due to parallelization. Having 2 processors does not mean that the computing time will halve because of the time required in dividing computation, exchanging the results by communication, and integrating the solutions to make the overall solution.

The simplest implementation of the parallel RRT is called the *OR parallelization*. Here each of the parallel nodes runs the RRT algorithm. Whenever a node can find a path from the source to the goal, it sends a message to the other nodes to stop computation because a path is found. The communication overheads are the least and the resultant computation time is the least of all the parallel runs, which does save time. However, this implementation effectively selects a node and a lot of computation done by the other computing nodes is wasted.

Another technique of parallel implementation is the *AND parallelization*. Here again, the problem is given to all nodes, each of which computes a solution under different parameter settings. However, the search does not stop when a node computes the path. Instead, the search stops when all the nodes compute the path. Then all paths are taken as the best of them is selected. The communication overhead is again minimal. The computation time is the maximum of all nodes, while the optimality is enhanced. To leverage over the additional time, each instance may even run the RRT* algorithm.

Typically, this is also a poor mechanism of working since again only one node's computation is used and the other nodes do not contribute anything to the solution. A better way to work is to fuse all trajectories to make a single best trajectory of the robot. Consider that one RRT finds a good path between source to a mid-way point A however has a rather poor path between A to the goal. Assume another RRT finds a good path between A to goal, but not from source to A. The situation is shown in Fig. 9.16. The integration is not so simple because point A is not discretized and therefore will not be at the same place at both trees. However, it is easy to search in a small neighbourhood of every node in one tree to connect to the other tree, and thus making a path benefitting from both trees.

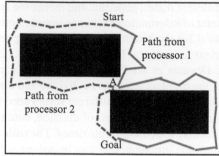

OR Parallelization: Use earliest computed path from any processor.

AND Parallelization: Use best path among all processor.

Fusion: Combine all paths and search for the best (Start → A from processor 1 and A→ Goal from processor 2)

FIG. 9.16   Parallel implementation of RRT.

9.6   Multi-tree approaches

A bi-directional RRT using 2 trees is better than a unidirectional RRT using only 1 tree, if both forward and backward simulations are possible. In the same discussions, a natural question arises that why the limit should be 2 trees, at source and goal only, and not generalized to many trees, rooted at strategized locations including source and goal. The multiple trees may help in dealing with *narrow corridors* as the corridor may be approached by multiple trees in different ways and any one of them may succeed in any direction to tackle the corridor and help the other trees. This also helps in *optimality* wherein the computations of different trees may be cumulatively used to make the resultant algorithm generate better paths as compared to a single tree. This is the genesis of the multi-tree approaches wherein multiple trees are worked upon to generate a path from source to goal.

The other motivation behind the algorithm is facilitating *parallel implementations*. In a multi-tree approach, each tree or section could potentially compute in parallel and exchange computed information in the form of a tree within some coordination cycles. Parallel implementations necessitate initialization and maintenance of multiple trees.

9.6.1   Local trees

A primitive mechanism of working with multiple trees is to extend the concept of bi-directional RRT to incorporate more than 2 trees. These trees are called *local trees* and are placed at strategic locations (Strandberg, 2004). Narrow corridors are excellent locations for the placement of these local trees. The narrow corridor sampling discussed in PRM generated samples inside difficult areas or narrow corridors. Here the same technique feeds the locations of the roots of the trees. Efforts may be made to limit the number of local trees and to maximize the diversity in their location so as not to have two trees close by in the same narrow corridor.

The growth happens just like the RRT algorithm by the generation of samples. Just like the bi-directional RRT, this technique makes a bias such that the trees grow towards each other. In the expansion of a tree, a q_{rand} from some other tree is selected because of the bias, which is used to attract the tree. This eventually results in the trees also growing towards each other, apart from growing throughout the space.

There will be many times when one tree *merges* with the other, just like the bi-directional RRT saw a merge of the two trees. In this case, however, it may not mean that the search is accomplished. When 2 trees merge, a single tree is made that replaces both trees, and the resultant algorithm runs with a reduced number of trees. To maintain the iterative nature of the algorithm, the event may itself be used to hunt for a new narrow corridor and for seeding a new local tree therein. The concept is shown in Fig. 9.17.

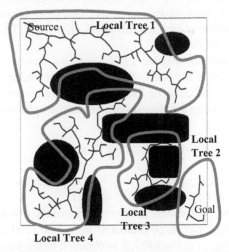

FIG. 9.17 Local Trees: Multiple trees start at difficult regions, source, and goal. The trees grow randomly with an affinity towards each other, and eventually merge to become one.

9.6.2 Rapidly-exploring random graphs

The *Rapidly-exploring Random Graph* (RRG) (Karaman and Frazzoli, 2011; Kala, 2013) is very similar to the RRT algorithm with the only difference that the constraint of a tree is not placed. The graph was relaxed to a tree previously because the source was fixed. The multi-tree approaches have arbitrarily placed roots acting as the source and minimizing the path cost from such an arbitrary root does not contribute towards minimizing the total cost from source to goal when all trees are merged. The constraint of graphs is hence placed again. This means that the data structure can admit cycles which will help in computing optimal paths between every pair of nodes while making the solution search a little more expensive. The RRG algorithm may hence be visualized as a *graph* ($RM < V,E>$) that explodes randomly while forming multiple connections including cycles as it traverses. The algorithm is suitable for multi-query approaches as well since it produces a roadmap as a graph similar to the PRM.

Based on the discussions on multi-tree implementations in general, the RRG could start from only the source; the source, and the goal; or multiple strategized points in the space. The multiple strategized points may be difficult regions like the narrow corridor. Alternatively, for multi-query approaches, if the sources and goals are known a priori, all of those could be the initial starting points. Like in the case of planning for multiple similar robots, for multi-robot motion planning problem, all the sources and goals of all robots can be taken as the initial starting points of the RRG algorithm. Alternatively, if there are important landmarks of interest where robots typically go to or start from, those are good candidates for online queries and therefore good starting points for the RRG algorithm. These constitute the initial vertices (V) of the RRG.

Initially, the algorithm only has a few vertices at strategized points taken as the initial seed points of the RRG. The RRG is exploded in all directions randomly in the same manner as the RRT. This is done by first generating a new sample q_{rand}; then finding the closest sample in the graph to q_{rand} called as q_{near}, that is Eq. (9.19).

$$q_{near} = \arg \min_{q_t \in V} d(q_t, q_{rand}) \qquad (9.19)$$

Finally, expanding q_{near} towards q_{rand} by a step size S generating the sample q_{new}. The expansion steps are the same as RRT.

However, in RRG, since there is no constraint of the tree and therefore more local edges from q_{new} may be possible and should be admitted. On accommodating any vertex (q_{new}) in the RRG, it is checked to the nearest neighbours as per the neighbourhood function $\delta(q_{new})$. The nearest-R neighbourhood function is given by Eq. (9.20).

$$\delta(q_{new}) = \{q \in V : d(q, q_{new}) < R\} \qquad (9.20)$$

The nearest-k neighbourhood function is also a valid strategy. The connection attempts by a local search algorithm Δ are made to all neighbours of q_{new} denoted by $\delta(q_{new})$. The local search algorithm Δ is typically taken as a straight line collision checker. The local collision checker checks for collisions for neighbours, $\Delta(q_{new}, q)$, $q \in \delta(q_{new})$ and all collision-free connections are added in the graph.

The goal-biased RRT was used to bias the search towards the goal. Similarly, the bias here may be used as a factor to strategize the RRG growth. It is assumed that, like the PRM, the connected components are stored as a disjoint set data structure, enabling querying whether two samples are in the same or different connected components and the number of connected components. There are two types of bias possible. Initially, the source and goal (for single query problems) will be parts of different connected components, and therefore a *goal bias* may be to attract the sub-graph containing source towards the sub-graph containing goal by sampling a random vertex of the goal sub-graph as q_{rand}. This requires that q_{near} is only searched for in the source sub-graph. The roles or sources and goals may be swapped like in the case of a bi-directional RRT.

The other bias is like the *connect connected components strategy* of PRM. Suppose the graph at any point of time consists of multiple connected components κ, which are otherwise disconnected from each other making the graph multiple independent sub-graphs. Let the sub-sets storing the vertices of the sub-graphs be $\{\kappa_i\}$. κ_i is a set of all vertices that make the connected component i, which are disconnected to all other sub-graphs. The aim is now to connect the different sub-graphs to account for connectivity. Therefore, one of the sub-graphs κ_i is specifically grown to be biased towards another sub-graph κ_j, by taking a sample in κ_j, that is $q_{rand} \sim \kappa_j$. The nearest sample is chosen in κ_i instead, that is Eq. (9.21).

$$q_{near} = \arg\min_{q_t \in \kappa_i} d(q_t, q_{rand}) \tag{9.21}$$

Thereafter q_{near} is propagated towards q_{rand} to generate q_{new}. This may eventually result in a connection between the two sub-graphs.

The KPECE implementation in RRT attempted to attract samples at the difficult areas or areas which are not judiciously represented by representation in the form of a multi-resolution grid. The same concept and implementation from RRT can also be directly extended to RRG.

The general concept is shown in Fig. 9.18A. The working on a synthetic map is shown in Fig. 9.18. The pseudo-code is given as Algorithm 9.11. Lines 1–8 are for initialization; wherein interesting points are chosen to initiate the RRG and added as vertices in line 1. The edge connections with the nearest neighbours are checked in lines 2–8. Lines 4 and 8 cater to the need to maintain a disjoint set of vertices like PRM, wherein every vertex is made a set initially an independent set, which is joined when an edge is added. Lines 12–14 are for implementing the connect strategy, wherein two disjoint graphs κ_i and κ_j are selected probabilistically as per strategy (say the source-goal pair has a higher probability than two other random ones, or the probability is proportional to the distance of closest vertices between them). The vertex q_{rand} chosen from κ_j (Line 13) attracts the closest vertex q_{near} in κ_i for expansion (Line 14). In the random strategy (Lines 16–17), a random vertex and the closest chosen vertex is taken. If the connection between q_{near} and q_{new} is successful (Lines 20–26), q_{new} is added to the graph as a vertex with an edge connection with q_{near} (Lines 21–22). The edges are assumed to be bi-directional. Similarly, edge connection with all neighbouring vertices is done and corresponding edges are added, maintaining the disjoint-set data structure (Lines 23–26).

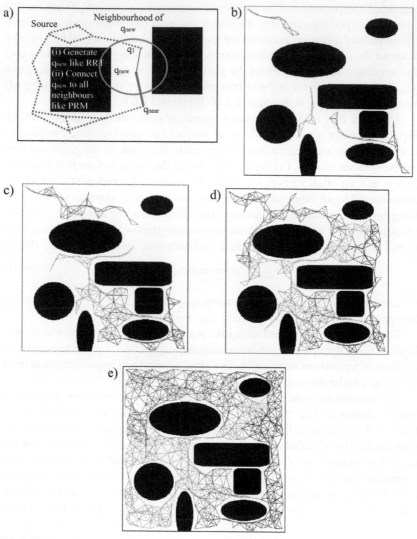

a)
Source Neighbourhood of
q_{new}
(i) Generate q_{new} like RRT
(ii) Connect q_{new} to all neighbours like PRM
$q1$
q_{new}
q_{near}

b)

c)

d)

e)

FIG. 9.18 Rapidly-exploring Random Graphs (RRG). Discard tree constraint to also admit graphs, (A) Notations, (B)–(E) RRG Expansion.

Algorithm 9.11 RRG (S,G).

Input: Source (*source*) and Goal (*G*). Configuration space definitions
including a collision checking mechanism
Output: Path from source to goal
▷ Initialization of vertices V and edges E ;
1   $V \leftarrow \{source, G, \text{interesting configurations like narrow corridors}\}$;
2   $E \leftarrow NIL$;   ▷ inital edges
3   **for** *all* $u \in V$ **do**
4      $MakeSet(u, \kappa)$;   ▷ disjoint set (like PRM)
5   **for** *all* $v \in \delta(v)$ **do**   ▷ edges to neighbours (like PRM)
6   **if** $\Delta(u, v)$ *is not null* **then**
7   add $\Delta(u, v)$ to E ;
8            $Union(\kappa(u), \kappa(v))$;

9   **while** *maximum running time* **do**
10    $r_{strategy} \sim U[0, 1]$;
11   **if** $r_{strategy} < p_{connect} \wedge size(\kappa) \geq 2$ **then**
       ▷ Expanding by connecting strategy ;
12   Select κ_i and κ_j as per strategy ;   ▷ 2 disjoint subgraphs
13   Select q_{rand} randomly from κ_j ;   ▷ from 2nd subgraph
14      $q_{near} \leftarrow argmin_{q_t \in \kappa_i} d(q_t, q_{rand})$;   ▷ from 1st subgraph
15   **else**
       ▷ Expanding by random strategy ;
16      $q_{rand} \leftarrow$ random sample as per sampling strategy ;
17      $q_{near} \leftarrow argmin_{q_t \in V} d(q_t, q_{rand})$;
18    $\lambda \leftarrow min(S, d(q_{near}, q_{rand}))/d(q_{near}, q_{rand})$;
19    $q_{new} \leftarrow (1 - \lambda)q_{near} + \lambda q_{new}$;
20   **if** $\Delta(q_{near}, q_{new})$ *is not null* **then**
21      $MakeSet(q_{new}, \kappa), union(\kappa(q_{new}), \kappa(q_{near}))$;
22   add q_{new} to V, add $\Delta(q_{near}, q_{new}), \Delta(q_{new}, q_{near})$ to E ;
23   **for** $v \in V$ **do**   ▷ edges to neighbours (like PRM)
24   **if** $\Delta(v, q_{new})$ *is not null* **then**
25   Add $\Delta(q_{new}, v), \Delta(v, q_{new})$ to E ;
26            $union(\kappa(q_{new}), \kappa(q_{near}))$;

27   **return** $GraphSearch(RM < \pi, E >, source, G)$;

9.6.3   CForest

The *Coupled Forest Of Random Engrafting Search Trees* (C-FOREST) (Otte and
Correll, 2013) algorithm is a parallel implementation of RRT* wherein each node
stores an RRT* tree which is continuously expanded with time in an RRT* formulation. The different mechanisms of parallel implementation discussed so far

carried coordination at the very end. All processors computed the path independently and the final results were integrated to make the resultant path from source to goal. The CForest algorithm carries *coordination* between the different trees running in different processors, so that each of them can be actively guided several times while they are searching for a path (and before the stopping criterion is exhausted). The different computing nodes exchange information during the running of the algorithm. The processor which has computed a good result immediately shares the result with the other processors.

The exchange of information between the processors happens through the best paths and path costs. Suppose a processor finds a path better than the current best, which is sent to all other processors. The best paths are added to the tree of the other processors. Let T_i be the tree at processor i and let the best path shared be $(\tau_{best} = \{q_{best, j}\})$, which consists of vertices $q_{best,j}$ in a sequential order, starting from the source, $q_{best,0} = Source$ and ending at the goal $q_{best,n} = G$. The best path in the same order $q_{best,j}$ is continuously sampled as q_{rand} for T_i and added into the tree using RRT* formulation. At the first iteration $q_{best,1}$ is taken as q_{rand}, then $q_{best,2}$ is taken as q_{rand} and so on. The best path discovered by one processor is then inserted in the RRT* of the other processor with the needed re-wirings of the trees. Some vertices, especially the source and goal, may already be available in the tree and therefore the same is not re-inserted. In such a manner, if one processor finds a better path, it is instantly incorporated by all the other processors so that they can also further reduce the cost. The concept is shown in Fig. 9.19.

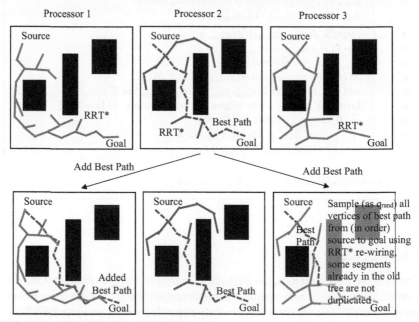

FIG. 9.19 CForest Transfer of best path.

The trees can become excessively large in terms of the number of vertices which severely affects the computation time. The CForest algorithm uses the concept of *pruning* like the Anytime RRT algorithm. Consider that the best path found so far has a cost L (length). In the case of RRT*, neither optimal cost from source is known nor optimal cost from the goal is known and therefore let $h_S(v)$ be the heuristic estimate of the cost from source and $h_G(v)$ be the heuristic estimate of cost from goal. While the current cost from source is known, since the algorithm runs RRT*, the costs will be iteratively reduced along with time. Both heuristics are defined as straight-line distances. If $h_S(v) + h_G(v) > L$, then the vertex is guaranteed not to give the shortest path and should therefore be pruned, including all decedents that it may have. The set of all v that satisfy the equation of acceptability, $h_S(v) + h_G(v) \leq L$, $v \in C^{free}$, constitute an ellipse in C^{free}. Any configuration outside the ellipse should be pruned and not admitted at any time during the search.

The other concept introduced by the algorithm is that of bounds. The paths from source to the goal are typically found in-between source and goal, however, this cannot be used as a strict constraint since many times the robot has to go at the back around an obstacle to connecting to the goal. However, if the best path so far is of cost L (length) and $d(S,G)$ is the Euclidian distance between source and goal, the maximum deviation from the straight-line path is $(L - d(S,G))$, which means the robot cannot move beyond a box of size $(L - d(S,G))/2$ around the axis-aligned box between source and goal. These bounds are used for the generation of q_{rand}. This is certainly not the tightest bound which will involve working within the ellipse; however, any sample generated will pass the same ellipse test before being used.

9.7 Distance functions

A necessity of some planning problems is to have a distance measure in the configuration space that ultimately translates to giving the path length. The Euclidian distance is a commonly used distance measure; however, it may need adaptations in different spaces. Imagine a planar rectangular robot operating in SE(2) configuration space with configuration space variables as X-coordinate, Y-coordinate, and orientation value. The robot goes from the pose $q_1(x_1, y_1, \theta_1)$ to the pose $q_2(x_2, y_2, \theta_2)$. A Euclidian distance function is given by Eq. (9.22).

$$d(q_1, q_2) = \sqrt{(x_1 - x_2)^2 + (y_1 - y_2)^2 + AngleDiff(\theta_1 - \theta_2)^2} \qquad (9.22)$$

The angle difference function $(AngleDiff(\theta) = abs(atan2(sin\ \theta, cos\ \theta)))$ is intentionally used to denote that the difference may be taken clockwise or anti-clockwise and needs to be reported in the closed range of 0 to π. So, the distance between $-\pi$ and π should be 0 instead of -2π. A problem here is that the XY positions can be arbitrarily large or small; however, the absolute difference of orientation will be strictly less than π. In any case, it is not advisable to arithmetically add quantities of different nature. The weighted addition therefore makes more sense, given by Eq. (9.23).

$$d(q_1, q_2) = \sqrt{(x_1 - x_2)^2 + (y_1 - y_2)^2 + w.AngleDiff(\theta_1 - \theta_2)^2} \quad (9.23)$$

Here the parameter w can be used to control the contribution of the orientation. It also accounts for the fact that small mobile robots that are easy to turn may have a small contribution of w, while large trucks that are hard to turn may have a large w. However, the parameter is practically hard to set.

A workaround is to only consider the displacement of a characteristic point in the robot. However, if the characteristic point is the centre of the robot, the distance function gives 0 if the robot only turns on its axis, which differential wheel drive systems can do. Hence multiple characteristic points may be taken. A better and parameter-less distance function is the *volume swept* distance function where the distance is the volume physically swept by the robot when it travels from one configuration to the other. The distance function can accommodate only orientation and only translation cases and is parameter-free. For 2D cases, the area swept has the same meaning.

An interesting case arises with the manipulator as well. Considering the manipulator has a limited turning angle range, the configuration is given by the joint angles. If the manipulator goes from the one configuration $q_1(\theta_{11}, \theta_{12})$ to the other $q_2(\theta_{21}, \theta_{22})$, the distance function using the Euclidian metric is given by Eq. (9.24).

$$d(q_1, q_2) = \sqrt{(\theta_{11} - \theta_{21})^2 + (\theta_{12} - \theta_{22})^2} \quad (9.24)$$

The angle difference function as above is not used because the orientation angles are assumed to be bounded that does not facilitate the cyclic property. However, the first link of a manipulator carries the second link on top of it and therefore it requires more effort to move the first link. To account for this, again a weight parameter (w) may be added making the distance function given by Eq. (9.25).

$$d(q_1, q_2) = \sqrt{(\theta_{11} - \theta_{21})^2 + w(\theta_{12} - \theta_{22})^2} \quad (9.25)$$

9.8 Topological aspects of configuration spaces

This concludes the coverage of the deliberative algorithms for motion planning. With a better understanding of the configuration spaces through several deliberative algorithms, let us conclude the discussions with another aspect of the configuration spaces. The configuration spaces have topological properties that are interesting to study. A generalization that has been applied in all the approaches so far is to say every bounded space is taken to be $\mathbb{R}^n$ configuration space.

More generally, we state that two spaces are *homeomorphic* to each other if one configuration space can be mapped into the other by a mapping function φ. Imagine a configuration space C_1 and another configuration space C_2. Assume that the mapping function φ maps C_1 to C_2, and similarly the inverse function φ^{-1} maps C_2 to C_1. If the mapping functions φ and φ^{-1} exist, it means that the planning problem could be mapped from C_1 to C_2 and solved in C_2 to yield a trajectory in C_2, which could then be transferred back into C_1. Imagine C_1 is $(-1,1)^2$ and C_2 is $(-10,10)^2$. Let us take the mapping function as $\phi(x,y) = (10x, 10y)$ and $\phi^{-1}(p,q) = (0.1p, 0.1q)$. Now planning and results can be transformed between spaces.

Two configuration spaces C_1 and C_2 are homeomorphic if one can be transformed into the other by applying infinitely small deformations infinite times making the mapping function φ. For φ and φ^{-1} to exist, it is necessary that φ is one-to-one and onto, or that φ is a bijection. Further, both φ and φ^{-1} should be continuous, otherwise, the trajectory definitions being continuous may not be transferable. As an example, the $\mathbb{R}^2$ can be transformed into a circle, ellipse, star, which are all homeomorphic to each other. A torus T^2 is not homeomorphic to $\mathbb{R}^3$. The little circle in the centre can be made infinitely small in the torus, but it cannot disappear to make $\mathbb{R}^3$. The torus T^2 is not homeomorphic to $\mathbb{R}^2$, since the torus cannot be cut through to make $\mathbb{R}^2$, as is done several times for visualization purposes including done by us in Chapter 6. Cutting loses the circular property hence represents a different problem.

The path has smoothness properties and homeomorphism does not ensure that a trajectory smooth in one space will also be smooth in the other space. This can be problematic as motion planning algorithms have smoothness guarantees to make for the trajectories, to ensure the traverse-ability of the path. *Diffeomorphism* is the same as homeomorphism which also enforces a smoothness constraint by stating that φ^{-1} should be smooth (or it must have derivatives of infinite degree possible). Two configuration spaces C_1 and C_2 are *diffeomorphic* if one can be transformed into the other by applying infinitely small *smooth* deformations infinite times making the mapping function φ. φ must be a bijection and φ^{-1} should be smooth. As an example, the $\mathbb{R}^2$ is not diffeomorphic to a circle, ellipse, star, etc. because there is no way to get a corner. The circle and ellipse are diffeomorphic forms of each other.

A common visualization technique is to cut through a torus to visualize that into a 2-D figure that can be easily seen on a piece of paper as was done in Chapter 6. This was possible because, around any arbitrary point, the torus was locally homeomorphic (and diffeomorphic) to $\mathbb{R}^2$. A configuration space C_1 may not be homeomorphic (and diffeomorphic) to another configuration space C_2; however, it may be possible to take a point q in C_1 and a neighbourhood around q, which is homeomorphic (and diffeomorphic) to some $\mathbb{R}^k$. The small space around q is called *locally homeomorphic* (and *locally diffeomorphic*). If such a local homeomorphism exists for all such points in the configuration space C_1 and all such local neighbourhood can be mapped into some real numbered space $\mathbb{R}^k$, then C_1 is said to have a *k-dimensional manifold*.

The torus has the same properties. Any point taken has a (large) neighbourhood which is locally diffeomorphic to $\mathbb{R}^2$. Even though mapping into $\mathbb{R}^2$ requires tearing down the torus, it could be torn down at any convenient place to plot in $\mathbb{R}^2$. One such mapping is called a chart. A *chart* is a mapping of a sub-space around c in C_1 into some $\mathbb{R}^k$ using a mapping function φ, such that the mapping is a diffeomorphism. The name is derived from the fact that even though the earth is spherical, the maps or charts are drawn as 2D, only because the earth at any point of the surface is a local diffeomorphism to $\mathbb{R}^2$.

The tearing down of a torus does not convey the cyclic property, and therefore it is necessary to make multiple such mappings covering redundant spaces in the torus to convey the message that the torus is cyclic. The earth can be approximately represented as a world map; however, it does not show that the earth is spherical. To convey the spherical shape, multiple such charts or mappings should be used. A collection of all such mappings constitutes the *atlas*. The name is derived from the fact that an atlas book has all maps of the earth.

The presence of multiple charts may though create a problem in planning. The source may be at some chart, while the goal may be at some other chart and planning may require working in several charts rather than the original space C. This creates a problem to travel between charts which is however possible since all mapping functions are known. It is a common exercise when programming angles, wherein hopping the angle π automatically switches to range beyond $-\pi$. Complex robots create numerous charts on different constraints. Planning in single configuration space is not possible due to the difficulties in representing such constraints. As an example, consider a humanoid walking. Now the configuration space encodes all possible joint angles and pose of the centre of the humanoid. It can model unrealistic configurations like the humanoid flying or at unstable positions. Modelling such constraints is much simple if a chart is taken, wherein (at least) one toe leg is fixed. This modelling rules out impossible cases and constraints are naturally represented in a chart. The complete humanoid is just an open kinematic chain (like a manipulator) with one toe fixed and the joint angle configurations constitutes the chart. A typical motion planning problem is to get all neighbouring configurations around the current configuration (q). For the chart with one leg fixed, this means moving all links of the kinematic chains by all combinations by a little amount. However, if working in the complete configuration space, this corresponds to simultaneously moving the centre of mass, arms, and legs in all combinations, while the motion of the centre of mass depends upon the motion of the legs creating constraints that most of the neighbours generated may not follow and thus a significant part of the calculations results in invalid states. However, the goal is not in the current chart, for which the humanoid will have to walk. Every step while walking is going from one chart to the other, eventually reaching the chart which has the goal.

Similarly, while pushing a box, the robot and the box make a complete composite system. However, the box itself cannot be controlled or moved. This breaks the problem into multiple charts, charts where the robot is free and charts where the robot is in contact with the box. When the robot is with the box, the forces govern the motion of the robot and box. This creates planning systems that continuously switch charts within an atlas. On a more general note, if a path exists between any two points of the manifold, then space is called *path-connected* or *connected*.

The 1D orientation has cyclic properties and it can be plotted in 2D as a circle. The 2D plot is a complete configuration space conveying the cyclic property. This does not need multiple charts making an atlas. The 2D plot is a submanifold representing orientation. However, care must be given while dealing with the 2D plot as all configurations must strictly lie on the circle. A planner may want to connect the top position ($x=0$, $y=1$) with the bottom ($x=0$, $y=-1$) using a straight-line connection, but a path is only possible by following the perimeter of the circle, representing valid orientations. The planner must constraint all paths to strictly lie on the circle. This may be easy in some planning algorithms and difficult for others. This leaves with two options when dealing with planning for manifolds. The first option is representing them in higher dimensions wherein they are completely represented, however, there are constraints that the positioning should follow to be always on a constrained surface within the higher dimension. To go from the top to the bottom of the circle as a classic search, the system may find neighbouring points (x,y) to the source ($x=0$, $y=1$), but out of infinite such points, only 1 point will be feasible that lies on the circle. The second option is to make multiple charts, which complicates planning due to the need of switching between the charts. In this example, taking the orientation as an angle makes travelling on the orientation axis as a simple problem; however, the logic that shooting out of π injects one into $-\pi$ due to the cyclic properties needs to be constantly added at all steps. Higher dimensional spaces have infinite charts with infinite switching points between charts making planning hard. Both are valid techniques to consider, and a choice is application-dependent.

Questions

1. Show the working of every RRT variant discussed for the maps as shown in Fig. 9.20.

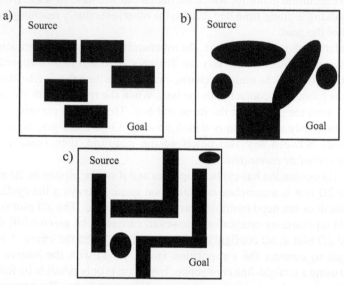

FIG. 9.20 Maps for question 1. (A) Map 1, (B) Map 2, (C) Map 3.

2. Explain the effect of using different sampling strategies (discussed in Chapter 8) for the generation of a random sample in the working of the RRT algorithm.
3. Compare and contrast between the techniques (a) kinodynamic planning using RRT in control space, (b) planning with RRT in configuration space with high clearance and thereafter locally optimizing the path, (c) Doing a coarser resolution planning using a graph search, followed by kinodynamic planning for every segment of the coarser path.
4. Explain the strategies to seed local trees in a multi-tree approach for different scenarios. Also, discuss the expected performance of the algorithm as the number of trees are increased. Also, explain mechanisms to intelligently determine the number and location of the trees for a single query planning problem.
5. A static map is to be used for motion planning. The map may get several queries at different times, each mentioning a source and goal. The application designer intends to use the RRT algorithm instead of the PRM algorithm for motion planning. However, the designer believes that the RRT run at time *i* could potentially use the computations of all past queries since the map is static. Give the modified RRT algorithm.
6. [Programming] Assume planning for a point robot in a 2D workspace. Read the map as bitmap images. Write a program to make a roadmap using every RRT variant discussed.

7. [Programming] In the above question, repeat the solution assuming the map as polygons instead of a bitmap.

8. [Programming] Instead of a point robot, solve for (a) a 2-link revolute manipulator, (b) a circular robot, (c) a rectangular robot, (d) Two robots as a circular disk of radius R. Solve for both polygon and bitmap workspace maps.

9. [Programming] Extend the RRT algorithm for the same problems to avoid a dynamic (polygon) obstacle that moves in a predefined trajectory. Ensure that in the simulated test case, the shortest path of the robot and the dynamic obstacle incurs collision.

10. [Programming] The RRT algorithm discussed so far only took path length as an objective. Modify the program to make the robot conscious of clearance and smoothness.

11. [Programming] The RRT algorithms discussed so far assumed that the map is fully known. Make a robot simulation for a mobile robot with a laser scanner. Make a noisy control for the same robot. Perform active SLAM using the RRT algorithm. The robot should do localization and mapping as it goes along. The RRT algorithm will try to make the robot go using short routes.

12. [Programming] Assume the robot needs to traverse through cliffs and troughs and the cost of taking a unit step depends upon the slope (costmap). Modify the RRT to work with such costmaps.

13. [Programming] Go through the Open Motion Planning Library. Study the implementation of the RRT algorithm. (a) Modify the code such that sometimes the RRT expands a single step and sometimes by multiple steps like RRT-Connect, (b) Modify the code such that the tree follows a combination of RRT and EST style expansions (c) Modify the code such that the tree largely focuses on expansions near areas with a high clearance, but sometimes also uses areas with a small clearance.

References

Akin, M., Bekris, K.E., Chen, B.Y., Ladd, A.M., Plaku, E., Kavraki, L.E., 2005. Probabilistic road-maps of trees for parallel computation of multiple query roadmaps. In: Robotics Research. The Eleventh International Symposium. Springer Tracts in Advanced Robotics 15. Springer, Berlin, Heidelberg, pp. 80–89.

Bekris, K.E., Chen, B.Y., Ladd, A.M., Plaku, E., Kavraki, L.E., 2003. Multiple query probabilistic roadmap planning using single query planning primitives. In: Proceedings of the 2003 IEEE/RSJ International Conference on Intelligent Robots and Systems. IEEE, Las Vegas, NV, pp. 656–661.

Bohlin, R., Kavraki, L.E., 2000. Path planning using lazy PRM. In: Proceedings of the 2000 IEEE International Conference on Robotics and Automation. IEEE, San Francisco, CA, pp. 521–528.

Bohlin, R., Kavraki, L.E., 2001. A randomized approach to robot path planning based on lazy evaluation. In: Rajasekaran, S., et al. (Eds.), Handbook on Randomized Computing. Kluwer Academic Publishers, Dordrecht, The Netherlands, pp. 221–253.

Devaurs, D., Siméon, T., Cortés, J., 2011. Parallelizing RRT on distributed-memory architectures. In: Proceedings of the 2011 IEEE International Conference on Robotics and Automation. IEEE, Shanghai, pp. 2261–2266.

Devaurs, D., Siméon, T., Cortés, J., 2013. Parallelizing RRT on large-scale distributed-memory architectures. IEEE Trans. Robot. 29 (2), 571–579.

Ferguson, D., Stentz, A., 2006. Anytime RRTs. In: Proceedings of the 2006 IEEE/RSJ International Conference on Intelligent Robots and Systems. IEEE, Beijing, pp. 5369–5375.

Ferguson, D., Stentz, A., 2007. Anytime, dynamic planning in high-dimensional search spaces. In: Proceedings of the 2007 IEEE International Conference on Robotics and Automation. IEEE, Roma, pp. 1310–1315.

Hsu, D., Latombe, J.C., Motwani, R., 1999. Path planning in expansive configuration spaces. Int. J. Comput. Geom. Appl. 9 (4–5), 495–512.

Hsu, D., Kindel, R., Latombe, J.C., Rock, S., 2002. Randomized kinodynamic motion planning with moving obstacles. Int. J. Robot. Res. 21 (3), 233–255.

Ichnowski, C., Alterovitz, C., 2012. Parallel sampling-based motion planning with superlinear speedup. In: Proceedings of the 2012 IEEE/RSJ International Conference on Intelligent Robots and Systems. IEEE, Vilamoura, pp. 1206–1212.

Kala, R., 2013. Rapidly exploring random graphs: motion planning of multiple mobile robots. Adv. Robot. 27 (14), 1113–1122.

Karaman, S., Frazzoli, E., 2011. Sampling-based algorithms for optimal motion planning. Int. J. Robot. Res. 30 (7), 846–894.

Karaman, S., Walter, M.R., Perez, A., Frazzoli, E., Teller, S., 2011. Anytime motion planning using the RRT*. In: Proceedings of the 2011 IEEE International Conference on Robotics and Automation. IEEE, Shanghai, pp. 1478–1483.

Kuffner, J., LaValle, S.M., 2000. RRT-connect: an efficient approach to single-query path planning. In: Proceedings of the 2000 IEEE International Conference on Robotics and Automation. IEEE, San Francisco, CA, pp. 995–1001.

LaValle, S.M., Kuffner, J.J., 1999. Randomized kinodynamic planning. In: Proceedings of the 1999 IEEE International Conference on Robotics and Automation. IEEE, Detroit, MI, pp. 473–479.

Lavalle, S.M., Kuffner, J.J., 2000. Rapidly-exploring random trees: progress and prospects. In: Algorithmic and Computational Robotics: New Directions. CRC Press, Boca Raton, FL, pp. 293–308.

LaValle, S.M., Kuffner, J.F., 2001. Randomized kinodynamic planning. Int. J. Robot. Res. 20 (5), 378–400.

Otte, M., Correll, N., 2013. C-FOREST: parallel shortest path planning with superlinear speedup. IEEE Trans. Robot. 29 (3), 798–806.

Plaku, E., Bekris, K.E., Chen, B.Y., Ladd, A.M., Kavraki, L.E., 2005. Sampling-based roadmap of trees for parallel motion planning. IEEE Trans. Robot. 21 (4), 597–608.

Raveh, B., Enosh, A., Halperin, D., 2011. A little more, a lot better: improving path quality by a path-merging algorithm. IEEE Trans. Robot. 27 (2), 365–371.

Strandberg, M., 2004. Augmenting RRT-planners with local trees. In: Proceedings of the IEEE International Conference on Robotics and Automation. IEEE, New Orleans, LA, pp. 3258–3262.

Şucan, I.A., Kavraki, L.E., 2009. Kinodynamic motion planning by interior-exterior cell exploration. In: Algorithmic Foundation of Robotics VIII. Springer Tracts in Advanced Robotics. vol. 57. Springer, Berlin, Heidelberg, pp. 449–464.

Urmson, C., Simmons, R., 2003. Approaches for heuristically biasing RRT growth. In: Proceedings of the 2003 IEEE/RSJ International Conference on Intelligent Robots and Systems. IEEE, Las Vegas, NV, pp. 1178–1183.

Chapter 10

Artificial potential field

10.1 Introduction

This chapter (and the few next) deals with the *reactive* mechanism for the motion of a robot, wherein the robot does not make long-term decisions by computationally expensive settings considering numerous possibilities, but instead directly maps the current sensory percept into a navigation decision with a small amount of algorithmic computation. Hence, the algorithms are computationally inexpensive; however, since most long-term plans are not considered, the plans are inherently laden with both optimality and completeness.

Most of the human navigation and planning is instinctive and reactive. One can casually walk around places while still texting a colleague or talking in person to another colleague, avoiding all static and dynamic obstacles. Since the human while walking focusses on texting and talking, they do not make long-term plans, considering a lot of possibilities, optimize plans, and then make a move. Instead, the navigation happens by subconsciously mapping whatever one sees with the eyes and ears as sensors, to the action signals given to the legs. This constitutes the *reactive* decision-making.

The reaction mechanism is sometimes inbuilt into humans (like crying when hungry), sometimes given to us by rules that constitute common sense (like not going out when it is raining), and sometimes learnt by practice (like the reaction of a good soccer player). In all cases, there is some mechanism by which a rule base or a *knowledge base* is made. An *inference engine* uses the knowledge base to decide the output action when given the inputs. The modelling of a knowledge base is different and depends on the tool used. As an example, the simplest systems use hard-coded rules that map the sensory percept into the actions that should be taken, and the knowledge base consists of rules in the form of if (antecedent) then (consequent). Similarly, many times the reaction of the system may be modelled by mathematical equations that are designed by a human. In such a case, the equations and their parameters constitute the knowledge base. In both cases, there is an element of learning possible by either learning just the parameters of the rules/equations or the complete rule/complete equation. The learning happens by measuring the performance, for which the real sensor readings are converted into a performance metric. The metric is used for the adjustment

Autonomous Mobile Robots. https://doi.org/10.1016/B978-0-443-18908-1.00009-1

of the knowledge base as per the modelling used. A characteristic is the use of neural networks as the knowledge base, wherein the sensory percept is given to the input neurons that are mapped by the neural network equations to the corresponding output. The weights and the architecture of the neural network are learnt. The concepts are shown in Fig. 10.1. The knowledge base constitutes the fundamental of how an action is taken and constitutes the *behaviour* of the system. Different humans have different behaviours that are characterized by a different mechanism by which they react to a situation.

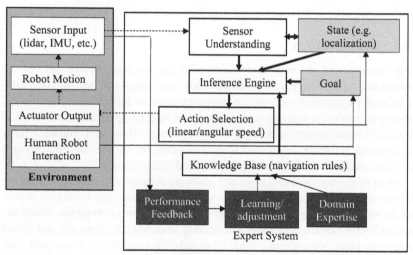

FIG. 10.1 Expert system design for reactive navigation of robots. The decisions are reactively made by knowledge base rules (like turn right if obstacle in front), which may be learned (if the robot went too close to the obstacle, the rule is adjusted).

So far, the maps have been inherently static. This was because, if the map changes anytime between the start of the planning and the end of the motion of the robot, the trajectory becomes invalid and needs to be re-computed. A better mechanism was to adapt the old trajectory and the old computed meta-data (e.g. roadmap, tree, cells, and search tree) to the changes in the environment. This was only applicable for small changes in the map. In this manner, the D*-Lite algorithm could quickly compute a new trajectory, in contrast to running a costly A* algorithm.

However, imagine a crowded marketplace with all the other moving people modelled as obstacles, and somehow you know the positions of each of them, and you intend to go to your favourite store. In highly crowded places, no trajectory will be possible that lets you reach the store, as there will be people all around. This does not mean that you cannot reach the store, as people just ahead of you will clear with time and you can move step by step. Further, since the number of obstacles (people) is large, the planning (and re-planning) time will also be very large. Most of this time is wasted since the dynamic obstacles (people) will clear with time to new positions, which will invalidate their old positions assumed for planning. If velocities of the other obstacles are known, it is possible to extrapolate their motion and account for the changing maps a priori. However, again, in such situations, the dynamics of the obstacles (people) are too complex to be accurately modelled to construct a good deliberative plan.

In such situations, the dynamic obstacles move at high speeds, have complex dynamics, have unclear intents, and may appear rather suddenly. The navigation is very conveniently handled by humans since they react to all such situations. A reactive algorithm does exactly what the humans do: move step by step at any instant of time.

This chapter aims to make a reactive navigation system wherein the sensory percept is mapped to the actuator actions and doing the same continuously in a loop moves the robot. The humans may have, at any instant of time, numerous obstacles, static and dynamic (other people, robots, etc.) around them. It is worth noting that the current reactive step only depends upon the distances, angles, and speeds to the nearest four to six obstacles (static obstacles and moving people), the distance to the goal and the angle to the goal. It does not depend upon the obstacles far away. This is further motivated by Fig. 10.2A. To understand the navigation behaviour, let us conveniently decompose it as a composite of two behaviours, first to avoid obstacle and second to move to the goal, shown in Fig. 10.2B.

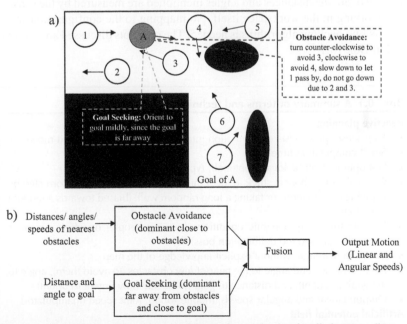

FIG. 10.2 Reactive decision-making. (A) The reactive motion is primarily based on distances, speeds, and angles of the nearest four dynamic obstacles (people, robots) or static obstacles, distance to the goal, and angle to goal. (B) Robot tries to balance obstacle avoidance and goal-seeking behaviour.

The *obstacle avoidance behaviour* ensures that the robot does not collide with any obstacle. If a robot is about to collide with an obstacle o, with the obstacle o at an angle α_o, the action should be to ask the robot to move at an angle $\pi + \alpha_o$. As an example, on getting an obstacle directly ahead, the best way is to step back; on getting an obstacle at the right, it is wise to step left; and so on. The closer a person gets to the obstacle, the more important it is to take an avoidance step. Therefore, obstacles close by have more effect and those far away have a lesser effect, meaning that the behaviour is primarily affected by the nearest few obstacles only. The decision is called *local* since it is only made based on the local sensory inputs.

The other behaviour is called *goal seeking*. If a robot is far away from the obstacles, it is imperative to walk towards the goal, and hence the angle to the goal as an input corrects the robot deviation. The robot may have to stop at the goal and may have to adjust its speed for which distance to the goal as an input is used. Furthermore, near the goal, it is more important to quickly correct any errors in terms of heading, rather than far away from the goal.

The fusion of the two behaviours depends upon the mechanism by which they are modelled or the reactive algorithm in use. In this chapter, the simplest case is taken, wherein a mathematical equation is taken by natural analogy, and the same is used for the actual motion of the robot. Another marked change from the earlier chapters is that reactive decision-making may be done directly in the workspace. So, the distances and angles mentioned are measured by the sensors while working in the workspace itself. No mapping to the configuration space by collision-checking algorithms is needed. The concepts discussed are summarized in Box 10.1.

Box 10.1 A summary of terms and techniques

Reactive planning
- Map sensor percept/state information into actuator outputs for a unit move
- Small computation time
- Not optimal (can be locally optimal), typically not complete
- Can make the robot get stuck in local minima. Sometimes maybe corrected by applying perturbations or taking a long random walk (biased towards a random direction)
- Suitable for highly dynamic conditions (moving people) or when dynamics cannot be guessed (no deliberation possible)
- Works in real-time, without explicit knowledge of the map
- Input: Distance and angle to the nearest few obstacles (to avoid them), angle to the goal (to orient), and distance to goal (to decide how urgent to orient)
- Output: Linear and angular speeds (mobile robot), joint speed (manipulator)

Artificial potential field
- Robot is positively charged and is attracted by negatively charged goal and repelled by positively charged obstacles

---

Box 10.1 A summary of terms and techniques—cont'd

- Attractive potential is quadratic to distance near goal and conic far away
- Repulsive potential is inversely proportional to the square of distance
- For multiple obstacles, repulsion may sum over all obstacles or take the nearest obstacle only. For a large obstacle, integrating over all interior points may not be feasible; the nearest point in the obstacle is preferable. For multiple obstacles, summation may be done.
- For a mobile robot, angular speed is set to attain total force direction, the component of the total force in the direction of motion changes linear speed.
- For a manipulator, the force gives the desired cartesian speed of the end-effector, converted into the joint speed using manipulator Jacobian. Alternatively, forces at links are converted into torques and then into joint speed.

Social potential field

- Every two people attract each other if they have a separation more than d_{soc} and repel each other if they have a separation less than d_{soc}.
- d_{soc} depends upon acquaintance between people and is modelled by person-specific attraction/repulsion coefficients and degrees.
- A person has separate coefficients for people in their group (intragroup) and people in other groups (intergroup).

Elastic strip

- The trajectory is modelled as an elastic strip, which can take any shape.
- The trajectory is repelled by obstacles (external force) and tries to contract to a straight line (internal force).
- At equilibrium, the forces cancel each other.

Adaptive roadmap

- A probabilistic roadmap where every edge is an elastic strip and responds to changing obstacles.
- Edges break on crossing tension due to obstacles. Vertices break in the absence of edges.
- Deleted vertices and edges are memorized and re-form when feasible.
- Connectivity is ensured by the Connecting Connected Component Strategy from the Probabilistic Roadmap algorithm.

10.2   Artificial potential field

The *Artificial Potential Field* (APF; Khatib, 1985; Hwang and Ahuja, 1992) is a method to navigate the robot from the source to the goal by using the analogy of an electrostatic potential field. Imagine a robot in any general position. The robot is assumed to be positively charged. The obstacles are assumed to be embedded with the positive charges as well. Since two like-charged particles repel each other, the robot will always be repelled by the obstacles. The repulsion becomes dominant when the robot is near the obstacle, in contrast to when the robot is far away from the obstacle. The goal is assumed to be strongly negatively charged. Therefore, the robot is always attracted to the goal. The presence of negatively charged goals and positively charged obstacles creates a *potential field*. The robot moves under the influence of the potential field, trying to go from a high potential area to a low potential area, and therefore attempts to reach the goal from the source while avoiding the obstacles on the way.

The force is given by a negative derivative of the potential field and models the virtual force acting on the robot to enable it to reduce its potential value in the greatest possible way. The free-body diagram of the robot shows the robot being attracted by the goal and repelled by each of the obstacles around it. This is shown in Fig. 10.3. The resultant force vector gives the desired direction of motion of the robot (Choset et al., 2005). The robot may not be able to move in the same direction due to the kinematic and nonholonomic constraints; however, it adjusts its linear and angular velocities and accelerations to go in the same direction.

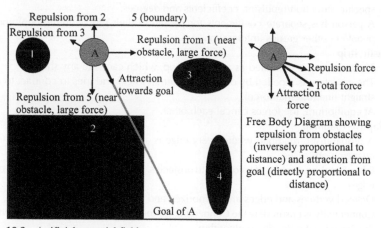

FIG. 10.3 Artificial potential field.

10.2.1 Attractive potential modelling

Imagine the robot as a positively charged particle at point q. It is subjected to an attractive potential (U_{att}^G) by the goal (G) and a repulsive potential (U_{rep}) by the other obstacles. The attractive potential (U_{att}^G) is assumed to be directly proportional to the distance of the robot to the goal $d(q,G)$, where G is the goal. Popular modelling is to make attractive potential proportional to the square of the distance given by Eq. (10.1), known as the *quadratic potential function*.

$$U_{att}^G = \frac{1}{2}k_{att}d^2(q,G) \tag{10.1}$$

Here k_{att} is the attraction constant associated with the attraction potential and d is the distance function. Correspondingly, the attractive force (F_{att}^G) is given by Eq. (10.2)

$$
\begin{aligned}
F_{att}^G &= -\nabla U_{att} = -k_{att}d(q,G)\nabla d(q,G) \\
&= -k_{att}q(q,G)\hat{u}(q-G) \\
&= -k_{att}(q-G)
\end{aligned}
\tag{10.2}
$$

Here $\hat{u}(q-G)$ is a unit vector in the direction $q-G$. Since, the potential is scalar and the position and goal are vectors, the derivative with respect to the distance generates a vector. The fastest mechanism to decrease the potential (or distance to goal) is to move towards the goal, which is directly incorporated in the derivation. The potential values are given by Fig. 10.4A, while the force as a result is given by Fig. 10.4B.

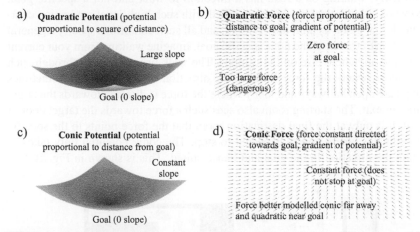

FIG. 10.4 Attractive potential (A) quadratic potential, (B) quadratic force, (C) conic potential, (D) conic force.

The problem with the quadratic potential function is that it becomes too large when the distance from the goal is large, which is when the robot is far away from the goal and starts its journey from the source. In such a case, a high attraction to the goal means that the robot goes close to the obstacles, which can be risky. The good aspect of the attractive force is that it becomes zero as the robot approaches the goal, and therefore, on reaching the goal, the robot (with the absence of momentum) successfully stops at the specific goal point.

To limit a large value of the force far away from the goal, the *conic potential* is used. The potential is directly proportional to the distance given by Eq. (10.3). The force is given by a derivate as shown in Eq. (10.4)

$$
U_{att}^G = k_{att}d(q,G)
\tag{10.3}
$$

$$F_{att}^G = -\nabla U_{att} = -k_{att}\nabla d(q, G)$$
$$= -k_{att}\hat{u}(q - G)$$
$$= -k_{att}\frac{(q - G)}{d(q, G)}$$

(10.4)

Fig. 10.4C shows the conic potential values for a synthetic scenario, while the forces because of it are shown by Fig. 10.4D. So, the conic potential models constant force acting towards the goal and therefore, the attraction force is constant. This is also visible from real-life wherein a person walks at a constant speed towards a room until the person comes close to the room and slows down in anticipation to stop. A car also moves at a constant speed until it is about to reach its destination and then slows to stop.

There is another major advantage associated with conic potential. Walking in a corridor has a general direction to follow but not a specific goal to reach. Similarly, walking on a road has a direction to walk and not a specific goal. Most of our practical navigation is filled with such high-level navigation commands that guide us until the very end, and all such conditions specify a general direction of traversal and not a specific goal. Imagine walking from your current room to anywhere else on the campus. The walk primarily goes through such corridors, where the force is along the direction of the corridors. Sometimes there are open hallways and lobby where the force is directed towards the target door or exit. The starting room also sees such a force towards the target door or exit. It is only at the final destination room that the force points to the specific coordinates where you're expected to stop. The conic potential only takes as input the direction to the goal and hence suited. This is shown in Fig. 10.5.

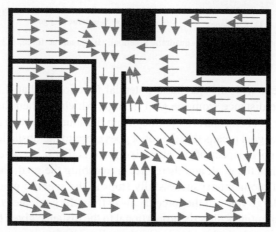

FIG. 10.5 Conic force modelling for real-life scenarios. Attraction force is constant in magnitude, towards the walking direction in corridor, towards the centre of entrance/exit/goal. This represents natural walking attraction.

The conic potential is useful far away from the goal (due to a constant magnitude) and the quadratic potential is useful near the goal (being zero at the goal), and therefore a simple rule is to fuse the two potential values, using conic potential at a distance of d_{att} or more, and quadratic potential near the goal at distances less than d_{att}. The fusion must ensure that the resultant function is continuous and has a derivative. The fused potential is given by Eq. (10.5), and the corresponding force is given by Eq. (10.6). The constants are added to make the potential and force functions continuous with a derivative:

$$U_{att}^G = \begin{cases} \dfrac{1}{2}k_{att}d^2(q,G) & \text{if } d(q,G) \leq d_{att} \\ d_{att}k_{att}d(q,G) - \dfrac{1}{2}k_{att}d_{att}^2 & \text{if } d(q,G) > d_{att} \end{cases} \tag{10.5}$$

$$F_{att}^G = \begin{cases} -k_{att}(q-G) & \text{if } d(q,G) \leq d_{att} \\ -d_{att}k_{att}\dfrac{(q-G)}{d(q,G)} & \text{if } d(q,G) > d_{att} \end{cases} \tag{10.6}$$

10.2.2 Repulsive potential modelling

The repulsive potential is used to prevent a robot from getting close to the obstacles and is modelled as being inversely proportional to the distance from the obstacle. The attractive potential is applied only by the goal; however, the repulsive potential is applied by each obstacle. Let the repulsive potential applied by a point obstacle o be given by U_{rep}^o. The potential is assumed to be inversely proportional to the square of the distance from the obstacle $d(q,o)$. Even though this methodology naturally fades away, the potential values away from the obstacles, an algorithmic limit is also applied to make the potential values 0 after a distance of d_{rep}. This limit is useful if the map is explicitly stored for static obstacles, in which case a proximity query to the nearest obstacles up to a distance of d_{rep} may be made, as computing repulsion for all obstacles in the map may be infeasible. Obstacles beyond d_{rep} are guaranteed to have a potential value of 0 and therefore do not affect the calculations. The potential function is hence adjusted to attain continuity around d_{rep}. The repulsive potential is given by Eq. (10.7):

$$U_{rep}^0 = \begin{cases} \dfrac{1}{2}k_{rep}\left(\dfrac{1}{d(q,o)} - \dfrac{1}{d_{rep}}\right)^2 & \text{if } d(q,o) \leq d_{rep} \\ 0 & \text{if } d(q,o) > d_{rep} \end{cases} \tag{10.7}$$

Here, k_{rep} denotes the constant associated with the repulsive potential and d is the distance function. The force is a derivative of the potential and is given by Eq. (10.8):

$$F_{\text{rep}}^0 = -\nabla U_{\text{rep}} = \begin{cases} -k_{\text{rep}} \left(\dfrac{1}{d(q,o)} - \dfrac{1}{d_{\text{rep}}} \right) \dfrac{-1}{d^2(q,o)} \nabla d(q,o) & \text{if } d(q,o) \le d_{\text{rep}} \\ 0 & \text{if } d(q,o) > d_{\text{rep}} \end{cases}$$

$$= \begin{cases} k_{\text{rep}} \left(\dfrac{1}{d(q,o)} - \dfrac{1}{d_{\text{rep}}} \right) \dfrac{(q-o)}{d^3(q,o)} & \text{if } d(q,o) \le d_{\text{rep}} \\ 0 & \text{if } d(q,o) > d_{\text{rep}} \end{cases} \qquad (10.8)$$

The cumulative repulsion is the repulsion from all obstacles o. It can be modelled in two ways. The first way to model the repulsion is to take the addition from all obstacles at a distance of d_{rep} or less (obstacles beyond d_{rep} do not contribute to repulsion). This is given by Eqs (10.9) and (10.10):

$$U_{\text{rep}} = \sum_o U_{\text{rep}}^o \qquad (10.9)$$

$$F_{\text{rep}} = \sum_o F_{\text{rep}}^o \qquad (10.10)$$

Fig. 10.6A shows the repulsion by such a technique. The forces are shown in Fig. 10.6B. As the force is too high, close to the obstacles, better visualization is by only showing the direction of the force and not its magnitude by drawing arrows of the same size, as shown in Fig. 10.6C.

a) Obstacles (infinite potential), large slope near obstacles

Repulsive Potential (inversely proportional to square of distance

b) Large force near obstacle boundaries, small force far away from obstacles

Repulsive Force (inversely proportional to cube of distance)

c)

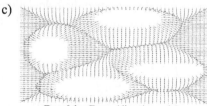

Repulsive Force (direction only)

FIG. 10.6   Repulsion from obstacles (A) potential, (B) repulsive force (C) repulsive force showing direction only.

The advantage of this approach is that the force values always have a smooth change as the obstacles come closer to the robot and the robot acts to avoid them. The problem with the approach is the need to model for each obstacle. Suppose a circular obstacle is directly ahead of the robot. Eqs (10.7) and (10.8) assume that the obstacle is a point, and a large obstacle will require integral over all the points in the obstacle, which is computationally hard and infeasible for many complex obstacles. However, the scheme is acceptable if the addition is done between different discrete obstacles. Representing every obstacle by the best (closest) point, which is the most active, solves the problem.

The second scheme takes the maximum overall obstacles to calculate the cumulative repulsive force (or potential). In the earlier discussion, when an obstacle was represented by the best point o, the fusion mechanism was to take the maximum overall force values for that obstacle. The repulsive potential and force are given by Eqs (10.11) and (10.12).

$$U_{\text{rep}} = \max_o \left(U_{\text{rep}}^o \right) \tag{10.11}$$

$$F_{\text{rep}} = \max_o \left(F_{\text{rep}}^o \right) \tag{10.12}$$

The problem with the approach is that the change in potential as the robot moves will no longer be smooth, as the robot changes from one obstacle to another. Imagine the robot going in a narrow corridor. The moment it is in the middle of the corridor, if it slightly deviates left, the direction of the force becomes right, and vice versa. The change in direction of force is sudden and large.

Consider a general workspace where a mobile robot is navigating. Consider a few obstacles, such as chairs, tables, etc., that the robot must surpass. If the robot adds the repulsions due to each of these obstacles, the sum methodology is being followed. For each of these obstacles (say a table), if the robot only considers the closest point in the table and calculates the repulsion due to it, the maximum modelling for the specific obstacle is being followed. Similarly, the robot may only take the boundary points of the table instead of all the internal points, in which case the solution is intermediate between sum (over all boundary points) and maximum (over a few internal points represented by a boundary point only).

10.2.3 Artificial potential field algorithm

The total potential and force account for both the attraction and the repulsive components, given by Eqs (10.13) and (10.14).

$$U = U_{\text{att}}^G + U_{\text{rep}} \tag{10.13}$$

$$F = F_{\text{att}}^G + F_{\text{rep}} \tag{10.14}$$

The robot moves under the influence of this force. The robot is said to be moving with a *gradient descent* of the potential values that is to move in the direction such that the reduction in potential is quick. This corresponds to moving towards the goal without an obstacle and moving opposite to the direction of the obstacle in the case of an obstacle in close vicinity. The resultant behaviour is a mixture of repulsions of all obstacles and attraction towards the goal. The resultant potential is shown in Fig. 10.7A. Since the potential becomes infinite near the obstacles, the potential values are sliced to largest value and a clearer figure is shown in Fig. 10.7B. The force values are similarly shown in Fig. 10.7C. The same figure is again clarified due to high force values near the obstacles, by only drawing for the direction and keeping the arrows of the same size, shown in Fig. 10.7D.

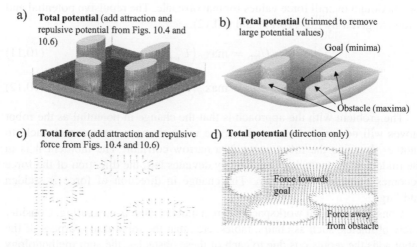

a) **Total potential** (add attraction and repulsive potential from Figs. 10.4 and 10.6)

b) **Total potential** (trimmed to remove large potential values)
Goal (minima)
Obstacle (maxima)

c) **Total force** (add attraction and repulsive force from Figs. 10.4 and 10.6)

d) **Total potential** (direction only)
Force towards goal
Force away from obstacle

FIG. 10.7 Total potential (A) attractive and repulsive potentials added (B) total potential, trimming large values. Slope large near obstacles, the overall curve inclines towards the goal. The robot path is tracing the slope in this landscape. (C) Total force, (D) total force showing direction only.

If the robot was holonomic, it could have moved in any direction. For such robots, the immediate speed can be given in the direction of the total force or the immediate direction of motion is $atan2(F_y, F_x)$. However, many robots have nonholonomic constraints that make it impossible for them to move sideways. The robots are typically moved by using linear velocity (v) and angular velocity (ω). The force is a virtual force that is exerted on the robot. The conversion of the virtual force into the control signals can be done by making the angular speed adjustments to move towards the resultant force vector given by Eq. (10.15).

$$\omega = k_\omega AngleDiff\left(atan2\left(F_y, F_x\right), \theta\right) \qquad (10.15)$$

Here θ is the orientation of the robot. The function *AngleDiff* denoting the angle difference is to imply that the output is bounded in the range $-\pi$ to π. An easy means is to use $AngleDiff(x,y) = atan2(\sin(x-y), \cos(x-y))$ that takes any unbound angle $x-y$ and bounds it to the range of $-\pi$ to π. The angular speed will be subjected to the minimum and maximum limits before being given to the robot. k_ω is the constant like the proportionality constant of a PID controller. If the force is already measured relative to the robot, the difference with θ will not be needed.

The linear speed of the robot is also adjusted by the projection of the force vector on the orientation of the robot. So, if the force is in the direction of the motion of the robot, the speed should increase, and if the force is opposite to the orientation of the robot, the speed should decrease, given by Eq. (10.16).

$$\Delta v = k_v(F \cdot u(\theta)) \tag{10.16}$$

Here $u(\theta)$ is a unit vector in the direction θ and Δv denotes the change in speed. k_v is a similar proportionality constant parameter. This mapping holds good only for mobile robots driven by a linear speed and an angular speed. Both speeds will have to be accounted for their maximum and minimum permissible values.

For manipulators, assume that the forces are calculated on the end-effector, which gives the desired cartesian speed of the end-effector. However, the manipulator moves according to the joint speed of the joints. The conversion between cartesian speed and joint speed is obtained by the *manipulator Jacobian*, which is a derivative of the forward kinematics model. The manipulator Jacobian converts the speed of the end-effector in the joint space to the speed of the end-effector of the manipulator. This does not account for the possible collisions with the arms of the robot. Another modelling, therefore, calculates the forces on all points of the manipulator and converts them into torques experienced by every link. The torques are then converted into the desired joint angular speeds. The problem here is that calculating repulsion from every obstacle in 3D to all links in 3D is computationally hard. The points of singularity (where the determinant of Jacobian becomes zero) act as additional obstacles.

The working of the algorithm on two synthetic scenarios is shown in Fig. 10.8. The robot was circular; however, had a heading direction and was controlled by linear and angular speed. The kinematic constraints of the limited velocity and acceleration for both linear and angular components were also accounted for.

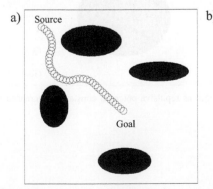
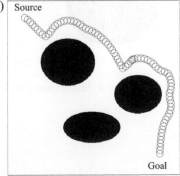

FIG. 10.8　Results using artificial potential field. (A) Map 1, (B) Map 2.

The algorithm has two major parameters, k_{att} and k_{rep}. The other parameters are either from the traditional control literature or limiting constants. The performance of the algorithm depends largely on the attraction and repulsion coefficients k_{att} and k_{rep}. Reducing k_{att} and increasing k_{rep} results in the repulsive potential dominating the attraction coefficient and therefore, the robot goes far away from the obstacles. This increases the clearance; however, inhibits the robot from entering narrow places, since the repulsion will be too strong and the attraction will be weak. An increase in clearance also results in an increase in the path length. This is shown in Fig. 10.9. On the contrary, increasing k_{att} and reducing k_{rep} result in the robot getting too close to the obstacles. In such a case, the path length may be reduced; however, the risks associated with the robot increase. The robot may accidentally slip and collide with the obstacles. This is shown in Fig. 10.10.

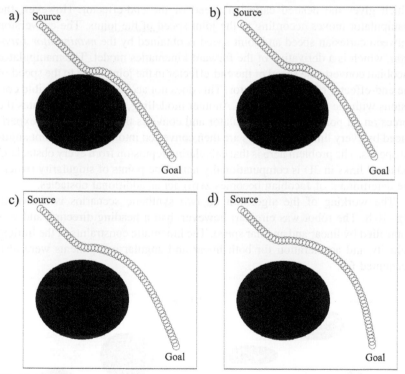

FIG. 10.9 Increase of clearance and path length as repulsive potential constant is increased is shown in (A–D).

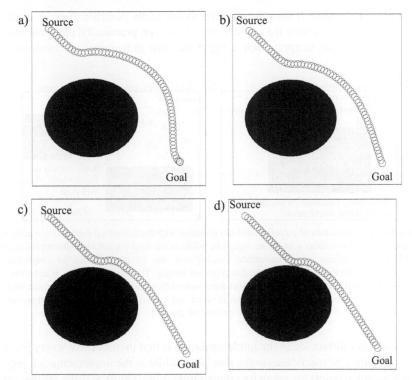

FIG. 10.10 Decrease of clearance and path length as attraction potential constant is increased is shown in (A–D).

Theoretically, two charged particles can never collide, since distance tending to zero implies that force tends to infinity and therefore the robot is repelled with the greatest force that cannot be balanced out by any means. However, practically the robots operate at control cycles with some frequency (say 10 Hz). This means that the control signals are applied after every 1/10 = 0.1 s. Within this duration, no change in the control signal is made and the robot continues to move with the same linear and angular speed. The robot possibly experienced a moderate force by an obstacle, which got sufficiently balanced out by the other obstacles and in the next step the robot was already in a dangerous condition of the collision, while the effect was due to the discretization of continuous motion into discrete control cycles. This is shown in Fig. 10.11A. Furthermore, the robots have kinematic and dynamic constraints, including braking limits, steering limits, etc. These disallow the behaviour that a free particle may exhibit. It may therefore be theoretically possible to avoid a collision by suddenly setting the speed to zero; however, practically, the robot may not be in a position to apply such strong brakes due to the physical constraints.

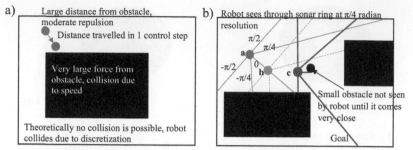

FIG. 10.11 Limitations of potential field (A) deviation with the theoretical model due to motion discretization. Robot faces a moderate repulsion, while in the next control cycle the repulsion is suddenly near infinity while immediate braking may not be possible due to kinematics. (B) Deviation with the theoretical model due to limited sensing. The robot traverses a, b and c when it suddenly sees the small obstacle. Repulsive force is directly calculated from the sonar/lidar readings, attractive force is calculated using localization and known goal coordinates (alternatively through intensity of electromagnetic signal emitted by the goal).

The other difference in the implementation is that theoretically, every point in every obstacle contributes to the repulsion, while in the implementation only a few sampled points are taken for computation. If proximity sensors are used to get distances to the obstacle, one may have limited sensors and any obstacle in between the sensors goes unnoticed. This is shown in Fig. 10.11B. A small obstacle can be continuously placed in the missed-out areas until the robot comes close to it. If the distances are taken from the global map, a discretization may be applied over the obstacles and distances. The noises due to this also make the robot behave differently than the theoretically expected actual motion.

The constants associated with the APF can be optimized for performance, such that the algorithm is locally optimal in terms of maintaining distances from the obstacles while navigating. However, being a reactive algorithm, the completeness and optimality cannot be ascertained. The algorithm can easily make the robot get stuck at local minima, and even if there is a path, the algorithm shall not find it. The biggest advantage of the APF algorithm is its real-time nature, wherein minimum computation is done at every step to move the robot.

10.3 Working of artificial potential field in different settings

So far, the working of the algorithm was theoretically studied. Now, we discuss the means to move the real mobile robots using the APF algorithm. Two common techniques are discussed. One wherein the robot has proximity sensing, and the other wherein a map of the world is known a priori.

10.3.1 Artificial potential field with a proximity sensing robot

The most common scenario is using robots with proximity sensors. A common (and budgetary) solution is using a sonar ring for proximity sensing. This is shown in Fig. 10.11B with a small ring of just five sensors. Here, each sonar sensor is located at a known angle (α) with respect to the heading direction of the robot. For a sonar ring of eight sonar sensors uniformly distributed with the 1st sonar sensor at the heading direction of the robot will have the sonar sensors at angles 0, $\pi/4$, $\pi/2$, $3\pi/4$, π, $-3\pi/4$, $-\pi/2$, and $-\pi/4$. The distance given by any sensor is used to calculate the magnitude of the repulsive force, while the direction of the repulsive force is opposite to that of the sensor at an angle of $(\pi+\alpha)$. Let the sonar ray at an angle α with respect to the robot heading record a distance of d_α. Reformulating Eq. (10.8) for the specific obstacle detected by the sonar ray at distance d_α and angle α gives the repulsion force formulated by Eq. (10.17), where the subscript R in the force calculation shows a representation of the force vector in the robot coordinate frame with the heading direction of the robot as the X axis.

$$
\left(F_{\text{rep}}^{\alpha}\right)_R = \begin{cases} -k_{\text{rep}}\left(\dfrac{1}{d_\alpha}-\dfrac{1}{d_{\text{rep}}}\right)\dfrac{-1}{d_\alpha^2}\nabla d_\alpha & \text{if } d_\alpha \leq d_{\text{rep}} \\[2ex] 0 & \text{if } d_\alpha > d_{\text{rep}} \end{cases}
$$

$$
= \begin{cases} -k_{\text{rep}}\left(\dfrac{1}{d_\alpha}-\dfrac{1}{d_{\text{rep}}}\right)\dfrac{1}{d_\alpha^2}[\cos\alpha \quad \sin\alpha]^T & \text{if } d_\alpha(q,o) \leq d_{\text{rep}} \\[2ex] 0 & \text{if } d_\alpha(q,o) > d_{\text{rep}} \end{cases}
$$

$$
= \begin{cases} k_{\text{rep}}\left(\dfrac{1}{d_\alpha}-\dfrac{1}{d_{\text{rep}}}\right)\dfrac{1}{d_\alpha^2}[\cos(\pi+\alpha) \quad \sin(\pi+\alpha)]^T & \text{if } d_\alpha \leq d_{\text{rep}} \\[2ex] 0 & \text{if } d_\alpha > d_{\text{rep}} \end{cases}
$$

$$(10.17)$$

It is assumed that the sensor is at the extrema of the robot and therefore, the distances reported are those from the end of the robot body. Hence, all repulsive forces can be conveniently calculated. The higher-end robots use lidar, which provides a higher resolution of the same data, wherein the angular resolution is of the order of half a degree with a remarkably high accuracy in distance computation.

The attraction force depends on the distance and direction of the robot from the goal. The goal is assumed to be specified in terms of the coordinates $G(G_x,G_y)$. Even if the goal is a well-known place 'entrance gate', the coordinates for the same on the map are known. The map is initially made using a SLAM technique, wherein all the goals, with their coordinates and names, are marked. This does not mean that the APF requires the map in the first place.

It only requires the goal's coordinates and its location. The location of the robot itself is available from a localization module, and the robot is known to be at pose q with position (x,y) and orientation θ. The attractive force is same as given in Eq. (10.6); however, the force should be in the robot coordinate frame with the X-axis as the heading direction of the robot to coincide with the coordinate frame used for the calculation of the repulsive force. Let $\theta_G = atan2(G_y - y, G_x - x)$ be the angle of goal from the robot in the world-coordinate frame. The attraction force is given by Eqs (10.18) and (10.19) and the total force is given by Eq. (10.20)

$$F_{att}^G = \begin{cases} -k_{att}(q - G) & \text{if } d(q, G) \leq d_{att} \\ -d_{att}k_{att}\dfrac{(q - G)}{d(q, G)} & \text{if } d(q, G) > d_{att} \end{cases} \qquad (10.18)$$

$$\left(F_{att}^G\right)_R = \begin{cases} k_{att}\, d(q, G)[\, \cos(\theta_G - \theta)\ \sin(\theta_G - \theta)]^T & \text{if } d(q, G) \leq d_{att} \\ d_{att}k_{att}[\, \cos(\theta_G - \theta)\ \sin(\theta_G - \theta)]^T & \text{if } d(q, G) > d_{att} \end{cases} \qquad (10.19)$$

$$(F)_R = \left(F_{att}^G\right)_R + \sum_\alpha \left(F_{rep}^\alpha\right)_R \qquad (10.20)$$

The steering control of the robot is now to rotate so as to point in the direction of the total force vector given by Eq. (10.21) and the speed is affected with a knowledge that the robot is heading at 0 rad in the robot coordinate axis system given by Eq. (10.22). The difference with respect to Eqs (10.15) and (10.16) is because the calculations are now in the robot coordinate system with the X axis heading at the orientation of the robot in the world coordinate frame (θ).

$$\omega = k_\omega atan2\left((F_y)_R, (F_x)_R\right) \qquad (10.21)$$

$$\Delta v = k_v (F_x)_R \qquad (10.22)$$

If the robot is budgetary, it may not have enough computing capability to find its location. Therefore, let us discuss a few schemes for a budgetary implementation. In the first scheme, the map may be accurately made by any technique (including another robot). Typically, a home or office environment can rent a high-end robot and run the SLAM algorithm on it to make an accurate map consisting only of static obstacles. On the same map, the coordinates of a few interesting places are computed, where the robot often has to go. For any one of these places, a user may ask the robot to go to a named place, whose coordinates are already known. Alternatively, the user may use the same map to compute the coordinates to which the robot should go. Budgetary encoders can be used to compute the robot's approximate position that is enough for the attractive force. Note that for APF, the robot only needs an approximate (and not exact) knowledge of its pose since obstacle avoidance is done by using onboard sensors and localization errors do not result in a collision. The robot may require re-localizations after a few go-to goal queries, by placing it at the home (charger) location whose coordinates are perfectly known, as the encoder-based localization incrementally adds errors.

A problem with the above approach is that incremental localization errors due to encoders lead to a difference in the final position reached by the robot and the goal. Commonly, robots are asked to reach popular places or people, both of which have a visual characteristic signature. Say the robot is to visit a person. The robot uses the encoder-based localization and known goal coordinates for attraction, while using a sonar sensor ring for repulsion. Once the robot reaches close enough to the person, face detection and recognition libraries identify the person at image coordinates (u,v), from the image taken from an onboard sensor in the heading direction of the robot. Say the camera has a Field of View (FOV) of $FOV_{min} = -\pi/3$ rad to $FOV_{max} = \pi/3$ rad, whose image is projected into an image of dimensions $(u_{max}, v_{max}) = 1280 \times 720$. The left-most pixel $(u = 0)$ corresponds to an angle of $FOV_{min} = -\pi/3$ rad, while the rightmost pixel $(u = u_{max} = 1280)$ corresponds to an angle of $FOV_{max} = \pi/3$ rad. The person's angular position (angle to goal) with respect to camera heading is thus (using linear interpolation) given by Eq. (10.23), where θ_G^R means that the angle to goal is measured with respect to the robot's coordinate axis with X axis as the heading direction.

$$\theta_G^R = FOV_{min} + \frac{u}{u_{max}}(FOV_{max} - FOV_{min}) \tag{10.23}$$

The distance from the person (goal) can also be calculated. Using the principles of camera projection (calibration matrix), the distance of the camera (robot) from the person (goal) is inversely proportional to the diagonal length (or length, or breadth) of the detected face in the image given by Eq. (10.24) for a camera with same focal length f for the X and the Y axis.

$$d(q,G) = f\frac{\text{diagonal length of face in real world}}{\text{diagonal length of face in image}} \tag{10.24}$$

Alternatively, a stereo camera can directly give the distance from the person as the perceived depth. The attraction to goal is given by Eq. (10.25).

$$(F_{att}^G)_R = \begin{cases} k_{att}\, d(q,G)\left[\cos\theta_G^R \quad \sin\theta_G^R\right]^T & \text{if } d(q,G) \le d_{att} \\ d_{att}k_{att}\left[\cos\theta_G^R \quad \sin\theta_G^R\right]^T & \text{if } d(q,G) > d_{att} \end{cases} \tag{10.25}$$

The other scheme assumes that most scenarios consist of corridors in which a constant force in the direction of the corridor may be assumed, doors in which the force may be at the centre of the door and so on. Vision sensors are used to detect the semantics and compute the orientation in the robot's coordinate axis system that the robot should align to. As an example, if the robot detects the door through which it needs to exit, it should rotate such that the door is at its centre whose calculations are as stated previously. If a corridor is detected, the intersection point of the left and right boundary lines in the image (extended till infinity) gives a point that should be at the centre, and hence the X-coordinate on the image is used to calculate the angular position and hence the attraction force. However, if a lobby has several exits, selecting the correct exit, and having an attraction towards the selected exit is a global planning problem that requires approximate knowledge of the pose. Therefore, even without accurate information about one's position, the APF may work. This is like the fact that we can walk around big campuses by moving in the direction of pathways, without explicitly knowing our position on a global map. When the goal is seen, the distance and direction to the goal are used to compute the actual attraction force.

In the last scheme, consider that the goal emits a characteristic sound or any electromagnetic signal like Wi-Fi, the intensity of the sound/electromagnetic signal received by the robot denotes its magnitude. The direction is computed by an array of receivers. The angle to the goal in the robot's coordinate frame is the direction corresponding to the receiver, which best catches the signal. Alternatively, if the signal strength increases as the robot travel, the robot is travelling in the correct direction and vice versa. The *swarm robotics* often uses such methods to compute the goal using budgetary sensors, and the topic is taken as a separate chapter.

10.3.2 Artificial potential field with known maps

The other use case is where a static map is known a priori, and the task is to move a robot from a given source to a given goal. A typical way of calculating the forces under laboratory settings is to set up a dummy robotics arena with a small robot and an overhead camera. The overhead camera looks at the map from a bird's-eye view and classifies all pixels as obstacles and nonobstacles. This makes the map. The overhead camera also looks at the robot and identifies it by its colour. The direction of motion of the robot can be computed by taking a derivative of its position with time, which is also the orientation of the robot. Since the map and obstacles are always known, the virtual proximity sensors can be embedded and used to calculate the distances from the image, just like one would have done in simulations.

If the image is not perfectly orthographic, the distances in the image will not correspond to the physical distances. An orthographic projection can be obtained by multiplying the image matrix by a transformation matrix. Let I be the image taken by a camera located at some point in a planar map, with small obstacles to make it planar. Such an image is shown in Fig. 10.12. The orthographic transformation is done by selecting a few points $P_I = \{(u_i, v_i, 1)\}$ in the image whose transformation in the real planar world $P_G = \{(x_i, y_i, 1_i)\}$ is known. Here, $(u, v, 1)$ denotes the position in the image plane while $(x, y, 1)$ denotes the position in the real world. The '1' in both terms is for homogenization. The subscript I means the image coordinate frame and G means the global coordinate frame. Assume there are n points. The transformation matrix is computed by Eq. (10.26).

$$[P_G]_{3 \times n} = [T_I^G]_{3 \times 3} \times [P_I]_{3 \times n} \tag{10.26}$$

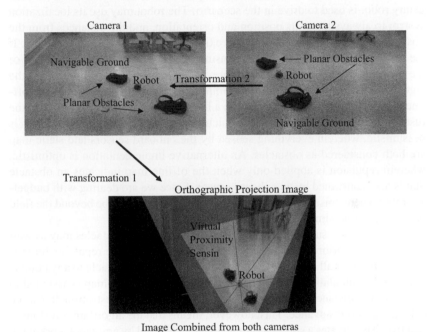

Camera 1

Camera 2

Navigable Ground

Planar Obstacles

Robot   Transformation 2   Robot

Planar Obstacles

Navigable Ground

Transformation 1

Orthographic Projection Image

Virtual Proximity Sensin

Robot

Image Combined from both cameras

FIG. 10.12 Potential field in a small indoor environment. Two cameras take two images and detect the robot (for calculating attractive force), ground, and obstacles using simple colour filtering techniques. The orientation is taken as a time derivative of position. Virtual proximity sensing gives the repulsive force.

Here T_I^G is the 3×3 transformation matrix, which transforms the image coordinate frame to the global coordinate frame. It is not a conventional 4×4 matrix and does not take the calibration matrix since the mapping is assumed to be planar only in the XY plane. The computation of T_I^G can be done by pseudo-inverse matrix computation ($[T_I^G]_{3 \times 3} = [P_G]_{3 \times n} \times pseudoInverse$ ($[P_I]_{3 \times n}$)), RANSAC algorithm, optimization (to minimize the error function $[T_I^G]_{3 \times 3} = \arg \min \sum_i \|[x_i y_i 1]^T - [T_I^G]_{3 \times 3} \times [u_i v_i 1]^T\|_2^2$), or any other technique. Once the matrix T_I^G is known the image is converted into the orthographic image by multiplication of every point by T_I^G and putting the results into an empty canvas, filling in the empty spaces by interpolation. A single overhead camera may have a restricted vision, in which case multiple such external cameras can be used, and their information can be fused. Stitching one camera's image to the other requires calibration of the transformation matrix across the two cameras. The specific result of such a transformation from two cameras is shown in Fig. 10.12.

A more realistic implementation that does not require an overhead camera is when one robot is used to make a rich map of the world, and thereafter a budgetary robot is used to drive in the scenario. The robot may use its localization system to always know its position and orientation, and the distances from the obstacle can be given by the virtual proximity sensors by querying the map. If the robot does have some proximity sensing, it can be appended to the sensing of the virtual proximity sensors. Therefore, new obstacles can be handled by the algorithm. This is a typical algorithm implementation in robotics where an offline computed static map is fused with a local map sensed by the robot, and the fused map is used for all algorithmic calculations. In this case, the assumption is pessimistic, wherein everything sensed by the onboard sensors and static map are both considered as obstacles. An alternative implementation is optimistic, wherein repulsion is applied only when the offline map indicates an obstacle that is also confirmed by an onboard sensor. Since we are dealing with budgetary robot where noise can be large and several obstacles may be beyond the field of vision, the pessimistic implementation is desirable.

If the map is assumed to be static, the distances from obstacles may as well be computed a priori. Consider the formulation wherein the repulsive force is the maximum for all obstacles, case in which the nearest obstacle to any point is taken in the calculation of the repulsive potential. Since the map is assumed to be static, the distances of all points in the map to the nearest obstacle can be calculated well in advance. This, however, means that the algorithm is no longer reactive. This creates a sort of roadmap that can be used by any robot working in the arena from any source to calculate the force to any goal in a multi-query motion planning approach. The algorithm specifically is known as the *Bush Fire Algorithm* discussed earlier.

The algorithm can be coded in a variety of ways. The easiest way to discretize the space into grids. The grid (i,j) will eventually store the distance from the nearest obstacle. Initially, the distances are assumed to be 0 for obstacles and infinity for nonobstacles, given by Eq. (10.27).

$$d_o(i,j) \leftarrow \begin{cases} 0 & \text{if } W(i,j) = \text{obstacle} \\ \infty & \text{if } W(i,j) = \text{free} \end{cases} \tag{10.27}$$

Here $d_o(i,j)$ denotes the estimated distance from the nearest obstacle for the grid (i,j) that is iteratively improved. The arrow denotes an initial estimate, which will change. At each iteration, a free grid investigates the neighbouring grids and attempts to compute a shorter distance to the nearest obstacle through the neighbouring grid. Suppose (k,l) is the neighbouring grid off the grid (i,j). The current estimated distance of (k,l) from the nearest obstacle is $d_o(k,l)$ while (k,l) is at a distance $d((i,j),(k,l))$ from (i,j). Therefore going from nearest obstacle to (k,l) and then to (i,j) will incur a cost of $d_o'(i,j) = d_o(k,l) + d((i,j),(k,l))$, where the prime is additionally used in the notations to denote a candidate and best out of all candidates shall be chosen and $d((i,j),(k,l))$ is the distance between (i,j) and (k,l). The update for every grid is given by Eq. (10.28).

$$d_o(i,j) \leftarrow \begin{cases} 0 & \text{if } W(i,j) = \text{obstacle} \\ \min_{(k,l) \in \delta(i,j)} (d_o(k,l) + d((k,l),(i,j))) & \text{if } W(i,j) = \text{free} \end{cases}$$

(10.28)

Fig. 10.13 shows the results of the Bushfire algorithm using the four-neighbourhood and eight-neighbourhood functions.

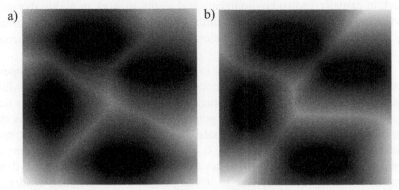

FIG. 10.13 Precalculation of distance to the nearest obstacle for calculating repulsive force (A) using eight neighbours, (B) using four neighbours. Brighter regions are away from obstacles.

Alternatively, the same distances may be achieved by using a Uniform Cost Search. The search will not start from a single source, but with all grids that have a 0 value, that is, all the boundary points. This propagates a wave simultaneously from all boundary points and stops when the fringe is empty. At that step, every node has the least cost from the nearest obstacle.

Assume that the robot knows its current position (x,y). The repulsive potential can hence be computed by injecting in the least distance value $d_o(x,y)$. The angle to the obstacle can be derived by looking at the neighbour that contributes the least distance (x_o,y_o), given by Eq. (10.29).

$$(x_o, y_o) = \arg \min_{(k,l) \in \delta(i,j)} (d_o(k,l) + d((k,l),(i,j)))$$

(10.29)

The resultant force is given by Eq. (10.30), where the force is only contributed by the nearest obstacle.

$$F_{\text{rep}} = \max_o F_{\text{rep}}^o = \begin{cases} -k_{\text{rep}} \left(\dfrac{1}{d_o(x,y)} - \dfrac{1}{d_{\text{rep}}} \right) \dfrac{-1}{d_o^2(x,y)} \nabla d_o(x,y) & \text{if } d_o(x,y) \le d_{\text{rep}} \\ 0 & \text{if } d_o(x,y) > d_{\text{rep}} \end{cases}$$

$$= \begin{cases} k_{\text{rep}} \left(\dfrac{1}{d_o(x,y)} - \dfrac{1}{d_{\text{rep}}} \right) \dfrac{1}{d_o^2(x,y)} \dfrac{[(x,y) - (x_o,y_o)]^{\text{T}}}{d((x,y),(x_o,y_o))} & \text{if } d_o(x,y) \le d_{\text{rep}} \\ 0 & \text{if } d_o(x,y) > d_{\text{rep}} \end{cases}$$

(10.30)

The fixed goal position is assumed to be known tocalculate the attractive force. The calculations in this case are all in the world coordinate axis and correspond to the ones of Section 10.2 and are therefore not repeated.

10.4 Problems with potential fields

The APF seems to be a good choice of the algorithm due to its ability to tackle obstacles in real-time, lack of a need of a preknown map, ability to work in partially known environments and ability to work in highly dynamic environments. The parameters can further be tuned to get good local optimality. However, the APF is not a solution to all problems due to a few inherent problems.

The major problem with the APF is its lack of completeness. This happens because a gradient descent on the potential values is used to navigate while the approach will result in getting stuck at the local minima. This means that the robot will come to a halt at a point, which is not the goal. This is the region where the attractive and repulsive forces cancel each other out, and therefore the resultant force is 0. One such situation is shown in Fig. 10.14A and B. This may be an exceptional condition; however, there are far too many local optima in a complex and realistic map rendering the APF useless as the only algorithm, unless it is fused with a more powerful deliberative planning technique.

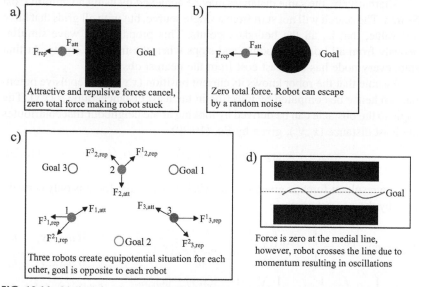

FIG. 10.14 Limitations of potential field (A–C) show equipotential regions Situation in (A) can only be avoided by persistent random motion (biased towards a randomly chosen direction) and then resuming potential field. In (C) $F_{i,att}$ denotes the attraction force by robot i and $F^j_{i,rep}$ denotes the repulsion force faced by robot i by robot j. (D) Oscillations in the narrow corridor.

Equipotential regions are more common in dynamic environments with dynamic obstacles. Imagine three robots need to go from one place to another. Each robot is attracted by the opposing goal and is repelled by the other two robots. This creates an equipotential region wherein the force values for all robots cancel out and the three robots get stuck because of each other. It may be noted that the environment was obstacle-free and yet the potential field does not produce a path. This is shown in Fig. 10.14C.

Since the APF can get stuck in multiple places, there needs to be some simple trick to get the algorithm working even if the robot is stuck. A clever case was shown in the previous example with three robots, which can be circumvented by simply adding a small noise in every step that continuously breaks the symmetry. However, this does not mean that adding noise removes all problems, as most of the problems of equipotential regions will still be prevalent.

A better technique is to identify when the robot is stuck, which is when it is nearly static, and far from the goal. In such a case, random perturbations can be applied to get the robot unstuck. The random perturbations can help overcome small local optima, like the one shown in Fig. 10.14B. The perturbations can be initially small, and the magnitude can be increased with time to get unstuck.

A more practical way of getting the robot out is by taking a random direction and moving the robot with a bias towards a chosen direction, while avoiding obstacles, if any, on the way (Kala, 2018). After a long walk, the robot effectively comes into a new area, from which it can start the potential field approach. If only a small walk is taken, the robot will likely end up in the same optima again. Even after taking a long walk, there is no guarantee that it will not make the robot get further stuck, which may even be worse than optimum. However, by continuously making such random motions, the robot may eventually be lucky enough to get to the goal, although its optimality will be highly compromised. There is a nonzero probability of getting unstuck and thus the setting is, theoretically, probabilistically complete.

Another problem with the approach is with the narrow corridors. Imagine a robot operating at a narrow region, shown in Fig. 10.14D. The goal is at one end of the corridor and the robot is repelled by the two boundary walls. If the robot is at the centre facing the end of the corridor, the repulsions from the two boundaries cancel each other out and the robot effectively moves only by the attraction from the goal. However, if the robot is even a little deviated from the centre, it feels a stronger repulsion from one boundary until it reaches the centre. However, the robot has kinematic constraints or a momentum, because of which it cannot correctly orient itself at the centre and will continue to take a few steps until the rectification is applied, in which time it slightly deviates towards the other direction. This continues and the robot oscillates inside the corridor instead, shown in Fig. 10.14D. A similar problem is associated with the Proportionate (P) controller which is likely to incur oscillations, which are corrected by the use of the PID controller instead. The APF alone behaves similarly to the P-controller without the integral and derivative terms and faces the same problem as the P-controller.

10.5   Navigation functions

The navigation function is a related concept that models the algorithm in such a way that there are no local optima. There is a single optimum at the goal, resulting in a valid path from any point in the space to the goal, and the trajectory traced because of the same path is smooth and collision-free. It is imperative that parameter tuning and adoption of the basic equations itself cannot result in the completeness property and making navigation functions requires either suitable assumptions to be placed in the environment so that geometrically potential values could be defined that guarantee the absence of a local optimum, or the same values could be deliberatively computed for any space and stored for all grids. Once such a computation is done, the robot chooses the path as per the precomputed values and traces its way to the goal. The path will be free of local optimum because all possible optima have already been directed towards the global optimum by the computations done.

In the simplest case, a grid is given, and the aim is to make a navigation function that maps every grid with an immediate direction of motion. Note that this will not strictly be a navigation function because of the absence of smoothness. One could run a wave from the goal outwards on all points, like a backward Uniform Cost Search algorithm. Specifically, the implementation that uses neighbouring grids and unit integral costs for all neighbouring edges is called the *Wave-Front planner*. The wave gives the distance of all nodes to the goal. The navigation function is then, for every grid, the direction to the neighbouring grid with the least cost. The local optima themselves point to neighbouring grids, which have been a better cost, as has already been computed by the wave propagation.

10.6   Social potential field

The social potential field is a socialistic extension of the APF (Reif and Wang, 1999; Gayle et al., 2009). Imagine a robot operating in a home or office environment with multiple other human workers that operate in the same space and act as obstacles for the robot. For the convenience, let us assume that the robot needs to operate just like human beings and therefore its decision-making and navigation etiquette should be like those of humans. This capability is added by modelling the potential field to direct the robot to follow social norms.

10.6.1   Force modelling

Imagine two people moving in a hallway towards a goal at the other end. The APF would model this as the two people repelling each other, the walls repelling both the people and the common goal attracting both the people. This will mean that the two people will be found walking with the maximum possible distance between each other, as shown in Fig. 10.15A. However, practically this is not observed in real life, wherein the two people will walk sufficiently close to each other, because of the affection that they have for each other. Therefore, the first person is attracted by the other person. However, the attraction will not mean the two people bumping into each other. Hence, it will be right to say that they also repel each other so as not to cause a collision.

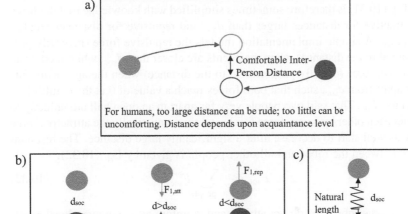

FIG. 10.15   Social potential. Every group of people attract and repel each other (A) motivation to keep socialistic distance, (B) force modelling. (C) Spring that models the same relation.

The force that exists between the two persons is hence special. It is attractive if the two are separated by a large distance (say more than d_{soc}) and repulsive when they are separated at a distance less than d_{soc}. The concept is shown in Fig. 10.15B. Here, d_{soc} is the distance that people typically like to keep with each other. The value of d_{soc} is the distance when the attractive and the repulsive forces are equal. Let q be the position of the robot and let o be the position of the other person. The modelling of the force applied by the person o to the robot is given by Eq. (10.31)

$$F_{soc}^o = \left(\sum_{k=1}^{L} \frac{c_{o,q}^{(k)}}{d^{\sigma_{o,q}^{(k)}}(q,o)} \right) \frac{(o-q)}{d(q,o)} \tag{10.31}$$

Instead of taking the inverse cube of the distance, the force takes the sum of L inverse powers, which is the generalization of the repulsion force case. The force is further generalized such that the force constant and power are dependent upon the relation between the person o and robot q, modelled as $c_{o,q}^{(k)}$ and $\sigma_{o,q}^{(k)}$. The k in the superscript is not a power but denotes the kth constant. The parameters depend upon the degree of affinity between the participating entities, or how they socially feel for each other. d is the distance function. The force is modelled such that it is positive when the distance $d(q,o)$ is less than d_{soc} resulting in an attraction, and negative when the distance $d(q,o)$ is larger than d_{soc} resulting in a repulsion.

Eq. (10.31) is therefore sometimes simplified with knowledge that the force is attractive for distances larger than d_{soc} and repulsive for distances smaller than d_{soc}. A simple implementation models the repulsive force (inversely proportional to the distance) when the agents are closer than d_{soc}, while models an attractive force (directly proportional to the distance) when the agents are further apart from d_{soc}, such that both forces reach a value of 0 at the equilibrium distance d_{soc}. That said two people very far apart in an open hall are unlikely to attract each other by a very large force, and thus realistically the attractive force may as well start to decrease after a significantly large distance. The resultant force between the robot and all other people is given by Eq. (10.32).

$$F_{soc} = \sum_{o=\text{people}} F_{soc}^o \qquad (10.32)$$

The other obstacles do not attract, and therefore have a conventional repulsion only model. Similarly, the goal only attracts. For corridor like situations, it may point to the end of the corridor, or the next door to take, and so on. The attraction is modelled as previously. The robot moves under the influence of all such forces and the resultant total force is given by Eq. (10.33).

$$F = \sum_{o=\text{people}} F_{soc}^o + \sum_{o=\text{obstacles}} F_{rep}^o + F_{att}^G \qquad (10.33)$$

A similar model uses a spring analogy with a natural length of d_{soc}. Smaller distances lead to compression, and hence an opposing force is applied. A larger distance leads to expansion, and hence the opposite force. The model is shown in Fig. 10.15C. The spring analogy is useful to understand in different contexts as well. Many times, soldiers are asked to parade in a rectangular formation, where every soldier attempts to place itself at the centre of the front, back, left, and right soldiers, which is a spring model. Deviations are immediately corrected by natural forces. Many times, the groups may be arranged in a pyramid or circular or other formations that can be modelled by virtual springs. Cloth simulation itself models cloth molecules bound to the four neighbouring molecules subjected to internal and external forces in a spring framework.

10.6.2 Groups

The value of the socialistic distance between the two people is not constant and depends upon the type of people. If the two are close friends, they walk close to each other with a small value of d_{soc}. However, if the two just know each other, the value will be a little higher. If they do not have any acquaintances, the value will still be higher. If the two are enemies, they might only repel each other and walk with the maximum possible separation with a large socialistic distance between them.

In a more general sense, the people belong to different *groups*, like a group of friends, a group of colleagues, etc. Each member of the group has a strong affinity for the members of the same group as friends stick together. The groups may have a varying degree of affinity or repulsion towards members of the other groups. This is clear from walking in marketplaces. If you see a small family coming towards you, you tend to walk left or right of the family. You do not dissect the family and walk in the middle. Even if you attempt to dissect the family, you will see the family adjusting leftward or rightward such that they do not break. This happens because the members of the family have a strong attraction to each other. Because of this their attraction coefficient and degrees are strong, and their equilibrium distance is small. Hence, dissecting the members would mean a high magnitude of repulsion, meaning that even if they dissect, it will not be until you have come awfully close to them. On the other hand, even if a member deviates slightly in response to your coming, the other members because of the strong attraction will move. Similarly, since you strongly repel with the family, you will probably adjust yourself because of this repulsion. Therefore, it is again difficult to walk in-between a chain of school children going one after the other unless someone externally stops the chain. The cases are shown in Fig. 10.16.

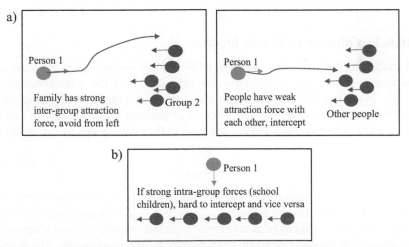

FIG. 10.16 Social potential in groups. Every group has a strong intragroup attraction in contrast to intergroup attraction and vice versa. (A) Person 1 avoids group 2 from the far left due to a strong intragroup attraction. However, person 1 walks in-between unrelated people due to a weak attraction between them. (B) Person 1 cannot intercept the chain of people due to a strong intragroup attraction.

On the other hand, if the same positions were occupied by the same people without any acquaintance, it is easier to walk in between. The people themselves repel each other and maintain a large equilibrium distance between themselves. On encountering you, they feel repulsion at opposite ends and move outwards, creating enough space for you to walk through. Similarly, if a chain had to be formed, like at the entrance of a building with a single queue of entry, breaking the chain between people not related to each other would be a lot easier.

The socialistic attraction and repulsion between people were controlled using coefficients and degrees, which were unique between every pair of people. This is now easier since the coefficient and degrees are assumed to be the same for every pair of groups. To limit the number of variables further, every group may be assumed to have the same coefficients/degrees for the members of the same group; and the same coefficient/degree for members of the different group. So, the modelling reduces to modelling *intra-group* attraction and repulsion coefficients and *inter-group* attraction and repulsion coefficients only.

Another assumption made here is that an individual cannot be a member of two groups or partial members of different groups, which otherwise only changes the parameters involved. The problem is not the modelling part of it, the problem is that membership degrees further introduce hard-to-determine and hard-to-set parameters that unnecessarily complicates the system. In real life, you are normally with the group that you went with. However, it is possible to have a *merging* of groups, *splitting* up of groups, and *re-groupings* depending upon the events and context.

10.6.3 Other modelling techniques

A modelling technique for social potential, widely studied in the literature of pedestrian modelling, incorporates the velocities, and the orientations also while modelling for the repulsive and attractive potentials (Helbing and Molnár, 1995). Assume that the robot is at position q and intends to reach the goal G. A unit vector (e) in the direction to the goal is given by Eq. (10.34).

$$e = \frac{G - q}{d(q, G)} \tag{10.34}$$

Here d is the distance function. The desired velocity is to move directory towards the goal and hence given by $v_{max}e$, where v_{max} is the maximum possible speed. The robot will not be able to move with this velocity because it may not be oriented in the same direction or there may be other constraints. The acceleration is applied to the robot such that the velocity is obtained after time T, in which case the acceleration can be given by Eq. (10.35).

$$F_{att}^G = \frac{v_{max}e - v}{T} \tag{10.35}$$

This constitutes the attraction force of the robot. It may be noted that the velocity is a vector here, while the robots are driven at linear and angular speeds. Hence, there is no direct mapping of the acceleration with the control inputs and it requires a small controller for the conversion of the acceleration vector into the linear and angular speeds.

The robot is repelled by all the other people and robots in the arena. For convenience, assume that the robot is not a member of any group, and that all other people and robots have the same degree of affinity to the robot. Therefore, the repulsion force is simply inversely proportional to the distance and the direction is opposite to the direction of the vector point from the robot to the other agent (o). The repulsive force is given by Eq. (10.36).

$$F_{rep}^o = k_{rep} \exp\left(-\frac{d(q,o)}{\sigma}\right) \frac{q - o}{d(q,o)} \tag{10.36}$$

Using exponentially decaying repulsive force is like the polynomial variants used earlier. The inverse polynomial (like the inverse cube) modelling used earlier tends to infinity as distance tends to zero. The repulsion force function becomes overly sensitive as the robot comes closer to the obstacle boundary. It is common in experiments with a modest control frequency to face small repulsions at the current time step, while facing excessively large repulsions at the next time step. Hence, it becomes very difficult to practically control the robot using the inverse polynomial (like inverse cube) functions, which could also lead to collisions. The exponential model in many cases realistically models the situation, as it only gives sufficiently large values and not infinity. The values moderately increase as the robot goes towards the obstacles. Even with static obstacles, Eq. (10.36) is a preferable implementation for the control of the mobile robot using the APF algorithm.

A problem with Eq. (10.36) is that it does not take into consideration the orientation of the robot with respect to the obstacle, which in this case is another person. The obstacle directly ahead is more fatal than the obstacle at the side. While doing brisk walking, if you see a person 2 m ahead, you are bound to take immediate action. However, if the person is 2 m at the side or even 0.5 m at the side, you barely care about it. This is called as the *elliptical model* for obstacles, wherein the obstacles in the walking direction or the goal direction are far more important, and the importance of those obstacle loses when they become more sideways. This is shown in Fig. 10.17. This effect is incorporated into the repulsive force, which is now given by Eq. (10.37).

$$F_{rep}^o = k_{rep} \exp\left(-\frac{d(q,o)}{\sigma}\right) \frac{q - o}{d(q,o)} \frac{1 + \cos \varphi_o}{2} \tag{10.37}$$

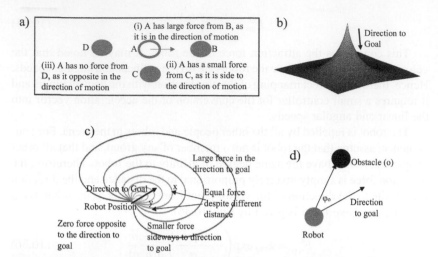

FIG. 10.17 Elliptical potential. (A–C) The magnitude of force for different spatial positions. (D) Calculation of the angle to goal φ_o.

Here φ_o is the angle between the direction to the goal of the robot to the line joining the robot to the obstacle (o). φ_o denotes the radial position of the obstacle to the robot, given in Fig. 10.17D. d is the distance function. The modelling is the same with the exception that discount factor of $(1+\cos\varphi_o)/2$ has been additionally applied to minimize the effect of obstacle at the back in contrast to the ones at the front. The angle is given by Eq. (10.38).

$$\cos(\varphi_o) = \frac{(o-q)\cdot e}{d(o,q)} \tag{10.38}$$

A few interesting cases are discussed. Suppose φ_o is 0, meaning that the obstacle is directly ahead of the goal (intended motion angle) of the robot. This causes the term to become the maximum 1 and has the full impact of the obstacle in terms of the distance. However, if φ_o is either $-\pi/4$ or $\pi/4$, the discounted term becomes $(1+1/\sqrt{2})/2$, which is much smaller. As you go further on either side, φ_o becomes $\pi/2$ or $-\pi/2$. In both cases, the discounted term is $1/2$, meaning that the force is active for only half of its current value. At the other extremity, when φ_o becomes π, the term becomes 0, meaning that there is no repulsive force. This is evident that any obstacle behind us is not perceived by us and therefore we take no action on the same basis. It is rare to look at the vehicles directly behind when driving a car to decide the current control action.

In many theories, it is proposed to use a blend of the two modelling techniques, the one that discounts the obstacles behind the ones that carry no such discount. A fused model integrates the two models with a contribution of λ without discount and $(1-\lambda)$ with discount, where $0\leq\lambda\leq1$ controls the relative contribution of the two models, given by Eq. (10.39).

$$F^o_{\text{rep}} = k_{\text{rep}} \exp\left(-\frac{d(q,o)}{\sigma}\right) \frac{q-o}{d(q,o)} \left(\lambda + (1-\lambda)\frac{1+\cos \varphi_o}{2}\right) \qquad (10.39)$$

For person as the obstacle, there will be simultaneously an attraction between the two, given by Eq. (10.40)

$$F^o_{\text{att}} = k_{\text{att}} \frac{o-q}{d(q,o)} \qquad (10.40)$$

The static obstacles like walls cause only repulsions and no attractions. Similarly, it is possible to have other attraction sites rather than a goal and other people. Say a person wants to peek out of the window or see the lab while passing by, etc. Such attractions are also modelled for. The resultant force is given by Eq. (10.41).

$$F = \sum_{o=\text{people}} \left(F^o_{\text{att}} + F^o_{\text{rep}}\right) + \sum_{o=\text{obstacles}} F^o_{\text{rep}} + \sum_{o=\text{attractions}} F^o_{\text{att}} + F^G_{\text{att}} \qquad (10.41)$$

10.7 Elastic strip

In the previous chapters, numerous techniques were presented for the planning of a robotic trajectory from a source (S) to a goal (G) making the trajectory function $\tau:[0,1] \to C^{\text{free}}$, with the constraints, $\tau(0)=S$, $\tau(1)=G$, $\tau(s)\in C^{\text{free}}$, $0 \leq s \leq 1$. However, the trajectory is only valid till the environment remains static. This is a big assumption to make and will not happen in most realistic scenarios. The only algorithm which could adapt the trajectory to small changes in the environment was D*, which as well only worked for small changes in the map. This stresses the need to build re-planning algorithms that adjust the trajectory of the robot as the environment changes along with time, without requiring the need to re-plan the trajectory by the parent algorithm. The elastic strip is one such method.

10.7.1 Environment modelling

The *elastic strip* algorithm models the trajectory of the robot as an elastic strip (Brock and Khatib, 1998, 2002, 2008; Yang and Brock, 2006; Sud et al., 2007). The elastic strip has a unique property that once stretched and held at some control points, it can be deformed into any possible shape. Therefore, the elastic strip can represent any trajectory. On top of this, the elastic strip is always smooth. Even if an elastic strip is stretched by applying a force at some point, it demonstrates a smooth curve along the impacted point. This means that the trajectories represented will additionally get smooth by such modelling. Further, the elastic strip, if unstretched, deforms back to the original shape, which is the straight-line distance between the endpoints. When all forces and holding points are removed, the elastic strip gets back into a straight-line between the extremities.

The visualization is done in both configuration space and workspace. In the workspace, the robot is represented as a complete mass; while the configuration space models a space with every axis as a configuration variable and a point in the space represents all configuration variable values of the robot, which can be mapped to the workspace by physically setting all robot configurations to the corresponding value.

Let us visualize the robot in the workspace first when placed at any specific configuration, shown in Fig. 10.18A. To assess the free space around the obstacle in the workspace, the robot is filled with balls of the largest possible size, such that the entire ball is obstacle-free excluding the robot itself. Since the robot may be a complex body like a mobile manipulator, it is eminent that a single such ball may not be enough to completely pack the robot and model the free space around it and therefore, multiple such balls are used. Typically, imagine the skeleton structure of the robot. Start placing large balls around the entire skeleton, till the complete robot is packed. Let the robot be at the configuration q, which represents a complete mass of the robot, in which the balls are placed at locations $\{p(q)\}$. Each ball is given by the region of space given by Eq. (10.42).

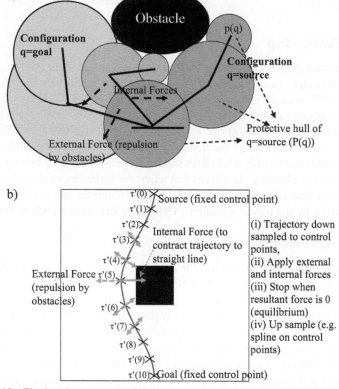

FIG. 10.18 Elastic strip concept. (A) Safe navigation tunnel shown by all shaded regions (drawn for three configurations in trajectory only), with the shaded regions corresponding to one configuration (manipulator) called the protective hull. In navigation, the robot can move anywhere freely within the safe navigation tunnel. The trajectory is adapted to changing obstacles shown in configuration space in (B) and workspace in (A) using internal and external forces. In workspace, the forces are converged into torques for motion of the manipulator.

$$B_\rho(p(q)) = \{a : d(a, p(q)) < \rho, a \in W^{\text{free}}\} \tag{10.42}$$

Here ρ is the radius of the ball and d is the distance function. It must be stressed that the radius is not constant; however, is the maximum possible radius as per the available free space. The complete set of balls together envelope the robot and constitute the *protective hull* of the robot. The robot may move freely to strictly lie in entirety within the protective hull, and it shall still be collision-free. The protective hull is given by Eq. (10.43)

$$P(q) = \bigcup_p B_\rho(p(q)) \tag{10.43}$$

Now visualize the trajectory as a continuous sequence of configurations. This constitutes placing similar balls at each configuration in the robot trajectory. The *safe navigation tunnel* constitutes the protective hulls of all configurations in the robot trajectory ($q = \tau(s), 0 \leq s \leq 1$), given by Eq. (10.44).

$$V_\tau = \cup_{q \in \tau} P(q) = \bigcup_s B_\rho(\tau(s)) \tag{10.44}$$

This represents a tunnel of space within which the robot can easily move without concerning about any kind of a collision.

10.7.2 Elastic strip

Let us discretize the trajectory to a few sampled points only, taking a sampled set of configurations $\{\tau'(s)\}$. For large enough sampling, the sampled trajectory τ' nearly represents the actual trajectory. The samples $\tau'(s)$ become the control points and the sampled trajectory is now called as the *elastic strip*. By moving the control points the trajectory can be modified to take different shapes in reaction to the change in environment. In terms of the configuration space, the motion of the robot shall always be collision-free if the trajectory up-sampled from $\tau'(s)$ is collision-free in the configuration space. Upsampling here means using interpolation and curve smoothing techniques with the control points taken as $\tau'(s)$. These techniques not only smooth the curve from the control points but also represent a continuous trajectory from a discrete set of control points. Using spline using $\{\tau'(s)\}$ as control points is a good technique.

In the workspace, the elastic strip representing real robot masses shall always be collision-free if the resultant volume swept by the robot in terms of the collision-free bounding balls is always a subset of the safe navigation tunnel of the robot. The navigation tunnel states how much the trajectory can be modified, still guaranteeing being collision-free. The relation is given by Eq. (10.45).

$$V_\tau \subseteq V_{\tau'} = \bigcup_s B_\rho(\tau'(s)) \tag{10.45}$$

Imagine the elastic strip as shown in Fig. 10.18B, represented in the configuration space. The control points of the strip are explicitly shown. The first and last control points are the source and the goal, which cannot move with the application of different forces. The strip faces an *internal force* applied by the neighbouring control points. The strip has a property that, in the absence of any external force, it contracts to the smallest size, which is the straight-line distance between the source and the goal. Hence, every control point tries to pull the neighbouring control points towards it to get the shortest distance.

The elastic strip is unable to contract because of the application of *external forces*. The external forces are experienced from the obstacles around, which apply a repulsion to let the control points, lie as far as possible. Because of such a force, the control points cannot shrink to make the trajectory lie inside the obstacle. As the distance of a control point from the obstacle tends to zero, the repulsion tends to infinity, and the control point is moved in the opposite direction. The internal and external forces act simultaneously on every control point and move the control point, which in turn adapts the trajectory. The control point stops its motion when the internal and external forces cancel each other out.

Suppose the obstacle shown in Fig. 10.19 moves towards the trajectory. Since the distance of the nearest control points to the obstacle decreases, the repulsion force increases and therefore, equilibrium is broken. Since the repulsion dominates, the control point moves further away and therefore, the trajectory moves further away in response to the obstacle. After moving sufficiently away, the repulsion from the obstacle reduces as the distance is now larger. At the same time, a larger stretch is being applied to the control point, and hence the internal force also increases to overcome the repulsion. The trajectory hence comes into equilibrium at a lesser distance from the obstacle as before.

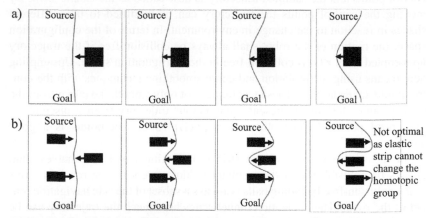

FIG. 10.19 Working of elastic strip. (A) The trajectory deforms (first figure to last) on the motion of the obstacle. As the obstacle retracts, the trajectory regains its shape (last figure to first). (B) Lack of optimality by continuous local adaptations.

In the same figure consider that the obstacle moves far away from the trajectory. In such a case the repulsive force reduces because of an increase in the distance with the control point and therefore the internal force becomes dominant and contracts the trajectory moving it towards the obstacle. As the control point moves towards the obstacle, the trajectory length reduces, at the same time increases the repulsion due to reducing the distance from the obstacle. The trajectory again stops at the equilibrium.

10.7.3 Modelling

More formally, let the control point be $\tau_t'(s)$ at any general time t. The sub-script of time (t) is added since now the trajectory would be changing due to forces along with time. Let $\tau_t'(s-1)$ and $\tau'(s+1)$ be the previous and next control points, respectively. The internal force is by $\tau_t'(s-1)$ and $\tau_t'(s+1)$ pulling $\tau_t'(s)$ towards itself, given by Eq. (10.46).

$$F_{\text{int}}(\tau_t'(s)) = k_{\text{int}} \left(\frac{d(\tau_0'(s), \tau_0'(s-1))}{d(\tau_0'(s), \tau_0'(s-1)) + d(\tau_0'(s), \tau_0'(s+1))} (\tau_t'(s+1) - \tau_t'(s-1)) \right.$$

$$\left. - (\tau_t'(s) - \tau_t'(s-1)) \right)$$

(10.46)

The equation may be observed as a summation of two relative forces at the previous and next control points. Here, k_{int} is the constant commonly referred to as the contraction gain constant of the strip, and d is the distance function. The typicality of implementation is that the distances mentioned $d(\tau_0'(s), \tau_0'(s-1))$ are measured from the original (τ_0') and not the modified trajectory (τ_t') after moving the control points. Hence, the proportional distances mentioned become constant, and Eq. (10.46) states that in the absence of any external force, the trajectory would re-orient itself to have the same proportion of distance as the original trajectory. This means that the force only depends on the curvature and not the actual distances. Eq. (10.46) should hence be read as the force applied to $\tau_t'(s)$ is a constant (k_{int}) times the deviation of the vector from $\tau_t'(s)$ to $\tau_t'(s-1)$ (or the vector $\tau_t'(s) - \tau_t'(s-1)$) from a constant proportion of the vector from $\tau_t'(s+1)$ to $\tau_t'(s-1)$ (or the vector $\tau_t'(s+1) - \tau_t'(s-1)$), where the constant proportion is given by proportionate distance of $\tau_t'(s)$ from $\tau_t'(s-1)$ in the original trajectory (τ_0') that the system tries to attain using the internal forces.

At equilibrium, in the absence of any external force, it is expected that the modified trajectory (τ_t') can be re-collapsed to a straight line, and the control points shall be distributed in the same ratio at which they were sampled in the original trajectory (τ_0'). At equilibrium without external forces, for fixed $\tau_t'(s-1)$ and $\tau_t'(s+1)$ the internal force is zero and we have, Eq. (10.47)

$$\tau_t'(s) = \tau_t'(s-1)$$

$$+ \frac{d\left(\tau_0'(s), \tau_0'(s-1)\right)}{d\left(\tau_0'(s), \tau_0'(s-1)\right) + d\left(\tau_0'(s), \tau_0'(s+1)\right)} \left(\tau_t'(s+1) - \tau_t'(s-1)\right)$$

$$(10.47)$$

Eq. (10.47) represents a point on the line connecting $\tau_t'(s-1)$ and $\tau_t'(s+1)$ at a constant proportion. Any deviation of $\tau_t'(s-1)$ causes an internal force. If a uniform sampling is used on the original trajectory, all distances in Eq. (10.47) become the same $(d(\tau_0'(s), \tau_0'(s-1)) = d(\tau_0'(s), \tau_0'(s+1)))$ and the ratio becomes 1/2. The control point $\tau_t'(s)$ faces a force to place itself at the mid-point of $\tau_t'(s-1)$ and $\tau_t'(s+1)$, given by Eq. (10.48), which forces the control points to align in a straight-line.

$$F_{\text{int}}\left(\tau_t'(s)\right) = k_{\text{int}}\left(\frac{1}{2}\left(\tau_t'(s+1) - \tau_t'(s-1)\right) - \left(\tau_t'(s) - \tau_t'(s-1)\right)\right)$$

$$= k_{\text{int}}\left(\frac{\left(\tau_t'(s+1) + \tau_t'(s-1)\right)}{2} - \tau_t'(s)\right) \qquad (10.48)$$

In general, the constant proportion means that the initial sampling can be done at a nonuniform rate, wherein more samples are placed as control points around some critical regions and a lesser number of samples are placed elsewhere for computational considerations.

The external force can be modelled like the repulsive force of the APF. The potential is still inversely proportional to the distance from the obstacle (o), however, in this specific modelling, the inversion is taken by subtracting the distance from a constant representing the maximum force, such that a reduction in distance results in an increase in the force value, given by Eq. (10.49).

$$V_{\text{ext}}^o\left(\tau_t'(s)\right) = \begin{cases} \frac{1}{2}k_{\text{ext}}\left(D_{\max} - d\left(\tau_t'(s), o\right)\right)^2 & \text{if } d\left(\tau_t'(s), o\right) \leq D_{\max} \\ 0 & \text{if } d\left(\tau_t'(s), o\right) > D_{\max} \end{cases} \qquad (10.49)$$

Here $D_{\max}$ is the maximum distance after which the potential is trimmed to 0 and d is the distance function. The force is given by the derivative of the potential, shown in Eq. (10.50).

$$F_{\text{ext}}^o\left(\tau_t'(s)\right) = \begin{cases} k_{\text{ext}}\left(D_{\max} - d\left(\tau_t'(s), o\right)\right)\frac{\tau_t'(s) - o}{d\left(\tau_t'(s), o\right)} & \text{if } d\left(\tau_t'(s), o\right) \leq D_{\max} \\ 0 & \text{if } d\left(\tau_t'(s), o\right) < D_{\max} \end{cases}$$

$$(10.50)$$

The total external force can be modelled as the sum of the maximum of the external forces for all obstacles. In case the maximum terminology is used, only the nearest obstacle needs to be computed and used in Eq. (10.50) for calculating the total external force. The total force is given by Eq. (10.51).

$$F\big(\tau_t'(s)\big) = F_{\text{int}}\big(\tau_t'(s)\big) + F_{\text{ext}}\big(\tau_t'(s)\big) = F_{\text{int}}\big(\tau_t'(s)\big) + \max{}_o F_{\text{ext}}^o\big(\tau_t'(s)\big) \quad (10.51)$$

The control point moves with the same force and converges when the internal and external forces cancel each other out, giving the resultant trajectory optimized as per the changing environment.

10.7.4 Discussions

It must be noted that the elastic strip mechanism of adapting the trajectory to the changing environment can only find a better trajectory as per the changed map in the same *homotopic group*. So, if the map changes large enough so that the homotopic group is no longer optimal, the trajectory will not be optimal as well. If the obstacle moves significantly, the trajectory may be highly sub-optimal, locally reacting to the changing obstacle. Therefore, by itself, it is not a good strategy to avoid all dynamic obstacles, which must be done by using the artificial potential field. This is shown in Fig. 10.19, wherein the trajectory after enough deformations no longer represents the optimal path.

A caveat in the discussions above was that the distances, potentials, forces, and motion were done in the configuration space. However, most of the discussions in the chapter revolved around using sensors for distance measurement, meaning that the calculations are in the workspace. This was also the case in the APF for manipulators, wherein the distances were assumed in the workspace for the end-effector, which gives the resultant force and hence the velocity of the end-effector in the workspace. The end-effector cartesian velocity is converted into the end-effector joint speed by using the Jacobian. Alternatively, the APF can use the forces into different links and joints and convert them into the artificial forces at the joint level resulting in joint speeds.

Let us visualize working in the workspace where all virtual forces are applied. The issue is how to visualize control points in the workspace. The balls representing the free space are placed at the spines representing the skeletal structure of the robot. The forces act upon the control points which are now the points in the workspace and hence represent the different parts of the body in the spine of the robot/manipulator. The internal and external forces together constitute the mechanism of motion of the control point in the workspace, which is translated into the joint movements by the Jacobian around the point. The resulting sum of all forces gives the motion to be made at every point, equivalent to moving the control point in the configuration space. The complete trajectory is such sampled configurations or a sampled set of virtual robots that move by application of virtual forces in the workspace, within the protective tunnel.

10.7.5 Adaptive roadmap

The elastic strip must not be the solution when the changes in the map cause a change in the homotopic group of the path. There must be a way to store more paths from source to goal in different honomotic groups, such that due to the movement of an obstacle, if the current path becomes suboptimal, still another alternative path may already be there that may now be optimal. This section meets the same expectation. The elastic strip as well models only for the single query motion planning problem. The generic approach is also extendible to a multi-query approach wherein there are potentially several agents which raise a source-goal motion planning query, or when the motion planning query is raised by the agent numerous times due to optimality in the changing map context, giving the agent the luxury to plan outside the current homotopic group as well.

The *adaptive roadmap* (Yang and Brock, 2010; Gayle et al., 2007) may be visualized as a roadmap consisting of vertices (V) and edges (E), wherein the vertices and edges react to the internal and external forces and therefore move as the obstacles move. Given an initial roadmap, the modelling is extended to move every edge as an elastic strip and similarly move every vertex as a variable control point of the elastic strip. As the obstacle approaches the edges, the edges incur tension and move away, while the tension of the edges at the other end is released with a propensity to move closer.

An interesting phenomenon is when an obstacle moves by a large amount, and in such a case, it is likely to highly deform the roadmap. The edges are assumed to have a maximum permissible tension beyond which they *break down* and cease to exist. This is also the case with a real elastic strip, where stretching it beyond a point causes the strip to break. When the internal force of the elastic strip becomes too high, the strip breaks. This deletes the edge from the roadmap. Similarly, the vertex will get removed when all edges associated with it are removed.

If the roadmap continues in this manner, adapting to the changes in the map, slowly the vertices and edges will keep getting removed from the roadmap resulting in a loss of connectivity affecting optimality and completeness both. Therefore, it is necessary to *add vertices and edges* to counterbalance the removal. The simplest and intuitive strategy is to memorize the vertices deleted. The same configurations are constantly checked for collision. If there is no collision, it means an obstacle has passed-by and therefore the vertices may be re-inserted into the roadmap. The strategy preserves the roadmap against obstacles passing by, wherein the vertices are temporarily deleted only for the time the obstacle passes.

The second strategy deals with losing many important edges. This will eventually result in a lack of completeness and connectivity. Therefore, it is necessary to take measures to induce connectivity back into the roadmap by a deliberative technique. The *connect connected component* strategy of Probabilistic Roadmap selects two roadmaps, which are not connected and uses RRT for the connection. This strategy is frequently used for connecting difficult subroadmaps in Probabilistic Roadmaps. This is also used frequently in the adaptive roadmap to get connectivity. Further, the connection attempts to the nearest samples are always checked and if two near samples are visible to each other, an edge is added. The concepts are shown in Fig. 10.20.

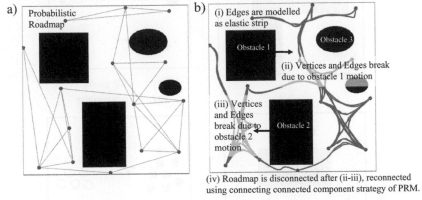

FIG. 10.20 Adaptive roadmap is shown in (B) for the probabilistic roadmap shown in (A). As obstacles move, edges break when tension increases more than a threshold; vertices break if all edges break. Deleted vertices/edges are re-inserted whenever possible.

It must still be stressed that if the environment has many obstacles constantly moving, maintaining the roadmap at a fast pace will be difficult, and computationally expensive. This is especially when the number of vertices and edges are too high. This must be considered before using adaptive roadmaps in any scenario. It is especially important to have a limited number of vertices and edges in the roadmap to enable computations as the map changes.

Questions

1. For the maps shown in Fig. 10.21, show the output of the Artificial Potential Field.

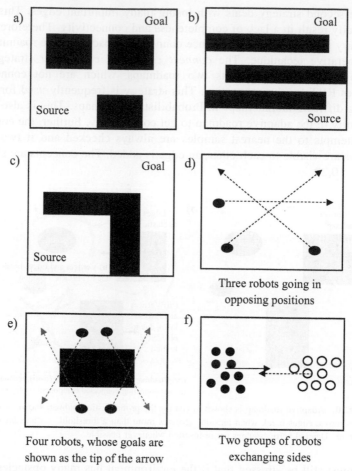

FIG. 10.21 Maps for questions. (A) Map 1, (B) Map 2, (C) Map 3, (D) Map 4, (E) Map 5, and (F) Map 6.

2. Comment on the similarities and differences between roadmap approaches, cell decomposition approaches and potential field algorithm, clearly stating in which context is each of these algorithms more appropriate. Also, suggest means to use a combination of such algorithms to create a better overall algorithm.

3. Give several examples where the Artificial Potential Field algorithm gets the robot stuck at a local optimum (equilibrium), including dynamic maps. For each scenario, think of an algorithm that can enable the robot, escape the local optima. Hence, think of different ways by which a selection or fusion of different algorithms may be done to make a better algorithm.

4. Imagine a hall filled with many robots and centralized system models an adaptive roadmap for every robot. Illustrate the computational feasibility of increasing the number of robots. Accordingly illustrate what kinds of obstacles are better modelled as moving obstacles in the adaptive roadmap, and what obstacles may as well be reactively avoided.

5. The adaptive roadmap can be too slow if many obstacles are moving, while the elastic strip can severely compromise on optimality when an obstacle sufficiently moves. Discuss any intermediate approach that can combine both benefits. Also, comment on how the initial elastic strip would be practically made in such an approach.

6. Some social etiquettes are commonly displayed by humans like overtaking on the right (or left in some countries), enabling a person at a back in a hurry overtake, letting people come in first (over people going out), not intercepting a group, etc. Observe and write all such behaviours. For every behaviour model a social force and formally write the applicable condition mathematically when the force is applied.

7. A bunch of people exit a building from the main exit and another bunch goes in. The number of people going in is far more than the number of people getting out. Assuming a Social Potential Field modelling explains the motion that you expect to happen. Also, explain the same scenario when the number of people in the two bunches is equal and the change with increasing the total number of people. Suppose the social potential of a VIP is modelled such that the repulsive potential is of a large magnitude. How does the answer change if a few VIPs are coming out, and no VIP is going in?

8. [Programming] Make a program that makes a simulated robot move in a defined map using potential field algorithm. The robot should use proximity sensors operating on (a) geometric map with polygon obstacles and (b) grid map given as a bitmap image.

9. [Programming] Extend the above program to model multiple robots going from source to goal. Use suitable strategies to detect deadlock. In case of a deadlock, make every robot take a few random steps biased towards a strategized direction. If the deadlock is still not resolved, increase the randomization step size and number of random steps.

10. [Programming] Extend the above program to model groups of agents that move using the potential field. The agents of the same group must have the same goal. Experiment with different densities of different groups. Display situations when the groups need to split to account for other moving groups. For faster results, use the k-d tree to store robot positions for faster proximity query to compute forces.

11. [Programming] Extend the program to simulate the behaviour that every two robots maintain a specific prespecified distance while crossing each other. Using multiple robots, simulate for situations when such a distance cannot be met.

12. [Programming] Use an algorithm to compute the path of a robot from a source to the goal. Represent the path using an elastic strip. Make the robot trace the path using any noisy control algorithm. Make obstacles move as the robot moves and adapts the path as per elastic strip. Teleoperate the obstacles to create adversarial situations and see how the robot reacts.

13. [Programming] Extend the above program to the adaptive roadmap. Particularly experiment for situations when the edges break and are re-formed.

14. [Programming] Make a map of a home or office environment. Mark different areas of interest where people move. Mark a normal navigation direction for every corridor for modelling attractive force. Make multiple simulated people move as per their typical routines, including waiting time. Study the areas of congestion.

15. [Programming] Consider two groups going as shown in Fig. 10.21F. Model all social forces such that in a decentralized manner (i) the members of the first group stay with each other, while members of the other group split to avoid a collision; (ii) the members of the second group stay with each other, while members of the other group split to avoid a collision; (iii) members of both groups split to avoid a collision; (iv) both groups align themselves on left and do not mingle with each other.

References

Brock, O., Khatib, O., 1998. Elastic strips: real-time path modification for mobile manipulation. In: Shirai, Y., Hirose, S. (Eds.), Robotics Research. Springer, London, pp. 5–13.

Brock, O., Khatib, O., 2002. Elastic strips: a framework for motion generation in human environments. Int. J. Robot. Res. 21 (12), 1031–1052.

Brock, O., Khatib, O., 2008. Elastic strips: a framework for integrated planning and execution. In: Experimental Robotics VI. Springer, London, pp. 329–338.

Choset, H., Lynch, K.M., Hutchinson, S., Kantor, G., Burgard, W., Kavraki, L.E., Thrun, S., 2005. Principles of Robot Motion: Theory, Algorithms, and Implementations. MIT Press, Cambridge, MA.

Gayle, R., Sud, A., Lin, M.C., Manocha, D., 2007. Reactive deformation roadmaps: motion planning of multiple robots in dynamic environments. In: 2007 IEEE/RSJ International Conference on Intelligent Robots and Systems. IEEE, San Diego, CA, pp. 3777–3783.

Gayle, R., Moss, W., Lin, M.C., Manocha, D., 2009. Multi-robot coordination using generalized social potential fields. In: Proceedings of the 2009 IEEE International Conference on Robotics and Automation. IEEE, Kobe, pp. 106–113.

Helbing, D., Molnár, P., 1995. Social force model for pedestrian dynamics. Phys. Rev. E 51, 4282.

Hwang, Y.K., Ahuja, N., 1992. A potential field approach to path planning. IEEE Trans. Robot. Autom. 8 (1), 23–32.

Kala, R., 2018. Routing-based navigation of dense mobile robots. Intell. Serv. Robot. 11 (1), 25–39.

Khatib, O., 1985. Real-time obstacle avoidance for manipulators and mobile robots. In: Proceedings of the 1985 IEEE International Conference on Robotics and Automation. vol. 2. IEEE, St. Louis, Missouri, pp. 500–505.

Reif, J.H., Wang, H., 1999. Social potential fields: a distributed behavioral control for autonomous robots. Robot. Auton. Syst. 27 (3), 171–194.

Sud, A., Gayle, R., Andersen, E., Guy, S., Lin, M., Manocha, D., 2007. Real-time navigation of independent agents using adaptive roadmaps. In: Proceedings of the 2007 ACM Symposium on Virtual Reality Software and Technology. ACM, Newport Beach, CA, pp. 99–106.

Yang, Y., Brock, O., 2006. Elastic roadmaps: globally task-consistent motion for autonomous mobile manipulation in dynamic environments. In: Robotics: Science and Systems. vol. 36. RSS, Philadelphia, PA.

Yang, Y., Brock, O., 2010. Elastic roadmaps—motion generation for autonomous mobile manipulation. Auton. Robot. 28, 113.

Chapter 11

Geometric and fuzzy logic-based motion planning

11.1 Introduction

This chapter aims to make a reactive function which can accept the current percept of the robot and make an immediate move of the robot without incurring high computational costs. Since the decisions shall be reactive, the navigation algorithm will be able to deal with dynamic obstacles, static obstacles, and suddenly appearing obstacles. This constitutes most of the navigation done by humans, wherein the humans subconsciously see the things around them to make the next step, and this step-by-step navigation results in the intelligent navigation of humans. The problem, however, is that the approach neither guarantees optimality nor completeness. Specifically, the chapter continues the discussions of the artificial potential field, which also makes reactive decisions for the navigation of the robot using a specific rule to solve the problem. This chapter gives more options for reactive decision-making.

Even though the problem of motion planning is to be solved for a variety of systems, including manipulators, drones, protein folding, and computational bio-informatics systems, the book has taken a special interest towards the navigation of mobile robots in the workspace. Let us focus on this specific use case.

The first intuition is to move as per the *geometry* of the environment. Looking back, the discussions over motion planning started with the *visibility graphs* to get to the corners to all obstacles and travel using the same as edges, thus minimizing path length. Instead of a graph search, one may as well take the tangent to the obstacle directly ahead and avoid the obstacles, one at a time, which is done in this chapter. Similarly, the *Voronoi graphs* stressed the notion of taking the centre of the obstacles to maximize clearance, which can also be reactively achieved. The graph search in both cases provides optimality and completeness, which will be absent in the reactive paradigms. However, the reactive paradigm enables handling dynamic and suddenly appearing obstacles and needs a small computational time and memory. The choices may not be limited to just the highest clearance (significantly compromising the path length) and the smallest path length (severely compromising the clearance).

Autonomous Mobile Robots. https://doi.org/10.1016/B978-0-443-18908-1.00004-2

It may be taken as a mix of the two depending upon the specific heuristic. Knowledge of the obstacle geometry or locally interpreting the obstacle geometry gives the luxury of taking reactive decisions.

The notion can be extended up to another layer of abstraction. The *rules* of navigation have so far been limited to the specifics allowable by the model. The artificial potential field was generalized to any force modelled by the notion of social potential, each force modelling a rule, but every rule must be a force. Similarly, the geometric approaches have so far been restricted to the model based on the geometrical assessment. At a higher level of abstraction, the notion is to design any possible rule and to further have a multitude of options to fire rules, aggregate rules, and interpret the results. This is done using a *fuzzy inference system* representing the motion behaviour of the agent. The different concepts are summarized in Box 11.1.

Box 11.1 A summary of terms and techniques used

Velocity obstacle approach

- Compute a set of all immediate feasible velocity vectors such that no collision is possible within a time horizon.
- Select a feasible velocity of the largest magnitude from the set, that leads to the goal within a small angular deviation from the straight line to the goal.
- Assume all other obstacles are circular with known velocities. Static obstacles are handled by fusion with a deliberative planner. Static obstacles are different from dynamic ones. One can follow a slow-moving vehicle in front, but not a broken-down vehicle resulting in an indefinite wait.
- Only consider kinematically feasible velocities.
- Typically aim at tangents to obstacles in line to goal.

Velocity obstacle variants

- **Global search**: In an A*-like search algorithm, for expansion of a node, select multiple velocities by multiple selection strategies (action function of A*). Each selected velocity makes a child in the search tree.
- **Reciprocal velocity obstacles:** Only do half the obstacle avoidance trajectory correction, the other half shall be done by the obstacle.
- **Hybrid reciprocal velocity obstacles:** If an initial attempt is to avoid an obstacle from the left, prefer avoiding the same obstacle from the left in the future iterations to avoid oscillations.
- **Optimal reciprocal collision avoidance:** Optimize the choice of velocity assignment, optimizing for the chosen side of avoiding obstacles.

Vector field histogram

- Consider a local grid around the robot, every cell applies a force based on distance and certainty.
- Create a polar plot of forces, smoothen by moving average.
- Binarize the plot. 1 or hill denotes high obstacle density at the angle, 0 or valley denotes a small obstacle-density (or no obstacles).
- Select the widest valley, also closer to angle to goal.

Box 11.1 A summary of terms and techniques used—cont'd

- Select an angle in the chosen valley, that offers large clearance and small deviation from angle to goal.
- Move towards the chosen direction.

Vector field histogram (VFH) variants

- **VFH+**: Inflate all obstacles by the size of the robot accounting for the radial occupancy of the robot. Prefer selection of the same corridor between the same obstacles as used previously, to avoid oscillations.
- **VFH***: In an A*-like search algorithm, for expansion of a node, select multiple angles (valleys) in the binarized polar histogram (action function of A*). Each selected angle makes a child in the search tree.

Other geometric approaches

- Radially scan all obstacles within a threshold distance, within the range governed by kinematic feasibility and visibility limits.
- Identify the radial range within which no obstacle exists.
- Aim to go to the spatial/radial bisector of the range (obstacle avoidance angle).
- Bias towards the goal for getting goal-seeking by taking a weighted addition of the obstacle avoidance angle and angle to goal for the immediate motion of the robot. Alternatively, limit selection of obstacle avoidance angle within some threshold of deviation from angle to goal.
- As another strategy, select an angle that has the largest obstacle clear distance, accounting for the size of the robot for obstacle avoidance.

Fuzzy logic

- **Fuzzy set**: An element may partially belong to a fuzzy set to the extent given by the membership degree.
- **Membership function**: Function that maps an input to the degree to which it belongs to a fuzzy set.
- **Fuzzy logic**: Calculate outputs using fuzzy rules operating on fuzzy sets.
- Divide inputs and outputs into fuzzy sets.
- **Fuzzification**: Convert applied crisp input into fuzzy input (calculate membership degree of different fuzzy sets).
- **Fuzzy inference of a rule**: (i) Evaluate antecedent of the rule from fuzzy inputs using fuzzy arithmetic over AND, OR, and NOT operators. (ii) Calculate the fuzzy output using the fuzzy implies operator over the antecedent fuzzy output and the consequent membership function.
- **Aggregation**: Summate multiple fuzzy outputs into one fuzzy output.
- **Defuzzification**: Convert a fuzzy output into a crisp (real number) output.

11.2 Velocity obstacle method

The first method of study is a geometric calculation for the immediate velocity (as a vector, which is different from linear speed as a scalar). Imagine a robot at any general point in the workspace. Imagine a dynamic obstacle in the same workspace. Intuitively, if this obstacle were in front of you, you would have either avoided it by moving to its left or by moving to its right. The same intuition was used in the Visibility Map with the difference that the map was static. The assumption is to have all circular dynamic obstacles.

11.2.1   An intuitive example

Consider that you intend to cross a road, which does not have a pedestrianized crossing area. Suppose a vehicle is coming on the road. Now you have two options to time yourself so that you can cross the road before the vehicle comes into your crossing area; and second to wait for the vehicle to go away from the crossing area and then start crossing the road. Let us put things into perspective by using some synthetic numbers. Consider the situation given in Fig. 11.1A. For simplicity (that is to make the problem one-dimensional), the person is assumed to have a zero radius; and the vehicle is assumed to have a zero width and a nonzero length. The person is further constrained to walk ahead in the identified crossing area only. If you move fast, you would be able to cross the road before the vehicle, and you would be safe. If you walk very slowly, you would be crossing after the vehicle has already left. Let us calculate the speeds, which are guaranteed to have a collision using two corner cases. The

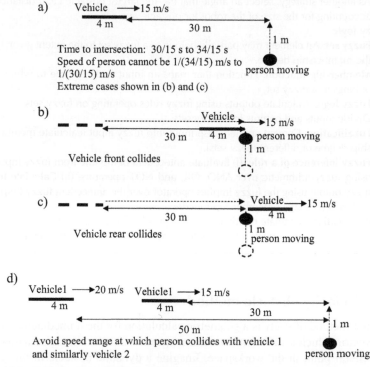

FIG. 11.1   Basic intuition of velocity obstacles. Find the speed range of the person at which an accident happens (velocity obstacle), and hence the person should avoid those speeds.

first corner case is when your edge touches the front of the vehicle (Fig. 11.1B). The time taken by the vehicle to cover the distance is given by $\frac{distance}{speed}=\frac{30\,m}{15\,m/s}=2$ s. The same time of 2 s is taken by you, so the speed is $\frac{distance}{time}=\frac{1\,m}{2\,s}=0.5$ m/s.

In the other corner case, you walk slow enough to let the vehicle just pass. The distance covered by the vehicle is 34 m, which will take a time of $\frac{distance}{speed}=\frac{34\,m}{15\,m/s}=2.27$ s. The speed so that you cover 1 m in 2.27 s is $speed=\frac{distance}{time}=\frac{1\,m}{2.27\,s}=0.44$ m/s. The scenario is shown in Fig. 11.1C. A speed range of [0.44,0.5] lands you into a collision, and therefore, these speeds act as an obstacle. This is called the *velocity obstacle* (VO). Alternatively, the speeds (positive only) $(0,0.44)\cup(0.5,\infty)$ are called the safe speeds wherein no collision occurs. You must select the one closest to your most comfortable speed, provided it is outside the collision-prone speed.

To make matters more interesting, let us insert another vehicle shown in Fig. 11.1D. Calculating both corner cases again, the quickest you can go must be when the vehicle reaches the near-collision site at $\frac{50\,m}{20\,m/s}=2.5$ s, in which case your speed is $\frac{1\,m}{2.5\,s}=0.4$ m/s. The slowest for a collision is with the vehicle advancing to reach at $\frac{54\,m}{20\,m/s}=2.7$ s, with your speed as $\frac{1\,m}{2.7\,s}=0.37$ m/s. The inter-vehicle collisions are intentionally neglected since they have a 0 width and can go close to each other. Now, the speeds [0.37,0.4] are also collision prone. The unsafe speeds are hence the union of the two $[0.37,0.4]\cup[0.44,0.5]$ and correspondingly the safe speeds are $(0,0.37)\cup(0.4,0.44)\cup(0.5,\infty)$. Again, the speed selection will be comfort-driven; however, it must be from the safe range denoting going before both vehicles, going in between both vehicles, and going after both vehicles. The procedure described is a complete working of the VO mechanism of motion planning; however, this will be extended to two-dimensions from one-dimension, and hence velocity will replace speed.

Such a mechanism of avoiding obstacles is seen in numerous situations of everyday life. If you need to overtake a slow-walking person ahead, you aim to walk at their right-hand tip to overtake on the right (left-side walking etiquette countries). A ship, on being informed about an obstacle glacier, calculates its heading to just avoid the obstacle glacier.

11.2.2 Velocity obstacles

The VO method (Fiorini and Shiller, 1998; Wilkie et al., 2009) is a mechanism to calculate the velocities that in the future will cause a collision and hence must be avoided. Alternatively, it is a method that suggests velocities that can be selected by the robot because they are safe and collision-free. The assumption is that the robot and all the obstacles continue to move at the same speed. This is not a limitation because, unlike deliberative methods, in reactive methods, a new move is calculated at every instant. Therefore, if the other obstacles change their velocities, the robot also reacts by selecting a new velocity that is guaranteed to be collision-free in the short run. Only the nearest few obstacles are considered. This is natural because if you are asked to select only one velocity and move with the same continuously and forever, no velocity may guarantee a collision-free motion.

The environment is assumed to be consisting of dynamic circular obstacles only (other humans, robots, etc.). If the obstacle is not circular, then the calculations will be hard to perform. Dynamics is not a hard assumption since static obstacles are a special case of dynamic obstacles with a velocity of 0. However, if you are behind a dynamic obstacle, either the obstacle will clear away giving you space to move, or the obstacle will move ahead so that you can continue the motion. However, if the obstacle is static, neither will happen. In traffic, it is acceptable to place your vehicle behind another vehicle moving very slowly and you still move; however, placing your vehicle behind a damaged vehicle waiting for service members is not suggestive.

Let ρ be the radius of the circular robot placed at position q in a 2D workspace. Let ρ_o be the radius of the circular obstacle placed at o, initially assumed to be stationary. For simplicity of calculations, let us make the robot point sized, which is by inflating all other obstacles by the radius ρ as is a usual process in the construction of the configuration space of circular robots. If the robot moves even at the thinnest speeds towards the obstacle o, it is bound to collide. Therefore, the colliding velocities are given by any magnitude with a direction towards the circular coverage of the obstacle of radius $\rho + \rho_o$, as shown in Fig. 11.2A. Notice that the shape is that of a cone and is therefore called the *collision cone* of ρ_o. The collision code is given by Eqs (11.1), (11.2):

$$CC(o) = \{v : \exists_\lambda \, q + \lambda v \cap R(o) \neq \varnothing\} \tag{11.1}$$

$$R(o) = \{q : d(q, o) < \rho + \rho_o\} \tag{11.2}$$

Here, $R(o)$ denotes the set of points within the circular obstacle located at o or a set of all points q whose distance from o is less than the sum of radii $\rho + \rho_o$. The equation is easier to solve analytically since the minimum and maximum angle is known and the velocity is any magnitude with the angle.

Now assume that the obstacle is moving with a speed v_o, which can be measured by numerous sensing technologies by the robot. There is no change in the formulation required except for interpreting the term velocity by relative velocity or to say that the relative velocity makes the collision-cone representing an obstacle. The VO is thus given by Eqs (11.3), (11.4) and Fig. 11.2B:

$$VO(o) \ominus v_0 = CC(o) \tag{11.3}$$

$$VO(o) = CC(o) \oplus v_o \tag{11.4}$$

The Minkowsky sum ($\oplus$) is used because v_o needs to be added to all the elements of the set $CC(o)$. The velocity vector taken by the robot has a tail at the robot position, and the head should never be at a grey region denoting the VO.

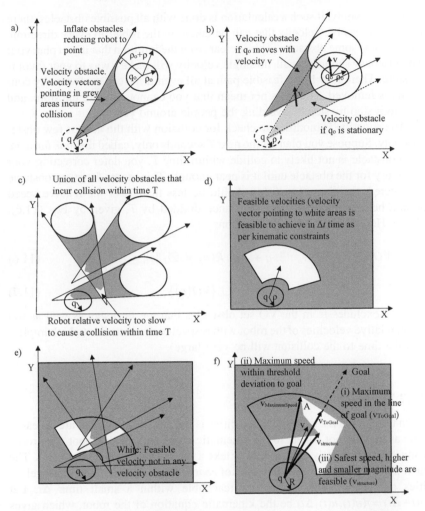

FIG. 11.2 Velocity obstacle method. (A) Velocity obstacle is any magnitude directed within the extreme tangents, shown in grey. (B) Account for obstacle speed, the relative velocity cannot be directed towards the obstacle, hence the absolute velocity obstacle is added with the speed of the obstacle. (C) The velocity obstacle is the union over velocity obstacle of every robot. (D) Accounting for kinematic constraints. (E) White regions, accounting for kinematics and velocity obstacles (F) Selection of velocity (for a different scenario). *A* is unsafe because a little change in magnitude causes a collision.

The set of velocities that acts as VOs includes not only the obstacle o but all such obstacles. The resultant VO is hence the union of all VOs given by Eq. (11.5) and shown in Fig. 11.2C:

$$VO = \cup_o VO(o) \tag{11.5}$$

Unfortunately, if such a calculation is done with all possible obstacles, there will be no feasible velocity that can be taken by the robot. In every direction, there will be some obstacles. This is clear from the intuition that if you plan your way in a busy shopping mall, no single velocity will make a way to your favourite store. There may be no feasible path at all even if the constant velocity constraint is removed. This does not mean that you cannot move to the store, and you do it step by step by avoiding the people around you.

The equation is modified to check for collision with the nearest few obstacles only. Suppose you plan for the next T seconds only, called the *time horizon*. If an obstacle is not likely to collide within time T, you defer correcting your trajectory for the obstacle until it is near enough. This means for every obstacle considered, the *time to collision* should be less than T, or the relative speed should be less than the current distance divided by T, given by Eqs (11.6), (11.7). Here, d is the distance function:

$$VO(o) = \left\{ v + v_o : \exists_\lambda\, q + \lambda v \cap R(o) \neq \varnothing, \|v\| \leq \frac{d(q, q_o)}{T} \right\} \qquad (11.6)$$

$$VO = \cup_o \{VO(o)\} \qquad (11.7)$$

This excludes from the VO set obstacles that are far away as well as too small relative velocities of the robot with respect to the obstacles (which implies that the time to the collision will be very large).

11.2.3 Reachable velocities

Even though the list of feasible velocities is infinite, most of them are unreachable as they do not adhere to the kinematic constraints of the robot. The robot cannot have infinitely large speeds or take a sudden turn of over 90 degrees. The next phase is hence to calculate a set of *reachable velocities*. Only those velocities are accommodated that are reachable within a small time Δt. Let $q(t+\Delta t) = f(q(t), u(t), \Delta t)$ be the kinematic equation of the robot, which gives the pose $q(t+\Delta t)$ of the robot at time $t+\Delta t$ given the pose of the robot $q(t)$ and a control signal, $u(t)$ applied uniformly till time Δt. The limits on the control signals are known, which gives a set of possible velocities. Furthermore, the velocities themselves have a maximum and minimum constraint, which further limits the set of reachable velocities.

For simplicity let us consider that the robot is operative with a linear speed $0 \leq s \leq s_{max}$ and an angular speed $0 \leq \omega \leq \omega_{max}$. The speeds are controlled by using accelerations for the linear term $a_{s,min} \leq a_s \leq a_{s,max}$, and for the angular term $a_{\omega,min} \leq a_\omega \leq a_{\omega,max}$. This means that in Δt time, the linear speed may be taken from the set $[s(t)+a_{s,min}\Delta t, s(t)+a_{s,max}\Delta t]$, which will need to be trimmed further to ensure adherence to the minimum and maximum speed limits to $[\max(s(t)+a_{s,min}\Delta t, 0), \min(s(t)+a_{s,max}\Delta t, s_{max})]$. Similarly, with the angular speed. This formulation is using linear and angular speeds, while the VO uses the formulation of a velocity vector. The relation is given by Eqs (11.8), (11.9).

$$\begin{bmatrix} v_x(t) \\ v_y(t) \end{bmatrix} = \begin{bmatrix} s(t) \cos(\theta(t)) \\ s(t) \sin(\theta(t)) \end{bmatrix} \tag{11.8}$$

$$\dot{\theta}(t) = \omega(t) \tag{11.9}$$

This translates the limited linear and angular speeds into the corresponding velocity vectors. Here, $(v_x(t), v_y(t))$ is the velocity vector, $\theta(t)$ is the orientation, and $\omega(t)$ is the angular speed at time t. The robot may choose from any velocity in the set of reachable velocities, and it will be kinematically feasible. The set of reachable velocities is shown in Fig. 11.2D.

Overall, we have a set of velocities called a VO that acts as obstacles and another set of velocities called reachable velocity (RV) that is feasible. The set difference between the two is called the *Reachable Avoiding Velocities* (RAV), which denote the velocities that are both feasible and avoid obstacles. This is shown in Eq. (11.10) and shown in Fig. 11.2E.

$$RAV = RV - VO \tag{11.10}$$

Here the − is the set difference operator.

11.2.4   Deciding immediate velocity

The calculations of RAV give a good set of candidate velocities, and any may be selected from the set. However, one velocity from the set needs to be selected. So far, the discussions were over obstacle avoidance behaviour. This subsection accounts for the goal and adds *goal-seeking* behaviour to select a velocity that will be given to the controller. *Heuristic measures* are used to decide the immediate velocity.

The first strategy called the *ToGoal* strategy aims to minimize the distance objective and, therefore, takes the velocity with the highest magnitude in the direction to the goal. This is shown in Fig. 11.2F. This strategy is to keep walking in a straight line towards the goal while avoiding obstacles by pausing and adjusting the speed. As a result, the shortest distance is obtained, however, this does not mean the shortest time because the robot may be going painfully slow or waiting for a prolonged time for the obstacles to clear. The strategy is given by Eqs (11.11), (11.12):

$$S_{ToGoal}(RAV) = \arg \max_{v \in RAV, \theta_v = \theta_G} \|v\| \tag{11.11}$$

$$\theta_v = \operatorname{atan2}(v_y, v_x), \theta_G = \operatorname{atan2}\left(g_y - q_y, g_x - q_x\right) \tag{11.12}$$

Here, θ_v is the direction component of the velocity, $\|v\|$ is the magnitude of the velocity, and θ_G is the angle to goal (g_x, g_y).

The other strategy is the opposite, which tries to maximize the speed so that the robot is always moving, called the *MaximumSpeed* strategy. So, the velocity with the highest value of speed (magnitude) is taken. However, this can mean going away from the goal. For this reason, an angular deviation from the goal constraint is added. The velocity selection is shown in Fig. 11.2F. The robot is always seen moving at high speeds, while may occasionally take elongated paths. The parameter controlling the angular threshold is important to set and controls the tradeoff between optimizing for speed and optimizing for distance. The strategy is given by Eq. (11.13):

$$S_{MaximumSpeed}(RAV) = \arg \max_{v \in RAV, \, abs(AngleDiff(\theta_v, \theta_G)) < threshold} \|v\| \qquad (11.13)$$

Here, the *threshold* is the maximum allowed angular deviation from the direction to goal and *AngleDiff* is the function that takes the difference between the two angles constrained to the range $[0, \pi]$.

The other common strategy assesses the *structure* of the VO to compute a safe velocity to travel with. The safety is that even if there are errors in setting the velocity to the desired value, the system should still be safe. The velocity here consists of the angular component denoting the direction and the magnitude component denoting the speed, and the safety is only applied to the latter. This means if the magnitude becomes a little high or low, it should still be possible to move safely. The velocity selection is shown in Fig. 11.2F. A large set from small to large speeds around the chosen value is collision-free and, therefore, is safe. The most unsafe region is in A where a little deviation also lands up the robot in a collision-prone state.

Alternatively, another strategy is to search for the best velocity near the preferred velocity given by $v_{pref} = \frac{s_{max}(G-q)}{d(q,G)}$. Here, s_{max} is the maximum linear speed, G is the goal, q is the current position, and $d()$ is the distance function. $\frac{(G-q)}{d(q,G)}$ is a unit vector towards the goal. The search produces the closest velocity in RAV.

It must be restressed here that the calculations are done at every instant and, therefore, the selected velocity will be overridden with a new one very soon. This also adapts to the changing velocities of the other obstacles and new obstacles that later come within the time horizon. Further, all strategies assume that all obstacles are dynamic. In the case of a static obstacle, the calculations may be made assuming that the obstacle eventually moves, which will not be the case. Therefore, the VO method must always be hybridized with deliberative planning that generates small subgoals for the VO method to achieve. These are not the only strategies, and the strategies may as well not be fixed and may change with time.

11.2.5 Global search

Since there are a good number of heuristics, it is tempting if all of them can be tried, and then the best of them may be taken. Even a single heuristic may not be the best throughout the navigation, so more generally, it is tempting if every permutation and combination of heuristics, one each for small segments of robot motion, can be used to make all possible moves, and then the best sequence of heuristics be so accepted as the final solution. Because of the computational complexity, the search is global, and the approach is deliberative. Furthermore, it is assumed that the obstacles either keep the same velocities throughout or their velocity profile is known a priori. If the obstacles change their velocities, then re-planning will have to be done in near real-time.

For the notion of discussions, let us call the heuristic or strategies as discussed previously as the *action function*. Let the functions be given by $S = \{S_1, S_2, S_3 \ldots\}$. Each such action function takes a reachable avoidance set of velocities (RAV) and selects a particular speed from the set. Not all action functions may be possible for all the RAV and the applicability also depends upon the shape of the RAV.

The *search-based techniques*, like the Uniform Cost Search, start with the source as the root of the search tree and expand the nodes by application of action functions making edges. As a graph search discretization of both states (vertices) and actions (edges) is necessary. As a tree search (without closed data structure) the actions (edges) should be discrete and not the states (vertices).

For the application of search, a discrete set of actions is needed, which is available from the already discretized action functions representing interesting heuristics. The search tree so produced is shown in Fig. 11.3. If Uniform Cost Search is used and every selected velocity is applied for ΔT time using time as the cost metric, the search will appear as a Breadth-First Search. Since heuristics (in action functions) are being used, the search will still not be optimal. However, practically, this is a much better option than using an exhaustive state-space search, which is extremely computationally expensive. The action functions so designed intelligently to prune the possible speeds for a speedup.

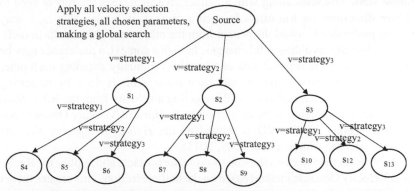

FIG. 11.3 Global search using the velocity obstacle method. At every step, different velocity selection strategies are used as the action functions, which makes a search tree. The search goes very similar to the graph-based search in control spaces.

The notion of using a *reactive technique* as the *action function* to make a global search may seem a specific implementation as such, however, the concept can be extended to any reactive technique if it chooses between some discrete possibilities. This was not discussed in the artificial potential field because the technique never made a choice between discrete options.

11.2.6   Variants

This subsection discusses several relevant advances of the classic VO method to make it better for working. Consider two robots going towards each other. A problem that happens is that the first robot assumes that the other would walk straight and takes the extreme measure of aiming at its extremity, while the other robot does the same thing. However, it is evident from the situation that the two robots would avoid each other by moving in the opposite direction, and each robot needs to only move by half the distance to avoid the other robot. If this was a traffic scenario, each driver knows that the other driver will do its half and only moves by half the distance. Similarly, on seeing someone walking towards you, you only take a small step aside, knowing that the other person will do the other half. This is called the *Reciprocal Velocity Obstacle* (van den Berg et al., 2008, 2011a), wherein the VO for each robot is adjusted to only account for half the corrective motion. The VO approach, at every step, realizes that the other robot has moved far away and corrects its velocity as a reaction. However, this causes too many changes in the velocity and incurs problems in control. Furthermore, the initial velocities are directed too far from usual. The Reciprocal Velocity Obstacle method avoids these problems.

A little assumption here is that the robot knows, which side to avoid the persons, which is possible in most cases as the person more on the left of the robot would be avoided rightward. The corresponding correction for the other person is obtained by symmetry. However, the actual motion, especially when done amidst static obstacles along with a deliberative plan, sees the robots need to change directions far too often. In such a context, if a robot rotates, it may be now preferable to avoid the person from the other side. This results in oscillations when the 'avoiding side' changes. In such a context, a preference may be given to stick to the original side of two robots mutually avoiding each other since some corrections have already been made. This is done by preferring the velocity on the desired side, while selecting a velocity from the RAV. Such an approach of applying some correction by Reciprocal Velocity Obstacle and applying scaling by the VO is termed as the *Hybrid Reciprocal Velocity Obstacle Method* (Snape et al., 2009, 2011).

Another method guesses the side to avoid the obstacle and applies the Reciprocal Velocity Obstacle method based on the half correction. This creates the set of candidate velocity spaces to be filled with half-planes produced by every obstacle, from which a velocity must be selected by additionally checking for the kinematic constraints. This is solved using a Linear Programming approach and the algorithm is termed as the *Optimal Reciprocal Collision Avoidance* (van den Berg et al., 2011b). It should be noted that optimality here is only for the local decision-making and not the optimality of the global path.

11.3   Vector field histogram

The *vector field histogram* (VFH, Borenstein and Koren, 1991) improves upon numerous limitations of the artificial potential field method and adds a notion of geometry. Imagine running a potential field algorithm on a synthetic map, which has some corridor that the robot must enter. It is highly possible that the repulsion of the corridor is too high as compared to the attraction, and the robot instead prefers to keep away. The potential field approach may make the robot go away from the obstacle often. A geometric algorithm, on the contrary, would have identified an opening and enabled the robot to enter. The geometric approaches discussed so far assumed perfect knowledge of the environment. So, the exact positions and speeds were known in the case of the VO method. Some sensor noise can be accounted for. However, practically the robot will face tricky situations, wherein it is not sure whether a space is clear or there appears to be a small obstacle, whether two obstacles have a navigable space in between or space is not enough to go through, etc. It may be better to go through an obstacle-free room with a longer path than a narrow corridor, with uncertain information about obstacles. In all such situations, richer modelling of uncertainty and noisy sensor readings needs to be accounted for. The VFH algorithm models uncertainties and selects immediate speeds, which are clearer, have higher certainty, and use geometry to enable get into difficult regions.

11.3.1   Modelling

Assume that the map is available as a grid map, as a global map, or the local map made by the onboard sensors. The VFH is a local reactive planning technique, and even if the global map is available, it is trimmed to get the local map around the robot only of size $w \times w$. Since sensor uncertainties are being talked about, every grid has a certainty $c(x_o, y_o)$ denoting the certainty that the grid is occupied by an obstacle. This is typicality of working with mobile robots, wherein *grid maps* are a convenient mechanism of storing the map and probabilistic fusion techniques are used to get the certainty of a grid being an obstacle based on multiple sensor readings for the same position.

Every grid in the local map exerts a repulsive force like the potential field, which is called the obstacle contribution, with nearby obstacles having a higher contribution. The obstacle contribution is inversely proportional to the distance from the robot. The modelling is the same as the potential field, with two differences. First is the inclusion of the certainty factor, and hence, if a grid is known with absolute certainty to be an obstacle, the force is high; and correspondingly, if the cell is known not to be an obstacle with absolute certainty, the force is 0. Second, is the modelling of the force. The inverse cube mechanism of modelling is nature-inspired, but, for a digital world with discretization losses and sensor uncertainties, the function becomes too sensitive close to the obstacles. The contribution is taken under a threshold to a maximum value with an additive inverse. The contribution for a grid (x_o, y_o) with the robot at the grid (x_r, y_r) is given by Eqs (11.14)–(11.16).

$$\|F_o\| = [c(x_o, y_o)]^2 \left(a - b\, d((x_o, y_o), (x_y, y_r))\right) \qquad (11.14)$$

$$\theta_o = \operatorname{atan2}(y_o - y_r, x_o - x_r) \qquad (11.15)$$

$$F_o = \|F_o\|[\cos\theta_o, \sin\theta_o]^T \qquad (11.16)$$

Here, $\|F_o\|$ is the magnitude of the contribution, and θ_o is the direction of the contribution from obstacle o at (x_o, y_o). a and b are constants, while d is the distance function. The contribution is intentionally modelled as a magnitude and a direction separately, as per the working requirements of the algorithm. The constants a and b are chosen to have a uniform decrease of the force along with distance. Since the window size is $w \times w$ with the robot at the centre, the maximum distance is $d_{\max} = \sqrt{2}(w - 1)/2$ and, therefore, a is taken as the maximum contribution value $F^{\max}$, and b is taken such that $(a - b\, d_{\max}) = 0$. The forces are shown in Fig. 11.4.

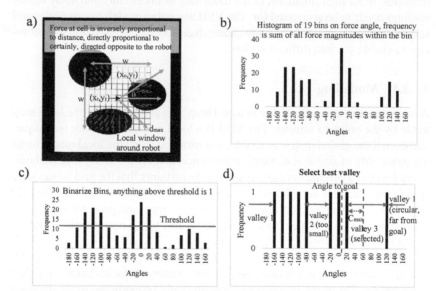

FIG. 11.4 Vector field histogram. (A) Force is calculated at every cell inside the local window (force reaches 0 at the extreme distance $d_{\max}$). (B) Circular histogram, plotted linearly for the sake of simplicity, smoothened by moving average for handling noise. (C) Thresholding, values above threshold are obstacles and vice versa. (D) Selection of best valley, wide enough (threshold to $2C_{\max}$, primary criterion) and near goal (secondary criterion).

11.3.2 Histogram construction

The aspect of geometrical analysis starts from here. The purpose is not to vector add the contributions as forces and move, but to geometrically assess the contribution values and identify interesting areas to move into. For decision-making, an angular *histogram technique* is used to convert the data into a finite number of discrete histograms. The histogram is a technique used to convert data of any dimensionality into 1D data, which can be used for visualization and decision-making. In this approach, the entire angular range $[0, 2\pi)$ is divided into α bins denoted by h_1, h_2, h_3... h_α. Each bin h_i covers an angular range of $[(i-1)2\pi/\alpha, i2\pi/\alpha)$ with a total coverage of $2\pi/\alpha$ radians. Every contribution calculated is added to the corresponding bin as per the angular value with a magnitude as of the contribution. This makes the *polar obstacle density* represented by histograms. This is given by Eq. (11.17), denoting that the ith histogram sums the magnitude of the contribution of all obstacles that lie within the angular range of the ith bin, or $(i-1)\frac{2\pi}{\alpha} \leq \theta_0 < i\frac{2\pi}{\alpha}$.

$$h_i = \sum_{o:(i-1)\frac{2\pi}{\alpha} \leq \theta_0 < i\frac{2\pi}{\alpha}} ||F_o|| \qquad (11.17)$$

The histograms are made from data, which was assumed to be noisy, and therefore, the histograms may themselves be noisy. It is, therefore, a common practise to *smoothen* the histograms, which removes noise as its effect gets distributed among neighbours, however, this also results in a loss of data. A bright light may be seen as a noise at a particular small spot in an image, which may be blurred out by taking a weighted addition with the neighbouring grids of the image. The size of the smoothing window determines the tradeoff between loss of data and smoothing out. For a window of size $2k+1$, with k elements on the left, k elements on the right, and 1 element at the centre, the smoothened histogram is given by Eq. (11.18):

$$h_i' = \frac{\sum_{j=-k}^{k} h_{((i+j+\alpha-1) \bmod \alpha)+1}}{2k+1} \qquad (11.18)$$

The equation uses a moving average with the difference that the coordinates are circular, and therefore, left of 1 lies α and right of α lies 1.

As a further step towards the reduction of the data for decision making, the histograms are *binarized* into 0 and 1, with a 0 denoting that the angular range covered by the bin is relatively obstacle-free and the obstacles are far away, hinting the robot to visit those areas. A value of 1 on the other hand denotes that the bin has obstacles nearby, and the obstacle density is high, and such areas must, therefore, be avoided. The binarized histogram is given by Eq. (11.19), where Th denotes the threshold of binarization.

$$h_i'' = \begin{cases} 0 & h_i' \leq Th \\ 1 & h_i' > Th \end{cases} \qquad (11.19)$$

11.3.3 Candidate selection

The histogram so represented is in the form of hills and valleys, with the hills denoting forbidden areas and the valleys denoting good areas. The criterion of the selection of the valley and the specific angle inside the valley is based on two objective functions. The first is that the robot would prefer navigating in a straight line towards the goal to have the least path length possible, however, since the straight line to the goal may not be feasible due to obstacles, the *angular deviation* between the chosen direction and the straight-line distance to the goal should be the least. This is given by Eq. (11.20), where θ is the chosen direction of motion of the robot.

$$\theta = \arg\min_{\theta \in h_i, \, h_i = 0} abs(AngleDiff(\theta, \theta_G)) \qquad (11.20)$$

The other objective function is *angular clearance*, meaning that the robot should walk in the centre of the widest valley. This helps to combat uncertainties in sensing and control. However, a threshold of C_{max} is placed and angular clearance of more than this is unnecessary. The two objectives are contradictory because increasing clearance may mean to have a large deviation from the goal.

To make decisions using the two objectives, multiple techniques can be used. Here, we'll use a technique, wherein the first objective (primary) would be maximized up to a threshold and the second objective (secondary) would be used as a tiebreaker. Here, clearance is the first criterion, which is optimized up to C_{max}. All valleys, which have a distance of $2C_{max}$ or more are taken. Out of them, the valley closest to the angle to goal θ_G is taken. In case, none of the valleys has such a high clearance, the widest valley is taken.

Thereafter, the task is the selection of a candidate angle inside this valley. Let the selected valley have θ_L and θ_R as the two endpoints. If the valley is narrow, the selected angle to motion is that maximizes the clearance given by $\theta = (\theta_L + \theta_R)/2$. If the valley is wide, and the valley is to the right (anticlockwise) of θ_G, the selected angle will be $\theta = \theta_R + C_{max}$. On the other hand, if the valley is on the left (clockwise), the selected angle will be given by $\theta = \theta_L - C_{max}$. Another case is when the direct angle to the goal is clear and has sufficient clearance, in which case the angle to goal θ_G is selected. The cases are shown in Fig. 11.5.

Since it is a local technique, the angle selection may be done to have some maximum deviation from the goal to avoid the robot getting back. The angular speed is set proportional to the angular deviation between the VFH calculated direction of motion and current orientation in a PID control mechanism. Similarly, the linear speed is inversely proportional to the obstacle density value as well as the angular error. This ensures that the robot slows when many obstacles are nearby and when a huge turn needs to be performed.

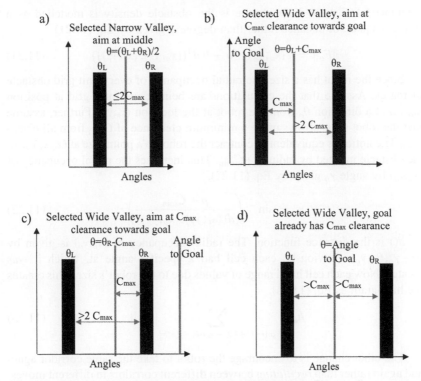

FIG. 11.5 Selection of target angle of motion in vector field histogram. Given a valley, an angle of motion is selected such that the angular clearance is maximized (up to C_{max}, primary criterion) and deviation from the goal is minimized (secondary criterion).

There are two important thresholds in the approach. The first binarizes the polar obstacle density, and the second classifies valleys as wide and narrow. Luckily, both thresholds are relative and can be set by the specific picture of the polar obstacle density and the number of valleys. Particularly, if the robot is detected as stuck, it is important to set the parameters to enable it to enter narrow corridors and go closer to the obstacles. Therefore, there is a bias to classify a histogram as obstacle-free in binarization, and the C_{max} is set to a low value to encourage the robot to select narrower valleys. However, in open areas, the thresholds can be conveniently set to high values. That said, it is important to adapt the behaviour for the proper running of the algorithm.

11.3.4   VFH+

A major limitation of the VFH is that the size of the robot is not accounted for. The VFH+ algorithm (Ulrich and Borenstein, 1998) primarily mends the calculations assuming the robot is a disk of radius ρ and further requires a minimum clearance of C_{min}. A little change is the obstacle density is modelled as a (minus) square of distance rather than degree 1 given by Eq. (11.21).

$$\|F_o\| = [c(x_o, y_o)]^2 \left(a - b\, d^2((x_o, y_o), (x_r, y_r))\right) \qquad (11.21)$$

Since the robot has a size, the radial occupancy of every unit grid obstacle increases. Assume that the calculations are being done for a grid at position (x_o, y_o) at a direction θ_o from the robot at the location (x_r, y_r). Further, assume that the robot must always make a minimum clearance of C_{min} from all obstacles. The notion is equivalent to contract the robot to a point size at (x_o, y_o) with the obstacle inflated by radius $\rho + C_{min}$. This increases the radial occupancy of (x_o, y_o) by angle γ_o given by Eq. (11.22).

$$\gamma_o = \sin^{-1}\left(\frac{\rho + C_{min}}{d((x_r, y_r), (x_o, y_o))}\right) \qquad (11.22)$$

$d()$ is the distance function. The radial occupancy of (x_o, y_o) is given by $[\theta - \gamma_o, \theta + \gamma_o]$. Previously, each cell had a specific angle at, which it was located. Now each cell has a range of values due to the robot's size. This creates the histogram as Eq. (11.23).

$$h_i = \sum_{o:(i-1)\frac{2\pi}{\alpha} \le \theta_o + \gamma_o \wedge \theta_o - \gamma_o < i\frac{2\pi}{\alpha}} \|F_o\| \qquad (11.23)$$

The other change is to encourage the robot to take the same corridor again and again rather than *oscillating* between different corridors in different moves. This is done by changing from one threshold that binarizes the polar obstacle density into two thresholds (high threshold and low threshold) that classify the obvious cases. An intermediate case is given the same classification as was used in the previous time step. This maintains corridor information and encourages the selection of the same corridor again.

11.3.5   VFH*

The VFH algorithm makes a choice of the valley to enter, which is a discrete choice, based on the metrics and parameter values. This motivates the attempt to try all valleys and different parameter settings for a preknown map, and later to take the best overall path produced. This is a deliberative and global planning technique using graph search primitives. Search-based techniques are applied over all valleys. The notion is that if a reactive technique makes some discrete choices, then a search can be easily applied over all the discrete choices, with the discrete choices making the action function of the search. The robot starts from the source. Each valley makes a child in the expansion of the source. The children are similarly expanded, with every valley discovered making a new child. The discussion is the same as the VO method, and hence, not repeated.

The only aspect that deserves discussion is the action function. Consider a state with the robot at position q. By calculating obstacle density, and polar obstacle density by histograms, and binarizing the histograms several candidate valleys are selected. Each valley becomes a candidate state and is added as a child. The number of children is the number of candidate valleys identified by the VFH. The simulation is done for ΔT steps with the same velocity selection to produce the specific child. The heuristic functions can be designed to make an A* algorithm do an efficient search. The resultant algorithm is called the VFH* algorithm (Ulrich and Borenstein, 2000).

11.4   Other geometric approaches

This section assesses geometry and uses complimentary heuristics for the navigation of the robot.

11.4.1   Largest gap

The VO method calculates the smallest distance by making the two robots kiss and avoid each other. Hence, it is required to study the equivalent method that maximizes clearance (Sezer and Gokasan, 2012). The obstacles and the robot are still considered as circular; however, the velocities of the obstacles are not considered. The calculations hence become simple. Consider that the robot sees several circular obstacles around. The robot is condensed as a point, and hence the obstacles are inflated by ρ, the radius of the robot. Suppose an obstacle at (x_o, y_o) with the angle to the robot as θ_o. To avoid the obstacle, the robot must travel aiming a tangent to the obstacle. The angular position of the anticlockwise (left) tangent (θ_{oL}) and clockwise (right) tangent (θ_{oR}) is given by Eqs (11.24), (11.25):

$$\gamma_o = \sin^{-1}\left(\frac{\rho_o + \rho}{d((x_r, y_r), (x_o, y_o))}\right) \tag{11.24}$$

$$\theta_{oL} = \theta_o - \gamma_o, \theta_{oR} = \theta_o + \gamma_o \tag{11.25}$$

The notations are shown in Fig. 11.6A. Consider the view from the robot, with different circular obstacles all around. Consider that the robot is oriented at an angle θ. The robot has *kinematic constraints* that specify the minimum and maximum turn that the robot can take at a small ΔT time. This means the immediate velocity of the robot should be angled between θ_{KL} and θ_{KR} where the subscript K stands for the kinematic constraints, and L and R are used to denote left and right, respectively. Let the current heading of the robot be θ. As per the current speed, the robot may not be able to make a turn of more than θ_K. Thus, the minimum orientation it can maintain at the next time instance is $\theta_{KR} = \theta - \theta_K$, while the maximum orientation it can maintain is $\theta_{KL} = \theta + \theta_K$.

Similarly, there are constraints into what the robot can sense because of the placement of the sensors at the front and side only, giving *view constraints* θ_{VL} and θ_{VR}. A lidar sensor can nearly sense the complete range without view constraints, baring the little region where the laser senses the robot body itself excluded from the sensing rage. A camera, on the contrary, has a much smaller field of view within which the robot can see, implying a higher constraint. The field of view of a ring of sonar sensors considers all the sonar sensors and the radial resolution of sensing depends upon the number of sonar sensors in the ring. A sonar sensor ring in this mechanism is a budgetary imitation of

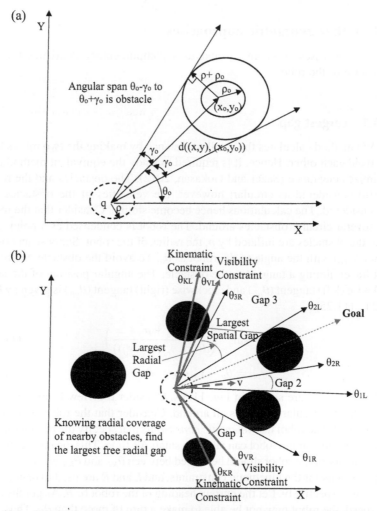

FIG. 11.6 Calculation of the largest gap. (A) Angular coverage of the obstacle with respect to the robot. (b) Finding the largest radial or spatial gap (within kinematic and visibility constraints) for motion. To get the goal-seeking behaviour, use either a weighted average of selected angle with the angle to the goal; or limit the selected angle within a threshold of the straight line to the goal.

the lidar sensor. The immediate motion of the robot can hence be in the range $[\max(\theta_{KR},\theta_{VR}),\min(\theta_{KL},\theta_{VL})]$. Note that the angles are being worked over, which have the cyclic property, and therefore, the range of values would be taken from $\max(\theta_{KR},\theta_{VR})$ till $\max(\theta_{KR},\theta_{VR})+2\pi$ or any convenient range to ascertain the cyclic nature. Working with angles is a big challenge due to the cyclic nature of angles, due to which difference, sum, and comparison operations need to be bounded to the working range.

The angular coverage of an obstacle acts as an obstacle since the robot cannot directly be heading towards the obstacle. Between the angle ranges under consideration, there will be multiple obstacles and free spaces between obstacles. Let the free spaces be given by $[\max(\theta_{KR},\theta_{VR}),\theta_{1R}], [\theta_{1L},\theta_{2R}], [\theta_{2L},\theta_{3R}]\ldots[\theta_{nL},\min(\theta_{KL},\theta_{VL})]$. This is shown in Fig. 11.6B. The angular ranges of obstacles may overlap and may not be continuous as per the notations shown here. The radial gap between obstacles is given by $[\theta_{1R} - \max(\theta_{KR},\theta_{VR})], [\theta_{2R} - \theta_{1L}], [\theta_{3R} - \theta_{2L}]\ldots [\min(\theta_{KL},\theta_{VR}) - \theta_{nL}]$. The heuristic of going through the largest gap states to find free space between obstacles that have the maximum gap. Suppose the maximum gap is recorded between $[\theta_{i-1L},\theta_{iR}]$, which is the angular space between the $(i-1)$th and ith obstacle. To move by the maximum radial gap, the robot takes the direction $(\theta_{i-1L}+\theta_{iR})/2$. A limitation here is that the calculations are done radially only. It is possible to calculate the actual distances from trigonometry; however, the calculations are more involved.

The limitation of the calculations made is that the goal did not appear in the calculations of the largest gap, or the absence of goal-seeking behaviour. There are two remedies for the same. The first is to limit the maximum deviation from the angle to the goal. This means that the robot always travels approximately towards the goal. Correspondingly the largest gap only within the range is taken. Second is to select the action as a weighted addition of the largest gap and angle to the goal. The latter ensures a bias of walking in general towards the goal. The technique gives the preferred angle to move. The steering or the angular speed of the robot is used to attain the preferred angle to move.

11.4.2 Largest distance with nongeometric obstacles

This subsection presents another heuristic, which is to move in the direction in which the robot can travel the maximum without incurring any collision, with any static or dynamic obstacle (Kala, 2018). The heuristic is to choose the direction, with the largest *longitudinal clearance* (clearance measured only in the specified angle). This subsection adds static obstacles of any general shape while still neglecting the velocities of the dynamic obstacles. The heuristic of maximization of the distance is done with a threshold (d_{max}) and any radial angle with more than the threshold is taken as just the same. While walking around, one observes obstacles everywhere in all directions. Some obstacles, like furniture, plants, and other people, are very close (less longitudinal clearance), and one cannot walk in the same direction. Some other obstacles like doors, walls making corridors, etc. are at a distant. One cannot currently see through the doors and corridors because of occlusions; however, when one comes near these doors and corridors, one would be able to see beyond. Thus, if the obstacles (doors, corridors, etc.) are far away (beyond d_{max}), it is considered safe to travel towards any of them. The longitudinal clearance at any general angle is given by Eq. (11.26).

$$d(\theta) = \min_{[x\,y]^T + l[\cos(\theta)\sin(\theta)]^T \in C^{obs},\, l \leq d_{max}} l \qquad (11.26)$$

Here, (x,y) is the position of the robot being planned. The equation represents the distance of a virtual proximity sensor at direction θ. A point at a distance of l at the proximity sensor ray oriented at angle θ with respect to the global coordinate frame has coordinates $[x\,y]^T + l[\cos(\theta)\sin(\theta)]^T$. The smallest l that incurs a collision is the reading of the proximity sensor.

Radial clearance is used as a secondary criterion (tiebreaker) when multiple heading directions have the same longitudinal clearance (especially when they enjoy a clearance of threshold). Radial clearance is the radial range, clockwise and anticlockwise, that has the same longitudinal clearance value. It measures the angular range within, which the robot may operate to be equally safe. The longitudinal clearance selects places where obstacles are far away, while the radial clearance ensures that angular deviations, while operating the robot do not harm the safety of the robot. The concept for a point robot is shown in Fig. 11.7A and B.

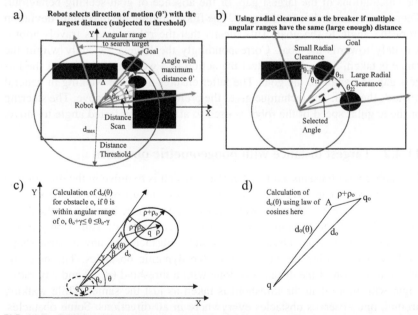

FIG. 11.7 Selection of target direction of motion: (A) Search for angle corresponding to the largest distance, within an angular deviation of Δ from the direction of goal (θ_G) to get goal-seeking behaviour. (B) Calculation of angle with the largest radial clearance out of angles with the largest (threshold) distance value (C and D) Calculation of distance from a robot at a scan angle θ.

The model only covers obstacle avoidance and not goal-seeking behaviour. The goal-seeking behaviour is conveniently obtained by limiting the search heading within a maximum threshold of the straight line to the goal, shown in Fig. 11.7. The chosen heading angle will, therefore, be within an angular threshold of the straight-line motion to goal. It must be stressed that this technique cannot solve for complex maps on its own and needs to be fused, with a global deliberative planner, like all other approaches of the chapter.

The maximum longitudinal clearance possible is given by Eq. (11.27):

$$d' = \max_{\theta \in [\theta_G - \Delta,\, \theta_G + \Delta]} d(\theta) \qquad (11.27)$$

Here, Δ is the maximum deviation allowable from the goal and θ_G is the angle to goal.

Now let us use the radial clearance as a tiebreaker. The directions within the maximum longitudinal clearance is recorded will typically be in ranges like $[\theta_{11}, \theta_{12}]$, $[\theta_{21}, \theta_{22}]$..., given by Eq. (11.28).

$$\Theta = \left\{ \langle \theta_{i1}, \theta_{i2} \rangle : d(\theta_j) = d' \forall \theta_{i1} \le \theta_j \le \theta_{i2} \right\} \qquad (11.28)$$

The largest gap is given by Eq. (11.29) and the chosen heading direction is given by Eq. (11.30).

$$[\theta_{j1}, \theta_{j2}] = \arg \max_{\langle \theta_{j1}, \theta_{j2} \rangle \in \Theta} AngleDiff(\theta_{j2}, \theta_{j1}) \qquad (11.29)$$

$$\theta' = (\theta_{j1} + \theta_{j2})/2 \qquad (11.30)$$

The robot is ultimately controlled by using a linear and angular speeds. The angular speed uses the difference between the current orientation of the robot and the chosen heading direction as the angular error. The linear speed is directly proportional to the distance in the chosen direction, with a large distance implying a higher speed. This is given by Eq. (11.31) and (11.32):

$$\omega = K_{P1} AngleDiff(\theta', \theta_R) \qquad (11.31)$$

$$v = K_{P2} d(\theta') \qquad (11.32)$$

Here, K_{P1} and K_{P2} are the proportionality constants θ_R is the current heading of the robot, θ' is the chosen heading direction. The equations will be subjected to kinematic constraints, including acceleration limits, maximum velocity limits, and other nonholonomic constraints.

Let us re-look at Eq. (11.26). The hard part is the calculation of the distance. Eq. (11.26) assumes that the problem can be mapped into a configuration space consisting of both the static obstacles as well as the other robots as the dynamic obstacles. The difference is because the geometry of the static obstacles is not known, and, like the potential field, the static obstacle information is either available as a local grid map or by using proximity sensing. Let us perform the calculations first for the case of dynamic obstacles with known circular geometry. The calculations for an angle θ (measured in the global coordinate frame) that intersects an obstacle (x_o, y_o) at an angle β (measured from the straight line joining the robot with the obstacle as the base) are given in Eqs (11.33)–(11.36). The notations are shown in Fig. 11.7C and D.

$$d_o(\theta) = \begin{cases} d_o \cos(\beta) - \sqrt{d_o^2 \cos^2(\beta) - \left(d_o^2 - (\rho + \rho_o)^2\right)} & \text{if } \theta_o - \gamma_o \le \theta \le \theta_o + \gamma_o \\ \infty & \text{otherwise} \end{cases}$$

$$(11.33)$$

$$\gamma_o = \sin^{-1}\left(\frac{\rho_o + \rho}{d_o}\right) \tag{11.34}$$

$$\beta = \theta - \theta_o \tag{11.35}$$

$$d_o = d((x_r, y_r), (x_o, y_o)) \tag{11.36}$$

Here, d is the distance function. The derivation of the distance is done by using the cosine law of trigonometry. Consider two sides of the triangle of length a and b, θ as the angle between the two sides, and c as the length of the third side. As per cosine law, Eq. (11.37):

$$c^2 = a^2 + b^2 - 2ab \cos\theta \tag{11.37}$$

The case of static obstacles can be taken as the same, with the radius of the obstacle ρ_o taken as 0 and the calculations repeated for every obstacle in the grid map or every point in the proximity sensing scan of any proximity sensor like lidar. The resultant distance value is hence the minimum overall possible obstacle, given by Eq. (11.38):

$$d(\theta) = \min_o (d_o(\theta), d_{\max}) \tag{11.38}$$

Here, o is iterated over all the static and dynamic obstacles.

11.5 Fuzzy logic

At a higher level of abstraction, intuition is that it should be possible to instruct the robot what to do in what situations, which is permanently stored as a navigation rule set in the robot. The robot uses a collection of such rules to navigate. This is done by using fuzzy rules computed by using fuzzy logic (Ross, 2010; Shukla et al., 2010a, b).

11.5.1 Classic logical rule-based systems

In every part of life, for any decision-making, you consider a set of rules that helps you make decisions, and the motion planning of a robot should not be an exception. The entire navigation behaviour is a simple set of rules like '*if you see an obstacle leftwards, then turn right*', '*if the goal is leftwards, then turn left'*, etc. If these rules are followed by the robot, it should be able to react to a variety of situations. The traditional rule-based expert system is a collection of such rules. A rule is in the form *'if (condition) then (action)'*. Here, the 'if' part is also called the antecedent, and the 'then' part is also called the consequent. A rule fires if the condition specified by its antecedent is true. The condition can have any Boolean operator like AND, OR, and NOT. A typical example is *'if the obstacle on the forward-left is near and the goal is far and the speed is not small then slow down, turn sharp right'* that consists of three base conditions *obstacle on the forward-left is near*, *the goal is far* and *the speed is not small*, connected by a AND proposition along with a NOT in the third condition. Only one rule fires for any input. If more than two rules have their preconditions met, then most specific rule, or most recently used rule is given a priority.

The general inputs to the problem of reactive motion planning are the distances, angles, and speeds of nearby obstacles, the distance, and angle from goal. Let us curtail the number of inputs to make the minimum rule base for the navigation. Hence, the input is taken as the direction and distance from the nearest obstacle, the angle to goal, and the distance to the goal. Furthermore, divide every input and output variable that can take as input a range of values into a set of discrete sets. Let us take only 1 such variable, *distance to the nearest obstacle* with (normalized) range [0,1] divided into just 3 sets, *near* [0,0.05), *far* [0.05,0.15) and *distant* [0.15,1]. These sets participate in every rule. It can be said that *distanceToObstacle = near ∪ far ∪ distant*, where *distanceToObstacle* is the set of all possible distance inputs. *Near*, *far*, and *distant* are the three disjoint subsets such that the intersection between any two of them is an empty set, or *near ∩ far* = ∅, *far ∩ distant* = ∅ and *near ∩ distant* = ∅, so every recorded distance value points to just one of the sets and participates in all rules of the particular set only. Suppose the input *angleToObstacle* be divided into the sets *farClockwise*, *Clockwise*, *antiClockwise* and *farAntiClockwise*. Similarly, suppose the input *angleToGoal* is also subdivided into the sets *farClockwise*, *clockwise*, *Straight*, *antiClockwise,* and *farAntiClockwise*. Suppose the input *distanceToGoal* be divided into the sets *near* and *distant*.

Thereafter, the rules are written. There are two distinct mechanisms to write the rules. The first is to populate the rule table with all possible cases (antecedent only) considering all possible sets for every input variable. A general rule is thus given by Eq. (11.39):

> if *distanceToObstacle* is {*near, far, distant*} and
>
> *angleToObstacle* is {*farClockwise, clockwise, antiClockwise,*
>
>   *farAntiClockwise*} and
>
> *distanceToGoal* is{*near distant*} and *angleToGoal* is {*farClockwise,*
>
>   *Clockwise, straight, antiClockwise, farAntiClockwise*} then···

$$(11.39)$$

Here, { } denotes the choice of set. Keeping the inputs in the same order, all possible rules are written (in shorthand) by Eq. (11.40)

$$\underset{\text{distanceToObstacle}}{\{near, far, distant\}} \times \frac{\{farClockwise, clockwise, antiClockwise, farAntiClockwise\}}{\text{angleToObstacle}}$$
$$\times \underset{\text{distanceToGoal}}{\frac{\{near, distant\}}{}} \times \frac{\{farClockwise, clockwise, straight, antiClockwise, farAntiClockwise\}}{\text{angleToGoal}}$$

(11.40)

The { } denote sets, × denotes the cross product of sets and the name of the input is written underneath. The cross-product will give all possible rules and a total of $3 \times 4 \times 2 \times 5 = 120$ rules. The consequent may be populated by a human expert with the support of a simulation system so that the human can see the robot performance, look at the rules that fired for wrong behaviour and correct the rules. Some of the rules of the rule table are shown in Table 11.1.

TABLE 11.1 Rule table. All possible rules are made by a combination of sets of all inputs. The consequent (output) participating set is written by human expertise.

RULE NUMBER	INPUT				OUTPUT
	Distance to obstacle	Angle to obstacle	Distance to goal	Angle to goal	Consequent (angular velocity) populated by a human expert
1.	near	farClockwise	near	farClockwise	farAntiClockwise
2.	near	farClockwise	near	Clockwise	farAntiClockwise
3.	near	farClockwise	near	Straight	farAntiClockwise
4.	near	farClockwise	near	anticlockwise	farAntiClockwise
....	....	....	....	....	....
120.	distant	farAntiClockwise	distant	farAntiClockwise	anticlockwise

Alternatively, the other mechanism of writing down the rules is as you would instruct a child to move. So, you write some generic rules, simulate the system, look at the robot performance, look at the rules that fired (if any) and either mend the current rules, write down new rules, delete some old rules, or write specific cases of the old rules by adding more variables, etc. A rule now may not have all the variables like '*if the obstacle is in front-left and distance to the obstacle is very near, turn steep right*' may have nothing to do with the goal. However, if the robot is seen to slow down just in front of the goal for a small obstacle away, then the rule may be mended with the goal information as well.

The problem with the classic logic system is that only one rule can be active. The robot's behaviour suddenly and sharply changes when the applicable membership set changes. Consider the *distanceToObstacle* consists of 3 sets, *near* [0,0.05), *far* [0.05,0.15), and *distant* [0.15,1]. If the *distanceToObstacle* changes from 0.051 (far) to 0.049 (near), the fired rule and thus the applicable behaviour (speed) sharply changes; while in the large range of far [0.05,0.15), the behaviour is the same.

11.5.2 Fuzzy sets

The *Fuzzy Logic* implements a more flexible logic system, wherein an element could be a member of multiple sets at the same time with varying degrees of membership. So, if the input *distanceToObstacle* is divided into three fuzzy sets: *near*, *moderate*, and *far*; it is said that every input is a member of each of the three sets, however, the membership degree will vary. Let some recording by the robot witness distance to the nearest obstacle as 90 cm. It is said that the input belongs to *near* to a membership degree of 0.8, *moderate* to a degree of 0.4, and *far* to a membership degree of 0. Correspondingly, the nonfuzzy classic sets are called the *crisp* sets.

Let d_o denote the input *distanceToObstacle* with fuzzy sets *near, far,* and *distant*. $\mu_{near}^{d_o}: \mathbb{R} \rightarrow [0,1]$ denotes the membership function of an input to the fuzzy set *near*. $\mu_{near}^{d_o}(d_o)$ is 1 if d_o has a complete membership of the set *near*, 0 if d_o has no membership of the set near and all other cases are intermediate. Fig. 11.8A shows the division of input into discrete sets. Fig. 11.8B shows the division as the fuzzy set. The boundaries are vague and there is no clear demarcation between the sets. So, especially near the boundary areas, the elements appear to be belonging to both the sets. Let us extend Fig. 11.8B into another dimension that shows the membership degree. The three sets are shown in Fig. 11.8C.

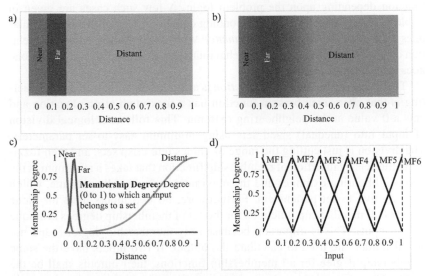

FIG. 11.8 Concept of fuzzy sets. (A) shows an input divided into crisp sets. (B) The sets are fuzzy in (B) and hence the boundary is not distinctly visible. The figure is extended by adding a new dimension in (C). (D) Decomposition of input into triangular fuzzy membership functions. Decomposition requires a specification of boundaries only with no other parameters, keeping the meaning of sets intact.

The mathematical representation of x belonging to set X with membership $\mu_X(x)$ is given by $\mu_X(x)/x$. Here, the numerator is a membership degree in the range $[0,1]$ and the denominator is the real input x, however, the '/' is just a separator and not a division operator. Correspondingly, the entire set for a discrete case is represented by Eq. (11.41). For the sake of discretization, the inputs are assumed to be discrete only.

$$\mu_{near}^{d_o} = \left\{ \frac{1}{0} + \frac{0.87}{0.01} + \frac{0.58}{0.02} + \frac{0.29}{0.03} + \frac{0.11}{0.04} + \frac{0.03}{0.05} + \dots \frac{0}{1} \right\} \quad (11.41)$$

Here, the term '+' is also just a separator and not the arithmetic addition operator. It can be seen as the input increases (denominator values), the membership value of the set near (numerator values) keeps decreasing, showing the real-life meaning that as the distances increase, the (membership of) nearness decreases. The case of a continuous variable is nearly the same with the '+' replaced by an integral (only as a separator and not as a real integration operation), given by Eq. (11.42).

$$\mu_{near}^{d_o} = \left\{ \int_{d_0=0}^{0} \frac{\exp\left(\frac{-d_o^2}{2\sigma^2}\right)}{d_o} \right\} \quad (11.42)$$

The membership function $\mu_X(x)$ gives the membership degree of the input x to the set X. There are numerous mechanisms to model this membership function depending upon the problem logic. A few such cases are given in Table 11.2, along with their corresponding equations. One of the most popular ones is the *Gaussian membership function* which beautifully captures the relation of a peak denoting the typical behaviour and the membership falling as one goes away from the peak.

The *triangular membership function* is a similar function, and a typical setting is that every membership function has a maximum at one location and gets a 0 value at the neighbouring extrema. This follows a logical division of input into (unequal) sized sets using minimum easy-to-set parameters. The division is just cutting the range of values into crisp sets, and then fuzzifying the sets by insertion of membership function that takes a minimum at the two neighbouring extremes and the maximum at the middle extreme, which has intuitive to set parameters. Furthermore, no two inputs have the same value of all membership functions (or the set of membership degrees is unique for an input). This is important because only membership degrees (and not actual inputs) are used to calculate the output. If two inputs have the same membership degree for all membership functions, their outputs shall be the same. The triangular decomposition of a synthetic input is shown in Fig. 11.8D.

TABLE 11.2 Commonly used membership functions.

S. no.	Name	Equation	Graph
1.	Triangular membership function	$\mu_X(x) = \begin{cases} 0 & x \leq a \\ \dfrac{x-a}{b-a} & a < x \leq b \\ \dfrac{c-x}{c-b} & b < x < c \\ 0 & x \geq c \end{cases}$	
2.	Gaussian membership function	$\mu_X(x) = \exp\left(-\dfrac{(x-\mu)^2}{2\sigma^2}\right)$	
3.	Generalized bell-shaped membership function	$\mu_X(x) = \dfrac{1}{1+\left\|\dfrac{x-c}{a}\right\|^{2b}}$	
4.	Sigmoidal membership function	$\mu_X(x) = \dfrac{1}{1+\exp(-a(x-b))}$	

Continued

TABLE 11.2 Commonly used membership functions.—cont'd

S. no.	Name	Equation	Graph
5.	Trapezoidal membership function	$$\mu_X(x) = \begin{cases} 0 & x \leq a \\ \dfrac{x-a}{b-a} & a \leq x \leq b \\ 1 & b \leq x \leq c \\ \dfrac{d-x}{d-c} & c \leq x \leq d \\ 0 & x \geq d \end{cases}$$	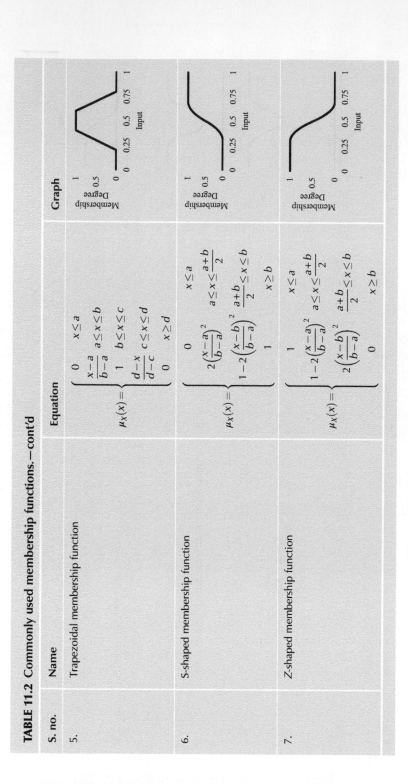
6.	S-shaped membership function	$$\mu_X(x) = \begin{cases} 0 & x \leq a \\ 2\left(\dfrac{x-a}{b-a}\right)^2 & a \leq x \leq \dfrac{a+b}{2} \\ 1-2\left(\dfrac{x-b}{b-a}\right)^2 & \dfrac{a+b}{2} \leq x \leq b \\ 1 & x \geq b \end{cases}$$	
7.	Z-shaped membership function	$$\mu_X(x) = \begin{cases} 1 & x \leq a \\ 1-2\left(\dfrac{x-a}{b-a}\right)^2 & a \leq x \leq \dfrac{a+b}{2} \\ 2\left(\dfrac{x-b}{b-a}\right)^2 & \dfrac{a+b}{2} \leq x \leq b \\ 0 & x \geq b \end{cases}$$	

Consider the problem of motion planning of a robot. The membership functions for all inputs and outputs are given by Fig. 11.9. The inputs are real numbers that are used to calculate the membership degree for all sets. The process of conversion of crisp inputs into membership of all fuzzy sets is called *fuzzification*. Let the membership functions be $\mu_{near}^{d_o}$, $\mu_{far}^{d_o}$ and $\mu_{distant}^{d_o}$, where the subscript is the membership function name, and the super-script is the input or output variable name The fuzzification of a synthetic *distanceToObstacle* $= 0.03$ is given by Eqs (11.43)–(11.45), assuming suitable values of the membership function parameters.

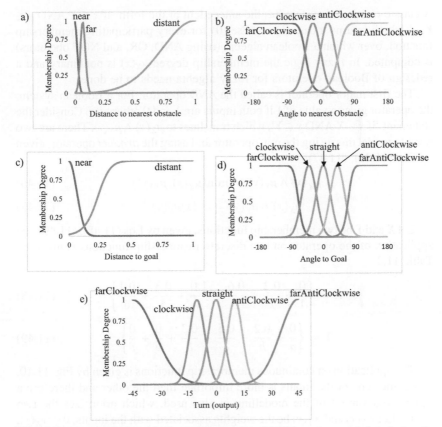

FIG. 11.9 Membership functions for motion planning. (A) Distance to nearest obstacle (input), (B) angle to nearest obstacle (input), (C) distance to goal (input), (D) angle to goal (input), (E) turn (output).

$$\mu_{near}^{d_o}(0.03) = \exp\left(\frac{-0.03^2}{2(0.01915)^2}\right) = 0.29 \qquad (11.43)$$

$$\mu_{far}^{d_o}(0.03) = \exp\left(\frac{-(0.03 - 0.066)^2}{2(0.0207)^2}\right) = 0.22 \qquad (11.44)$$

$$\mu_{distant}^{d_o}(0.03) = \exp\left(\frac{-(0.03 - 1)^2}{2(0.2797)^2}\right) = 0.0024 \qquad (11.45)$$

11.5.3 Fuzzy operators

A classic rule-based system has the antecedent of the form 'if $x \in X$ AND $y \in Y$...' which returns a true (1) or false (0) for every participating membership function, over which a Boolean algebra (using AND, OR, and NOT operators) is computed. In fuzzy logic the membership degree $\mu_X(x)$ is nonbinary and a redesign of Boolean operators for fuzzy algebra needs to be done.

The first major operator is called the *AND* operator. For a Boolean system, the operator gives a value of 1 if both inputs are 1 and 0 otherwise. Consider the statement 'if $x \in X$ AND $y \in Y$', which translates to $\mu_X(x) \wedge \mu_Y(y)$. There are two ways to model this, using the min operator and using the *product* operator, given by Eqs (11.46), (11.47):

$$\mu_X(x) \wedge \mu_Y(y) = \min(\mu_X(x), \mu_Y(y)) \qquad (11.46)$$

$$\mu_X(x) \wedge \mu_Y(y) = \mu_X(x)\mu_Y(y) \qquad (11.47)$$

Let X and Y be two membership functions, given by Eqs (11.48), (11.49), the application of the operator on two discrete membership functions is given by Table 11.3.

$$X = \left\{\frac{0}{1} + \frac{0.2}{2} + \frac{0.6}{3} + \frac{1.0}{4} + \frac{0.3}{5} + \frac{0.1}{6}\right\} \qquad (11.48)$$

$$Y = \left\{\frac{0}{a} + \frac{0.2}{b} + \frac{0.5}{c} + \frac{0.3}{d} + \frac{0}{e} + \frac{0}{f}\right\} \qquad (11.49)$$

The application on continuous membership functions is given by Fig. 11.10. Sometimes one of the inputs is more important than the other and therefore a generalized variant of the modelling is also used, which prioritizes the two inputs. Let $w_X(x)$ and $w_Y(y)$ be the weights associated with the inputs, $0 \le w_X(x)$, $w_Y(y) \le 1$. The generalized modelling is shown as Eqs (11.50), (11.51):

$$w_X\mu_X(x) \wedge w_Y\mu_Y(y) = \min(w_X\mu_X(x), w_Y\mu_Y(y)) \qquad (11.50)$$

$$w_X\mu_X(x) \wedge w_Y\mu_Y(y) = \mu_X(x)^{w_X}\mu_Y(y)^{w_Y} \qquad (11.51)$$

It is worth verifying with all operators that when the input is crisp true (1) and false (0), the operators should return the same output as the classic Boolean operator. The working of the fuzzy logic is like that of probabilistic

TABLE 11.3 Application of AND operator ($X \land Y$). a/b represents that the membership degree of b is a.

(a) Min modelling $\mu_{X \land Y}(x,y) = \min(\mu_X(x), \mu_Y(y))$

$X \land Y$

Membership values of Y (μ_Y)	Membership values of X (μ_X)					
	0/1	**0.2/2**	**0.6/3**	**1/4**	**0.3/5**	**0.1/6**
0/a	0/(1,a)	0/(2,a)	0/(3,a)	0/(4,a)	0/(5,a)	0/(6,a)
0.2/b	0/(1,b)	0.2/(2,b)	0.2/(3,b)	0.2/(4,b)	0.2/(5,b)	0.1/(6,b)
0.5/c	0/(1,c)	0.2/(2,c)	0.5/(3,c)	0.5/(4,c)	0.3/(5,c)	0.1/(6,c)
0.3/d	0/(1,d)	0.2/(2,d)	0.3/(3,d)	0.3/(4,d)	0.3/(5,d)	0.1/(6,d)
0.0/e	0/(1,e)	0/(2,e)	0/(3,e)	0/(4,e)	0/(5,e)	0/(6,e)
0.0/f	0/(1,f)	0/(2,f)	0/(3,f)	0/(4,f)	0/(5,f)	0/(6,f)

(b) Product modelling $\mu_{X \land Y}(x,y) = \mu_X(x)\mu_Y(y)$

$X \land Y$

Membership values of Y (μ_Y)	Membership values of X (μ_X)					
	0/1	**0.2/2**	**0.6/3**	**1/4**	**0.3/5**	**0.1/6**
0/a	0/(1,a)	0/(2,a)	0/(3,a)	0/(4,a)	0/(5,a)	0/(6,a)
0.2/b	0/(1,b)	0.04/(2,b)	0.12/(3,b)	0.2/(4,b)	0.06/(5,b)	0.02/(6,b)
0.5/c	0/(1,c)	0.1/(2,c)	0.3/(3,c)	0.5/(4,c)	0.15/(5,c)	0.05/(6,c)
0.3/d	0/(1,d)	0.06/(2,d)	0.18/(3,d)	0.3/(4,d)	0.09/(5,d)	0.03/(6,d)
0.0/e	0/(1,e)	0/(2,e)	0/(3,e)	0/(4,e)	0/(5,e)	0/(6,e)
0.0/f	0/(1,f)	0/(2,f)	0/(3,f)	0/(4,f)	0/(5,f)	0/(6,f)

a)

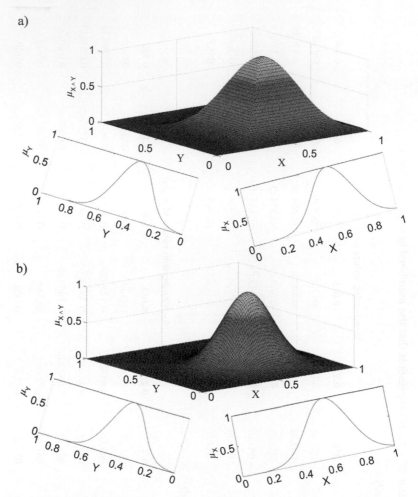

b)

FIG. 11.10 Graphical illustration of fuzzy X AND Y (A) using $\mu_{X \wedge Y}(x,y) = \min(\mu_X(x), \mu_Y(y))$ (B) using $\mu_{X \wedge Y}(x,y) = \mu_X(x)\mu_Y(y)$.

reasoning, where the membership degrees denote the probabilities. That said there is a major difference between probabilities and fuzzy sets. If $P(A) = 0.8$, it means that A occurs with a probability of 0.8. Based on luck, either A will occur or not occur. So, if $P(tomorrow\ is\ raining) = 0.8$, it means either it will rain tomorrow (more likely) or not. However, in fuzzy systems, the events occur partially. So, if $\mu_A = 0.8$, it also means that A occurs by 80%. Hence, $\mu_{tomorrow\ is\ raining} = 0.8$ also denotes a partial rain.

If an event has two preconditions, both required and independent, with probabilities $P(X)$ and $P(Y)$, it is natural that the probability of the event is $P(X)P(Y)$, which is product modelling. Imagine that the event happens by a succession of two related subconditions, which require human effort. Assume that a person who can solve for the harder subcondition can also solve for the easier one, and therefore the probability of the event is governed by the probability of solving for the harder subcondition and is given by the minimum of the two membership values.

The OR is the next vital operator of Boolean logic. The two popular means to model the OR is by max and *sum*. This is given by Eqs (11.52), (11.53).

$$\mu_X(x) \vee \mu_Y(x) = \max(\mu_X(x), \mu_Y(x)) \tag{11.52}$$

$$\mu_X(x) \vee \mu_Y(x) = \min(\mu_X(x) + \mu_Y(x), 1) \tag{11.53}$$

The generalized variant of the operator is given by Eqs (11.54), (11.55)

$$w_X\mu_X(x) \vee w_Y\mu_Y(x) = \max(w_X\mu_X(x), w_Y\mu_Y(x)) \tag{11.54}$$

$$w_X\mu_X(x) \vee w_Y\mu_Y(x) = \min(w_X\mu_X(x) + w_Y\mu_Y(x), 1) \tag{11.55}$$

The threshold in sum ensures that the output is within the possible membership range and does not go beyond 1. The modelling gives the same output as the Boolean logic when the input is crisp (0 or 1). The formulation is again simple to understand. The probability of X or Y, both independent, is the sum of the two probabilities. Again, if an event happens by either of two related subconditions, it is imperative that now a person will solve the easier one only and therefore the membership is modelled by the max operator. The application of the operator on the same discrete sets is given in Table 11.4. In the case of continuous membership functions, the output is given in Fig. 11.11.

The $\neg$ operator is relatively simple to model. It is given by Eq. (11.56).

$$\neg\mu_X(x) = 1 - \mu_X(x) \tag{11.56}$$

The notion from probability is also that if an event happens with probability $P(A)$, it does not happen with probability $1 - P(A)$. Consider the membership function given by Eq. (11.57), the application of not is given by Eq. (11.58):

$$X = \left\{\frac{0}{1} + \frac{0.2}{2} + \frac{0.6}{3} + \frac{1.0}{4} + \frac{0.3}{5} + \frac{0.1}{6}\right\} \tag{11.57}$$

$$\neg X = \left\{\frac{1}{1} + \frac{0.8}{2} + \frac{0.4}{3} + \frac{0}{4} + \frac{0.7}{5} + \frac{0.9}{6}\right\} \tag{11.58}$$

In the case of continuous membership function, the application of the operator is shown in Fig. 11.12.

Another importation operator is the implication operator. Even though it can be derived by AND, OR, and NOT, it is wise to also model it up explicitly. The modelling is by using min or product, just like AND, which is given by Eqs (11.59), (11.60).

TABLE 11.4 Application of OR operator. a/b represents that the membership degree of b is a.

(a) Max modelling $\mu_{X \vee Y}(x,y) = \max(\mu_X(x), \mu_Y(y))$

$X \vee Y$

		Membership values of X (μ_X)					
		0/1	**0.2/2**	**0.6/3**	**1/4**	**0.3/5**	**0.1/6**
Membership values of Y (μ_Y)	**0/a**	0/(1,a)	0.2/(2,a)	0.6/(3,a)	1/(4,a)	0.3/(5,a)	0.1/(6,a)
	0.2/b	0.2/(1,b)	0.2/(2,b)	0.6/(3,b)	1/(4,b)	0.3/(5,b)	0.2/(6,b)
	0.5/c	0.5/(1,c)	0.5/(2,c)	0.6/(3,c)	1/(4,c)	0.5/(5,c)	0.5/(6,c)
	0.3/d	0.3/(1,d)	0.3/(2,d)	0.6/(3,d)	1/(4,d)	0.3/(5,d)	0.3/(6,d)
	0.0/e	0/(1,e)	0.2/(2,e)	0.6/(3,e)	1/(4,e)	0.3/(5,e)	0.1/(6,e)
	0.0/f	0/(1,f)	0.2/(2,f)	0.6/(3,f)	1/(4,f)	0.3/(5,f)	0.1/(6,f)

(b) Sum modelling $\mu_{X \vee Y}(x,y) = \min(\mu_X(x) + \mu_Y(y), 1)$

$X \vee Y$

		Membership values of X (μ_X)					
		0/1	**0.2/2**	**0.6/3**	**1/4**	**0.3/5**	**0.1/6**
Membership values of Y (μ_Y)	**0/a**	0/(1,a)	0.2/(2,a)	0.6/(3,a)	1/(4,a)	0.3/(5,a)	0.1/(6,a)
	0.2/b	0.2/(1,b)	0.4/(2,b)	0.8/(3,b)	1/(4,b)	0.5/(5,b)	0.3/(6,b)
	0.5/c	0.5/(1,c)	0.7/(2,c)	1/(3,c)	1/(4,c)	0.8/(5,c)	0.6/(6,c)
	0.3/d	0.3/(1,d)	0.5/(2,d)	0.9/(3,d)	1/(4,d)	0.6/(5,d)	0.4/(6,d)
	0.0/e	0/(1,e)	0.2/(2,e)	0.6/(3,e)	1/(4,e)	0.3/(5,e)	0.1/(6,e)
	0.0/f	0/(1,f)	0.2/(2,f)	0.6/(3,f)	1/(4,f)	0.3/(5,f)	0.1/(6,f)

a)

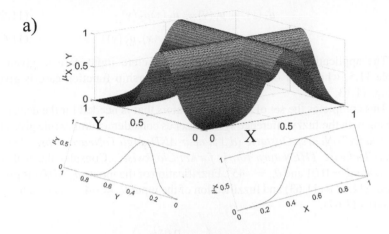

b)

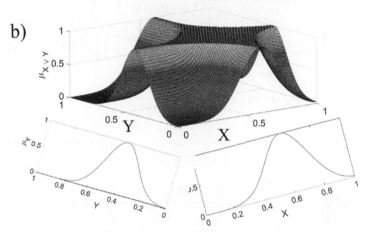

FIG. 11.11 Graphical illustration of fuzzy X OR Y (A) using $\mu_{X \vee Y}(x,y) = \max(\mu_X(x), \mu_Y(y))$ (B) using $\mu_{X \vee Y}(x,y) = \min(\mu_X(x) + \mu_Y(y), 1)$

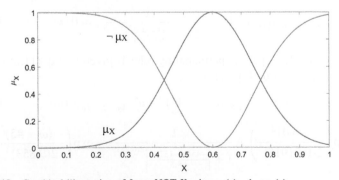

FIG. 11.12 Graphical illustration of fuzzy NOT X using $\mu_X(x) = 1 - \mu_X(x)$.

$$\mu_X(x) \rightarrow \mu_Y(y) = \mu_X(x)\mu_Y(y) \tag{11.59}$$

$$\mu_X(x) \rightarrow \mu_Y(y) = \min(\mu_X(x), \mu_Y(y)) \tag{11.60}$$

The application of the operator on the discrete input case is given in Table 11.5, while that of the continuous membership function case is given in Fig. 11.13.

This completes the set of Boolean operators that is needed for the decision-making with the fuzzy inference system. Let us consider a dummy rule given by '*if distanceToNearestObstacle (d_o) is near AND angleToNearestObstacle (θ_o) is farClockwise THEN turn (ω) is farAntiClockwise*'. Consider the real-life inputs are $d_o = 0.01$ and $\theta_o = -45°$. Fuzzification of the inputs $d_o = 0.01$ is given by Eqs (11.61)–(11.63) and fuzzification of the input $\theta_o = -45°$ is given by Eqs (11.64)–(11.67).

$$\mu_{near}^{d_o}(0.01) = \exp\left(\frac{-0.01^2}{2(0.01915)^2}\right) = 0.87 \tag{11.61}$$

$$\mu_{far}^{d_o}(0.01) = \exp\left(\frac{-(0.01 - 0.066)^2}{2(0.0207)^2}\right) = 0.02 \tag{11.62}$$

$$\mu_{distant}^{d_o}(0.01) = \exp\left(\frac{-(0.01 - 1)^2}{2(0.2797)^2}\right) = 0.001 \tag{11.63}$$

$$\mu_{farClockkwise}^{\theta_o}(-45) = \frac{1}{1 + \left|\frac{-45 - (-137.1)}{98.5}\right|^{2(6.28)}} = 0.70 \tag{11.64}$$

$$\mu_{clockkwise}^{\theta_o}(-45) = \exp\left(\frac{-(-45 - (-20))^2}{2(20)^2}\right) = 0.46 \tag{11.65}$$

$$\mu_{antiClockkwise}^{\theta_o}(-45) = \exp\left(\frac{-(-45 - 20)^2}{2(20)^2}\right) = 0.01 \tag{11.66}$$

$$\mu_{farAntiClockkwise}^{\theta_o}(-45) = \frac{1}{1 + \left|\frac{-45 - 140}{93.7}\right|^{2(4.475)}} = 0.00 \tag{11.67}$$

The fuzzy rule as per the participating rules is given by Eq. (11.68), which resolves to as given by Eq. (11.69)

$$\mu_{near}^{d_o}(0.01) \wedge \mu_{farClockkwise}^{\theta_o}(-45) \rightarrow \mu_{farAntiClockkwise}^{\omega}(\omega) \tag{11.68}$$

$$\exp\left(\frac{-0.01^2}{2(0.01915)^2}\right) \wedge \frac{1}{1 + \left|\frac{-45 - (-137.1)}{98.5}\right|^{2(6.28)}} \rightarrow \exp\left(\frac{-(\omega - 45)^2}{2(9.853)^2}\right)$$

$$\tag{11.69}$$

TABLE 11.5 Application of implication operator. a/b represents that the membership degree of b is a.

(a) Min modelling $\mu_{X\to Y}(x,y) = \min(\mu_X(x), \mu_Y(y))$

$X \to Y$		Membership values of X (μ_X)					
Membership values of Y (μ_Y)		0/1	0.2/2	0.6/3	1/4	0.3/5	0.1/6
	0/a	0/(1,a)	0/(2,a)	0/(3,a)	0/(4,a)	0/(5,a)	0/(6,a)
	0.2/b	0/(1,b)	0.2/(2,b)	0.2/(3,b)	0.2/(4,b)	0.2/(5,b)	0.1/(6,b)
	0.5/c	0/(1,c)	0.2/(2,c)	0.5/(3,c)	0.5/(4,c)	0.3/(5,c)	0.1/(6,c)
	0.3/d	0/(1,d)	0.2/(2,d)	0.3/(3,d)	0.3/(4,d)	0.3/(5,d)	0.1/(6,d)
	0.0/e	0/(1,e)	0/(2,e)	0/(3,e)	0/(4,e)	0/(5,e)	0/(6,e)
	0.0/f	0/(1,f)	0/(2,f)	0/(3,f)	0/(4,f)	0/(5,f)	0/(6,f)

(b) Product modelling $\mu_{X\to Y}(x,y) = \mu_X(x)\mu_Y(y)$

$X \to Y$		Membership values of X (μ_X)					
Membership values of Y (μ_Y)		0/1	0.2/2	0.6/3	1/4	0.3/5	0.1/6
	0/a	0/(1,a)	0/(2,a)	0/(3,a)	0/(4,a)	0/(5,a)	0/(6,a)
	0.2/b	0/(1,b)	0.04/(2,b)	0.12/(3,b)	0.2/(4,b)	0.06/(5,b)	0.02/(6,b)
	0.5/c	0/(1,c)	0.1/(2,c)	0.3/(3,c)	0.5/(4,c)	0.15/(5,c)	0.05/(6,c)
	0.3/d	0/(1,d)	0.06/(2,d)	0.18/(3,d)	0.3/(4,d)	0.09/(5,d)	0.03/(6,d)
	0.0/e	0/(1,e)	0/(2,e)	0/(3,e)	0/(4,e)	0/(5,e)	0/(6,e)
	0.0/f	0/(1,f)	0/(2,f)	0/(3,f)	0/(4,f)	0/(5,f)	0/(6,f)

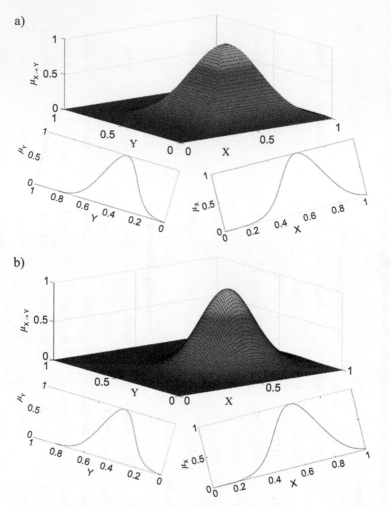

FIG. 11.13 Graphical illustration of fuzzy X IMPLIES Y (A) using $\mu_{X\to Y}(x,y) = \min(\mu_X(x), \mu_Y(y))$ (B) using $\mu_{X\to Y}(x,y) = \mu_X(x)\mu_Y(y)$.

$$\min(0.87, 0.70) \to \exp\left(\frac{-(\omega - 45)^2}{2(9.853)^2}\right)$$

$$0.70 \to \exp\left(\frac{-(\omega - 45)^2}{2(9.853)^2}\right)$$

$$\min\left(0.70, \exp\left(\frac{-(\omega - 45)^2}{2(9.853)^2}\right)\right)$$

The fuzzy output is shown in Fig. 11.14.

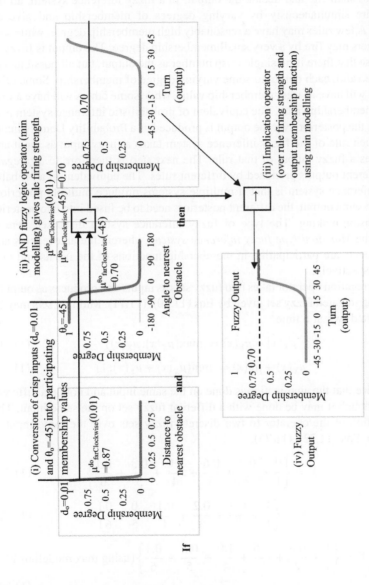

FIG. 11.14 Implication of a rule. The crisp inputs are fuzzified and passed through modelled fuzzy operators, which gives a fuzzy output for the rule.

11.5.4 Aggregation

The fuzzy inference system is a collection of many such rules. The notion of the rule is not that of the classic rule-based system, wherein only one of the numerous rules shall fire and decide the output. In a fuzzy inference system, all the rules fire simultaneously by varying degrees of membership and give an output. A few rules may have a reasonably high membership degree, while several others may fire by a very small membership degree. The output is fuzzy in the sense that there is no single crisp number as the output, but all possible outputs are valid, each output to some varying degree of membership. Some crisp outputs will have a larger membership value, while some others will have a very small membership value. The equivalent of probabilistic inference system is to say that the posterior over the output is produced as a Probability Density Function. Each rule of the fuzzy inference system takes a fuzzy input as input and produces a fuzzy output of that rule. The next step is therefore to *aggregate* the different outputs produced by different rules. The equivalent in a probabilistic inference system is that if multiple systems produce multiple posteriors over the same output, the different posteriors need to be fused into one posterior for decision-making. The type of fuzzy inference system discussed here is called the *Mamdani type fuzzy inference system*, wherein the antecedents and consequents are participating in membership functions of the inputs and outputs, respectively.

Aggregation operator takes two fuzzy sets as inputs and produces as output a single aggregated fuzzy set given by Eqs (11.70), (11.71). Multiple sets may be aggregated two at a time.

$$\mu_X(x) + \mu_Y(x) = \max(\mu_X(x), \mu_Y(x)) \tag{11.70}$$

$$\mu_X(x) + \mu_Y(x) = \min(\mu_X(x) + \mu_Y(x), 1) \tag{11.71}$$

Notice that the aggregation is done on the same input and not on a different input, though it may be done with a different fuzzy set on the same input. The application of the operator to two discrete fuzzy sets over the same input is given as Eqs (11.72)–(11.75).

$$X = \left\{ \frac{0}{1} + \frac{0.2}{2} + \frac{0.6}{3} + \frac{1.0}{4} + \frac{0.3}{5} + \frac{0.1}{6} \right\} \tag{11.72}$$

$$Y = \left\{ \frac{1}{1} + \frac{0.3}{2} + \frac{0.2}{3} + \frac{0.1}{4} + \frac{0}{5} + \frac{0}{6} \right\} \tag{11.73}$$

$$X + Y = \left\{ \frac{1}{1} + \frac{0.3}{2} + \frac{0.6}{3} + \frac{1.0}{4} + \frac{0.3}{5} + \frac{0.1}{6} \right\} \text{(using max modelling)} \tag{11.74}$$

$$X + Y = \left\{ \frac{1}{1} + \frac{0.5}{2} + \frac{0.8}{3} + \frac{1.0}{4} + \frac{0.3}{5} + \frac{0.1}{6} \right\} \text{(using sum modelling)} \tag{11.75}$$

For a continuous membership function, the application of the operator is given as Fig. 11.15. The rules may be weighted, meaning that some rule has a higher weight as compared to the other. Consider that the problem is to move a robot from a source to a goal. If the robot is sufficiently far from the obstacle, a robot may still be asked to orient itself to avoid the obstacle, so that it does not have to come near to the obstacle to make a sharp avoidance steer. However, this rule is of lesser importance than another rule, which orients the robot towards the goal. To make the system prefer orienting towards the goal, rather than avoiding a far obstacle, the weights can be specified to the rules. Similarly, if some rules are for obstacle avoidance and some for goal-seeking, the weights can be used to denote the importance of different rules to balance the two opposing behaviours. The aggregation for rules with weights is given by Eqs (11.76), (11.77).

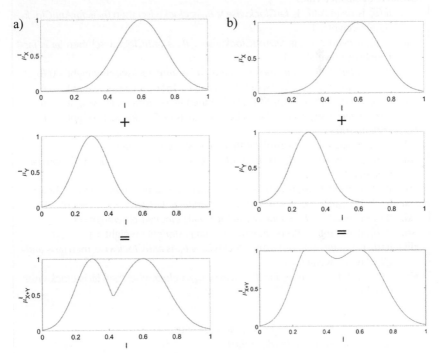

FIG. 11.15 Graphical illustration of fuzzy aggregation (A) using $\mu_{X+Y}(x)=\max(\mu_X(x),\mu_Y(x))$ (B) using $\mu_{X+Y}(x)=\min(\mu_X(x)+\mu_Y(y),1)$.

$$w_X\mu_X(x) + w_Y\mu_Y(x) = \max(w_X\mu_X(x), w_Y\mu_Y(x)) \tag{11.76}$$

$$w_X\mu_X(x) + w_Y\mu_Y(x) = \min(w_X\mu_X(x) + w_Y\mu_Y(x), 1) \tag{11.77}$$

Here, w_X and w_Y denotes the weights of the rules, $0 \le w_X, w_Y \le 1$.

The application of the rules is shown in Box 11.2. Aggregation on the specific example is explicitly shown in Fig. 11.16. The complete inference process is also shown in Fig. 11.17.

Box 11.2 Calculation of fuzzy output.

A. Notations:
d_o: Distance to nearest obstacle
θ_o: Angle to nearest obstacle
θ_G: Angle to goal
d_G: Distance to goal
ω: Turn

B. Rules:
Obstacle avoidance rules
   i. If (d_o is *near*) $\wedge$ (θ_o is *farClockwise* $\vee$ θ_o is *clockwise*) then (ω is *farAntiClockwise*): weight$=1$
   ii. If (d_o is *near*) $\wedge$ (θ_o is *farAntiClockwise* $\vee$ θ_o is *antiClockwise*) then (ω is *farClockwise*): weight$=1$
   iii. If (d_o is *far*) $\wedge$ (θ_o is *farClockwise*) then (ω is *antiClockwise*): weight$=0.6$
   iv. If (d_o is *far*) $\wedge$ (θ_o is *clockwise*) then (ω is *farAntiClockwise*): weight$=0.6$
   v. If (d_o is *far*) $\wedge$ (θ_o is *farAntiClockwise*) then (ω is *clockwise*): weight$=0.6$
   vi. If (d_o is *far*) $\wedge$ (θ_o is *antiClockwise*) then (ω is *farClockwise*): weight$=0.6$
Goal seeking rules
   vii. If (d_G is *near*) $\wedge$ (θ_G is *straight*) then (ω is *straight*): weight$=1$
   viii. If (d_G is *near*) $\wedge$ (θ_G is *farAntiClockwise*) then (ω is *farAntiClockwise*): weight$=1$
   ix. If (d_G is *near*) $\wedge$ (θ_G is *antiClockwise*) then (ω is *antiClockwise*): weight$=1$
   x. If (d_G is *near*) $\wedge$ (θ_G is *farClockwise*) then (ω is *farClockwise*): weight$=1$
   xi. If (d_G is *near*) $\wedge$ (θ_G is *clockwise*) then (ω is *clockwise*): weight$=1$
   xii. If (d_G is *distant*) $\wedge$ (θ_G is *straight*) then (ω is *straight*): weight$=1$
   xiii. If (d_G is *distant*) $\wedge$ (θ_G is *farAntiClockwise* $\vee$ θ_G is *antiClockwise*) then (ω is *antiClockwise*): weight$=1$
   xiv. If (d_G is *distant*) $\wedge$ (θ_G is *farClockwise* $\vee$ θ_G is *clockwise*) then (ω is *clockwise*): weight$=1$

C. Fuzzification:
Inputs: $d_o=0.01$, $\theta_o=-45°$, $d_G=0.2$ and $\theta_G=60°$.
Distance to nearest obstacle (d_o) $= 0.01$

$$\mu_{near}^{d_o}(0.01) = \exp\left(\frac{-0.01^2}{2(0.01915)^2}\right) = 0.87$$

$$\mu_{far}^{d_o}(0.01) = \exp\left(\frac{-(0.01 - 0.066)^2}{2(0.0207)^2}\right) = 0.02$$

$$\mu_{distant}^{d_o}(0.01) = \exp\left(\frac{-(0.01 - 1)^2}{2(0.2797)^2}\right) = 0.001$$

Box 11.2 Calculation of fuzzy output—cont'd

Angle to nearest obstacle (θ_o) = $-45°$

$$\mu_{farClockkwise}^{\theta_o}(-45) = \frac{1}{1 + \left|\frac{-45-(-137.1)}{98.5}\right|^{2(6.28)}} = 0.70$$

$$\mu_{clockkwise}^{\theta_o}(-45) = \exp\left(\frac{-(-45-(-20))^2}{2(20)^2}\right) = 0.46$$

$$\mu_{antiClockkwise}^{\theta_o}(-45) = \exp\left(\frac{-(-45-20)^2}{2(20)^2}\right) = 0.01$$

$$\mu_{farAntiClockkwise}^{\theta_o}(-45) = \frac{1}{1 + \left|\frac{-45-140}{93.7}\right|^{2(4.475)}} = 0.00$$

Distance to goal (d_G) = 0.2

$$\mu_{near}^{d_G}(0.2) = \exp\left(\frac{-0.2^2}{2(0.0594)^2}\right) = 0.0035$$

$$\mu_{distant}^{d_G}(0.2) = \frac{1}{1 + \left|\frac{0,2-1}{0.773}\right|^{2(7.25)}} = 0.38$$

Angle to goal (θ_G) = $60°$

$$\mu_{farClockwise}^{\theta_G}(60) = \frac{1}{1 + \left|\frac{60-(-155)}{85.5}\right|^{2(6.19)}} = 0.00$$

$$\mu_{clockwise}^{\theta_G}(60) = \exp\left(\frac{-(60-(-40))^2}{2(20)^2}\right) = 0.00$$

$$\mu_{straight}^{\theta_G}(60) = \exp\left(\frac{-(60-0)^2}{2(20)^2}\right) = 0.01$$

$$\mu_{antiClockwise}^{\theta_G}(60) = \exp\left(\frac{-(60-40)^2}{2(20)^2}\right) = 0.61$$

$$\mu_{farAntiClockwise}^{\theta_G}(60) = \frac{1}{1 + \left|\frac{60-155}{85.5}\right|^{2(6.19)}} = 0.21$$

Turn (ω) (output variable)

$$\mu_{farClockwise}^{\omega}(\omega) = \exp\left(\frac{-(\omega-(-45))^2}{2(12)^2}\right)$$

$$\mu_{clockwise}^{\omega}(\omega) = \exp\left(\frac{-(\omega-(-10))^2}{2(4)^2}\right)$$

Continued

Box 11.2 Calculation of fuzzy output—cont'd

$$\mu^{\omega}_{straight}(\omega) = \exp\left(\frac{-\omega^2}{2(4)^2}\right)$$

$$\mu^{\omega}_{antiClockwise}(\omega) = \exp\left(\frac{-(\omega - 10)^2}{2(4)^2}\right)$$

$$\mu^{\omega}_{farAntiClockwise}(\omega) = \exp\left(\frac{-(\omega - 45)^2}{2(12)^2}\right)$$

D. Calculation of rules
Step 1: Rules with inputs

   i. $\mu^{d_o}_{near}(0.01) \wedge (\mu^{\theta_o}_{farClockwise}(-45) \vee \mu^{\theta_o}_{clockwise}(-45)) \rightarrow \mu^{\omega}_{farAntiClockwise}(\omega)$ +

   ii. $\mu^{d_o}_{near}(0.01) \wedge (\mu^{\theta_o}_{farAntiClockwise}(-45) \vee \mu^{\theta_o}_{antiClockwise}(-45)) \rightarrow \mu^{\omega}_{farClockwise}(\omega)$ +

   iii. $\mu^{d_o}_{far}(0.01) \wedge \mu^{\theta_o}_{farClockwise}(-45) \rightarrow \mu^{\omega}_{antiClockwise}(\omega) \times 0.6$ +

   iv. $\mu^{d_o}_{far}(0.01) \wedge \mu^{\theta_o}_{clockwise}(-45) \rightarrow \mu^{\omega}_{farAntiClockwise}(\omega) \times 0.6$ +

   v. $\mu^{d_o}_{far}(0.01) \wedge \mu^{\theta_o}_{farAntiClockwise}(-45) \rightarrow \mu^{\omega}_{clockwise}(\omega) \times 0.6$ +

   vi. $\mu^{d_o}_{far}(0.01) \wedge \mu^{\theta_o}_{antiClockwise}(-45) \rightarrow \mu^{\omega}_{farClockwise}(\omega) \times 0.6$ +

   vii. $\mu^{d_G}_{near}(0.2) \wedge \mu^{\theta_G}_{straight}(60) \rightarrow \mu^{\omega}_{straight}(\omega)$ +

   viii. $\mu^{d_G}_{near}(0.2) \wedge \mu^{\theta_G}_{farAntiClockwise}(60) \rightarrow \mu^{\omega}_{farAntiClockwise}(\omega)$ +

   ix. $\mu^{d_G}_{near}(0.2) \wedge \mu^{\theta_G}_{antiClockwise}(60) \rightarrow \mu^{\omega}_{antiClockwise}(\omega)$ +

   x. $\mu^{d_G}_{near}(0.2) \wedge \mu^{\theta_G}_{farClockwise}(60) \rightarrow \mu^{\omega}_{farClockwise}(\omega)$ +

   xi. $\mu^{d_G}_{near}(0.2) \wedge \mu^{\theta_G}_{clockwise}(60) \rightarrow \mu^{\omega}_{clockwise}(\omega)$ +

   xii. $\mu^{d_G}_{distant}(0.2) \wedge \mu^{\theta_G}_{straight}(60) \rightarrow \mu^{\omega}_{straight}(\omega)$ +

   xiii. $\mu^{d_G}_{distant}(0.2) \wedge (\mu^{\theta_G}_{farAntiClockwise}(60) \vee \mu^{\theta_G}_{antiClockwise}(60)) \rightarrow \mu^{\omega}_{antiClockwise}(\omega)$ +

   xiv. $\mu^{d_G}_{distant}(0.2) \wedge (\mu^{\theta_G}_{farClockwise}(60) \vee \mu^{\theta_G}_{clockwise}(60)) \rightarrow \mu^{\omega}_{clockwise}(\omega)$

Step 2: Substitute membership values

   i. $0.87 \wedge (0.70 \vee 0.46) \rightarrow \exp\left(\frac{-(\omega - 45)^2}{2(12)^2}\right)$ +

   ii. $0.87 \wedge (0.00 \vee 0.01) \rightarrow \exp\left(\frac{-(\omega - (-45))^2}{2(12)^2}\right)$ +

   iii. $0.02 \wedge 0.70 \rightarrow \exp\left(\frac{-(30 - 40)^2}{2(20)^2}\right) \times 0.6$ +

   iv. $0.02 \wedge 0.46 \rightarrow \exp\left(\frac{-(\omega - 45)^2}{2(12)^2}\right) \times 0.6$ +

Box 11.2 Calculation of fuzzy output—cont'd

---

v. $0.02 \wedge 0.00 \rightarrow \exp\left(\frac{-(\omega-(-10))^2}{2(4)^2}\right) \times 0.6 +$

vi. $0.02 \wedge 0.01 \rightarrow \exp\left(\frac{-(\omega-(-45))^2}{2(12)^2}\right) \times 0.6 +$

vii. $0.00 \wedge 0.01 \rightarrow \exp\left(\frac{-\omega^2}{2(4)^2}\right) +$

viii. $0.00 \wedge 0.21 \rightarrow \exp\left(\frac{-(\omega-45)^2}{2(12)^2}\right) +$

ix. $0.00 \wedge 0.61 \rightarrow \exp\left(\frac{-(\omega-10)^2}{2(4)^2}\right) +$

x. $0.00 \wedge 0.00 \rightarrow \exp\left(\frac{-(\omega-(-45))^2}{2(12)^2}\right) +$

xi. $0.00 \wedge 0.00 \rightarrow \exp\left(\frac{-(\omega-(-10))^2}{2(4)^2}\right) +$

xii. $0.38 \wedge 0.01 \rightarrow \exp\left(\frac{-\omega^2}{2(4)^2}\right) +$

xiii. $0.38 \wedge (0.21 \vee 0.61) \rightarrow \exp\left(\frac{-(\omega-10)^2}{2(4)^2}\right) +$

xiv. $0.38 \wedge (0.00 \vee 0.00) \rightarrow \exp\left(\frac{-(\omega-(-10))^2}{2(4)^2}\right)$

Step 3: Use min ***for*** $\wedge$, max ***for*** $\vee$, ***min for*** $\rightarrow$, **+ *for aggregation multiplied by rule weight,***

$$\mu(\omega) = \min\left(0.70, \exp\left(\frac{-(\omega-45)^2}{2(12)^2}\right)\right) + \min\left(0.01, \exp\left(\frac{-(\omega-(-45))^2}{2(12)^2}\right)\right)$$

$$+ \min\left(0.02, \exp\left(\frac{-(30-40)^2}{2(20)^2}\right)\right) \times 0.6 + \min\left(0.02, \exp\left(\frac{-(\omega-45)^2}{2(12)^2}\right)\right) \times 0.6$$

$$+ \min\left(0.00, \exp\left(\frac{-(\omega-(-10))^2}{2(4)^2}\right)\right) \times 0.6 + \min\left(0.01, \exp\left(\frac{-(\omega-(-45))^2}{2(12)^2}\right)\right) \times 0.6$$

$$+ \min\left(0.00, \exp\left(\frac{-\omega^2}{2(4)^2}\right)\right) + \min\left(0.00, \exp\left(\frac{-(\omega-45)^2}{2(12)^2}\right)\right)$$

$$+ \min\left(0.00, \exp\left(\frac{-(\omega-10)^2}{2(4)^2}\right)\right) + \min\left(0.00, \exp\left(\frac{-(\omega-(-45))^2}{2(12)^2}\right)\right)$$

$$+ \min\left(0.00, \exp\left(\frac{-(\omega-(-10))^2}{2(4)^2}\right)\right) + \min\left(0.01, \exp\left(\frac{-\omega^2}{2(4)^2}\right)\right)$$

$$+ \min\left(0.38, \exp\left(\frac{-(\omega-10)^2}{2(4)^2}\right)\right) + \min\left(0.00, \exp\left(\frac{-(\omega-(-10))^2}{2(4)^2}\right)\right)$$

$$\mu(\omega) \approx \min\left(0.70, \exp\left(\frac{-(\omega-45)^2}{2(12)^2}\right)\right) + \min\left(0.38, \exp\left(\frac{-(\omega-10)^2}{2(4)^2}\right)\right)$$

E. Defuzzification:

$$output = \frac{\sum_{\omega=-45°,}^{45°} \mu(\omega)\omega}{\sum_{\omega=-45°,}^{45°} \mu(\omega)} = 26.8°$$

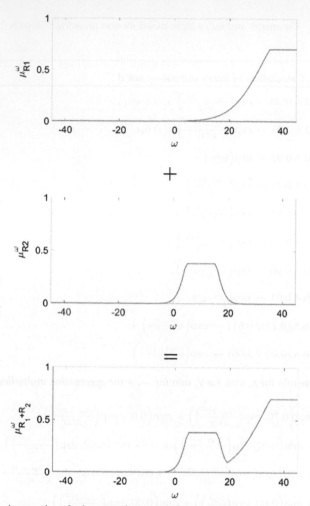

FIG. 11.16 Aggregation of rules to produce output for the example

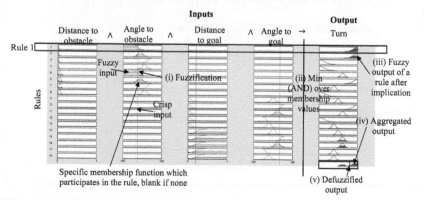

FIG. 11.17 Inference for an input. The rules are written in strict AND format (splitting OR into multiple rules). The participating membership function is drawn for every rule (row). The vertical lines give the crisp input whose membership is computed as fuzzy input. The min across the columns of a row denotes AND operation, which undergoes implication by min, thus slicing the participating output membership function. All outputs are aggregated and then defuzzified.

Each term of the expression represents a rule. It may be seen that even though there were many rules, only 2 rules effectively fire. Even though the rule base may be large, at any point of time only a small proportion of the rules fire and contribute towards decision-making.

11.5.5 Defuzzification

The aggregation of all rules produces a valid aggregated output, which is fuzzy, and therefore, every possible real number is the output with an associated membership degree. This output can however not be fed inside the robot, which only takes crisp real numbers. The conversion of a fuzzy output into a crisp output is called *defuzzification*. The membership degree denotes the weightage or likelihood of the output which is considered to defuzzify the output.

The commonly used technique of defuzzification is the *centroid defuzzification*. Consider a membership function plotted against an input. The area under the membership function is taken to be a body and the aim is to calculate the centre of gravity or the centre of this area. In simple words, the method takes a weighted addition using membership degree as the weight. For a discrete set, the defuzzified output is given by Eq. (11.78):

$$x = \frac{\sum_x \mu_X(x) x}{\sum_x \mu_X(x)} \tag{11.78}$$

Consider a discrete set given by Eq. (11.79), the defuzzified output is given by Eq. (11.80):

$$X = \left\{ \frac{0}{1} + \frac{0.2}{2} + \frac{0.6}{3} + \frac{1.0}{4} + \frac{0.3}{5} + \frac{0.1}{6} \right\} \tag{11.79}$$

$$x = defuzzify(X) = \frac{0(1) + 0.2(2) + 0.6(3) + 1.0(4) + 0.3(5) + 0.1(6)}{0 + 0.2 + 0.6 + 1.0 + 0.3 + 0.1} = 3.77 \tag{11.80}$$

For a continuous membership function, the only change is that the summation becomes integral, given by Eq. (11.81).

$$x = \frac{\int_x \mu_X(x) x dx}{\int_x \mu_X(x) dx} \tag{11.81}$$

Consider the membership function given in Fig. 11.18, the defuzzified output is shown in the same figure. There are other methods possible for defuzzification, like the *Bisector* method, which chooses the crisp value which divides the fuzzy output into two equal areas. This behaves like the centroid. Imagine driving a car with an obstacle ahead. A friend gives you some rules for which the obstacle should be avoided from the right. Another friend gives you some other rules for which the obstacle should be avoided from the left. Instead take the centroid or bisector, which takes the mean of the two decisions which bangs the vehicle into the obstacle. Hence, sometimes it is better to go with one expert for which you have the maximum degree of confidence. This is modelled by the *maximum defuzzification* method, which selects the output corresponding to the largest membership degree. In case of multiple such options, the middle of an area, the smallest of an area or largest of an area are good options to consider.

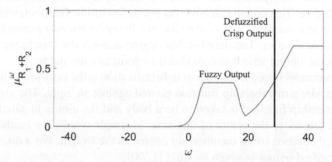

FIG. 11.18 Defuzzification operation. Conversion of a fuzzy (membership function) output into a crisp (real number) output. Here, taken as the centroid assuming the area under the curve as a solid body.

11.5.6 Designing a fuzzy inference system

Consider the problem of robot motion planning using the already discussed formulation (Kala et al., 2010). The robot uses virtual proximity sensors at $\pi/2$ rad clockwise and anticlockwise to scan the nearest obstacle and get the distance from the nearest obstacle. The robot keeps a small linear speed, and the angular speed is taken as the output of the fuzzy inference system. The path traced by the robot for a couple of synthetic scenarios is shown in Fig. 11.19.

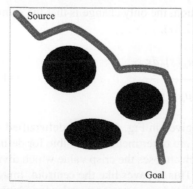
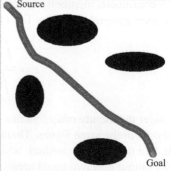

FIG. 11.19 Simulation results for motion planning using fuzzy logic.

The robot comfortably obtained the goal while avoiding the obstacle. However, it is important to consider the general design philosophy of the fuzzy inference system, which can be done in the following steps that enable writing down the rules using human expertise.

1. Identify the inputs and outputs, preferably having the least number of variables without losing information or having important inputs unexpressed. Deciding on the inputs is like feature engineering in machine learning approaches.
2. Break down all variables into a small number of membership functions. If every membership function effectively covers a large part of the input space the system has a high *generalization* and a small effect of noise. Generalization is the ability of a system to give answers to the inputs that were not foreseen by the human expert in the design phase.
3. Write the rules. The first way to write down the rules is to populate all possible antecedents considering all combinations of membership functions of all inputs and writing consequents using a human expert. The mechanism used in the example was, on the contrary, visualizing a robot going from source to goal and visualizing all possible situations, and translating the same into rules.
4. Simulate using designed rules. If the robot collides or makes a poor move, note the inputs and the rules fired and mend the rules, add rules, or mend the parameters of the membership functions. The biggest advantage of the fuzzy inference system is its *interpretability* or that the output can be traced to the active rules and membership functions, which can be modified to get the desired output.
5. In case it is not possible to get the desired behaviour, add membership functions to get more expression ability of writing rules for the specific behaviours at specific inputs. The addition of membership function must though be done cautiously, as it increases possible rules which may be hard to completely model by hand. Too many variables make a complex system that is hard to design by hand.

A smaller number of membership functions and a smaller number of rules result in a simpler system. If the output (*turn*) is plotted against all the possible inputs, a fuzzy system that has a smaller number of membership functions and rules has a smaller number of crests and troughs that nicely blend. The system is said to have a *high bias* and *low variance*. Excessive membership functions and rules enable very rough output spaces with numerous high spikes (crests and troughs). As a result, a small noise moves the input in the input space and results in a dramatic change in the output value, which makes a sensitive and nonrobust system. The system is said to have *high variance* and *low bias*.

The desired output has a characteristic surface across inputs. If a lower number of membership functions and rules are used, resulting in higher bias, the system will not have enough expression potential to imitate the desired output curve and thus the performance will be poor. The system is said to be *under-fit*. On the contrary, if too many rules and membership functions are specified, it is highly possible that the human is unable to write the consequences accurately or the human makes errors. The system is operating at a high variance and low bias condition. The system is said to be *over-fit*. Hence, the approach in design is always to start from a high bias and low variance condition, to slowly increase the complexity by adding membership functions and rules, and to stop in time when the system has just enough rules and membership functions to model the problem. The stopping condition is called the *ideal-fit*.

11.6 Training

The other major issue is to be able to train the system, assuming that training data is available consisting of the inputs as well as the outputs. The training is responsible for setting all the (trainable) parameters associated with the membership functions of both input and output as well as the (trainable) antecedents and consequents of the rules. The term trainable is associated because in a Gaussian membership function, typically the user decides whether only the mean should be optimized or both the mean and variance. The former injects a reasonable heuristic controlling generalization and thus makes training easier, while the latter lets the training algorithm play with the complete system. Similarly, in a triangular membership function, a user may impose symmetry which will reduce the number of parameters from 3 to 2, which heuristically is a sign of a good design also limiting training time. Similarly, the antecedents may be an exhaustive list of permutations of all possible membership functions, in which case they may not be trained.

This chapter discusses some mechanisms to train or automatically construct a fuzzy inference system from a training data. The discussions are further continued in Chapter 12 for training as a neural network.

11.6.1 Gradient-based optimization

So far, the fuzzy system was trained by trial and error, which may not be the best way out. There are several training methodologies. The first methodology uses the *gradient-based optimization*, where an iterative correction in parameter values is a local derivative of the error function. This, however, requires the output from the fuzzy system as a mathematical closed-form solution with real number weights. Therefore, let us keep the rules as fixed, the number of membership functions as fixed, and the shape of the membership functions as fixed. The parameters of the membership functions are trained. Gradient-based training is covered in the next chapter.

It is possible for the machine to tweak the rules as well by using *gradient-free optimization approaches*. The tweaking of rules happens by changing the participating membership functions of antecedents or consequents and checking if the resultant system after training has a lesser loss, in which case it is accepted as the current best and tweaked even further. Since training data is available, the rules also need not be guessed. They are imperative from the training data itself. Assuming that the antecedents are fixed, the consequents can be guessed by finding all the training data for which the antecedent is largely true. The corresponding targets are noted and mapped to the best consequent. If the rule table is an exhaustive list of all antecedents possible, this can be done for all rules to populate the consequents automatically.

11.6.2 Direct rule estimation

An exhaustive list of antecedents has the problem that it assumes all types of inputs are possible. This is rarely the case. Machine learning techniques identify the types of inputs or regions in the input space that never occur. A camera never takes an image in which every pixel has a contrasting colour as compared to the neighbours. A lidar scan is never such that every scan reading is one opposite extremity of the previous scan reading. Hence it is unreasonable to consider all possible inputs.

Consider the input for the dummy problem is (4,6) with output 7, which represents a single training data. This can be conveniently written as a fuzzy rule '*if input₁ is like 4 and input₂ is like 6, then output is like 7*'. Here, '*like 4*' represents a Gaussian membership function with a mean 4 and variance σ (user-defined parameter). One rule is designed for every data in the training data set, which gives a rich set of rules. There is no further training required. The closer a new test input gets to 4, the higher is the membership degree; while the further away is the test input from 4, the membership value slowly degrades to 0. σ denotes the degree of generalization. If σ is 0, the input only fires for the exact input 4, and not even for close inputs like 3.9. This means the rule will rarely be active (making a high variance system). A high value of σ will always fire the rule and result in under-fitting (making a high-bias system), wherein output roughly becomes the mean of all rule outputs. A high value of σ has smoothing effects in the input-to-output mapping, where the output is smoothened by many neighbouring inputs from the training data.

This manner of writing rules is called *linguistic rule writing*, wherein the training data set is read as English statements and translated directly into fuzzy rules. Linguistic hedges can be added as well. As an example, *very much* is a linguistic hedge where *very much like* 4 is modelled as a square of the membership function, given by $\left(\exp\left(-\frac{(x-4)^2}{2\sigma^2}\right)\right)^2$. Similarly saying *a little like* 4, the hedge *a little* can be modelled as a square root, given by $\left(\exp\left(-\frac{(x-4)^2}{2\sigma^2}\right)\right)^{0.5}$.

The problem with this approach is that all training data is taken as rules, which will make many rules, thus making testing on new input computationally very heavy. It is reasonable to assume the smoothness of the resultant output over the input. Hence, the inputs may be *clustered* in the input space and every cluster head makes one such rule.

Questions

1. In the scenarios given in Fig. 11.20, give the output of the artificial potential field, VO method, VFH, and largest gap approach. The robot is shown by a black circle. Explain the calculation when the robot is mid-way to the goal. In scenarios involving multiple robots, each robot has the same speed, and the arrow shows the goal of the corresponding robot. Compare and contrast the methods. Please enlist the assumptions involved in all of them. Which of the approaches is the best to solve the problem of reaching the goal and why? If no method can solve a problem, also suggest if other algorithms and techniques can solve the problem and how?

FIG. 11.20 Sample scenarios for question 1.

2. Two robots are facing directly at each other. The source of the 1st robot is the goal for the 2nd robot and vice versa. Both robots use the VO method to reach to their respective goals. Suggest possible trajectories of the robots if the robots use the velocity selection strategy as the largest speed within an angle α to the goal. Suggest for two cases of tie-breaking: (i) both robots prefer clockwise direction if two velocity vectors are equally preferable and (ii) one robot prefers clockwise direction while the other prefers anticlockwise direction if two velocity vectors are equally preferable. Show snapshots of velocity calculation at a few locations in the trajectory.

3. Common upon the optimality and completeness of VFH algorithm. If not complete, give a situation where the algorithm fails, showing the trajectory. Explain how the algorithm can be modified to maintain a large clearance from the obstacles? Also, explain how the algorithm can be modified to avoid other robots on the way?

4. Comment on how the VFH algorithm handles narrow corridors. Also, comment on how the parameters can be modified to enable the algorithm to take more narrow corridors and vice versa. How the parameters can be

modified to enable the algorithm to maintain larger clearances from obstacles and vice versa. Comment upon the difference in ways by which VFH and Voronoi diagram handle narrow corridors, explaining which approach is better and why?

5. Consider a circular robot at position (x,y) and radius ρ, and a static circular obstacle at position (x_o,y_o) with a radius ρ_o. The robot is currently at an orientation of θ with respect to the coordinate axis system. Assume that the goal is somewhere behind the obstacle in the line from the current position of the robot to the goal.

   (a) Calculate the immediate direction of motion of the robot, such that it kisses the obstacle as it moves.

   (b) What should be the angular range of values of the angle to goal such that the robot motion using the solution to (a) is optimal?

   (c) Calculate the immediate direction of motion of the robot, such that it maintains a distance of Δ from the obstacle as it moves.

   (d) Consider that there are 2 obstacles at positions (x_{o1},y_{o1}) and (x_{o2},y_{o2}), each with a radius ρ_o. The goal lies in-between these obstacles. Calculate the immediate direction of motion of the robot, such that the robot maximizes the distance between the obstacles. Assume enough space between the obstacles that allow the robot to pass by.

   (e) Consider the same situation in (d). Now the robot must go with the direction of maximization of the gap if the gap is less than 2Δ, and by maintaining a distance of Δ if the gap is more than 2Δ. In the case of the latter, the equations must also mention which of the 2 obstacles is used for the computation of the angle. Assume enough space between the obstacles that allow the robot to pass by.

6. A robot needs to use the VO method, VFH, and largest gap approach. The robot is not allowed to use any external sensor (the ones installed outside the robot like on the ceiling). Explain the sensors that the robot will require and how can the robot practically get the desired inputs for each of the mentioned algorithms.

7. The VO approach only considers collisions till a time horizon T. Explain how the approach performs if this horizon is indefinitely large for both high-density and low-density scenarios. Similarly, the largest gap approach only considers the nearest few obstacles. Explain what happens if all obstacles are considered, while a gap is defined as the obstacle-free region with a visibility till the map boundary. Similarly, explain the performance of the VFH approach if the local map is nearly as large as the global map.

8. Consider the fuzzy sets $A = \{0.2/1 + 0.8/2 + 0.6/3 + 0.5/4 + 0.2/5\}$, $B = \{0.1/a + 0.8/b + 0.5/c\}$, and $C = \{0.1/x + 0.8/y\}$. Calculate the output of $A \wedge B$, $A \vee B$, $\neg A$, $A \rightarrow B$, $A \wedge B \vee A \wedge C$, $(\neg A) \wedge B \vee A \wedge (\neg C)$, $A \wedge B \rightarrow (\neg C)$.

9. Consider the rule $A \wedge B \to C$. The membership functions are given by $\mu_A(a) = \exp\left(\frac{-(a-0.5)^2}{2(0.1)^2}\right)$, $\mu_B(b) = \exp\left(\frac{-(b-0.3)^2}{2(0.5)^2}\right)$, and $\mu_C(c) = \exp\left(\frac{-(c-0.3)^2}{2(0.5)^2}\right)$. Calculate the fuzzy output corresponding to the crisp inputs $a = 0.1$ and $b = 0.4$.

10. Consider the fuzzy inference system shown in Box 11.2. Calculate the crisp output for the input $d_o = 0.2$, $\theta_o = 30°$, $d_G = 0.6$ and $\theta_G = -10°$.

11. [Programming] Make a fuzzy logic system that orients the robot in the direction to goal. Extend the same system to make the robot get to the goal without obstacles. Extend the same system to make the robot get to the goal with obstacles. Extend the same algorithm to locally optimize the parameters and rules, to have a better performance on chosen scenarios by chosen metrics.

12. [Programming] Make a fuzzy logic system that enables two robots to avoid each other, each going to its own goal. Add rules to ensure that if the two robots are nearly head-on to each other, avoiding each other on the left is socially preferable.

13. [Programming] Write a program to make a robot escape static obstacles of any arbitrary shape using the VFH algorithm. Account for the robot being circular with radius ρ. Examine the performance with a change of parameters for different kinds of maps, including narrow corridors.

14. [Programming] Extend the program using VFH to account for dynamic obstacles. Examine the performance for increasing speed of the obstacle.

15. [Programming] Write a program that uses VO to make n circular robots avoid each other. Teleoperate one robot to behave nonidealistically and check the performance of the algorithm.

16. [Programming] Write a program that does a radial scan and makes the robot avoid the other robots with the largest clearance. Use a suitable technique to add goal-seeking behaviour. Extend the program to account for static obstacles and the size of the robot.

17. [Programming] Extend the previous program to account for the social etiquette that if two robots are head-on, they prefer to avoid each other from the left side.

18. [Programming] At any situation, numerous reactive algorithms can be used to compute the immediate motion of the robot. Make a robot operate by using an ensemble of reactive algorithms using (a) fusion of outputs of all reactive techniques used and (b) selection of a random reactive algorithm at each step. Using these reactive algorithms as the action function, also makes a global search algorithm.

References

Borenstein, J., Koren, Y., 1991. The vector field histogram-fast obstacle avoidance for mobile robots. IEEE Trans. Robot. Autom. 7 (3), 278–288.

Fiorini, P., Shiller, Z., 1998. Motion planning in dynamic environments using velocity obstacles. Int. J. Robot. Res. 17 (7), 760–772.

Kala, R., 2018. Routing-based navigation of dense mobile robots. Intell. Serv. Robot. 11 (1), 25–39.

Kala, R., Shukla, A., Tiwari, R., 2010. Fusion of probabilistic A* algorithm and fuzzy inference system for robotic path planning. Artif. Intell. Rev. 33 (4), 275–306.

Ross, T., 2010. Fuzzy Logic With Engineering Applications. John Wiley & Sons, West Sussex, UK.

Sezer, V., Gokasan, M., 2012. A novel obstacle avoidance algorithm: follow the gap method. Robot. Auton. Syst. 60 (9), 1123–1134.

Shukla, A., Tiwari, R., Kala, R., 2010a. Real Life Applications of Soft Computing. CRC Press, Boca Raton, FL.

Shukla, A., Tiwari, R., Kala, R., 2010b. Towards Hybrid and Adaptive Computing: A Perspective. Springer-Verlag, Berlin, Heidelberg.

Snape, J., van den Berg, J., Guy, S.J., Manocha, D., 2009. Independent navigation of multiple mobile robots with hybrid reciprocal velocity obstacles. In: 2009 IEEE/RSJ International Conference on Intelligent Robots and Systems. IEEE, St. Louis, MO, pp. 5917–5922.

Snape, J., Berg, J.V.D., Guy, S.J., Manocha, D., 2011. The hybrid reciprocal velocity obstacle. IEEE Trans. Robot. 27 (4), 696–706.

Ulrich, I., Borenstein, J., 1998. VFH+: reliable obstacle avoidance for fast mobile robots. In: Proceedings of the 1998 IEEE International Conference on Robotics and Automation vol. 2. IEEE, Leuven, Belgium, pp. 1572–1577.

Ulrich, I., Borenstein, J., 2000. VFH*: local obstacle avoidance with look-ahead verification. In: Proceedings of the IEEE International Conference on Robotics and Automation vol. 3. IEEE, San Francisco, CA, pp. 2505–2511.

van den Berg, J., Lin, M., Manocha, D., 2008. Reciprocal velocity obstacles for real-time multi-agent navigation. In: 2008 IEEE International Conference on Robotics and Automation. IEEE, Pasadena, CA, pp. 1928–1935. 2008.

van den Berg, J., Snape, J., Guy, S.J., Manocha, D., 2011a. Reciprocal collision avoidance with acceleration-velocity obstacles. In: 2011 IEEE International Conference on Robotics and Automation. IEEE, Shanghai, pp. 3475–3482.

van den Berg, J., Guy, S.J., Lin, M., Manocha, D., 2011b. Reciprocal n-body collision avoidance. In: Pradalier, C., Siegwart, R., Hirzinger, G. (Eds.), Robotics Research. Springer Tracts in Advanced Robotics, vol. 70, Springer, Berlin, Heidelberg, pp. 3–19.

Wilkie, D., van den Berg, J., Manocha, D., 2009. Generalized velocity obstacles. In: 2009 IEEE/RSJ International Conference on Intelligent Robots and Systems. IEEE, St. Louis, MO, pp. 5573–5578.

Chapter 12

An introduction to machine learning and deep learning

12.1 Introduction

Attributed to the advancements of machine learning, it is imperative to think that the problem of motion planning can be conveniently solved using neural networks alone, wherein a neural network takes the sensor percept (and goal information) values as input and the motor commands as output, thus freeing the human programmer from writing any planning program. Neural networks have lately solved the most complex problems and attained huge success. They can detect and recognize humans under challenging settings, identify complex objects in different lighting conditions and poses, understand complex text documents, and identify sentiments or emotions.

The neural networks (Haykin, 2009; Shukla et al., 2010) imitate the human brain containing biological neurons. The *biological neuron* gets electrical signals through the dendrites from several other neurons that get attenuated. The attenuated signals are added up and further get modified (passed through an activation function). This forms the output that is transmitted to the other neurons by axons. The human brain has billions of neurons that are connected and exchange signals. The neurons keep forming new connections, breaking other connections, and changing the strength between neurons. Humans perceive the world from sensory organs like eyes, which produce electrical signals to the receptive biological neurons. The information is processed in a layered manner with the initial layers doing the basic data processing, producing richer information to be used by the next layers. Consider vision as an example. The raw input constitutes the pixels. The initial layers interpret the lines and the next few layers interpret the corners made from two intersecting lines, while the following layers interpret the basic shapes made from lines and corners. The following layers interpret complex shapes constituted from spatial relations between simpler shapes. Finally, the actions are performed by the hands which are controlled by the muscles which are again connected to the neurons. The initial layers have the information in the rawer format, while the higher layers have more abstract and informative features for decision-making.

Autonomous Mobile Robots. https://doi.org/10.1016/B978-0-443-18908-1.00022-4

12.2   Neural network architecture

Let us first understand the architecture of the neural network and then understand a learning mechanism.

12.2.1   Perceptron

The perceptron (or an artificial neuron) is an imitation of the biological neuron. The perceptron takes input as real numbers. The inputs are multiplied by weights given by Eq. (12.1):

$$v = \sum_{i=1}^{n} w_i x_i + b \tag{12.1}$$

Here w_i is the weight of the ith connection, x_i is the ith input, and n is the number of inputs. b is called the *bias*, which is a constant added to the input. The output v is called the *induced field* of the neuron. The bias may also be visualized as the weight (w_0) associated with an additional input (x_0) such that the input is always kept as 1, giving $w_0 = b$ and $x_0 = 1$, (Eq. 12.2):

$$v = \sum_{i=0}^{n} w_i x_i \tag{12.2}$$

Let w be $(n+1) \times 1$ vector given by $[w_0 \ w_1 \ w_2 \ ... \ w_n]^T$. Similarly, let x be $(n+1) \times 1$ vector given by $[x_0 \ x_1 \ x_2 \ ... \ x_n]^T$. The equation is also expressed by Eq. (12.3):

$$v = w \bullet x = w^T x \tag{12.3}$$

Representing as a matrix multiplication allows using library optimizations, multi-core and GPU parallelization. If the output to I independent inputs needs to be computed, the inputs can be represented as $x_{(n+1) \times I}$ matrix with every column representing one input, whose outputs would still be given by $(w_{(n+1) \times 1})^T x_{(n+1) \times I}$, with every column representing the output of the corresponding input.

The output is passed through a step activation function given by Eq. (12.4):

$$y = \phi(v) = \begin{cases} 1 & \text{if } v > 0 \\ 0 & \text{if } v \leq 0 \end{cases} \tag{12.4}$$

Here ϕ is the *activation function*. The complete perceptron is shown in Fig. 12.1.

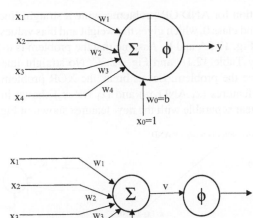

FIG. 12.1 Model of Perceptron. The inputs get multiplied by weights and a summation is taken at the perceptron ($v = \sum_{i=1}^{n} w_i x_i + b$), which passes an activation function to produce the output ($y = \phi(v)$). Bias (b) is an extra input added, modelled as w_0 attached to a fix input $x_0 = 1$, to make the resultant model $y = \phi(w^T x)$.

12.2.2 XOR problem

To visualize the working of a perceptron, let us take a simple binary classification problem. If the features are plotted in a 2-dimensional space, the representation is as shown in Fig. 12.2 with every data as a point that has the target label as 1 or 0. A line that divides the two classes is called the *decision boundary*. Equation ($w_0 + w_1 f_1 + w_2 f_2 = 0$) denotes a straight line as a decision boundary. The weights control the slope of the decision boundary, and the bias translates the decision boundary.

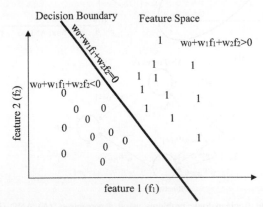

FIG. 12.2 Feature space. The classification requires separating the 0 s and 1 s. The decision boundary represents the weighted addition of the perceptron. A threshold activation function labels the points at one side of the decision boundary as 1 and vice versa.

To find a solution for AND OR problems, write a straight-line equation that separates class 1 and class 0, which gives the weight and bias values of the perceptron, as shown in Fig. 12.3A and B. Assume that the problem is to learn an XOR function, given by Table 12.1A and Fig. 12.3C. No straight line as a decision boundary can solve the problem. Let us break the XOR problem by extracting two higher-order features (x_1 AND x_2) and (x_1 OR x_2) shown in Table 12.1B. The problem is linear separable with the new features shown in Fig. 12.3C and F.

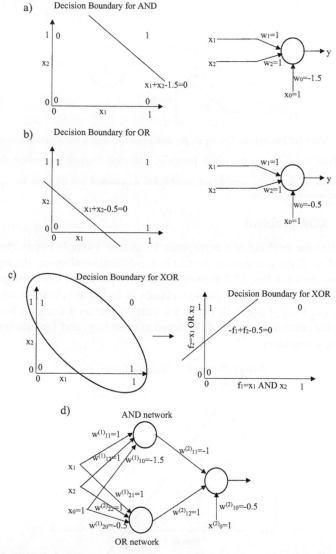

FIG. 12.3 Solving Boolean logic problems using a perceptron. (A) AND (B) OR (C) XOR (D) XOR network architecture. In XOR a single perceptron cannot solve the problem and hence AND and OR solutions are used as features, which appear as a new layer of perceptrons. The super-scripts in weights denote the layer and the sub-scrips in weights denote the target and preceding neuron number.

TABLE 12.1A XOR problem.

x_1	x_2	y
0	0	0
0	1	1
1	0	1
1	1	0

TABLE 12.1B XOR problem in terms of AND and OR.

x_1	x_2	x_1 AND x_2	x_1 OR x_2	y
0	0	0	0	0
0	1	0	1	1
1	0	0	1	1
1	1	1	1	0

This suggests that even if a problem is linearly non-separable with some inputs, it is possible to transform the problem into some higher-level features that make the problem linearly separable. Such features shall be computed by the learning algorithm. The problem could be solved in such a layered manner only because the two activation functions (step function) were non-linear. If the activation functions are linear ($y = \phi(v)$), the inputs will get multiplied by weights as they pass along the network; the resultant output will still be described as $y = w^T x$ for some weight vector w. This does not solve the problem of the need for non-linearity.

12.2.3 Activation functions

The role of the activation function is to distort and transform the input to produce the output. The popular choices are linear ($\phi(v) = v$, (b)), piecewise linear (Eq. 12.5), rectified linear unit (ReLU, $\phi(v) = \max(v,0)$), sigmoid (Eq. 12.6), and hyperbolic tangent (Eq. 12.7). All activation functions are shown in Fig. 12.4.

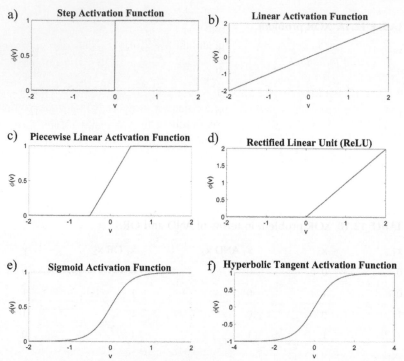

FIG. 12.4 Activation Functions (A) Step, (B) Linear, (C) Piecewise Linear, (D) Rectified Linear Unit (ReLU), (E) Sigmoid, (F) Hyperbolic Tangent.

$$y = \phi(v) = \begin{cases} 1 & if\ v \geq \dfrac{1}{2} \\ v + \dfrac{1}{2} & if\ -\dfrac{1}{2} \leq v \leq \dfrac{1}{2} \\ -1 & if\ v \leq -\dfrac{1}{2} \end{cases} \qquad (12.5)$$

$$y = \phi(v) = \frac{1}{1+e^{-av}} = \frac{1}{1+e^{-aw^T x}}, \quad a > 0 \qquad (12.6)$$

$$y = \phi(v) = a\tanh(bv), \quad a > 0, b > 0 \qquad (12.7)$$

Note that a single perceptron with a non-linear activation function will not be able to solve the non-linear XOR problem. The effect of the non-linear activation function only affects the threshold on which classes are classified; the inputs are still multiplied with the weights into a single number used for decision-making.

12.2.4   Multi-layer perceptron

Artificial neural networks (ANN) consist of a network of neurons that are arranged in a layered manner called the *multi-layer perceptron*. The first layer is a passive layer (does not do any data processing) which is called the *input layer* that receives the problem input. The last layer of the network is called the *output layer*. In between the input and the output layers, there are numerous layers called *hidden layers*. They are called so because the target output that these layers try to produce is unknown.

Each neuron has a bias term as per the modelling philosophy. This can be conveniently modelled by adding a new neuron in the preceding layer whose input is always fixed to 1 and whose weight is the bias value. Each layer of the neural network takes input from the preceding layer and multiplies the inputs by the weights stored at the connections, passing the output through the activation function to add non-linearity, which forms the input to be fed into the next layer. A complete neural network is shown in Fig. 12.5A. The weights are taken as matrices with $w^{(k)}$ denoting the weight matrix between layer $k-1$ and layer k. The parenthesis () denotes that the number is a superscript and not power. The size of the weight matrix is (number of neurons in layer $k-1$) $\times$

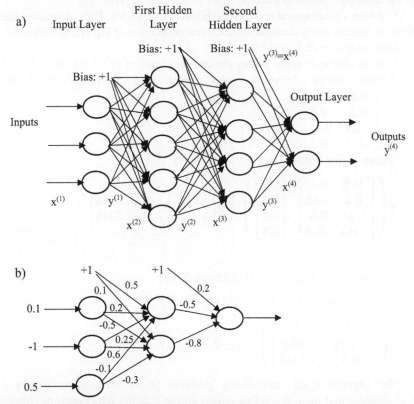

FIG. 12.5 (A) Multi-Layer Perceptron. Every circle represents a perceptron which performs a weighted addition followed by the application of activation function. Every perceptron has an additional constant input (+1) with a weight corresponding to the bias. The intermediate layers are called hidden layers. (B) An Illustrative Example using Multi-Layer Perceptron.

(number of neurons in layer k). Every column of the weight matrix denotes the inbound weights of a neuron. As data transfer from one layer to another, they get multiplied by this weight matrix. Let $x^{(k)}$ be the output of layer $k-1$, which forms the input of the layer k. The weighted addition of all neurons is given by Eq. (12.8):

$$v^{(k)} = w^{(k)T}x^{(k)} \tag{12.8}$$

The weighted addition $v^{(k)}$ represents a column vector of size (number of neurons in layer k) × 1, denoting the output of every neuron in layer k. If all neurons of layer k have the same activation function, the output of the layer is given by Eq. (12.9):

$$y^{(k)} = \phi\left(v^{(k)}\right) = \phi\left(w^{(k)T}x^{(k)}\right) \tag{12.9}$$

The notation is read as applying the activation function $\phi(.)$ to all the elements of the column vector, which produces a column vector as the output of the layer $y^{(k)}$, which is also the input to the next layer $x^{(k+1)}$.

Consider a small neural network as shown in Fig. 12.5B. The calculations are given as follows using the sigmoidal activation function of Eq. (12.6) with $a=1$.

Input Vector $= [0.1 \ -1 \ 0.5]^T$

Input to first layer (after adding the constant input of 1) $x^{(1)} = [1 \ 0.1 \ -1 \ 0.5]^T$

Weight vector between input layer and first hidden layer:

$$w^{(1)} = \begin{bmatrix} 0.5 & 0.1 \\ 0.2 & -0.5 \\ 0.25 & 0.6 \\ -0.1 & -0.3 \end{bmatrix}$$

Output of first hidden layer $y^{(1)} = \phi\left(\left(w^{(1)}\right)^T x^{(1)}\right) =$

$$\phi\left(\begin{bmatrix} 0.5 & 0.1 \\ 0.2 & -0.5 \\ 0.25 & 0.6 \\ -0.1 & -0.3 \end{bmatrix}^T \begin{bmatrix} 1 \\ 0.1 \\ -1 \\ 0.5 \end{bmatrix} \right) = \phi\left(\begin{bmatrix} 0.22 \\ -0.7 \end{bmatrix} \right) = \begin{bmatrix} 0.5548 \\ 0.3318 \end{bmatrix}$$

Input to output layer $x^{(2)} = [1 \ 0.5548 \ 0.3318]^T$

Output of output layer $y^{(2)} = \phi\left(\left(w^{(2)}\right)^T x^{(2)}\right) =$

$$\phi\left(\begin{bmatrix} 0.2 \\ -0.5 \\ -0.8 \end{bmatrix}^T \begin{bmatrix} 1 \\ 0.5548 \\ 0.3318 \end{bmatrix} \right) = \phi(-0.3428) = 0.4151$$

The *regression* or *curve-fitting problem* is to draw a surface in a high-dimensional input space that covers all the training data points as closely as possible while being smooth as well, as shown in Fig. 12.6. The classification problems regress the probability distribution with the target probabilities for the training data as 1 or 0. Some designs add a *softmax* activation function to the output layer to adhere to the constraints that the probability output should always be between 0 and 1, and all probabilities should add up to 1. Consider y_i to be the output of an output neuron (corresponding to class label i) that passes through the softmax activation function to yield the output z_i given by Eq. (12.10):

$$z_i = \frac{\exp(y_i)}{\Sigma_i \exp(y_i)} \tag{12.10}$$

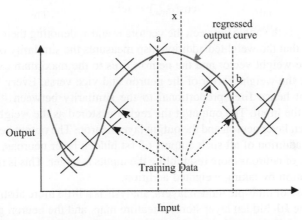

FIG. 12.6 Regression problem. Given training data items shown by crosses, the problem is to regress a smooth curve that nearly fits all data well. The output for an unknown input x can be computed by interpolation around known a and b. By memorizing enough such points, the complete curve can be approximately interpolated (universal approximation).

12.2.5   Universal approximator

A neural network with a single hidden layer and enough neurons in the hidden layer can act as a *universal approximator*. The network can approximately learn any possible mapping function or learn any surface between the inputs and outputs. This means that the neural networks can be used to solve any problem, assuming that there are enough data that completely represent the characteristics of the problem, there is sufficient time to learn, and enough neurons. Because of this property, typically, the classic neural networks only have a single hidden layer.

The ability of neural networks to act as universal approximators is not surprising. Imagine a complex surface to regress shown in Fig. 12.6. Memorize large areas of space, leaving out the spaces where the transition is smooth. Using this one can predict the areas which are not in the memory (say x) by interpolation from nearby known areas (say a and b). If you get a part wrong for which a training sample was available, you memorize that by adding it to your memory. By constantly expanding your memory, you would have learned enough to sufficiently estimate the surface.

The neural networks do a similar thing. Let w be the weight vector of a single neuron and x be the input given to it. Let us assume that the input is normalized so that $||x|| = 1$, or $x_0^2 + x_1^2 + x_2^2 + \ldots = 1$; and the weights are also normalized or $||w|| = 1$ or $w_0^2 + w_1^2 + w_2^2 + \ldots = 1$. Both w and x become vectors that lie on a unit sphere, and hence angular similarity is a good measure between the vectors. The cosine similarity denoting the concept is given by Eqs (12.11), (12.12).

$$w.x = w^\mathrm{T}x = ||w||.||x|| \cos(\theta_{w,x}) \qquad (12.11)$$

$$\cos(\theta_{w,x}) = w^\mathrm{T}x \qquad (12.12)$$

Here $\theta_{w,x}$ is the angle between the vectors w and x, denoting their similarity. This means that the weighted addition also measures the similarity of the input vector to the weight vector and the neuron fires to the maximum extent if the input is like the weight vector of the neuron and vice versa. Every 1st hidden layer weight hence fires proportional to the similarity between its encoded weight and the input. The output at the region is stored as the weight between the 1st hidden layer neuron and the output layer neuron. The output neuron takes a weighted addition of all such outputs of 1st hidden layer neurons, where the contribution of neurons more resembling the inputs is larger. This is the same as an interpolation by taking weighted addition.

In the case of multiple hidden layers, analysis is a little more abstract. Every neuron of the kth hidden layer stores a feature map, and the neuron gets highly activated if the input matches the feature map. The neurons of the $(k+1)$th layer also store feature maps built using the features of kth layer as input. As an example, the feature corner was defined on lines; and the feature square was defined on a spatial placement of corner features. Every layer independently learns a feature mapping. This ultimately translates to the desired output.

12.3 Learning

The model of the neural network can solve any problem and imitate any function subjected to correct values of weights, bias, and the correct neural network architecture. The essence of the neural network is to be able to *learn* the best weight and bias values. Assume that the training data available consist of $D = \{(x,t)\}$ where D is the data set, x is the input, and t is the target. The data set is divided typically into 3 parts, a training part (say 60%), a validation part (say 20%), and a testing part (say 20%).

Any system is never trained and tested on the same data set, which necessitates to keep a different training and testing data set. Even a lookup table can give an error of 0 on any input from the training data, which does not mean that the lookup table is a good machine learning system. The goodness of a machine learning system is its ability to *generalize* or give correct outputs to data which it has not seen before. The *training data* are used to set the weights and the bias values. The training only fixes the weights and biases and not the architecture and other hyper-parameters like the type of activation function. The architecture (number of hidden layer neurons) and other hyper parameters are fixed by trial and error, by trying multiple possible values and selecting the best architecture that has the least validation error on the *validation data set*. Using the training data for training weights and biases and architecture will not detect overfitting in architecture that will be discussed later. The *testing data* are to report the final accuracy. Nothing can be changed once the testing data are applied to the system, because if any parameter is changed noting the testing accuracy, erroneously that parameter would get trained on the testing data by human training in the form of noting the testing accuracy and changing the parameter value.

12.3.1 Bias-variance dilemma

Given several training points, many curves can sufficiently regress the surface. Consider the problem in Fig. 12.7. The training data and validation data are separately shown. The training error is the difference in the output value shown on the curve and the target value shown as the data. Similarly, for the validation error. Since training aims to reduce the training error, both Fig. 12.7A and B have a small training error and therefore both are good candidates. The training error of Fig. 12.7B is slightly lower than Fig. 12.7A. However, we follow *Occam's razor* that states if two systems equally solve the problem, the least complex one is preferable. Fig. 12.7B is wrong because if the validation data are seen, the error is large. The corresponding cases for a classification problem are shown in Fig. 12.8.

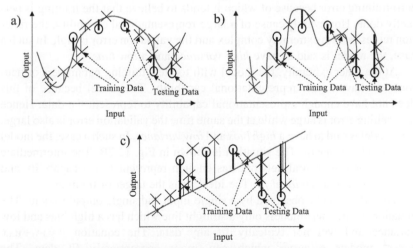

FIG. 12.7 Bias Variance Dilemma for a regression problem. The training data are shown by crosses and testing data by circles. As the model complexity increases, the curve can represent complex shapes leading to overfitting and for very small model complexity, the curve is overly simple (under-fit). The aim is to find the ideal fit. (A) Ideal fit, ideal bias, ideal variance, moderate training error, moderate testing error. (B) Overfit, low bias, high variance, nearly zero training error, high testing Error. (C) Underfit, high bias, low variance, high training error, high testing error.

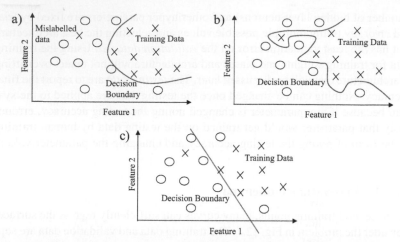

FIG. 12.8 Bias Variance Dilemma for a classification problem. The training data are shown by crosses and testing data by circles. (A) Ideal fit, ideal bias, ideal variance, moderate training error, moderate testing error. (B) Overfit, low bias, high variance, nearly zero training error, high testing error. (C) Underfit, high bias, low variance, high training error, high testing error.

The representational capacity of the neural network is given by the number of learnable parameters (weights and biases) or the number of neurons. A neural network, having too many weights and biases, has a large representational capacity and can represent complex shapes. Such a neural network is prone to *over-fit* the data as shown in Fig. 12.7B. The neural network gets nearly a zero-training error because of which it tends to believe that the training is perfectly done. However, because of a large representational capability, the function represented is extremely complex and the validation error is high. In such a case, the model is said to have *high variance* and a *low bias*.

At the other extremity, the model with too few training parameters or neurons will have a little representational capability. The model because of this does not have enough representational capability to represent the data. Hence the training error is large while at the same time the validation error is also large. The model is said to have a *high bias* and *low variance*. In such a case, the model is said to be *under-fit*. The situation is shown in Fig. 12.7B. The intermediate case is when the neural network has enough representational capability and hence is termed as an *ideal fit*. The ideal fit is the target of training.

The effect can be observed by a single input and single output system. The equation $y = w_0 + w_1 x$ models only a straight line which has a high bias and low variance and will not typically fit any data. The equation $y = w_0 + w_1 x + w_2 x^2$ models a curve which can more appropriately fit data. The equation $y = w_0 + w_1 x + w_2 x^2 + w_3 x^3$ also has the capability of modelling a crest and trough. As the degree is increased, the number of crests and troughs can also be increased. This increases model complexity, increases variance, and reduces bias.

To achieve ideal-fit, neural network architecture is first set to have very few neurons. The training is completed which shows a high bias and low variance leading to an underfit characterized by a large training error and a large validation error. The problem is resolved by increasing the number of neurons. As the number of neurons is increased, the network slowly increases the representational capability leading to an increase in variance and reduction of bias, characterized by a falling training error and a falling validation error. If the neurons are continuously increased, the network tends towards over-fit, which is visible by a large validation error while a very small training error. The theoretical trends of increasing the number of neurons are shown in Fig. 12.9A. The increase in neurons hence stops when the validation error starts to increase, leading to the curve going from underfitting to overfitting. This is called *early stopping*.

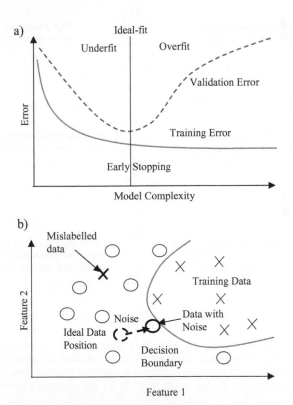

FIG. 12.9 (A) Early Stopping. Starting with low complexity, slowly add neurons (increase model complexity) till validation error reduces. Any further increase tends to overfit. (B) Noisy and mislabelled data in the training set. Limiting complexity makes the system reject noises and mislabelled data.

The same effect can also be observed with the model complexity denoted by the square of the norm of all the weights, $||w||^2$. Consider that a neuron $g(x)$ is added on top of a neural network $f(x)$. This adds model complexity, however, consider the norm of the weights of the added neurons is taken to be small. In such a case, the neurons fire to a small degree and only produces a small change in the output. The neuron partially contributes to representational capability or model complexity. Since the weight of the neuron is small, it can only partially distort $f(x)$ to produce $g(x)$ shown in Fig. 12.10. If the neuron can take high values of weights, it can sufficiently contribute to the model complexity and vice versa. Hence an aim of the training is also to reduce $||w||^2$, which is called *regularization*, because the reduction of $||w||^2$ will reduce the model complexity, which will reduce variance and increase bias and thus avoid overfitting. If the term is not taken, there is no way that the training realizes that it has over-fit the data as the training only has access to the training data.

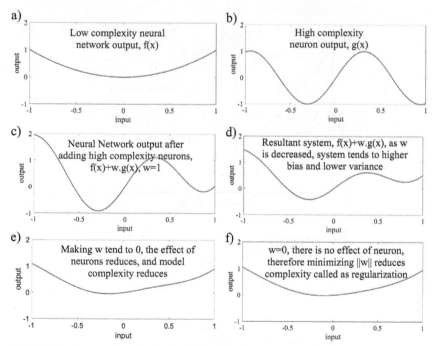

FIG. 12.10 Concept of regularization. Minimizing norm of weight $||w||$ avoids overfitting by increasing bias. (A) Low complexity neural network output. (B) High complexity neural network output. (C–F) Neural network output after adding high complexity neurons with weight 1 shown in (C), high weight shown in (D), low weight shown in (E), and zero weight shown in (F).

Consider that a few data samples get corrupted by *noise* that make a class label move a little in the input space, which mixes the two classes, and therefore drawing a clear decision boundary is neither possible nor desirable. It may be possible to still draw a decision boundary, however, will require too many turns making the boundary unsmooth which typically may mean a high validation and testing error. The learning algorithm should be able to figure out that there is noise and therefore should not over-learn and try to segregate the noise. In case of a regression problem as well, the noise makes the data move in the input space and therefore multiple readings around the same ideal point also appear as a noise distribution in the training data. A system that attempts to get a 0 error will also try to model this noise to best fit the data, thus over-training. On the contrary, a good learning algorithm should be able to see that the change in output values around close points is noise and only the genuine data should be learned and not the noise. Avoiding over-fitting avoids learning noise. The cases are shown in Fig. 12.9B.

A larger problem is *mislabelled data* for the classification problem. This occurs when a human annotator wrongly labels a class. This appears in the input space as a single instance of a class label that is far away from the other data of the same class. A learning algorithm should be able to figure out that the class is wrongly labelled and filter it out. This will happen when smoothness is enforced, and the curve is stopped to take very complex shapes. The same problem also appears in regression problems when the input is coming from a sensor, which suddenly breaks down and therefore gives random readings, which may be far away from the normal readings of the sensor. This appears as an *outlier* in the input space and the learning algorithm should detect this and avoid learning the outlier.

12.3.2   Back-propagation algorithm

Let us now discuss the training. First, let us design a loss function that would be minimized because of the training. The loss function is given by Eq. (12.13).

$$loss(w) = \frac{1}{2} \sum_i \sum_d \left(t_{i,d} - y_{i,d}(x_d, w)\right)^2 + \frac{1}{2}\gamma \|w\|^2 \tag{12.13}$$

Here $y_{i,d}$ is the output of the ith output neuron of the neural network as a function of the neural network weights (w) and the dth training input (x_d). The bias terms are already taken as additional weights. $t_{i,d}$ is the target available from the supervised training data for the input x_d for the ith output neuron. i iterates over all the output neurons. d iterates through all the data in the training data set only. The first term is called the *sum of square error*. The extra ½ is added which will aid in the optimization process. The second term is the *regularization parameter*, $\|w\|^2$ which avoids the network from over-fit. The parameter γ controls the contribution of regularization in the training process and is called the regularization constant.

The aim is now to adjust the weights in such a way that the loss function is minimized, which is an optimization problem. The problem is solved by *gradient descend*. Let the error space or weight space or parameter space be given in

Fig. 12.11A. The optimization or learning process is to find the global minima shown in Fig. 12.11A. Let us start with a random point in this space and keep moving till the error reduces. The gradient descent moves the solution in the direction that leads to the largest reduction in the error function (steepest slope), which is given by the gradient of the loss function. The step size of movement is proportional to the derivate value. Far away from the minima, it is preferable to take large steps to quickly come to the minima. Near the minima, large steps have the risk of over-stepping the minima and therefore the step size should be small. At the minima, the step size should be zero, as any step taken will deviate the weights from the locally best value. The learning rule is hence given by Eqs (12.14), (12.15).

$$\Delta w(t) = \eta \frac{\partial loss(w)}{\partial w} \tag{12.14}$$

$$w(t + 1) = w(t) + \eta \frac{\partial loss(w(t))}{\partial w(t)} \tag{12.15}$$

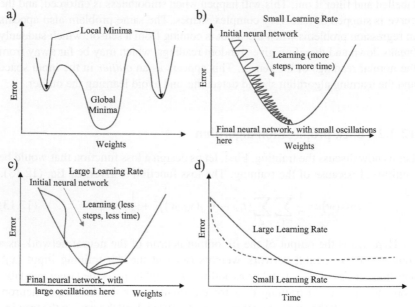

FIG. 12.11 Learning in Neural Networks (A) Error Plot (error v/s all weight values) (B)–(D) Effect of learning rate.

Here η is the *learning rate* which controls the proportionality constant of the step size or how fast the learning happens. A large learning rate can mean that the algorithm takes large step sizes and therefore learns fast, however is highly likely to step over the local optima and get to the other side. With a large learning rate, the algorithm will oscillate around the minima with a very large error, while the time needed to come to the oscillation stage will be very small. On the contrary, if the learning rate is small, the algorithm will take very small steps which will take a long time to reach the minima. The risk of overshooting is small, and the algorithm oscillates within a small error. The two cases are shown in Fig. 12.11B and C. The error plot with time is also shown in Fig. 12.11D. The learning rate is hence made adaptive called the *adaptive learning rate*, which typically starts with a high value and decays along with time.

It must be mentioned that the gradient descend is a *local search* algorithm only and the search is highly likely to get stuck at a local minimum. If the search process is stuck at the minima, there is nothing that the algorithm does to come out of the local minima and get to the global minima.

The application of the gradient descent technique to a multi-layer perceptron is called the *back-propagation algorithm*. The algorithm computes the error of the output neurons by comparing the output with the target. This error is propagated from the output layer to the input layer in a backward manner, going from a higher layer to a lower layer. Every layer receives the errors and uses the same to adjust the layer's weights. Thus, the resultant algorithm has two stages, a forward pass in which the input is applied to the input layer and the output is received at the output layer, and a backward pass in which the error is computed at the output layer and back propagated to the input layer. The conceptual diagram is shown in Fig. 12.12A.

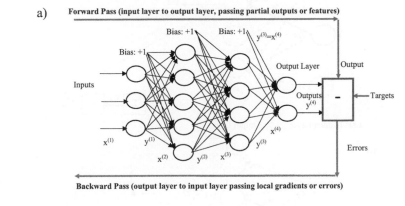

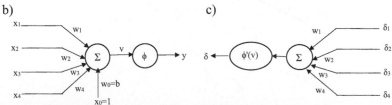

FIG. 12.12 Back-Propagation Algorithm. (A) The algorithm loops through a forward calculation of output, calculation of error, and backward calculation of local gradients (used to calculate weight corrections). Forward calculation (weighted addition and activation function) of a neuron is shown in (B) and correspondingly, the inverse calculation (weighted addition of local gradient, multiplication by a derivative of the activation function of the local field v computed in the forward pass) is shown in (C).

Let $w_{ij}(t)$ be a weight from a neuron j to a neuron i. The change of weight of the neuron by the same rule is given by Eqs (12.16), (12.17).

$$\Delta w_{ij}(t) = \eta \frac{\partial loss\big(w_{ij}(t)\big)}{\partial w_{ij}(t)} \tag{12.16}$$

$$w_{ij}(t+1) = w_{ij}(t) + \eta \frac{\partial loss(w(t))}{\partial w(t)} \tag{12.17}$$

The problem now is taking the derivative of the loss function which consists of the mean square error and the regularization parameter, given by Eq. (12.18).

$$\Delta w_{ij}(t) = \eta \frac{\partial loss\big(w_{ij}(t)\big)}{\partial w_{ij}(t)} = \eta \frac{\partial}{\partial w_{ij}(t)} \left(\frac{1}{2} \sum_k \sum_d e_{k,d}^2 \right) + \frac{1}{2}\eta\gamma \frac{\partial \|w\|^2}{\partial w_{ij}(t)} \tag{12.18}$$

Here $e_{k,d}$ is the error of the kth neuron for the dth data in the training data set. As a special case consider that the weight w_{ij} is from the last hidden layer to the output layer, in which case the error $e_{k,d}$ is given by Eq. (12.19). The notations are shown in Fig. 12.13A.

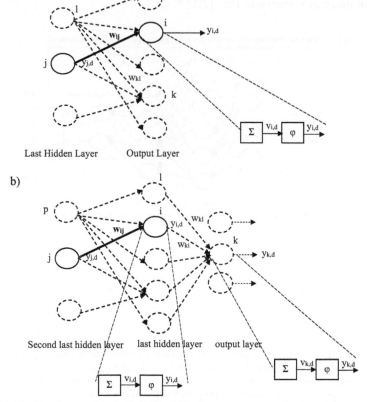

FIG. 12.13 Notations used in the derivation of the Back-Propagation Algorithm. (A) Last hidden layer to output layer weights. (B) Other layer weights.

$$e_{k,d}^2 = \left(t_{k,d} - y_{k,d}(x_d, w)\right)^2 \tag{12.19}$$

Here the subscript d stands for the training data set. The error is the square error over the dth data for the ith output neuron. The derivation uses the chain rule of differentiation. The weight w_{ij} only affects the output neuron i and no other neuron and therefore $\partial e_{k\neq i,d}/\partial w_{ij}=0$. Further, solving for this specific error function for the specific case of w_{ij} being a weight from the last hidden layer to the output layer, gives Eq. (12.20), obtained from Eqs (12.18), (12.19).

$$
\begin{aligned}
\Delta w_{ij}(t) &= \eta \frac{\partial}{\partial w_{ij}(t)} \left(\frac{1}{2} \sum_d e_{i,d}^2 \right) + \frac{1}{2} \eta \gamma \frac{\partial \|w\|^2}{\partial w_{ij}(t)} \\
&= \eta \sum_d \frac{1}{2} \frac{\partial}{\partial w_{ij}(t)} e_{i,d}^2 + \eta \gamma w_{ij}(t) \\
&= \sum_d \left(\eta e_{i,d} \frac{\partial}{\partial w_{ij}(t)} e_{i,d} \right) + \eta \gamma w_{ij}(t) \\
&= \sum_d \left(\eta e_{i,d} \frac{\partial}{\partial w_{ij}(t)} \left(t_{k,d} - y_{i,d}(x_d, w) \right) \right) + \eta \gamma w_{ij}(t) \\
&= \sum_d \left(-\eta e_{i,d} \frac{\partial}{\partial w_{ij}(t)} \left(y_{i,d}(x_d, w) \right) \right) + \eta \gamma w_{ij}(t)
\end{aligned}
$$

$$\tag{12.20}$$

The output $y_{i,d}$ is given by Eqs (12.21), (12.22).

$$y_{i,d}(x_d, w) = \phi(v_i) \tag{12.21}$$

$$y_{i,d}(x_d, w) = \phi\left(\sum_l w_{il} y_{l,d} \right) \tag{12.22}$$

Here w_{il} is the weight from neuron l to neuron i (which is assumed to be the output layer). y_{ld} is the output of neuron l for the data item d. Taking the derivative $\partial y_{i,d}/\partial w_{ij}$ as $\partial y_{i,d}/\partial v_i \, \partial v_i/\partial w_{ij}$ using the chain rule gives Eq. (12.23).

$$\Delta w_{ij}(t) = \sum_d \left(-\eta e_{i,d} \phi'(v_i) \frac{\partial}{\partial w_{ij}(t)} \left(\sum_l w_{il} y_{l,d} \right) \right) + \eta \gamma w_{ij}(t) \tag{12.23}$$

Here $\phi'(v)$ is the derivative of $\phi(v)$ with respect to v. The partial derivative of all weights w_{il} is zero for all l except $l=j$, because the weight w_{ij} only affects itself. This gives Eq. (12.24).

$$
\begin{aligned}
\Delta w_{ij}(t) &= \sum_d \left(-\eta e_{i,d} \phi'(v_i) \frac{\partial}{\partial w_{ij}(t)} \left(w_{ij} y_{j,d} \right) \right) + \eta \gamma w_{ij}(t) \\
&= \sum_d \left(-\eta e_{i,d} \phi'(v_i) y_{j,d} \right) + \eta \gamma w_{ij}(t)
\end{aligned}
$$

$$\tag{12.24}$$

Here the term $\delta_{i,d} = e_{i,d}\phi'(v_i)$ is called the *local gradient* denoted as $\delta_{i,d}$. The term stores the local gradient associated with the ith neuron, which is supplied to all the neurons of the preceding layer as an intermediate result. The local gradients are propagated back as the error terms which gives the computational benefit to the algorithm. Without the local gradients, all the weights would be affected by each other, and computing the new updated value of every weight would result in a lot of computation. The local gradient stores partial reusable results to reduce computation. The local gradient is technically defined as the partial derivative of the error term with respect to the local induced field for the particular neuron ($\delta_{i,d} = \partial e_{i,d}^2/\partial v_i$). The definition is such that the weight correction (without regularization) becomes a product of the local gradient ($\delta_{i,d}$) with the learning rate (η) and the input to the particular neuron ($y_{j,d}$) from the weight w_{ij} in the backward pass. It will be worth verifying that if the gradient of the error with respect to the weight ($\partial e_{i,d}^2/\partial w_{ij}$) needs to be the local gradient ($\delta_{i,d}$) times the input to the neuron (y_j) from the weight w_{ij} as per the aim, the local gradient (δ_i) should be the derivative of the error with respect to the local-induced field ($\partial e_{i,d}^2/\partial v_i$). The term is defined as Eq. (12.25).

$$\delta_{i,d} = \frac{\partial e_{i,d}^2}{\partial v_i} = e_{i,d}\phi'(v_i) \tag{12.25}$$

Thus, re-deriving the gradient and using the definition of local gradient, for an output neuron the changes in weight are given by Eqs (12.26), (12.27).

$$\Delta w_{ij}(t) = \frac{1}{2}\sum_d \frac{\partial e_{i,d}^2}{\partial w_{ij}} + \eta\gamma w_{ij}(t) = \sum_d \frac{1}{2}\frac{\partial e_{i,d}^2}{\partial v_i}\frac{\partial v_i}{\partial w_{ij}} + \eta\gamma w_{ij}(t) \tag{12.26}$$

$$\Delta w_{ij}(t) = \sum_d \left(-\eta\delta_{i,d}y_{j,d}\right) + \eta\gamma w_{ij}(t) \tag{12.27}$$

The harder case is to compute the weight adjustment for the hidden layer neurons. Consider a weight $w_{ij}(t)$ from a neuron j at the second last hidden layer to a neuron i at the last hidden layer. The computations are hard to do because the target of the neuron is not known as the earlier case and therefore the error cannot be directly computed. The notations are shown in Fig. 12.13B.

The weight change is again the derivative of the loss function given by Eq. (12.28). The derivation is the same with the difference that now every output neuron is affected by the weight $w_{ij}(t)$ and therefore all the output neurons are considered. A change in $w_{ij}(t)$ affects the output of neuron i, which is the last hidden layer neuron, whose output is given to all the output neurons by the corresponding weights.

$$\Delta w_{ij}(t) = \eta \frac{\partial loss\left(w_{ij}(t)\right)}{\partial w_{ij}(t)} = \eta \frac{\partial}{\partial w_{ij}(t)} \left(\frac{1}{2} \sum_{k} \sum_{d} e_{k,d}^2\right) + \frac{1}{2}\eta\gamma \frac{\partial \|w\|^2}{\partial w_{ij}(t)}$$

$$= \sum_{d} \left(\eta \sum_{k} \frac{1}{2} \frac{\partial}{\partial w_{ij}(t)} e_{k,d}^2\right) + \eta\gamma w_{ij}(t)$$

$$= \sum_{d} \left(\eta \sum_{k} e_{k,d} \frac{\partial}{\partial w_{ij}(t)} e_{k,d}\right) + \eta\gamma w_{ij}(t)$$

$$= \sum_{d} \left(\eta \sum_{k} e_{k,d} \frac{\partial}{\partial w_{ij}(t)} \left(t_{k,d} - y_{k,d}(x_d, w)\right)\right) + \eta\gamma w_{ij}(t)$$

$$= \sum_{d} \left(-\eta \sum_{k} e_{k,d} \frac{\partial}{\partial w_{ij}(t)} \left(y_{k,d}(x_d, w)\right)\right) + \eta\gamma w_{ij}(t)$$

$$(12.28)$$

The output $y_{k,d}$ of the output layer neuron k is given by Eqs (12.29), (12.30).

$$y_{k,d}(x_d, w) = \phi(v_k) \tag{12.29}$$

$$y_{k,d}(x_d, w) = \phi\left(\sum_{l} w_{kl} y_{l,d}\right) \tag{12.30}$$

Again taking $\partial y_{k,d}/\partial w_{ij}$ as $\partial y_{k,d}/\partial v_k \, \partial v_k/\partial w_{ij}$ using the chain rule gives Eq. (12.31).

$$\Delta w_{ij}(t) = \sum_{d} \left(-\eta \sum_{k} e_{k,d}\phi'(v) \frac{\partial}{\partial w_{ij}(t)} \left(\sum_{l} w_{kl} y_{l,d}\right)\right) + \eta\gamma w_{ij}(t) \tag{12.31}$$

The output of neuron k is affected by all incoming weights w_{kl}. However, the weight w_{ij} only affects the output of the neuron i, which in turn only affects the weights w_{ki} and any weight $w_{kl,l\neq i}$, therefore, has a partial derivative of 0 with respect to w_{ij}. Incorporating the relation gives Eq. (12.32).

$$\Delta w_{ij}(t) = \sum_{d} \left(-\eta \sum_{k} e_{k,d}\phi'(v) \frac{\partial}{\partial w_{ij}(t)} \left(w_{ki} y_{i,d}\right)\right) + \eta\gamma w_{ij}(t)$$

$$= \sum_{d} \left(-\eta \sum_{k} e_{k,d}\phi'(v) w_{ki} \frac{\partial}{\partial w_{ij}(t)} y_{i,d}\right) + \eta\gamma w_{ij}(t) \tag{12.32}$$

$$= \sum_{d} \left(-\eta \frac{\partial y_{i,d}}{\partial w_{ij}(t)} \sum_{k} w_{ki} e_{k,d}\phi'(v)\right) + \eta\gamma w_{ij}(t)$$

Taking the definition of the induced local field from Eq. (12.25), gives Eq. (12.33).

$$\Delta w_{ij}(t) = \sum_{d} \left(-\eta \frac{\partial y_{i,d}}{\partial w_{ij}(t)} \sum_{k} w_{ki} \delta_{k,d}\right) + \eta\gamma w_{ij}(t) \tag{12.33}$$

The term $\partial y_{i,d}/\partial w_{ij}$ is familiar from the same case as the output neuron and is obtained with the knowledge that y_{id} is given by Eqs (12.34), (12.35). The notations are shown in Fig. 12.13.

$$y_{i,d} = \phi(v_i) \tag{12.34}$$

$$y_{i,d} = \phi\left(\sum_p w_{ip} y_{p,d}\right) \tag{12.35}$$

Taking derivative by the chain rule gives Eq. (12.36).

$$\Delta w_{ij}(t) = \sum_d \left(-\eta\varphi'(v_i)\frac{\partial}{\partial w_{ij}(t)}\left(\sum_p w_{ip}y_{p,d}\right)\sum_k w_{ki}\delta_{k,d}\right) + \eta\gamma w_{ij}(t) \tag{12.36}$$

Out of all weights w_{kp}, the only non-zero partial derivative weight will be of w_{ij} as previously, given by Eq. (12.37).

$$\Delta w_{ij}(t) = \sum_d \left(-\eta\varphi'(v_i)\frac{\partial}{\partial w_{ij}(t)}\left(w_{ij}y_{j,d}\right)\sum_k w_{ki}\delta_{k,d}\right) + \eta\gamma w_{ij}(t)$$

$$= \sum_d \left(-\eta y_{j,d}\varphi'(v_i)\sum_k w_{ki}\delta_{k,d}\right) + \eta\gamma w_{ij}(t) \tag{12.37}$$

Using the definition of a local gradient gives Eq. (12.38).

$$\delta_{i,d} = \varphi'(v_i)\sum_k w_{ki}\delta_{k,d} \tag{12.38}$$

This leads to a much simpler relation of weight change of the hidden layer neurons as shown in Eq. (12.39). Not surprisingly, this relation is the same as that of the output layer neuron weights or rather any other weights, which is because of the manner of defining the local gradient.

$$\Delta w_{ij}(t) = \sum_d \left(-\eta\delta_{i,d} y_{j,d}\right) + \eta\gamma w_{ij}(t) \tag{12.39}$$

This shows the mechanism of back propagating an error from a forward layer to the preceding layer. In the forward propagation, every neuron gives its output to the neurons of the forward layer which get multiplied by the corresponding weights. This can be visualized as an outward flow of information simultaneously to many neurons of the forward layer. The backward phase is exactly the opposite wherein the local gradient or δ values from a forward layer come back to the previous layer neuron, visualized as the exact opposite flow of data. This is shown in Fig. 12.12B and C. The local gradient values get multiplied by the same weights when they come and reach the neuron. The values are multiplied additionally by the derivative of the activation function on the local induced field of the forward propagation.

There is an interesting trick with the derivative of the activation function $\phi'(v)$. Consider that the activation function is sigmoidal, given by Eq. (12.40), for a constant a.

$$y = \phi(v) = \frac{1}{1 + e^{-av}} \tag{12.40}$$

The derivative of the sigmoidal function is given by Eq. (12.41).

$$\phi'(v) = \frac{ae^{-av}}{(1 + e^{-av})^2}$$

$$= a.\frac{1}{1 + e^{-v}}\left(1 - \frac{1}{1 + e^{-v}}\right) \tag{12.41}$$

$$= \phi'(v) = a.v(1 - v)$$

This gives a simple term to express the derivative of the sigmoid activation function. Plugging this relation in Eq. (12.25) and Eq. (12.38) gives Eqs (12.42), (12.43).

$$\delta_{i,d} = \begin{cases} ae_{i,d}y_{i,d}(1 - y_{i,d}) = a(t_{i,d} - y_{i,d})y_{i,d}(1 - y_{i,d}) & \text{if } i \text{ is the output neuron} \\ ay_{i,d}(1 - y_{i,d})\sum_k w_{ki}\delta_{k,d} & \text{if } i \text{ is not the output neuron} \end{cases}$$

$$\tag{12.42}$$

$$\Delta w_{ij}(t) = \sum_d \left(-\eta \delta_{i,d} y_{j,d}\right) + \eta \gamma w_{ij}(t) \tag{12.43}$$

Similarly, for the hyperbolic tangent function, given by Eq. (12.44), the derivative is given by Eq. (12.45).

$$y = \phi(v) = a \tanh(bv) \tag{12.44}$$

$$\phi'(v) = ab \sec^2(bv)$$

$$= ab(1 - \tan^2(bv))$$

$$= ab(1 - \tan(bv))(1 + \tan(bv)) \tag{12.45}$$

$$= ab(1 - y/a)(1 + y/a)$$

$$= (b/a)(a - y)(a + y)$$

This also gives a simple term to express the derivative. Plugging this relation in Eqs (12.25), (12.38) gives Eqs (12.46), (12.47).

$$\delta_{i,d} = \begin{cases} (b/a)e_{i,d}\left(a - y_{i,d}\right)\left(a + y_{i,d}\right) = (b/a)\left(t_{i,d} - y_{i,d}\right)\left(a - y_{i,d}\right)\left(a + y_{i,d}\right) & \text{if } i \text{ is the output neuron} \\ (b/a)\left(a - y_{i,d}\right)\left(a + y_{i,d}\right)\sum_k w_{ki}\delta_{k,d} & \text{if } i \text{ is not the output neuron} \end{cases}$$

$$\tag{12.46}$$

$$\Delta w_{ij}(t) = \sum_d \left(-\eta \delta_{i,d} y_{j,d}\right) + \eta \gamma w_{ij}(t) \tag{12.47}$$

12.3.3 Momentum

A problem with the back-propagation algorithm is slow training. To accelerate the training, a term called *momentum* is added. The momentum tries to move the weight search in the same direction as that of the previous iteration. Consider that the current weight correction is negative, while the erstwhile weight correction was also negative, and the addition becomes more negative that accelerates the rate of decrease of weight or accelerates learning. Similarly, for positive weight corrections.

On the contrary, consider that the search process is near the minima and momentum adds to the step size thus resulting in the search overshooting the minima. Say that the weight reduction was going on and in the next iteration, the search has a positive weight correction. However, the previous weight correction was negative, which means the momentum tries to resist a weight change by applying acceleration in the opposite direction. Now momentum slows down the search or reduces the step size. As the weights oscillate around minima, momentum continuously resists the change by a reduction of step size that enables converge deeper into the minima.

Thus, momentum accelerates motion towards the minima if there is no change in the sign of the motion direction, while the momentum retards the motion by reduction of step size if the weight correction direction constantly changes signs shown in Fig. 12.14A. The application of momentum is shown by Eq. (12.48).

$$w_{ij}(t + 1) = w_{ij}(t) + \Delta w_{ij}(t) + \alpha \Delta w_{ij}(t - 1) \tag{12.48}$$

α is the proportionality constant that determines the contribution of momentum in the learning process.

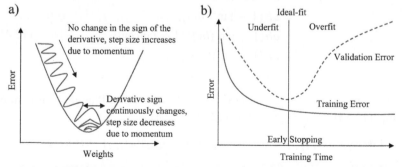

FIG. 12.14 (A) Effect of Momentum in Learning. Momentum adds a proportion of the last weight change to the current weight change. As a result, learning accelerates till the sign of weight change remains the same and decelerates when the sign of weight change continuously changes. (B) Early Stopping. The training is stopped when the validation error starts to increase, even if the training error is reducing, to stop overfitting. Weights increase from small initial values to large values, thus tending to overfit.

12.3.4 Convergence and stopping criterion

One iteration of learning characterized by the variable t is called an epoch. Along with time, the error becomes constant, and the learning is said to have *converged*. The *stopping criterion* determines when the training should be stopped. The neural networks are initially assigned with small random weights starting from a high bias and low variance condition. The weights must not be all zero and not all the same, in which case the neurons in each layer maintain symmetry, learn the same weight values, learn the same latent features, and hence are equivalent to a single neuron only. As time goes on, the model complexity increases due to the increase in the weight values and therefore the training error reduces. A stopping criterion called *early stopping* is typically used to avoid overfitting as the model learns. As the model increases, its complexity and learning starts from an under-fit to an ideal fit, the validation error reduces. However, as the network goes towards over-fit, the validation error starts to increase. The learning is stopped at this point when the validation error starts to increase. The effect is shown in Fig. 12.14B.

12.3.5 Normalization

Another visualization of the weight space is to draw it as contours. The contours are a convenient way to visualize 3D data in 2D. Consider the loss function because of two weights drawn in Fig. 12.15. The contours show the points

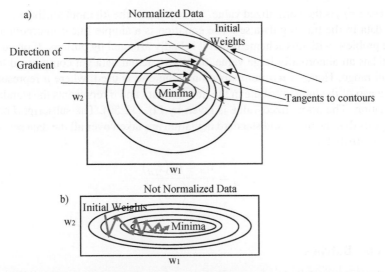

FIG. 12.15 Effect of normalization. (A) shows the error plot as a contour with normalized weights. The derivatives nicely point to the minima requiring fewer steps. (B) shows the same function with unnormalized weights. The contours are highly elliptical, and the derivatives are sensitive, requiring significantly more time.

which have the same loss function joined by curves. The spacing between the curves denotes the slope of the loss function curve. The weights of the neural network at any point in time represent a point in this space. The gradient of steepest descend is orthogonal to the contour curves. The algorithm works by taking steps in the steepest direction and thus proceeds towards the minima as shown in Fig. 12.15. If the inputs and outputs are normalized between -1 and 1, the weights typically lie between -1 and 1 and the activation functions are most sensitive within this range.

A particular problem happens when there is no normalization. Suppose the data associated with the weight w_1 are of a high range (say of the order of 10^5) as compared to the data associated with weight w_2 (say of the order of 10^{-5}). In such a case, w_1 should ideally take a small value and w_2 should take a very large value, otherwise, in a weighted addition the first input will dominate. Now, the slope is sensitive, and the algorithm will end up taking a greater number of steps and thus the training will take a long time. In another sense, the weights are initialized to small random values. It will take a long time for w_1 to continuously increase and for w_2 to continuously decrease.

There are two common techniques to normalize the input. If a strict criterion of -1 to 1 needs to be followed, the max-*min normalization* is the best technique. The normalized variable value is given by Eq. (12.49)

$$x'_{i,d} = -1 + 2\frac{x_{i,d} - min_d(x_{i,d})}{max_d(x_{i,d}) - min_d(x_{i,d})} \tag{12.49}$$

Here $x'_{i,d}$ is the normalized value. $x_{i,d}$ represents the ith input attribute of the dth data in the training data set. The equation is a simple linear interpolation. The problem with this scheme is that it is badly affected by outliers. If one data item has an abnormally large value, the entire actual data get compressed to a small range. Hence a $\mu\,\sigma$ normalization technique is used, where μ represents the mean of the input attribute in the training data and σ represents the standard deviation. The normalized value is given by Eq. (12.50). The subscript d only suggests that the mean and standard deviation are taken over all the data sets for the ith attribute.

$$x'_{i,d} = \frac{x_{i,d} - \mu_d(x_{i,d})}{\sigma_d(x_{i,d})} \tag{12.50}$$

12.3.6   Batches

The learning modelled so far is called *batch learning*, in which a batch is made consisting of all the items of the training data set and the error of the entire batch is used to compute the correction in weights. So, every step is applied only after computing the error on all items in the batch, which can take too long on large data sets. However, there is one loss function, and all steps lead towards the minima. At the same time, it is highly likely to converge into the local minima rather than the global minima.

The converse learning philosophy is called *sequential learning*, wherein one data instance is applied, which is a small calculation. One step error function is measured, and weight corrections are applied on one instance of data. At the next iteration, a new data instance from the training data is chosen. The learning is fast. However, every data have their target and hence every iteration is a completely new error function. However, learning does happen as different data items are a snapshot of the same problem. The other major advantage of this update rule is that the chances of escaping the local minima and going towards the global minima are high since some data items may model the problem differently and take a step away from the local minima and move towards the global minima.

Mini-batch learning takes advantage of both techniques. Here the learning is done in batches, however, the batch size is not the complete training data size but is much smaller as specified by the user. The batch size is such that it is reasonably fast to compute the error of the batch, and the batch is large enough to represent the complete problem. Randomly sampled training data make a mini-batch. Thus, every iteration is with a new mini batch. For this reason, the resultant training algorithm is also called the *Stochastic Gradient Descend*. Unlike sequential learning, the error function is always close to the ideal complete batch error function, and thus the algorithm largely proceeds towards its minima; and unlike batch learning, it does not take too long to take one step. Because of a random sampling of mini batches, there is some possibility to avoid the local minima. The different methods are summarized in Table 12.2. Note that increasing learning rate does not solve the slow learning problem as it may largely always overshoot the minima. It is better to use smaller data to estimate the gradient for faster results while sticking to a smaller learning rate.

TABLE 12.2 Different learning methods in terms of batch size.

Property	Batch learning	Sequential learning	Mini batches
Working	Every iteration works on the mean square error of the entire data set	Every iteration works on mean square error of 1 data instance only	Every iteration works on the mean square error of some sampled data instances
Computation time per iteration	Highest	Least	Moderate

Continued

TABLE 12.2 Different learning methods in terms of batch size—cont'd

Property	Batch learning	Sequential learning	Mini batches
Loss Function	Every iteration has a single loss function	Loss function per iteration can change dramatically due to change in data	Reasonably stable estimate of the loss function, some changes per iteration
Possibility to get stuck at local optima	Highest	Least	Moderate
Number of iterations for convergence	Least	Highest	Moderate

12.3.7 Cross-validation

Consider the problem where there is limited overall data available. To solve the problem, the data need to be split into training, validation, and testing parts, leaving extremely small pieces for each part. Further, the resolution of validation and testing will be poor to make decisions. If there are only 10 validation data items for binary classification, you only get classification accuracy in multiples of 10%. Further, some data will be easy, and some will be difficult, and the performance largely depends upon whether the difficult examples go into the training data or the validation data. The testing data are a problem requirement; however, the data need to be shared between training and validation, which is a problem with limited size data sets.

The *cross-validation* scheme divides the data into multiple parts and uses some parts for training, which are cross-validated by the other parts used as the validation data. A shuffling of data ensures that the same data which act as the training part are later used for cross-validation and vice versa. The most common technique for cross-validation is called the *k-fold cross-validation*. In this technique, the training data (including validation data) are divided into k-parts, where k is an algorithm parameter. k-1 parts are used for training, and 1 part is used for validation. The process is repeated in the entirety of k-times with the validation part changed at every iteration and the rest of the parts used for training. The training and validation errors are averaged for the k-runs. It is imperative that based on the validation data, the architecture or training parameters would be changed and the process repeated. The concept is shown in Fig. 12.16.

Consider the extreme case where $k = n$, where n is the number of training data items, or every part has just 1 data item. At the first iteration, the machine

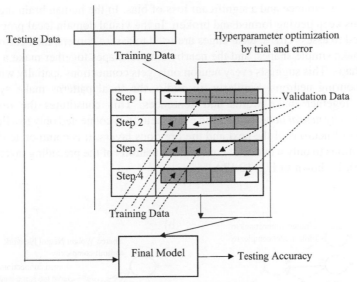

FIG. 12.16 Cross-validation: The training data are divided into k-folds, 1 used for validation, and the other $k-1$ for training. The training happens k times and the results are averaged. Effectively k chunks are used for validation and $k-1$ for training, dissimilating the need of dividing a small data set into training and validation.

is trained on $k-1=n-1$ parts with $n-1$ data items overall. This is the maximum training data size possible. The validation data are however 1 item. This process is repeated $k=n$ times and the training and validation errors are averaged out. Overall, the validation data size becomes n, while every training happens with $n-1$ training data items. This stops the reduction of data items because of the division between training and validation sets. While the scheme looks promising, there is another problem. Consider T as the time for training the machine once. In the previous case, every trial and error on the architectural and learning parameters could happen in T time. Now the complete cycle takes n rounds (in general k rounds) and hence nT time (in general kT time). This is a significant time and therefore the computational time is large. For this reason, the value of k is usually limited. The value should be large enough to give enough training examples for the machine to learn, while small enough to facilitate a conveniently small training time. A typical value of $k=10$ is chosen for most average-sized data sets.

12.4 Limited connectivity and shared weight neural networks

The neural networks discussed so far had a *fully connected* architecture, wherein every neuron of a layer is connected to all neurons of the previous layer and all neurons of the next layer. This creates a problem as even if a single neuron is added, the model complexity significantly increases, leading to a significant increase in variance and a significant loss of bias. In the human brain the connections keep getting formed and broken. In the visual domain local properties are used to assess edges, local edges are used to assess corners, the nearby corners make simple shapes, and the nearby simple shapes together make a complex shape. This suggests every neuron only gets connections spatially with the neighbouring neurons. Similarly, in audio, the local patterns make syllabi, which together make words and sentences. This constitutes the *limited-connectivity neural network* wherein every neuron is connected only to a limited number of neurons of the next and the previous layers. It is common to define connections to only a bunch of consecutive neurons of the preceding layer. The network is shown in Fig. 12.17A.

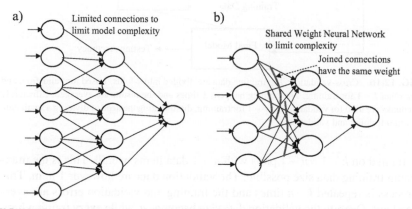

FIG. 12.17 (A) Limited Connectivity Neural Network. Every neuron of the hidden layer is locally connected to only 3 neurons of the preceding layer. This limits complexity. (B) Shared weight neural network. The same trained feature mapping represented by the weights is used spatially multiple times.

Another mechanism is called the *weight-sharing neural network*. The same image filter can find the corner in any part of the image, while the same signal filter can find syllabi at any part of the speech signal. The interesting patterns of the input can be detected by any template among all the input. This motivates the need to have neurons that share weights. The same filter when applied to two parts of the image are interpreted as two different neurons because of the spatial difference, but with the same weight because they signify the same filter. Two connections sharing a weight means that the numerical value of the weight shall be the same for the two weights. Hence any update on one weight would automatically update the other weight. The network is shown in Fig. 12.17B.

12.5 Recurrent neural networks

The neural network discussed so far was a directed acyclic graph, meaning that there were no cycles. So the output was only governed by the current input. Imagine a network that has direct connections to neurons in the preceding layer or the same layer. This forms cycles in the neural network. Such a neural network is called the *recurrent neural network*. Because of the cycles formed, the output of the recurrent neural network not only depends upon the current input but also on the previous inputs and outputs of the neurons. Consider the neural network shown in Fig. 12.18A. Suppose the input applied is [[0.5 –0.5], [0.8 0.3]] for the two neurons for 2 iterations. The internal state is taken as 0 for all neurons. The output of the network for the first two-time steps are shown as follows using the sigmoidal activation function:

External inputs at $t=1$, [0.5 –0.5]

$$\text{Weight } w^{(1)} = \begin{bmatrix} w_{c,bias} & w_{d,bias} \\ w_{ca} & w_{da} \\ w_{cb} & w_{db} \\ w_{cc} & w_{dc} \\ w_{cd} & w_{dd} \end{bmatrix} = \begin{bmatrix} 0.5 & 0.1 \\ -0.3 & 0.8 \\ 0.2 & -0.5 \\ 0 & 0.6 \\ 0.2 & 0 \end{bmatrix}$$

$$\text{Weight } w^{(2)} = \begin{bmatrix} w_{e,bias} \\ w_{ec} \\ w_{ed} \end{bmatrix} = \begin{bmatrix} -0.3 \\ -0.1 \\ 0.2 \end{bmatrix}$$

Inputs to first hidden layer at $t=1$, $x^{(1)}(1) = [1\ 0.5 -0.5\ 0\ 0]^T$ (bias, a, b, c, d)

Output of first hidden layer at $t=1$, $y^{(1)}(1) = \phi\left(\left(w^{(1)}\right)^T x^{(1)}(1)\right) =$

$$\phi\left(\begin{bmatrix} 0.5 & 0.1 \\ -0.3 & 0.8 \\ 0.2 & -0.5 \\ 0 & 0.6 \\ 0.2 & 0 \end{bmatrix}^T \begin{bmatrix} 1 \\ 0.5 \\ -0.5 \\ 0 \\ 0 \end{bmatrix}\right) = \phi\left(\begin{bmatrix} 0.25 \\ 0.75 \end{bmatrix}\right) = \begin{bmatrix} 0.5622 \\ 0.6792 \end{bmatrix}$$

Output of output layer at $t=1$, $y^{(2)}(1) = \phi\left(\left(w^{(2)}\right)^T x^{(2)}(1)\right) =$

$$\phi\left(\begin{bmatrix} -0.3 \\ -0.1 \\ 0.2 \end{bmatrix}^T \begin{bmatrix} 1 \\ 0.5622 \\ 0.6792 \end{bmatrix}\right) = \phi(-0.2204) = 0.4451$$

External Input at $t=2$, [0.8 0.3]

Inputs to first hidden layer at $t=2$, $x^{(1)}(2) = [1\ 0.8\ 0.3\ 0.5622\ 0.6792]^T$ (bias, a, b, c, d)

Output of first hidden layer at $t=1$, $y^{(1)}(2) = \phi\left(\left(w^{(1)}\right)^T x^{(1)}(2)\right) =$

$$\phi\left(\begin{bmatrix} 0.5 & 0.1 \\ -0.3 & 0.8 \\ 0.2 & -0.5 \\ 0 & 0.6 \\ 0.2 & 0 \end{bmatrix}^T \begin{bmatrix} 1 \\ 0.8 \\ 0.3 \\ 0.5622 \\ 0.6792 \end{bmatrix}\right) = \phi\left(\begin{bmatrix} 0.4558 \\ 0.9273 \end{bmatrix}\right) = \begin{bmatrix} 0.6120 \\ 0.7165 \end{bmatrix}$$

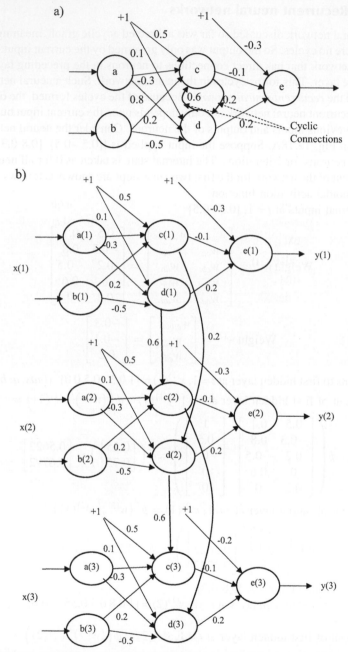

FIG. 12.18   Recurrent Neural Network (A) An example network (B) Unfolding with time. Unfolding is done for 3-time steps only and the unfolded network represents a shared weight network. The corresponding weights are shared. The clamping between the shared weights is not shown.

Output of output layer at $t=2$, $y^{(2)}(2) = \phi\left(\left(w^{(2)}\right)^T x^{(2)}(2)\right) =$

$$\phi\left(\begin{bmatrix} -0.3 \\ -0.1 \\ 0.2 \end{bmatrix}^T \begin{bmatrix} 1 \\ 0.6120 \\ 0.7165 \end{bmatrix}\right) = \phi(-0.2179) = 0.4457$$

In this manner, the recurrent neural networks can take in *sequences as input* and generate *sequences as output*. Whenever an input is applied an output is supplied. When the next input is applied, the output of the hidden neurons in the previous time step is also taken as additional (recurrent) inputs. The output is thus dependent on the first input as well as the last output of hidden neurons. More generally, the network decisions are based on all past inputs, which gives the ability to learn and deal with sequential data. This forms the basis of human decision making as well. If you see a person just ahead of you, you may decide to overcome the person by going through the right of the person. However, if you have observed the person long enough, you may know that the person is walking sideways and will therefore automatically clear by the time you get near. Hence you walk straight instead. The humans use the past percept to guess the intent of the dynamic obstacles or people around, as well as to get the latent information of their motion.

The recurrent neural networks are particularly important in robotics as the robotic decisions are sequential. Consider that the robot can sense the distances (and angles) from the obstacles around and the distance (and angle) from the goal. The recurrent neural networks can model the derivative of the distances which gives the speed towards goal and the relative speed of the robot with respect to the obstacles, and similarly model the double derivatives for accelerations. The recurrent neural networks represent a highly dynamic network that can model higher-order derivatives of any order. The modelling will though be limited by the model complexity.

The problem now is learning. The recurrent connections make it difficult to derive the learning rule for the recurrent neural networks. The recurrent neural networks learn an input sequence to produce an output sequence. For simplicity let the input sequence be $x(1)\ x(2)\ x(3)\ x(4)\ x(5)$, where $x(i)$ is not a real number but a vector of inputs given to the network at the time i. Let the corresponding target sequence be $t(1)\ t(2)\ t(3)\ t(4)\ t(5)$. For simplicity let the neural network have just 1 hidden layer which has cyclic connections, along with the input layer and the output layer. The output at the time i is given by the inputs at time 1 to i. The network for the sequence can thus be represented by a 5-layer network by *unfolding in time*, where each layer has input at the time i, produces output at the time i. The weights between input and hidden layers and the ones between the hidden layer and output layer are the same when unfolded with time. The cyclic weights in the hidden layer are represented as weights between the $i-1$th layer representing time $i-1$ and ith layer representing time i, for every such time. Physically there is only one weight which has been shown multiple times as a representation in unfolding. The weights are not independent but need to be the same which is the same concept as the *shared weight neural network*.

The learning is now a simple application of the back-propagation algorithm in a multi-layered manner with the difference that since the weights are shared, the updates are applied simultaneously to all the shared connections to ensure the same value. Such an application of the back-propagation algorithm is known as *Back-propagation with time*. A normal implementation of the back-propagation algorithm only iterates from a forward layer to the previous layer. However, in this implementation, the algorithm is iterating with time, going from a larger time instance to a lower time instance which was symbolically depicted as going to the previous layers by the notion of unfolding. The unfolding is shown in Fig. 12.18B.

12.6 Deep learning

The neural networks with a single hidden layer and a limited number of neurons can work with small data sets, learn in small computation times, and avoid the risk of overfitting. Deep learning is a technique that has revolutionized numerous application domains including robot navigation. *Deep Learning* (Goodfellow et al., 2016; LeCun et al., 2015; Schmidhuber, 2015) lets the neural network have deep architectures with many hidden layers and many neurons in these layers, and therefore the conventional neural networks are termed as the shallow networks. The technology is driven by the dramatic increase in the data for any learning problem, implying that the *data set size limitations* may no longer be valid. Specifically, in robotics, a robot operating manually generates data at the rate of tens to hundreds of hertz depending upon the sensor. Therefore, manually driving a robot over a few days of human driving can mean a very large amount of data amounting to terabytes with billions of records. The *processing speed* is also no longer a major concern with many companies providing cloud solutions and powerful GPUs that process data in parallel. The *risk of over-fitting* is eliminated because of the availability of a massively large data.

Traditionally, *feature engineering* techniques were used for reducing high-dimensional input data into low-dimensional features without much of a loss of information. The features are selected by humans and humans tend to take not the best decisions owing to numerous biases. Simultaneously making efforts towards turning the architecture, tuning the parameters, and selecting the correct set of features is not possible and humans tend to select sub-optimal features as a result. It may even be possible that the best feature for a specific problem is not yet discovered. Because there is an option to have numerous neurons, it is not required to do feature-engineering. The deep learning system can take all the inputs as a high-dimensional input vector and discover interesting meta-features by looking at the training data, without requiring human intervention.

The Back-Propagation algorithm is used which is prone to get stuck at *local minima*. However, through experiments, it is observed that using any deep learning technique and a stochastic gradient descend (wherein mini-batches are made by random sampling of data), the local minima is not a problem given a large volume of quality data. In other terms, the data quality and quantity matter more than the local minima. It is again observed that the data quality and quantity are superior to the *architectural parameters* as well. Hence, the architecture is kept fixed, as available from the best practices and the computational availability, and not set by trial and error.

12.7 Auto-encoders

The first architecture of deep neural networks is called *auto-encoder*. The aim is to make a machine learning system that will do feature extraction on its own, that is to convert the high-dimensional input into a low-dimensional output by learning.

Suppose a high-dimensional input vector (x) is reduced to a lower-dimensional vector of feature $(F_{bottleneck})$ by a mapping function $(F_{bottleneck} = f_{enc}(x))$ called a feature extractor or *encoder*. The best mapping functions keep all the unique information while throwing away all redundancies. If there is no loss of information, it should be possible to get the original input back using the inverse function, $x = f_{enc}^{-1}(F_{bottleneck})$, called the *decoder*.

Since the aim is to learn the feature extraction process, let us model a neural network with one input layer where the high-dimensional input (x) will be applied, one hidden layer which will produce the features $(F_{bottleneck})$ as output, and one output layer which will reconstruct the original input from features, thus producing the original input as the output (x_{dec}). The input to hidden layer mapping is called the encoding process $(F_{bottleneck} = f_{enc}(x))$, while the hidden layer output to output layer mapping is called the decoding process $(x_{dec} = f_{dec}(F_{bottleneck}))$. The auto-encoder is thus a neural network with a single hidden layer, which has the same input and target. The output of the hidden layer is the feature vector $(F_{bottleneck})$ and the layer is also called the *bottleneck layer*. The neural network is hence given by Eqs (12.51), (12.54). The architecture is shown in Fig. 12.19.

$$v_1 = w_{1,enc}^T x \tag{12.51}$$

$$F_{bottleneck} = f_{enc}(x) = F_1 = \phi(v_1) \tag{12.52}$$

$$v'_1 = w_{1,dec}^T F_{bottleneck} \tag{12.53}$$

$$x_{dec} = f_{dec}(F_{bottleneck}) = \phi(v'_1) \tag{12.54}$$

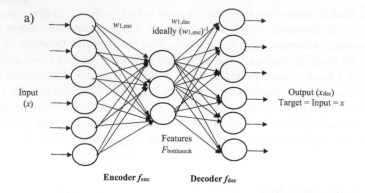

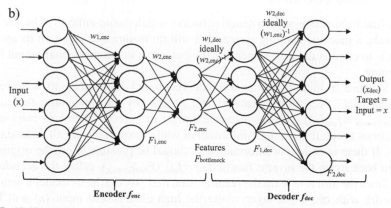

FIG. 12.19   Autoencoder. The network aims to produce the same output as the input and thus attempts to compress the original input into features (output of the bottleneck layer) $F_{bottleneck}$, such that the data are compressed and can still be re-constructed into the original input. (A) Autoencoder using 3-layers, (B) Stacked Autoencoders using multiple encoding and decoding layers.

The sub-script 1 associated the variables refer to the first layer (of encoder or decoder). Subscript *enc* refers to the encoder and *dec* refers to the decoder. The ' denotes the decoding variables. φ is the activation function. $w_{1,enc}$ and $w_{1,dec}$ are the weights of the 2 layers (encoder and decoder). The loss function is called as the re-construction loss given by Eq. (12.55).

$$loss(w) = \frac{1}{2}\sum_i \sum_d (x_{i,d} - f_{dec}(f_{enc}(x_d)))^2 + \frac{1}{2}\gamma\|w\|^2$$

$$= \frac{1}{2}\sum_i \sum_d (x_{i,d} - x_{dec,i,d})^2 + \frac{1}{2}\gamma\|w\|^2 \qquad (12.55)$$

Here $x_{i,d}$ is the dth input vector while i iterates through all the elements of the input vector. $x_{dec,i,d}$ is the output of the neural network. γ is the regularization parameter that aims to reduce the norm of all weights combined or $\|w\|$. After the network has learned, the decoding layer or the output layer is removed to get the features $F_{bottleneck}$. This network represents the encoder that maps the inputs to feature values learned by the data.

Since we are studying deep learning, let us take deeper networks for both the encoder and decoder. The concept is called a *stacked auto-encoder*. Here a series of encoding layers are applied one after the other. The input is subjected to the first layer which gives features $F_{1,enc}$. Thereafter another layer is added that further reduces the features from $F_{1,enc}$ to $F_{2,enc}$. Similarly, more layers are stacked on top of each other to make still further reduced features. The last layer (say kth layer) is called the bottleneck layer that gives the most reduced features ($F_{bottleneck}$), given by Eq. (12.56), where f_{enc} denotes the encoder part of the network.

$$F_{bottleneck} = f_{enc}(x) = \phi\left(w_{k,enc}^T\left(....\phi\left(w_{2,enc}^T\phi\left(w_{1,enc}^T x\right)\right)\right)\right) \qquad (12.56)$$

This constitutes the encoding phase of the auto-encoder. The same number of layers are added in the reverse order for the decoding process. The output is received at the last output layer, giving the reconstructed output as Eq. (12.57).

$$x_{dec} = f_{dec}(F_{bottleneck}) = \phi\left(w_{k,dec}^T\left(....\phi\left(w_{2,dec}^T\phi\left(w_{1,dec}^T F_{bottleneck}\right)\right)\right)\right) \quad (12.57)$$

During training the output is compared to the input to measure the reconstruction loss. The learning so far was *unsupervised feature extraction* and not a typical supervised classification or a regression problem. For supervised classification, the features extracted by the auto-encoders from the bottleneck layer are used as an input and connected to the final output layer which gives the class posterior or the regression output. A way to visualize the complete network is thus all stacked layers of the auto-encoder, followed by the supervised layers(s) that ultimately give the output. The overall design is shown in Fig. 12.19.

An interesting question is whether the stacking was required in the auto-encoder when even a single layer reduction of data into features was possible. The stacking ensures that every layer does a primitive task of getting rid of some redundant features as visible in that layer. Hence, the function to be learned is as simple as possible. The same analogy is used when a complex task is done in a layered and hierarchical manner and seldom is such a task attempted to be solved in a single step. The simplicity of the function enables reducing training time, reducing the risk of overfitting, and reduces the risk of local minima, while enabling richer transformations across multiple layers. For similar reasons, the stacked auto-encoder outperforms a single layer neural network despite the latter being a universal approximator.

An element of robustness is also added to the network. This means that even if the inputs were corrupt or some inputs were missing, the network should be able to regenerate noise-free inputs. This is done by intentionally corrupting some inputs by adding Gaussian noise to the inputs in the input layer. However, the output is the correct and uncorrupted input. An auto-encoder trained in this mechanism is called the *de-noising stacked auto-encoder*, where the keyword de-noising stresses on the network's capability to remove noise. Another variant instead completely removes an input by setting a value of 0. Now the auto-encoder has the capability of recovering missing data. This technique is also referred to as the *dropout*.

Another notion added is called *sparsity*. In the human brain, even though there are a huge number of neurons, most of them are inactive most of the time. In a multi-layer perceptron as well, every input only succeeds in firing a small number of hidden neurons. Every auto-encoder tries to learn a template stored in the form of the weight connections. An input should only fire a very few such neurons or templates and the other neurons should be inactive with an output of nearly 0. This constraint is additionally enforced by penalizing neurons if they fire and hence have output far away from 0. Let $\widehat{\rho}_i$ be the average activation of neuron i, averaged over all the data in the training data set. The aim is to keep the mean as small as possible around 0, say a small number ρ, given by Eqs (12.58), (12.59).

$$loss_{sparse}(w) = \sum_i \rho \, log \frac{\rho}{\widehat{\rho}_i} + (1 - \rho) \log \frac{1 - \rho}{1 - \widehat{\rho}_i} \tag{12.58}$$

$$loss(w) = \beta \, loss_{sparse}(w) + \frac{1}{2} \sum_i \sum_d (x_{i,d} - x_{dec,i,d})^2 + \frac{1}{2}\gamma\|w\|^2 \tag{12.59}$$

The term is derived from the Kullback-Leibler divergence (KL divergence) and can be conveniently written as $loss_{sparse}(w) = KL(\rho \| \widehat{\rho}_i)$. Here ρ is a small constant called the sparsity parameter. β controls the tradeoff between the mean square error and sparsity.

Any data recorded have noise. In other words, multiple samples of the same data are different due to noise. We say that the data have natural variations, which is a good aspect to model in the auto-encoders as well, making the *Variational Autoencoders* (VAE, An and Cho, 2015). Let the encoder now output a feature distribution (also called a latent distribution) instead of a feature vector (also called a latent vector), modelling the possible natural input variations. Assuming the output distribution is Gaussian in nature, the distribution can be represented by a mean ($\mu_{bottleneck}$) and standard deviation ($\sigma_{bottleneck}$). The output is denoted as $N(\mu_{bottleneck}, \sigma_{bottleneck})$ where $N()$ is the normal distribution function. The encoder network is now given as $[\mu_{bottleneck}, \sigma_{bottleneck}] = f_{enc}(x)$. The decoder still takes a real feature vector (latent) as input. For decoding a sample is taken from the distribution $F_{bottleneck} \sim N(\mu_{bottleneck}, \sigma_{bottleneck})$ and used for decoding, which gives the reconstructed input as previously, $x_{dec} = f_{dec}(F_{bottleneck})$. The loss function needs to be modified to learn a distribution. The aim is to minimize the KL-divergence between the output distribution of the encoder and a unit normal distribution (denoted by $N(0,1)$ representing a mean 0 and standard deviation 1). In other words, we force the encoding process to create latents (features) that represent a unit-normal distribution around the origin in the latent space. If the data come from different classes, the data will be distributed at different regions of the latent space, overall making a unit normal distribution. The loss function is given by Eqs (12.60)–(12.63) and shown in Fig. 12.20.

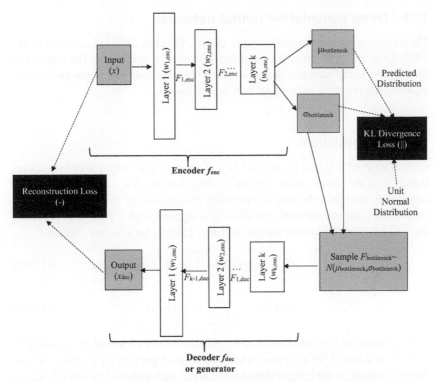

FIG. 12.20 Variational Autoencoders. The encoder outputs a distribution of features characterized by a mean and variance. The latent feature is sampled and decoded. The network learns on the reconstruction loss and the KL divergence of the output distribution with a unit normal distribution (having mean 0 and variance 1).

$$\left[\mu_{bottleneck,i}, \sigma_{bottleneck,i}\right] = F_{enc}(x_i) \tag{12.60}$$

$$F_{bottleneck,i} \sim N\left(\mu_{bottleneck,i}, \sigma_{bottleneck,i}\right) \tag{12.61}$$

$$x_{dec,i} = f_{dec}(F_{bottleneck,i}) \tag{12.62}$$

$$loss(w) = \frac{1}{2} \sum_i \sum_d (x_{i,d} - x_{dec,i,d})^2$$
$$+ KL\left(N\left(\mu_{bottleneck,i}, \sigma_{bottleneck,i}\right) \middle\| N(0,1)\right) + \frac{1}{2}\gamma\|w\|^2 \tag{12.63}$$

Here $KL(a\|b)$ is the KL divergence between the distributions a and b. An advantage now is that the decoder network acts as a *generator*. Take a random sample from a unit normal distribution, $r \sim N(0,1)$ and pass it to the decoder, generating a reconstructed data sample $F_{dec}(r)$. This represents a synthetically generated, but realistic-looking data for the trained problem. Without the variational factor, it was not possible to generate synthetic data since the real data are only found in small unknown fragments of the entire possible latent space (or feature space).

12.8 Deep convolution neural networks

The deep convolution neural network (CNN, Hu et al., 2014; Lawrence et al., 1997; Li et al., 2015) uses the *convolution* operation taken from the signal and image processing literature as the activation function, which empowers them to be able to work well with signal and image data.

12.8.1 Convolution

Convolution of a signal by another signal produces as output a third reaction signal. In other words, when a signal is subjected to a filter, by convolving the signal with the filter, the output represents the reaction of the signal to the specific filter. The convolution operation of a signal f with a filter g is given by Eq. (12.64) for continuous signals and Eq. (12.65) for discrete signals.

$$(f*g)(t) = \int_{\tau=-\infty}^{\infty} f(t-\tau)g(\tau)d\tau = \int_{\tau=-\infty}^{\infty} f(\tau)g(t-\tau)d\tau \qquad (12.64)$$

$$(f*g)(t) = \sum_{\tau=-m}^{m} f(t-\tau)g(t) \qquad (12.65)$$

The equation is written as a mathematical construct wherein $g()$ is assumed to be zero centred with a range from $-m$ to m, representing a signal of size $2m+1$. Instead, if $g()$ is represented as an array with indices 1 to $2m+1$, a constant will have to be added.

Practically, $f()$ will have a finite size, which affects computation at the two ends of the signal wherein the access to $f(t-\tau)$ may become out of bound. There are two ways to handle this. The first is to only compute the response when the indices are valid, which means that size of the response signal will be less than that of the original signal by a factor of $2m$. The other mechanism is to do *padding* of extra zeros at the two extremes so that the indices are always valid. This produces the response signal of the same size as the input signal. An illustrative example of 1-D convolution is shown in Fig. 12.21.

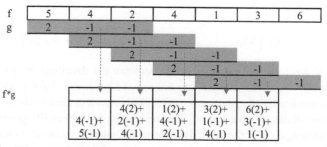

FIG. 12.21   1D convolution operation.

The discussions can be extended into 2D as well, which is our target application since the input is a 2D image taken by the camera. The extension is given by Eq. (12.66).

$$(f*g)(u,v) = \sum_{\tau_u=-m}^{m} \sum_{\tau_v=-m}^{m} f(u-\tau_u, v-\tau_v)g(\tau_u, \tau_v) \tag{12.66}$$

An illustrative example of 2-D convolution is shown in Fig. 12.22. The concept can also be extended to 3-dimensional or higher-dimensional tensors. There are numerous examples of higher-dimensional inputs. An image is itself 3-dimensional with the 3rd dimension being the channel (R,G,B). A grayscale video represents 3 dimensions with 2 spatial dimensions and 1 dimension of time.

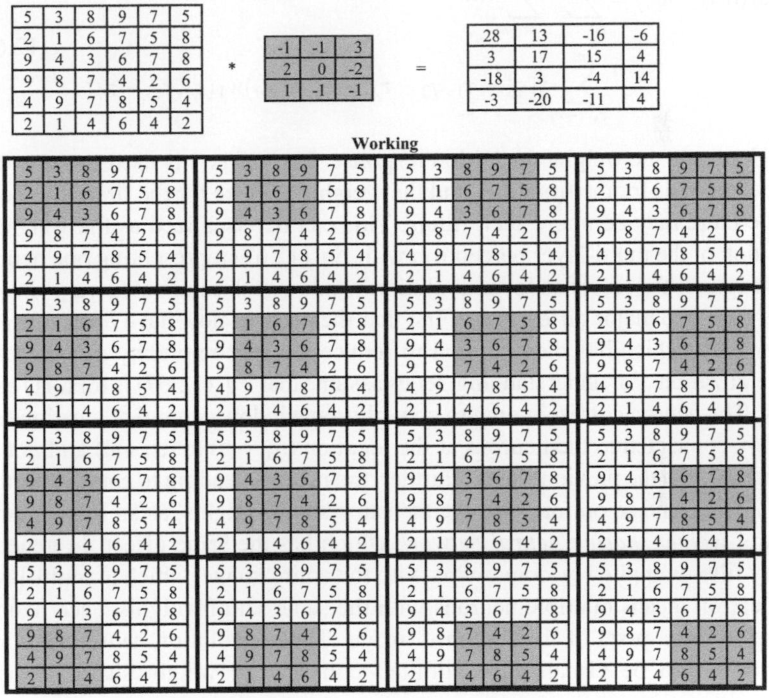

Input:

5	3	8	9	7	5
2	1	6	7	5	8
9	4	3	6	7	8
9	8	7	4	2	6
4	9	7	8	5	4
2	1	4	6	4	2

*

Kernel:

-1	-1	3
2	0	-2
1	-1	-1

=

Result:

28	13	-16	-6
3	17	15	4
-18	3	-4	14
-3	-20	-11	4

Working

3(-1) + 4(-1) + 9(3) + 6(2) + 1(0) + 2(-2) + 8(1) + 3(-1) + 5(-1)	6(-1) + 3(-1) + 4(3) + 7(2) + 6(0) + 1(-2) + 9(1) + 8(-1) + 3(-1)	7(-1) + 6(-1) + 3(3) + 5(2) + 7(0) + 6(-2) + 7(1) + 9(-1) + 8(-1)	8(-1) + 7(-1) + 6(3) + 8(2) + 5(0) + 7(-2) + 5(1) + 7(-1) + 9(-1)
7(-1) + 8(-1) + 9(3) + 3(2) + 4(0) + 9(-2) + 6(1) + 1(-1) + 2(-1)	4(-1) + 7(-1) + 8(3) + 6(2) + 3(0) + 4(-2) + 7(1) + 6(-1) + 1(-1)	2(-1) + 4(-1) + 7(3) + 7(2) + 6(0) + 3(-2) + 5(1) + 7(-1) + 6(-1)	6(-1) + 2(-1) + 4(3) + 8(2) + 7(0) + 6(-2) + 8(1) + 5(-1) + 7(-1)
7(-1) + 9(-1) + 4(3) + 7(2) + 8(0) + 9(-2) + 3(1) + 4(-1) + 9(-1)	8(-1) + 7(-1) + 9(3) + 4(2) + 7(0) + 8(-2) + 6(1) + 3(-1) + 4(-1)	5(-1) + 8(-1) + 7(3) + 2(2) + 4(0) + 7(-2) + 2(1) + 4(-1) + 7(-1)	4(-1) + 5(-1) + 8(3) + 6(2) + 2(0) + 4(-2) + 8(1) + 7(-1) + 6(-1)
4(-1) + 1(-1) + 2(3) + 7(2) + 8(0) + 4(-2) + 7(1) + 8(-1) + 9(-1)	6(-1) + 4(-1) + 1(3) + 8(2) + 7(0) + 9(-2) + 4(1) + 7(-1) + 8(-1)	4(-1) + 6(-1) + 4(3) + 5(2) + 8(0) + 7(-2) + 2(1) + 4(-1) + 7(-1)	2(-1) + 4(-1) + 6(3) + 4(2) + 5(0) + 8(-2) + 6(1) + 2(-1) + 4(-1)

FIG. 12.22 2D convolution operation.

Such convolution of images representing the *f* signal with specially designed filters representing the *g* signal is common in different applications. Removing noise, beautifying images, image editing operations, sound editing operations, sound enhancement, etc. are convolution filters. Since convolution can do such enhancement to image and sound signals, it is interesting to use them for the information processing in neural networks.

Let us model the neural network. A filter on the application of a signal (of any dimensions, images, videos, etc.) will produce a signal of the same size (in case of a zero-padding) or slightly smaller size (in case of no such padding). The filter acts as weights and the input acts as the original signal, modelled as Eq. (12.67). The representation is for a general *n*-dimensional input.

$$s_1(q_1 q_2 \cdots q_n) = \sum_{\tau_1=-m}^{m} \sum_{\tau_2=-m}^{m} \cdots$$

$$\sum_{\tau_n=-m}^{m} x(q_1 - \tau_1, q_2 - \tau_2, \cdots, q_n - \tau_n) w_1(\tau_1, \tau_2, \cdots \tau_n) \tag{12.67}$$

Here s_1 is the weighted addition response. The suffix 1 is used to show that the response is for the first filter and the weights (w_1) are for the first filter only, while there may be many such filters. This can be visualized as the weighted addition of the multi-layer perceptron with Eq. (12.67) representing one neuron that captures the response for convolution window with centre at $(q_1, q_2, ..., q_n)$. There are two major differences. First that the neurons now have *local connectivity only*, which has the potential to reduce model complexity and increase generalization. For understanding, let us assume that Eq. (12.67) is for a 1D case only. The neuron corresponding to convolution window centred at location (q_1) is only connected to the neurons at the location ($q_1 - m$) to ($q_1 + m$) and not all the neurons in the previous layer. Further, the neurons leading to computation of $s_1(1)$, $s_1(2)$, and so on for 1D share the same set of weights, as an implementation of *weight sharing neural network* which has the same effect. In general, every filter is applied by convolving with the entire image, which in multi-layer perceptron means multiple neurons with the same set of weights, each neuron representing a pixel of the response image. The convolution neural networks can thus also be visualized as limited connectivity weight sharing neural networks.

The response is subjected to an activation function. A problem when using the back-propagation algorithm on a multi-layer perceptron is *vanishing gradient*. In the back-propagation algorithm, the local gradients are propagated from the output layer to the input layer in a backward direction, while a weighted addition of the local gradients is taken. Typically, the activation functions used are sigmoid or hyperbolic tangent which gives outputs between -1 and 1, which is also the typical range of derivatives and weights. Hence, as we back-propagate the local gradients, we continuously make multiplications by fractional numbers and reduce the magnitude of the local gradient. The weight change is proportional to the local gradient. Hence the layers near the output layer learn very fast while the layers near the input learn very slowly. As one travels from the output to the input layer, the *learning keeps reducing* drastically. Practically, after 2 layers there is no significant learning. Hence there is a need to take an activation function that induces non-linearity so as not to make the entire deep neural network linear, however also does not have a fractional gradient.

Another problem associated with the activation functions is that they have a near-linear performance on a small region within some limits, and in all other regions, their output is saturated or constant between -1 and 1 (or 0 and 1). This is called the problem of *saturation* wherein the output of the neuron is nearly constant for high or low input values, and therefore learning becomes difficult. A large change in input in such regions produces no change in the output. Since the learning uses the gradient of the activation function, a negligible change in weight takes place since the gradient is nearly 0. Sigmoid and tangent hyperbolic are two typical activation functions and both face this problem. This problem must also be solved. A common activation function is the *Rectified Linear Unit (ReLU)*, which is given by Eq. (12.68).

$$\phi(x) = \max(0, x) \tag{12.68}$$

The activation function is non-linear because of the max function, while the derivative is 1 for all positive inputs and 0 otherwise. The activation function hence avoids the vanishing gradient and saturation problems. The ReLU activation function does no learning when the input is negative. A neuron that had negative response to all inputs will never fire and will never learn. Sometimes, it is better to get some output and have some weight corrections, which is done by a Leaky ReLU given by Eq. (12.69).

$$\phi(x) = \max(\alpha x, x) \tag{12.69}$$

Here $\alpha < 1$ is a small positive number. For positive inputs, $x > \alpha x$ and the function behaves same as ReLU. For negative inputs ($x < 0$) the first term dominates ($\alpha x > x$), which gives a small (negative) output which was 0 for ReLU.

12.8.2   Pooling and subsampling

Each filter produces a signal of roughly the same size as the input. *Pooling* or *subsampling* is used to reduce the amount of data. For 1D, pooling of size w means that consecutive w items of the signal shall be replaced by a single value. Similarly, in 2D, pooling of $w \times w$, means that a block of the same size shall be replaced by a single value. Pooling is similar to resizing an image. Typically, a pooling size of 2×2 for images (and similarly 2 for signals) is good enough. The process is also called *downsampling* where the size is called the *sampling ratio*. The pooling discussed here is such that no two pooling windows overlap each other. *Pooling with overlapping windows* is also an acceptable technique, where the overlap size is an additional parameter.

There are 3 common types of pooling: *Max-Pooling*, where a maximum overall value is used as the replacement value; *Min-Pooling*, where a minimum of all values is used as a replacement; and *Average-Pooling* where an average of all values is used instead. Max-Pooling is the most common pooling strategy. Each filter tries to get a particular pattern from the preceding signal and very few spatial areas find the pattern and result in an output. The max-pooling tries to maximize the output of the active areas. In contrast, the min-pooling only results in an output when all the areas covered by the pooling window were active. This diminishes most of the outputs. The max-pooling operation is shown in Fig. 12.23.

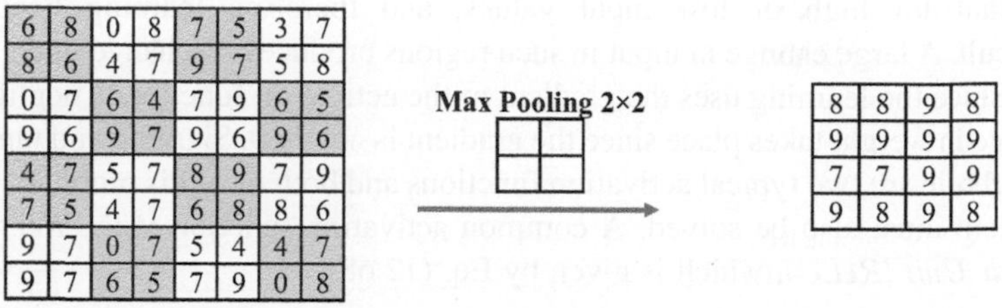

6	8	0	8	7	5	3	7
8	6	4	7	9	7	5	8
0	7	6	4	7	9	6	5
9	6	9	7	9	6	9	6
4	7	5	7	8	9	7	9
7	5	4	7	6	8	8	6
9	7	0	7	5	4	4	7
9	7	6	5	7	9	0	8

Max Pooling 2×2

8	8	9	8
9	9	9	9
7	7	9	9
9	8	9	8

FIG. 12.23 Max pooling operation with a 2 × 2 window with a stride 2. Take the maximum of all elements in a 2 × 2 window, shift window by stride (2).

Another mechanism of getting low-dimensional output from a filter is called *stride*. In convolution, a filter window is convolved with a signal to get the response of one data item in the signal, and the filter window moves by 1 step to calculate the convolution response of the next data in the signal. With a stride, the filter takes steps of magnitude given by the stride length. So, after convolving the filter centred across a data item from the signal, the filter moves several steps to deliberately miss calculating responses against some part of the signal data. Hence the output response is of a smaller size as compared to the input by a factor of the stride length.

A single filter may not extract all interesting aspects of the signal. Hence, multiple such filters are used. The response of k filters shall be (without subsampling) k such signals. Thus, the application of k filters adds a new dimension to the input. In the case of grayscale images, the data will become 3 dimensional with the additional dimension denoting the filters and the size of the 3rd dimension as the number of filters. Further stacking layers could do a 3D convolution and 3D pooling. However, typical implementations only carry pooling in 1 or 2 of the 3 dimensions. The number of dimensions does not go increasing with the number of layers as the filter outputs are stacked upon in the last dimension for every layer.

The continuous application of convolutions in a layered manner only highlights interesting aspects of the signal. A typical problem is either classification or regression which is not done by these layers and is only done in a *fully connected* multi-layer perceptron network. Finally, the inputs are *flattened* to a single dimension only. This is done by iterating through all the data in all the dimensions in any specific order and appending in a single dimensional array. This forms the input to one (or more) fully-connected layers called the *dense layers*. The final layer is the output layer that gives the output. The complete architecture of the use of multiple layers is shown in Fig. 12.24.

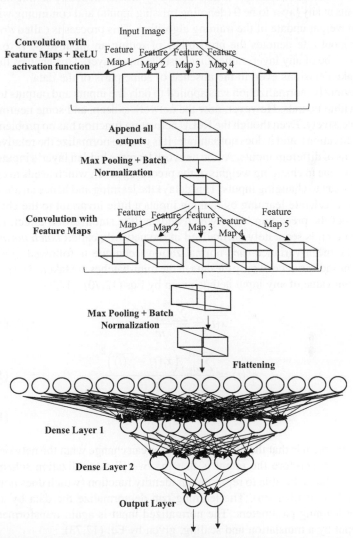

FIG. 12.24 Convolution Neural Network: Convolution with multiple feature maps that extract features. ReLu function (max(x,0)) avoids vanishing gradient problem (only layers close to output learn), still adding non-linearity. Max Pooling reduces data, while batch normalization accelerates learning. Finally, the inputs are flattened into multiple dense or fully-connected layers as a multi-layer perceptron.

12.8.3 Training

The architecture only provides a model, and the bigger problem is now to train the network. The back-propagation algorithm is suitable for the training purpose. Mini batches selected stochastically from the data are hence used, which is the process of *Stochastic Gradient Descend*.

The notion of robustness is also added by probabilistically setting some of the inputs at any layer to be 0 (denoting missing inputs) and continuing with the general weight update of the training algorithm. This process is called *dropout*. The dropout rate denotes the probability of setting an input to be zero. Here input may be of any layer, rather than just the input layer. This enables the system make decisions even in the absence of some parts of the data.

Previously, normalization was applied to only the inputs and outputs to keep them within bounds. However, now the network is deep, and some neurons can fire excessively. Even though the ReLU activation function has no problem with large activations, and it does not saturate, it is good to normalize the relative contributions of different inputs. As the network learns, a hidden layer's inputs keep changing due to changing weights of the preceding layer, which needs to continuously adapt to changing inputs. This delays the learning and hence an attempt is made to accelerate learning by making inputs a little invariant to the changing weights of the preceding layer, which adds a little stabilization effect. Hence, every neuron is separately normalized in the network, called *batch normalization*. The mean-variance ($\mu\sigma$) normalization technique is followed, however, the mean and variance of only the data in the mini-batches are taken. The normalized input value of any input is thus given by Eqs (12.70)–(12.72).

$$\overline{x(t)} = \frac{\sum_{i=1}^{m} x_i(t)}{m} \tag{12.70}$$

$$\sigma^2(t) = \frac{\sum_{i=1}^{m} \left(x_i(t) - \overline{x(t)}\right)^2}{m} \tag{12.71}$$

$$x'_i(t) = \frac{x_i(t) - \overline{x(t)}}{\sigma(t)} \tag{12.72}$$

The problem is that the normalized values can change what the network represents, and therefore the intent is that the overall normalization scheme for ideal data should be able to represent an identity function (which does not carry any batch normalization). The learning can de-normalize the data by adding suitable learning parameters. The normalized input is again transformed to a new input by a translation and scaling, given by Eq. (12.73).

$$x''_i(t) = \gamma(t)x'_i(t) + \beta(t) \tag{12.73}$$

Here $\gamma(t)$ and $\beta(t)$ are parameters that are learned in the training process and hence the parametrization is with time. If $\gamma(t)$ takes the value of the standard deviation and $\beta(t)$ takes the value of the mean, while the mini-batch mean and variance are indicative of the mean and variance of the entire data set, the normalization function does become an identity. The learning is a standard back-propagation with mini-batches. The derivatives of the entire network can be obtained by using the chain rule.

A little problem is when performing testing in which case no mini batch is available and without batch normalization, the system represents a different function. The problem can be solved by replacing the mini-batch mean and variance with that of the entire data set computed during the training process or explicitly after training with a similar process but the weights frozen for changes.

The layered architecture resembles nature. Complex object recognition happens by taking pixel inputs, understanding lines from pixels, understanding corners from lines, then at a higher layer understanding simple shapes, then at a higher layer understanding complex shapes, and so on. The corner detector shall remain the same irrespective of the place in the image where the corner is found.

12.9 Long-short term memory networks

The study is now focussed upon the recurrent domain. The recurrent neural networks enable taking sequences of data as input and therefore are excellent agents to model highly dynamic temporal systems. This is of our interest because the robot motion is a complex temporal system with temporal sequences of data coming from all the on-board sensors.

12.9.1 Problems with recurrent neural networks

The problem with the traditional recurrent neural networks is that they tend to derive the output at any time primarily based on the most recent data in the input sequence. It can hence be said that these networks have a *short-term memory* to memorize the recent percept stored in the form of the outputs of the recurrent connections, while they tend to forget the older inputs in the sequences. A part of it is also true with humans. We look at the recent remaining work and try to complete it. We look at the immediate crisis and solve it. Even in the case of navigation we extrapolate the recent observations of dynamic obstacles and decide on how to navigate.

However, there may be situations wherein some knowledge of the past data is required. Our decisions are affected by the success and failures reasonably long time back. While turning to initiate overtaking of a vehicle, you may lose sight of the vehicle being overtaken for some time, which does not mean that you start retreating as if there was no vehicle being overtaken.

It may be inquisitive why a typical recurrent neural network primarily makes decisions based on the most recent information only. Imagine a recurrent neural network of a single hidden layer, which can be unfolded with time to visualize many hidden layers, one for each time step, trained in a back-propagation style. The network will incur the problem of *vanishing gradients*, wherein the layer at the output will learn fast, while the initial layers will virtually not learn. This means that the inputs at the first few layers are not important since the corresponding weights are not learning. It may be argued that unlike vanishing gradient, all layers in a recurrent neural network have the same shared weights. Hence, a more accurate answer is, taking the inspiration from vanishing gradients in back-propagation algorithm, the change in weight due to the latter layers is high while the change in weights due to the initial few layers is low. Further, imagine that the inputs and outputs are normalized, and the weights are typically between −1 and 1, and the activation functions are adjusted to be most active within the range. The input from the first few layers of the unfolded network gets multiplied by too many fractional weights and activations to lose value, unlike the forward layers whose inputs are directly (or after a layer or so) fed to produce the output.

12.9.2 Long-short term memory networks

The manner of remembering long-term items in a sequence is by frequently *forgetting*. Intuitively, if somehow, we forget a little of our immediate past, it leaves memory for the more historic events to stay intact. The new memory does not erode the old one, as the new memory is restricted by intentionally forgetting a little of the immediate past input. The network is hence called the *Long-Short Term Memory* (LSTM) network (Gers et al., 1999; Greff et al., 2017). The term implies that the network has a short-term memory of the immediate past events for decision making; however, at the same time, the network also has a long-term memory for decision making.

Imagine information (recurrent connection outputs) coming from the past and at every step, it is modified by some data fed as input. Let the new information be the weighted addition of the old information and the new input, while the weights are dependent upon the content (or relative importance) of the new input and old information. If the new input is multiplied by zero, the old information would not get modified, or the network completely forgets the new input. If the old information is multiplied by a zero weight, the new information is entirely the new input, and the network is said to have forgotten the old information.

Let us now discuss the architecture of the LSTM. Imagine an unfolded recurrent neural network. The input layer is represented at the bottom, the output layer is represented at the top and the unfolded recurrent layers are represented horizontally. A unit layer is called a *cell* that takes external inputs, inputs from the previous time cells in a recurrent framework, produces outputs, and passes information and outputs to the cells ahead in time. The *cell state* is defined as the information that flows over time in this network (as recurrent connections) with the information content having a value of $c(t)$ at time t. The cell state would be affected by inputs and outputs of the different cells, as we go over the network (or more concretely in time over the temporal sequences). Similarly, the network passes the output $y(t)$ from the previous time to the next time as a recurrent connection. The conceptual architecture is shown in Fig. 12.25A.

The first and most interesting modification to be done is *forgetting* implemented by a *forgetting gate*. The external input $x(t)$ applied at the time step t, along with the output at time step $t - 1$, called $y(t - 1)$, which represents the input due to the recurrent connections is affected by the weights and a normal weighted addition takes place, along with the addition of a bias. However, sigmoidal activation is applied. The sigmoidal function results in outputs between 0 and 1 and this hence controls the level by which the old information should be

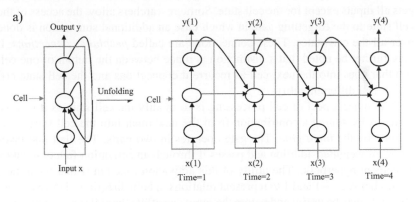

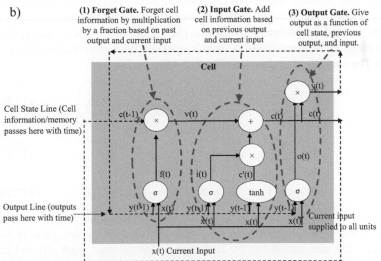

FIG. 12.25 (A) Unfolding a Recurrent Network with time: Recurrent neural network takes timed sequence as input, producing a sequence as output, with a cyclic architecture. A single circle represents a complete layer of the multi-layer perceptron, (B) LSTM Architecture.

forgotten. The forgetting itself happens by multiplication of the old information with the forgetting degree computed by the new input. The gate is shown in Fig. 12.25B. The modelling of the gate is given by Eqs (12.74)–(12.75)

$$f(t) = \sigma \left(\sum_a w_{f,a} x_a(t) + \sum_b w_{f,b} y_b(t-1) + b_f \right) \tag{12.74}$$

$$v(t) = f(t)c(t-1) \tag{12.75}$$

Here σ is the sigmoid activation function. $v(t)$ is the cell state after forgetting (but before being affected by the input). The first term takes weighted addition over all the external inputs $x(t)$ and the second over all the recurrent connection inputs $y(t-1)$. b_f is the bias term. It may be observed that the forgetting gate gets all inputs except for the cell state. Some researchers allow the access of the cell state to the forgetting gate in which case an additional summation is done over all the cell states. These connections are called *peephole connections*. It may further be noted that a distinction is made between the output of one cell $y(t)$ that goes into the next cell as recurrent connections and the cell state $c(t)$ which is a different entity.

The other aspect of cell processing is to modify the cell state as it travels, which is by adding a contribution from the new input into the cell state. This is done by the *input gate*. The gate operates in two parts. The first one takes a typical weighted addition and passes it through an activation function, taken as tangent hyperbolic. The utility of this activation function is that it can take values between -1 and 1 to represent relations in both directions. However, not all inputs may be useful and before the input can affect the cell state, it must pass a check whether and how much of it deserves to be used and how much of it must be instead immediately forgotten. Hence, another sigmoid is applied, whose range is between 0 and 1, which operates on a weighted addition of inputs. The modelling is shown as Eqs (12.76)–(12.78).

$$i(t) = \sigma \left(\sum_a w_{i,a} x_a(t) + \sum_b w_{f,b} y_b(t-1) + b_i \right) \tag{12.76}$$

$$c'(t) = \tanh \left(\sum_a w_{i,a} x_a(t) + \sum_b w_{f,b} y_b(t-1) + b_{c'} \right) \tag{12.77}$$

$$c(t) = v(t) + i(t)c'^{(t)} = f(t)c(t-1) + i(t)c'(t) \tag{12.78}$$

Here $i(t)$ is the importance of the new weight in the scale of 0 to 1, maintained by the sigmoid function. $c'(t)$ is the contribution of the input in the new state. The summation has the first term as the external input $x(t)$ and the second term as the recurrent connections $y(t-1)$, with $b_{c'}$ as the bias. The contribution $c'(t)$ on being added to the forget value $v(t)$ makes the new cell state $c(t)$. The new cell state is thus the weighted addition of the old cell state $c(t-1)$

with a weight $f(t)$ and the new transformed input $c'(t)$ with a weight $i(t)$. Again, it is possible to take peephole connections and include the terms from the cell state $c(t-1)$ as well.

The last step is to produce the output of the neuron to be given as the output of the current time step. Note again that the cell state is different from cell output. Both cell state and cell output need to be calculated and passed between unfolded layers. The output is a function of the cell state that passes through the activation function, which is taken as tangent hyperbolic to get a range of -1 to 1. However, the sigmoid is still applied based on the input to select the relevant content of the state relevant to the output and to suppress the rest. The gate is modelled as given by Eqs (12.79), (12.80).

$$o(t) = \sigma \left(\sum_a w_{i,a} x_a(t) + \sum_b w_{f,b} y_b(t-1) + b_o \right) \qquad (12.79)$$

$$y(t) = o(t) \tanh(c(t)) \qquad (12.80)$$

Here $o(t)$ is the intermediate output because of activation and weighted addition on external inputs $x(t)$ and recurrent connections $y(t-1)$; which on multiplication with the activation function over the cell state $c(t)$, produces the final output $y(t)$. The network is trained like the recurrent neural network as back-propagation through time.

12.10   Adaptive neuro-fuzzy inference system

The neural networks have a well-formulated mechanism of training and hence numerous problems are first represented as a neural network and thereafter the derivations of the back-propagation algorithm are directly applied. The fuzzy inference system of Chapter 11 also falls into the category wherein the entire mathematical formulation is cast into a neural network so that a neural network like training can be applied. The network is called as the adaptive neuro-fuzzy inference system (Rutkowski, 2004).

The type of fuzzy inference system discussed here will be the *Takagi-Sugeno type fuzzy inference system*. Here the antecedents are the participating membership functions on the inputs like the Mamdani type; however, the consequent is a function over the inputs. The genesis is that the antecedents still decide the relevance of a rule, however the rule itself produces a prospective crisp output. This makes it easier to learn the system. A typical rule is given by Eq. (12.81).

if distanceToObstacle (d_o) *is near and angleToObstacle* (θ_o) *is farClockwise then turn* $(\omega) = f(d_o, \theta_o)$ $\qquad (12.81)$

Here $f()$ is any general function. The type-0 Takagi-Sugeno system has a degree 0 equation as the consequent and hence the rule is given by Eq. (12.82).

> *if distanceToObstacle (d_o) is near and angleToObstacle (θ_o) is*
> *farClockwise then turn (ω) = 45°*   (12.82)

Similarly, a type-1 system has a degree 1 rule given by Eq. (12.83).

> *if distanceToObstacle (d_o) is near and angleToObstacle (θ_o) is*
> *farClockwise then turn(ω) = $A\,d_o + B\,\theta_o + C$*   (12.83)

$A, B,$ and C are constants to be learned. The antecedent of every rule still has fuzzification, application of fuzzy operators to determine the firing strength of a rule, say (α_r) for the rule r. Since the output of a rule is crisp, the aggregation gives the final output of the inference system, given by Eq. (12.84).

$$y = \frac{\sum_r \alpha_r f_r(x)}{\sum_r \alpha_r} \qquad (12.84)$$

Here α_r is the firing strength of rule r and $f_r(x)$ is the consequent of rule r with input vector x. The output is based on the observation that every rule gives a likely output $f_r(x)$ with likelihood α_r. For simplicity, let us take a type-0 fuzzy system.

The complete neural network representing the fuzzy inference system is shown in Fig. 12.26. Let us very quickly re-visit the step-by-step working of the fuzzy inference system and re-formulate the problem in terms of a neural network. The first step is to apply membership functions to all the inputs possible. The input is represented as an input layer of the neural network. Every input is mapped to k number of neurons in the forward layer, where k is the number of membership functions per input. So, there are a total of $I \times k$ neurons in this layer, where I is the number of inputs. For convenience, assume that a Gaussian membership function is used. Let the weight associated with the mth membership function of the ith input be $w^{(1)}_{i,\,m} = 1$ that joins ith input to the neuron representing the mth membership function of this input. The neural network is imagined to be fully-connected and therefore let us join this neuron to all other input neurons with a fixed weight of 0. The bias is the mean of the Gaussian (b^i_m), and the activation function becomes the Gaussian membership function equation. The output of a neuron representing the membership of input named i and the membership function named m with membership degree $\mu^i_m(x)$ is given by Eq. (12.85). The output of the complete layer can be conveniently written as Eq. (12.86).

$$\mu^i_m(x_i) = e^{-\frac{(x_i - b^i_m)^2}{2\sigma^2}} = \phi(x_i) = \phi\left(w^{(1)}_{i,m} x_i\right) = y^{(1)}_{i,m} \qquad (12.85)$$

$$y^{(1)} = \phi\left(\left(w^{(1)}\right)^T x\right) \qquad (12.86)$$

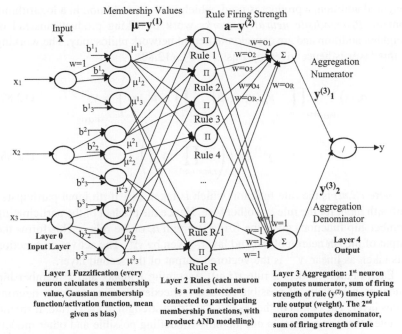

FIG. 12.26 Neuro-Fuzzy Inference System Architecture. Type-0 Takagi-Sugeno fuzzy inference system is modelled as a neural network to use the Back-Propagation training algorithm on a data set.

Here i is one of the input neurons of the input layer, m is one of the associated membership functions, x is the complete input vector, x_i is the ith component of the input vector corresponding to input i, σ^2 is the variance of the membership function, which is an algorithm parameter, ϕ is the associated membership function and $w^{(1)}$ is the weight vector taken as 1 for the specific input (i) and membership function (m) and 0 for all others. $y_{i,m}^{(1)}$ refers to the output of one neuron of the first hidden layer of the neural network corresponding to the neuron associated with input i and membership function m. $y^{(1)}$ is the vectored output of the complete layer.

Thereafter, the next step of fuzzy reasoning is to apply the rules, typically written using ANDs among the antecedents. To get differentiability, the product modelling of AND is used. A typical single fuzzy rule consists of one membership function taken from all inputs. The membership degrees are already available from the first layer. A rule is represented by one neuron. The neuron takes input from one membership function per input (whose weight is set as 1) and not from other membership functions (whose weight is set as 0). Here instead of the weighted addition, a product is used, which is like an addition in a logarithmic domain. *Probabilistic neural networks* work by taking products instead of weighted addition and are common in neural network philosophy. The working of this layer is hence given by Eqs (12.87), (12.88).

$$\alpha_r(x) = \prod_{i,m,r(i,m)=1} \mu_m^i(x) = \phi\left(\prod_{i,m,r(i,m)=1} \mu_m^i(x) \right) = y_r^{(2)} \qquad (12.87)$$

$$y^{(2)} = \phi\left(\prod_{i,m,r(i,m)=1} y_{i,m}^{(1)} \right) \qquad (12.88)$$

Here $r(i,m)$ is the rule function which is taken as 1 if ith input participates with mth input for the rule r, 0 otherwise. Hence the product of the participating membership functions gives the firing strength α_r of the rule r. This forms the output of the rth neuron of the 2nd layer, given by $y_r^{(2)}$. The activation function ϕ is taken as linear. $y^{(2)}$ is the vectored output of the second layer.

Each neuron models a rule $r(i,m)$. Assuming I inputs and k membership functions per input, all possible rules are encoded, leading to $I \times m$ neurons in this layer. The output of the layer is the firing strength of every rule. It should be noted that the product is not the only modelling possible and other models can also be encoded in a neural framework. Further, typically the generalized variants of the different models would be taken that accommodates weights to be learned by the Back-propagation algorithm.

Since a type-0 Takagi-Sugeno fuzzy inference system is being modelled, the next step is simply to take a weighted addition of all the outputs, modelled as Eq. (12.89).

$$y = \frac{\sum_r \alpha_r o_r}{\sum_r \alpha_r} \qquad (12.89)$$

Here o_r is the typical output that the rule r represents.

However, it is difficult to model this relation in the neural framework. Hence, let the numerator be calculated independently and the denominator independently in 1 layer. The third layer hence has 2 neurons, 1 for the numerator, and 1 for the denominator. The numerator is given by Eq. (12.90).

$$y_1^{(3)} = \sum_r \alpha_r o_r = \sum_r \alpha_r w_{r,1}^{(3)} = \left(w_1^{(3)} \right)^T y^{(2)} = \phi\left(\left(w_1^{(3)} \right)^T y^{(2)} \right) \qquad (12.90)$$

Here $y_1^{(3)}$ is the output of the first neuron of the 3rd layer. $w_1^{(3)}$ is the weight vector of the third layer to the 1st neuron, which in actuality represents the consequent of the rule. The activation function ϕ is taken as linear. It may be noted that a type 0 inference system was taken for the sake of ease of modelling. It is possible to model any degree by taking inputs from the input layer.

The second neuron is modelled in a similar manner given by Eq. (12.91).

$$y_2^{(3)} = \sum_r \alpha_r = \sum_r \alpha_r w_{r,2}^{(3)} = \left(w_2^{(3)}\right)^T y^{(2)} = \phi\left(\left(w_2^{(3)}\right)^T y^{(2)}\right) \quad (12.91)$$

Here $w_2^{(3)}$ represents the weights of the 3rd layer for the 2nd neuron and the weights are taken as 1, which are frozen and non-trainable. The activation function is again identity.

The final layer is simply the division of the two, given by Eq. (12.92).

$$y = y^{(4)} = \frac{y_1^{(3)}}{y_2^{(3)}} \quad (12.92)$$

The network so formed can be trained as a multi-layer perceptron. The modelling of every neuron is different and will therefore require computation of the derivatives, however, the generics remain the same. In such a case, the network learns itself based on the supplied inputs and outputs.

Questions

1. Consider the limited connectivity neural network given in Fig. 12.27. Calculate the output for the given inputs, error, and weights after 1 iteration of the back-propagation algorithm. Assume parameters not given in the question.

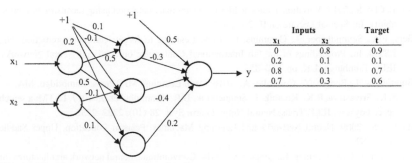

	Inputs	Target
x_1	x_2	t
0	0.8	0.9
0.2	0.1	0.1
0.8	0.1	0.7
0.7	0.3	0.6

FIG. 12.27 Neural Network for question 1.

2. The robot is attempting to learn images of daily household items. The robot records a lot of images on an everyday basis, which is somehow labelled at the rate of 1000 images per day. Compare and contrast between sequential and batch learning in this scenario, showing which one is preferable. In sequential learning, the robot is likely to forget the old images if unseen for a long time (known as the stability v/s plasticity dilemma in learning). Suggest ways to avoid forgetting the historical image data.

3. In the example of question 2, consider that the robot takes a lot of images of objects, but those are unlabelled. Suggest a strategy by which the robot could benefit from a limited labelled data available as a fixed dataset, and unlabelled data that the robot captures on an everyday basis. The robot may ask the human to annotate a small amount of data. This is known as semi-supervised and active learning in machine learning.

4. In the example of question 2, consider that the robot has 10 objects. Model this 10-class classification problem as multiple binary classification problems. Also, suggest what happens when one of the objects is extremely rare in contrast to the others (data imbalance problem in learning)?

5. [Programming] Take popular data sets from the UCI machine learning repository and solve the same using different machine learning techniques. Study the effect of changing hyper parameters and reason for the change in results.

6. [Programming] Using CNN, solve some standard problems on standard data sets, e.g. MNIST handwritten digit data set, NIST Special Database 19, Imagenet, Pascal Visual Object Recognition Challenge, Annotated Faces in the Wild, Face Detection Dataset and Benchmark, and Sports 1 M data set.

7. [Programming] Using LSTM, solve some standard problems on standard data sets, e.g. TIMIT Acoustic-Phonetic Continuous Speech Corpus, CoNLL 2000 chunking, and CoNLL 2003 Named Entity Recognition.

References

An, J., Cho, S., 2015. Variational autoencoder based anomaly detection using reconstruction probability. In: Special Lecture on IE 2.1, pp. 1–18.

Gers, F.A., Schmidhuber, J., Cummins, F., 1999. Learning to forget: continual prediction with LSTM. In: Proceedings of the 9th International Conference on Artificial Neural Networks. IEEE, Edinburgh, UK, pp. 850–855.

Goodfellow, I., Bengio, Y., Courville, A., 2016. Deep Learning. MIT Press, Cambridge, MA.

Greff, K., Srivastava, R.K., Koutník, J., Steunebrink, B.R., Schmidhuber, J., 2017. LSTM: a search space odyssey. IEEE Trans. Neural Netw. Learn. Syst. 28 (10), 2222–2232.

Haykin, S., 2009. Neural Networks and Learning Machines. Pearson Education, Upper Saddle River, NJ.

Hu, B., Zhengdong, L., Hang, L., Qingcai, C., 2014. Convolutional neural network architectures for matching natural language sentences. In: Advances in Neural Information Processing Systems. NIPS, Montreal, Canada, pp. 2042–2050.

Lawrence, S., Giles, C.L., Tsoi, A.C., Back, A.D., 1997. Face recognition: a convolutional neural-network approach. IEEE Trans. Neural Netw. 8 (1), 98–113.

LeCun, Y., Bengio, Y., Hinton, G., 2015. Deep learning. Nature 521, 436–444.

Li, H., Lin, Z., Shen, X., Brandt, J., Hua, G., 2015. A convolutional neural network cascade for face detection. In: Proceedings of the IEEE Conference on Computer Vision and Pattern Recognition. IEEE, Boston, MA, pp. 5325–5334.

Rutkowski, L., 2004. Flexible Neuro-Fuzzy Systems: Structures, Learning and Performance Evaluation. Kluwer Academic Publishers, New York.

Schmidhuber, J., 2015. Deep learning in neural networks: an overview. Neural Netw. 61, 85–117.

Shukla, A., Tiwari, R., Kala, R., 2010. Towards Hybrid and Adaptive Computing: A Perspective. Springer-Verlag, Berlin, Heidelberg.

Chapter 13

Learning from demonstrations for robotics

13.1 Introduction

The use of machine learning for modelling the navigation behaviour of the robot enables the robot to adapt any behaviour based on the mechanism to train the system. Humans learn as they perform any task including navigation. Even as an adult, if you fall near a construction site, it is imperative that you take extra vigilance and caution the next time you visit any construction site. Here, you have learned that your past action was poor and replaced it with a better action that makes you walk cautiously. Similarly, if you tend to go too near the obstacles at blind turns and get hit sometimes, you realize the importance of keeping a clearance. The notion of learning was also prevalent in the Artificial Potential Field, wherein the performance in terms of the path lengths and clearances experienced could be used to fine-tune the attractive and repulsive coefficients on the fly. In this chapter, the notion of learning is not just for small parameter optimization, but to learn a large behaviour overall.

Even though the neural network can model any possible behaviour, the real challenge is to be able to learn such behaviour. The simplest mechanism that does not use learning is to hard code the rules of the Fuzzy Inference System or any similar model. The rules state that an obstacle on the left motivates a rightward turn and vice versa, and the general navigation is to orient towards the goal. The parameters associated with such rules and weights can be adapted based on performance. The notion of learning in this chapter is more involved and uses no pre-fed behaviour. Hard-wired behaviours are hard for human designers to initially write, and human designers can tend to be lazy in the comprehensiveness of the rules, which can be fatal for the robot when dealing with novel situations. The notion of learning is to enable a robot to learn its navigation function based on the available data.

Learning is of three types: supervised learning, unsupervised learning, and reinforcement learning. In *supervised learning,* the data are in the form of input along with their target values. In case of motion planning, some historic record

Autonomous Mobile Robots. https://doi.org/10.1016/B978-0-443-18908-1.00002-9

is available that has the sensor percept of the robot (input) along with the optimal action taken by the robot at the time (output). In *unsupervised learning*, the inputs are available but not the target. Hence, the aim is primarily to cluster and classify the inputs based on their spatial occupation in the input space. It gives frequent patterns and trends in the input. The robot cannot act if no target is given in the training. However, the notion is to cluster the input and get different types of inputs, which can be annotated by a human expert. Unlike supervised learning where human annotator is used to annotate a significantly large data set, the notion in unsupervised learning is just to ask for annotations pertaining to a few frequently occurring patterns. Clustering of inputs is a commonly used unsupervised learning technique, using human to annotate the target of the cluster heads only. In *reinforcement learning*, the input is given without the target; however, a *reward function* is given which tells the robot how good or bad it performed. For every type of input, the robot learns to take action to maximize the reward. It tends to repeat actions for every state, which results in a good reward and minimizes the use of actions that result in poor performance. The motivations of learning, including adapting parameters on the fly based on performance and taking actions to maximize clearance on collision, all pointed to reinforcement learning wherein the performance measure came from the environment, measured by the sensors. This chapter is only for the supervised learning of the robot. Reinforcement learning is taken as the next chapter.

13.2 Incorporating machine learning in SLAM

Before we take on the harder problem of using machine learning for navigation, let us take a simpler problem of simultaneous localization and mapping (SLAM) and see the use of machine learning for improving the same algorithm. The concepts are summarized in Box 13.1.

Box 13.1 Summary of the Visual Place Recognition concepts

Visual place recognition:
- Extract global features of the current (query) image
- Lookup for the closest registered place images in the feature space (using *k*-d tree)
- If the current image matches the registered place image, perform loop closure

Reducing false positives. Only accept places if between the candidate place and query image
- Global features match well
- Local (pixel) features using feature detectors and descriptors match well
- Place is within the uncertainty region of the robot pose as per its estimate

Using better global features
- Use deep learning techniques like CNN
- Use sequence of images for CNN

Box 13.1 Summary of the Visual Place Recognition concepts—cont'd

- Use images at multiple resolutions
- Extract temporal features like using LSTM

Handling sequential nature of the problem

- Suppose the places are consecutively numbered. The robot will visit places in the same sequence. An out-of-turn detection is a false positive to be eliminated.
- Place trajectory (place ID v/s time) needs to be smooth that can be optimized against observations (matching scores), or smoothness enforced using techniques like a Particle Filter.

Patch features

- Intermediate between local features (pixels) and global features (global image descriptors)
- Encode patches or blocks of image into a feature descriptor.
- Two images match if the patch descriptors across several patches match with the same spatial consistency (position with similar translation, rotation, and scaling between the images).

Handling multiple domains

- Images do not match against different domains (e.g., morning and evening)
- **Spatial temporal maps:** One map for every domain. Detect domain and load the most relevant map (features).
- **Cross domain image matching:** Make a deep learning network that can match images from different domains.
- Convert one domain into the other

Using semantics in SLAM

- **Filtering:** Only do correspondence matching between pixels that have the same semantic label (e.g., door pixels match with door pixels only).
- **Reconstruction loss:** Use reconstruction loss as the difference between the projected and observed shape, size, and pose of the regular semantic object, like the 3D bounding box specifications should match.
- **Semantic prior:** For localization, load a map of semantics only as the feature map is memory intensive. The semantic map may be made manually or by an expensive robot. In test run, the robot localizes by minimizing the nonsemantic reprojection error (doing another SLAM) and reprojection error of the registered semantics.
- **Occupancy map prior:** In the same method as above, may add occupancy map that requires less memory, where the robot only optimizes for poses free of obstacles. Such positions can be taken from a trajectory database.

Handling dynamic objects

- Filter out detected dynamic objects and do not add them into the map or use them for localization.

SLAM as a learning problem

- Train a deep neural network with an image sequence as input and a robot pose as output.
- Or, train a deep neural network with an image sequence as input and visual odometry as output.
- In the training data set inject artificial dynamic objects, changes in common object positions, and patches resembling occlusions.

13.2.1 Visual place recognition

The overall notion of SLAM is that the robot makes a map as it travels. On subsequent runs, the robot may use the same map for localization (and maybe to further improve the map by accounting for new and changed obstacles). Localization is possible using visual odometry (using landmarks from the map), given the initial position of the robot. Typically, the robot is always aware of its position. It is origin at first, while thereafter the position is always derived from the last position. In case the robot shuts down, the position on wakeup is the same as the position on shut down. However, this is a risky logic considering many times the robot may be physically moved, while being powered off. *Global localization* is used to get the initial (approximate) position of the robot. It is also commonly called as the problem of *Visual Place Recognition*. Sometimes, the robot can get increasingly wrong with time when using visual odometry alone and therefore may require an external *relocalization* (Kenye et al., 2020), which is the same problem. Consider that the robot map consists of some known set of places (rooms in the case of indoors and different prominent places in the case of outdoor). Different sensor readings (pictures) are taken from different angles for every place. Thereafter, a classifier is used that maps the images as input into place label as output. It could be done by using visual bag of words, deep learning, or other approaches.

Another mechanism for implementing visual place recognition is to extract global features from all the images of the places. The visual bag of words technique uses visual words as global features. The images may be fed into a convolution neural network (CNN)-based model whose last layer gives the global features. Taking the gradient of each pixel represents the oriented gradient local feature, which can be binned into histograms by dividing the orientation angles into bins to create a 1D global feature called as the *Histogram of Oriented Gradients*. The global features are extracted from the query image. The nearest place image is searched whose label is a solution to the place recognition problem (or the majority label among the k-nearest place images in the feature space).

The place recognition gives a place comprising of a range of poses that the robot can be in. The specific pose still needs to be computed. Place recognition specifically solves the problem of *loop closure detection* discussed in Chapter 2. Loop closure over the key-frame pose graph is used to get the specific pose, discussed as the problem of *loop closure correction* in Chapter 2. From a machine learning perspective, there is a severe risk of false positive or a wrong place being tagged against the live image. There are three possible techniques to limit the false positives. (i) The place and the query images are checked for local consistency using the local feature matching algorithm. Feature detectors and feature descriptors are used on both the images and the point features are matched to make a correspondence function. The geometric consistency of the feature matching is done. Loop closure detection is said to have taken place only when the image pair matches. (ii) Further, another trick is that a *sequence of images* is checked for loop closure instead of the current image only. If multiple sequential images detect a positive loop closure at the same place, only then the loop closure correction is initiated. (iii) A robot may know the approximate pose or the range of possible poses within which it lies, which can be used to limit the possible place matches. A GPS may give an indicative location and only places within the GPS error are considered. Previous place matching may accurately point to a place, and the candidate places may only be within an uncertainty level time the subsequent time of motion.

A factor to consider is the size of the place. There is no hard limit on how large or small a place should be. Small places mean many classes, which are difficult to learn; but the global localization gives them a fairly good position on the map and vice versa. Knowing the place does not completely solve the localization problem. Visual place recognition for loop closure detection followed by loop closure correction is a good way to solve the problem at the global and local levels. However, an assumption is that the pose key-frame graph representing the map is highly accurate, against which the vehicle is localized using loop closure detection and correction. The map can be made with a richer sensor suite by a robot operating in ideal conditions or made with an active involvement of a human who corrects the map as it goes wrong, like manually triggering the loop closures and smoothing the roads as per the visual appeal.

13.2.2 Sequential and deep learning-based place recognition

The CNNs are actively being used for solving vision problems and visual place recognition is no exception. The CNNs can extract global features for matching the query and registered place images. An aspect of the visual place recognition problem is that it is sequential in nature. The robot takes a live sequence of images, while the place registration itself happens with a sequence of images witnessed by the robot at the time of place registration. The algorithms that match one image sequence (video) with another image sequence (video) are more useful. CNN is used to extract spatial features. The spatial features with time may be pooled to extract temporal features by another CNN or by using a separate recurrent model like the LSTM. Sample architecture is shown in Fig. 13.1. Sometimes, the features are also extracted at multiple scales (resolutions) and merged (concatenated or aggregated), highlighting different aspects at different resolutions (Lin et al., 2017).

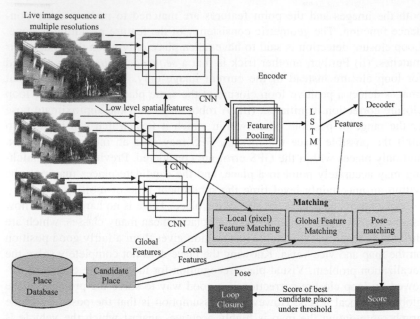

FIG. 13.1 Visual place recognition. Spatial and temporal features and extracted using CNNs and LSTMs at multiple resolutions, which are used for global matching of the candidate registered place and the current image sequence. Local (pixel level) feature matching and pose matching (is the registered place pose near/within the radius of uncertainty of the pose estimate of the robot) is used to refine the results. If a place matches, it is used for loop closure.

Another way in which the sequentially can be added is as a candidate-scoring metric. Suppose a vehicle is travelling on a road with a fixed route, while every few metres of the route makes a place. The problem is to compute the place output for every image seen by the vehicle as it travels the same route. Suppose $d(t,p)$ is the similarity between the image taken by the robot at time t and place p. The places are continuous as per definition (say, the distance from the source in the fixed route), while the vehicle also takes a continuous sequence of images with time. As the vehicle travels, it will visit from the first place to the last place along with time, making the place output a smooth trajectory as a function of time. This is shown in Fig. 13.2. Suppose the n sequential place outputs are p_1, p_2, p_3, and so on till p_n. A smooth sequence is an indication of the correct output, while a nonsmooth entry indicates an incorrect output. The nonsmooth entry can thus be replaced by another candidate place that best matches the region in consideration. Equivalently, a candidate place is correct if the place as a sequence correctly matches the observed sequence as a whole. This will be subsequently verified by using correspondence matching with local features and being within an expected place region as per the previous localizations.

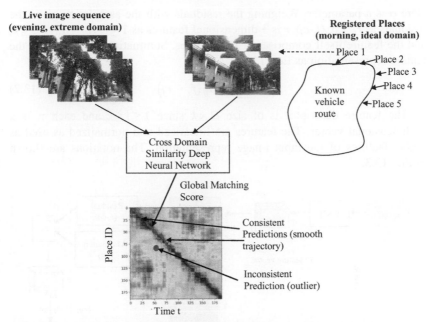

FIG. 13.2 Predicting consistent place trajectories. The places are numbered sequentially in the route of the vehicle and hence the place prediction output must be a smooth trajectory with time. The place trajectory may be optimized, locally searched with smoothness constraint, or constructed by filtering techniques (like a Particle Filter). Constructing the entire similarity matrix is time consuming and can only be done locally, which constraints the method of use and the resolution of operation.

Let us specifically discuss one popular architecture called the *Network Vector of Locally Aggregated Descriptors* (NetVLAD, Hausler et al., 2021; Arandjelovic et al., 2016). Let the image be of size $m \times n$, which is supplied to a trainable CNN model to extract features from every pixel. Each CNN filter extracts and adds the filter response of the pixel, for all pixels, repeated in multiple layers. Let the output be of size $D \times m \times n$. Read the output of the CNN as $m \times n$ feature descriptors, 1 for each pixel, each of size D, or $f_1, f_2, f_3, \dots f_{m \times n}$. Assume that a data set of several descriptors is already available, which can be clustered to get the typical features. Let there be k clusters, each cluster having a cluster centre c_j as a D dimensional vector, or the clusters are $c_1, c_2, c_3, \dots c_k$. Let the residual of feature of the ith pixel in the query image with the cluster j be given by $(f_i\text{-}c_j)$, $1 \le i \le m \times n$, $1 \le j \le k$, where the residual is a D dimensional vector. Let the association of ith pixel feature with cluster j be α_{ij}, given by Eq. (13.1)

$$\alpha_{ij} = \frac{\exp\left(\frac{\|f_i - c_j\|_2^2}{\sigma^2}\right)}{\sum\limits_{j=1}^{k} \exp\left(\frac{\|f_i - c_j\|_2^2}{\sigma^2}\right)} \tag{13.1}$$

Here σ is a parameter. Weighing the residuals with the association gives the residuals as $r_{ij} = \alpha_{ij}(f_i - c_j)$. r is 3-dimensional features as $1 \leq i \leq m \times n$, $1 \leq j \leq k$, and the feature itself comprises of D attributes. Summating across i gives the feature representation as Eq. (13.2).

$$v_j = \sum_{i=1}^{m \times n} \alpha_{ij}(f_i - c_j) \qquad (13.2)$$

The feature descriptor is of size $D \times k$ since $1 \leq j \leq k$ and each v_j is a D dimensional vector. The features are linearized and normalized as used as global features of the input image representation. The notations are shown in Fig. 13.3.

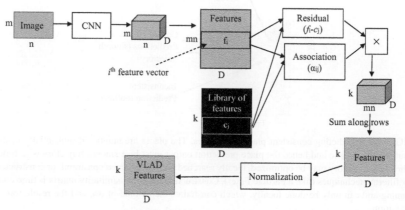

FIG. 13.3 Computation of the NetVLAD features.

So far, the focus was either on global features where the complete image is encoded into a feature vector (like passing through a CNN), or local features where point features on the image pixels were extracted, or a combination of both. Another level to work on is the *patch features*, where the image is decomposed into several patches, and each patch is assigned a global feature value (Hausler et al., 2021). Imagine a famous landmark at either the outdoor city, or a unique artefact inside the hall. In any image taken by the robot, the landmark will be a small part of the entire image, but more than a pixel. Let us say that the major chunk of the landmark is encoded as a patch. Now if only the patch matches between the candidate place image and the query image, it is a good indicator of the real place. Sometimes the landmarks may be too large and will have multiple patches that encode them, each patch being relatively unique in the data set. Not all patches may be visible due to occlusion or noise. The patch features are thus intermediate to local and global features, larger than pixel and smaller than the complete image. The matching between the candidate place image and query image does a correspondence match between the patches in both images. The type and location of patches in both images are considered for correspondence matching. The mutual errors in the patches, the geometric consistency of matched patches, and the number of matched patches is useful for scoring the candidate place. Since the feature extractor is a CNN, which is learnable, it learns to extract better features that increase distinctiveness between the patches.

As per the definition, there are two challenges to consider. First is the selection of the patches. The data set will have several possible patches that can be considered in the registered list of patches, while a patch with no uniqueness is an unlikely candidate. Since the uniqueness of a patch to differentiate places is hard to define, it is reasonable to use cluster heads as the patches against a very large data set of patches. The second is to find the best match (location and size) for a patch in the query image. The object detection libraries already move candidate windows to get the best object bounding box that best matches against the registered template, and the same techniques are applicable here. Here we can move the candidate bounding box of the patch (in terms of location and scale) until it has the best matching feature vector close to the registered patch.

13.2.3 Handling multiple domains

A problem with loop closure detection or place recognition is that a visual map may be made by using a camera in the day conditions (*ideal domain*), while the same map is used to localize the robot using loop closure in some extreme conditions like low light, evening, night, rain, etc. (*extreme domain*). The types of features visible in ideal and extreme conditions are different and therefore both global and local feature matching techniques are likely to fail (Yadav and Kala, 2022).

One of the solutions to the problem is to use *spatial–temporal maps*. So far, every landmark has been associated with a position and signature. Now the landmarks may have different signatures for different time, light, and weather conditions (different domains). For different domains, different features may get highlighted that may not be observable in other domains. Consider time as a special case. The feature signature may show a smooth change in the descriptor value with time, as light becomes dark. Sometimes it is possible to regress the feature descriptor value with time. In any case, each domain has a different map, and the localization module identifies the domain and loads the most specific map for the domain.

The challenging conditions, like evening, are hard because it is difficult to perform SLAM in the first place to record place pose against a registered place image that can later be used for loop closure. Suppose the vehicle is on a run in the evening domain. An option is now available to find the closest place image in the ideal conditions for a query image in the evening conditions for place recognition. Previously, we were maintaining separate maps for every domain. Now the aim is to have *cross-domain* place recognition, where the query image is in one condition (domain) and the candidate place is in another condition (domain). Consider a deep neural network that takes two images (or a sequence of images) and gives an output whether the two images (or image sequence) are of the same place (with a similarity 1) in different domains, or the images are of different places (with a similarity 0) in different domains. This is shown in Fig. 13.2. The network can be trained by image pairs sampled from the data sets having multiple runs in the morning (ideal domain) and the evening (extreme domain).

There is a problem with the specific architecture. Previously, global features of the query image were used to search for the closest registered place in the feature space in logarithmic time. Now there are no global features but a deep neural network that reports cross-domain image similarity. This means that a new query image in the extreme domain (evening) must be matched with all registered candidate place images in the ideal domain, which is computationally infeasible considering that deep neural networks are being used. A trick is to suppose that the previous pose (and hence the place) was known. The robot travels a known route. So, the next pose will be a little ahead in the route and equivalently, the next candidate places to consider will be among the next few in the route. Only such candidate places are chosen and the best matching candidate place among them is the current place of the robot.

Another alternative is to use a particle filter or any tracking algorithm. The algorithm predicts the pose (and hence the next place) of the robot and uses observation as the cross-domain similarity between the query image and the registered image of the candidate particle place from the deep neural network. It is possible to lose track. Using a small set of globally distinctive landmarks that can be checked against the current image is recommendable. If the algorithm loses track, it continuously gives the wrong output. However, if the current image matches a global landmark, the algorithm knows the place and hence the pose.

13.2.4 Benefiting from semantics

Let us now incorporate learning-based techniques into the working of the SLAM algorithm. A SLAM algorithm takes an image sequence as input and computes the pose of the robot and the landmark map as a pose-graph optimization problem. The images taken by the robot's camera constitute raw pixels used for solving the problem of SLAM. The pixels are affected by several factors of variations including light and weather conditions. A lot of deep-learning architectures are now available to classify the pixels into a set of semantic labels (like furniture, human, etc.) and to further detect the objects visible in the raw image and draw them as 2D or 3D bounding boxes. This can be done by using either both RGB and depth images or a combination of both images. Lidar information may sometimes be available in addition to help in the semantic segmentation.

There are several ways in which this semantics can be used to make a better SLAM algorithm. The first use case is for *filtering* wrong correspondence matches. Every landmark now has a semantic label in addition to its position and signature. When doing correspondence matching between a live frame and the key frame, the semantic label is also checked. A correspondence matching is said to have taken place between the two features only when their semantic class is the same. This helps to remove the wrong matches.

So far, the pose-graph optimization only attempted to minimize the reprojection error. Consider that the camera looks at the front face of a chair distant ahead (similarly for all faces). It is possible to triangulate and regress a regular shape like a rectangular plane in 3D representing the front face of the chair (similarly for all faces) at the key frame. The live frame can now also use the loss function as the difference in the position, shape, and size of the faces of the chair between the current frame and the key frame, in addition to the standard reprojection loss (Yang and Scherer, 2019).

A common problem with SLAM algorithms is that the maps are extremely large and often cannot be stored on the system. Hence, they must be either stored in clouds or downloaded on the fly as the robot runs, or some nonuseful features (landmarks) need to be restricted from being admitted into the map. It is, however, easy to store *semantic maps*, which may be made initially by robots with high-precision SLAM capabilities and used for localization of the low-cost robots. Feature (landmark) maps are heavily affected by light conditions; while in contrast, the semantics are relatively invariant to such factors of variations. Another use case of semantics is thus to do visual odometry in the absence of semantics, but to additionally account for the known position and shape of the semantics when detected (Agrawal et al., 2022). The error function for optimization is the reprojection error of visual odometry and the semantic reprojection error. The *semantic reprojection error* projects the registered semantics' position, shape, and size and matches with the observed position, shape, and size. If a semantic object has multiple occurrences, place recognition can be used to solve the correspondence matching problem.

Similarly, while the point features can be too many to store and be prone to factors of variations, an *occupancy map* or *topological map with shape information* is generally available. For the roads, the shape and structure are available from satellite navigation data, including the number of lanes. For indoor robots, lidar-based robots can be rented to make an occupancy map using SLAM. A new run of a robot using only the camera can now localize based on this occupancy map. The structure of the road or the corridor boundaries in the map are projected into a camera and the difference between the projected and the observed road or corridor shape is used to compute the error and calculate the pose. Similarly, the robot in both cases should be in the navigable region of the road; or in a pose from an earlier run of the robot, while all possible robot trajectories are available from a trajectory database. As the robot runs, the trajectory database is populated, removing redundancies.

13.2.5 Dealing with dynamic objects

The landmarks used for SLAM are assumed to be static as the pose of the robot is computed based on the registered position of a static landmark. Consider the moving people captured by the robot. The robot will detect features on the moving people, which cannot be used for localization. The standard localization procedure is designed to remove outliers and a few landmarks can be easily filtered out by the PnP and RANSAC approaches. However, there is a problem if the moving people constitute a reasonable proportion of the features. Therefore, such people are detected at the semantic (pixel) level or the level of the bounding box and filtered out (Kenye and Kala, 2022). No feature on such people is registered.

Another interesting object is parked vehicles. The parked vehicles are good for pose-graph optimization of the current run representing good static features, but cannot be used for other runs as loop closures, since it is highly likely that these vehicles would have moved to some other place. Other semantic-level features to be potentially filtered out are the ones in the sky.

13.2.6 Localization as a machine learning problem

In the same vein, the problem of matching the sensor percept (images, lidar scans, etc.) to the location can also be taken as a regression-based machine learning problem. The localization benefits from visual odometry computed using consecutive images. To model the same, the recurrent models are of greater interest. The model takes the sensor percept as input and gives the pose as output. The major problem with the machine learning approach is the generation of the training data since a supervised machine learning technique is used. In a controlled environment, some labels of positions may be supplied by humans and others interpolated from the human-supplied input. This is, however, not a very practical approach. Therefore, the robot needs to initially make a map under ideal settings, maybe with higher precision sensors and odometry systems for a high precision SLAM. Thereafter, the sensors may be scaled down to the ones available and extra sensors removed for localization. The scaled-down version becomes the input while the position given by high-precision SLAM becomes the target for supervised learning. The SLAM often fails due to moving objects, people, and temporary entities (parked vehicles, water bottles, etc.) on the map. For robust learning, artificial objects (patches in the image, laser scans) are added to all inputs in the training data to imitate actual objects. An alternative approach eliminates moving people, vehicles, and other temporarily looking entities from the image (replacing with a black patch) before doing localization, which is an equally justifiable methodology.

Similarly, a common problem with localization is that different lightning conditions have different visual appearances. The domain adaptation and transfer learning techniques in the machine learning literature can be used to adapt the model represented by learning in one domain (conditions) into the one of another domains (or conditions), used to convert training data from one domain to another, or used to compute the most resembling data from the other domain to be used for localization. The machine learning tricks can work when a rich map exists in one domain, but the map in the other domain is not equivalently rich.

The visual odometry relied on sensitive operations like feature extraction and matching. The features may not be worthy and may not exactly match in appearance as stored in the map, which affects performance. It is worthwhile to think if machine learning can learn a generic visual odometry as well, which takes sensor inputs interleaved with some time and gives the transformation. This is possible with the better results if the test map looks like the one used for training. The advantage is that computer simulations with virtual sensors can be used over maps hand designed to look like actual test maps to give an initially trained model. There is, therefore, no shortage of synthetic data for training. The synthetic data will have the targets given by the simulator. Factors of variation can be added in the simulator itself. A greater discussion on the aspects of machine learning will be done later in the book.

13.3   Dealing with a lack of data

To use deep learning, we need a large amount of data to learn a huge amount of model parameters, which may not be available for new problems. This is practical with motion planning as well. If a robot needs to be able to shoot its sensory percept into a deep learning network and get linear and angular speeds as output, it needs to be understood that the number of inputs possible is too large and it will require a lot of data. Furthermore, especially for this problem, the quality of the human-generated data is reasonably poor due to the high variations possible within people and the laziness of the humans to navigate the robot idealistically.

There are a few common solutions to the problem. A common technique to synthetically multiply the data by a factor is called *data augmentation*. If the problem was visual recognition to identify an object in the image, it is known that the recognition system should be rotation-invariant, translation-invariant, symmetric across all axes, and prone to resistance to noise. This can be used to multiply the number of images by applying rotations, translations, and flipping to the images. Further, noise models can be used to add noise to images. All these produce more images and increase the data set size.

The other common technique is called *transfer learning*. The technique makes use of natural observations that knowing English makes it easier to learn French, knowing C++ makes it easier to learn JAVA, etc. In all such cases, you already know some basics, which form a part of the problem to be learned. Like a person knowing English is familiar with the concepts of words, sentences, grammar, etc., and all these are also needed in French. *Transfer learning* involves the transfer of knowledge from the source problem to a related target problem. Hence, the target problem already knows some important aspects from the source problem and learns the rest. This reduces the requirement of data.

In deep learning, a convenient mechanism of transfer learning is *pretraining* (Erhan et al., 2010). In this technique, the deep learning network is first learned for a related problem over which a lot of data are available. Hence, the network is well-learned. This network is then attempted to be reused for the target problem in hand. Since the problem may have a different output dimension, the last (output) layer is removed from the source-learned network, and a new layer is added with random weights that match the output specifications of the target problem. Since the weights are randomly set, it is important to first get them into reasonable values. So, training is done on the target data with all weights fixed, except for the ones associated with the last layer. Even though the size of the data set is small, there is only one final layer to be learned. The network is further trained with the new target data with all weights trainable. This is shown in Fig. 13.4.

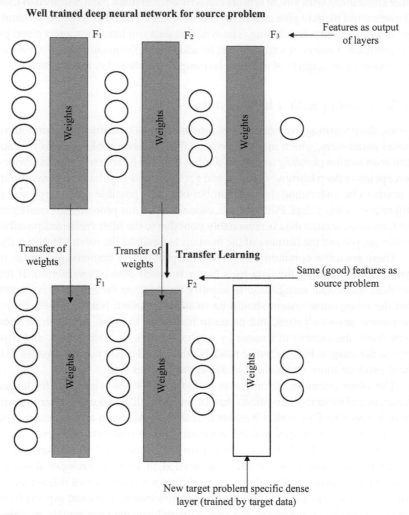

FIG. 13.4 Transfer learning: a well-trained network for a source problem is available. For the target problem with a small amount of data, the last layer of the network is replaced with a new layer, trained by the target data. Other layers can also be retrained.

Suppose the target problem was object recognition for a home application with specific objects, for which a large data set was not available. However, since a lot of data are available for general object recognition, this is regarded as the source problem and a deep-learning network is trained on it. The features extracted in the final layer represent interesting features that are generic enough for any object recognition with some bias towards the specific data set objects. However, the generics are reusable and can be transferred to any new problem. This is attempted by using the neural network on the source problem until the prefinal layer. In such a case, the network has generated good features that can be used for object recognition in a new problem. Hence the deep neural network only acts as a feature extractor, which is followed by a multilayer perceptron for the actual classification.

Imagine the robot is human-like, which sees from the video camera and uses the images to do the actual navigation. Now we do not have enough data to train such a system. However, there are many well-trained networks on images that are already available. These networks are trained on clouds by big institutions and given to the community. These represent training with big data sets and many training iterations. Now if the robot is to be trained, the learning starts from this well-trained network, because of which the robots know a lot about vision. Further training is done to specialize in the problem domain.

13.4   Data set creation for supervised learning

Let us now attempt to solve the problem of robot navigation using supervised techniques. The supervised training requires training data with inputs and targets. Getting such data is not a big problem. Video cameras already track the people from shopping malls to airports, which can generate a rich trajectory of every person over a huge space. Given the information of all the people and the static obstacles captured by such cameras, any simulation library can reconstruct the entire scene and place a virtual robot in place of the humans and record the readings as seen by the robot to make the inputs. In such a mechanism, one can potentially collect a huge volume of data as the sensors are active around the clock and give continuous data streams. The good aspect about this approach is that even though there is a little knowledge of the objective function that humans optimize when walking, there are some unexplained *social norms* in human navigation like interpersonal separation, giving right of way to a person coming in, overtaking from the right, enabling a fast person to overtake by stepping left, etc. All these are hard to mathematically study. However, since the tracked humans already exhibit behaviours that adhere to such social norms, the machine learning system will naturally learn and exhibit those to have an overall navigation system that adheres to the social norms. There are research challenges to project 2D image data into 3D, detecting and projecting obstacles in 3D, tracking the deformable human body in 3D, to handle handoff when the person moves outside the field of view of one camera into the field of view of another camera, handling occlusions, and to calibrate all cameras to report readings into a single-coordinate frame. All these problems can only be approximately solved, which limits the performance.

An alternative way to capture similar data is to use the advances in *virtual reality*, wherein a virtual world can be set up and a person can be given a virtual reality headset. The virtual reality is enabled by using a high-quality simulator that already has libraries to capture human motion and blend the world accordingly. Furthermore, the world as perceived by a robot in the position of the human can also be easily recorded using the same libraries. This avoids the hassles of embedding video recorders at numerous places and enables recording high-quality data in the comfort of a laboratory; however, the quality of data is compromised. The human may be moving nicely, but the virtual avatars in the surroundings as animated computer characters may be problematic in their motion, which captures a different and nonrealistic reaction of the human. Both approaches have a limitation that the different people themselves vary in their motion behaviour. Human navigation depends upon mood, personality, sex, age, health condition, context (like getting late for a meeting or doing a casual walk), etc. Learning one system for trajectories for all behaviours means confusing the system about the output since the labelled output for the same sensory percept input will be different. In the same situation a human may prefer overtaking the human ahead, while another human may instead prefer following. Furthermore, human kinematics and dynamics are different from those of the robot. Even if humans have social norms, the robot itself is a different class and may have different norms.

The other mechanism of data capture is by taking a physical robot and making humans *teleoperate* it, which is moving the robot with a keyboard or joystick. The teleoperation ensures that the motion of the robot is feasible and sound. However, the problem is getting actors (as obstacles) and trainers for the experiment, which may be costly or motivated enough to record genuine data to the best of their capability. The quality of learning is entirely dependent upon the quality of the data, and therefore it is important to record a lot of data, which may not be practically feasible. The robots especially operate at small speeds and the trainer may lose patience early in the experiment.

The simulations are an easy mechanism for taking high-quality data and so are for collecting data for supervised-learning systems. A simulator can be used for teleoperation by humans instead of a physical robot. The best aspect is that the simulations can be replicated across computers, which means that the simulation can be gamified and launched as a game for numerous people to play online and have their data recorded. A game may have numerous agents, each controlled by a separate person, and numerous such games may run in parallel. This multiplies the data. However, how you drive cars in games is different from how you do it in reality. Similarly, is the case of inter-personal relationships. The person will feel no force if a virtual agent collides but will never risk that in real life. Hence, the quality of data is compromised.

So far, the discussions have focussed on generation of a large pool of data using humans. A problem with the specific problem of motion planning is that mostly people travel straight at speed sufficiently small enough to avoid collision from anybody ahead or close to preferred speeds otherwise. Hence, even if a significantly large quantity of data is recorded, more than 90% of the data will be similar. There are *rare cases* like avoiding a group of oncoming people, overtaking a group of slow-moving people, have a person crossing another person aggressively while running, a person just in front suddenly changing goal, group-splitting, etc. Even if a lot of data are recorded, such rare data would be found in small numbers, typically insufficient to learn. The stress is therefore also on the creation of systems that can record a large volume of data automatically without human intervention, from which rare behaviours can be mined for rigorous training. Alternatively, if the rare behaviours are exhaustively known, it is possible to have automated systems to create data of situations where such rare behaviours are compulsorily displayed, to increase the quantity of rare data.

A good way to generate the data is from a *well-established planner* and to train the supervised system over the same data, in which case the supervised system imitates the same well-established planner. A simulator is loaded with several robots, each controlled by the well-established planner. The simulator automatically generates data useful for learning. All reactive algorithms discussed so far were prone to get stuck in local optima or get trapped in situations even after fusing with the deliberative algorithms. Typically, a set of parameters (and rules) are handcrafted that navigate the robot in some situation. However, the robot faces problems with the same parameters (and rules) in some other situation. It is possible to get better parameters (and rules) for the new situations, which however, worsens the performance in the previous situations. Getting one set of parameters (and rules) for all situations is extremely cumbersome and sometimes impossible. The parameters (and rules) of the reactive algorithms are thus extremely contextual. Sometimes a reactive algorithm may not be able to solve for a specific source-goal pair, in which situation another algorithm may have a better relevance. Machine learning on the contrary requires the human to pool in all contextual data and the machine itself learns to extract the context and activate the relevant neurons that maps the inputs to outputs representing the context and is thus better capable to handle the problem.

There is a disadvantage since the limitations of the well-established planner will also be learned. This is controllable to some extent. Suppose data are being generated from an algorithm that fails to make the robot reach the goal is some situation. Now instead of populating the data into the data table, it is possible to try a different parameter combination or a different algorithm that successfully makes the robot reach the goal, and to populate the same data into the data table. Sometimes, it is possible to curate the data in such situations where the motion planning algorithms fail. All situations where the motion planning algorithms fail are noted down. A human is used to break the complex motion that cannot be achieved into sub-goals and ask the reactive motion planning algorithm to reach the sub-goals sequentially. If the failure was because of multiple robots blocking each other, the human prioritizes them and gives relevant sub-goals to break the blockage. The curated data are fed into the data table. This is also a good technique to pretrain the machine learning system to get the basics of the navigation, which a child learns in the initial stages. High-quality human data can be fed later once such basics are learned. This type of learning is called *transfer learning*, where the machine is first learned on a problem for which there are a lot of data; and thereafter the learned model is transferred, and the machine is further learned on the specific problem with a limited high-quality data.

Reinforcement learning is the other extreme case, wherein no human intervention is needed. This removes all the limitations noted earlier. The data neither depend on the limitations of the tracking algorithms on real-life data nor suffer from the limitations of an artificial human reaction to an artificial simulator. As an added advantage, since the scenarios of training are randomized by letting the robot make good and bad moves, it also enters situations that humans would have completely avoided at the first place. The robot also gets trained on those difficult situations. Like a human will never go into the pavement to train what a self-driving car should do in such a situation, but reinforcement learning may do the same. In return, the reinforcement learning system also learns how to act when the car, for some reason, happens to be on the pavement. Suppose a model is trained using supervised learning, which is used to autonomously drive the robot. The initial percept of the robot (input) will match the ones seen in the training data and the robot will have similar motion commands as seen in the training data. However, since training is rarely perfect, the robot will deviate a little. The subsequent percept will hence be slightly deviated from the ones in the training data, for which there is typically a less training data and hence the output will be approximate, causing the robot to further deviate. Subsequently, the robot is likely to be in novel places, which were not present in the training data. The problem of motion planning is sequential, where the current actions affect the future actions. Errors in supervised learning may constantly deviate the robot from the common paths with large training data. This is unlike episodic problems like recognizing the human in front where an error in recognizing the current human does not cause errors in recognizing the future humans. Getting into novel situations (because of cumulative errors in the past or noise) is the biggest limitation of the supervised-machine learning and reinforcement learning techniques have a clear advantage.

The biggest advantage is that the system is automated, and learning happens to maximize the cumulative reward received. Learning can take place in parallel cores and on multiple machines autonomously for days, without requiring any human intervention. If the robot's behaviour needs to be modelled to act in a particular way, the robot can be additionally rewarded for behaving in such a manner. However, again there are problems to be considered in the reinforcement learning of the behaviour. (i) The performance entirely depends on the reward function, which is like an objective function. It is a rarity to get a good reward function straight-away. (ii) The trajectories are based on reward, so they do not naturally capture the social etiquette conveyed by humans. The robot does not behave like humans, unless this is modelled and set as a reward function, which is difficult. (iii) The reinforcement learning is based on the maximization of reward and therefore, the system particularly does not learn when the positive rewards are deep (good rewards come after making too many bad moves) and scanty. This imposes a big constraint on the design of the system. It is possible to have a system train for days together with no behaviour learned. The *passive reinforcement learning* technique uses a human (or another motion planning algorithm) to operate the robot, and the reinforcement learning technique learns from such data. This differs from the active reinforcement learning technique where the learner also selects the action eliminating the need of any human (or another motion planning algorithm). The human intervention to drive the robot may appear as a limitation, but it helps when the positive rewards are deep wherein the human-driven robot (or driven by another motion planning algorithm) is appropriately driven and thus gets enough positive rewards for learning. The passive stage can be followed by an active stage on a reasonably well-trained model.

The debate between supervised and reinforcement learning technique is big and important. Currently, the domain of indoor robot navigation is largely dominated by reinforcement learning techniques due to the absence of high-quality data to train or realistic data-capturing systems. There are middle grounds as well, like initial training using a supervised technique and using that as the initial model for reinforcement learning that constitutes the *transfer learning*. The supervised learning trains the model, while the reinforcement learning data are used to perform subsequent training of the same model. Similarly, the reinforcement learning reward function can be initially designed based on the supervised data and later use reinforcement learning technique to train the overall system. Extracting a reward function from a learnt model or from human trajectories constitutes the *inverse reinforcement learning technique*. The discussions are summarized in Table 13.1.

TABLE 13.1 Comparison of different data capturing methods for learning-based motion planning.

Method	Magnitude of data	Quality of data	Captures human socialistic etiquette?	Captures robot kinematic constraints?	Other limitations
Tracking humans in crowd at public places	+++ High as recording of natural motion happens continuously	Different people will react differently to the same situation	+++ Yes	– No	Tracking across multiple cameras and 3D projection from 2D images is a hard research problem and prone to losses
Virtual reality setup with simulations	+ Needs volunteers, who may work in parallel	Different people will react differently to the same situation	++ Limited as human reaction can be different from real life	– No	Moving other people randomly generates non-practical multi-robot behaviours, while moving them socially by humans will require too many volunteers
Teleoperation of a real robot	– Needs volunteers and a real robot	Different people will react differently to the same situation	+ Limited as human reaction can be different from real life	++ Yes	Real robots can be very slow and with only a single real robot data collection can be very slow.
Teleoperation on a simulator	++ Needs volunteers, who may work in parallel without a physical presence	Low as human will be lazy to correctly teleoperate the robot	+ Limited as human reaction can be different from real life	+ Yes	Poor visual feedback can make people react non-idealistically. Limitations on the motion of other humans as above
Using a motion planning algorithm for robot motion in a simulator	+++++ Simulators can generate data autonomously in parallel without human aid	Limited to the goodness of the motion planning algorithm used	– No. Some social etiquette can be added as an objective function	+ Yes	The robot can, at best, only imitate the motion planning algorithm and inherits its behavioural limitations
Reinforcement learning on a simulator	+++++ Simulators can generate data autonomously in parallel without human aid	Most of the time the robot explores with poor actions, which hence takes a long time	– No. Some social etiquette can be added as a reward function	+ Yes	Difficulty in designing a good reward function. Reward function with scanty deep positive rewards (good rewards only after numerous bad moves) are hard to learn *

*Reinforcement learning is detailed in Chapter 14. Supervised data can be used to solve the problem of the design of a reward function as inverse reinforcement learning. Supervised data can also be used to pretrain a model to be further trained by reinforcement learning. Supervised data can also be used for passive reinforcement learning.

13.5   Some other concepts and architectures

Let us discuss a few concepts very briefly that would help us to make better architectures for motion planning. Previously, a CNN was used to process data in a layered format, where each layer applied some convolutions to extract features followed by pooling and batch normalization, to be given to the forward layers. CNN with *skip connections* allows some features to pass from an earlier layer to a (not immediately) subsequent layer, skipping a few layers in between. This helps to pass selected few good features without distorting them in between.

The networks so far were learning to extract latent features (say f_i at layer i) from the previous layer (f_{i-1}) to the next layer (f_i), where every layer had more abstract features than the previous. The residual connections (Wu et al., 2019) instead make a block of layers learn the difference in features between the preceding (say f_{i-1}) and the subsequent (may not be the immediately next) layer (say f_{i+1}). The layer block learns the residual function $r_{i+1} = f_{i+1} - f_{i-1}$. These connections take input from the preceding layer (f_{i-1}) to calculate the residual (r_{i+1}). The residual r_{i+1} is added to the feature vector of the preceding layer to be given to the subsequent layers ($f_{i+1} = f_{i-1} + r_i$). The architecture is shown in Fig. 13.5A.

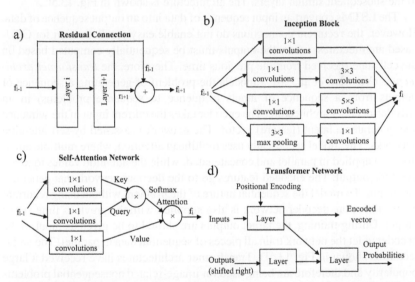

FIG. 13.5 Some important architectures. (A) Residual connections. (B) Inception block. (C) Self-attention network. (D) Transformer architecture. The layer consists of a multihead attention network, feed forward and batch normalization layers. The outputs are subjected to a softmax in addition.

The networks aim to create deep architectures to add nonlinearity. However, the deeper architectures also incur computational expenses and difficulty in learning due to vanishing gradients. The *inception networks* (Szegedy et al., 2015, 2017) aim to have the same representational capability by using wider networks instead of deeper networks. The network uses several convolution filters with varying sizes on the same input along with the max pooling operation, which is used to get the output. The 1×1 convolutions are applied to limit the dimensionality of the data. Different sizes of the filters highlight different features due to their different scales. The architecture is shown in Fig. 13.5B.

The CNNs and LSTMs treat every data as equal, while the weights are adjusted to model abstract features. Different inputs may have different importance in different contexts. As an example, as the robot moves, different laser rays will point to the most dangerous obstacles and those inputs are more important than the other laser ray inputs. The important parts of the laser image change as the robot moves. The *attention networks* (Zhang et al., 2019; Vaswani et al., 2017) multiply the inputs by an attention score to amplify their importance. The attention score is calculated based on a key embedding and a query embedding of the input, while the score is normalized by using a softmax function. The attention score may be based on a global or local context. The self-attention networks model a layer to learn the importance of the specific input as per the current context. The input goes through the processing as per the modelling to produce a value. The values are multiplied by the attention value that can amplify or dimmish the output. This forms the input to the subsequent similar layers. The architecture is shown in Fig. 13.5C.

The LSTMs convert an input sequence of data into an output sequence of data. However, the recurrent connections do not enable easy parallelization for GPU-based architectures, where the outputs must be sequentially computed based on the cell state received from the previous time. Therefore, the *transformer architecture* (Vaswani et al., 2017) models the problem of generating a sequence of outputs from a sequence of inputs (sequence to sequence problems) in an encoder-decoder architecture. The encoder takes the current input in the sequence and generates a latent (feature) vector. The network is assisted by self-attention networks. The model specifically uses multihead attention, where multiple attentions are applied in parallel and concatenated, while the attention values together yield the output. The encoded features go to the decoder network that generates the output. To model the sequential nature of the problem without using recurrent connections, the decoder network is also supplied with the previous time step's output. During training, the target outputs are shifted to be given as input to the decoder to let the network train all pieces of sequential data in parallel. The architecture is shown in Fig. 13.5D. Transformer architectures have received a large popularity and therefore are being used for image-related nonsequential problems as well. In such cases, the images are read as a sequence of sub-image blocks, read sequentially from top to bottom and left to right, making a visual transformer architecture (Dosovitskiy et al., 2020; Han et al., 2020).

13.6   Robot motion planning with embedded neurons

The problem of robot motion planning is now solved using the general principles of neural networks. The simplest approach involves representing the entire configuration space of the robot as a neural network by plugging in neurons in different parts of the configuration space (Yang and Meng, 2003). Every neuron is connected to the neighbouring neurons, which thus forms a neural network. The aim of the problem solved by embedding neurons in this manner is not to solve the problem in a typical supervised-learning technique; however, the aim is to model the neural dynamics in such a network in a manner that the robot naturally avoids the obstacles and moves towards the target because of the information exchanged between the neurons in such a network.

The inspiration for such modelling of the problem is again by the biological analogy. In the brain, every neuron has a spatial placement and the information stored in the network is in the form of the potential that it maintains. The difference between the potential values of two connected neurons gives rise to a transfer of electrical signals between the neurons that resembles information sharing. Since the biological brain is a complex network, there is a complex transfer of information within cycles and complex topologies that constantly happens to denote the dynamics of the human brain. Some of the neurons are directly connected to the sensory percept, which derives their inputs from the external senses that continuously drive the neural dynamics. If the sensory percept is kept constant for a time, the constant dynamics and transfer of information between the neurons will happen until the system converges, which constitutes the stable state of the biological neural network.

Using the same analogy, consider the neural network embedded in the configuration space, with every neuron connected to the neighbouring neurons. For a circular robot, the network is shown in Fig. 13.6. The obstacles act as the external inputs which continuously pass information in the network about the presence of obstacle and therefore the robot must keep away from such configurations. This can be viewed as the neurons in or near obstacles clamped to a high-potential value. Similarly, the goal also acts as an external input, which tries to convey the information in the network that the configuration is very desirable, and the robot must try to attain it. The goal can be similarly visualized as a neuron clamped to a small potential value. If the map does not change, the neural network dynamics continuously changes the potential values of all the neurons, in which the obstacle information and dynamic information as potential values are continuously migrated into the neural network. The network eventually converges with the potential value of every neuron denoting its desirability in terms of being away from the obstacle and near the goal.

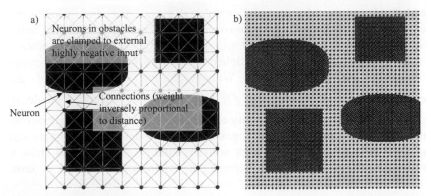

FIG. 13.6 Motion planning using embedded neurons. Neurons transmit information about obstacles (negative potential) or targets (positive potential), within the network. The robot moves towards the neighbouring neuron with the highest potential value. (A) Sparse view. (B) Dense view.

If the map however changes, some neurons not inside obstacles may now denote obstacles, and some neurons earlier in obstacles may now be free. This instantaneously changes the external inputs and the values of the clamped neurons, which are readily propagated to the other neurons in the network, to stabilize in a new convergent state. This points to the neural network modelling to readily adapt to any changing environment, which is needed for motion planning for dynamic environments.

Consider a neuron representing the configuration q. The dynamics of the neuron is given by Eq. (13.3):

$$\frac{d\xi(q)}{dt} = -A\xi(q) + (B - \xi(q))\left([I(q)]^+ + \sum_{q'} w(q, q')[\xi(q')]^+\right)$$
$$- (D + \xi(q))[I(q)]^- \tag{13.3}$$

The term $I(q)$ denotes the external input supplied directly to the neuron at q. The input is given by Eq. (13.4):

$$I(q) = \begin{cases} E & \text{if } q \text{ is target} \\ -E & \text{if } q \text{ is obstacle} \\ 0 & \text{otherwise} \end{cases} \tag{13.4}$$

Here, E is a large constant representing that the target is highly desirable and the obstacle is highly undesirable. The term $[I(q)]^+$ will only be active for the target, which is called an *exhalatory input*. Similarly, the term $[I(q)]^-$ will only be active for the obstacle and is called the *inhibitory input*. The other terms denote the neural activity. The summation over q' as per representation is over all the neurons in the configuration space; however, only the neighbouring ones have a sufficiently large weight and the summation can be conveniently read as over the neighbouring neurons only. The term $\left([I(q)]^+ + \sum_{q'} w(q, q')[\xi(q')]^+\right)$ adds the desirability of a neuron by summing all the positive parts, including the external inputs and weighted addition over the neighbouring neurons, and hence a large value denotes a favourable neuron, which stresses on gradual increase of the potential. The term is multiplied by $(B - \xi(q))$ where, B is the upper bound of the neural activity. Without B, the second term of the equation $\left([I(q)]^+ + \sum_{q'} w(q, q')[\xi(q')]^+\right)$ acts as a decay rate with the decay proportional to the positive activity of the external inputs and neighbouring neurons. The inclusion of B changes the activity such that for targets and neurons near the target with high activity initially, the term will show a fast decay until the activity becomes B, wherein the change is zero.

Similarly, if the initial neural activity was small, the activity will rise to B and then the change will be zero, or the activity will be B. If the neuron is not associated with any positive connections, the term will not contribute to the dynamics. The exhalatory effect is counterbalanced by the inhibitory term. The inhibitory external input $[I(q)]^-$ contributes to the dynamics by multiplication with a similar term $(D + \xi(q))$, where D denotes the lower bound of the neural activity and the discussions are similar. Not taking neighbouring neurons is a convention that the robot just needs to avoid the obstacles and hence the penetration of inhibition in the neural network may not be high.

The weights of the network are not learned in a backpropagation style, however, they are set heuristically and in this case, the heuristic is clear that two neurons close to each other affect each other dramatically and hence have a high weight. The high weight also enables the neurons to share similar equilibrium potential value. On the contrary, two far-away neurons are not connected or are said to be connected with a weight of zero. The weight function is modelled as Eq. (13.5).

$$w(q, q') = \begin{cases} \dfrac{W}{d(q, q')} & \text{if } q' \in \delta(q) \\ 0 & \text{otherwise} \end{cases} \tag{13.5}$$

Here W is a constant. $\delta()$ is the neighbourhood function defined by the neurons at a distance of less than a threshold (η), or $\delta(q) = \{q' : d(q, q') < \eta\}$.

Each neuron of the neural network represents a configuration, while the potential associated with the neurons denotes the cumulative desirability of the configuration. Hence, at any point in time, the robot chooses the next configuration or neuron to travel to by selecting the best neuron or configuration among the neighbours. It is assumed that a controller exists that can get the robot from its current configuration towards the next configuration. The next configuration would also dynamically change as the robot gets close enough, wherein it again chose among the best neighbours. The dynamics can be written in such a manner that there are no local optima. The selected target neuron or configuration q_t by the robot currently at configuration q is hence given by Eq. (13.6)

$$q_t = \arg\max_{q' \in \delta(q)} \xi(q') \tag{13.6}$$

The path planned by a robot is shown in Fig. 13.7. Fig. 13.7A, C, E show the path in the configuration space (along with neurons) planned by jumping to the highest value neighbouring neuron at every point of time. Fig. 13.7B, D, F show the path traced by a circular robot in the workspace, at every time moving towards the highest value neuron in the neighbourhood.

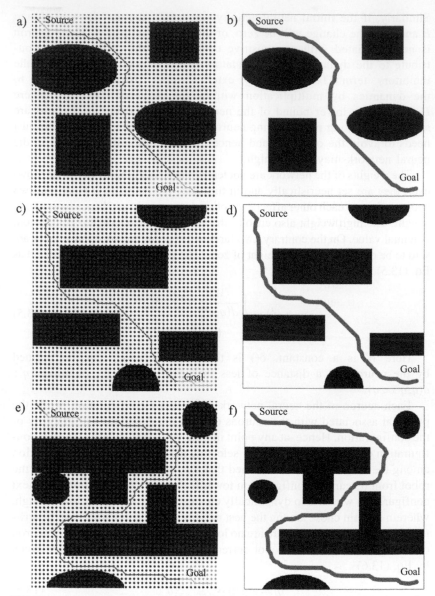

FIG. 13.7 Robot motion planning using embedded neurons. (A, C, E) Planned path in the configuration space by jumping to the nearest highest potential neuron. (B, D, F) Reactive motion of a robot constantly going to highest value neighbouring neuron.

It will indeed be curious to note if this mechanism of motion is doing anything different from the potential field, which also reactively gets the same properties satisfied. In the potential field as well, the repulsion is applied from all obstacles, and attraction is applied to the goal, while a cumulative summation of both is used for the actual navigation of the robot.

The manner of discussion of the potential field in this book was as a reactive planner, which lacks completeness and global optimality; however, the embedded neurons ensure completeness, which depends on the dynamics modelling of the network. More specifically, the embedded neurons act as a *navigation function*, giving a direction of motion in the configuration space for every possible robot position. The major difference is that in the embedded neurons, the information about the target traverses all possible candidate paths and reaches all the neurons representing a sampled configuration space. However, in the potential field, a direct straight-line distance to the goal was modelled, that may not be a possible path. Furthermore, the neural networks enable a far greater possibility of modelling the dynamics of the network as a recurrent neural network, contrary to the potential field restricted to a single attractive and repulsive equation.

Another analogy in the same effect can be taken from the wireless sensor networks and the computer networks routing algorithms. The data packet can cover the entire geographical world and travel autonomously from your computer system to a remote web server without knowing the complete map. In both the cases, the environment or the network route itself is dynamic and can change very often. Typically, the neighbouring nodes are connected in complex topologies. The routing happens by routing tables at every system, which in this case means that every node selects the best outbound destination for a data packet. The network discovery happens when every node propagating its local knowledge of the network. In the specific case of robot motion planning, the neural network imitates the computer network or the wireless sensor network, and the local knowledge of the obstacle and targets is propagated across the neural network for pathfinding. The only difference is that the neurons discussed here are virtual, unlike the physical networks.

The application discussed here did not incorporate kinematic constraints. It may be curious to know what happens if two neighbouring configurations are not possible by any controller like a sideways-walk. The primitives are the same as cell-decomposition. Only two easy to connect configurations are added as neighbours and hence for differential wheel drive robot, two configurations that only or primarily make the robot travel sideways shall not be connected. This produces paths that are possible to control by the low-level control algorithm.

13.7 Robot motion planning using behaviour cloning

It is intuitive to solve the robot navigation problem in a supervised-learning paradigm. It is assumed that the machine has access to a large trajectory data set, which will be used for learning. The trajectories are now called as *demonstrations*, where the word is derived from the fact that a human demonstrates a play of the navigation game (trajectory) while the machine is expected to learn from several such demonstrations. The problem is therefore called as *Learning from*

Demonstrations. Another nomenclature calls the problem as *Imitation Learning* where the name is derived from the fact that a human is playing the navigation game observed by the machine, while the machine is expected to *imitate* the human game play.

Given some demonstrations, there are two things that the machine can do. The first is to learn a model (say a deep neural network) which given the sensor input gives the control (velocity) output. This is specifically called as *behaviour cloning*. The name is derived from the fact that the human game play is an exhibition of behaviour, and we try to create a machine learning system as a clone of the same human behaviour. This is done by generating several demonstrations using human and the same data to train the machine.

The other way of doing imitation learning is to learn what the human is attempting to achieve. This is done by modelling a reward function from the demonstrations. The machine tries to learn a reward function, which the human is attempting to maximize. This is called as *inverse reinforcement learning*. Thereafter, given the reward function, the machine itself plays the game to maximize the reward. Inverse reinforcement learning techniques shall be taken as the next chapter in the book alongside reinforcement learning.

The behaviour cloning problem will be solved by using deep neural networks. The deep neural networks in this section can be taken as a black box, which gives the architecture and meta-parameters of the learning algorithm, takes a training data set as input and give the optimized neural network as the output. The optimized neural network can then be used to solve the motion planning problem. The challenges from an application perspective are only to be discussed now. This section discusses the modelling of inputs and outputs for different types of data. The section also discusses the creation of the data set and the challenges in the same.

A typical design principle in every problem being solved with deep learning in the modern times is to have as many inputs as possible that help in decision making, while making strategies to get a large pool of labelled data, assuming availability of an abundance of compute typically provided by public cloud services. However, care must be given for the specific problem of behaviour cloning as sometimes more inputs may lead to a drop in the performance due to the problem of *causal confusion* (de Haan et al., 2019). As an example, imagine that the robot is trained to give linear and angular velocity outputs using the camera's image as input. Mostly the robot will travel straight and make a few turns. Now imagine if the current speed is also added as an input. Mostly, the robot will be using the same speed as previously for large durations of time. Therefore, if the current speed is directly used as an output taking nothing from the image, the training error will be small, which appears to be a good network. The problem is a high correlation between the current speed as input and next speed as output that resists the machine to take anything from the image data. The concepts are summarized in Box 13.2.

Box 13.2 A summary of terms.

Using learning for navigation

- Reactively avoids static and dynamic obstacles
- Uses raw sensor readings as input and produces motor commands as output
- Does not rely on hard-wired rules/model written with a human bias. Any input/output mapping model can be learned.

Variants of learning-based motion planning

- A trajectory (data set) collected by any means is called a *demonstration*. Equivalently, the problem is called *Learning from Demonstrations* or *Imitation Learning*.
- **Behaviour cloning:** Learns a machine that maps demonstration inputs to target outputs.
- **Inverse reinforcement learning:** Extracts a reward function from demonstrations.

Types of learning

- **Supervised:** Inputs and target outputs are given. E.g., a human teleoperates a robot. The model learns a mapping of lidar readings and goal direction (inputs) to robot linear and angular speeds (outputs).
- **Unsupervised:** Only inputs are given, while learning tries to cluster or model frequent input patterns. E.g., robot sees images of numerous objects. It will not know the object names, but after learning, it can give one image per object to an expert for labelling.
- **Reinforcement:** Inputs and performance judged by a reward function are given. E.g., robot makes a few steps and is rewarded for reaching closer to the goal or punished for hitting an obstacle.

Challenges in behaviour cloning

- Most of the data are straight and not important. Challenging data like requiring a sudden turn is rare; such data need to be mined out and amplified during training.
- Data on rare situations and rare behaviours may not be present. It needs to be either mined from larger data sets, or data sets curated to create situations involving rare behaviours.
- The robot will eventually drift into regions not foreseen in training due to training errors and noise. The sequential nature of the problem keeps adding drift. Data are thus iteratively collected and aggregated, moving the robot by the trained system, and using targets for such inputs from new demonstrations (or targets collected from the expert).
- Demonstrations are imperfect and target outputs may not be correct. Training should filter them out.
- Robot kinematics and dynamics are different from those of a simulated robot or moving people. Models should be pretrained on simulators and models adapted on the real robots.
- A robot needs to be trained to avoid moving people, which requires modelling the social behaviours of real humans. This may not be possible as the human reaction to robots cannot be predicted. The machine may be trained to imitate preknown behaviours like overtaking from right, crossing, following, etc.
- The same situation may have multiple correct targets. Sometimes a human follows a person, while other times it overtakes the person in the same situation. The models may hence be made to give a multimodal probability distribution of actions as output.

13.7.1 Goal seeking using raw sensor inputs

The first and foremost problem from a design perspective is deciding the inputs and outputs of the neural network. Since the robots are being kinematically controlled and the aim is to make a controller that kinematically moves the robot, the linear and angular speeds are acceptable as outputs. For reactive systems, the inputs are typically and conveniently taken as the distances from obstacles, directions to obstacles, distance to the goal, and direction to goal, while typically the nearest 3–5 obstacles are considered. The obstacles can be taken in a clockwise direction and within a field of view and distance threshold for the sake of similarity to the methods discussed so far. The problem with such inputs is that it needs the segmentation of obstacles from raw sensor readings by the robot.

The alternative used here is hence to place a ring of sonar sensors on the robot and take those as the inputs to denote the obstacles, along with the distance and direction to the goal. Richer and higher dimensional input data employ a (real or virtual) lidar scan or video camera, alongside the distance and direction to goal.

Care must be given when placing the virtual lidar scan though. The number of inputs will be large, which may mean the necessity of having a large amount of training data, large training time, and the need to operate with a large model complexity network. The lidar scan may alternatively be sampled down to make the number of inputs as small. Unless deep learning is used, feature engineering approaches need to be applied for dimensionality reduction of a complete laser scan. Typically, *principle component analysis* (PCA) is an unsupervised dimensionality reduction technique that transforms the original data into a new coordinate axis system by applying transformations. The transformations are such that the variance along the first few axes is maximized, while that along the last few axes is minimized. The axes are now numbered or ordered as per their importance. Since the variance along the lower axis is very small, they do not carry much information as all the data have the same value and hence those could be dropped. The technique is only possible when there is a natural redundancy in the data or some features can be obtained as a weighted combination of the others, which is what the PCA exploits. PCA is a typical way to get rid of the number of dimensions of input data.

Attributed to the advancements in deep learning, let us model a CNN that takes the lidar readings as input. The first layer uses a 1D convolution to extract the features. The CNN here extracts the spatial features through several convolution layers. The features are flattened and given to a dense layer for deciding the linear and angular speed. The dense layers also use the goal information (distance and angle to the goal with respect to the current heading of the robot) for deciding the outputs.

Instead of a laser, another navigation mechanism gaining popularity imitates human motion with inputs from the eyes. The setup uses a camera as input, along with the relative distance and direction to the goal. The monocular cameras do not sense depth and hence the use of a depth camera or stereo camera can give additional information to the network for learning. CNN is a good choice for both vision and lidar inputs.

The CNN does not take sequences as input, which may be important to steer a vehicle. This can be done by an LSTM, which however does not have the richness of taking convolution and extracting those as features. Hence, a hybrid network is typically the choice wherein the CNN acts as a feature extractor. The output is fed into an LSTM, which handles the sequences on such an input with problem-specific-rich features. The LSTM extracts the temporal features, while the CNNs extract the spatial features. The hybrid architecture is shown in Fig. 13.8.

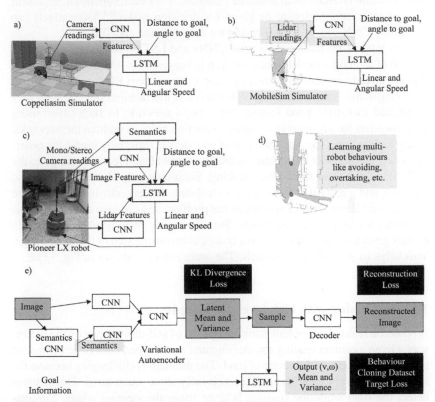

FIG. 13.8 Behaviour cloning. (A) Using a vision camera. (B) Using a lidar sensor. (C) Using a camera and lidar sensor. (D) Learning multirobot behaviours. (E) Using a variational auto-encoder to get features that are an input to the Behaviour Cloning LSTM model.

The lidar input sees the world in 2D through the laser rays and good to extract the obstacle occupancy in the 2D world. The cameras can look at the visual features like an oncoming person that should be avoided differently than a static obstacle, the semantics of the obstacles that can affect the avoidance strategies, the terrain, and the nature of the road that can affect navigation, and higher obstacles that are above the lidar's sensing plane. It is possible to use one CNN to extract the lidar features, another CNN to extract the image features, and finally give a concatenation of both features to the LSTM. The LSTM also receives the goal information for computing the output.

While theoretically the CNNs can extract semantic features and people's information and use the same for decision making, powerful semantic and object detection libraries exist that can convert any image into semantics of the pixels (road, obstacle, human, etc.) or detect bounding boxes of the objects (humans, furniture, etc.). This is a good information for decision making. It is possible to add the semantic information as additional channels in the image. A standard image is three channels, consisting of the red, green, and blue colour channels. Additional semantic channels, 1 for each semantics, represent the semantic information to be given for decision making. Alternatively, an additional CNN can be used for handing the semantic specific input.

The models are built upon several CNNs and LSTMs on images and lidars. Instead of training a model from scratch, it is suggested to use pretrained models that are already capable of extracting good features from RGB or lidar images. This constitutes transfer learning. Alternatively, the labelled training data may be limited to extract good features for a large network. In such cases, using auto-encoders for unsupervised feature extraction helps to pretrain the networks. Using architectures like variational auto-encoders is especially useful for generalization. The encoder architecture is used to extract latent features that go to the LSTM or dense layer for decision making. Simultaneously, the features are also decoded, and a reconstruction error is calculated. The encoder is now trained to extract good latent features specific to the problem so as to give a small error on the target linear and angular speeds. Simultaneously, the encoder is trained to extract generic features that help in a noiseless reconstruction of the input image. This helps to generalize the models. The architecture is shown in Fig. 13.8E.

13.7.2 Driving with high level commands

A related and simpler problem is the navigation of self-driving cars on the motorway. Given a camera reading, the steering and linear speed need to be controlled to keep the vehicle on the centre of road. The problem is challenging because the road may not be straight and may occasionally show sharp turns. The cameras detect the road while the centring is done using the steering wheel. Simultaneously the vehicle must avoid collision with the vehicle in front and stop in good time. The camera has a visual input and therefore, the CNN is a good candidate to extract features and regress the desired steering angle and linear speed.

The problem is slightly complicated for the scenarios having multiple vehicles. In such cases, the vehicle must additionally decide whether to change lanes or drive in the current lane. It is neither advisable to drive behind a very slow vehicle, or to frequently change lanes. The decision may be made by using simple heuristics or machine learning. Overtaking requires lane changes with good speed controls. In either case, the machine learning system needs to execute the decision, controlling the steering and linear speed. Similarly, in the case of an intersection, the vehicle must first decide whether it has the right to access the intersection using machine learning or heuristic measures. If it does, the onboard learning-based controller must execute the decision. The controller output depends on whether the vehicle will take the left exit, the right exit, any other exit, or the straight road.

The overall problem is the same as discussed previously. However, there is no specific distance to goal and angle to goal in this problem. The problem instead has goal as some high-level commands like go straight on the road, change to the left lane, change to the right lane, overtake, take the left exit, stop before intersection, etc. The decisions may come from heuristics or another machine learning system. The goal information is hence supplied as a vector that encodes a high-level command. Say there are k goal action commands a_1, a_2, .. a_k. The input is a 1-hot vector encoding where the input is a k-dimensional vector with 1 against the selected action and 0 for all remaining.

The same problem happens with the indoor autonomous mobile robots as well, where high level commands like walk through the corridor, enter the next room on the right, avoid the person in front from the right only, do not enter the left room at any cost, etc. can be supplied.

Let us revisit using linear and angular speeds as output. Imagine that there is a small obstacle straight ahead of the robot. Now the robot can avoid the obstacle by going from the left or the right side, both being equally good means for obstacle avoidance. Several humans on being asked to teleoperate will use different sides to avoid the obstacles. The machine gets very different target outputs for the same inputs, which can lead to an ambiguity and difficulty in learning. Therefore, the output is more correctly a *multimodal probability density function* of linear velocity and angular velocity, where every subjective action of the human is probabilistically represented as a modality. In this case both negative and positive angular speeds will have (nearly the same) maxima in the output probability density function. Similarly, humans will disagree in the distance to keep while avoiding the obstacle. Different humans will have small to large clearances, while the machine can be made to learn the distribution modelled by having the output as a probability density function. The outputs are both mean and standard deviations of linear and angular speed (or both mean and a covariance matrix). The inputs are given to the network to get a probability density function as output. A sample is taken from the probability density function to be given to the robot as the current velocity. The loss function is the probability of observing the velocities in the demonstration target (to be maximized).

13.7.3 Goal seeking avoiding dynamic obstacles

A challenge is when the robot operates amidst moving people. Moving people have social conventions which governs their future motion. As an example, people give way to a person in a hurry, people avoid each other from the left (right in certain countries), give priority to an incoming person in a room, etc. The robot must be able to predict the future trajectory of each human considering the social conventions and use the same for motion planning. This divides the problem into a prediction problem and a navigation problem. The prediction system predicts the future trajectory of each human in the scenario, while the same algorithm can be used to predict the future position of the robot itself (called the *dreamer*). Thereafter, given a future position of the robot, the navigation problem is just to learn a controller as the predicted position acts as an immediate goal that the robot must attain as the next step (called the *controller*). Together the two systems make the *Dreamer-Controller architecture*. It is possible to learn both the problems together as one machine learning system as well. The problems of prediction and navigation are thus similar. Prediction gives the next position as output, while navigation gives as output the next control command. The position and control commands are related to each other by the robot kinematics.

Since the people behave differently than static obstacles, let us model them separately. Assume that the robot knows the position and speeds of the other humans. The robot continuously detects and tracks people in the robot coordinate frame and knowing the pose of the robot, we can transform the people's position and speeds in the world coordinate frame. Both camera and lidars or a combination of both can be used for detection and tracking.

Let us model the problem as a different encoder-decoder network (Kothari et al., 2022). The encoder takes the trajectories of all people as input and encodes them into a vector. The decoder however does not reconstruct the input. The decoder is used to construct the future trajectories. The loss function of the encoder-decoder network is the error between the future trajectories predicted by the decoder with the target ground truth.

The input to the encoder-decoder is now the historic trajectory (including positions and speeds) of a variable number of people, while the output is to predict the next position of the robot. The historic trajectory of all agents consists of a temporal data of positions. Let us encode or embed the trajectory into a feature vector, called the single human *trajectory embedding*. As an example, pass the historic trajectory into a LSTM (equivalently CNN for a fixed history size), whose output is the embedding. Alternatively let the model be an encoder-decoder network, whose bottleneck layer gives the feature embedding of the trajectory. The logged trajectories were in the world coordinate frame while the system is being modelled in the robot coordinate frame. Therefore, all positions are transformed into the robot coordinate frame with the robot as the origin and the heading direction of the robot as the X-axis, before passing through the trajectory embedding network. Similarly, the velocity vectors are taken as relative to the robot velocity.

The information is separately available for each human, which should now be *socially pooled* to get the input. Social pooling assimilates the information of all neighbouring people that affect motion into an input tensor fed into the network. Humans mutually affect each other's motion. Therefore, pooling of information is needed to model the inter-human interactions. Let us cast the individual human's information into a local grid of size $m \times m$ around the robot (with the X-axis of the grid as the heading direction of the robot) and discretize the grid space into cells. A cell may be free, may have a moving person (encoded by an embedding), or may have a static obstacle. The cell stores the embeddings for each of the case. Static obstacle and free cells are given a constant embedding. The input is now the grid of embeddings. The grid embedding encodes the spatial interaction features. Passing the grid embeddings into a CNN extracts the interaction features among the agents. Passing the grid embedding into a LSTM can get the temporal interaction features useful for decision making.

The pooled embeddings after LSTM are given to the decoder. The decoder may be a dense layer to determine the output (decoded) future trajectory. The layer is also supplied the goal information. Alternatively, the pooled embeddings can be given to an LSTM decoder model along with the goal information to determine the output. The output is typically a probability distribution of the next position represented by a mean and a covariance matrix for a single modal case. The loss function is the probability of seeing the target output as per the output probability density function, which needs to be maximized. The architecture is shown in Fig. 13.9A.

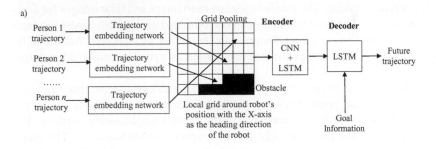

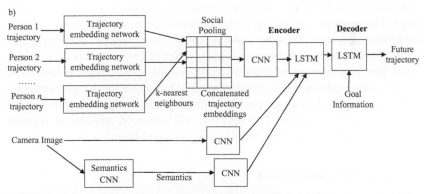

FIG. 13.9 Trajectory prediction. The spatial temporal trajectories of all humans are embedded by passing through a network, the trajectories (positions and speeds) taken relative to the robot. (A) Pools the trajectories as a grid (along with modelling obstacles and free segments), while (B) concatenates the trajectories. (B) Additionally takes the image (or occupancy map) along with its semantics. CNN and LSTM are used for encoding. The encoded vector is decoded to produce the future trajectory, also using the relative goal coordinates (or a high-level command like take a left turn).

Let us discuss another mechanism of solving the same problem. Suppose the robot is operating at a relatively wide place and therefore the static obstacles need not be considered. If the robot always moves towards the goal in such wide areas, it is unlikely to come in the way of static obstacles. Let the robot's trajectory be only affected by the k-nearest neighbours. The positions are converted into a weight vector using the elliptical model discussed in the Artificial Potential Field approach. A person nearby has a higher weight in contrast to a far-off person, and a person in the direction of goal has a higher weight in contrast to a person behind. Using these heuristics, the top k neighbours are considered. The relative position and speed embeddings of these people are pooled. The pooling is by a direct concatenation of the trajectory embeddings. Alternatively, the information may be aggregated by applying max pooling. The pooled vectors thereafter go through a CNN and LSTM network for the extraction of temporal features. The encoding is given to the decoder to predict the future position as a probability density function.

The above mechanisms better model the agent interactions and use the same for decision making. The previous section used image and lidar images for decision making. The two techniques can be combined giving both inputs for decision making. The relevant features are extracted and pooled. Additionally, semantic features can be extracted and pooled as well. The combined architecture is shown in Fig. 13.9B.

The problem of prediction of future trajectories can also be visualized as a spatial temporal graph. Each person (and the robot) is a vertex of the graph. Every person's vertex is connected to the nearest neighbouring vertices. The vertices store the properties like position, speed, embeddings, etc. The graph has a temporal aspect as the properties change with time. The *Graph Neural Networks* (GNN, Wu et al., 2020) can be used to process the information of this graph. The problem is given the historic information of this graph, to calculate the future graph state that stores each person's future position. The graph convolution neural network propagates the properties of vertices (and edges) to calculate new properties as an implementation of the convolution operator. The output is the properties transformed by another weight vector that gives the output properties. The GNNs are popular tools to process graphs and can be used for predicting the motion. The network will however, requires a separate controller module to drive the robot to the predicted pose.

13.7.4 Generation of data

The data can come from a variety of ways that have already been discussed. We focus on the most obvious method, which is to employ many users to teleoperate robots in simulation to generate the data for training. At every instance of time, the simulator is used to note the inputs from the virtual sensor on the virtual robot, or to measure the distances and angles of the obstacles with respect to the robot, based on the chosen inputs. The outputs are already available with the simulator which is the command velocity applied by the human teleoperator. The maps are easy to produce in simulators. The inputs and outputs at every point of time are logged into a data file, which grows fast and produces a lot of data for learning. Care must be given that different people will move the robot differently and hence the learned model will be intermediate between all the training provided. The change in robot driving behaviour by different people will create difficulties in learning and possibilities of overfitting. The data collection setup is shown in Fig. 13.10.

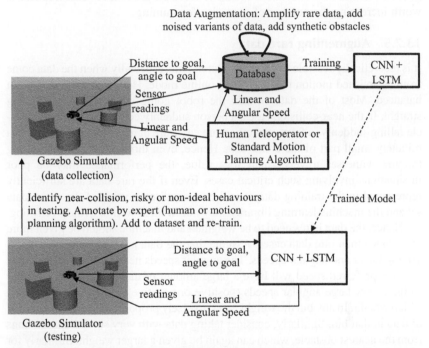

FIG. 13.10   Data collection and testing framework. Data are iteratively collected, using an expert to give targets on trajectory segments that the learned model gets wrong. Rare data (like sharp turns, near-collision) are amplified during training. Instead of CNN, using classic features or down sampling lidar data are also possible to make easier learning models. SONAR ring replaces lidar for budget robots.

Humans are likely to wear out and drive the robot randomly and thus not generating enough data. Therefore, let us say another method is utilized to record the data. This is to use established motion planning algorithms to autonomously move all robots in the simulation scenario. Given that the robot navigates well, the recordings of the virtual sensors on the robot are used as inputs and the robot action as output to populate the data file. The data file is then used to learn the robot behaviour.

It is important to ensure that all possibilities have been taken by generating many scenarios, maps, sources, and goals. If the robot needs to operate among other robots or humans, multiple behaviours like following, getting aside, overtaking, getting overtaken, etc., are possible and each of them needs to be explicitly recorded. Since it is a machine learning approach, the robot can make only good decisions if some similar input was seen by it in the past. Whatever be the method, an input and output file is generated. Typically, it would be broken into training, testing, and validation data sets. After training, the network is applied in a new scenario with a new pair of source and goal. Ideally, if the training was good, the neural network should be able to solve the problem. In case it fails, it is worth increasing the number of scenarios for training.

13.7.5   Augmenting rare data

It is theoretically feasible to generate a lot of data, especially when the data come from established motion planners in an online mode. However, the data are not balanced. Most of the data contain the robot going straight or nearly going straight. In the near-collision cases, a person suddenly coming in front, an obstacle falling suddenly in front, and other rare cases are hard to find and constitute a minutely small part of the entire data. Hence, even though the machine during training witnesses a very small loss value, the performance may be poor in situations involving such critical cases. Even if the rare data are sufficiently represented in the training data, it will be a small proportion of the entire data set and the machine learning libraries may not fit them well to avoid overfitting.

Hence, the data items need to be weighted and machine should be given more importance than rare data cases. Assume that the data are filtered in bins representing linear and angular speeds. The angular speeds near zero and linear speed near the preferred speed will have a large amount of data, and hence a smaller weight. Very large angular speeds (positive or negative) will be rare and hence a high weight. In any bin the weight is set inversely proportional to the magnitude of data in that bin. Similarly, consider taking data with very small lidar readings from the nearest obstacle, which can again be given a larger weight. Similarly for the angle to goal, where large angular deviations to the goal are rare. The rare data can be augmented or supplied repeatedly to the machine learning algorithm.

Sometimes the machine needs to learn from rare situations, which may not be present or easy to record. Such data are curated where the scenario specifications are such that the rare behaviours are generated. As an example, consider a simulator is used to automatically move and generate data, while an additional robot is tele-operated or given a fixed trajectory to create a risky situation that the robot must learn to avoid. Sometimes rare data may be mined from extremely large data sets of several runs, while still the rare data will be a minute fraction of the entire data set.

13.7.6   Dealing with drift

Suppose a large data set is recorded that is used to learn a model for behaviour cloning. The model is used to move a robot. The model will not perfectly generate the same outputs as used for training, due to which the robot will deviate a little. The deviation will cause a slightly different future position that may trigger a new sensor perception not available in the data set. Due to model generalization, the network will still be able to give a nearly correct output; however, the robot will deviate more. This will cause another input that is relatively unrepresented. The robot starts to drift with time, eventually leading to dangerous states not available in the training data set, in which the model is unreliable.

Hence, the training happens iteratively through *iterative data aggregation* (DAGGER, Ross and Bagnell, 2010; Ross et al., 2011). The model is trained on a large data set, which is used to test the robot. During testing, the robot generates new sequences of sensor percept. New demonstrations are requested from the human (or motion planning library used for data generation) to these sensor inputs if the robot appears to be going wrong. The robot is driven by the model and as soon as the robot starts to deviate, the human is asked to take over and drive the robot and the data are added to the database. If a motion planning algorithm is used to generate data, it can give the target to all new percept of the drifting robot.

Another mechanism to generate data is to continually add drift as the human drives the robot for data generation. Drift is added by adding a large noise vector to the human-supplied target velocity to move the robot. Sometimes random exploration is applied by giving random velocities to the robot, which gets the robot into novel situations, and the human drives and takes the robot out, which is added to the data set.

13.7.7 Limitations

This section presents several ways to perform Behaviour Cloning to train a model that can autonomously operate a robot. However, it is important to note that the navigation models can be very complicated to learn as the robot can get into several novel situations. It is not possible to generate data for every possible situation using the methods and techniques described here. As a result, the robot may still drift and get into a novel situation. Once the robot starts to go wrong, it continues to have a poor future trajectory. Hence, Reinforcement Learning is subsequently used to let the robot explore and autonomously discover novel situations. The robot simulates and generates its data and learns from such simulations to further improve the models. The Reinforcement Learning of the same problems is taken up as a separate chapter.

Questions

1. Common upon the optimality and completeness of learning-based motion planning. If not complete, give indicative situations where the algorithm fails. Comment on the algorithm's ability to handle narrow corridors.
2. Different people have different walking strategies. Similarly, different robots may have different navigation strategies. Suggest ways to implement this concept using learning-based motion planning.

3. Using specific examples from learning-based motion planning, explain why dropout is important in a convolution neural network and similarly why denoising important in autoencoders?

4. The CNN extracts spatial features, while the LSTM extracts the spatial and temporal features. Explain the process of extraction of temporal features in LSTM and mention a few informal temporal features that are useful for navigation in general.

5. [Programming] Here we attempt to solve the localization problem using machine learning. You should use transfer learning from well-established models for all questions. (i) Make an XY coordinate axis system at home and mark a few coordinates. Take photos from a camera at a known height and with different measurable orientations (say multiples of $\pi/4$ degrees) multiple times a day in different lighting conditions. Make a localization system that takes an image as input and prints the (x,y,θ) coordinates. Learn this as a regression and classification problem both. (ii) Additionally, make a learning function that calculates the similarity score of two images. The score is 0 if the images were of the same (x,y,θ) taken at different times of the day, while the score is 1 for all other pair of images. Given an image, calculate the (x,y,θ) as the image that has the least distance from all images in the data set. (iii) Additionally, take a sequence of images, at every time stepping on at one of the neighbouring marked coordinates. Make an LSTM network that predicts the pose (x,y,θ) of all images in the sequence, given the pose of the starting position. (iv) Record a video to make a continuous image stream, walking through marked coordinates only. Use interpolation to tag output of all intermediate frames in-between marked coordinates. Use LSTM to predict the pose for a data set with multiple rectilinear trajectories and walking speeds. (v) Train the network robustly by applying multiple noises to images and adding snapshots of synthetic objects in the image to see if the results improve.

6. [Programming] Solve all motion planning problems of Chapter 11 using machine learning using the same inputs. Additionally, take input as the readings of a virtual lidar along with distance and direction to goal. Ensure to collect enough data to facilitate learning and solve easier problems before the harder ones.

References

Agrawal, A., Agrawal, D., Arora, M., Mahajan, R., Beohar, S., Kenye, L., Kala, R., 2022. SLAM and map learning using hybrid semantic graph optimization. In: 2022 30th Mediterranean Conference on Control and Automation, pp. 731–736.

Arandjelovic, R., Gronat, P., Torii, A., Pajdla, T., Sivic, J., 2016. NetVLAD: CNN architecture for weakly supervised place recognition. In: Proceedings of the IEEE Conference on Computer Vision and Pattern Recognition, pp. 5297–5307.

de Haan, P., Jayaraman, D., Levine, S., 2019. Causal confusion in imitation learning. In: Advances in Neural Information Processing Systems 32.

Dosovitskiy, A., Beyer, L., Kolesnikov, A., Weissenborn, D., Zhai, X., Unterthiner, T., Dehghani, M., Minderer, M., Heigold, G., Gelly, S., Uszkoreit, J., 2020. An Image is Worth 16x16 Words: Transformers for Image Recognition at Scale. arXiv preprint. arXiv:2010.11929.

Erhan, D., Bengio, Y., Courville, A., Manzagol, P.A., Vincent, P., Bengio, S., 2010. Why does unsupervised pretraining help deep learning? J. Mach. Learn. Res. 11, 625–660.

Han, K., Wang, Y., Chen, H., Chen, X., Guo, J., Liu, Z., Tang, Y., Xiao, A., Xu, C., Xu, Y., Yang, Z., 2020. A Survey on Visual Transformer. arXiv preprint. arXiv:2012.12556, 2(4).

Hausler, S., Garg, S., Xu, M., Milford, M., Fischer, T., 2021. Patch-netvlad: multiscale fusion of locally-global descriptors for place recognition. In: Proceedings of the IEEE/CVF Conference on Computer Vision and Pattern Recognition, pp. 14141–14152.

Kenye, L., Kala, R., 2022. Improving RGB-D SLAM in dynamic environments using semantic aided segmentation. Robotica 40 (6), 2065–2090.

Kenye, L., Palugulla, R., Arora, M., Bhat, B., Kala, R., Nayak, A., 2020. Relocalization for self-driving cars using semantic maps. In: 2020 Fourth IEEE International Conference on Robotic Computing, pp. 75–78.

Kothari, P., Kreiss, S., Alahi, A., 2022. Human trajectory forecasting in crowds: a deep learning perspective. IEEE Trans. Intell. Transp. Syst. 23 (7), 7386–7400.

Lin, G., Milan, A., Shen, C., Reid, I., 2017. Refinenet: multi-path refinement networks for high-resolution semantic segmentation. In: Proceedings of the IEEE Conference on Computer Vision and Pattern Recognition. IEEE, pp. 1925–1934.

Ross, S., Bagnell, D., 2010. Efficient reductions for imitation learning. In: Proceedings of the Thirteenth International Conference on Artificial Intelligence and Statistics, pp. 661–668.

Ross, S., Gordon, G., Bagnell, D., 2011. A reduction of imitation learning and structured prediction to no-regret online learning. In: Proceedings of the Fourteenth International Conference on Artificial Intelligence and Statistics, pp. 627–635.

Szegedy, C., Wei, L., Jia, Y., Sermanet, P., Reed, S., Anguelov, D., Erhan, D., Vanhoucke, V., Rabinovich, A., 2015. Going deeper with convolutions. In: 2015 IEEE Conference on Computer Vision and Pattern Recognition. IEEE, Boston, MA, pp. 1–9.

Szegedy, C., Ioffe, S., Vanhoucke, V., Alemi, A.A., 2017. Inception-v4, inception-resnet and the impact of residual connections on learning. In: Thirty-First AAAI Conference on Artificial Intelligence.

Vaswani, A., Shazeer, N., Parmar, N., Uszkoreit, J., Jones, L., Gomez, A.N., Kaiser, L., Polosukhin, I., 2017. Attention is all you need. In: Advances in Neural Information Processing Systems 30.

Wu, Z., Shen, C., van den Hengel, A., 2019. Wider or deeper: revisiting the resnet model for visual recognition. Pattern Recogn. 90, 119–133.

Wu, Z., Pan, S., Chen, F., Long, G., Zhang, C., Philip, S.Y., 2020. A comprehensive survey on graph neural networks. IEEE Trans. Neural Netw. Learn. Syst. 32 (1), 4–24.

Yadav, R., Kala, R., 2022. Fusion of visual odometry and place recognition for SLAM in extreme conditions. Appl. Intell. 52, 11928–11947.

Yang, S.X., Meng, M.Q., 2003. Real-time collision-free motion planning of a mobile robot using a neural dynamics-based approach. IEEE Trans. Neural Netw. 14 (6), 1541–1552.

Yang, S., Scherer, S., 2019. CubeSLAM: monocular 3-D object SLAM. IEEE Trans. Robot. 35 (4), 925–938.

Zhang, H., Goodfellow, I., Metaxas, D., Odena, A., 2019. Self-attention generative adversarial networks. In: International Conference on Machine Learning. PMLR, pp. 7354–7363.

Chapter 14

Motion planning using reinforcement learning

14.1 Introduction

This chapter gets realistic and models the environment as *stochastic* or the systems where uncertainties prevail, and the robot often enters a condition that it did not expect. More formally, in a *deterministic system*, if the robot applies an action a from a state s, it reaches state s' given by $s' = a(s)$, where a is the action function. So, taking a step right from state (2,3) landed the robot in state (3,3). Similarly, the action *pick book* perfectly had the state where the book was in the hand. Optimizing the trajectory assumed that the control algorithms would be able to nicely trace the trajectory.

In a *stochastic system*, if the robot is at state s and applies action a, it cannot be determined in which state the robot would end up. There is only a probability that the robot lands at the expected state s'. The robot could end up in either of the numerous states possible. Formally, it is said that the robot lands in state s' with a probability $P(s'|s,a)$ on the application of action a from state s. Here, P is the probability density function for a given state and action. As it is the probability, we have Eq. (14.1):

$$\sum_{s'} P(s'|s,a) = 1 \tag{14.1}$$

The term $P(s'|s,a)$ may sometimes be more conveniently written as $T(s,a,s')$, called the transition probability density function. An assumption made here is that the probability of reaching s' only depends upon the current state and the current action and not the historical states and actions. This is called the *Markovian assumption*. If s_t is the state at time t and a_t is the selected action at time t, the assumption is defined as Eq. (14.2):

$$P(s_{t+1}|s_0 a_0 s_1 a_1 s_2 a_2 \ldots s_t a_t) \approx P(s_{t+1}|s_t a_t) \tag{14.2}$$

Say that the robot is on a slippery floor. The robot cannot predict how much the wheels would slip, and therefore, moving forward could move one, two, or

Autonomous Mobile Robots. https://doi.org/10.1016/B978-0-443-18908-1.00016-9

669

three units forward depending upon the slip or misorientation. The book picked by the robot could slip from its grippers, leading to a different state. As per the Markovian assumption, if the robot is at a particular state and attempts to go ahead, the slippage only depends on where the robot is currently at and what action is currently applied. It does not depend upon the path through which the robot reached the current state or the actions applied previously.

Most robots plan a trajectory, and the control algorithms trace the trajectory well. Keeping some *contingency* in planning (or contingency planning) in the form of extra clearances ensures that the control algorithms work well in real life. It may thus be interesting to question the need for a new set of algorithms for handling stochasticity. Imagine a robot in a room slips by a few metres. Now, the control algorithm may still correct it, but the robot may not have been planned with clearances of such a high degree. This means that the control correction can cause collisions. Furthermore, the robot can slip even while correcting itself, causing more deviation. In such cases, the robot should be re-planned with the current position as the source. However, the robot can slip while executing the new plan as well. Similarly, imagine that the book falls while picking up at a place which is known. Now, the system is suddenly in a new state and will require a new plan.

An intuitive way to work over the problem is thus to know the best possible action for every possible state of the robot. This is called the *policy* (also called a *strategy*) or a *policy function* $\pi:S \to A$ that maps every state in the state space S to the best action to be taken in the state out of all possible actions A. S denotes the state space (a set of all states) and A denotes the action space (a set of all actions). $\pi(s)$ is thus the best action at a state s. The problem under these settings changes from a path search to searching a policy. The policy can be computed offline and can be referred to by the robot when acting in the real world. This is like the fact that the robot plans a path offline and acts upon it online.

Another form of uncertainty is when the environment has agents that act as an adversary to the robot or the agent being planned, commonly found in games. A classic example is a *pursuit-evasion* problem where a pursuer tries to catch an evader as soon as possible, while the evader tries to inhibit or delay being caught. The future action of the agent depends upon the next action of the adversary that is neither known nor may be easy to predict leading to uncertainties.

The chapter starts with decision-making under stochastic settings. This is used as a foundation for making a navigation system using *Reinforcement Learning* (RL). The RL is prevalent in humans learning to ride a bike, learning to play video games, learning how to walk, etc. Nobody tells the humans exactly what to do in what situations. The humans just do things randomly. They get an appreciation or punishment as feedback called *reward*. The humans try to do things that result in appreciation and minimize the actions for, which they eventually get a punishment. The terms are summarized as Box 14.1.

Box 14.1 A summary of different terms and techniques

- **Stochastic systems:** The outcome of an action is dependent upon probabilities (robot may slip and travel more than expected).
- **Markovian assumption:** Outcome (s') only depends upon the current state (s) and current action (a), and not the previous states and actions, resulting in probability distribution $T(s,a,s') = P(s'|s,a)$. If the robot moves ahead, the slippage only depends upon where the robot is and not on the path used by the robot to reach the current place.
- **Adversarial agent:** An agent whose utility is the additive inverse of the utility of the agent being planned, who thus tries to take opposing actions. E.g., games like tic-tac-toe.
- **Cooperative agent:** An agent who shares the same utility function as the agent being planned, who thus tries to take supporting actions. E.g., moving as a swarm.
- **Policy (π):** Action with the most expected returns (utility) for a state (s).
- **Reward (R):** A feedback given by the environment on making any move, telling the immediate goodness of the move. E.g., losing battery on moving ahead (negative), reaching a goal (positive), getting too close to an obstacle (negative), moving towards the goal (positive), deviating away from goal (negative), the score of a game, etc.
- **Utility (U):** Total sum of all rewards that the agent expects to get from the current state s till the terminal state (end of the game). Suppose starting from (0,0), moving by an optimal sequence, the robot acquires rewards -1 (loss of battery), then -1 (loss of battery), then -10 (too close to obstacle), and then 100 (goal achieved). The immediate reward at (0,0) is -1, but the utility (indicated by a single observed trajectory alone) is 88. The actual value is averaged for infinite such runs when moving using an optimal policy.
- **Discounting (γ):** Immediate rewards are preferable to the future rewards of the same value, and thus the future rewards are discounted. In the sequence above, indicative utility becomes $(-1)+\gamma(-1)+\gamma^2(-10)+\gamma^3(100)$. Smaller values of γ enable early convergence but restrict the effect of a reward in the utility calculation to only its surrounding states, affecting optimality.
- **Reinforcement learning:** Learning by acting in the real/simulated world, with stochastic/adversarial/cooperative elements, trying to maximize the total rewards earned, without supervision (someone who tells the correct action at every step).
- **Deep reinforcement learning:** Using deep neural networks to model utility and/or policy function of reinforcement learning.
- **Inverse Reinforcement Learning:** Extract a reward function from demonstrations generated by an expert (human or an existing motion planning algorithm).
- **Avoid deep rewards:** If positive rewards come only after too many steps, the chances that initial random motion reaches the positive reward are slim, and the algorithm may never discover it. Prefer small rewards for moving towards the goal.
- **Design of a reward function:** Bad reward function means bad learned behaviours. Tweaking reward function parameters may not be easy as every tweak requires solving the computationally intensive reinforcement learning problem. Inverse Reinforcement Learning techniques may help.

Continued

Box 14.1 A summary of different terms and techniques—cont'd

- **Pretraining from demonstrations (Imitation Learning)**
  - Train policy network using states and actions available in the demonstration dataset.
  - Train value network as a passive reinforcement learning, using demonstrations for action selection and temporal difference for learning.
  - Alternatively, value (target) of a state *s* is the total discounted cumulative reward till the terminal in the demonstration. Train value network for all *s* in the demonstrations and computed targets

---

14.2 Planning in uncertainty

Imagine a robot operating under stochastic conditions. A natural way to model the problem is now by using game trees with the MAX nodes and chance nodes (without any opponent). However, the game may end up in a loop with the robot slipping to one of the states it had been in before. An instinct is, therefore, to model graphs instead of trees. Correspondingly, a better solution technique is also used (Russell and Norvig, 2010).

14.2.1 Problem modelling

Let us assume a simple problem where a point robot operates on an $m \times n$ sized grid map. The robot needs to reach a specific goal state. The robot moves using the action up, down, right, and left. The robot wheels often slip, and therefore, the actions do not always result in the desired state. On attempting moving right, the robot actually moves right with a probability of 0.8. On the contrary, even though the robot has attempted moving right, it ends up moving up with a probability of 0.1 and down with a probability of 0.1. Similarly, on attempting moving left, the robot actually moves left with a probability of 0.8, while it moves up with a probability of 0.1 and down with a probability of 0.1. The robot similarly, on attempting moving up ends up actually taking up with a probability of 0.8 and ends up taking a left with a probability of 0.1 and a right with a probability of 0.1. In case of attempting moving down, the robot actually moves down with a probability 0.8, left with a probability 0.1, and right with a probability of 0.1. Colliding with walls or obstacles makes the robot stay at the same place. In general, let $T(s,a,s') = P(s'|s,a)$ denote the probability of reaching state s' by application of action a from state s, considering slippage and obstacles both.

The agent receives a *reward* by the environment on making every step. The reward may be a big payoff upon reaching the terminal state, or small rewards on the way. The reward can be positive, stating something good for the agent, or negative, stating that the agent has done something not desirable. Let $R(s)$ be the reward received by the agent upon leaving (or landing) at state s. The rewards could equivalently be an association with an action $R(a)$, state–action pair

$R(s,a)$, or the state–action pair along with the state that the agent lands on $R(s,a, s')$. For the sample problem, let the rewards be -0.1 for each step (denoting the loss of battery while moving), $+100$ for reaching the goal, and -10 for colliding with an obstacle or wall (to tell the robot that colliding is bad). The goal is more generally the *terminal state* where the simulation stops and the reward achieved on reaching the goal is more generally the *terminal reward*.

14.2.2 Utility and policy

Let us first informally define the utility as the total summation of all rewards achieved by the robot when moving by an optimal policy. We will use the notation $R(\tau)$ to denote the summation of all rewards for a trajectory τ. Consider a map given in Fig. 14.1A. Let us simulate the motion of the robot till it reaches the goal. The actions will not always result in the expected state because of stochasticity. Such simulations under the influence of stochastic actions are called *Monte Carlo Simulations*. Consider the simulation shown in Fig. 14.1A. The total rewards collected are an indicator of the utility. The robot makes five normal steps (that accumulates $-0.1 \times 5 = -0.5$ reward), 1 step where a collision occurs due to stochasticity (that accumulates $-10 \times 1 = -10$ reward), and 1 step to reach the goal (that accumulates $1 \times 100 = 100$ reward). The total summative reward is 89.5.

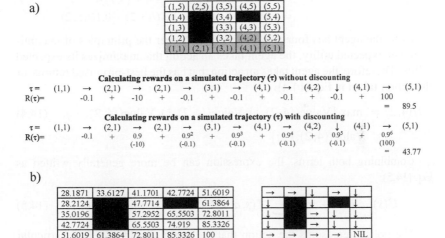

FIG. 14.1 Estimation of utility. (A) Monte Carlo simulation to calculate the summative rewards of a trajectory. The actions are stochastic and sometimes the robot slips. The robot gets -0.1 on moving, -10 on colliding, and $+100$ for reaching the goal. The total rewards indicate the utility of $(1,1)$. (B) Utilities and Policies computed for the dummy problem by value iteration.

From Fig. 14.1A, starting from $(1,1)$ with the best policy, the agent accumulates a reward of 89.5, indicating that the utility of $(1,1)$ should be 89.5. However, the different runs of the same simulation will have different returns because of the stochastic actions, which cannot mean different values of the utilities of the state. Let us now more formally define utility. The *utility* of a state is the total cumulate reward that the agent is expected to get, till the end of the simulation, when moving by an optimal policy. Let the utility function be given by $U:S \rightarrow \mathbb{R}$, such that the utility of a state s is given by $U(s)$. Let the agent take action a from state s to stochastically reach state s'. Let $T(s,a,s')$ be the probability of reaching state s' from state s on the application of action a.

To formulate the expression for utility, consider an example. Say the utility of state (4,2) is being calculated. Assume that the utilities for all the other states are well known (a common assumption to model recursively defined solutions). Suppose the agent takes the action down and lands at state (4,1) with a probability of 0.8, (3,2) with a probability of 0.1, and (5,2) with a probability of 0.1. In all cases, it takes a -0.1 reward to make the action. Subsequently, the cumulative reward from (4,1) to the goal is $U(4,1)$ and similarly for others. So, if the action down results in going from (4,1) to (4,2), there is an immediate reward of -0.1 and subsequently a total cumulative reward of $U(4,2)$, resulting in a total cumulative reward from (4,1) as $-0.1 + U(4,1)$. Similarly, for (3,2) and (5,2). This is classically called a *lottery*, and the calculations are called expected returns of the lottery. The value of the action down is given by Eq. (14.3):

$$
\begin{aligned}
\text{Value}((4,2), \downarrow) &= P((4,1)|(4,2), \downarrow)(-0.1 + U(4,1)) \\
&\quad + P((3,2)|(4,2), \downarrow)(-0.1 + U(3,2)) \\
&\quad + P((5,2)|(4,2), \downarrow)(-0.1 + U(5,2)) \\
&= -0.1 + 0.8U(4,1) + 0.1U(3,2) + 0.1U(5,2)
\end{aligned}
\tag{14.3}
$$

Now, the agent has four actions possible. As per the principles of maximization of expected utility, the agent takes an action that maximizes its expected utility. Therefore, the agent takes an action that has the best-promised returns, or the action taken is Eq. (14.4):

$$
\begin{aligned}
U(4,2) = \ &\max \{ \text{Value}((4,2), \downarrow), \text{Value}((4,2), \uparrow), \text{Value}((4,2), \rightarrow), \\
&\text{Value}((4,2), \leftarrow) \}
\end{aligned}
\tag{14.4}
$$

Combining both terms, the expression can be more generally written as Eq. (14.5):

$$
U(s) = R(s) + \max_a \sum_{s'} T(s, a, s') U(s'), s \neq \text{terminal state}
\tag{14.5}
$$

The policy represents the action that the agent should take at a particular state, which is given by Eq. (14.6):

$$
\pi(s) = \arg\max_a \sum_{s'} T(s, a, s') U(s'), s \neq \text{terminal state}
\tag{14.6}
$$

The reward is modelled on a state s and the agent gets the same on leaving a state. Equivalently it could be modelled based on a landing state (s'), or state–action pair, or based on state (s) action (a), and landing state (s') vector by placing the reward inside the summation with $U(s')$. More generally, say the reward is modelled as $R(s,a,s')$. The equations in such modelling are given as Eqs (14.7) and (14.8):

$$U(s) = \max_a \sum_{s'} T(s, a, s')(R(s, a, s') + U(s')), s \neq \text{terminal state} \quad (14.7)$$

$$\pi(s) = \arg\max_a \sum_{s'} T(s, a, s')(R(s, a, s') + U(s')), s \neq \text{terminal state} \quad (14.8)$$

It is important here to differentiate between reward and utility. The reward is immediate feedback given by the system. It tells immediately what is good or bad. The utility is the total expected sum of all rewards that the agent will receive in the future. A state may have a negative reward but a positive utility, implying that even though landing on the state may be immediately poor, but eventually, it goes through a path that has high returns at the end. As an analogy, studying may be painful with a negative reward, but the utility of studying is highly positive as it enables getting a good job and research results in the future. Similarly, the robot will lose battery immediately on moving, which is poor but will eventually get a big terminal reward on reaching the goal.

14.2.3 Discounting

Consider the environment shown in Fig. 14.2, where the numbers represent the rewards associated with the states. A natural policy is thus given in Fig. 14.2. Since all rewards are positive, the agent can continue to collect them and therefore does not even move to the goal. The utilities of all states are thus infinite. Most of the methods to solve problems are iterative. While the infinite answer is correct, a problem with the iterative methods is that they will start with a guess and increase the guessed value until infinity, which will take infinite corrective steps.

To solve the problem, a notion of *discounting* is applied. The immediate rewards are more desirable to a human as compared to future rewards. So, getting a few thousand dollars now is preferable to getting the same money (with interest) at a future date. Hence, the value of future rewards is diminished by multiplying them with a value γ, where γ is the discount factor $(0 < \gamma \leq 1)$.

Rewards				Policy				Utility			
0.1	0.1	0.1	Terminal +100	→	→	←	Terminal NIL	∞	∞	∞	Terminal +100

FIG. 14.2 Discounting. The summation of rewards is discounted by a factor γ. Without discounts, the total cumulative reward and hence the utility is infinite, which is limited by discounting.

Consider the same run as shown in Fig. 14.1A. Let $\gamma = 0.9$. After discounting, the cumulative reward of the trajectory may be approximated as Eq. (14.9).

$$\begin{aligned} R(\tau) &= (-0.1) + \gamma(-10) + \gamma^2(-0.1) + \gamma^3(-0.1) + \gamma^4(-0.1) + \gamma^5(-0.1) + \gamma^6(100) \\ &= (-0.1) + 0.9(-10) + 0.9^2(-0.1) + 0.9^3(-0.1) + 0.9^4(-0.1) \\ &\quad + 0.9^5(-0.1) + 0.9^6(100) = 43.77 \end{aligned}$$

$$(14.9)$$

The utility function is now formally given as Eq. (14.10) and the policy function is given by Eq. (14.11) for the case when $R(s)$ is the reward on leaving state s. Note that the terminal utility is fixed and cannot be calculated by this equation, which is fixed at the terminal reward.

$$U(s) = R(s) + \gamma \max_a \sum_{s'} T(s, a, s') U(s'), s \neq \text{terminal state} \qquad (14.10)$$

$$\pi(s) = \arg\max_a \sum_{s'} T(s, a, s') U(s'), s \neq \text{terminal state} \qquad (14.11)$$

For the reward function $R(s,a,s')$, the utility and policy are given by Eqs (14.12) and (14.13).

$$U(s) = \max_a \sum_{s'} T(s, a, s')(R(s, a, s') + \gamma U(s')), s \neq \text{terminal state} \quad (14.12)$$

$$\pi(s) = \arg\max_a \sum_{s'} T(s, a, s')(R(s, a, s') + \gamma U(s')), s \neq \text{terminal state}$$
$$(14.13)$$

The formulation is shown in Fig. 14.3. For any problem with all positive rewards bounded by R_{max}, the utility can never be more than $\sum_i \gamma^i R_{max}$, which is a geometric progression bound to $R_{max}/(1-\gamma)$ and, hence, the algorithms converge.

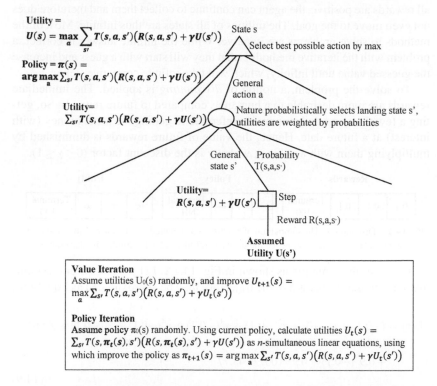

Value Iteration
Assume utilities $U_0(s)$ randomly, and improve $U_{t+1}(s) = \max_a \sum_{s'} T(s, a, s')(R(s, a, s') + \gamma U_t(s'))$

Policy Iteration
Assume policy $\pi_0(s)$ randomly. Using current policy, calculate utilities $U_t(s) = \sum_{s'} T(s, \pi_t(s), s')(R(s, \pi_t(s), s') + \gamma U(s'))$ as n-simultaneous linear equations, using which improve the policy as $\pi_{t+1}(s) = \arg\max_a \sum_{s'} T(s, a, s')(R(s, a, s') + \gamma U_t(s'))$

FIG. 14.3 Calculating utilities. The game tree shows the formulation of utilities and policies. Since there will be cycles, the calculation is done using value iteration and policy iteration algorithms.

More practically, a lower discount factor means that the algorithm converges quickly, mostly looking at short-term rewards only. At the same time, higher discount values encourage the agent to look at the long-term future rewards for decision-making, delaying convergence. If the discount value is small, the discount term γ^t becomes small after a few steps. So, only a few future states affect the policy of state s. States near the terminal states converge soon, which soon cause the convergence of their neighbours and so on. If γ is large, a state must consider all possible future state values for deciding its optimal policy, which takes time. Consider two extreme examples. In the first, the discount factor is $\gamma = 0$. Now the agent is simply a greedy agent that takes every move based on the best reward and no learning is needed. In the other extreme case, consider that discount is $\gamma = 1$. If there is a high positive goal at the very end, it sufficiently affects all states at the very start and the algorithms will take time to compute the best utility and value.

14.2.4   Value iteration

The problem is to compute the utilities $U(s)$ and policy $\pi(s)$, given the rewards $R(s)$ and the transition probability (also called the *model T(s,a,s')*). Eq. (14.12) has n states and n equations (1 for each state), and therefore, it may appear to be solvable for the simultaneous equations using linear equation solvers to get an answer straight away. However, the equations are nonlinear due to the presence of the max function and, hence, need to be solved iteratively. The equations are called Bellman in nature, as they encode a recursive relation between the utilities of the state variables. Every state's utility value depends upon the other states.

The first algorithm to solve the problem is called the *value iteration*. Here, we first guess the initial values of the utilities as arbitrary real numbers (they will eventually get their actual values). This can be used to break the recursion in Eq. (14.12) by rewriting as Eq. (14.14).

$$U_{t+1}(s) = \max_a \sum_{s'} T(s, a, s')(R(s, a, s') + \gamma U_t(s')), s \neq \text{terminal state}$$

$$(14.14)$$

Here, $U_t(s)$ is the utility function at time t, which is improved with time. The improvement is casually stated without proof. Intuitively, the magnitude by which the constraint is violated is called the error shown by Eq. (14.15), which reduces with iterations.

$$\text{error}_t = \sum_s \left(U_t(s) - \max_a \sum_{s'} T(s, a, s')(R(s, a, s') + \gamma U_t(s')) \right)^2 \quad (14.15)$$

Now the improved utilities can again appear on the right hand of Eq. (14.14) to get still better values. Iteratively, the guess is improved. The pseudo-code of the approach is thus given by Algorithm 14.1. The working of the algorithm on the sample problem produces a result shown in Fig. 14.1B.

Algorithm 14.1 Value iteration

---

1   $U[s] \leftarrow rand \forall s \neq$ terminal state ;   ▷ random guess of utility
2   $U[s]$=terminal reward $\forall s$ =terminal state ;   ▷ fixed utilities
3   **while** *true* **do**
4     $U'[s] \leftarrow U[s] \forall s$;
     ▷ calculate new utilities U using old utilities U' ;
5   **for** *all* $s \in S$ **do**
6   **if** $s \neq$ *terminal state* **then**
7        $U[s] \leftarrow max_a \Sigma_{s'} T(s,a,s')(R(s,a,s') + \gamma U'(s'))$
8   **else** $U[s]$ =terminal reward;
9     $e \leftarrow \Sigma_s (U[s] - U'[s])^2$;
10   **if** $e <$ *acceptable convergence error* **then** break;
11   return U ;

14.2.5 Policy iteration

The problem with value iteration is that it requires a lot of time to converge for sufficiently large state spaces. The algorithm is, hence, improved. Now, let us break the same nonlinearity, however, by assuming a fixed policy $a = \pi_t(s)$ for all states, which eliminates the max term. Here, π_t is the policy at time t. The modified equation is given by Eq. (14.16):

$$U_t(s) = \sum_{s'} T(s, \pi_t(s), s')(R(s, \pi_t(s), s') + \gamma U_t(s')), s \neq \text{terminal state}$$

$$(14.16)$$

Now, for n states, there are n linear equations, which can be solved by any linear equation package solver. This gives the utility values for the guessed policy. This step is also called *policy evaluation*. Next, given these utilities, let us calculate a better policy for the given utility values. This is called *policy refinement*, where the improved policy is given by Eq. (14.17).

$$\pi_{t+1}(s) = \arg \max_{a} \sum_{s'} T(s, a, s')(R(s, a, s') + \gamma U_t(s')), s \neq \text{terminal state}$$

$$(14.17)$$

The better policy is used to calculate utilities, which results in a still better policy. This goes on and the policy improves. The algorithm has discrete choices of policy, and the utilities always follow the Bellman equation. Therefore, the algorithm converges very quickly. The pseudo-code is shown as Algorithm 14.2.

Algorithm 14.2 Policy iteration

---

1   $\pi[s] \leftarrow rand(A) \forall s \neq terminal\,state$;   ▷ guess a random policy
2   $\pi[s] \leftarrow NIL \forall s = terminal\,state$;
3   **while** *true* **do**
4     $U \leftarrow PolicyEvaluation(\pi)$ by solving n linear equations,
      $U(s) = \Sigma_{s'} T(s, \pi(s), s')(R(s, \pi(s), s') + \gamma U(s'))$, preferably by
      using an iterative approximate solver like Value Iteration ;
5   **for** $s \in S$ **do**   ▷ Policy Refinement using U
6   **if** $s \neq terminal\,state$ **then**
7         $\pi'[s] \leftarrow argmax_a \Sigma_{s'} T(s, a, s')(R(s, a, s') + \gamma U(s'))$
8   **else** $\pi'[s] = NIL$;
9   **if** $\pi[s] = \pi'[s] \forall s$ **then** break;   ▷ no improvement
10    $\pi[s] = \pi'[s] \forall s$;
11 **return** U

The major problem is that solving n linear equations is a computationally heavy algorithm for large n, naively done incurring a complexity of $O(n^3)$. However, the values are only used to discretely improve the policy, and therefore, accurate values may not be necessary. The values need to be accurate enough till a point that the *arg max* over actions remains the same used for policy refinement. A variant approximately calculates the utilities by using an iterative algorithm like value iteration with a fixed policy, instead of solving n linear equations. For a fixed policy, a few iterations of value iteration will only require a complexity of $O(n)$ practically, as a state's utility value depends only on a few (constant) others. This is called the *modified policy iteration algorithm*. The algorithm may be further generalized. The algorithm starts with random utility and policy values. Then, the algorithm heuristically selects whether value refinement needs to be done or policy refinement, and the specific states over, which the refinement needs to be done. If a value refinement is chosen, for a fixed policy, the new utility value for the state is estimated. If a policy refinement is chosen, the new policy value for the state is computed for fixed utility values. Powerful heuristics exist to choose between value or policy refinement at every iteration, and the selection of states for refinement.

14.3   Reinforcement learning

The aim is to solve the hard problems of making reactive controllers for robots under stochastic, cooperative, and adversarial settings or a mixture of these. Examples include a robot that often slips on its way because of poor wheels, a robot playing a game like tennis against a human, a robot picking up items just by looking at the image of the item, robots moving as a swarm overcoming obstacles, etc. All these problems would be now solved by using Reinforcement Learning (RL).

There are three major problems due to, which the currently discussed methods cannot be used. The first problem is large number of states possible, which makes the previous approaches computationally infeasible. The second is the absence of the transition probability (model) in the problem specification. The transition probabilities will not be given for a real system, which must be computed by the robot itself.

The third is that the states are continuous, while they are currently assumed to be discrete. In the easiest case, whatever the robot looks at is the state, like the lidar scan of the environment around it, the video input from the stereo camera, etc. The aim is to make a system where the sensor percept can be given as input that gives the best action as an output, as done in the machine learning approaches. So, the state space is all possible images that can ever be seen by the robot, or all possible lidar readings that can come. Even a single iteration of this state space is infeasible.

Now, instead of exactly solving the problem, the emphasis is on *learning*. If a new video game is given, humans try random moves to get feedback on what is good and what is not. This helps them to eventually become great players by practicing without supervision, trying to maximize the goods, and ultimately achieving the desirable things as per the game. The humans may slip on slippery floors, but that sends them feedback to correct themselves, and avoid slipping anytime in the future. The teams learn to coordinate and distribute responsibilities between themselves based on historical good and bad experiences. In all these examples, the agent gets rewards (score, the player gets hit, the player gets killed, fallen, team fights, etc.), and the agent tries action sequences that maximize the cumulative rewards (total score, safely reached the destination, timely project completion, etc.). Sometimes an agent may have to take actions that give a lower (negative) reward but lead to a better outcome (e.g., hiding for a video game monster to pass by, turning back on a longer route avoiding a slippery floor, firing a noncooperative team member, etc.). A move may appear good immediately (quickly reaching towards the destination, quick completion of a project module without discussions, etc.), but may turn out to be bad later (player got killed, fallen, incompatible solution of a project module that increases the time of all other modules, etc.). In neither of the examples, does a person supervise and state what the best action was in every situation. In neither of these examples, explicit computation of policy for all states is done. Only a tiny fraction of all possible states may be ever visited.

There are two types of RL. The first is called *Active Reinforcement Learning*. Here, the agent makes decisions in real life, while learning based on the outcomes of the decisions. The other type of learning is called the *Passive Reinforcement Learning*, where the agent learns from the observations, however, the actual motion and selection of action is done by some intelligent human being. As an example, playing a video game is active RL, but seeing your sibling play the game is passive RL.

14.3.1 Passive reinforcement learning

For the ease of discussion, let us first only consider the problem of passive RL. The system needs to learn the transition probabilities ($T(s,a,s')$ or model) as well as the utilities ($U(s)$), from which a policy can be derived. An intuitive way is to learn the transition probabilities by observing many sequences and then using the probability definition given by Eq. (14.18):

$$T(s, a, s') = N(s, a, s')/N(s, a) \tag{14.18}$$

Here, $N(s,a,s')$ is the number of times that the transition from s to s' through action a was observed, while $N(s,a)$ is the number of times action a was selected from state s. Similarly, the utility $U(s)$ can be an average of summative rewards of all runs from state s to the terminal state. This mechanism, however, lacks the Bellman nature of the update and thus will require a significantly large amount of data.

Let us, therefore, makè a better learning mechanism that uses the Bellman nature of the equation and does not use the transition probability (model), eliminating a need to learn it. Let $U(s)$ be the currently estimated utility of the agent at state s. Suppose the agent is at state s and applies an action a to reach a state s' with a reward $R(s,a,s')$. The aim is now to learn a better estimate for $U(s)$ with this observation. The agent has observed reaching from s to s' with a reward $R(s,a,s')$, and subsequently to the terminal state with a cumulative reward of $U(s')$, implying that the utility at s should be $R(s,a,s')+\gamma U(s')$. However, the current estimate for the same quantity is $U(s)$. The error of $R(s,a,s')+\gamma U(s') - U(s)$ needs to be adjusted. The error function is also called the *temporal difference error* since the prediction after one step is used to compute the Bellman error. As per the learning technique, a fraction of the correction is made, where the fraction is controlled by the learning rate η. The learning is called *temporal difference learning*. The resultant equation becomes Eq. (14.19):

$$U(s) \leftarrow U(s) + \eta(R(s, a, s') + \gamma U(s') - U(s)), s \neq \text{terminal state} \tag{14.19}$$

The learning rate has the same effect as supervised learning. High values may lead to too many fluctuations, making the model very plastic. A low value converges to a stable point; however, the convergence can be slow. In contrast to supervised learning, the model is incremental (apply learning equation for every piece of data) rather than as a batch (apply learning equation for a batch of data).

The other interesting thing to note in the equation is the absence of the transition probabilities. The learning equation is invoked many times for every state s. Nature automatically selects s' as per the natural probabilities. Consider that the human always selects the down action from state $s = (4,2)$ and in the entire play of the human state (4,2) appeared 10,000 times. Now, it is good to believe that approximately 8000 times the robot would land at state $s' = (4,1)$ and (4,1) would be used to calculate the target utility. Similarly, the target utilities will be calculated 1000 times each for the states $s' = (3,2)$ and $s' = (5,2)$. The repeated application of the learning equation is equivalent to the cumulative addition of the corrections, as also discussed for the supervised learning when making an equivalence between incremental learning, learning by mini-batches, and batch learning. So the correction from state $s' = (4,1)$ happens 8000 times, while that from $s' = (3,2)$ and $s' = (5,2)$ only happens 1000 times each. This is nearly equivalent to the multiplication of the probabilities. Nature selects s' by their natural probabilities to eliminate a state and take undue dominance for contributing as the target in the error equation.

14.3.2 Q-learning

Now, let us focus on active learning, where no human is there to select the best action at every time. The problem can be studied in two parts. The first is given an action selection, to learn the observed behaviours. This problem has already been solved as Passive RL and the same method is applicable. The second problem is to make an action selection mechanism, given some utility estimates. The problem is shown in Fig. 14.4.

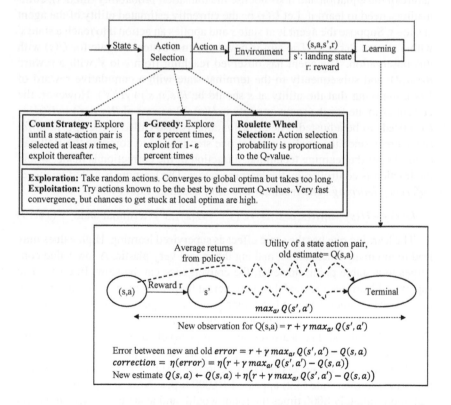

FIG. 14.4 Q-learning. An action is selected, executed, and learned. The learning is model-free (no $T(s,a,s')$). Nature selects s' by their natural probabilities, to call correction by their natural proportions, imitating a natural multiplication of probabilities. The probabilities are explicitly multiplied in model-based approaches.

The specific technique studied here is called Q-learning. Let the utility of state s for a specific given action a be denoted by $Q(s,a)$. This is a slightly dented definition, where the utilities are defined on a state–action pair, and hence, the new function Q. Utility of a state s, a term as used previously, is given by Eq. (14.20).

$$U(s) = \max_a Q(s,a) \tag{14.20}$$

The learning equation may now be modified from Eqs (14.19) to (14.21).

$$Q(s,a) \leftarrow Q(s,a) + \eta(R(s,a,s') + \gamma \max_{a'} Q(s',a') - Q(s,a)), s \neq \text{terminal state} \tag{14.21}$$

The equation uses temporal difference error to perform temporal difference learning. The formulation is specifically known as State (s) Action (a) Reward (R) landing State (s') next Action (a') framework, or *SARSA*, denoting the terms used in the equation.

The important aspect is the *action selection*. In the active RL nomenclature, the robot needs to decide, which action to take at every instant of time and learn from the same observation. There are two philosophies of action selection. The first philosophy always selects the best action as per the current knowledge. This is a *greedy* action selection measure. It is called *exploitation* because the agent exploits the currently available knowledge to assume that it is the best thing possible. The action selection is given by Eq. (14.22):

$$\pi_{\text{exploitative}}(s) = \arg\max_{a'} Q(s,a') \tag{14.22}$$

The other philosophy to select the action is exactly the opposite strategy, where the agent selects a random action, out of the available actions A. This is called *exploration* because the agent gives no importance to the current knowledge of the Q-values and simply explores by taking random actions. This is given by Eq. (14.23):

$$\pi_{\text{explorative}}(s) = \text{rand}(A) \tag{14.23}$$

Now, it is important to argue whether exploration or exploitation strategy is better. Let us first study the concept by a human analogy. Imagine that you need to go to a good restaurant to dine out. Now, there are two ways, in which a restaurant could be selected. First is to try a new restaurant that just opened and none of your colleagues has tried it out or a few colleagues tried out the restaurant with a very average review. This is exploration. The chances of failing and having a poor experience are high, but you get greater knowledge about the place. Maybe, the colleagues who earlier went to the place had been extremely unlucky to have an average experience, while the place was good. Alternatively, you may go to your usual place which is well-tried and tested. This is exploitation. The chances of getting a good experience are high, but the best experience may come from a place you have not tried so far and are too restrictive to experiment with something new.

Like humans, in learning, more exploration leads to trying everything numerous times, and therefore, the utility estimates are good. However, it takes a severely long for the system to converge. On the contrary, being exploitative readily converges the system, however, the chances of missing out on good options are high and the policy will mostly be suboptimal.

More realistically, consider a greedy agent. It will always try the same route to reach the terminal state and learn new observations. The observations change with time, which as well changes the greedy policy. Eventually, the policy converges. Now, a better route may have a poor initial policy due to bad initialization, but the greedy agent does not try it out and misses out on the better path, even though the convergence is very quick. On the contrary, if the agent is completely explorative. It knows a good enough path but will keep exploring new paths. As a result, it will reach states, which are far off the path (like walking too much in the opposite direction), states, which are not good as per the Q-values, and states, which have already been visited many times and are known to be bad. The agent wastes most of its time in such states, rather than learning from the better states and converging. This causes no convergence or slow convergence.

A mixture of exploration and exploitation is the correct way to go, to explore enough to unveil the optimal policy while exploiting enough for convergence after a reasonable amount of time. There are numerous ways by, which a balance between exploration and exploitation can be achieved.

A simple strategy is to use exploration if a state has been visited under a threshold time and to exploit thereafter. This is like saying that every restaurant must be tried five times, after which it makes sense to only try the best one. The problem here is that the parameter is hard to set since a change in the utility for some states may affect a state reasonably far away. In our example, a restaurant may borrow a part of the ingredients from some other restaurant, which may instead borrow some staff from some other. So, a great restaurant may have a poor performance for a long time only because of poor ingredients, an error that is eventually corrected. The parameter heavily depends upon the problem, reward function, and training duration. The value may be smaller near the goal, while larger further away, while the value may need to be increased or decreased based on the observed convergence trend.

Better modelling is called the ϵ-greedy method. Here, the agent explores for ϵ proportion of times, while exploits in the other $1 - \epsilon$ proportion of the time. ϵ is a small constant that balances between exploration and exploitation. The action selection is given by Eq. (14.24).

$$\pi_{\epsilon-\text{greedy}}(s) = \begin{cases} \text{rand}(A) & \text{with probability } \epsilon \\ \text{argmax}_a Q(s, a) & \text{with probability } 1 - \epsilon \end{cases} \tag{14.24}$$

Another variant gets rid of the parameters and states that the probability of selection of an action is directly proportional to its Q-value. The normalization may be done by functions like the softmax function. The *Roulette Wheel*

Selection (RWS) is a convenient way to stochastically select an action, given the probabilities. The probability of selection of an action a is given by Eq. (14.25).

$$P_{RWS}(a) = \frac{\exp(Q(s, a))}{\sum_a \exp(Q(s, a))}$$

(14.25)

The working of the algorithm is explained by Algorithm 14.3.

Algorithm 14.3 Q-learning

---

1   $Q(s, a) \leftarrow random \forall s, a$;   ▷ no. states × no. actions
2   **while** *stopping criterion* **do**   ▷ loop on episodes
3      $s \leftarrow$ initial/random state ;
4   **for** $t = 1$ *to maximum sequence length* **do**   ▷ steps in episode
         ▷ Step 1: Action selection;
5   **if** *rand* $< \epsilon$ **then** $a \leftarrow rand(A)$;   ▷ explorative
6   **else** $a \leftarrow argmax_a Q(s, a)$;   ▷ exploitative
7   implement action a in a simulated environment and get reward r and new state s' ;
         ▷ Step 2: Learning;
8         $Q[s, a] \leftarrow Q[s, a] + \eta(r + \gamma max_{a'} Q(s', a') - Q(s, a))$;
9         $s = s'$;
10   **if** $s =$*terminal state* **then** break;
11 return Q

14.4 Deep reinforcement learning

This section solves the other problem associated with the modelling so far, that the continuous states were taken as discrete variables. Now, the states would be continuous high-dimensional inputs, which may be images seen by the robot, laser scan readings of the robot, detailed information of the map, etc.

14.4.1 Deep Q-learning

The major change now is that the Q-table is now a neural network (Mnih et al., 2015). The neural network imitating the Q-table takes the states as input and produces the Q-value for all actions as output. The network is thus given by Fig. 14.5. The neural network could be a multilayer perceptron to deal with low dimensional inputs, a convolution neural network to deal with rich images and signals, or an LSTM to deal with sequential data.

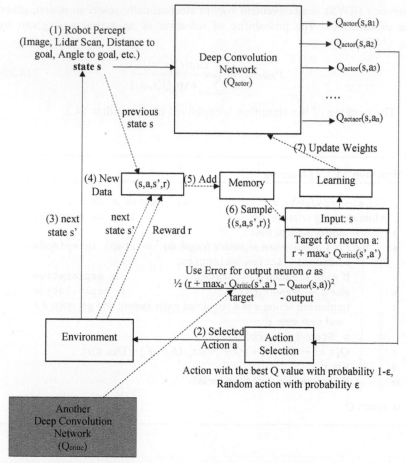

FIG. 14.5 Working of Deep Q Learning. Sensor percept (1) is used to calculate Q-values for action selection using an ε-greedy policy (2). The environment gives next state s' (3) and reward (r), which is added to the memory (4–5) for learning by calculating the targets using the critic network (6–7). The critic network copies the actor network with some frequency. Using a separate actor and critic network stabilizes training.

The learning now involves changing the weights of the neural network. For the gradient descend-based approach to work, it is important to specify the error function, whose derivative can be back-propagated. For the robot to go from state s to state s' by the application of action a using the reward $R(s,a,s')$, the error function for a single observation is given by Eq. (14.26).

$$\text{error} = \frac{1}{2}\left(R(s,a,s') + \gamma \max_{a'} Q(s',a') - Q(s,a)\right)^2 \qquad (14.26)$$

This is the same *temporal difference* error equation. Unlike a normal mean square error function associated with the neural networks, the neural network model $Q(s,a)$ appears twice in the equation. The first one is taken as a constant by injecting the actual value of state (s') observed and maximizing for all actions. The second one appears in the computation of the gradient. The conventional neural network learning works. The neural network libraries can be used with the state as input and the expression $R(s,a,s') + \gamma \max_{a'} Q(s',a')$ as the target value. The learning is incremental or data-by-data rather than by using batches. An off-the-shelf machine learning library can incrementally learn from an input and target pair. State s makes the input. However, the target is known for only one of the neurons representing action a, out of several possible actions, each action making an output neuron. To eliminate the other neurons to learn, a trick is to set their target as the current output ($Q(s',a'), a' \neq a$), which causes a zero error, and hence, the network does not learn from those neurons. Alternatively, the gradients can be modified to work only for the selected output neuron representing action a, whose target is known.

In Eq. (14.26), we travel 1 step using action a to reach a state s' in the simulations and use the observed error to compute the one-step temporal difference error. Let us extend it to k-steps instead. Suppose starting from a state $s = s_0$, the simulation takes k-steps using a fixed policy. Let the observed state–actions rewards in the run be $s_0 = s, a_0, r_0, s_1, a_1, r_1, s_2, a_3, s_3, \ldots a_{k-1}, r_{k-1}, s_k$, where s_i is the state at the ith step and a_i is the selected action in the ith step (say using the policy $\pi_{\epsilon\text{-greedy}}(s_i)$) that stochastically lands the agent into state s_{i+1} with a reward $r_i = R(s_i, a_i, s_{i+1})$. The total rewards received are $r_0 + \gamma r_1 + \gamma^2 r_2 \ldots \gamma^{k-1} r_{k-1}$. From s_k, the future rewards till terminal are expected as $U(s_k) = \max_{a'} Q(s_k, a')$. This suggests a k-step lookahead error as Eq. (14.27):

$$\text{error} = \frac{1}{2} \left(\sum_{i=0}^{k-1} \gamma^i R(s_i, a_i, s_{i+1}) + \gamma^k \max_{a'} Q(s_k, a') - Q(s,a) \right)^2 \quad (14.27)$$

A major difference with the supervised learning of the neural network is that here the target $R(s,a,s') + \gamma \max_{a'} Q(s',a')$ is not constant and changes at every instant, unlike supervised learning, where the target is fixed and supplied in the dataset. Every time the weights of the neural network representing $Q(s,a)$ are modified, the target changes due to a change in the term $Q(s',a')$, which is obtained from the same neural network. This causes the learning to be unstable and is a major factor of concern. A solution to the problem is called *actor-critic learning*. Here, there are two networks instead of a single network. One network is called the actor-network, which is the network that is trained as and when an observation is available from a move of the agent. The second network is called the critic network, and it acts as an oracle to give a target value to the agent (actor-network). It is assumed that the critic always knows the correct Q-values, and the agent uses it to calculate the target value $R(s,a,s') + \gamma \max_{a'} Q_{\text{critic}}(s',a')$, where the suffix critic denotes the use of the critic neural network. Initially, the actor and critic are randomly set. Even though, the actor learns against a random critic, it ends up learning something valuable, which satisfies the Bellman nature of the equation. After a few iterations, the actor has gained intelligence, and the same is directly transferred to the critic by copying the same set of weights and biases. The actor now starts to learn against this improved critic. The concept is also shown in Fig. 14.5.

Another problem in the system so far is that the agent learns incrementally from every observation. This is known to cause problems that led to the use of mini-batches in supervised learning. The different observations mean different functions, which guide the network in different directions altogether. The problem is amplified in RL since the consecutive frames have a high correlation to each other, which may result in nearly forgetting some aspects of the problem. In incremental learning, the robot does not see past data again in the future. It is, hence, likely to forget the learning from such past data in the future, which can be fatal if the data contain something important and rare.

A better way is thus to use mini-batches. The mini-batches are sampled by a large dataset storing all past experiences. In RL, the technique is specifically known as *experience replay*. The name stresses the fact that some historical experiences stored in the dataset are replayed so that the robot does not forget them. Unlike supervised learning, in RL the dataset is in the form of (s, a, s', r), where r is the reward achieved while going from s by an application of action a to state s'. It is immaterial to store exact target or Q-values as targets because the targets themselves change as the network parameters change. The robot now does two things simultaneously. The first is *experience gathering*, which is to select an action as per the current utility estimates using a mixture of exploration and exploitation. This constantly feeds new experiences in the experience dataset. The other task of the robot is *learning from the experiences*. For this, the robot samples a few experiences $\{(s, a, s', r)\}$ from the experience dataset. The actual targets for (s, a, s', r) are calculated as $r + \gamma \max_{a'} Q_{\text{critic}}(s', a')$ from the current network parameters. This is used to train the actor-network.

The working of the overall approach is given as Algorithm 14.4. The algorithm starts by randomly initializing an actor-network and a critic network, both taken as deep neural networks, in lines 1–2, while *Memory* (line 3) is the dataset that will be created as the robot moves and learns. Line 5 loops through the episodes, at every episode resetting the agent to the source (line 6). In every episode, the robot moves through a maximum sequence length (line 8) or till it reaches the terminal (line 24). Lines 9–10 implement an ε-greedy policy to select an action, which is implemented in the environment to collect the reward and the next state (lines 11–13). The dataset (memory) is appended with the current state (s), current action (a), reward (r), and the next state (s') shown in line 14. Lines 15–21 prepare a small, sampled dataset for training the actor-network as an implementation of experience replay, by noting the target of every sampled input (s) in lines 18–20, which is trained in line 22. Line 23 copies the learned actor to the critic with a set frequency of Δ.

Algorithm 14.4 Deep Q-learning

---

1 $Q_{actor} \leftarrow$ random neural network that takes robot percept as input and returns the Q-value for each action ;
2 $Q_{critic} \leftarrow Q_{actor}$;
3 $Memory \leftarrow Empty$;   ▷ all training data
4 $c \leftarrow 0$;
5 **while** *stopping criterion* **do**   ▷ loop on episodes
6   reset robot to initial/random state ;
7     $s \leftarrow$ current sensor observation ;
8   **for** $t = 1$ *to maximum sequence length* **do**
9   **if** *rand* $< \epsilon$ **then** $a \leftarrow rand(A)$;
10   **else** $a \leftarrow argmax_a Q_{actor}(s, a)$;
11   implement action a in a simulated environment and get reward r with robot at new state ;
12       $s' \leftarrow$ sensor observation at the new state;
13       $r \leftarrow$ observed reward ;
14   add (s, a, s', r) in $Memory$;
15       $B \leftarrow$ mini batch from $Memory$;
16       $T \leftarrow Empty$;   ▷ data used for training actor
17   **for** $(s, a, r, s') \in B$ **do**
         ▷ get target for neuron a using critic ;
18   **if** s' *is not at terminal* **then**
19             $target_a \leftarrow r + \gamma max_{a'} Q_{critic}(s', a')$
20   **else** $target_a \leftarrow$ terminal reward;
21   add $(s, target_a)$ to T ;   ▷ input,target of neuron a
22   train Q_{actor} using T ;   ▷ experience replay
      ▷ with frequency Δ, reset critic to actor ;
23   **if** $mod(c, \Delta) = 0$ **then** $Q_{critic} \leftarrow Q_{actor}$;
24   **if** *robot is at terminal state* **then** break;
25       $s \leftarrow s'$;
26       $c \leftarrow c + 1$

27 return Q_{actor}

14.4.2 Policy gradients

So far, we were learning the utilities, or the Q-values, while the policy mapped every state to the best possible action using the computed Q-values. Such a policy is also called a deterministic policy. We first slightly redefine policy, now also called the *stochastic policy*. Policy $\pi(s)$ is a probability density function giving the probability of execution of every action. Good actions have a higher probability and vice versa. The aim will now be to learn the policy. Under the new settings, the states are given as input to a neural network, which outputs the probability of selection of every action and thus closely represents the classification problem with the difference that the learning shall be using RL. Therefore, let $\pi(a \mid s, W)$ be the probability of selection of action a given state s and weights and biases of the neural network W.

Suppose the agent makes a series of moves under a policy and the traced trajectory is given by $\tau = [s_0, a_0, s_1, a_1, s_2 \ldots a_{T-1}, s_T]$. Let the total cumulative rewards be given by Eq. (14.28):

$$R(\tau) = \sum\nolimits_{t=0}^{T} \gamma^t R(s_t) \tag{14.28}$$

The aim is to find a policy that maximizes the total summative reward for all such runs, given by Eqs (14.29) and (14.30):

$$J(W) = \mathbb{E}_{\tau \sim \pi(\tau \mid W)} R(\tau) = \sum\nolimits_{\tau} \pi(\tau \mid W) R(\tau) \tag{14.29}$$

$$W^* = \arg \max_W J(W) = \arg \max_W \mathbb{E}_{\tau \sim \pi(\tau \mid W)} R(\tau) = \arg \max_W \sum\nolimits_{\tau} \pi(\tau \mid W) R(\tau) \tag{14.30}$$

Here, J is the total summative rewards that need to be maximized. $\mathbb{E}$ is the expectation over τ over the probability density function π that generates it. The probability of tracing the trajectories is conditioned on the policy, which in turn is conditioned on the weights (W) of the neural network. W^* are the optimal neural network weights.

To maximize the function, let us go the gradient ascent way, which is to iteratively add the derivative of the objective function to the weight vector. This requires the gradient of the objective function, given by Eq. (14.31):

$$\nabla_W J(W) = \nabla_W \left(\sum\nolimits_{\tau} \pi(\tau \mid W) R(\tau) \right)$$

$$= \sum\nolimits_{\tau} \nabla_W \pi(\tau \mid W) R(\tau)$$

$$= \sum\nolimits_{\tau} \pi(\tau \mid W) \frac{\nabla_W \pi(\tau \mid W)}{\pi(\tau \mid W)} R(\tau) \tag{14.31}$$

$$= \sum\nolimits_{\tau} \pi(\tau \mid W) \nabla_W \log(\pi(\tau \mid W)) R(\tau)$$

Here, step 2 divides and multiplies the expression by $\pi(\tau \mid W)$ to get the equation in the desired form. The result is interesting because the derivative of the probability has been reduced to a derivative of the log probability. An interesting aspect is that the probability of the trajectory would be a multiplication of small numbers, which will become small and create the problem of vanishing gradient as the derivative is used for weight updates. The log-likelihood solves the problem. Using the Markovian assumption, we have the probability of observing a trajectory given by Eq. (14.32) and its log given by Eq. (14.33):

$$
\begin{aligned}
\pi(\tau|W) &= \pi(s_0 a_0 s_1 a_1 \ldots s_T|W) \\
&= P(s_0)\pi(a_0|s_0)P(s_1|s_0 a_0)\pi(a_1|s_1)P(s_2|s_1 a_1)\ldots P(s_T|s_{T-1}a_{T-1}) \\
&= P(s_0)\prod_{t=0}^{T-1}\pi(a_t|s_t)P(s_{t+1}|s_t a_t)
\end{aligned}
$$

$$(14.32)$$

$$
\log \pi(\tau|W) = \log P(s_0) + \sum_{t=0}^{T-1}\log \pi(a_t|s_t) + \sum_{t=0}^{T-1}\log P(s_{t+1}|s_t a_t)
$$

$$(14.33)$$

Combining Eqs (14.31), (14.28), and (14.33) gives Eq. (14.34):

$$
\nabla_W J(W) = \sum_\tau \pi(\tau|W)\nabla_W\left(\log P(s_0) + \sum_{t=0}^{T-1}\log \pi(a_t|s_t)\right) \quad (14.34)
$$

$$
+ \sum_{t=0}^{T-1}\log P(s_{t+1}|s_t a_t)\right)\left(\sum_{t'=0}^{T}\gamma^{t'} R(s_{t'})\right)
$$

Removing the terms from the derivative that do not depend upon the weight (W) gives Eq. (14.35):

$$
\begin{aligned}
\nabla_W J(W) &= \sum_\tau \pi(\tau|W)\left(\nabla_W \sum_{t=0}^{T-1}\log \pi(a_t|s_t)\right)\left(\sum_{t'=0}^{T}\gamma^{t'} R(s_{t'})\right) \\
&= \sum_\tau \pi(\tau|W)\left(\sum_{t=0}^{T-1}\nabla_W \log \pi(a_t|s_t)\right)\left(\sum_{t'=0}^{T}\gamma^{t'} R(s_{t'})\right) \\
&= \sum_\tau \pi(\tau|W)\left(\sum_{t=0}^{T-1}\nabla_W \log \pi(a_t|s_t)\left(\sum_{t'=0}^{T}\gamma^{t'} R(s_{t'})\right)\right) \\
&= \mathbb{E}_{\tau\sim\pi(\tau|W)}\sum_{t=0}^{T-1}\nabla_W \log \pi(a_t|s_t)\left(\sum_{t'=0}^{T}\gamma^{t'} R(s_{t'})\right)
\end{aligned}
$$

$$(14.35)$$

The first summation is over time, inside which there is another summation on time. At time t (inside the first summation), the gradient computed should not affect the past gradients because of causality, while the gradient can affect all future rewards. The consequence of an action affects future actions, but not past actions. Adding causality gives Eq. (14.36):

$$
\nabla_W J(W) = \mathbb{E}_{\tau\sim\pi(\tau|W)}\sum_{t=0}^{T-1}\nabla_W \log \pi(a_t|s_t)\left(\sum_{t'=t}^{T}\gamma^{t'-t} R(s_{t'})\right) \quad (14.36)
$$

The important aspect of the derivation is that the transition probabilities $P(s_{t+1}|s_t, a_t)$ are eliminated and the learning is now model-free, otherwise, there were no easy means to compute these probabilities. Defining the derivative in such a manner by the expectation means that the derivative could be sampled by a few trajectories and the average derivative used for the calculations. This is similar to the drop of the probabilities in Q-learning and replacing them with a few samples stochastically selected by nature in the Monte Carlo simulations. The derivative is thus given by Eq. (14.37):

$$\nabla_W J(W) = \frac{1}{N} \sum_{\tau} \sum_{t=0}^{T-1} \nabla_W \log \pi(a_t|s_t) \left(\sum_{t'=t}^{T} \gamma^{t'-t} R(s_{t'}) \right) \tag{14.37}$$

Here, N is the number of sample trajectories drawn. The weights are updated iteratively as Eq. (14.38):

$$W(n+1) = W(n) + \eta \nabla_W J(W) \tag{14.38}$$

Here, η is the learning rate and n is the learning iteration number. The regularization parameter (norm of the weights or $\|W\|_2^2$) will be added in the loss function (and hence the weight term in the derivative in addition), which is not explicitly shown.

The implementation of policy gradient using a custom supervised machine learning library is easy. Observe Eqs (14.37) and (14.38). They model a neural network for a classificatory problem. Simulate a trajectory and calculate its discounted total cumulative reward ($\sum_t \gamma^t R(s_t)$). If the cumulative reward is high, the corresponding probabilities should be increased, and vice versa. Thus, the loss function is the cumulative reward times the log probabilities $\log \pi(a_t|s_t)$, where the log probabilities are obtained by adding a log to the neural network output. Automatic differentiation is available in most libraries and can be used to calculate the derivative of this loss function as a log of network output times cumulative reward and backpropagate it into the network.

This manner of implementation of the algorithm is called the *Monte Carlo policy gradient* because the trajectories are sampled by a Monte Carlo simulation and the weight is updated using the policy gradient. The algorithm in such an implementation is commonly called the REINFORCE algorithm.

One of the major limitations so far is that the learning happens based on a complete trajectory from the initial state to the terminal state, while the Deep Q learning approaches could learn based on offline generated data consisting of the state–action-landing state pair along with the observed reward function. So far, the agent cannot learn from every step, while learns only upon termination of the trajectory at the terminal state. Let us modify the algorithm. The last term ($\sum_{t'=t}^{T} \gamma^{t'-t} R(s_{t'})$) of Eq. (14.37) is the cumulative reward of trajectory r, which can be approximated by the utility of starting state–action pair $Q_{\text{critic}}(s_t, a_t)$. Consider the learning equation as Eq. (14.39):

$$\nabla_W J(W) = \frac{1}{N} \sum_{\tau} \sum_{t=0}^{T-1} \nabla_W \log \pi_{\text{actor}}(a_t|s_t) Q_{\text{critic}}(s_t, a_t) \tag{14.39}$$

A unique algorithm uses one network to estimate the Q-values in the learning equation, called the *critic*, while uses another network to learn the action selection policies, called the *actor*. The subscript critic denotes that the

Q-values are based on the weights of the critic neural network. Similarly, the policy is based on the weights of the actor. The resultant algorithm is called the *actor-critic algorithm*. The learning of the critic is based on the ideas discussed in deep Q-learning. The values are used in the weight update equation of the actor. The actor and critic networks need to be updated one after the other. Observing one state–action-landing state pair (s_t, a, s_{t+1}). For the actor, the neural network outputs are taken $(\pi_{\text{actor}}(a_t| s_t))$, transformed by log $(\log \pi_{\text{actor}}(a_t| s_t))$, and multiplied to the Q-value obtained from the critic network $(Q_{\text{critic}}(s_t, a_t))$. This makes the loss function. Automatic differentiation libraries are used to compute the derivative to update the actor's weights by back-propagation. The critic, on the contrary, uses the loss function discussed in deep Q-learning.

A problem in the algorithm is high variance. The agent moves as it generates the data, and a little change in the selection of action at one time instant can lead to a different observation sequence, which is learnt. In the optimization of objective functions, it is always safe to subtract constants that do not affect the results (optimal weights). Therefore, the derivative is changed to Eq. (14.40):

$$\nabla_W J(W) = \frac{1}{N} \sum_\tau \sum_{t=0}^{T-1} \nabla_W \log \pi_{\text{actor}}(a_t|s_t) \left(Q_{\text{critic}}(s_t, a_t) - V(s_t)\right) \quad (14.40)$$

Here, the function $V(s_t)$ is also called the *baseline* or a *value function*. Even though the function depends upon state s_t, it does not depend upon the weights W and, will therefore, be a constant that does not affect the optimization. The specific term $Q(s_t, a_t) - V(s_t)$ is referred to as the *advantage,* and the advantage function, therefore, reports the difference between the Q-value and the baseline.

The methods are now offline. The methods can generate a lot of data as (s_t, a, s_{t+1}) pairs along with the reward function $R(s_t, a, s_{t+1})$. The data are populated on a table and a standard machine learning library can learn from the data by extracting samples from the data in the form of experience replay. In parallel, an actor populates the data tables. To parallelize learning, it is possible to use several actors to learn asynchronously in parallel and to timely synchronize their updates, called the Asynchronous Advantage Actor Critic (A3C; Mnih et al., 2016). The updates by different actors can be synchronized, making the Advantage Actor Critic (A2C) method.

14.4.3 Deterministic policy gradients

Previously, the policy was taken as stochastic in nature, where the assumption was that action a_t is selected from a state s_t with a probability $\pi(a_t|s_t)$, making the policy function. The derivations involved accounting for all possible actions a_t with the associated probabilities. Typically, the policy function $\pi(a_t|s_t)$ is a neural network that takes state s_t as input and produces a probability density function of actions as output with one output neuron reserved for every action. For this, the actions need to be discretized. Now, let us make the policy function deterministic in nature, or the system selects a specific action a_t from state s_t, making the policy

$a_t = \pi(s_t)$, while no other action or probability is considered during learning or acting in the world. The actions may as well be continuous in nature.

Let us re-derive the policy gradient for the case of a deterministic action, called the *Deterministic Policy Gradient* (Silver et al., 2014). While the actions do not involve probabilities, the state transition is still stochastic in nature, which means that there is a probability of the agent being at state s time t given by $P_t(s)$, given infinite runs from the optimal policy π. Let $P_0(s)$ be the probability density function of the system at time 0 (typically the probability is 1 for the known source and 0 for all remaining states for systems, where the agent can only start from an accurately known source state). The agent at state s will select an action deterministically, $a = \pi(s)$, which will have a utility of $Q(s,a)$ or $Q(s,\pi(s))$. The loss function is thus given by Eq. (14.41):

$$J(W) = \sum_{s,t} P_t(s)Q(s, \pi(s)) \tag{14.41}$$

Taking a derivative of Eq. (14.41) gives Eq. (14.42), where the derivative of the Q-function is with respect to action a, and the derivative of the policy function is with respect to the weights W as per the chain rule, for the specific action $a = \pi(s)$.

$$\nabla_W J(W) = \sum_{s,t} P_t(s)\nabla_W Q(s, \pi(s))$$

$$\nabla_W J(W) = \sum_{s,t} P_t(s)\nabla_a Q(s, a)\nabla_W \pi(s)|_{a=\pi(s)} \tag{14.42}$$

Given several runs using the policy, nature will naturally sample the states as per the natural probabilities, which means that the derivative can be defined in terms of expectations given by Eq. (14.43).

$$\nabla_W J(W) = \mathbb{E}_{s \sim P_t(s)}\left(\nabla_a Q(s, a)\nabla_W \pi(s)|_{a=\pi(s)}\right) \tag{14.43}$$

Now, we take two neural networks. The first one for the Q-values and the second one for the policy π. The algorithm is off-policy, meaning that an agent shall explore and add data to the data tables, which will be used to simultaneously learn the Q-values and policy both. For the Q-values, the temporal difference error suffices as used previously for the value function. For the policy network π the change in weight is given by Eq. (14.44):

$$W(n + 1) = W(n) + \nabla_a Q(s, a)\nabla_W \pi(s)|_{a=\pi(s)} \tag{14.44}$$

The *Deep Deterministic Policy Gradient Algorithm* (DDPG; Lillicrap et al., 2015) uses a deep neural network implementation of both the actor (policy network) and the critic (Q-value network) while disallowing large changes in any of the networks. The *Distributed Distributional DDPG* (D4PG; Barth-Maron et al., 2018) algorithm additionally uses several actors along with adding weights to the data for being selected for the experience replay called prioritized experience replay. The *prioritized experience replay* samples difficult data more in contrast to the easy data to let the machine learn more from such data. The temporal difference error on the data is a good metric to decide the sample importance. The RL techniques can dramatically change the policy at every iteration of the algorithm, while many algorithms try to restrict the same. The *Trust Region Policy Optimization* (TRPO; Schulman et al., 2015) is a widely used policy gradient algorithm that enforces the updated and old policy to have a KL-divergence less than a threshold, while the *Proximal Policy Optimization* (PPO; Schulman et al., 2017) clips the difference between the old and the new policy that discourages sudden changes in the policy. Another common problem associated with RL is overestimation of the Q-value function that makes some states appear to have a larger than real value. The *Twin Delayed Deep Deterministic* (TD3; Fujimoto et al., 2018) network uses two value approximations using two separate networks and uses a minimum of both estimates to restrict overestimation. Furthermore, the Q-value network is learnt at every iteration, while policy learning is delayed (only to be learnt at some interleaved iterations), to make it stable to learn the Q-value network in contrast to learn it against a constantly changing policy function.

14.4.4 Soft actor critic

Another interesting RL algorithm deserves a mention. This is called the *Soft Actor Critic* (SAC; Haarnoja et al., 2018). The algorithm uses three separate neural networks for three different aspects. The first network is the Q-value function that models $Q(s,a)$. In other words, the network takes a state as input and gives as output the value for each action. The second network is a soft value function that computes the value (utility) of an input state $V(s)$. It takes the state as input and gives the value of the state as output. Even though the value can be derived from the Q-values ($V(s) = \max\limits_{a'} Q(s, a')$), the SAC method implements a separate network for the same reasons as the actor-critic networks of Q-learning, which is to stabilize training. Both of these networks together constitute the critic networks. The actor is the third network that computes the stochastic policy $\pi(a|s)$. The network takes a state as input and gives as output the probability of selection of every action.

The Q-value network $Q(s,a)$ is trained on the temporal difference error. However, the critic value is taken from the value network and not the same Q-value network. The error for a step from state s to state s' by the application of an action a and receiving a reward $R(s,a,s')$ is thus given by Eq. (14.45), where $V(s')$ is obtained from the value network and W_Q are the weights of the Q-value network.

$$J_Q\left(W_Q\right) = \frac{1}{2}\left(R(s, a, s') + \gamma V(s') - Q(s, a)\right)^2 \qquad (14.45)$$

The value network $V(s)$ is itself trained on a loss function as the residual between the value $V(s)$ and the estimated Q-value $Q(s,a)$ with action a sampled by using the policy network. $\pi(a|s)$ represents the probability of selection of action a at state s, which has a Q-value of $Q(s,a)$. The value of the network should ideally be $\mathbb{E}_{a \sim \pi(a|s)}Q(s, a)$. Additionally, an aim is to maximize the entropy of the network. Several policies may have similar values, while in such cases maximization of entropy ensures a lesser chance of ambiguity. Accounting for the residual and entropy, the error function is thus given by Eq. (14.46), where W_V is the weights of the value network.

$$J_V(W_V) = \frac{1}{2}\left(V(s) - \mathbb{E}_{a\sim\pi(a|s)}(Q(s,a) - \log\pi(a|s))\right)^2 \tag{14.46}$$

The policy is explicitly available as the policy network $\pi(a|s)$. The same policy is also represented as the Q-value network. The probability of selection of an action a at state s or $\pi(a|s)$ is proportional to $Q(s,a)$, and thus the policy is a softmax applied to the Q-value network or $\pi(a|s) = \text{softmax}(Q(s,a))$. The policy network is thus trained to closely imitate the Q-value network. The policy network is trained to minimize the KL-divergence loss over the Q-network. The loss function is given by Eq. (14.47), where $KL(x||y)$ denotes the KL-divergence between the distributions x and y, and W_π are the weights of the policy network.

$$J_\pi(W_\pi) = KL(\pi(a|s) \,||\text{softmax}(Q(s,a))) \tag{14.47}$$

14.5 Pretraining using imitation learning from demonstrations

RL starts from randomly initialized value and policy networks, and the robot takes random actions to explore the world. The state space is extremely complex with infinite possible state values. RL faces problems when the positive rewards are too deep, or the terminal state is too deep. In such a case, whatever the robot does, it receives a reasonably negative similar reward for all actions. The robot keeps wandering in every iteration. If the terminal state is deep, it is highly unlikely that the robot will ever reach the terminal state. The robot must complete at least one full episode to discover the terminal state and the terminal reward that can eventually be propagated. Sometimes it is possible to make a reward function where a robot is rewarded for getting nearer to the goal, rather than getting at the goal, avoiding deep rewards. In the case of a deep reward, the effect of that reward needs to travel the entire state space. This will take a severely long time. First, the neighbours of the goal will get converged, followed by their neighbours, and so on. The convergence towards the area of the start of the robot may take a long time.

An alternative is therefore to pretrain the network before applying RL. Suppose many demonstrations are available, where a demonstration is a game played by an intelligent human or an expert algorithm (like another motion planning algorithm for the navigation of robots). The policy and value networks are *pretrained* from such demonstrations. The policy network takes the state as input and generates the probability distribution of actions as output. The network can be trained from demonstrations as a standard supervised learning paradigm for classification problems. The demonstration data record expert actions that act as a target for the policy network training.

The value network is interesting because no target is available. There are two ways in which the network can be trained. The first is to do *passive reinforcement learning* over the demonstrations, where the standard Q-learning is used to train the network, however, the action selection is from the demonstrations instead of policies like ε-greedy. This is like populating the dataset with state–action-landing state and reward values and learning the Q-value network from such pairs using experience replay. The data are populated from demonstrations instead of the robot explorations.

Another technique to pretrain the value network is to explicitly calculate the value for all states in the demonstration dataset. Suppose one demonstration records state–reward–action pairs as $s_0, a_0, r_0, s_1, a_1, r_1, s_2, a_3, s_3, \ldots a_{n-1}, r_{n-1}, s_n$, where s_i is the state at the ith step and a_i is the action in the ith step that stochastically lands the agent into state s_{i+1} with a reward $r_i = R(s_i, a_i, s_{i+1})$. For state s_i, the cumulative summation of rewards till the terminal is $r_i + \gamma r_{i+1} + \gamma^2 r_{i+2} \ldots \gamma^{n-1-i} r_{n-1} = \sum_{j=i}^{n-1} \gamma^{j-i} r_j$. This is the indicative value (utility) of state s_i, and therefore, used as a target for the training of the value network.

14.6 Inverse Reinforcement Learning

The modelling so far assumed the presence of a reward function, which may not always be the case. For some problems, no reward function may exist, while for some others the reward function must be constructed against many reward indicators while the parameters for a weighted addition of the indicators may be hard to set intuitively.

The problem of *Inverse Reinforcement Learning* (IRL) is to extract an unknown reward function from a sequence of expert demonstrations or having access to an expert's policy. Let us assume that several demonstrations are available from the human or an existing intelligent algorithm. Let us say that each demonstration has come from an expert policy (π_E), which is also the best possible policy for the encountered states. Let the demonstration τ be given as a sequence of state–action pairs, or $s_0, a_0, s_1, a_1, \ldots a_{n-1}, s_n$, that shows a part of the expert policy for the observed state–action pairs. The problem is to find a reward function $R(s)$, such that the expected cumulative sum of rewards for the expert policy is always higher than the expected cumulative sum of rewards for any other policy (π) since the expert is assumed to be the optimal policy, or Eq. (14.48).

$$\forall_\pi \mathbb{E}_{s_t \text{ obtained from } s_{t-1} \text{ using } \pi_E(s_{t-1})} \left(\sum_t \gamma^t R(s_t) \right)$$

$$\geq \mathbb{E}_{s_t \text{ obtained from } s_{t-1} \text{ using } \pi(s_{t-1})} \left(\sum_t \gamma^t R(s_t) \right) \tag{14.48}$$

14.6.1 Inverse Reinforcement Learning for finite spaces with a known policy

First, let us take a discrete world containing a small number of states, like a grid world (Ng and Russell, 2000). Furthermore, assume that the policy $\pi(s)$ is perfectly known. Since the number of states is finite, the policy is typically available as a table $\pi_{S\times1}$ of size $S \times 1$, where S is the size of the state space. Also assume that the transition model $T(s,a,s')$ is known. Assume $T(s,a,s')$ is stored as a matrix of size $S \times S$ for every action policy a denoted by $P_{a,S\times S}$. Let $V_{S\times1}$ denote the value of all states as a $S \times 1$ array. Let $R_{S\times1}$ denote the reward of all states as a $S \times 1$ array. As per the definition of value, we get Eq. (14.49), where action $a1$ is the best action obtained by using the policy function $\pi(s)$.

$$V_{S\times1} = R_{S\times1} + \gamma P_{a1,S\times S}V_{S\times1}$$
$$V_{S\times1} = (I_{S\times S} - \gamma P_{a1,S\times S})^{-1}R_{S\times1} \tag{14.49}$$

Here, $I_{S\times S}$ is the $S \times S$ identity matrix. As per the definition we have the constraint that the value calculated by using the optimal action policy $a1$ should be better than the value calculated by using any other action other than $a1$, or Eq. (14.50).

$$\forall_{s,a\neq a1}\ R(s) + \Sigma_{s'}P_{a1}(s,s')V(s') \geq R(s) + \Sigma_{s'}P_a(s,s')V(s') \tag{14.50}$$

Rewriting the relation as a matrix (and eliminating the rewards from both the left and right sides) gives Eq. (14.51), where the inequality needs to hold for all the elements of the array.

$$P_{a1,S\times S}V_{S\times1} \succcurlyeq P_{a,S\times S}V_{S\times1}, \forall a \neq a1 \tag{14.51}$$

Substituting the value of $V_{S\times1}$ from Eq. (14.49) gives Eq. (14.52).

$$P_{a1,S\times S}(I_{S\times S} - \gamma P_{a1,S\times S})^{-1}R_{S\times1} \succcurlyeq P_{a,S\times S}(I_{S\times S} - \gamma P_{a1,S\times S})^{-1}R_{S\times1}\forall a \neq a1$$
$$(P_{a1,S\times S} - P_{a,S\times S})(I_{S\times S} - \gamma P_{a1,S\times S})^{-1}R_{S\times1} \succcurlyeq 0\forall a \neq a1 \tag{14.52}$$

Any reward function that satisfies the constraint of Eq. (14.52) is a valid solution, while the reward function can be obtained by using Linear Programming techniques. There will be infinitely many solutions. Therefore, let us take the solution that maximizes the margin between the best policy $a1$ demonstrated by the expert, and the second-best policy that selects the second-best action apart from $a1$. The second-best action $a \neq a1$ is the one that maximizes its utility, which in turn minimizes the objective measure of Eq. (14.52) since the component of action a is with a negative sign. The objective measure should be a scaler, and therefore, the summative margin across all states s is taken. The reward function is constrained to a maximum absolute value of $R_{\max}$ for all states. A regularization term for the reward is also added to the objective measure to encourage smaller rewards making simpler solutions. The reward R is obtained by maximizing the objective function in Eq. (14.53), where α is the regularization constant.

$$R = \arg\max_R \min_{a\neq a1} \sum_s (P_{a1}(s) - P_a(s))_{1\times S}(I_{S\times S} - \gamma P_{a1,S\times S})^{-1}R_{S\times1} - \alpha\|R\|$$

such that $(P_{a1,S\times S} - P_{a,S\times S})(I_{S\times S} - \gamma P_{a1,S\times S})^{-1}R_{S\times1} \succcurlyeq 0\forall a \neq a1$

$$|R(s)| \leq R_{\max}\forall s$$

$$\tag{14.53}$$

14.6.2 Apprenticeship learning

We now discuss the solution for the case, where there can be infinitely many state variables called Apprenticeship Learning (Abbeel and Ng, 2004; Zhifei and Joo, 2012). Let us assume that the reward is a linear weighted addition of the state attribute features with the weight vector W_r. Let us further constraint W_r so as not to take very large values, and hence, $||W_r|| \leq 1$. Assume $\phi(s)$ are the features extracted from state s. The reward function is given by Eq. (14.54):

$$R(s) = W_r^T \phi(s) \qquad (14.54)$$

Using Eq. (14.54) in Eq. (14.48), we get Eq. (14.55).

$$\forall_\pi \, \mathbb{E}_{\pi_E}\left(\sum_t \gamma^t W_r^T \phi(s_t)\right) \geq \mathbb{E}_\pi\left(\sum_t \gamma^t W_r^T \phi(s_t)\right)$$

$$\forall_\pi \, W_r^T \mathbb{E}_{\pi_E}\left(\sum_t \gamma^t \phi(s_t)\right) \geq W_r^T \mathbb{E}_\pi\left(\sum_t \gamma^t \phi(s_t)\right) \qquad (14.55)$$

$$\forall_\pi \, W_r^T \mu(\pi_E) \geq W_r^T \mu(\pi)$$

$\mathbb{E}_{\pi_E}(\sum_t \gamma^t \phi(s_t))$ samples state s_t using the expert policy $a_{t-1} = \pi_E(s_{t-1})$ on the previous state s_{t-1} (or calculates the quantity from the demonstration sequences directly). $\mathbb{E}_\pi(\sum_t \gamma^t W_r^T \phi(s_t))$ does the same for a generic policy π, such that the results hold for all such policies π.

Here $\mu(\pi_E) = \mathbb{E}_{\pi_E}(\sum_t \gamma^t \phi(s_t))$ is called the *feature expectation*, which is the discounted averaged feature vectors for several runs of the robot when operating with the expert policy π_E. If π_E contains demonstrations τ as a sequence s_0, a_0, $s_1, a_1, \ldots a_{n-1}, s_n$, the value may be given by $\mu(\pi_E) = \sum_t \gamma^t \phi(s_t)$, while the results must hold good for all such demonstrations τ. Similarly, $\mu(\pi)$ is the feature expectation of policy π, which is the discounted averaged feature vector for several runs of the robot when operating with the policy π. The term $\mu(\pi)$ should be valid for all policies π. It is obviously not possible to enumerate for the infinitely many policies π to check the validity of the equation. Let us parametrize the policy as a function of weights W_π and work for a general policy π only.

There are infinitely many weight vectors (W_r) that will satisfy the inequality, including the naïve case of having a zero reward for all states. Let us, therefore, add a constraint that the expert policy should be significantly better by value than any nonexpert-guessed policy. The problem objective function, therefore, becomes, Eq. (14.56):

$$W_r = \arg\max_{W_r, ||W_r|| \leq 1} W_r^T \mu(\pi_E) - W_r^T \mu(\pi)$$

$$W_r = \arg\max_{W_r, ||W_r|| \leq 1} W_r^T (\mu(\pi_E) - \mu(\pi)) \qquad (14.56)$$

Let us further find the best policy π, assuming a reward function $R(s) = W_r^T \phi(s)$. This is a RL problem that attempts to maximize the cumulative summation of rewards, given by Eq. (14.57):

$$W_\pi = \arg\max \Sigma_t \gamma^t R(s_t) = \arg\max \Sigma_t \gamma^t W_r^T \phi(s_t)$$
$$W_\pi = \arg\min \; - W_r^T \mu(\pi) \tag{14.57}$$

The two objective functions of Eqs (14.56) and (14.57) are formally incomplete. Eq. (14.56) uses an unknown policy π to be obtained from Eq. (14.57). Eq. (14.57) has an unknown weight W_r to be obtained from Eq. (14.56). More formally, combining Eqs (14.56) and (14.57), gives Eq. (14.58).

$$W_r = \arg\max_{W_r, \|W_r\| \le 1} \min_{W_\pi} W_r^T (\mu(\pi_E) - \mu(\pi)) \tag{14.58}$$

Eq. (14.58) represents a game between the policy network π with weights W_π and reward network $R(s)$ with weights W_r. The reward network aims to maximize the objective measure, while the policy network aims to minimize the same objective measure.

A problem is now to solve the maximization and minimization simultaneously. Let us break Eq. (14.58) into an outer problem to find the reward (IRL problem) given a policy and an inner problem to find the policy (the RL problem) given a reward. The two problems are solved iteratively and interchangeably. This is like iteratively playing the game, where the first player (RL problem) plays to get a policy, the second player (IRL problem) uses the same policy to compute the reward, and the reward is used to further improve the policy and get a new policy, which in turn is used to give an updated reward. This goes on and on.

Let us first take a random policy π_0, where the subscript denotes the iteration 0. For a constant policy, solve Eq. (14.56) to get the weights $W_{r,1}$, where the subscript 1 denotes the iteration. Now, assuming the reward function as $R_1(s) = W_{r,1}^T \phi(s)$ solve Eq. (14.57) to get a new policy π_1. Now, we have two policies (π_0 and π_1), and let us take the better of the two. This is used to get the updated reward function weights $W_{r,2}$. At any iteration i, we have the previous policies π_0, π_1, π_2, and so on till π_{i-1}. The updated reward function weight is given by Eq. (14.59), where the minimizer is used to get the best of the i available policies.

$$W_{r,i} = \arg\max_{W_r, \|W_r\| \le 1} \min_j W_r^T \left(\mu(\pi_E) - \mu(\pi_j) \right) \tag{14.59}$$

Thereafter, using the reward function $R_i(s) = W_{r,i}^T \phi(s)$, a new policy π_i is calculated as a standard RL. The policy is used to generate the feature expectations $\mu(\pi_i)$ by executing several sequences using the policy and averaging for the observed feature values. This goes on. Eventually the improvements in the feature expectations $\min_j W_r^T \left(\mu(\pi_E) - \mu(\pi_j) \right)$ will become extremely small and the algorithm terminates. The algorithm is shown as Algorithm 14.5.

Algorithm 14.5 Apprenticeship learning

---

1 Initialize a random policy π_0 ;
2 Calculate feature expectations of expert $\mu(\pi_E)$ (from demonstrations) ;
3 Calculate feature expectations $\mu(\pi_0)$ (estimate from runs using π_0) ;
4 $i \leftarrow 1$;
5 **while** *stopping criterion* **do**
   ▷ Inverse Reinforcement Learning ;
6   $W_{r,i} = argmax_{W_r, \|W_r\| \leq 1} min_{0 \leq j \leq i-1} W_r^T (\mu(\pi_E) - \mu(\pi_{i-1}))$;
7   **if** $min_{0 \leq j \leq i-1} W_r^T (\mu(\pi_E) - \mu(\pi_{i-1})) \leq \epsilon$ **then** break;
   ▷ Reinforcement Learning ;
8   Calculate optimal policy π_i using reward function $R(s) = W_{r,i}^T \phi(s)$;
9   Calculate feature expectations $\mu(\pi_i)$ (estimate from runs using π_i) ;
10   $i \leftarrow i + 1$;

   ▷ Best mixture policy ;
11 Calculate $\lambda = argmin_{\lambda, 0 \leq \lambda_j \leq 1, \sum_{j=0}^{i} \lambda_j = 1} \|\mu(\pi_E) - \sum_{j=0}^{i} \lambda_j \mu(\pi_j)\|$;
12 return $\pi(s) = \sum_{j=0}^{i} \lambda_j \pi_j(s)$

The output of the algorithm is not a single policy, however, a collection of policies $\pi_0, \pi_1, \pi_2, \ldots \pi_n$ for n iterations. The best policy amongst the policies needs to be finally picked. Let the best policy be given by a mixture (weighted addition) of the available policies, with the policy π_j having a weight λ_j, $0 \leq \lambda_j \leq 1, \Sigma_j \lambda_j = 1$. The weights λ_j are optimized. The mixture should be as close as possible to the expert, or Eq. (14.60):

$$\lambda = \underset{\lambda, 0 \leq \lambda_j \leq 1, \sum_{j=0}^{n} \lambda_j = 1}{\arg\min} \left\| \mu(\pi_E) - \sum_{j=0}^{n} \lambda_j \mu(\pi_j) \right\| \tag{14.60}$$

An interesting structure in the problem is that it can be reformulated as a quadratic programming problem which is very common with the Support Vector Machine (SVM) solvers. Let $\mu(\pi_E) - \mu(\pi_j)$ be denoted by a margin t_j. The problem is now finding the reward function weight W_r that maximizes the margin t_j. Out of all margins in the computed policies π_j, let the smallest margin be t. Therefore, the expert is always better than all policies by a margin of at least t, or $W_r^T(\mu(\pi_E) - \mu(\pi_j)) \geq t$ or $W_r^T \mu(\pi_E) \geq W_r^T \mu(\pi_j) + t$. Since the weights are constrained, $\|W_r\| \leq 1$. The aim is now to maximize t, which gives the quadratic programming problem as Eq. (14.61) that can be solved by using SVM solvers.

$$W_{r,i} = \arg\max_{W_r} t$$

$$\text{such that } W_r^T \mu(\pi_E) \geq W_r^T \mu(\pi_j) + t \tag{14.61}$$

$$\|W_r\| \leq 1$$

14.6.3 Maximum entropy feature expectation matching

The problem of IRL is to find a reward function such that the expert demonstrations have a better cumulative reward value than any other policy. The problem is that there are infinitely many reward functions, including a function with a zero reward for all states, that adheres to the specification. Therefore, let us have a different objective measure, which is to maximize the entropy of the policy network (Hussein et al., 2021a,b). Increasing the entropy makes a clearer action selection where one of the actions takes a very high probability value, while the remaining actions have a near-zero probability value. The second objective is that the policy should match the expert or that the feature expectations of the policy network are the same as the expert.

Let us now have features of a state–action pair instead of only the state. The feature vector of state s and action a is given by $\phi(s,a)$. The expectation from the expert policy is given by Eq. (14.62) and a general policy π by Eq. (14.63).

$$\mathbb{E}_{\pi_E}(f) = \sum_s \sum_a P(s)\pi_E(a|s)\phi(s,a) \tag{14.62}$$

$$\mathbb{E}_{\pi}(f) = \sum_s \sum_a P(s)\pi(a|s)\phi(s,a) \tag{14.63}$$

Here $P(s)$ is the visitation of s or the probability of visiting s. The maximization of entropy, subjected to equivalence of the feature expectations between the expert and policy π is given by Eq. (14.64).

$$W_{\pi} = \arg\max_{W_{\pi}} - \sum_s \sum_a P(s)\pi(a|s)\log\pi(a|s) \tag{14.64}$$

such that

$$\mathbb{E}_{\pi_E}(f) - \mathbb{E}_{\pi}(f) = 0$$

$$\sum_a \pi(a|s) = 1$$

The solution to the equation has the form given by Eq. (14.65)

$$\pi(a|s) = \frac{1}{Z(s)} \exp\left(W_r^T \phi(s,a)\right) \tag{14.65}$$

Here $Z(s) = \sum_a \exp(W_r^T \phi(s,a))$ is the normalization factor. The equation is intuitive since $W_r^T \phi(s,a)$ represents the reward of taking action a at state s. Rewards are exponentially preferable to their value that models the action selection probabilities.

14.6.4 Generative adversarial neural networks

Let us briefly de-tour to a different problem than IRL. The problem is to generate synthetic data that look realistic. Some real data samples are given to understand the problem domain, based on which the machine should learn how to create synthetic data. As an example, consider that the robot is taken on a tour and the robot's camera takes real images of the environment from a camera. However, an algorithm now needs a large pool of images that will be too slow to record from the real robot.

A *generator* takes a random noise vector $(z \sim P_z)$ from a distribution P_z and generates synthetic images as output. For convenience let the generator be a CNN that convolves and upsamples the data in a layered manner, starting from a noise vector and ending at the synthetic image. Let the distribution P_z be a unit normal distribution with mean 0 and standard deviation 1 for all attributes in the noise vector. Let the generator be given by $G(z)$. The problem is now to learn the generator model, given some real samples.

Let us make a *discriminator* that takes an image as input and gives as output the probability that the input is a real image from the real dataset. For convenience let the discriminator be a CNN. Let us denote the discriminator as $D(I)$. To train the discriminator, take the real dataset images associated as the positive examples. Use the generator to generate several synthetic images as the negative examples. The discriminator is thus trained using Eq. (14.66), commonly called the *binary cross-entropy loss*, where W_D is the weight of the discriminator.

$$W_D = \arg\max_{W_D} \mathbb{E}_{I \sim \text{real data}} \log D(I) + \mathbb{E}_{z \sim P_z} \log(1 - D(G(z))) \qquad (14.66)$$

The loss function tries to increase the log probability of the positive samples while tries to reduce the output of the negative samples (or increase the log probability of the sample being negative).

The generator is now trained to fool the discriminator. The generator aims to generate images that the discriminator cannot distinguish and thus misclassifies them. The loss for a random vector $z \sim P_z$ that generates the image output $G(z)$ is given by $\log(1 - D(G(z)))$ that needs to be minimized. The generator is thus given by the loss function in Eq. (14.67), where W_G is the weight of the generator.

$$W_G = \arg\min_{W_G} \mathbb{E}_{z \sim P_z} \log(1 - D(G(z))) \qquad (14.67)$$

Eq. (14.66) trains the discriminator against a generator that is not explicitly shown in Eq. (14.66), while Eq. (14.67) trains the generator against the discriminator that is not explicitly shown in Eq. (14.67). More formally, combining both equations, gives Eq. (14.68).

$$W_D, W_G = \arg\max_{W_D} \min_{W_G} \mathbb{E}_{I \sim \text{real data}} \log D(I) + \mathbb{E}_{z \sim P_z} \log(1 - D(G(z)))$$

$$(14.68)$$

Eq. (14.68) represents a two-player game between the discriminator and the generator. The discriminator attempts to best segregate the samples as real and synthetic, while the generator tries to reduce the same value. Both the networks are trained in an iterative and interchangeable manner. With time, the generator becomes more competent to generate synthetic images that are hard to discriminate from the real images, while simultaneously the discriminator attempts to perfect itself to raise the bar for the generator. The network is called the *Generative Adversarial Network* (GAN; Goodfellow et al., 2020).

Overall, two networks are obtained after training. A generator that can generate synthetic but realistic-looking images. At the same time a discriminator can distinguish if the input image is real or synthetic. The concept is shown in Fig. 14.6A.

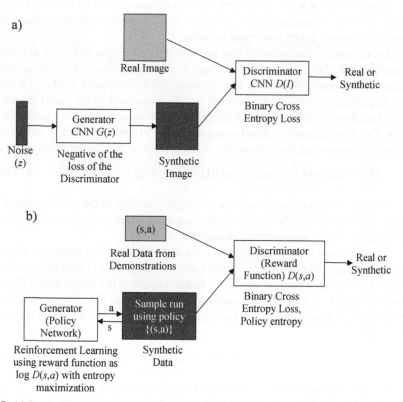

FIG. 14.6 (A) Generative Adversarial Network. The network trains a generator to generate synthetic images from noise and a discriminator to differentiate between the real and synthetic images. The generator uses a loss function as the negative of the discriminator's loss. (B) Generative Adversarial Imitation Learning. The generator is the policy network that is used to generate trajectory sequences given as synthetic data to the discriminator. The discriminator acts as a reward function to train the generator (policy) network using Reinforcement Learning.

14.6.5   Generative Adversarial Imitation Learning

Let us use the GAN working methodology for solving the IRL problem (Ho and Ermon, 2016; Fahad et al., 2020) known as *Generative Adversarial Imitation Learning* (GAIL). It is assumed that some demonstration trajectories are available from an expert. Let us first make a discriminator $D(s,a)$ that takes a state–action pair (s,a) as input and predicts whether the state–action pair is from the expert demonstration trajectory or a synthetically generated trajectory. The discriminator can be trained by taking all state–action pairs in the demonstration dataset as the positive examples, and random runs generated from a random policy as the negative examples.

The generator is the policy network $\pi(a|s)$. The generator attempts to generate state–action pairs such that the discriminator cannot distinguish them from the demonstration data. The loss function for the discriminator is similar to GAN. The demonstration samples are taken as positive, while several negative samples are generated using the policy network (generator). The discriminator weights are given by Eq. (14.69).

$$
W_D = \arg\max_{W_D} \min_{W_\pi} \mathbb{E}_{(s,a)\sim\text{demonstrations}} \log D(s,a) + \mathbb{E}_{(s,a)\sim\pi} \log(1 - D(s,a))
$$
$$
- \alpha H(\pi)
$$

$$(14.69)$$

Here $H(\pi)$ is the causal entropy function which is added for regularization using the constant α.

The training of the policy network (generator) needs a reward function. The closer is a state to the demonstration trajectory, the better is its feature expectation and higher should be its reward value. The reward function is the misclassification done by the discriminator for the samples generated by the policy. The reward function can thus be conveniently taken to be Eq. (14.70):

$$
R(s,a) = \log D(s,a) \qquad (14.70)
$$

The reward function is used to compute a better policy as a standard implementation of RL, while the RL also tries to maximize the entropy $H(\pi)$. The discriminator is thus analogues to the reward function. The game is now played between the reward function (discriminator) and the policy function (generator). The policy network (generator) aims to generate a better policy that generates trajectory traces similar to the expert demonstrations with similar feature expectations. The reward function (discriminator) on the contrary aims to distinguish between the expert demonstrations and the sequences generated by the policy network, drawing new samples of trajectories from the generator. The concept is shown in Fig. 14.6B.

14.7 Reinforcement learning for motion planning

Chapter 13 presented several architectures, scenarios, inputs, and outputs to perform behaviour cloning for the navigation of the mobile robot. This chapter discussed the mechanism to train any deep learning model to represent the value and policy networks using a RL paradigm. The same architectures as discussed in Chapter 13 can now be trained by using RL techniques. A sample architecture is shown in Fig. 14.7. The architecture uses the Pioneer LX robot equipped with a lidar scanner and a camera. The camera images are enhanced by extracting semantic information. The CNNs are used to extract and concatenate the features. The goal information (distance and angle to goal) is additionally supplied. The features representing the states go to two parallel networks, one predicts the value and the other predicts the policy. The policy is used to get the linear and angular speed and drive the robot. The environment gives the rewards for populating the memory dataset, from which data are sampled for experience replay. The value network (critic) trains by using the temporal difference error. The policy network (actor) is trained by using the Deterministic Policy Gradient algorithm that uses the value network as a critic. Let us specifically discuss two other implementations.

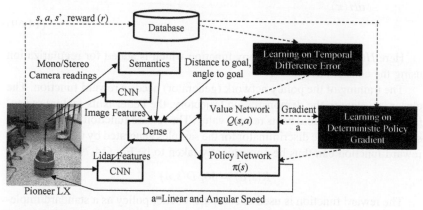

FIG. 14.7 Pioneer LX training by using Deep Reinforcement Learning. CNNs are used to extract features that are given to both the value network and the policy network. The lidar and image readings along with the goal information make the state.

14.7.1 Navigation of a robot using raw lidar sensor readings

Reinforcement learning is done for a robot equipped with a laser scanner. The robot always knows its position in the world due to the presence of a highly accurate localization system. The state (inputs) is the raw laser scan readings, distance to goal, and direction to goal. The laser scan readings pass through a CNN for feature extraction and the goal information is added to the dense layer. The network outputs the Q-values for all actions.

The discussions were on discrete actions, while the linear and angular velocities make continuous actions. Therefore, let us discretize the action space. The
linear and angular speeds are discretized into a small number of values and each
combination of linear and angular speed makes a discrete action.

The major challenge is the design of the reward function. The robot is given
small positive rewards for moving towards the goal and small negative rewards
for moving away from the goal. The projection of the motion direction on the
straight line to the goal is used to compute the reward. The robot is also given
small negative rewards for being too close to the obstacle. Clearance is used
for the same. The robot is given a highly positive reward on reaching the goal
and a highly negative reward for colliding. Every step has an additional penalty
to encourage the robot to reach the goal early. The robot is also trained for multi-
robot behaviours, where additional rewards are awarded for maintaining social
etiquettes, like avoiding an oncoming robot preferably from the left and overtaking a robot preferably from the right. It should be noted that the weight constants
of the rewards have a huge impact on the performance of the final system. It is
hard to set these weight constants. A change in the values means that the robot
will have to re-learn, which will take a long time.

The robot can be left for training in a simulation setup for prolonged times.
However, for initial accelerated training (with a bias), first, an algorithm that is
known to perform well is used to move the robot and RL is used in a passive
move only. A well-tuned model of artificial potential field is used for the purpose. Once the model is trained enough, it is asked to learn in an active move.
The Pioneer DX robot learning itself in a simulation using the Gazebo simulator
can be seen in Fig. 14.8. The map is a depiction of the Centre of Intelligent
Robotics at IIIT Allahabad.

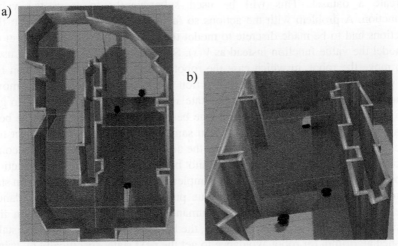

FIG. 14.8   Pioneer DX robots learning in a simulated Gazebo environment using Reinforcement
Learning. (A) Orthographic view. (B) Oblique view.

14.7.2 Navigation of a robot amidst moving people in a group

Another use case is that of a robot trying to avoid moving people or other robots. The problem is interesting because it represents a *Multi-Agent Reinforcement Learning* framework where the need is to move multiple agents at the same time. To make matters simple, let us assume that there are only two agents, a robot, and a human, while the aim is to move the robot alongside a dynamically moving human (Chen et al., 2017a). The state vector is now the joint state of the robot as well as the human along with the goal information. Let the robot and human be circular and holonomic, which can move at any direction at any time. Therefore, the orientations are not important. Let us encode the state information in the robot coordinate frame with the origin at the robot body and the X axis points from the robot body towards the robot goal. Let us further use the polar coordinates to encode the states. The robot state will be origin as the position with a velocity vector (v_{xr}, v_{yr}) and radius ρ_r, where r denotes the robot. The human state in the polar coordinates will be the distance between the robot and human $d(r,h)$, the angle of the vector from human to robot (θ_h) in the robot coordinate frame, speed (v_{xh}, v_{yh}) relative to the robot and human radius ρ_h. Here, h denotes the human. The goal will be at the X-axis at a distance $d(r,g)$, which the robot aims to reach at the preferred speed of v_{pref}. Here, g refers to the goal. The joint state thus becomes $s = [s_r, s_h, s_g] = [(v_{xr}, v_{yr}, \rho_r), (d(r,h), \theta_h, v_{xh}, v_{yh}, \rho_h), (d(r,g), v_{pref})]$.

The robot is given a reward of +1 on reaching the goal. The reward on colliding is -0.25. The robot is penalized on being closer to the obstacle by less than 0.2 with the penalty as $-0.1 + c/2$ for the clearance c. A very small reward (say -0.007) is given for every step to encourage the robot to reach the goal early.

Let us first make an algorithm that makes the robot (and human) move to create a dataset. This will be used subsequently to learn the value function. A problem with the actions so far in deep Q-learning was that the actions had to be made discrete to model the Q-value function $Q(s,a)$. Let us model the value function instead as $V(s)$. Suppose an ε-greedy policy is used to move the robot, in which case the robot must choose a random action for ε proportion of times and the best action for the remaining time. To move the robot by the best policy from a state s, several actions a are used to get to target states s' with values $V(s')$. The best action a is the one with the best value function. Now the algorithm can sample several actions a to select the best one, without directly discretizing the action. However, state s' has a component of the human's state, which cannot be obtained from the selected action applicable for the robot only. A naïve implementation is to keep the human stationary for the one step used for value prediction. However, the avoidance behaviour can have the robot and human close by, in which situation the assumption is hard to make. Therefore, the human is moved by using an established algorithm for robot avoidance to get the next state, specifically by using the Optimal Reciprocal Collision Avoidance (ORCA; van den Berg et al., 2010) algorithm. The same is used to get the human component of the next state.

The best action selection is thus given by $a = \text{argmax}_a R(s, s') + \gamma V(s')$, where state $s' = [s'_r, s'_h, s'_g]$ contains a join of the robot state (s'_r), human state (s'_h), and goal state (s'_g) at the next frame. s'_r is obtained by applying action a on the previous robot state s_r. s'_h is obtained by applying ORCA motion planner on the human at s_h. s'_g is the goal information calculated from the updated robot state.

The network is trained using the Temporal Difference error on the value function. The problem is challenging and therefore an initial training is done by moving the robot and human using the ORCA simulator. The trajectories as a joint state are noted and the cumulative reward till the terminal is used as the value of the joint state. This is used to pretrain the value network using demonstrations or imitation learning. Thereafter, the robot moves by using RL.

Now let us extend the solution to the multiple robot case (or the multihuman case). Currently, the value function is for one robot and one human only. To use the same value network, let us say that the robot only tries to avoid the most affecting human, out of several humans in the scenario. This limits the problem to finding the most affecting human and using the same value function as used previously. The most affecting human is the one that most limits the value of the robot. In other words, the robot plays a 1-step game with all other humans. The robot selects the best move or the action that maximizes its value. The nature however plays one move, which is to move one human. ORCA motion planner is used to make the humans move. Nature aims to reduce the value of the robot as an adversary. Let s_i be the join of states considering human i. State s_i is thus $s_i = [s_r, s^i_h, s_g]$, where s_r is the robot state, s^i_h is the state of the ith human, and s_g is the goal state. The action selection is thus modified to Eq. (14.71).

$$a = \text{argmax}_a \min_i R\left(s_i s'_i\right) + \gamma V\left(s'_i\right) \tag{14.71}$$

Here state $s'_i = [s'_r, s'^i_h, s'_g]$ contains the states considering the selected human i. s'_r is obtained from s_r by the selected action a. s'^i_h is obtained by applying ORCA motion planner on the human i at s^i_h. s'_g is the same goal s_g in the updated robot's coordinate frame.

Now we can place multiple humans in the scenario. Let us now modify the reward function to incorporate social etiquettes. The first etiquette is that the robot must never break a group of humans going together. In other words, the robot must not be within the convex hull formed by a group of humans and simultaneously must not be within a circular coverage of the human group. A negative reward is given if this constraint is violated. If the robot is a part of a group, it is given a penalty if someone else comes within the group, or if the group does not maintain a cohesion (small distance of the members to each other). Similarly, a larger reward is given if the group of robots reach the goal, in contrast to a smaller reward if only one robot of the group reaches the goal. This stops the robots from blocking other members of the group to reach their goal (Vallecha and Kala, 2022). The other conventions include passing, crossing, or overtaking people from the correct side. As an example, overtaking should be done on the right while passing people should happen from the left side. If such a norm is violated, a penalty is added (Chen et al., 2017b).

14.8 Partially observable Markov decision process

In all the approaches so far, it was always assumed that the state of the agent and everything in the environment is always known or the problem was solved as a Markov Decision Process. Another class of problem argues that the robot may not know its position, the entire environment used for decision-making or some important information may be missing. In such a case we do not have a clear state s used in the formulations. However, there is a probability density function $P(s)$ over all states in the state space S. This is also called the *belief* of the agent about its state represented by $bel(s)$. Such systems are modelled as a *Partially Observable Markov Decision Process* (POMDP), where the term Partially Observable highlights the fact that the state is not known in absolute terms, but as a probability density function; while Markov Decision Process highlights the Markovian assumption that the probability of the landing state only depends upon the current state and the current action.

The formulation of POMDP is thus not difficult and can be modelled by using the belief instead of the true knowledge of the state. This translates to taking a weighted addition over all states s, weighted by the probability density function, exactly as done in the calculation of the expected utility. However, now the system is doubly probabilistic. The first probability is transition probability (model) as used before or $T(s,a,s')$. The states here are again as a belief or a probability density function given by $P(s)$. This necessitates double integration of every calculation and the POMDPs are thus regarded as extremely computationally expensive for most problems. The POMDPs very beautifully model the uncertainties and calculate the policy that is best suited with a limited amount of information. The information may be updated as the agent moves and gathers more data about the environment.

Questions

1. Consider a 2×2 grid with no obstacles, in which the robot can take up, down, right, or left actions. The cost of taking every step is 1 while the terminal utility is +100. The agent can collide with the walls. On collision, the reward is -20. The Q-table is initially kept with +100 value for everything. Learning rate is 0.4. The robot is initially at (1,1). It takes action up and reaches (1,2). It takes another action up and collides with the wall. It takes action right, however, stays at (1,2) and collides with the wall. It takes another action right and reaches (2,2), the terminal state. (i) Show the

updated Q-table after all actions. (ii) The agent is again re-initialized to (1,1). Explain the choice of action taken by the agent along with the policy used for the same. (iii) Simulate the agent for 3 steps and show the learnt Q-table. Assume suitable parameter values.

2. Give one example of model-free reinforcement learning and one example of model-based reinforcement learning. Explain the pros and cons of using a model-free approach in contrast to a model-based approach.

3. Compare and contrast between the algorithms: A* algorithm, Q-learning, value iteration, and ExpectiMax game tree (with chance and max nodes) for planning the robot in a grid. State which algorithm is better in what context and why?

4. Explain what the effect on optimality and computation time is when solving for game trees (i) by adding $\alpha\beta$ pruning, (ii) by increasing the cutoff depth, (iii) by uniformly scaling all utility values (without chance node) while using $\alpha\beta$ pruning, (iv) by uniformly scaling all utility values with chance node while using $\alpha\beta$ pruning. Can the technique be used when MIN decides not to play optimally?

5. [Programming] Consider a robot operating in a grid of size $m \times n$ with obstacles. The robot can be facing any of the eight directions: N, S, E, W, NE, NW, SE, and SW (all directions have their usual meanings). There are only three actions possible: turn clockwise, turn anticlockwise, and move forward. Each turn normally makes the robot obtain the next direction. Each move forward normally makes the robot take a step ahead and land at the nearest grid. So, a move forward command, while the robot is at (1,1) and facing NE, will make the robot move to (2,2). However, the actions are stochastic. So, taking 1 step N can result in taking 1 step N with a probability of 0.8, taking 1 step NE with a probability of 0.1, and taking a step NW with a probability of 0.1. Similarly, with the other cases. The action of turning is not stochastic. Each action has a penalty of −0.2, while collision has a penalty of −10. The robot needs to reach the exit location (m,n) which has a high positive reward. The discount factor is taken as 0.9. Compute the policy using value iteration, policy iteration, and reinforcement learning. Simulate the robot using the computed policy. See the effect of the different algorithms on increasing map size.

6. [Programming] Make an intelligent player for a game of Othello. Play the game for m moves considering both the players play optimally. The terminal utility is the difference in the number of disks of the players. The evaluation function for nonterminal states is given by the difference in the number of corners occupied (highest weight), the difference in the number of moves possible (moderate weight), and the difference in the number of disks (lowest weight). The terminal score must always be preferred to utility estimates. Study the effect of different cutoff depths. Also, solve for three players with no alliances and any two players forming an alliance.

7. [Programming] Experiment with different deep reinforcement learning techniques using Open AI Gym framework and Atari games.
8. [Programming] Simulate a robot operating by using virtual lidar sensors. The robot always knows its position. Using deep reinforcement learning, demonstrate the behaviours (i) going to the goal without obstacles, (ii) obstacle avoidance while navigating ahead in a straight corridor, (iii) motion to the goal avoiding obstacles, (iv) two robots walking in parallel in a corridor without obstacles (they should neither be too far nor too close), (v) two robots walking in parallel in a corridor with obstacles, (vi) a group of robots going together, avoiding obstacles. Use actions as different combinations of linear and angular speeds. Assume that each robot knows the relative position to the surrounding robots. Show mechanisms to increase/decrease clearance with obstacles by tuning the reward function. Compare the behaviours with the ones obtained from the potential field.
9. [Programming] In question 8, suggest heuristics for (i) using a deliberative path to guide a robot running using Reinforcement Learning, and (ii) using data from hand-drawn trajectory/teleoperated trajectory to make a reward function, which is used for reinforcement learning of the robot, and (iii) using a deliberative algorithm to trace a trajectory, while passively learning an initial reinforcement learning model.

References

Abbeel, P., Ng, A.Y., 2004. Apprenticeship learning via inverse reinforcement learning. In: Proceedings of the Twenty-First International Conference on Machine Learning, p. 1.

Barth-Maron, G., Hoffman, M.W., Budden, D., Dabney, W., Horgan, D., Tb, D., Muldal, A., Heess, N., Lillicrap, T., 2018. Distributed distributional deterministic policy gradients. arXiv preprint arXiv:1804.08617.

Chen, Y.F., Liu, M., Everett, M., How, J.P., 2017a. Decentralized non-communicating multiagent collision avoidance with deep reinforcement learning. In: 2017 IEEE International Conference on Robotics and Automation. IEEE, pp. 285–292.

Chen, Y.F., Everett, M., Liu, M., How, J.P., 2017b. Socially aware motion planning with deep reinforcement learning. In: 2017 IEEE/RSJ International Conference on Intelligent Robots and Systems. IEEE, pp. 1343–1350.

Fahad, M., Yang, G., Guo, G., 2020. Learning human navigation behavior using measured human trajectories in crowded spaces. In: 2020 IEEE/RSJ International Conference on Intelligent Robots and Systems, pp. 11154–11160.

Fujimoto, S., Hoof, H., Meger, D., 2018. Addressing function approximation error in actor-critic methods. In: International Conference on Machine Learning. PMLR, pp. 1587–1596.

Goodfellow, I., Pouget-Abadie, J., Mirza, M., Xu, B., Warde-Farley, D., Ozair, S., Courville, A., Bengio, Y., 2020. Generative adversarial networks. Commun. ACM 63 (11), 139–144.

Haarnoja, T., Zhou, A., Abbeel, P., Levine, S., 2018. Soft actor-critic: off-policy maximum entropy deep reinforcement learning with a stochastic actor. In: International Conference on Machine Learning. PMLR, pp. 1861–1870.

Ho, J., Ermon, S., 2016. Generative adversarial imitation learning. In: Advances in Neural Information Processing Systems 29.

Hussein, M., Crowe, B., Petrik, M., Begum, M., 2021a. Robust maximum entropy behavior cloning. arXiv preprint arXiv:2101.01251.

Hussein, M., Crowe, B., Clark-Turner, M., Gesel, P., Petrik, M., Begum, M., 2021b. Robust behavior cloning with adversarial demonstration detection. In: 2021 IEEE/RSJ International Conference on Intelligent Robots and Systems. IEEE, Prague, Czech Republic, pp. 7858–7864.

Lillicrap, T.P., Hunt, J.J., Pritzel, A., Heess, N., Erez, T., Tassa, Y., Silver, D., Wierstra, D., 2015. Continuous control with deep reinforcement learning. arXiv preprint arXiv:1509.02971.

Mnih, V., Kavukcuoglu, K., Silver, D., Rusu, A.A., Veness, J., Bellemare, M.G., Graves, A., Riedmiller, M., Fidjeland, A.K., Ostrovski, G., Petersen, S., Beattie, C., Sadik, A., Antonoglou, I., King, H., Kumaran, D., Wierstra, D., Legg, S., Hassabis, D., 2015. Human-level control through deep reinforcement learning. Nature 518, 529–533.

Mnih, V., Badia, A.P., Mirza, M., Graves, A., Lillicrap, T., Harley, T., Silver, D., Kavukcuoglu, K., 2016. Asynchronous methods for deep reinforcement learning. In: International Conference on Machine Learning. PMLR, pp. 1928–1937.

Ng, A.Y., Russell, S., 2000. Algorithms for inverse reinforcement learning. In: ICML. vol. 1, p. 2.

Russell, S., Norvig, P., 2010. Artificial Intelligence: A Modern Approach. Pearson, Upper Saddle River, NJ.

Schulman, J., Levine, S., Abbeel, P., Jordan, M., Moritz, P., 2015. Trust region policy optimization. In: International Conference on Machine Learning. PMLR, pp. 1889–1897.

Schulman, J., Wolski, F., Dhariwal, P., Radford, A., Klimov, O., 2017. Proximal policy optimization algorithms. arXiv preprint arXiv:1707.06347.

Silver, D., Lever, G., Heess, N., Degris, T., Wierstra, D., Riedmiller, M., 2014. Deterministic policy gradient algorithms. In: International Conference on Machine Learning. PMLR, pp. 387–395.

Vallecha, M., Kala, R., 2022. Group and socially aware multi-agent reinforcement learning. In: 2022 30th Mediterranean Conference on Control and Automation. IEEE, Athens, Greece, pp. 73–78.

van den Berg, J., Guy, S.J., Lin, M., Manocha, D., 2010. Optimal reciprocal collision avoidance for multi-agent navigation. In: Proceedings of the IEEE International Conference on Robotics and Automation. IEEE, Anchorage (AK), USA.

Zhifei, S., Joo, E.M., 2012. A review of inverse reinforcement learning theory and recent advances. In: 2012 IEEE Congress on Evolutionary Computation. IEEE, Brisbane, QLD, Australia, pp. 1–8.

Chapter 15

An introduction to evolutionary computation

15.1 Introduction

Optimization has proven to be useful in a variety of applications. Optimization is also valuable for numerous robotics problems, including the specific problem of motion planning (Kala et al., 2011, 2012; Kala and Warwick, 2014). This chapter formally discusses optimization algorithms that can be used to solve the problem of motion planning. This chapter covers all the theoretical paradigms of optimization with a special reference to the specific problem of motion planning.

The specific optimization technique discussed in this chapter is called the evolutionary approach. Evolutionary algorithms solve a large variety of problems in numerous problem domains including optimization. It represents a class of algorithms motivated by the natural evolution process of different species. Evolution makes the species fitter with generations. The individuals making the population pool of the species collectively improve with generations. It should similarly be possible for a pool of trajectories to continuously evolve and reach a better objective value.

The chapter discusses the most popular evolutionary algorithms: the genetic algorithms (GAs), particle swarm optimization (PSO), and differential evolution (DE). The chapter also discusses the means to make the algorithms faster using memetic computation and solve the problem with time constraints as a local search.

15.2 Genetic algorithms

GA (Davis, 1987; Goldberg, 1989) is one of the most basic and widely used techniques. The algorithm imitates natural evolution, specifically imitating the mating process of natural animals to produce a newer generation.

Autonomous Mobile Robots. https://doi.org/10.1016/B978-0-443-18908-1.00011-X

15.2.1 An introduction to genetic algorithms

The GA is an inspiration from the natural evolution process which consists of a population of individuals. The individuals of a population mate between themselves and further have characteristics developed by them on their own to make a new generation of the population. Evolution follows Darwin's principles of *survival of the fittest*, which means that only those individuals that are the most adaptive to change survive, while the others die in the process. The fitter individuals mate more with the other individuals and produce more offsprings, while the weaker ones are unable to mate and therefore die. So, every new population primarily consists of the characteristics of the fitter individuals and new characteristics developed by them which makes the new population better than the previous one. If the environment changes, the individuals adaptive to change or the fitter ones as per the new environment are the ones who survive.

The same analogy is used to construct an artificial process that continuously improves the solution to a problem. An *individual* is a solution to the problem. The individual in the specific format as applicable to the problem is called the *phenotype*. So, broadly speaking, the complete trajectory from source to goal represents the phenotype in the problem of robot motion planning. The *genotype* represents the same problem as seen by the GA. Typically, the GA is used on a string or array of real numbers, which becomes the genotype. The conversion of a phenotype to a genotype is called the *individual representation strategy* and is a part of the problem design process. Every item in the genotype is called the *gene*. This also represents the natural analogy, wherein the entire human is composed of genes and chromosomes.

As an example, for the problem of learning a neural network, the complete neural network is the phenotype; while a list of all weights, biases, and connections as a continuous string is the genotype. Similarly, for the problem of optimizing the visit schedule of places, or the travelling salesman problem, the actual routes and places constitute the phenotype, while the array of places represents the genotype. This specific problem comes under the general class of combinatorial optimization problems. For the problem of robot motion planning, the trajectory constitutes the phenotype. Let us say that the trajectory passes through a list of *via points* that make the genotype, while the trajectory itself is straight-line connections between the via points.

The *population* is defined as a collection of individuals. The individuals constituting a population pool undergo numerous modifications by the application of *genetic operators*. The genetic operators are applied to one, two, or more individuals of the population pool to produce new offspring. A set of all offspring produced by all the genetic operators constitutes the candidate population for the next generation. However, not all individuals undergo modification by genetic operators. In alignment with Darwin's theory of survival of the fittest, the fitter individuals create more offspring by being selected multiple times. The *fitness function* tests the goodness of an individual.

This function is specified as a part of the problem design process. For the problem of neural network learning, the mean square error (and regularization term) denoting the loss function is the fitness function. For the travelling salesman problem, the length of the final path is the fitness function. For the problem of robot motion planning, the fitness function is a mixture of path length, smoothness, and clearance. Formally, fitness means a good attribute and therefore the fitness function must be maximized. In all the examples, the function was to be minimized, and therefore a more appropriate term is the *cost function*. The book will use the two terms interchangeably since one can be obtained by the negation of the other.

Selection decides which individuals (and in what quantity) will be used for the application of genetic operators. Selection decides the trade-off between the selection of the best individual too many times and selection of even weak individuals. Selection of best individuals too many times dominates the new population by a single individual that may lead to deletion of key characteristics not present in the individual. However, since most population comes from the fittest set of individuals, the resultant population has a high fitness leading to early convergence. On the contrary, selection of even weak individuals makes a weak new generation because of weak parents leading to slow convergence. However, at the same time this can preserve key characteristics found in currently unfit individuals till a time the characteristics further develop leading to high fitness.

The two typical genetic operators used are crossover and mutation. *Crossover* represents the mating process of the individual of a population to create children. In the primitive form, two individuals of a population pool exchange genes or characteristics and produce two children, with the children typically having genes from either of the two parents. Since the actual evolution is synthetic, the crossover definition is generalized to any k parents producing m children, by exchanging characteristics.

The other common operator also inspired by the natural evolution process is *mutation*. In the mating process, many times there is a copying error in the transfer of genes and therefore the children also develop some new characteristics not present in the parents. This process is called a mutation. In the mutation operation, some noise is added to the genes of an individual, which adds new characteristics.

Overall, the application of all the genetic operators combined creates a new candidate population. In the natural process, the old population dies off and the new population completely replaces the older population. This is called one *generation* of the algorithm. The newer generations replace the older ones, and the process goes on generation by generation. Here, one generation represents the typical notion of iteration. A *stopping criterion* decides the number of generations or time that the algorithm executes for. The concept may even be generalized. The old population and new candidate population may be combined, and the best half population may be selected. This represents a competition between the old and the new candidate population pool. In another case, every new child produced by a parent competes with its parent for survival, and the fitter of parent and child is accepted for the new population.

A convention is to keep the population size as fixed. This makes sense because the algorithm operates iteratively for an indefinite time, depending upon the stopping criterion. If the new population is larger, then the number of individuals constantly increases causing an explosion. If the new population is smaller, there can be an extinction. A summary of all operators and terms is given as Box 15.1.

Box 15.1 Different terms used in genetic algorithm.

Individual: A representation of a solution that solves the problem, while the algorithm tries to get the best individual or the optimal solution to the problem.

Phenotype: Individual in the format suitable to the problem.

Genotype: Individual in the format suitable for optimization, e.g., array of real numbers, with the information that needs to be optimized encoded.

Population: A collection of individuals.

Genetic operators: Operations that take one or more individuals called parents to produce one or more individuals called children.

Fitness function: A function to decide the goodness of an individual.

Selection: A function that probabilistically selects the best individuals for genetic operations or to carry forward to the next generation.

Crossover: A genetic operator that produces children by the exchange of genes between parents.

Mutation: A genetic operator that introduces errors or adds noise in the parent to produce the child.

Generation: Creation of a population pool for the next time step by selection of individuals, application of genetic operators, and selection of the individuals from the parents and children.

Stopping criterion: Condition meeting which the optimization process must stop.

Exploration: Intent to explore or dispense the population into a larger hypervolume in the optimization space. High exploration increases convergence time and increases chances to get into the correct global optima region, but not closer into the global optima.

Exploitation: Intent to aggregate the individuals within a smaller hypervolume in the optimization space. High exploitation gets faster convergence, but the chances of convergence into a local optimum are higher.

The working of GA is now discussed in detail. For the same, let us take a typical *optimization problem* given by Eqs. (15.1)–(15.3).

$$x^* = argmin_x f(x) \qquad (15.1)$$

$$LB_i \leq x_i \leq UB_i \qquad (15.2)$$

$$g(x) \leq 0 \qquad (15.3)$$

The problem is defined as finding the best value of the vector $x^* = [x_1^* \ x_2^*$ $x_3^* ... x_n^*]^T$, which minimizes a given function $f(x)$ called as the cost (fitness) function, while being subjected to the lower and upper bounds $LB_i \leq x_i \leq UB_i$ and some general constraints, say $g(x) \leq 0$. For simplicity, let us neglect the constraints, even though the specific problem of motion planning has the constraint of the trajectory being obstacle-free. Let us take a synthetic optimization problem, called sphere function, of dimension 3, given by Eqs. (15.4)–(15.6).

$$x^* = argmin_x f(x) \qquad (15.4)$$

$$f(x) = \sum_{i=1}^{3} x_i^2 \qquad (15.5)$$

$$-1 \leq x_i \leq 1, 0 \leq i \leq 3 \qquad (15.6)$$

15.2.2   Real-coded genetic algorithms

The first problem is to represent the individual in the form of a genotype. The evolutionary algorithms vary dramatically in the mannerism of representation of the individual. In this chapter, we will only use the representation called real-coded representation and the corresponding GA called the *real-coded genetic algorithm*. A solution to the optimization problem can be represented by three real numbers representing the values of x_1, x_2, and x_3. A sample individual along with fitness value is shown in Fig. 15.1. In this problem, the genotype and the phenotype are the same and therefore do not require an explicit conversion.

A *binary-coded GA* encodes the individuals as bit strings. Every real number can be converted into a binary number, which is what the technique does. A problem is that the length of a genetic individual needs to be fixed beforehand. Therefore, every real number is given a fixed number of bits which limit the precision in the genotype.

Another representation is *tree-based*. This representation is suited to store computer programs, algebraic expressions, formal models, recursively defined design, etc. This is commonly used in *genetic programming* (Koza, 1992; Angeline, 1994), a branch of evolutionary computing dedicated to the evolution of computer programs, using genetic operators. Consider an algebraic expression tree in which the nonleaf nodes denote the operators and leaf nodes denote the inputs or real numbers. The regression of the expression represented in this tree on a dataset is also machine learning on a dataset using evolution. The expression denotes the phenotype while its tree-representation makes a genotype. The loss function is the mean square error (and regularization) between the expression output on the inputs in the dataset and the target in the

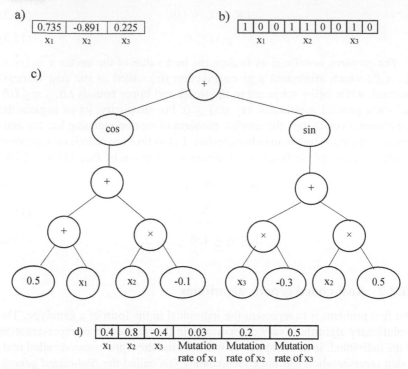

a)

0.735	-0.891	0.225
x_1	x_2	x_3

b)

1	0	0	1	1	0	0	1	0
x_1			x_2			x_3		

c)

d)

0.4	0.8	-0.4	0.03	0.2	0.5
x_1	x_2	x_3	Mutation rate of x_1	Mutation rate of x_2	Mutation rate of x_3

FIG. 15.1 Individual representation techniques: (A) shows a real-coded representation; (B) shows a binary-coded individual, wherein every real number is converted into a binary number with a limited number of bits; (C) shows a tree-based representation, typically for a machine learning problem to regress training data for some expression computed by genetic programming; (D) shows the technique wherein mutation rate is embedded into the individual as used in evolutionary strategy. The complete representation using covariance matrices for mutation is not shown to limit complexity.

dataset. A related branch is called *evolutionary programming*, which does not use the crossover operator to evolve the programs.

Another technique also injects in hyper-parameters of the algorithm as the genes of the individual, so that the hyper-parameters are also simultaneously optimized. The technique is called *evolutionary strategies* (Bäck et al., 1991; Hansen and Ostermeier, 1996) and this individual representation makes the algorithm self-adaptive. The mutation parameter is hard to set for the given problem, as shall be discussed later. So, the mutation rate for every gene can be embedded into the individual which can be optimized. The techniques are shown in Fig. 15.1.

A population is a collection of individuals. The population of individuals is shown in Table 15.1. The *population size* is the number of individuals in the population pool and is an algorithm parameter. The *initial population* is randomly generated in general. The initial population has a lot of effect on the algorithm performance. This is another place where *heuristics* are used. Every problem has its heuristics that guess good initial solutions to be inserted. If no heuristic is known, or if heuristics can only generate very few solutions, the other individuals are randomly generated.

TABLE 15.1 A sample population.

Individual no.	Individual			Fitness $(f(x) = \sum_{i=1}^{3} x_i^2)$ (smaller is better)	Rank	Expectation of being selected (7-rank)/21
x_1	0.5	0.8	−0.3	$(0.5)^2 + (0.8)^2$ $+(-0.3)^2 = 0.98$	6	1/21
x_2	0.3	−0.2	0.8	$(0.3)^2 + (-0.2)^2$ $+(0.8)^2 = 0.77$	4	3/21
x_3	−0.7	0.3	−0.4	$(-0.7)^2 + (0.3)^2$ $+(-0.4)^2 = 0.74$	3	4/21
x_4	0.5	−0.2	−0.3	$(0.5)^2 + (-0.2)^2$ $+(-0.3)^2 = 0.38$	1	6/21
x_5	0.6	−0.5	0.2	$(0.6)^2 + (-0.5)^2$ $+(0.2)^2 = 0.65$	2	5/21
x_6	0.4	0.8	0.1	$(0.4)^2 + (-0.8)^2$ $+(0.1)^2 = 0.81$	5	2/21

15.2.3 Selection

Given a population, the first task is to select some individuals that will incur genetic operation. Every individual has an expectation value, which denotes the likelihood of the individual to be selected for the genetic operations. Given the expectation value of n individuals, the population size, the selection operator selects k individuals. The expectation is proportional to the fitness value (higher is better), which is normalized to represent selection probabilities. This is called as the *fitness-based scaling*, given by Eq. (15.7).

$$E(x_i) = \frac{f(x_i) - \min_j f(x_j)}{\sum_i (f(x_i) - \min_j f(x_j))} \tag{15.7}$$

Here, $E(x_i)$ is the expectation of x_i to be selected. The translation is applied to get them between 0 and 1 with a summation of 1 as in the case of probabilities, considering that the fitness values may be negative.

The fitness-based scaling has a major problem that a very few individuals initially have a significantly better fitness value and hence get selected many times, while the comparatively weaker ones die off very quickly. This cases a premature convergence for the algorithm. Imagine in a natural evolution process some individuals developed the ability to hunt and some others the ability to run. Instantly, the individuals who could run survived, while the ones who could hunt were extinct due to immature hunting skills. Now nobody knows hunting. For the longer run, hunting may have been a better skill. It is hence important for a few individuals who know how to hunt to survive for a few generations till which they develop their skills enough to dominate the population pool by fitness. However, if slow running is taken as a skill, retaining it for too long may not be beneficial. Hence, there is a trade-off between selecting fitter individuals many times and selecting enough different individuals to retain good genes.

A better technique is hence called *rank-based scaling*, wherein the expectation value is (inversely) proportional to the rank of the individual. This is given by Eq. (15.8).

$$E(x_i) = \frac{n - rank(x_i) + 1}{\sum_i (n - rank(x_i) + 1)} \tag{15.8}$$

Here, n is the population size while the subtraction denotes the relation that lower is the numeric rank, higher is the expectation value. The expectation values for the synthetic population pool are given in Table 15.1.

There are numerous techniques of selection, some of the prominent ones are discussed. One of the most basic techniques is called the *roulette wheel selection*, in which a roulette when is spun and the individual pointed by the static arrow is selected. The portion of the circumference of the wheel for every individual is proportional to its expectation value. To select k individuals, the wheel is rotated k times. This is shown in Fig. 15.2A. A naïve implementation of the roulette wheel for k spins can have a complexity of O(nk), which will have a complexity of $O(n^2)$ if k is $O(n)$. A better implementation is using a binary search instead. The wheel total circumference is taken as 1, in which case the portion of circumference for every individual is its expectation value. After the wheel is spun, the arrow points to a random position $r \sim U[0,1]$ in the wheel, obtained by a uniform distribution of 0 and 1. The positional coverage of every individual is obtained by a cumulative sum. This indexes the individual on its occupancy within a scale of 0 to 1. Now a binary search over the population pool over the coverage values can be used to get to the individual within the range in a logarithmic time, giving the complexity $O(k \log n)$ and for a system with k of the order of n, given by $O(n \log n)$. This process is also shown in Fig. 15.2B. The algorithm is shown as Algorithm 15.1.

The roulette wheel has a problem that an individual is likely to get selected numerous times because of its greater coverage over the wheel. A little modified variant is called the *stochastic universal sampling*. Here, the same roulette wheel is used. However, the wheel has k selectors or arrows. Here, k is the number of individuals to be selected. With one spin, all k individuals are selected. The generic implementation is the same as the roulette wheel with the first pointer at a random position given by the random number (r). The ith pointer is the same

a)

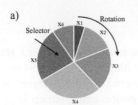

b)

Random Number between 0 and 1, say 0.8 — Binary Search

Individual No.	Expectation Value	Expectation Value on Wheel (inclusive)	
		Start (inclusive)	End (exclusive)
x_1	1/21 = 0.05	0.00	0.05
x_2	3/21 = 0.14	0.05	0.19
x_3	4/21 = 0.19	0.20	0.38
x_4	6/21 = 0.29	0.38	0.67
x_5	5/21 = 0.24	0.67	0.90
x_6	2/21 = 0.10	0.90	1.00

c)

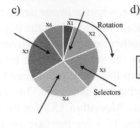

d)

Random Number between 0 and 1, say 0.8
Number of individuals to select = 4
Step Size: 1/selectors = 1/4 = 0.25
Pointers locations:

(0.8) mod 1 =0.8mod 1 =0.8,	(0.8+1×0.25) mod 1 =1.05 mod 1 =0.05	(0.8+2×0.25) mod 1 =1.3 mod 1 =0.3	(0.8+3×0.25) mod 1 =1.55 mod 1 =0.55

Individual No.	Expectation Value	Expectation Value on Wheel (inclusive)	
		Start (inclusive)	End (exclusive)
x_1	1/21 = 0.05	0.00	0.05
x_2	3/21 = 0.14	0.05	0.19
x_3	4/21 = 0.19	0.19	0.38
x_4	6/21 = 0.29	0.38	0.67
x_5	5/21 = 0.24	0.67	0.90
x_6	2/21 = 0.10	0.90	1.00

FIG. 15.2 Selection (A and B) Roulette wheel selection. Radial occupancy is the expectation value of the individual. The wheel is spun k times for k selections. (C and D) Stochastic universal sampling. The wheel is spun once for k selections, with k uniformly placed selectors.

Algorithm 15.1 Roulette wheel selection.

Input: Population (x_i) with expectation values ($E(x_i)$), number of individuals to select (k)
Output: k selected individuals

1 **for** i *from 1 to n* **do** ▷ calculate positions on wheel
2   **if** $i = 1$ **then** $StartPosition(x_i) \leftarrow 0, EndPosition(x_i) \leftarrow E(x_i)$;
3   **else**
4        $StartPosition(x_i) \leftarrow EndPosition(x_{i-1})$;
5        $EndPosition(x_i) \leftarrow EndPosition(x_{i-1}) + E(x_i)$

6 $selectedIndividuals \leftarrow \emptyset$;
7 **for** i *from 1 to k* **do** ▷ number of individuals to select
8     $r \sim U[0, 1]$; ▷ selector's position
9     $startSearch \leftarrow 1, endSearch \leftarrow n$; ▷ binary search r
10   **while** $startSearch \leq endSearch$ **do**
11        $mid \leftarrow (startSearch + endSearch)/2$;
12   **if** $r \geq StartPosition(x_{mid}) \wedge r < EndPosition(x_{mid})$ **then**
13           $selectedIndividuals \leftarrow selectedIndividuals \cup \{x_{mid}\}, break$;
14   **else if** $r \geq EndPosition(x_{mid})$ **then**
15           $startSearch \leftarrow mid + 1$
16   **else if** $r < StartPosition(x_{mid})$ **then**
17           $endSearch \leftarrow mid - 1$

18 **return** $selectedIndividuals$

position r with an offset $(i-1)/k$ representing the position of pointer i, starting from $i = 1$ as the first pointer. The position of the ith pointer is thus $r + (i-1)/k$. The calculations follow circular arithmetic and hence $1 + \varepsilon$ ($\varepsilon < 1$) represents the circular position after 1 complete round and ε motion extra, which is just the position ε. Another mechanism to implement the same algorithm, considering k is of the order of n, is to get all k individuals in a single linear iteration over the population size. So, an iterator iterates over all individuals in the population pool with their circular positions noted, and another iterator iterates over all the pointers with their known positions. Knowing the selection made by the ith pointer, for the $(i + 1)$th pointer, we search from the individual pointed by the ith pointer, in the forward direction till the individual is found. The selection is shown in Fig. 15.2. The algorithm is given as Algorithm 15.2 for the implementation using a binary search and Algorithm 15.3 for the case of a linear search.

Algorithm 15.2 Stochastic universal sampling by binary search.

Input: Population (x_i) with expectation values $(E(x_i))$, number of individuals to select (k)

Output: k selected individuals

1 **for** i *from 1 to n* **do**
2   **if** $i = 1$ **then** $StartPosition(x_i) \leftarrow 0, EndPosition(x_i) \leftarrow E(x_i)$;
3   **else**
4         $StartPosition(x_i) \leftarrow EndPosition(x_{i-1})$;
5         $EndPosition(x_i) \leftarrow EndPosition(x_{i-1}) + E(x_i)$

6 $selectedIndividuals \leftarrow \emptyset$;
7 $r_1 \sim U[0,1]$;   ▷ first selector's position
8 **for** i *from 1 to k* **do**
9     $r_i \leftarrow r_1 + (i-1)/k$;   ▷ i^{th} selector's position
10   **if** $r_i > 1$ **then** $r_i \leftarrow r_i - 1$;   ▷ circular, mod 1 arithmetic
11     $startSearch \leftarrow 1, endSearch \leftarrow n$;   ▷ binary search r_i
12   **while** $startSearch \leq endSearch$ **do**
13         $mid \leftarrow (startSearch + endSearch)/2$;
14   **if** $r_i \geq StartPosition(x_{mid}) \wedge r_i < EndPosition(x_{mid})$ **then**
15             $selectedIndividuals \leftarrow$
            $selectedIndividuals \cup \{x_{mid}\}, break$;
16   **else if** $r \geq EndPosition(x_{mid})$ **then**
17             $startSearch \leftarrow mid + 1$
18   **else if** $r < StartPosition(x_{mid})$ **then**
19             $endSearch \leftarrow mid - 1$

20 **return** $selectedIndividuals$

Algorithm 15.3 Stochastic universal sampling by linear search.

Input: Population (x_i) with expectation values $(E(x_i))$, number of
individuals to select (k)

Output: k selected individuals

1 **for** i *from 1 to n* **do**
2 $\quad$ **if** $i = 1$ **then** $StartPosition(x_i) \leftarrow 0, EndPosition(x_i) \leftarrow E(x_i)$;
3 $\quad$ **else**
4 $\quad\quad$ $StartPosition(x_i) \leftarrow EndPosition(x_{i-1})$;
5 $\quad\quad$ $EndPosition(x_i) \leftarrow EndPosition(x_{i-1}) + E(x_i)$

6 $pointerLocations \leftarrow \emptyset$;
7 $selectedIndividuals \leftarrow \emptyset$;
8 $r \sim U[0,1]$;
9 **for** i *from 1 to k* **do**
10 $\quad$ $p \leftarrow r + (i-1)/k$;
11 $\quad$ **if** $p > 1$ **then** $p \leftarrow p - 1$;
12 $\quad$ $pointerLocations \leftarrow pointerLocations \cup \{p\}$;

$\quad$ ▷ cur : pointer on individuals, binary search for 1st ;
13 $cur \leftarrow RouletteWheelSearch(r, pointerLocations(1))$;
14 $p \leftarrow 1$; $\qquad\qquad\qquad$ ▷ counter of pointers (selectors)
15 **while** $p \le k$ **do**
$\quad$ ▷ select if p^{th} pointer points to cur^{th} individual ;
16 $\quad$ **if** $pointerLocations(p) \ge$
$\quad\quad StartPosition(x_{cur}) \wedge pointerLocations(p) < EndPosition(x_{cur})$
$\quad\quad$ **then**
17 $\quad\quad$ $selectedIndividuals \leftarrow selectedIndividuals \cup \{x_{cur}\}$;
18 $\quad\quad$ $p \leftarrow p + 1$;
19 $\quad$ **else** $\qquad$ ▷ move cur by 1 following circular arithmetic
20 $\quad\quad$ $cur \leftarrow (cur \bmod populationSize) + 1$

21 **return** $selectedIndividuals$

A simpler and faster technique is called the *Tournament Selection*, which imitates a standard tournament of sports. A match is played between 2 random individuals. A fitter individual (as per the fitness or expectation value) is more likely to win a match as compared to a lesser fit individual. The winning probability of the fitter individual is kept as t, while that of the weaker individual is kept as $1 - t$. This happens in sports wherein it is common to see weaker teams win as well. Here, t is an algorithm parameter. A variant of tournament selection introduces a parameter S, called the *tournament size,* and selects S random individuals who play a tournament. All S individuals are taken in sorted order from the best fitness to the worst. If a fitter individual wins, it is said to have been selected; however, if a weaker individual wins, it enters competition with the next competitor. k tournaments are played for the selection of k individuals. The complexity is thus $O(k)$. If k is $O(n)$, the complexity becomes $O(n)$.

For the expectation values in Table 15.1, consider that tournament size is 3 and 3 randomly sampled individuals are x_2 with expectation value $3/21$, x_5 with expectation value $5/21$, and x_6 with expectation value $2/21$. The first match is played between the best 2 individuals x_5 and x_2. Say, unluckily, the fitter individual x_5 lost. Now, x_2 plays with the next fittest individual, x_6. This time the fitter individual wins and hence x_2 is selected. This process will repeat k times to select k individuals. The approach is presented as Algorithm 15.4.

Algorithm 15.4 Tournament selection.

Input: Population (x_i) with expectation values ($E(x_i)$), number of individuals to select (k), probability of winning of the fitter individual (t)

Output: k selected individuals

1   $selectedIndividuals \leftarrow \emptyset$;
2   **while** $size(selectedIndividuals) < k$ **do**
3      $x \leftarrow$ randomly sampled S individuals ;   ▷ for tournament
4   sort (descending) x by expectation values ;   ▷ order of play
5   **for** i *from 1 to S-1* **do**   ▷ matches
6         $r \sim U[0,1]$;
            ▷ with high probability t fitter wins and is selected; with probability $1 - t$ weaker wins and competes with the next ;
7   **if** $r < t$ **then**
8            $selectedIndividuals \leftarrow selectedIndividuals \cup \{x_i\}$, break
9   **else if** $i = S - 1$ **then**   ▷ if the last individual wins, it is selected
10           $selectedIndividuals \leftarrow selectedIndividuals \cup \{x_S\}$

11   return $selectedIndividuals$

15.2.4 Crossover

The *crossover* operator mates two individuals to produce two offspring. The children carry the characteristics of the parents only. If some characteristics or genes are better in one parent, and some other characteristics or genes are better in the other parent, the child produced could potentially combine all good characteristics or genes to be significantly better. This happens in natural evolution wherein a child may take the best characteristics of both parents.

The first type of crossover is the *1-point crossover*, wherein a crossover point is selected randomly between 0 and the length of the individual. The first child keeps the genes of the first parent till the crossover point and the second parent after the crossover point. The second child takes the genes of the second parent till the crossover point and the first parent after the crossover point. The technique is shown in Fig. 15.3A. A problem is that the good characteristics may only be found somewhere in-between the first parent. Hence, the *2-point crossover* generates 2 crossover points. The first child is primarily formed by the first parent with the genes in-between the crossover points taken from the second parent, and vice versa for the second child. This technique is shown in Fig. 15.3B.

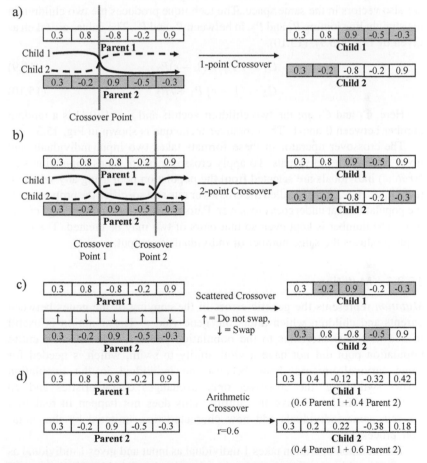

FIG. 15.3 Crossover techniques (A) 1-point crossover, genes after the crossover point are swapped, (B) 2-point crossover, genes between the crossover points are swapped, (C) scattered crossover, swapping of genes happens by random numbers, (D) arithmetic crossover, the children are weighted addition of parents, weighted by random numbers.

It is worth generalizing the notion even further to have any number of crossover points. This technique is called the *scattered crossover* technique. For every gene *i*, with a probability of 0.5, the first parent's gene goes to the first child and the second parent's gene goes to the second child. Otherwise, the two genes are swapped and given to the children, meaning the first child gets the corresponding gene from the second parent and vice versa. This can be done by generating a random string of ↑ and ↓, with ↑ denoting no exchange and ↓ denoting an exchange of genes. This is shown in Fig. 15.3C.

The above crossover techniques hold good for both binary and real-coded genetic algorithms. A technique specifically for the real-coded genetic algorithm is called the *arithmetic crossover*. Imagine the parents (P_1 and P_2) as vectors in a high-dimensional space. Further imagine that the children to be formed are also vectors in the same space. The technique produces the two children in the straight line joining P_1 and P_2, in between P_1 and P_2. The values are taken as given by Eqs. (15.9), (15.10).

$$C_1 = r\,P_1 + (1 - r)P_2 \tag{15.9}$$

$$C_2 = (1 - r)\,P_1 + rP_2 \tag{15.10}$$

Here, C_1 and C_2 are the two children vectors and $r \sim U[0,1]$ is a random number between 0 and 1. The crossover technique is shown in Fig. 15.3D.

The crossover operator in these formats takes two input individuals and gives two output individuals. To apply crossover, *xrate* times population size (*xrate.n*) individuals are selected from the population pool using any selection technique. Here, *xrate* is called the crossover rate and specifies the proportion of the population that undergoes crossover. Pairs of two individuals undergo crossover. The number is kept even so that pairs of two may be created. The technique produces the same number of individuals as output.

15.2.5 Mutation

Mutation represents the genetic errors in the copying of the genes between parents and children which alters the gene values. The mutation is useful to add new characteristics to the population pool. Imagine that the entire population pool did not have a vital ability to swim, which is needed for some survival process. Now because no individual in the population knows swimming, the crossover only exchanges characteristics and no future individual will have the ability. This does not happen in real life, wherein some individuals add new characteristics, imitated by the mutation process.

The mutation operation takes 1 individual as input and gives 1 individual as output. For a binary representation, the operation is carried by flipping one or more gene values, by which a 0 becomes 1 and vice versa. Every gene value is flipped with a probability given by the *mutation rate*.

If the individuals are represented as real numbers, the technique is called the *Gaussian mutation*, in which a small Gaussian noise with mean 0 and variance

σ^2 is added to the gene value. Here, σ controls the amount by which the individual is expected to be distorted called the mutation rate. The new value is hence given by Eq. (15.11).

$$c_i = x_i + r, r \sim G(0, \sigma) \qquad (15.11)$$

Here, c_i is the ith gene of the child (c) and x_i is the ith gene of the parent (x). G $(0, \sigma)$ is the Gaussian distribution with mean 0 and deviation σ. The mutations are also shown in Fig. 15.4.

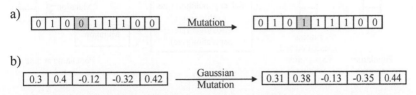

FIG. 15.4 Mutation techniques. (A) Mutation by flipping bits for binary-encoded individuals. (B) Mutation by adding a Gaussian noise.

15.2.6 Other operators and the evolution process

The exact evolution process varies with design. A common implementation is to select *xover* rate times population size parents for crossover. The individuals produced from crossover also undergo mutation and are added to the new population. This leaves a space for (1-*xover* rate) times population size individuals, which are selected by the selection operator, undergo mutation, and added to the new population. The process is shown in Fig. 15.5. After evolution, the solution is the best individual in the population pool.

A little problem with evolution is that, even though the best individual will get selected multiple times and make numerous offspring, it is possible that all offspring thus made will undergo an unfavourable mutation, resulting in a drop in the fitness of the best individual. This problem is solved by making an *elite* operator that directly transfers the best few individuals from the old population pool to the new population pool and the number of individuals transferred is called the *elite count*.

In another implementation, the children population pool is made by applying genetic operations to the parents. The parent population pool of size n and the children population pool are combined to make a combined pool of parents and children. From this population, the best n individuals are selected that make the new population pool. This selection strategy is called the *top strategy* of selection. It is possible to use other discussed strategies as well.

The algorithm in this manner can go on and on, iteratively improving the solution. A stopping criterion is used to decide when to stop the evolutionary process. The choice of stopping criterion includes maximum running time, the maximum number of generations, externally supplied by the user

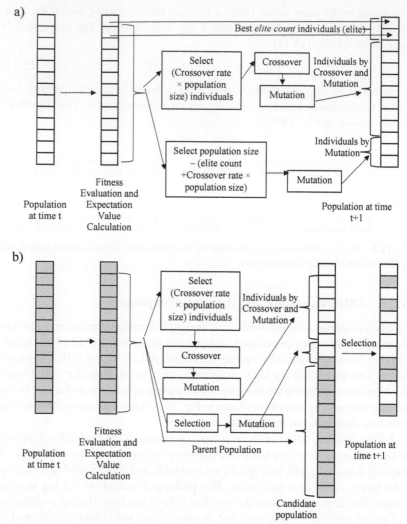

FIG. 15.5 Genetic algorithm process. (A) Make a new population by elite individuals of the last population, selected individuals that undergo crossover + mutation, and selected individuals that undergo mutation. (B) Previous and candidate populations are united and the top 'population size' individuals are selected as the next generation.

as an external signal, minimum improvement per generation, maximum stall time (maximum time till which the solution does not further improve), maximum stall generation (maximum generations till which the solution does not further improve), etc. The overall algorithm in one such flavour is given as Algorithm 15.5.

Algorithm 15.5 Genetic algorithm.

---

1 initialize *population*(0) by random individuals or using problem specific heuristics ;

2 $t \leftarrow 0$;

3 **while** *stopping criterion* **do**   ▷ generations

4   *population*($t + 1$) $\leftarrow \emptyset$;

5   Calculate fitness and expectation values ;

6   Add best k individuals to *population*($t + 1$) ;   ▷ elite operator
    ▷ Select and do crossover on *xrate*% individuals ;

7     $xIndiduals \leftarrow Select(population(t), (xrate \times popSize))$;

8   **for** $(x_1, x_2) \in xIndiduals$ *in pairs of 2* **do**

9         $c_1, c_2 \leftarrow Crossover(x_1, x_2)$;

10         $c_1 \leftarrow mutate(c_1), c_2 \leftarrow mutate(c_2)$;

11   add c_1, c_2 to *population*($t + 1$) ;

    ▷ Fill remaining population slots by mutation ;

12     $MutateIndividuals \leftarrow$
    $Select(population(t), popSize - (k + xrate \times popSize))$;

13   **for** $x_1 \in MutateIndividuals$ **do**

14   add $mutate(x_1)$ to *population*($t + 1$)

15     $t \leftarrow t + 1$;

16 return best individual in *population*(t)

15.2.7   Analysis

It is interesting to visualize the evolution process. Let us plot the fitness function $f(x)$ over a high-dimensional space, keeping every gene as an axis and the fitness value as one axis. This is called the *fitness landscape*. This for a synthetic function is drawn in Figs. 15.6 and 15.7 (for a 1-input problem) and by a contour (for a 2-input problem) in Figs. 15.8 and 15.9. Every individual in the population pool along with its calculated fitness value represents a point in this space. Initially, the population is spread throughout this space. Let us draw a rectilinear box that covers the entire population pool, which for convenience is called the *schema* of the population pool. So initially the schema will be roughly the entire space given by $[LB_1, UB_1] \times [LB_2, UB_2]$. It may be noted that the general schema theory is for the binary encoding of the population pool, in which case all the individuals can have a value of just 0 or 1 in every axis, and the schema hence represents a high-dimensional cube with every individual at a corner of the cube.

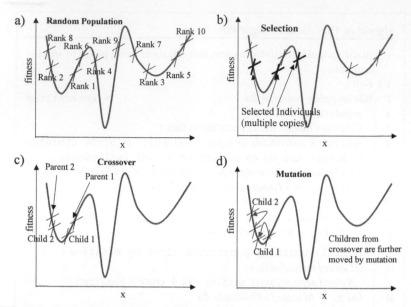

FIG. 15.6 Graphical representation of different operators in the fitness landscape. (A) Random population. (B) Selection. (C) Crossover. (D) Mutation.

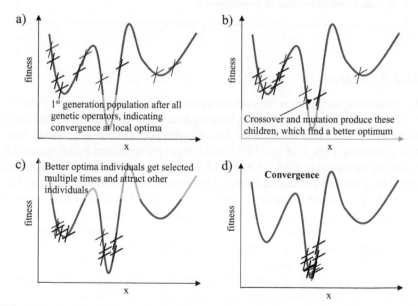

FIG. 15.7 Working of the genetic algorithm in the fitness landscape for the initial population in Fig. 15.6A. (A) Population after 1st generation. (B) Crossover and mutation at the next generation. (C) Selection at a further generation. (D) Eventual convergence at global optima.

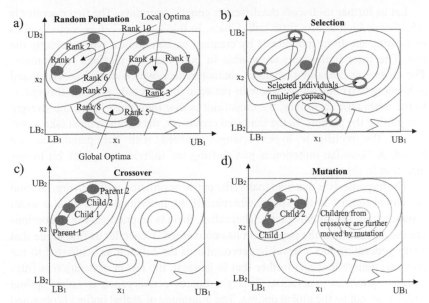

FIG. 15.8 Graphical representation of different operators in the fitness landscape contour. (A) Random population. (B) Selection. (C) Crossover. (D) Mutation.

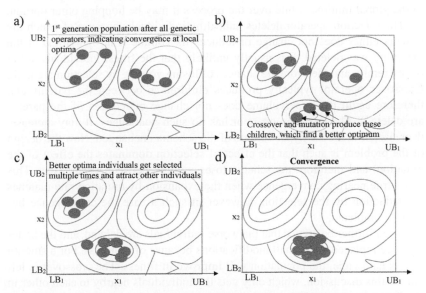

FIG. 15.9 Working of the genetic algorithm in the fitness landscape contour for the initial population in Fig. 15.8A. (A) Population after 1st generation. (B) Crossover and mutation at the next generation. (C) Selection at a further generation. (D) Eventual convergence at global optima.

Let us further re-discuss the different genetic operators. The first operator is selection, which selects fitter individuals. In this space, it can be visualized as replicating the fitter individuals by creating carbon copies, while deleting the rest. The fitter individuals are visible in both the 3-D plot and the contour. The crossover operator exchanges characteristics, which should be visualized as keeping two children anywhere in-between the selected parents in this space. In the arithmetic crossover, the constraint is that they must lie on the straight line joining the parents. In the scattered crossover, the children must take a corner of the rectilinear hyper-rectangle formed with the parents as the bound. A Gaussian mutation is just shifting the individuals a little bit in any direction in this space.

The hypothesis is that eventually, the entire population will converge around the global optima in this space, wherein all the individuals will be at a small distance from the global optima. Thereafter, there is no practical use of continuing the algorithm as the population has converged. Hence let us first argue that every generation has a lower hypervolume of the schema as compared to the previous generation; and further that in general, the newer generation is fitter than the previous generation. This guarantees convergence to some optima which may not be the global optima. The guarantee of global optima is obtained from the fact that there is a nonzero probability of applying a mutation to an individual such that it escapes the current optima, and lands near the global optima. So, given infinite time, the individual will escape all minima and go to the global minima, while over the process it may be hopping other optima.

The selection operator deletes unhealthy individuals, because of which the schema either does not change or reduces its volume if an individual at the extremity was deleted. It is usually unlikely that the fittest individuals are at the extremity and therefore the chances of losing volume by selection are high. Crossover operator typically produces children in-between the parents and therefore again the hypervolume does not change or reduces (if the parents are deleted). The mutation operator has no such guarantees. It may increase, decrease, or make the hyper volume of the schema unchanged. Now the design of the problem is such that the effect of selection dominates the effect of the mutation, and therefore the schema loses volume between generations. This is not true nearing convergence when the expansion by mutation rate matches the reduction by the selection, however in such a case the convergence has nearly completed.

It can further be argued that the worse individuals are deleted, and the better ones are replaced with small modifications because of mutation. For a smooth fitness landscape, the population will improve in fitness. The crossover is left out for this discussion, which only gets the individuals nearby to each other in the fitness landscape. This makes the new population better and thus indicating a step closer to optima.

The genetic operators come in two varieties. The first is *exploitative*, which tries to contract the schema at every step and thus exploit the current individuals with the assumption that the global optima lies within and is well-represented by the current population. The other operators are *explorative* which try to expand the schema and explore outside and inside the schema by moving around. Every evolutionary technique attempts to maintain a trade-off between the opposing factors of exploration and exploitation, that governs the performance.

High exploitation very quickly converges the population to a small space. However, an individual near global optima may currently have a worse fitness than an individual near a local-optima. In such cases, due to high exploitation, the individual near global optima may get deleted, wrongly converging at the local optima. Extremely exploitative settings could lead to no exploration around the current optima as well, and the algorithm may not make steps towards the current local optima.

On the other hand, too much exploration may overcome exploitation and the algorithm may never converge. Even at convergence, the individuals may be reasonably far away. However, this also ensures that eventually the global optima will be explored as the algorithm does a lot of exploration around, before converging or exploiting. While the exploration leads to a greater chance to escape local optima and get to the global optima, it also significantly delays convergence thus increasing computation time. At the same time, high exploitation gets very quick convergence, typically at the cost of premature convergence at the local optima.

The crossover rate is an exploitative setting and increasing the same increases exploitation, while the mutation rate is an explorative setting that increases exploration. Higher number of individuals for a fixed number of generations means better exploration, giving more time for the search, which is good and increases the chance of getting to the global optima. However, increasing the number of individuals for a fixed stopping time means that the number of generations shall reduce, which will increase exploration and thus making it unlikely that convergence happens soon. This is an explorative setting.

15.3   Particle swarm optimization

To add a little variety in the discussion over the evolutionary approaches, another technique called the particle swarm optimization (PSO) is studied (Kennedy and Eberhart, 1995; Poli et al., 2007). The technique imitates the food search process carried out by several organisms, here called the particles.

15.3.1   Modelling

The problem is to give velocities to these particles, such that they are directed largely towards the global optima, while also ensuring better coordination to explore the space and escape the local optima. Imagine the fitness landscape and the particles searching the global optima. Each particle knows the goodness of the current position; past positions visited by it; and the goodness of the current and past positions of other particles communicated by them.

A particle i in the PSO has a position $(x_i(t))$ and a velocity $(v_i(t))$ at time t. The position is used to compute the fitness value, while the velocity is used to move the particle with time. Both quantities are n-dimensional vectors, where n is the number of genes. The position of the particle i at time $t+1$ is thus derived from its position at time t by Eq. (15.12). The notations are shown in Fig. 15.10.

$$x_i(t + 1) = x_i(t) + v_i(t + 1) \tag{15.12}$$

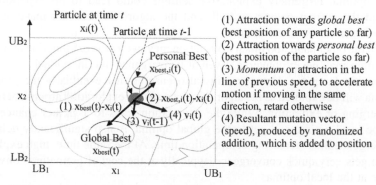

FIG. 15.10 Notations used in particle swarm optimization.

The velocity is set by three strategies. Assuming that every particle is near an optimum, it would like to go and explore more in that area to establish whether the optima is global optima or local optima and to get the best fitness value within the optima. For this, the particle must consistently move towards the best position that it has seen so far, known as the *personal best*, and denoted by $x_{best,i}(t)$ for particle i. Upon reaching near its personal best, the particle may roam around by randomization to find better fitness nearby. This velocity is modelled as Eq. (15.13):

$$v_i^{(1)}(t + 1) = c_1 r_1 \left(x_{best,i}(t) - x_i(t) \right) \tag{15.13}$$

Here, $x_{best,\ i}(t) - x_i(t)$ denotes the intended magnitude and direction of motion further scaled by a constant c_1, and r_1 is added to induce some randomization in the process. It represents a random number between 0 and 1.

The second strategy is an attempt to explore global optima. All particles share their positions and fitness value. Let $x_{best}(t)$ be the position that records the best fitness value for all particles at any point in time called the *global best position*. This position attracts more particles to reach $x_{best}(t)$ and thereafter explores the surroundings. The velocity due to this strategy is given by Eq. (15.14). The terms are like the previous strategy.

$$v_i^{(2)}(t+1) = c_2 r_2 (x_{best}(t) - x_i(t)) \qquad (15.14)$$

It is not advisable for all particles to march towards $x_{best}(t)$ in which case there will be a premature convergence. Some particles may be close to the global optima but may neither represent $x_{best}(t)$ nor may be close towards $x_{best}(t)$. $x_{best}(t)$ may be currently the local optima. Here, the first strategy of moving towards personal best comes to the rescue. On the other hand, a particle may be in a poor position and maybe wasting time at the current optima. It may be better to proceed to the global best position to explore within the same area, which when repeated for many such particles may result in a better optimum on the way.

Finally, the particle may be going in some suitable direction, while the other strategies force the particle to come towards some other position. A *momentum* term is added that enables a particle to continue exploring in the same direction as previously by taking its previous velocity direction as the current velocity direction. The main aim of momentum is like the back-propagation algorithm in neural networks, which is to accelerate the particle if it is constantly moving in the same direction and to retard the motion if it is constantly fluctuating directions. Here, direction means the polarity or sign of the velocity in the dimension. Consider the motion of a particle at the jth dimension. If the particle is constantly travelling on the positive jth axis, it means that the preferred position lies on the positive side and the algorithm accelerates the motion by giving large positive speed. On the contrary, if the particle keeps changing its direction between the positive jth axis and the negative jth axis, it means that the particle is oscillating around an optima and momentum reduces the speed to enable the particle to go deep into the minima. The strategy is given by Eq. (15.15):

$$v_i^{(3)}(t+1) = \omega v_i(t) \qquad (15.15)$$

Here, ω is an algorithm parameter denoting the weight of the strategy.

There are numerous strategies to choose from. One way to handle such a situation is to stochastically select one strategy every iteration, where every strategy has a probability of being selected. This is called the *selection approach*. The other manner of handling the situation is to take a weighted summation over the outputs of all strategies, weighted by random numbers. This is called the *fusion approach*. Sometimes, one strategy due to the weights and randomness will dominate, while at other times, some other strategy will dominate. Since there is an inherent need for randomization which flickers the particles around the best positions and aids in exploration, the fusion strategy is adopted, with the resultant velocity and position given by Eqs. (15.16), (15.17):

$$v_i(t+1) = \omega v_i(t) + c_1 r_1 (x_{best,i}(t) - x_i(t)) + c_2 r_2 (x_{best}(t) - x_i(t)) \qquad (15.16)$$

$$x_i(t+1) = x_i(t) + v_i(t+1) \qquad (15.17)$$

A problem with the approach is that the speed across any dimension may occasionally become large, in which case the particle starts making big jumps. The particle can jump over the global optima. The maximum speed across all dimensions is hence fixed to a magnitude of v_{max}. Let $v_{ij}(t+1)$ be the speed of particle i at dimension j at time $t+1$. The speed must follow the constraint given by $-v_{max} \le v_{ij}(t+1) \le v_{max}$. Similarly, for dimension j, the position variables are bounded to a lower bound LB_j and upper bound UB_j. Hence, constraint $LB_j \le x_{ij}(t+1) \le UB_j$ must be imposed. The pseudo-code is given in Algorithm 15.6.

Algorithm 15.6 Particle swarm optimization.

1   $x_{best} \leftarrow NIL$;   ▷ global best
2   **for** *i from 1 to population size* **do**
3      $x_i \leftarrow$ random individual or by using problem specific heuristic ;
4      $v_i \leftarrow$ random velocity ;
5      $x_{best,i} \leftarrow x_i$;   ▷ personal best for i
6   **if** $x_{best} = NIL \lor f(x_i) < f(x_{best})$ **then** ▷ update global best using fitness function f
7        $x_{best} \leftarrow x_i$;

8   **while** *stopping criterion* **do**
9   **for** *all individuals i* **do**
10   generate random vectors r_1 and r_2 ;
11        $v_i \leftarrow \omega v_i + c_1 r_1(x_{best,i} - x_i) + c_2 r_2(x_{best} - x_i)$;
12        $x_i \leftarrow x_i + v_i$;
          ▷ update personal and global best ;
13   **if** $f(x_i) < f(x_{best,i})$ **then** $x_{best,i} \leftarrow x_i$;
14   **if** $f(x_i) < f(x_{best})$ **then** $x_{best} \leftarrow x_i$;
15   Limit too large (positive or negative) velocity in any dimension;

16 **return** x_{best}

15.3.2 Example

Consider the problem of optimizing a spherical function. Consider a sample initial population given in Table 15.2. The personal best of every individual is the current position only. The global best is the best position among all individuals. For simplicity, let us consider $\omega = 0.8$, $c_1 = 2$, and $c_2 = 2$. The random numbers are randomly inserted as $r_1 = [0.4, 0.3, 0.5]^T$ and $r_2 = [0.5, 0.3, 0.4]^T$. The new velocity and position of x_1 is given by Eqs. (15.18), (15.19).

TABLE 15.2 Sample initial population for particle swarm optimization.

Particle	Position (x_i)	Velocity (v_i)	Fitness $(f(x) = \sum_{i=1}^{3} x_i^2)$ (smaller is better)	Personal best $(x_{best,i})$	Global best (x_{best})
1	[0.5 −0.4 0.4]	[0.1 −0.05 0.1]	0.57	[0.5 −0.4 0.4]	[0.3 0.3 −0.2]
2	[−0.3 0.4 −0.1]	[0.08 −0.1 0.05]	0.26	[−0.3 0.4 −0.1]	
3	[0.2 −0.3 0.4]	[0.1 −0.1 −0.04]	0.29	[0.2 −0.3 0.4]	
4	[0.3 0.3 −0.2]	[−0.01 0.04 0.09]	0.22	[0.3 0.3 −0.2]	
5	[0.6 0.8 −0.4]	[0.05 −0.03 0.02]	1.16	[0.6 0.8 −0.4]	

$$v_1(1) = 0.8[0.1, -0.05, 0.1]^T + 2[0.4, 0.3, 0.5]^T \cdot \left([0.5, -0.4, 0.4]^T - [0.5, -0.4, 0.4]^T\right)$$
$$+ 2[0.5, 0.3, 0.4]^T \cdot \left([0.3, 0.3, -0.2]^T - [0.5, -0.4, 0.4]^T\right)$$
$$= 0.8[0.1, -0.05, 0.1]^T + 2[0.4, 0.3, 0.5]^T \cdot [0, 0, 0]^T + 2[0.5, 0.3, 0.4]^T \cdot$$
$$[-0.2, 0.7, -0.6]^T = [0.08, -0.04, 0.08]^T + [-0.2, 0.42, -0.48]^T$$
$$= [-0.12, 0.38, -0.4]^T$$

$$(15.18)$$

$$x_1 = [0.5, -0.4, 0.4]^T + [-0.12, 0.38, -0.4]^T = [0.38, -0.02, 0]^T \qquad (15.19)$$

Here, '·' represents the element-wise product and is applied to save space. The fitness of the particle now becomes $0.38^2 + (-0.02)^2 + 0^2 = 0.1448$, which is better and hence this also updates the personal best in Table 15.2 (not shown). This is also better than the global best and therefore the global best is updated in Table 15.2 (not shown). This can be done for all the particles in the population pool to make the next-generation population.

15.3.3 Analysis

It is imperative to imagine the running of the algorithm in the fitness landscape. The direct imagination is a bunch of particles moving step by step in the fitness landscape. This can hence be better visualized by the contours representing the fitness landscape of a two-dimensional optimization problem. For any individual i, let the global best and personal best be negatively charged entity, and the particle itself be a positively charged entity. The global and personal best attract the particles by a force proportional to the distance, a randomization makes the forces fluctuating with time. Without randomization, the particle would converge at an equidistant region between the personal best and global best.

Let us discuss the case when the attraction to personal best continuously dominates due to randomizations. The particle constantly moves towards the personal best, however, upon reaching the personal best it overshoots due to momentum with some fluctuation from the global best term. If the new position is better, the personal best is updated. By constantly approaching towards, overshooting, fluctuating near, and updating the personal best, the particle slowly converges to its local optima. It is possible that the local optima is later found to be better than the global optima, and this results in updating the global best and attracting all other particles.

Because of randomization, there is a nonzero probability that the global best term continuously dominates, and the particle escapes its optima and goes towards the currently known global best in multiple steps. Once the particle reaches a point better than its personal best and the personal best updates, the personal best also points to the optima region of the global best. While still being attracted to the global best, the particle may find a better global best and updates the same. Eventually, a convergence happens.

It is imperative to think about the betterment that PSO gives to the GA. The main parameter in the GA is the *mutation rate* which controls the magnitude by which an individual is moved in the fitness landscape. A fixed mutation rate in GA is problematic. Initially, when the individuals in the fitness landscape are far away and the schema hyper volume is large, you need to make large steps while selecting the optima collaboratively, escaping the local optima, and moving nearer to the global optima. For this, a higher mutation rate is better. When convergence happens, the algorithm should take the smallest steps so as not to overshoot the optima from any direction. This suggests a smaller mutation rate. Many theories and postulates help in adapting the mutation rate parameter in GA.

In PSO, the equivalent of the mutation rate is the velocity term, which is the magnitude by which the particle moves in the fitness landscape. Initially, the diversity is high when the particles are spread far apart and therefore the difference between the personal and global best is high, which gives a high velocity. However, at the later iterations, all the individuals nearly converge and therefore the difference vectors from personal and global best are small, which gives a small velocity. Hence, the adaptation of mutation rate is inbuilt in the PSO, which makes it a powerful algorithm.

The other point of difference is that the mutation in GA is completely random and therefore the individuals can take a step anywhere in the space. However, PSO has intelligently placed strategies that make the mutation a lot more directed, constraining the motion of the particles and leading to an early convergence, while making the most of the exploration.

15.4 Topologies

The PSO has a major problem, which is one global attractor forcibly attracts all the particles. The personal best adds a resistance; however, the dominance by a single global best causes a premature convergence towards the current best. In terms of GA, this can be imagined as a crossover in which 1 parent is always the fittest individual and a mutation with a heavy bias towards the current best.

PSO uses an intelligent mechanism to handle this effect by arranging all the individuals in a *topology*. The topology is in the form of a graph wherein the vertices represent the individuals and any two individuals sharing an edge are said to be called neighbours. The natural intuition is that every decision in life primarily uses local information of the neighbours only. As an example, local celebrities in sports and education attract youth in the region. With the growing globalization, it may appear that such neighbourhood-based decisions are not being taken, however the locality still defines much of our perspectives.

This means every particle has a different global best which is found by consulting only the neighbouring particles as per the topology, and hence the individuals are differently directed. A particle only takes the global best out of all particles in its neighbourhood. It cannot consult all particles in the population to find the global best. For analysis only, consider that there is one particle x which is currently the global best out of all particles in the population. The neighbouring nodes of x, say $\delta(x)$, have access to the real global best x. Eventually, there will be a convergence in the PSO represented in the subgraph consisting of x and all its neighbours, say $\{x \cup \delta(x)\}$. At convergence, the particles representing the subgraph $\{x \cup \delta(x)\}$ are all near the global best. Hence, the neighbouring nodes of this subgraph, say $\{\delta(y) : y \in x \cup \delta(x)\}$, get access to the global best. The convergence of these new nodes is delayed which gives a chance to find a better global best. If a better global best is found, it starts to attract the neighbouring particles. If there is no change in the global best, the convergence happens in a layer-by-layer manner like the Breadth-First Search.

Any undirected graph is a candidate for a topology. However, two topologies are widely used. The first is the *star topology* which has a central particle with direct connections to all other particles, while the other particles are only connected to the central particle, shown as Fig. 15.11. The other is the *ring topology*, in which case the particles form a ring with every particle connected to the one forward and backward in the ring, shown in Fig. 15.11. Assuming that a global-optima has been identified as one particle, which does not change, the convergence eventually happens in a layered mechanism, with the neighbours getting converged followed by their neighbours representing a breadth-first search. Now a star topology has a maximum distance of two between any two particles and therefore the global optima soon influence the centre, which in turn influences all other particles and thus shows a fast convergence. On the contrary, the maximum distance possible between any two nodes in a ring is $n/2$ for n particles. This means that convergence from one particle to reach to the diagonally opposite particle shall take a lot of time, delaying convergence. The effect of the risk of getting into a local-optima is the opposite. A particle currently in a poor region, however, somewhere near the global optima in a star topology very quickly gets affected by the converged population of the centre. However, in a ring topology, it gets more time to explore till convergence hits. Hence convergence is fastest in the star topology, however, so is the risk of converging at the local optima.

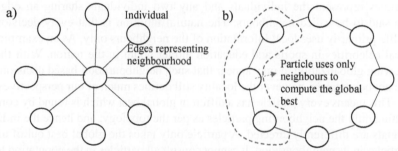

FIG. 15.11 Topologies used in particle swarm optimization. Different particles have different global best (found from neighbours only), which delays convergence and increases the chance of converging into global optima. (A) Star topology, wherein individuals are close-by and result in faster convergence. (B) Ring topology, wherein information from one individual takes time to reach the other end of the ring resulting in delayed convergence and higher chance to get to global optima.

The notion of topologies exists in GA as well. The only difference is that every vertex of the topology graph represents a subpopulation instead of just 1 individual. The subpopulations evolve in isolation without consultation with the neighbourhood. However, the best few individuals from one subpopulation transfer to the other subpopulation in the neighbourhood and thus affect the functioning. Maintaining different subpopulations stops premature convergence in GA since different subpopulations have different optima. The coordination is done by the exchange of individuals. The outgoing migration of individuals from one population is called emigration, controlled by the emigration rate; while the number of individuals coming in is called immigration controlled by the immigration rate. The overall model of GA with such a topology is called the *island population method*, denoting isolated islands of subpopulations that evolve, affected by emigration and immigration.

15.5 Differential evolution

A major advantage in the use of PSO is that the mutation rate can adapt from a high value at the start to reasonably small values at the end. The trick was the use of vectors for mutation which are based on the difference of positions of individuals. This motivates the design of an algorithm using difference vectors to fix the magnitude and direction of mutation. This gives the name *differential evolution* (DE), or the evolutionary technique using the differential of position vectors as mutation (Storn and Price, 1997; Das and Suganthan, 2010).

15.5.1 Mutation

Let us draw 3 individuals from the population pool. As the first strategy, let them be random individuals, called the *random mutation strategy*. Let the individuals be x_{r1}, x_{r2}, and x_{r3}. The differential serves as the mutation vector, given by $(x_{r2} - x_{r3})$, which applied to the individual x_{r1} gives the mutated individual (v_i) shown by Eq. (15.20). The mutated individual so formed is called the *mutant*.

$$v_i = x_{r1} + F(x_{r2} - x_{r3}), x_{rk} \sim U(population), k = \{1, 2, 3\} \qquad (15.20)$$

Here, F is a constant that amplifies the mutation vector and enables algorithmic control over the mutation extent. v_i is the resultant mutant vector. The suffix i mean the mutation applied to get the ith vector in the new population. In all terms, the parametrization on time or generation is avoided for clarity.

Mutation applied in this manner moves the candidate individual randomly and the only utility above GA is that the mutation rate is adaptive due to the use of a differential vector. A more directed strategy is called the *best mutation strategy*, which applies the mutation to the best individual to create a candidate vector. The aim is to explore more near the best-known position so far in pursuit of it being the global optima. The mutant individual is given by Eq. (15.21).

$$v_i = x_{best} + F(x_{r1} - x_{r2}), x_{rk} \sim U(population), k = \{1, 2\} \qquad (15.21)$$

Here, x_{best} is the best individual in the current population. The random strategy causes random motions in the fitness landscape which can take a long time to converge. On the contrary, the best mutation strategy always applies the mutation vector to the best individual giving rise to the candidate vector nearby the best individual, thus risking premature convergence to a local-optima.

The last strategy is called the *target to best mutation strategy*. Here also, the best individual is taken for mutation, but in the determination of the mutation vector instead. The mutation vector is taken as a vector that points towards the best individual. The overall mutation is thus given by Eq. (15.22).

$$v_i = x_{r1} + F(x_{best} - x_{r2}) + F(x_{r3} - x_{r4}), x_{rk} \sim U(population), k = \{1, 2, 3, 4\}$$
$$(15.22)$$

The different mutation strategies are shown in Fig. 15.12. The topologies and their applications are the same in DE like the PSO. Hence, the best is only taken from the neighbourhood which reduces the risk of convergence to local optima.

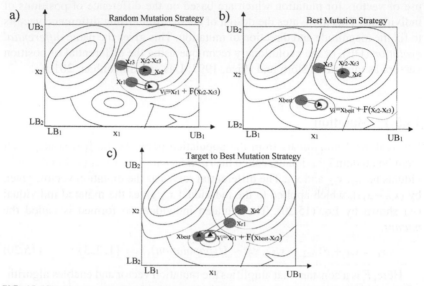

FIG. 15.12 Mutation in differential evolution. (A) Random mutation strategy. (B) Best mutation strategy. (C) Target-to-best mutation strategy.

Another generalization done in the mutation is to take multiple difference vectors. The process so far used just 1 difference vector as mutation. For 2 and 3 vectors, the modelling is shown by Eqs. (15.23), (15.24).

$$v_i = x_{r1} + F(x_{r2} - x_{r3}) + F(x_{r4} - x_{r5}) \tag{15.23}$$

$$v_i = x_{r1} + F(x_{r2} - x_{r3}) + F(x_{r4} - x_{r5}) + F(x_{r6} - x_{r7}) \tag{15.24}$$

15.5.2 Recombination

The GA had a crossover coupled with the mutation as an important genetic operator. The same operator is called the *recombination* operator in DE. The utility of the operator is slightly different in DE. If the mutant vector is taken as the new individual, it may be close to the target, for nonrandom strategies leading to premature convergence. Further, since the vectors are random, a competent individual may not contribute to any mutant vector, thus rending itself useless and deleted. Hence the parent x_i, which is the ith vector in the current population pool is used to give some genes or characteristics while the others come from the mutant vector.

The recombination is done between the parent vector x_i in the current population and the mutant vector v_i. The individual produced after recombination is called a *candidate vector* because it is a candidate made to be inserted into the new population pool. The term *crossover rate (xrate)* is also re-defined, which now means the proportion of the candidate vector that comes from the mutant vector. There are two recombination strategies popular, although any mechanism of crossover can be used here. The first is called the *binomial crossover strategy.* This strategy is like the scattered crossover, wherein for every gene j, the probability of the gene coming from the mutant is taken as the crossover rate. So, by expectation, crossover rate times individual length genes come from the mutant. A problem however is that it is probabilistically possible that the mutant does not contribute anything, and the parent is unaltered in the generation of the candidate vector. A small constraint is hence added that the mutant should contribute at least one gene and is implemented by copying a random gene $r \sim U(size\ of\ individual)$ from the mutant to the candidate. The binomial crossover is given by Eq. (15.25).

$$u_{i,j} = \begin{cases} v_{i,j} & \text{if rand} \leq xrate \vee j = r \\ x_{i,j} & \text{if rand} > xrate \vee j \neq r \end{cases} \tag{15.25}$$

Here, $u_{i,j}$ is the jth gene of the ith candidate vector, $v_{i,j}$ is the jth gene of the ith mutant vector, and $x_{i,j}$ is the jth gene of the ith individual. *rand* is a function that returns a random number between 0 and 1.

The other operator is called the *exponential recombination.* This is like the 2-point crossover with the difference that the space between the two crossover points is such that (probabilistically) crossover rate percent of genes come from the mutant. The first crossover point has no restriction. The length of the individual to be taken from the mutant starting from the first crossover point is taken as C, where probabilistically C must be *xrate.n* and n is the length of the individual. However, it should be able to take small values to the maximum possible value of n. An implementation is given by Algorithm 15.7.

Algorithm 15.7 Exponential recombination.

Input: Individual x_i, Mutant v_i
Output: individual u_i
1   $C \leftarrow 0$;   ▷ length of genes taken from mutant
2   **do**
3 | $C \leftarrow C + 1$;
4   **while** *rand* $\leq x_{rate} \wedge C \leq n$;
     ▷ C is probabilistically *xrate*×(individual size);
5   $u_i \leftarrow x_i$;
6   $j \sim U[0,\text{individual size}]$;   ▷ start position in mutant
7   **for** k *from 1 to C* **do**   ▷ copy C genes
8 | $u_{i,j} \leftarrow v_{i,j}$;
9 | $j \leftarrow (j \text{ modulo } n) + 1$;   ▷ mutant is circular
10   return u_i

rand is a function that returns a random number between 0 and 1. Given the first crossover point from where copying must happen and the length of copying, it is a naïve mechanism to do the copy of the mutant vector into the parent vector to generate the candidate vector. However, the individual is taken as circular in nature, and iterating after the last gene, you get to the first gene. So, iterating after the *n*th gene position gives the 1ˢᵗ gene position, popularly implemented in computing by the modulo function. The recombination operators are shown in Fig. 15.13.

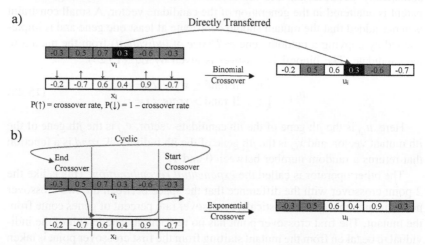

FIG. 15.13 Recombination in differential evolution. (A) Binomial recombination. (B) Exponential recombination.

15.5.3 Algorithm

The last operator is *selection*. So far, the parent x_i was used to generate the child u_i. There needs to be a mechanism to make the newer population better than the old one. Unlike GA, the DE does not apply any discretion in the individuals that undergo evolution. The selection strategy in DE makes the candidate compete with the parent and the fitter of the two is used in the new population. This is given by Eq. (15.26).

$$x_i(t+1) = \begin{cases} u_i(t) & \text{if } f(u_i(t)) \geq f(x_i(t)) \\ x_i(t) & \text{if } f(u_i(t)) < f(x_i(t)) \end{cases} \tag{15.26}$$

The notations are for a maximization problem since the term fitness is used. For the minimization problem, the sign of inequality shall change. The complete algorithm is given as Algorithm 15.8.

Algorithm 15.8 Differential evolution.

---

1 **for** *i from 1 to population size* **do**
2     $x_i \leftarrow$ random individual or by using problem specific heuristic ;
3 **while** *stopping criterion* **do**
4   **for** *all individuals i* **do**
5        $v_i \leftarrow$ mutant vector as per strategy ;
6        $u_i \leftarrow recombination(x_i, v_i)$;
7   **if** $f(x_i) < f(u_i)$ **then** $\triangleright$ accept child if better
8           $x_i \leftarrow u_i$

9 **return** best individual in population

The DE has choices regarding the mutation strategy, number of difference vectors, and crossover strategy. The algorithm has its naming convention. The general template is DE/*x*/*y*/*z*, where *x* represents the mutation strategy, *y* represents the number of difference vectors and *z* represents the recombination strategy. *z* or recombination strategy may occasionally be missed out. So, DE/*rand*/1/*bin* states a DE with a random mutation strategy, 1 difference vector, and a binomial recombination strategy. Similarly, DE/*best*/1/*exp* means a DE with the best mutation strategy, 1 difference vector, and exponential recombination operator.

15.5.4 Example

Consider the same problem of optimizing a spherical function. Consider the population shown in Table 15.3. To produce the individual for the next generation from x_1, consider random mutation strategy with chosen vectors 2, 4, and 5. Let $F = 0.8$. The mutant vector is given by Eq. (15.27).

$$v_1 = x_2 + F(x_4 - x_5)$$

$$[-0.3, 0.4, -0.1]^T + 0.8\left([0.3, 0.3, -0.2]^T - [0.6, 0.8, -0.4]^T\right)$$

$$[-0.3, 0.4, -0.1]^T + [-0.24, -0.4, 0.16]^T$$

$$[-0.54, 0, 0.06]^T$$

(15.27)

TABLE 15.3 Sample initial population for differential evolution.

Particle	Position (x_i)	Fitness ($f(x) = \sum_{i=1}^{3} x_i^2$) (smaller is better)
1	[0.5 −0.4 0.4]	0.57
2	[−0.3 0.4 −0.1]	0.26
3	[0.2 −0.3 0.4]	0.29
4	[0.3 0.3 −0.2]	0.22
5	[0.6 0.8 −0.4]	1.16

Consider binomial recombination strategy between $x_1 = [0.5, -0.4, 0.4]^T$ and $v_i = [-0.54, 0, 0.06]^T$, say it gives $u_1 = [0.5, 0, 0.06]^T$. Now the fitness value of x_1 is, $f(x_1) = 0.5^2 + (-0.4)^2 + 0.4^2 = 0.57$, while that of u_1 is, $f(u_1) = 0.5^2 + 0^2 + 0.06^2 = 0.2536$. As u_1 is better, x_1 takes a new value of u_1. This is repeated for all individuals and done for multiple iterations of the evolution.

15.5.5 Self-adaptive differential evolution

A major limitation of the DE algorithm is the need to set parameters like F, crossover rate, number of difference vectors, type of mutation, etc. In general, this is a limitation of most algorithms that perform well only for good parameter values, while the parameters are dependent upon the type of input (fitness function in this case) that is not known during the design phase. A trial and error mechanism of parameter setting can take too long and a human may not always do the same religiously. The *self-adaptive* variants of the algorithms aim to constantly adapt the parameters with time such that the resultant algorithm performs the best. The parameters constantly adapt as the input (fitness function) changes or the context (convergence status) changes with time. In other words, the algorithm can itself select good parameter values as per the operating conditions.

Consider that k sets of parameters appear to be promising by the designer, each parameter-set distinct and diverse from the others. However, the designer is unsure which setting will perform the best. Each set differs in terms of the type of crossover, mutation, etc. Let the first parameter setting be s_1, second be s_2, third be s_3 and so on till s_k. The DE consists of a population that undergoes genetic operations (dependent upon the parameters) to produce a new population. Under the new setting there are k sets of parameter values, each set will produce a different new population. Let us probabilistically select the a parameter setting s_i with a probability $p_i(t)$ at time t to produce a child, $\sum_{i=1}^{k} p_i(t) = 1$.

For a population of n individuals, the parameter setting s_i would be expected to produce $n.p_i(t)$ children. Initially, the probabilities are all equal, or $p_i(0) = 1/k$.

Now different parameter settings will be good or bad at different times. A parameter setting may be good initially but may become poor later, in which case its probability should be initially high which should be gradually reduced. Similarly, a parameter setting may become very good at the last stages of the algorithm, in which case its probability should be increased at the end. The probability of selection of s_i, given by $p_i(t)$, is proportional to the current goodness of s_i. At generation t, let the parameter set s_i be invoked to generate $n_i(t)$ children, $\sum_{i=1}^{k} n_i(t) = n$. The generation of a child is called as successful if the generated child is fitter than the parent. On the contrary, the generation of a child is called unsuccessful if the parent is fitter than the generated child. Let $n_{is}(t)$ be the number of successful children produced by the parameter set s_i, while $n_{if}(t)$ be the number of failed children, $n_{is}(t) + n_{if}(t) = n_i(t)$. The success rate of the parameter set s_i is the percentage of successful children generated by it, given by Eq. (15.28).

$$sr_i(t) = \frac{n_{si}(t)}{n_i(t)} = \frac{n_{si}(t)}{n_{si}(t) + n_{fi}(t)} \tag{15.28}$$

The probability of selection of s_i at any time is proportional to its current success rate, given by Eq. (15.29), where a small constant (ε) is added to restrict the probability of selection of any parameter set to become zero, in which case it will never get selected again and have no chance to improve later.

$$p_i(t+1) = \frac{sr_i(t) + \varepsilon}{\sum_{i=1}^{k} sr_i(t) + \varepsilon} = \frac{\dfrac{n_{si}(t)}{n_{si}(t) + n_{fi}(t)} + \varepsilon}{\sum_{i=1}^{k} \dfrac{n_{si}(t)}{n_{si}(t) + n_{fi}(t)} + \varepsilon} \tag{15.29}$$

In this manner, a parameter set will be selected more number of times if it generates fitter individuals, while the probability and hence the number of selections will drop as and when the parameter set starts to produce weaker individuals. The concept is summarized in Fig. 15.14. The concept is generic to any algorithm and parameter set. The *Self-Adaptive Differential Evolution* (SADE) (Qin and Suganthan, 2005) specifically uses $k = 2$ parameter sets, which are $s_1 = DE/rand/1/bin$ and $s_2 = DE/target-to-best/2/bin$.

The other parameters, specifically F and crossover rate (*xrate*), are also important and hard to set since they represent continuous choices and not discrete choices that can be divided into a small number of sets. The first change made for such parameters is to sample the parameter value from a distribution at every few iterations rather than setting them as fixed values. Sampling from a normal distribution of a set mean and standard deviation ensures that some iterations experience a very small value, while some others experience a very large value and different ranges of values are tried in the evolutionary process.

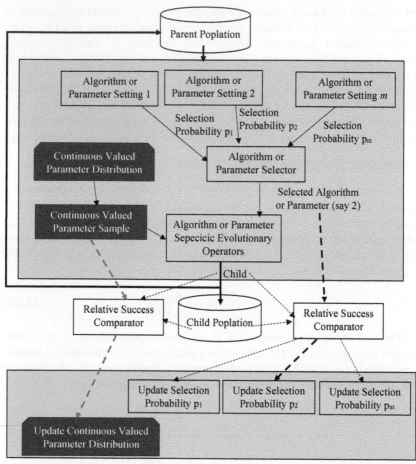

FIG. 15.14 Self-adaptive evolutionary algorithms/evolutionary ensembles. The algorithm maintains sets of parameters (or different evolutionary algorithms) with every parameter set (or evolutionary algorithm) producing a part of child population proportional to its selection probability. The parameter sets (or evolutionary algorithms) generating more successful or fitter individuals have their selection probabilities increased, making the overall algorithm self-adaptive. For continuous parameter values, samples are drawn from a distribution and the distribution learnt from the performance of the drawn samples.

Now every sampled parameter setting generates a few individuals over time whose success (and failure) rate can be measured. Say a sampled value ($xrate_1$) of the parameter $xrate$ was used for some generations that gave some success rate, while later another sampled value ($xrate_1$) for some generations gave a lower success rate. Now $xrate_1$ appears to be a better value and hence the mean of the distribution may be shifted towards $xrate_1$. Here, the distribution itself learns with time using the observed samples. The distribution at every time is adjusted to better represent the samples that gave a higher success rate. This learning of the distribution makes the algorithm self-adaptive in nature.

15.5.6 Evolutionary ensembles

The ideas of the SADE can be extended to other evolutionary algorithms (and algorithms beyond evolution). Let us therefore generalize the same (Wu et al., 2018, 2019). Every evolutionary technique takes a population pool, applies several operators and generates a new population pool. Different evolutionary algorithms have different strengths and weaknesses, and it is unknown which evolutionary technique may perform the best. Furthermore, the choice of the evolutionary algorithm that performs the best may change with time as convergence happens. Therefore, the proposal is to use multiple evolutionary algorithms in parallel making an *evolutionary ensemble*, instead of just a single evolutionary algorithm.

In general, in computing if there is an uncertainty about which of the k algorithms will perform the best, it is possible to execute them in parallel and select the one that performs the best. However, algorithms like evolution are limited by time. k-parallel executions may increase the overall computation expense by a factor of k, while the result may not be better than a single algorithm executed k times more. The more time an algorithm is given, the better are the results.

Therefore, evolutionary ensembles must strategize the investment of time between the evolutionary algorithms used. An algorithm performing well at the moment should be given more time, while an algorithm not currently productive must be limited by time. Therefore, at any time, the entire population of n individuals is subdivided into k subpopulations, where every algorithm s_i is given a slice of $n_i(t) = p_i(t) \cdot n$ individuals. $p_i(t)$ is the proportion of individuals given to the algorithm s_i, $\sum_{i=1}^{k} p_i(t) = 1$. Each algorithm s_i takes its share of individuals, uses its specific operators and adds the same to the new population pool. The proportion of population $p_i(t)$ is adapted with time. The new population pool contains individuals from all algorithms. An algorithm generating fitter individuals is given a large proportion, while the algorithm generating relatively weaker individuals is given a smaller proportion.

An added problem with the evolutionary ensembles is to stop premature convergence, previously solved by the use of topologies. An algorithm may appear very interesting because of its ability to give the fittest individuals, however, in the long run it may lead to a premature convergence, an attribute that does not appear in the success metrics that decide the proportion of population given to the algorithm. A solution is naturally to use the evolutionary ensembles within the framework of topologies and other diversity preservation techniques (to be discussed in Chapter 16). More importantly, the topology itself is a parameter suitable for adaptation by looking at the convergence of the different segments of the algorithm and the likelihood of convergence at a local optima. The neighbourhood size of a node, and the selection of the neighbours could thus be adapted as well.

15.6 Local search

The searches discussed so far are called the global searches as they aim to get to the global optima using a population of multiple individuals. However, this is not practical for applications wherein the results need to be computed very quickly. In such a case, a population of only a single individual is used that primarily aims to get deep into the local optimal only and to do so very quickly. This type of search is called the *local search*. Numerous problems need a local search. Examples include adapting the robot path to the changing environment, which needs to be done quickly as the robot would otherwise be at a standstill; and needs to be further done faster than the change in the environment itself. The postprocessing of the path in grid search and cell-decomposition approaches is also a local search, where the global optimality has already been taken care of. The gradient descend algorithm, used to make the back-propagation algorithm, is also a local search, where the gradient is additionally available.

15.6.1 Hill climbing

The most basic local search algorithm is called *hill climbing*. The name is derived from the fact that in a fitness landscape with a maximization aim, the local maxima is a hill, in which a single individual must always walk up the hill, thereby eventually reaching the peak. The initial individual is randomly generated. Thereafter mutation is applied to get a new candidate individual. If the candidate is weaker than the individual, it represents going downwards and the move is not accepted. However, if the candidate is better, the candidate becomes the individual. So, the individual always moves up and eventually reaches the peak, which is the local optimal value. The pseudo-code is given by Algorithm 15.9. The process is shown in Fig. 15.15.

Algorithm 15.9 Hill climbing.

1 $x \leftarrow$ initial random individual ;
2 **while** *stopping criterion* **do**
3      $x' \leftarrow x+$ Gaussian noise ;   ▷ mutated individual
4   **if** $f(x') > f(x)$ **then** $x \leftarrow x'$;   ▷ accept if better
5 **return** $x, f(x)$

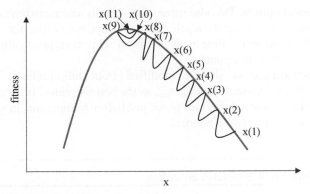

FIG. 15.15 Hill climbing algorithm: the individual is mutated at every time step, and the better of the parent or mutated individual is accepted. The problem is assumed to be a maximization problem.

15.6.2 Simulated annealing

A problem with the hill climbing is that it shall never get away from the local optima, even if it was faintly small. The *simulated annealing* adds a clause to the hill climbing algorithm that a downward motion may also be accepted with a small probability, in a hope that the same will result in getting out of the local optima and getting towards the global optima. The probability is kept sufficiently large to start off and soon dies off to nearly zero, which is the convergent phase and the algorithm primarily gets into its current optima. This is modelled by the annealing process in metallurgy wherein a metal is first heated and then cooled.

Assume that the current individual x is mutated to give the candidate individual x', which has a lower fitness ($f(x') < f(x)$). The probability by which a downhill motion can be accepted (x' is accepted) is given by Eq. (15.30).

$$P(accept\ x'|f(x') < f(x))) = \exp\left(-\frac{f(x) - f(x')}{T(t)}\right) \quad (15.30)$$

Here, $T(t)$ is the current temperature inspired by the annealing process. It acts as a parameter that controls acceptance probability. $T(t)$ constantly decreases with time like the annealing process. Initially (at high temperatures) the acceptance probability is high and therefore the algorithm more openly accepts downward moves. However, as time increases, the temperature $T(t)$ becomes low, which significantly decreases the acceptance probability. Similarly, if the new individual is closely competent than the current one, the probability to accept the new individual is good. However, if the new individual is bad, characterized by a large difference in their fitness values, it is unlikely to get selected because of a low accepting probability.

The acceptance probability is always nonzero, which means that there is a nonzero probability that the algorithm will take a series of downward moves to escape the local optima. This also means that there is a nonzero probability that the algorithm will eventually reach the global optima, or the algorithm will reach the global optima as time tends to infinity. However, practically the algorithm rarely escapes an optimum.

The pseudo-code is slightly modified from hill-climbing, given by Algorithm 15.10. Explicitly storing x_{best} as the best individual is for the protection that x may accept a downward move just before termination, in which case the best observed should be returned.

Algorithm 15.10 Simulated annealing.

---

1 $x \leftarrow$ initial random individual ;
2 $x_{best} \leftarrow x$;   ▷ best individual ever
3 $t \leftarrow 0$;
4 **while** *stopping criterion* **do**
5   calculate temperature $T(t)$;
6      $P(t) = exp\left(-\frac{f(x)-f(x')}{T(t)} \right);$ ▷ probability of accepting a worse individual
7      $x' \leftarrow x+$ Gaussian noise ;   ▷ mutated individual
8   **if** $f(x') > f(x)$ **then** $x \leftarrow x'$;   ▷ accept if better
9   **else if** *rand* $< P(t)$ **then**   ▷ or accept worse by chance
10        $x \leftarrow x'$;
11   **if** $f(x) > f(x_{best})$ **then** $x_{best} \leftarrow x$;
12      $t \leftarrow t + 1$;
13 return $x_{best}, f(x_{best})$

15.7 Memetic computing

So far, two types of optimizations have been discussed, global optimization and local optimization. The *global optimization* intends to find the global optima, for which the techniques make use of a multiindividual approach and incur heavy computation costs. On the other hand, in *local optimization*, the techniques in principality converge to a local-optima only, while giving a very quick performance. It is particularly interesting to note that the pros and cons of the two techniques are contrasting and hence a popular computing paradigm is to *fuse* the two techniques, which gives rise to memetic computing. The trade-offs are given in Table 15.4. *Memetic computing* combines the global search and local search to take the best advantage of the two techniques, producing an algorithm lesser likely to get stuck at a local optimum.

TABLE 15.4 Global v/s local search.

S. no.	Property	Local search	Global search
1.	Computation Time	**Small**	Large
2.	Optimality	Typically, converges to local optima	**Typically, converges to global optima**
3.	Number of individuals	1	Many

Better property values are shown in bold.

The specific manner of fusion is a matter of design and specific to the application. Some typical architectures are discussed. A popular architecture is called the *master-slave architecture*. Here, the global optimization acts as a master and the local optimization acts as the slave. Every individual is subjected to a local search often in the optimization process, say after every few generations.

All evolutionary algorithms had a dilemma between exploration and exploitation. The dilemma is between exploring a local-optima further that delays convergence; or instead converge quickly to what currently appears as the global optima. The memetic algorithm makes a quick search around the current optima of every individual by a local search. Within a few iterations, a good indicative idea can be obtained about the goodness of an optimum. Hence even if selection deletes an individual, it is more certain that it was a local-optima only as its value could not be appreciably improved by the local search.

The architecture can be best visualized as the motion of individuals in the fitness landscape. The local search is a super-power given to the algorithm that places every individual close to their local optima, which is exercised by the master GA to assess the goodness of the individuals. Now if an individual continues to have a poor fitness value despite moving towards its optima by the local search algorithm, it is reasonable to assume that the particle was at the local optima and worthy of being deleted.

Assume T is the total computation time available for the optimization, and there are n individuals in the population pool running for G generations. Further, assume that the genetic operators take no time to compute. Let the local search be applied with a time threshold of T_L. As a poor implementation let all individuals undergo the local search for every generation. In such a case the total computation time approximately becomes $T = nGT_L$, not considering the time to sort individuals in the global optimization. If the total computation time is fixed, then either there can be a large value of T_L and subsequently small values of n and G which makes the resultant algorithm largely local. On the other hand, a large value of n and G and correspondingly a small value of T_L makes the algorithm largely global. Hence, parameter control enables switching between extremely global to extremely local settings.

Local search is more computationally expensive than a naïve fitness evaluation and therefore must be sparingly used. Every crest and trough (hills and valleys) of the fitness landscape is called a modality. Evolution is difficult because of the presence of multiple modalities that threat the algorithm being stuck at the local optima. Often, many individuals close by represent the same modality (same hill or valley), and local search gets them close to each other, still closer to the local-optima. In such a situation, different individuals are approaching the same optima using the local search, while even if a single individual approached closer to the local optima, it would have been enough to guide the global search about the fitness of that modality. Hence the local search becomes wasteful, especially towards the later generations. It is better to *cluster* the population and apply local search on only the best individuals within the cluster. This increases the time needed to cluster the individuals in the first place. For problems with computationally heavy fitness function, the clustering overhead is small as clustering does not need fitness evaluation. As an example, using GA to optimize the neural network has fitness function as the loss function which iterates through all the data, which is computationally heavy. Similarly, in trajectory planning, the complete trajectory needs to be ascertained for being collision-free, which is again computationally heavy. Another heuristic is that the local optimization is only done for some sampled generations. The sample size of generations and the number of clusters may themselves be adaptive parameters.

There is a greater problem if the fitness function is computationally heavy. Every local search calls the fitness function multiple times, which even done for a few individuals in every few iterations of optimization can be computationally expensive. Hence another architecture of memetic computing uses a local search to make a *good initial population pool for the problem*. The initial population has a strong effect on the algorithm performance and a good initial population can reduce the computation time significantly. So, starting from a heuristic or random population, all the individuals are subjected to a local search. This initially predicts the goodness of the optima represented by every individual, which is a one-time use of local search, and the dynamics thereafter is just by the GA.

A related problem is that the GA is only effective till all individuals come to the same optima region. Thereafter, a local search is much faster. The GA takes every single individual to the same optima region, while the local search needs to take only 1 such individual deep into the optima which adds speed. If the problem is unimodal, constraint-free with a gradient, the local search is directed in the direction of the gradient which is fast. Another architecture makes use of the same property. Here, the local search is used as a *postprocessing technique* of the global search. The global search is run for some time. Thereafter, the best individual is extracted and subjected to local search.

The different architectures are summarized in Fig. 15.16. Finally, these architectures are not exclusive, and it is possible to mix different architectures or create new architectures on top of these.

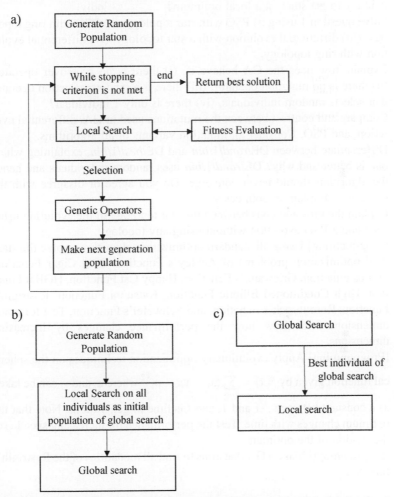

FIG. 15.16 Memetic algorithm architectures. (A) Master-slave architecture, wherein a slave local search is applied for some sampled individuals/generations alongside fitness evaluation of the global search. (B) Initial population generation architecture, wherein the local search is used to generate better initial population of the global search. (C) Postprocessing architecture, wherein the local search further optimizes the best global search individual.

Questions

1. Explain the working of GA, PSO, and differential evolution for two generations to solve the minimization problem $f(x_1,x_2,x_3) = abs(x_1.x_2.x_3)$ where '.' is the arithmetic multiplication operator and $abs()$ is the absolute function. You may approximate your calculations and use any randomization. Assume a small random initial population. Specify the operators used. Is the function unimodal or multimodal? Under what conditions the solution is likely to get stuck at a local optimum?

2. Solve question 1 using (i) PSO with star topology, (ii) PSO with ring topology, (iii) differential evolution with a star topology, (iv) differential evolution with ring topology.

3. Explain how does the GA behave if, (i) there is no crossover operator, (ii) there is no mutation operator, (iii) there is a random selection operator that selects random individuals, (iv) there is only 1 individual?

4. Compare and contrast between the mutations used in GA, differential evolution, and PSO. Consider all mutation variants of all algorithms.

5. Differentiate between DE/*rand*/1/*bin* and DE/*best*/1/*bin*, explaining which one is better and why? DE/*rand*/1/*bin* uses random individuals and hence the algorithm should never converge. Do you agree or disagree with the statement? Explain in both cases.

6. Explain the pros and cons between the star topology in the evolution using PSO and a PSO evolution without using any topology.

7. [Programming] Using all standard optimization algorithms, solve the standard optimization problems of Ackley's Function, Bent Cigar Function, Discus Function, Griewank's Function, Happy Cat Function, HGBat Function, High Conditioned Elliptic Function, Katsuura Function, Rastrigin's Function, Rosenbrock's Function, and Schwefel's Function. Test for many dimensions and show how the performance changes by increasing dimensions.

8. [Programming] Apply evolutionary optimization techniques on the spherical function given by $f(x) = \sum_{i=1}^{n} (x_i - v_i t - a_i)^2$ where a_i and v_i can be taken as a constant between -1 and 1, and t is time or generations. Note that the optimum changes with time. Test the performance for increasing speed v_i of the motion of the optimum.

9. [Programming] Make a GA that aims to find all modalities of the Rastrigin's function.

References

Angeline, P.J., 1994. Genetic programming and emergent intelligence. In: Complex Adaptive Systems archive Advances in genetic programming. MIT Press, Cambridge, MA, pp. 75–97.

Bäck, T., Hoffmeister, F., Schwefel, H.P., 1991. A survey of evolution strategies. In: Proceedings of the Fourth International Conference on Genetic Algorithms. Morgan Kaufmann Publishers, San Mateo, CA, pp. 2–9.

Das, S., Suganthan, P.N., 2010. Differential evolution: a survey of the state-of-the-art. IEEE Trans. Evol. Comput. 15 (1), 4–31.

Davis, L., 1987. Handbook of Genetic Algorithms. Van Nostrand Reinhold, New York.

Goldberg, D.E., 1989. Genetic Algorithms in Search, Optimization, and Machine Learning. Addison-Wesley, Reading.

Hansen, N., Ostermeier, A., 1996. Adapting arbitrary normal mutation distributions in evolution strategies: the covariance matrix adaptation. In: Proceedings of the IEEE International Conference on Evolutionary Computation. IEEE, Piscataway, New Jersey, pp. 312–317.

Kala, R., Warwick, K., 2014. Heuristic based evolution for the coordination of autonomous vehicles in the absence of speed lanes. Appl. Soft Comput. 19, 387–402.

Kala, R., Shukla, A., Tiwari, R., 2011. Robotic path planning using evolutionary momentum based exploration. J. Exp. Theor. Artif. Intell. 23 (4), 469–495.

Kala, R., Shukla, A., Tiwari, R., 2012. Robotic path planning using hybrid genetic algorithm particle swarm optimization. Int. J. Inf. Commun. Technol. 4 (2-4), 89–105.

Kennedy, J., Eberhart, R., 1995. Particle swarm optimization. In: Proceedings of the International Conference on Neural Networks, vol. 4. IEEE, Perth, WA, Australia, pp. 1942–1948.

Koza, J.R., 1992. Genetic Programming: On the Programming of Computers by Means of Natural Selection. MIT Press, Cambridge, MA.

Poli, R., Kennedy, J., Blackwell, T., 2007. Particle swarm optimization: an overview. Swarm Intell. 1 (1), 33–57.

Qin, A.K., Suganthan, P.N., 2005. Self-adaptive differential evolution algorithm for numerical optimization. In: Proceedings of the 2005 IEEE Congress on Evolutionary Computation. IEEE, Edinburgh, UK, pp. 1785–1791.

Storn, R., Price, K., 1997. Differential evolution—a simple and efficient heuristic for global optimization over continuous spaces. J. Glob. Optim. 11 (4), 341–359.

Wu, G., Shen, X., Li, H., Chen, H., Lin, A., Suganthan, P.N., 2018. Ensemble of differential evolution variants. Inf. Sci. 423, 172–186.

Wu, G., Mallipeddi, R., Suganthan, P.N., 2019. Ensemble strategies for population-based optimization algorithms—a survey. Swarm Evolut. Comput. 44, 695–711.

Chapter 16

Evolutionary robot motion planning

16.1 Introduction

The problem is to move a robot from a start configuration to a goal configuration with an added constraint that the robot should not collide with any obstacle. The motion, if planned in control spaces, may have more kinematic and dynamic constraints. Traditionally, all hard problems that cannot be solved by exhaustive searches are often solved by optimization. *Optimization* aims to get the best possible solution to a problem, judged by an objective measure while satisfying some bounds and constraints. Optimization is used in numerous hard problems like machine design, product design, art design, scheduling, and planning. This includes problems from operational research, like the delivery of items to places as a travelling salesman problem; vehicle routing; deciding timetables of academic universities, airlines, and trains.

It is hence evident that a major tool used for solving the problem of motion planning is optimization. The aim is to define the problem formally with an objective function and leave it to an optimization algorithm to find the best trajectory possible. The optimization algorithm does thousands of simulations and gives the best trajectory. The chapter remodels the problem of motion planning to be fit for the optimization approaches. This chapter primarily presents the use of heuristics to accelerate the search for the best trajectory.

So far, numerous algorithms have been discussed in the book. It would be interesting to note the pros, cons, and specific use cases of the algorithm. The algorithm belongs to the deliberative technique, and the deliberative algorithms are better candidates for discussion. The search-based techniques were computationally expensive and thus not useful for problems of large dimensionality. They further added a hard-to-set resolution parameter. Cell-decomposition techniques used preprocessing to produce a roadmap; however, the utility was practically limited to small dimensional problems planned in the workspace (sometimes in geometric workspaces) only. Neither of these approaches could be used to solve the problem of planning a high degree of freedom manipulation

Autonomous Mobile Robots. https://doi.org/10.1016/B978-0-443-18908-1.00020-0

arm. Sampling-based approaches were a respite that could scale very well to high-dimensional problems while exploring (sampling) only a part of the configuration space. The single query variant or RRT was particularly interesting. All techniques required the use of a postprocessing technique of *local search* to produce smooth trajectories, which is an optimization problem. Thus, the approaches discussed here blend with all other deliberative approaches for final trajectory generation.

Now, it is imperative to consider whether having a deliberative search followed by a local optimization makes a better system, or instead, the optimization could be used to solve the complete problem in the first place. There is no clear answer. The strong evolutionary techniques discussed in this chapter can solve the global optimization problem, which is to mine out the best path having numerous possible homotopic groups of trajectories and to optimize it further locally on the run. However, at the same time, the overall problem of motion planning has stringent time constraints to be adhered to. Hence, if the sampling-based approaches can give a good enough initial seed trajectory to be quickly locally optimized, it would be a better solution. If it is possible to do a global optimization for the specific problem within the time bounds, optimization will represent a better solution.

Sampling-based approaches and optimization approaches have a different approach to solve the problem of robot motion planning. The differences can be understood with the analogy that if you were required to solve the shortest path from your university to home, you could do so in multiple ways. Evolution creates a database of all the paths taken by you previously and suggests new paths based on cross-bred and noised paths, whose cost is experienced and added to the database. Sampling, on the contrary, stores every segment visited separately as a graph. The problem is that the network of roads is infinitely large, and we can only explore a minute fraction of it. Along with time, the graph is expanded, and finally, a single graph search is made to get the path. In learning, every day, we witness a new experience and incorporate that into the best path so far. Good experiences modify the best-known path.

More concretely, optimization samples complete trajectories, using the information of all the previous sampled trajectories; while sampling-based techniques sample subgraphs, or vertices and edges. Depending on the map, and problem, the two techniques could have their pros and cons. Like a complex map will be hard to optimize since a single trajectory connecting the source and goal will be hard to find. On the contrary, if the trajectory can be optimized by a few turns alone, it may be better to combine search and local optimization as one approach that is done by global optimization.

RRT and sampling-based approaches unnecessarily incur too many zigzags (with each edge representing a turn or a zigzag path), which must be later locally optimized. The real robots face simpler maps for most of the time of operation, while they need smooth paths as any turn or zigzag produces inconvenience to the humans cooperating in the same space. Humans anticipate the intent and future path of other people and robots to make their plans, while if a person or robot constantly zigzags and changes direction, the human cannot anticipate where the person or robot is going and what their future trajectory would be. Each zigzag or turn makes it difficult to control the robot, and the difference between the planned and actual position becomes large. Hence, a robot should have smooth and simple paths for straightforward queries, better facilitated by optimization than sampling-based approaches.

Comparing optimization with the reactive algorithms is more straightforward since every difference between the deliberative and reactive algorithms can be named straight away. However, robots today have a high computing capability, in which case local optimization is easy and can be done on the fly. This means if the robot needs to overcome small obstacles around moving at moderate speeds, a small amount of continuous online optimization will suffice. The robot initially plans and traces a trajectory, while the trajectory is continuously adapted by local optimization as the robot and obstacles move. Initially, the robot may only consider a small planning horizon or may consider obstacles till a small window. As the robot moves, it considers the obstacles further away that later come within the current planning horizon, or the planning window. The terms and techniques are summarized in Box 16.1.

Box 16.1 A summary of terms and techniques

Modes of operation of evolutionary motion planning
- Postprocessing of paths by deliberative algorithms like A*, Cell Decomposition, RRT, etc., as a local search for smoothness, simplification, shortening, clearance, etc.
- Complete deliberative offline computation
- Like reactive algorithms, computing on the fly as the robot moves, adapting to changing environment, the build up of uncertainties, and changing source.

Evolution v/s sampling-based motion planning
- Sampling-based motion planning produces complex paths with many turns, often requiring postprocessing.
- Evolution does not work when it is not possible to get initially good/feasible trajectories, possible in complex maps.
- Practically, most common scenarios are simple for which evolution results in smooth and simple paths.

Evolutionary motion planning process
- **Individual representation:** Via points through which the robot travels. The via points are smoothened by splines.

Continued

Box 16.1 A summary of terms and techniques—cont'd

- **Fitness function:** Path Length, Clearance, length of the path in obstacle, smoothness (angle between path segments/via points), complexity (number of via points).
- **Multiresolution fitness evaluation:** Resolution of collision/clearance computation is poor at start and high at later generations.
- **Incremental evolution:** Initial population is allowed a lower number of via points, while the limit is iteratively increased till the maximum value.
- **Operators:** Crossover, parametric mutation, shorten (structural mutation), add point (structural mutation), repair trajectory (make it feasible), and insert random individuals.
- **Dynamic maps:** Fitness function adapts trajectory to changing map. Diversity preservation or maintaining trajectories at each of the competing homotopic groups ensures completeness when an obstacle blocks the current way.
- **Control space:** The individual is a sequence of control commands and duration of time for which every such command is applied. Heuristics is useful to get initial good solutions, like the use of other deliberative and reactive algorithms.

Diversity preservation

- Preserve diversity among individuals to resist convergence into a local optimum.
- **Fitness sharing:** Fitness function is modified to incentivize individuals in less crowded areas.
- **Crowding:** Children replace closest individuals in a sampled population.
- **Restricted mating:** Crossover only between close individuals, unless there is a large difference in fitness values.

Multiobjective optimization

- Get diverse solutions such that no individual dominates (better on all objectives) any other. Selection of a single individual can be done by multicriterion decision-making, modelling weights of objectives based on manually selected items by the user in training data or selecting individuals with the best tradeoff (knee analysis, either objective does not dramatically change from a small change in the other).
- **NSGA:** Use evolution using Pareto rank (primary) and objective space diversity (secondary) as the fitness function.
- **MOEA/D:** Use weighted addition of objectives to get one solution on the Pareto front. Use multiple such weights to get a complete Pareto front.

16.2 Diversity preservation

Let us first discuss a few theoretical concepts that will be helpful in the design of optimization algorithms for the problem of motion planning. A major problem with every evolutionary technique is that an aim to converge the population deletes some important genes or some interesting individuals altogether, which leads to these techniques getting converged in a local optimum. The solution to this problem is also nature-inspired, and called diversity preservation. Traditionally, cultures evolve and breed locally, which enables them to carry the traditional knowledge and values of a vital asset. If a culture is encouraged to evolve with all others, it is possible that every individual uses the best knowledge available globally and throws away the traditional knowledge. The local breeding restriction may initially seem unreasonable, but this helps the traditional knowledge base to develop. Ultimately, the traditional knowledge is either on par with or better than the global best-known knowledge or it is ascertained that the traditional knowledge is inferior with no scope for improvement. It is not true that cultures are completely independent. Some individuals may still have cross-cultural effects through travel, study abroad, or likewise, which are some skills brought from outside to within the culture. If the world were a global village from the outset, the specialty of every single culture that exists today would be lost. Even with the increasing effects of globalization, universities, countries, and institutions lay a lot of stress on the diversity of their workplace, to take people who may not be the best suited immediately but provide diversity, which can eventually become a strong asset.

The evolutionary process is modified to encourage *diversity* within the fitness landscape (Sareni and Krahenbuhl, 1998; Lozano et al., 2008). This means that efforts are made such that the individuals representing new places are encouraged to continue improving themselves, rather than being deleted for a poor fitness value. This also means that the children are constrained to be near the parents so that the new children are not different from the parents and try to develop further locally, rather than have large migrations. Furthermore, the individuals are not encouraged to have a crossover with far individuals but should have a crossover within the locality to eliminate the risk of the children being formed far away from the parents. Different algorithms have different means of attaining diversity, which is discussed in detail.

16.2.1   Fitness sharing

The first mechanism of diversity preservation mends the fitness function in such a way that the individuals that represent new areas (where there are not enough individuals) are rewarded by an increase in their fitness value (for a maximization problem). This creates *evolutionary pressures* for the motion of individuals in the fitness landscape to avoid crowded areas and to venture into less congested areas. Convergence still happens, when the absolute fitness is poor and the incentive of being in solitude is not good enough to sustain.

In this approach, all individuals *share* their positions and hence the name fitness sharing. The reformulated fitness function (larger is better) is given by Eq. (16.1).

$$f'(x) = \frac{f(x)}{\sum_i \text{share}(d(x, x_i))} \tag{16.1}$$

The denominator is specifically called the *crowding distance* because it measures the crowding at a place and is a metric used to denote the diversity of the solution x. Here $f(x)$ is the problem-specified fitness function, and $f'(x)$ is the reformed fitness function. Let $d(x, x_i)$ be the distance between the individuals x and x_i. Since distances are being taken, there is a need to normalize all axes. *share*() is a monotonically decreasing sharing function given by Eq. (16.2).

$$\text{share}(d(x, x_i)) = \begin{cases} 1 - \left(\dfrac{d(x, x_i)}{d_{max}}\right)^\alpha & \text{if } d(x, x_i) \le d_{max} \\ 0 & \text{if } d(x, x_i) > d_{max} \end{cases} \tag{16.2}$$

Here d_{max} is the maximum distance within which the sharing happens. It is also called the *sharing radius* or *niche radius*. A value of 0 after d_{max} means that for calculating the fitness of x, a proximity query can be used to get all individuals till a distance of d_{max} as the rest of the individuals do not contribute, that can be implemented in a logarithmic time if the population is stored as a k-d tree. d_{max} is an algorithm parameter. The subtraction with 1 is to implement the relationship that an increase in distance decreases the sharing value. α is another constant that controls the scale of the overall function.

The reformulated function is such that if an individual has only a few individuals in its vicinity, its sharing value for all individuals combined is small, and therefore the fitness is increased. The further away the other individuals are, the lesser is their effect, the lower is their sharing value, and the higher is the reformulated fitness. On the other hand, the closer the individuals are in the vicinity, the higher is the sharing value and its sum, and the smaller is the fitness value.

Consider the negative spherical function ($f(X) = 3 - (x_1^2 + x_2^2 + x_3^2), -1 \le x_i \le 1$) with an aim to find the largest value of the function. The individuals, along with their modified fitness value as shown in Tables 16.1a and 16.1b. The evolution happens with this modified fitness function.

TABLE 16.1a Fitness sharing on a sampled population.

Individual	Position	Fitness	Total share $\sum_i \left(1 - \left(\frac{d(x_i, x_j)}{d_{max}}\right)^{\alpha}\right)$, $d_{max} = 0.75$, $\alpha = 1$	Effective fitness (fitness/share)
x_1	[0.5 −0.4 0.4]	2.43	$\left(1 - \left(\frac{0}{0.75}\right)^1\right) + \left(1 - \left(\frac{0.3162}{0.75}\right)^1\right) = 1.5784$	1.5395
x_2	[−0.3 0.4 −0.1]	2.74	$\left(1 - \left(\frac{0}{0.75}\right)^1\right) + \left(1 - \left(\frac{0.6164}{0.75}\right)^1\right) = 1.1781$	2.3258
x_3	[0.2 −0.3 0.4]	2.71	$\left(1 - \left(\frac{0.3162}{0.75}\right)^1\right) + \left(1 - \left(\frac{0}{0.75}\right)^1\right) = 1.5784$	1.7169
x_4	[0.3 0.3 −0.2]	2.78	$\left(1 - \left(\frac{0.6164}{0.75}\right)^1\right) + \left(1 - \left(\frac{0}{0.75}\right)^1\right) + \left(1 - \left(\frac{0.6164}{0.75}\right)^1\right) = 1.3562$	2.0498
x_5	[0.6 0.8 −0.4]	1.84	$\left(1 - \left(\frac{0.6164}{0.75}\right)^1\right) + \left(1 - \left(\frac{0}{0.75}\right)^1\right) = 1.1781$	1.5618

Fitness function $f(X) = 3 - (x_1^2 + x_2^2 + x_3^2)$, $-1 \leq x_i \leq 1$. Evolution happens using 'effective fitness' to incentivize individuals at sparsely occupied areas for diversity preservation (maximization problem assumed).

TABLE 16.1b

Distance	x_1	x_2	x_3	x_4	x_5
x_1	0	1.2369	0.3162	0.9434	1.4457
x_2	1.2369	0	0.9950	0.6164	1.0296
x_3	0.3162	0.9950	0	0.8544	1.4177
x_4	0.9434	0.6164	0.8544	0	0.6164
x_5	1.4457	1.0296	1.4177	0.6164	0

Distances between individuals.

16.2.2 Crowding

Crowding is a technique that attains diversity by resisting individuals to crowd at a place. The Genetic Algorithm (GA) selects individuals which undergo genetic operations to produce a candidate population, which may be made to compete with the parent population to make the next generation population. For this technique of diversity preservation, let us generate the candidate children for the new population one by one. So, two parents are selected for the crossover, which produces two children, and both undergo mutation to produce two candidate individuals of the next generation. In a typical implementation, these two individuals would have occupied two spots in the new population. However, in crowding these individuals, they instead *replace* two individuals from the population pool. The strategy that decides which candidate individual replaces which individual in the population pool is decided by the *crowding strategy*.

The simplest crowding strategy is that each child replaces the *closest individual* in the population pool. In such working, the child is bound to be close to the replaced individual, and therefore the reduction of diversity is limited. Let us call the replacing individual the child and the individual being replaced the parent (which may not be the one used to produce the child). It is ensured that the child is near the parent, leading to a limited reduction of diversity. This step is repeated $n/2$ times to create the new population pool. Since crossover produces two children, a variant replacement strategy is called *deterministic crowding*. Here the two closest individuals (called parents, not the individuals used to produce the children) in the population pool are computed to the children, and based on the distances, a decision is made whether the first child replaces the first parent, or the first child replaces the second parent, and the converse for the second child. Furthermore, the replacement only happens if the child is fitter than the parent, resulting in the children competing with the parents for a position. A stochastic implementation uses Tournament Selection rather than just taking the better of the parent and the child. In such a case, a worse child can also replace the parent with some small probability. The technique is called the *Restricted Tournament Crowding*.

There should be some parameter control over the extent to which crowding happens. The aim is also to add some stochasticity in the process. Hence only *CF* randomly sampled individuals are taken as candidates for replacement, over which the search is done to get the best parent. *CF* is called the *crowding factor*, which is an algorithm parameter denoting the extent of crowding. The overall approach is summarized in Fig. 16.1.

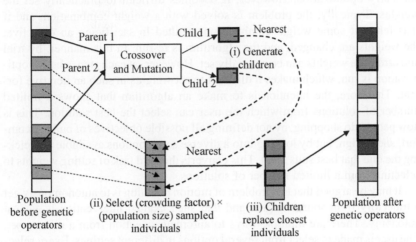

FIG. 16.1 *Crowding strategy.* The parents are selected to make children. However, the children replace the closest individuals in a sampled population.

16.2.3 Other techniques

Other techniques also enable diversity preservation. The simplest mechanism of diversity control is a restricted crossover, wherein only two individuals that lie within a distance threshold can crossover. The selection operator is accordingly modified. Here, the threshold of distance becomes important, and the parameter may be varied as the evolution grows and the population converges. Another technique introduces the fitness function in decision-making whether the individuals can mate or not. Here, the fitness and distance ratio between the individuals is used as a factor to assess the mating probability between the individuals.

16.3 Multiobjective optimization

Practical problems do not have a single objective function to be optimized; however, they have multiple objective functions. Often the objective functions are conflicting in nature, wherein optimizing one objective function generates a solution that is poor for the other objective value. Like in the problem of motion planning, the path length and path clearance are conflicting objectives. The intention is to have the smallest path length and the maximum clearance; however, minimizing path length unintentionally also minimizes the clearance, and maximizing clearance unintentionally increases the path length.

The problems with multiple objectives were so far dealt with by a weighted average method, which seemed a reasonable technique to establish tradeoffs and carry optimization. However, it becomes difficult to practically set the weights. Typically, the problem is solved with a weight combination, and if it is felt that some weight combination resulted in sacrificing an objective, the weights are changed, and the algorithm is re-run. In this manner, by trial and error, the weights can be manually set. However, at each step, a heavy optimization is run, which makes it difficult to set weights in such an iterative format. Therefore, the intention is to make an algorithm that shows a limited number of solutions from which the user can select the best solution. This is how people do shopping, by not defining all possible objectives of budget, comfort, and design, but by looking into a diverse set of options available and picking the one that best suits them. This converts the hard task of setting weights to selecting from a limited number of solutions.

It may be argued that the problem of motion planning is to autonomously get a trajectory from source to goal and that such manual intervention is against autonomy. There are multiple ways to automate selection from a diverse set. The user is made to select from several options in different settings. Every selection suggests the preference of one solution in contrast to the other by the person. This populates a preference database. The weights are learned from many such entries in the database, such that the trained model imitates the user selection preferences. More generally, the field of *multicriterion decision-making* deals with a selection among a set of choices based on multiple criteria by setting utilities against each criterion. Since a lot of user data can be made available, using the same data, the modelling can be done as a training procedure. Later, the trained utility function can be used to automatically select the best trajectory.

Alternatively, by looking at the limited solutions, it can be ascertained which solutions have the best tradeoff by using automated means, which is a good strategy if no such user data of choice is available. A popular mechanism here is *knee selection*, which selects the solution that is the most central in terms of all objectives and whose sensitivity in terms of all objectives combined is small.

16.3.1 Pareto front

A multiobjective problem has multiple objective functions. Suppose the objective functions are $f_1(x), f_2(x), f_3(x), \dots f_m(x)$, which must all be minimized. m is the number of objective functions. Unlike previous approaches, now the problem is being handled as a minimization problem and hence the term objective function instead of fitness function. There will be no single solution to the problem, as one solution cannot simultaneously have the minima for all objectives, as per the notion of contradictory objectives.

A solution x_1 is said to dominate the solution x_2 if x_1 is at least as good as x_2 on all objective functions and x_1 is better than x_2 on at least one objective function. This is represented by $x_1 \prec x_2$, defined as Eq. (16.3):

$$x_1 \prec x_2 \text{ iff } f_i(x_1) \le f_i(x_2) \forall i \wedge \exists_j f_j(x_1) < f_j(x_2) \tag{16.3}$$

The individuals x_1 and x_2 are called *nondominated* if neither x_1 dominates x_2 nor x_2 dominates x_1, defined as Eq. (16.4).

$$x_1 \sim x_2 \text{ iff } x_1 \not\prec x_2 \wedge x_2 \not\prec x_1 \tag{16.4}$$

Let us take an example. Suppose you want to buy a laptop and you are uncertain between two options, A and B. For simplicity, say you only care about the processing speed and price. Option A has a higher processor and a lower price as compared to option B. Now, there is no reason why you buy the second laptop (option B), as it is worse in both objectives. Here, it is said that option A dominates option B. In the other and more practical case, consider that option A has more processing power but is expensive, while option B has the lesser processing power and is cheaper. Now, it is evident to have a problem in the selection since you get something but also lose something. Here, the two laptops are called nondominated, and the decision needs to be made by a human expert.

A solution x is said to be nondominated in the population (X) if it is not dominated by any other individual in the entire population pool. This is given by Eq. (16.5). All laptops in shops are typically nondominated; otherwise, they will not get buyers.

$$\text{Nondominated}(x) \text{ iff } \nexists_{x_i \in X} \, x_i \prec x \tag{16.5}$$

To better visualize the concepts, consider a multiobjective optimization problem given by Eqs (16.6)–(16.8):

$$f_1(x) = (1 - x_1)^2 + (x_2)^2 \tag{16.6}$$

$$f_1(x) = (x_1)^2 + (1 - x_2)^2 \tag{16.7}$$

$$-1 \le x_1 \le 1, \, -1 \le x_2 \le 1 \tag{16.8}$$

Consider a random population given in Table 16.2 and Fig. 16.2A. Now the population cannot be plotted in a single fitness landscape as there are multiple objective functions. So, let us plot two spaces instead. The first space is called the *decision space*, wherein each dimension of the input optimization problem is an axis and an individual in the problem (x) is hence a point in this space. This space does not cover the objective functions, and therefore the objectives are separately plotted in another space called the *objective space*. A point in this space is given by $F(x) = [f_1(x), f_2(x), ...f_m(x)\,]^T$. Every point in the decision space (x) corresponds to some point in the objective space $(F(x))$ and vice versa.

TABLE 16.2 Sample random population for multiobjective optimization.

Individual	Decision space $[x_1\ x_2]$	$f_1(x) =$ $(1 - x_1)^2 + (x_2)^2$	$f_2(x) = (x_1)^2 +$ $(1 - x_2)^2$	Objective space $[f_1\ f_2]$
x_1	[−0.3 0.7]	2.18	0.18	[2.18 0.18]
x_2	[0.5 −0.1]	0.26	1.46	[0.26 1.46]
x_3	[−0.6 −0.3]	2.65	2.05	[2.65 2.05]
x_4	[−0.5 0.9]	3.06	0.26	[3.06 0.26]
x_5	[−0.2 0.8]	2.08	0.08	[2.08 0.08]
x_6	[0.5 0.1]	0.26	1.06	[0.26 1.06]
x_7	[0.9 0.2]	0.05	1.45	[0.05 1.45]
x_8	[0.3 0.7]	0.98	0.18	[0.98 0.18]
x_9	[0.1 −0.3]	0.90	1.70	[0.90 1.70]
x_{10}	[−0.4 −0.7]	2.45	3.05	[2.45 3.05]

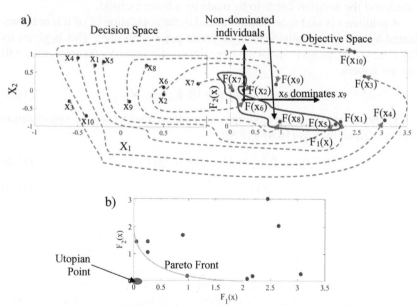

FIG. 16.2 (A) Decision Space and Objective Space for the population in Tables 16.1a and 16.1b. x_6 dominates x_9 since x_9 lies on the top-right space of x_6 (for a minimization problem). x_5, x_6, x_7, and x_8 are not in the top-right corner of any individual and hence constitute nondominated individuals. (B) Pareto Front. Set of individuals such that no two individuals dominate each other. Utopian point shows the optimal values of all objective functions, here (0,0).

Now consider the objective space, with both objectives for the minimization problem. If a solution x_2 lies in the top-right corner of x_1, then by definition, x_1 dominates x_2. If two solutions x_1 and x_2 are such that neither of them is in the top-right corner of the other, then the solutions are said to be nondominated by each other. If a solution x_2 is in the top-right corner of any other individual in the entire population, it is said to be dominated by that solution. All solutions which do not lie in the top left corner of any other solutions are said to be nondominated. Fig. 16.2B shows the dominance of the different solutions and explicitly the set of nondominated solutions.

A set of all nondominated solutions constitutes the *Pareto front* of the problem. The Pareto front is given by Eq. (16.9):

$$\text{Pareto front}(X) = \{x \in X : \nexists_{x_i \in X} \; x_i \prec x\} \tag{16.9}$$

Since the objectives are continuous, the Pareto front will consist of infinite points. The front will always have a convex shape, as shown in Fig. 16.2B. For any other shape, some point will lie on the top right of another solution and will become dominated. A special point is called the *utopian point* (U^*), which lies at the minimum objective value of all objectives, given by Eq. (16.10):

$$U^* = [\min(f_1(x)), \; \min(f_2(x)), \ldots \min(f_m(x))]^T \tag{16.10}$$

The set of all nondominated solutions is good solutions that can be presented to the user for selection. However, the problem is that these are far too many (or infinite) points that a user may not be able to go through, and since a human interface is involved, the number of points needs to be limited. Therefore, sample out a set of *diverse* solutions from the Pareto front, the points being large enough to represent the Pareto front while being small enough for human understanding. The corresponding points in the decision space are taken, converted into the phenotype, and presented for human selection. This is the output of multiobjective optimization.

Consider the Pareto front in Fig. 16.2B. This represents a knee, and the central point of the knee is usually a good solution to the problem if a human is not available to train a utility function. Consider solutions on the X-axis (similarly Y-axis). By going towards the central knee, the objective represented by the X-axis (similarly Y-axis) significantly improves with no major reduction in the objective value represented by the Y-axis (similarly X-axis). The solutions represented at the extremity are thus a poor choice, and the central knee solution is a good candidate.

16.3.2 Goodness of a Pareto front

In a single objective optimization, the fitness function itself acts as a metric that can be used for comparing algorithms. Multiobjective algorithms use different metrics to assess algorithm and solution quality. *Hypervolume* is a common technique that states the volume of space in multiple dimensions covered by all individuals in the objective space, with a better algorithm having a larger hypervolume. The *Generational Distance (GD)* metric takes a complete and dense Pareto front and tries to see how much of it is imitated by the computed Pareto front. The dense Pareto front may be constructed offline by computationally intensive algorithms. It is the distance of every point in the solution in the objective space to the nearest point in the dense Pareto front. The metric is given by Eq. (16.11).

$$GD(X) = \frac{\sum_{i=1}^{|X|} \min_{p_j \in P} d(F(x_i), p_j)}{|P|} \qquad (16.11)$$

Here $|X|$ is the population size, and X is the population. x_i iterates through all individuals of the population, which in the objective space is given by $F(x_i)$. P is the set of individuals that represent the Pareto front, and p_j is an individual from the Pareto front. The metric is shown in Fig. 16.3A.

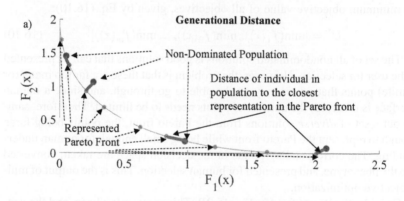

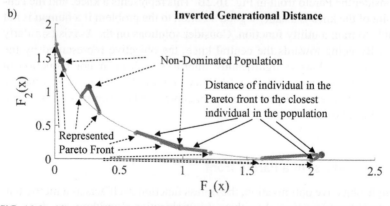

FIG. 16.3 *Metrics used to assess MultiObjective Optimization Algorithm.* (A) Generational Distance, from every individual in the population, distance to closest point in the Pareto front and (B) Inverted Generational Distance, from every point in the Pareto front, the distance to the closest individual in the population.

Unfortunately, the generational distance can get a 0 value if all the individuals are the same at one point on the Pareto front. It does not measure diversity. An inverted metric is called the *Inverted Generational Distance* (*IGD*). Here the distance between all points in the Pareto front is found with the nearest point in the solution set. The metric is given by Eq. (16.12). The metric is also shown in Fig. 16.3B. The outer loop is now on the dense Pareto front.

$$IGD(X) = \frac{\sum_{j=1}^{|P|} \min_{x_i \in X} d(F(x_i), p_j)}{|P|} \tag{16.12}$$

16.3.3 NSGA

The first algorithm to solve the problem is the *Non-Dominated Sorting Genetic Algorithm* (NSGA; Deb et al., 2002). The algorithm runs the GA for the optimization process, including selection, crossover, mutation, elite, and all the specifics remaining the same as the GA. The only difference is that the GA works on a single objective function. More specifically, in rank-based scaling, the objective function sorts all the individuals, after which the selections are done based on the ranks alone. To do sorting, the only program module to be written is a *comparator* that compares two individuals and decides which one is better among the two.

The primary criterion in comparison or sorting is taken as the *Pareto rank*. The nondominated solutions have a rank of 1 and represent the best solutions. Let us delete the Pareto front solutions temporarily to compute the rank. There will be a new set of solutions in the objective space that now make the Pareto front. These are called the rank 2 solutions. Deleting these solutions as well gives a new Pareto front, called rank 3 solutions. This goes on till all solutions have their ranks calculated. This is shown in Fig. 16.4A.

Too many solutions have the same Pareto rank. Therefore, a secondary criterion is added, which is *objective space diversity*. This also meets the requirements of the multiobjective optimization algorithm to generate diverse solutions (in the objective space) to represent the Pareto front. The notion of diversity is directly taken from the previous discussions, with the difference that earlier diversity was strictly in the decision space. Now the concepts are the same, however, applied in the objective space. More concretely, the diversity in decision space is still important for the same reasons as discussed earlier, and a better mechanism is to take diversity measures in both decision space and objective space.

A typical measure of diversity is the *crowding distance* described earlier, which acts as the secondary criterion. The problem with the crowding distance metric is that it calculates the distance between all individuals and passes it through the sharing function, giving a high complexity of $O(mn^2)$ where n is

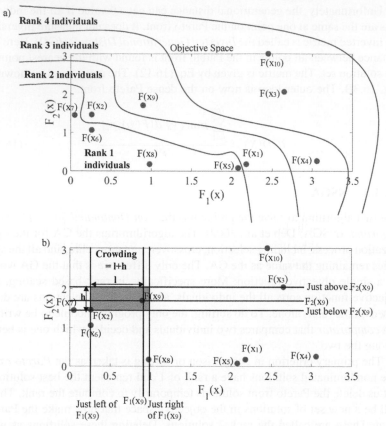

FIG. 16.4 *Multiobjective Optimization using NSGA (A) Pareto ranking.* A smaller rank indicates a better individual. First, nondominated individuals are ranked 1. After removing rank 1 individuals, the nondominated individuals in the resultant pool are ranked 2, and so on. (B) *Crowding Distance in Objective Space for x_9*. Distance between the left and right individual in each axis, summed for all axes. The Pareto rank (lower is better) is the primary criterion and crowding distance (higher is better) is the secondary criterion that acts as a fitness function for multiobjective evolution.

the number of individuals and m is the number of objectives. A faster variant introduced in NSGA-II is to calculate the distance independently in all objective axes and to only consider the individual in the specific objective axis at the left and right. In every axis, the distance between the left and right individual in that axis is a measure of the diversity, summated for all axes. Consider a problem with three objectives and visualize the population as a set of points in the 3D objective space with axes called as X, Y, and Z. To calculate the diversity of an individual x, first project all points in the X-axis (look at the X-axis values). Look at the left and right neighbours of x (which are the individuals with a just-lower and just-higher fitness value corresponding to the X-axis as compared to x). This makes the length of a cuboid. Now project all points in the Y-axis and look at the up and down neighbours of x in the Y-axis. This gives the width of the cuboid. Similarly, the projection on the Z-axis will give the height of the cuboid. The sum of the sides of the cube (hypercube in a general case) is the diversity measure, shown in Fig. 16.4B.

To simultaneous compute the diversity of all individuals, sort the individuals in every objective axis, one axis at a time. After every sort, store the side length of the hypercube (distance between the left and right neighbours for the axis) corresponding to every individual. For one axis the sorting takes $O(n \log n)$ time, while for m axes, the time required is $O(mn \log n)$. Given the side-lengths of the hypercube, the sum of sides (diversity) can be calculated in $O(nm)$ time for n individuals and m objectives. The total complexity is thus $O(mn \log n)$.

Another major betterment included in NSGA-II is the calculation algorithm for the Pareto ranks. The naïve approach calculates dominance between two individuals in $O(m)$ time, where m is the number of objectives. Calculating the dominance of one individual takes $O(nm)$ time by iterating through all individuals. This means it takes $O(n^2 m)$ time to calculate the Pareto front, and the process is repeated for all Pareto ranks, which could go up to n, giving a complexity of $O(n^3 m)$. A trick is to store the number of (and list of) individuals that every individual dominates and is dominated by. The Pareto front is a list of candidates who are not dominated by any one available from the metric. We start by giving a rank 1 to all individuals (say x) whose metric value of the number of individuals that dominate x is zero. Thereafter, these individuals are temporarily deleted. Every time an individual (say x) is deleted from the Pareto front, the list of individuals that it (x) dominated (say $\{y\}$) is retrieved that was explicitly stored. The metric of the number of individuals that dominate y is reduced by 1 since one of the individuals that dominated y (specifically x) is now deleted. Again, all individuals (say x) whose metric value is zero (for the metric number of individuals that dominate x) are given rank 2 and the process goes on.

The overall algorithm compares two individuals first on their rank, and for all individuals with equal rank, the diversity measure is used as a secondary criterion. The comparator is given by Eq. (16.13)

$$x_1 < x_2 \text{ iff } \text{rank}(x_1)$$
$$< \text{rank}(x_2) \vee (\text{rank}(x_1) = \text{rank}(x_2) \wedge \text{diversity}(x_1) < \text{diversity}(x_2))$$

$$(16.13)$$

16.3.4   MOEA/D

To add flavour to the discussions, a reasonably different approach is discussed, called the *Multi-Objective Evolutionary Algorithm by Decomposition* (MOEA/D; Zhang and Li, 2007). A classic way to solve the multiobjective problem is to take a weighted addition of the objective values. In the rest of the text, $z = F(x)$ will denote a variable in the objective space, while x will denote a variable in the decision space. The objective function is given by Eq. (16.14), and the solution by Eq. (16.15).

$$g_w(z) = w_1 z_1 + w_2 z_2 + \ldots + w_m z_m = w \cdot z \qquad (16.14)$$

$$z_w^* = \text{argmin}_z g_w(z) = \text{argmin}_z w \cdot z \qquad (16.15)$$

Here $g_w(z)$ represents the cumulative objective function being optimized, which is parametrized on the weight w, and hence the subscript. The typicality of the notation is that the input parameter $z = F(x)$ is not directly controllable but can be controlled only by altering the vector x, which is a point in the decision space. $z = F(x) = [f_i(x), f_2(x) \ldots f_m(x)]^T$ is a point in the objective space (with axis as the objective functions $f_1(x)$, $f_2(x)$, $\ldots$ $f_m(x)$) and $w = [w_1, w_2, w_3 \ldots w_m]^T$ is the weight vector. z_w^* is the optimal solution that minimizes the function $g_w(z)$.

Since optimization is translation invariant. Therefore, the function is translated to the utopian point ($U^* = [u_1^*, u_2^*, u_3^* \ldots u_m^*]^T$), giving the modified function as Eqs (16.16) and (16.17). This modelling is commonly referred to as the Tchebycheff Approach.

$$
\begin{aligned}
z_w^* &= \text{argmin}_z g_w'(z) \\
&= \text{argmin}_z w_1 (z_1 - u_1^*) + w_2 (z_2 - u_2^*) + \cdots w_m (z_m - u_m^*) \qquad (16.16) \\
&= \text{argmin}_{F(x)} w \cdot (z - U^*)
\end{aligned}
$$

$$u_i^* = \min_x f_i(x) \qquad (16.17)$$

The equation $w_1(z_1 - u_1^*) + w_2(z_2 - u_2^*) + \ldots w_m(z_m - u_m^*) = 0$ represents the equation of a line passing through the utopian point $U^* = (u_1^*, u_2^*, \ldots u_m^*)$ in the objective space. The value of $w_1(z_1 - u_1^*) + w_2(z_2 - u_2^*) + \ldots w_m(z_m - u_m^*)$ represents the projection made by the point at this line with the origin shifted to U^*. The other way to look at the equation is that it represents the dot product between the vectors w and $F'^{(x)} = [z_1 - u_1^*, z_2 - u_2^*, \ldots z_m - u_m^*]^T = z - U^*$, where w is normalized, in which case the dot product is the projection of $z - U^*$ on w. Since w is fixed, let us alter x, which in turn alters $z = F(x)$ to yield the optimal solution of the objective function, $z_w^* = F(x_w^*)$. The concept is shown in Fig. 16.5. Now $z_w^* = [z_1^*, z_2^* \ldots z_m^*]^T$ represents one point in the Pareto front; since if an individual $z = [z_1, z_2 \ldots z_m]^T$ in the objective space was possible and dominated z_w^*, then by definition of dominances, Eq. (16.18) must be true.

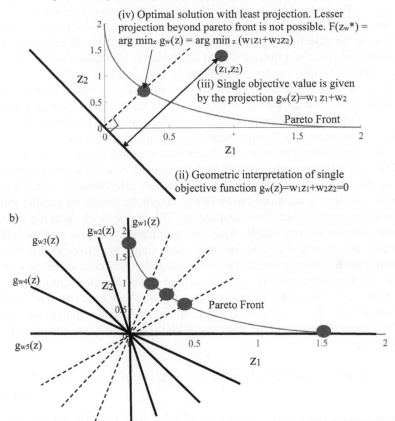

a)
(i) Multi-objective optimization problem reduced to a single objective optimization problem by using weighted addition $g_w(z) = w_1z_1 + w_2z_2$

(iv) Optimal solution with least projection. Lesser projection beyond pareto front is not possible. $F(z_w^*) =$ arg $\min_z g_w(z) =$ arg $\min_z (w_1z_1 + w_2z_2)$

(z_1, z_2)

(iii) Single objective value is given by the projection $g_w(z) = w_1 z_1 + w_2$

Pareto Front

(ii) Geometric interpretation of single objective function $g_w(z) = w_1z_1 + w_2z_2 = 0$

b)
$g_{w2}(z)$ $g_{w1}(z)$

$g_{w3}(z)$

$g_{w4}(z)$

Pareto Front

$g_{w5}(z)$

FIG. 16.5 *Multiobjective Evolutionary Algorithm by Decomposition (MOEA/D).* (A) Finding one point on the Pareto front by using one weighted addition of objective functions. (B) Taking numerous such positive normalized weights to get the Pareto front (normals shown by dashed lines).

$$z_i \leq z_i^* \forall i$$
$$\Rightarrow w_1z_1 + w_2z_2 + \cdots + w_mz_m \leq w_1z_1^* + w_2z_2^* + \cdots + w_mz_m^*$$
$$\Rightarrow w_1(z_1 - u_1^*) + w_2(z_2 - u_2^*) + \cdots w_1(z_m - u_m^*)$$
$$\leq w_1(z_1^* - u_1^*) + w_2(z_2^* - u_2^*) + \cdots w_m(z_m^* - u_m^*) \tag{16.18}$$
$$\Rightarrow w \cdot (z - U^*) \leq w \cdot (z^* - U^*)$$
$$\Rightarrow g_w(z) \leq g_w(z^*)$$

This means z would yield a better value of the objective $g_w(z)$, which nullifies the assumption that z_w^* was the optimal solution for $g_w(z)$.

This is the basic intuition behind the algorithm. By continuously changing the weight vector w, we can get more points in the Pareto front, up to the point wherein the Pareto front has a good number of well-represented weight vectors. The other requirement of the problem is that the solutions should have a large diversity in the objective space. A diversity of solutions in the objective space ultimately translates to a diverse weight set, which is otherwise random. Generating diverse weight vectors is a reasonably easy problem to solve.

The name of the algorithm is derived from the fact that the problem is solved by finding n diverse weight vectors $W = \{w^{(i)}\}$. Each vector w_i represents a single objective optimization problem, the solution of which gives one point in the Pareto front. Thus, the multiobjective optimization problem is *decomposed* into n single-objective optimization problems, which represent reasonably easier problems.

The algorithm however uses more computationally efficient techniques than solving the n single-objective optimization problems independently, which is to state the fact that the solutions can be shared among the otherwise parallel and independent optimizations for coordination. This happens by checking if an optimal solution for one single-objective optimization problem (one weight vector) is also potentially good for some other single-objective optimization problem (another diverse weight vector) or the individual creation in the evolution of one problem turns out to be better for some other problem instead.

Let $W = \{w^{(i)}\}$ be a set of diverse weight vectors, where each vector $w^{(i)}$ is of the form $w^{(i)} = [w_1^{(i)}, w_2^{(i)}, w_3^{(i)} \ldots w_m^{(i)}]^T$. Each weight captures the optimal solution for the weighted addition of the objective, with the objective function given by Eq. (16.19):

$$g_{w_i}(z) = w_1^{(i)}\left(z - u_1^*\right) + w_2^{(i)}\left(z - u_2^*\right) + \ldots + w_m^{(i)}\left(z - u_m^*\right) = w^{(i)} \cdot (z - U^*)$$

(16.19)

Each weight $(w^{(i)})$ is associated with an individual (x_i), which is the best individual it has found so far for the minimization of $g_{w_i}(z)$ (or its translated variant). The population pool is the set of all such individuals $X = \{x_i\}$.

Even though the algorithm runs in a typical GA style, there is no single objective function, but every individual has its objective function. So, the selection is random. The genetic operators are applied on two parents, typically doing a crossover followed by mutation to produce a child. The child can replace any individual in the population if the child has better fitness. Remember again that this problem represents a unique evolution, where a separate fitness function is associated with every individual, and the ith individual has the fitness function $g_{w_i}(z)$ where $z = F(x)$. The utopian point $U^* = [u_j^*]$ is not initially known and is also updated continuously.

The solution however has a major problem of complexity. Every new individual generated does a comparison with all other individuals for possible replacements, giving a complexity of $O(n^2)$ for n such individuals constituting a generation. The aim is to reduce this complexity. The weights share a spatial relationship or there is a distance function $d(w^{(i)}, w^{(j)})$ that shows the similarity between two weights $w^{(i)}$ and $w^{(j)}$. The notion is similar weights have similar fitness and vice versa, or if $w^{(i)}$ is similar to $w^{(j)}$, then the solution of $g_{w_i}(z)$ represented by x_i is similar to the solution of $g_{w_j}(z)$ represented by x_j. So, to make a computationally lesser intensive solution, say two individuals nearby to x_i mate to produce a new individual (y_i), which by genetic operator functioning will also be in the vicinity of x_i. Previously, y_i was a candidate for replacing all individuals. However, now it is known that only individuals near x_i are likely to be affected and are checked for replacement.

The generation of initial, random diverse weights is a naïve problem and assumed to be solved. Every weight calls the nearest k weights as its neighbours. The neighbourhood function is given by Eq. (16.20):

$$\delta(i) = \left\{ j : \text{rank}_{d(w^{(i)}, w^{(j)})} w^{(j)} \leq k, j \neq i \right\} \qquad (16.20)$$

The equation should be read as the neighbourhood of i denoted by $\delta(i)$ is the set of all j ($j \neq i$) that have a rank less than or equal to k using the ranking function as the distance between weights i and j, or $d(w^{(i)}, w^{(j)})$.

Let us attempt generating a candidate individual near $w^{(i)}$. This can be done by sampling two parents from the neighbourhood of i, that is $j \sim U(\delta(i))$ and $k \sim U(\delta(i))$, $k \neq j$. The parents x_j and x_k now undergo crossover and mutation and generate a child y_i (or two children, taken one after the other), which is expected to be close to x_i. In the GA the individuals that undergo crossover and mutation are selected based on fitness or rank. However, since the problem is decomposed with a separate weight vector $(w^{(i)})$, making an objective function (g_{w_i}) for every individual (x_i), a new child (y_i) is generated near every individual (x_i) by a crossover and mutation of nearby parents.

Now the child y_i can potentially replace all other individuals by being better as per their objective measure. However, for computational reasons, since y_i is expected to be closer to x_i, it is expected that y_i is likely to be better only to the neighbours of i. All individuals near x_i, given by the neighbourhood function, are checked for a possible replacement. If better, y_i replaces the current representative. The individual y_i is first projected to objective space $z = F(y_i)$, which is used to calculate the objective function corresponding to weight $w^{(j)}$, which is $g_{w_i}(z) = w^{(j)} \cdot (z - U^*)$. For all neighbours, $j \in \delta(i)$, replacement of the current representative is given by Eq. (16.21), where $\leftarrow$ denotes that the new value will overwrite the old value.

$$x_j \leftarrow \begin{cases} y_i & g_{w_j}(F(y_i)) < g_{w_j}(F(x_j)) \\ x_j & \text{otherwise} \end{cases} \qquad (16.21)$$

Numerous metrics used for the comparisons of the algorithms require the complete Pareto front, and hence the same is constructed by pooling in all solutions while only keeping the nondominated ones, commonly called as the *external population*. This population is only for the final metric assessment and does not contribute to the normal working of the algorithm. The pseudo-code is shown as Algorithm 16.1. In the algorithm, the notation of the utopian point (u_i^*) is replaced by u_i', since the optimal value is not known and is iteratively improved instead.

Algorithm 16.1 MOEA/D

Input: objective functions $F(x) = [f_1(x), ...f_m(x)]^T$, population size n
Output: non-dominated solutions $x_1, x_2, ..x_n$, pareto front (E)
1   $E \leftarrow \emptyset$;   ▷ external population, entire Pareto front
2   make a diverse set of weights W ;
3   **for** *i from 1 to population size* **do**
4 | ▷ weight neighbourhood function, nearest k ;
      | $\delta(i) = \{j : rank_{d(w_i, w_j)}(w_j) \le k, j \ne i\}$

5   $X \leftarrow \emptyset$;   ▷ population
6   **for** *i from 1 to population size* **do**
7 | add random individual x_i to X corresponding to weight $w^{(i)}$

8   **for** *i from 1 to number of objective functions* **do**
9 | $u'_i = min_{x \in X} f_i(x)$;   ▷ guessed Utopian point

10   **while** *stopping criterion* **do**
11 | **for** *i from 1 to population size* **do**
      | ▷ x_i updates by random neighbours indexed p_1 and
      | p_2 (corresponding parents $parent_1$ and $parent_2$) ;
12 | $p_1 \sim U(\delta(i))$, $parent_1 = x_{p_1}$;
13 | $p_2 \sim U(\delta(i))$, $parent_2 = x_{p_2}$;
14 | $child \leftarrow Mutate(Crossover(parent_1, parent_2))$;
15 | **for** *j from 1 to number of objective functions* **do**
16 | | $u'_j \leftarrow min(u'_j, f_j(child))$;   ▷ update Utopian point
      | ▷ check if $child$ is a better solution for
      | neighbouring weights only ;
17 | **for** $j \in \delta(i)$ **do**
      | | ▷ $F(x) = [f_1(x), ...f_l(x), ...f_m(x)]$ is a point in
      | | objective space. x_j stores best solution of
      | | $g_{w_j}(F(x)) = \sum_{l=1}^{m} w_l^{(j)}.(f_l(x) - u'_l)$, update if better ;
18 | | **if** $g_{w_j}(F(child)) < g_{w_j}(F(x_j))$ **then** $x_j \leftarrow child$;

      | ▷ delete all $e \in E$ dominated by any $x \in X$;
19 | **for** $e \in E$ **do**
20 | **if** $\exists_{x \in X} x \prec e$ **then** $E \leftarrow E - \{e\}$;
      | ▷ add all $x \in X$ not dominated by any $e \in E$;
21 | **for** $x \in X$ **do**
22 | **if** $\forall_{e \in E} e \nprec X$ **then** $E \leftarrow E \cup \{x\}$;

23 return X,E

Line 1 initializes the external population. Over time, individuals keep getting generated from the evolutionary process. Lines 19–20 watch if all the individuals in the external population are still nondominated. If any individual e gets dominated by a new individual, it is deleted from the external population. Similarly, lines 21–22 see if one of the new individuals is nondominated to all individuals in the external population, which is added to the external population. Line 2 makes a one-time generation of diverse weights (W), which are unchanged. Lines 3–4 compute the neighbours of each weight (δ). Lines 5–7 start with a random population (X). In principle, every individual x_i corresponds to a weight $w^{(i)}$. Lines 8–9 make the best guess for the optimal value of the utopian point. Lines 15–16 maintain the best estimates for the utopian point for all dimensions (input objective functions) as new individuals are generated.

Lines 10–18 make the main logic of the algorithm. A potential child is produced corresponding to every weight indexed at i. As per principles, every weight $w^{(i)}$ corresponds to an individual x_i, denoting the best individual for the objective function made by that weight. Neighbouring weights have neighbouring individuals. Hence, parents are sampled from the neighbouring weights of i (lines 12–13) by first sampling the indexes in the weight neighbourhood function (δ) and then getting the individual at the same index. The parents produce a child (line 14). Lines 17–18 judge if the new child, for any weighted fitness function, is better than the previously best-known individual. If found better, it is regarded as the best individual for the specific weight and stored in the population.

The NSGA algorithm discussed previously has issues when the number of objectives becomes large, wherein even the initial population generates solutions that are nondominated to each other, and the optimization process becomes very computationally heavy. The requirements of the number of solutions to judiciously represent the Pareto front is exponential to the number of objectives. Hence, the third version of NSGA (NSGA-III) uses some of the concepts discussed here to solve these problems (Deb and Jain, 2014; Jain and Deb, 2014).

16.4   Path planning using a fixed size individual

The focus now shifts to utilizing the principles of evolutionary computation for the specific problem of robot motion planning, or finding a trajectory $\tau:[0.1] \to C^{\text{free}}$ from the source (or $\tau(0)$) to the goal (or $\tau(1)$), such that all intermediate points are collision-free, or $\tau(s) \in C^{\text{free}} \; \forall \, 0 \le s \le 1$. This section models the problem in such a manner that any optimization toolbox from any library can be used to solve the problem.

The first major challenge is to represent a continuous trajectory of the robot from this phenotypic representation to a genotype, wherein it represents a string of real numbers. Consider a trajectory that is broken down into straight line segments, and every segment is connected. The motion from the source to the goal is said to have been decomposed into several via *points* that encode the trajectory. Imagine going from a source to a goal with a square obstacle in between. The optimal path is going from the source to a corner of the obstacle and thereafter to the goal, where the corner of the obstacle is a via point. In high-dimensional and nongeometric spaces, such via points are not obvious and may be larger in number. The optimization technique is supposed to compute the optimal sequence of via points. The trajectory in the phenotype and genotype is thus shown in Fig. 16.6. It may be nice to note that the source and goal points are fixed and not a part of the genotype. After optimization, it is expected that every via point will be optimally located near a turn that the robot is supposed to make. The genotype is given by Eq. (16.22).

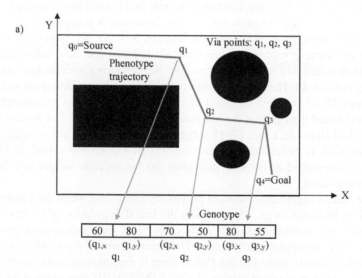

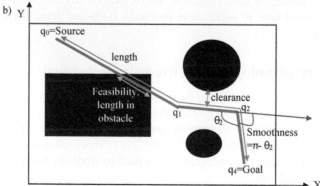

FIG. 16.6 *(A) Individual Representation and (B) Fitness Function.* The phenotype is the trajectory as straight lines joining via points. The genotype is all via points. The fitness function consists of length, clearance (minimum/average over trajectory), smoothness (angle, maximum/average over trajectory), complexity (number of via points), and feasibility (length of the trajectory in obstacle).

$$q = [q_1, q_2, q_3 \ldots q_m] \tag{16.22}$$

Here q_i is a configuration with the dimensionality as that of the configuration space, and m is the number of via points. The dimensionality of the genetic search space is thus the number of dimensions of the configuration space, times the number of via points. The lower and upper bound as needed for the library implementation is typically available from the knowledge of the robot's configuration. The phenotype represents straight line segments between every pair of points in the genotype, given by Eq. (16.23).

$$\tau(q) = [(1 - \lambda)q_i + \lambda q_{i+1}, 0 \leq \lambda \leq 1, 0 \leq i \leq m] \tag{16.23}$$

Here, q_0 is the source configuration, and q_{m+1} is the goal configuration. Both of these are not present in the genotype. The trajectory is made by sequentially appending every segment after the other. $(1 - \lambda)q_i + \lambda q_{i+1}$ represents the equation of a straight line between q_i and q_{i+1}.

The other important input to be given in any library implementation of optimization is the fitness function. For simplicity, the fitness function is taken as the path length. However, a constraint in the problem is the path feasibility, or that the path must never collide with an obstacle. So far, the optimization was assumed to be constrained free. There are numerous techniques for handling constraints. The technique used here is called *penalization*, wherein the fitness function of the problem is modified, and a term is added in proportion to the constraint violation or the magnitude of the trajectory inside obstacles.

This requires the collision-checking of the line segment. The collision checking used is called incremental collision-checking, wherein a traversal is made in the entire line-segment iteratively with some resolution. The collision-checking of a configuration is done with the samples taken at this resolution. Each collision-checking corresponds to a query to the collision-checking algorithms, which perform intersection tests. If any configuration is found to be collision-prone, the distance between it and the next segment is assumed to be collision-prone. Consider the line segment (q_i, q_{i+1}), with a collision checking resolution Δ (distance between consecutive points checked for collision). The number of points to be checked for collision is thus $d(q_i, q_{i+1})/\Delta$, where d is the distance function. The points checked for collision are thus given by Eq. (16.24):

$$Q = \{q_{i,j}\}$$
$$= \left\{ (1 - \lambda_j)q_i + \lambda_j q_{i+1} \mid \lambda_j = \frac{j\Delta}{d(q_i, q_{i+1})}, 0 \leq j \leq \frac{d(q_i, q_{i+1})}{\Delta}, 0 \leq i \leq m \right\} \tag{16.24}$$

Here, i iterates through via points and j through points to be checked for collision. Care must be given in the implementation that λ_j is always between 0 and

1 for validity and the extreme cases $\lambda_j = 0$ and $\lambda_j = 1$ are always checked. The fitness function is thus given by Eqs (16.25)–(16.27).

$$\text{fit}(q) = \text{length}(q) + \alpha\,\text{penalty}(q) \tag{16.25}$$

$$\text{length}(q) = \sum_{i=0}^{m} d(q_i, q_{i+1}) \tag{16.26}$$

$$\text{penalty}(q) = \sum_{i=0}^{m} \sum_j \Delta \cdot I\left(q_{i,j} \notin C^{\text{free}}\right) \tag{16.27}$$

$\alpha \gg 1$ is the penalty constant, q_0 is the source, and q_{m+1} is the goal. $I()$ is the indicator function that gives 1 if the condition is true and 0 if the condition is false. q_{ij} is a point in the path segment from q_i to q_{i+1}. Fitness function creates evolutionary pressures for individuals to get better along with generations. The penalty added encourages individuals to move via points such that the segment within the obstacles keeps reducing, up to a point where the trajectory becomes feasible. Similarly, for a feasible trajectory, there are evolutionary pressures that aim to move the via points closer to each other to continuously minimize the path length, up to the point where a reduction in path length makes the trajectory infeasible.

The initial population consists of random individuals. Each individual is a concatenation of free samples ones after the other to make the genotype. It is possible to use good sampling techniques like the one used in Probabilistic Roadmaps for the generation of random individuals. Real Coded Genetic Algorithm with default implementations of scaling, selection, crossover, mutation, and elite is enough to optimize the population. Post optimization, the best individual is taken and given to the controller for navigation.

16.5 Path planning using a variable sized individual

A major problem with the approach discussed was that the number of via points, m, was fixed by a human and denoted the maximum number of turns allowed for the robot, with every via point pointing to a robot's turn. This is difficult to do as some problems are simple, wherein even a single turn is enough; however, some other problems require the robot to turn around several obstacles. Thus, this section makes the number of via points as variable. A problem now is that the individual size of the genotype is the number of via points times the dimensionality of the configuration space, which is now a variable number. Hence, the genetic operators need to be modified to account for the problem of a variable size individual (Kala et al., 2009a,b, 2012). A glimpse of the population pool is given in Fig. 16.7.

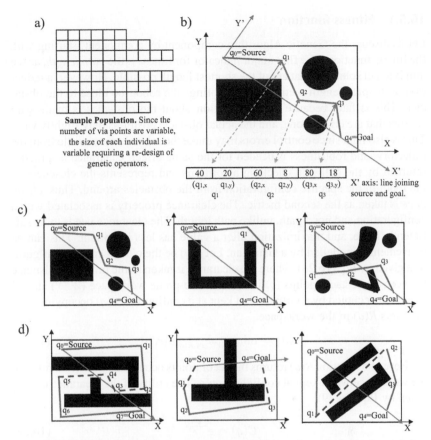

FIG. 16.7 Heuristic Individual Representation. (A) A sample variable-sized population. (B) Individual Representation. Assuming the robot moves roughly ahead, $X'Y'$ axis (line joining source and goal) is used for representation, and points are sorted as per X' axis values. (C) Examples where the robot path has points sorted in X' axis. (D) Examples where robot path has places (shown by the *dotted line*) where points are not sorted by X' axis values.

The typicality of motion planning is that in most cases, the humans walk such that the projection of every step towards the goal is positive, irrespective of whether the obstacle framework is convex or concave. This is seen in everyday life, wherein it is a rarity to take a step 'back' on an optimal route. A few examples are given in Fig. 16.7. If this notion is followed, the via points can be conveniently represented by a transformed coordinate axis system, such that the X-axis is the longitudinal axis joining the source and goal and the Y-axis is perpendicular to the X-axis. The genotype in such a system always has points that are sorted by their X values. The coordinate axis system is shown in Fig. 16.7. This forms a dominant heuristic that can be used to accelerate the search. However, care must be given that the heuristic may not always hold.

16.5.1 Fitness function

Let us discuss the Genetic Algorithm component by component, starting with the fitness function. The first major indicator for fitness is the *path length,* as the aim is to get solutions that are of the shortest length possible. However, a reduction in the path length metric typically brings the robot too close to the obstacles. This is fatal because the information about the obstacles is made with sensors that are error-prone, and the actual obstacle positions may slightly vary. This is also fatal since control errors may make the robot deviate a little from the trajectory, and robustness is needed for the same effect. Clearance is a metric returned by the collision-checking algorithms and represents the closeness of the robot in the specific configuration from the obstacles around. Thus, *clearance* is taken as the second metric. The clearance property is associated with a configuration and not a path, unlike path length. The clearance needs to be sufficiently large, and maximization after a value has less important. This can be conveniently modelled by a Gaussian. Let $c(q)$ be the clearance at a configuration given by Eq. (16.28), where the clearance is taken as the minimum distance between all obstacle points (o) in the obstacle-prone workspace (W^{obs}) and all points (r) occupied by a virtual robot kept at a configuration (q) occupying a set of points $R(q)$ in the workspace.

$$c(q) = \min_{o \in W^{obs}, r \in R(q)} d(r, o) \qquad (16.28)$$

Here, the function $R(q)$ returns the set of points occupied by a virtual robot in the workspace when kept at configuration q. The utility of the clearance $C(q)$ is given by Eq. (16.29).

$$C(q) = e^{-\frac{c^2(q)}{\sigma^2}} \qquad (16.29)$$

The clearance of a path may be defined as the *minimum clearance* out of all configurations in the path or as the *average clearance* of the path. Consider the trajectory as straight-line segments $[q_i, q_{i+1}]$, where every segment is further divided into points $[q_{i,j}]$ as per some resolution where collision checking takes place. The clearances by minimum and average are given by Eqs (16.30) and (16.31).

$$C(\tau) = \min_{i,j} C(q_{i,j}) \qquad (16.30)$$

$$C(\tau(q)) = \frac{\sum_i \sum_j C(q_{i,j})}{\sum_i \sum_j 1} \qquad (16.31)$$

The denominator in Eq. (16.31) is to denote averaging by dividing by the number of terms summated in the numerator. Attempt to maximize the average clearance means that, in general, the trajectory will be away from the obstacles. However, it may be possible that at some segments, the trajectory goes close to the obstacles. This problem can be solved by attempting to maximize the minimum clearance. However, it does not guarantee a general ability to maximize the clearance around general obstacles, and the algorithm only looks at the point where minimum clearance occurs.

A starting point for many programmers doing motion planning is to take grid maps, much like the ones used throughout the book. In such a case, the collision-checking becomes a constant time grid map query; however, clearance is required to be computed. There are a few tricks to use. The first is to make a virtual lidar that emanates a few virtual rays around in all directions in parallel and accepts the reading that first hits the obstacle and terminates all other rays when this happens. In such a case, the clearance calculation is computationally expensive and needs to be done for all points for all queries for every individual and generation. Using a hash table to store the configuration values with the computed value can avoid repeated lookups. Furthermore, the clearance values of nearby configurations follow the triangle law of inequality, and hence, the clearance bounds may as well be estimated for a configuration, for which a clearance to a nearby point is already computed. Alternatively, all obstacles may be inflated by a small size to give clearance values, like in the computation of Voronoi using the Bushfire algorithm. Since only small clearance values matter, the expansion may be done for a few iterations only. This makes one heavy computation expense but calculates clearances for all configurations at once. The techniques have no practical usage to high-dimensional robots or when operating on a map with geometric (polygonal or polyhedral) obstacles, for which a collision-checking library must be used.

Another popular and important metric is *smoothness*. The robots are typically not able to make sharp turns. The control errors become large if the turns are not smooth. The curvature traceable by the robot is inversely proportional to its speed. So only a slow robot can trace sharp turns. The smoothness in this problem is denoted by the angles between the line segments, with a larger angle denoting a smoother curve. It is normalized by the length of the smallest line segment to denote the importance of the angle. The angle is a property of every three consecutive points, and therefore, smoothness is not measured around the source and goal points. This is shown in Fig. 16.6. The smoothness of a via-point is given by Eqs (16.32) and (16.33), while the smoothness of the complete path is the average smoothness of all via points.

$$\text{smooth}(q_i) = \frac{\pi - \theta_i}{\min(d(q_i q_{i-1}), d(q_i q_{i+1}))} = \frac{\pi - \cos^{-1}\left(\frac{(q_{i-1}-q_i)\cdot(q_{i+1}-q_i)}{d(q_{i-1}q_i)\, d(q_{i+1}q_i)}\right)}{\min(d(q_i q_{i-1}), d(q_i q_{i+1}))}$$

$$(16.32)$$

$$\text{smooth}(\tau(q)) = \frac{\sum_i \text{smooth}(q_i)}{\sum_i 1}$$

$$(16.33)$$

Another factor that contributes to fitness is the number of via points. It is possible to characteristically place many via points and get good trajectories. However, every via point adds genes to the population pool that exponentially increasing the search space volume and make the optimization a lot harder. This increases the convergence time, and the optimization is prone to getting stuck in local optima. Therefore, the aim is to produce simple paths with a small number of via points only, unless they significantly improve the other metrics. This acts as a *regularization* factor seen in neural networks. Correspondingly, this factor is called *complexity*, or the number of via points. Too many via points can produce a zigzag like a path that must be avoided. It is better to have a smaller number of turns and simpler paths. Every turn incurs a potential error in the controller to make the turn, which must be limited.

Even though a reduction in the number of points leads to a reduction in path length, the metrics of path length and number of via points can lead to different paths. Imagine a problem having a maze-like structure. Going through the maze incurs a small path length but incurs many turns. On the contrary, one may prefer to take a longer route away from the complete maze, thus, walking away and then, at the maze boundary, incurring just a few turns and a simple path. Both paths are competent as per their metrics.

The last factor contributing to fitness is feasibility constraint, which is the length of the segment inside obstacles and is taken the same as before. The resultant fitness function is given by Eq. (16.34):

$$
\text{fit}(q) = w_{\text{length}} \frac{\text{length}(q)}{d(S,G)} - w_{\text{clearance}} C(q) + w_{\text{smooth}} \frac{\text{smooth}(q)}{\text{smooth}_{\text{max}}}
$$
$$
+ w_{\text{complex}} \frac{\text{complex}(q)}{\text{complex}_{\text{max}}} + \alpha \cdot \text{penalty}(q) \tag{16.34}
$$

Here all factors are normalized to lie on a small scale around 1, which enables easy setting of the parameters. $d(S,G)$ is the straight line distance between source and goal. Clearance is already normalized and hence neglected. Smoothness and complexity can be arbitrarily large and therefore convenient values ($\text{smooth}_{\text{max}}$ and $\text{complex}_{\text{max}}$) are used for normalization. The different weights (w_{length}, $w_{\text{clearance}}$, w_{smooth}, w_{complex}, and α) denote the relative contributions of the factors.

16.5.2 Multiresolution fitness evaluation

The fitness function computational expense is controlled by the resolution parameter Δ. Very high-resolution settings (low Δ) mean taking too many intermediate points, which significantly increases the computational expense (Kala et al., 2011). However, high-resolution settings enable to get accurate values of the path feasibility and clearances. On the other hand, poor values of the resolution settings can mean checking too few points for collisions and clearances. Poor resolution speeds up the algorithm significantly and can result in many collision-prone trajectories being reported as collision-free. The two cases are shown in Fig. 16.8.

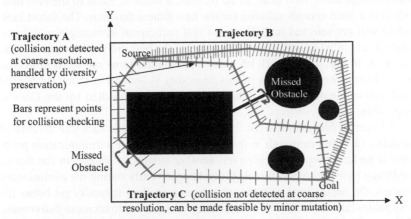

FIG. 16.8 Multi-Resolution Collision Checking. The coarser resolution gives fast results but misses detecting collisions. Trajectory C can be made feasible by a small mutation. Diversity preservation ensures enough different trajectories avoiding convergence near trajectory A, which appears best initially and infeasible later with increase in resolution.

It is known that initially, the population is random and diverse. The individuals vary dramatically in their fitness values. Hence, it is acceptable to have a poor resolution setting to get approximate fitness values. However, at the later iterations, the trajectories are very competent, and the aim is to get the best trajectories among very competent trajectories. Correspondingly, the algorithm later operates at a high-resolution setting.

The first mechanism to make the parameter adaptive is to make it deterministically vary from the initial value (Δ_{max}) to some convergent value (Δ_{min}), which here denotes the highest resolution, till time T, beyond which it is kept as a constant. This is called a *deterministically adaptive* setting of the parameter. For the resolution parameter, this is given by Eq. (16.35).

$$\Delta(t) = \begin{cases} \Delta_{max} - \dfrac{t}{T}(\Delta_{max} - \Delta_{min}) & \text{if } t \leq T \\ \Delta_{min} & \text{if } t > T \end{cases} \qquad (16.35)$$

The factor $\Delta(t)$ was the distance between consecutive points checked, and hence, a smaller distance means a higher resolution. This makes a major change in the evolutionary algorithm, denoting evolution under a constantly changing fitness function since the fitness function may change with time as the resolution is increased.

The other mechanism is to make the parameter *adaptive*, wherein the parameter setting happens by noting the position and fitness values of the individuals in the population pool. Diversity is a convenient measure to judge convergence. Upon noticing that the diversity of the population has dropped, the parameter value can be made to point to high-resolution settings.

The discussions here are generic for any problem with a dynamic fitness function. The first type of dynamics in the fitness function is when there is a small change along with time. In such cases, a mutated value of the best individual is a good enough solution for the new fitness function. The fittest individual will get selected multiple times and participate in multiple mutations, while it is highly likely that a mutated copy is the best individual. Consider trajectory C in Fig. 16.8, which on close examination turns out to be collision-prone. Now a weak mutation of this trajectory is enough to represent a valid (and near-optimal) trajectory. This can be naturally handled by the Genetic Algorithm.

A larger problem is when the fitness landscape changes such that the current modality (a valley or trough in the fitness landscape for a minimization problem) is no longer optimal. However, another trough (or valley) in the fitness landscape now becomes optimal. If this happens early during the evolutionary process, the populations near the other trough (valley) naturally get better fitness values later and create an evolutionary pressure to attract more individuals. However, if this happens after the convergence has already taken place, it is a threat since all individuals near the other trough (what would later turn out to be the optimal region) have already died.

Luckily for this problem, and like other problems, the increase of resolution to the maximum value takes place in some time T, after which the fitness landscape is constant with reference to the resolution parameter. The solution is to avoid convergence till this time, and can be conveniently handled by *diversity preservation*. To understand the effect of losing diversity, consider trajectory A in Fig. 16.8. At poor resolutions, it appears to be on a good trajectory. However, if convergence happens near this trajectory, eventually, when the resolution is increased, it will be discovered that it is infeasible and there are no nearby trajectories that are feasible.

16.5.3 Diversity preservation

Diversity preservation has multiple uses for the specific problem. From the perspective of a conventional evolutionary algorithm, it enables to escape of the local optima and converges to the global optima. From the perspective of a multiresolution setting, it enables to escape the situation wherein a trajectory may later turn out to be infeasible at higher resolution with no feasible trajectory nearby, or a situation wherein an alternate modality (or a homotopic group in the case of trajectories) may turn out to be better under high-resolution settings. From the perspective of dynamic environments also, where obstacles change with time, diversity preservation helps to maintain optimality when a different distant trajectory turns out to be better later.

The diversity preservation can be done by any one of the methods discussed earlier. However, it is important to note that now the trajectory represents a variable-length genotype, and the diversity preservation mechanisms use a distance function between genotypes as an input. No standard distance function exists which can take the distance between two vectors of unequal length. A distance function is hence devised.

Consider the aim is to find the distance between two trajectories $\tau(q^A) = [q_i^A]$ and $\tau(q^B) = [q_i^B]$, both represented as a sequence of configurations. Consider that every segment of $[q_i^A, q_{i+1}^A]$ of the first trajectory is divided into smaller segments (also used for collision checking and clearance), $[q_i^A, q_{i+1}^A] = \{q_{i,j}^A\}$. The distance is found from a smaller length trajectory to a larger length trajectory and is given as the distance between every unit segment point making up the first trajectory and the closest segment in the second trajectory. This is given by Eq. (16.36).

$$
d\left(\tau(q^A), \tau(q^B)\right) = \begin{cases} \dfrac{\displaystyle\sum_i \sum_j \min_{k,l}\left(d\left(q_{i,j}^A, q_{k,l}^B\right)\right)}{\displaystyle\sum_i \sum_j 1} & \text{if length}\left(\tau(q^A)\right) \leq \text{length}(\tau(q^B)) \\[3ex] \dfrac{\displaystyle\sum_k \sum_l \min_{i,j}\left(d\left(q_{i,j}^A, q_{k,l}^B\right)\right)}{\displaystyle\sum_k \sum_l 1} & \text{if length}\left(\tau(q^A)\right) > \text{length}(\tau(q^B)) \end{cases}
$$

$$(16.36)$$

The denominator is used to represent averaging by dividing by the number of additions done in the numerator. The distance is also explained in Fig. 16.9. The algorithm to compute the closest point from the first trajectory $q_{i,j}^A$ to the

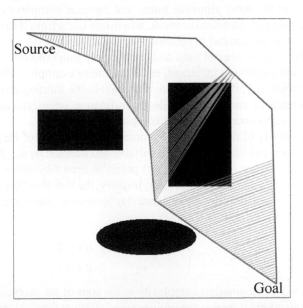

FIG. 16.9 *Calculating the distance between two unequal trajectories.* Diversity preservation needs a distance function between trajectories. From all points in the shortest trajectory, the closest point in the second trajectory is taken and average Euclidian distance between all such pairs is used.

second trajectory $q^B_{k,l}$ does not require a complete search of the second trajectory and can be done by a simultaneous traversal of both trajectories and searching the neighbours only. Correspondingly, the complete trajectory can be parametrized by the ratio of length between 0 and 1, in which case the distance is simply the summation over the corresponding indices, given by Eq. (16.37).

$$d\big(\tau(q^A), \tau(q^B)\big) = \frac{\int_{s=0}^{1} d(\tau(q^A,s), \tau(q^B,s))ds}{\min(\text{length}(\tau(q^A)), \text{length}(\tau(q^B)))} \tag{16.37}$$

Here $\tau(q^A,s)$ denotes the configuration on the trajectory q^A at the parametrization s.

16.5.4 Incremental evolution

The number of via points in the population can be variable, which enables the representation of a large variety of trajectories and gives the ability for the optimization tool to mine the best of them. However, the strategy of setting the number of via points in the initial and the subsequent population pools is another important perspective of the algorithm.

Biological inspiration is attained from the fact that life started from the unicellular organisms in the simplest forms and attained complex forms much later. This is termed as *incremental evolution,* wherein solutions to problems start evolution from the most primitive forms and increase complexity later. This is like the training of the neural network, where one starts from the least number of neurons while the number of neurons is slowly increased; design of the fuzzy systems where the number of rules and membership functions are lowest at the start which are iteratively increased, etc. In all these examples, after attaining sufficient complexity, there is a minor reduction in the training error on a subsequent increase of complexity while the validation error typically starts to increase on a subsequent increase of complexity.

The complexity of this problem is denoted by the number of via points. The maximum number of via points is kept variable and denoted by $o_{\max}(t)$. No individual may ever have more than $o_{\max}(t)$ points at time t by initial population generation or any other genetic operator. Initially, the aim is to keep the parameter as small as possible, which is iteratively increased along with time. The factor is given by Eq. (16.38).

$$o_{\max}(t) = \begin{cases} o_1 + \dfrac{t}{T}(o_2 - o_1) & \text{if } t \leq T \\ o_2 & \text{if } t > T \end{cases} \tag{16.38}$$

Here $o_1 = 1$ is the smallest complexity at the start of the problem, and o_2 is the maximum complexity that the problem may attain at time (generation) T. It can be kept arbitrarily large after a long time. The ability to take complex forms does not mean the individuals in the population will take such forms. Complex individuals will become unfit after attaining more than sufficient complexity, resisting the creation of more complex forms by evolutionary pressures.

The utility of such modelling is important to understand. Consider a reverse mechanism of starting from a high complexity and reducing the complexity with time. Now the parameter of maximum complexity is important and cannot be arbitrarily large. Here, initially, the population will get flooded with complex individuals, representing complex search space, which will be wasteful if the solution is simple. On the contrary, consider that the modelling allows simple to complex individuals to be represented as a pool. Again, this requires the specification of the largest complexity, which is an important parameter, and the complexity distribution of individuals varying from low to high complexity solutions. Taking on too many complex solutions may be wasteful. The incremental nature is better because the maximum complexity need not be specified accurately. It may be argued that the time wasted in the initial generations when the complexity needed to represent solutions to the problem was more than that allowed by the parameter. However, smaller complexity individuals are simpler and take lesser time for fitness evaluations and the time wasted is lesser than the other cases, with the advantages intact.

16.5.5   Genetic operators and evolution

The genetic operators need to be reformulated as per the problem of having a variable-length genotype (Xiao et al., 1997; Kala et al., 2011). All operators are summarized in Fig. 16.10. The first major operator is *crossover*. The smaller parent is made of the size of the larger parent by replication of some of its genes randomly, producing individuals of the same length, in which a scattered crossover technique can be carried out. In the children, if the same gene is found twice, one of them is deleted.

The other parameter is a *mutation*. Now there are two types of mutations possible. The first mutation is called the *parametric mutation* that moves the via points around in the configuration space by adding a small Gaussian noise. This most closely represents the generic Gaussian mutation as discussed previously. Since Gaussian mutation is used, with a small probability, the points are also likely to be moved significantly, mostly the deviation of points being small. By such deviations, the via points can slowly get to their optimal values by constant small motions. Also, an infeasible trajectory can be converted into a feasible trajectory by this operation because the fitness value is proportional to the magnitude of infeasibility, which by mutations can be slowly reduced.

The other form of mutation is called a *structural mutation*. This mutation technique changes the structure of the trajectory. The first variant is called the *shorter operator*. This operator deletes a via point to shorten the trajectory. By the triangle law of distance, the deletion of via points will certainly shorten the trajectory, which may not always be likely considering the effects of feasibility, and other metrics.

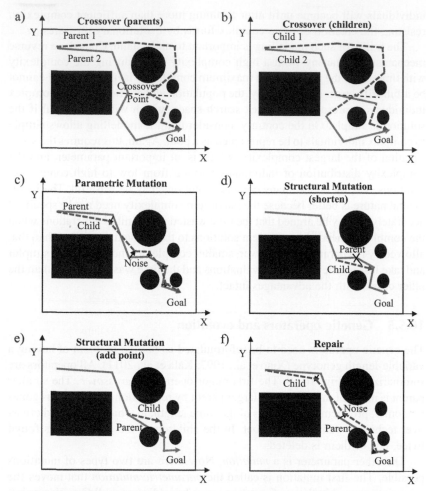

FIG. 16.10 Different genetic operators. (A and B) Crossover, segments of trajectories are exchanged. Adjustment is made to account for individuals of different lengths. (C) Parametric mutation, apply a Gaussian mutation to all via points. (D) Shorten by deletion of a random point. (E) Add point, add a new point nearby two selected points. (F) Repair, add noise to an infeasible trajectory to make it feasible.

The opposite operator is called the *add point*, which instead selects a pair of neighbouring via points and adds a new via point somewhere in-between the two. Keeping the via point in-between the neighbouring points is a heuristic specific to the problem. It may not strictly be in-between the hypercube (box) shown in the figure however may occasionally lie somewhat outside the hypercube, keeping geometrically close to the neighbouring points.

A *repair* operator is only applied to individuals which are infeasible. The operator, in a memetic style, applies multiple iterations of mutations to get a feasible individual. An individual may become infeasible on an increase of resolution, which may be made feasible by a few small mutations alone. Rather than waiting for the evolutionary process to generate feasible trajectories using evolutionary pressures, this operator quickly repairs all such trajectories.

It is possible to inject heuristics in the application of the different genetic operators instead of doing a naïve implementation. Consider the repair operator. The trajectory represented by the genotype is simulated for the calculation of the feasibility. This gives the pair of via points between which there is an infeasibility. These via points are good candidates for the application of noise as mutation to impart feasibility. Furthermore, consider the configuration where a collision happens. The retraction direction is the direction to move the configuration so as to make it feasible and is given by the collision-checking algorithms. It is hence advisable to apply mutation to the neighbouring via points as per the knowledge of the retraction direction. Consider the configuration inside the obstacle that is causing infeasibility. Move the configuration in the retraction direction to make the configuration feasible, and move further apart to add sufficient clearance. This is similar to the obstacle-based sampling in the Probabilistic Roadmap technique. Adding this point in between the neighbouring via points is also likely to make the trajectory feasible. Similarly, if clearance needs to be maximized, the specific configuration of the minimum clearance and the neighbouring via points are known, and so is the direction that will increase the clearance. Move the neighbouring via points in the same direction to increase the clearance value as an implementation of the mutation operator.

The *insert* operator is a new operator that adds new random individuals to the population pool of the maximum complexity possible. When the complexity of the individuals is lesser than the required complexity, the individuals are infeasible, and their optimization has little value. Adding new random individuals of maximum complexity quickly feeds feasible individuals of the maximum complexity, which guides the evolution thereafter. Even when complexity is further increased, this operator adds more individuals that represent different higher-complexity individuals which are fundamentally different from others and may have a better fitness in case a more complex individual turns out to have better fitness. For still higher generations, this factor contributes to the safety for dynamic maps, wherein after convergence the map changes and some other modality becomes better. Heuristics are encouraged for the generation of random individuals, like the generation of random individuals from a roadmap computed offline, an RRT-like algorithm, etc.

The genetic operators are applied in different proportions to make a candidate population from a parent population, where the proportion of time for which a genetic operator is applied is an algorithm parameter. A few percentages of solutions are contributed by every operator. It is suggestive to look at the success metric of every operator to assess its utility and thus set the contribution parameter. This makes the parameter adaptive.

The initial population generation is done by the generation of random individuals. Since the complexity is mostly limited, random via points are added in-between the source and goal. A better technique is to generate paths quickly heuristically by using other algorithms studied in the other chapters; however, the choice of the algorithm must be such that a lot of computation time is not wasted. Operating other algorithms in low-resolution settings is suggestive. As an example, a Probabilistic Roadmap Technique can be used to construct a roadmap, from which a diverse set of trajectories in different homotopic groups represents a good initial population.

The evolutionary process is probabilistically optimal, or the probability of generation of an optimal solution tends to one as time tends to infinity, and probabilistically complete, meaning that the probability of finding a solution tends to one as time tends to infinity.

16.6 Evolutionary motion planning variants

We now highlight numerous limitations of the evolutionary planning discussed so far and mend them, making the algorithm more suitable.

16.6.1 Evolving smooth trajectories

A major problem is that the approach generates line segments that are connected via points, making the trajectory nonsmooth at the via points. The evolution of the smooth curves involves the representation of a genotype that encodes a smooth curve. Assume the genotype still be to several via points; however, the phenotype is a *spline curve* that uses the genotype via points (along with source and goal) as the control points. This is shown in Fig. 16.11A.

By varying the via points, the smooth curve shape changes. The splines also have a parameter called the degree of the spline, with a lower degree meaning smooth splines (the spline may be far from the control points); at the same time, the higher-degree splines may have a nonsmooth curve (but the curve is close to the control points). The parameter may be chosen.

Under the new setting, the only change to be applied is in the fitness evaluation. In every evolutionary technique, the fitness function takes a genotype as input, converts it into the phenotype, and assesses the goodness of the phenotype that is returned as the fitness value. For the modified problem, the fitness function takes the list of via points (genotype) as input, adds source and goal at the beginning and end, respectively, uses a spline curve smoothening to get a smooth trajectory, and calculates the fitness (path length, clearance, smoothness, complexity, feasibility, etc.) of the smoothened trajectory that is returned as the fitness value.

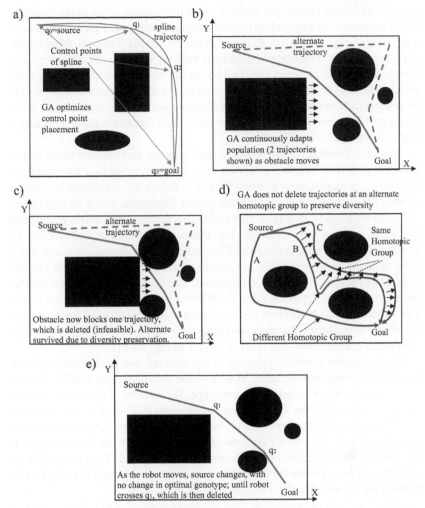

FIG. 16.11 Genetic Algorithm for dynamic maps. (A) Using spline smoothening. (B and C) Using diversity preservation. (D) Preservation of homotopic groups. (E) Adapting genotypes as the robot moves (source changes).

16.6.2 Adaptive trajectories for dynamic environments

Algorithms discussed so far in the book had means to adapt the path as the environment changes or obstacles move, appear, or disappear. Like D* algorithm, elastic strip, and adaptive roadmaps.

GA does not need any other means to constantly adapt the path to the changing map. The mapping modules constantly sense the change in the map, which is queried by the evolutionary algorithms during optimization. The change in map changes the fitness landscape, while the new individuals adapt to the changing fitness landscape, by responding to the changing evolutionary pressures (Shukla et al., 2009; Kala et al., 2010; Kala and Warwick, 2014). The only thing important is to maintain diversity as the optimal homotopic group of trajectories may change. For the trajectory in Fig. 16.11B, as the obstacle moves, the trajectory quickly adapts as per the evolutionary process. For the trajectory in Fig. 16.11C, the new obstacle completely blocks the old population of trajectories. It is therefore important to maintain diverse trajectories such that an alternate trajectory is readily available.

That said, eventually, convergence shall happen, and the converged trajectory may suddenly become infeasible due to a new obstacle. An intelligent mechanism to handle such a situation is to maintain all competent homotopic groups of trajectories and not let a competent homotopic group of trajectories die by the deletion of all trajectories within that group. Trajectories in different homotopic groups pass through the obstacles differently (clockwise v/s anticlockwise). A trajectory from a different homotopic group may not be going in-between the obstacles, which are now blocked, and survival of such a trajectory in the evolutionary process is important. The harder part is to judge whether two trajectories belong to the same or different homotopic groups. This can be approximately done by finding corresponding points (by relative length parametrization) and ensuring that the lines connecting the corresponding points in the trajectories are collision-free. This is shown in Fig. 16.11D.

The last aspect involves a change of the map not due to the moving obstacles, but due to the moving robot, in which case the source gets modified. Since the source was never a part of the genotype, the motion of the robot does not cause any change in the optimal genetic individual (except a little mutation to combat control noise). However, a change becomes important when the robot crosses a via point, in which case it should be deleted. The situation is shown in Fig. 16.11E. The shortened operator already has this effect. To better enable the robot motion to be accounted for, the first control point is explicitly checked for deletion for all individuals in the population pool.

16.6.3 Multiobjective optimization

So far, the different objectives were solved by using a weighted addition. Hence, weights become important parameters to set. A better technique is to use multiobjective optimization (Ahmed and Deb, 2011). The number of objective functions is small (path length, clearance, smoothness, and complexity), and the functions are conflicting with each other (making path length better makes clearance worse, and similarly for the different pairs), which makes it ideal for multiobjective optimization.

An important aspect is to handle the infeasible solutions in this approach. If the infeasibility is added as an axis, the algorithm will produce a Pareto front that has many infeasible solutions. The infeasible solutions are therefore given a poor objective value for all objectives, to be dominated and deleted.

The Pareto front cannot be used by the robot to get to the goal, since only one trajectory and not a set of trajectories are required. The selection of one trajectory can happen by either using measures like a knee analysis of the Pareto front, which gives the best tradeoff between the objectives. Alternatively, the human user can be presented with several scenarios and corresponding Pareto front trajectories in the learning phase. The human is then asked to choose the best trajectory. The weights can be learned from such choices of the human over several different scenarios.

16.6.4   Optimization in control spaces

Control space-based planners are increasingly getting an interest due to the complexity of the robots and the inbuilt mechanism of control space trajectories to deal with the kinematics and other constraints inherent to the robot. The optimization also marks an interesting perspective for optimization.

The primary problem is still the genotypic representation, wherein the trajectory is represented from a phenotypic representation into a genomic sequence. The concept of configuration space via point cannot be used, as there may be no control signal possible that makes the robot reach one point from the other.

Consider that a constant signal u_1 is applied for a fixed duration of time Δt. Subsequently, a signal u_2 is applied for another fixed duration of time Δt and so on. Let the maximum time that the robot may take to reach the goal be T and therefore the maximum control signals needed will be $m = \text{ceiling}(T/\Delta t)$. All these are unknown control signals, which are optimized by an optimization toolbox. The genotype hence is represented by an ordered sequence of control signals, $x = [u_1, u_2, ..., u_m]$, where each control signal has a dimensionality as per the kinematics of the robot.

The advantage of the scheme is that all concerns about smoothness and uncontrollability of the trajectory have been accounted for, and the time to reach the goal is a good objective function, along with feasibility and some resistance to errors using a small clearance. The problem is that the initial random control signal sequence may never make the robot reach the goal in the first place. Alternatively, the robot may reach the goal reasonably early and the rest of the control sequence, when stimulated, may make the robot move away from the goal. Hence, if the robot reaches near enough to the goal, the rest of the control sequences are neglected. A penalty is added in proportion to the closest point reached by the robot to the goal. The fitness function is thus given by addition of the time to reach the goal, clearance, large penalty times the segment of trajectory inside obstacles, and small penalty times the distance between the closest point in the trajectory to the goal; all metrics calculated by neglecting any trajectory after reaching the goal.

To calculate the fitness function, the robot needs to be simulated. Let $q_{i+1} = K(q_i, u_{i+1}, \Delta t)$ be the kinematic equation of the robot that gives the state at $i+1$ when a constant control signal u_{i+1} is applied to the state q_i for

Δt amount of time. Starting from the source (q_0), the first control sequence u_1 is applied for Δt time that gives the trajectory of the first segment along with the ending point $q_1 = K(q_0, u_1, \Delta t)$. q_1 is used as the starting point for the next segment with control signal u_2, giving the trajectory and the ending point as $q_2 = K(q_1, u_2, \Delta t)$. This goes on till all the control signals are in the sequence. As an example, consider a planar differential wheel drive robot with state $q_i(x_i, y_i, \theta_i)$ consisting of the position (x_i, y_i) and orientation (θ_i); and control signal $u_i(v_i, \omega_i)$ consisting of the linear (v_i) and angular speed (ω_i). The kinematics of the robot is given by Eqs (16.39)–(16.41).

$$
\begin{bmatrix} x_{i+1} \\ y_{i+1} \\ \theta_{i+1} \end{bmatrix} = K\left(\begin{bmatrix} x_i \\ y_i \\ \theta_i \end{bmatrix}, \begin{bmatrix} v_{i+1} \\ \omega_{i+1} \end{bmatrix}, \Delta t \right)
$$

$$
= \begin{bmatrix} \cos(\omega_{i+1}\Delta t) & -\sin(\omega_{i+1}\Delta t) & 0 \\ \sin(\omega_{i+1}\Delta t) & \cos(\omega_{i+1}\Delta t) & 0 \\ 0 & 0 & 1 \end{bmatrix} \begin{bmatrix} x_i - ICC_{x,i+1} \\ y_i - ICC_{y,i+1} \\ \theta(t) \end{bmatrix} + \begin{bmatrix} ICC_{x,i+1} \\ ICC_{y,i+1} \\ \omega_{i+1}\Delta t \end{bmatrix}
$$

$$\tag{16.39}$$

$$
ICC_{i+1} = \begin{bmatrix} x_i - R_{i+1}\sin(\theta_i) \\ y_i + R_{i+1}\cos(\theta_i) \end{bmatrix} \tag{16.40}
$$

$$
R_{i+1} = \frac{v_{i+1}}{\omega_{i+1}} \tag{16.41}
$$

Here ICC_{i+1} is the centre of the circle around which the robot turns, which has a radius of R_{i+1}. The optimizer aims to find a sequence of linear and angular speeds that drives the robot optimally till the goal.

The assumption of using a uniform control sequence for a Δt duration of time is wasteful. To get fine dextrous motion, the time resolution, Δt, must keep small. At the same time, this will increase the number of variables in the genotype ($m = \text{ceiling}(T/\Delta t)$) and thus make evolution a lot harder. A better technique is to thus make the time-resolution variable for every step and to have a variable number of such steps. The individual may now be represented as a variable-length sequence $x = [u_i, \Delta t_i]$, stating that the control u_i was applied for Δt_i duration of time, while both the quantities are optimized. This generates a complete control sequence, which is given to the robot and the robot moves as per the same sequence. The fitness function is taken as the time (and maybe a small factor for clearance), coupled with penalties for infeasibility (length of infeasible path) and not reaching the goal (distance from the closest point to goal). The trajectory beyond reaching the goal is trimmed.

The search space in such a case is large and the problem is best solved by using heuristics for the generation of the initial population. For simple robots, reactive planning techniques can be used for some fractions of time which give indicative control actions. Alternatively, some deliberative algorithms may be tried to get an approximate path, and the initial control signals generated to lie close to the same path. This resembles the *optimal control* problem. Furthermore, it is possible to generate some configuration space roadmap, which can be used further as a heuristic means by the control space optimization algorithm.

Using heuristics for the initial population generation is essentially running other algorithms presented in the book in their local formats. The algorithms give a path as output. The path is traversed by any control technique. The control signals are encoded to make a genotype, which is further optimized by optimization. The original path may face problems specific to the algorithm. The control sequence obtained from the control algorithm over this path may also end up in many near-collisions in the traced trajectory. However, optimization will mend all such errors and result in an optimal sequence of the control signal to move the robot.

16.7 Simulation results

The algorithm was tested in simulation settings for synthetic maps. The results are demonstrated for different algorithmic settings. The first discussions involve optimizing a single objective function. The results for a synthetic map are shown in Fig. 16.12. The results of optimization with smoothening using spline curves produce trajectories, as also shown in Fig. 16.12.

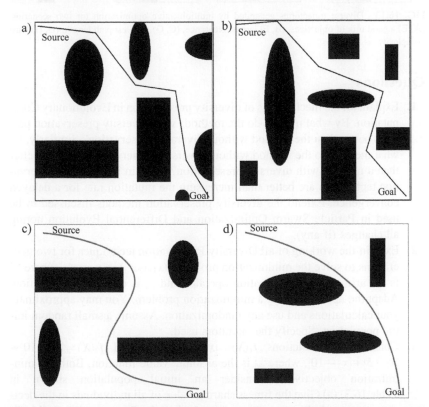

FIG. 16.12 Simulation results. (A and B) Genetic Algorithm using variable via points. (C and D) Genetic Algorithm using spline smoothing.

The approach was also discussed in the context of multiobjective settings, which produces a Pareto front of diverse trajectories. Some diverse trajectories for the two scenarios are shown in Fig. 16.13.

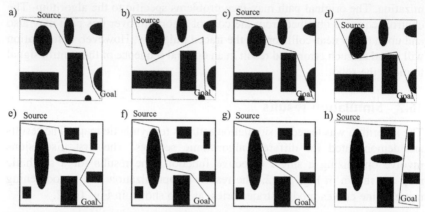

FIG. 16.13 Diverse trajectories generated by multiobjective optimization for two scenarios: (A, E) a good mix of objectives, (B, F) largest clearance, (C, G) smallest path length, (D, H) least complexity.

Questions

1. Explain all the mechanisms of diversity preservation in Evolutionary Computation. By what metrics do the methods with diversity preservation perform better than the method without diversity preservation? Similarly, by what metrics do the method without diversity preservation perform better than a method with diversity preservation? Explain if diversity preservation techniques are better than increasing the mutation rate for a delayed convergence. Extend the diversity preservation methods discussed to be used in Particle Swarm Optimization and Differential Evolution noting all changes (if any).

2. Explain the working of all Diversity Preservation techniques for two generations to solve the minimization problem $f(x_1, x_2, x_3) = |x_1.x_2.x_3|$ where '.' is the arithmetic multiplication operator and $|.|$ is the absolute function. Adapt the approaches for a minimization problem. You may approximate your calculations and use any randomization. Assume a small random initial population. Specify the operators used.

3. Consider the functions $f_1(X) = |x_1| + |x_2| + |x_3|$ and $f_2(X) = |x_1 - 10| + |x_2 - 5| + |x_3 - 10|$, where $|.|$ is the absolute value function. Both are minimization objectives. Consider an initial population shown in Table 16.3. (a) Give the fitness sharing value of all individuals in the decision space and objective space, (b) calculate the Pareto ranks of all individuals, (c) give the output of two generations of NSGA and MOEA/D,

(d) propose a mechanism to consider diversity in both the decision and objective spaces in NSGA and MOEA/D.

TABLE 16.3 Initial population for question 3.

	x_1	x_2	x_3
Individual 1	0	0	0
Individual 2	5	0	10
Individual 3	0	5	0
Individual 4	10	0	10
Individual 5	10	5	10

4. Suggest whether MOEA/D is better, equivalent, or worse than an isolated optimization of n objective functions.
5. Explain the effects of increasing the number of objective functions in the NSGA and MOEA/D algorithms. Also, explain the effects of disproportionately scaling the different objective functions on the NSGA and MOEA/D algorithms.
6. Consider the optimization of the trajectory using a Genetic Algorithm. Explain the tradeoff between high-resolution settings and low-resolution settings; explaining which setting is more effective in what context? Suggest metrics to adapt the resolution parameter. Is there a need to apply post-processing that is usually done with the outputs of the A* algorithm and PRM? Why or why not?
7. Consider the optimization of the trajectory using a Genetic Algorithm. For each of the following scenarios, state whether a typical implementation of a Genetic Algorithm is most likely to solve the scenario or not. (a) An environment with too many small convex obstacles cluttered in the map, (b) an environment with a small number of nongeometric obstacles, (c) a environment with concave obstacles, (d) an environment with a narrow corridor, (e) a complex maze, (f) dynamic environment with slowly moving obstacles, (g) dynamic environment with obstacles moving very quickly, and (h) dynamic environment with a suddenly appearing obstacle.
8. Compare and contrast evolutionary robot motion planning, elastic strip, and adaptive roadmap approaches.
9. Numerous deliberative and reactive algorithms for motion planning have already been discussed. Suggest how can these algorithms be incorporated as heuristics into the problem of evolutionary robot motion planning. Some designs should incorporate more than one algorithm.

10. [Programming] Add diversity preservation techniques to all standard optimization problems detailed in question 6 of Chapter 15.
11. [Programming] Solve the standard multiobjective optimization problems of ZTD and DTLZ using NSGA and MOEA/D.
12. [Programming] Solve the problem of evolutionary robot motion planning using evolutionary computation with smoothing and diversity preservation. The algorithm should optimize the trajectory as the robot moves. Show the performance of the algorithm with a few other robots moving to make a dynamic map.
13. [Programming] *n* robots need to go from their source to goal. Plan each robot separately using evolutionary algorithms, treating all other robots as dynamic obstacles.
14. [Programming] Use Probabilistic Roadmap to get a few competing trajectories, preferably in different homotopic groups. Use these trajectories as the initial population for evolutionary motion planning.
15. [Programming] Using evolutionary algorithms, plan the trajectory of a two-link planar revolute manipulator with both geometric (polygon-shaped) and nongeometric (grid-map) obstacles.
16. [Programming] Control a robot using different deliberative and reactive algorithms, along with a naïve controller. Using these sequences as the initial population, optimize the control sequence using evolutionary computation.
17. [Programming] Construct an approximate, deliberative path to the goal using sampling-based techniques. Supply points at a distance as subgoals that a robot should reach. Make a local evolutionary algorithm that continuously adapts a trajectory to the nearest subgoal ahead for a continuous motion of the robot. Add a few dynamic obstacles. How does the design change, considering that the local map in consideration is reasonably simple?

References

Ahmed, F., Deb, K., 2011. Multi-objective path planning using spline representation. In: 2011 IEEE International Conference on Robotics and Biomimetics. IEEE, Karon Beach, Phuket, pp. 1047–1052.

Deb, K., Jain, H., 2014. An evolutionary many-objective optimization algorithm using reference-point-based nondominated sorting approach, part I: solving problems with box constraints. IEEE Trans. Evol. Comput. 18 (4), 577–601.

Deb, K., Pratap, A., Agarwal, S., Meyarivan, T., 2002. A fast and elitist multiobjective genetic algorithm: NSGA-II. IEEE Trans. Evol. Comput. 6 (2), 182–197.

Jain, H., Deb, K., 2014. An evolutionary many-objective optimization algorithm using reference-point based nondominated sorting approach, part II: handling constraints and extending to an adaptive approach. IEEE Trans. Evol. Comput. 18 (4), 602–622.

Kala, R., Warwick, K., 2014. Heuristic based evolution for the coordination of autonomous vehicles in the absence of speed lanes. Appl. Soft Comput. 19, 387–402.

Kala, R., Shukla, A., Tiwari, R., 2009a. Fusion of evolutionary algorithms and multi-neuron heuristic search for robotic path planning. In: Proceedings of the 2009 IEEE World Congress on Nature & Biologically Inspired Computing. IEEE, Coimbatore, India, pp. 684–689.

Kala, R., Shukla, A., Tiwari, R., Rungta, S., Janghel, R.R., 2009b. Mobile robot navigation control in moving obstacle environment using genetic algorithm, artificial neural networks and A* algorithm. In: Proceedings of the IEEE World Congress on Computer Science and Information Engineering. IEEE, Los Angeles/Anaheim, pp. 705–713.

Kala, R., Shukla, A., Tiwari, R., 2010. Fusion of probabilistic A* algorithm and fuzzy inference system for robotic path planning. Artif. Intell. Rev. 33 (4), 275–306.

Kala, R., Shukla, A., Tiwari, R., 2011. Robotic path planning using evolutionary momentum based exploration. J. Exp. Theor. Artif. Intell. 23 (4), 469–495.

Kala, R., Shukla, A., Tiwari, R., 2012. Robotic path planning using hybrid genetic algorithm particle swarm optimization. Int. J. Inf. Commun. Technol. 4 (2–4), 89–105.

Lozano, M., Herrera, F., Cano, J.R., 2008. Replacement strategies to preserve useful diversity in steady-state genetic algorithms. Inf. Sci. 178 (23), 4421–4433.

Sareni, B., Krahenbuhl, L., 1998. Fitness sharing and niching methods revisited. IEEE Trans. Evol. Comput. 2 (3), 97–106.

Shukla, A., Tiwari, R., Kala, R., 2009. Mobile robot navigation control in moving obstacle environment using genetic algorithms and artificial neural networks. Int. J. Artif. Intell. Comput. Res. 1 (1), 1–12.

Xiao, J., Michalewicz, Z., Zhang, L., Trojanowski, K., 1997. Adaptive evolutionary planner/navigator for mobile robots. IEEE Trans. Evol. Comput. 1 (1), 18–28.

Zhang, Q., Li, H., 2007. MOEA/D: a multiobjective evolutionary algorithm based on decomposition. IEEE Trans. Evol. Comput. 11 (6), 712–731.

Chapter 17

Hybrid planning techniques

17.1 Introduction

So far, numerous algorithms have been presented that solve the problem of motion planning for robotics. It is evident to question the availability of numerous algorithms from different schools of learning for a single problem. The choice of algorithm is dominantly driven by the availability of computing infrastructure, the dynamics of the environment ranging from static to highly dynamic environments, the knowledge of the environment, whether it is fully observable or partially observable, the presence or absence of a high-quality rich sensor suite, etc. There is, unfortunately, no single algorithm that completely solves the problem exhaustively. This is the reason that motion planning algorithms have been dominant in the literature since the beginning of robotics and continue to be published in leading avenues.

If a problem cannot be perfectly solved by a single algorithm, while several good candidate algorithms exist; it is imperative to investigate the limitations of an algorithm and to get complimentary algorithms that overcome the limitations. The fusion of the two algorithms should hence be a more sophisticated algorithm that adds the advantages of the two complementary algorithms and removes the limitations of the individual algorithms. This chapter does not introduce any new fundamental algorithm; however, it tries to tailor the different algorithms in a hybrid architecture such that the resulting algorithm is better in performance with lesser limitations than all the other algorithms.

The notion or attempts to fuse different modules to make a sophisticated algorithm is not new in computing. From a similar domain of machine learning, while it is known that many different machine learning techniques exist, the *ensemble algorithms* use numerous techniques at the same time and fuse their output for better decision-making. The motivation is taken from *committee-based decision-making*, in which a decision is not made by a single expert, but by several experts who are individually good and diverse (differ from each other and hence get unique points for collective decision-making). Such fused machine learning algorithms, namely ensemble learning, bagging, boosting, and random forest, are known to give a better performance as compared to a standalone machine.

Autonomous Mobile Robots. https://doi.org/10.1016/B978-0-443-18908-1.00018-2

Similarly, the evolutionary algorithms are known to be global optimizers, which are often fused with a local search algorithm in a *memetic architecture*. This forms a fusion of a global search, which is lesser prone to get stuck in a local-optima but computationally expensive; with a local search, which very quickly gets to a local-optima but will rarely get out of a local-optima. Using the same philosophy, neural networks can be easily trained by using evolutionary algorithms forming the branch of neuro-evolution. Similarly, fuzzy systems can be trained by using powerful evolutionary algorithms. The fusion of neural and fuzzy systems with an evolutionary algorithm is a hybrid architecture wherein each algorithm operates in a different domain.

In the same manner, in robotics, using multiple redundant sensors is a fusion that enables making better decisions regarding the obstacles. At the same time, using complementary sensors to record the same information in different domains also enables one to get a better understanding of the obstacles since different sensors have different noise models and may be accurate in different conditions. As an example, the sonar sensors may report nearer obstacles while high-frequency ultrasonic sensors may report obstacles further away. The radar at the front of a self-driving car may detect a close cut-in, which may even be detected by a 3-D lidar at the top of the vehicle. The video camera may be used for localization in the morning, while the localization at night may be primarily by using lidars. The lidar may detect the distance to the obstacle for keeping safety distances, while the camera may detect the obstacle type to decide how to combat the obstacle.

The fusion in all such formats has already been discussed and studied in different segments of the book. The fusion studied in this chapter is focused on motion planning algorithms, including all decision-making done by the robot.

17.2 Fusion of deliberative and reactive algorithms

Fundamentally, the motion planning algorithms may be divided into two categories, deliberative algorithms, which incur heavy computation in pursuit of getting the best possible trajectory; and reactive algorithms, which directly map the sensor percept into a motion command as a reaction to the situation. The demarcation between algorithms is not crisp but may be fuzzy as some reactive algorithms may show more deliberative characteristics than others, while some algorithms may lie nearly at the boundary. The fusion is largely between the deliberative and reactive algorithms.

17.2.1 Deliberative planning

The deliberative algorithms plan the best possible trajectory that avoids all the obstacles as the robot moves from the source to the goal. The algorithms typically (heuristically) try out all possible trajectories and select the best possible trajectory. The graph search-based algorithms represent the entire map as grids and discretize the edges, and thus do optimal graph search. The cell decomposition-based algorithms as well represent the map by discrete cells and perform a graph search to get the trajectory. The roadmap-based approaches instead summarize the environment as a roadmap first and then use the roadmap for a graph search. The Probabilistic Roadmap technique iteratively grows such a roadmap, while the tree-based techniques represent the same roadmap as a tree for a single source and single goal problem. The optimization represents multiple trajectories in parallel and optimizes them, in pursuit to get the best trajectory. All such approaches form the deliberative mechanism of working.

The deliberative algorithms construct a trajectory. The control algorithms help in tracing the trajectory. Therefore, the robot must largely know the map for the deliberative algorithms and the robot must always know its position in the map, which is needed for the identification of the source as well as for the control algorithms to calculate the position error. This requires the robot to have a large memory as well as a rich set of sensors to enable localization and mapping.

The deliberative algorithms are typically *computationally expensive* as they deal with searching in a set of many trajectories. Therefore, the algorithms typically also require a rich computing infrastructure and rich availability of memory. In all roadmap-based approaches, the query for a source and goal pair was called online, however, the overall algorithm is still tagged under the computationally expensive category because the online query comes with a big assumption that the map does not change. If the map changes, the query on the new map will be typically offline since a new roadmap will have to be made.

The advantage of the deliberative algorithms is that they are *complete*, meaning that the algorithms find a trajectory if one exists. The completeness may have a resolution clause attached to it (resolution complete, like the grid search) or a probabilistic clause attached to it (probabilistically complete, like Probabilistic Roadmap). Similarly, the algorithms are *optimal* (with the notion of resolution optimal and probabilistically optimal). This forms a big advantage as the robot takes good moves and *does not get itself stuck* requiring human help.

The deliberative algorithms largely assume a good knowledge of the map. It is possible to use the deliberative algorithms on partially known environments, in which the deliverable algorithms can do reasoning for the best path as per the limited availability of knowledge. As more knowledge is made available as the robot moves, the plans are extended or re-planned. The robots thus should have a rich sensory suite. Therefore, deliberation is a rarity in robot swarm algorithms, wherein the robot rarely knows its position or has any knowledge of the map.

17.2.2   Reactive planning

The reactive planning paradigm believes in immediate decision-making to move the robot as quickly as possible. The algorithms directly map the sensor percept and state information into an immediate control action of the robot. The algorithms do not first construct a trajectory and then make the robot trace the trajectory. The potential field approach directly maps the proximity sensor readings and location information into the current motion of the robot. The geometric approaches also analyze the obstacle corner information to quickly compute the immediate motion direction. The neural and fuzzy approaches also use the sensor percept as input to a formulation that gives the motion commands as output.

The reactive algorithms have a *low computation time*. The algorithms do not require a complete map and can make decisions from the current percept alone. Therefore, the algorithms may work only with a small number of proximity or similar sensors, without requiring a high memory, sensor set, and computation power. The algorithms can work with *(very) partially known environments*. Because the decisions are primarily based on the current sensor percept, the algorithms can work well with *highly dynamic environments* and suddenly appearing/disappearing obstacles.

As a loss, typically, the reactive algorithms are neither complete nor optimal. Therefore, a robot can often get stuck in simple situations and the robot will not come out until help is provided by the humans or the map naturally changes. However, many characteristic situations have simple operating rules and strategies and are naturally free of any trapping conditions.

17.2.3 The need for fusion and hybridization

Most of the navigation of humans uses deliberation. For example, to drive to a different city for vacation by satellite navigation uses deliberative algorithms. While going from one room to the other, one already knows the map, and the steps are calculated by deliberation. Similarly, for the robots, most of the navigation is done by using deliberation alone. Most offices and homes have a largely static map consisting of walls, doors, couches, plants, etc.

Imagine a robot going from one room of the house to the other while a housemate is also walking. Now the deliberative algorithm already has a trajectory, which is invalidated due to the new person walking. The trajectory now needs to be reconstructed by using deliberative algorithms, which may take a few milliseconds to seconds. However, as the new trajectory is computed, the housemate further moves and immediately nullifies the new trajectory. The method can only work if the rate of change in the environment is sufficiently slower than the rate of re-planning, which is clearly not the case with the use of deliberative algorithms. Similarly imagine a robotic manipulator arm trying to hold a glass, while a person is constantly moving his hands, changing the map at a fast pace. Again, the trajectory of the robot needs to constantly be re-computed faster than the motion of the hand, which is hard to do with the use of deliberative algorithms.

On the contrary, now the reactive paradigm sounds better since the robot will continuously make small moves that avoid moving (and static) obstacles. However, consider a simple situation, wherein the robot needs to go from one room to the other. The robot will try to go straight to the goal, which is blocked by the wall. Turning a little left has the goal on the right motivating a right turn;

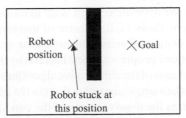

FIG. 17.1 Robot getting stuck when operated by a reactive algorithm. The robot needs to go from one room to the other. The reactive algorithms fail to make the robot deviate by a large amount from the straight-line motion to the goal and hence make the robot get stuck.

while going a little rightward will feature the goal on the left, motivating a leftward turn. Because there is a local optimum, the robot is very likely to get stuck or show oscillations around some point. The situation is shown in Fig. 17.1. This means that the algorithm performs poorly in practical scenarios.

Let us re-look at the pros and cons of the two algorithms. The deliberative algorithms are (near) complete, (near) optimal, and not do not make the robot get stuck. However, the deliberative algorithms require a large computation time; cannot work with dynamic obstacles and suddenly appearing obstacles; generally, assume knowledge of the map and location; and generally, require a large sensory, computing, and memory infrastructure. On the contrary, the reactive algorithms, require little computing time which is ideal for dynamic and suddenly appearing/clearing obstacles; require no knowledge of the map and can work with a small number of sensors, memory, or computing infrastructure. However, the reactive algorithms are neither optimal nor complete and often make the robot gets stuck in situations.

The reactive algorithms only require a position relative to the goal and obstacles, which is easier to compute using the onboard sensors in contrast to the absolute position in the map, which requires a highly accurate global localization system onboard. Imagine the Artificial Potential Field algorithm. The sensors (e.g., lidar) make a local instantaneous map around the current robot position, which is used to calculate the repulsion force that does not require a knowledge of the global pose of the robot. The attraction to goal may be calculated by an approximate knowledge of the robot's pose till the robot comes close enough to the goal and the goal is visible, in which case again the onboard sensors see the goal relative to the current position of the robot and the attraction force can be calculated without knowing the global pose of the robot. On the contrary, imagine that the robot's deliberative trajectory was calculated by using the Probabilistic Roadmap technique. Control algorithms are used to execute the trajectory. The controllers continuously calculate the error between the current global robot pose and the expected pose as planned in the deliberative trajectory, which is given as feedback to the control system. A few centimetres of error in the localization to determine the global pose near the obstacles can now make the control algorithm deviate the robot by a few centimetres away from the planned trajectory, which is enough to cause a collision.

The two algorithms are opposite of each other in terms of the pros and cons, which are summarized in Table 17.1. Because of this tradeoff, the deliberative algorithms can handle static and semistatic situations very well, however, the dynamic obstacles like other people were best handled by the reactive algorithms. The conflicting pros and cons of the deliberative algorithms motivate the fusion of the two algorithms to make a single algorithm that has the advantages of both algorithms, while it minimizes the disadvantages of the two algorithms (Kala et al., 2009; Kala, 2013; Tiwari et al., 2012). Typically, the intuition is to use deliberative algorithms for the static obstacles. However, in case of dynamic obstacles (like moving people and moving other robots), change of map, suddenly appearing/clearing obstacles, etc., the reactive algorithms are used. In the case of a large map change like blockage of a route due to fallen furniture, cleaning in progress, lift not working, sudden human traffic, etc., the reactive algorithms are used till the deliberative algorithm can figure out a new plan. While using satellite navigation, the route is decided by deliberative algorithms. If a road on the route gets blocked due to factors like accidents or gets too much sudden traffic, the navigation system actively re-plans the route using deliberative algorithms to avoid such a road. In the case of highly dynamic obstacles like moving people, the algorithm largely relies on the reactive algorithm to combat the dynamic obstacle.

TABLE 17.1 Pros and cons of the deliberative and reactive algorithms.

Property	Deliberative algorithm	Reactive algorithm
Completeness	*(Near-)complete*	Not complete
Optimality	*(Near-)optimal*	Not optimal
Stuck at local optima	*No*	Yes
Computation time	Large	*Small*
Computation complexity	Large	*Constant*
Memory requirements	High	*Small*
Dynamic obstacles	No	*Yes*
Suddenly appearing obstacles	No	*Yes*
Works in partial environments	Limited use when most of the environment is unknown	*Yes*
Assumption of absolute robot position	Yes	*No. Works on position relative to goal and obstacles*
Sensors	High sensor requirements	*Low sensor requirements*

Better algorithm's property in italics.

The trajectory correction algorithms like elastic strip, adaptive roadmap, and D* algorithm give an interesting ground wherein the deliberative algorithm trajectories are locally corrected to deal with the dynamics. This forms a boundary case between the deliberative and reactive algorithms. Even though the trajectory can be locally corrected for moving obstacles, the correction involves iterating over the entire trajectory, which can be computationally expensive. If the obstacle moves at a speed faster than such a correction, the correction can never be done. Furthermore, if too many obstacles move, the updates will be very costly. The approach assumes full knowledge of the map and some motion dynamics of the obstacles. These are not purely reactive behaviours.

17.3 Fusion of deliberative and reactive planning

The basic intention is to fuse a deliberative planning algorithm that ensures that the robot does not get stuck and aims at optimality and completeness, primarily considering the static obstacles; and the reactive algorithm that handles the dynamic and suddenly appearing obstacles including moving people. The first fusion architecture uses the deliberative algorithm to get a trajectory considering the static obstacles only (Kala et al., 2010a). The resolution may be kept lower to get a timely trajectory. The robot will not ultimately be following this exact trajectory and it is acceptable to deviate a little from the optimal trajectory during the reactive phase. This further solves the problem of a large computation time associated with the deliberative algorithm as the search settings are low resolution. However, it may be noted that low-resolution searches can yield paths in a different homotopic group, which can severely lack optimality. The output of the deliberative algorithm is a trajectory as shown in Fig. 17.2.

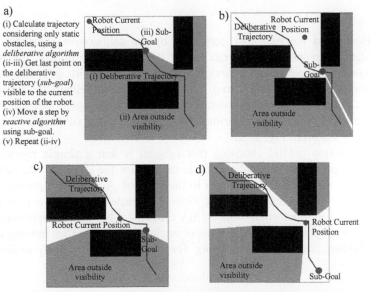

FIG. 17.2 Fusion of Deliberative and Reactive Planning using the last visible point in the deliberative trajectory as the subgoal. (A–D) Show the subgoal moving as the robot traverses.

The trajectory computed by the deliberative algorithm will not be able to handle new obstacles and dynamic obstacles and therefore, the robot is not moved by the deliberative trajectory. However, the robot is moved by a reactive algorithm that takes care of all types of obstacles. The problem with the reactive algorithm is the lack of completeness and optimality. Therefore, the reactive algorithm is not given the goal, but a subgoal chosen in the deliberative trajectory. The reactive algorithm only tries to get to the subgoal. As soon as the robot reaches close enough to the subgoal, the subgoal is changed to a point that is more distant, until the subgoal lies at the goal itself. The deliberative trajectory thus provides a series of subgoals to be sequentially (approximately) visited by the robot.

This mechanism of working is used every day in problem-solving. A difficult problem is usually broken into multiple subproblems to be solved sequentially, and efforts are made to solve one subproblem after the other. A big project is typically first decomposed into tasks, and all tasks are solved one after the other. While working over the current task, all efforts are towards that task only.

Similarly, reaching the goal is a difficult problem, which is decomposed into getting from one *subgoal* to the other. In the first approach, the subgoal is conveniently chosen as the last point visible in the deliberative trajectory, considering only the static obstacles present at the time of deliberative planning and not the dynamic obstacles. The subgoal selection is thus given by Eq. (17.1):

$$\text{subgoal} = \tau_d(s), s = \arg\max_s \lambda\tau_d(s) + (1 - \lambda)q \in C^{\text{free}} \forall 0 \leq \lambda \leq 1 \quad (17.1)$$

Here, τ_d is the trajectory computed by the deliberative algorithm with any general point given by $\tau_d(s)$. q is the current position of the robot. $\lambda\tau_d(s) + (1 - \lambda)q$ represents a straight-line joining $\tau_d(s)$ with q. If this line is collision-free, $\tau_d(s)$ is visible to q. The last point in the deliberative trajectory visible to the current robot's position q is the chosen subgoal. Subgoal is the goal given to the reactive algorithm. The selection of the subgoal is diagrammatically shown in Fig. 17.2.

The notion behind choosing this subgoal is that a simple mechanism exists to get to the subgoal, which is to wait for the dynamic obstacles to clear with time and move straight to the subgoal. A reactive algorithm will take a few turns in response to the moving obstacle but overall is less likely to get stuck in this formulation. At any time, the robot has the option to first reach the subgoal by waiting for the obstacles to clear and then moving using the deliberative trajectory waiting for the obstacle to clear, eventually reaching the goal. The robot could detect that it is stuck or oscillating by keeping track of the configurations visited. In case, the robot is stuck it will be in the same configuration for a long time, while if it is oscillating, it would be repeatedly visiting the same configuration. In such a case it is acceptable to move directly towards the subgoal by waiting for the obstacles to clear. After reaching close enough to the subgoal the reactive algorithm may continue. For the same reasons, to ensure completeness, a valid constraint is that the robot must always have visibility of at least one point in the deliberative trajectory considering the static obstacles only. If a move leads to a configuration such that no point in the deliberative trajectory is visible, the robot may instead decline to make such a move. This is because if the move is made and later the robot detects being stuck, it has no knowledge of how to connect to the deliberative trajectory. Note that the strategy to get unstuck discussed here assumes that the other agents will intelligently clear with time. Sometimes two robots could wait for each other to clear indefinitely. Such deadlocks shall be handled later in the chapter.

Another variant places the subgoal at a little distance from the current position of the robot. While doing a project, it is common to enlist all the work and to plan to complete some of the modules in the next few days as the first subgoal. This gets the focus clear to work on a few of the several modules. However, as the days proceed and some of the modules are completed, new modules, next in the order of priority are added in the immediate subgoal to be worked over in the next few days. As the modules complete, the subgoal is populated with more work, until the entire project is complete.

Similarly, this strategy first shows a subgoal in the deliberative plan that is a little distance ahead. However, as soon as the robot makes the first few steps, the subgoal is moved further back into the deliberative trajectory. As the robot continues to reach nearer to the subgoal, the subgoal is pushed further back, until the robot reaches the final goal. In general, the subgoal is always ahead of the current position of the robot by a distance η. The subgoal selection is given by Eqs (17.2) and (17.3).

$$\text{subgoal}(t + 1) = \tau_d(s(t + 1)) \tag{17.2}$$

$$s(t + 1) = \arg\max_{s'} s' : d(q, \tau(s')) \leq \eta, s(t) \leq s' \leq 1 \tag{17.3}$$

Here, η is the threshold distance that should be always maintained between the robot and the subgoal, q is the current robot position and d is the distance function. As soon as the robot reaches a distance less than η, the subgoal is shifted away. The subgoal is not changed after it reaches the goal itself. The selection of the subgoal is diagrammatically shown in Fig. 17.3.

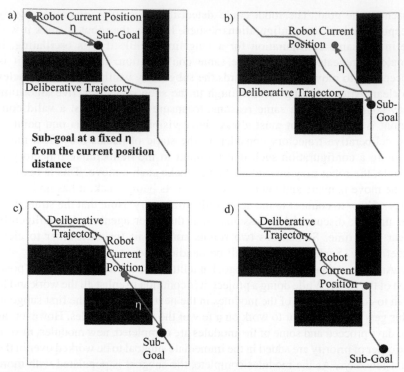

FIG. 17.3 Fusion of Deliberative and Reactive Planning using subgoals at a fixed distance of the current position. (A–D) Show the subgoal moving as the robot traverses.

If η is too small, the robot essentially follows a deliberative trajectory with no use of reactive planning. This can be fatal as the subgoal may be on top of a dynamic obstacle and will make the robot wait for too long for such an obstacle to clear. If η is too large, the algorithm is primarily reactive and suffers from all limitations of the reactive algorithms. For the sake of completeness, sometimes it is wise to set a constraint that the subgoal must always be visible to the robot considering only the static obstacles only. Hence, a new subgoal selection only moves the subgoal further such that the visibility from the robot is maintained. Furthermore, the robot is only allowed to make a move to a configuration such that the visibility to the current subgoal is retained, considering the static obstacles only. By choosing a subgoal as the last visible point in the deliberative trajectory, the optimality and completeness are under check, while the utility of the reactive algorithm is maximized, which is thus a better approach than taking a parameter η.

So far, the discussions have revolved around the notion that dynamic obstacles like moving people and other robots are handled by the reactive algorithm, while static obstacles are already accounted for in the deliberative trajectory. If a new static obstacle arrives or leaves the map, the reactive algorithm will still be able to handle the situation as it does for the dynamic obstacle. However, consider the case that a new static obstacle *completely blocks* the way of the robot, and the robot should now move from a different homotopic group of trajectories altogether. The situation is shown in Fig. 17.4 with the obstacle as the closed gate. Imagine the subgoal is after the closed gate. After the gate closes, the subgoal remains the same. There is no direct way to reach the subgoal, and the robot is stuck. It is important to keep the deliberative algorithm running for handling such changes in the map which are beyond the paradigm of the reactive algorithm. A deliberative algorithm is supposed to *reroute the robot* and take it from a new route. Similarly, if a blockage previously present gets cleared, the deliberative algorithm is supposed to reroute the robot. Another interesting situation is when there is a lot of congestion at an area, wherein a rerouting is necessary to be done by the deliberative algorithm. The navigation systems for cars give a route that avoids congested areas, while the cars are actively rerouted to avoid congestion as the congestion changes between the roads.

Gate closes midway, deliberative algorithm senses
change of map and re-plans the path

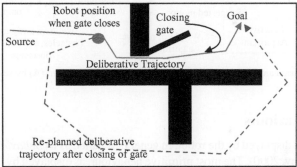

FIG. 17.4 Rerouting by the deliberative algorithm. Rerouting is done by the algorithm when the map changes (or congestion level changes, like rerouting on seeing congestion from a car). The robot will not exactly follow the deliberative path.

So far, the robot was only following a reactive algorithm for navigation. Another architecture uses an additional fusion of deliberative and reactive algorithms apart from the subgoal approach. Here, both algorithms are used to constantly give the output commands to the robot. The output of a reactive algorithm is naturally a motion command or a command which can be directly transformed into a motion command. The deliberative algorithm stores a trajectory initially and uses an online control algorithm to trace the trajectory. If the robot is far away from the obstacles, the deliberative algorithm's output is primarily used. If the robot is near the obstacles, the reactive algorithm's output is used. The fusion may be by *selection*, wherein the fusion module selects either the motion command from the deliberative algorithm or the one from the reactive algorithm based on the distance from obstacles. Alternatively, the fusion may be by *addition*, in which a weighted addition of both outputs is done, where the weight is largely in favour of the reactive algorithm for small distances from the obstacles and towards deliberative algorithm for large distances. The fusion architecture is shown in Fig. 17.5.

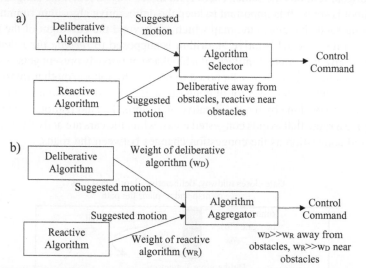

FIG. 17.5 Fusion architecture of Deliberative and Reactive Algorithm: (A) by selection, (B) by aggregation.

17.4 Behaviours

The reaction displayed by the robot, in general, is called as its *behaviour* (Brooks, 1986; Murphy, 2000). The most primitive behaviour, as displayed by the humans, is *reflexes*, which are natural to the humans and is a direct response to the sensory percept. The humans are born with such behaviours. Similarly, in robots, many of the behaviours like the potential field are a rule or a mapping that is hard-coded as software or hardware inside the robot, and the robot itself merely performs the reflex. Some reflex behaviours have a response even after the percept is no longer existent and constitute a little longer-term response, called as *taxes*. The birds and insects being attracted by light, sun or magnets are such behaviours. A unit behaviour that an agent is born with is called as the *innate behaviour*. Sometimes agents display a sequence of such innate behaviour called as the innate sequence. Some innate behaviours require memory to store the state. Like in case of the potential field, the location of the robot was based on some memory state. Similarly, the bug algorithms required some memory to store the direction of boundary following.

A more interesting higher-level behaviour is called as the *reactive behaviour*, which involves some component of *learning*. The robot driven by a potential field, wherein the coefficients are constantly changed for performance is a reactive behaviour. Similarly, the neural and fuzzy networks continuously adapt the architecture and weights and constitute learning making a reactive behaviour. On the same lines, the deliberative algorithms are called as a *conscious behaviour*, wherein a rich computation and reasoning are performed.

The robots may have not just one but multiple behaviours and may chose the behaviour depending upon the state, current percept, or any other external criterion. Such conditions that trigger behaviours are called as the *innate release mechanism* in the behavioural sciences. In a potential field, the proximity of obstacles acts as an innate release mechanism for the unit behaviour of obstacle avoidance, while goal proximity acts as an innate release mechanism for the unit behaviour of goal seeking. The potential field is a fusion of these two behaviours.

It is evident that a robot shows multiple behaviours like avoiding obstacles (other vehicles), going to goal, following a corridor (road), following a person (car), overtaking a person (car), drifting to allow a person (car) overtake, following the map boundary (patrolling), waiting for a person to come inside before going out, etc. Each application has multiple behaviours that are triggered based on the need. Each behaviour is associated with a precondition or release mechanism that states whether the behaviour should be displayed or not at a particular point of time (Fig. 17.6A). A simple mechanism is to have all behaviours fire *concurrently* when their precondition holds good. This is shown in Fig. 17.6B.

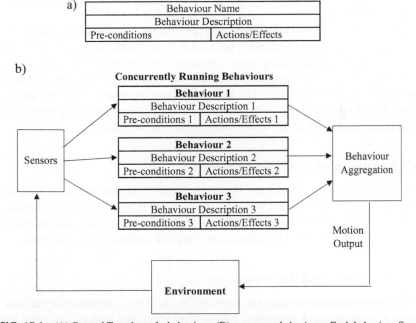

FIG. 17.6   (A) General Template of a behaviour, (B) concurrent behaviours. Each behaviour fires based on preconditions. A single behaviour may dominate (likely), two behaviours may cancel their effects (one says left, other says right), or the behaviours may reach an equilibrium (drifting one side, will dominate the other).

Every behaviour is associated with several sensors that give information about the world as perceived by the robot. The information also comes from the state of the robot. The state stores information accumulated over time like the position calculated by using the localization algorithm that integrates information over time. The map itself is made on the fly by integrating obstacle information over time. For a reactive paradigm, the sensors are mapped to the action directly by the behaviours, unlike the deliberative paradigm incurring a lot of layered processing. Therefore, the *behaviour schema* is said to have a *sensor schema*, that specifies the sensors and sensor processing functionalities, and a *motor schema*, specifying the motion commands. The behaviour schema is *instantiated* for every robot, wherein specific behaviour properties are specified. Two robots may have the same schema, but different specific properties like the minimum distance to keep from the obstacles and speed preferences. This follows the object-oriented programming paradigm of defining schemas and instantiating for every behaviour separately. Imagine two robots with the same obstacle avoidance behaviour schema. To make the obstacle avoidance behaviour, both robots instantiate the obstacle avoidance behaviour schema with different attributes or parameter values. The first robot may have a large repulsion constant from the obstacle, while using the same schema, the second robot may have a small repulsion constant from the obstacle. The displayed behaviour for the same schema will show the first robot as too cautious of the obstacles, while the second robot as relatively aggressive, not fearing going near the obstacles.

In case of concurrent behaviours, it is possible that one behaviour *dominates* the other and the most dominating behaviour is displayed (like obstacle avoidance dominates goal seeking near obstacles), or one behaviour *cancels* the effect of the other (like an obstacle demanding the robot to turn right and goal asking the robot to turn left). The behaviours may typically reach an *equilibrium*, like the robot staying in the middle of the corridor as turning at either side will incur an attempt to move the robot at the opposite side.

17.5   Fusion of multiple behaviours

At any point of time a robot may display multiple behaviours and other complementary modules and services which aid the decision-making of the robot (Brooks, 1986; Murphy, 2000). This section talks about the different architectures that facilitate having multiple such behaviours and program modules in one software.

17.5.1   Horizontal decomposition

From a programming perspective, a natural inclination in the design of the solutions is to decompose a problem into several steps that happen one after the other. Assignments are usually broken down into sequential steps. Similarly, mathematical problems are solved step-by-step with later steps building over the previous ones.

Similarly, the overall robotics problem can largely be decomposed horizontally with the sensors at one end and actuators at the other end. The foremost step is taking the sensor readings. Thereafter, the sensor readings are processed for noise removal. The signal so generated undergoes processing to extract information, typically done by image processing, signal processing and computer vision techniques. The information is used for both determining the world knowledge model and the state. The knowledge model includes the grid map used for navigation along with the semantic knowledge of the world (like the position of key items where the robot may have to go to pick items, shape of the objects to pick, primitive knowledge of the machines to operate like pulling a drawer, etc.), the facts the world (like whether the object to grasp is clear or has obstacles, whether the robot will be able to go under an obstacle, over an obstacle or will have to avoid an obstacle, etc.). Many of these are required for task planning to be discussed later in the book. The state of the robot including the pose for navigation and manipulator position is also calculated from the same information. The information in the form of state and world model is then used for planning or decision-making. The planning gives the motion command that the robot must invoke from its controllers. The controllers operate the end level motors. The complete decomposition is shown in Fig. 17.7A.

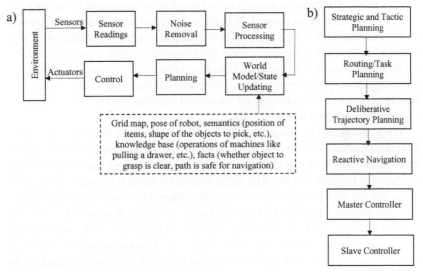

FIG. 17.7 Decomposition of behaviours: (A) horizontal decomposition, (B) vertical decomposition.

17.5.2 Vertical decomposition

From a planning perspective, the most revered architecture is called as the *vertical decomposition* of the problem. Organizations have a vertical hierarchy, and every top layer management decomposes the problem to be done by a lower-level management, the results of which are received and fused by the top-level management. Upon receiving a complex problem, humans tend to break it down into levels of hierarchy and solve the same. Similarly, is the case of the robot.

The robot decision-making is largely a vertical hierarchy. Let us first discuss it with an example of the self-driving cars (Kala and Warwick, 2013). A self-driving car is given a mission which asks the vehicle to solve complex problems like visiting a set of mission sites with some preferences, some ordering constraints, and some Boolean constraints. Say the car needs fuelling from any fuel station before it is impossible to move; while it needs to go to a mall for shopping, deliver a gift to a friend and get dinner from a restaurant, preferably just before heading home. For a vehicle to decide in which order should the places be visited constitutes the *strategic planning* and *tactic planning* of the vehicle. Thereafter, given an order of sites to visit, the vehicle must find a route, including a specification of the roads and intersections to take. This is the problem of *routing*. The route cannot be easily navigated since there are other vehicles and the self-driving car needs to make decisions whether to overtake some vehicle, to follow it, to allow a vehicle to overtake, to pass the vehicle from the left or right, etc. Also, given a decision, the spaces to be left from the vehicles and lanes needs to be decided. This constitutes the problem of *deliberative trajectory planning*. The trajectory needs to be traced in realization of the other vehicles and lane restrictions around. A vehicle that constantly adapts and corrects for the uncertainties with time uses a *reactive navigation*. The decisions of a reactive navigator are typically used by a *master control* algorithm to infer the preferred motion into linear and steering velocity. The master control algorithm signals are given to a *slave controller* that translates the same into the voltages to be applied on the individual wheel and steering motors. This constitutes a hierarchy as shown in Fig. 17.7B.

For a mobile robot, the statements and hierarchy largely remain the same. Imagine a humanoid operating in a service sector needs to carry repair services at several homes in any order, for which the exact repair service operation will be known at the time of visit. The highest-level problem is to order all the operations that the robot needs to do one after the other. This constitutes the levels of strategic planning and tactic planning. Every service will require the robot to arrange a few items here and there, perform actions to acquire information, get all preconditions of the operations met, etc. This constitutes the problem of *task planning*. The task planning approaches shall be discussed later in the book. The task planning gives a smaller level problem to the robot, which is typically a classic motion planning problem, requiring either the mobile base, or the manipulator, or both to go from one configuration to the other. This is called as the *deliberative geometric motion planning problem*. All algorithms discussed so far largely carry a geometric motion planning. Naturally, there will be reactive algorithms fused with deliberative ones to solve the geometric motion planning in a hybrid architecture. The next hierarchy is thus *reactive geometric motion planning problem*. The term geometric is specifically added since task planning is also a deliberative planning approach. The term denotes planning in a continuous configuration space of some unknown geometry. The *master control algorithms* convert the desired motion into the control signals to be given to the robot. The *slave control algorithms* operate each actuator (motor) of the robot to give the desired voltages and signals to operate the physical motor.

In all cases, it should be remembered that all modules will require access to sensors, location, map, and any input as may be given by the human from time to time. These are commonly maintained and accessed by all the modules, giving a little horizontal touch to the overall architecture. It is imperative that there is no strict rule whether the decomposition should be horizontal or vertical, and a mixture of both is the architecture to be typically used in robotics.

17.5.3 Subsumption architecture

Imagine a robot can operate by displaying multiple behaviours, each of which may be more suited for some or the other operation. The preceding section gave the concept of using the behaviours concurrently and summating over the outputs. Typically, this is not a big problem since, at every point of time, very few behaviours operate as per the precondition and typically just one behaviour dominates. So, obstacle avoidance as a behaviour dominates near the obstacles and goal seeking dominates far away. However, there are problems that different outputs could cancel each other, or a genuine behaviour could be wrongly dominated by another behaviour.

Subsumption architecture is another architecture for the behavioural motion of the robot (Brooks, 1986). The motive here is to have multiple concurrent behaviours, however, the behaviours are said to be in a *layered architecture*. The layering denotes a prioritization scheme associated with the individual behaviours. In a subsumption architecture, a higher layer behaviour can *subsume* or *supress* the output of a lower-level behaviour. So, the behaviours do not only add up and cancel each other out. On the contrary, the higher layer behaviours take a dominance and is effective in deciding the navigation of the robot, without concerning the indicative output of the lower layer. The lower layer behaviour may still be displayed when the higher layer behaviour is inactive or does not have its precondition met. In such a case the control is largely left over the lower-level behaviour. The ideology of the subsumption architecture comes from creating purely reactive behaviours. As per the same motivation, the layers cannot encode the internal state, which is an aspect primarily associated with the higher-order living beings and the deliberative algorithms.

To understand the working of the subsumption architecture, let us take a small example. Imagine a robot is doing an inspection of a critical facility and reports any suspicious activity to the security office. The robot keeps wandering around in any random order to catch suspicious people. The most basic behaviour that the robot needs to display is its linear velocity control, which stops the robot from colliding with any obstacle around. The behaviour takes the distance from the front obstacle by a laser scanner or sonars and sets the speed low enough to avoid collision. In the worst case, when the robot is close to the obstacle, the linear speed will become and remain zero.

Thereafter, a higher order behaviour at the first level, is that the robot must be able to display obstacle avoidance. This is a simple implementation of the potential field algorithm, wherein the repulsive forces from the sonars or lidar scan is used to calculate the repulsive force and hence the steering to avoid the obstacles. This is directly transmitted as the angular speed of the robot.

A still higher behaviour is to make the robot wander around randomly. This happens by asking the robot to move towards a random direction, which acts as a goal direction for the robot. The wandering ensures that the robot keeps moving in pursuit of visiting new places for security. The output of the wandering behaviour overrides the generic obstacle avoidance behaviour. However, the wandering also may cause a collision and therefore it is necessary for wandering to have a similar obstacle avoidance.

The wandering is just a random motion of the robot, which it does when it sees nothing interesting. It is possible that the robot has seen something suspicious and therefore needs to walk towards the suspicious looking entity for a closer examination. This means that the robot should no longer move aimlessly but be directed towards the suspicious entity in the environment. This subsumes the random goal of the wandering behaviour and gives a new directed goal. The precondition is checked by the vision cameras onboard. In the process of examination and moving towards the suspicious entity, if the robot is clear about the entity being unsafe or hazardous, it needs to give a signal to the security office, which forms a next higher-level behaviour.

The overall architecture discussed is given by Fig. 17.8. From the figure it may be observed that the subsumption architecture does not do a strict vertical or horizontal decomposition of the problem in hand, however, enables the different behaviours or modules to intermix with each other. The concurrent running of the behaviours can have problems when the behaviours cancel each other's effect. Hence the subsumption architecture allows prioritization, and a higher-level behaviour takes a priority. Not all behaviours will always be active and in the passive state of a higher-order behaviour, a lower order behaviour continues to work. The fusion is not only at the level of the final linear and angular velocities, but also at any other level like selection of immediate goal, selection of intent, selection of inputs for the next behaviours, etc.

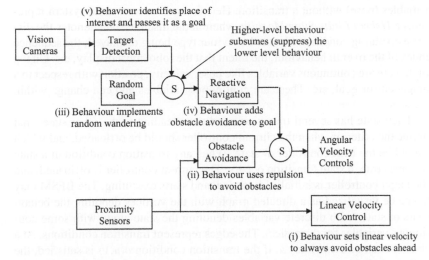

FIG. 17.8 Subsumption Architecture. A higher-order behaviour (like an identified target) takes priority and subsumes (suppresses) a lower order behaviour (like random wandering).

17.6 Behavioural finite state machines

Another interesting philosophy that enables programming for different behaviours applicable in different times is using *Behavioural Finite State Machines* (BFSM; Curtis et al., 2016). In this philosophy, at any point of time, only one behaviour is possible, while the invoked behaviour changes along with time. The program hence keeps switching controllers that drive the robot or keeps switching the behaviour to be displayed. Different behaviours may be more appropriate at different times. Unlike a rule table, wherein the precondition is used as a criterion to select the applicable behaviour, the Finite State Machine can represent sequences of transitions that govern the behaviour to be invoked. This makes it an appropriate tool to model a linear sequence of controller changes, cyclic sequence of controller changes, controller changes abrupted/ split based on a qualifying criterion, etc. The notion is that the preconditions that trigger a state or behaviour change is bound to the current operating state and not as a global set of rules. This makes it possible to better control the transitions.

Every behaviour or controller of the robot is represented by a *state*. The state mentions its own set of sensors, actions, decision-making modules and other requirements. It models the behaviour where the robot moves by interacting with the environment and obstacles around. In a classic Finite State Machine, the state of the system cannot change unless a transition occurs, while the BFSM has controllers as state that move the robot causing a change of the robot's state variables (pose) without a transition. Hence, more correctly, the system represents a *Hybrid Finite State Machine* wherein the discrete state denotes the discrete operating condition like the behaviour type being displayed, the sequence index of the overall behaviour, the intent that the robot must display, etc. On top of it, there are continuous variables like pose, velocity, position with respect to a neighbour or goal, etc. These are continuous variables and can change within the state.

Each state has several transition conditions leading to the other states that define the criterion when the current controller should be offloaded, and which should be the next controller or state. When any transition condition in a state becomes true, a state transition occurs. The current controller is offloaded, and the target controller is initialized, loaded, and starts executing. The BFSM may hence be visualized as a directed graph with the vertices denoting the behaviours or state (with discrete variables denoting the state along with some continuous variables for dynamics). The edges represent transition conditions. At a state (vertex or behaviour) u, if the transition condition $c\langle u, v \rangle$ is satisfied, the state (or behaviour) changes to v. Here $c\langle u, v \rangle$ denotes the condition associated with the transition from state (behaviour) u to state (behaviour) v.

Sometimes timing information is used to denote some state transitions, like a random motion behaviour will be displayed for 2 s, or the robot should follow straight motion for 5 s, etc. In all such cases it may be assumed that a global time is accessible by all the states. This further indicates that the BFSM is more correctly a *Timed Hybrid Finite State Machine*, wherein timing information for all states is naively available.

The transitions may not always be deterministic. In many cases the humans may show stochastic transitions to their behaviours. As an example, on getting late for a meeting, it is stochastic whether to cancel the 'take coffee' behaviour before 'go to meeting' behaviour. The humans may stochastically decide whether to overtake a vehicle or not. Upon qualifying, a transition condition may thus have an effect that is stochastic, and the target may vary based on randomness. The probability distribution of the effect is specified as per the modelling. As an example, on seeing a person ahead walking slowly, the robot may invoke the 'follow the person for 5 s' behaviour with a probability of 0.25, 'overtake from the left side' behaviour with a probability of 0.25, and 'overtake from the right side' behaviour with a probability of 0.5.

Let us take a small example to start with. A robot needs to visit several places in the same order. The robot already has a generic 'motion to goal avoiding obstacle' behaviour. The realization of the sequence of behaviour is given by Fig. 17.9A. A common mistake when programming BFSM with a graph search background is to make goals as vertices and going to goal behaviours as actions. In the BFSM, the convention is opposite, and 'going to goal' behaviours are vertices and 'reaching goal conditions' are edges denoting the transition criterion.

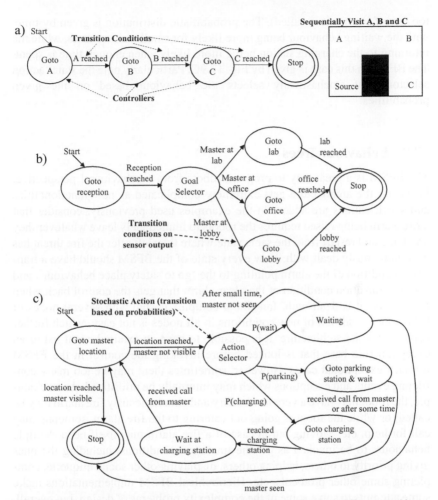

FIG. 17.9 Behavioural Finite State Machine (BFSM) representing different problems: (A) sequential problem, (B) problem with sensors, (C) stochastic action selection.

In another small example, consider that the robot needs to go to the reception, where it will know the master's current location. The robot then goes to the master, interacts, and then comes out. The FSM is given by Fig. 17.9B. The goal selector state uses an external sensor whose output determines the transition condition. In another example, consider that the robot needs to wait for the master, however, the robot waits for a designated time. Thereafter, the robot must choose between the behaviours of 'continuing waiting for the master' (assuming the master is still around), 'go to the charging station' (assuming the master left with a caveat that the robot may return when called back) or 'go to a nearby parking station' (thus neither blocking the way in front of the office nor being too far off in case it is called). The probabilistic distribution is given by time, with the waiting behaviour being more likely for small waiting times, while the returning to the charging station being more likely when enough time is spent. The BFSM in this case is given by Fig. 17.9C. Particularly note the state 'action selector' that stochastically selects the behaviour based on the given probabilities.

17.7 Behaviour Trees

The BFSM is a great way to program robots from a behavioural perspective. However, the architecture gets extremely complicated as new functionalities and specifications are added. In the examples used previously, consider that a fire alarm being raised requires the robots to immediately leave whatever they are doing and restart from the same state where they left after the fire threat has been judiciously dealt with. Now every state of the BFSM should have a transition condition of the alarm pointing to the 'go to safety place behaviour', and another transition condition of the clear alarm that gets the control back when the fire threat is dealt with. Imagine a complex design with several nodes or behaviours. Addition of new conditions in all nodes is not easy, which further complicates the architecture. It is possible to have an external event list or an exception rule base that is looked before invoking the control in the BFSM for such exceptional cases. However, sometimes there may be too many conflicting events and exceptions which may internally be prioritized. As an example, the robot may have a very low battery and simultaneously fire alarm may be called; or while the robot is going out catering to the fire alarm, someone may call for help, etc. Moving out in case of a fire alarm may itself need multiple behaviours to be invoked as per the transition conditions, including the ones giving priority to others, helping others, displaying other social etiquette, completing some other protocols, etc. Hierarchical BFSM implementations make some attempts to solve some of the complexity problems of design, but overall, the system becomes too cumbersome to handle. A BFSM may have several states and several transition conditions that are hard to maintain as the design complicates. Theoretically addition of one new node could have transitions to/from all others requiring a compatibility with all of them.

A better way to organize behaviours is thus to store them hierarchically as a tree. This gives rise to *Behaviour Trees* (Colledanchise and Ögren, 2018; Colledanchise et al., 2019). The behaviours are stored as leaves. The behaviours are hierarchically grouped to model skills or complex behaviour set. Therefore, as one transcends from top to bottom of the tree, the behaviours start getting concrete and specific, while at the lower depths the nodes represent more abstract or complex behaviours. Every complex behaviour at a depth is refined at the next depth, ultimately leading to a specific behaviour at the leaves. Such a hierarchical representation of behaviours enables to easily add skills or update the behaviour flow by working at the correct nodes and branches and abstracting out the details of the other branches and nodes.

Let us first re-define behaviours to suit the modelling. Every behaviour takes a sensor percept or inputs from the state to control the next step of the robot. The behaviours however take a unit control step only on receiving a system *tick*. Consider the obstacle avoidance behaviour that perceives the distance from the obstacles and makes a step in a direction relatively clear of obstacles. The behaviour will only make the robot move by one step if the behaviour receives a system tick. In the absence of a tick the behaviour will be passive. Consider the twin behaviours of obstacle avoidance and goal seeking. At any time if only the obstacle avoidance behaviour gets the tick, the robot displays only obstacle avoidance with no goal seeking and vice versa. If obstacle avoidance keeps getting ticks for 10 s, followed by goal seeking for 5 s, and thereafter obstacle avoidance for 10 s, the robot avoids obstacles for the first 10 s. At the pose after 10 s, the robot starts goal seeking; while after another 5 s, the goal seeking is disabled, and the obstacle avoidance behaviour drives the robot for another 10 s. The decision to invoke (send tick) or not to invoke (do not send tick) the behaviour is programmed in the tree.

The behaviours also send the status to the parent node that helps the parent node to decide whether to send or block the tick. Typically, the behaviours will send the status as *running* which means that the behaviour can be further displayed subjected to the receipt of ticks. However, sometimes a behaviour may have completed, and subsequent ticks are thus wasteful. In such a case a *success* status is sent by the node to its parent. Consider the goal seeking behaviour as an example. Once the robot lands at the goal, the behaviour sends success as a status to intimate the parent a successful termination of the behaviour. However, sometimes the behaviour may fail instead and thus getting the subsequent ticks is useless. In such a case the behaviour sends the *fail* status to its parent. Consider the same goal-seeking example. If the goal is not visible or goal coordinates cannot be computed, the behaviour will fail as the robot cannot do anything. Similarly, obstacle avoidance behaviour may get the robot stuck at the local minima and the robot may be unable to move, further, in which case the behaviour returns a failure.

Now let us design the nonleaf nodes that drive the behaviour invocation logic by controlling the system ticks. Every nonleaf node may or may not get a tick from the parent. In case a node does not get a system tick, it is passive and cannot do any computation. Upon receiving a tick from the parent, the node must decide which child (or multiple children simultaneously) need to be transmitted the system tick, depending upon the status (running, failure, success) of the children. The choice is only for one control step in which the robot takes a unit action. At the next control cycle, a new decision is made.

A common requirement is that the robot performs one action out of several possible actions. This is done by the *selector node* (?). The node is given a set of children in the order of preference, and the node attempts to execute one of them till the node gives a success status message. Say, the master asks the robot to (in the order of preference) either get cake from the refrigerator, or get snacks from the snack bar, or order pizza. Now the robot attempts to get cake. If it is not possible to fetch cake (as there is none in the refrigerator), the robot will attempt to get snacks. If that is not possible as well (snack bar closed), the robot will order pizza. The selector node attempts to execute the first child. Once the first child sends the success status message, the purpose of the selector node is over, and the selector node sends the success message to its parent. However, it may be possible that the first child eventually fails. Now the selector node sends the tick to the second child and continues execution of the same. If any child fails, the selector node starts the execution of the next child. Once any child returns a success, the selector node reports a success to its parent. If all children fail, the selector node reports a failure to the parent node. Typically, in most control cycles, the selector node will have a node that reports a running status, asking for more ticks in the future cycle. Consequently, the selector node shall send running as the status to its parent.

Very commonly, the robots need to sequentially perform actions, like visiting a sequence of places or picking up (and placing) a sequence of items. The *sequence node* ($\rightarrow$) models the same behaviour by sending ticks sequentially to all children. The node first sends the ticks to the first child. Eventually the first child will send a success status showing that the operation is successfully accomplished. Now the sequence node starts sending the tick to the second child. This goes on till the last child accomplishes the tasks and sends a success signal. After all children record a success status, the sequence node has completed the asked behaviour and the sequence node sends a success status to its parent. However, at any time, if any node fails, the sequence cannot be completed and hence the sequence node responds to the parent with a failure message. In most control cycles, the current node receiving the tick would still be running, in which case the sequence node sends a running status to its parent.

Sometimes, a robot may display several behaviours concurrently and in parallel. This is modelled by the *parallel node* ($\|$). As an example, the robot may be using cameras to look at suspicious people around as a behaviour, while concurrently another behaviour may be handling the motion of the mobile base of the robot. The parallel node, upon receiving a tick from the parent, sends the tick to all the children simultaneously. All children behaviours are invoked. If all nodes return a success status (maybe at different times), the parallel node sends a success status. Generalizing the success condition, the parallel node returns a success if at least m of n children send a success status. Using the same generalization, if more than $n - m + 1$ nodes return a failure, it is not possible that the node will eventually terminate successfully, and the parallel node sends a failure to its parent. Commonly several (more than $n - m + 1$) nodes will send a running or success status (with less than m nodes returning a success status) and therefore the parallel node will intimate the running status to its parent, hoping to get the ticks in the subsequent control cycles.

A major difference with the behaviours in this section in contrast to the other sections is the presence of the *precondition* which is not modelled so far in Behaviour Trees. A behaviour typically executes only if the invoking precondition is true. This is done by adding *conditional nodes*. The conditional nodes do not perform any action. They only take sensor percept and state information to check if a condition is true or false. In case the condition is true, the node returns success, while in case the condition is false the node returns a failure. As an example, say the goal seeking behaviour is only to be performed if the goal can be seen (precondition). The goal seeking action with goal visibility as a precondition is thus performing the sequence (goal visible) → (go to goal). In case the condition 'goal visible' fails, the sequence node returns a failure (in the current control cycle only) without calling the 'go to goal' behaviour. In any future system tick, if the goal visible becomes true, the go to goal behaviour will be sent a tick by the sequence node.

Similarly, let us also model the *postcondition* or effects of the behaviours. In general, a behaviour must only be executed if the postcondition has not already been met. Many behaviours check the validity of the postcondition and (if valid) return success instead of performing the action. In such behaviours checking the postcondition explicitly is not required. Many other behaviours will explicitly need the postcondition to be checked. Consider that a manipulator must place the glass at a specific location before pouring in water. The glass can be picked up from the current location and placed at the desired location. Here the postcondition is the presence of glass at the desired location. If this postcondition is not checked before executing the action and the postcondition is already valid, the manipulator will pick and place the glass at the same place. Sometimes behaviours have no success conditions and postconditions thus need to be explicitly written. As an example, a robot carrying items around an office may be waiting for the person to press the 'go' button, until which the robot will wait. Waiting as a behaviour has no success condition that comes from an external sensor in this program logic. The postconditions are modelled by using a conditional node. Consider the selector node (glass at desired pose) ? (pick and place glass). If the glass is already at the desired pose, the selector node will return a success without executing the 'pick and place glass' behaviour.

The different nodes are shown in Fig. 17.10. Note that the designer is free to propose and program new types of nodes as per need. An above example required the robot to 'follow the person for 5 s' behaviour with a probability of 0.25, 'overtake from the left side' behaviour with a probability of 0.25, and 'overtake from the right side' behaviour with a probability of 0.5. This requires a stochastic selector node that randomly selects a node with the given probability, while once a node is selected, the node should have the memory to always select the same node in the subsequent ticks. Custom nodes can be created based on probabilities, timers, costs, rewards, etc. Sometimes the nodes need to re-order children or dynamically change the behaviour hierarchy based on conditions known at runtime.

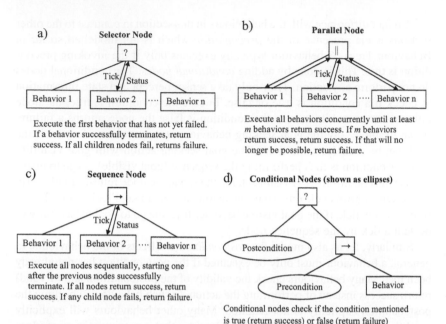

a) **Selector Node**

Execute the first behavior that has not yet failed. If a behavior successfully terminates, return success. If all children nodes fail, returns failure.

b) **Parallel Node**

Execute all behaviors concurrently until at least *m* behaviors return success. If *m* behaviors return success, return success. If that will no longer be possible, return failure.

c) **Sequence Node**

Execute all nodes sequentially, starting one after the previous nodes successfully terminate. If all nodes return success, return success. If any child node fails, return failure.

d) **Conditional Nodes (shown as ellipses)**

Conditional nodes check if the condition mentioned is true (return success) or false (return failure) without invoking any physical action

FIG. 17.10 Behaviour Trees. Each behaviour (leaf) receiving a system tick takes a unit control step as per modelling. Invoked behaviours return status (successfully terminated, failed, running). Each node sends ticks to one or more children. (A) Selector node (?), (B) parallel node (∥), (C) sequence nodes (→), (D) conditional nodes. The precondition conditional node ensures the behaviour is never executed unless the precondition is true. The postcondition conditional node ensures that the behaviour is never executed after the postcondition has been met.

Let us make the Behaviour Trees to model some interesting scenarios. As the first scenario consider that the robot must sequentially visit four places (say A, B, C, and D). However, if the master calls, the robot must abandon this task temporarily and go to the master to hear the work. Thereafter the robot must go to the place asked by the master. In case the master is not present, the robot must wait. At any time, if the battery is critically low (<20%), the robot must go for charging and continue charging till the battery is 100%. The solution is shown in Fig. 17.11A. The sequence of visiting five places is a simple sequence node. It is assumed that the 'Goto Goal' behaviour returns a success on reaching the goal. The 'master called' is added as a precondition for the master calling behaviour, while a wait behaviour (with 'master present' postcondition) is added to let the robot wait till the master comes. The charging is the most important behaviour and is thus added first guarded by the precondition 'charge <20%'. The behaviour terminates when the last item in the sequence (wait) terminates with the postcondition of a 100% charge. In this example, at any time only one leaf behaviour is called, while the specific behaviour called changes along with time depending upon the success or failure of the conditional nodes (that use external sensors) and the behaviours.

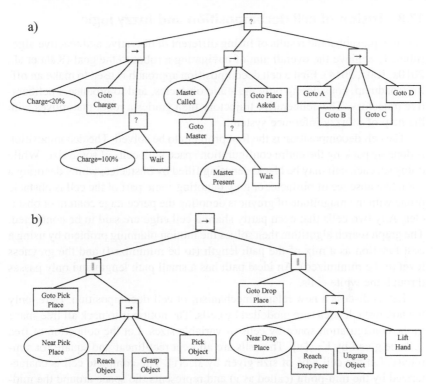

FIG. 17.11 Behaviour Tree for the example. (A) The robot generally intends to visit places A, B, C, and D sequentially in the same order. However, whenever the master calls for an urgent work, the robot must go to the master, wait for the master (if the master is not currently available), take instruction and go to the place pointed by the master. At any stage, a low charge immediately triggers going to the charger and getting charged till the battery is 100%. (B) The robot needs to pick and place an object, in a sequence. For the picking the robot attempts to parallelly go to the picking site and pick the object while the robot is moving. Picking only initiates when the robot is near enough as a precaution. The robot will not execute drop sequence until the item is safely picked. Similarly for drop.

As another example, consider that a mobile manipulator must pick an item from one location and drop it at another location. While the mobile manipulators generally first operate the mobile base and then the manipulation arm. Ambitiously consider that the manipulation arm can prepare to pick up the object in anticipation such that the manipulation arm and mobile base can work in parallel. The manipulation arm however must only be operated if the mobile base is near enough to the picking location, otherwise the motion of the mobile base can incur collisions. The solution is given in Fig. 17.11B. Picking and dropping are similar problems, solved as a sequence. Both subproblems operate the mobile base and the arm in parallel. The arm is guarded by the 'near place' precondition, followed by the picking actions in a sequence.

17.8   Fusion of cell decomposition and fuzzy logic

We now present some results of fusing different deliberative and reactive algorithms to achieve the overall aim of navigating a robot to the goal (Kala et al., 2010a; Kala, 2014). First a cell decomposition approach is used to make an offline roadmap, considering only the static obstacles, and to produce a deliberative trajectory. The deliberative trajectory then guides a robot that moves with the help of a fuzzy inference system.

The cell decomposition is the first problem to be solved. The decomposition is done by packing the entire configuration space with cells of some size. While doing so, each cell may be black, denoting filled by obstacles; white, denoting a complete absence of obstacle; or grey denoting some part of the cell is obstacle prone with the magnitude of greyness denoting the percentage content of obstacles. Any two cells that even partly share a cell edge are said to be connected. The graph search algorithm then solves the motion planning problem by using a cost function as a mix of the path length (to be minimized) and the greyness level (to be minimized). An ideal path has a small path length and only passes through the white cells.

Let us discuss a new related mechanism of cell decomposition. Here, only the nonobstacle areas are modelled by cells. The notion is to pack all free space in the configuration space by cells of variable sizes. Let the collection of free cells be given by $V = \{v\}$. The cells are all kept rectilinear and square for simplicity. Let the cell v be of size given by $size(v)$. The complete cell is characterized by the mid-point (called as v) and represents the space around the mid-point upto a size of $size(v)/2$ along every axis in the positive side and $size(v)/2$ along the negative side, giving the cell volume as Eq. (17.4)

$$volume(v) = \times_d [v_d - size(v)/2 v_d + size(v)/2]   \quad (17.4)$$

Where $\times_d$ means the cross product in the dimension d. v_d represents the dth dimension value of the mid-point. For a 2-D space, this corresponds to Eq. (17.5) with (v_x, v_y) denoting the centre of the cell.

$$area(v) = [v_x - size(v)/2 v_x + size(v)/2] \times [v_y - size(v)/2 v_y + size(v)/2]$$
$$\quad (17.5)$$

Since only free spaces are being modelled, the entire space bounded by all cells v must be in the free configuration space, or $volume(v) \subseteq C^{free}$ or $q \in volume(v) \Longrightarrow q \in C^{free}$. For bounded spaces, it also means that the complete cell must obey the bounds of every axis of the space, or Eq. (17.6)

$$LB_d \leq v_d - size(v)/2 \leq v_d + size(v)/2 \leq UB_d   \quad (17.6)$$

where LB_d is the lower bound of the dth dimension and UB_d is the upper bound of the dth dimension.

The cells must additionally be nonoverlapping, or the volume intersection of any two distinct cells (v_i and v_j) should always be an empty set (ø), given by Eq. (17.7).

$$\forall_{v_i, v_j, v_i \neq v_j} volume(v_i) \cap volume(v_j) = \emptyset \qquad (17.7)$$

There are limitations applied on the size of the cell, or $sizeMin \leq size(v_i) \leq sizeMax$. The cells cannot be so small (less than $sizeMin$) that it becomes computationally difficult to process the cells in online planning. Too small cell size makes it possible to have too many cells that makes the graph search computationally heavy. Similarly, if the cell size is too large (larger than $sizeMax$), there will be a severe loss of optimality. Specifically, the centre of the cell is used as a characteristic point for the robot. If the cell is large, the robot may end up keeping too much clearance which affects optimality. The setting of the factor $sizeMax$ is in cognizance of the desired maximum clearance and the acceptable deviation in path length.

The placement of cells in the free space is done to minimize the total number of cells needed to fill the entire free space, or minimize $|V|$, where $|V|$ is the set size of all cells V. At the same time, the cells together must cover the entire free space, or as much as possible, that is the objective is to maximize $\cup_{v \in V} volume(v)$. The cell placement algorithm works in a greedy manner and starts fitting cells in free space all around starting from the maximum size. If the cells of maximum size can no longer be filled in, a little smaller size is tried in the entire configuration space. If that also fails, the size is further reduced. Iteratively, the entire configuration space is filled with cells. The reduction of size takes place only till the minimum accepted size. The generic algorithm is given by Algorithm 17.1.

Algorithm 17.1 Cell decomposition (C^{free}).

Input: Free configuration space, typically available as a grid map
Output: Cells

1 $V \leftarrow \emptyset$;   ▷ list of cells
2 $used \leftarrow \emptyset$;   ▷ space already filled by cells
   ▷ fill cells in a greedy manner ;
3 **for** $size$ *from* $sizeMax$ *down to* $sizeMin$ *in steps* Δ **do**
4   **for** *cells with centre* v *and size 'size' such that*
       $volume(v) \subseteq C^{free}, volume(v) \cap used = \emptyset, LB_d \leq$
       $v_d - size(v)/2 \leq v_d + size(v)/2 \leq UB_d$ **do**   ▷ unused cells
5          $c \leftarrow clearance(V_i)$;
6          $V \leftarrow V \cup < v, size, c >$;
7          $used \leftarrow used \cup volume(v)$;

The *used* data structure stores the parts of the configuration space which have already been covered by some cell and therefore should be avoided by the other cell. The data structure is implemented as a hash table of dimensionality same as the configuration space. In the case of 2-D, it represents a map like the grid map, and stores the area covered by cells.

The cells are connected by a neighbourhood function. The neighbourhood function (δ) states any two cells v_i and v_j to be neighbours if they are at a distance η times the maximum grid size, that is Eq. (17.8)

$$\delta(v_i) = \left\{v_j : d(v_i, v_j) \leq \eta \, sizeMax\right\} \tag{17.8}$$

Here d is the distance function. η is a constant with a small value implying limited connectivity and a large value implying large connectivity, in which case the clearance may not be guaranteed. Every neighbouring cell is checked for collision by a straight-line motion between the characteristic points and a feasible motion adds the edge.

The robot uses this roadmap for online planning. First, the source and goal are added and connected by the same neighbourhood function. The A* algorithm is used to search for a path. The cost function is given by path length and clearance. The tradeoff between the two is controlled by another factor. The same algorithm also acts as the re-planning algorithm if the map changes and the same is informed to the robot. This changes the deliberative path as and when required.

The reactive algorithm is used as the Fuzzy Logic controller. Broadly, the algorithm steers the robot to avoid the obstacles around, which is handled by the obstacle avoidance rules; and simultaneously to steer the robot towards the goal, which is handled by the goal-seeking rules. The fusion of the two algorithms takes place by providing the reactive algorithm the subgoals, which are points taken from the path of the deliberative trajectory. In this approach, the last visible point by the robot at the deliberative trajectory is used as the subgoal of the reactive algorithm. Only the static map is used for visibility.

The results are shown in two cases. In the first case, the cell decomposition is shown in Fig. 17.12A and B. The deliberative trajectory is shown in Fig. 17.12C. All figures show the configuration space. The robot moves in the workspace. The first decision (also to be continuously made) is the choice of the subgoal. Fig. 17.12D shows all configurations visible to the robot at the source. The last point visible within the deliberative trajectory is taken as the subgoal. The motion of the robot is shown in Fig. 17.13. As the robot moves ahead, further points in the trajectory become visible to the robot and therefore the subgoal also moves ahead. This stops when the subgoal becomes the goal, and the subgoal can move no further. The same set of figures for another scenario are shown in Figs. 17.14 and 17.15.

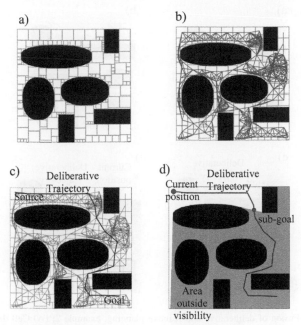

FIG. 17.12 Fusion of deliberative and reactive planning, example 1: (A) cell decomposition by placing free space with cells of different sizes, (B) ioining adjacent cells by edges, (C) computing a deliberative trajectory, (D) selection of 1st subgoal based on visibility.

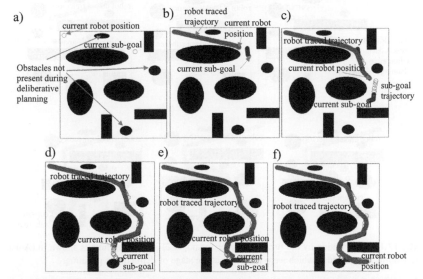

FIG. 17.13 Navigation by the fuzzy navigator (example 1) along with time is shown in (A–F). Subgoal is the last point visible in the configuration space (workspace is plotted) in the deliberative trajectory. Subgoal comes near to the current position near the corner due to low visibility.

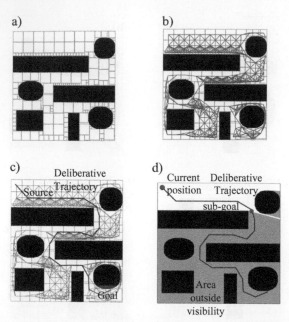

FIG. 17.14 Fusion of deliberative and reactive planning, example 2: (A) Cell decomposition, (B) joining adjacent cells by edges, (C) computing a deliberative trajectory, (D) selection of 1st sub-goal based on visibility.

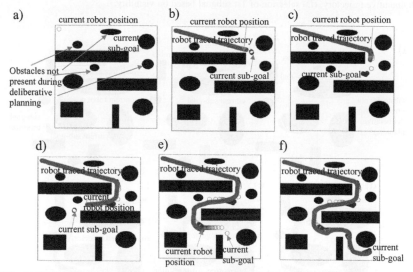

FIG. 17.15 Navigation by the fuzzy navigator (example 2) along with time is shown in (A–F). New obstacles are added which were unknown at the stage of deliberative planning.

17.9 Fusion with deadlock avoidance

As another choice, let us take the Probabilistic Roadmap as the deliberative planning mechanism (Kala, 2018). The sampling technique uses a hybrid of different samplers to generate all homotopic group of paths in a small computation time.

The roadmap is made offline, which is used for the online working of the reactive algorithm. For the reactive algorithm, the geometric planner is considered. The planner works on the principle of geometries of the obstacle and other robots as sensed by the onboard laser sensors. The gaps are identified. The largest gap, which is also closest to the angle to goal and has obstacles at the furthest distance is chosen for the motion. Here also fusion happens using the subgoals. However, the subgoal is fixed to a point in the deliberative path after a threshold distance to the current position of the robot. To complicate the problem, let us keep numerous robots in the scenario that act as dynamic obstacles to each other.

17.9.1 Deadlock avoidance

When there are multiple robots, it is common for the robots to get stuck even though they may be operating with sophisticated algorithms. As an example, consider the situation in Fig. 17.16A. The robots need to go to the opposing ends and are blocked by the other two robots. The solution is that the robots take a step back, which is not what the reactive planning algorithms do.

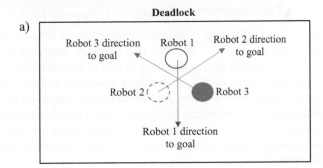

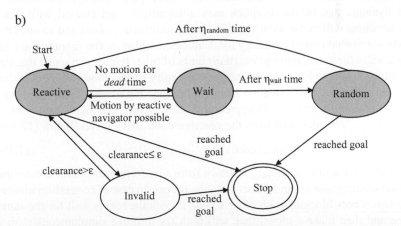

FIG. 17.16 Deadlock: (A) deadlock caused by three robots by a reactive planner, each robot cannot move until the other two move. (B) BFSM for deadlock avoidance. The robot, on detection of a deadlock (no motion for some time), first waits, and then moves in a random direction (deadlock avoidance). Reactive → Wait → Random is the normal operation cycle. If clearances are very small in any state, adjustments are made by the reactive methodology.

The robot normally is in a state wherein it moves by the reactive navigation system (guided by a deliberative trajectory through subgoals). The blockage situation discussed is formally called the *deadlock*, which requires mechanisms for deadlock detection and subsequently deadlock avoidance. In the simplest case, *deadlock detection* is done by monitoring the robot and when the robot is unable to move for some time, a deadlock is said to have occurred. This excludes very minor changes in the position of the robot to counter small changes. The deadlock is said to have occurred if the robot cannot sufficiently move for the past *dead* amount of time. For *deadlock avoidance*, the algorithm first makes the robot wait, and thereafter makes the robot take random steps.

17.9.2 Modelling behavioural finite state machine

Suppose $T(a, b)$ denotes the transition condition from state a to the state b in the BFSM. Deadlock detection is the transition condition from the reactive to the wait state, when the robot cannot move for *dead* time, modelled as Eq. (17.9)

$$T(\text{reactive}, \text{wait}) = \forall_{t-\text{dead} \leq s \leq t, \ s>0} u_{\text{reactive}}(s) = 0 \qquad (17.9)$$

Here $u_{\text{reactive}}(s)$ is the input of the reactive navigator at time s, consisting of a linear and angular velocity. The *dead* is a constant, implying if there is no motion for a *dead* unit of time, a deadlock is said to have occurred. The parameter is important to set. Typically, a deadlock avoidance in high congestion situation results in the robot entering another deadlock, and only after a series of deadlock avoidance, the robot can navigate to pass the congestion areas. In such a case, a high value wastes too much time with the robot not doing anything. Small values may wrongly call a situation a deadlock and make the robot take unwanted movements.

In many cases, considering that the other robots participating in the deadlock are dynamic agents, the deadlock may automatically get cleared with time. Some robots will move, clearing the way for some other robots and so on. This is the *wait* behaviour as a strategy for deadlock avoidance. The robot in the wait state waits for a maximum of *maxWait* units of time. It may so happen that the deadlock seems to have cleared before the expiry of this time and the reactive motion continues. The transition from wait to reactive states is hence governed by the possibility to make an appreciable move (there is a significant control input $u_{\text{reactive}}(t)$ calculated from the reactive controller), given by Eq. (17.10).

$$T(\text{wait}, \text{reactive}) = (u_{\text{reactive}}(t) > 0) \qquad (17.10)$$

The exact waiting time η_{wait} is drawn from a uniform distribution to have the actual waiting time as random between the robots. In case of congestion involving two robots blocking each other's way, if both the robots wait for the same time and then make a move, they will both try to move simultaneously (in a random direction) and may cause another deadlock. However, if the robots wait for random times, one of the robots with a lesser waiting time will take a random step (by the next strategy) and will clear the way for the other robot. This also happens practically when a person entering a door and a person leaving the door come together simultaneously, one waits and the other steps away showing courtesy towards the other person. If both people step away at the same time, no one will be able to get in or out. This could happen indefinitely if the waiting time is constant, wherein both will step away at the same time, knowing that the other person has stepped away, step in at the same time, then step out after waiting for the same time, and so on.

maxWait another parameter, which has the same tradeoffs. Keeping a large value makes robots wait for too long without moving, while small values mean that the robots have a strong urge to go to the behaviour of avoiding deadlock even though a deadlock may have already cleared. The transition conditions are given by Eqs (17.11) and (17.12):

$$T(\text{wait}, \text{random}) = (t_{\text{wait}} \geq \eta_{\text{wait}}) \tag{17.11}$$

$$\eta_{\text{wait}} \sim U(0, \ maxWait) \tag{17.12}$$

Here t_{wait} is the time that the robot has spent in the wait behaviour. $U(0, maxWait)$ is a uniform distribution between 0 and *maxWait*. η_{wait} is the amount of time that the robot decides to spend in the waiting behaviour.

If the deadlock does not clear, it is important to make movements to get the robot moving. This is done by switching to a *random motion behaviour*, wherein the robot makes random moves. The robot displays this behaviour uninterrupted for η_{random} time. Like the η_{wait} variable, the time duration of displaying the random motion behaviour is taken as a random time subjected to a maximum of *maxRandom*. It is important not to terminate this behaviour any earlier because it may appear that the deadlock has cleared as it is possible to make the robot move, while the robot if let loose to the reactive planner can end up in the same deadlock again. It is important to significantly move the robot. With the same notion, *maxRandom* cannot be too small to risk the robot getting to the same deadlock again and it cannot be too large to force the robot to make unnecessary moves for a prolonged time. After the random motion, the robot continues to move by the reactive planner. If the deadlock is not cleared, the same set of transitions will invoke the behaviour again. The transition conditions are given by Eqs (17.13) and (17.14):

$$T(\text{random}, \text{reactive}) = (t_{\text{random}} \geq \eta_{\text{random}}) \tag{17.13}$$

$$\eta_{\text{random}} \sim U(0, \text{maxRandom}) \tag{17.14}$$

Here η_{random} is the time for which the robot displays the random motion behaviour, which is taken from a uniform distribution between 0 and *maxRandom.* t_{random} is the time that the robot has spent in the random motion behaviour.

The random movement may be done by taking a target random vector and making a move towards the chosen vector at the maximum speed in the same direction, avoiding collisions. The random direction is chosen so that it is possible to move appreciably in the same direction. This behaviour is invoked for some time.

An interesting case happens that because of the dynamics of the different robots, two robots may not collide but may get closer to each other. Imagine that one robot sees the other robot at a distance and makes a step forward towards the second robot and the second robot also simultaneously makes a step forward towards the first. In this case, both robots will have a lesser distance than anticipated. The distances can also reduce due to noise, modelling errors, sensing errors, control errors, etc. In such a case, it is important to take appropriate action to maintain a sufficient distance between the robots. This is called the *invalid state* wherein corrective measures are applied to resume normal functioning. The transitions are given by Eqs (17.15) and (17.16):

$$T(\text{reactive, invalid}) = (\text{clearance} \leq \varepsilon) \tag{17.15}$$

$$T(\text{invalid, reactive}) = (\text{clearance} > \varepsilon) \tag{17.16}$$

Here ε is a small constant implying the least needed clearance.

At any instance of time, if the robot reaches the goal, the *stop* state has occurred. The transitions are given by Eqs (17.17)–(17.19):

$$T(\text{reactive, stop}) = (q = \text{Goal}) \tag{17.17}$$

$$T(\text{invalid, stop}) = (q = \text{Goal}) \tag{17.18}$$

$$T(\text{random, stop}) = (q = \text{Goal}) \tag{17.19}$$

As a summary, the set of states discussed for the BFSM are given by $S = \{Reactive, Wait, Invalid, Random, Stop\}$. The robot starts from the reactive state, wherein it is guided by the reactive navigator. The stop is the exit state when the robot reaches the goal. The different states and the transition conditions are shown in Fig. 17.16B.

17.9.3 Results

The working of the algorithm is shown for two scenarios. In the first scenario, the roadmap is shown in Fig. 17.17. The motion of the robots is shown in Fig. 17.18. The same set of figures for another scenario is shown in Figs. 17.19 and 17.20. In both cases, the robots are generated at the four corners and they need to go to the opposite corners. The robots easily avoid each other on the way. The robots enter deadlock situations at many points in time. The deadlock is handled by the same scheme.

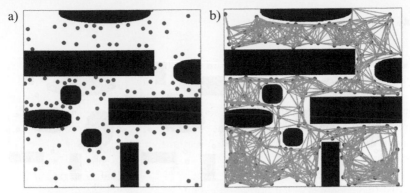

FIG. 17.17 Roadmap used for scenario 1: (A) vertices of PRM, (B) edges of PRM.

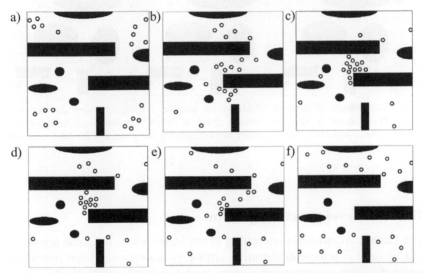

FIG. 17.18 Results of deadlock avoidance for scenario 1 along with time are shown in (A–F). The position of robots at different times is shown. The robots go from the four corners to the opposite four corners, causing blockages on the way cleared by the deadlock avoidance scheme.

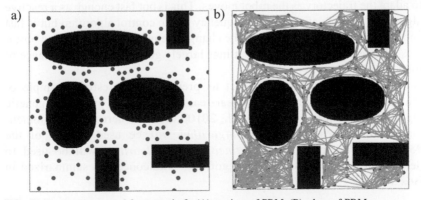

FIG. 17.19 Roadmap used for scenario 2: (A) vertices of PRM, (B) edges of PRM.

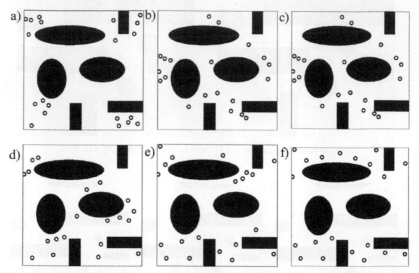

FIG. 17.20 Results of deadlock avoidance for scenario 2 along with time are shown in (A–F). The position of robots at different times is shown. The robots go from the four corners to the opposite four corners, causing blockages on the way cleared by the deadlock avoidance scheme.

17.10 Bi-level genetic algorithm

The Genetic Algorithm is interesting in implementation. The same algorithm can be used in a more deliberative setting by acting as a global optimizer, while the same algorithm can also be used as a local search which is near-reactive. The near-reactive nature is not a loss since it has already been discussed that the trajectory correction algorithms like the elastic strip and genetic correction of the trajectory can be done at a reasonable frequency to adapt the trajectory as the robot moves. This is not fast enough as a reactive algorithm working at nearly the frequency of the control algorithm but is fast enough to make the robot react to the changing environment or moving obstacles. Hence the implementation settings largely governed the performance of the algorithm.

To demonstrate this nature, let us present an interesting example of the fusion of a deliberative and a near-reactive algorithm by using a genetic algorithm at both levels (Kala et al., 2010b, 2011). Thereafter, the nomenclature of the *coarser genetic algorithm* will be used instead of the deliberative algorithm and the *finer genetic algorithm* will be used to denote the near-reactive genetic algorithm. The concept is summarized in Box 17.1.

Box 17.1 A summary of bi-level Genetic Algorithm based motion planning.

Motion Planning with dynamic obstacles
- If the obstacle trajectory is known, any forward simulation algorithm can be used on a spatial-temporal map
- If obstacle position is not known, it needs to be guessed by:
  - (i) Assuming obstacle is static: simple, poor performance
  - (ii) Assuming constant (current) speed forever in the future: simple (used here)
  - (iii) Guessing intended target and trajectory to target by heuristics: hard
  - (iv) Learning/Predicting the behaviour using LSTM and similar models: hard

Overall approach
- Continuous trajectory optimization as the robot moves
- Done at two levels to suit complex maps

Deliberative planning
- Coarser Genetic Algorithm
- Works on a lower resolution
- Considers only static obstacles (predicted obstacle trajectory is unreliable in long-term)

(Near-)Reactive planning
- Finer Genetic Algorithm
- Works on a finer resolution
- Computation time-limited by giving a small local map around the current robot position
- Predicts future dynamic obstacle trajectory making a spatial-temporal map
- For fitness evaluation, simulate a candidate trajectory of the robot and simultaneously all other obstacles with guessed trajectories. Compute collision, length, clearance, etc.

Integration of deliberative and reactive planning
- A point in the deliberative trajectory outside the local map of the finer Genetic Algorithm (subgoal) acts as the goal for the finer Genetic Algorithm.

17.10.1 Coarser genetic algorithm

First, let us concentrate on the coarser Genetic Algorithm. The previous examples used a roadmap-based approach and therefore it was possible to process the map offline considering the static obstacles only. The Genetic Algorithm does not work by producing an offline roadmap. This algorithm runs in two stages. In the first stage, it quickly tries to find the deliberative trajectory. The initial movement may be delayed a little which appears as the robot taking a while to think about the motion, which also happens with humans. Thereafter, the algorithm continuously adapts the deliberative trajectory for any changes that it may be able to spot.

The design of a genetic algorithm-based motion planning is already discussed, and the same principles are used. Since some results must be readily available and the algorithm must not take long to adapt to any map changes, the resolution of collision checking is typically low. This is a valid mechanism considering that the coarser genetic algorithm only provides a reference for the finer genetic algorithm and the motion of the robot is by the finer genetic algorithm alone.

A different variant is also presented. The coarser genetic algorithm works on a static map available offline, which is summarized in a lower resolution. This is done by placing cells all around the map and noting the greyness level as the percentage of obstacle content in the cell. The cells may be re-binarized by making any cell with a reasonable obstacle content as black. This step is the same as reducing the resolution of the input configuration space. The Genetic Algorithm uses incremental complexification. The number of via points in the path of the robot defines the problem complexity, which is increased along with time, subjected to a maximum limit.

The coarser Genetic Algorithm keeps optimizing the coarser trajectory as the robot moves. New static obstacles or removal of static obstacles are handled by the coarser Genetic Algorithm during optimization. The goal is always fixed, but the source keeps changing. The robot at any point in time considers the best individual to be used as a guide for motion planning. When an individual crosses a via point, it is deleted from the genetic individual. The crossing is detected by the sign of the projection, given by Fig. 17.21A–B.

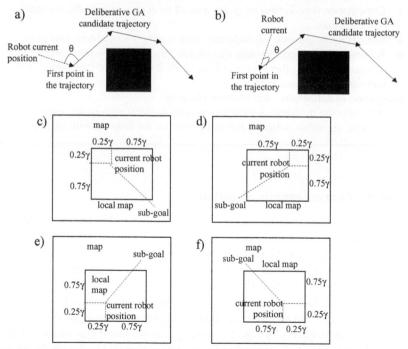

FIG. 17.21 Bi-level Genetic Algorithm Planning Concepts (A and B) If a robot crosses a via point, it is deleted from the genetic individual, identified by the angle θ. (C–F) The local map is given to the finer genetic algorithm, as a fixed size around the robot's current position. The subgoal is a point just outside the local map in the trajectory of a coarser Genetic Algorithm. The finer Genetic Algorithm optimizes the trajectory from the current position to the subgoal.

The coarser Genetic Algorithm is not fed with dynamic obstacles that continuously move. This is because by the time the robot would come near the goal, the dynamics of the obstacles near the goal would change dramatically which is unknown. Since the robot does not know the obstacle's trajectory, it will have to make a guess, which can only be accurate for a short time. Dynamic obstacles are thus not considered for global optimization. From the Velocity Obstacle method to the potential field algorithm, the dynamic obstacles are considered only when near the robot.

17.10.2 Finer genetic algorithm

The finer genetic algorithm is responsible for reacting to the suddenly appearing obstacles, account for the moving obstacles nearby, and working at a finer level optimization of the trajectory. The optimization needs to take place in near-real-time and therefore it is important to make restrictions for faster performance. The coarser Genetic Algorithm restricted the resolution. The finer Genetic Algorithm restricts the region of the space where optimization is carried while working on the highest possible resolution. This means that out of the entire space, a small segment of size γ around the robot's position in each dimension is taken and given to the finer Genetic Algorithm. The complexity is limited as there can only be very few via points in the small segment carved out.

The robot primarily moves towards the subgoal and rarely moves back. Hence most of the map (heuristically 75% of the space) is towards the subgoal and 25% is away from the subgoal. A local map of volume γ^d is given to the finer Genetic Algorithm, where d is the number of dimensions. The local map for all the possible cases in 2-D space for a point robot is given in Fig. 17.21C–F.

The fusion of the coarser and finer Genetic Algorithm is again using subgoals, where a subgoal taken in the deliberative path of the coarser Genetic Algorithm acts as a goal for the finer Genetic Algorithm. The subgoal is taken as a point in the coarser Genetic Algorithm trajectory at a fixed distance from the current position of the robot. The subgoal is taken outside the local map window for optimization.

17.10.3 Dynamic obstacle avoidance strategy

In this section, the obstacles are taken to be dynamic. Typically, the dynamic obstacles are of two types, the ones whose trajectories are known before-hand and the ones whose trajectories are not known. The example of the former includes multirobot motion planning with communication, in which the robots communicate their precise trajectories to all others. If the trajectory of the obstacle is known, the avoidance strategies can be precalculated. As an example, if it is known that the person in front on the pavement will go to a particular shop and will turn, the person walking in the opposite direction can account for the turn of the person beforehand. This can be valuable information for planning.

However, if no information about the trajectory of the obstacles is available, planning becomes harder. In such cases, it is important to guess the possible trajectories of the obstacles which may be other robots, people, pets, or other dynamic agents. The problem is called *intent detection*. Intent detection is a hard problem. One means to solve the problem is to guess the goal location by knowing the interesting locations in the map and the direction that the obstacle is mostly looking at. So, the person may be going to the exit, vending machine, etc., which can be guessed given the motion. Thereafter, knowing the goal, the simplest trajectory that connects to the goal is used. This however converts the problem into a multirobot motion planning problem without communication, which is another hard problem. A recent approach to solve the trajectory prediction problem and the intent detection problem is by using *Long Short-Term Memory Networks* which can regress the motion behaviour and extrapolate the motion of the person.

As the simplest case, the intended direction of motion of the person can be guessed by the maximum direction that the person faces for the past window of time. The humans mostly travel straight towards the subgoal and the trajectory is only temporarily affected only by the obstacles around as the humans turn to avoid them. Similarly, the preferred speed of navigation of the person is the one that the person displayed in the recent past window of time. The finer genetic algorithm only does the short-term obstacle trajectory prediction. The assumed motion of the obstacle is given by a constant preferred speed (v_o) and a constant direction of motion (θ_o), both available from the recent past window of observation. The motion is given by Eq. (17.20).

$$o(t) = o(0) + v_o t [\cos \theta_o, \ \sin \theta_o]^T \tag{17.20}$$

Here $o(0)$ is the current position of the obstacle, θ_o is the guessed direction of motion of the obstacle and v_o is the preferred speed of motion of the obstacle.

The finer genetic algorithm makes a time-varying map and uses the same for motion planning. While traversing a candidate trajectory, a simulation of the robot moving at the candidate trajectory and the obstacles moving as per the guessed trajectory is made. If the robot does not collide with any static obstacle or any dynamic obstacle with guessed motion, the candidate trajectory is called feasible, and the corresponding fitness is used as a quality measure.

17.10.4 Overall algorithm and results

Since static obstacles may appear suddenly, it is important to add enough random individuals at every time. Otherwise, if an obstacle suddenly appears close to the robot, all individuals may turn out to be infeasible which cannot be quickly mended.

The fusion of the two algorithms is interesting to observe. The finer Genetic Algorithm makes a finer trajectory with the advantage of avoiding suddenly appearing obstacles, adapt for dynamic obstacles, and make a precise fine trajectory for local optimality. At the same time, the coarser Genetic Algorithm selects the homotopic group to consider, the coarser obstacle avoidance strategy, and the manner (clockwise/anticlockwise) of avoiding each obstacle.

The algorithm is summarized in Fig. 17.22. The working of the algorithm for two scenarios is shown in Fig. 17.23. In both cases, new obstacles were added after the robot moved when the robot was close enough to the obstacles. Such obstacles are explicitly marked in the figures.

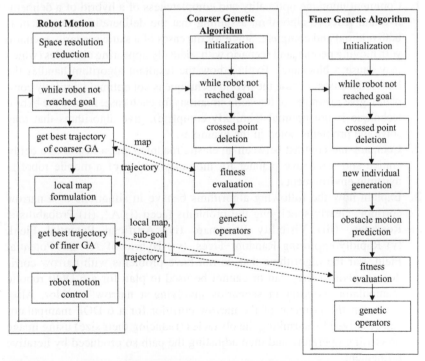

FIG. 17.22 Architecture of the bi-level Genetic Algorithm. The coarser GA works on the lower resolution map and the finer GA works on a local map (cut out around the current robot position). The coarser GA supplies a distant subgoal (outside the local map), which acts as the goal of the finer GA. The finer GA predicts the position of all dynamic obstacles, avoiding them.

 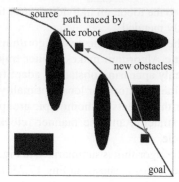

FIG. 17.23 Results of planning using a bi-level Genetic Algorithm. New obstacles are added visible to only finer Genetic Algorithm.

Questions

1. Comment upon the optimality and completeness of a hybrid of a deliberative and reactive algorithm, assuming that the deliberative component is both optimal and complete. Solve for the cases of a static scenario, scenario with moving people, and scenario with suddenly appearing obstacles (which may cause a blockage). Explain how the resultant algorithm handles the individual cases. In case the hybrid technique is not optimal (similarly complete), give examples of the same situations for each case. In case the hybrid technique is not optimal (similarly complete), give algorithms that may solve the problems noted in the examples.

2. Compare and contrast between an elastic roadmap and a fusion of a deliberative and a reactive planner for motion planning of a mobile robot in dynamic environments.

3. Explain how the following algorithms behave in situations of a narrow corridor, clearly specifying the problems faced (i) A*, (ii) Probabilistic Roadmap, (iii) Visibility Roadmap, (iv) Artificial Potential Field, (v) Rapidly-exploring Random Trees, (vi) Hybrid of A* and Potential Field. For the algorithms that do not have problems with narrow corridors, explain if they can or cannot be used to plan for a 6 DOF robotic manipulator working in scenarios involving a narrow corridor. Also, explain if the problem of the narrow corridor for a 6 DOF manipulator can be solved by shrinking the obstacles (reducing their size) using image processing operators and then adjusting the path so produced by iterative refinement.

4. A robot needs to go from the current position to the corner of an office. Give the algorithmic framework for the following cases. In all cases explain the sensors and actuators used for every robot and where are they used in the proposed algorithm. (i) No obstacles, (ii) Only static obstacles (complex framework),

(iii) Static obstacles (complex framework) and moving people around, (iv) Static obstacles (complex framework), moving people around, a group of robots of the same type. (v) No obstacles, with a group of robots of the same type forming a chain/line, (vi) Static obstacles and moving people around, with a group of robots of the same type forming a chain/line. Ensure to distinguish between static obstacles and moving people appropriately. (vii) No obstacles, with a group of robots of the same type forming a circle. The circle patrols iconic places in the office by vising all places one after the other in a sequence. Any abnormality means one of the robots in the circle goes to the security office nearby and reports the abnormality, continuing the sequence thereafter.

5. Consider different behaviours displayed by self-driving cars, including changing lanes, overtaking, entering/exiting an intersection, stopping before pedestrian zones, etc. Suggest controllers for all behaviours along with the applicable precondition. Suggest a software architecture to organize all behaviours. Try to model the behaviours using all possible architectures to establish a tradeoff. Repeat the solution to navigation in indoor environments alongside humans, displaying the social etiquette of getting aside on seeing a person coming, cooperating for overtaking, etc.

6. [Programming] Take a large map representing an office environment. Plan and execute the path of a robot using a hybrid of any deliberative and any reactive algorithm of your choice. Extend the solution to have multiple robots in the same arena, which must be reactively avoided. Furthermore, let each robot move by sequentially visiting several named positions in named rooms of the office environment repeatedly in a loop.

7. [Programming] Program a two-link revolute manipulator with some static and a dynamic obstacle that passes by, as the manipulator travels towards the goal.

8. [Programming] A robot is typically parked near the master bedroom and goes to the master whenever called, where the robot gets the instruction that it needs to execute. The instructions are either to get snacks or look for mails (if found fetch them back). If asked for snacks, the robot either gets biscuits from the kitchen (with a probability of 0.8) or fruits from the dining table (with a probability of 0.2). The robot must always go to the charging station whenever its battery levels are down and must inspect the entrance for security every hour (unless the robot is busy catering to the master). Model the robot behaviours as a BFSM and Behaviour Tree and simulate the robot. Use multiple random robots that wander to create dynamic obstacles.

9. [Programming] Use multiple deliberative algorithms under multiple parameter settings to give a trajectory each and fuse the trajectories into a single trajectory. The overall algorithm should behave like an iterative algorithm, where the trajectory improves with time. Similarly, fuse multiple reactive algorithms for the motion of the robot. Make the robot move from such a hybrid framework.

References

Brooks, R., 1986. A robust layered control system for a mobile robot. IEEE J. Robot. Autom. 2 (1), 14–23.

Colledanchise, M., Ögren, P., 2018. Behavior Trees in Robotics and AI: An Introduction. CRC Press.

Colledanchise, M., Parasuraman, R., Ögren, P., 2019. Learning of behavior trees for autonomous agents. IEEE Trans. Games 11 (2), 183–189.

Curtis, S., Best, A., Manocha, D., 2016. Menge: a modular framework for simulating crowd movement. Collect. Dyn. 1, 1–40.

Kala, R., 2013. Multi-robot motion planning using hybrid MNHS and genetic algorithms. Appl. Artif. Intell. 27 (3), 170–198.

Kala, R., 2014. Navigating multiple mobile robots without direct communication. Int. J. Intell. Syst. 29 (8), 767–786.

Kala, R., 2018. Routing-based navigation of dense mobile robots. Intell. Serv. Robot. 11 (1), 25–39.

Kala, R., Warwick, K., 2013. Multi-level planning for semi-autonomous vehicles in traffic scenarios based on separation maximization. J. Intell. Robot. Syst. 72 (3–4), 559–590.

Kala, R., Shukla, A., Tiwari, R., 2009. Fusion of evolutionary algorithms and multi-neuron heuristic search for robotic path planning. In: Proceedings of the 2009 IEEE World Congress on Nature & Biologically Inspired Computing. IEEE, Coimbatore, India, pp. 684–689.

Kala, R., Shukla, A., Tiwari, R., 2010a. Fusion of probabilistic A* algorithm and fuzzy inference system for robotic path planning. Artif. Intell. Rev. 33 (4), 275–306.

Kala, R., Shukla, A., Tiwari, R., 2010b. Dynamic environment robot path planning using hierarchical evolutionary algorithms. Cybern. Syst. 41 (6), 435–454.

Kala, R., Shukla, A., Tiwari, R., 2011. Robotic path planning using evolutionary momentum-based exploration. J. Exp. Theor. Artif. Intell. 23 (4), 469–495.

Murphy, R.R., 2000. Introduction to AI Robotics. MIT Press, Cambridge, MA.

Tiwari, R., Shukla, A., Kala, R., 2012. Intelligent Planning for Mobile Robotics: Algorithmic Approaches. IGI Global, Hershey, PA.

Chapter 18

Multi-robot motion planning

18.1 Multi-robot systems

In the future, any working environment may not deploy a single robot, but a team of robots that will work in coordination with each other. Multiple robots can solve overall problems more efficiently. Imagine having a long list of tasks to be done by a robot, like getting numerous items from different places, inspecting some remote sites, and carrying a load of raw materials from one place to another. In all cases, having a large fleet of robot means that the robots can divide all work among themselves and solve the problem more efficiently. This saves a lot of execution time. In many problems, the work can only be solved by a team of robots, like moving furniture with the weight of the furniture more than the payload of the robot or guarding a facility.

Different robots are manufactured with different requirements. Many cases arise where the robots complement each other to achieve the desired task. As an example, a robot operating on a deserted land with a large payload uses an aerial drone for vision, localization, and mapping. Industrial facilities use a limited number of manipulators that carry the loading and unloading on mobile robots to move a lot of materials around the facility.

If n robots together solve a problem, it may initially appear that the efforts get reduced by a factor of n. However, this assumes that it is possible to divide the problem equally among robots which may not always be true. As an example, for the problem to inspect k sites around, a robot may be too far off, and the nearby robots will have to inspect additional sites. A robot may not have the expertise to do some tasks, which must be done by the other robots instead. If three items need to be assembled into one, a robot may have to wait after getting the first item for the other two to be carried by the other robots for assembly.

On a broader note, the problem is now to *divide the entire task* into multiple robots, which requires sophisticated Artificial Intelligence algorithms. Some algorithms may guarantee the optimality of task division; however, they may incur a computational time over the time they save (as compared to naïve or greedy strategies). As an example, the optimal solution may save 1 min over a greedy plan; however, computing the optimal solution may

Autonomous Mobile Robots. https://doi.org/10.1016/B978-0-443-18908-1.00014-5

require 5 min in an online setting, and thus the benefits do not meet the costs incurred. Some algorithms may instead take a little time, but significantly lose on the optimality. The optimization algorithms with an iterative formulation of the solution give a trade-off, but impose an additional problem of trading off between the planning time and execution time. The concepts are summarized in Box 18.1.

Box 18.1 A summary of some terms.

Multiple Robots
- Can divide larger work among themselves
- Solve problems that cannot be solved by one robot (lifting heavy items collectively)
- Complement each other's capabilities (aerial robot uses aerial vision to guide a ground mobile robot)
- May co-exist for different applications, owned by different users

Task Division Algorithms
- Divide a larger work into multiple robots
- Optimal division may neither be equal nor *n* robots may divide the execution time by *n*
- E.g., for delivering letters, a robot may only deliver to a distant person, while another robot may deliver to a large number of people close by

Coordination
- Mitigate errors that emerge as the robots act
- If a robot solves its part earlier than anticipated, it takes a part of another robot's work
- If a robot fails, its part of work is redistributed
- Two robots carrying a piece of furniture together need to adjust for each other's errors to move synchronously.

Centralized Multi-robot Systems
- One offboard system takes all robots' inputs, the offboard system plans, and passes the actions to be performed by all robots.
- No issues with reaching a consensus.
- Centralized Systems may elect a robot as a leader
- Large complexity

Decentralized Multi-robot Systems
- Every robot takes information from neighbours, assesses, and acts accordingly.
- Robots may have conflicting plans (both want the same bottle simultaneously) and reaching a consensus in the absence of a leader may be difficult.
- With communication, robots can transmit information for rectification of anomalies
- Without communication, a robot has to guess what the other robot may do based on what it sees, and act accordingly. Wrong guesses (a robot turns left to make way for another, who instead was about to stop) can be fatal.
- Small complexity

Box 18.1 A summary of some terms—cont'd

Applications
- Collective SLAM
- Multi-robot task planning
- Multi-robot control, making interesting shapes, and dancing

Multi-robot Motion Planning
- Individual robots, when planned without considering other robots, may have trajectories that are collectively collision-prone.
- The coordination system detects and resolves such conflicts, asking some robots to deviate.
- The problem is simultaneously planning the path and speed assignment for all the robots.
- Many optimal plans have lower than preferred speeds (e.g., need to wait for a robot to pass by).

18.1.1 Coordination

Even if the work is divided among the robots, it cannot be broken down into the problem of multiple robots solving their part of the problem. It requires *coordination* to be done between the robots. Coordination is necessary at multiple phases. First, if a robot executes its part of the job earlier than anticipated, it relieves some other robots operating slowly. Second, if a robot fails due to any reason, it requires the other robots to share its job. Thereafter, many operations may be performed jointly by robots. While moving a piece of furniture using two robots, if one of the robots is going slowly, the other must adjust, and if one robot gets disturbed, the other robot must adjust to balance the forces and torques, keeping the furniture balanced.

Since there are coordination issues, there must be information interchange and communication between the robots. In many cases, communication is *direct* and formal in the form of a wireless connection between the robots. Communication protocols are needed to reliably exchange information at high rates. All robots exchange information about their state. The robots communicate their capabilities and preferences, based on which a task allocation is centrally done and communicated back to the robots using formal communication channels. If a robot cannot perform its job, it communicates, and other robots are deputed. This imitates human behaviour when people talk while carrying an item together; players use gestures and voice to state what should be done in a football match; and in a game of tennis (doubles), the players would shout about who needs to hit the ball.

In many other cases, communication cannot be facilitated by a direct exchange of information; however, the robots can still look at each other for coordination. This is called *indirect communication*. As an example, a robot senses the pose of a heavy box to calculate the forces to be applied, which adjusts for the forces by the other robot. Similarly, if two robots come in each other's way in navigation, they look at the direction of motion of each other and adjust their trajectories without explicit communication.

Coordination, in general, takes care of the adjustment in the decision-making by one robot attributed to the other robots, including to account for the uncertainties associated with the other robots. If many people are simultaneously operating in a common arena or are working on a common problem, a person expects the other person to act in a certain way. However, the other person may not be able to act in the same way, which creates an ambiguity in the system. Coordination is the process to remove such ambiguities by reaching a consensus. Coordination becomes much harder in the absence of communication when a guess must be made about the intent of the other person for corrections. The problem arises when the intent is wrongly guessed and is observed by an unanticipated action of the other robot, which must be dynamically resolved.

For larger teams, the overall problem is divided into smaller parts, where each part is a sufficiently small sub-problem; however, the division itself takes a longer time. Reaching a consensus with many individuals is a hard problem even with human teams. If the situation changes often and in real-time, the changes need to be quickly adapted. Good human teams are known to be robust to such disturbances as they can coordinate and align themselves quickly. In a game of soccer, the situation changes throughout the match; however, the teams adapt themselves and assume or alter responsibilities as per what they see. Similarly, the robots constantly align themselves by the changing circumstances, which is difficult since a consensus must be reached directly or indirectly.

18.1.2 Centralized and decentralized systems

The multi-robotic systems are classified into two categories: centralized and decentralized. *Centralized systems* have a single system that does all decision-making. It assimilates all information from all robots in real-time, decides the action that every robot needs to take, and transmits the decision to the individual robots. The robots are typically low processing units, who primarily upload their percept and execute the decision communicated. This requires the need to have one high computing and high memory system with reliable and fast communication because the assimilated data from all robots can be enormously large. Many algorithms thus have a limitation on the number of robots that can be considered in a centralized manner. The advantage of the centralized system is that all robots obey one system and therefore a consensus is always reached.

Mostly, an offboard computer system is used for doing central computations. The robots can also *elect* a leader from among themselves that acts as a central leader. The election protocol asks the robots to bid for the position of the leader and the best bid among all is designated as the leader. The bid can be in terms of computational availability, available battery, resources, or sometimes just a random number. The bids are distributed throughout the robot network and the best bid robot wins the election. Once, the leader is elected it takes all decisions for all the robots, until re-election is necessary for situations like the fall of the leader due to breakdown, battery below a critical level, or that some other robot is now more suitable to be the leader.

The other type of system is the *decentralized system*. Here, all the robots make their own decisions rather than relying on a central master. The robots consider their information and that of the neighbours. So, the information is limited, and therefore, a timely computation is possible. This also means that the systems are scalable to many robots. However, the robots may take opposing decisions, and coordination among them becomes a hard problem. As inputs from all the robots are not collectively considered, it is possible to make suboptimal decisions using the partially available information.

18.1.3 Applications

There are numerous examples in robotics where multiple robots are used. Collective mapping uses multiple robots to collectively explore and map (SLAM) a remote site. In multi-robot vision systems, the robots share their percept like self-driving cars can collectively log the traffic levels, and road conditions and perform SLAM for better routing, localization, or decision-making (using the ahead vehicle's vision to decide the feasibility of overtaking). Similarly, the aerial robots are good for an overhead vision for the ground robots.

Patrolling and surveillance are better done with multiple robots, eliminating blind spots. Painting services, construction, and forming interesting shapes for the visual appeal are interesting multi-robot problems. Multi-robot dancing, where the robots not only coordinate between themselves but also with humans and music is also an interesting application.

18.2 Planning in multi-robot systems

The purpose of this chapter is to focus on a single aspect of the multi-robotic systems, which is planning. The planning itself is at the level of task planning, motion planning, or trajectory control. At the level of task planning, the overall robotic system is given a task to solve, which is divided between the robots. Each robot does its part by active coordination with each other, continuously adapting for the changing situation like a robot falling or otherwise being unable to complete the task. The aim is that the overall task should be efficiently solved, even if the distribution of work between the robots may not be equal because of their capabilities. The task planning paradigm is given in the later chapters.

At the control level, the problem is that a team of robots share a control system that adapts the motion of one robot accounting for the others, such that the overall team moves as desired. The typical use cases are to move the robots together as a pyramid, circle, or chain or to make interesting shapes. The error function is taken as a cumulative error of all robots. Adjustments are done to every robot to minimize the overall error. This chapter does not focus on such systems.

18.2.1 Problem definition

The focus of the chapter is *multi-robot motion planning*, which is an extension of the classic problem of motion planning into the use of multiple robots. Consider that there are n robots that together operate in the same environment and co-exist. Each robot does something as per the needs of its master, or many robots together solve a problem. Each robot has a configuration it starts from called as the source (S_i), where the subscript i stands for the ith robot. The robots also have a goal denoted by G_i. The problem is to find a trajectory from source to goal for every robot, such that no robot collides with any obstacle and no two robots collide with each other. Let $\Gamma = \{\tau_i\}$, be the trajectories for all robots with the trajectory of robot i given by $\tau_i : \mathbb{R}_{\geq 0} \to C_i^{free}$. Here, C_i^{free} is the free configuration space of robot i. The robot's trajectory must start from the source, or $\tau_i(0) = S_i$ and end at the goal or $\tau_i(t) = G_i$, $t \geq T_i$. Here, T_i is the time at which the robot reaches the goal and t is the time. Here, it is assumed that the robot stays at the goal forever. In some definitions, the robot ceases to exist on reaching the goal, like reaching the end of the road under observation, entering a house, etc. The robot must not collide with any static obstacle, or $\tau_i(t) \in C_i^{free}$. Furthermore, no two robots must collide with each other at any point in time, or Eq. (18.1):

$$R_i(\tau_i(t)) \cap R_j(\tau_j(t)) = \varnothing \forall t, i \neq j \tag{18.1}$$

Here, $R_i(q)$ is the function that gives all points occupied by the robot i in the workspace when placed at the configuration q. The equation states that at any time, the configurations must be such that the volume of workspace occupied by any two robots must be non-intersecting (or the intersection must be a null set). The problem is summarized in Fig. 18.1A.

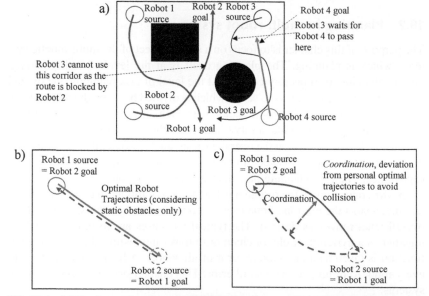

FIG. 18.1 Multi-robot motion planning. (A) Multiple robots avoiding collisions with each other. (B and C) The robots deviate from the shortest path to goal, demonstrating coordination.

18.2.2 Metrics

The metric to assess the quality of a trajectory can be more varied in the case of a multi-robot motion planning problem. The first metric is the average path length, for all the robots. This is given by Eq. (18.2):

$$length(\Gamma) = \frac{\sum_{i=1}^{n} \|\tau_i\|}{n} \tag{18.2}$$

Here, $\|\tau_i\|$ denotes the length of the trajectory τ_i and n is the number of robots The average traversal time is similarly given by Eq. (18.3):

$$time(\Gamma) = \frac{\sum_{i=1}^{n} T_i}{n} \tag{18.3}$$

The clearance is another factor that can be considered. The clearance for the case of a single robot came with the notion of taking a minimum or average over a trajectory, both with their pros and cons. In the same way, the clearance for multiple trajectories may be by taking a minimum or average over robotic trajectories. Here the average method is used, given by Eq. (18.4):

$$clearance(\Gamma) = \frac{\sum_{i=1}^{n} clearance(\tau_i)}{n} \tag{18.4}$$

Here, $clearance(\tau_i)$ is the clearance of the trajectory τ_i as per the choice of modelling. Similarly, for smoothness, given by Eq. (18.5).

$$smoothness(\Gamma) = \frac{\sum_{i=1}^{n} smoothness(\tau_i)}{n} \tag{18.5}$$

The overall measure may be a weighted addition of all measures.

18.2.3 Coordination

Assume that every robot does not consider the other robots and therefore plans an optimal trajectory for itself, given by $\{\tau^*_i\}$. If the plan is executed there may be collisions between the robots. Consider the simplest case shown in Fig. 18.1B. Each robot travels straight to the goal, causing collisions on the way. So, the robots must now alter their trajectories or slow-down to avoid a collision. The altered plan is shown in Fig. 18.1C. It is said that the robots have coordinated among themselves to achieve a collaborative plan which is collision-free. The magnitude by which a robot coordinates is the magnitude by which it is happy to accept a worse cost, given by Eq. (18.6).

$$coordination(\tau_i) = cost(\tau_i) - cost(\tau_i^*) \tag{18.6}$$

Here, $cost(\tau_i)$ is the cost incurred in executing the trajectory τ_i as per any suitable cost function. In multi-robot motion planning, the overall cost should be small, while no single robot should be asked to reduce the cost by a large factor. The total coordination considering all the robots may be given by maximum coordination shown by a robot, or by the sum of coordination of all robots. For the case of sum, the total coordination is given by Eq. (18.7).

$$coordination(\Gamma) = \sum_i coordination(\tau_i) \tag{18.7}$$

So, for every robot, Eq. (18.6) should be minimized, while overall Eq. (18.7) should also be minimized.

18.2.4 Importance of speeds

While defining the problem for a single robot, the focus was path planning (where the continuum of configurations to visit is the output) and not trajectory planning (where the speed needs to be specified as well). For static environments, it was confidently stated that the robot path can be suffixed with the heuristically computed speed information giving the trajectory. In the simplest case, the speed can be assigned with the knowledge of the curvature of the robot and the maximum safe operating speed of the robot. At any point, the speed is inversely proportional to the curvature, limited by the maximum speed.

However, in multi-robot motion planning, the definition is directly on trajectory planning by the inclusion of speed. As different robots need to plan to avoid each other, they must know at exactly what time will they be reaching a particular configuration or volume of workspace. To avoid each other, the robots must plan their speeds as well. Consider the case shown in Fig. 18.2. There are multiple ways by which the robots A and B can avoid each other. In the first case (Fig. 18.2A), A goes straight to the goal quickly and B takes a curve to avoid A. In the second case (Fig. 18.2B), B takes the opposite curve to avoid A while A slows down a bit to let B pass through first. In the other case (Fig. 18.2C), A goes as usual to the goal, while B waits for A to pass through. All cases have some implication on speeds which must be worked out to make the trajectories safe. Especially in the last case, B's slowing down, and waiting is an adjustment of the speed to avoid a collision. This is also the best path as per the shortest average path metric.

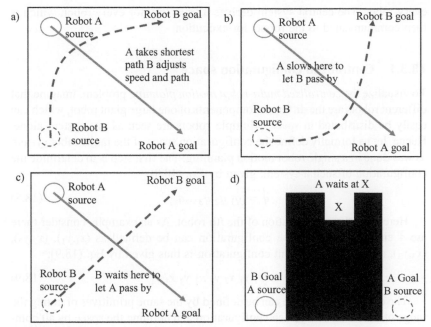

FIG. 18.2   Importance of speed in multi-robot motion planning. Different plans shown in (A)–(D) may be optimal by different metrics. The planner must plan both path and speed for optimality. In (D) no solution is possible until the planner models the waiting at 'X' behaviour.

In some cases, it may not be possible to make the robots come out with even a single collaborative plan unless the robots agree to reduce the speeds. Take an example in Fig. 18.2D. There are 2 distinct plans possible: to ask the first robot to wait till the second robot passes and to ask the second robot to wait till the first robot passes. The planning algorithms must account for a reduction in speed and waiting options. In any case, the optimal behaviour may ask a robot to slow down. The multi-robot motion planning is an amalgamation of both problems to be solved simultaneously, the *multi-robot path planning*, devising the paths to be taken by the robots, and *multi-robot speed assignment*, dealing with speed assignment to every segment of the path.

18.3   Centralized motion planning

The problem is first solved by using a centralized technique. In a centralized multi-robot motion planning problem, the map is known by a single leader or a central system, which may be made collectively by the robots and/or external sensors. The central system also knows the source and goal configuration of every robot. The central system derives a trajectory for every robot, which is then communicated to the robots for execution.

18.3.1 Centralized configuration space

To visualize the *centralized multi-robot motion planning* problem, imagine that different robots are the different components of one large giant robot, which can easily be distributed in space. Multiple robots are seen as one system alone. Hence, let us formally define the configuration space of the multi-robot system as was done in a single robot motion planning. The first step is to determine the configuration of such a system, given by Eq. (18.8):

$$q = [q_1 \ q_2 \ q_3 ... q_n] \tag{18.8}$$

Here, q_i is the configuration of the *i*th robot. As an example, consider there are 4 circular robots whose configuration can be defined as (x_1,y_1), (x_2,y_2), (x_3,y_3), and (x_4,y_4). The joint configuration is thus given by Eq. (18.9):

$$q = [x_1 \ y_1 \ x_2 \ y_2 \ x_3 \ y_3 \ x_4 \ y_4] \tag{18.9}$$

The configuration space can be defined by the same primitives of noting the range of values taken by each configuration and defining the space by all combinations of values taken. This comes out to be the cross-product of the configuration spaces of the individual robots, given by Eq. (18.10):

$$C = C_1 \times C_2 \times C_3 \times ... \times C_n \tag{18.10}$$

Considering the same example, the configuration space of the first robot is $\mathbb{R}^2$, and similarly to the other three. The resultant configuration space is thus given by Eq. (18.11):

$$C = \mathbb{R}^2 \times \mathbb{R}^2 \times \mathbb{R}^2 \times \mathbb{R}^2 = \mathbb{R}^8 \tag{18.11}$$

Even though the physical workspace occupied by every robot is the same, the $\mathbb{R}^8$ configuration space means that every combination of positions among all robots is possible.

The other important task is to classify the configuration space into free configuration space and obstacle-prone configuration space, done by the collision checking algorithms. Since the entire multi-robot system is taken as one system, a collision between the robots is analogous to self-collision discussed previously. It can be mathematically formulated as Eqs (18.12), (18.13):

$$q \in C^{obs} \text{iff} \left(\exists_i q_i \in C_i^{obs} \right) \vee \left(\exists_{i,j,i \neq j} R_i(q_i) \cap R_j(q_j) \neq \varnothing \right) \tag{18.12}$$

$$q = [q_1 \ q_2 \ q_3 ... q_n] \tag{18.13}$$

The equation states that for a joint configuration to be collision prone, either any of the robots collides with an obstacle, or two robots collide with themselves, making the intersection between the volume occupied in the workspace non-null. The free configuration space is simply the compliment over the entire configuration space or $C^{free} = C \backslash C^{obs}$. The free configuration space can also be explicitly modelled as Eqs (18.14), (18.15):

$$q \in C^{free} iff \left(\forall_i : q_i \in C_i^{free} \right) \wedge \left(\forall_{i,j,i \neq j} R_i(q_i) \cap R_j(q_j) = \varnothing \right) \tag{18.14}$$

$$q = [q_1 \, q_2 \, q_3 ... q_n] \tag{18.15}$$

The equation states that for a configuration to be free, all robots must themselves be collision-free, and no two robots must collide with each other.

The source is taken as a joint of sources of all robots $S = [S_1 \, S_2 \, S_3 ... S_n]$, where S_i is the source of robot i. The goal is taken as the joint of goals of all robots, $G = [G_1 \, G_2 \, G_3 ... G_n]$, where G_i is the goal of robot i.

The multi-robot motion planning has been reduced to the planning of a single joint system, for which any of the deliberative approaches which work in high-dimensional spaces can be used.

18.3.2 Centralized search-based planning

Consider the A* algorithm. Notionally, every axis of the joint configuration space is discretized into cells and every cell represents a vertex of the graph. The A* algorithm for a 2-D map typically took a actions, making the robot go in a directions. Typical choices were up, down, right, and left for 4 actions, and for 8 actions the choices included the diagonals. In the case of multiple robots, every pair of actions taken by every robot is worth considering, and therefore the actions possible are the cross-product of all actions of all robots.

It may be possible that a specific robot reaches its goal earlier, which must wait. A robot may have to wait for another robot to cross. A robot may have to slow down for optimality. The importance of considering speeds has already been detailed. This also emphasizes that a 'no action' (denoted by $\varnothing$) at a particular time step should be an option for the robots. Let A_i be the possible actions that can be taken by the robot i. The possible actions for all robots are hence given by Eq. (18.16):

$$A = (\{A_1 \cup \varnothing\} \times \{A_2 \cup \varnothing\} ... \times \{A_n \cup \varnothing\}) - (\varnothing, \varnothing, \varnothing, ..., \varnothing) \tag{18.16}$$

The action of the multi-robot system is the cross product of all actions of all robots, with a difference of one action wherein no robot does anything. These are all possible actions. Some of these actions may not be feasible. To check for feasibility, a simulation is done wherein every robot moves with the chosen action. If the entire simulation results in no collision with any robot, either with a static obstacle or between the robots, the action is called feasible. A simulation or a joint action may be infeasible even if the configuration before and the configuration after the simulation may both be feasible. As an example, if two circular robots at (1,1) and (1,2) exchange position, the action is infeasible, while the configuration before and after the simulation are both feasible.

The other primitives remain the same. The source is the join of all sources. The goal is the join of all goals. The configuration space is discretized along every axis to yield vertices. The actions or the edges are a join to the neighbours as also given explicitly by the action function. The edge cost function is based on the metric used. If the total path length is being optimized, it shall be the sum of distances from all actions. On the contrary, if time is being minimized, each edge cost shall be the time taken by the largest time-consuming action. The heuristic function can be obtained for the relaxed problem, considering no other obstacles. For the total path length metric, the sum of Euclidian distances to goals for all robots is a valid heuristic, where the relaxation is that no robot considers any other robot or obstacle on the way. For the total travel time heuristic, the largest Euclidian distance on the maximum travel speed of the robot is a valid heuristic.

It may be noted that speed assignment is not done here. Speed assignment will mean that the problem is solved in the state space and not the configuration space. Increasing or decreasing speeds are valid actions to be taken for a state-space solver.

Let us study the problem using an example in which 4 circular robots must go to their own goals. Let (S_{x_i}, S_{y_i}) be the X and Y coordinates (configuration) of the source of the ith robot. Similarly, let (G_{x_i}, G_{y_i}) be the X and Y coordinates (configuration) of the goal of the ith robot. The joint source, goal, and any general configuration are given by Eqs (18.17)–(18.19):

$$S = \left[S_{x_1}, S_{y_1}, S_{x_2}, S_{y_2}, S_{x_3}, S_{y_3}, S_{x_4}, S_{y_4}\right] \tag{18.17}$$

$$G = \left[G_{x_1}, G_{y_1}, G_{x_2}, G_{y_2}, G_{x_3}, G_{y_3}, G_{x_4}, G_{y_4}\right] \tag{18.18}$$

$$q = \left[x_1, y_1, x_2, y_2, x_3, y_3, x_4, y_4\right] \tag{18.19}$$

Let the actions of any robot be given by up ($\uparrow$), down ($\downarrow$), right ($\rightarrow$), and left ($\leftarrow$). The action at any time taken by the planner is denoted by (a_1, a_2, a_3, a_4), where a_i denotes the action taken by robot i. The total actions (edges) possible for any state are given by Eq. (18.20):

$$A = \{\uparrow, \downarrow, \rightarrow, \leftarrow, \varnothing\}^4 - (\varnothing, \varnothing, \varnothing, \varnothing)$$
$$= \{(\uparrow, \uparrow, \uparrow, \uparrow, \uparrow), (\uparrow, \uparrow, \uparrow, \uparrow, \downarrow), (\uparrow, \uparrow, \uparrow, \uparrow, \rightarrow), (\uparrow, \uparrow, \uparrow, \uparrow, \leftarrow), \tag{18.20}$$
$$(\uparrow, \uparrow, \uparrow, \uparrow, \varnothing) \cdots (\varnothing, \varnothing, \varnothing, \leftarrow)\}$$

Consider the metric used to assess the plan is the time when all robots take to reach their goal. Every simultaneous action by every robot takes a unit step. Hence, the edge cost to go from q_a to q_b by an action a is given by

Eq. (18.21) and the historic cost function (for all configurations but source) is given by Eq. (18.22):

$$w(q_a, a, q_b) = 1 \tag{18.21}$$

$$g(q_b) = g(q_a) + 1 \tag{18.22}$$

The heuristic function is given by the time needed by the slowest robot to reach the goal (Eqs 18.23, 18.24). The total cost function is the sum of historic and heuristic components, given by Eq. (18.25):

$$h(q_b) = \max_i \; d\big((x_{b_i}, y_{b_i}), (G_{xi}, G_{yi})\big) \tag{18.23}$$

$$q_b = \begin{bmatrix} x_{b_1} \; y_{b_1} \; x_{b_2} \; y_{b_2} \; x_{b_3} \; y_{b_3} \; x_{b_4} \; y_{b_4} \end{bmatrix} \tag{18.24}$$

$$f(q_b) = g(q_b) + h(q_b) \tag{18.25}$$

Note that the speed is not taken as a factor in both the historic and heuristic costs since the costs are scale-invariant. Because only 4 rectilinear directions are being the Manhattan distance function is suitable.

If the total path length was being minimized, the edge weight is given by Eqs (18.26)–(18.28):

$$w(q_a, a, q_b) = cost(a_1) + cost(a_2) + cost(a_3) + cost(a_4) \tag{18.26}$$

$$a = (a_1, a_2, a_3, a_4) \tag{18.27}$$

$$cost(a_i) = \begin{cases} 1 & if \; a_i \neq \varnothing \\ 0 & if \; a_i = \varnothing \end{cases} \tag{18.28}$$

The heuristic function is now given by the total path length, or Eqs (18.29), (18.30):

$$h(q_b) = \sum_i d\big((x_{b_i}, y_{b_i}), (G_{xi}, G_{yi})\big) \tag{18.29}$$

$$q_b = \begin{bmatrix} x_{b_1} \; y_{b_1} \; x_{b_2} \; y_{b_2} \; x_{b_3} \; y_{b_3} \; x_{b_4} \; y_{b_4} \end{bmatrix} \tag{18.30}$$

The search tree is too large as the branching factor is $a^n - 1$, where n is the number of robots (4) and a is the number of actions per robot (including no action), which turns out to be $5^4 - 1$. This cannot be shown in the paper. However, a little segment of the search tree is shown in Fig. 18.3. The Uniform Cost Search has a complexity of $O(b^{1 + \lfloor C^*/\varepsilon \rfloor})$, where b is the branching factor (number of actions), C^* is the cost of the optimal solution, and ε is the minimum improvement in edge cost during search. For multi-robot motion planning $b = a^n - 1$, giving complexity as $O(a^{n(1 + \lfloor C^*/\varepsilon \rfloor)})$. The problem is *exponential in terms of the number of robots* and the solution is unsuited

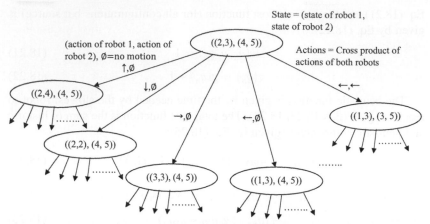

FIG. 18.3 Centralized search based planning. Joint configuration space is divided into grids. If using diagonals, the time of action by every robot will be different ($\sqrt{2}$ for diagonal, 1 for rectilinear) and hence a robot will have to travel slower/wait. To solve, actions may be defined as a motion for Δ time in the same direction with a continuous state space.

for many robots. Typically, only 2–3 simple robots can be planned by using such a technique.

Now let us extend the solution to account for 8 actions, instead of the regular 4 actions. The modelling can be simply extended to allow the $\sqrt{2}$ costs of the diagonals. Consider the action set at any particular point of time as planned by the planner to be ($\downarrow,\uparrow,\rightarrow,\nearrow$), meaning that the first robot is asked to go down, second to move up, third to move right, and the last to go top right. The first 3 robots have a smaller distance (1) to travel and hence reach the new configuration earlier, while the last robot has a larger distance to travel ($\sqrt{2}$) and reaches its new configuration later. The configuration is at any time, meaning that the first three robots will have to wait for the last robot. The other robots cannot execute their next action while the fourth robot is yet to complete the execution of its current action. This is because the entire multi-robot system goes from one configuration to another by one joint action set, and a subsequent joint action set from the trajectory can only be used when the execution of the previous one is over.

This would not have been a problem with state-space planners incorporating speeds, wherein the states are not discretized and the position at the end of a unit of time could be used. The state space-based approach works on a tree search and not a graph search and the revisited states are readmitted. This can lead to higher complexity, but solves the problem of the robots unnecessarily waiting. The complexity is a severe issue, which is already high for a multi-robot system.

It is also possible to use only the X and Y coordinate value of the robot, but to not discretize the vertices and to expand like a tree of state space. So, the configuration is the position after a unit time in a continuous domain, without any discretization. Since the states are continuous, revisiting a state means that it is re-inserted in the queue as a new node. This also does not make one robot wait for the other robot to complete the current action.

18.3.3 Centralized probabilistic roadmap based planning

The Probabilistic Roadmap approach can also be directly applied to the joint configuration space. Consider the example of 3 robots going from their source to goal. Let us take a random vertex, given by $q = [x_1, y_1, x_2, y_2, x_3, y_3]$. The decision is to decide whether the configuration is free or not. This can be done by placing the virtual robots at (x_1, y_1), (x_2, y_2), and (x_3, y_3). If none of the robots collide with a static obstacle or themselves, it is a free vertex and can be admitted. After the vertices, the algorithm needs to check for edges, for which the first definition is a distance function to get the neighbouring vertices, for which a Euclidian function is sufficient. The collision checking for an edge happens by simulating a motion of all 3 robots simultaneously. If neither of the robot collides with a static obstacles and the robots do not collide with themselves, the edge is added.

Let us take a specific example of an edge. Consider an edge from the joint configuration (10,48,35,10,50,25) to the joint configuration (20,38,25,50, 50,25). The antecedent vertices of the three robots are (10,48), (35,10), and (50,25) while their destination vertices are (20,38), (25,50), and (50,25). In terms of collision checking, it is easy to construct a line that samples the configurations lying on this line and to check each of them for collision. The continuous collision techniques can do the same thing faster. Let us visualize the motion in the simulation. This is shown diagrammatically in Fig. 18.4. The start time and end time of all robots for this simulation are fixed. Meaning, they speed themselves proportionally to the distance that they need to travel. Robot 3 has no distance to travel and stays static in the entire simulation. Robot 2 has a long distance and has a maximum relative speed. Robot 1, which has a small distance to travel moves much slower. The relative speed (and not absolute speed) is already considered in this approach.

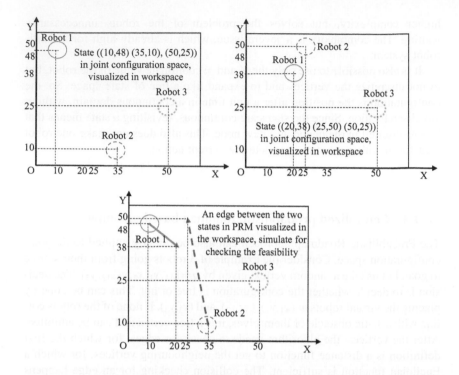

FIG. 18.4 Centralized probabilistic roadmap planning. In the given edge, because of a difference of lengths, robot 1 travels slowly, robot 2 travels very fast and robot 3 is static. If the simulation is collision-free, the edge is added to the roadmap.

The total hypervolume of the space remains to be exponential on the number of robots, which means that the requirement of vertices and edges will be large. Previously, it was argued that the expected number of samples drawn for a robot being planned in a space of d-dimensions is $(R/\rho)^{fd}$, where ρ is a small constant (length of hypercube in which the sample may be moved without affecting connectivity), R is the normalized length of each axis of the space, f is the number of difficult samples in space and d is the number of dimensions. For n robots, this term naturally becomes $(R/\rho)^{nfd}$, which significantly limits the number of robots that can be planned in a centralized manner.

18.3.4 Centralized optimization-based planning

The optimization-based technique was an interesting mechanism of deliberative motion planning of a robot. Here, let us extend the same approach to multiple robots. The primary task in optimization was the representation of a genotype, which was done by the insertion of a variable number of via points in between fixed source and goal. The via points hence become the parameters to optimize the trajectory and they form the genotype. In the case of multiple robots, such a genotype is replicated for every robot, and a set of all via points makes the genotype. Since the problem is multi-robot motion planning, for already discussed issues, it is important to account for the speed of the robot, otherwise, the system would not be able to model for the slowing and waiting for behaviours. Hence, every segment (in between the via points) also mentions the relative speed at which the robot moves in that segment. The complete modelling is shown in Fig. 18.5A. The number of via points may be different for different robots and different across the individuals that make the population pool.

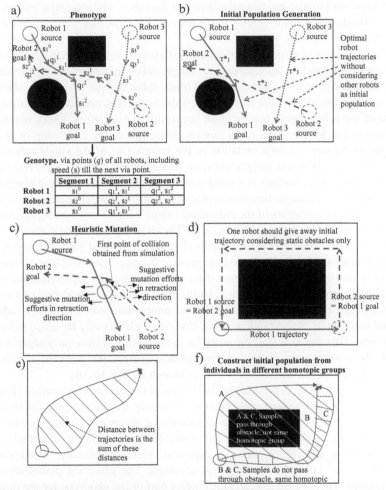

FIG. 18.5 Centralized Evolutionary Motion Planning. (A) Individual representation as via points and speeds. (B) Initial population using trajectories without considering other robots (and their mutated variants). (C) Heuristic mutation of via points surrounding collision/close proximity site, in the retraction direction. (D) The initial population of (B) leads to no solution because mutations of static obstacle trajectories do not make any robot take the longer route. (E) Distance function between trajectories for diversity preservation. (F) Generation of diverse initial trajectories from different homotopic groups.

The population is randomly generated in the same mechanism as for a single robot. The only difference is that the trajectories of all robots must be generated to make a single individual. The fitness evaluation involves a simulation of all the robots going by the represented trajectories and the metrics are computed. Additionally, the inter-robot collisions are checked in such a simulation. The collisions are added as a penalty in the cost function. The genetic operators largely remain the same for multi-robot motion planning as the single robot motion planning. Crossover is applied for the corresponding trajectories of the robots in the joint individuals. The parametric mutation not only moves the via points but also modifies the speed of the robot. The structural mutation, like a single robot, adds/deletes via points in the trajectory corresponding to any robot.

The working of any evolutionary algorithm can be significantly improved by supplying good initial solutions. A single-robot motion planning problem is a lot easier to solve as compared to a multi-robot motion planning problem. For most practical cases the single robot trajectories may not non-colliding and none of the robots may come in close proximity to each other, which makes a valid solution for the multi-robot motion planning case. If a collision or close proximity happens, a little mutation to the trajectories of the single robots at either the path or speed assignment may be enough to get rid of the collision or close proximity, leading to a good plan for the multi-robot motion planning problem. Hence, a clever heuristic is placed in the *initial population generation*. First, the best trajectories of the individual robots without considering the other obstacles are computed, denoted as $\{\tau^*_i\}$. The trajectories may be computed by using evolution or any other algorithm. Typically, the best path is computed, and a maximum permissible speed assignment is done. The collection of trajectories is additionally encoded from the current phenotypic representation to a genotypic representation as required by the algorithm. This is added as one individual of the initial population. Mutated variants of this plan are added as additional individuals. Some random individuals may additionally be added to have diversity. This forms a starting point for running the evolutionary algorithm. Now the evolutionary algorithm can quickly correct for only the conflicts and make the plan collision-free. This is shown in Fig. 18.5B.

Another heuristic is possible in the design of the *mutation operation* (Kala, 2014b). The genotype separately encodes all robotic trajectories, and the fitness evaluation involves a simultaneous simulation of all the robotic trajectories. In the worst case, if two robots collide in the simulation of a genetic individual, it is known which trajectories caused a collision. More specifically, it is known across which via points the collision occurred. The same via points should be mutated. This is used to select the correct part of the genotype for the mutation operation. The heuristic can be extended to find out the retraction direction of the collision and to apply the mutation in the same direction to both robots. If the first robot was rightward to the second robot while colliding, the neighbouring via points of the first robot are mutated by pushing rightwards, and the neighbouring via points of the second robot are mutated by pushing leftwards. In other words, consider the situation just before collision (or the situation just before a close proximity or a near-collision) between robot i and robot j. First, imagine robot i as a static obstacle just before collision and imagine the robot j's trajectory as per its current representation in the genetic individual as an elastic strip. Robot i applies an external repulsion force used to correct robot j's trajectory, which serves as a mutation operation for robot j. Now, imagine robot j just before collision as a static obstacle and imagine robot i's trajectory as an elastic strip. Now robot j applies a repulsion force to change robot i's trajectory, which is the mutation operation for robot i. Only colliding or near-colliding pairs of robots may be chosen for such a heuristic mutation. Similar arguments can also be given to adding smoothness across the most unsmooth part of the trajectory or adding clearance to the least clear parts of the trajectory associated with any robot. This is shown in Fig. 18.5C.

Using the initially generated good trajectories severely boosts the algorithm performance. However, consider the situation as shown in Fig. 18.5D. The best trajectories use the same area, and no soft mutation can avoid the collision. One of the robots needs to *reroute* during the evolutionary process, for which no good heuristic lies. This makes the problem significantly more difficult as the initial population is deceptive, trying to bias the evolution towards the best initial trajectory of the individual robots. A better implementation is hence not to inject a single good trajectory for a robot, but a diverse set of trajectories for a robot coming from different homotopic groups.

The diversity preservation measures require a distance function to be designed between trajectories. This is given by Eq. (18.31) and shown in Fig. 18.5E.

$$d(\tau_1, \tau_2) = \int_{s=0}^{1} d(\tau_1(s), \tau_2(s))ds \approx \sum_{s=0, steps\ \Delta}^{1} d(\tau_1(s), \tau_2(s))\Delta \qquad (18.31)$$

Another implementation that facilitates the injection of initial good trajectories is to have the initial trajectories from a different *homotopic group of trajectories*. Computation of whether two trajectories belong to the same homotopic group is a hard problem, especially in high-dimensional and non-geometric spaces. However, it is possible to sample out points within trajectories to approximately compute if they belong to the same of different homotopic groups. This is given in Fig. 18.5F. The computation of a competing set of trajectories may be done by building a roadmap. For high-dimensional spaces, the Probabilistic Roadmap is a suitable approach. The traversal of the roadmap generates trajectories, out of which the best trajectories belonging to different homotopic groups may be taken. The individuals of the centralized optimization now have a mix of homotopic group combinations as different robot trajectories and can model for situations in which a poor initial robot trajectory is not taken and a trajectory from another homotopic group is mutated to give the best trajectory plan.

18.4 Decentralized motion planning

In decentralized motion planning, every robot is planned independently. Therefore, there are limited options with every robot and the complexity is not exponential to the number of robots. On the contrary, the complexity is typically linear to the number of robots, since a single source single goal motion planning algorithm is applied for all the robots. However, if the robots are planned independently, the trajectories generated may be collision prone. In centralized planning, all the robots were considered at a time and the planner had the luxury of considering an alternate pair of collision-free trajectories. In the case of the decentralized system, first, the robots plan their trajectories. If a pair of trajectories are found to be collision-prone, then the *coordination system* does the task of locally modifying the trajectories or taking suitable actions such that the collision is resolved. The emphasis is on the coordination system to devise strategies for resolving collisions. This is practical with everyday decision-making in organizations. Multiple teams work over different modules of a problem to design solutions. If the solutions cannot be integrated and are not compatible, the senior management then does the task of resolving conflicts.

It is interesting to assess the relative pros and cons of centralized and decentralized motion planning systems. The centralized systems work in the joint spaces and therefore the planners have the liberty to generate all possible combinations of trajectories. The solutions are therefore generally both complete and optimal with any restriction and connotations put by the individual planning algorithms like resolution optimality and probabilistic optimality. However, the centralized system works in a massive search space which is exponential to the number of robots. The solutions can therefore only be applied for a few robots due to computational complexity constraints.

The decentralized systems on the other hand work individually for every robot. Hence, a robot may need to suboptimally take a different route or take a trajectory from a suboptimal homotopic group, which it would not know at the time of initial planning. Initially, every robot attempts to take the best trajectories and aims to locally correct the trajectories for a collision. Sometimes the robots need to travel by using trajectories reasonably different from the optimal ones, going through a different homotopic group, to avoid collisions, which the decentralized systems may not be able to compute. Hence, the systems lack optimality. The completeness is similarly not guaranteed. However, the computational complexity is typically linear to the number of robots. This is a significant advantage. The centralized systems cannot be used for more than 2–5 robots. Hence, typically decentralized solutions are used for motion planning of multi-robotic systems. The differences are also summarized in Table 18.1.

TABLE 18.1 Differences between centralized and decentralized motion planning.

S. No.	Property	Centralized motion planning	Decentralized motion planning
1.	Working	Planned in a joint space of all robots	Each robot is planned independently, coordination algorithms locally resolve collisions
2.	Computation Complexity	Exponential to the number of robots	Linear to the number of robots, plus additional complexity for coordination
3.	Space Complexity	Exponential to the number of robots	Linear to the number of robots
4.	Completeness	Complete, subjected to connotations of the search algorithm	Typically, not complete for a linear complexity case
5.	Optimality	Optimal, subjected to connotations of the search algorithm	Not optimal
6.	Number of robots	Works for 2–5 robots only	Works for a very large number of robots

Communication plays a major role in the planning of a decentralized system. If the robots have direct communication between themselves, they can exchange information, and the coordination manager then resolves the differences and communicates the resolved trajectories to the robot. This makes it much easier to plan for the individual robot. In the case of planning using a decentralized scheme without direct communication, the robots start executing their trajectories and only realize about a prospective collision when it is eminent. Thereafter, the robot must guess the intent of the other robot to correct its trajectory.

18.4.1   Reactive multi-robot motion planning

The reactive approaches only take a unit action for a robot, which avoids both static and dynamic obstacles. Therefore, approaches with the same advantages and disadvantages can solve the multi-robot motion planning problem. It does not matter whether the dynamic obstacles are moving robots or moving people.

In case the robots have direct communication between them, a robot can communicate with the other robot what it expects the other robot to do. It is up to the other robot to implement the suggested action or not, but the algorithm does take that into account in decision-making. There are numerous practical examples of such communication in everyday human life. While driving, if there is an emergency vehicle, it communicates using the hooter that the right of way should be given. Similarly, if two people block each other's way in an intersection or door, they communicate to state who goes first.

The ideas can be extended to the robots as well. If a robot wishes to overtake, it sends a message that the robot in front should drift sideways (Kala and Warwick, 2012, 2013a, 2015a). This acts as an additional force (potential field), rule (fuzzy logic), or objective measure (optimization) using communicated information. Similarly, a robot can communicate the preferred side (left or right) of avoidance to a robot. These cases are shown in Fig. 18.6A and B. In the working of the reactive approaches now imagine that a robot sees another

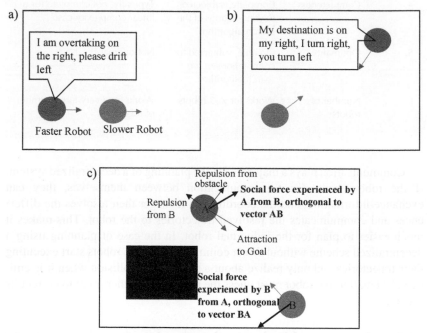

FIG. 18.6   Decentralized Multi-robot Motion Planning (A) and (B) Role of Communication in Reactive Motion Planning with multiple robots. One robot suggests an action for the other robot, which may be incorporated by taking additional input, implemented by adding new forces (potential field), rules (fuzzy logic), or objective measures (optimization) in the reactive planner. (C) Orthogonal Social Force in Potential-based motion planning. If two people are nearly head-on, a force is applied in the orthogonal direction to increase the separation with time. The force is on the left on A, if A is likely to avoid B from its left (either because A is already relatively on the left, or A's goal is relatively on the left). In the case of a tie when two people are head-on, in some countries left avoidance is preferred and vice versa.

robot (dynamic obstacle) directly ahead, or for other reasons, it is not clear whether the avoidance will be from the left or right. The robot makes a decision (say avoiding from the right) and communicates the same to the other robot. Now it is eminent that the other robot may also have such a delimma and may be considering an opposing plan, as possible in Fig. 18.6. The receipt of such a message means that the algorithm incentivizes decision-making that adheres to the preferred side of avoidance communicated by the robot. Hence, the robot decides to take an acceptable plan. As another example consider a robot is fairly confident that the robot (dynamic obstacle) in front will go straight and therefore avoiding the same robot from the left is feasible. However, if the robot in front has a sudden leftward drift, it poses a great threat to the robot at back. Hence, it is better to, for security reasons, give a communication signal to the front robot asking it not to turn left. In such a case the robot at the back will expect the robot in front not to turn left, unless it is extremely important for other reasons in which case the robot at the back will adjust its reactions instantaneously.

When using the potential field algorithm, there is a problem of local minima when two robots face each other. The two robots would repel each other, pushing each other back, and not move aside to avoid each other. Humans typically follow a principle of to-left (or to-right) in such situations. Hence, an additional social force is modelled that drifts the robots on their left (or right). The force is only applicable when both robots are nearly in front of each other. The force is applied in the direction at which the robot is originally present (left or right, with a tie-breaking as per social norms). If the robot A being planned is relatively left of another robot B (obstacle), a left avoidance is preferred to be applied on A, and vice versa, as shown in Fig. 18.6C. By symmetry, the other robot (B) achieves an opposite force. A person may have its goal nearby and therefore may have to avoid a person on the right to minimize the deviation from the goal, even if it was relatively on the left. This is another factor that helps to decide whether two people avoid each other on their left or right. Sometimes there is a nearly perfect symmetry with two people facing each other head-on. In such situations, social norms come into play. Some countries have a to-left walking rule and each robot should drift on the left, while some other countries prefer walking on the right and the same may be preferred. The force is higher in front of the robot (angle 0 between the robots) and weaker at the side (angle $\pi/2$ or $-\pi/2$). The force is shown in Fig. 18.6C.

If the robots do not have a direct communication, the reactive planner may continue to have the same inputs and rules with the only difference that now the intent of the other robot will have to be guessed for (Kala and Warwick, 2013b). The trajectory tracking and prediction algorithms do the same.

The static obstacles and dynamic obstacles are differently modelled since they act differently. In other words, the velocity of the motion of the obstacle plays a significant role to decide the immediate motion of the robot (Kala, 2014a; Kala and Warwick, 2015b). Consider the situation in Fig. 18.7. If the obstacle is static, the robot will normally avoid it by steering left or right. If the obstacle is dynamic, going away from the robot with a lower speed, the robot has a far larger time to avoid it. If the obstacle is dynamic going away at a much larger speed, the robot will not even consider it, since it cannot overtake the obstacle. On the contrary, if the obstacle is dynamic and coming towards the robot, the robot will have to turn much swiftly to avoid the obstacle.

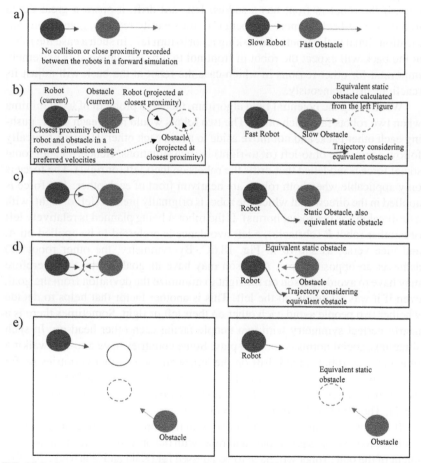

FIG. 18.7 Importance of incorporating obstacle speed in decision-making for obstacle avoidance is shown by several examples in (A)–(E). Both robot and obstacle are simulated by preferred (or current) speeds and closest proximity/collision is drawn with a solid lined circle denoting robot and dashed circle denoting obstacle. This is the equivalent static obstacle corresponding to the dynamic obstacle. As the relative speed of the robot increases, the turn gets sharper.

To get some equivalence between the static and dynamic obstacles, let us guess the trajectory of the dynamic obstacle (another robot). This is a (somewhat) difficult machine-learning problem. Therefore, for the time being, let us assume that the obstacle continues to move with the same speed with time, till it collides with another, maybe static, obstacle. To complete the simulation, let the robot being planned, be assumed to be moving with the current speed. The simulation done for a small-time step records the smallest distance (or collision) between the robot and all other obstacles (or other robots). Let o' be the position at which the moving obstacle o and the robot being planned by their extrapolated trajectories are the closest (or suffer a collision). It is convenient to replace the moving obstacle o with a single equivalent static obstacle o' for planning.

Consider Fig. 18.7. If the robot and obstacle o are moving towards each other at the same speeds, the collision will be in the middle. When planning the robot, it is good to imagine the obstacle in the middle itself and steer the robot accordingly. Similarly, if the obstacle is moving at a lower speed away from the robot, the collision will happen much later, and the equivalent static obstacle is placed at the same position. Any general situation is shown in Fig. 18.7E. The obstacle as per this definition can be used to define all inputs for navigating the robot by a reactive paradigm.

Let us now revisit the reactive approaches. The first reactive approach was the artificial potential field. The changes to account for multiple robots are (i) The robot listens to the communication messages received from the other robots. All such messages are converted into additional forces in the suggestive direction. (ii) The robot detects other robots (dynamic obstacles) directly ahead which need to be avoided from the left or right. The robot makes a decision and applies a force in the same direction. The force is orthogonal to the direction to the robot. Two robots facing each other only repel each other longitudinally and an external lateral force is required to steer them away from each other. The chosen direction should not oppose a message received from the same robot and should also preferably not oppose a decision made at the previous time steps. At the same time, the robot communicates this decision to the robot in front, asking it to drift at the other side. (iii) Previously the spatial position of the obstacle and the angular position of the obstacle from the direction to goal of the robot (elliptical model method) were considered to compute the force. Now the speed of the dynamic obstacle is also added as a model parameter. The algorithm extrapolates the trajectory of the obstacle and the robot and the position of the obstacle in the extrapolated trajectories is taken as an equivalent static obstacle corresponding to the dynamic obstacle. If the term force is replaced by a rule, the modelling is valid for the fuzzy logic-based system as well. The concept is shown in Fig. 18.8A.

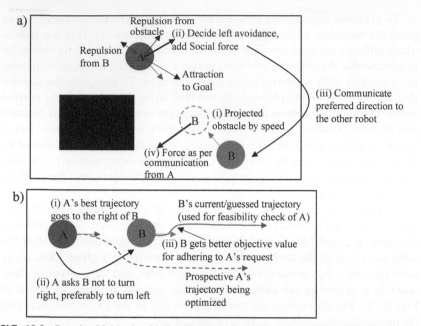

FIG. 18.8   Reactive Multi-robot Motion Planning (A) Artificial Potential Field. A projects B's motion (i) to calculate repulsion. Since it is head-on, A decides to drift left (ii) and communicates this to B (iii), and who adds the opposite force (iv). (B) Local Optimization. A computes a trajectory that goes to the right of B and can be a potential risk (i). A requests B to stay left (ii) which is implemented in B's optimization objective measure (iii).

A lot of avoidance of static and dynamic obstacles is possible by continuous optimization of the trajectory, as the obstacles and robot move. The major change that is of interest catering to the fact that there are multiple moving robots is (i) Each prospective trajectory needs to be assessed for feasibility. Previously the obstacles were static and thus a simulation was done of the prospective trajectory and if the simulated robot did not collide with any obstacle, the trajectory was regarded as feasible. Now, the algorithm needs to avoid collision with the dynamic obstacles. In case of a direct communication, it can be assumed that every robot publishes the current trajectory that it is following, and every robot simulates its prospective trajectories against the ones already published by the other robots. In the absence of direct communication, the robot guesses the trajectory of the other robots using machine learning and similar approaches. The most basic implementation is that the other robots continue to move with the same speed till they collide with a static obstacle. (ii) The robot listens to the communication messages received from the other robots, and adherence of (at least) an initial part of the trajectory is given a better objective function value as an incentive. (iii) The robot analyses the current trajectory and observes the other robots being avoided from the left or right. If a robot in close proximity is being avoided from one side or a robot in the extrapolation of trajectories comes directly ahead that is being avoided from one side, a message is given to that robot to not drift at the chosen side of avoidance.

18.4.2   Congestion avoidance in decentralized planning

The reactive planner as discussed will be assisted by a deliberative planner using a fusion of deliberative and reactive planning techniques. The deliberative planner of one robot does not know about the deliberative planner of the other robot. Hence, the deliberative planners schedule too many robots in an area causing congestion. This is not a completely new problem in robotics. In case of traffic scenarios, while planning a route, people do not consider other people and often clog the traffic of city centres causing congestion. Once congestion is caused, it is difficult to clear. A little deviation does not clear the congestion.

Congestion avoidance is a complex problem. The congestion avoidance systems are of two types, *anticipatory congestion avoidance systems*, which can look at the motion of different agents, predict in advance that congestion may happen and the area where the congestion is likely to happen, and take actions a priori such that congestion does not occur. The algorithms predict the area that is likely to get congested at a time and ask the robots visiting that area to reroute avoiding that area at that time a priori. The other system is the *congestion clearance system,* which detects that congestion at a region has already taken place and takes measures to resolve the congestion after it has already occurred. They typically stop new robots from entering that area, while asking the robots in the congestion area to reroute by taking alternative ways if available.

The anticipation of congestion ideally happens by simulating the entire system with knowledge of the intents of each robot, and if the simulation results in congestion at an area, it is communicated to the robots that the area should be avoided. The deliberative planners can then use a cost function that penalizes going through the areas which are collision prone as per the a priori simulation. The assumption is that the intent or the goal location of every agent is known, which is a fair assumption with communication.

A simpler way of implementing congestion control is using heat maps (Kala and Warwick, 2015c; Kala, 2018a). The *heat map* shows the current occupancy in the workspace. It is assumed that every agent emits heat. At the spatial level, the heat is large closer to the agent, while the heat fades away quickly as one goes away from the agent. At the temporal level, the heat values die off with time. So, if a person suddenly disappears from a place, the heat value slowly fades away with time to a zero value. The larger the number of people in an area, the larger the value of the heat map.

Motion planning on heat maps is one example of *costmap-based planning*. In a costmap-based planning, there is not a single binary map that designates whether the configuration is free or not; however, there is a cost associated with every region. The cost of the obstacles is infinite. The aim is to compute the least cost trajectory that passes through the least cost areas. The examples of costmap-based planning include navigating in an uneven surface where the unevenness is the cost, navigating on a multi-terrain surface where the cost of navigation is different for different terrains, and navigating in a hazardous environment in which every area has a hazard value denoting the cost. A sample heat map is given in Fig. 18.9. Any roadmap-based technique, including Probabilistic Roadmaps, can be used to construct a roadmap around the heat map. The cost of an edge from q_1 to q_2 is now given by Eq. (18.32):

FIG. 18.9 Congestion avoidance by using a Heat map. Every individual transmits heat with an exponential loss with distance, and decays with time for temporal cohesion. The deliberative planner avoids regions with high heat.

$$w(q_1, q_2) = \int_{\lambda=0}^{1} C(\lambda q_1 + (1 - \lambda)q_2)d\lambda \qquad (18.32)$$

Here, $C()$ is the heat map value that accounts for congestion. The equation goes through the entire edge and computes the cost by adding costs of all intermediate points. The integral may be replaced by a discrete summation as well. In case, there is an obstacle on the way, the cost associated with that point will be infinite and the edge will cease to exist.

18.4.3 Prioritization

In decentralized motion planning, every robot plans its trajectory. However, the individual trajectories may be such that when simultaneously executed incur collisions. The decentralized planning comes in the same situation for the resolution of collisions. A convenient to-use technique is prioritization. In *prioritization*, every robot has a priority value (Bennewitz et al., 2001, 2002). Let the priority of robot i be given by π_i. In this scheme, a higher-priority robot does not consider any lower-priority robot during motion planning and it is the sole responsibility of the lower-priority robot to avoid a collision with a higher-priority robot. So, when the coordination manager investigates a prospective collision, it rules that the higher-priority robot will have the liberty to retain its trajectory and asks the lower-priority robot to re-plan. It is important to mention that the robots have communication and therefore all planning and re-planning happens at the start only and not when the robots start making a motion. First, a conflict-free plan is found and then it is executed.

Considering that lower-priority robots will have to re-plan multiple times, on accounting for a collision with a higher-priority robot. The easiest implementation is to plan the robots in the order of their priorities. First, the highest-priority robot is planned. Once the trajectory of the highest-priority robot is fixed, then the second-highest-priority robot is planned. The second highest-priority robot plans to avoid static obstacles as well as the highest-priority robot only. This goes on and, in the end, the lowest-priority robot is planned. The lowest-priority robot avoids all the other robots, knowing their trajectories that have already been fixed. The planning procedure is shown in Fig. 18.10.

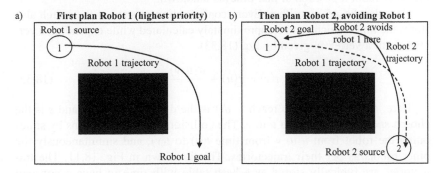

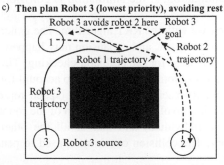

FIG. 18.10 Prioritization coordination scheme. Robots are planned in the order of priority. Lower priority robot avoids the higher priority ones. (A) Planning for robot 1. (B) Planning for robot 2 avoiding robot 1. (C) Planning for robot 3 avoiding robots 1 and 2.

Let the robots be indexed in the order of priority with 1 being the highest-priority robot and n being the lowest-priority robot. While planning robot i, the trajectories of all robots with a lower index are already known, since they were planned earlier. Let the trajectory for robot j be τ_j, $j < i$. This is a single robot motion planning algorithm with additional obstacles with pre-known trajectories $\{\tau_j\}$. Consideration must be given in planning that it is important to consider speeds and taking multiple speed options should be facilitated by the planning system.

Furthermore, since the trajectory of the other robots is known and is time-dependent, collision-checking is only possible by doing a forward simulation of the trajectory of the robot being planned and that of the other robots as per their trajectories. This is a *simulation-based collision-detection*, which replaces the conventional collision detection with static obstacles discussed previously (Kala and Warwick, 2011a,b). The simulations can only be done in a forward mechanism and not a backward mechanism. So forward simulation algorithms like A* algorithm, Rapidly-exploring Random Trees (RRT), optimization, etc., can be used. However, algorithms like bidirectional A* algorithm, bidirectional RRT, etc., cannot be used as a backward simulation from goal towards the source is required, while the time that the robot reaches the goal (and hence the other robot's position at that time) is unknown.

Consider the A* algorithm. Every configuration or state of the search also stores the time, which can be approximately calculated while expanding a vertex v from a vertex u, given by Eq. (18.33).

$$t(v) = t(u) + \frac{d(u, v)}{s} \qquad (18.33)$$

Here, $t(v)$ is the time to reach v, $d()$ is the distance function and s is the chosen speed in going from u to v. The collision checking happens by simulating the robot from u to v from time $t(u)$ to $t(v)$, and simultaneously the other robots as per their trajectories. This is shown in Fig. 18.11. The trajectories are typically stored as a hash table with time to have a constant lookup for the simulation. If the simulation is collision-free, the vertex v is admitted.

Similar is the case with the RRT algorithm. The generation of random vertex, selection of near vertex, and expansion may happen as usual. However, a similar simulation is used for the collision-checking. Time is as an additional state variable or an additional axis in search to account for slowing behaviours. Genetic Algorithm represents the trajectory (speed representation included). The fitness evaluation involves the simulation of the robot as per the candidate trajectory (genetic individual) and that of all the other robots as per their planned trajectory. The collision checking is used to penalize collision-prone trajectory.

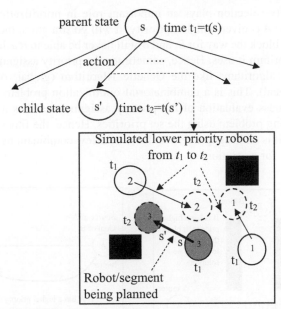

parent state s time $t_1 = t(s)$

action $\cdots$

child state s' time $t_2 = t(s')$

Simulated lower priority robots
from t_1 to t_2

t_1

2

t_2

t_2

2

1 t_2

t_2

3

s' s 3

1

1

t_1

t_1

Robot/segment
being planned

FIG. 18.11 Planning robots considering priorities using the A* algorithm. While planning a robot, consider candidate segment (from optimization, RRT, etc.) from s at t_1 to s' at t_2. Simulate all lower priority robots within the time, if feasible, accept the candidate (add edge). Backward simulation algorithms (bidirectional RRT, backward search) cannot be used.

The working of the overall algorithm is shown by Algorithm 18.1. Here, $Plan(S_i, G_i, C_i, \Gamma)$ denotes single robot motion planning for the robot i from source S_i to goal G_i using its configuration space C_i and the previously planned trajectories contained in Γ.

Algorithm 18.1 Prioritization based multi-robot motion planning.

Input: Sources (S_i), Goals (G_i), Configuration space (C_i) of all robots
Output: trajectories of all robots $\Gamma = \{\tau_i\}$

1 $\Gamma \leftarrow \emptyset$;   ▷ set of planned trajectories
2 Assign priority Π for all robots ;
3 **for** *all robots i in the order of priority* Π **do**
4 $\quad$ $\tau_i \leftarrow Plan(S_i, G_i, C_i^{obs}, \Gamma)$;   ▷ plan robot i, avoiding all robots in Γ
5 $\quad$ **if** $\tau_i = NULL$ **then** return $NULL$;
6 $\quad$ add τ_i to Γ ;
7 return Γ ;

The priority selection plays an important role in prioritization. Consider Fig. 18.12A. If *A* is given a higher priority, it will go straight to the goal. However, this will block the way for *B*, and *B* will never be able to reach the goal for whatever algorithm it uses. Hence, sometimes the priority assignment may be optimized by algorithms like the Genetic Algorithm (global) or Simulated Annealing (local). This is a combinatorial optimization problem. The caveat is that the fitness evaluation of the priority assignment solves a multi-robot motion planning problem using the set priorities. Hence, the fitness evaluation is expensive. It is still suggestive to at least try a few combinations of priorities before setting for a solution.

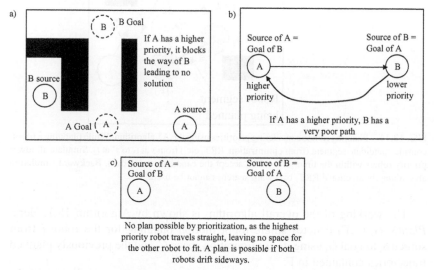

FIG. 18.12 Importance of priority assignment is shown through examples in (A)–(C). Performance significantly depends upon priorities, whose assignment may be optimized.

In prioritization, the highest-priority robot gets the best trajectory while the lowest-priority robot gets the worst trajectory. The approach is not optimal. Consider the situation in Fig. 18.12B. The highest-priority robot does nothing to cooperate (getting aside) with the lower-priority robot, sacrificing optimality. It is therefore suggestive to use some post-processing, local optimization, or modelling like the elastic strip for multiple robots so as to locally correct such trajectories. The approach is not complete as shown using a related problem in Fig. 18.12C.

18.5 Path velocity decomposition

Another mechanism of multi-robot motion planning problem is *path velocity decomposition* (Kant and Zucker, 1986). The multi-robot motion planning problem asks for a solution to the twin problems of path planning of multiple robots and speed assignment to multiple robots. In this approach, first, only the path planning problem is solved in a purely decentralized manner. Each robot plans its path considering only the static obstacles and no other robot. Let the path for the ith robot be given by $\tau_i[0,1] \to C_i^{free}$, such that the robot is at configuration $\tau_i(s) \in C_i^{free}$ at a relative distance of s ($0 \le s \le 1$) from the source, computing which is a classic robot motion planning problem. Note that the parametrization is by a relative distance from the source and not time.

Thereafter, the paths of all the robots are fixed. It is possible that tracing the path with the desired speeds may result in a collision. Hence, the speeds for every segment of the trajectories are set such that the resultant plan is collision-free and the best for the fixed paths. The speed assignment algorithm has the liberty to change the speed of any segment of any robot.

Let the speed assignment of the robot i be given by $\sigma_i : \mathbb{R}_{\ge 0} \to [0,1]$, which gives the path parametrization for any time t. The path parametrization can be given to the computed path to get the corresponding configuration. So, at time t, the robot i is at a relative distance of $\sigma_i(t)$ from its source, and its configuration is given by $\tau_i(\sigma_i(t))$. The aim is to find the optimal speed setting σ_i for every robot. This problem will be solved in a centralized setting, but it is a single-dimensional problem for every robot and therefore the dimensionality of the overall problem is limited to the number of robots.

Let us understand the speed setting function σ_i in a better way. Consider the path of one robot shown in Fig. 18.13A. The parametrization of the trajectory is over the relative distance (0 to 1), which is convertible into the real distance by a

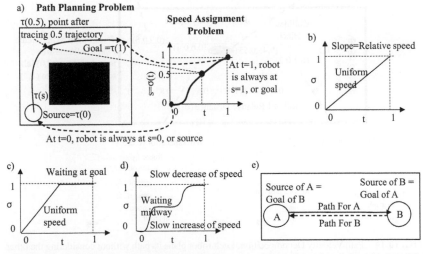

FIG. 18.13 Speed assignment problem. (A) First, the path is planned and then the speed is assigned with meanings given in (B)–(D). (E) An example showing that the approach is not complete as no combination of speed assignment can avoid collisions.

multiplication of the trajectory length. Similarly, the time is scaled to lie between 0 and 1. The slope of the figure hence shows the relative speed, which is convertible into the real speed by a multiplication of the trajectory length on the total time. Fig. 18.13B shows the robot moving at a uniform speed over the entire trajectory length. Fig. 18.13C shows the robot moving at a uniform speed and the robot then stays at the goal for the remaining time. Fig. 18.13D shows the robot traversing half the trajectory, then waiting for a while, and then traversing the other half of the trajectory. The initial velocity is zero and increases slowly, while the final reaching velocity is zero and decreases slowly. Note that the curve shall always increase in terms of the distance as the robot can only move forward in the trajectory. The curve will also always increase in terms of time as the robot cannot be at two places at the same time.

Now consider 2 robots instead of just 1. The trajectory of the robots is shown in Fig. 18.14. Now consider the relative speed assignment among the robots given in Fig. 18.14. The figure shows the speed assignment of the two robots at the two axes. The X-axis denotes $\sigma_1(t)$ and the Y-axis denotes $\sigma_2(t)$, the speed mapping function of the two robots at any general time t. The origin is $\sigma_1 = 0$

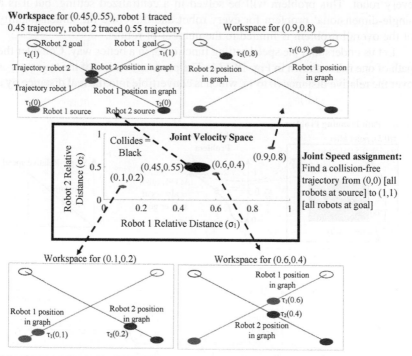

FIG. 18.14 Path Velocity Decomposition. Each robot plans its path without considering the other robots. Relative speed assignment problem is solved as a conventional robot motion planning problem that can be solved by any motion planning algorithm. In the speed-assignment space (centre), white denotes relative positions of robots with no collision and vice versa.

and $\sigma_2 = 0$, which corresponds to the robot configuration $\tau_1(0)$ and $\tau_2(0)$, meaning both the robots are at the origin. The other extreme position $\sigma_1 = 1$ and $\sigma_2 = 1$ denote the robot configurations $\tau_1(1)$ and $\tau_2(1)$, meaning both the robots are at the goal. Similarly, $\sigma_1 = 1$ and $\sigma_2 = 0$, denote the first robot at goal and the second at the origin, while $\sigma_1 = 0.5$ and $\sigma_2 = 0.5$ denote both robots are mid-way to their goal at the fixed trajectory. Note that since the initial trajectories are collision-free, any combination of σ_1 and σ_2 will not have any collision with static obstacles; however, there may be a collision between the robots. This means the problem is to find a trajectory in this space from $(0,0)$ to $(1,1)$, such that the joint positions taken by the robots at any point in time are collision-free.

Let us now generalize the problem to n robots. For n robots, consider the paths as $\tau_1, \tau_2, \tau_3 \ldots \tau_n$. The speed assignment space is denoted by $\Sigma = [0,1]^n$. Any general point in this space is the relative positioning of the robots in their trajectories, given by $\sigma = (\sigma_1, \sigma_2, \sigma_3, \ldots \sigma_n)$ denoting the robots at position $\tau_1(\sigma_1)$, $\tau_2(\sigma_2)$, $\tau_3(\sigma_3) \ldots \tau_n(\sigma_n)$. Let us now divide the entire space into collision-prone areas (Σ^{obs}) and collision-free areas (Σ^{free}). The robots will collide with each other at a joint position σ if Eq. (18.34) holds.

$$\sigma \in \Sigma^{obs} \text{iff} \ \left(\exists_{i,j,i \neq j} R_i(\tau_i(\sigma_i)) \cap R_j(\tau_j(\sigma_j)) \right) \neq \varnothing) \qquad (18.34)$$

The equation states that $\sigma = (\sigma_1, \sigma_2, \sigma_3, \ldots \sigma_n)$ is collision-prone if a pair of robots exist such that the volume occupied by the two of them is overlapping. $R_i(q)$ is the volume occupied by the robot i in workspace when placed at configuration q. Correspondingly, the free space is given by $\Sigma^{free} = \Sigma \backslash \Sigma^{obs}$, which can also be explicitly modelled as Eq. (18.35):

$$\sigma \in \Sigma^{free} \text{iff} \ \left(\forall_{i,j,i \neq j} R_i(\tau_i(\sigma_i)) \cap R_j(\tau_j(\sigma_j)) \right) = \varnothing) \qquad (18.35)$$

The problem is to find a trajectory $\lambda : [0,1] \to \Sigma^{free}$ in this space, such that it starts with all robots at their source $\lambda(0) = (0,0, \ldots 0)$, ends with all robots at their goal, or $\lambda(1) = (1,1, \ldots 1)$ such that all intermediate positions do not incur collision between any two robots, or $\lambda(s) \in \Sigma^{free}$, $0 \leq s \leq 1$.

Surprisingly, speed assignment is another classic robot motion planning problem that can be solved by any technique already studied in the book. The constraints are that the motion must always move forward along every axis, which makes the problem a lot easier for each algorithm discussed. Even though there may be several robots, most pairs of robots will be non-interacting which do not come into each other's space even remotely and therefore do not affect each other's trajectory. The choice of the algorithm should be such that if two pairs of robots are not interacting, the algorithm should not waste too much time in deciding a mutually agreeable plan between them by considering all possible pairs of assignments. Optimization is a good technique for the same reason. Alternatively, some algorithms may *cluster* robots such that any pair of robots from different clusters do not interact with each other, and thereafter, the speed assignment problem is solved for every cluster of robots separately. Breaking a large planning problem into multiple independent planning problems is a powerful tool that significantly reduces the complexity.

The approach is neither optimal nor complete. Consider a simple scenario shown in Fig. 18.1E). No matter what the speed assignment is, the robots shall not be in a position to avoid each other.

18.6 Repelling robot trajectories

In the path velocity decomposition approach, the path computation was decentralized, while a joint velocity was optimized in a centralized architecture. Similarly, the centralized evolutionary planning theoretically worked on a joint state space of all robots, optimizing speed and path both; however, practically the approach selects near-colliding individual robot trajectories and optimizes them to be collision-free. The simulation gives points where the modification needs to be applied. Both approaches were optimization-based.

The discussions are further extended to the use of heuristics for coordination. Earlier the reactive methods (or the ones that apply heuristic rules for the navigation of the robot) were applied to the problem of multi-robot motion planning. Now, they will be applied to the problem of multi-robot coordination. So, it is assumed that the initial trajectory of all the individual robots is available (or the trajectory of other robots can be guessed), planned by considering static obstacles only. The problem is to coordinate their motion and to make the same collision-free.

The solution can be visualized in an abstract space called the *trajectory space* (Kala, 2018b), wherein every point represents a trajectory of the robot or any obstacle (static or dynamic). Space contains the position and time information of all the robots and obstacles. Now, the objective is to make the trajectories attract and repel each other as an implementation of a potential inspired solution to the coordination problem.

Every robot, for coordination, intends to keep itself as close as possible to its initial trajectory made without considering the other robots. Therefore, in a trajectory sense, it is said that the optimal single robot trajectory for a robot, planned without considering the other robots, *attracts* the current trajectory of the robot at any point in time. Similarly, the robot trajectory needs to avoid the static and dynamic obstacles, which in this space is said that the trajectory of the robot at any point of time is *repelled* by the trajectories of the other robots and static obstacles. This is shown in Fig. 18.15A.

In single robot motion planning, the paths were planned, while the speed was set as the maximum permissible value accounting for the obstacles in front (and curvature). The same principles are used to generate a simulated motion of the robots. At any time, if a simulation of all current robot trajectories needs to be done, only the path is taken. The speeds are set by noting the readings of a virtual proximity sensor at the robot and the known curvature. The curvature limits the speeds to the ones admissible as per the kinematics. The virtual proximity sensor sets speed such that no two robots collide, stopping the robot if necessary. This is shown in Fig. 18.15B–F. This reduces the problem to solve for a path for every robot which is collision-free as per the speed setting heuristic function. Note that the robots shall never collide in such a simulation, although they may end up stopping and waiting for each other to move endlessly.

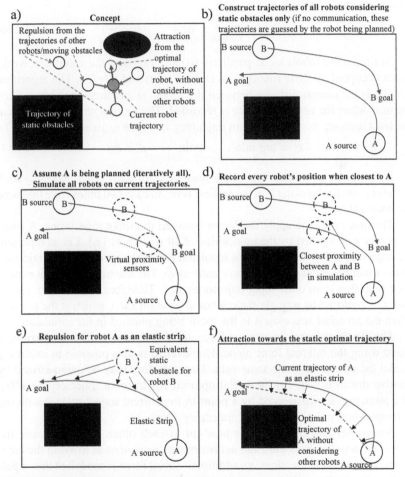

FIG. 18.15 Repelling Robotic Trajectories. (A) A robot's trajectory is repelled by other robots' trajectories and attracted towards its optimal trajectory constructed considering static obstacles only. Starting from a static optimal trajectory (B), simulate all robots setting speed based on distance from the front obstacle and curvature (C), to get the closest proximity (D), used for repulsion as an elastic strip (E), and attraction towards the static optimal trajectory (F). Elastic strip deforms using forces in (E) and (F). Steps (C)–(F) are iteratively repeated for all robots.

The attraction and repulsion of trajectories depend on the attraction and repulsion forces, which are dependent on the distance function. Hence, let us define the *distance function* between any two trajectories of any two robots (or static obstacles, which is a constant position for any general time). The distance between a trajectory of the robot being planned τ with the trajectory of any other robot/obstacle τ_i, is the closest distance recorded in the simulation of the robots along the trajectories using the speed setting heuristic in the workspace. This is given by Eq. (18.36):

$$d(\tau, \tau_i) = \min_{t} \min_{o \in R(\tau(t)), o_i \in R(\tau(t))} d(o, o_i) \tag{18.36}$$

At time t, the robots are at positions $\tau(t)$ and $\tau_i(t)$, and the volume of workspace occupied by these robots is given by $R(\tau(t))$ and $R_i(\tau_i(t))$. The algorithm first finds the closest distance between the robots at any time, thereafter, finds the time when the nearest distance is recorded. The robot with trajectory τ is at the least distance from a robot with trajectory τ_i at time c_i given by Eq. (18.37):

$$c_i = \arg\min_{t} \min_{o \in R(\tau(t)), o_i \in R(\tau(t))} d(o, o_i) \tag{18.37}$$

It should be noted that the robots may wait for each other endlessly and, therefore, the tie-breaking is at the earliest time, in case two times have the same distance value.

The task now is to apply the forces between the robot trajectories. A robot trajectory τ is repelled by the trajectories of all the robots $\{\tau_i\}$. Let us represent the trajectory at the workspace. The repulsion is done by modelling the trajectory of the robot being planned (τ) as an *elastic strip*. The elastic strip cannot experience repulsion from other robot trajectories $\{\tau_i\}$. Therefore, the ith robot trajectory τ_i is *replaced* by a single static robot at position $\tau_i(c_i)$, which is the position when the ith robot was closest to the robot being planned in the simulation.

The elastic strip has an internal force and can maintain the least disruptive shape using the internal force alone. However, it is also possible to explicitly model the attraction in the same vein. The trajectory may be parametrized by relative distance $\tau(s)$. Let the initial optimal single robot trajectory be $\tau^*(s)$. The attraction is experienced by a point in the current trajectory $\tau(s)$ with the corresponding point in the initial trajectory $\tau^*(s)$.

Sometimes two robots can be head-on to each other. In such a case, the robots are repelled backward and neither makes a move so as to avoid the other by the side. In such situations, an additional social force is modelled that repels both robots to move sideways and is taken as an external force for the elastic strip. The force is leftward for a robot being planned if it lies leftward of the obstacle robot and vice versa. In case of a tie or both sides of avoidance being nearly equally competent, social conventions are taken. In many countries avoiding people is on the left and a preference to the left side may be given.

The algorithm is iterative. First, every robot plans its trajectory. Thereafter, a simulation is done to determine the closest position between every pair of robots. Every robot experiences a repulsion by the other robots, at a position when they are the closest as per their trajectories. Every robot adapts its trajectory to account for the other robots. Thereafter, in the next iteration, a new simulation is done. Typically, the collisions are resolved, and the robots move a little away from each other. In the new simulation, again the robots note the new closest points and adjust their trajectories. This goes on until no robot can improve its trajectory. The communication is not a strict requirement because if the trajectory of the other robots is not known, it shall be guessed and used to calculate the repulsive forces.

18.7 Co-evolutionary approaches

The path velocity decomposition made an interesting system where the path computation was decentralized, and the velocity computation was centralized. Another interesting methodology is optimization. The centralized solution to the problem of multi-robot motion planning has already been discussed. However, the approach works on a centralized space of all possible combinations of robot trajectories. It should be possible to tell the optimization technique that the individual robots have a near-independent trajectory and only need a little coordination among them to resolve the conflicts if they happen to develop a conflicting set of trajectories.

18.7.1 Motivation

Cooperative evolution (or *co-evolution*) is another interesting evolutionary algorithm, which models the behaviour wherein multiple species evolve in cooperation with each other (Potter and de Jong, 2000). The species together make an eco-system. The characteristic of the ecosystem is that all the species must live harmoniously helping each other and deriving help from each other. A good ecosystem features species that develop complementary skills that are of value to the members of some other species. A species of the eco-system does not only live for itself or develop characteristics for its own, but for the entire ecosystem.

In any evolutionary technique, the fitness evaluation provides the evolutionary pressures for the individuals to react to. The focal point of the cooperative evolution is that an individual is given a high fitness not for what goodness it has; however, for the magnitude, it cooperates well with the members of the other species to enable the entire ecosystem to have a good value. The cooperative evolution does not measure the fitness of an individual of a species in terms of its merit; however, it assigns the fitness value of the entire ecosystem with individuals from all species. Many times, a good-looking individual may be given poor fitness if it fails to cooperate well with the other individuals of the other species in the ecosystem.

Let us discuss the notion with a few examples. Imagine a large project is to be made by a team, which represents an ecosystem. The project is decomposed into components and the staffing problem is solved for the different components of the project. Each component here represents the species. Now an individual of a team is not assessed on how he/she solves that component but in terms of how the component gets well in the entire project. Many times, extremely talented people are not good employees to take because it is difficult to explain to them the requirements, or they fail to change the module in the realization of the entire project. A lack of cooperation causes the individual's fitness value to be underrated. Similarly, many times the components may be developed well, but the components integrated may not make a great product. Say a sensor of a robot requires heavy battery power, which requires a large battery size and therefore an increase in the robot payload. The sensor may be accurate and of low cost, but its utility in the actual robot is limited.

Similarly, for the evolution of a neural network, every neuron (species) of the neural network (ecosystem) aims in finding interesting features, converting a non-linear problem into a linearly separable problem. The neuron features (species) evolved may be individually good, but collectively bad as they do not complement each other. The fitness of a neuron (species) is based on how it cooperates with the other neurons to make an overall good neural network (ecosystem), and not based on its merit. This encourages the neurons to develop complementary features, not discovered by any other neuron.

Consider the problem of navigation of a robot based on a set of rules. The quality of the rule (species) is based on how it cooperates with the other rules making an overall good navigation system (ecosystem), and not on its own ability to navigate the robot. This encourages the system to evolve complementary rules for situations not covered by any other rule.

In the problem of multi-robot motion planning as well. The trajectories of the individual robots make a species, and the ecosystem is all the trajectories of all the robots combined. The metric used to assess the quality of a trajectory is based on its cooperation with the other robot trajectories to make the plan collision-free and with sufficiently high clearance.

18.7.2 Algorithm

To understand the algorithm, let us take a synthetic problem, to optimize the expression given by Eqs (18.38), (18.39)

$$f(X) = (x_1 - 0.5)^2 + (x_2 - 0.25)^2 + 10x_1^2x_2^2 + (x_3^2 - 0.1)^2 + 20x_3^2x_4^2 + x_1^2x_3^2$$

$$(18.38)$$

$$-1 \leq x_1, x_2, x_3, x_4 \leq 1 \tag{18.39}$$

If the last term $(x_1^2x_3^2)$ is removed, it would be simple to solve the problem as the objective can be decomposed into two completely independent objective functions $(x_1 - 0.5)^2 + (x_2 - 0.25)^2 + 10x_1^2x_2^2$ and $(x_3^2 - 0.1)^2 + 20x_3^2x_4^2$. The two can be solved independently and the results can be combined. However, the presence of the last term signifies that no decomposition is possible, which constrains the two functions to coordinate with each other to take values that are collectively good. The complete individual as a genotype is given by $[x_1 \ x_2 \ x_3 \ x_4]$, representing the ecosystem. Let us break up this individual into 2 parts representing species, the first species being $[x_1 \ x_2]$ and the second being $[x_3 \ x_4]$. The two species (represented by their population) are evolved separately.

In general, consider any general problem. It is known that the complete problem (or ecosystem) can be decomposed into n components (or species) using some heuristics. This division and generation of an initial random (or as per some heuristic technique) population are shown in Fig. 18.16A. Even though the representation of all species is separate, they will have to interact with each other to evolve.

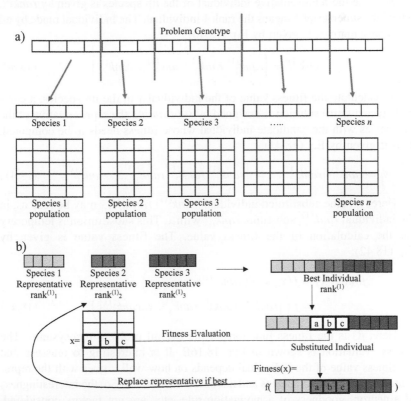

FIG. 18.16 Cooperative co-evolution (A) Division of a problem into species. Every genotype is divided into multiple parts called species, such that each species can be evolved nearly independently. (B) Fitness Evaluation by substitution.

Imagine a Genetic Algorithm (or any other evolutionary algorithm) that needs to be executed for the optimization of all the species population. The crossover, mutation, elite, and all other genetic operators do not require any collaboration and they may be done independently by all the species. The fitness evaluation requires a collaboration between the species, which is also because no fitness function of the individual species exists; however, the fitness function exists for the ecosystem.

To get a fitness measurement, an individual from a species needs to collaborate with the other individuals from the other species. Therefore, let us take a representative solution (the one with the best fitness) from all species. It may seem recursive that the fitness calculation requires a knowledge of the best individual, which itself is based on a fitness measure. To solve the recursion, the initial representatives are taken randomly. Once some representatives (even if random) are available, it is possible to calculate the fitness value of all individuals in all species. A better-fitness individual replaces the representative of the corresponding species.

Suppose the representative individual of the ith species is given by $rank_i^{(1)}$, where the superscript 1 means the rank 1 individual. The individual made by all the representatives is given by Eq. (18.40):

$$rank^{(1)} = \left[rank_1^{(1)} \ rank_2^{(1)} \ rank_3^{(1)} ... rank_n^{(1)} \right] \qquad (18.40)$$

To calculate the fitness value of the individual x in the ith species, a complete (ecosystem) solution is designed by *substituting* the representative of the ith species with the candidate individual whose fitness needs to be computed. This is given by Eq. (18.41).

$$y_i = rank^{(1)} | rank_i^{(1)} : x = \left[rank_1^{(1)} \ rank_2^{(1)} \ rank_3^{(1)} ..x..rank_n^{(1)} \right] \qquad (18.41)$$

Here, y_i is the substituted individual. $rank^{(1)} | rank_i^{(1)} : x$ may be read as, in the individual $rank^{(1)}$, substitute $rank_i^{(1)}$ with x. This substitution is temporary for the calculation of the fitness value. The fitness value is given by Eq. (18.42):

$$
\begin{aligned}
fitness(x) = f(y_i) &= f\left(rank^{(1)} | rank_i^{(1)} : x \right) \\
&= f\left(\left[rank_1^{(1)} \ rank_2^{(1)} \ rank_3^{(1)} ..x..rank_n^{(1)} \right] \right) \qquad (18.42)
\end{aligned}
$$

Here, $f()$ is the fitness function for the overall problem (ecosystem). The fitness evaluation is shown in Fig. 18.16B. It is interesting to reassess that the fitness value of the individual depends on how well it goes with the representatives of the other species and not based on its merit. So, the interestingness of a neuron, goodness of a navigation rule, etc., are not factors considered. However, the overall fitness function in terms of neural network loss, the merit of the navigation system, etc., is used. Once the fitness of all the individuals is computed, the representatives can be changed as Eq. (18.43), assuming a minimization problem is being solved.

$$rank_i^{(1)} \leftarrow argmin_{x_i} fitness(x_i) \tag{18.43}$$

Consider the same example of 2 species for a synthetic function. Let the current representatives be given by $[-0.1 \ 0.6]$ and $[0.3 - 0.5]$. The fitness evaluation is shown in Table 18.2.

TABLE 18.2 Fitness evaluation by substitution in cooperative co-evolution.

(a)

S. No.	Individual of species 1	Rank 1 representative	Individual after substitution	Fitness value
1.	$[-0.1 \ 0.5]$	$[-0.1 \ 0.6 \ 0.3 \ -0.5]$	$[-0.1 \ 0.5 \ 0.3 \ -0.5]$	0.7635
2.	$[0.2 \ -0.4]$	$[-0.1 \ 0.6 \ 0.3 \ -0.5]$	$[0.2 \ -0.4 \ 0.3 \ -0.5]$	0.8952
3.	$[0.9 \ -0.4]$	$[-0.1 \ 0.6 \ 0.3 \ -0.5]$	$[0.9 \ -0.4 \ 0.3 \ -0.5]$	2.2665
4.	$[0.3 \ 0.1]$	$[-0.1 \ 0.6 \ 0.3 \ -0.5]$	$[0.3 \ 0.1 \ 0.3 \ -0.5]$	*0.3947*
5.	$[0.1 \ -0.5]$	$[-0.1 \ 0.6 \ 0.3 \ -0.5]$	$[0.1 \ -0.5 \ 0.3 \ -0.5]$	1.0635

(b)

S. No.	Individual of species 2	Rank 1 representative	Individual after substitution	Fitness value
1.	$[-0.3 \ 0.1]$	$[-0.1 \ 0.6 \ 0.3 \ -0.5]$	$[-0.1 \ 0.6 \ -0.3 \ 0.1]$	*0.56378*
2.	$[-0.2 \ 0.3]$	$[-0.1 \ 0.6 \ 0.3 \ -0.5]$	$[-0.1 \ 0.6 \ -0.2 \ 0.3]$	0.61308
3.	$[0.3 \ 0.8]$	$[-0.1 \ 0.6 \ 0.3 \ -0.5]$	$[-0.1 \ 0.6 \ 0.3 \ 0.8]$	1.96238
4.	$[0.1 \ -0.9]$	$[-0.1 \ 0.6 \ 0.3 \ -0.5]$	$[-0.1 \ 0.6 \ 0.1 \ -0.9]$	1.67802
5.	$[-0.5 \ 0.6]$	$[-0.1 \ 0.6 \ 0.3 \ -0.5]$	$[-0.1 \ 0.6 \ -0.5 \ 0.6]$	1.5051

Let the representative best individual be $[-0.1 \ 0.6 \ 0.3 \ -0.5]$ with $[-0.1 \ 0.6]$ as representative of species 1 $(x_1 \ x_2)$ and $[0.3 \ -0.5]$ as the representative of species 2 $(x_3 \ x_4)$. The fitness function is $f(X) = (x_1 - 0.5)^2 + (x_2 - 0.25)^2 + 10x_1^2 x_2^2 + (x_3^2 - 0.1)^2 + (x_4^2 + 0.2)^2 + 20x_3^3 x_4^4 + x_1^2 x_3^2$. The genes corresponding to the individual whose fitness is being calculated are shown in bold. The best fitness is shown in italics. The new representatives are shown in italics. (a) Species 1 (b) Species 2. The new best specie is $[0.3 \ 0.1 \ -0.3 \ 0.1]$.

The fitness evaluation can be used to assign ranks to be used for finding the expectation for the selection of an individual in the evolution of any species. The selection, crossover, mutation, and other genetic operators are unchanged and are done independently for all species. This produces a new population pool for all species. The working of the algorithm is given by Algorithm 18.2.

Algorithm 18.2 Co-evolutionary optimization.

1 **for** *all species i* **do**
2 $\quad\lfloor$ generate random population $X_i = \{x_i\}$
3 **for** *all species i* **do** $\qquad\qquad \triangleright$ `initially random representatives`
4 $\quad\lfloor$ $rank_i^{(1)} \leftarrow$ random individual from X_i ;
5 **while** *stopping criterion* **do**
6 $\quad$ **for** *all species i* **do**
7 $\quad\quad$ **for** *all individuals* x_i *in species i* **do**
8 $\quad\quad\quad\lfloor$ $fitness(x_i) \leftarrow f(rank^{(1)}|rank_i^{(1)} : x_i)$;
9 $\quad\quad$ $rank_i^{(1)} \leftarrow argmin_{x_i} fitness(x_i)$; $\qquad \triangleright$ `new representative`
10 $\quad\quad$ generate new generation population by genetic operators ;

An issue is that fitness is being evaluated based on a single representative alone. The single representative may not be good, which can adversely affect the fitness evaluation of the individuals. Hence, instead of rank 1 individual making a representative, the best k individuals are made the representatives. The jth representative is taken as Eq. (18.44):

$$rank^{(j)} = \left[rank_1^{(j)} \; rank_2^{(j)} \; rank_3^{(j)} ... rank_n^{(j)} \right], 1 \le j \le k \qquad (18.44)$$

The substitution is carried for all the representatives, instead of just 1. The fitness associated with an individual x in the ith species for the rank j is given by Eq. (18.45), and the total fitness is considering all the ranks from 1 to k given by Eq. (18.46):

$$fitness_j(x) = f\left(rank^{(j)}|rank_i^{(j)} : x \right) = f\left(\left[rank_1^{(j)} \; rank_2^{(j)} \; rank_3^{(j)} ...x...rank_n^{(j)} \right] \right) \qquad (18.45)$$

$$fitness(x) = \sum_{j=1}^{k} fitness_j(x) \qquad (18.46)$$

18.7.3 Master-slave cooperative evolution

The co-evolutionary technique finds good individuals for all the species. However, the solution to the overall problem (ecosystem) needs to be constructed. Typically, it is acceptable to take the best representatives of all the individuals as the solution to the problem and return the $best^{(1)}$ individual. However, it may be possible to have some other permutation of individuals which get an overall better objective, which should be computed. The same ideology may be used during the optimization as well. Combining all rank 1 solutions to make a rank 1 representative is a heuristic, while better combinations may exist. Therefore, as the optimization goes along, it may be valuable to get the best combination of individuals. The problem of finding good species is different from the problem of finding a good combination of existing species that make the best ecosystem. Here, we will use different algorithms to solve both problems simultaneously.

The task of finding the best combination of individuals is done by a separate algorithm called the *master algorithm*. Finding the best combination of the individual of species that together make the best solution to the overall problem (ecosystem) is a simple combinatorial optimization problem that can be conveniently solved by a simple Genetic Algorithm. The optimization is run in parallel to the co-evolutionary optimization over the species, which is called the *slave algorithm*. The master genetic algorithm is used to find the best combination of individuals from the slave. The master algorithm does not itself have an individual which stores a real encoded vector representing a solution. It instead only stores the pointers to the slave individual solutions, denoting the combination of slave individuals from different species that form a master solution. The slave algorithm runs a co-evolutionary optimization and stores the real encoded vectors representing the species population.

The master and the slave optimizations run in parallel. The two optimizations are not independent but related to each other. The master genetic algorithm stores pointers to the individuals of the slave population. So, the master cannot exist without the slave. Similarly, the slave algorithm draws representatives from the master or more generally uses the master population for fitness evaluation.

A population showing the individuals of both the master and slave population is shown in Fig. 18.17A. The master Genetic Algorithm is a conventional Genetic Algorithm doing combinatorial optimization. The most interesting part is the fitness evaluation. To compute the fitness of the master individual, the individuals of the slave being pointed to are retrieved. These individuals are then joined together which makes a complete individual that can be passed to the fitness evaluation of the overall problem. The mutation operator changes the pointer from one slave individual to another. The crossover operator, typically a scattered crossover, exchanges the pointers of the two parents to get the two children. The operators are shown in Fig. 18.17B and C.

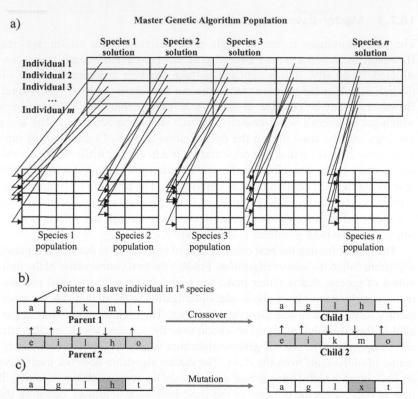

FIG. 18.17 Master-Slave Cooperative Coevolution. (A) Master population selects the best combination of species individual. Every master population gene stores a pointer to one individual from the corresponding species. The best few master individuals act as representatives for slave population fitness evaluation. (B) Scattered crossover (C) Mutation (change to a different individual from the same species). Each letter represents a pointer to a slave individual shown as actual pointers earlier.

The slave is a co-evolutionary population that is evolved as discussed excluding the fitness evaluation. The best k individuals used previously are the best (fittest) k individuals of the master genetic algorithm. Substitution-based fitness evaluation is used. Let z_j be the individual of the master algorithm with rank j, which is composed of individuals from all the species, or $z_j = [z_1^{(j)} z_2^{(j)} z_3^{(j)} ... z_n^{(j)}]$. $z_i^{(j)}$ is the ith specie solution in the individual z_j. More specifically $z_i^{(j)}$ is the ith gene of the master population that points to a slave individual whose genes are retrieved for fitness evaluation. The fitness value of individual x in slave species i is given by Eqs (18.47), (18.48).

$$fitness_j(x) = f\left(z_j | z_i^{(j)} : x\right) = f\left(\left[z_1^{(j)} z_2^{(j)} z_3^{(j)} ..x..z_n^{(j)}\right]\right) \qquad (18.47)$$

$$fitness_j(x) = \sum_{j=1}^{k} fitness_j(x) \qquad (18.48)$$

Another fitness function now measures the average fitness of the individuals in which the individual x in slave species i participates. Therefore, k best individuals from the master population are searched for, whose ith specie individual points to x. The average fitness of these individuals is taken. If the individual x is good, it will be found in the better master individuals and vice versa. If including or pointing to x improves a master individual, more individuals would point to x. This is called the *best-k* fitness function.

18.7.4   Analysis of co-evolutionary algorithm

The co-evolutionary algorithm very interestingly breaks down the problem into species to be evolved in cooperation with each other. To understand, consider the simplest fitness function, which has the two species independent of each other, given by Eqs (18.49), (18.50):

$$f(X) = (x_1 - 0.5)^2 + (x_2 - 0.25)^2 + 10x_1^2 x_2^2 + (x_3^2 - 0.1)^2 + 20x_3^2 x_4^2 \quad (18.49)$$

$$-1 \leq x_1, x_2, x_3, x_4 \leq 1 \qquad (18.50)$$

Let the *best*$^{(1)}$ solution at any point of time be $(x_1, x_2, x_3, x_4) = (b_1, b_2, b_3, b_4)$, with (b_1, b_2) as the best representative of the first specie and (b_3, b_4) as the best representative of the second specie. The fitness function for any individual (x_1, x_2) of the first species at a particular point of time hence becomes Eq. (18.51).

$$f(X) = (x_1 - 0.5)^2 + (x_2 - 0.25)^2 + 10x_1^2 x_2^2 + (b_3^2 - 0.1)^2 + 20b_3^2 b_4^2 \quad (18.51)$$

The value (b_3, b_4) does not contribute anything in the ranking of the individuals of species 1, which governs their selection probability. The effect of changing representatives thus does not affect the working of species 1 and the evolution is independent. The co-evolution was not told that species 1 would be able to evolve independently, co-evolution figured that out. Similarly, for the second species, the fitness function of a general individual (x_3, x_4) becomes Eq. (18.52):

$$f(X) = (b_1 - 0.5)^2 + (b_2 - 0.25)^2 + 10b_1^2 b_2^2 + (x_3^2 - 0.1)^2 + 20x_3^2 x_4^2 \quad (18.52)$$

Again, the choice of the representative of the first species does not determine the rank of the individual of the second species, and hence, changing representatives has no role to play in the evolution of the first species. Even though both evolutions are independent, it may be noted that the number of fitness evaluations for a fixed number of individuals and generations would double, because the fitness function needs to be called separately for every species. This is typically a problem when fitness evaluation is time-consuming.

Let us take the opposite case, wherein a generic problem is divided into too many species. Now, the fitness function of a general individual of a species depends dramatically on the representatives from other species. As soon as a species is towards convergence, the representatives change and destabilize the species. The species now must adjust to a changed fitness function due to the changed representatives. In cooperative evolution, the fitness of one species depends on all the other species and therefore they keep destabilizing themselves. Eventually, some species will dominate and converge, thus fixing its representative, which will become a reason for some other dominating species to fix its representative and converge. A consensus is thus eventually reached. However, the convergence will be much slower due to a wrong division, besides the fact that the fitness evaluation is separately done for every species thus multiplying the evaluation requirements with the number of species. Evolution under a dynamic fitness function is a harder problem attempted by co-evolution with changing representatives.

Overall, co-evolution is a good method with faster convergence, if the species are *nearly independent* and need to cooperate for smaller terms only. The master algorithm itself consumes a lot of time, especially when the fitness evaluation is time-consuming. However, the algorithm does a good job of getting better representatives identified and better representatives can highly boost the convergence. If better representatives can be computed quickly, the species will have a faster convergence.

18.7.5 Motion planning using co-evolution

We have already solved the problem of motion planning of a single robot by optimization, including adding the speed of every segment as a gene in the individual. The motion planning using co-evolution is a minor extension of the same problem (Kala, 2012, 2020). The master-slave approach is used, wherein the master is a Genetic Algorithm solving the best selection of individuals from the slave population.

The slave is a coevolutionary algorithm. The entire problem is decomposed into species with every robot's trajectory representing one species. The representation of the individual species is as used previously and is also shown in Fig. 18.18. The individual is modelled as a collection of via points, with each via point having the configuration and speed to travel that segment. The master connects to the slave using the same architecture as discussed previously. The fitness function is done for the entire plan, using the metrics of average path length, average traversal time, average clearance, and smoothness. The penalization scheme is used to penalize solutions that are collision prone. The genetic operators are the same as discussed in the case of a single robot.

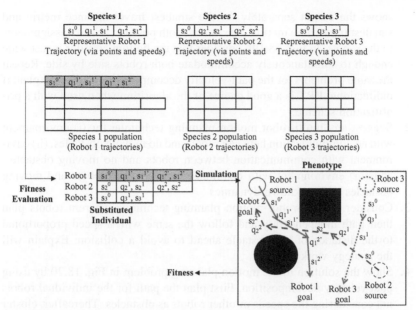

FIG. 18.18 Multi-robot Motion Planning using Co-evolution. Each robot's trajectory is a species, evolved like the single robot motion planning problem. Fitness evaluation by substitution is explicitly shown. The concept is extendable to a master-slave architecture.

Questions

1. Using a prioritization scheme, simulate the priority-based robot avoidance for the cases shown in Fig. 18.19. Give options when navigating with different speeds and different priority assignments. For Fig. 18.19E and F,

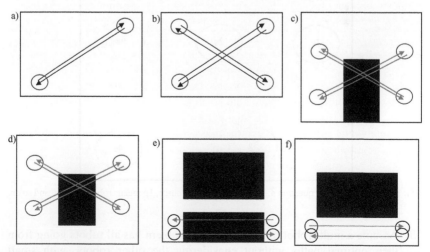

FIG. 18.19 Problems for question 1. Tails show the sources of the robots and heads show the goals of the robots. Consider all possible priority assignments. (A) Problem 1, (B) Problem 2, (C) Problem 3, (D) Problem 4, (E) Problem 5, (F) Problem 6.

shows the output separately for the smallest travel distance metric and smallest travel time metric, considering both possible priority assignments to the robots. In Fig. 18.19F consider that the bottom corridor is just wide enough to simultaneously accommodate both robots side by side. Repeat the same problems for the path-velocity decomposition scheme. Explain if bidirectional RRT is a good choice of the algorithm when used with a prioritization scheme.

2. Suggest any multi-robot motion planning technique for (i) environment with no communication between robots and no moving obstacles, (ii) environment with communication between robots and no moving obstacles, and (iii) environment with communication between robots and moving obstacles with unknown dynamics.

3. Consider a multi-robot motion planning technique wherein robots plan their path independently and follow the same with a speed proportional to the distance of the obstacle ahead to avoid a collision. Explain will the strategy work or not.

4. Show the solution to the motion planning problem in Fig. 18.20 by using path velocity decomposition. First plan the path for the individual robots not considering the speeds or other robots as obstacles. Thereafter, cluster the robots such that any pair of robots from different clusters cannot affect each other's motion. Thereafter, for every cluster show the speed assignment search space. 1, 2, 3, and 4 are robot numbers. s and g represent source and goal.

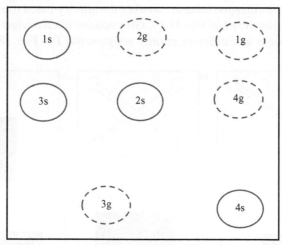

FIG. 18.20   Problem for question 4. 1s means robot 1's source, 1g means robot 1's goal, and so on.

5. Assume a multi-robot motion planning problem has all robots going from their optimal paths without considering the other robots, with small

deviations and speed adjustments. For such problems, suggest heuristics to make centralized grid search, centralized Probabilistic Roadmap, and centralized optimization-based approaches faster.

6. Optimize the function in Eqs (18.38), (18.39) using co-evolution with genetic algorithm and particle swarm optimization. Take a small random population and solve for 2 generations.

7. The motion of an agent in a grid map from a known source grid to a known target grid is in the form of motion commands. E.g., command '$\rightarrow \rightarrow \downarrow \rightarrow \downarrow\downarrow \rightarrow$' asks the agent to go 2 steps right, 1 step down, one step right, 2 steps down, and 1 step right. Show how Evolutionary Algorithms can be used to compute a good path for the agent. Describe how 5 agents can be planned using cooperative evolution. Explain the pros and cons of evolutionary algorithms in contrast to the A* algorithm for the problem for the cases of 1 agent and n agents.

8. Two robots are facing directly at each other. The source of the 1st robot is the goal for the 2nd robot and vice versa. Both robots use Artificial Potential Field method to reach to their respective goals. There is no direct communication between the robots. Suggest possible trajectories of the robots if the agents use an orthogonal social force such that (i) Both agents prefer anticlockwise (left) direction avoidance if they are head-on, and (ii) One agent prefers clockwise (right) direction while the other prefers anticlockwise (left) direction if they come head on.

9. [Programming] Consider that there are 2 robots. Both robots can only move rectilinearly, without diagonal movements. Plan the robots such that they meet at one common point using the A* algorithm.

10. [Programming] Plan multiple robots in a home or office scenario using centralized techniques with different deliberative algorithms. Study the effect of the increase in computation time due to increasing the number of robots. See if a reasonable performance can be achieved by varying the resolution parameter and whether it is a suggestive method for dealing with a large number of robots using centralized techniques. In the simulation, also add some new dynamic obstacles and use reactive techniques to avoid those obstacles.

11. [Programming] Plan multiple robots in the home or office scenarios using prioritization. Ensure that the robots exercise options to reduce their speeds. Test on scenarios in which waiting by one robot is essential. Try different priority combinations and select the best among them. Modify the final trajectories to add cooperation among the robots using local optimization or post-processing, where a higher-priority robot gets aside to give more space to the lower-priority robot.

12. [Programming] The chapter discussed path velocity decomposition by fixing the path and optimizing the speeds. Extent the approach to modify path and speed profile iteratively and interchangeably for the coordination of

multiple robots. First, the speeds should be optimized for fixed paths, and then the paths should be optimized for fixed speeds. This process goes on.

13. [Programming] Take a large map representing an office environment. Plan and execute the path of a robot using a hybrid of any deliberative and any reactive algorithm of your choice. Extend the solution to have multiple robots in the same arena, which must be reactively avoided. Further, let each robot move by sequentially visiting several named positions in named rooms of the office environment repeatedly in a loop. Most of the simulated people simultaneously visit a few key areas (like canteen, reception, etc.), causing congestions. Make the deliberative algorithm congestion conscious, which actively re-plans the robot taking it through low-congestion areas. The congestion may be actively stored and updated as a heat map. Use heuristic means to create congestion and see how the simulated people attempt to avoid congestion.

14. [Programming] Take a large map representing an office environment. Keep doors, corridors, and places of interest (like the cafeteria, washrooms, desks, etc.) (i) Plan and execute the path of multiple robots using any algorithm between the places of interest. Make an algorithm that guesses the immediate destination of the other robots (door, desk, etc.), only looking at the past motion. (ii) Make an algorithm that estimates the short-term trajectory of the robots using heuristics. (iii) Make an algorithm that estimates the short-term trajectory of the robots using machine learning. (iv) Plan every robot using the estimated trajectories of the other robots, using the RRT algorithm. The algorithm will only consider the static obstacles and nearby dynamic obstacles with their guessed trajectories. The trajectory shall be repaired continuously. The robots cannot communicate directly.

References

Bennewitz, M., Burgard, W., Thrun, S., 2001. Optimizing schedules for prioritized path planning of multi-robot systems. In: Proceedings of the 2001 IEEE International Conference on Robotics and Automation. IEEE, Seoul, South Korea, pp. 271–276.

Bennewitz, M., Burgard, W., Thrun, S., 2002. Finding and optimizing solvable priority schemes for decoupled path planning techniques for teams of mobile robots. Robot. Auton. Syst. 41 (2–3), 89–99.

Kala, R., 2012. Multi-robot path planning using co-evolutionary genetic programming. Expert Syst. Appl. 39 (3), 3817–3831.

Kala, R., 2014a. Navigating multiple mobile robots without direct communication. Int. J. Intell. Syst. 29 (8), 767–786.

Kala, R., 2014b. Coordination in navigation of multiple mobile robots. Cybern. Syst. 45 (1), 1–24.

Kala, R., 2018a. Routing-based navigation of dense mobile robots. Intell. Serv. Robot. 11 (1), 25–39.

Kala, R., 2018b. On repelling robotic trajectories: coordination in navigation of multiple mobile robots. Intell. Serv. Robot. 11 (1), 79–95.

Kala, R., 2020. Robot mission planning using co-evolutionary optimization. Robotica 38 (3), 512–530.

Kala, R., Warwick, K., 2011a. Multi-vehicle planning using RRT-connect. Paladyn J. Behav. Robot. 2 (3), 134–144.

Kala, R., Warwick, K., 2011b. Planning of multiple autonomous vehicles using RRT. In: Proceedings of the 10th IEEE International Conference on Cybernetic Intelligent Systems. IEEE, Docklands, London, pp. 20–25.

Kala, R., Warwick, K., 2012. Planning autonomous vehicles in the absence of speed lanes using lateral potentials. In: Proceedings of the 2012 IEEE Intelligent Vehicles Symposium. IEEE, Alcalá de Henares, Spain, pp. 597–602.

Kala, R., Warwick, K., 2013a. Planning autonomous vehicles in the absence of speed lanes using an elastic strip. IEEE Trans. Intell. Transp. Syst. 14 (4), 1743–1752.

Kala, R., Warwick, K., 2013b. Motion planning of autonomous vehicles in a non-autonomous vehicle environment without speed lanes. Eng. Appl. Artif. Intell. 26 (5–6), 1588–1601.

Kala, R., Warwick, K., 2015a. Motion planning of autonomous vehicles on a dual carriageway without speed lanes. Electronics 4 (1), 59–81.

Kala, R., Warwick, K., 2015b. Reactive planning of autonomous vehicles for traffic scenarios. Electronics 4 (4), 739–762.

Kala, R., Warwick, K., 2015c. Intelligent transportation system with diverse semi-autonomous vehicles. Int. J. Comput. Intell. Syst. 8 (5), 886–899.

Kant, K., Zucker, S.W., 1986. Toward efficient trajectory planning: the path-velocity decomposition. Int. J. Robot. Res. 5 (3), 72–89.

Potter, M.A., de Jong, K.A., 2000. Cooperative coevolution: an architecture for evolving coadapted subcomponents. Evol. Comput. 8 (1), 1–29.

Chapter 19

Task planning approaches

19.1 Task planning in robotics

The book so far has discussed numerous planning techniques for a variety of scenarios and applications. All the approaches so far were *geometric*, wherein a configuration space or workspace was available, whose geometric properties were either directly available or derived by the planning algorithm. The planning took place in a continuous space. The book now covers a different planning technique called *task planning* (Edelkamp et al., 2006; Galindo et al., 2008; Gerevini et al., 2009). The problem is a task that must be solved by performing several actions in the correct sequence and each action requiring some navigation or manipulation. The states and actions will typically be discrete in nature and can be specified by a few Boolean variables alone. While the action itself will require navigation of the mobile base or the operation of the arm, both of which are complex geometric problems; the actions manipulate a subset of a fixed set of objects between a fixed set of poses and therefore the actions have a discrete effect. The states can take one of the discrete forms. Since objects are being manipulated, the state of the system is the pose of the different objects in the system, with each object restricted to one of the discrete choices. The state of the system can often be described by a set of Boolean variables making a discrete state.

Let us first consider a few planning situations with our understanding of geometric planning. Imagine a cluttered work-desk that needs to be cleaned, which involves moving the items to their designated place. Naturally, if a book is kept over a file, the book needs to be moved first, before moving the file. Similarly, if the books need to be stacked in one place, the users will typically have a preference to keep the commonly referred book at the top. Similarly, it is not possible to first keep a stack of books and then clean the base of the table. Every book moved, every item picked up or placed, every item cleaned, etc., make up a unit step.

Imagine that the robot has several files to process. In each case, the file should be collected from a place, it must be passed through several people in some order of preference, and finally it is submitted to some office. Even though most processes are now online, it is not uncommon to see hard copy of papers and files go through offices. Again here, the robot must first sequence the pickups, drop-offs, and movement of files in between. The robot may itself have constraints on the number of

Autonomous Mobile Robots. https://doi.org/10.1016/B978-0-443-18908-1.00013-3

909

files that can be simultaneously carried, the availability of people signing files, etc. In all such cases, the larger problem is to sequence, schedule, and plan several simpler steps. After all the operations, the system reaches its final desired state.

The problem is still a planning problem, and therefore it is important to first define a *state*, which is a representation that completely encodes the condition of the system at any general point of time, as the operations are done in a sequence that transforms a source condition to the desired goal condition. The state should be minimally encoded, only encoding things that change with time and with some actions as the plan is executed from a source to a goal.

Here, the problem is attacked at a higher layer of abstraction, and therefore all the geometric planning problems within are assumed to be naively solvable. So, moving a book from one place to another is a complex geometric planning problem that, however, is considered solvable by a unit move action. This makes a minimal encoding a lot easier, which would have otherwise meant specifying the pose of every entity including the robot with complex constraints (like the books cannot be floating in the air without support).

19.2 Representations

Let us now model the problem using a logical representation (Russell and Norvig, 2010) or as a set of Boolean variables. The modelling needs to be done for the states making the vertices of the search graph and actions making the edges of the search graph.

19.2.1 Representing states

Let us consider the same problem used in the example. The overall problem consists of several entities. Say a book B_1 is one such entity. Considering that the problem is being solved at a higher level of abstraction, the book may be at a limited number of places like on top of the table, on top of another book B_2, on top of a file F, in the robot's hand, etc. The state is thus discrete. For generality, let us encode the state in a way such that there are only Boolean state variables. Say the book B_1 is currently on top of the table, encoded as $TopTable(B_1) = true$, where $TopTable(B_1)$ is a Boolean variable. More specifically, $TopTable$ is a function that takes one entity as input and returns true, if the entity is on top of the table, false otherwise. Similarly, $Top(X,Y)$ returns, if X is placed on top of Y. $InHand(X)$ returns, if X is in the hand of the robot for any entity X. The complete problem is encoded using several such entities and functions.

$Moveable(X)$ is a domain function that is true for all the entities that can be picked, placed, and manipulated by the robotic hand. Most programming languages have data types like int, character, etc., and every function call ensures that the input is of the same data type. In logical programming, the entities have different domains, which remain constant throughout the planning process. However, the classic planning systems have no means to specify the domains of the different entities. The domains are thus given as an additional fact in the state representation. Since the facts will never change, the domain variables are often neglected in the representation of the state.

The state can thus be conveniently represented by several Boolean variables, using Boolean functions for ease of modelling. The problem with the functions is that the number of Boolean variables becomes far too high, even though the number of functions may be very few. Take the single function $Top(X,Y)$ as an example that returns true, if X is on top of Y. If the problem is modelled by n entities that include all the items on the table, the robot, etc., the number of Boolean variables by the top function alone is n^2, one Boolean variable with every pair of entities.

Every variable can take any of the three values: *true, false,* or *unknown.* The value unknown is more curious. The systems can either be fully observable, wherein all the information about the entire system that affects decision-making is available; or partially observable, wherein not all the information needed for decision-making is known. In case of a partially observable system, some information is not available. Some actions uncover more detail about the truth of the unknown variables but incur an operational cost as well. In the case of geometric planning, the map may not be known, which is typically built as the robot moves around.

In the case of a partially observable system, the state variable encodes all the Boolean variables along with their truth value. The presence of a Boolean variable $TopTable(B_1)$ means that the book B_1 is placed on the table, while the presence of a $\neg TopTable(B_1)$ denotes that the book is not kept on the table. The Boolean variables whose state is not known are not specified in the state. Say neither $TopTable(B_1)$ nor $\neg TopTable(B_1)$ is specified in the state. It means that it is unknown whether B_1 is on top of the table or not. Only once an action is applied that sees B_1 on the table or confirms its absence from the top of the table, either $TopTable(B_1)$ or $\neg TopTable(B_1)$ may be inserted into the state. Till then, the world is partially observable and does not know the truth of $TopTable$ (B_1). A state variable is given in Fig. 19.1. Such a representation is called an

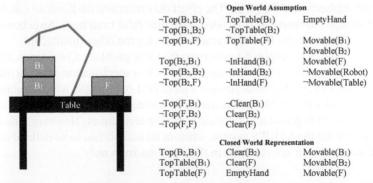

Open World Assumption		
$\neg Top(B_1,B_1)$	$TopTable(B_1)$	EmptyHand
$\neg Top(B_1,B_2)$	$\neg TopTable(B_2)$	
$\neg Top(B_1,F)$	$TopTable(F)$	$Movable(B_1)$
		$Movable(B_2)$
$Top(B_2,B_1)$	$\neg InHand(B_1)$	$Movable(F)$
$\neg Top(B_2,B_2)$	$\neg InHand(B_2)$	$\neg Movable(Robot)$
$\neg Top(B_2,F)$	$\neg InHand(F)$	$\neg Movable(Table)$
$\neg Top(F,B_1)$	$\neg Clear(B_1)$	
$\neg Top(F,B_2)$	$Clear(B_2)$	
$\neg Top(F,F)$	$Clear(F)$	

Closed World Representation		
$Top(B_2,B_1)$	$Clear(B_2)$	$Movable(B_1)$
$TopTable(B_1)$	$Clear(F)$	$Movable(B_2)$
$TopTable(F)$	EmptyHand	$Movable(F)$

FIG. 19.1 State representation for the problem. Every variable is Boolean/logical. In an open-world assumption, all variables are mentioned as true or false. Variables not mentioned may have an unknown value. In the closed-world assumption, only the true variables are mentioned. Movable(X) are domain variables that do not change their value and are normally omitted.

open-world assumption, since, it is assumed that the world of knowledge is open and there are entities whose status is not known.

For fully observable systems, the value of all Boolean variables is known a priori. In such a case, the state only mentions the positive variables in the system. If a variable is true, it is denoted as usual in the state. However, if a variable is false, it is omitted from the state. So, the presence of $TopTable(B_1)$ means that book B_1 is placed on the table. The absence of $TopTable(B_1)$ means that book B_1 is not placed on the table. The reason for doing so is that the number of false variables in any state is generally much larger than the number of true variables. In the same function, if $TopTable(B_1)$ is true, the book cannot be on any other entity and all those variables are false, or $\neg Top(B_1,B_1)$, $\neg Top(B_1,B_2)$, $\neg Top(B_1,F)$, $\neg Top(B_1,Robot)$, etc. Instead of specifying a large number of false variables, the false variables are omitted altogether. Such a representation is said to follow the *closed-world assumption,* where the name is derived from the fact that the world is closed, or the truth of every entity is known a priori. The *Planning Domain Definition Language (PDDL)* is the classic mechanism to represent the world knowledge and the state encoding. The language is similar and an extension of the language used for the development of the *Stanford Research Institute Planning System (STRIPS).*

19.2.2   Representing actions

The *actions* denote the edges of the graph search in such a state space. An action connects one state to another showing a valid transition between the states. The *path* is the sequence of steps taken to reach the goal state from the given source state. The actions are also specific to the mannerism of encoding the state. Every action is associated with an *action name,* a *precondition* that must be true for the action to be invoked, and an *effect* of the action. The precondition specifies the conditions, which must be met for the specific action to be invoked. The planner must make sure that no action is taken from a state, which does not satisfy the precondition. The effect states what the action accomplishes and indicates the target state after the application of the action. The effect only mentions the Boolean variables, which change (become true from false or become false from true). An action only changes a subset of the Boolean variables, leaving the others unaffected.

Let us take an example of action from the same problem. Consider an action $Pick(X)$, wherein the robot picks up an entity. The specification of the action is given by Box 19.1a. The notations use a lot of first-order logic, which enables us to easily model the problem. The use of first-order notations like for all ($\forall$) or there exists ($\exists$) is possible for classic planning techniques. However, this was not possible for the STRIPS system, wherein all actions had to be defined using a 0th order logic or all variables in their ground truth only.

Box 19.1a Action schemas using first-order logic. Instead of this modelling, the modelling in (b) will be used for all problems.

Pick(X)
Precondition: $(\forall_Y \neg\text{InHand}(Y)) \wedge (\forall_Z \neg\text{Top}(Z,X)) \wedge \text{Moveable}(X)$ *//Empty hand and nothing on top of X*
Effect: $\text{In-Hand}(X) \wedge (\forall_Y \neg\text{Top}(X,Y))$ *//Hand holds X and X on top of nothing*

Place(X,Y) *// Place X (currently in hand) over Y*
Precondition: $\text{InHand}(X) \wedge (\forall_Z \neg\text{Top}(Z,Y)) \wedge \text{Moveable}(X) \wedge \text{Moveable}(Y)$ *//X in hand and Nothing on top of Y*
Effect: $\neg\text{InHand}(X) \wedge \text{Top}(X,Y)$ *//X not in hand and X on top of Y*

PickTable(X) *// Pick X (currently on the table)*
Precondition: $(\forall_Y \neg\text{InHand}(Y)) \wedge (\forall_Z \neg\text{Top}(Z,X)) \wedge \text{TopTable}(X) \wedge \text{Moveable}(X)$ *// Empty hand and nothing on top of X*
Effect: $\text{In-Hand}(X) \wedge \neg\text{TopTable}(X)$ *//Hand holds X and X not on top of the table*

PlaceTable(X) *// Place X (currently in hand) over the table*
Precondition: $\text{InHand}(X) \wedge \text{Moveable}(X)$
Effect: $\neg\text{InHand}(X) \wedge \text{TopTable}(X)$

The problem can be converted to a 0th order logic by taking additional auxiliary variables. Let *Clear*(X) encode whether X has anything on top of it and *EmptyHand* denote whether the robot's hand is empty or not. The use of such auxiliary variables is common to avoid the use of 1st order logic, which can be computationally expensive for search. The modified modelling is shown in Box 19.1b.

Box 19.1b Action schemas using zeroth order logic. This modelling is used in all subsequent examples.

Pick(X,Y) *//Pick X (currently placed on top of Y)*
Precondition: $\text{EmptyHand} \wedge \text{Clear}(X) \wedge \text{Top}(X,Y) \wedge \text{Moveable}(X) \wedge \text{Moveable}(Y)$
Effect: $\neg\text{EmptyHand} \wedge \text{In-Hand}(X) \wedge \text{Clear}(Y) \wedge \neg\text{Clear}(X) \wedge \neg\text{Top}(X,Y)$

Place(X,Y) *// Place X (currently in hand) over Y*
Precondition: $\neg\text{EmptyHand} \wedge \text{InHand}(X) \wedge \text{Clear}(Y) \wedge \text{Moveable}(X) \wedge \text{Moveable}(Y)$
Effect: $\text{EmptyHand} \wedge \neg\text{InHand}(X) \wedge \text{Top}(X,Y) \wedge \neg\text{Clear}(Y) \wedge \text{Clear}(X)$

PickTable(X) *// Pick X currently on the table*
Precondition: $\text{EmptyHand} \wedge \text{Clear}(X) \wedge \text{TopTable}(X) \wedge \text{Moveable}(X)$
Effect: $\neg\text{EmptyHand} \wedge \text{InHand}(X) \wedge \neg\text{Clear}(X) \wedge \neg\text{TopTable}(X)$

PlaceTable(X) *// Place X currently in hand over the table*
Precondition: $\neg\text{EmptyHand} \wedge \text{InHand}(X) \wedge \text{Moveable}(X)$
Effect: $\text{EmptyHand} \wedge \neg\text{InHand}(X) \wedge \text{TopTable}(X) \wedge \text{Clear}(X)$

The STRIPS technique, working on a closed-world assumption, makes it easier to programme the action functions. The effects of action only mention the state variables that either change from true to false or vice versa. In the closed-world assumption, only the true state variables are mentioned. Hence, the state variables that change from false to true by the application of action are added to the effect. Such variables constitute the *add list*. Similarly, the variables that change from true to false are deleted, making the *delete list*. In the STRIPS technique, using a PDDL-based representation, the effects are thus an *add list* and a *delete list*, shown in Box 19.1c. The set arithmetic is used to first delete the elements in the delete list and then add the ones in the add list. An action a applied to the state s thus creates the state s' given by Eq. (19.1).

$$s' = (s - DeleteList(a)) \cup AddList(a) \text{ iff } s \text{ satisfies } PreConditions(a) \quad (19.1)$$

Box 19.1c Action schemas in closed world assumption.

Pick(X,Y) *//Pick X (currently placed on top of Y)*
Precondition: EmptyHand $\wedge$ Clear(X) $\wedge$ Top(X,Y) $\wedge$ Moveable(X) $\wedge$ Moveable(Y)
Add List: InHand(X), Clear(Y) *//These variables do not appear with NOT*
Delete List: EmptyHand, Clear(X), Top(X,Y) *//These variables appear with NOT*

Place(X,Y) *// Place X (currently in hand) over Y*
Precondition: ¬EmptyHand $\wedge$ InHand(X) $\wedge$ Clear(Y) $\wedge$ Moveable(X) $\wedge$ Moveable(Y)
Add List: EmptyHand, Top(X,Y), Clear(X)
Delete List: InHand(X), Clear(Y)

PickTable(X) *// Pick X currently on the table*
Precondition: EmptyHand $\wedge$ Clear(X) $\wedge$ TopTable(X) $\wedge$ Moveable(X)
Add List: InHand(X)
Delete List: EmptyHand, Clear(X), TopTable(X)

PlaceTable(X) *// Place X currently in hand over the table*
Precondition: ¬EmptyHand $\wedge$ InHand(X) $\wedge$ Moveable(X)
Add List: EmptyHand, TopTable(X), Clear(X)
Delete List: InHand(X)

The application of a dummy action for both representation techniques (open-world and closed-world) is shown in Fig. 19.2. In the implementation of the action $Pick(X,Y)$ the variable X is *substituted* by entity B_2 and Y by B_1. More specifically, the pick is an *action schema*, which can be substituted by any entity to create a specific action. For n entities, the pick schema gives n^2 actions, all of which can be obtained by substituting every input variable with every combination of entities. However, not all the actions will be applicable as per the precondition. As an example $Pick(Robot,Table)$ will not be applicable since neither *Robot* is movable, nor *Table* are movable. Similarly, initially $Pick(B_1,B_2)$ will not be applicable since B_1 is not initially on top of B_2. Therefore, while a problem may have very few action schemas, the number of actions obtained after all possible substitutions may be extremely large making planning very difficult.

a)

¬Top(B₁,B₁)	**Top(B₂,B₁)**	¬Top(F,B₁)	TopTable(B₁)	¬InHand(B₁)	¬Clear(B₁)	EmptyHand
¬Top(B₁,B₂)	¬Top(B₂,B₂)	¬Top(F,B₂)	¬TopTable(B₂)	¬InHand(B₂)	**Clear(B₂)**	
¬Top(B₁,F)	¬Top(B₂,F)	¬Top(F,F)	TopTable(F)	¬InHand(F)	Clear(F)	

Pick(B₂,B₁)
Pre-condition: EmptyHand ∧ Clear(B₂) ∧ Top(B₂,B₁) ∧ Moveable(B₂) ∧ Moveable(B₁)
Effect: ¬EmptyHand ∧ In-Hand(B₂) ∧ Clear(B₁) ∧ ¬Clear(B₂) ∧ ¬Top(B₂,B₁)

¬Top(B₁,B₁)	**¬Top(B₂,B₁)**	¬Top(F,B₁)	TopTable(B₁)	¬InHand(B₁)	**Clear(B₁)**	¬EmptyHand
¬Top(B₁,B₂)	¬Top(B₂,B₂)	¬Top(F,B₂)	¬TopTable(B₂)	**InHand(B₂)**	**¬Clear(B₂)**	
¬Top(B₁,F)	¬Top(B₂,F)	¬Top(F,F)	TopTable(F)	¬InHand(F)	Clear(F)	

b)

Top(B₂,B₁)	TopTable(B₁)	TopTable(F)	**Clear(B₂)**	Clear(F)	**EmptyHand**		

Pick(B₂,B₁) //Pick X currently placed on top of Y
Pre-condition: EmptyHand ∧ Clear(B₂) ∧ Top(B₂,B₁) ∧ Moveable(B₂) ∧ Moveable(B₁)
Add List: InHand(B₂), Clear(B₁)
Delete List: EmptyHand, Clear(B₂), Top(B₂,B₁)

~~Top(B₂,B₁)~~	TopTable(B₁)	TopTable(F)	~~Clear(B₂)~~	Clear(F)	~~EmptyHand~~	InHand(B₂)	Clear(B₁)

FIG. 19.2 Actions. Implementation of the Pick(B_2,B_1) action by the substitution $X = B_2$ and $Y = B_1$ in Pick(X,Y). (A) Open world assumption modelling. (B) Closed world assumption modelling. Preconditions in the parent state and effects in the child state are shown in bold.

In an open world, an action may enable discovering the status of a variable and thus adding information. Like an action, the open gripper discovers that after opening, there is nothing in the hand, even though the robot may be initially holding an object. Problems have tautologies or implication rules that are always true. *Logical inference engines* take the current knowledge of the state and aim to extract more information about the state of some variables that can be inferred by the application of the implication rules. In this way, the logical inference and the power of search combine to make the classic planning techniques.

Let us take an initial state and apply a series of actions to see how the state changes with time. This becomes the foundation for a search. The sequence is only done for the closed-world assumption for the shortage of space and only variables true in a state are mentioned. A sample run in the application of the functions is shown in Fig. 19.3.

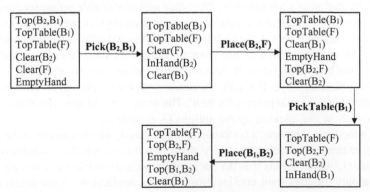

FIG. 19.3 A sample run. Random actions are applied from the source state to get new states using a closed world assumption. Actions are given in Box 19.1b.

19.3 Backward search

Since the problem has already been enumerated by the definition of states that make the vertices and actions that make the edges, the search techniques can now be applied to get a solution.

19.3.1 Backward search instead of forward search

The most appropriate thing to do will be to have a forward search like Breadth-First Search. Consider the same problem of moving several entities on the table. Consider the problem modelled using a closed world assumption as Eqs (19.2), (19.3).

$$
\textbf{Source} : Top(B_2, B_1) \land TopTable(B_1) \land TopTable(F) \land Clear(B_2) \land
$$
$$
Clear(F) \land EmptyHand \land TopTable(O_1) \land Top(O_2, O_1) \land \qquad (19.2)
$$
$$
TopTable(O_3)...
$$

$$
\textbf{Goal−Test} : TopTable(F) \land Top(B_1, F) \land Top(B_2, B_1) \qquad (19.3)
$$

In the goal test, all the positive conditions mentioned must mandatorily be satisfied or proven to be true and all the negative conditions mentioned must mandatorily be false. For all other state variables, it does not matter whether they are true or false. The state will be said to have satisfied the goal test irrespective of the truth value of variables not specified in the goal-test. The goal specified in Eq. (19.3) wants F on top of the table, B_1 on top of F and B_2 on top of B_1. It does not care how the other objects (O_1, O_2, O_3, etc.) are distributed. So a valid goal state may have either $Top(O_2, O_1)$ or $\neg Top(O_2, O_1)$ and the truth of this variable does not affect the validity of the goal state.

The problem is simple, and the robot just needs to pick up book B_2, place it temporarily on the table, pick up book B_1 and place it on the file F, and finally pick and place the book B_2 on book B_1. Several other entities (O_i, like pens, papers, bottles, etc.) are arbitrarily added.

Proceeding as a conventional forward search, an initial source condition is given. Now the first thing to do is expand the source into the child nodes by applying all applicable actions. However, there is a clear problem with a naïve implementation of such searches. The *place* and *placeTable* action schemas do not result in any applicable action (as the hand is empty). However, the *pick* and *pickTable* action schemas can be substituted by several entities to pick up any of them. All entities, which do not have anything over them, can be picked up. Hence, the branching factor is the number of entities that are at the top, which can be arbitrarily large. For action schemas with k variables and n entities, the branching factor could potentially be n^k. The search should hence be able to find out somehow that picking up the entities O_i is useless.

To facilitate the same, a *backward search* is used, which starts from the goal and goes towards the source. Luckily, the goal is invariant to the condition of the entities O_i, which means that the backward search starts from a slim state featuring only the important entities. However, the application of the action may not be naïve. The need is now to find out all states s, which on the application of any action schema with any substitution, give the goal state.

The problem of a large number of actions prevails. Consider expanding from the goal. From the goal in a backward direction, the action $PickTable(O_1)$ is a valid expansion that produces a state s as a child (s on the application of the action leads to the goal). However, the action is not called relevant, because in the forward simulation, the system already satisfied the goal condition at state s after which O_1 was picked to reach the goal. Performing the last action of picking O_1 is not useful.

In general, an action in a backward search is called *relevant*, if it enables reaching at least one *subgoal* of the problem. The different logical conditions mentioned in the goal specification are called subgoals, and all subgoal conditions together make the goal-condition. The problem has three subgoals $Top(B_2, B_1)$, $Top(B_1,F)$, and $TopTable(F)$. This means only those actions are relevant, which enables getting the subgoal $Top(B_2,B_1)$ (or the ones which change $\neg Top(B_2,B_1)$ to $Top(B_2,B_1)$), and similarly for the other subgoals. Note that the goal test can have negative conditions, which must be converted from a true to a false by the action after substitution. Actions that have $Top(B_2,B_1)$ or $Top(B_1,F)$ in the effects are of interest. The subgoal $TopTable(F)$ is already satisfied by the source and is therefore of less interest.

The only action with a substitution that can get the effect $Top(B_2,B_1)$ is $Place(X,Y)$ for the substitution $X=B_2$ and $Y=B_1$. It is good to ascertain that the action does not give an additional effect that is contradictory to another subgoal, in which case the action will not be applicable.

In a forward search, to expand one state (s) by an action (a), all the preconditions are checked (s satisfies $Precondition(a)$) and if feasible, the effects are added in the new state vector (deleting from delete list, adding from the add list and equivalently for the open-world case). In a backward search, the methodology shall be opposite. First, it is checked, if the effects that the action produces are either in the state vector (goal to start with) or are unspecified (in which case the target does not care about the value of a variable). In such a case, the expansion is done, and the preconditions are added to the state vector. Consider that an expansion of state s is done by an action a to produce state 's' (or in the forward path s is created by applying action a to s'). Provided that the action is relevant, the state s' will be given by Eq. (19.4).

$$s' = (s - Effects(a)) \cup PreConditions(a) \text{ iff } s \text{ satisfies } Effects(a) \wedge$$
$$a \text{ achieves at least } 1 \text{ SubGoal of } s \qquad (19.4)$$

An expansion is shown in Fig. 19.4. (step i). The arrows in the expansion are reversed, since it is a backward search. The arrow-head points towards the specific subgoal(s) being achieved by the action, while the tail is the expanded state. The subgoals satisfied by the source are of lesser interest, since the subgoals satisfy the initial condition (although may have to be first negated and then re-achieved to meet the problem constraints). The overall search happens using any search technique and stops when all the state variables mentioned are satisfied by the source.

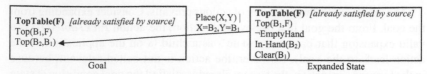

FIG. 19.4 First expansion of the backward search. Search happens by selecting actions that achieve at least one subgoal (subgoal present in effects), without negating the existing ones. Since the search is backward, the arrow is reversed. Head points to the subgoal(s) satisfied. Tail points to the expanded state. Expanded state deletes achieved subgoals and adds preconditions of the selected action. Subgoal(s) satisfied by source are shown in *bold*.

Continuing with a depth first search mechanism of search, let us expand the newly obtained state. All expansions are shown in Fig. 19.5A. One subgoal is $Top(B_1,F)$, which can be obtained by action schema $Place(X,Y)$ with the substitution $X=B_2$ and $Y=B_1$. The subgoal is deleted, while the preconditions of $Place(B_2,B_1)$ are added (step i). In the resultant state, one unmet subgoal is

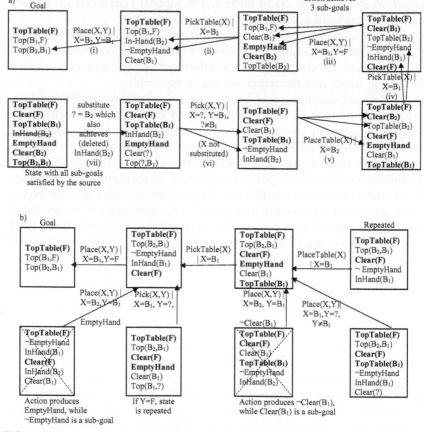

FIG. 19.5 Backward search. (A) Selecting actions that always succeed. In step (vi), the value of X is unsubstituted, noting the constraint. A substitution in step (vii) results in the source state and the search stops. (B) Search by selecting actions that eventually fail. Some actions negate an available sub-goal and are not expanded. Domain variables are not shown.

In-Hand(B_1), which can only be obtained by the action *Pick*(X) with the substitution $X=B_1$ (step ii), which also achieves ¬*EmptyHand*. In the new state, let us identify *Top*(B_1,F) as the subgoal to achieve next, which is done by the *Place*(X,Y) action with the substitution $X=B_1$ and $Y=F$ (step iii), which also achieves subgoals *Clear*(B_1) and *EmptyHand*. Next, to meet the subgoal *InHand*(B_1), the action *Pick*(X) is applied with the substitution $X=B_1$ (step iv). One of the ways to achieve *TopTable*(B_2) is by the action *PlaceTable*(X) with $X=B_2$, which is attempted (step v).

Now *Clear*(B_1) is an unmet subgoal, which can be achieved by *Pick*(X,Y) with $Y=B_1$ and an unknown (un-substituted) value of X (step vi). The only constraint is that X cannot be B_1, because such a substitution will produce ¬*Clear*(X) or ¬*Clear*(B_1) as an effect while *Clear*(B_1) is already a subgoal to achieve. An action cannot negate another unsatisfied subgoal. For substituting X, a lot of assignments will have to be tried and most of them may later be found as infeasible. This has the same problems as the forward search wherein many actions by different substitutions exist, and it is not feasible to try all of them. Any value of X with the constraint that it does not violate an achieved subgoal is a correct substitution. The variable is left unsubstituted. As the search continues, the variable can get more constrained.

Now the resultant state has an unsubstituted variable. If the variable is replaced by B_2 (step vii), a state is achieved that is already satisfied by the source, and the search can hence stop. This also achieves the last subgoal *InHand*(B_2).

The example discussed selected actions that led to a valid solution, which is a matter of chance. The branches may fail and necessitate backtracking. One small expansion is shown in Fig. 19.5B. Only the most interesting expansions are shown due to space constraints. The branch with action *Place*(X,Y) with substitution $X=B_2$ and $Y=B_1$ fails because an effect of this action is *EmptyHand*, while ¬*EmptyHand* is an unachieved subgoal. The unsubstituted branches will be expanded further (not shown); however, there is no direct substitution that achieves all subgoals. Every unsubstituted variable has constraints to ensure that it does not take a value that negates an unachieved subgoal.

19.3.2 Heuristics

Heuristics play an important role in determining the nodes to expand and to get a solution faster. The role of the heuristic function is to speed up the search, while optimality can be guaranteed with the help of an admissible heuristic function created using the relaxed problem method. The states were restricted to Boolean variables while the continuous variables could have been more generic. The utility of the Boolean constraint has a payoff in the design of good heuristic functions that work for any problem represented in the PDDL format, without needing the problem designer to write problem-specific heuristic function as was the case so far.

One of the obvious relaxations is to neglect the preconditions of all actions or to have all actions with *no preconditions*. Another assumption is that the sequence of actions does not cancel each other's effects. Thus, in the effects, only the state variables in the goal are used. The problem now is that several actions directly obtain several subgoals. The problem is to select some of these actions, which together achieve all subgoals. This is the *set cover problem*, wherein all subgoals need to be covered by several sets (actions), while every action covers a subset of them. The actual solution has an exponential time complexity, but good greedy solutions exist (which may not always guarantee optimality).

To study this, let us take a small synthetic example. After removing the preconditions and only noting the literals that exist in the goal as effects, the action set for a synthetic problem is given by Eqs (19.5)–(19.8).

$$\text{Action 1 : Precondition : NIL, Effects : } a \wedge c \wedge \neg e \qquad (19.5)$$

$$\text{Action 2 : Precondition : NIL, Effects : } c \wedge \neg d \qquad (19.6)$$

$$\text{Action 3 : Precondition : NIL, Effects : } a \wedge b \qquad (19.7)$$

$$\text{Action 4 : Precondition : NIL, Effects : } b \wedge \neg d \qquad (19.8)$$

Assume the goal is $a \wedge b \wedge c \wedge \neg d \wedge \neg e$. It can be reached in two steps as a result of the set cover of action 1 and action 4.

Alternatively, keeping the preconditions intact, the *negative effects of the actions are dropped* in the relaxation method. This again means that all actions make the system reach closer to the goal and no action cancels the effect of the other. The optimal action sequence is again not a polynomial solution, but again greedy solutions can be used.

Finally, another relaxation assumes that the overall problem can be *decomposed* into achieving several subgoals. The cost of computing a single subgoal is not large, which can be done by backtracking the variables needed. If the subgoals are assumed to be independent, all of them need to be achieved and thus the sum of all costs to attain all subgoals is the heuristic. This is called the *subgoal independence* assumption that states all subgoals are independent in nature. Achieving one subgoal does not automatically achieve another subgoal or achieving one subgoal does not make the problem of achieving other subgoals easier. Every subgoal needs to be achieved by an independent action set. Suppose the goal is to pack five boxes. The subgoal independence assumption states that an independent action sequence will have to achieve each of the subgoals (packing of each of the boxes), and a sequence of actions will not get the related subgoals achieved together. So, packing one box does not get some of the subproblems of packing of the 2nd box solved automatically. However, practically a single action can achieve multiple subgoals and thus the maximum of all costs is an admissible heuristic.

Assume that the goal is $a \land b \land c$. Assume that it takes 2 steps to get a independently, 3 steps to get b independently, and 0 steps to get c independently (already in the source). The admissible heuristic is $\max(2,3,0) = 3$ since the action sequence to get b may also bet a and c. Summing will get a heuristic of $2+3+0 = 5$, if it is assumed that getting a subgoal does not simultaneously get others.

Due to the very large size of the state space, optimality may not always be the criterion to select an algorithm. Focus is often to take algorithms that give reasonable solutions at a reasonable computation time, even if it be by compromising on optimality. Thus, it is acceptable to take heuristics that do not guarantee admissibility.

19.4   GRAPHPLAN

Let us discuss another interesting planning algorithm that uses a new data structure called a planning graph. The motivation is to first make a heuristic function, which is then extended as a complete planning algorithm called GRAPHPLAN.

19.4.1   Planning graphs

Let us first develop a heuristic function that can be used by the planning algorithms. The aim is to determine the number of steps that it takes to reach the goal in any particular state. The ideal solution is using a search for the same, which results in exploring a very large number of states with an exponential complexity using a search tree. To reduce this complexity, let us apply all actions and get newly expanded states; however, to make the complexity linear instead of exponential, all children states are merged into one state. This forms the basis of the *planning graphs*.

Let us take an example and calculate the heuristic value of the source state. Due to limitations of space, let us take a very small problem given by Fig. 19.6. Taking the source state, apply all actions; however, keep appending the literals produced into just one variable to avoid an exponential growth. The expansion is shown in Fig. 19.6. The source state is denoted as S_0 or the state at level 0. All possible actions applied are shown as A_0 or the action set at level 0. This makes a new state called S_1 or state at level 1. The problem is said to be solved in levels, with the source as level 0 giving rise to the next state at level 1 by the application of actions at level 0. The actions at level 1 are then applied to the level 1 state to produce a level 2 state and so on. The complete planning graph is shown in Fig. 19.7.

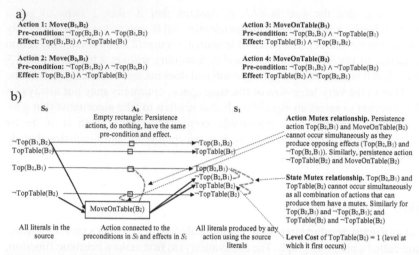

a)

Action 1: Move(B₁,B₂)
Pre-condition: ¬Top(B₂,B₁) ∧ ¬Top(B₁,B₂)
Effect: Top(B₁,B₂) ∧ ¬TopTable(B₁)

Action 2: Move(B₂,B₁)
Pre-condition: ¬Top(B₁,B₂) ∧ ¬Top(B₂,B₁)
Effect: Top(B₂,B₁) ∧ ¬TopTable(B₂)

Action 3: MoveOnTable(B₁)
Pre-condition: ¬Top(B₂,B₁) ∧ ¬TopTable(B₁)
Effect: TopTable(B₁) ∧ ¬Top(B₁,B₂)

Action 4: MoveOnTable(B₂)
Pre-condition: ¬Top(B₁,B₂) ∧ ¬TopTable(B₂)
Effect: TopTable(B₂) ∧ ¬Top(B₂,B₁)

b)

FIG. 19.6   Planning graphs. S_i denotes the state level i with all possible attributes that can be created by any combination of level $i-1$ actions. A_i denotes action level, showing the possible actions that can be applied. (A) Problem description. (B) 1st expansion.

The state at any level is similar to a *belief state*, which encodes all the possible configurations that the robot could be at the time. Of course, the robot will be at a configuration that is a valid subset of the configurations represented. So both $Top(B_2,B_1)$ and $\neg Top(B_2,B_1)$ are present at level 1 state (S_1) because it is possible (by different actions) to reach both the configurations. However, after 1 step, the robot will either be in a position where $Top(B_2,B_1)$ is valid or $\neg Top(B_2,B_1)$, and not both.

To produce a state at one level from the previous level, walk through all possible actions. For every action, check if all preconditions exist at the previous level, in such a case action is applicable to be applied. The action is added at the action layer shown by a boxed name. Arrows are added from all preconditions in the previous layer to the names of the actions. All effects (if not already inserted) are inserted in the next layer. Arrows are added from the action name to the effects in the next layer.

In artificial intelligence, the no-operation action has a great relevance. The action does as the name suggests, that is nothing. With every variable hence an additional action is defined, which has the same precondition and effect. Since the precondition and the effect is the same variable, there is practically no change in the application of such an action. These are called *persistence actions*. The full name of the action is the variable name followed by the persistence action. However, in the expansion, it is denoted by a small box with no name. So the $Top(B_2,B_1)$ persistence action is denoted by a small box with no name written inside. The action has a precondition $Top(B_2,B_1)$ and effect $Top(B_2,B_1)$. A line from $Top(B_2,B_1)$ in the previous state layer to the action

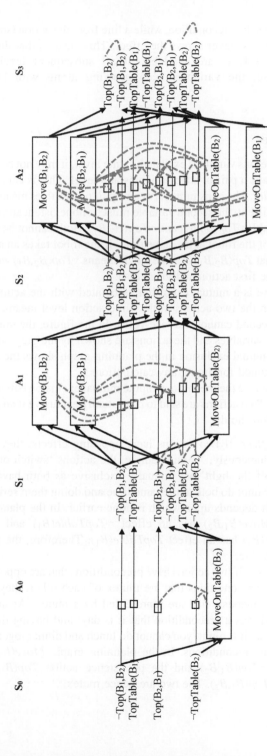

FIG. 19.7 Output of the planning graphs algorithm.

(small box) denotes the precondition, while a line from the action (small box) to $Top(B_2,B_1)$ in the next layer shows the effect. This ensures that the variables once present at a level are available at all subsequent levels and are never deleted. So, the variables keep growing along with layers in a planning graph.

19.4.2 Mutexes

Even though numerous variables exist at the state level, it may not be possible to simultaneously have a pair of state variables at a level. This is called a *mutual exclusion constraint*. The constraint between a pair of variables means that if the first variable is existent, the second variable must be nonexistent and vice-versa. Consider the variables $Top(B_2,B_1)$ and $\neg Top(B_2,B_1)$. Both cannot be used in the state description of the robot at the same time. If the robot takes an action from the source such that $Top(B_2,B_1)$ is achieved, it means $\neg Top(B_2,B_1)$ would not be achieved, after the first action and vice versa.

Similarly, there is a mutual exclusion associated with the actions. Mutual exclusion between the two actions in the same action level means, if the first action fires, the second cannot and vice-versa. In the figure, the mutual exclusions between the variables and the actions are shown by the grey-dashed lines. The presence of mutual exclusion in the planning graph shows the constraints that may be accounted for during any calculations.

Let us first describe the means to compute mutual exclusion between actions at an action level. Two actions are said to be mutually exclusive if any one of the following conditions hold:

- *Inconsistent effect*: If two actions have opposing effects, they cannot be invoked simultaneously. As an example, the actions 'switch on the light' and 'switch off the light' are mutually exclusive as both have opposing effects. You cannot do both at the same time and doing them serially means the final effect depends upon which is done earlier. In the planning graph, the actions $Move(B_1,B_2)$ has an effect $\neg TopTable(B_1)$ and the action $MoveOnTable(B_1)$ has an effect $TopTable(B_1)$. Therefore, the actions are mutually exclusive.
- *Competing needs*: If two actions have preconditions that are opposite to each other, or have preconditions that are mutex of each other, they cannot be invoked simultaneously and are represented by a mutex. As an example, if having lunch has a precondition that it is day; and having dinner has a precondition that it is night, you cannot do lunch and dinner together as they have opposing preconditions. In the planning graph, $Move(B_2,B_1)$ has a precondition $\neg Top(B_1,B_2)$ and the persistence action $Top(B_1,B_2)$ has a precondition $Top(B_1,B_2)$. The two are hence mutex.

- *Interference*: If the effect of one action negates the precondition of the other, or vice versa, they cannot be invoked together and must have a mutex relationship. As an example, two actions A_1 and A_2 that place a book on the same shelf (that has a capacity of one book only) have interference. If A_1 happens first, the shelf is no longer available to accommodate another book by A_2. The precondition of A_2 is negated by the effect of A_1, and vice versa. The result depends upon the sequence of applying the actions and the second action will always become nonapplicable. In the planning graph, the actions $Move(B_2,B_1)$ has an effect $Top(B_2,B_1)$ while the action $MoveOnTable(B_1)$ has a precondition $\neg Top(B_2,B_1)$. The effect of one is opposite to the precondition of the other and the two actions are thus mutex.

The mutex can similarly be defined between any two variables at the same level. Two variables are mutex to each other if all possible combinations of actions to achieve them are mutex to each other. For two variables to co-exist at a level, both must be produced by a regular or persistence action from the previous level, 1 action producing the first variable and another producing the second variable. Out of all combinations of actions that can produce the two variables (1 action for each variable) some of the combinations of actions will have a mutex relationship meaning that they cannot be fired together or cannot produce the two variables together. If there is at least one nonmutex action combination to produce the two variables, it may be possible to have both the actions fired and thus, producing both the variables without the mutex relationship between the variables. However, if no such nonmutex action combination exists at the previous action level, the variable pair will have a mutex relationship. In the planning graph, level 1, the only way to get $Top(B_2,B_1)$ is by its persistence action. Similarly, the only way to get $TopTable(B_2)$ is by the action $MoveOnTable(B_2)$. However, the two actions are mutex to each other and therefore their effects are also mutex to each other.

The planning graph is expanded from one level to the next, computing the applicable actions, and mutexes at every time. After some time, the planning graph will *level off* and show no change. This is because the persistence action will transfer all variables to the next level. The actions, which were previously applicable, will remain to be applicable. Because no new state is formed (graph has levelled off), the nonapplicable actions remain to be nonapplicable. If a state was not reached till the graph levelled off, it will not be found. The mutexes in states typically reduce as more action combinations produce the same state variable.

19.4.3   Planning graphs as a heuristic function

Now the planning graph is used to calculate the heuristic value. A goal has several subgoals, each of which must be met. The *level-cost* of a subgoal is the cost at which the subgoal first exists, which is also the least number of actions required to get the subgoal. Therefore, if a subgoal does not exist in the planning graph, it is nonattainable and the problem has no solution. The admissible heuristic is the *maximum level cost* of all subgoals. The heuristic, even though admissible, leads to small estimates. Another heuristic is the *sum of all level-costs*. The heuristic assumes subgoal independence. The heuristic assumes that a series of actions only leads to one subgoal and the subgoals are independent. In reality, an action may lead to multiple subgoals and thus the heuristic is not admissible and does not guarantee optimality.

Mutex relations also play a role in determining the heuristic value. If two subgoals at a level have a mutex relationship, they cannot occur simultaneously. The *set-level* heuristic is the first level at which all subgoals are found and none of them have a mutex with each other. The heuristic is admissible and results in optimal plans.

Consider the goal as $Top(B_1,B_2) \land TopTable(B_2)$. The first subgoal $Top(B_1, B_2)$ is first found at level 2 and the second subgoal $TopTable(B_2)$ is first found at level 1. The max level heuristic is thus $max(2,1)=2$ and the sum-level heuristic is $2+1=3$. At level 2 both subgoals are present without a mutex relationship with each other. This means that the set level heuristic is also 2. The calculation of the heuristic values is shown in Tables 19.1 and 19.2.

TABLE 19.1 Level cost of literals in the planning graph. First level at which the literal occurs, refer Fig. 19.7.

Literal in subgoal	Level cost
$Top(B_1,B_2)$	2
$TopTable(B_2)$	1

TABLE 19.2 Calculation of the heuristic value using the planning graph.

S. no.	Heuristic name	Admissibility	Heuristic value
1.	Maximum level heuristic	Admissible	$max(2,1)=2$
2.	Sum level heuristic	Assumes all subgoals are independently achieved, not admissible	$2+1=3$
3.	Set level heuristic	Maximum level with all subgoals without mutex, admissible	2

19.4.4 GRAPHPLAN algorithm

The planning graph is an interesting data structure. An attempt is thus made to extract a solution from it, rather than just to use it as a heuristic measure. The planning algorithm expands a planning graph level by level. It is not possible to find any solution till all subgoals are seen with no mutex between each other or until the set-level heuristic value. Therefore, the expansion happens as usual.

After all subgoals without mutex are found, a solution may be possible. Hence an attempt is made to extract a solution from the planning graph. To extract a solution, the problem is to apply several actions out of all actions modelled in the planning graph, at every layer. While applying any action, care must be taken not to apply two actions in the same layer, which have a mutex relationship to each other. Imagine a planning graph with some subset of actions selected to be applied. Now start from the source state and apply all the selected actions at the 0th action level, which will give the state at the 1st level. Since only nonmutex actions were selected, the outcome will be the same irrespective of the order in which the actions are applied at the 0th action level. From the state hence obtained in the first level, apply all selected actions whose preconditions are met, to give the state at 2nd level. This goes on. If the final simulation results in the goal state, the plan is possible.

The problem is hence to select a few (nonmutex) actions and to check whether it leads to the goal. This is a Boolean constraint satisfaction problem (CSP). A CSP has several variables, each of which can take one of the several values out of all possible values from the domain. The problem is to assign values to the variables, such that the constraints are satisfied. For this specific problem, the actions in the different layers act as variables. Each action can be selected (1) or not-selected (0), where 0 and 1 act as values. The constraint is that no two actions with a mutex can be selected and that the forward application of actions should lead to a goal and have the preconditions met. The fundamental way of solving a CSP is by backtracking where iteratively variables are selected and different value assignments are tried as different branches, such that the constraints are always satisfied. If no value selection is possible, the algorithm backtracks to a different branch. Efficient CSP solvers can be used to find a solution. The solvers use advanced heuristics to try iterative assignments and correct wrong assignments. The solvers can also detect absence of a solution reasonably earlier than trying all possible variable-value combinations.

CSP is not the only means to extract a solution. Searching for a solution within the planning graph can be seen as another search problem, with the constraint that only actions visible in the planning graph should be used. The forward or the backward search with good heuristics may be used. For the same reasons as stated earlier, better heuristics are possible with the backward search rather than the forward search. The typical means is to first select subgoals that were visible at the end; however, out of all actions that give the subgoal, the one that has the easiest precondition is preferred.

If the solution extractor finds a solution in the planning graph, the solution is returned. If a solution cannot be found by the solution extractor, it may be possible to get a solution by expanding the planning graph by another layer. A new layer of expansion is done, and the extraction of the solution is re-attempted. The algorithm keeps adding new layers and attempts extraction of a solution from the expanded planning graph. This makes the GRAPHPLAN algorithm. The pseudo-code is given by Algorithm 19.1. An interesting thing is a condition when the algorithm detects a failure and stops, which could otherwise go infinitely. After the graph has levelled off, there may still be a possibility to get a plan at the higher levels and this cannot be used as the sole criterion for breaking. While extracting a solution, a track is kept of which subgoals could not be found. The number of such subgoals typically reduces. However, if the same set of subgoals cannot be consistently found, as reported by the solution extractor, then it is said that no solution exists.

Algorithm 19.1 GRAPHPLAN.

---

1 $P \leftarrow Source$;   ▷ Planning graph
2 **while** *true* **do**
3      $P \leftarrow Expand(P)$;
4   **if** *all sub-goals found in P with no mutex* **then**
        ▷ ExtractSolution extracts a solution as a
        constraint satisfaction problem. Each action can
        be 1 (applied) or 0 (not applied), with mutex and
        other constraints ;
5          $solution \leftarrow ExtractSolution(P, goal)$;
6   **if** *solution* $\neq NIL$ **then** return *solution*;
7   **if** *P has levelled off and no new sub-goals can be satisfied by the solution extractor* **then**
8   return NIL

19.5 Constraint satisfaction

A mechanism that very recently got a lot of popularity is to model the planning problem as a CSP and to use powerful CSP solvers for getting a plan.

19.5.1 Constraint satisfaction problems

In CSP, the aim is to find a single state that satisfies several constraints. The problem is modelled as a set of *variables* that can take several *values* chosen from a prespecified *domain*. The variables, however, share constraints between themselves, and therefore all possible value assignments to the variables may not represent a valid solution due to the breaking of one or more constraints. Any solution that assigns all variables some values while keeping the constraints satisfied is taken as the output.

A large number of problems are CSPs. The sudoku game is played on a 9×9 board and asks the players to place numbers 1–9 in each cell. The constraints are that no digit must be repeated in the individual rows, no digit must be repeated in the individual columns, and no digit must be repeated in the 9 subsquares of size 3×3 in the board. Here each cell is a variable (81 variables) and values can be digits from the domain 1 to 9. Another classic example is the timetable scheduling problem, where each day-time slot needs to be assigned an activity. The constraints are that two activities cannot be scheduled for the same class together, an instructor cannot take two activities at the same time, and a room cannot have two activities at the same time. The 8-queens problem is another CSP where each queen's position is a variable and the different cells that it can take are the values. The constraints are that no two queens must occupy the same cell or cut each other.

The classic mechanism to solve the problem is by *backtracking*. The backtracking method takes a partially solved problem state representing the vertices of the state-based search. Actions are represented by taking an unassigned variable and assigning all possible values adhering to the constraints. There are numerous interesting heuristics to be considered that make backtracking powerful and computationally efficient. Firstly, at any level, only one variable needs to be selected. The other variables may be selected subsequently in the next levels. The selection of the variable follows the principle that the variable with the least number of remaining values and the variable, which is the most constrained must be selected first, which relieves most of the constraints at the next level. Thereafter, the values, which are less constraining are tried first to make it highly possible to get a solution in the same branch and eliminate the risk of backtracking. Other heuristics also aim to decompose the problem into independent or partially independent subproblem that makes it computationally efficient. Other heuristic techniques try to predict a failure in the current branch and backtrack right away. There is rich literature on the use of heuristics for the CSPs.

The other mechanism to solve the problem is by using *local search algorithms*, which especially work well in the case of many variables. The solutions first assign values to the variables without considering the constraints. Thereafter, the solutions identify the constraints, which are violated and adjust the value assignments to the variables to reduce the number of constraint violations. Iteratively reducing the number of constraints leads to a constraint-free solution. Again, good heuristics enable us to decide the good variables whose values must be replaced and good new values for such variables. A popular heuristic is to choose the variables which have the minimum conflicts with the other variables.

19.5.2 Modelling planning as a CSP

The CSPs find application in many areas and have therefore gained huge research interest. Over the years, very powerful heuristic algorithms have been designed that efficiently solve CSPs for a very large number of variables. This hence paves the way to model problems as CSP to use these solvers.

The planning problem is hence modelled as a CSP. A major difference between the classic CSP and planning is the temporal nature of the planning problem. In CSP, there is no dimension of time, and everything exists at the same time. However, in temporal planning, the system changes its state at every instant of time. Literals that are initially true may become false and then again true as so on. To model the problem, imagine a planning problem and a solution path for the problem. A literal may be true at some point in time and false at another, in the path from source to goal. Hence, the time annotation is added on top of the literal. Let $X^{(t)}$ denotes that X is true at time t and similarly let $\neg X^{(t)}$ denote that X is false at time t.

Consider the problem of moving items on the desk. The variable $\neg Top^{(0)}(B_1,B_2)$ means that book B_1 is not on top of book B_2 initially at time 0. Similarly, $Top^{(4)}(B_1,B_2)$ means that book B_1 is on B_2 at time 4. Similarly, let us consider the actions. The actions follow a similar convention with $A^{(t)}$ denoting that the action A is applied at time t. So, $Pick^{(3)}(B_1,B_2)$ denotes that the action $Pick(B_1,B_2)$ was applied at time 3. The modelling is summarized as Box 19.2.

Box 19.2 Planning as a constraint satisfaction problem (CSP).

Variables: all literals and actions in their ground state at every time, e.g. $Top^{(2)}(B_2,B_1)$ [Is B_2 on top of B_1 at time 2?], $Pick^{(2)}(B_2,B_1)$ [Is action $Pick(B_2,B_1)$ applied at time 2?], etc.

Domain of every variable: {0 (false), 1 (true)}.

Constraints:

- **Variable constraint**: If a variable is positive (similarly negative) at time t, either action is applied at time $t-1$ which makes it positive (similarly negative); or the variable is positive (similarly negative) at time $t-1$ and no action is applied at time t-1 that makes it negative (similarly positive).

- **Precondition constraint**: If an action is positive at any time t, all positive (similarly negative) variables in the precondition must be positive (similarly negative) at time t.

- **Action exclusion constraint (optional)**: Only 1 action is true at any time step.

- **Goal constraints**: All variables specified at goal as positive (similarly negative) must be positive (similarly negative) at a specific time t.

Solution: Add goal constraints for time $t=1$. Return if a solution is found by a custom CSP solver, else add goal constraints for time $t=2$ and similarly iterate on t.

19.5.3 Modelling constraints

The states and actions are heavily constrained. Every positive action necessitates that its precondition at the previous time step be applicable, and its effects should be true at the next time step. Let us hence model these constraints between the variables. The first problem is that action modelling only stated the literals that changed their state without mention of the literals that remain unchanged. Now, it is possible to have every combination of state value at every time step as a new variable and even a single action constrains all of them. So, if $Pick^{(3)}(B_1,B_2)$ is the only action applied at time 3, the state of any general object O may remain unchanged, but since it is an independent variable, this needs to be specified as a constraint to the algorithm. Having $TopTable^{(3)}(O)$ and $\neg TopTable^{(4)}(O)$ is impossible, if $Pick^{(3)}(B_1,B_2)$ was the only action applied. This is called the *frame problem*. One solution is to state $Pick^{(3)}(B_1,B_2)$ leaves all variables (like $TopTable^{(3)}(O)$) in the same state as they were before the application of the action. However, this specification repeated for every action-variable pair is time and space consuming. Here $Pick^{(3)}(B_1,B_2)$ is one of the a actions (produced from a small number of action schemas by every substitution possible) and $TopTable^{(3)}(O)$ is one of the b ground literals (produced by every possible substitution). The number of constraints (that every rule leaves every unaffected literal in the same previous state) is $a.b$, which will be practically very large.

A better solution to the problem is to model constraints between the value a literal can take between time t and time $t+1$. Consider that a variable $X^{(t+1)}$ is true at time $t+1$. Now there are only two possible ways for it to happen. First, that $X^{(t+1)}$ changed from false to true from time t to $t+1$, in which case it is important to apply at least one action that changes X from false to true at time t. Let a disjunction of such actions be given by $ActionsPositive^{(t)}(X)$. This is given by Eq. (19.9)

$$ActionsPositive^{(t)}(X) = \vee_{A:X\in Effects(X)}A^{(t)} \qquad (19.9)$$

The other way to produce $X^{(t+1)}$ is that it was preexisting from time t (and therefore need not be produced). However, no action that has an effect of $\neg X$ should be applied at time t. Let $ActionsNegative^{(t)}(X)$ denote the conjunction of all actions that can produce $\neg X$. This is given by Eq. (19.10)

$$ActionsNegative^{(t)}(X) = \wedge_{A:\neg X\in Effects(A)}A^{(t)} \qquad (19.10)$$

Adding the two conditions, we get the constraint shown in Eqs (19.11), (19.12)

$$X^{(t+1)} \Longleftrightarrow ActionsPositive^{(t)}(X) \vee X^{(t)} \wedge \neg ActionsNegative^{(t)}(X) \qquad (19.11)$$

$$X^{(t+1)} \Longleftrightarrow \vee_{A:X\in Effects(A)}A^{(t)} \vee X^{(t)} \wedge \neg\left(\wedge_{A:\neg X\in Effects(A)}A^{(t)}\right) \qquad (19.12)$$

The equations state that $X^{(t+1)}$ is true if either of the actions was applied that produces X at time t; or X was already true at time t (that is $X^{(t)}$) and neither of the actions was applied at time t that produces $\neg X$.

Consider the variable $Top(B_1, B_2)$ at time 2. The only way to obtain $Top(B_1, B_2)$ is by either applying action $Place(B_1, B_2)$ at time 1 or by asserting that $Top^{(1)}(B_1, B_2)$ was already obtained at time 1, but it should not be moved at time 1 by application of action $Pick(B_1, B_2)$. The constraint is given by Eq. (19.13).

$$Top^{(2)}(B_1, B_2) \Leftrightarrow Place^{(1)}(B_1, B_2) \vee Top^{(1)}(B_1, B_2) \wedge \neg Pick^{(1)}(B_1, B_2)$$

$$(19.13)$$

These constraints bind the actions to their effects. However, it is also important to bind the constraint that an action is only performed at time t when all its preconditions are met at time t. This means if the action A is applied at time t (or $A^{(t)}$), then its preconditions must be true at time t, or Eq. (19.14)

$$A^{(t)} \Longrightarrow Precondition\left(A^{(t)}\right)$$

$$(19.14)$$

Taking the small 2 object example as used in planning graphs, the constraint for one such action can be represented as Eq. (19.15)

$$Place^{(t)}(B_1, B_2) \Longrightarrow \neg EmptyHand^{(t)} \wedge InHand^{(t)}(B_1) \wedge Clear^{(t)}(B_2) \quad (19.15)$$

Another constraint is that the goal must be finally achieved called the *goal constraint*. The goal is assumed to be true exactly at time t, wherein the constraint is added on top of all the subgoals to have a state as per the goal specification at time t.

The constraints so far allow having multiple actions applied simultaneously in parallel at any time. If a strictly linear path is needed where there is only a single action at every time step, the *action exclusion constraints* are added that ensure only 1 action is true at any time step. Alternate modelling is to take a variable $AppliedAction^{(t)}$ that states which action will be applied at that time. The domain of the variables includes all possible actions.

19.5.4 Getting a solution

An assumption is that the solution depth or exact time t when the goal is achieved is known in advance. This is certainly not the case. The algorithm proceeds like the GRAPHPLAN algorithm. First, the constraints are modelled till $t = 0$. This makes a CSP, which is then solved for a solution. If the solution cannot be computed, the problem is modelled as a CSP until time $t = 1$, and again solved by a CSP solver to see, if a solution is possible or not. The problem depth limit is iteratively increased with a query to the CSP solver. Once a solution is found, it is returned.

It is further important to note that all the action schemas must be converted into all possible actions by substituting all applicable variables. This must be done for every instance. All these actions act as separate variables for the working of the CSP solver. The state at time $t = 0$ is known and therefore, those variables can be taken as constants. All variables are modelled separately for all instances of time. This, however, does not include variables that have no temporal dependency, domain variables, or those that are constants.

19.6   Partial order planning

A good ability of humans that enables them to solve complex problems is the ability to decompose a large problem into smaller problems that can be solved parallelly. Humans can work through all the subproblems to develop partial solutions. Once the partial solutions are ready, they can be integrated to formulate the solution to the main problem. The problem solving becomes easy because each subproblem achieves some subgoals and different subproblems can be executed in parallel. This limits the computational complexity.

19.6.1   Concept

Using the same motivation, *partial order planning* (POP) works by identifying the partial problems that can be solved independently. The entire problem is thus decomposed into multiple partial problems, which can be solved to get partial solutions. The partial solutions are then knitted together to make a solution to the overall problem as a sequence of steps. The hard part is to identify the partial problems and how the different partial plans interact with each other.

As an example, consider the problem is to clean two work-desks by moving items in the individual work-desks. Now, a human may clean first work-desk followed by the second work-desk, clean the second work-desk followed by the first work-desk, or interchange between the two desks at any time. The two are independent problems. The algorithm should be able to identify and attack such independence from the problem specification.

For problem-solving using POP, the general methodology is to identify all the *open conditions,* which must be met by an action or an action sequence to plan. An open condition models a *flaw* in the current plan that does not achieve something necessary for the problem's solution. Initially, we need to achieve all the subgoals and thus all subgoal conditions are open and need to be satisfied. Because currently the subgoals are not satisfied, the different subgoals are the flaws that need to be corrected. The manner of satisfying an open condition (removing a flaw) is to apply an action that has the same effect as the open condition. However, the application of action means that it must be applicable, or all its preconditions must be satisfied at the first instance. This adds up open conditions to be satisfied or increases the flaws, which may be done by applying more actions. Once all open conditions are satisfied by the source, the problem is solved.

To denote the search graphically, let us denote every action by a rectangle. The name of the action is written inside the rectangle. The preconditions are written on top of the rectangle and the effects are written at the bottom of the rectangle. For completeness, two special actions are added. The *source action* has no preconditions but only has an effect, which is the source state. Similarly, the *goal action* has a precondition as the goal but has no effects. This allows placing source and goal in the same POP representation. It may be observed that the source and goals were states, but are now converted to actions. It may also be observed that so far, the states were boxed, and the actions were drawn as arrows. However, now the actions are instead boxed.

A *causal link* is placed between the actions to denote, which action must be compulsorily performed before another action. A causal link is denoted as an arrow from the action that should be done before to the one that should be done after. Many actions are added to the solution to remove a flaw or close an open condition. Suppose the action A_1 is added to achieve the precondition p for an action A_2. Hence, p is the effect of A_1 and precondition of A_2. It is said that A_1 *achieves* the precondition p for A_2 with the causal link from between the effect p in A_1 and precondition p in A_2. This is represented by $A_1 \prec_p A_2$. In case no strict relation of achieving the precondition exists, the link may just be demoted by $A_1 \prec A_2$ stating that for some unspecified reasons, A_1 must be done before A_2.

19.6.2 Working of the algorithm

To show the working of the algorithm, consider the same problem that the robot must move several items. It is assumed that the robot can hold an infinite number of items in its hand (or the robot can attach any number of items to itself). The problem is shown in Fig. 19.8. The nomenclature is shown in Fig. 19.9 and the solution to the problem is shown in Fig. 19.10. The search starts with the source and the goal as the only two actions. The source is the first action to be performed and the goal is the last action to be performed. Hence, there is a causal link from source to every action (including goal) and a causal link from every action (including source) to the goal. These links shall not be shown for clarity.

Initially, the preconditions of the goal action are open. Let us pick any one precondition and achieve it. $TopTable(B_2)$ condition can be closed by adding a function $PlaceTable(B_2)$ and adding the causal links (step *i*). It opens the precondition $InHand(B_2)$, which is satisfied by the action $Pick(B_2,B_1)$, opening new preconditions that are already satisfied by the source (step *ii*). Similarly, the other open condition for the goal is $Top(B_1,B_2)$, which can be closed by the action $Place(B_1,B_2)$, opening new preconditions $Clear(B_2)$ and $InHand(B_1)$ shown as step *iii*. $Clear(B_2)$ is already satisfied by the action $PlaceTable(B_2)$ of step *i*, while $InHand(B_1)$ is satisfied by the action $PickTable(B_1)$ as step *vi*. Steps (*iv-v*) are currently omitted. The new open conditions $Clear(B_1)$ and $TopTable(B_1)$ are already satisfied by the action $Pick(B_2,B_1)$ of step *ii* and the source respectively, and similarly for steps *vii–xii*.

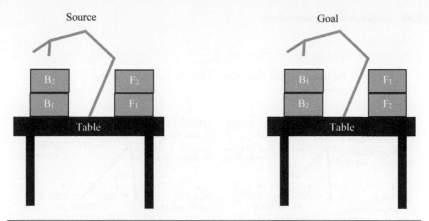

Source (closed world modelling): TopTable(B₁) ∧ Top(B₂,B₁) ∧ TopTable(F₁) ∧ Top(F₂,F₁) ∧ Clear(B₂) ∧ Clear(F₂) ∧ Moveable(B₁) ∧ Moveable(B₂) ∧ Moveable(F₁) ∧ Moveable(F₂)

Goal: TopTable(B₂) ∧ Top(B₁,B₂) ∧ TopTable(F₂) ∧ Top(F₁,F₂)

Action Schemas
Pick(X,Y) //Pick X (currently placed on top of Y)
Pre-condition: Clear(X) ∧ Top(X,Y) ∧ Moveable(X) ∧ Moveable(Y)
Effect: InHand(X) ∧ Clear(Y) ∧ ¬Clear(X) ∧ ¬Top(X,Y)

Place(X,Y) // Place X (currently in hand) over Y
Pre-condition: InHand(X) ∧ Clear(Y) ∧ Moveable(X) ∧ Moveable(Y)
Effect: ¬InHand(X) ∧ Top(X,Y) ∧ ¬Clear(Y) ∧ Clear(X)

PickTable(X) // Pick X currently on the table
Pre-condition: Clear(X) ∧ TopTable(X) ∧ Moveable(X)
Effect: InHand(X) ∧ ¬Clear(X) ∧ ¬TopTable(X)

PlaceTable(X) // Place X currently in hand over the table
Pre-condition: InHand(X) ∧ Moveable(X)
Effect: ¬InHand(X) ∧ TopTable(X) ∧ Clear(X)

FIG. 19.8 Graphical representation of the problem for partial order planning.

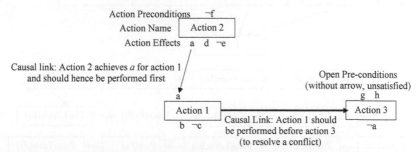

FIG. 19.9 Nomenclature of partial order planning. Boxes are actions with preconditions written above and actions written below. Arrows indicate precedence constraint, which may be as one action achieves the precondition of the other.

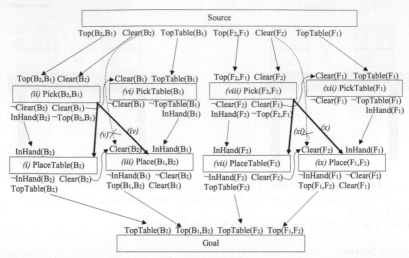

FIG. 19.10 Partial order planning. The search finds subplans that can be executed in parallel. The search starts with the source and the goal. More actions are added to close (add a link) preconditions that are not yet met. The search may be needed to backtrack (delete a previously satisfied precondition).

The overall solution is more clearly represented by Fig. 19.11A, after removing the preconditions and the effects. It can be seen that the algorithm naturally found a parallel element of the problem and represented it automatically, which reduces the search efforts. The plan is still as a graph and needs to

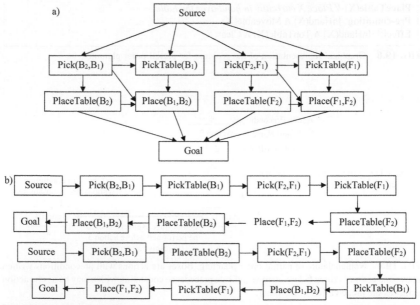

FIG. 19.11 Depiction of parallelism in the problem. (A) Action graph, simplified version of Fig. 19.10. (B) Two possible linear paths to be given to the robot.

be *linearized* to get the final solution. Any traversal of the POP graph as long as it accepts the causal link constraints is a valid solution. Two such traversals are shown in Fig. 19.11B. Algorithms like topological sort can be used for the linearization.

19.6.3 Threats

There is one caveat at the algorithm though. Consider an action A_1 achieves the precondition p for an action A_2 ($A_1 \prec_p A_2$). Consider another branch wherein an action A has an effect $\neg p$. Since A is in parallel to A_1 and A_2, any serialization is possible including A then A_1 then A_2, A_1 then A then A_2 or A_1 then A_2 then A. Consider one such serialization as A_1 then A then A_2 shown in Fig. 19.12. Now, p is a precondition of A_2; however, the application of action A negates the precondition by producing $\neg p$.

This situation is called a *threat*. An action A is a threat to an action A_2, since the application of A negates a precondition of A_2. Every threat must hence be identified and corrected. There are two ways to correct a threat. The first is to apply action A after A_2, which is called *promotion* of the threat causing action. This is given by Eq. (19.16). In this case, the negated precondition will not matter to A_2, because the action has already been performed.

$$Promotion : A_1 \prec_p A_2 \prec A \text{ iff } \neg p \in \text{effect}(A) \tag{19.16}$$

The other option is to perform A first followed by A_1 and finally A_2. In such a case, A will negate the precondition of A_2 by producing $\neg p$. However, A_1 will again result in the precondition p, thus making A_2 feasible. The strategy is called the *demotion* strategy because the threat action A is demoted to be done before A_1. This is given by Eq. (19.17)

$$Demotion : A \prec A_1 \prec_p A_2 \text{ iff } \neg p \in \text{effect}(A) \tag{19.17}$$

Consider the same problem in Fig. 19.10. The bold lines represent the resolution of the threats. *Source* achieves $Clear(B_2)$ for the action $Pick(B_2,B_1)$ or $Source \prec_{Clear(B_2)} Pick(B_2,B_1)$. However, action $Place(B_1,B_2)$ produces $\neg Clear(B_2)$ and is a threat. $Place(B_1,B_2)$ cannot be demoted before the source as the source is always the first action. Therefore $Place(B_1,B_2)$ is promoted after $Pick(B_2,B_1)$ or $Source \prec_{Clear(B_2)} Pick(B_2,B_1) \prec Place(B_1,B_2)$, shown as a causal link (step *iv*). Similarly, $PlaceTable(B_2)$ achieves $Clear(B_2)$ for $Place(B_1,B_2)$ and $Pick(B_2,B_1)$ is a threat, which is resolved by demotion ($Pick(B_2,B_1) \prec PlaceTable(B_2) \prec_{Clear(B_2)} Place(B_1,B_2)$). Consider step *v*. $Clear(B_2)$ was purposefully not achieved from the source. Suppose source achieves $Clear(B_2)$ for $Place(B_1,B_2)$, or suppose $Source \prec_{Clear(B_2)} Place(B_1,B_2)$. Now $Pick(B_2,B_1)$

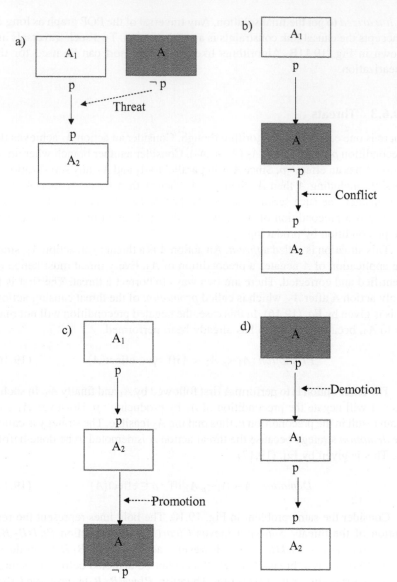

FIG. 19.12 Threats. (A) Action A_1 achieves the precondition p for action A_2. The action A produces $\neg p$, which violates the precondition. Therefore, A cannot lie in-between A_1 and A_2 as shown in (B). It must be either promoted after A_2 as shown in (C) or demoted before A_1 as shown in (D).

is a threat as it produces $\neg Clear(B_2)$. Demotion before *Source* is not possible to resolve the threat as nothing lies before source. If promotion is attempted, a causal link from $Place(B_1,B_2)$ to $Pick(B_2,B_1)$ will be added (or *Source* $\prec_{Clear(B_2)}$ $Place(B_1,B_2) \prec Pick(B_2,B_1)$).

This will cause a *cycle* in the POP graph $(Place(B_1,B_2) \prec Pick$ $(B_2,B_1) \prec_{Clear(B_1)} PickTable(B_1) \prec_{InHand(B_1)} Place(B_1,B_2))$ and cyclic graphs cannot be linearized to get a solution. If both promotion and demotion fail, it is said that the threat is nonresolvable, and therefore adding the link is not possible. Cycles may get discovered after a few steps and the algorithm may have to backtrack.

The overall algorithm is rather interesting. From the point of view of conventional graph search, every vertex is a partial plan as seen in Fig. 19.10. The vertices, therefore, have several POP action nodes with open conditions, closed conditions and causal links. The edges correspond to taking one open condition and satisfying it by the application of some action. At a vertex being expanded, select an open condition, and close it in every way possible by applying all possible actions, every action makes a branch or an edge of the expanded node. Backtracking is the search algorithm used. If the search identifies that a threat is unresolvable or identifies a cycle in the causal links, the algorithm backtracks. The algorithm, thus searches in a plan space, where a (partially constructed) POP graph at any point of time is a vertex, unlike the approaches seen so far where a set of literals represented the vertices. The selection of the open condition uses heuristics to speed up the process. The easiest heuristic is to close conditions that do not cause too many new open conditions (unless already satisfied by the source).

Let us also see what happens when we place a constraint that the hand can only carry one item at a time, by the introduction of *EmptyHand* in preconditions and effects of actions. The solution is given in Fig. 19.13. The solution is the same as Fig. 19.10, except steps *v*, *viii*, *ix*, and *x*. In step *v*, $Pick(B_2,B_1)$ achieves $\neg EmptyHand$ for $PlaceTable(B_2)$ and therefore $Place(B_1,B_2)$ is a threat that is promoted. Similarly in step *viii*, $PickTable(B_1)$ produces $\neg EmptyHand$ for $Place(B_1,B_2)$ and therefore $PlaceTable(B_2)$ is a threat that produces *EmptyHand*. The threat is resolved by demotion of $PlaceTable(B_2)$ before $PickTable(B_1)$. Similarly in step *ix*, the *source* achieves *EmptyHand* for $Pick(B_2,B_1)$ and therefore $PickTable(B_1)$ is a threat that is promoted. Step *x* does not satisfy *EmptyHand* from *source* for $PickTable(B_1)$ as the following threat from $Pick(B_2,B_1)$ can neither be demoted (cannot lie before source) nor promoted (adding causal link from $PickTable(B_1)$ to $Pick(B_2,B_1)$ causes the cycle $PickTable(B_1) \prec Pick(B_2,B_1) \prec_{Clear(B_1)} PickTable(B_1)$).

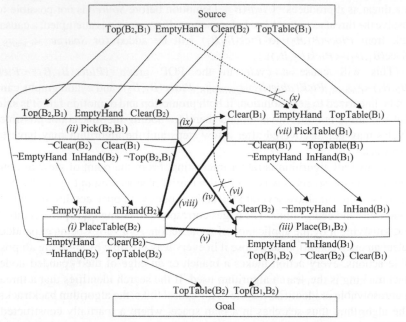

FIG. 19.13 Resolving threats. Source achieves *Clear* (B_2) for *Pick* (B_2,B_1) (shown as *ii*), however, *Place*(B_1,B_2) (shown as *iii*) has an effect ¬*Clear* (B_2) and is thus a threat. It cannot be demoted before the start. It is therefore promoted after *Pick* (B_2,B_1) by the causal link (*iv*). In (*x*) and (*vi*), if the deleted branch is used, the conflict resolution will result in a cycle in the graph, which has no linearization.

19.7 Integration of task and geometric planning

The book so far discussed two types of planning. The first one discussed for numerous chapters is *geometric planning*. Here the space is mostly continuous, and planning is done to get a trajectory so that the robot reaches one configuration from the other. The manipulator moving in its environment to pick a book, the mobile robot going from one room to the other, are all examples of geometric planning. On the other hand, this chapter motivated and designed upon a *task planning* paradigm wherein the robot was given a higher-level task that could conveniently be represented by a set of Boolean variables. The problem was presented using Boolean propositional variables and the problem was solved using the classic planning techniques.

The geometric and task planning approaches are linked to each other (Cambon et al., 2009; Srivastava et al., 2014). In general, the task planning approaches work at a higher layer of abstraction. The solutions to task planning ultimately require a geometric solver to plan a trajectory and then control a robot over the trajectory as a solution to the task. The task planner hence, gives a sequence of goals to the geometric planner, while the geometric planner responds when the goal is reached so that the next action (goal) in the sequence can be given to the geometric planner. A failure to reach the goal can also be given as a response by the geometric planner to the task planner to initiate replanning. So, picking up books and files are sequenced by the task planner and the trajectory for the same is computed by the geometric planner. Actions like $PickTable(B_1)$ were casually used by the task planner, while the same needs a geometric solver used by the robot.

A more intensive interest calling for the integration of task and geometric planning is when the task planning uses geometric planning to compute the feasibility of an action. Consider the same problem of moving several objects on the table. Currently, the modelling assumes that the table can hold any number of objects. This is however not true as the table has a capacity. The capacity cannot be defined in terms of the number of the objects, as every object has a different size. Hence, a geometric planner is needed to compute the feasibility of whether an object picked up can be placed somewhere in the table, and if yes, suggest a place of placing the object. These are geometric problems solved by a geometric planner. The planning state now has a Boolean component corresponding to the task planning problem (e.g. $TopTable(B_1)$) and a geometric component for the geometric planning problem (e.g. specifying the pose of B_1 on the table). Both components of the state representation are updated as the task planning works. A task may be feasible because of the preconditions but may be geometrically infeasible. So, the feasibility conditions of $PlaceTable(B_1)$ may hold considering the Boolean variables, but there may be not enough free space to actually place B_1 on the table, making the action geometrically infeasible. Similarly, many times there may be space on the table, but carrying the physical object may not be feasible due to a large number of cluttered obstacles, threatening a collision. This is also taken care of by geometric planning. The task planner continuously uses the geometric planner to assess the feasibility of the actions.

The task planning approaches largely disregarded optimality and mostly took all actions at equal cost. In case optimality is of a concern and the different actions have different costs, it needs to be determined how much physical effort will take place to act or to determine the cost of the actions. In the example, it needs to be known how much time will be required to move a book from the first pile to the table. Carrying a heavy book may have a larger cost as compared to carrying a light file. This is again a geometric planning problem, and the geometric planning now gives the cost function for the task planning search, along with checking for feasibility.

19.8 Temporal logic

The task planning approaches so far have enabled the specification of any goal with multiple Boolean operators. The discussions largely assumed that the subgoals are separated by AND operators, while the system must achieve all the subgoals. The OR operators can be similarly handled, while the NOT operators are already modelled.

This, however, does not enable a plan for more realistic scenarios wherein the humans typically have a *temporal* specification of the goal. Say the problem is to clean the table *before* cleaning the room and thus an action sequence that cleans the room first followed by cleaning the table results in the same state, which is however not admissible. Similarly, the specification that the sugar should be put in the coffee *after* the coffee is poured has a temporal clause attached.

The formal language that considers the temporal aspect is called *temporal logic*. The temporal logic is an extension to the Boolean logic and includes some additional operators to handle the temporal specifications. The task planning can now have Boolean and temporal logic operators and is therefore more generally known as *mission planning*. As an example, the human could instruct the robot to '*get groceries after cleaning the working desk; while also cleaning the room whenever free; however, when the master calls for the robot to get some water, it must do so leaving all work first*'. This system and mission specification can be easily written in the form of a formal language using temporal logic using which the robot does the planning.

The section first discusses model verification (Baier and Katoen, 2008; Fisher, 2011), which will be used to design a reactive controller or a navigation system from the specifications (Kress-Gazit et al., 2009; Lahijanian et al., 2012; Svorenova et al., 2015). Thereafter, the section discusses the specific languages to specify the missions and the operating rules. The languages can be extended to deal with probabilities (Fu et al., 2016; Schillinger et al., 2018), multiple robots (Kantaros and Zavlanos, 2018; Ulusoy et al., 2013) and using sampling-based approaches (Karaman and Frazzoli, 2009; Bhatia et al., 2010) for speedup.

19.8.1 Model verification

Model verification is inspired by means to provide guarantees to the modern complex software systems that the software suite will do exactly what is expected out of it, and that it will not do anything which is specified as hazardous. The testing techniques work by guessing for the conditions that can go wrong or trying them randomly, to ascertain the guarantees on working. However, this does not guarantee that nothing bad will happen and the goal of the software shall be met. In model verification, the computer programs are represented as formal *models*, which are *verified* to always hold certain important properties, no matter what the sequence of events or inputs is. It is also verified that the software suite eventually satisfies certain desirable results that are expected out of the software suite. The model verification techniques typically have two possible outputs: the model is guaranteed to hold all the specified properties for any possible input, or at least one *counterexample* wherein the software does not meet the properties and does not pass the verification.

In this approach, we will model the properties that we need to be guaranteed (as an example, the robot must eventually do everything desired, the robot must ensure the coffee is done before the sugar, the robot must never go to a dangerous area, etc.) at one end. We also state the possible moves or actions of the robot and the possible transitions that the system can have based on the qualifying conditions. Our problem statement demands to get one path or action sequence or a reactive control policy that satisfies the model, while the model verification system checks whether all paths satisfy the model. A naïve way to approach the problem is to model the inverse system, which is the system that does not satisfy the properties. This is done by putting a not in front of the property specification. Thereafter, the transition system is given, specifying the valid actions or transitions possible by the robot or system. The model verification system tries to find whether all paths satisfy the input *negated model* (equivalently all paths that *do not* satisfy the model), and if a single path does not satisfy the negated input model (equivalently, satisfies the model), the verification fails, and the counterexample is given as the output. This counterexample satisfies the model and is therefore a valid controller.

The term used here is specifically reactive controller or policy and not linear path. Consider that the robot should immediately go when the master calls, leaving all current actions. This has a reference to an external sensor input, which is unknown during the planning and therefore, the output is a policy and not a path. A policy specifies what the robot needs to do given its state and every possible the external input. In this example, a call from the master can come at any time. The policy must account for this and have an action ready if the master's call is received at any step. Sometimes the actions may be nondeterministic. So, executing an action can make the robot reach either of the possible target states. Here as well, the robot does not know beforehand which state it will reach. However, the robot must plan such that whichever state it reaches, the next action must be readily available from the policy.

The easiest way to model the system is by using a *Finite State Automaton* (FSA) represented by (Q,Σ,q_0,A,δ). Q is a set of all possible states. q_0 is the initial state. A is a set of accepting states. Any execution or simulation stops, if the system reaches any of the terminal states. Σ is the set of input alphabets that governs the path that the system takes. The transition function (δ) gives the next state s' of the system based on the current state s and input alphabet a, or $s' = \delta(s,a)$.

The *Nondeterministic Finite State Automaton* (NFSA) represented by (Q,Σ, Q_0,A,δ) adds the nondeterminism to the model. Here there may be more than 1 initial state denoted by Q_0. Similarly, a transition may lead to any one of the many states, or $\delta(s,a)$ may have more than one output value.

The automaton may have one or more initial states. Starting from the initial state there are many paths possible. The path depends upon the external conditions as well as nondeterminism in the automaton. Starting from any source, following any path is called the *execution* of the path. The execution stops on reaching any *terminal state*. An execution is valid only if it ends at the terminal state and starts at the initial state.

Imagine having a string that is read from left to right, character by character. Starting from the initial state, you read the next character from the string and make the same transition as per the transition function. The string is *accepted* by the automaton if making the transitions ends up in the terminal state. The set of all strings accepted by the automaton is called the *language* for the automaton.

A path must satisfy all possible properties as desired by the system. Moreover, the path also depends upon the inputs and the nondeterminism. So, for a system to guarantee all properties, all possible paths must satisfy all properties. A collection of all possible paths is called the *trace*.

19.8.2 Model verification in robot mission planning

Any robotic system can be modelled as a transition system, which defines from which state (s), the robot can go to which robot state (s') on the application of which action (a). Any transition system (TS) is characterized by the tuple (Q,A,I,δ,AP,L). Here Q is a set of states, A is the set of actions, I is the set of initial states, δ is the action function. Application of action a on state s produces the state $s' = \delta(s,a)$. AP is additionally a set of *atomic propositions*. The atomic propositions may be understood from the binary encoding used in the task planning, where the truthness (or not) of the variable is represented as an atopic proposition variable. L is a labelling function. More generally, the states themselves may not be visible; however, the labels may be visible. Like in the localization problem, the state (position) is not visible, but the sensor readings are visible as labels.

The essence of model checking comes from parallel and concurrent systems, wherein multiple threads run in parallel, requiring guarantees. The same notion is also very relevant to robotics. The parallel systems share *resources* that may be with only one or a finite number of systems at a time. This necessitates the need to communicate between the parallel systems to make the resource allocation exclusive or within the available limit. The communication is normally done via shared variables or formal communication channels.

It is important to guarantee that the modelling is *deadlock-free*, or not all threads are waiting for a resource at the same time. Imagine that two robots are cleaning a desk, which requires a cleaning lotion and a cloth. Unfortunately, there is one unit of each. To avoid both robots picking up the cloth, a reservation strategy is used (like a *semaphore*) where the robots must first reserve through a reservation system. Only a successful reservation acknowledgement entitles the robots to use the resource. Imagine the first robot reserves and picks the cloth, while the other reserves and picks the cleaning lotion. Now both robots will be waiting infinitely for the other item, resulting in a deadlock. The verification system must identify this and urge modelling that is deadlock-free (like the cloth cannot be reserved until the cleaning lotion is reserved).

The text so far talked about various properties that must be satisfied by the system. Let us formalize these a little. There are three important kinds of properties: safety, liveness, and fairness, which must be ensured and verified by the system.

The *safety properties* specify whatever is dangerous for the robot and must never be done. These properties specify the conditions which must never be true for any path or trace of any execution. The language represented by the model must ensure that every state in every path possible is safe. As an example, the robot must never go to a dangerous area; the robot must never cause an action that breaks an item, etc.

The *liveness property* specifies the conditions which must eventually be reached by the system. Safety properties do not ask the system to achieve anything specific and therefore an idle system is also a safe system. However, we want that eventually something should be achieved by the robot which is specified by the liveness properties. As an example, the robot must eventually reach the goal, the robot must eventually clean the table and the room, etc. The liveness properties may ask the system to eventually reach some criterion, to reach some criterion repeatedly often, and sometimes to reach a criterion in a finite time without any possibility of getting stuck on the way. Each of these specification measures has a different level of strictness that impacts verification.

The last category of properties is *fairness*. Imagine two robots working in a working place and both want to clean the desk by reserving the cleaning equipment. However, the system keeps favouring the first robot and always allocates the equipment to it whenever asked. As a result, the second robot may never be able to meet its objectives. This is not fair, which is checked by the fairness properties. The fairness is of three varieties. The first simply mentions that without any conditions, every process must be given a turn infinitely often. The second variant is called strong fairness and states that every process that is enabled infinitely often must also get its turn infinitely often. This notion relies on the condition that for every process requesting for its turn, the request must be met. The last modelling is called weak fairness and states that every process which is continuously enabled from some time instant should also get its turn infinitely often.

The notions of visiting some states infinitely often give rise to a more expressive modelling instead of the nondeterministic automaton. The model is extended to work with languages of infinite size. The *Büchi Automaton* (BA) is one such model that accepts or rejects languages of *infinite length*. Imagine a robot is asked to visit a gate infinitely often. Now the robot will keep moving or have a plan of an infinite length. In this plan, the robot should be at the gate after every few times. The model used for the representation now needs to be able to work with such infinite length paths. Alternatively, the policy output should represent an infinite length sequence that has the robot visiting the gate infinitely often. The *Nondeterministic Büchi Automaton (NBA)* is represented by $(Q, \Sigma, Q_0, \delta, F)$, where Q is the set of all states, Σ is the input alphabets, Q_0 is the set of initial states, δ is the transition function, and F is the set of terminal states.

Typically, the properties desired by the system are all represented as an *NBA*. The automaton can accept any transition made, such that the sequence is accepted, if all the modelled properties are satisfied. The NBA is itself made by looking at the desired properties that may not model the context of robot actions. In other words, the NBA may model transitions that are not physically attainable by any action of the robot. The *transition system* (TS) models the specific robot actions that only hold the transitions which are attainable. The TS may, however, not have a representation of acceptable properties. Thus, an automaton is made that accepts sequences that meet the desired properties like the NBA, but only models valid transitions like the TS. This is done by taking a product of the NBA with the TS, used for model checking. Using a negated NBA representing the desired properties and computing a counterexample of the NBA gives a valid solution to the problem.

The *Deterministic Büchi Automaton* (DBA) is, as the name suggests, a deterministic variant of the NBA. In terms of expressiveness, the deterministic variants of the automaton are less expressive. Another variant is the *Generalized Nondeterministic Büchi Automaton* (GNBA) that adds expressiveness by enabling the algorithm to visit several finishing states infinitely often.

19.8.3 Linear temporal logic

The language used to specify the properties to hold or specify the mission that the robot must solve while doing only the desirable things is the *linear temporal logic* (LTL). The LTL uses the Boolean operators (and, or, not) to specify the mission, but adds the *temporal operators* over these to specify the temporal constraints. The following are the commonly used temporal operators in LTL. The graphical representation of the operators is shown in Fig. 19.14.

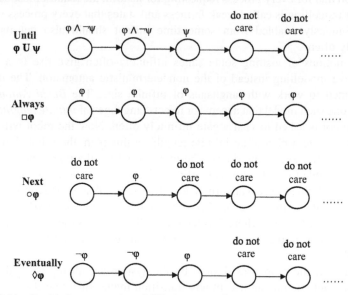

FIG. 19.14 Graphical representation of different temporal operators.

- *Until (U):* It is the most basic temporal operator. A formula $\varphi U \psi$ is true if the formula φ holds up to the point (until) the formula ψ holds in the path. All temporal operators can be derived by suitably writing the until operator.
- *Always* ($\square$): It is a unary operator. $\square\varphi$ is true if and only if all states in the path satisfy the formula φ.
- *Next* ($\circ$): It suggests something to hold from the next time instant onwards. $\circ\varphi$ is true if the formula φ holds from the next time onwards.
- *Eventually* ($\Diamond$): It is used to denote the properties, which may hold some time in the future. As an example, the goal must be eventually reached. $\Diamond\varphi$ states that the formula φ must hold good sometimes in the future.

The eventually operator ($\Diamond\varphi$) is equivalent to *true* $U\varphi$ and the always operator (φ) is equivalent to $\neg\Diamond\neg\varphi$, which makes it easier for deduction using until operator alone. All Boolean expressions can be expressed using $\wedge$ and $\neg$ operators alone. This means that the only basic operators in LTL are $\neg$, $\wedge$, U, and $\circ$. To verify whether a string s satisfies a formula φ, the semantics is given by Eqs (19.18)–(19.26) are used.

$$s \models true \tag{19.18}$$

$$s \models a \ iff \ a \in A_0 \tag{19.19}$$

$$s \models \phi_1 \wedge \phi_2 iff \ s \models \phi_1 \wedge s \models \phi_2 \tag{19.20}$$

$$s \models \phi_1 \vee \phi_2 iff \ s \models \phi_1 \vee s \models \phi_2 \tag{19.21}$$

$$s \models \neg\phi iff \ s \not\models \phi_1 \tag{19.22}$$

$$s \models \phi_1 U \phi_2 iff \ \exists_i s(i:) \models \phi_2 \wedge s(j) \models \phi_1 \forall 0 \le j < i \tag{19.23}$$

$$s \models \circ \phi \ iff \ s(1:) \models \phi \tag{19.24}$$

$$s \models \Diamond\phi \ iff \ \exists_i s(i:) \models \phi \tag{19.25}$$

$$s \models \phi iff \forall_i \ s(i) \models \phi \tag{19.26}$$

Here $s(i)$ denotes the ith term and $s(i:)$ denotes the substring from position i.

The formula $\varphi_1 U \varphi_2$ requires φ_2 to be true somewhere in the future. Another similar operator *weak until* (denoted by W) is satisfied if additionally, φ_1 is true throughout the sequence. Another operator release (denoted by R) is used in LTL. $\varphi_1 R \varphi_2$ models that φ_2 is true until φ_1 becomes true, while after φ_1 becomes true, the requirement of φ_2 to be true is released.

The operators can be used to represent several scenarios, like asking the robot to clean 3 rooms can be represented by $\Diamond Clean(r_1) \wedge \Diamond Clean(r_2) \wedge \Diamond Clean(r_3)$. The expression to buy sugar and either tea or coffee is given by $\Diamond Buy(Sugar) \wedge (\Diamond Buy(Tea) \vee \Diamond Buy(Coffee))$. The expression to clean 3 rooms until the battery is low is given by $(\Diamond Clean(r_1) \wedge \Diamond Clean(r_2) \wedge \Diamond Clean(r_3))$ $U batteryLow$.

There are a few notations that are commonly used in LTL. *Sequencing* of φ_1, φ_2, and φ_3 is given by $\Diamond(\varphi_1 \wedge \Diamond(\varphi_2 \wedge \Diamond\varphi_3))$. Asking the robot to do a task *infinitely often*, like the security of a premise is given by $\Box\Diamond\varphi$. On the contrary, asking a robot to do something *eventually forever* is given by $\Diamond\Box\varphi$. Sometimes the robot may need to respond to a request like go to master when called, open the door on a doorbell, etc. The general syntax of a *request-response* system is given by $\Box(request \rightarrow \Diamond response)$. The system would naturally have a very large number of such specifications, which must be all satisfied by the robot as a part of the same mission. All such specifications are given separated by the $\wedge$ operator.

The LTL is converted into an NBA. The inverse formula is used to compute a counterexample as the solution. The question then is whether a transition system (TS) representing the robot transitions satisfies the LTL. The product of TS and NBA is used to hunt for a counterexample. Since the sequence is infinite, a finite prefix is returned as a counter-example. The common libraries for model verification are SPIN, Tulip, NuSMV, etc.

The bottleneck in the LTL is that the conversion to the NBA and thereafter the verification is an exponential time operation. The model checker typically only finds a counterexample which does not guarantee optimality.

19.8.4 Computational tree logic

The LTL-based model checking is based on linear sequences or paths. The notion is extended to a nonlinear data structure using *Computational Tree Logic* (CTL). Specifically, the representation is in the form of a tree and the notion is to emphasize the branching over which the verification is carried out. The major difference with the LTL (besides a tree representation) is that the CTL uses existential path qualification ($\exists$) and universal path qualification ($\forall$) operators in the specification. This means, given a tree, a *TS* satisfies the CTL formula $\exists\varphi$, if any branch satisfies φ. Similarly, a *TS* satisfies the CTL formula $\forall\varphi$, if all branches satisfy φ. In LTL, all paths should satisfy the given formula. The model checking, hence, happens by using techniques of simulation and bisumation. A significant improvement is that the CTL makes the verification polynomial in time, instead of the LTL, which is exponential in time. However, care must be taken because LTL and CTL formulas do not convey the same meaning and are not equivalent. Fairness is another difference. In LTL, it can be specified as a straight formula given to the system and no special verification techniques are needed. In CTL though, special techniques need to be devised for fairness checking.

The CTL uses two different formulas called the state formula and the path formula. The *state formula* concerns the specific properties that the state should possess. The *path formula* concerns the properties that are applicable to a path in the tree. Due to distinction, the CTL is described by the Backus-Naur Form Grammar. The parenthesis is omitted for clarity and is otherwise included in the specification. Only minimal base operators are used as others can be derived by these operators. The state formula is given by Eq. (19.27), while the path formula is given by Eq. (19.28).

$$\phi_{state} := true \,|\, a \,|\, \phi_{state} \,|\, \phi_{state} \wedge \phi_{state} \,|\, \neg \phi_{state} \,|\, \exists \phi_{path} \,|\, \forall \phi_{path} \tag{19.27}$$

$$\phi_{path} := \phi_{state} U \phi_{state} \,|\, \circ \phi_{state} \tag{19.28}$$

19.8.5 Other specification formats

LTL and CTL are popular means to represent tasks, missions, and properties. Other formats are briefly discussed. The CTL* representation has a syntax that is a mixture of LTL and CTL. It has the property to arbitrarily specify the operators without distinguishing the state and path properties like the LTL, while also uses the for all and there exists qualifiers, like the CTL.

The *Timed Linear Temporal Logic* allows specifying time constraints in the specification. Like a property must be true within k units of time, or something should be true after k units of time, etc. The time constraints are shown by a superscript. The specification $\Diamond^{\leq k} \varphi$ means that φ must be true within k units of time.

The *Probabilistic Computational Tree Logic* further adds the notion of probabilities to model the stochastic systems. It enables specifying the minimum probability of satisfying a constraint, given that the sequence that the robot follows is stochastic. The probability specifications are given as the subscript using the probability function to denote the probabilities. So, $P_{>p}(\Diamond \varphi)$ denotes that the probability of φ eventually holding should be greater than a threshold p and $P_{>p}(\Diamond^{\leq k} \varphi)$ denotes that the probability of φ holding within k steps should be greater than a threshold p. The probability calculations use the Markov chains (discussed in Chapter 14). There can be constraints on the expected rewards on the solution as well, called the *Probabilistic Reward Computational Tree Logic* (PRCTL). The reward is typically shown as a subscript. So, $\Diamond_{\leq r} \varphi$ shows that φ must hold well with a reward less than r.

Questions

1. Solve the standard AI problems (Flat Tire Problem, Blocks World Problem, Air Cargo Problem) by using Backward Search, Planning Graphs, and POP.
2. Solve the following problems using POP
   (i) Source: $e \wedge f$; Goal: a
      Action a_1: Precondition: f, Effect: d
      Action a_2: Precondition: e, Effect: b
      Action a_3: Precondition: $b \wedge d$, Effect: a
   (ii) Source: $e \wedge f$; Goal: a
      Action a_1: Precondition: f, Effect: d.
      Action a_2: Precondition: e, Effect: $b \wedge \neg f \wedge \neg d \wedge \neg c$.

Action a_3: Precondition: e, Effect: $b \wedge \neg f \wedge \neg c$.
Action a_4: Precondition: e, Effect: $b \wedge \neg f \wedge \neg d$.
Action a_5: Precondition: $b \wedge d$, Effect: a

3. Compare and contrast the different task planning algorithms, trying to establish the utility of each of these in different settings.

4. For the following problem definition, calculate the max-level heuristic, set-level heuristic, and sum-level heuristic. Show the complete planning graph with mutex relations.

Source: a, Goal: (i) Case I: $a \wedge \neg c$ (ii) Case II: $a \wedge b$.
Action: A_1: Preconditions: $a \wedge \neg b$, Effects: c.
Action: A_2: Preconditions: $b \wedge c$, Effects: $\neg c$.
Action: A_3: Preconditions: a, Effects: $\neg b$.

References

Baier, A., Katoen, J.P., 2008. Principles of Model Checking. MIT Press, Cambridge, MA.

Bhatia, A., Kavraki, L.E., Vardi, M.Y., 2010. Sampling-based motion planning with temporal goals. In: Proceedings of the 2010 IEEE International Conference on Robotics and Automation, pp. 2689–2696.

Cambon, S., Alami, R., Gravot, F., 2009. A hybrid approach to intricate motion, manipulation and task planning. Int. J. Robot. Res. 28 (1), 104–126.

Edelkamp, S., Jabbar, S., Nazih, M., 2006. Large-scale optimal PDDL3 planning with MIPS-XXL. In: 5th International Planning Competition Booklet, 2006.

Fu, J., Atanasov, N., Topcu, U., Pappas, G.J., 2016. Optimal temporal logic planning in probabilistic semantic maps. In: Proceedings of the 2016 IEEE International Conference on Robotics and Automation. IEEE, Stockholm, pp. 3690–3697.

Galindo, C., Fernandez-Madrigal, J., Gonzalez, J., 2008. Multihierarchical interactive task planning: application to Mobile robotics. IEEE Trans. Syst. Man Cybern. B Cybern. 38 (3), 785–798.

Gerevini, A., Haslum, P., Long, D., Saetti, A., Dimopoulos, Y., 2009. Deterministic planning in the fifth international planning competition: PDDL3 and experimental evaluation of the planners. Artif. Intell. 173 (5–6), 619–668.

Fisher, M., 2011. An Introduction to Practical Formal Methods Using Temporal Logic. Wiley, West Sussex, UK.

Kantaros, Y., Zavlanos, M.M., 2018. Temporal logic optimal control for large-scale multi-robot systems: 10400 states and beyond. In: Proceedings of the 2018 IEEE Conference on Decision and Control. IEEE, Miami Beach, FL, pp. 2519–2524.

Karaman, S., Frazzoli, E., 2009. Sampling-based motion planning with deterministic μ-calculus specifications. In: Proceedings of the 48h IEEE Conference on Decision and Control (CDC) Held Jointly With 2009 28th Chinese Control Conference. IEEE, Shanghai, pp. 2222–2229.

Kress-Gazit, H., Fainekos, G.E., Pappas, G.J., 2009. Temporal-logic-based reactive mission and motion planning. IEEE Trans. Robot. 25 (6), 1370–1381.

Lahijanian, M., Andersson, S.B., Belta, C., 2012. Temporal logic motion planning and control with probabilistic satisfaction guarantees. IEEE Trans. Robot. 28 (2), 396–409.

Russell, S., Norvig, P., 2010. Artificial Intelligence: A Modern Approach. Pearson, Upper Saddle River, NJ.

Schillinger, P., Bürger, M., Dimarogonas, D.V., 2018. Simultaneous task allocation and planning for temporal logic goals in heterogeneous multi-robot systems. Int. J. Robot. Res. 37 (7), 818–838.

Srivastava, S., Fang, E., Riano, L., Chitnis, R., Russell, S., Abbeel, P., 2014. Combined task and motion planning through an extensible planner-independent interface layer. In: Proceedings of the 2014 IEEE International Conference on Robotics and Automation, Hong Kong, pp. 639–646.

Svorenova, M., Cerna, I., Belta, C., 2015. Optimal temporal logic control for deterministic transition systems with probabilistic penalties. IEEE Trans. Autom. Control 60 (6), 1528–1541.

Ulusoy, A., Smith, S.L., Ding, X.C., Belta, C., Rus, D., 2013. Optimality and robustness in multi-robot path planning with temporal logic constraints. Int. J. Robot. Res. 32 (8), 889–911.

Schulman J., Ryan, M., Duan, Abbeel, P., V. 2018. Gradient estimation using stochastic computation graphs...

Schulman J., Ryan, in Reinforcement multi-robot teams. Int. J. Robot. Res. 37(2), 818–838.

Schulman, S., Peng, B., Kang, L., Chhatr., K., Levine, S., Abbeel, P. 2011. Combined task and motion planning through an extensible planner-independent interface layer. For proceedings of the 2014 IEEE International Conference on Robotics and Automation, Hong Kong (pp. 639–646).

Srivastava, M., Otte, C. 2015. Optimal coupling logic for a branch-and-bound-based motion system with probabilistic penalties. IEEE Trans. Autom. Control 60(6), 1528–1541.

Huang, A., Smith, S.L., Tang, X., Rus, D. 2012. Optimality and robustness in multi-robot path planning with temporal logic constraints. Int. J. Robot. Res. 32(8), 889–911.

Chapter 20

Swarm and evolutionary robotics

20.1 Swarm robotics

So far, it was assumed that the robots have reasonable computing, memory, and sophisticated sensors, which enabled localization, planning, and storing of a complete map. *Natural Swarms* of primitive organisms can display complicated behaviours even though each organism has a very primitive sensory capability and a limited brain. The *fish of schools* can easily navigate over a long distance in harmony with each other, while the school reacts when the predator comes and re-assembles when there is no predator. The *ant colonies* can very effectively find sweet items kept at home and coordinate with each other to get the entire colony towards the item of interest. In the *flocking of birds*, the birds do not have complex localization techniques with them nor can the birds communicate with words. However, they are collectively able to migrate thousands of miles at the same time of the year. The *chirping of birds* is in complete harmony, which may require hours of lectures and practice sessions with human experts if done by humans. The *bacterial colonies* can locally exchange information and attack the human cells or food items and grow in the same body.

The natural swarms exhibit emergent properties. A system has an *emergent property* when it is much better than the sum of all its parts. Every ant in the colony may not be able to find food for itself, which an ant colony can find. A single bird may not be able to migrate and flock, but collectively the birds correct and navigate each other. A single neuron can barely do anything, but a network of neurons in the human brain is responsible for all human intelligence. Teams typically do not only divide the work and solve themselves but discuss in multiple meetings to get more understanding. As a result, the solution developed is better than what every member of the team would have made in isolation and integrated at the end. Such complex behaviours have led to numerous global optimization methods: Ant Colony Optimization, Particle Swarm Optimization (PSO), and many other interesting algorithms.

The same inspiration in this chapter is adopted to robotics giving rise to the field of *swarm robotics* (Tan and Zheng, 2013; Navarro and Matía, 2013;

Autonomous Mobile Robots. https://doi.org/10.1016/B978-0-443-18908-1.00019-4

Brambilla et al., 2013). A single robot with sophisticated sensors and computing capabilities may do a lot of jobs. However, the single robot will be expensive. For many tasks, it may be better to have thousands of robots that together make a swarm. The robots individually have extremely low computing, memory, battery, and sensory requirements. Each robot itself only does a very primitive job. However, collectively the robots become a powerful system because of their emergent properties. A robotic swarm can solve a complex task by using thousands of very primitive robots, at the same time being highly efficient and affordable as compared to a single robot. *Swarm intelligence* uses motivation from natural swarms to make intelligent algorithms, which may not be limited to robotics.

The swarms derive their intelligence based on the information exchange done locally by a robot with the neighbouring robots. This interaction denotes the information flow within the swarm network, and all intelligence is distributed and built up in this network. No single node manages the thousands of robots in a swarm; however, every robot makes its decision based on its neighbours in a purely decentralized manner.

20.1.1 Characteristics of swarm robotics

Let us look at the important characteristics of the swarm. Swarms have thousands of robots, and hence the decision-making must be purely *decentralized*. At no time, any robotic agent or a central server can have the information from all the robots and use the same for decision-making. Navigation and decision-making are purely *reactive*. The robots make a move based on the robots in the near vicinity alone. The robots may be aided by *local sensing* to get information about the robots in the vicinity. In some cases, there may be a *local communication* which can be used by a robot to communicate and exchange formal messages with the neighbouring robots only.

The solutions are completely *distributed,* and no single robot solves a big chunk of the problem, while there is no central system that can be used to integrate the part solutions solved by the robots. All computation and intelligence generated therein are distributed across the swarm network. The robots work in *parallel* to each other. This gives the biggest efficiency of the robotic swarm system in contrast to a single robot system. Human intelligence is due to many biological neurons working in parallel and not attributed to the speed of a single neuron. Even in terms of geographical occupancy, the robots distribute themselves in the area and work in parallel, which is better than a single robot visiting different places of interest. The swarms typically take very little power, and the overall solutions are also energy efficient.

All the robots are *low cost*. This restricts putting any expensive sensor, actuator, processing board, or memory unit onboard the swarm. The solutions developed should also be *scalable* to a high level. Meaning that the number of robots may be significantly increased to get better performance with no change in the algorithm. The swarms are *homogeneous*. Different robots should be the same in all perspectives and should be running the same control algorithm. This makes it possible to add robots on the fly and get better results. Some works are now creating a hybrid swarm like a swarm of robots that fly and provide a vision for the ground robots; a swarm of robots that does manipulation, while another is more suited to transport the objects from one place to another, etc. The *flexibility* is another added criterion, meaning that the same swarm, with a change of the controller, should be suitable for doing different kinds of work. The same swarm can thus be deployed across a wide set of applications.

The swarm robotics must hence not be confused with multirobotic systems, wherein the number of robots ranges to a few tens. The robots can share much information like knowledge base and map across the network while also solving problems with a linear or polynomial complexity like priority-based motion planning. Each robot is an expensive unit, which can typically localize itself easily and has access to a central dynamically changing map. That said, many times, the line between the multirobots and swarms can become fuzzy, especially with the sensing and computing technologies being more affordable, meaning that the swarms can have a larger set of sensors otherwise to be used only in multirobotics.

20.1.2   Hardware perspectives

The most interesting aspect of the swarm is its hardware, wherein there is little possibility of giving any sophisticated sensor or computation unit (Dorigo et al., 2004). The robots have a highly limited battery and *power system*. Actuators consist of motors that are primarily used for navigation, preferably differential wheel drive systems with possibilities of spherical or caster wheels for support. Many times, the robots also have *clippers* that enable them to clip an object of interest. The robots may also have small *manipulators* to pick up small items. The robots may also want to connect to each other. This typically happens by making one robot *clamp* to the other. The robots reach close enough, and the clamp is operated by the follower robot onto the leader robot. The computing is limited to low-power embedded boards.

The sensors are limited. In a swarm, the most basic behaviour for a robot is to reach a specific goal, which needs to be done without having a localization system. The goal is said to emit a signal, which can be detected by the robot, the sensors for which are placed on the robot. The most primitive system has the goal that emits light of a certain frequency (colour). The LEDs are suitable as emitters. The robot has *light sensors*, which detect the light. A ring of light sensors is placed around the body of the robot to get the direction of the goal (light). The robot knows which sensor detected the light and moves in the same direction. The direction of the goal (light) is taken as the direction of the sensor (with respect to the robot's heading direction) that captured the light, which is the angular error used to give the angular velocity. The setting of the linear velocity in such a motion is such that no collision occurs. An *ultrasonic* or *sonar sensor* is used to detect the obstacles ahead. The linear velocity is proportional to the distance from the front obstacles. The setting of the different sensors is shown in Fig. 20.1A.

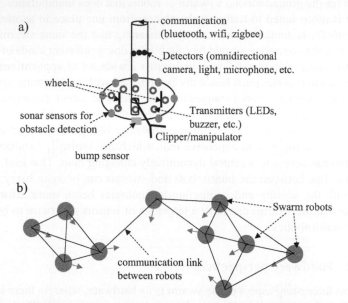

FIG. 20.1 Swarm robotics. (A) A typical swam robot. The robots communicate by transmitting signals (like LEDs) and detecting them. Sonars enable obstacle avoidance. The clipper is to hold the object or for robots to form a chain. (B) Mobile ad hoc network formed by moving robots.

The other option to reach the goal is to take sound instead of light. The transmitter is a buzzer (speaker), while the receiver is a microphone. Unlike LEDs, the sound is omnidirectional and will be received by all microphones at the rim of the robot body. The differential of the sound from the different microphones can be used to indicate the direction of the goal (sound), while the amplitude can be used to detect the distance from the goal. The fundamental is if the left ear hears a sound louder than the right, the buzzer (speaker) is more on the left side relative to the person. Alternatively, it is possible to use a single microphone also to get to the goal. The robot knows the current amplitude of the sound. It also knows the amplitude of the sound at the previous step. If the sound got louder, the robot moved towards the goal and vice versa. By taking a differential of the sound along with time, the direction can again be estimated. Luckily, there is no shortage of carrier waves and magnets, Wi-Fi signals, radio raves, etc., may also be used.

For any robot, the generic behaviour not only includes a motion to the goal but also avoids other obstacles on the way. The obstacle avoidance is done by the *ultrasonic or sonar sensors* packed around the robot, the same as the artificial potential field method. The *bump or tactile sensors* are also useful; however, the robot collides with an obstacle for getting feedback about the obstacle using such sensors.

Some behaviours require the robots to go towards the other robots. As an example, the robots may need to form a chain, assemble at the centre, make groups, make a pattern, etc. This requires a robot to get attracted to the other robots, as well as repel to avoid a collision, the same as the social potential field. Whenever one robot needs to attract another robot, it transmits signals like light, sound, wave at a characteristic frequency, etc. The other robot has the receiver that gets attracted to the robot-like any other goal. The sonar sensors ensure that a physical collision does not happen. This necessitates every robot to have transmitters like LEDs, buzzers, etc. Another interesting aspect is that different lights can be used to attract/repel different types of robots. Similarly, different frequencies of sounds can be used to attract/repel different types of robots. Imagine that there are two types of robots, and they need to cluster in space. Now the robots have red or blue lights. A robot with red light is repelled by the blue light and attracted to the red light. So, whenever a red robot's sensor sees a blue light, a force is applied in the other direction, while if it sees a red light, a force is applied in the same direction. This moves the robots.

The last module associated with the swarm robots is for *communication*. A robot can connect to the near robots using formal communication channels. The typical choices are wifi, Bluetooth, or Zigbee. As every robot is connected to the neighbours, together the entire network of robots forms a mobile ad-hoc network that keeps changing its topology as the swarm moves. This is shown in Fig. 20.1B. Such a communication network ensures that a message can be eventually broadcasted to all robots over multiple hops with different robots.

20.2 Swarm robotics problems

Let us now solve some interesting behaviours using the principles of swarm robotics. In all cases, the fundamental algorithm used is the *social potential field*. The robots sense some goals and some obstacles relative to themselves, which are converted into forces. The robots calculate the resultant desired vector of motion and move accordingly. Furthermore, many times the main behaviour is supported by some other behaviours, modelled as *Probabilistic Behavioural Finite State Machine*. The robot has a state which governs the invoked controller (behaviour). Based on some conditions, the state changes, and the controller (behaviour) corresponding to the new state is loaded. The change may be stochastic.

20.2.1 Aggregation

In *aggregation,* the robots come close and collect in one place. The aggregation is achieved by making every robot attract to each other. Every robot emits a sound or light or any carrier wave to broadcast information about its existence. Every robot also has sensors that sense the other robots by the same carrier wave. The basic algorithm is a social potential field. A robot may not detect anything. In such a case, it moves in a random direction.

The robots may end up making many clusters shown in Fig. 20.2A. If there was a central system, the robots could have directed themselves towards the centre of mass of the entire robotic system. However, here the problem must be solved locally only and in a decentralized manner. The implementation is hence a Probabilistic Behavioural Finite State Machine with two states, random motion and social potential field motion. The transition from the social potential field to random motion takes place if there is no sensory percept or randomly with a small probability. The motion continues to the social potential field after some time, provided a sensory percept was available. This is shown in Fig. 20.2B.

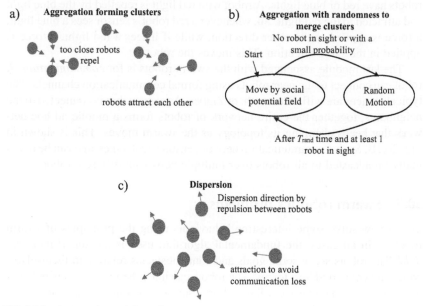

FIG. 20.2   Aggregation and dispersion behaviours. (A) Aggregation, (B) aggregation with a behavioural finite state machine controller for the robot, (C) dispersion.

Assume that the robots are stuck in multiple clusters. The probability that a robot leaves its cluster and reaches another cluster by random motion is nonzero. The joint probability of all robots in a cluster leaving their cluster and joining a single cluster is also nonzero. Subsequently, the robot may leave the single cluster, but there will be a single cluster to re-join. So, probabilistically, eventually, the robots converge at one cluster.

20.2.2 Dispersion

The *dispersion* behaviour is opposite to that of aggregation. In dispersion, the robots are expected to go as far apart from each other as possible. Now the robots repel each other. If the robots go far, they may end up losing communication and may end up being completely disconnected. Therefore, the robots only move when the communication signal strength from the robots is high enough. If the signal strength is too poor, the robot instead attempts to go towards the other robots. The attraction and repulsion are shown in Fig. 20.2C.

An interesting behaviour that combines both aggregation and dispersion is *clustering*. In clustering, robots of a kind aggregate in one place, while the entire cluster of robots of different kinds disperses off. This is implemented by using emitters of different frequencies. So, every type of robot has a different colour of light or a frequency of sound. The robot not only receives the signals but also identifies the frequency associated with the signal. This enables the robot to decide whether the force experienced will be attractive in nature or repulsive. Robots of different clusters can repel each other indefinitely, and therefore, some restriction may be made to limit the repulsion.

20.2.3 Chaining

The chaining behaviour enables the robots to go as a chain from one place to the other. The behaviour first forms a chain and thereafter the chain of robots travel from one place to the other. To form a chain, the first requirement is to have a leader. The leader is *elected*, which becomes the start of the chain. Assuming all robots are connected by a communication channel forming a distributed mobile ad-hoc network, each robot transmits its bid to the neighbouring robots along with its ID and broadcasts the best bid along with the corresponding robot ID that it has received so far. If a robot receives a better bid from some other robot, it updates its knowledge of the best bid. After sufficient time every robot knows the best bid. The robot with the best bid ID then declares itself as the leader and becomes the start of the chain. The bid here may be battery left, capability, or just a random number.

The leader now starts attracting other robots by transmitting signals. Let the rear of the leader robot emit light as an attraction. Every robot gets attracted to the leader, however, only one robot will reach the leader robot first. When the robot is close enough it attempts to form a connection. Let the chain be with a physical connection, which is made by the clamp of the robot reaching the leader being invoked. If the clamp is successful, the robot gets physically connected. The clamp is applied over the emitting light, which means that the light is no longer visible to any other robot. The joined robot detects that a clamp is successful, and this robot now plays the role of the last robot in the chain. Therefore, it starts emitting light. The clamping mechanism is shown in Fig. 20.3A. If the clamp is unsuccessful, the robot uses some randomization to give some other robot attempts to clamp. Otherwise, the robots may block each other. Randomization is also added throughout the motion for the same reason. Similarly, all robots add each other at the end of the chain. The mechanism is shown in Fig. 20.3B.

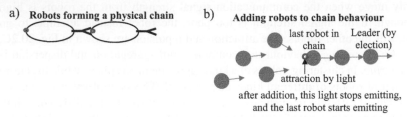

FIG. 20.3 Chaining behaviour. (A) Formation of a physical chain by clamping. (B) Robots attempting to be added at the end of the chain by attraction to the light.

Once the chain is formed, the motion of the leader is governed by the artificial potential field algorithm. All the robots follow the leader because of the physical clamp, which moves the entire chain. The robots have proximity sensors like sonar to keep themselves away from obstacles, which gives the repulsive force. The attraction force is due to the physical clamp.

It is not necessary to have a physical clamp to implement the chain. When a robot is close enough, the leader (or the last robot in the chain) is requested to register the robot as a follower. The leader (or the last robot in the chain) may simultaneously get multiple requests but responds to only one robot, which is attributed as the follower. The last robot in the chain now changes the colour (or frequency of the signal), stating that it is no longer the source of attraction, and the newly joined robot starts emitting the signal for attraction. The leader still moves by the artificial potential field method. However, each follower detects the leader with its sensors and feels an attractive force towards the leader.

20.2.4 Collective movement

The collective movement enables a swarm of robots to move collectively as a group. Imagine a swarm of robots that is already aggregated and arranged as a group. Every robot senses the surrounding robots. Normally, the robot would take an orientation which is the average orientation of all the surrounding robots (Reynolds, 1987). The magnetometers or other sensors can be used to approximately get the orientation of the robots that are broadcasted. If a robot is far away from the swarm, it needs to first join the swarm to exhibit the collective motion. If the distance of a robot to the nearest robot is larger than a threshold, it tries to move towards the other robot to join the swarm. Similarly, if the distance between two robots is less, it is a prospective threat of collision, and the robots exhibit a repulsive force to each other. The behaviour of the robot can be summarized in Fig. 20.4. A variant of the algorithm also asks the robot to collectively move in a direction (like towards a goal). In such a case, every robot feels an attractive force towards the goal in addition to the other forces.

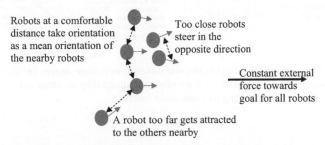

FIG. 20.4 Collective motion to goal behaviour. Robots attempts to orient in the same direction as the nearby robots. A robot is repelled by robots too close, and a robot far away attempts to join the group by attraction towards the nearest robots.

20.2.5 Olfaction

Olfaction is a problem in swarm robotics wherein the swarm of robots tries to search for a single or all sources of odour. The practical applications include sensing gas leaks, chemical leaks, finding dangerous mines, etc. The robots are expected to find the odour source and to either call for human help to resolve the issue or a robot may resolve the same itself.

Let us first consider that there is a single odour source. The robots have olfaction sensors that sense the intensity of the odour. An interesting implementation uses the PSO based algorithm for implementation. Assume every robot is connected using a mobile ad hoc network. The robots relay their positions and odour values. Now, every robot feels an attractive force towards the global best particle, its own local best, and has some momentum term. The probabilistic addition of the forces, like a PSO, gives a direction of motion, which the robot aims to get to.

In case there are multiple odour sources, the robots need to ensure that they, as a team, find all such sources. The robots continue to go by an attraction towards the odour source, which is possible since every robot has an odour sensor that causes an attractive force. However, once a robot reaches close to the odour, it indicates that the source has been found and other robots must find other sources. This happens with the robot emitting repulsive forces in the form of a signal. A random motion behaviour is also added in case a robot does not feel a strong enough odour. The working and modelling of the algorithm are shown in Fig. 20.5.

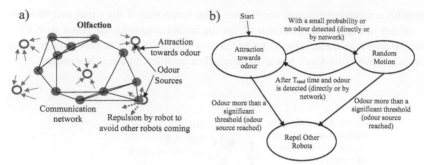

FIG. 20.5 Olfaction behaviour. (A) Robots collectively search odour sources. (B) Robot controller as a behavioural finite state machine. Each robot gets attracted by the odour and once reached, repels the other robot from locating the same odour source.

20.2.6 Shape formation

The problem of shape formation asks the robots to make interesting shapes. Here the swarm of robots arranges themselves to make shapes like 'ROBOTICS', make an emblem, etc. The problem is typically solved by making the robots feel the attraction towards the shape. Furthermore, there is a repulsion among robots. The repulsion ensures that the robots distribute themselves equally within the shape. When the robots are equidistant, the repulsive forces cancel each other out. The attraction is kept large to attract the robots from outside to inside, while the attraction inside the shape is constant.

Randomization enables the robots to escape the local optima. Imagine the task was to make the shape 'ROBOTICS' wherein every character is disconnected from the other. The robots at the letter 'R' may not feel a strong force to go to the letter 'O' and there may be too many robots at 'R'. Therefore, randomization is added to enable some robots to escape.

A related problem is that of collective transport. Here, the robots first surround the object of interest. Thereafter, the robots exert a force to move the object of interest. To surround the object, assume that the object emits attractive signals, while the robots repel each other to distribute themselves around the perimeter of the object. The carrying of the object happens by a small clipping done by every robot or any other pushing/pulling mechanism. The robots during carrying feel an attractive force towards the goal, along with repulsion from each other and the obstacles.

20.2.7 Robot sorting

Some applications may need to sort the robots in a specific order like their battery availability (Kumar et al., 2020; Zhou et al., 2016). The best and worst robots know their extreme positions and reach the same. The basic logic is that every robot tries to place itself in-between two robots with a higher and lower order. A naïve implementation can result in the robots scattered forming a tree instead of a line between the maxima and minima.

Therefore, a distributed Breadth-First Search is done from the first robot, which produces a Breadth-First Tree. This can be done in a distributed manner. Another search is done from the last robot, which produces another Breadth-First Tree. Preference is made to select the same parent-child edges that make the first robot rooted Breadth-First Tree. If a node lies in the path from the first robot to the last robot in the path produced by both trees, it is said to be on the main path connecting the first and the last robot. The other robots attempt to join the main path. Each robot moves towards its parent as per the Breadth-First Search Tree. In the navigation, if a robot locates two robots with a lower and higher-order value, the robot injects itself into the main path and tries to place itself geometrically at the centre of the two robots. Similarly, if a robot in the main path queries and finds that it is not in a sorted sequence, it detaches itself from the main path. It moves in the relevant direction (towards the smallest or largest order robot, depending upon the robot's order compared to the neighbours) to insert itself in a better place. The problem is described in Fig. 20.6.

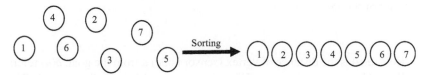

FIG. 20.6 Robot sorting behaviour.

20.2.8 Seeking goal of a robot

A traditional mechanism of robot motion planning uses routing algorithms from the mobile wireless ad hoc network literature. In this mechanism, several wireless sensor nodes are dropped in the entire arena in which the robot operates. The sensors become the vertices of a roadmap. If two sensors can communicate, they are said to be connected by an unweighted edge. The network is shown in Fig. 20.7. Each sensor tries to emit the distance from the goal, along with the location and ID of the sensor. The goal is a special sensor that emits a 0-value. Every other sensor listens to the values of the neighbouring sensors and emits a value 1 larger than the smallest valued neighbour. This is as per the graph search formula given by Eq. (20.1)

$$d(u) = \min_{v \in \delta(u)} d(v) + 1 \tag{20.1}$$

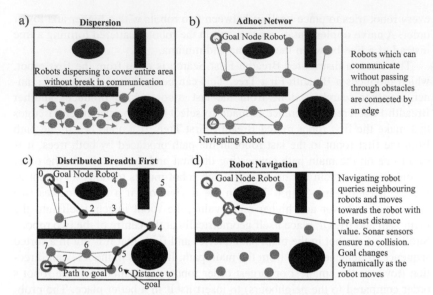

FIG. 20.7 Goal seeking behaviour. (A) Robots dispersing, (B) robots forming an ad-hoc network, (C) distributed breadth-first search (every robot in the network estimates its distance to goal based on neighbours and transmits the same), (D) robot navigation by moving towards the network robot with least reported distance.

Here, $d(u)$ is the distance of vertex (sensor node) u from the goal. $\delta(u)$ refers to all neighbours of the vertex u. The d value is emitted across the network. The algorithm is a *distributed Breadth-First Search*, wherein the global graph is not available, and each sensor node only knows its neighbours. The search is backward in nature since the distance wave propagates from the goal with a distance value of 0.

The robot travelling to the goal only listens to all the neighbouring sensor nodes. It walks towards the neighbouring sensor node with the smallest d value. As it travels along, it may get a better neighbouring sensor node with a smaller value, and the robot walks towards it. In this manner, the robot reaches the goal. *Randomization* is used in case the robot is not connected to any other sensor node or is getting very weak signals from the other sensor nodes. The approach provides a *globally shortest path* subjected to the resolution of the placement of the sensor nodes. The approach *adapts* to changing maps and changing the placement of the sensor nodes. The navigation of a robot is shown in Fig. 20.7.

The sensor nodes can be replaced by robotic swarms that use the dispersion technique to spread throughout the arena. The dispersion is adaptive to adding more robots or changing maps. The wandering robot is additionally given proximity sensors to avoid obstacles and other robots.

20.3 Neuro-evolution

In this section, we study the problem of learning a neural network using an evolutionary technique instead of the Back Propagation Algorithm (Kala et al., 2010, 2011). The Back Propagation Algorithm is an example of a *local search algorithm*, and more specifically, a *gradient-based local search algorithm*. In the gradient-based local search, the direction of the optima is known, which can be used to quickly get to the minima. Therefore, the search is even faster than local search algorithms like hill-climbing and simulated annealing, in which random attempts are made in pursuit of going to the local optima.

However, the local search algorithms are prone to get stuck at the local optima and do not guarantee to reach the global optima. The global search algorithms take much longer but have a greater chance of escaping the local optima and getting to the global optima. It is now possible to efficiently learn the neural networks using global search methods.

The other problem associated with the Back Propagation Algorithm is that it only trains the network weights (and biases) and not the network architecture (number of hidden layer neurons, the connectivity of the hidden layer neurons). A common assumption made is a *fully connected neural network*, wherein every input layer neuron is connected to every hidden layer neuron; however, this may not form the optimal structure. The motivation here is that learning of the weights, architecture, and any important parameter should happen automatically, not requiring human input. This gives rise to the domain of neuro-evolution, wherein the neural network is learned by using an evolutionary algorithm. The concepts are summarized as Box 20.1.

Box 20.1 Neuro-evolution concepts.

Back Propagation Algorithm
- Performs gradient-based local search only, prone to get stuck at local optima
- Does not learn the architecture (number of hidden layer neurons, which connection should be present)
- Fast learning

Neuro-evolution
- Use genetic algorithm to learn a neural network
- Leans architecture, weights and bias values
- Combine with Back Propagation Algorithm as a local search, making a memetic architecture

---

Continued

Box 20.1 Neuro-evolution concepts—cont'd

Neuro-evolution procedure

- **Representation**: Encode a neuron by noting weights between the input to hidden, and hidden to the output layer (and whether the connection is present). An individual is a variable number of such encoded neurons.
- **Crossover**: Adapted to work with a variable-length genotype.
- **Mutation**: Structural (add/delete connection/neuron) and Parametric (add noise to weights and biases)
- **Fitness function**: mean square error and regularization (model complexity, norm of weights, number of neurons, number of connections).
- **Incremental**: Start with a few neurons and incrementally let the neural networks add neurons.
- Diversity preservation needs a distance measure between encoded networks X and Y which have a variable length. The distance can be (i) Distance between output vectors produced by X and Y to many random input vectors, or (ii) Aligning neurons in Y to closest matching neurons in X. Distance between X and Y is the Euclidian distance between aligned neurons plus a penalty proportional to the difference in the number of neurons.

Evolutionary robotics

- Apply genetic algorithm to train a recurrent neural network that acts as a controller for a robot.
- Network takes sensor percept as input and returns actuator commands (linear and angular speeds) as output.
- Fitness function is the behaviour specific performance as observed by the sensors/environment.

20.3.1 Evolution of a fixed architecture neural network

First, let us solve the problem using the easiest means possible, which is to have a fully connected neural network with a fixed architecture. Imagine a neural network as shown in Fig. 20.8A. The first task is to convert the phenotypic neural network representation into a real vector-encoded genotypic representation, which can be used by the off the shelf optimization libraries. Since the neural network performance only depends upon the weights (and biases), encoding the weights (and biases) in any fixed order gives the genotype taking values typically between −1 and 1. The initial population is generated randomly.

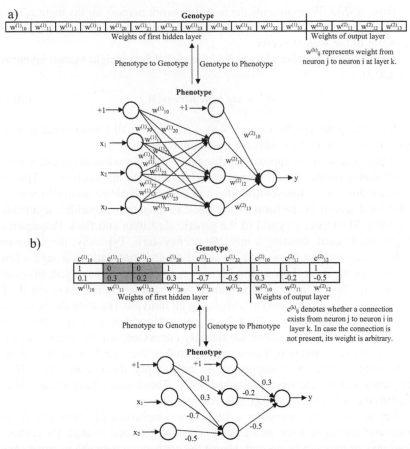

FIG. 20.8 Phenotype and genotype for the evolution of a neural network. (A) Fixed architecture, fully connected. All weights and biases are encoded. Genetic algorithm is used to optimize the genotype, thus giving the best neural network not prone to get stuck at local optima. 0 is used to represent bias. (B) Not fully connected.

For the fitness evaluation, the genotype (a string of real numbers) is converted into a phenotype (neural network with all weights set as per any library implementation). The training data $\{x_i\}$ is then passed to the neural network with weights W, which gives the output $\{y_i = ANN(x_i|W)\}$. The target for every input is known $\{t_i\}$. The fitness function is thus taken as the regularized mean square error, as was also adopted in the neural network, given by Eq. (20.2).

$$fitness(\theta) = \sum_i \|ANN(x_i|W) - t_i\|^2 + \lambda\|W\| \tag{20.2}$$

Here, $ANN(x_i|W)$ represents the neural network outputs for the input x_i for a given weight vector W. λ is the regularization parameter, and $||W||$ is the Euclidian norm of the weight vector.

The genetic algorithm aims to find the global best weight vector, given by Eq. (20.3).

$$W^* = \arg \min_W fitness(W) \tag{20.3}$$

Since the genotype W is encoded as a real vector, all conventional genetic operators can be used to yield the best weight vector.

A problem with the approach is that every fitness evaluation of every individual shall pass the entire training data set, making the approach slow. The fitness function is very time-consuming, and the same is called multiple times. A good local search in the form of the Back Propagation Algorithm is already available. Therefore, a hybrid of the genetic algorithm and Back Propagation Algorithm is used, creating a *memetic architecture*. Typically, the common architectures include (i) passing the initial random population through a few iterations of the Back Propagation Algorithm, which are then evolved; (ii) passing the final best solution after evolution through a few iterations of the Back Propagation Algorithm; and (iii) subjecting all individuals during the evolution through a local search using Back Propagation Algorithm in every iteration. In (ii), the local and global search are strongly connected, and the local search is applied at every iteration. However, this makes the search extremely time-consuming. Hence, it is suggestive to only apply local search at a little frequency and for clustered individuals only. These issues have already been discussed in Chapter 15 as memetic computing.

Let us make the approach a little more sophisticated. Now the fully-connected neural network architecture is discouraged and instead, the connections may or may not be present. Hence the connection between neurons j and i have genes c_{ij} denoting whether a connection exists ($c_{ij} = 1$) or not ($c_{ij} = 0$). The gene also encodes the actual weight w_{ij}, which is only applicable if the weight exists. The genes hence get double with no major change in the evolution. A sample individual is given in Fig. 20.8B. The fitness function now may also be used to regularize based on the number of connections, or Eq. (20.4)

$$fitness(W, C) = \sum_i ||ANN(x_i|W, C) - t_i||^2 + \lambda_1 ||C.W|| + \lambda_2 ||C|| \tag{20.4}$$

Here the individual consists of weight terms (W), linearized into a vector form for ease; and connection terms C), linearized into a vector form for ease. λ_1 and λ_2 are the regularization constants. The overall evolution can be naively done since the representation is a linear string. It may be noted that half of the individual (weights) has a real encoded representation, while the other half (connections) has a binary representation. Correspondingly the crossover and mutation operators may be different for the two sets of genes.

20.3.2 Evolution of a variable architecture neural network

A major parameter now is the maximum number of neurons. Even though evolution is free to limit complexity by breaking connections. However, setting a high number of neurons increases the genotype length, which exponentially increases the search space. A better representation is hence taken, which is to allow the representation techniques to have a variable number of neurons, which can be obtained by keeping a variable length of the genotype. Let us encode a neuron first. The number of hidden layers is fixed to one. A neuron of this hidden layer can be encoded by noting all the weights from the input layer to the neuron, and all weights from the neuron to the output layer, given in Fig. 20.9A. A neural network is a collection of such neurons, and therefore, the genes of all such neurons can be appended one after the other to make a neural network, shown in Fig. 20.9B.

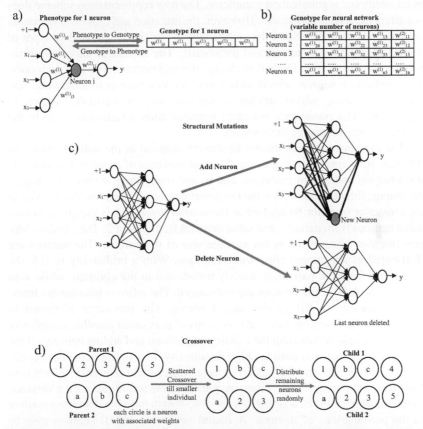

FIG. 20.9 Variable architecture neuro-evolution. (A) Single neuron encoding, (B) neural network as a variable number of such neurons. The population is an array of such individuals. (C) Structural mutation operators, (D) crossover with variable sized parents.

To leverage the possibility of having a variable number of neurons, the concept of *incremental complexification* is used. The same concept was also used in the evolution of robot trajectories. Evolution will start with the simplest neural networks with just 1 (or very less) neurons. As the generations increase, the neural networks will be allowed to take more complex forms by allowing them to add more neurons to increase their size, limited to $C_{max}(t)$, given by Eq. (20.5)

$$C_{max}(t) = C_{max}^{\infty}\left(1 - \exp\left(-\frac{t^2}{\sigma^2}\right)\right) \tag{20.5}$$

Here, C_{max}^{∞} is the maximum number of possible neurons and σ is an algorithm parameter. After the networks reach a reasonable size, there is no incentive to increase the size as the increased size is penalized by regularization without affecting the mean square error.

The fitness function is as used previously, with the number of neurons taken as an additional regularization parameter. The new representation scheme does not affect the selection operator. However, the mutation and crossover operators are now to be adjusted. Since the architecture is variable, there are two types of mutation operators, structural and parametric. The *structural mutation* operators are applied to an individual to change the architecture. The two such operators and *add a neuron*, which adds a new random neuron to a network; and *delete a neuron*, which deletes a neuron from a network, shown in Fig. 20.9C. The *parametric mutation* operator adds a Gaussian noise to the weight vectors to change their values.

The crossover operator cannot be directly applied as the individuals are of different sizes corresponding to the different numbers of neurons. Crossover is not a helpful operator in evolution, and Evolutionary Programming techniques discourage their use. However, the crossover operator can still be modelled. Let the crossover operator be applied at the neuron level, that is the children take some neurons from parent 1 and some neurons from parent 2. The children take size (number of neurons) as the average size of the parents. The neurons are distributed as a scattered crossover technique. With a probability of 0.5, the *i*th neuron from the parents is directly transferred to the children, while with a probability of 0.5, the neurons are exchanged. The leftover neurons are transferred randomly to fill up the vacant places. The procedure is shown in Fig. 20.9D. An *insert operator* adds neurons of maximum possible complexity by selecting individuals from the existing population and adding neurons. If the individual becomes too unfit for high complexity, it shall die in the next cycle.

Diversity preservation is added to save weak neural networks, which may later turn promising. Every diversity preservation technique uses a distance function, which must be adapted to a variable-length individual and be invariant to the permutations of neurons. A neural network with 10 neurons may be exactly the same as another neural network with the same 10 neurons with the neurons written in the reverse order. A direct Euclidean distance metric of the genotype string would however report a very large distance. To compute distances between the neural networks, first, the neurons are aligned in a greedy manner. Thereafter, the Euclidian distance is taken for the corresponding neurons. Say ith neuron of the first network is the closest to the jth neuron of the second network, and the Euclidean distance between this matching pair is computed, summated for all such pairs.

Another problem however is that the number of neurons may not match. A heavy penalty is given if a neuron cannot be paired because of the difference in architectures. Alternatively, a more formal way to find the distance is to subject both the networks to random data of a large size and get their output vector. The difference in the neural response (output vector) of the two networks is used as an indicator of the difference in the neural networks. If the networks are similar, their response will also be similar. However, this method can be extremely time-consuming. The overall working of the algorithm is shown in Algorithm 20.1.

Algorithm 20.1 Neuro-evolution.

1 generate a population of initial random neural networks with a variable number of neurons ;
2 apply a few iterations of back propagation algorithm as a local search ;
3 convert the networks into genotype ;
4 **while** *stopping criterion* **do**
5   convert into phenotype and calculate fitness (loss on training data, regularization) ;
6   apply diversity preservation aware selection ;
7   apply crossover operator ;
8   apply structural and parametric mutation operators ;
9   apply other genetic operators ;
10   make the next generation population ;
11   with some probability, apply back propagation algorithm to some clustered networks ;

12 get the best genotype in the population ;
13 apply a few iterations of back propagation algorithm ;
14 return best network

A popular architecture is the *Neuro-Evolution using Augmented Topologies* (NEAT, Stanley and Miikkulainen, 2002), which models every connection instead of the neuron. The system uses innovation number to trace the neuron connections as they get modified by genetic operators. A new connection is given a new number. However, when the same connection goes to a child gene or gets modified, it carries its innovation number, which enables the algorithm to detect its origin. This information is used to carry crossover between the same origin connections or genes only and to estimate the distance for diversity preservation called speciation.

20.3.3 Co-operative evolution of neural networks

Cooperative evolution enables the division of an entire problem into species or components, which results in a better performance of the evolution, given that the species are nearly independent of each other. In a neural network, every neuron attempts to extract good features, which can make a highly nonlinear problem linear. If multiple techniques independently find good neurons or features, each one of them may likely find the same best feature, which does not make the best neural network. In a neural network, the different neurons should be able to complement each other, and a neuron should solve for data items that another neuron is not able to do effectively. Therefore, the neurons need to evolve in cooperation with each other.

Cooperative evolution was discussed in Chapter 18. The technique broke an individual (or population) into multiple components, each carrying some genes. Each component was called a species, which together make an ecosystem or the complete individual. The species (or components) evolved in cooperation with each other. So, an individual of a specie was given a better fitness if it cooperated with the other species and enabled the ecosystem to have a higher fitness.

Previously each species had a different subproblem. However, here each species tries to evolve a solution to the same problem, which is to find good neurons. Therefore, instead of taking a population pool for each species, it is reasonable to take an overall single population pool of neurons that evolve through co-evolution. Taking one population for all species changes the fitness evaluation mechanism. Sampling is done to select k neurons, where k is the desired number of neurons in the hidden layer. For a variable architecture, the number of neurons may be varied randomly. The sampled neurons are used to make a complete neural network, which is assessed for fitness as previously (mean square error and regularization). This gives an indicative measure of the fitness of the neurons that were sampled. This is indicative only, as a participating neuron could be poor, however, the overall neural network gets a good fitness because the other neurons were good. Similarly, a neuron may be good but may get a poor fitness because of the absence of a neuron that solves for the complementary data items. These errors can be resolved by doing sampling multiple times. This allows every neuron to pair up with many neurons and get an indicative fitness, which is representative of its real value.

The fitness of the neuron is the average fitness of all the neural networks made by sampling, in which the neuron was a part. It may be stressed that a neuron gets a better fitness value not because of its merit alone, but because of its ability to cooperate with the other neurons and provide something on top of what the other neurons can solve. Once the fitness evaluation is done, the other genetic operators can be implemented in the usual manner. Diversity plays an important role. The overall process is shown in Fig. 20.10A and Algorithm 20.2.

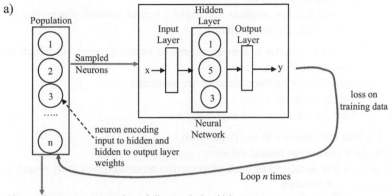

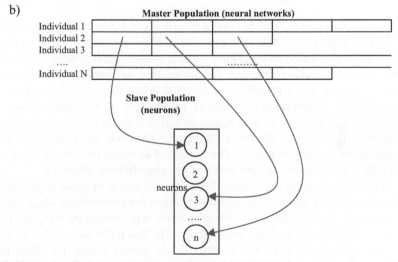

FIG. 20.10 Cooperative evolution of a neural network. (A) Cooperative evolution of neurons. The neurons cooperate to develop complementary features. Only the fitness evaluation process is explicitly shown. The final network comprises of the best few neurons. (B) Hierarchical cooperative evolution of a neural network. The master population has pointers to a variable number of slave population neurons forming a neural network (only 1 individual shown). The master finds the best neural network, while the slaves find the best neurons.

Algorithm 20.2 Cooperative neuro-evolution.

---

1   generate initial random neuron population ;
2   **while** *stopping criterion* **do**
3   **for** *some iterations* **do**
4   sample a few neurons and make a neural network phenotype ;
5   calculate the loss function ;
6   fitness of a neuron = average fitness of neural networks in which the neuron participated ;
7   apply diversity preservation aware selection ;
8   apply genetic operators ;
9   make the next generation population ;
10 get best few neurons in the population ;
11 make a phenotype neural network ;
12 apply a few iterations of back propagation algorithm as a local search ;
13 return best network

The co-evolution is extended into a master-slave architecture, wherein the master is a genetic algorithm that aims to select the best individuals from the slave species population and thus form the best overall solution, while the slaves evolve in a co-evolution paradigm. The slaves used the master for fitness evaluation. The master is a genetic algorithm that points to a variable number of slave neurons, representing the variable architecture of the neural network. The master does not have its gene sequence and only stores pointers to the slave population. The fitness evaluation is the loss function over the training data and regularization as used previously. The manner of handling crossover and mutation with a variable architecture problem has already been discussed.

The slave is a single population of neurons, instead of multiple populations representing every species. The genetic operators are conventional since every neuron is encoded as a sequence of real numbers representing the weights. The fitness function is obtained by three methods. The first is the *substitution*. In this method, k best neural networks as stored in the master genetic algorithm are taken, and a neuron is substituted with the neuron from the slave population whose fitness is to be calculated. The average fitness of the neural networks so obtained by substitution indicates the fitness of the slave neuron.

Further, to ensure that the neuron adds value, the k best neural networks from the master population are taken in which the neuron (whose fitness is being calculated) is a part. In these networks, the neuron is deleted and the difference that it makes in the fitness of these neural networks is taken. If the fitness dramatically deteriorates, it means that the neuron has a remarkably high value and is indispensable to the working of the neural network. On the contrary, if there is no change in fitness, it means that the neuron was not individually adding anything.

Furthermore, another indicative measure is the average fitness of the best k neural networks that the neuron (whose fitness is being calculated) is a part of. The overall fitness is a weighted addition of each of these measures. The overall architecture is given in Fig. 20.10B. Two popular architectures using the co-evolutionary concepts in neuro-evolution are *Symbiotic Adaptation in Neuro-Evolution* (SANE, Moriarty and Mikkulainen, 1996), which uses a single population of neurons for neuro-evolution; and *CovNet* (Garcia-Pedrajas et al., 2003), which uses a 2-layer master-slave architecture.

20.3.4   Multiobjective neuro-evolution

Another concept studied was the multiobjective evolutionary algorithms, wherein there are multiple objectives instead of just one. This is applicable for the current problem as well wherein the *model loss* (mean square error) over the training data set and *regularization* are two different objective functions that were so far combined into one by the application of a weighted addition. For multiobjective optimization, the fitness functions must ideally be anticorrelated. Improving one fitness function should deteriorate the other and vice versa.

The model loss over the training data and the regularization term are anticorrelated. To improve the model loss over the training data, the networks should have a higher model complexity and should be prone to overfit with a large variance, which happens by increasing the number of neurons and connections. On the contrary, if the number of neurons is limited in nature, the network is underfit and has a high bias, which increases the model loss.

The multiobjective optimization produces a Pareto front of individuals, each representing a neural network. The problem in multiobjective optimization is the need to do multicriterion decision-making to select one network from the Pareto front. However, multiple neural networks make an *ensemble*. In an ensemble, multiple neural networks instead of just one are taken. The same input is given to all the networks. Each network produces an output. The final output is the integral of the outputs of all the networks. This is shown in Fig. 20.11A. Making decision using ensembles or committees instead of a single expert is a common practice for humans in everyday life.

An ensemble is good if the individual members are all highly accurate, while the individual members are also as diverse as possible. These are conflicting requirements as the highly accurate systems usually agree with each other and therefore indicate a smaller diversity. This is another important aspect of decision-making using ensembles with humans. The corporate decisions typically involve diverse members with different backgrounds. Each member puts a unique perspective on the table that is used for decision-making. Similarly, academic evaluations may have evaluators from different departments or institutions for diversity.

There are numerous ways to increase diversity while making the individual neural networks still accurate. One of the primary measures is by having different members (neural networks) train on different data. Each network is exceptionally good in the part of data that it is trained on. Since the overall problem is the same, each network also has a fair accuracy on the other parts of the data not given to it for its training, the accuracy being somewhat higher than a random neural network. Such a network is called the weak learner, which represents that the accuracy of such a neural network is not competent to the one trained on the entire data (or the strong learner), while the drop in accuracy leads to a significant increase in diversity. Since the performance in several parts of the input data set not shown in training will be on a lower side, the networks will disagree implying diversity. A change in the architecture or hyperparameters also adds to the diversity, but the diversity is not very high using such measures.

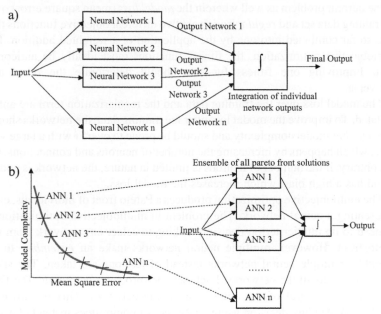

FIG. 20.11 Multiobjective neuro-evolution. (A) Ensemble architecture. The ensemble performs best when each network has a low individual loss and a high diversity, (contradictory objectives). (B) Multiobjective evolution using model complexity and mean square error as objectives. The objectives are contradictory (high complexity implies lower training error and vice versa). Pareto front neural networks have a high diversity and low individual loss, making a good ensemble.

For the problem of multiobjective neuro evolution, the individual neural networks in the Pareto front are known to have high accuracy; otherwise, they would have been dominated. At the same time, they also have a high diversity since the multiobjective optimization increases the diversity in the objective space. The networks vary from a low size/high error to a high size/low error, making a Pareto front. Therefore, an ensemble of these networks is ideal for problem-solving, and there is no need to select a single network. All the neural networks in the Pareto front are taken and used in the ensemble. The input is given to all the networks in the Pareto front. The outputs from all these networks are collected and integrated. This is shown in Fig. 20.11B.

20.4 Evolutionary robotics

The problem is now to evolve the controllers of the robots, which makes a reinforcement learning mechanism of learning the controllers for robots (Harvey et al., 1997; Nelson et al., 2009; Floreano and Nolfi, 2000). The controllers are modelled as a recurrent neural network. The control of any robot depends not only on the current percept but also on the state of the robot, which can be calculated by integrating information of all past percepts and actions. As an example, obstacle avoidance depends upon the relative speed of the obstacle, which cannot be estimated from the current percept of the proximity sensors; however, it can be estimated by using a difference of the consecutive readings. Similarly, during overtaking, an obstacle may not be currently visible, but information about it may be available from the past percept. Recurrent neural networks can model all higher-order derivatives, integrals, etc., and can therefore model any state variable automatically.

The neuro-evolutionary principles are used for the evolution of the controllers. The controllers take the sensory percept as input (e.g. reading of the light sensor, microphone, and camera) and produce the controller output (e.g. typically linear speed, angular speed, gripper action, and light display). So far, rules have been used for the navigation of the robot or robot swarm (represented by algorithms like the social potential field). In evolutionary robotics, the notion is to have the robot learn its behaviour.

An interesting approach used previously was Deep Q-learning as a reinforcement learning technique. Evolutionary robotics is also a reinforcement learning technique. The first difference is that in evolutionary robotics, a population of controllers is simultaneously evolved, while the Q-learning only updates a single controller. The difference is like the Back Propagation Algorithm and neuro-evolution. Furthermore, the Q-learning approach learns from a reward at every step. However, the evolutionary robotics techniques directly take a summated reward or the overall performance for a simulation run and learn the same.

20.4.1 Fitness evaluation

While the overall principles of neuro-evolution have already been taken, the step that deserves special attention is fitness function evaluation. First, the genotype is converted into a phenotype which is a complete recurrent neural network. The controller is loaded onto the robot. The robot's sensors take live input, which forms the input to the recurrent neural network. The output is taken and supplied to the actuators. This moves the robot and its state changes. The run is done iteratively for some time. As the robot runs, the overall metric of performance is calculated. The process is shown in Fig. 20.12A.

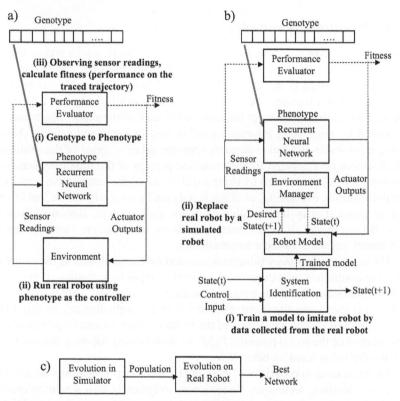

FIG. 20.12 Evolutionary robotics. (A) Every individual represents a recurrent neural network, whose fitness is measured based on the sensory percept (indicating safety distance from obstacles, curvature, velocity maintained, distance moved towards the goal, etc.) of a robot using the same neural network as a controller. (B) The robot is replaced by a simulated robot to get faster results. (C) After simulations, the final population is further leaned on the real robot.

Suppose the robot needs to avoid obstacles as it travels. The fitness function is given by Eqs (20.6), (20.7). The function can also be written as a product of terms, or Eq. (20.8):

$$fit(X) = \sum_{t=0}^{T} \left(\frac{v_L(t) + v_R(t)}{2} - w\, abs\left(\frac{v_R(t) - v_L(t)}{2R\omega_{\max}} \right) - \eta \min_i \phi(d_i(t)) \right)$$

$$(20.6)$$

$$\phi(d) = \exp\left(-\frac{d^2}{\sigma^2}\right) \tag{20.7}$$

$$fit(X) = \sum_{t=0}^{T}\left(\frac{v_L(t) + v_R(t)}{2}\right)\left(1 - abs\left(\frac{v_R(t) - v_L(t)}{2R\omega_{max}}\right)\right)^{w}\left(1 - \min_{i}\phi(d_i(t))\right)^{\eta} \tag{20.8}$$

Here the robot is executed for T steps, and the fitness is summed over the performance on all T steps. $v_L(t)$ is the speed of the left wheel at time t, $v_R(t)$ is the speed of the right wheel at time t, R is the radius of the robot, $d_i(t)$ is the distance of the ith proximity sensor, assuming k such sensors are in the circumference of the robot. $\varphi()$ is the penalty function that penalizes if the distance is very small. The function takes a value of nearly 0 if the distance is large enough and takes a higher value otherwise. ω_{max} is the maximum angular speed. The overall fitness function needs to be maximized. The first term of the function suggests the robot travel as much as possible (otherwise, even stopping is a valid behaviour). The second term penalizes making turns. Sharp turns do not denote good motion. The term has a corresponding weight attached, w. The third term penalizes small distance values. η is a constant denoting the weight given to the penalty term.

Let us add goal seeking to make a reactive controller that takes the robot to the goal while avoiding obstacles. The fitness function is given by Eq. (20.9):

$$fit(X) = \sum_{t=0}^{T}\left(\frac{v_L(t) + v_R(t)}{2} + \lambda\cos(\theta_G(t)) - w\,abs\left(\frac{v_R(t) - v_L(t)}{2R}\right)\right. \tag{20.9}$$
$$\left. - \min_{i}\phi(d_i(t))\right)$$

Here θ_G is the angle of the goal relative to the robot. It is assumed that either the goal is visible to the robot for the calculation of the angle or the position of the robot can be calculated by using encoders or other systems. The second term encourages the robot to orient as close as possible to the goal and avoids the robot from turning away from the goal. λ is the associated constant to the term.

As another example, suppose the robot needs to follow a boundary like patrolling a site. There is a boundary wall always on the left of the robot. The robot needs to always keep itself at a constant distance (e) from the same wall. To detect the boundary, there are proximity sensors. Now the fitness function is given by Eqs (20.10), (20.11):

$$fit(X) = \sum_{t=0}^{T}\left(\frac{v_L(t) + v_R(t)}{2} + \mu(\min_{i}d_i(t))\right) \tag{20.10}$$

$$\mu(d) = \eta \exp\left(-\frac{(d - e)^2}{\sigma^2} \right) \tag{20.11}$$

The sensor ray that hits the wall is the one that has the smallest distance reading, considering that the robot is reasonably close enough to the wall and all other obstacles or buildings are far away. The distance to wall is thus $\min_i d_i(t)$. The fitness function penalizes small and large distance values from the distance to be maintained.

20.4.2 Speeding up evolution

An issue with the fitness function discussed is that the evolution is done on the real robot. Imagine that a population is initialized, and the fitness needs to be computed for all individuals. The first individual's phenotype is loaded on the robot and the robot is moved to a new position and state, which gives the fitness function of the first individual. The second individual's phenotype is then loaded, which starts from where the first individual ends. Hence the different individuals do not start from the same initial state of the robot. Some good networks may start from a poor starting point of the robot and get penalized. It is good to design a fitness function with this constraint in mind.

The major problem of evolution is that it is extremely time-consuming. To get a good fitness measure, the threshold time (T) for which the robot is executed with one controller needs to be reasonably large. However, this means that the evaluation of all controllers from all generations can take too long. This is even painstaking when no human intelligence has been used, and the initial individuals are random causing mostly random motion of the robot. Hence the threshold time is kept *incremental*. During the initial generations, when the population is mostly random, even a small time is enough to judge the individuals. However, much later, the individuals become sophisticated and move the robot well. Now it is important to run the robot by different individuals for a longer time to better judge which is the best system. The threshold time parameter (T) increases with an increase in the number of generations.

A better acceleration is by first simulating the robot on simulation software. One can run thousands of robots in parallel in simulations. However, the best controller may not run well on the real robot. The simulation systems account for some sensor and actuator noise; however, the noise models may not be realistic and depend upon the operational area, faulty placement of the sensor, friction, battery performance, and wheel mounting error. The kinematics and dynamics of the robot can be different than modelled. The constants of mass, length, and dimensions may not be the same. Hence the final evolution happens in the real robot to adapt to such changes. However, the starting population is competitive due to the use of simulations.

In some cases, the kinematics and dynamics of the robot may not be known. This makes it impossible to simulate such robots. It may also be possible that the kinematics and dynamics are known; however, the parameters are not accurately known, or the modelling errors may be largely due to the operating conditions. In such cases, the *system identification* technique is used to learn the model of the robot to be used for simulation. The robot is operated by a human operator for a prolonged time in diverse random motions. Using highly accurate external systems, the configuration at every instant of time, the speeds at every instant of time, and the control inputs at every instant of time are noted down. This creates a big database of the system state $q(t)$, velocities $\dot{q}(t)$, and control input $u(t)$ at every point of time. The state at the next instant $q(t + \Delta t)$ is also known from the time-stamped data set. A supervised machine learning technique learns the kinematics $f()$ given by Eq. (20.12):

$$q(t + \Delta t) = f(q(t), u(t), \Delta t) \qquad (20.12)$$

This gives the kinematics of the robot, and the robot is simulated using such a model. It must be noted that the overall problem of evolutionary robotics is to learn the controller in a way that no training data is available. Simulation-based evolutionary robotics requires a kinematic function, which is here proposed to be a secondary learning problem. The problem of learning the controller and learning the kinematics are solved completely independently, the former primarily using Reinforcement Learning while the latter using supervised learning on a human-created data set of robot operation. The resultant evolution process is shown in Fig. 20.12B and C.

20.4.3 Evolution of multiple robots and swarms

Let us extend the solution to multiple robots. If the robots have communication between themselves, the problem can be modelled as making a central controller that controls each of the other robots. The central controller takes the sensor readings of all the robots and gives the control signal to all the robots. The resultant recurrent neural network is shown in Fig. 20.13A.

a) **Centralized Controller**

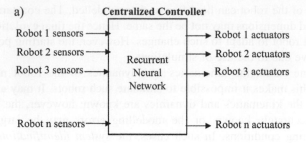

Robot 1 sensors ⟶
Robot 2 sensors ⟶
Robot 3 sensors ⟶

Recurrent Neural Network

⟶ Robot 1 actuators
⟶ Robot 2 actuators
⟶ Robot 3 actuators

Robot n sensors ⟶

⟶ Robot n actuators

b) **Decentralized controller without direct communication**

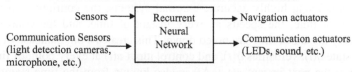

Sensors ⟶

Communication Sensors (light detection cameras, microphone, etc.)

Recurrent Neural Network

⟶ Navigation actuators

Communication actuators (LEDs, sound, etc.)

c) **Decentralized controller with direct communication**

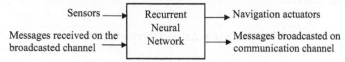

Sensors ⟶

Messages received on the broadcasted channel ⟶

Recurrent Neural Network

⟶ Navigation actuators

Messages broadcasted on communication channel

d) **Decentralized controller with direct communication**

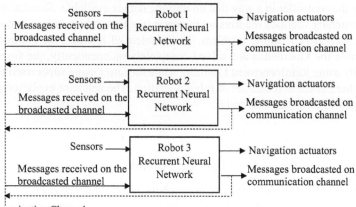

Sensors ⟶
Messages received on the broadcasted channel ⟶

Robot 1 Recurrent Neural Network

⟶ Navigation actuators

Messages broadcasted on communication channel

Sensors ⟶
Messages received on the broadcasted channel ⟶

Robot 2 Recurrent Neural Network

⟶ Navigation actuators

Messages broadcasted on communication channel

Sensors ⟶
Messages received on the broadcasted channel ⟶

Robot 3 Recurrent Neural Network

⟶ Navigation actuators

Messages broadcasted on communication channel

Communication Channel

FIG. 20.13 Evolutionary multirobotics. (A) Centralized architecture where one central controller controls all robots. (B) Using specialized sensors and actuators to communicate without direct communication. (C and D) Using a direct communication channel where every robot can broadcast messages to one/all other robots.

The centralized system is ideal, but learning it will require too much effort. The neural networks face the problem of the curse of dimensionality, where the magnitude of training data and time required for training exponentially increases with the increase in the number of input parameters. Furthermore, every robot sharing its every detail with every other robot is a cumbersome approach and not suited for swarms. Let us solve the problem in a decentralized manner. Each robot has its sensory percept and its own recurrent neural network controller. A robot can exchange information with the nearest robots using a communication channel. What precisely is transferred in this signal is not known or modelled; however, the machine learns to transfer good signals. The input is the sensor percept and the signals from the nearest robots, while the output is the actuator signals and the signal to be given to the other robots. This is shown in Fig. 20.13B.

In the case of multiple robots, each robot tries to solve the same problem, and therefore, a single controller phenotype is replicated for all robots for the fitness evaluation. Co-evolution is useful if the robots need distinct controllers for cases like heterogeneity. If the robots do not have a formal communication channel, they communicate by some communication media (light, sound, or other signals), which are typically binary, which is learnt. This is shown in Fig. 20.13C and D.

20.4.4 Evolution of the goal-seeking behaviour

The classic problem of robot motion planning, wherein a robot needs to go to a specific goal, is a hard problem from a learning perspective. The two behaviours of obstacle avoidance and goal-seeking are contradictory to each other. The goal-seeking asks the robot to steer towards the goal, which may mean going close to the obstacles; while obstacle avoidance asks the robot to steer away from the nearest obstacle, which may be away from the goal. Goal-seeking behaviour without obstacles and obstacle avoidance behaviours separately are relatively easy to learn. We learn these easy behaviours separately. We only focus on learning obstacle avoidance behaviour, since the goal-seeking is too simple. The output of this behaviour is then used in another network, which fuses the two behaviours to produce a final output. When learning the hybrid behaviour, the robot already has the capabilities of the elementary behaviours and only the fusion needs to be learned. The generic architecture is shown in Fig. 20.14A. The architecture-specific for goal-seeking with obstacles is shown in Fig. 20.14B.

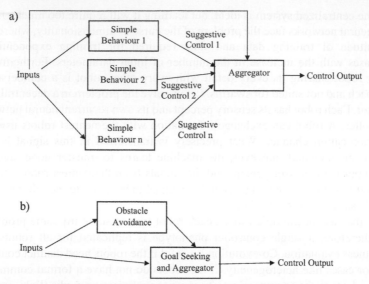

FIG. 20.14 Speeding up evolutionary robotics. (A) Individual simple behaviours are separately learnt, then the complex behaviour is learnt. (B) Goal seeking and avoiding obstacle behaviours are learnt separately. The network can be pretrained using available algorithms like artificial potential field to generate training data.

The network forms the policy network of Chapter 14 on Reinforcement Learning. For the reasons motivated in Chapter 14, *pretraining* the network using data generated from an established algorithm or demonstration data set is suggestive.

20.5 Simulations with ARGOS

The theoretical foundations of swarm motivate the use of a simulation platform to implement the concepts. A popular swarm simulation library is *ARGoS* (*Autonomous Robots Go Swarming*). The library (Pinciroli et al., 2012) implements everything as a plugin that can be easily loaded and unloaded into the simulation library. To speed up computation, the entities also define bounding boxes for fast collision checks. To make dynamics fast, every robot has its physics engine. Mobile robots are kinematically driven, while aerial robots have more intensive dynamics settings. Dynamics is also separated in different spaces, and two robots that do not interact are simulated separately by dynamics. The simulation is multithreaded with different threads for sensing and control, computing actions, and physics. The simulator in this mechanism has a discrete-time implementation.

Sensing and actions are local. The loop functions are used to take actions with global information, say to change the map based on some global condition, make a robot dead based on some global criterion, or log performance metric based on a global criterion. Similarly, the QT functions can be used to adapt the visual output, which can also be used to display information for debugging or visual aesthetic purposes. These functions are separately executed to not limit the performance with many robots.

The library has its roots in the swarmnoid project (Dorigo et al., 2004), which provides three unique robots. The first robot is the *foot bot*, which has all the features quoted for the swarms in this chapter. The footbots can also clamp to each other forming chains or other structures. The other type of bots is the *hand bots*, which incorporate manipulation units and can be individually or collectively be used for dextrous manipulation. There is limited sensing or mobility. The mobility can be provided by attaching hand bots to foot bots. The unique feature of these bots is that they can hold items to be used for vertical climbing like insects and even attach to the ceiling using magnets or suction. The last type is the *eye bots*, which are aerial vehicles with sensing. Their primary use is to provide aerial sensing to direct foot bots and hand bots. To get an extended vision, multiple such eye bots are used in coordination. These bots can also get attached to the ceiling.

Fig. 20.15 shows a swarm of robots that aggregate, besides being attracted to a light source. The robots are repelled by the obstacles. The robots use proximity

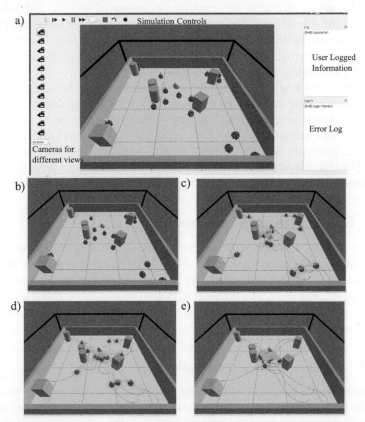

FIG. 20.15 Aggregation using the ARGoS simulator with an external light. (A) Simulator setup. (B–E) Robots aggregating.

sensors to detect obstacles, a light sensor to detect the light source, and an omnidirectional camera to detect LEDs from the other robots, while using the differential wheel drive to move and LEDs as the additional actuator. The standard deviation is printed as the simulation happens to get a metric assessment; whose logic is maintained as a loop function and written in the GUI through the logging utilities. Also, the trajectory trace is shown, which are drawn as QT functions and stored in the loop functions.

The ARGoS configuration file defines as an XML the overall settings like the number of threads, experiment duration, ticks (iterations) per second, random seed, control algorithms (with their actuators, sensors, and parameters), loop functions (along with parameters), arena (boundary, obstacles, robots, entitles like light, etc.), a physics engine for each entity, media (information exchange between robots and the media associated with the exchange), visualization library (cameras), QT functions, etc. The robots and obstacles can be distributed by distribution functions within a defined area.

The controller is defined as a derived class of a generic controller class, whose main logic is written in the step function. The step function adds forces from all sources, calculates the preferred direction of motion, and sets the speeds of the two wheels. The sensors, actuators, and parameters defined in the XML are available as variables initialized in the *init* function. The constructor, init, reset, destroy, and destructor methods can be used for initialization and cleanup, like initially turning on the LEDs.

Similarly, the loop function operations are written by extending the relevant class and overriding the *PostStep* function, also using the constructor, destructor, init, and reset as appropriate. The loop functions have access to space. The 'get space' and 'search entities by type' are called to get pointers to all foot-bots to extract positions. The QT utility overrides the *DrawInWorld* function to draw. The class maintains a reference to the loop function to get the trajectories calculated in loop functions.

20.6 Evolutionary mission planning

Another domain where evolution meets robotics is for complex mission planning, task planning, scheduling or the *operational research* problems. Typical solutions use classic AI techniques and temporal logic that have already been discussed in Chapter 19. The problem with both approaches is that the optimal solution search is exponential hence motivating the use of evolutionary techniques.

20.6.1 Mission planning with Boolean specifications

A classic problem is the Travelling Salesman Problem (TSP). In a robotic context, the problem is defined by asking one or multiple robots to do many tasks. Each task requires the robot to visit a mission site. The robot thus needs to visit all the mission sites or places. This can be solved by using evolutionary computation, which provides probabilistic optimality and thus applicable for very large problem sizes.

The TSP takes a cost matrix as input, denoting the cost of going from one mission site (place) to the other. The TSP solver then computes a sequence of visiting all mission sites which has the least travel cost. Imagine that the robot is being planned to visit mission sites in a home or office environment. Populating every entry of the cost matrix is a classic robot motion planning problem. Therefore, the entire map is converted into a roadmap using a Probabilistic Roadmap technique. The cost of traversal between every pair of mission site is computed. Thereafter the cost matrix is fed to an evolutionary TSP solver, which is a combinatorial optimization problem with very good heuristics.

In the case of multiple robots, the problem can be solved as a centralized optimization problem by an evolutionary problem. The solver simultaneously optimizes the assignment of mission sites to robots and the route of each robot. The genotype is a sequential list of mission sites (as a combinatorial optimization problem) along with the robot assigned to the site (as a parametric optimization problem). Some mission sites may require special skills by the robot that only a few robots have, which acts as a constraint (Kala, 2016).

Alternatively, *clustering* is a very powerful heuristic for the problem to be solved in a *decentralized* manner. For *n* robots, the mission sites to be visited are first clustered into *n* parts, each part to be assigned one robot. This breaks one multirobot TSP problem into *n* single robot TSP problem. Each TSP is solved independently (without considering the source as the robot assignment is not done), giving the sequence of visiting the mission sites. In the last step, the problem of assigning the robots to clusters is taken, which is another evolutionary combinatorial optimization problem. The genotype represents the cluster assigned to every robot. The fitness function takes the TSP solution of every cluster and, in a greedy manner, adds the source of the assigned robot to the route. The traversal of all robots is used for the fitness computation (sum of distance travelled by all robots, time taken by the longest working robot, etc.). Alternatively, another heuristic first clusters the mission sites and then uses a greedy strategy to assign the robots to clusters and then solves the TSP for every cluster independently by optimization.

The *Generalized TSP* problem allows visiting any one (or a few) of a group of sites while visiting all groups. This allows us to model situations wherein the robot can get an item asked (like water) from any of the 2–3 places where it is available. The optimization now simultaneously optimizes the order of visit of mission sites, and the choice within the mission site (like which place is used to get water), which are jointly represented in the genotype.

Consider the robot needs to get several items, along with water from either of the two places. But upon reaching the designated place, the robot finds that there is no availability of water. There is a solution to the problem by going to the other site that provides water. More generally, every mission site has a known probability of being usable for the mission. This is a policy planning problem that plans for every possibility of the availability of all mission sites. The representation of the policy is itself exponential in terms of the number of mission sites. The humans, on the contrary, make intuitive decisions by assuming all the mission sites with a probability less than a threshold will not be available and vice versa, where the threshold is the personal optimism of the person. In case all options related to a site have a value less than the threshold, only the one that has the maximum probability is assumed to be present. The humans act on such an intuition, till they find out a mission site which they assumed to be available is not available and re-optimize the complete plan with the new knowledge.

Many others tradeoff between knowledge of the probability as the additional cost, optimizing between the total length of the tour and the probability that the mission is satisfied (Bhardwaj and Kala, 2019). A solution is to use a genetic algorithm for optimization using the fitness function as the weighted addition of the total length of the tour (to be minimized) and the probability that the mission is solved (to be maximized), considering the probability of success of the individual operations based on their availability. The genetic algorithm optimizes the sequence of mission sites to be visited, also selecting multiple mission sites per group to increase probability (even if only one of them is needed as per the mission). In the above example, the individual may contain, apart from a permutation of the other places, either one or both sources of water anywhere as a part of the sequence. If water was at two nearby locations, both may be planned to be visited as that does not increase the cost much; if a site nearby has a small probability and a site far away has a large probability, it is wise to visit the nearby site, as getting water nearby may dramatically reduce the traversal cost; and if the water is available at two places far off from each other, only the one with the maximum probability may be tried. Planning to visit more places increasing the probability of success (since it is unlikely that water will not be found at multiple locations planned), while also increases the tour length. Note that the computation of the probability may also be exponential to the number of mission sites, which needs to be heuristically done. Upon reaching a site, the robot gets to know whether the resource is available or not, and in both cases, the robot is re-planned with the new knowledge.

More generalization may be added to have any Boolean expression solved as a TSP (Kala et al., 2018), wherein the sequence of mission sites visited must satisfy the given input Boolean expression. As an example, the mission specification asking the robot to either get coffee from any machine and cookies from the favourite shop or get pizza from either of the joints; and to compulsorily show a card design to any 2 of the 4 teammates; can be converted into a Boolean specification.

Any Boolean expression can be given as an infix expression and converted into a postfix expression. Verifying whether a sequence of places visited satisfies the Boolean expression can be done by solving the Boolean expression, replacing every mission site visited by true and mission sites not visited by false. The evolutionary solver then finds the sequence that satisfies the Boolean expression and has the smallest cost. In combinatorial optimization, any permutation of all mission sites is a feasible solution, which is a property that the solver previously used. However, now any general string represents the geno-type, and getting a single feasible sequence that satisfies the generic Boolean expression may be a hard problem. Attacking any heuristic in the structure of the problem is important. As an example, visualize the generic Boolean expression as a tree. In the case of an AND operator, both branches need to be visited. In the case of an OR operator, only one of the branches randomly needs to be visited. In the case of a NOT operator, the branch is not to be visited. Random traversals may be done in a random order to get a list of mission sites, whose random permutation may be a feasible solution. This mechanism is used to make the initial population pool.

In the case of multiple robots, using any Boolean expression, the decentra-lized heuristic of clustering can still be used (Dumka et al., 2018). Any negation-free Boolean expression can be converted into a disjunctive normal form (OR of ANDs, e.g. $a \wedge b \wedge c \vee d \wedge e \wedge f \vee g \wedge h$). The input is a general Boolean expression, and the robot must find the shortest path sequence such that the generic Boolean expression is satisfied (with every visited place taken as true and the unvisited place taken as false). The expression is converted into OR of ANDs. A choice needs to be made between any one of the ANDs ($a \wedge b \wedge c, d \wedge e \wedge f$, or $g \wedge h$), and hence they are solved (optimized) in parallel, and the one returning the best solution is taken. Each of these is a multirobot TSP that can be solved using the clustering heuristic in parallel. The mission sites are clustered, and every cluster is assigned to one robot to visit all the sites in the same cluster.

20.6.2 Mission planning with sequencing and Boolean specifications

The temporal constraints like visiting several mission sites after some other group can be equivalently modelled and optimized by evolutionary techniques. Let us model a simple temporal operator called *then* ($\rightarrow$), where $a \rightarrow b$ means that first solve for a and then solve for b, like make coffee and then add sugar. Again, the aim is not only to get the lowest cost sequence or plan, but also a sequence or plan that adheres to the Boolean and temporal constraints, which are handled by constraint handling in optimization.

In the easiest case, consider that the robot is asked to solve several Boolean expressions sequentially. Let a *subtask* be a specification consisting of Boolean expressions only, and a *task* be a sequence of subtasks that must be solved by the robot in the same order. Let the robot be given several such tasks, each of which may be done in parallel that constitutes a *mission* (Kala, 2020). As an example, multiple users ask the robot to (i) either get coffee from any machine and cook-ies from the favourite shop or get pizza from either of the joints; *then* return to the user to deliver the items; *then* take the letter; and *then* put the letter in the mailbox; (ii) show the design to any 4 of 5 teammates and *then* meet the head; and (iii) inspect all places for security. The total specification is a mission con-sisting of three tasks. Consider the first task. Here 'either get coffee … joints'; 'return to the user to deliver the items'; 'take the letter'; and 'put the letter in the mailbox' are the subtasks that should be sequentially solved. Let us break down the mission into tasks and subtasks, given by:

- [Task 1 or T_1]
  - [task 1 subtask 1 or s_{11}] either get coffee from any machine and cookies from the favourite shop or get pizza from either of the joints;
  - [task 1 subtask 2 or s_{12}] *then* return to the user to deliver the items;
  - [task 1 subtask 3 or s_{13}] *then* take the letter; and
  - [task 1 subtask 4 or s_{14}] *then* put the letter in the mailbox;
- [Task 2 or T_2]
  - [task 2 subtask 1 or s_{21}] show the design to any 4 of 5 teammates and
  - [task 2 subtask 2 or s_{22}] *then* meet the head;
- [Task 3 or T_3]
  - [task 3 subtask 1 or s_{31}] inspect all places for security.

The individual is represented as a sequence of mission sites visited and operations done by the robot. The verification of whether a string satisfies the mission is polynomial. Consider the tasks $\{T_i\}$, where task i has subtasks $[s_{ij}]$ to be solved sequentially. In the example, we state mission (M) is to solve all three tasks, or $M = T_1 \wedge T_2 \wedge T_3$. The first task consists of four subtasks sequentially, or $T_1 = s_{11} \rightarrow s_{12} \rightarrow s_{13} \rightarrow s_{14}$; while task 2 has 2 subtasks or $T_2 = s_{21} \rightarrow s_{22}$; and task 3 consists of 1 subtask, or and $T_3 = s_{31}$. For verification if a string x satisfies the task T_i, extract the smallest prefix substring $x(:,l_1)$ till index l_1 that satisfies the first subtask x_{i1}, or $l_1 = \arg \min_{x(:,l) \models s_{i1}} l$. The remaining string $x(l_1 + 1,:)$ starting from index $l_1 + 1$ is used for the verification of the rest of the tasks. Again, find the smallest prefix substring $x(l_1 + 1, l_2)$ that satisfies the subtask s_{i2}, or $l_2 = \arg \min_{x(l_1 + 1, l) \models s_{i2}} l$. Continue this till the last subtask. The subtask j is satisfied by string $x(l_{j-1} + 1, l_j)$ given by $l_j = \arg \min_{x(l_{j-1} + 1, l) \models s_{ij}} l$. If there are remaining subtasks that cannot be satisfied, the verification fails.

Verification of each subtask is a Boolean expression evaluation with all mission sites visited replaced with true and the rest with false. A string s satisfies the mission if it simultaneously satisfies all tasks $\{T_i\}$, or $s \models M$ iff $\forall_i s \models T_i$. The fitness is calculated as the total time of traversal by the robot. A penalty is added proportional to the number of tasks and subtasks not satisfied. In the above example, the genotype 'show design to team member F, inspect place P_2, get coffee from machine 1, show design to team member B, inspect place P_3, get a cookie from store 2, go to the user, take the letter from the user, go to the mailbox, show design to team member D, show design to the team member I, go to the head, inspect place P_1' is a valid sequence and the cost of going through the sequence is computed from the cost matrix. The evolution finds the lowest cost feasible string.

There is a trick in modelling the mission as a collection of tasks (all of which must be solved), while a task is a sequence of subtasks, subtasks being Boolean expressions. Each subtask making a task is nearly independent. Suppose the robot is asked to 'show the design to any 4 of 5 teammates and then meet the head'. The robot may solve the Boolean expression 'show the design to any 4 of 5 teammates' and 'meet the head' independently and execute the solutions in the same order. However, if a teammate sits very close to the head, it is expected to meet that teammate at the last. Hence there is some dependency between the subtasks. The solution to a task may hence be made by a *co-evolution* (Kala, 2020), with every subtask as a species. Similarly, each task is nearly independent and may be represented as different species. The number of species is the total number of sub-tasks across all tasks in the mission. The representation is shown in Fig. 20.16.

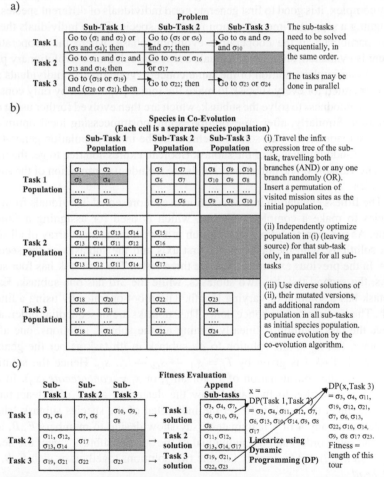

FIG. 20.16 Evolutionary mission planning. (A) Problem description format. (B) Different species in co-evolution. (C) Fitness evaluation. If x, y, z is a solution to task 1 and a, b is a solution to task 2, possible solution to (task 1 and task 2) are (all super-sequences) $[x\,y\,z\,a\,b], [x\,y\,a\,z\,b], [x\,y\,a\,b\,z], [x\,a\,y\,z\,b], [x\,a\,y\,b\,z], [x\,a\,b\,y\,z], [a\,x\,y\,z\,b], [a\,x\,y\,b\,z], [a\,x\,b\,y\,z], [a\,b\,x\,y\,z]$. Dynamic programming is used to compute the lowest cost super-sequence, for a given source and cost matrix.

Typicality here is that many of the subtasks may be independent of all others, while there may be many such subtasks. Therefore, an initial genetic algorithm finds, in parallel, diverse solutions of all subtasks of all tasks (without considering the prior subtasks or the ones ahead) and uses the same as a part of the initial population (Kala, 2019). Consider the subtask 'show the design to any 4 of 5 teammates'. Some solutions of this subtask are computed without considering the previous or the next subtasks in the sequence and added as a population. The other individuals are either random or mutated versions of these solutions. So far, in co-evolution, an individual of a species was awarded a good fitness based on the extent to which it cooperated with the other representatives of the other species to make an overall good solution. However, if the problem is very complex, it is good to first generate good individuals of different species by designing a separate fitness function for every specie. These individuals thereafter participate in the cooperative evolution phase, where the cooperative nature is evaluated, and the individuals adapt through the co-evolutionary process. As an implementation of the same concept, good diverse individuals are first generated for every subtask to make the initial population by only considering the goodness to solve the subtask, which are then evolved further using co-evolution. Similarly, after co-evolution, some postprocessing local optimizations are made on the best solution achieved. The initial population generation uses the random traversal of the subtask Boolean expression tree to get the mission site variables. The initial individuals are a random permutation of the same variables.

The fitness evaluation in co-evolution uses represented individuals from all species to make a complete solution, which is used for assigning a fitness value. As per representation, the solution is a two-dimensional array of all subtask solutions, with the tasks as the first axis and the subtasks as the second axis. In the previous example, there are three tasks. The first task has four subtasks, the second task has two subtasks, while the 3rd has one subtask. Each subtask has a solution of varying length. The robot operates by using a linear path. The linearization is hence done. The subtasks are in sequential order, and hence the task solution is linearly joining all the subtask solutions, one after the other. Let x_{ij} be the solution to jth subtask in ith task as per the genetic individual. Task i is given by $T_i = s_{i1} \rightarrow s_{i2} \rightarrow \dots s_{in}$. Hence the solution of task i is the concatenation of x_{ij} for all j, or $x_i = concatenate_j(x_{ij})$. In the example, a solution to the task 'show the design to any 4 of 5 teammates and *then* meet the head' is to get the individual of the species subtask 'show the design to any 4 of 5 teammates' (say show design to members F, B, and D in the same order) and the individual of the species subtask 'meet the head' (say visit *head*) and to linearly join the two solutions together, or visit F, B, D, and *head*.

The different tasks can be solved in parallel by the robot, much like the humans simultaneously work for multiple projects in parallel, doing the steps of the individual projects interchangeably. In the problem, there are 3 tasks that the robot is to solve interchangeably. If the specification has no NOT operator, the fusion of the solutions of all tasks into a solution of a mission can be done by using Dynamic Programming (Kala, 2018). The problem of combining solutions to two tasks into one solution that solves both tasks is like finding the lowest cost super-sequence that has both task solutions as subsequence, with the transition costs given by the cost matrix. Say task 2 requires the robot to visit F, B, D, and *head* in the same order; while task 3 requires the robot to visit places P_2, P_1, and P_3 in the same order (for inspection). Any super-sequence that has subsequences '$F, B, D, head$' and 'P_2, P_1, P_3' is a valid solution to tasks 2 and 3 combined. Say the Dynamic Programming finds the combined solution as $F, P_2, B, P_1, P_3 D$, *head*, which has the lowest cost out of all possible super-sequences.

All tasks solution may be fused, two at a time. Dynamic Programming now linearizes the individuals for the fitness calculation. Say the solution of task i be x_i. The solution to the mission $M = T_1 \wedge T_2 \wedge ...T_m$ is given by fusing pairs of task solutions in a prespecified order (that is optimized, say o) by using Dynamic Programming. o_i is the index of the task to be fused at the ith step, or x_{o_i} is the ith string to be fused by using Dynamic Programming. Mission solution is given by $x = x_{o_1} +_{DP} x_{o_2} +_{DP} ...+_{DP} x_{om}$, where $x_{o_1} +_{DP} x_{o_2}$ uses Dynamic Programming to fuse strings x_{o_1} and x_{o_2} into the lowest cost super-sequence. x_{o_i} is the solution to the task to be fused at the ith step as per a prespecified/optimized order o.

The fitness evaluation first checks for the feasibility of the subtask solutions ($x_{ij} \vDash s_{ij}$), then combines the subtask solutions into task solutions by linearly appending them ($x_i = concatenate_j(x_{ij})$), and then uses task-wise Dynamic Programming to get a linear list of mission sites visited by the robot ($x = x_{o_1} +_{DP} x_{o_2} +_{DP} ...+_{DP} x_{om}$), which is used to compute the tour length or the fitness value.

20.6.3 Mission planning with generic temporal specifications

Let us generalize to any possible expression using the Boolean operators (AND, OR, NOT) and the temporal operator (THEN). Not taking other possible temporal operators is to make a verification system with a polynomial complexity. A generic instruction given to the robot is shown in Fig. 20.17 and should be read as 'Go to both σ_1 and σ_2; then either σ_3 and σ_4, or σ_5 and σ_6. Thereafter go to either σ_7 and σ_8, or σ_9 and σ_{10}, and compulsorily visit σ_{11} followed by σ_{12}' (Kala, 2023).

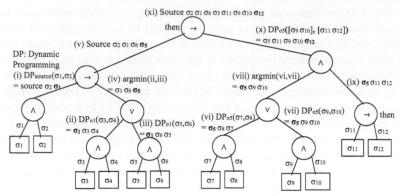

FIG. 20.17 Generation of a greedy solution for mission planning using expression trees. Perform a postorder expression tree traversal. $\wedge$: Merge by dynamic programming, $\vee$: take best of both solutions, then ($\rightarrow$): append the solutions. $DP_X(y,z)$ denotes finding the lowest cost super-sequence of y and z using source x. Each node also supplies a source to the next node, shown as bold (unchanged for all nodes, except *then*). The cost matrix is not shown (computed as a classic robot motion planning problem). The solution is not optimal.

Let us use evolutionary computation to find an optimized string that satisfies the specification. First, let us make a greedy solution to the problem. The tree contains AND, OR, and THEN operators. In the case of an OR operator, only one of the branches needs to be visited, while the choice of the branch to be visited needs to be optimized. A greedy heuristic is to solve both branches independently but use the one that gives a smaller cost (irrespective of how that affects the solution to the subsequent specifications). In the case of an AND operator, both branches may give their solutions, which may be integrated using a Dynamic Programming approach. In the case of a THEN operator, the solution of the first branch may be appended before the solution of the second branch. A greedy solution is therefore a postorder traversal of the tree. Every operator recursively solves the children and gets the children solutions. Depending upon the choice of the operator, the solution of the parent is made using the knowledge of the solution of the children. As a base case, the leaves ask the robot to visit one mission site, and the solution is the mission site alone (placed after the current source in the postorder tree traversal).

However, such a greedy solution may not be optimal. The greedy heuristic of taking the smaller cost solution out of the branches is replaced with optimizing the choice of branches for all OR nodes. Similarly, the Dynamic Programming heuristic does not consider the effect on the subsequent specifications, and therefore, it is replaced with optimizing the order of merging the solutions of both branches. The AND node gets two linear sequences consisting of only the leaf nodes (mission sites). Hence every mission site is associated with a priority (that is reused by every AND node). The AND node first computes the solution of all branches and then merges the solutions of both branches in the order of their priority. The algorithm is similar to merge-sort where the merge step gets two sorted sequences and merges them as per the item value. The AND operator gets two sequences in which the operations need to be performed, and the sequences are merged into one sequence as per the optimized priority value.

The tree now represents the genotype as shown in Fig. 20.18. Even though the genotype is a tree, the optimization is parametric that only consists of the priorities of the leaves and the choice of all OR nodes. The architecture of the tree shall always remain constant, and optimization is for a few parameters of the tree nodes. The greedy strategy is used to give the 1st individual in the population pool a biased start. The greedy solution sequence is converted into a genotype by noting which branch of the OR node was active and setting the index of every mission site in the sequence as its preference.

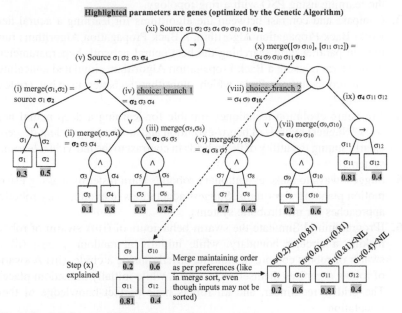

FIG. 20.18 Mission planning using expression trees. Individual representation and fitness evaluation for genetic algorithm are shown. The greedy solution is used as one of the individuals in the genetic algorithm for faster results.

The genetic algorithm thereafter computes the optimized genotype tree, which is linearized to get a solution. Since only the OR choices and priorities of the leaf nodes are being optimized, the conventional genetic operators of mutation and crossover are used. For fitness evaluation, the genotype is written on top of the mission expression tree, noting the choices of the OR operators and leaf preferences. A postorder traversal of the tree is then performed. For the OR nodes, only the chosen branch as per the genetic individual is chosen. For the AND node, the children sequences are merged as per the priority of the leaves in the genetic individual. In the case of the THEN operator, the solutions are appended. The linearized solution of the root node is used to calculate the cost.

Questions

1. Compare and contrast between the algorithms: Q-learning, deep Q-learning, grid search, evolutionary robotics, and social potential field for the motion of a robot to goal in static and dynamic environments. State which algorithm is better in what context and why?

2. Using neuro-evolution, learn the XOR function for a population of 10 random individuals for a fixed architecture neural network with two neurons in the hidden layer. Show the calculations for two generations. Also, show the learning using PSO with a ring topology.

3. Compare and contrast between the algorithms for learning a neural network: Back Propagation Algorithm, n Back Propagation Algorithms running in parallel, grid searching the best neural network hyperparameters while learning using a Back Propagation Algorithm, simulated annealing, and neuro-evolution. State which algorithm is better in what context and why?

4. Are neuro-evolution techniques suitable for learning a deep neural network? Suggest hybrids of neuro-evolution and stochastic gradient descent, clearly stating the utility of the hybrids in contrast to the individual learning systems.

5. Suggest ways to make evolutionary robotics learn faster by making use of motion planning algorithms. Suggest adaptations to evolutionary robotics approaches for multirobotic systems.

6. [Programming] Simulate the swarm behaviours of (i) A swarm of robots making a circular boundary, while initially at random places; (ii) A swarm of robots compresses and lie strictly inside a circle, (iii) A swarm of robots organizes themselves in a grid, while initially at random places. The grid is rectilinear, and all robots have perfect knowledge of their orientation.

7. [Programming] A swarm has four types of robots that transmit lights of different colours. Simulate the self-organizing behaviour of the robots, where robots of different colours gather in one place, while the clusters of different robots are far apart.

8. [Programming] In the classic game of pursuit-evasion, there is a pursuer who tries to catch the evader; while there is an evader who tries to escape the pursuer. The game ends when the pursuer catches the evader. (i) Simulate one pursuer and one evader using swam robots as the agents. Describe the sensors used, actuators used, and the navigation rule set. Assume an obstacle-free environment. Describe the trajectory of the

pursuer and evader in such a case, assuming the pursuer has a higher maximum speed in contrast to the evader. (ii) Assume that there are obstacles, and the evader does not move. The evader is hidden somewhere within the obstacles. Describe the algorithm if the pursuer knows the location of the evader, assuming a complex obstacle framework. The map is fully known. (iii) Assume that the pursuer does not know the location of the evader and will only know the location once the pursuer sees the evader. Describe the change in the algorithm. The map is however fully known. (iv) In question (ii), assume that the evader can move. Both pursuer and evader always know all positions. Describe the search-based algorithm used for both pursuer and evader. (v) Extend the solution in (iii) to incorporate multiple robots as pursuers and evaders. Any pursuer can catch any evader. All evaders must be caught. (vi) Using evolutionary robotics, learn the best behaviours for both pursuer and evader.

9. [Programming] A swarm of robots is randomly placed. Make rules such that the swarm of robots make a sorted chain (virtual links) in a decentralized manner. Simulate the chain to reach a goal, avoiding obstacles. Assume one of the robots has its battery phased out and therefore the robot cannot move. The rest of the chain must abandon this robot and continue the motion. Simulate the chain of robots splitting into two chains using a decentralized approach.

10. [Programming] (i) Simulate the behaviour where swarm patrols and guards a celebrity, surrounding him/her while maintaining a distance R from the celebrity. Only local communication may be assumed. (ii) Simulate the behaviour where the robots hover around the celebrity. The robots must move as celebrity moves. (iii) Assume that there are multiple celebrities. Simulate the behaviour when the robots divide themselves among celebrities and move. The celebrities may come close to each other and pass each other with proximity; however, the swarm of robots must always maintain a distance of R from all celebrities.

11. [Programming] (i) Solve the multirobot olfaction problem for the cases of a small number of robots and a swarm of a very large number of robots. (ii) For the same problem, consider that the location of the odour sources is available; however, the robots must reach the odour sources for repair. Solve the same problem for the cases of a small number of robots as a centralized solution as well as a swarm of robots. For the former consider the cases when the number of odour sources is smaller/larger than the number of robots. (iii) For the case of a small number of robots, also show the solution for (ii) if the robots operate at a preknown map with obstacles.

12. [Programming] Using evolutionary robotics, learn the behaviour where groups of robots go from one place to another, while avoiding the other groups and static obstacles.

References

Brambilla, M., Ferrante, E., Birattari, M., Dorigo, M., 2013. Swarm robotics: a review from the swarm engineering perspective. Swarm Intell. 7 (1), 1–41.

Bhardwaj, A., Kala, R., 2019. Sensor based evolutionary mission planning. In: 2019 IEEE Congress on Evolutionary Computation. IEEE, Wellington, New Zealand, pp. 2697–2704.

Dorigo, M., Tuci, E., Groß, R., Trianni, V., Labella, T.H., Nouyan, S., Ampatzis, C., Deneubourg, J.L., Baldassarre, G., Nolfi, S., Mondada, F., Floreano, D., Gambardella, L.M., 2004. The swarm-bots project. In: International Workshop on Swarm Robotics. Springer, Berlin, Heidelberg, pp. 31–44.

Dumka, S., Maheshwari, S., Kala, R., 2018. Decentralized multi-robot mission planning using evolutionary computation. In: Proceedings of the 2018 IEEE Congress on Evolutionary Computation. IEEE, Rio de Janeiro, Brazil, pp. 662–669.

Floreano, D., Nolfi, S., 2000. Evolutionary Robotics: The Biology, Intelligence, and Technology of Self-Organizing Machines. MIT Press, Cambridge, MA.

Garcia-Pedrajas, N., Hervas-Martinez, C., Munoz-Perez, J., 2003. COVNET: a cooperative coevolutionary model for evolving artificial neural networks. IEEE Trans. Neural Netw. 14 (3), 575–596.

Harvey, I., Husbands, P., Cliff, D., Thompson, A., Jakobi, N., 1997. Evolutionary robotics: the Sussex approach. Robot. Auton. Syst. 20 (2–4), 205–224.

Kala, R., 2023. Mission planning on preference-based expression trees using heuristics-assisted evolutionary computation. Appl. Soft Comput. 136, 110090.

Kala, R., Shukla, A., Tiwari, R., 2010. A novel approach to classificatory problem using grammatical evolution based hybrid algorithm. Int. J. Comput. Appl. 1 (28), 61–68.

Kala, R., Janghel, R.R., Tiwari, R., Shukla, A., 2011. Diagnosis of breast cancer by modular evolutionary neural networks. Int. J. Biomed. Eng. Technol. 7 (2), 194–211.

Kala, R., 2016. Sampling based mission planning for multiple robots. In: Proceedings of the 2016 IEEE Congress on Evolutionary Computation. IEEE, Vancouver, BC, Canada, pp. 662–669.

Kala, R., Khan, A., Diksha, D., Shelly, S., Sinha, S., 2018. Evolutionary mission planning. In: Proceedings of the 2018 IEEE Congress on Evolutionary Computation. IEEE, Rio de Janeiro, Brazil, pp. 1–8.

Kala, R., 2018. Dynamic programming accelerated evolutionary planning for constrained robotic missions. In: Proceedings of the IEEE Conference on Simulation, Modelling and Programming for Autonomous Robots. IEEE, Brisbane, Australia, pp. 81–86.

Kala, R., 2019. Evolutionary planning for multi-user multi-task missions. In: 2019 IEEE Congress on Evolutionary Computation. IEEE, Wellington, New Zealand, pp. 2689–2696.

Kala, R., 2020. Robot mission planning using co-evolutionary optimization. Robotica 38 (3), 512–530.

Kumar, U., Banerjee, A., Kala, R., 2020. Collision avoiding decentralized sorting of robotic swarm. Appl. Intell. 50 (4), 1316–1326.

Moriarty, D.E., Mikkulainen, R., 1996. Efficient reinforcement learning through symbiotic evolution. Mach. Learn. 22 (1-3), 11–32.

Navarro, I., Matía, F., 2013. An Introduction to Swarm Robotics. ISRN Robotics 2013: 608164.

Nelson, A.L., Gregory, J.B., Doitsidis, L., 2009. Fitness functions in evolutionary robotics: a survey and analysis. Robot. Auton. Syst. 57 (4), 345–370.

Pinciroli, C., Trianni, V., O'Grady, R., Pini, G., Brutschy, A., Brambilla, M., Mathews, N., Ferrante, E., Di Caro, G., Ducatelle, F., Birattari, M., Gambardella, L.M., Dorigo, M., 2012. ARGoS: a modular, parallel, multi-engine simulator for multi-robot systems. Swarm Intell. 6 (4), 271–295.

Reynolds, C., 1987. Flocks, herds, and schools: a distributed behavioural model. Comput. Graph. 21 (4), 25–34.

Stanley, K.O., Miikkulainen, R., 2002. Evolving neural networks through augmenting topologies. Evol. Comput. 10 (2), 99–127.

Tan, Y., Zheng, Z.Y., 2013. Research advance in swarm robotics. Defence Technol. 9 (1), 18–39.

Zhou, Y., Goldman, R., McLurkin, J., 2016. An asymmetric distributed method for sorting a robot swarm. IEEE Robot. Automat. Lett. 2 (1), 261–268.

Chapter 21

Simulation systems and case studies

21.1 Introduction

Robotics, unlike other domains, requires expensive robots whose behaviours are programmed and learned. This causes a fundamental difference with the other domains. First, a research group needs expensive robots for full-stack robotics research and development, which may not be affordable. There may be too many users with a single robot, necessitating sharing and long unavailability. Imagine making software with only a single computer system across your entire working lab. Second, it is impossible to test for adverse settings and for the robustness of the algorithms using the real robot, as the chances of damaging the robot are high. Laboratories may not even have all possible cases for the operation of the robot, and researchers may often program using heuristics that work in their working environment or laboratory only. In some cases, for environments like roads, hills, and underwater; it may be expensive to program the robot directly for the environment with much higher risks of damage. Third, many robots learn their behaviour by experiences. Some others rely on drawing examples from the real environment for performance evaluation and adaptation. For all such cases, relying on a single robot can make learning and adaptation slow.

It is therefore desired that the initial research and development be done on the simulated environments. The simulated robots do not cost (except for software license cost for some software), can be repaired from damage by restarting the simulation, allow the user to design complex to simple scenarios, and millions of robots can be simulated in parallel under random or adversarial setting for faster learning and adaptation. A problem is that the simulators may not be able to simulate the real behaviour, and it is possible to have excellent runs on the simulations with the robot still failing in real life. Therefore, a current development pattern is to do research and development on simulations and then adapt the models to a real robot.

The simulations are broadly classified into three categories. The *microscopic simulation systems* see every entity separately and model their behaviour

Autonomous Mobile Robots. https://doi.org/10.1016/B978-0-443-18908-1.00017-0

1001

and interaction. The simulations discussed so far in the book were all microscopic, wherein the individual robots and their coordination were modelled. Specifically, they were an *agent-based microscopic simulation*, where each entity was modelled as an agent that reads from sensors and acts with actuators. The simulation systems explicitly model the pose, velocity, and other parameters of all the individual agents. The microscopic simulations typically have a complexity of $O(n \log n)$, where the logarithmic term is typically attributed to the proximity queries. Such systems can therefore be slow if there are too many simulated entities.

The *macroscopic simulation systems* model a distribution of entities overall rather than simulating every entity directly. The practical examples are simulation of flowing water, blowing air, and diffusion. It is hard to model the motion and interaction of each water and air molecule. Instead, physics formulas and differential equations are applied at a macroscopic level. The simulation systems typically model for density functions, average speeds, average flow, and such macroscopic parameters. Interestingly, the exact number of entities may not be explicitly modelled though may be derived from the density functions. The problem with the macroscopic simulation is that it becomes very difficult to model atypical behaviours caused by a single individual because it is unmodelled, like a traffic accident, travelling on the wrong side, and single-agent overtaking.

The *mesoscopic simulation systems* attempt to combine the pros and cons of both types of simulation systems by modelling the agents as a group, while also allowing them to play with some specific agents. They combine the concepts of both simulation types.

21.2 General simulation framework

Imagine that you need to simulate the behaviour of a robot doing some operations in a physical environment. This section makes a walkthrough of the different programming and design modules involved. The emphasis is on the agent-based microscopic simulation only. The concepts are summarized in Box 21.1.

Box 21.1 Simulations

Simulation advantages:
- Robots do not cost
- No loss from robot damage
- User can design complex scenarios not found in real life
- Numerous robots can be simulated in parallel
- Faster data generation for learning
- **Cons:** Real robots may behave differently than the simulated ones, requiring algorithmic adaptations

---

Box 21.1 Simulations—cont'd

Simulation types:
- **Microscopic:** Simulate each entity separately that behaves based on its neighbours
- **Macroscopic:** Simulate distributions of entities like spatial/temporal change of density, average speed, and flow (e.g. simulation of diffusion)
- **Mesoscopic:** Combination of microscopic and macroscopic

Simulation elements:
- **Object hierarchies:** A hierarchical relation between objects showing which object is 'attached' to which other, such that a motion of the parent object also moves the child. The spatial relation between two objects in a hierarchy may (or may not) be controlled by a motor that can be programmed.
- Simulated proximity sensors use ray-tracing algorithms and vision sensors project elements on a virtual camera. Sensors and actuators also simulate specified noises.
- **Kinematic chain:** A specific object hierarchy of links of the manipulator facilitating the computation of the forward kinematics (compute end-effector pose given joint angles) or inverse kinematics (compute joint-angles given desired end-effector pose).
- **Graphics library:** Project 3D entities into a 2D image representing a user-controlled virtual camera. Popularly, OpenGL.
- **Dynamics/physics engine:** Compute the motion of each entity based on entity properties, forces, and torques. Popularly, OpenDE, Bullet, Vortex, etc.
- **Collision/proximity checking:** Check if the 3D entity descriptions have an intersection (collision). If not, return the closest proximity. Used by planning algorithms on a virtual (ghost) robot. Popularly, FCL.
- **Transformation:** Library that stores a transformation between every entity (robot, object, sensor, actuator). Popularly, TF.
- **Mapping library:** Maintains a 2D or 3D map of the robot's world, which may be made by the onboard robot sensors. Popularly, GMapping.
- **Planning library:** Answers to motion planning queries. Popularly, OMPL.
- **Control library:** Moves the robot based on a reference trajectory.

Writing simulation programs
- Every entity is handled by a program. All such programs run in parallel.
- Inputs/outputs are taken from/given to networking streams, shared memory, or simulated system calls, facilitating the sharing of information between programs including simulated sensors and actuators by the simulator.
- A program typically has an initialization function (called once at the start), a control loop (what continuously happens), and a termination logic (called once at the end).

21.2.1 Graphics and dynamics

The first concept is to physically model everything in the simulated world that is seen on the simulation screen. Everything is assumed to be an abstract *object*, which has a shape, size, colour, pattern, position, and orientation. The object may be obstacles, a robot, a manipulator, a surface, a camera, sensors, or anything. The objects are arranged in *object hierarchies*. This is done for two reasons. First, even a simple-looking water bottle with a nice curvy shape needs to be rendered on the screen and therefore stored as simple render-able shapes. Every complex shape can be recursively broken down as a collection of simpler shapes, which have a fixed spatial relation to the parent. Second, imagine moving a table. Not only a table will move, but all the objects over the table will also move, but the inverse is not true. This means that the objects on the table are connected to the table by a link. This can be specified as an object hierarchy with the table as the parent connected to all objects as the children with a link storing the transformation from the parent frame to the child frame. At the time of rendering, the children pose is derived from the parent pose and the (maybe fixed) transformation between the parent and child. Therefore, moving the parent moves the children; however, moving the children changes the transformation between the parent and child, which does not move the parent. Similar is the case of the manipulator. If the first link moves, the entire hand moves, which is specified by making the second link as a child of the first link, the third link as the child of the second link, and so on. This is also called the *kinematic chain*. Every link stores the transformation between the parent and child, which may change as the manipulator moves. Once an object is grasped, it is added as a child of the last link. The object and object hierarchies may be made by design and drag-and-drop utilities, without the need to program. In nongraphical libraries, they need to be specified by XML files, which is harder.

The simulations use a variety of techniques to model individual objects. For some common objects, the models are available from CAD and other 3D design software. The robots also have their designs readily available to the manufacturers. Some objects may be 3D scanned by 3D scanners. The others may be hand-drawn and added as hierarchies. Some objects may have a spatial relationship that changes. For example, when a manipulator rotates, the spatial relationship between the links changes. In some cases, the relation is constant. As an example, if a tabletop is sitting on four legs, the legs have the same spatial relationship to the top. Similar is the case of a passive joint of a manipulator with no motor. The spatial relation may change if there is a motor fitted on the link. The motor can be programmed, which is the task of the control algorithm. The entire environment is a collection of many object hierarchies.

The sensors are also simulated entities. A typical sensor is the proximity sensor, which can be imitated using the *ray-tracing algorithms* to get the actual reading over which white noise is added. The ray-tracing algorithm finds the first intersection of a ray with an object with a defined geometry. The other typical sensor is a video camera, which is a reflection of all objects in the world when passed by a calibration or projection matrix. The actuators are similarly simulated by taking the true reading from the programming scripts and adding noise.

The display of all the simulated entities with defined geometries is done by the *graphics library*. Currently, *OpenGL* is a powerful and popular graphics library. The library takes all objects with their display properties as input and gives the output as a 2D image that would be seen by a virtual camera at the view, using the calibration matrix of the virtual camera.

Another important aspect of the simulation is a *physics engine* or a *dynamics engine*. The engine also uses all objects along with their models, however, considers the physics rather than a visual display. The engine models every primitive entity and computes all real forces and torques on the model. This is then used to compute the immediate motion of the object. The engine, therefore, relies on the dynamic properties of the object like mass, the centre of mass (or distribution of mass), the moment of inertia, coefficient of restitution, and spring constant. The current popular dynamics engines are *VORTEX*, *OpenDE*, and *Bullet*. It must be noted that physics differs from the problem of control studied previously. In control, we compute a signal which is given to the motor that applies a torque for motion. The physics uses the same torques to compute the effects on the motion. Physics can also model situations like a collision with a wall, in which case giving a linear velocity does not move the vehicle as it physically cannot. Instead, the damage is simulated. The physics can also model situations like falling and breaking objects, where an uncontrollable motion happens.

21.2.2 Modules and services

The simulation thereafter requires several libraries, typically available as services. A prominent library among them is the *collision and proximity checking library*. The library represents all objects internally in a different mechanism, like an oct-tree or *k*-d tree. The same is used to query for proximity and collision by the motion planning algorithms. Typicality is that even though the two links of a manipulator are connected, it is not called a collision while touching an object boundary is called a collision. A manipulator sees a book as an obstacle before grasp and does not touch it, while after grasping it touches the book and it is not regarded as a collision. The tags specify what can collide and what cannot (because it is a part of the same body and connected directly). In the case of an object being grasped, the object after grasp is called an *affixment*. A popular library currently is the *Flexible Collision Checking Library* (FCL).

Every sensor and actuator of the robot has its coordinate axis system. Further, every robot has its coordinate axis system. The world is associated with a global coordinate axis system. Some relations between the coordinate axis systems are constant, like that between the robot and its camera; while some others are dynamic, like between a mobile robot and the world. The *transformation library* gives a utility to dynamically compute a transformation between any coordinate system to any other coordinate system. The TF transformation library is a common choice that represents all coordinate axis systems as a tree with the edges storing the transformation between the coordinates. The transformations are updated at every instant.

Manipulation is a common requirement for the simulations. Simulations see a fleet of robots picking and placing items from one place to the other. Each manipulator chain is defined as a kinematic chain. The *forward kinematics* is directly available as a service or library implementation. The service is typically provided directly by the manufacturer. The service takes all joint angles as input to compute the pose of the end-effector. Similarly, *inverse kinematics* is typically provided as a library implementation. The grasping programs recognize the object of interest, localize the object, compute the grasping positions, and thus compute the pose of the end-effector. The joint angles for the end-effector are computed by the inverse kinematics libraries taking the pose of the end-effector as an input.

21.2.3 Planning and programming

Motion planning is a service provided such that the manipulation unit or the mobile robot unit can directly reach the place of interest, even if there are obstacles on the way. The *Open Motion Planning Library* (OMPL; Şucan et al., 2012) is a popular library. The library started with sampling-based algorithms, however, slowly integrated several other algorithms. The library does only the deliberative planning and may thus require aid from a reactive planner implemented separately. The library typically makes use of a *mapping service* that provides the maps in a format that can be used by a collision-checker for planning. The library produces an output that is processed by a *control algorithm*, typically provided by the manufacturer. A manipulator can get its joint position directly from the internal encoders; however, a mobile robot needs to always solve the localization problem. The localization server is actively running in the background for the same purpose. A SLAM library further allows the creation of maps.

The story so far includes graphics created by drag and drop and displayed by the graphics libraries, dynamics computed by the physics engine, and a lot of services running in the background. The task is now to program the robot. In any simulation, each entity works in parallel in a separate thread. The entities may be robots, simulated sensors, or items like fans constantly rotating. Therefore, each entity is programmed and controlled by an independent script, executed as a separate thread or process. Each script is an infinite loop where the input is taken (either by a network stream simulating the sensor or a simulated system call), and the output is computed and given (either by writing to a network stream simulating message to the actuator or a simulated system call). This is called the *control loop* of the program. The scripts typically have an initialization part, the main control logic loop, and a part called if the process is terminated. Programming is done for each entity separately, calling the necessary services if needed. A simultaneous play of all entities makes a simulation.

It is interesting to comment on the aspects of parallelism in simulation. Executing scripts in parallel means that the different scripts will need synchronisation that increases programming effort but models the system more realistically. Taking a single robot as an example, numerous behaviours run in robotics with different frequencies, starting from a deliberative re-planning at a small frequency to a controller at a high frequency. This is implemented by using parallel scripts for each module. Furthermore, assume that a module works by using image and lidar data. Now image processing may take time and the lidar data will be wasted while the system is doing image processing sequentially. Since multiple sensors work at multiple frequencies, sequentially reading all data means that the resultant frequency will be lesser than the smallest frequency sensor. Parallel scripts facilitate deciding as per the latest available data from every sensor, while the sensors read their data in parallel.

21.3 Robot Operating System

To give the discussion on simulation a more applied look, let us discuss one of the most popular frameworks, which is the *Robot Operating System* (ROS). Like the operating system gives a platform for all hardware and software modules to interact, share and use resources; the ROS gives a framework for the robots, sensors, and actuators to share information, communicate with each other and carry the logical operations. ROS is not an operating system and is a framework typically installed on a Linux operating system. It is also not a simulator, although many simulators use ROS for interaction with programming modules, services, and even real robots. The concepts are summarized in Box 21.2.

Box 21.2 Robot Operating System (ROS) concepts

- **Node:** A computing program that sends/receives information from other nodes. Multiple nodes execute in parallel exchanging information.
- **Topic:** A named channel over which messages of a specific type are exchanged.
- **Message:** The physical data flowing over a topic
- **Message type:** Data type of the physical data flowing over a channel
- **ROSTCP:** Communication protocol for sending/receiving data
- **Publisher:** A node that writes data on a topic
- **Subscriber:** A node that receives data from a topic.
- A node may be a publisher to some topic and a subscriber to another topic. A node may write to a topic, which is received by many subscribers simultaneously, making it better than every subscriber taking the data from the publisher on a one-on-one basis.
- **Call-back function:** A function registered by the subscriber that is automatically called by the ROS master when data is received on the subscribed topic in an event-driven paradigm. In this mechanism the subscriber nodes do not need to actively ping the topics to check for new data.

Continued

Box 21.2 Robot Operating System (ROS) concepts—cont'd

- **Communication queue:** A queue that stores all data to be written to a topic/read from a topic if a publisher/subscriber cannot immediately read/write data. On exceeding queue capacity, the oldest data is lost. The queue gets populated when the subscriber cannot process data at the rate with which it is written, and vice versa.
- **Launch file:** An XML file to execute multiple nodes simultaneously.
- **Service:** Utilities that take a request, process it, and produce a response.
- **Parameter server:** A server that allows creating, amending, or reading the values of parameters across the ROS framework.
- **ROS master:** Server that coordinates communication between all nodes, topics, services, and parameter server.
- **Namespaces:** A hierarchical organization of all node/topic names in the format /xx/yy/zz.
- **Re-mapping:** Renaming a topic name used in a node for a specific execution of the node. Useful when a control program reads from the topic *pose* and writes to the topic *cmd_vel*, while the user intends to re-use the same controller for robot 1 that should read from /*robot1*/*pose* and write to /*robot1*/*cmd_vel*.
- **ROS graph/RQT graph:** Graph with nodes as vertices. Two nodes (u, v) are connected by a directed edge t if u publishes to topic t and v subscribes to topic t. The graph shows information exchange between the nodes.
- **RQT-Console:** ROS specific console to analyse logs and information.
- **RQT-Plot:** Plotting utility from live ROS topic streams
- **RVIZ:** Graphical utility for displaying paths, images, maps, etc.
- **Stage and Gazebo:** Typical simulators used in ROS
- **Unified Robot Description Format (URDF):** Format to describe robot shape, size, or composition properties
- **TF library:** Library that manages transforms between entities. Enables transforming between any two coordinate axis systems. The dynamic transforms are published by nodes (e.g. that do localization), while the static transforms can be specified in the launch files.
- **Action library:** A library that receives requests for 'actions' that take time to execute; requests for changing, querying status, or cancelling actions as they execute; and requests to call a callback function when the action is completed.
- **ROS bag:** Enables recording data flowing in selected topics which can be played back later. E.g. recording data published by real sensors which can be played back later for online SLAM programming as and when needed.
- **Navigation stack:** Motion planning standard in ROS consisting of modules for global mapping (Gmapping), local mapping by onboard sensors, localization (Adaptive Monte Carlo Localization), global planner (Movelt), local planner (Dynamic-Window Approach), global goal, local goal, base controller, and associated message types (move base).

21.3.1 Understanding topics

In this theme, let us discuss a simple situation, wherein a mobile robot needs to navigate and go to a goal of interest. Let us discuss the system from a real hardware perspective. Let the robot have a stereovision system, lidar sensors, and other proximity sensors that interact with their device drivers to the program running in the operating system. In ROS, every program is called a *node*. ROS has a lot of nodes that run in parallel simultaneously. The sensor information needs to be given to the localization and mapping nodes. Now all sensors have a different frequency of working. Further, the same set of sensors provides input to localization, mapping, planning, and emergency stop nodes, which are otherwise independent in terms of functioning.

Instead of every node making sequential calls to all the physical sensors, in ROS, the sensors operate by independent scripts or programs (called nodes) that continuously read and *publish* the data over the networking channels. The nodes that need the sensor values read from the same networking channel and are thus said to have *subscribed* to the channel. Consider the localization module as an example. The output of the lidar sensor node is an input to the localization node. The lidar sensor node is called the *publisher* as it publishes the lidar reading values continuously. The localization node is called the *subscriber* as it gets the lidar reading values from the channel. The networking channel is more formally called a *topic*. In ROS, all information is exchanged over such topics.

The physical information which is sent from a publisher to one or more subscribers is called the *message*. A message is a physical packet that is transferred over the channel. The message has a *message type* associated with it, which may be an int, string, array, or any custom type. The messages are encoded into binary and decoded as per the type of the message. Since the publisher and subscriber may be executed in different computer systems or different programming languages, the communication happens by the *ROSTCP protocol*, which is like the TCP/IP protocol. The relation between the publisher and subscriber is shown in Fig. 21.1A and B.

The lidar data also goes to the mapping, planning and emergency stop nodes and each of them is a subscriber to the same topic. Hence, a data generated by one sensor and published to one topic is accessed by multiple subscribers. Imagine writing a naïve program with no data sharing. Programs for mapping, planning, and emergency stop will individually access and collect data from the same lidar sensor. Hence a reading supplied to the emergency stop will be different from the one supplied to the mapping system as both programs have individually read the sensor readings at different times. ROS enables giving the same piece of data to all subscribers to a topic that maximizes the data utility and prohibits any node to block the sensor by reading for a prolonged time and not giving up the resource.

An interesting observation is that the localization node gives the output as the pose, which is an input to the planning node. The localization node is thus

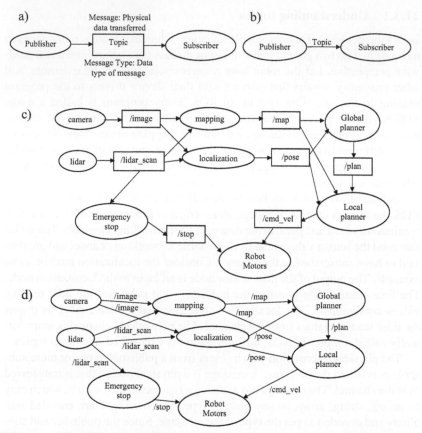

FIG. 21.1 ROS graph. Each ellipse represents a computational node, and a rectangle represents a topic over which communication happens. All nodes run in parallel, exchanging information using this network. The publisher publishes messages over a topic, to which the subscriber is subscribed. The messages automatically go to the subscriber by an automated call to its registered callback function. (B, D) are a simplified view of (A, C).

the publisher and the planning node is the subscriber for the topic *pose*, where messages of the type SE(2) (position and orientation) go. The localization node is thus a subscriber to the lidar reading topic and simultaneously a publisher to the pose topic. Many nodes simultaneously act as a publisher to some topic and a subscriber to some other topic. Most nodes or programs need inputs and give outputs. In ROS, the inputs typically come from the relevant network channels, or a node needs to subscribe to the topics from where it needs the inputs. Similarly, the outputs are written on network channels, or a node publishes a topic for which it has computed the output. This publisher–subscriber relation between all nodes can thus be represented as a graph where all programs or nodes act as the vertices, and a directed edge called *t* exists from a node *a* to a node b if a is a publisher to the topic t and b is a subscriber to the same topic t. This is called the *ROS graph* or *RQT Graph*. A sample ROS graph for the mobile robot problem is shown in Fig. 21.1C and D.

The manner of working of ROS is described by the *event-driven programming* paradigm. Imagine writing a subscriber program of localization for which the lidar and video data are available from the network streams. One way is to constantly ping the network stream for new data and to collect the latest piece of data. This needs to be sequentially done for both lidar and video and is hence wasteful. Since most nodes require data, constantly pinging topics can lead to a waste of computation. A more efficient manner of handling the situation is that the localization node registers itself as a subscriber with the *event manager*, to the topic lidar reading once during the program run. The localization node also supplies a *callback function*, which is a function that will be automatically called when a new message is available at the channel. Now, the localization node need not constantly ping to check for the presence of more data. It is the duty of the event manager to call the callback function and supply the latest data. Similarly, the lidar node registers as a publisher with the event manager. Once the lidar is ready with new data, it notifies the event manager that new data is available and is published at the channel. It is now said that the event is triggered. The event manager instead invokes the callback function of all the registered subscribers and supplies the new data.

The publisher publishes the data over a topic, which is consumed by the subscriber. The publisher and subscriber work at different frequencies, which also varies with time. Assume that at some instant the publisher publishes data at a higher rate than the subscriber can consume. The extra data is stored in a *queue*, waiting for it to be consumed. The queue has a size and if the subscriber cannot consume the data while the queue itself is full, the older data will be permanently deleted and lost. This will be the case if the publisher constantly has a higher rate of supplying the data in contrast to the rate at which the subscriber can consume the data. Sometimes such a deletion of data may be desirable. Imagine if the video takes some time to process while the camera can publish videos at a high rate. The subscriber may decide based on the latest frame it gets; however, it may be 2 s old with all the frames produced in those 2 s in the queue. It is better to throw away all old frames and instead work with the latest frame. Conversely, many times data should not be deleted. In the same example, consider that the video was being used to detect the signboard. Missing a few frames can mean that the robot misses the signboard altogether, even if the robot was unfortunately stuck in the processing of a single old frame and detection otherwise was fast enough. Consumption of data by the subscriber dequeues the data, while publishing of a new data calls the enqueue over the queue.

If the publisher is slow to produce the data, there is no queue, while any existent queue will diminish. Some subscribers wait for the new data, some others are programmed to make decisions with the partially available data from other sensors, and some others use the old data for decision-making. In the example, the localization node is more likely to make decisions only using the lidar data as it comes at a higher frequency; however, once the video data is available its information is used once for correction at a smaller rate.

21.3.2 Services and commands

Services are like nodes that operate on a request-and-response basis. The services take the request and compute the output which is returned as the response. The services themselves do nothing till the next request is received. ROS also provides a *parameter server*. The different modules can register, access, and change the different parameters. The parameters should not be used for communication, and their use is ideal for storing the constant values and studying their effects. The ROS master manages the topics, nodes, services, and parameter server.

The programs and tutorials are beyond the focus of the book. A lot of help is available from the ROS tutorials, which is a recommended starting point. ROS is an easy install. The framework uses the catkin build system, which is convenient for compiling the programs. A good starting point is to write a dummy publisher that publishes integers and a dummy subscriber that subscribes to the same integers to understand the concepts discussed here. The ROS master server is executed by calling *roscore*. It is a server that keeps running in the background. The ROS programs are executed by "*rosrun* <name of package> <name of program>", which is a convenient way than calling the executables directly.

Multiple nodes can be called at once by using an XML file specification with the command "*roslaunch* <name of package> <name of XML launch file>". This is specifically convenient with robots that need to invoke several sensors, actuators, and numerous other nodes simultaneously. The manufacturers provide a launch file that simultaneously invokes all modules of the robots. The file also allows to name the different nodes, suppress their output, or pass input argument values.

Practically, ROS master would have too many nodes. Namespaces are used for the ease of organizing the nodes. A typical problem is having a topic called *pose* with 10 robots of different kinds. Invoking all robots using the manufacturer's launch file can mean all of them simultaneously publish on the same topic called *pose*, which is a variable name conflict. Therefore, every topic also has a namespace that is hierarchically organized. The namespace hierarchies are separated by using the/operator. So, */robot1/pose* refers to the topic *pose* in the namespace of *robot1*. Similarly, */groupA/moduleB/robotC/pose* means the pose of *robotC* in *moduleB*, which is itself in *groupA*.

Let us consider a different problem. Imagine writing a good algorithm that makes a mobile robot escape static and dynamic obstacles and reach a goal. This reactive controller is implemented as a ROS node, which listens to the lidar readings (say topic */robot1/lidar*) and pose (say topic */robot1/pose*) and publishes to the velocity topic (say */robot1/cmd_vel* read as robot1's command velocity). Now an organization has five such robots that continuously provide the same service simultaneously. It is possible to write five different programs, one for each robot, but it is always good to try if the same program can be called five times for easy code maintenance. The only difference between the programs is in the name of the topic. */robot1/lidar* becomes */robot2/lidar*, */robot1/pose* becomes */robot2/pose*, and */robot1/cmd_vel* becomes */robot2/cmd_vel* when executing the program written for robot 1 for running robot 2, and similarly for the other robots. A technique to run a single program multiple times for different robots is called *re-mapping*. ROS can re-map a topic name to another topic name in the execution of the program. This is done by calling "*rosrun <package> <program> <old name1>:=<new name1> <old name2->:=<new name2>...*". The same can be done in a launch file in a different syntax by writing *<remap from="old name" to="new name" />* inside a node specification.

It is particularly good to mention a few tricks to quickly inspect ROS programs. If making a publisher, it is good to see what the publisher is sending on the topic, which can be done using "*rostopic echo <topic name>*" that prints everything on the channel. Similarly, if testing the subscriber, "*rostopic pub <topic name> <message type> <message data>*" can be used to send the data to the channel, which can give the relevant inputs to the subscriber being tested for.

Imagine that a new robot is either purchased in real or in a simulated environment. If the robot software is already installed, it will come with some launch files that invoke the different functionalities of the robot. Let us assume that the robot has been invoked. The first interesting thing is to see the functionalities as provided by the manufacturer. "*rostopic list*" can be used to list all topics that the nodes of the robot publish to or subscribe from, called after invoking the robot. The details of any topic can be found by using "*rostopic info <name of topic>*". The information will display the message type dealt by the topic, the publishers, and subscribers associated with the topic. "*rosmsg show <message type>*" can be used to get more details about a message type. Similarly, the services can be listed by "*rosersvice list*" and the details can be shown by "*rosservice type <name>*". The service can even be called by using "*rosservice call <name> <arguments>*". The commands corresponding to the nodes are "*rosnode list*" to get all nodes and "*rosnode info <name>*" to get information about a node.

There are a few GUI tools readily available in ROS that make it easier to make sense of the message passing between different ROS nodes. Foremost, *RQT-Graph* shows the different nodes, node groups, and topics. It enables an easier understanding of the message exchanges in the entire ROS framework. The *RQT-Console* can be used to have GUI access to all the logs created by different nodes including the nodes being programmed. The nodes can be sorted, filtered, or analysed using RQT-Console. The *RQT-Plot* can be interfaced with any topic to visually display the data like proximity sensor output and speed of the robot. The graphs are thus made on the go for easy visualization. RVIZ is the graphic utility that can be used to display several things including images, depth images, point clouds, data streams, and moving robots.

21.3.3 Navigation

Like every framework, the robot including all functionalities, shapes, actuators, and sensors needs to be specified, along with the spatial relation (fixed or controlled) between the entities as a transformation matrix. ROS uses *Unified Robot Description Format* (URDF), an XML specification for the same. *ROS control* enables modelling of the robot and to interface the controller in the ROS framework. Similarly, the calibration utilities may be used to calibrate the different sensors. The most popular planning library is currently the *Open Motion Planning Library* (OMPL), which is interfaced with ROS, especially for the manipulation scenarios by the *MoveIt* library. Any robot needs to additionally specify properties to be usable for planning using MoveIt, like the identification of kinematic chains, which can be done using the tools provided by MoveIt. Luckily, when working on the most popular robots on hardware or simulation, the ROS-supplied packages have all these descriptions already available and interfaced. These can be downloaded and installed as standard ROS packages, typically either supplied by the manufacturers or the open-source community.

The robot is often given multiple things to do like go to a place and pick up something. This is possible to do using a topic (over which the task is sent, and an acknowledgment is received) or a service (where the request is the task to do, and the response is obtained after the task is done with the success status of the task). The problem with just sending the task over the topic is that the acknowledgment status shall be received as a callback and the program logic thus gets a little messed up for primitive tasks like asking a robot to sequentially do a few things. The problem with the service is that the program may want to do something in the time when the robot is accomplishing a task, like tracking for changes in the map and updating the requirements.

The *action library* is a library that makes programming such *actions* easier. Actions are things to be done by the robot which shall take some time. The action server is given the specification and the control of the program is not kept but given back. The task can be updated or cancelled by the program by sending adequate requests. The action server can be queried any number of times for status like successful completion, abortion of the task, or change of task. The action server can call a callback function upon successful completion or an unsuccessful completion.

Imagine a utility where a robot is serving by going to multiple places as per the user's request in an office. A user wants the robot to go to the classroom, which is given as a request to a service, and the robot returns a successful response once it has reached the classroom given as a response. However, while the robot was navigating, the professor instead urgently wanted the robot to meet him/her. In a request-response service, such cancellation is not natural program logic. Conversely, consider that the robot listens to a topic to get the goal, and continuously moves towards the same goal. Now consider that the robot must sequentially visit several places, which (in a procedural sense) is merely five lines program of the form "Goto place *i*". However, the program will have to give the first place, wait till that place is reached (known at the callback function), and then give the second place. Action library is a cleaner way to implement such long-term actions, where (if needed, nearly) sequential lines could be given for the robot to achieve, or the action could be continuously monitored and amended, as necessary.

ROS itself is a framework that enables the different libraries, modules, and programs to communicate with each other over topics. It does not provide simulation facilities like a physics engine, creating a simulated robotic world, and creating virtual sensors and actuators. The two popular simulators often interfaced with ROS are *Stage* and *Gazebo*. The classic simulator of Player-Stage had two modules, the player who controlled the robotic simulated agent, and the stage that provided the simulation framework over which the player worked. The ROS interface provides the means to control the robot over program logic. The simulation part is still provided by Stage. The stage does only 2D simulations. *Gazebo* is a more heavy-weight and realistic simulator that uses a physics engine to simulate the actual dynamics of the system. Gazebo needs its specific modelling of the robot for simulating dynamics, which can be easily available for popular robots that come with their packages for Gazebo.

Each sensor gives readings in its frame of reference, which may need to be transformed. As an example, the robot may sense obstacles in its frame of reference, but the mapping module will require the same in the world coordinate frame of reference. Similarly, the mapper may give the map in the world coordinate frame of reference, but the robot reactive decision-making may need to convert it into its coordinate frame of reference. The *TF library* handles all the transformations. There are two types of transformations, static and dynamic. Static transformations have the same transformation in the entire run, like the transformation between the camera of the robot and the base of the robot. The static transformations can be published by the launch files. The dynamic transformation is the one that changes, like the transformation between the robot pose and the world. The localization library publishes such transforms dynamically. Any data communicated over ROS, if it has a coordinate frame attached to it, the reference system is typically specified in the message header. This makes it easier to interpret and convert the reference frames using the TF library.

The robots primarily deal with vision data, for which OpenCV is often used. Similarly, for handling depth images, point clouds, and other 3D images, Point Cloud Library (PCL) is a popular choice. Both are easily interfaced with ROS. Typicality when working with Simultaneous Localization and Mapping

(SLAM) algorithms is that the input is a real-world stream of images, other sensor readings, and control inputs, while the output is a map with the pose of the robot at every instant of time. While all these can be simulated, it is preferable to work with real-world data, which otherwise requires a real robot with real sensors. It is luckily possible to *record and playback data* in ROS associated with each sensor and control sequence. This is done by using ROS bag files. The files can record data over mentioned topics, and in playback, the same sequence of data is played back over the same topics. So, SLAM algorithm developers do not take the robot on the run on every change in the design of the algorithms, they instead only re-play the bag files and re-run the SLAM algorithmic nodes to take the played data as input.

Much of the discussions in the book were for navigation, and ROS has its standards for the same known as the *navigation stack* that allows using any library for any of the components of navigation with seamless integration. This is contradictory to the classic mechanism wherein the navigation software suite for one robot would only work with that robot. The navigation stack defines the different modules and the topics over which the modules interact. The foremost is the odometry of the robot, giving the robot pose. The odometry is calculated using the sensor messages of the different sensors onboard. A popular implementation is the *Adaptive Monte Carlo Localization* (AMCL). The *map server* is responsible for providing the map for the planning module. The global map server uses a binary costmap (obstacle and no obstacle areas) which can be given as a binary image with pixel-to-distance mapping ratio. If the map is not available for an area, the *Gmapping* is a popular package providing SLAM utilities. The *local map* also accounts for the additional obstacles detected by the onboard proximity sensors like moving people who are not a part of the static map. The module then publishes the map over the same named topic. The *base controller* is responsible for moving the robot, which is done by sending the velocity commands over the *cmd_vel* (command velocity) topic of the robot. Finally, the planning module is specified in two heads, a *global planner* and a *local planner*. The OMPL is a popular global planner, while the Dynamic-Window Approach (DWA) is a popular reactive local planner. The planning typically assumes that the robots are circular of the mentioned radius, and obstacle inflation is used for planning the robot. The inflated map denoting the configuration space is separately published. Global planning gets the goal information from the *goal* topic. The planner automatically computes the *local goal* to be given to the local planner. Both global and local planners also publish their plans on the same named topics. The overall navigation framework is called the *move base* as developed by the community behind the navigation stack.

21.3.4 Example

Let us study the basic concepts of ROS using a small example. We will make use of the simplest robot provided by ROS, which is called a turtlebot. The robot will be simulated over a custom simulator provided by ROS. First, we will move one simulated turtle bot by using keyboard teleoperation. Thereafter, we will move a second turtlebot which automatically takes the first turtlebot as the goal and moves towards it. Similarly, a third simulated turtlebot will follow the second turtlebot, and a fourth turtlebot will follow the third turtlebot.

Let us first program a generic motion to goal behaviour for one turtlebot. Suppose the robot pose of the robot is available at the topic /*pose*, and the turtlebot can be moved by writing the velocity commands to the topic /*cmd_vel*. Further, assume that the goal is available from the topic /*goal*. The programming is done in C++ for ease of understanding. The class hence has one publisher for velocity commands and two subscribers for pose and goal. The subscribers have their callback functions. The constructor mentions the topics, queue size, and message types of all publishers and subscribers. The subscribers additionally have the callback function specification. The subscriber gets new data which is used to compute and send velocity commends. The main is thus empty, expect initialization as a ROS node. The function *spin*() ensures that the program does not terminate, which will forcefully terminate the publishers and subscribers. Function spin is specifically callback aware and ensures that the callbacks are made active when needed. The class is shown in Box 21.3.

Box 21.3 Overall class architecture

```
#include "ros/ros.h"
#include "turtlesim/Pose.h"
#include "geometry_msgs/Twist.h"
class turtleMover
{
    ros::NodeHandle nh;
    ros::Publisher velPublisher;
    ros::Subscriber poseSubscriber;
    ros::Subscriber goalSubscriber;
    turtlesim::Pose pose;
    turtlesim::Pose goal;
    void poseCallback(const turtlesim::Pose& newPose);
    void goalCallback(const turtlesim::Pose& newGoal);
    public:turtleMover();
};
turtleMover::turtleMover()
{
    //register publisher with message type geometry_msgs::Twist named cmd_vel
    with queue size 10
    velPublisher=nh.advertise<geometry_msgs::Twist>("cmd_vel",10);
    //register pose subscriber. On receiving a message turtleMover::poseCall-
    back is automatically called
    poseSubscriber=nh.subscribe("pose",10,&turtleMover::poseCallback,this);
```

Continued

Box 21.3 Overall class architecture—cont'd

```
        goalSubscriber=nh.subscribe("goal",10,&turtleMover::goalCallback,this);
    //register goal subscriber
        goal.x=-1;goal.y=-1;goal.theta=0; //current goal (overwritten when a
    message at /goal is received)
    }
    int main(int argc, char **argv){
        ros::init(argc,argv,"turtleMover"); //register node with all re-mappings given at
    command line
        turtleMover mover; //constructor registers all publishers and subscribers
    //spin disallows the program to terminate unless ROS exits.
    //also allows transfer of control to registered callback functions when a
    message is received
        ros::spin();
        return 0;
    }
```

The main logic is implemented through subscribers. The goal subscriber simply updates the goal. The pose subscriber however calculates the velocity command. The linear velocity is proportional to the distance to the goal. The robot should ideally be facing the goal, and hence the angular speed is proportional to the angular deviation from the goal. The *atan2* function ensures that the angles are within the range of $-\pi$ to π. The velocity is specified in the robot's coordinate axis frame. As the robot moves forward on its X-axis, the linear velocity is specified in the X-direction, while rotation is around the Z-axis. The robot itself limits the speed in case an out-of-bound speed is supplied. Because of kinematic constraints, the robot cannot translate along the Y- and Z-axis or rotate around the X- and Y-axis, and the simulator simply discards those values. The program is given in Box 21.4.

Box 21.4 Subscriber functions. The constants kp1 and kp2 should be substituted with the desired parameter values

```
    //automatically called on receiving a new goal on /goal topic
    void turtleMover::goalCallback(const turtlesim::Pose& newGoal)
    {
        goal=newGoal;
    }
    //automatically called on receiving a new pose on /pose topic
    void turtleMover::poseCallback(const turtlesim::Pose& newPose)
    {
        pose=newPose;
        if(goal.x>0) //valid goal
        {
            double dis=sqrt(pow(goal.x-pose.x,2)+pow(goal.y-pose.y,2));
```

Box 21.4 Subscriber functions. The constants kp1 and kp2 should be substituted with the desired parameter values—cont'd

```
        double v=kp1*dis;
        double goalAngle=atan2(goal.y-pose.y,goal.x-pose.x);
        double angleError=goalAngle-pose.theta;
        angleError=atan2(sin(angleError),cos(angleError)); //limit to range (-pi, pi]
        double omega=kp2*angleError;
        geometry_msgs::Twist vel;
        vel.linear.x=v; //liner speed is in the robot's X direction
        vel.linear.y=0;
        vel.linear.z=0;
        vel.angular.x=0;
        vel.angular.y=0;
        vel.angular.z=omega; //angular speed is in the robot's Z direction
        velPublisher.publish(vel);
    }
}
```

The code is used to move all but the first turtlebot and hence let us make a launch file to move all those turtlebots. The launch file calls the same code three times with different names keeping the package name and program name the same. The goals, positions, and velocity topics are different for each robot which are handled by remappings. The interesting aspect is that the position of turtlebot 1 becomes the goal for turtlebot 2 and so on. The launch file is shown in Box 21.5.

Box 21.5 Launch file. The same controller is called three times for three different robots with a change in name and topic names (done by re-mapping)

```
<launch>
    <node pkg="turtleExperiments" name="turtle2" type="turtleMover">
        <remap from="cmd_vel" to="/turtle2/cmd_vel" />
        <remap from="pose" to="/turtle2/pose" />
        <remap from="goal" to="/turtle1/pose" />
    </node>
    <node pkg=" turtleExperiments " name="turtle3" type="turtleMover">
        <remap from="cmd_vel" to="/turtle3/cmd_vel" />
        <remap from="pose" to="/turtle3/pose" />
        <remap from="goal" to="/turtle2/pose" />
    </node>
    <node pkg=" turtleExperiments " name="turtle4" type="turtleMover">
        <remap from="cmd_vel" to="/turtle4/cmd_vel" />
        <remap from="pose" to="/turtle4/pose" />
        <remap from="goal" to="/turtle3/pose" />
    </node>
</launch>
```

TABLE 21.1 Executing the ROS simulation program.

Terminal	Command	Utility
Terminal 1	roscore	ROS server keeps running in the background
Terminal 2	rosrun turtlesim turtlesim_node	Server and GUI for turtlebots simulator
Terminal 3	rosservice call /spawn "x: 2.0 y:=2.0 theta: 0.0 name='turtle2'"	Spawn 3 robots as the simulator initially gives only one robot by default. Tab is used to auto-populate the message type shown in quotes. Repeat to add the desired number of robots
Terminal 3	rosrun turtlesim turte_teleop_key	Move the 1st turtlebot by using a keyboard
Terminal 4	roslaunch turtleExperiments launchTurtles.launch	Call the launch file to automatically move the other robots
Terminal 5	rostopic list	Inspect available topics
	rostopic info /turtle2/cmd_vel	Inspect all available topics one by one
	rosmsg show geometry_msgs/Twist	Inspect all message types used in all topics
	rostopic echo /turtle2/cmd_vel	Inspect messages flowing in all available topics one by one. Ensure that the robots are moving by teleoperating the first
	rostopic pub /turtle2/cmd_vel -r 1 geometry_msgs/Twist "linear: x: 0.5 y: 0.0 z: 0.0 angular: x: 0.0 y: 0.0 z: 1.0"	First, stop the launch file execution. Then make a turtle move by publishing speed at the rate of 1 Hz. The linear speed is given in the X coordinate and angular speed in the Z coordinate as per the coordinate axis system of the robot
	rosservice list	Inspect available services
	rosservice info /spawn	Inspect all services one by one
	rosservice type /spawn	Inspect all services one by one
	rosnode list	Inspect available nodes
	rosnode info /turtlesim	Inspect all available nodes one by one
	rosrun rqt_graph rqt_graph	Inspect the ROS graph
	rosrun rqt_console rqt_console	Inspect RQT console. Rerun all programs including the simulator
	rosrun rqt_ plot rqt_plot	Select topic as /turtle2/pose and see the plot

The compilation is through catkin. The execution is done using the *rosrun* and *roslaunch* commands for the different nodes. The execution is done to launch (in different terminals) the ROS server, the turtlebot simulator, the keyboard teleoperation (already provided in the simulator), and the launch file created. The simulator provides a ROS service to spawn robots to the needed four robots. The execution is summarized in Table 21.1. The trace of the robots is shown in Fig. 21.2A and B. The corresponding ROS graph is shown in

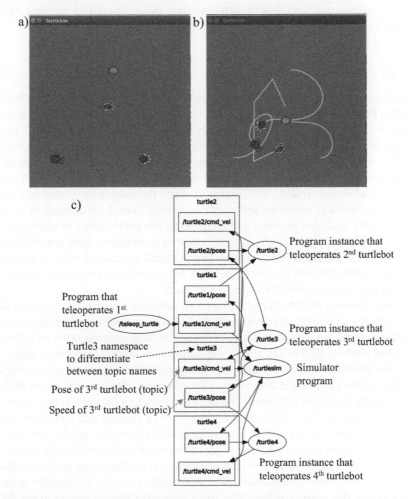

FIG. 21.2 Simulation with four turtlebots. The first turtlebot is teleoperated, second turtlebot follows first turtlebot, third turtlebot follows second turtlebot, and fourth turtlebot follows third turtlebot. (A) Initial configuration, (B) final configuration, (C) ROS graph. Each turtlebot has a velocity topic (*/turtle{i}/cmd_vel*), published by controller node (*/turtle{i}*) and subscribed by the simulator node. Each turtlebot also has a pose topic (*/turtle{i}/pose*), published by the simulator node (*/turtlesim*) and subscribed by the controller node (*/turtle{i}*) and as goal by the following robot controller node (*/turtle{i+1}*).

Fig. 21.2C. Here the controllers of the four turtlebots, the keyboard teleoperation utility, and the simulator are the nodes shown as ellipses. The topics and their interaction with the nodes are shown in rectangles. The topics associated with each turtlebot are grouped by the name of the turtlebot.

21.4 Simulation software

In this section, we very briefly discuss some other simulation systems and libraries that are often used in robotics.

21.4.1 Motion planning libraries

The focus of the book was motion planning and therefore we start with the *Open Motion Planning Library* (OMPL). Even though the library was discussed in the context of ROS, it can be used separately as a standalone library as well. The library primarily has sampling-based motion planning techniques inbuilt along with several scenarios to test the algorithms. Slowly other planners are being integrated into the same library. The library is modular, open-source, and has easy-to-understand codes. This makes it easy to make new planners with collision-checking, samplers and controls already inbuilt. It is easy to read the currently available planners and samplers and to use the same to create your own. The library is particularly useful because of its benchmarking capabilities. Once a novel planner is made, it can be compared against all other available planners with different parameter settings by automated scripts that directly give graphs against several metrics predefined.

The extension of the library, *MoveIt* also deserves a re-mention. Since it is an extension of OMPL, it provides all the features of OMPL including benchmarking. The library is particularly of value for manipulation tasks including pick and place. A GUI utility allows simulation over several predefined scenarios. One of the most difficult aspects of grasping is to compute a grasp configuration or the pose of the end-effector for grasping. Humans can orient their hands easily for different types of objects, but computing the same is not so simple for the robot. The library detects the object of interest and calculates the grasp configuration. Thereafter, inverse kinematics is used to get the goal configuration. The motion planning gets the manipulator to the goal configuration. Sometimes it is better to motion plan to a pregrasp configuration and then take the robot to the grasp configuration. The library also provides a controller to get the robot to the grasp position. The robot then uses the gripper to grasp the object. The gripper may be a vacuum, electrical using fingers, or using friction pads to reduce the chance of slippage. The vacuum is the most suited as it works even if there are small errors. However, there is a severe limitation on the shape and weight of the object that the vacuum can pick. After grasp, the object becomes a part of the manipulator and is added to the object hierarchy, and the object can no longer be called as colliding with the manipulator end-effector. The robot first goes to a postgrasp location and thereafter to the pre-specified drop configuration. The inverse kinematics is again used to convert the drop pose into the joint angles, and motion planning is used to compute a trajectory to the drop location.

21.4.2 Crowd simulation with Menge

Crowd simulation is another unique application of simulation and motion planning. Crowd simulation is useful for planning indoor infrastructure to foresee congestion buildup, chances of a stampede, and overall user experience within a facility. This can enable authorities to decide on the number of exits, number of lifts, number of staircases, limit attendees of events, re-schedule events, decide reporting time for events, etc. Imagine a football field has expanded the number of seats, crowd simulation enables us to decide how quickly will the crowd disperse after the match. Similarly, crowd simulation enables us to determine if a building can be timely vacated in case of fire. Another interesting application of crowd simulation is animations in movies. Animating armies in war, animating swarms of animals, or animating background characters are all examples of crowd simulation. It may be noted that in the simulations so far, the aim was to make the robots operate most efficiently. However, in crowd simulation, the aim is to make the simulated humans operate as humans do. Therefore, the models first observe a huge number of trajectories of the humans and thereafter regress the reactive planner parameters to imitate humans most closely. The simulation is done using circular models; however, graphics are used to cast a walking human gait over the simulated behaviour of the circular agents.

A popular library for crowd simulation is *Menge* (Curtis et al., 2016). The crowd simulation can be best described by a heterogeneous swarm. There are several groups of agents with different behaviours, while within a group the agents have the same behaviour. Menge models each behaviour by a Probabilistic Behavioural Finite State Machine. The deliberative and reactive planning is provided to enable motion to goal to be specified as a single behaviour. The probability functions can be used in behaviour selection. The library further allows specifying external events like fire, in which case all the agents behave differently.

21.4.3 Vehicle simulation with Carla

Another application of motion planning and navigation is a self-driving car. It may initially appear that the simulation is the same as any mobile robot with specific kinematic models; however, there is typicality associated with planning vehicles operating in traffic scenarios. At the highest level of abstraction, the planning is *route planning*. A well-defined road-network graph is already available, which can be searched for a route to minimize length, minimize expected travel time, minimize the number of traffic lights, etc. In route planning, it is common that if all vehicles take the best route, the congestion is bound to grow by a big margin on popular roads, and therefore *congestion control* mechanisms must be placed. The congestion is controlled by avoiding areas with a lot of traffic. This, however, can only cause a slowdown in the further development of congestion and cannot stop the development of congestion in the first place. In anticipatory congestion avoidance, the systems simulate and anticipate the roads that will cause a lot of congestion in the future and avoid taking such roads in anticipation.

At a finer level, the planning has several specific scenarios like moving on a road, intersection, diversion, or parking. The normal road consists of behaviours like overtaking, lane change, and vehicle following. There are strong heuristics to plan each behaviour and select between behaviours. The intersections without traffic lights are particularly hard wherein deciding the right of way is a difficult problem. The vehicle needs to additionally decide when to enter the intersection and when to wait.

A popular library for simulating cars is *Carla* (Dosovitskiy et al., 2017). *Torcs* and *Apollo* are also used for similar use cases. Carla specifically has rich graphics. It gives a near-real experience of virtual cities which makes it possible to work with virtual vision cameras for learning and planning the vehicle motion.

21.4.4 Other simulators

There are numerous simulators available that differ in the application scenario, pros, and cons. Every year numerous new simulators are made, and capabilities are added to existing ones. The list of popular simulators is very dynamic. Here let us very briefly mention a few robotic simulators. CoppeliaSim (Rohmer et al., 2013) is particularly easy to start simulations. It can also integrate with ROS. CoppeliaSim provides drag-and-drop utilities to create an environment. Even robots can be dragged and dropped without caring about the background drivers. The scenarios can also be inspected by object hierarchies. Scripts can be attached to all robots and components by using Lua scripts. The library has numerous tutorials and examples to head start.

Two simulators already discussed, Player Stage and Gazebo, also deserve a re-mention. Both are also stand-alone simulators without using ROS. Gazebo can also be used with the Player-Stage framework where the player component provides means to move around the robots. The Stage does simple 2D simulations, while Gazebo is used for 3D dynamics.

Another interesting simulator is *MuJoCo*. The simulator is particularly good for dynamics with manipulators. The controllers can be provided with the simulations. *Drake* (Tedrake, 2019) is another simulator good for simulating robots, which also draws its strength in accurately modelling the robot dynamics. *Unity*, another simulation framework, has recently gained much attention due to the high graphic value wherein the simulations look realistic. The simulator is particularly common for game-playing scenarios. Finally, *Webots* is another general-purpose simulator that can be used for various scenarios.

21.5 Traffic simulation

Traffic simulation scales up the simulation to be able to simulate the entire traffic at a city level. The road-network data, including a specification of roads, intersections, lanes, parking spots, and speed limits, are readily available. The number of vehicles flowing at different parts of the city is also commonly available from the transportation authorities, measured by using means including detectors at different roads and intersections. As the vehicles pass by, loop detectors vibrate, which is used to count the vehicles. The aim is thus to be able to realistically simulate the actual traffic data. The traffic simulation systems enable transportation units and governments to make key decisions, including the construction of new roads, the construction of new flyovers, deciding traffic light operating policies, deciding speed limits, and deciding lane usage rules. In all such conditions, the transportation authority simulates the traffic under different settings, and the setting resulting in the most efficient traffic is selected. The utility of the construction of new roads and flyovers is essentially based on the improvement per unit of money spent. The safety factors also become critical in making such decisions which are studied based on the simulations.

21.5.1 Kinematic wave theory

First let us discuss the simplest scenario, which is a small road that has several vehicles. Let us focus on a single-lane road only, as the simplest scenario. The road is observed for limited space and for some duration of time. This limited space-time observation of the road is referred to as the *observation window* (Immers and Logghe, 2002; Maerivoet and de Moor, 2004). A synthetic observation window is shown in Fig. 21.3A. The lines represent the different vehicles that pass through the observation window in time and space. Note that the lines shall never intersect as two vehicles can never be at the same time at the same place, which is the condition of a collision. Let us assume that the rear of the vehicle is used to measure the position of the vehicle, which is the rear bumper. Let us define a few terms for the observation window. The first is the *space gap* between the vehicles, which is the distance that a vehicle keeps with the vehicle in front. The space gap is the difference in the positions of two consecutive vehicles minus the vehicle length. Interestingly, the same gap can also be defined in terms of time, called the *time gap*. The time gap is the time that a vehicle will take to collide with the vehicle in front if the vehicle in front does not move. Both gaps can be seen in the space-time observation window. The *average speed* of the vehicle is the total distance travelled by the vehicle in the observation window by the total time taken to travel the said distance. These variables can be observed for any traffic system. One of the ways is to send service vehicles with GPS receivers that record the data that can be used for computing the variables. The CCTV cameras also facilitate the occupancy and flow of information. The roads have loop detectors, which are devices that send signals as vehicles pass by, which also record the same data.

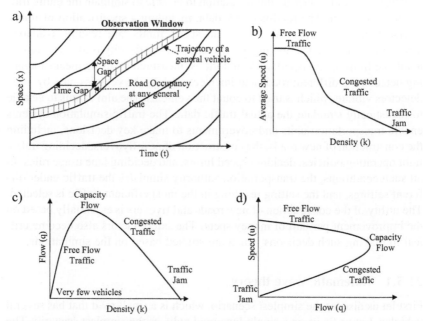

FIG. 21.3 Macroscopic traffic simulation (A) observation window for a single lane road (B–D) fundamental diagrams. The diagrams show a relation between the different fundamental macroscopic variables in different phases of traffic.

This is the first example where we discuss the macroscopic study of a simulation system. Therefore, let us define three macroscopic entities. The first is *density* (k), which is the total number of vehicles per unit space in the observation window. Density is a spatial concept that can be studied at any instance of time. Since time is varying, the average density over time is of interest. Alternatively, density can also be defined as the total time spent by all vehicles in the observation window, divided by the area of the observation window, which mathematically gives the same expression. The other interesting macroscopic variable is traffic *flow* (q). The flow is the total number of vehicles that pass a region per unit of time. The entity is temporal and may therefore be averaged for space. Like density, the definition becomes the total distance travelled by all vehicles within the observation window, divided by the area of the observation window. The last interesting macroscopic variable is the *average speed (v)*. Here, the reference can be the average speed of all vehicles at any instance of time (space-based definition) or the average speed of all vehicles that cross a region at any time within the observation window (time-based definition). The space-based and time-based definitions may have different numeric values within the same observation window. For discussions here, we restrict to the average over the space and not time. To use time average, the harmonic mean instead of the arithmetic mean is used.

The three macroscopic variables are related to each other by the *fundamental equation* of traffic flow, leading to the kinematic wave theory. The relation is given by Eq. (21.1):

$$\text{Flow} = \text{Average density} \times \text{Speed}$$
$$q = k\,v$$

(21.1)

21.5.2 Fundamental diagrams

Given an observation window, let us now plot the three macroscopic entities to get the relation between them in different traffic conditions. These are called *fundamental diagrams* as they explain the fundamental behaviour of the traffic. By observation of traffic, the generic nature of plots between the three quantifies is shown in Fig. 21.3B–D. First, let us discuss the relation between density and speed. As density increases, the speed reduces. It is intuitive to find a reason for the trend. The drivers are used to driving at the top allowed speeds on freeways with the entire freeway with themselves. As density increases, they have vehicles ahead of them at smaller distances and they are conscious of them. They desire to keep safe distances with the vehicles ahead. Traffic cannot have a large density of traffic with all the vehicles moving at high speeds even though possible in simulation. Higher speeds require maintenance of larger safety distance to overcome uncertainties. In actual traffic, a human driver would be cautious about a low distance with the vehicle ahead and would speed down as this distance becomes small, to a value accommodable as per the current distance. The driver cannot trust that the vehicle ahead will continue to travel at high speeds. The extreme case is when the density becomes extremely large, wherein the speed almost reaches zero and a *traffic jam* is said to have occurred. Hence traffic controllers ensure to keep only a limited number of vehicles on road.

The other relationship is between density and flow. A density with nearly zero value sees vehicles moving at top speed, but the flow is still small as the number of such vehicles is small. The road infrastructure is underutilized. This region is also called the free-flow region, where the vehicles freely move without affecting each other much. As density increases, the flow increases due to an increase in the number of vehicles, up to a point called the *capacity flow*. This is the maximum flow possible. As the number of vehicles is further increased, the density increases, which reduces the average speed of the vehicles, which further reduces the flow. Here, the vehicles are high in number, but the flow is small as the vehicles move with a smaller average speed. This region is called the *congested flow*. On further increasing the density, the vehicles stop moving due to a traffic jam, and therefore the flow is reduced to zero.

Considering the fundamental equation, the last plot is self-explanatory. The plot between flow and speed shows that a zero-flow value can either be because of free-flow traffic with very few vehicles or with traffic jams in which case the speed is small. The capacity flow on the contrary sees reasonable speeds with the highest flow value. Other results are intermediate. As speed increases, the flow initially increases from the congestion to the capacity flow region but then starts to reduce from the capacity flow region to the free-flow region.

Every transportation management system aims to keep the flow as close as possible to the capacity flow. A higher density leads to congestion while a lower density is underutilization of expensive infrastructure. The problem is that users do not consider flow metrics before joining the transportation network. Therefore, the density over the entire network keeps increasing. Now to avoid congestion developing at certain parts of the road network, the vehicles must be diverted to other areas. The traffic lights ensure that the flow in certain regions is under the capacity flow. The traffic lights reduce the rate of traffic getting into the roads if the current traffic is more than the capacity flow. However, to get some regions under capacity flow, they increase the density at other areas by reducing the outflow rate, while the inflow rate may be high. A better mechanism to ensure that the density is limited is by encouraging re-routing, wherein longer routes are suggested by the navigation guidance systems to avoid congestion on some roads. In worst cases, when all the users enter the transportation network and the traffic cannot be re-routed, congestion is bound to happen, and traffic lights can only do a little to avoid that. Users must be encouraged to travel at later parts of the day with a smaller traffic density as a last resort.

21.5.3 Kinematic waves

Imagine a segment of the road. The outflow rate is the rate at which the vehicles leave the segment, while the inflow rate is the rate by which the vehicles enter the segment. In the ideal case, both the rates are the same leading to a stationary traffic condition. For some reason, if the outflow rate is small while the inflow rate is large, the vehicles start *queuing* one after the other. The queuing rate is the difference between the inflow and outflow rates. The queuing goes from one end of the road and increases with time towards the other end. This is called a *kinematic wave*. Specifically, the situation will send a *shock wave* as more and more vehicles get into the jammed condition. The easiest example is a traffic light. If the light turns red, the vehicles all start queuing.

The opposite scenario is also interesting. Imagine that the outflow is larger than the inflow. This causes a reduction in the queue, which is another kinematic wave. The queues can be reduced to zero leading to free-flow traffic. Thereafter, the outflow rate will be reduced to the inflow rate with no congestion. Imagine the traffic light turns red to green, it causes the same effect.

These are two clear waves. The traffic can have multiple such waves from different sources that go at different speeds and intersect, add, or cancel each other. An uphill slope on a road means that the vehicles will have to slow down on approaching the uphill, which sends a similar wave. Narrowing of the road has the same effect. Broadening of the road has the opposite effect, wherein the outflow can be higher and the traffic queues normally clear. Two roads merging means that the merged road may not have the bandwidth to accommodate traffic from both merging lanes and this will limit the outflow. The traffic jams may propel backward. If a road diverges into two segments, the opposite behaviour may be observed, although the vehicles may need to slow due to the turning of the road. The *stop and go* waves are common wherein vehicles need to stop, go for a small distance, and stop again.

21.5.4 Intelligent Driver Model

After observing the macroscopic behaviours and patterns, we attempt to do agent-based microscopic modelling of one vehicle for the simulation. In this section, we discuss the *Intelligent Driver Model* (Treiber et al., 2000), which models the behaviour of one vehicle as it tries to follow another vehicle. Hence only the linear speed control is of interest as the vehicle displays a straight-line vehicle-following motion only. The characteristic of such a behaviour is that once simulated, it should adhere to the fundamental equation and diagrams. Suppose a vehicle has a preferred speed of v_{max}. Suppose the vehicle is moving at a speed v. If there is no traffic on road, the vehicle may show acceleration to attain the best speed proportional to the speed difference with the preferred speed, given by Eq. (21.2)

$$\dot{v}_{acc} = a\left(1 - \left(\frac{v}{v_{max}}\right)^{\delta}\right) \tag{21.2}$$

Here, a is the proportionality constant and δ is a constant that adapts as per the driving behaviour of the driver. The effect will be an increase in speed. The acceleration will be higher when the speed difference is more and will reach a value of 0 when the vehicle is at the preferred speed.

Let s^* be the preferred distance that the driver intends to keep with the vehicle in front. Let the current distance from the vehicle in front (space gap or distance from the front of the vehicle to the rear of the vehicle in front) be s. The driver would tend to decelerate with a profile given by Eq. (21.3)

$$\dot{v}_{\text{dec}} = -a \left(\frac{s^*}{s} \right)^2 \tag{21.3}$$

As s becomes large, the deceleration nearly becomes 0, showing no real gain in deceleration. However, as the distance from the vehicle tends to 0, the deceleration tends to infinity, showing a strong tendency to decelerate. The resultant acceleration is thus given by Eq. (21.4), using both the terms together.

$$\dot{v} = \dot{v}_{\text{acc}} + \dot{v}_{\text{dec}} = a \left(1 - \left(\frac{v}{v_{\text{max}}} \right)^{\delta} - \left(\frac{s^*}{s} \right)^2 \right) \tag{21.4}$$

The safety distance (s^*) is given by Eq. (21.5)

$$s^* = s_0 + s_1 \sqrt{\frac{v}{v_{\text{max}}}} + T_{\text{react}} v + \frac{v \Delta v}{2 \sqrt{ab}} \tag{21.5}$$

The preferred distance is simple to understand. There is a constant s_0, which is the least distance that the vehicle would like to always have from the vehicle in front under all circumstances. Thereafter, a higher vehicle speed means a larger caution and thus a need to keep more distance. This is a normal reaction of the drivers and is modelled by the ratio v/v_{max}.

Let T_{react} be the reaction time of the driver. This means upon spotting anything unusual (that requires immediate stopping), by the time that the driver reacts and starts application of the brakes, the vehicle would have already moved $T_{\text{react}} v$ distance. The final term uses the speed difference with the vehicle in front Δv into the equation. If the vehicle ahead is at the same speed as the vehicle being modelled, there is no possibility of a collision and therefore the term becomes 0. If the vehicle ahead is at a larger speed in comparison to the vehicle being planned, collision is never possible, and the term instead becomes negative. If the vehicle ahead has a smaller speed, it is necessary to apply brakes in time otherwise collision is bound to happen. The term takes care of the same as well encouraging a larger safety distance. a and b are the acceleration and braking coefficients used for the modelling. The different notations are shown in Fig. 21.4.

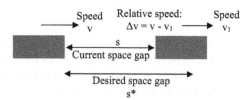

FIG. 21.4 Intelligent Driver Model. The model sets the current acceleration based on the distance from the vehicle in front.

21.5.5 Other behaviours and simulation units

So far, the assumption was only a single lane in a small road segment. Let us now add other traffic elements to traffic simulation. In the case of multiple lanes, an additional behaviour is a *lane change*. If a driver is in a lane with a slow-moving vehicle in front, it is advisable to change lanes. Out of all lanes (basically the one on the left, the current lane, or the right lane), the choice of the lane is based on metrics that evaluate the best lane. *Time to collision* is a reasonable metric, calculated by assuming the preferred speed of the vehicle being planned. So, a lane that has the most distant vehicle and the fastest vehicle ahead is used as the choice of lane. Many places have speed limits on different lanes, in which case the choice of the lane is additionally based on the preferred driving speed. *Overtaking* is another behaviour that happens by changing to the overtaking lane, completing the overtaking, and returning to the original lane. The time to collision is again a good metric to decide the need for overtaking. The feasibility needs to be checked as well. The mergers and intersections without traffic lights are also an interesting scenario, wherein a vehicle needs to decide whether to use the intersection or wait for the other vehicles.

The other interesting simulation element is *traffic lights*. The traffic lights coordinate the motion of vehicles to avoid collision by stating the right of way at the local level. At the same time, the traffic lights coordinate global traffic management to avoid congestions at the global level. The current traffic lights in many places are nonintelligent and operate in cycles. The two common operating styles are to allow traffic between every road pair in cycles or to allow traffic to go in any direction from a specific road and the specific road is cycled. The pedestrian traffic lights can also be incorporated. Modern traffic lights have loop detectors that detect the vehicle flow. Therefore, when there is no traffic on a road, the traffic light change happens without waiting for the end of the planned duration. The simulators are used to decide the traffic light operating policy by simulating all policies with their parameters. If the traffic lights operate in isolation, they may end up directing too much traffic to congestion-prone areas. The interconnected traffic light models are used to link the traffic lights to avoid congestion globally. The duration of one traffic light is based on the traffic at the other traffic lights and their durations.

Traffic simulators also allow simulating accidents, reservation of lanes for public buses, reservation or closure of roads, pedestrian movements, multi-modal transportation networks involving trams, trains, buses, etc., having multiple kinds of vehicles and traffic, and other situations. The traffic simulations use a road network graph to simulate traffic. The Openstreet Map is a good data source regarding the road-network graph. The vehicles are generated using macroscopic modelling, wherein the vehicles are specified using the *origin-destination matrix* (OD) matrix. The three-dimensional matrix specifies how many vehicles travel between every origin-destination pair for every block of time. Such matrices are available for several regions, made approximately based on observed data.

A good simulator for traffic is Simulation of Urban Mobility (SUMO; Lopez et al., 2018). The user can draw any road network graph or load data from other sources. Similarly, the demand can be manually generated using either microscopic or macroscopic variables or loaded from external sources. The simulator allows programming of every traffic element for research. *AutonoVi* (Best et al., 2018) is another emerging traffic simulation system.

21.6 Planning humanoids

To end the discussions, let us take the hardest planning problem, which is the planning of a humanoid robot (Dalibard et al., 2013; El Khoury et al., 2013; Harada et al., 2008, 2009). The humanoids may have over 40 degrees of freedom and it is, therefore, a challenge to plan them. The humanoids have a mobile base which is actuated by two legs modelled as an inverted pendulum, have hands that act as the manipulator, and the humanoids can show complex behaviours like bend and walk, squat, and make complex shapes to avoid obstacles. It would be unfair to compare the walking gait of the humanoids with a wheeled mobile robot, even though both are used to go from one position to the other. This is because stability is difficult with humanoids, while a wheeled mobile robot base is always stable. The advantage is that the humanoids can step over obstacles, take stairs, and walk on uneven terrains, which is impossible in the case of a wheeled mobile robot.

21.6.1 Footstep planning

Let us take the problem that a humanoid must go to and get an object of interest. The problem is broken down into a navigation problem of a mobile base and a manipulation problem of the hand. Using the recognized object along with its pose, the grasping library gives the desired pose of the end-effector. The inverse kinematics library further gives the position of all joint angles of the manipulator and the desired pose of the mobile base. For navigation, let us assume the humanoid has a circular base. Any motion planning technique can be used to get a trajectory of the base to the computed goal pose. Control algorithms are used to make the humanoid trace the trajectory. The humanoids come with libraries preimplemented for gait, which includes moving forward a sufficient step length, making a turn of a small angle, and sometimes mixtures of both motions. The mobile robot trajectory can thus be broken into small segments of linear motion forward and small turns, till the goal is reached. Once the base reaches the goal, the motion planning and control of the manipulator is done, followed by grasping.

The kinematics of a wheeled mobile robot allows it to navigate curves, while the kinematics of the humanoid is different. Specifically, the planner would like to make the best use of the kinematic capability of the robot and include some knowledge of the same in making a trajectory of the mobile base, while always ensuring stability. Therefore, the approach is improved by a better algorithm. The planner is called the *footstep planning* algorithm for a humanoid and the planner plans where the humanoid should keep its footsteps. The walking of any biped is based on two legs and at every point in time, either of the legs is fixed to the ground while the other one is moving. The leg fixed at the ground is said to be at the *stance phase* of walking, while the moving leg is said to be at the *swing phase* of walking. The legs keep switching between the stance phase and the swing phase in a synchronized manner. It may be noted that in running and jumping, both legs are in the air, but such motions are not discussed in humanoids because of a lack of guarantee of stability.

The footstep planner plans for the placement of each foot at every step of motion. Each foot has a position and orientation and therefore represents a SE(2) configuration space, while both legs combined is $(SE(2))^2$ configuration space. There are however constraints between the two legs since the span of both the legs and their orientations are both limited. We have already seen planning for such constraints in any forward simulation algorithm, including graph searches, RRT, and optimization. First, the two feet are at the source configuration. Thereafter, the first foot is moved within the vicinity of the second by finding a neighbouring configuration. The configuration should be collision-free, reachable by keeping the second leg fixed, and stable so that the robot does not risk tipping. For all (sampled) such placements of the first foot, the (sampled) placements of the second foot are found with the same constraints, this time keeping the first leg fixed.

A PRM style planning is also possible by finding random joint configurations of both legs, such that the collision, stability, and reachability constraints are satisfied. A local planner is then used to compute the edge. The local planner here is a custom library implementation that moves the robot from one pose to the other. In the simplest case, the planner rotates the robot in the line of motion, moves the robot straight using small steps, and rotates the robot in the goal direction keeping both legs at the desired orientation. The local planner is obtained from the *walking pattern generators*, which move the robot using different commonly occurring walking patterns. Take a humanoid robot at the origin and move it multiple times to random goals by using teleoperation on an obstacle-free map. Populate a database with all the motions made, called the motion database. Given a source and goal, try different patterns already stored in the motion database to see if they make the robot reach the goal. If any pattern can make the robot move to the desired configuration, the samples are connected. Working is shown in Fig. 21.5.

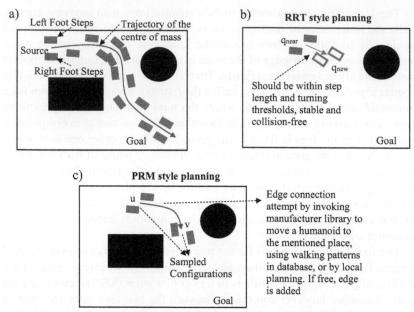

FIG. 21.5 Footstep planning. (A) The humanoid plans its footsteps to avoid collision and have stability. (B) In forward simulation algorithms like RRT, every forward simulation step should ensure the constraints on turns and step size. (C) Probabilistic Roadmap based planning.

21.6.2 Whole-body motion planning

There is a problem with footstep planning. The humanoids can work on uneven surfaces, step over obstacles, take stairs, etc., while the planner so far has used only binary two-dimensional modelling of the workspace. Planning in the complete space is called the *whole-body motion planning* of the humanoid robot. The advantage is that it allows the robot to bend and walk to avoid an obstacle overhead, walk over obstacles, step over obstacles, take stairs, orient hands to avoid a collision, etc. in the 3D workspace.

Let us first make an obvious solution to the problem, which is doing *kinodynamic planning* in the control space. Every robot can be given a control sequence, and a simulator can be used to get the state obtained after applying the control sequence if the simulator has the kinematic modelling of the robot. The problem is to find a control sequence that when applied to the simulation takes the robot to the desired goal configuration. Any forward simulation algorithm like optimization, RRT, or graph search can be used as already discussed in earlier chapters, replacing the kinematic equation with a simulation in the simulator. As an example, optimization assumes many via points and represents the control sequence between every consecutive via points as the genotype. The RRT instead applies sampled control sequences to a parent, which become the tree edges and uses a simulator to get the state after the application of the control sequence which becomes the child nodes of the parent.

The concepts can be translated into a configuration space with a forward simulation as well. The configuration space has every joint angle of the humanoid. To visualize, let us isolate one joint angle (say of the ankle) which will take some trajectory as shown in Fig. 21.6A. Now the aim is to find joint trajectories that guarantee stability, no collision, high clearance, high smoothness, and a small travel time. For the RRT algorithm (similarly for others), the conventional way of generation of child joint angles from the parent joint angles can be used. However, the resultant edge representing a joint trajectory will have to be given to a simulator to ensure that the segment is collision-free and stable.

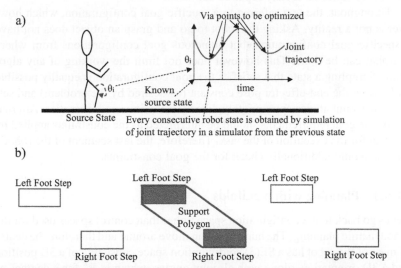

FIG. 21.6 Whole-body motion planning. (A) Each joint has a trajectory sampled by using via points (or planned by expansion like RRT). The (part) trajectory is given to a simulator to get the state at the next via point, stability, collision, and cost metrics. Preknown good sequences can be loaded from a motion database. (B) The Centre of Mass and Zero Moment Point should always be within this polygon for stability.

It needs to be again stressed here that *stability* is of prime importance. Given a configuration, the stability can be derived by several indicators. Foremost, the centre of mass projected on the ground should be contained within a polygon bounded by the two legs, called the *support polygon*. This ensures that the robot never gets a torque due to its weight which tumbles it down. Similarly, the *Zero Moment Point* (ZMP), which is a point where the moment due to the motion and gravity cancel each other out, should also be within the same polygon. This is shown in Fig. 21.6B. The simulator may have unstable intermediate states, but such states should not be used in the actual run of the humanoid which can be sensitive and fatal for the robot. Using an algorithm, if an intermediate state is found which is unstable, it needs to be either completely discarded or corrected by getting to the nearest stable state. Consider RRT as an example. If a state s is being simulated by a sampled control sequence u which lands the robot at an unstable state s_{new}, the control sequence could be iteratively locally optimized such that the landing state is stable.

The overall optimization for kinodynamic planning can take too long. Therefore, a *motion database* is used which consists of the motion of the humanoid in numerous known situations. The library is biased to load precomputed sequences. The sequences may become collision-prone due to differences in the query scenario and the scenario over which the sequence was generated. However, this gives a good starting point for further optimization. The motion may many times be locally repaired and used.

Throughout, the text referred to a specific goal configuration, which however is not a reality. Asking the robot to go and grasp an object does not have a specific goal configuration, but numerous goal configurations from where grasping can be done. This however does not limit the working of any algorithm. Sampling a state that satisfies the goal configuration is equally possible by keeping the end-effector pose constant (as desired by the problem) and setting other joint angles appropriately, like by using inverse kinematics. Alternatively, the goal requirements should be interpreted as the constraints applied by the goal for the execution of the task. Therefore, the last segment of the robot's motion should additionally check for the goal constraints.

21.6.3 Planning with manifolds

Let us go back to the configuration space rather than control spaces used for the kinodynamic planning. The humanoid can move around and therefore the centre of mass of the robot has a SE(3) configuration space consisting of a 3D position and a 3D orientation, along with all joint angles, which is $\mathbb{R}^n$ for n degrees of freedom. Note that a real number rather than a torus is used since each joint angle is now assumed to be limited in orientation. However, only a very few of these configurations are valid. The robot cannot fly in the air, go such that either of the body parts is inside the ground, or be unstable. The robot should neither collide with obstacles, nor self-collide. The robot configuration should be stable. Such highly constrained configuration space should be viewed as a high-dimensional space, in which only a line adheres to the constraints. As an example, if a 3D space (x, y, z) has an equality constraint $x^2 + y^2 + z^2 = 1$, the constraint-free space is just a sphere and not a complete space.

Since the space is highly constrained, let us try to eliminate the main constraints (Hauser and Latombe, 2009). Let us fix one of the feet of the humanoid anywhere on the ground. Now the configuration reduces since one of the legs is constrained not to move. The complete humanoid is now just a few open kinematic chains. The stability and collision constraints exist, but they are significantly less constraining, leaving a significant part of the reduced configuration space as free. The reduced configuration space is called a *manifold*. Similarly, let us fix both legs of the humanoid for an even reduced configuration space, which is another manifold.

The manifolds are joined to each other. Consider two feet of the robot as constrained to lie at a specific position, as shown in Fig. 21.7. This manifold is a special case of two manifolds, one in which only the left foot is constrained and another in which only the right foot is constrained. Therefore, it is possible to travel from the manifold of the left foot fixed to the manifold with both feet fixed to the manifold with the right foot fixed. The traversal is natural since all manifolds are parts of the same joint configuration space. Unfortunately, there are infinite such manifolds.

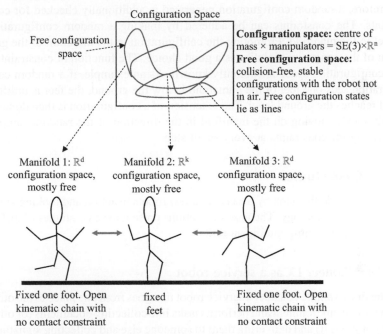

FIG. 21.7 Manifolds. Planar samples some manifolds (vertices of the probabilistic roadmap) with some configurations within the manifold and attempts to connect neighbouring manifolds by moving the robot (edges of the probabilistic roadmap), generating more manifolds. The manifold of both feet fixed connects the adjoining manifolds with one foot fixed.

As the first step, let us sample out some of the infinite manifolds by keeping both legs fixed at sampled locations, keeping the other joint angles at random values. These become analogues to the vertices of a Probabilistic Roadmap (PRM). Note that even if both feet are fixed, there are still infinite configurations possible with the other joints. The connections between the manifolds are a simpler motion planning problem. In this motion, the humanoid can move all the joints to stabilize and avoid collisions. Imagine a specific configuration with both feet fixed in the associated manifolds. The attempt is to make a motion plan that can take the humanoid from this configuration to a configuration of another manifold, where again both feet are fixed at a different location. These are analogues to the edges of a PRM. The motion can be obtained by a local planning technique that moves the humanoid using primitive motion or pattern databases. This motion has multiple new manifolds discovered at every step, which are added to the roadmap.

Alternatively, let us consider an RRT-like expansion. The key here is to expand a state in the full configuration space in some direction. A naïve expansion in the full configuration space can give configurations which are not in the manifold and thus infeasible. There will be too many such configurations. Therefore, a random configuration generated is additionally checked for constraints. The constraints can be adhered by making a random configuration travel to the manifold. Promoting the configuration in the direction of the gradient of the constraint violation is a good choice. The function of constraint in the configuration space is explicitly known. As an example, if a random configuration sees one feet of the humanoid below the ground, the feet is uplifted till it reaches the ground level. The extension of a configuration is then done by continuously moving on the manifold in the direction of the random sample, correcting for constraints at every small step.

21.7 Case studies

Let us conclude the book by observing a few robots in action and looking at the background technology. The next subsections present the case studies of different robots performing intelligent tasks.

21.7.1 Pioneer LX as a service robot

As the first example consider a service robot that goes from one place to the other of the office environment and performs tasks like collecting documents (or other items of interest) and delivering them to someone else and collecting signatures on documents by interacting with the humans. The robot used for this purpose is the Pioneer LX robot. The robot first makes a map using the SLAM algorithm. Thereafter different places of interest are marked on the map like places where different people sit. The robot has an autonomous navigation capability. In other words, if a goal pose or a goal name is relayed to the robot, it can autonomously move to the desired location avoiding people and static obstacles on the way. This happens by using the ROS Navigation Stack implementation.

Thereafter, the robot collects several tasks from several users as input (Bhardwaj and Kala, 2019; Kala, 2020, 2021). The tasks are given as an expression with AND, OR, and Sequence operations. In the case of AND, all subtasks will have to be done in any order. In the case of OR, only one of the subtasks will have to be done. In the case of a sequence, the subtasks must be done in the same order as specified. As an example, a user may ask the robot to get signatures of five lab members. Another user may want the robot to go to a friend, collect keys and deliver the keys back. Another user may simultaneously want the robot to get coffee from any of the outlets and deliver the same. The robot must cater to the needs of all these users.

A mission planner is used to schedule the unit operations. A typical unit operation is visiting a human by going to its goal place. The robot then detects and recognizes the human of interest. If found, the robot tells the task-specific lines to the human and waits for the human to respond. Once the necessary operations are done by the human, the robot starts executing the next unit operation in the planned sequence. The overall conceptual diagram is shown in Fig. 21.8. Fig. 21.8 also shows the robot in action.

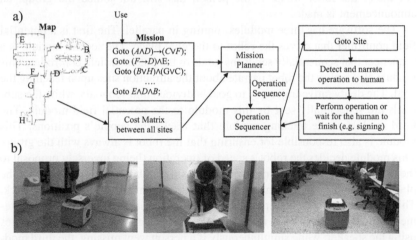

FIG. 21.8 Pioneer LX as a service robot. Users add tasks to the mission which is solved by the mission planner and executed by the sequencer. The robot interacts with the users like asking them to sign a document. (A) Concept diagram, (B) one run of the robot.

21.7.2 Pioneer LX as a tourist guide

Let us take a related application of the same robot. The robot must now take a group of people on a guided tour of a facility (Malviya et al., 2020; Malviya and Kala, 2022). The navigation of the robot is sequentially visiting several goals or tour attraction sites. Upon reaching any goal (attraction site), the robot is expected to explain the group members about that site and then go to the next site. An application-level requirement is that the robot must slow down when the people are far behind and pause the tour if a group member has temporarily left the tour. If a group member is missing for a long time, the robot must make a suitable announcement.

The robot is now additionally equipped with a rear-looking web camera. The robot detects the humans who are a part of the guided tour and tracks them. The tracking is done in the 3D real world using the observations as the detected bounding box in the image. For doing the same, the motion model of the humans considers a Social Potential Field method. The motion model assumes that the people get attracted and repelled by each other, while also being attracted and repelled by the robot. The observation model uses the pose of the robot from the localization system, calculates the pose of the camera, knowing the constant transformation between the robot and camera, and then projects the humans on the image plane.

The robot thus knows the position estimates of all the members of the group. The robot sequentially visits the attraction sites using the Artificial Potential Field algorithm. If a member is not seen for a little while, the robot slows to let the person catch up and re-join the group. Upon seeing the person still not there, the robot waits. If the person has still not joined the group, an announcement is made.

The software has three modules running in parallel. The first is a potential field controller that drives the robot to the goal place. The second module is a sequencer that sequentially sends the goals to the potential field controller. This module also makes the robot narrate about the attraction sites upon reaching the goal. The deliberative trajectory to goal is divided into subgoals, while the subgoals are sequentially sent by this module to the potential field module. The third module is the tracking module that tracks the people's positions. This module is also responsible for ensuring that the robot is always with the group. This module decides the robot's speed of operation (being inversely proposal to the time for which a person in a group is missing), robot's decision to wait at the place, and making an announcement requesting the missing person to join back. This module sends the preferred speed to the potential field module. On implementing, the second module made the robot reach a goal and explained the sites using the speaker; while simultaneously if a person went missing, the third module made the relevant announcements using the speaker. Both modules simultaneously used the speaker. Two modules share the speaker as a resource, and thus, a semaphore-style reservation strategy is used to avoid both modules from simultaneously accessing the speaker and making announcements.

It was observed that the robot had to take sharp turns several times due to congested corridor spaces. Upon taking a sharp turn, the visual contact between the robot and the person would be lost as the person would be outside the field of view of the turning robot, and the robot would assume that the person is missing, which is a false positive. The robot would slow down and eventually wait for the person. This made the robot wait at a sharp tuning point, blocking the way and disallowing the people to move behind the robot. Hence a heuristic is added that the robot must never slow down, stop, wait, or make announcements when making a sharp turn. If the robot makes a sharp turn, it is assumed that the person is following. Similarly, after the robot makes a sharp turn, the people following the robot would later make the same turn, at which time also it is assumed that the people are following the robot.

The tracker will lose track of the people when such sharp turns happen. In several other cases, the track will be lost like the person quitting the group and later re-joining. A new track is initialized whenever the person is detected again. The area of the face is used to get the depth from the monocular camera, which is then used to get the X and Y coordinates in the camera coordinate frame and thereafter (knowing the robot's pose and the person's height), get the position of the person in the world coordinate frame. The concept is shown in Fig. 21.9. A run of the robot is also given in Fig. 21.9.

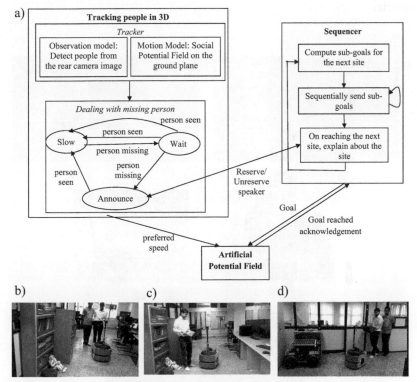

FIG. 21.9 Pioneer LX as a tourist guide. (A) The robot moves to all sites, explaining about the sites. If a person is not visible, the robot slows, waits, and then makes an announcement (B) robot explaining about the site, (C) robot stops and announces as people are missing, (D) the robot is about to encounter a sharp turn and anticipates a loss of visibility of the human. The robot will assume that the person is present even though the person will not be visible.

21.7.3 Chaining of Amigobot robots

Let us now focus our attention on low-cost robots. The Amigobot robot is considered that does not have a lidar sensor. The robot is equipped with sonar sensors for proximity. Localization happens by using wheel encoders. The aim is first to make the robot autonomously go from a source to a goal. The occupancy map of the arena is made by the Pioneer LX robot and supplied to the Amigobot robot. The initial position is supplied by the human. The robot plans a path to the goal using the A* algorithm. The robot is driven by the Artificial Potential Field algorithm, supplying a distant point in the deliberative trajectory as the subgoal. Each sonar ring detects obstacles and applies repulsion from the same. The robot knows its pose using wheel encoders and the constant goal position, which is used to calculate the attraction to the goal. A run is shown in Fig. 21.10A. Note that the robot encountered a new obstacle which was not in the map. However, using sonar, the robot could detect and avoid the obstacle.

FIG. 21.10 Robots in action (A) An Amigobot moving using a fusion of A* algorithm with the Artificial Potential Field algorithm on a map made by the Pioneer LX robot. (B) A chain of Amigobots moving by using the elastic strip algorithm. (C) Baxter robot picking an object. Deep learning libraries detect the object and predict the grasp rectangle, which is used to compute the end-effector's pose. (D) A Jackal mobile base with a Kinova Gen3 Lite arm doing mobile manipulation using the lidar sensor and a top mounted RGBD camera. (E) NaO robots dancing, programming by a graphical utility that lets the user select poses and pose sequences to be followed by the robot.

To make the run interesting, let us add more Amigobots to make them travel in a chain (Apoorva et al., 2018). The first (leader) Amigobot drives to the goal using the above-mentioned algorithm. Each following Amigobot follows the Amigobot ahead in the chain. The nonleader Amigobots are driven by the elastic strip algorithm. The sonars detect obstacles and provide repulsions. If a robot reaches too close to its immediate leader or its immediate follower, a repulsion is applied to stop the robots from colliding. An elastic strip algorithm is used to compute the internal force to make the robots maintain a chain formulation. The run is shown in Fig. 21.10B.

21.7.4 Pick and place using the Baxter robot

The next case study is using manipulators. The robot is expected to pick up an object of interest. The robot has an Intel Real Sense RGBD (RGB and depth) camera attached to the torso that provides a table-top view of the table. The live camera image is passed through an object detection and recognition library. The algorithm searches for the desired object on the table. The next task is to get a grasping rectangle that shows how the object should be grasped. This is another deep-learning library that takes the image as input and gives the grasp rectangle configuration as an output (Shukla et al., 2021, 2022). To grasp, we should know the position and orientation of the gripper. Since the object is on the table-top, the only configurations needed are the 2D position on the tabletop and the 1D orientation of the gripper. The centre of the bounding box is taken as the position of the end-effector. The depth component of the RGBD image is used to get the Z-axis value of the centre, while the camera calibration matrix is used to get the X- and Y-axis values of the end-effector's position. The position is transformed from the camera to the robot's coordinate frame. The orientation of the gripper is obtained from the orientation of the bounding box. The inverse kinematics is used to get the joint angles. Each joint is controlled to get to the desired joint angle. Thereafter the gripper is closed, and the object is lifted. The robot's execution is shown in Fig. 21.10C.

The mobile base and the manipulator together make a mobile manipulator. This example will use the Kinova Gen3 Lite manipulator mounted on top of a Jackal mobile base. The setup uses an Intel Real Sense camera along with a SICK lidar. In the combined setup, first the mobile base moves to the place of interest. ROS Navigation stack is used. Adaptive Monte Carlo Localization (AMCL) is used for localization while the Dynamic Window Approach (DWA) planner is used. Thereafter, the manipulator uses the same algorithm as above to calculate the pose of the end-effector. However, now the manipulator has its own body along with the obstacles in the environment with which a collision can happen. Therefore, the depth image taken from the depth camera is published as an obstacle map. The manipulator uses the MoveIt library (that uses the Open Motion Planning library) to plan the manipulator and execute the same. The robot is shown in Fig. 21.10D.

21.7.5 Dancing NaO robots

As the last example, let us consider the humanoid robot NaO. The robot moves using 2 ft and can grasp objects using the two hands. The robot has a vertical stereovision system where one camera is at the forehead and another at the mouth. The eyes are socialistic which have LED to display the robot's mood. The robot has four microphones for sound localization. The ears are fitted with speakers to let the robot talk. The body is fitted with tactile sensors.

Let us discuss a dancing application using the robots. The robots are programmed for dancing by using a graphical utility. The dance moves feature the robot taking characteristic poses in a sequential manner that is synchronized with music. It is hard to specify the pose sequence manually. Therefore, the joint positions are specified by a graphical library that lets the programmer move a virtual robot. The programmer moves each joint to the desired pose and saves the pose as the next in the sequence. The selected poses are sequenced to be attained at the desired beat of the music. When all the poses are taken by the robot, one after the other, it appears that the robot is dancing to the beat of the music.

Alternatively, it is possible to move the physical robot by hand to get the joint angles. This is called as the *teach-pendant* mode of programming, where a human holds the hands (and legs) of the robot to demonstrate a trajectory and the robot logs this trajectory. The trajectory is internally smoothened and optimized. The optimized trajectory is played back by the robot when asked. With the backdrop of music, this appears as the robot dancing. The dancing robots are shown in Fig. 21.10E.

Questions

1. Draw the fundamental diagrams of a transportation system with human-driven vehicles. Compare the diagrams with the following scenarios (a) Transportation system with only autonomous vehicles, (b) Transportation system with a mixture of autonomous and human-driven vehicles, (c) Transportation system with all human-driven vehicles but with aggressive drivers, (d) A group of people walking on a pavement, (e) A group of people moving by a travellator.
2. Consider the Intelligent Driver Model. Show how the constants are likely to be different between (a) human-driven vehicle, (b) autonomous vehicle, (c) human-driven vehicle with a fatigued driver. Simulate the traffic using the Intelligent Driver Model and the social potential field to comment upon the similarities and differences.
3. Since the ROS programs are C++ or Python codes, can they be executed directly from a command line? If not, why? If yes, explain if there is a difference between executing them directly from the command line and executing them by using roslaunch/rosrun. Also explain why two nodes

exchange data should using a topic when they can just have common shared variables in C++ or Python.

4. [Programming] Write ROS nodes and execute them from a launch file such that after execution the RQT graph is as shown in Fig. 21.11. The edges represent topics of integer data type. The topic names are not written for clarity. You are free to process the integers in the nodes by any function. Additionally, echo the contents of every topic on the console and in the RQT-plot. You may need to publish the 1st message manually in some cases.

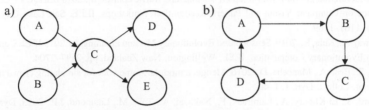

FIG. 21.11 Graphs for question 4. (A) Problem 1. (B) Problem 2.

5. [Programming] Using ROS, program 6 turtlesim robots to assemble at one place, without colliding with each other. The robots can communicate to each other. Also, program 2 groups of turtlesim, each group with three robots. The robots in a group attract each other, while the robots in different groups repel each other. The simulation should allow a seamless increase in the number of robots.

6. [Programming] Consider the simulations done previously for multirobot motion planning with each robot being moved using a combination of a deliberative and a reactive technique. Re-perform the simulations using ROS, such that the deliberative planner, reactive planner, program that computes the local goal, simulated sensors, simulated actuators and similarly all the other modules are ROS nodes or services.

7. [Programming] Using ROS on turtlesim, (a) Simulate one robot that traces a perfect square trajectory, stopping and turning on all four corners, (b) Simulate one robot that traces a square trajectory approximately, without stopping at the corners, (c) Simulate four robots that first form a chain and thereafter they move as a chain on the same trajectory as (b). Once a signal is received, all robots split into two chains, consisting of alternate robots. One chain moves clockwise and the other anticlockwise in the same trajectory as (b).

8. [Programming] Explore all available robots in ROS as per their website. Simulate some robots using Gazebo and the ROS navigation stack. Add obstacles in Gazebo and allow the robot to plan and move to a specified goal. Refer to the tutorials and documentation provided for each robot. Log data and use the same to train a machine learning system for motion planning. Test the performance on the same simulator.

9. [Programming] Explore all simulators discussed in the chapter by following their tutorials and documentation. Thereafter, comment upon the general characteristics of simulation as provided by most simulators.

References

Apoorva, Gautam, R., Kala, R., 2018. Motion planning for a chain of mobile robots using A* and potential field. Robotics 7 (2), 20.

Best, A., Narang, S., Pasqualin, L., Barber, D., Manocha, D., 2018. AutonoVi-Sim: atonomous vehicle simulation platform with weather, sensing, and traffic control. In: 2018 IEEE/CVF Conference on Computer Vision and Pattern Recognition Workshops. IEEE, Salt Lake City, UT, pp. 1161–1169.

Bhardwaj, A., Kala, R., 2019. Sensor based Evolutionary Mission Planning. In: 2019 IEEE Congress on Evolutionary Computation. IEEE, Wellington, New Zealand, pp. 2697–2704.

Curtis, S., Best, A., Manocha, D., 2016. Menge: a modular framework for simulating crowd movement. Collect. Dyn. 1, 1–40.

Dalibard, S., El Khoury, A., Lamiraux, F., Nakhaei, A., Taïx, M., Laumond, J.P., 2013. Dynamic walking and whole-body motion planning for humanoid robots: an integrated approach. Int. J. Robot. Res. 32 (9–10), 1089–1103.

Dosovitskiy, A., Ros, G., Codevilla, F., Lopez, A., Koltun, V., 2017. CARLA: an open urban driving simulator. In: Conference on Robot Learning. CoRL, Mountain View, CA, pp. 1–16.

El Khoury, A., Lamiraux, F., Taix, M., 2013. Optimal motion planning for humanoid robots. In: 2013 IEEE International Conference on Robotics and Automation. IEEE, Karlsruhe, pp. 3136–3141.

Harada, K., Morisawa, M., Miura, K., Nakaoka, S.I., Fujiwara, K., Kaneko, K., Kajita, S., 2008. Kinodynamic gait planning for full-body humanoid robots. In: 2008 IEEE/RSJ International Conference on Intelligent Robots and Systems. IEEE, Nice, pp. 1544–1550.

Harada, K., Morisawa, M., Nakaoka, S.I., Kaneko, K., Kajita, S., 2009. Kinodynamic planning for humanoid robots walking on uneven terrain. J. Robot. Mechatron. 21 (3), 311.

Hauser, K., Latombe, J.C., 2009. Integrating task and PRM motion planning: dealing with many infeasible motion planning queries. In: ICAPS09 Workshop on Bridging the Gap Between Task and Motion Planning. AAAI, Greece, pp. 34–41.

Immers, L.H., Logghe, S., 2002. Traffic flow theory. Faculty of Engineering, Department of Civil Engineering, Secion Traffic and Infrastructure, Kasteelpark Arenberg 40, 21.

Kala, R., 2020. Robot mission panning using co-evolutionary optimization. Robotica 38 (3), 512–530.

Kala, R., 2021. Multi-robot mission planning using evolutionary computation with incremental task addition. Intell. Serv. Robot. 14, 741–771.

Lopez, P.A., Behrisch, M., Bieker-Walz, L., Erdmann, J., Flötteröd, Y.P., Hilbrich, R., Lücken, L., Rummel, J., Wagner, P., Wiessner, E., 2018. Microscopic traffic simulation using SUMO. In: 2018 21st International Conference on Intelligent Transportation Systems. IEEE, Maui, HI, pp. 2575–2582.

Maerivoet, S., de Moor, B., 2004. Traffic flow theory, Santiago 11, 1.

Malviya, V., Kala, R., 2022. Trajectory prediction and tracking using a multi-behaviour social particle filter. Appl. Intell. 52, 7158–7200.

Malviya, V., Reddy, A., Kala, R., 2020. Autonomous social robot navigation using a behavioral finite state social machine. Robotica 38 (12), 2266–2289.

Rohmer, E., Singh, S.P.N., Freese, M., 2013. V-REP: a versatile and scalable robot simulation framework. In: 2013 IEEE/RSJ International Conference on Intelligent Robots and Systems. IEEE, Tokyo, pp. 1321–1326.

Shukla, P., Kumar, H., Nandi, G.C., 2021. Robotic grasp manipulation using evolutionary computing and deep reinforcement learning. Intell. Serv. Robot. 14, 61–77.

Shukla, P., Pramanik, N., Mehta, D., Nandi, G.C., 2022. Generative model based robotic grasp pose prediction with limited dataset. Appl. Intell. 52, 9952–9966.

Şucan, I.A., Moll, M., Kavraki, L.E., 2012. The open motion planning library. IEEE Robot. Autom. Mag. 19 (4), 72–82.

Tedrake, R., 2019. Drake: Model-Based Design and Verification for Robotics. https://drake.mit.edu.

Treiber, M., Hennecke, A., Helbing, D., 2000. Congested traffic states in empirical observations and microscopic simulations. Phys. Rev. E 62 (2), 1805.

Index

Note: Page numbers followed by *f* indicate figures, *t* indicate tables, and *b* indicate boxes.